Lecture Notes in Computer Science 10010

Commenced Publication in 1973
Founding and Former Series Editors:
Gerhard Goos, Juris Hartmanis, and Jan van Leeuwen

More information about this series at http://www.springer.com/series/7407

Adam Day · Michael Fellows
Noam Greenberg · Bakhadyr Khoussainov
Alexander Melnikov · Frances Rosamond (Eds.)

Computability and Complexity

Essays Dedicated to Rodney G. Downey
on the Occasion of His 60th Birthday

 Springer

Editors

Adam Day
Victoria University of Wellington
Wellington
New Zealand

Michael Fellows
University of Bergen
Bergen
Norway

Noam Greenberg
Victoria University of Wellington
Wellington
New Zealand

Bakhadyr Khoussainov
University of Auckland
Auckland
New Zealand

Alexander Melnikov
Massey University
Auckland
New Zealand

Frances Rosamond
University of Bergen
Bergen
Norway

Cover illustration: The figure on the cover are the steps to a Scottish Country Dance progression called *La Spirale*. This new dance was created by Rod Downey for the golden anniversary of Wellington's Johnsonville Scottish Country Dance Club. Kristin (Rod's wife) introduced Rod to Scottish Country dance thirty years ago and they have been dancing together ever since. Rod has written three books of dances and a fourth in progress. The links can be found at his website http://homepages.ecs.vuw.ac.nz/~downey/. Connections between mathematics and dance are described by Rod in a Mathreach interview at http://www.mathsreach.org/wiki/images/7/7d/06DanceofMathematics.pdf.

Photograph of the honoree on p. V: Archives of the Mathematisches Forschungsinstitut Oberwolfach

ISSN 0302-9743   ISSN 1611-3349 (electronic)
Lecture Notes in Computer Science
ISBN 978-3-319-50061-4   ISBN 978-3-319-50062-1 (eBook)
DOI 10.1007/978-3-319-50062-1

Library of Congress Control Number: 2016958514

LNCS Sublibrary: SL1 – Theoretical Computer Science and General Issues

Printed on acid-free paper

This Springer imprint is published by Springer Nature
The registered company is Springer International Publishing AG
The registered company address is: Gewerbestrasse 11, 6330 Cham, Switzerland

Rodncy G. Downcy

Foreword

In the friendly rivalry between ANZAC neighbours, Australia and New Zealand, the Aussies have a habit of claiming New Zealand exports as their own – actor Russell Crowe, legendary racehorse Phar Lap, and the incomparable dessert known as pavlova. But, for New Zealand's part, we will put all that aside and happily claim Professor Rod Downey, mathematician and theoretical computer scientist extraordinaire and thoroughly good bloke, born in Oz, but working most of his adult life at Victoria University of Wellington in New Zealand.

Having spent the last 17 years working for the Marsden Fund at the Royal Society of New Zealand, and for all of that time following Rod's research, proposal by proposal, report by report, and publication by publication, it is an honor to be invited to write this foreword. The Marsden Fund is arguably New Zealand's most prestigious funding, and Rod has been funded as Principal Investigator for 20 of Marsden's 22 years. This represents six Marsden Fund grants, a stunning achievement given that the average success rate for proposals is just 9%. There is no doubt that the Marsden Fund has invested well; over 20 years, Rod has averaged about ten publications per year, working with leading mathematicians around the world, and developing a whole new generation of researchers in the field.

In the early days of Marsden, Rod was funded for his pioneering work with Mike Fellows on parameterized complexity. In later years, algorithmic randomness has been a preoccupation. But, all along the way, he has explored widely within the fields of computability and complexity.

In 2010, after more than a decade of work, Rod and co-author Denis Hirschfeldt produced their magnum opus, an 855-page reference work on "Algorithmic Randomness and Complexity." But just three years later, a beaming Rod Downey walked into the Marsden office with *Fundamentals of Parameterized Complexity*, another Springer volume, this time of 763 pages and co-authored with Fellows. History will decide the true "greatest work."

Both volumes provide a hint of Rod's zest for life. On page 517 of *Algorithmic Randomness and Complexity*, the authors use a precious liquid, 1955 Biondi-Santi Brunello, to visualise the proof of Theorem 11.3.1. And in *Parameterized Complexity* (Springer 1999), Rod and Mike acknowledge the pivotal role of surfing and wine in the success of parameterized complexity.

Rod has also made a wider contribution to the mathematics and theoretical computer science fields. For three years, from 2009 to 2011, he was the government's appointee as Convenor of the Mathematical and Information Sciences panel of the Marsden Fund Council, demonstrating a phenomenal knowledge of diverse fields, and a refreshing approach to the governance aspects of the role. He has also been keen to engage with the wider public, explaining algorithmic randomness brilliantly on one occasion in the

Wellington Town Hall, and also contributing another wonderful lecture on Alan Turing in 2015, noting that, "Turing was a prodigy, a brilliant and original man who was terribly treated for being gay. His story is a study in ideas and social commentary."

Wellington is a small place, and you are liable to run into Rod and his family on their mountain bikes, or perhaps with his surfboard in tow, or even at the odd public performance of Scottish country dancing, for which he is an accomplished choreographer. But, above all, Rod stands out for his sense of humour and enjoyment of life, even in the face of adversity.

Rod is an outstanding mathematician, a lovely human being, and a wonderful New Zealander.

September 2016   Peter Gilberd
                                              Programme Manager of the Marsden Fund
                                                          Royal Society of New Zealand

Preface

This Festschrift is an appreciative worldwide scientific community's fitting recognition of the outstanding mathematical and theoretical computer science contributions made by Rodney Downey. It was presented at a celebratory "RodFest," an International Computability and Complexity Symposium that took place in the town of Raumati Beach outside of Wellington, with many of the authors gathered to wish him a happy birthday. The volume contains contributions from invited speakers and papers that present original unpublished research or expository and survey results in the following areas in which Rod Downey has had significant interests: Turing degrees, computably enumerable sets, computable algebra, computable model theory, algorithmic randomness, reverse mathematics, and parameterized complexity.

As Rod turns 60, there is much to celebrate. He has advanced our understanding of several areas of mathematics and computer science. His students and postdocs have gone on to become some of the finest scientists in the world. Rod's colleagues find him supportive and energetic, always with a wealth of new ideas and valuable insights. Rod's work with mathematical associations, organizing conferences and joint meetings, and bringing in world-class scientists has changed New Zealand's mathematical landscape.

Rod's mentoring often includes helmeted surfing in New Zealand's rocky waves and Scottish country dancing. Perhaps the body-boarding helps build the courage needed for mathematical research. Rod's wife, Kristin, introduced him to Scottish country dancing, and he is now a teacher and active in the dance community both locally and internationally. Rod has written three books and devised over 80 dances, accompanied by detailed notes and history, with some on YouTube. The cover picture on this Festschrift is of a popular dance progression, La Spirale, devised by Rod in 1998 and used in many dances.

Building on the work of Church, Gödel, Turing, and others, Rod has developed exciting interactions between computability, complexity, and randomness. This line of research applying algorithmic information theory/Kolmogorov complexity and the incompressibility method to recursion theory in the analysis of notions of randomness, still holding many open problems, is described in the survey by Hirschfeldt. Rod has received many honours and awards, including the Royal Society of New Zealand Hector Medal for outstanding, internationally acclaimed work in recursion theory, computational complexity, and other aspects of mathematical logic and combinatorics. The surveys and articles in this Festschrift are a fitting tribute to Rod's many significant and deep contributions.

Rod is cofounder, together with Mike Fellows, of the highly successful field of parameterized/multivariate complexity. Parameterized complexity is a two-dimensional refinement of classical complexity, where instance complexity is measured by both the size of the input and by a problem-specific parameter. With this fine-grained analysis, normally intractable problems can be solved efficiently provided the parameter is not too large. The story of Rod and Mike's collaboration is well told by Rod in Mike's Festschrift (Springer LNCS 7370) and in Mike's article herein.

In the area of mathematics popularization, Rod's lectures on Turing have drawn large crowds. He has also contributed to the Mathreach Project. In his video interview called "Complexity, Computation and a Bit of Fuzzy Logic," Rod describes the use of mathematics in areas as diverse as industrial smelting, computer chip manufacturing, and tumble drier sensors.

We would like to thank all the contributors to this Festschrift both for their scientifically interesting articles as well as for their enthusiasm to contribute. The positive and immediate response to our invitations made assembling this volume a joyful experience for us. We are very thankful to all the authors and to the many reviewers, who made the excellent articles even better, to Kristin Downey, who sourced many photos, and to Mateus de Oliveira Oliveira for help in many ways. We are indebted to Alfred Hofmann at Springer for his feedback, and to Springer for giving us the possibility to publish this Festschrift in their *Lecture Notes in Computer Science* series. We especially thank Anna Kramer and Ronan Nugent and all those at Springer for their gracious help and advice.

With this book we celebrate Rod's vision and achievements, and honor this eminent scientist who we are privileged to have as mentor, teacher, and friend.

Happy Birthday, Rod!

January 2017   Adam Day
Michael Fellows
Noam Greenberg
Bakhadyr Khoussainov
Alexander Melnikov
Frances Rosamond

Tribute to Rod Downey

Marston Conder, Distinguished Professor of Mathematics

Department of Mathematics, University of Auckland, Auckland, New Zealand
Former Co-director of the New Zealand Institute of Mathematics
and its Applications (the NZIMA)

It is a pleasure to write a brief tribute for Rod Downey on the occasion of his 60th birthday. I've known Rod for almost 30 years, since he came to New Zealand to take up his position at Victoria University. Rod has made some outstanding contributions to the mathematical sciences community in New Zealand, on many fronts. First and foremost these include his own excellent research, which won him the New Zealand Mathematical Society's annual Research Award in 1992, election as a Fellow of the Royal Society of New Zealand in 1996, and the Hector Medal of the Royal Society of New Zealand in 2011, as well as several other awards and honors both locally and internationally.

Rod is one of the founding co-directors of the New Zealand Mathematics Research Institute Inc. (otherwise known as the NZMRI), which is the body that has been running annual summer workshops for professional mathematicians and students since the summer of 1994/1995, and Rod has been an active participant, as well as an organiser or co-organiser of three of these meetings. Rod was also a Principal Investigator in the NZIMA (the mathematical sciences CoRE) from 2002 to 2012, and made many valuable contributions in the selection of programmes, visitors, and students supported by the NZIMA, driven by pursuit of excellence. He served as President of the New Zealand Mathematical Society from 2001 to 2003, and as Chair of the Mathematical and Information Sciences Panel of the Marsden Fund from 2009 to 2012, and has served as a member or chair of several other award committees for the Royal Society of New Zealand.

At a more personal level, some of Rod's impressive characteristics include the energy and passion he has for his subject and his research, his advocacy for discrete mathematics and theoretical computer science, and the support and mentorship he clearly offers to his students, postdoctoral fellows, and other young colleagues. These are legendary.

Rod Downey, a *Taonga* of Victoria University

Peter Donelan, Department Chair

Department of Mathematics, Victoria University of Wellington, Wellington,
New Zealand

Rod Downey joined the Department of Mathematics at Victoria University of Wellington in the southern summer of early 1986. At the time of his appointment, just three years out from his PhD, Rod had already amassed 18 publications, a raft of papers in preparation and a rapidly growing reputation. So his decision to move to New Zealand, adding to Victoria's proud tradition as a centre for research in logic, was greeted with considerable pleasure here and perhaps some dismay by his colleagues in the US. No doubt despite some attractive offers to move on in the ensuing years, it is to our continuing benefit that Rod has remained in this beautiful and mathematically conducive capital city. His mathematical achievements and reputation are surely unmatched in New Zealand.

The immense range and quality of Rod's research will be evident throughout this Festschrift. Its recognition locally has been manifold. Successive and rapid promotions saw Rod awarded a personal chair in 1995 and accede to the highest rank of Professor at Victoria in 2011. In both 2000 and 2008, Rod received University research excellence awards. Rod's influence on the development of mathematics at Victoria and indeed throughout New Zealand has been profound. He has attracted to the (now) School of Mathematics and Statistics and its precursors a succession of postdoctoral fellows and postgraduates who have gone on to become leaders in computability, complexity, and randomness across the globe, including here at Victoria in the forms of current faculty Noam Greenberg, Georgios Barmpalias, and Adam Day. They, collectively, are testimony to Rod's hugely committed and professional approach to mentoring young talent. Nationally, Rod has been President of the New Zealand Mathematical Society, a principal investigator of the NZ Institute of Mathematics and its Applications, executive member of the NZ Mathematics Research Institute and Chair of the Royal Society of New Zealand's Marsden Fund Panel for Mathematical and Information Sciences. In all those roles he has been able to raise the profile and standing of mathematics in New Zealand.

I would also like to acknowledge Rod as a committed and innovative teacher at Victoria. He has been responsible for many curriculum developments that have contributed to our strong undergraduate mathematics programme and he has been an effective adherent of the view that our students will benefit from early exposure to a reflective and deeply knowledgeable research mathematician. So Rod continues to inspire first-year algebra and logic classes and, equally, takes our best and brightest students to the research frontiers of computational complexity. As all those will know, who were fortunate enough to attend his packed public lectures on Turing's life and work during the centenary year, or who have attended any of the numerous plenary addresses and talks he has delivered over the years, Rod well deserves the ovations and invitations he receives.

Rod is a *taonga* of Victoria University, treasured by his colleagues as a unique, generous, and multitalented Scottish country dancer, surfer, wine connoisseur, mathematician, and friend.

Curriculum Vitae Rodney Graham Downey

Current Position
Personal Chair, Professor of Mathematics

Address
School of Mathematics, Statistics & Operations Research
Victoria University of Wellington
P.O. Box 600, Wellington, New Zealand
Telephone: (04) 463 5067 (Office) 463 5045 (Fax) (04) 4784948 (Home)
E-mail: rod.downey@vuw.ac.nz

Personal Information
Born: 20 September, 1957 in Queensland, Australia
Married: to Kristin Macdonald Downey, two children Carlton and Alex
Nationality: New Zealand and Australian
Hobbies: Surfing, Tennis, Scottish Country Dancing

Education
1975–1978 Undergraduate at University of Queensland, St. Lucia, Queensland, Australia B.Sc. with first class honours in Mathematics
1979–1982 Postgraduate at Monash University, Clayton, Victoria, Australia, Ph.D. in Mathematics (November 1982)

Experience
1982 Lecturer in Mathematics at Chisholm Institute of Technology (now Monash University), Caulfield campus, Caulfield East Victoria Australia
1983 (Spring) Visiting Assistant Professor at Western Illinois University, USA
1983–1985 Lecturer, National University of Singapore Kent Ridge, Republic of Singapore
1985–1986 Visiting Assistant Professor, University of Illinois Urbana-Champaign, USA
1986–1987 Lecturer in Mathematics, Victoria University of Wellington, New Zealand
1988–1990 Senior Lecturer in Mathematics, Victoria University of Wellington
1989 Member, Mathematical Sciences Research Institute, Berkeley, California, USA
1991–95 Reader in Mathematics, Victoria University of Wellington, New Zealand
1992 Visiting Scholar, Mathematics Department, Cornell University, Ithaca, NY, USA
1992 Member, Mathematical Sciences Institute, Cornell University, Ithaca, NY, USA
1993 Lee Kong Chiang Visiting Fellow, National University of Singapore
1995 Visiting Professor, Mathematics Department, Cornell University
1997 Visiting Scholar, University of Siena
1999 Visiting Scholar, University of Wisconsin, Madison
1999 Visiting Professor, National University of Singapore

2000 Visiting Professor, University of Notre Dame, Indiana
2001 Visiting Scholar, University of Chicago
2003 Visiting Scholar, University of Chicago
2005 Visiting Professor, University of Chicago
2008 Visiting Professor, University of Chicago
2010 Visiting Scholar, University of Chicago
2014 Visiting Scholar, University of Chicago
2008 Visiting Scholar, University of Madison, Wisconsin
2005, 2011, 2017 Member, Institute for Mathematical Sciences, Singapore
2011 Visiting Professor, Nanyang University of Technology, Singapore
2003 Inaugural MacLaurin Fellow, New Zealand Institute for Mathematics
  and its Applications (Center of Research Excellence)
2008–2010 James Cook Fellow, Royal Society of New Zealand
2009–2012 Chair MIS Panel and Member of Council, Marsden Fund
2012 Fellow, Isaac Newton Institute, Cambridge
1995- Personal Chair, Professor of Mathematics, Victoria University of Wellington

Prizes and Awards

1990 New Zealand Royal Society Hamilton Award for Science
1991 Foundation Fellow of the Institute of Combinatorics and its Applications
1992 New Zealand Mathematical Society Award for Research.
1994 New Zealand Association of Scientists Research Medal for the best
  New Zealand based scientist under 40
2007 Elected Fellow of the New Zealand Mathematics Society
2000 Vice-Chancellor's Award for Research Excellence
1996 Elected Fellow of the Royal Society (NZ)
2006 Invited Speaker, International Congress of Mathematicians
2007 Invited Speaker, International Congress of Logic, Methodology, and Philosophy
  of Science
2008 Elected Fellow of the Association for Computing Machinery (one of two
  New Zealanders)
2008–2010 James Cook Fellowship, Royal Society of New Zealand
2008 Victoria University of Wellington Award for Research Excellence
2010 Shoenfield Prize from the Association for Symbolic Logic
2011 Hector Medal, Royal Society of New Zealand. (New Zealand's oldest research
  medal)
2012 Fellow Newton Institute (Cambridge) for the Alan Turing Programme
2012 Elected Fellow American Mathematical Society. (Inaugural intake, one of 3
  New Zealand based)
2013 Elected Fellow of the Australian Mathematical Society
2014 European Association for Theoretical Computer Science/International
  Symposium in Parameterized Complexity and Exact Computation Nerode
  Prize (joint with Bodlaender, Fellows, Hermelin, Fortnow and Santhanam)

Grants

1979–1982 Commonwealth Postgraduate Research Award (Australia)

1983–1985 Research Grant (Singapore)

1986 Support Grant from U.S. National Science Foundation

1988–91, 92–95, 96–99 PI for three US/NZ Binational Cooperative Grants

1989 Support Grant from Mathematical Sciences Research Institute, Berkeley, USA

1992 Support Grant from Mathematical Sciences Institute, Ithaca New York, USA

1993 Support Grant from the Lee Foundation, National University of Singapore

1995 PI Research Grants continuously from The Marsden Fund for Basic Science

2005 AI on Catherine McCartin's Research Grant, Marsden Fund for Basic Science

1998–2004 AI on 2 Marsden Grants to support the *NZ Mathematical Sciences Research Institute,* of which I am one of the directors, along with Professors Marston Conder, David Gauld, Gaven Martin, and Vaughan Jones

2002–10 PI on the CoRE grant from the New Zealand Government for the *New Zealand Institute for Mathematics and its Applications*

1997 Support grant from the Italian Government

2003- AI on NSFC Grand International Joint Project Grant No. 60310213 "New Directions in Theory and Applications of Models of Computation" (China)

2005–2008 AI on Noam Greenberg's Marsden Grant

2008–2010 James Cook Fellowship, Royal Society of New Zealand

2011 PI on NSF Grant 1135626 with Charles Steinhorn for Travel Grants for 12 US based researchers to speak at the 12th Asian Logic Conference in Wellington, December, 2011. ($US 31K)

2014 PI on Randomness and Computation programme, Institute for Mathematical Sciences, June 2014. ($S 125K)

Postdoctoral Fellows Supervised (Current or Last Known Position Listed.)

1. Michael Moses (George Washington University)
2. Peter Cholak (University of Notre Dame)
3. Geoff LaForte (Western Florida University) (deceased)
4. Richard Coles (Telecom, UK)
5. Reed Solomon (University of Connecticut).
6. Walker White (Cornell University)
7. Denis Hirschfeldt (University of Chicago)
8. Evan Griffiths (New Zealand Risk Assessment Programme)
9. Wu Guohua (Nanyang Technological University, Singapore)
10. Joe Miller (University of Wisconsin, Madison)
11. Yu Liang (Nanjing University, China)
12. Rebecca Weber (Dartmouth)
13. Noam Greenberg (Victoria University)
14. Antonio Montalbán (Berkeley)
15. George Barmpalias (Wellington)
16. Laurent Bienvenu (CIRM Montpellier University)
17. Asher Kach (Google)

18. Dan Turetsky (Wellington)
19. Alexander Melnikov (Massey University, Albany)
20. Greg Igusa (Current)

Ph. D. Students Supervised

Wu Guohua (1999–2002) (Nanyang University of Technology)
Catherine McCartin (1999–2003) (Massey University)
Ng Keng Meng (Selwyn) (2006–2009) (Nanyang University of Technology)
Adam Day (2008–2011) (Wellington)
Michael McInerary (joint with Greenberg) (2013–2016) (Nanyang University Technology)
Katherine Arthur (current)
Day, Wu and McCartin won the *Hatherton Award* for the best paper arising from a PhD paper by a New Zealand based PhD. Day won the *Sacks Prize* for the best PhD in logic worldwide from the *Association for Symbolic Logic*. He was the first New Zealand graduate to get a Fellowship to the Miller Institute at Berkeley.

M. Sc. Student Supervised

Stephanie Reid (2003), John Fouhy (2003), Michelle Porter (2015), Katherine Arthur (2015).
All received A^+ masters with distinction.

Professional Service

- Managing Editor *Bulletin of Symbolic Logic*, 2004–2010. (full term)
- Editor *Journal of Symbolic Logic, 1999–2004*, Coordinating editor 2000–2004. (full term)
- Editor, *Theory of Computing Systems* (formerly *Math. Systems Theory*), 2006-
- Editor, *Archive for Mathematical Logic*, 2009-
- Editor, *Computability*, 2011-
- Co-director, New Zealand Mathematical Sciences Research Institute.
- Co-director, New Zealand Institute for Mathematics and its Applications.
- Vice-President, New Zealand mathematics Society 2000–2001.
- President, New Zealand Mathematical Society 2001–2003, immediate past president, 2004.
- Prizes committee, council, and Australasian committee Association for Symbolic Logic 2000-.
- Nominating committee and committee on plagarism Association for Symbolic Logic.
- Fellows' Committee Royal Society New Zealand 1999–2001. Hamilton Prize Committee, 2004.
- Marsden panel for Mathematical and Information Sciences 1997, 2002, 2003 (chair 2009–2011).
- Marsden Council 2009–2011.
- Royal Society Travel Grants Committee 2008–2010.

- New Zealand Mathematical Sciences Advisory Group 1999–2003.
- New Zealand representative on the International Mathematics Union 2001–2004.
- Fellows Selection Panel, Royal Society, 2005.
- Chair of the Steering Committee for *Computability, Complexity and Randomness* series 2003-.
- Steering Committee for *International Workshop of Parameterized Complexity and Exact Computation*, 2005–2009.
- Chair review committee, science faculty, University of Samoa.
- PC member of 28 computer science conference committees in the last 6 years.
- Judge, Alan Turing Research Fellowship Awards, 2012.
- Fellows panel Royal Society, 2015.
- Assessor, Rutherford Discovery Fellowships (2016)
- 2015- Council, Association for Symbolic Logic.

Areas of Interest: Algebra, Logic, Complexity theory

Publications

Thesis

Abstract Dependence, Recursion Theory and the Lattice of Recursively Enumerable Filters Thesis, Monash University, Victoria, Australia, (1982). J.N. Crossley, Supervisor.

Books

1. *Parameterized Complexity*, (with M. Fellows) Springer-Verlag, Monographs in Computer Science, 1999 xiii+533 pages.
2. *Algorithmic Randomness and Complexity* (with D. Hirschfeldt), Springer-Verlag, Computability in Europe Series No 1, December 2010. xxvi+855 pages.
3. *Fundamentals of Parameterized Complexity*, (with M. Fellows), Springer-Verlag, 2013, texts in computer science, ISBN 978-1-4471-5559-1, online http://link. springer.com/book/10.1007/978-1-4471-5559-1, xxx+763 pages.
4. *A Transfinite Hierarchy of Lowness Notions in the Computably Enumerable Degrees, Unifying Classes and Natural Definability*, (with Noam Greenberg) submitted, 172pp.

Books Edited

1. *Aspects of Complexity*, (with D. Hirschfeldt, editors), de Gruyter Series in Logic and Its Applications, Volume 4, 2001, vi+172 pages.
2. *Proceedings of the 7th and 8th Asian Logic Conferences*, (Chief Editor, with Ding Decheng, Tung Shi Ping, Qiu Yu Hui, Mariko Yasugi, and Wu Guohua, editors) World Scientific, 2003, viii+471 pages.

3. *Parameterized and Exact Computation: First International Workshop, IWPEC 2004, Bergen, Norway, September 14–17, 2004. Proceedings* (Rod Downey, Frank Dehne, Michael Fellows, editors) Springer-Verlag Lecture Notes in Computer Science, Vol 3162, Springer Verlag, 2004. 300 pages.

4. *Mathematical Logic in Asia: Proceedings of the 9th Asian Logic Conference,* (Rod Downey, Sergei S. Goncharov and Hiroakira Ono, eds) World Scientific, 2006, Singapore, viii+319 pages.

5. *Proceedings Fifteenth Computing: The Australasian Theory Symposium (CATS 2009),* Wellington, New Zealand. CRPIT, 94. (Downey, R. and Manyem, P., Eds.) ACS.

6. *Proceedings of the 10th Asian Logic Conference:* (with Joerg Brendle, Chong Chi Tat, Hirotaka Kikyo, Hiroakira Ono and Feng Qi), World Scientific, 2009.

7. *The Multivariate Algorithmic Revolution and Beyond, Essays Dedicated to Michael R. Fellows on the Occasion of His 60th Birthday,* Lecture Notes in Computer Science, Vol. 7370 Subseries: Theoretical Computer Science and General Issues (Bodlaender, H.L.; Downey, R.; Fomin, F.V.; Marx, D. (Eds.)) 2012, 2012, XXII, 506 p. 32 illus.

8. *Proceedings of the 11th Asian Logic Conference*, (with Rob Goldblatt, Joerg Brendle and Bungham Kim), 2013, World Scientific, 325 pages.

9. *Turing's Legacy,* Cambridge University Press, Lecture Notes in Logic, Cambridge University Press, 2014. (Featured in the 19th Annual ACM Computing Reviews Notable Books and Articles (2014).)

Journal Special Issues Edited

1. Special Issue of the *Annals of Pure and Applied Logic* Volume 138, Issues 1–3, Pages 1–222 (March 2006), devoted to the NZIMA Logic Programme (with Rob Goldblatt).

2. Special Issue of *Theoretical Computer Science*, devoted to *Parameterized Complexity and Exact Computation*, (with Mike Langston, and Rolf Niedermeier) Volume 351, Issue 3, Pages 295–460 (28 February 2006) Parameterized and Exact Computation

3. Special Issue of *Theory of Computing Systems* Exact Computation and Parameterized Complexity. Vol 41 No 3 (October 2007).

4. Two special issue of *The Computer Journal* devoted to Parameterized Complexity (with Mike Fellows and Mike Langston). Volume 58 Numbers 1 and 3, 2008, Oxford University Press.

5. Special issue of *Theory of Computing Systems*, Theory of Computing Systems, Vol. 52, Issue 1, 2013, Computability, Complexity and Randomness.

Electronic Article

1. *Algorithmic randomness,* (with Jan Reimann) for *Scholarpedia*, (Rodney G. Downey and Jan Reimann (2007) Algorithmic randomness. Scholarpedia, 2 (10):2574) http://www.scholarpedia.org/article/Algorithmic_Randomness

Papers

1. On a question of A. Retzlaff, *Z. Math. Logik Grund. der Math.*, **29** (1983) 379–384.
2. Abstract dependence, recursion theory and the lattice of recursively enumerable filters, *Bull. Aust. Math. Soc.*, **27** (1983) 461–464.
3. Nowhere simplicity in matroids, *J. Aust. Math. Soc. (Series A)* **35** (1983) 28–45.
4. Co-immune subspaces and complementation in V_∞, *J. Symbolic Logic*, **49** (1984) 528–538.
5. Perfect McLain groups are super-perfect, (with A.J. Berrick), *Bull. Aust. Math. Soc.*, **29** (1984) 249–257.
6. Bases of supermaximal subspaces and Steinitz systems, *J. Symbolic Logic*, **49** (1984) 1146–1159.
7. Decidable subspaces and recursively enumerable subspaces, (with C.J. Ash), *J. Symbolic Logic*, **49** (1984) 1137–1145.
8. Some remarks on a theorem of Iraj Kalantari concerning convexity and recursion theory, *Z. Math. Logik Grund. der Math*, **30** (1984) 295–302.
9. The universal complementation property, (with J.B. Remmel), *J. Symbolic Logic*, **49** (1984) 1125–1136.
10. A note on decomposition of recursively enumerable subspaces, *Z. Math. Logik Grund. der Math.*, **30** (1984) 456–470.
11. Automorphisms of supermaximal subspaces, (with G.R. Hird), *J. Symbolic Logic*, **50** (1985) 1–9.
12. Effective extensions of linear forms in a recursive vector space over a recursive field, (with I. Kalantari), *Z. Math. Logik Grund. der Math.*, **31** (1985) 193–200.
13. The degrees of r.e. sets without the universal splitting property, *Trans. Amer. Math. Soc.*, **291** (1985) 337–351
14. Sound, totally sound, and unsound recursive equivalence types, *Annals Pure and App. Logic*, **31** (1986) 1–22.
15. Splitting properties of r.e. sets and degrees (with L.V. Welch), *J. Symbolic Logic*, **51** (1986) 88–109.
16. Recursion theory and ordered groups, (with S. Kurtz) *Annals Pure and App. Logic.*, **32** (1986) 137–151.
17. Undecidability of $L(F_\infty)$ and other lattices of r.e. substructures, *Annals Pure and App. Logic*, **32** (1986) 17–26. (Corrigendum in *ibid* 48 (1990) 299–301.)
18. Bases of supermaximal subspaces and Steinitz systems II, *Z. Math. Logik Grund. der Math*, **31** (1986) 203–210.
19. Structural interactions of the recursively enumerable T- and W-degrees, (with M. Stob), *Annals Pure and App. Logic.*, **31** (1986) 205–236.
20. Orbits of creative subspaces, *Proc. Amer. Math Soc.*, **99** (1987) 163–170.
21. Subsets of hypersimple sets, *Pacific J. Math.*, **127** (1987) 299–319.
22. Maximal theories, *Annals Pure and App. Logic.*, **33** (1987) 245–282.
23. Degrees of splittings and bases of an r.e. vector space, (with J.B. Remmel and L.V. Welch), *Trans Amer. Math Soc.*, **302** (1987) 683–714.
24. Localization of a theorem of Ambos-Spies and the strong anti-splitting property. *Archiv. für Math Logik Grundlagenforschung*, **26** (1987) 127–136.

25. Δ_2^0 degrees and transfer theorems, *Illinois J. Math.*, **31** (1987) 419–427.

26. *T*-degrees, jump classes and strong reducibilities, (with C. Jockusch). *Trans. Amer. Math. Soc.*, **301** (1987) 103–136.

27. Automorphisms and recursive structures, (with J.B.Remmel), *Z. Math. Logik Grund. der Math.*, **33** (1987) 339–345.

28. Two theorems on truth table degrees, *Proc. Amer. Math. Soc.*, **103** (1988), 281–287.

29. Recursively enumerable *m*-degrees and *tt*-degrees II: the distribution of singular degrees. *Archive for Mathematical Logic*, **27** (1988), 135–148.

30. Intervals and sublattices in the r.e. weak truth table degrees, Part I: density, *Ann. Pure and Appl. Logic*, **41** (1989) 1–27.

31. Completely mitotic r.e. degrees (with T. Slaman), *Ann. Pure and Appl. Logic*, **41** (1989) 119–153.

32. D-r.e. degrees and the nondiamond theorem. *Bull. London Math. Soc.*, **21** (1989) 43–50.

33. Classification of degree classes associated with r.e. subspaces, (with J.B. Remmel), *Ann. Pure and Appl Logic*, **42** (1989) 105–125

34. Degrees bounding minimal degrees, (with C.T. Chong), *Math.Proc.Cambridge Phil. Soc.*, **105** (1989) 211–222.

35. Honest polynomial reductions, non relativizations and $P = NP$ (with W. Gasarch, S. Homer and M. Moses) *Proceedings of the 4th Annual Conference on Structures in Complexity Theory*, (1989), IEEE Publ. 196–207.

36. Recursively enumerable *m*-degrees and *tt*-degrees I: the quantity of *m*-degrees. *J. Symb. Logic*, **54** (1989) 553–567

37. On choice sets and strongly nontrivial self-embeddings of recursive linear orders. (with M.F. Moses) *Z. Math. Logik Grundlagen Math.*, **35** (1989) 237–246.

38. Intervals and sublattices in the r.e. weak truth table degrees, Part II: nonbounding, *Ann. Pure and Appl. Logic*, **44** (1989) 153–172

39. On hyper-torre isols, *J. Symbolic Logic*, **54** (1989) 1160–1166

40. A contiguous nonbranching degree, *Z.Math Logic Grundlagen Math.*, **35** (1989) 375–383.

41. On Ramsey-type theorems and their applications, *Singapore Math. Medley*, **17** (1989) 58–78.

42. Lattice nonembeddings and initial segments of the recursively enumerable degrees, *Annals Pure and Appl. Logic*, **49** (1990) 97–119.

43. Automorphism and splittings of recursively enumerable sets, (with M Stob) in *Proceedings of the Forth Asian Logic conference, CSK publication*, Tokyo (1990) 75–87.

44. Array recursive sets and multiple permitting arguments (with M. Stob and C. Jockusch) in *Proceedings Oberwolfach 1989*, Springer Verlag, *Lecture Notes in Mathematics 1990*, 141–174.

45. Notes on the $0'''$ priority method with special attention to density theorems, in *Proceedings Oberwolfach 1989*, Springer Verlag, *Lecture Notes in Mathematics 1990*, 111–140.

46. Superbranching degrees, (with J. Mourad) in *Proceedings Oberwolfach 1989*, Springer Verlag, *Lecture Notes in Mathematics 1990*, 175–186.

47. Minimal degrees recursive in 1-generic degrees, (with C.T. Chong), *Annals Pure and Appl. Logic*, **48** (1990) 215–225.

48. On complexity theory and honest polynomial time degrees, *Theoretical Computer Science*, **78** (1991) 305–317.

49. Jumps of hemimaximal sets, (with M. Stob), *Z. Math. Logik Grundlagen Math.*, **37** (1991) 113–120.

50. Recursive linear orderings with incomplete successivities (with M.F. Moses), *Trans. Amer. Math.Soc.*, **326** (1991) 653–668.

51. Automorphisms of the lattice of recursively enumerable sets: Orbits (with M. Stob), *Advances in Math.*, **92** (1992) 237–265.

52. Tabular degrees and α-recursion theory, (with C. Bailey), *Annals Pure and Applied Logic*, **55** (1992),205–236.

53. Splitting theorems in recursion theory, (with M. Stob) *Annals Pure and Applied Logic*, **65** (1)((1993) 1–106).

54. Orderings with α-th jump degree 0^{α} (with J.F. Knight), *Proc. Amer. Math. Soc.*, **114** (1992) 545–552.

55. On Π_1^0 classes and their ranked points, *Notre Dame J. of Formal Logic*, **32** (1991) 499–512.

56. An invitation to structural complexity, *New Zealand Journal of Mathematics*, **21** (1992) 33–91.

57. On co-simple isols and their intersection types (with T. Slaman), *Annals pure and Appl. Logic*, (Special issue in honour of John Myhill), **56** (1992) 221–237.

58. Fixed parameter intractability, (with M. Fellows), *Proceedings Structure in Complexity, Seventh Annual Conference, IEEE Publication*, (1992) 36–50.

59. Fixed parameter tractability and completeness, (with M. R. Fellows), *Congressus Numerantium*, **87** (1992) 161–187.

60. Automorphisms of the lattice of recursively enumerable sets: promptly simple sets (with P. Cholak and M. Stob), *Trans. American Math. Society*, **332** (1992) 555–570.

61. Fixed parameter intractability II, (with K. Abrahamson and M. F. Fellows) in *Proceedings Tenth Annual Symposium on Theoretical Aspects of Computer Science* (STACS'93)(Ed. G. Goos and J. Hartmanis), Springer-Verlag Lecture Notes in Computer Science, Vol 665 (1993) 374–385.

62. Parameterized computational feasibility, (with M. Fellows) in *Feasible Mathematics II* (ed. P. Clote and J. Remmel) Birkhauser (1995) 219–244.

63. Nondiamond theorems for polynomial time reducibility, *Journal of Computing and System Sciences*, **45** (1992) 385–395.

64. Computability Theory and Linear Orderings, in *Handbook of Recursive Mathematics* (ed Ershov, Goncharov, Nerode and Remmel) Vol 2, North Holland, (1998), 823–977.

65. Effective algebras and relational systems; coding properties (with J.B. Remmel), in *Handbook of Recursive Mathematics* (ed Ershov, Goncharov, Nerode and Remmel) Vol 2, North Holland, (1998), 977–1041.

66. Friedberg splittings of recursively enumerable sets (with M. Stob), *Annals Pure and Applied Logic*, **59** (1993) 175–199.
67. Degrees of inferability (with P. Cholak, L. Fortnow, W. Gasarch, E. Kinber, M. Kummer, S. Kurtz, and T. Slaman), *Proceedings of Colt '92 (Fifth Annual Workshop on Computational Learning Theory, 1992)*, 180–192.
68. On the Cantor-Bendixon rank of recursively enumerable sets, (with P. Cholak), *J. Symbolic Logic*, **58** (1993) 629–640.
69. Countable thin Π_1^0 classes, (with D. Cenzer, C. Jockusch, and R. Shore), *Annals Pure and Applied Logic*, **59** (1993) 79–139.
70. Array nonrecursive sets and lattice embeddings of the diamond, *Illinois J. Mathematics*, **37** (1993) 349–374.
71. Embedding Lattices into the wtt-degrees below $\mathbf{0}'$, (with C. Haught), *J. Symbolic Logic*, **59** (1994) 1360–1382.
72. Effectively and noneffectively nowhere simple subspaces, (with J.B. Remmel), *Logical Methods* (ed. Crossley, Remmel, Shore, and Sweedler) Birkhauser, Boston, 1994, 314–351.
73. Permutation and presentations (with P. Cholak), *Proc. Amer. Math. Soc.* **122** (1994) 1237–1249.
74. Every recursive boolean algebra is isomorphic to one with incomplete atoms, *Annals Pure and Applied Logic*, **60** (1993) 193–206.
75. Fixed-parameter tractability and completeness II: on completeness for W[1] (with M. Fellows), *Theoretical Comput. Sci.* **141** (1995) 109–131.
76. Recursively enumerable m- and tt- degrees III: realizing all finite distributive lattices (with P. Cholak), *J. London Math. Soc.*, (2) **50** (1994) 440–453.
77. Fixed-parameter tractability and completeness III: some structural aspects of the W-hierarchy (with M. Fellows) in *Complexity Theory: Current Research* (Ed. K. Ambos-Spies, S. Homer and U. Schoning) Cambridge University Press, (1993) 166–191.
78. Lattice nonembeddings and intervals in the recursively emumerable degrees, (with P. Cholak), *Annals Pure and Applied Logic*, **61**, (1993), 195–222.
79. Decidability and definability for parameterized polynomial time m-reducibilities, (with P. Cholak) *Logical Methods*, (ed. Crossley, Remmel, Shore, and Sweedler) Birkhauser, Boston, 1994, 194–221.
80. On irreducible m-degrees, *Rendiconti Seminario Matematico Dell'Universita e Del Politecnico di Torino*, (Rend. Sem. Mat. Univ. Pol. Torino), **51**, (1993) 109–112.
81. Parameterized learning complexity, (with M. Fellows and P. Evans) in *Proceedings of the Sixth Annual Conference on Computational Learning Theory*, ACM Press, New York, (1993) 51–57.
82. The parameterized complexity of some problems in logic and linguistics, (extended abstract) (with M. Fellows, B. Kapron, M. Hallett, and T. Wareham), in *Proceedings of the Workshop on Structural Complexity and Recursion Theoretical Methods in Logic Programming*, October 29, Vancouver, Canada (ed. Blair, H., V. Marek, A. Nerode, J. Remmel) Mathematical Sciences Institute Publ. (1993) 44–58. Final version in *Logic at St. Petersberg* (Ed. A. Nerode and Yu. Matiyasevich) Springer Verlag Lecture Notes in Computer Science, Vol 813, (1994) 89–101

83. Every low boolean algebra is isomorphic to a recursive one, (with C. Jockusch) *Proceedings Amer. Math. Society*, **122**, No. 3, November 1994, pp. 871–880.

84. Highness and bounding minimal pairs, (with S. Lempp and R. Shore), *Mathematical Logic Quarterly* **39** (1993) 475–491.

85. Fixed-parameter tractability and completeness I: basic results, (with M. Fellows), *SIAM J. Computing* **24** (1995), 873–921.

86. There is no plus-capping degree, (with S. Lempp), *Archive for Mathematical Logic*, **33** (1994) 109–119.

87. Fixed-parameter tractability and completeness IV: on completeness for W[P] and *PSPACE* analogues, (with K. Abrahamson and M. Fellows), *Annals Pure and Applied Logic*, **73**, (1995) 235–276.

88. On the structure of parameterized problems in *NP* (extended abstract), (with L. Cai, J. Chen and M. Fellows,) in *Proceedings Eleventh Annual Symposium on Theoretical Aspects of Theoretical Computer Science, 1994* (STACS'94) (Ed. E. Mayr and k. Wagner), Springer-Verlag, Lecture Notes in Computer Science, Vol 775, (1994) 509–520. Final Version in *Information and Computation*, Vol 123 (1995) 38–49.

89. A rank one cohesive set, (with Yue Yang), *Annals Pure and Applied Logic*, **68** (1994) 161–171.

90. The parameterized complexity of sequence alignment and consensus, (extended abstract) (with H. Bodlaender, M. Fellows, and H. Todd Wareham), in *Combinatorial Pattern Matching*, (5th Annual Symposium, CPM'94, Asilomar June 1994) (Ed. Maxime Crochemore and Dan Gusfield) Lecture Notes in Computer Science Vol 807 (1994) 15–30.

91. Parameterized Complexity Analysis in Computational Biology, (with H. Bodlaender, M. Fellows, M. Hallett, and H. Todd Wareham), in *IEEE Computer Society Workshop on Shape and Pattern Matching in Computational Biology*, held in conjunction with the 1994 IEEE Conference on Computer Vision and Pattern Recognition, June 1994. Final Version appeared in *Computer Applications in the Biosciences*, **11**, (1995), 49–57

92. The Parameterized Complexity of Short Computation and Factorization, (with L. Cai, J. Chen and M. Fellows), Proceedings of the *Sacks Symposium*, in *Archive for Mathematical Logic,* Vol. 36, No 4/5 (1997), 321–338.

93. The structure of honest polynomial *m*-degrees, (with B. Gasarch and M. Moses), *Annals of Pure and Applied Logic*, **70** (1994), 1–27.

94. The parameterized complexity of sequence alignment and consensus, (with H. Bodlaender, M. Fellows and H. Todd Wareham), *Theoretical Computer Science*, **147**, (1995) 31–54.

95. Lattice embedding below a non-low$_2$ recursively enumerable degree, (with R. Shore), *Israel Journal of Mathematics*, **94** (1996), 221–246.

96. Degree theoretical definitions of the low$_2$ recursively enumerable sets, (with R. Shore), *J. Symbolic Logic*, **60** (1995) 727–756.

97. Advice classes of parameterized tractability, (with L. Cai, J. Chen, and M. Fellows), *Ann. Pure and Applied Logic*, **84** (1997), 119–138.

98. On the structure of parameterized problems in *NP*, (with L. Cai, J. Chen and M. Fellows,), *Information and Computation*, **123** (1995), 38–49.

99. Array nonrecursive degrees and genericity, (with C. Jockusch and M. Stob) in *Computability, Enumerability, Unsolvability,* (ed. Cooper, Slaman, and Wainer), London Mathematical Society Lecture Notes Series Vol 224, Cambridge University Press (1996), 93–105.

100. (with S. Lempp and R. Shore) Jumps of minimal degrees below $0'$ *Journal of the London Mathematical Society,* **54,** No. 2 (1996), 417–439.

101. There is no fat orbit, (with L. Harrington) *Annals of Pure and Applied Logic,* **80** (1996), 227–289.

102. Parameterized circuit complexity and the W-hierarchy, (with M. Fellows and K. Regan), *Theoretical Computer Science.* Vol. 191 (1998), 97–115.

103. Minimal pairs in initial segments of the recursively enumerable degrees, (with M. Stob), *Israel Journal of Mathematics,* **100** (1997), 7–27

104. On presentations of algebraic structures, in *Complexity, Logic and Recursion Theory,* (A. Sorbi, ed.), Marcel Dekker, Lecture Notes in Pure and Applied Mathematics, Vol. 197 (1997), 157–206.

105. There is no degree invariant half jump, (with R. Shore), *Proceedings of the American Mathematical Society.* Vol. 125 (1997), 3033–3037.

106. Contiguity and distributivity in the enumerable degrees, (with S. Lempp), *Journal of Symbolic Logic,* Vol. 62 (1997), 1215–1240. (Corrigendum in *ibid* Vol 67 (2002) 1579–1580.)

107. On the universal splitting property, *Mathematical Logic Quarterly,* **43** (1997) 311–320.

108. Intervals without critical triples, (with P. Cholak and R. Shore), in *Logic Colloquium '95,* (Ed. Johann A. Makowsky and Elena V. Ravve), Lecture Notes in Logic, Vol. 11, Springer-Verlag (1998) 17–43.

109. The parameterized complexity of relational database queries and an improved characterization of $W[1]$, (with M. Fellows and U. Taylor), in *Combinatorics, Complexity and Logic,* Proceedings of DMTCS '96, (D. Bridges, C. Calude, J. Gibbons, S. Reeves, I Witten, Eds) Springer-Verlag (1996), 194–213.

110. Difference sets and computability theory, (with Z. Feuredi, C. Jockusch, and L. Rubel), *Annals Pure and Applied Logic.* Vol. 93 (1998), 63–72.

111. Undecidability results for low complexity degree structures, (with A. Nies), in *Proceedings Complexity Theory, 12th Annual Conference,* (1997) 128–132.

112. Splitting Theorems and the Jump Operator, (with R. Shore) *Annals Pure and Applied Logic,* Vol. 94 (1998), 45–52.

113. On computing graph minor obstruction sets, (with Kevin Cattell Michael J. Dinneen, Michael R. Fellows, and Michael A. Langston) *Theor. Comp. Science.* Vol. 233 (2000), 107–127.

114. Threshold dominating sets and an improved characterization of $W[2]$, (with M. Fellows) *Theoretical Computer Science* Vol. 209 (1998), 123–140.

115. Descriptive complexity and the W-hierarchy, (with M. Fellows and K. Regan), in *Proof Complexity and Feasible Arithmetic* AMS-DIMACS Proceedings Series Vol. 39, (1998) (P. Beame and S. Buss, ed.), 119–134.

116. Effective presentability of Boolean algebras of Cantor–Bendixson rank 1, with C. Jockusch, *Journal of Symbolic Logic,* Vol. 64 (1999), 45–52.

117. Infima in the recursively enumerable weak truth table degrees (with R. Blaylock and S. Lempp) *Notre Dame J. Formal Logic* Vol 38 (1997), 406–419.

118. The Parameterized Complexity of Some Fundamental Problems in Coding Theory, (with M. Fellows, G. Whittle and Alexander Vardy) *SIAM J. Comput.* Vol. 29 (1999), 545–570.

119. A note on the computability of graph minor obstruction sets for monadic second order ideals, (with B. Courcelle and Mike Fellows), *Journal of Universal Computer Science,* Vol. 3 (1997), 1194–1198

120. Enumerable sets and quasi-reducibility, (with G. LaForte and A. Nies), *Annals of Pure and Applied Logic* Vol. 95 (1998), 1–35.

121. Parameterized complexity: a systematic method for confronting computational intractability, (with M. Fellows and U. Stege) *Contemporary Trends in Discrete Mathematics,* (R. Graham, J. Krachovil, J. Nesetril, and F. Roberts, eds) DIMACS Vol. 49, American Mathematical Society, (1999) 49–100.

122. Computational tractability: the view from mars, (with M. Fellows and U. Stege) *Bulletin of the European Association for Theoretical Computer Science,* No. 69, (1999), 73–97.

123. Parameterized complexity after (almost) 10 years: Review and Open Questions, (with Mike Fellows), in *Combinatorics, Computation & Logic*, Springer-Verlag, 1999, 1–33.

124. Uniformly hard languages (with L. Fortnow.) in *Complexity Theory, Annual Conference, 1998*, 228–235. Journal Version in *Theoretical Computer Science,* 298(2):303–315, 2003

125. Initial segments of computable linear orderings. (with B. Khoussainov and R. Coles) *Order* Vol. 14 (1997–1998), 107–124

126. Index sets and parametric reductions, (with M. Fellows) *Archive for Mathematical Logic.* Vol. 40 (2001), 329–348.

127. On the proof theoretical strength of the Dushnik-Miller theorem for countable linear orderings, (with S. Lempp) in, *Recursion Theorey and Complexity,* (Arslanov and Lempp, eds) de Gruyter, 1999, 55–58.

128. A Δ_2 set of barely Σ_2 degree, (with G. LaForte and S. Lempp) *Journal of Symbolic Logic*, Vol. 64 (1999), 1700–1718.

129. A note on btt-degrees, *Rendiconti Seminario Matematico Dell'Universita e Del Politecnico di Torino,* (Rend. Sem. Mat. Univ. Pol. Torino), Vol 58 (2000), 449–456.

130. The Complexity of Irredundant Sets Parameterized By Size, (with M. Fellows and V. Raman), *Discrete Applied Math.* Vol. 100 (2000), 155–167.

131. Undecidability results for low complexity time classes, (with A. Nies) *Journal of Computing and Sys. Sci.* Vol. 60 (1999), 465–479.

132. Every set has a least jump enumeration, (with R. Coles and T. Slaman) *Journal London Math. Soc.* (2) Vol. 62 (2000), 641–649.

133. Computability, Definability and Algebraic Structures, in *Proceedings of the 7th and 8th Asian Logic Conferences*, World Scientific, 2003, 63–102.

134. Some Orbits for $\mathcal{E}$, (with Peter Cholak and E. Herrmann) *Annals Pure and Applied Logic* Vol. 107 (2001), 193–226.

135. Genericity and Ershov's Hierarchy, (with Amy Gale) *Mathematical Logic Quarterly* Vol. 47 (2) (2001), 161–182.

136. A Δ_2^0 set with no low subset in it or its complement, (with D. Hirschfeldt, S. Lempp, and R. Solomon) *Journal of Symbolic Logic* Vol. 66 (2001), 1371–1381.

137. Questions in computable algebra and combinatorics, (with J. Remmel) in *Computability Theory and its Applications*, (Cholak, Lempp, Lermann, and Shore eds.), Contemporary Math, Vol. 257, AMS Publ., 2000, 95–126.

138. Automorphisms of the lattice of Π_1^0 classes: perfect thin classes and anr degrees, (with P. Cholak, R. Coles and E. Herrmann) *Trans. Amer. Math. Soc.* Vol. 353 (2001), 4899–4924.

139. Presentations of computably enumerable reals, (with G. LaForte) *Theoretical Computer Science* Vol. 284 (2002), 539–555.

140. Maximal contiguous degrees, (with P. Cholak and R. Walk) *Journal of Symbolic Logic* Vol. 67 (2002), 409–437.

141. Randomness, computability, and density, with D. Hirschfeldt and A. Nies, SIAM J. Comput. Vol. 31 (2002), 1169–1183. (Extended abstract appeared in *Symposium for Theoretical Aspects of Computer Science, STACS'01* January, 2001. Lecture Notes in Computer Science, Springer-Verlag, (Ed A. Ferriera and H. Reichel, 2001, 195–201)

142. Parameterized Complexity for the Skeptic, in *Computational Complexity, 18th Annual Conference*, IEEE, 2003, 147–169.

143. The reverse mathematics of the Nielson-Schrier Theorem, (with D. Hirschfeldt, S. Lempp, and R. Solomon) in *Proceedings International Conference on Mathematical Logic Honouring Ershov on his 60th Birthday and Mal'tsev on his 90th Birthday*, (ed. Goncharov et. al.), Novosibirsk, 2002, 59–71.

144. Uniformity in Computable Structure Theory, (with D. Hirschfeldt and B. Khoussainov), *Algebra and Logic* Vol. 42, No. 5, 2003, 318–332.

145. Degree Spectra of relations on boolean algebras, (with S. Goncharov and D. Hirschfeldt) *Algebra and Logic* Vol. 42, No. 2, 2003, 105–111.

146. Computably Enumerable Reals and Uniformly Presentable Ideals, (with S. Terwijn) *Archive for Mathematical Logic* Vol. 48 (2002), 29–40.

147. Decomposition and infima in the c.e. degrees, (with G. LaForte and R. Shore) *Journal of Symbolic Logic* Vol. 68 (2003), 551–579.

148. Computability-theoretical and proof-theoretical aspects of partial and linear orderings, (with D. Hirschfeldt, S. Lempp, and R. Solomon) *Israel J. Math.* vol 138, 2003, pp 271–289.

149. Complementing Cappable Degrees in the Difference Hierarchy (with A. Li and G. Wu), *Annals of Pure and Applied Logic*. Vol. 125 (2004), 101–118.

150. The Kolmogorov Complexity of Random Reals, with Yu Liang and Ding Decheng *Annals of Pure and Applied Logic* Volume 129, Issues 1–3, (2004), 163–180

151. Some Computability-Theoretical Aspects of Reals and Randomness, in *The Notre Dame Lectures*, (Peter Cholak, ed.) Vol. 18 in the *Lecture Notes in Logic* series published by the Association for Symbolic Logic. (2005) 97–146.

152. Hahn's Theorem, Archimedian Classes and reverse mathematics, (with Reed Solomon), *Reverse Mathematics, 2001* Lecture Notes in Logic Vol 21 (ed. S. Simpson). Association for Symbolic Logic, A. K. Peters, Massachussetts (2005) 147–163.

153. Randomness and reducibility, (with G. LaForte and D. Hirschfeldt) extended abstract appeared in *Mathematical Foundations of Computer Science, 2001* J. Sgall, A. Pultr, and P. Kolman (eds.), Mathematical Foundations of Computer Science 2001, Lecture Notes in Computer Science 2136 (Springer, 2001), 316–327). final version in *Journal of Computing and System Sciences*. Vol. 68 (2004), 96–114.

154. Trivial Reals, with D. Hirschfeldt A. Nies and F. Stephan, extended abstract in *Computability and Complexity in Analysis* Malaga, (Electronic Notes in Theoretical Computer Science, and proceedings, edited by Brattka, Schröder, Weihrauch, FernUniversität, 294-6/2002, 37–55), July, 2002. Final version appears in *Proceedings of the 7th and 8th Asian Logic Conferences*, World Scientific, 2003, 103–131.

155. Cutting Up Is Hard To Do: the Parameterized Complexity of k-Cut and Related Problems, (with Vladimir Estivill-Castro, Michael R. Fellows, Elena Prieto-Rodriguez and Frances A. Rosamond) *Electronic Notes in Theoretical Computer Science,* Vol. 78 (2003), 205–218.

156. Invariance and noninvariance in the lattice of Π_1^0 classes, (with P. Cholak), *Journal of the London Mathematical Society* (2) Vol. 70 (2004), 735–749.

157. On Schnorr randomness, (with Evan Griffiths), extended abstract in *Computability and Complexity in Analysis* Malaga, (Electronic Notes in Theoretical Computer Science, and proceedings, edited by Brattka, Schröder, Weihrauch, FernUniversität, 294-6/2002, 25–36)July, 2002. Final version in *Journal of Symbolic Logic*. Vol 69 (2) (2004), 533–554.

158. On Kurtz randomness, (with E. Griffiths and S. Reid) *Theoretical Computer Science* Vol. 321 (2004), 249–270.

159. On Schnorr and computable randomness, martingales, and machines, with E. Griffiths and G. LaForte. *Mathematical Logic Quarterly*. Vol. 50, No 6 (2004), 613–627.

160. Completing pseudojump operators, (with R. Coles, C. Jockusch and G. LaForte) *Annals of Pure and Applied Logic*. Vol. 136 (2005), 297–333.

161. Degrees of d.c.e. reals, (with G. Wu and X. Zheng) *Mathematical Logic Quarterly*. Vol. 50 (2004), 345–350.

162. There are no low maximal d.c.e. degrees, (with Yu Liang) *Notre Dame Journal of Formal Logic*, Vol. 45 (2004), 147–159.

163. Some recent progress in algorithmic randomness, in *Proceedings of the 29 th Annual Conference on Mathematical Foundations of Computer Science, Prague, August 2004* (invited paper) (J. Fiala, V. Koubek and J. Kratochvil eds), Springer-Verlag, Lecture Notes in Computer Science, Vol. 3153 (2004), 42–81.

164. A basis theorem for Π_1^0 classes of positive measure with applications to random reals, (with J. Miller) *Proceedings of the American Mathematical Society*. Vol. 134 (2006), 283–288.

165. (with Catherine McCartin) Online problems, pathwidth, and persistence, *Proceedings IWPEC 2004*. Springer-Verlag Lecture Notes in Computer Science 3162, pp 13–24, 2004.

166. On Self-Embeddings of Computable Linear Orderings (with C. Jockusch and J. Miller) *Annals of Pure and Applied Logic* Volume 138, Issues 1–3 (2006), 52–76.

167. Bounded persistence pathwidth, (with Catherine McCartin) in *Proceedings of CATS 2005, Computing, The Australasian Theory Symposium,* Conferences in Research and Practice in Information Technology, Vol 41, pp 51–56, 2005.

168. Some New Directions and Questions in Parameterized Complexity, (with Catherine McCartin), *Developments in Language Theory, 2004*, (Edited by C. Calude, E. Calude, and M. Dinneen) Springer Verlag Lecture Notes in Computer Science Vol. 3340 (2004), 12–26. (Invited Paper)

169. Schnorr dimension (with W. Merkle and J. Reimann) preliminary version appeared in *Computability in Europe, 2005*, Springer-Verlag Lecture Notes in Computer Science Vol. 3526 (2005) 96–105. Final version in *Mathematical Structures in Computer Science* Vol. 16 (2006), 789–811.

170. Five Lectures on Algorithmic Randomness, in *Computational Prospects of Infinity, Part I: Tutorials* (Ed. C. Chong, Q. Feng, T. A. Slaman, W. H. Woodin and Y. Yang) Lecture Notes Series, Institute for Mathematical Sciences, National University of Singapore Vol 14, World Scientific, Singapore, 2008, 3–82.

171. Relativizing Chaitin's Halting Probability (with D. Hirschfeldt, J. Miller, and A. Nies) *Journal of Mathematical Logic.* Vol. 5 (2005), 167–192.

172. Arithmetical Sacks Forcing, (with Liang Yu) *Archive for Mathematical Logic*, Vol. 45, No 6. (2006), 715–720.

173. Totally $<\omega^\omega$-computably enumerable degrees and m-topped degrees, (with Noam Greenberg), in Proceedings of *TAMC 2006: Theory and Applications of Models of Computation,* (Jin-Yi Cai, S. Barry Cooper, and Angsheng Li, eds), Springer-Verlag Lecture Notes in Computer Science, Vol. 3959, (2006) 46–60.

174. Every 1-generic computes a properly 1-generic, (with Barbara Csima, Noam Greenberg, Denis Hirschfeldt, and Joe Miller), *Journal of Symbolic Logic.* Vol. 71, No. 4 (2006), 1385–1393.

175. Algorithmic randomness and computability, *Proceedings of the 2006 International Congress of Mathematicians,* Vol 2, *European Mathematical Society*, (2006), 1–26.

176. Lowness and Π_2^0 nullsets, (with Andre' Nies, Yu Liang and Rebecca Weber) *Journal of Symbolic Logic* Vol. 71 (2006), 1044–1052.

177. Prompt Simplicity, Array Computability and Cupping, (with Noam Greenberg, Joe Miller, and Rebecca Weber), *Computational Prospects of Infinity: Part II Presented Talks* (Ed. C. Chong, Q. Feng, T.A. Slaman, W.H. Woodin and Y. Yang) Lecture Notes Series, Institute for Mathematical Sciences, National University of Singapore Vol 15, World Scientific, Singapore, 2008, 59–79.

178. Lowness for Computable Machines, (with Noam Greenberg, Nenad Mihailovic', and A. Nies) *Computational Prospects of Infinity: Part II Presented Talks* (Ed. C. Chong, Q. Feng, T. A. Slaman, W. H. Woodin and Y. Yang) Lecture Notes

Series, Institute for Mathematical Sciences, National University of Singapore Vol 15, World Scientific, Singapore, 2008, 79–87.

179. Online promise problems with online width metrics, (with Catherine McCartin) *Journal of Computing and System Sciences*, Vol. 73, No. 1 (2007), 57–72.

180. Parameterized approximation problems, (with Mike Fellows and Catherine McCartin), in *International Workshop in Parameterized Complexity and Exact Computation, IWPEC, 2006*, (H. Bodlaender and M. Langston, Eds.), Springer-Verlag Lecture Notes in Computer Science, Vol. 4169 (2006), 121–129.

181. Calibrating randomness, (with Denis Hirschfeldt, Andre Nies, and Sebastiaan Terwijn) *Bulletin of Symbolic Logic*, Vol. 12 (2006), 411–491.

182. *The Sixth Lecture on Algorithmic Randomness, Logic Colloquium 2006*, (S.B. Cooper, H. Geuvers, A. Pillay, and J. Vaananen, eds) Lecture Notes in Logic, Cambridge University press/ASL Publ., 2009, 103–134.

183. Undecidability of the structure of the Solovay degrees of c.e. reals, (with D. Hirschfeldt and G. LaForte) *Journal of Comput. and Sys. Sci.* Vol. 73 (2007), 769–787

184. Subspaces of computable vector spaces, (with D. Hirschfeldt, A. Kach, S. Lempp, J. Mileti, and A. Montalbán). *Journal of Algebra*. Vol. 314 (2007), 888–894.

185. Ideals in computable rings, (with S. Lempp, and J. Mileti) *Journal of Algebra* Vol. 314 (2007), 872–887.

186. Bounded fixed-parameter tractability and reducibility, (with J. Flum, M. Grohe, and M. Weyer) *Annals of Pure and Applied Logic* Vol. 148 (2007), 1–19.

187. Parameterized Algorithms, (with Catherine McCartin) in *Handbook of Algorithms and Complexity, Second Edition* (M. Atallah and M. Blanton, eds), CRC press, Boca Raton, USA, 2009, pp 25-1 to 25-30.

188. Strong jump traceablilty I, the computably enumerable case (with Peter Cholak and Noam Greenberg) *Advances in Mathematics* Vol. 217 (2008), 2045–2074.

189. The upward closure of a perfect thin class, (with Noam Greenberg and Joe Miller), *Annals of Pure and Applied Logic*, Vol. 165 (2008), 51–58.

190. The Complexity of Orbits of Computably Enumerable Sets, (with Peter Cholak and Leo Harrington), *Bulletin of Symbolic Logic*. Vol. 14, No. 1 (2008), 69–87.

191. On the orbits of computably enumerable sets, (with P. Cholak and L. Harrington) *Journal of the American Mathematical Society* Vol. 21, No. 4 (2008), 1105–1135.

192. Slender Classes, (with A. Montalbán) *Journal of the London Mathematical Society*, Vol. 78, No. 1 (2008), 36–50.

193. Totally $< \omega$ computably enumerable degrees and bounding critical triples, (with Noam Greenberg and Rebecca Weber), *Journal of Mathematical Logic*, Vol. 7 (2007), 145–171.

194. Degree spectra of unary relations on $(\omega, \leq)$, (with Bhakhadyr Khoussianov, Joseph Miller and Liang Yu), in *proceedings of the International Congress in Logic, Methodology and Philosophy of Science, Beijing, 2007*. King's College Publications, London, 2009, 36–55.

195. (with A. Montalbán) The isomorphism problem for torsion-free abelian groups is analytic complete, *Journal of Algebra*. Vol. 320 (2008), 2291–2300.

196. Parameterized approximation of dominating set problems, (with M. Fellows, C. McCartin and F. Rosamond), *Information Processing Letters* Vol. 109(1) (2008), 68–70.

197. Turing degrees of reals of positive effective packing dimension, (with Noam Greenberg) *Information Processing Letters*. Vol. 108 (2008), 298–303.

198. The Computer Journal Special Issue: Foreword by Guest Editors, (with Mike Fellows and Mike Langston), *The Computer Journal*, Vol. 51, No 1. (2008), 1–6.

199. On Problems without Polynomial Kernels (Extended Abstract), (with Hans Bodlaender, Mike Fellows, and Danny Hermelin) in Proceedings of *Automata, Languages and Programming, 35th International Colloquium, ICALP 2008*, Reykjavik, Iceland, July 7–11, 2008, Proceedings, Part I: Tack A: Algorithms, Automata, Complexity, and Games. Lecture Notes in Computer Science 5125 Springer 2008, 563–574. Final version in *Journal of Computing and System Sciences*. Vol. 75(No. 8): 423–434 (2009)

200. Working with strong reducibilities above totally ω-c.e. degrees, (with George Barmpalias and Noam Greenberg) *Transactions of the American Mathematical Society,* Vol. 362 (2010), 777–813.

201. K-trivial degrees and 'the jump traceability hierarchy, (with George Barmpalias and Noam Greenberg), *Proceedings of the American Mathematical Society*. Vol. 137 (2009), 2099–2109.

202. Kolmogorov complexity and Solovay functions, (with Laurent Bienvenu) in *Proceedings STACS 2009* in *Leibniz International Proceedings in Informatics* (Susanne Albers and Jean-Yves Marion, Eds) Schloss Dagstuhl - Leibniz-Zentrum fuer Informatik, Germany, 2009, 157–158.

203. On computable self embeddings of computable linear orderings, (with Bart Kastermans and Steffen Lempp) *Journal of Symbolic Logic*, Vol. 74 (4) (2009) 1352–1366.

204. Space Complexity of Abelian Groups (with D. Cenzer, J.B. Remmel and Z. Uddin) *Archive for Math. Logic*, Vol. 48 (2009), 63–76.

205. Lowness for Demuth randomness, (with Keng Meng Ng) in *Mathematical Theory and Computational Practice*, (Proceedings CiE 2009) Ambos-Spies, Löwe, Merkle, eds, Springer-Verlag, LNCS 5635, (2009) 154–166.

206. Limits on jump inversion for strong reducibilities, (with Barbara Csima and Keng Meng Ng), *Journal of Symbolic Logic*. Vol. 76, No. 4 (2011), 1287–1296.

207. Effective packing dimension and traceability, (with Keng Meng Ng), *Notre Dame Journal of Formal Logic*. Vol. 51(2) (2010), 279–290.

208. Jump inversions inside effectively closed sets with applications to randomness, (with George Barmpalias and Keng Meng Ng), *Journal of Symbolic Logic*. Vol. 74, No. 2 (2011), 491–508.

209. Binary trees with few labelled paths, with Noam Greenberg, Carl Jockusch and Kevin Milans. *Combinatorica* Vol. 31, No 3 (2011), 285–303.

210. Extensions of embeddings below computably enumerable degrees, (with Noam Greenberg, Andrew Lewis, and Antonio Montalbán) *Transactions of the American Mathematical Society* Vol. 365 (2013), 2977–3018.

211. Algorithmic randomness and complexity, in *Randomness Through Computation*, (H. Zenil, Ed) World Scientific, Singapore, 2011, 223–241.

212. Decidability and computability of certain torsion-free Abelian groups, (with Asher Kach, Sergei Goncharov, Julia Knight, Oleg Kudinov, Alexander Melnikov, and Dan Turetsky) *Notre Dame Journal of Formal Logic.* Vol. 51 (2010), 85–97.

213. Euclidean functions of computable Euclidean domains, (with Asher Kach), *Notre Dame Journal of Formal Logic.* Vol. 52 (No 2) (2011), 163–172.

214. Limitwise monotonic functions and their applications, (with Asher Kach and Dan Turetsky), in *Proceedings of the 11th Asian Logic Conference, 2009.* (Ed. T. Arai, Q. Feng, B. Kim, G. Wu and Y. Yang) World Scientific, 2012, 59–87.

215. On the complexity of the successivity relation in computable linear orderings, (with Steffen Lempp and Guohua Wu) *Journal of Mathematical Logic.* Vol. 10 (2010), 83–99.

216. Strong jump traceability II: K-triviality. (with Noam Greenberg) *Israel Journal of Mathematics*, Vol. 191 (2012), 647–667.

217. Confronting intractability via parameters, (with Dimitrios Thilikos) *Computer Science Review.* Vol. 5 (2011), 279–317

218. Effective categoricity of abelian groups, (with Alexander Melnikov) *Journal of Algebra* Vol. 373, (2013), 223–248.

219. Computability Theory, Algorithmic Randomness and Turing's Anticipation, in *Turing's Legacy*, (R. Downey, Ed), Cambridge University Press, 2014. pp 90–123.

220. Computable Categoricity Versus Relative Computable Categoricity, (with Asher Kach, Steffen Lempp, and Dan Turetsky) *Fundamentae Mathematica.* Vol. 221 (2013), 129–159.

221. Characterizing Lowness for Demuth Randomness, (With Bienvenu, Greenberg, Nies and Turetsky), *Journal of Symbolic Logic.* Vol. 79 (2014), 526–56.

222. Lowness for Bounded Randomness, (with Selwyn Ng), *Theoretical Computer Science A*, Vol. 460 (2012), 1–9.

223. Pseudo-jump inversion, upper cone avoidance, and strong jump-traceability, (with Noam Greenberg) *Advances in Mathematics,* Vol. 237 (2013), 252–285.

224. Bounded randomness (with Paul Brodhead and Selwyn Ng), *Festschrift for the 60th Birthday of Cris Calude* (M. Dinneen, B. Khoussainov eds), Lecture Notes in Computer Science, Vol. 7160, Springer-Verlag, 2012, 59–70.

225. A Parameterized complexity tutorial, in *Proceedings, Language and Automata Theory and Applications, LATA 2012*, (ed A. Dediu and C. Martin-Vide) Springer-Verlag LNCS 7183 (2012), 38–56.

226. A basic parameterized complexity primer, *The Multivariate Algorithmic Revolution and Beyond, Essays Dedicated to Michael R. Fellows on the Occasion of His 60th Birthday,* Lecture Notes in Computer Science, Vol. 7370 (ed. Bodlaender, Downey, Fomin and Marx) Springer-Verlag LNCS 7370, 2012, 91–128

227. The birth and early years of parameterized complexity, *The Multivariate Algorithmic Revolution and Beyond, Essays Dedicated to Michael R. Fellows on the Occasion of His 60th Birthday,* Lecture Notes in Computer Science, Vol. 7370 (ed. Bodlaender, Downey, Fomin and Marx) Springer-Verlag LNCS 7370, 2012, 17–38.

228. Asymptotic density and computably enumerable sets, with Jockusch and Schupp. *Journal of Mathematical Logic* Vol. 13 (2013), 43 pages.

229. Completely decomposable groups and arithmetical categoricity, (with Alexander Melnikov), *Transactions of the American Mathematical Society*. Vol. 366 (2014), 4243–4266

230. Randomness, Computation and Mathematics, In *How the World Computes: The Turing Centenary Conference*, S.B. Cooper, A.Dawar, B. Löwe (eds.), CiE 2012, Lecture Notes in Computer Science 7318: Springer, Heidelberg, 2012, 162–181.

231. The complexity of computable categoricity, (with Asher Kach, Steffen Lempp, Andrew Lewis, Antonio Montalbán and Dan Turetsky) *Advances in Mathematics* Vol. 268 (2015), 423–66

232. Dynamic Dominating Set and Turbo-Charging Greedy Heuristics, (with Judith Egan, Michael Fellows, Frances Rodamond, and Peter Shaw) *Tsinghua Science and Technology*, Vol. 19 (2014), 329–337.

233. Resolute Sequences in Initial Segment Complexity, (with George Barmpalias) in *Proceedings of the 11th Asian Logic Conference*, (Ed. Downey, Brendle, Goldblatt, Kim), World Scientific, 2013, 1–23.

234. Exact pairs for the ideal of the K-trivial sequences in the Turing degrees, (with George Barmpalias), *Journal of Symbolic Logic*, Vol. 79, (2014), 676 –692

235. Asymptotic density and the Ershov hierarchy, with Carl Jockusch, Timothy McNicholl, and Paul Schupp *Mathematical Logic Quarterly* Vol. 61, Issue 3, (2015) 189–195.

236. Solovay functions and their applications to algorithmic randomness, (with L. Bienvenu, W. Merkle, A. Nies) *Journal of Computing and System Sciences* Volume 81, Issue 8, December 2015, 1575–1591.

237. Integer valued betting strategies and Turing degrees, (with Barmpalias and McInerney), *Journal of Computing and System Sciences* Volume 81, Issue 7, November 2015, Pages 1387–1412.

238. Iterated effective embeddings of abelian *p*-groups, (with Melnikov and Ng), *International Journal of Algebra and Computation*. Vol. 24 (2014) 1055–1084.

239. Courcelle's Theorem for Triangulations, (with Benjamin Burton) accepted *Journal of Combinatorial Theorey A*

240. Computability theory, complexity theory, reverse mathematics and algorithmic randomness, in *The Human Face of Computing*, C. Calude, ed., Imperial College Press, Advances in Computer Science and Engineering Texts, Vol. 9, 2015, 203–224.

241. The Finite Intersection Property and Genericity (with Diamondstone, Greenberg and Turetsky), *Mathematical Proceedings of the Cambridge Philosophical Society*. Vol. 160 (02), 2016, 279–297

242. Random strings and truth table degrees of Turing complete c.e. sets (with M. Cai, R. Epstein, S. Lempp and J. Miller) *Logical Methods in Computer Science*. 10 (3):15, 24 pp., 2014.

243. Minimal pairs in the c.e. truth table degrees, (with Keng Meng Ng) in *Proceedings of the 13th Asian Logic Conference*, ed X. Zhao, Q. Feng, B. Kim and L. Yu) World Scientific 2015, 53–67.

244. Abelian p-groups and the halting problem, (with Melnikov and Ng), accepted *Annals of Pure and Applied Logic*.
245. On Δ_2^0 categoricity of Equivalence Relations, (with Melnikov and Ng), *Annals of Pure and Applied Logic*, (2015), 166(9), pp. 851–880.
246. Generic Muchnik reducibility and presentations of fields, (with N. Greenberg and J. Miller), accepted *Israel Journal of Mathematics*.
247. Myhill-Nerode Methods for Hypergraphs, (with M. Fellows, S. Gaspers, F. Rosamond, R. van Bevern) *Algorithmica* Vol. 73(4):696–729, 2015
248. Any FIP real computes a 1-generic (with Peter Cholak and Greg Igusa) accepted, *Transactions of the American Mathematical Society*
249. Turing and Randomness, for Alan Turing in Perspective, volume of Cambridge University Press, 2016.
250. The members of thin and minimal Π_1^0 classes, their ranks and Turing degree s, (with Y. Yang and G. Wu) Annals of Pure and Applied Logic, Vol. 166 (2015), 741–754.
251. Avoiding effective packing dimension 1 below array noncomputable c.e. degrees (with Jonny Stephenson) submitted.
252. Multiple recurrence and algorithmic randomness, (with Satyadev Nandakumar and André Nies) submitted.
253. A minimal degree computable from a weakly 2-generic one, (with Satyadev Nandakumar), submitted.
254. A Friedberg Enumeration of Equivalence Structures, (with Melnikov and Ng) submitted.
255. Kobayashi compressibility, (with George Barmpalias), submitted.

Papers in Preparation

256. Decompositions of c.e. sets and degrees, with Steffen Lempp and Guohua Wu.
257. Lower bounds for the SJT hard sets, (with David Diamondstone, Noam Greenberg and Dan Turetsky)
258. On minimal wtt degrees and computably enumerable Turing degrees, (with Reed Solomon, and Keng Meng Ng) being written.
259. Degrees containing members of thin classes are dense and co-dense. (with Wu and Yang) being written.
260. A hierarchy of degrees, (with Noam Greenberg)
261. Splitting Theorems for computably enumerable degrees, (with Keng meng Ng), in preparation.
262. Index sets and Π_1^0 classes, (with Ng and Csima) being written.
263. Abelian Groups Categorical Relative to the Halting Problem, (with Melnikov and Ng)

Professional Societies

I am a member of the following societies American Math. Society (life member), Australian Math. Society, New Zealand Math. Society, European Association for Theoretical Computer Science, London Math. Society, Association for Symbolic

Logic, (Council) Combinatorial Mathematics Society of Australasia (life member), Royal Society of New Zealand, Wellington Mathematical Society, The Association for Computing Machinery (life member).

Other Information

I am a reviewer for Mathematical Reviews, Zentralblatt für Mathematik, and the Journal of Symbolic Logic. For these I have over 300 reviews including 14 book reviews. I am a referee for various journals such as the Journal of Symbolic Logic, Annals of Pure and Applied Logic, the Transactions of the American Math. Society, the Archive for Mathematical Logic, Theoretical Computer Science, Journal of Computer and System Sciences, Journal of Computing and System Sciences, SIAM Journal of Computing, Journal of Graph theory, and the Australasian Journal of Combinatorics. I have been on numerous programme committees for Computer Science and logic meetings.

I have given numerous invited addresses at international meetings and colloquia. For instance, here are some recent invited lectures:

- 2005 February Plenary Speaker at the UCLA Meeting for the opening of its Logic Center, sponsored by the ASL, NSF and UCLA.
- 2005 May. Invited speaker at University of Chicago for 4 lectures on algorithmic randomness, whilst Visiting Scholar.
- 2005 July-August. One of the only 2 invited Tutorial Speakers (in computability theory, the second month, the other being Ted Slaman at Berkeley) at the 2 month meeting Computational Prospects of Infinity, Singapore. Five Lectures on Algorithmic Randomness.
- 2005 September Invited Speaker at the 16th Australasian Workshop on Combinatorial Algorithms
- 2006 May. Plenary speaker at Theory and Applications of Models of Computation, Beijing.
- 2006 July. Tutorial Speaker at The European Logic Colloquium, Nijmegen, Holland. Three Lectures.
- 2006 August. Invited Speaker, International Congress of Mathematicians, Madrid.
- 2007 August. Invited Lecture, International Congress of Logic Methodology and Philosophy of Science, Beijing.
- 2007 December. Plenary Lecture, First Joint Meeting of the New Zealand Mathematical Society and the American Mathematical Society, Wellington.
- 2008 February, Tutorial speaker, NZIMA Algorithmics Meeting, Napier.
- 2008 March, Invited speaker, American Math. Society Special Session on Computability, Irvine.
- 2008 June, Invited Speaker, Logic Computability and Randomness, Nanjing, China.
- 2008 December, Invited Speaker, special session on algorithmics, NZMS/Aust MS annual Meeting, Christchurch.
- 2009 February, Royal Society Invited Speaker for Rutherford Foundation Dinner, Wellington Town Hall.
- 2009 May, Invited Speaker, Algorithmic Randomness Meeting, Madison.

- 2009 June, Plenary Speaker, Asian Logic Meeting, Singapore.
- 2010 May, Invited Speaker, Midwest Computability Seminar, University of Chicago.
- 2010 May, Plenary Speaker, 5th Logic, Computability and Randomness Conference, University of Notre Dame, USA.
- 2011 February, Plenary Speaker, 6th Computability and Randomness Conference, Cape Town, South Africa.
- 2011 July, Invited Speaker Computational prospects of Infinity II, National University of Singapore.
- 2012 January, Schloss Dagstuhl, Computability and Randomness
- 2012 February, Oberwolfach, Computability
- 2012 March, Tutorial Speaker, Language, Automata Theory and Applications, A Coruna, Spain.
- 2012, April, University of Leicester.
- 2012, April, Tutorial Speaker, British Computer and Theoretical Computer Science, University of Manchester. (London Mathematical Society Discrete Mathematics Keynote Speaker)
- 2012, June, Special Session Speaker, The Incomputable, Chichley Hall.
- 2012, June, Plenary Lecture, How The World Computes-The Turing Centenary Conference, CIE, Cambridge, UK.
- 2012, June, Data Reduction and Problem Kernals, Schloss Dagstuhl.
- 2012, July, plenary speaker, Computability and Randomness, Isaac Newton Institute, UK
- 2012, August, Plenary Speaker, Turing Memorial Programme, Palacio De La Magdalena, Santander, Spain.
- 2012, October, Alan Turing, the Birth of Computers and the Power of Mathematics, Public Lecture, Victoria University.
- 2012, November, Seminar on parameterized complexity, Cornell University.
- 2012, November, Seminar on the Finite Intersection property.
- 2012, November, Harvard/MIT logic seminar. Finite Intersection property.
- 2012, November, Plenary Lecture, Alan Turing Centenary Conference TURING 100, Boston University.
- 2012, December, Plenary Lecture, Midwest Computabililty Seminar, University of Chicago.
- 2013, January, Plenary Lecture: My Mathematical Encounters with Anil Nerode, Logical Foundations of Computer Science, San Diego.
- 2013, January, Parameterized complexity basics, Joint Meetings special session on incremental and multivariate computation.
- 2013, January, Effective Torsion-Free abelan groups, special session on Effective Mathmatics, Joint meetings, San Diego.
- 2013, April, Effectivity in Abelian Group Theory, Kobe University.
- 2013, May, What have I been thinking about in parameterized Complexity, Shonan Village Conference Center, Japan.
- 2013, May, Alan Turing and Randomness, Workshop on Information Theory and Randomness, invited Lecture, University of Tokyo.

- 2013, May, Integer Valued Randomness, Workshop on Information Theory and Randomness, invited Lecture, University of Tokyo.
- 2013, May, Recent Progress in Multivariate Algorithmics, Colloquium Lecture, University of Auckland.
- 2014, May, Integer Valued Randomness, Invited Lecture, Midwest Computability (Chicago)
- 2014, May, Effectivity in Abelian Group Theory, University of Notre Dame.
- 2014, Courcelle's Theorem for Triangulations, Invited Lecture, Subfactors in Mathematics and Physics, Maui.
- 2015, Alan Turing, Computing, Bletchley, and Mathematics, Public Lecture.
- 2015, Courcelle's Theorem for Triangulations, Invited Lecture, TAMC, Singapore.
- 2015, June Courcelle's Theorem for Triangulations, Invited Lecture, Computability, Probability and Logic, Radboud University, Nijmegen.
- 2015, June Computability in Mathematics-Turing's Legacy MATCH Kolloquium Lecture, University of Heidelberg.
- 2015, July, Alan Turing, Computing, Bletchley, and Mathematics, Singapore Public Library, Singapore (Public Lecture).
- 2015, April, The Life of π, CAPT Masters Lecture Singapore
- 2016, April, Parameterized Complexity, Chinese Academy of Sciences Colloquium, and Tsinghua University.
- 2016, June, Logic for Algorithms, University of Montpellier, colloquium lecture
- 2016, June, The Computational Power of Random Strings, Plenary Lecture, Luminy Conference Center, Computability, Complexity and Randomness.

I review for various granting bodies such as the New Zealand-U.S. Cooperative Science Foundation, and the United States National Science Foundation (both in Mathematics and Computer Science), European Research Council, Saudi Arabian Granting Agency, NSERC (Canada), EPSRC, the Irish NSF, South African NSF, and the Chinese NSF.
From 1987–1996 I edited the New Zealand Mathematical Society publication "Postgraduate Topics in Mathematics and Related Areas".
I have been a member of the Publications Committee of the New Zealand Mathematical Society, the Committee of the Wellington Mathematical Society, was a member of the Board of Governors of Newlands College from 1987–1989, and am currently a member of the Council and the Australasian Committee of the Association for Symbolic Logic. I served on the National Committee for Mathematics of the Royal Society from 1992–1995, and 2000-. In 1991 I was on the organising committee of the New Zealand Association of Mathematics Teachers Biennial Conference. In 2000, I organized the major NZMRI summer meeting in Kaikoura. In 2002, I am co-organizing the NZMRI meeting in New Plymouth. I co-organized the VIC 2004 meeting in Wellington, and 4 conferences on parameterized complexity and exact computation, such as Dagstuhl 2005. I have been on many (>50) conference committees for computer science conferences. I organized the Asian Logic Meeting in Wellington in December 2011. I co-organized the Dagstuhl "Computability" in 2017, and earlier ones in Oberwolfach 2012.

In 1997, 2001, 2004 I was on the Marsden Mathematical and Information Sciences panel. I chaired the panel from 2008–2011. I have served on the Royal Society of New Zealand Fellows Committee for Mathematical and Information Sciences, and as the Fellows representative on the New Zealand National Mathematical & Information Sciences Advisory Group.

My name appears in -Marquis "Who's Who in the World", and "Who's Who Aotearoa".

Contents

Contents   XLI

There Are No Maximal d.c.e. *wtt*-degrees 479
  Guohua Wu and Mars M. Yamaleev

A Rigid Cone in the Truth-Table Degrees with Jump 487
  Bjørn Kjos-Hanssen

Asymptotic Density and the Theory of Computability: A Partial Survey..... 501
  Carl G. Jockusch Jr. and Paul E. Schupp

On Splits of Computably Enumerable Sets........................... 521
  Peter A. Cholak

1-Generic Degrees Bounding Minimal Degrees Revisited.............. 536
  C.T. Chong

Nondensity of Double Bubbles in the D.C.E. Degrees................ 547
  *Uri Andrews, Rutger Kuyper, Steffen Lempp, Mariya I. Soskova,
  and Mars M. Yamaleev*

On the Strongly Bounded Turing Degrees of the Computably
Enumerable Sets... 563
  Klaus Ambos-Spies

Permutations of the Integers Induce only the Trivial Automorphism
of the Turing Degrees 599
  Bjørn Kjos-Hanssen

Algorithmic Randomness

On the Reals Which Cannot Be Random 611
  Liang Yu and Yizheng Zhu

A Note on the Differences of Computably Enumerable Reals 623
  George Barmpalias and Andrew Lewis-Pye

Effective Bi-immunity and Randomness............................ 633
  Achilles A. Beros, Mushfeq Khan, and Bjørn Kjos-Hanssen

On Work of Barmpalias and Lewis-Pye: A Derivation on the D.C.E. Reals ... 644
  Joseph S. Miller

Turing Degrees and Muchnik Degrees of Recursively Bounded
DNR Functions.. 660
  Stephen G. Simpson

Algorithmic Statistics: Forty Years Later 669
  Nikolay Vereshchagin and Alexander Shen

Memories, Complexity and Horizons

Cameo of a Consummate Computabilist

Robert Goldblatt[✉]

Victoria University of Wellington, Wellington, New Zealand
Rob.Goldblatt@msor.vuw.ac.nz

Rod Downey took up a lectureship in mathematics at the Victoria University of Wellington[1] in 1986. At the time he was a promising young researcher with a dozen papers to his name. Thirty years on, the dozen has mushroomed into 250-plus as he has developed into a world leader in his field who has made a profound contribution to the mathematical research environment both internationally and in New Zealand. This article briefly outlines his exceptional career.

1 Beginnings

Rod grew up in Brisbane, Australia. An interest in logic developed early: his high school offered the subject, but only to the bottom class. He opted to move down to take it. After graduating with first class honours in mathematics from the University of Queensland in 1978, he became a research student at Monash University in Melbourne. There he completed a PhD in 1982 with a thesis entitled *Abstract Dependence, Recursion Theory and the Lattice of Recursively Enumerable Filters*, supervised by John Crossley. Before settling in Wellington he held a number of postdoctoral positions, including at the National University of Singapore with Chong Chi Tat and at the University of Illinois at Urbana-Champaign with Carl Jockusch.

2 Research

Rod is a pre-eminent authority on many aspects of the theory of computability, including the structure of the computably enumerable degrees, computable algebra, and complexity theory. His work on such topics will be discussed by other articles in this volume; here we mention two major aspects.

Firstly, together with Mike Fellows he founded the field of *parameterized complexity* which has shown that apparently intractable computations can become feasible once fixed values are given to certain fundamental parameters, such as the size of the object to be computed. This has grown into an important branch of theoretical computer science, with its own conferences, special sessions, and special issues of journals. There are already several books on it by

---

R. Goldblatt—Thanks to Noam Greenberg and Gaven Martin for information and comments.

[1] Founded as Victoria College in 1897 in celebration of the Diamond Jubilee year of the reign of Queen Victoria.

© Springer International Publishing AG 2017
A. Day et al. (Eds.): Downey Festschrift, LNCS 10010, pp. 3–8, 2017.
DOI: 10.1007/978-3-319-50062-1_1

others, in addition to the pioneering and prodigious Downey & Fellows 530-page monograph from 1999 [1], and its completely rewritten and updated 760-page successor of 2013 [6]. These are surely some of the most cited academic works ever produced in New Zealand, with more than 3700 Google Scholar citations. Applications of the theory have appeared in numerous areas, e.g. computational biology, databases, information and coding theory, linguistics, circuit complexity, and approximation theory. Like Eilenberg and Mac Lane, or Jagger and Richards, the Downey-Fellows partnership has been a brilliantly successful and enduring one, in their case cemented by a shared devotion to surfing and good wine. Rod has recounted his own story of the origins of their work in [2].

Secondly, he initiated a comprehensive development of the study of *algorithmic randomness*, a subject that lies at the junction of computability theory, measure theory, and information theory. This resulted in another monumental (850-page) research monograph [7], co-authored with Denis Hirschfeldt, giving a unifying treatment of several related but historically separate approaches to the question of what makes a sequence of numbers or other symbols random. One connection to computability theory comes through the idea that a sequence can be viewed as being random if there is no short description for it, i.e. any computer program that generates a segment of the sequence must be as long, or as complex, as the segment itself. There has been an explosion of work in this area in recent years, led by Rod with an army of co-authors, former students and post-doctoral supervisees.

3   Collaboration and Supervision

Rod's working style is highly gregarious, involving extensive collaboration and mentoring. The MathSciNet database records over 100 different Downey co-authors. He has travelled the world in a seemingly endlessly fashion, acquiring airpoints Platinum Status for life as he pursues new contacts and research associates. A consequence has been the attracting to New Zealand of a steady stream of visitors who have contributed greatly to the research culture here.

Particularly unique is his record of supervision of post-doctoral fellows, demonstrating his special talent for training others to do advanced research. Many newly graduated PhD's in logic and computability from the USA and Europe have chosen to begin their post-doctoral careers gaining the experience and benefit of Rod's masterful guidance. He has used his long series of grants from the Marsden Fund to host 20 such fellows so far, maintaining a strong oversight of their work and high expectations that go beyond the mathematical. Their duties include showing up for Friday lunch at a downtown noodle shop, and trying out their supervisor's Scottish country dancing classes. When Rod's sons Carlton and Alex were small they thought 'post-doc' was synonymous with 'child-minder'.

Many of the post-docs have become long-term friends and research collaborators. Rod has worked assiduously to help them find positions around the world and move on to the next stage of their careers. Among them is Noam Greenberg who stayed on to take up a lectureship at Victoria and, in a similar vein to his mentor, has achieved rapid promotion to Professor. Another, George Barmpalias, has recently returned to Wellington as a lecturer in computational mathematics.

In several cases Rod interacted with students at other universities before they graduated, and gave them ideas for projects; some of these later became postdoctoral fellows with him. At Victoria itself, four students to date have completed PhD's under Rod's supervision and gone on to establish their own academic careers. A fifth has just completed. Three of those first four received the NZ Royal Society's Hatherton Award for the best scientific paper by a PhD student at any New Zealand university in physical, earth, or mathematical and information sciences. One of them, Adam Day, was also awarded the 2011 Sacks Prize from the international Association for Symbolic Logic for the best PhD thesis in logic of its year worldwide. After a couple of years as a Miller Fellow at UC Berkeley, Adam returned to a lectureship at Victoria.

These developments have ensured that Victoria's long tradition of high quality research in logic will continue well into the future.

4 Downey Descendants

Graduate Students
> Wu Guohua (PhD 2002),
> Catherine McCartin (PhD 2003),
> Stephanie Reid (MSc 2003),
> John Fouhy (MSc 2003),
> Ng Keng Meng (Selwyn) (PhD 2009),
> Adam Day (PhD 2011),
> Katherine Arthur (MSc, 2016),
> Michelle Porter (MSc 2016),
> Michael McInerney (PhD 2016).

Postdoctoral Fellows
> Michael Moses,
> Peter Cholak,
> Geoff LaForte,
> Richard Coles,
> Reed Solomon,
> Walker White,
> Denis Hirschfeldt,
> Evan Griffith,
> Wu Guohua,
> Joseph Miller,
> Yu Liang,
> Rebecca Weber,
> Noam Greenberg,
> Antonio Montalbán,
> George Barmpalias,

Laurent Bienvenu,
Asher Kach,
Daniel Turetsky,
Alexander Melnikov,
Gregory Igusa.

5 NZ Environment

Rod's time in New Zealand has coincided with a revolution in the local resourc-
ing of mathematical research. In his early years in the 1980's there was very little
support available and he struggled to find ways to get overseas and to bring in
visitors. This changed dramatically in the 1990's as the public scientific research
establishment underwent a radical re-structuring. Sir Ian Axford's vision and
enterprise led to the creation of the Marsden Fund[2], which for the first time provided grants for fundamental "investigator driven" research, open to nationwide
competition. A crucial factor was the recognition of the mathematical and information sciences (MIS) as a separate discipline area for allocation purposes. It
made all the difference that mathematicians now had their applications subject
to *peer* evaluation, rather than being treated as the poor relation by committees
made up of people from the laboratory sciences and other areas.

Rod became a director (unpaid) of a new NZMRI: New Zealand Mathematics Research Institute, along with a number of other prominent mathematicians
(Marston Conder, David Gauld, Vaughan Jones, Gaven Martin). The NZMRI
has been wonderfully effective in organising an annual series of summer meetings at exotic coastal locations, still on-going after more than 20 years, that has
brought some of the world's best mathematicians to New Zealand to lecture
on topics of current interest and interact with local researchers and graduate
students. As well as providing leadership and involvement in all these events,
Rod himself set-up and ran the meeting in Kaikoura in 2000 on the topic of
Computability, Complexity, and Computational Algebra, attracting more than
100 participants (see [5] for proceedings). In another capacity he was similarly
successful in organising the large 12th Asian Logic Conference at Victoria in
2011 [4].

Other local leadership contributions have including serving as President of
the NZ Mathematical Society and being a member and chair of the Marsden
Fund's MIS evaluation panel for several years. He also served on the governing
board of the NZ Institute of Mathematics ands its Applications (NZIMA), one
of the country's first Centres of Research Excellence, which existed during 2002–
2012 and did a great deal to distribute resources widely to the benefit of the
mathematics community.

---

[2] Named after Ernest Marsden, Professor of Physics at Victoria (1915–1922) and later
head of the Government's Department of Scientific and Industrial Research. As a
student at Manchester he had conducted the famous Geiger–Marsden experiments
that led Ernest Rutherford to conceive his nuclear model of the atom.

6   Editing and Expositing

This picture of a busy life would be incomplete without reference to the enormous contribution that Rod has made to the time-consuming process of evaluating and publishing scientific literature. He has been a member of uncountably many conference committees, and chair of many of them, including the international conference series *Computability, Complexity and Randomness (CCR)*, which he cofounded and which has held eleven conferences around the world since 2004. He has edited numerous books of conference proceedings and special issues of journals. The many functions he has performed for the Association for Symbolic Logic include being the coordinating editor of *The Journal of Symbolic Logic* and then *The Bulletin of Symbolic Logic* over a period of a decade altogether.

As a contribution to the Turing Centenary celebrations, Rod edited a collection of expert essays entitled *Turing's Legacy* [3]. He has given many popular public lectures on Turing's life and work: one in Wellington attracted an overflow audience. Other outreach activities having included talks on logic to a University of the Third Age audience, and on critical thinking to undergraduate college students in Singapore.

7   Recognition

Rod's achievements have brought him many honours and awards. Victoria University appointed him to a Personal Chair in 1995. The following year he was elected a Fellow of the Royal Society of NZ. In 2008 he became the only NZ mathematician to have been elected a Fellow of the Association for Computing Machinery (there is one other ACM Fellow in the country, a computer scientist). In 2012 he was one of three NZ-based mathematicians to be designated Inaugural Fellows of the American Mathematical Society. He is also a Fellow of the Australian Mathematical Society, and of course the NZ Mathematical Society.

The Royal Society of NZ has given him its Hamilton Prize for emerging researchers; a James Cook Fellowship providing support for two years of full-time research; and the Hector Medal, its highest award for research in the chemical, physical or mathematical sciences.

Rod was the first recipient of a MacLaurin Research Fellowship from the NZIMA. He has been continuously supported by large grants from the Marsden Fund since its inception, and is the only member of the mathematical sciences community to be granted continuous post-doctoral funding by Marsden.

He also holds the NZ Mathematical Society's Research Award, and the New Zealand Association of Scientists' Research Medal (for researchers under the age of 40). In 2006 he became the first NZ based mathematician to be invited to address the International Congress of Mathematicians. In 2010 he received the Shoenfield Prize, awarded once every three years by the Association for Symbolic Logic for outstanding expository writing in the field of logic, and in 2014 he was awarded the Nerode Prize by the European Association for Theoretical Computer Science for outstanding papers in the area of multivariate algorithmics.

8 Salutation

This sketch portrays a person of tremendous energy and drive, with a strong disposition to promote the interests of others as well as his own. These qualities perhaps help to explain how he was able to overcome an attack of cancer that might have overwhelmed a less robust character. His friends and colleagues congratulate him on his 60th birthday and wish him many more years of productive creativity and enjoyment of life. Those good wishes are extended also to his wife Kristin, who deserves recognition, if not sainthood, for her support of Rod and his work.

References

1. Downey, R.G., Fellows, M.R.: Parameterized Complexity. Monographs in Computer Science. Springer, New York (1999)
2. Downey, R.: The birth and early years of parameterized complexity. In: Bodlaender, H.L., Downey, R., Fomin, F.V., Marx, D. (eds.) The Multivariate Algorithmic Revolution and Beyond. LNCS, vol. 7370, pp. 17–38. Springer, Heidelberg (2012). doi:10.1007/978-3-642-30891-8_2
3. Downey, R. (ed.): Turing's Legacy: Developments from Turing's Ideas in Logic. Lecture Notes in Logic, vol. 42. Cambridge University Press and the Association for Symbolic Logic, Cambridge (2014)
4. Downey, R., Brendle, J., Goldblatt, R., Kim, B. (eds.): Proceedings of the 12th Asian Logic Conference. World Scientific, Singapore (2013)
5. Downey, R., Hirschfeldt, D. (eds.): Aspects of Complexity: Minicourses in Algorithmics, Complexity and Computational Algebra. de Gruyter Series in Logic and Its Applications, vol. 4. Walter de Gruyter, Beijing (2001)
6. Downey, R.G., Fellows, M.R.: Fundamentals of Parameterized Complexity. Texts in Computer Science. Springer, London (2013)
7. Downey, R.G., Hirschfeldt, D.R.: Algorithmic Randomness and Complexity. Theory and Applications of Computability. Springer, New York (2010)

Surfing with Rod

Michael R. Fellows[✉]

Department of Informatics, University of Bergen,
P.O. Box 7803, 5020 Bergen, Norway
michael.fellows@uib.no

Abstract. I wish Rod a happy birthday. Rod has offered a good historical account of our early days of collaboration in his paper, "The Birth and
Early Days of Parameterized Complexity" [D12]. One of the themes of our
life-long collaboration has been a shared passion for surfing, and many of
our best ideas were hammered out on surf trips. This contribution is best
viewed as a gloss on that earlier account by Rod, taking as an organizing
skeleton, our various surf trips and what we were thinking about, *with a
particular emphasis on open problems and horizons that remain, and some
reflections on the formation of our research community.*

1 Introduction

Happy birthday Rod! When I first heard about the festival, I was somewhat
incredulous: "Why is he doing this? He is only 50!!" (I probably needed an egg.)
Keep going, man! We've had a lot of fun!

This being a personal reminiscence about our collaboration, my first thought
was to write a history of the development of the central ideas of parameterized
complexity, recalling the historical context, which is a bit colorful, as many of
the key ideas were developed on surf trips.

But Rod, you have already done a lot of that in the entertaining and informative article [D12] about how the field developed in its early years.

So Plan B was to focus on our surf trips as the narrative skeleton of a colorful
history of (most of) the main ideas in PC. But that was not going to work as
I found it impossible to collate the many surf adventures with the intellectual
adventures.

So I have settled on Plan C: to discuss a small selection of key surfing adventures, and the ideas we were discussing then (so a limited historical window on
the field) *and especially the open questions that remain.*

I think we have surfed together at least a hundred times at more than 25
different surf spots, some entirely and perhaps deservedly obscure (Red Rocks,
Wairaka Rock), always talking about our parameterized complexity projects,
latest ideas, poetry, literature, the latest goings-on of the math-for-kids projects,
etc. Always another surf trip and new ideas. Long may this be so!

M.R. Fellows—Research supported by the Bergen Research Foundation, the Government of Norway through its Elite Professorship Program, and the Australian
Research Council.

A. Day et al. (Eds.): Downey Festschrift, LNCS 10010, pp. 9–18, 2017.
DOI: 10.1007/978-3-319-50062-1_2

2   About Surfing

What People Generally Don't Know About Surfing

What people who do not surf generally do not understand is how much time surfers spend driving around looking at the waves and not actually surfing.

This happened to me on my first trip to New Zealand. I had a lot of grant money and freedom thanks to the enlightened administration of Canada's NSERC at that time (early 1990's). NSERC was not in the business of herding the cats. There were essentially two principal rules:

(1) Thou shalt not spend your research money on snow tires for your car.
(2) Thou shalt not spend your research money on taking out the entire department to dinner at an all-you-can-eat place just because you have a research visitor.

I contributed a paper to a relatively obscure (to me, anyway) combinatorics conference ACCMCC in New Zealand. Rod has hilariously written about this adventure in his essay about the origins of parameterize complexity [D12]. I had all of my surf gear along: wetsuit, flippers, bodyboard. Down the stairs from the airplane, cattle class, and there came up down the other stairs from First Class, Ron Graham, and he said to me: "What are *you* doing here?!"

I was there to present some of my results on algorithmic aspects of the Graph Minor Theorem, and I was also there to surf.

I met Rod through a social pointer by Marston Conder. Rod and I immediately hit it off about the basic ideas of what we would now call FPT *versus* XP and agreed that this was a rich theme for mathematical investigation. After two bottles of Villa Maria (as has been recounted elsewhere), we had a rough plan about how to proceed, with this obviously natural research horizon. The tentative plan was to write a series of numbered papers (like Robertson and Seymour) and fold it into a small book. My recollection is that in our visionary state, the plan was something like:

(PC I) *basic complexity framework and results*
(PC II) *concrete complexity results*
(PC III) *applications.*

My recollection is that we got this far (tentatively) over the Villa Maria, without having the concrete mathematical results to fill in the three containers, even (PC I), although there was a start in the paper [AEFM89] and some then unpublished results building on that. The earlier work of Langston and me [FL87, FL88] laid the groundwork for (PC II) and (PC III). My recollection is that we had a confident vision that there were plenty of issues to support a general program.

I can't recall if we discussed surfing at that time. I think this shared passion turned up later in email or phone conversations. Or maybe Rod had prior commitments that weekend.

I had my gear and a rental car, and I had a couple of days to drive around looking for surf. I had a map. Castle Point looked promising, so I drove over there, over the scenic hills, slowly through large herds of sheep on the backroads at several points along the way. Castle Point looked good on the map, but when I got there, I was surprised. On the one side of the point, the wind was ferocious — like in a movie about Lawrence of Arabia in a sandstorm — if you stood there for 15 min the skin would be sandblasted off your legs. Ouch! The surf was huge and totally chaotic and there was nobody there. It was all rocky and horrible.

On the other side of the point, sandy beaches, and the wind was a bit less strong and offshore, but there were no waves, or only just tiny ones, and absolutely nobody there. I don't know what I was expecting. So I just drove back to Palmerston North.

Thus ended my first surf trip to New Zealand. Just drove around and looked at stuff and never got wet. As surfers do, more often than you would think.

Makara Point and Courage

Makara Point is close to Wellington and pretty good when it is on, but it is in a weird location and requires special conditions. Rod and I were standing in his office talking about parameterized complexity. His office has a wonderful view out across the harbor, and all of the sudden Rod yells, "It's happening!" and out window one could see this slanting black wall of rain approaching from the east. "We have to get moving!"

Then the Chairman of the Department knocked on the door and asked Rod if he would be attending the meeting in the afternoon. Rod said, "Sorry, I have a prior engagement."

The engagement was that we needed to rush off to Makara Point right now! Because the way it works is that when one of the storm fronts from the Antarctic wave machine hits, there are super-winds from the west which build up this big chaotic swell that can make it into Makara Point (unsurfably chaotic). Then, when the front moves through, as we were seeing, the winds reverse direction, and blow offshore, cleaning off the swell, and for about 1-2 h during the reversal there are good surf conditions, which is why you have to get moving! After some hours, it is gale force again, offshore this time, and it is no good again, as the reef is a long paddle out with too much fetch.

One time we drove around to the north from Makara where there is another obscure point. It is really nasty, like a lot of the surf breaks in geologically young New Zealand, just a rocky long shoreline in the middle of nowhere (maybe some sheep farms). Rod claimed this was sometimes a pretty good spot. Me: "You *surfed* this?! What, with Craig?" Rod: "No, alone." Me: "You surfed this *alone*!?".[1]

Rod's answer was, *"Look, if you don't have physical courage, you will never do great mathematics."*

---

[1] Most surfers do not surf alone, because it is a statistically proven fact that if you do not surf alone, your chances of being taken by a shark are improved by at least fifty percent.

I buy that! This attitude has infused the research community of parameterized complexity that we have nurtured, to a significant extent. We have tutored many young researchers into surfing (meaning body-boarding, which you can have fun with in 15 min, and is arguably superior to stand-up surfing).[2]

Surf Sufism

Rod and I would probably agree on the following mystical tenets:

- Surfing conforms well with mathematical science, where research is generally *not* a team sport. If there is a multi-author paper in the mathematical sciences, it is generally "1 + 1 + 1" rather than "3". Biologists and Physicists should choose rugby as their courage mascot, since research there is typically done by surprisingly large teams. In the mathematical sciences, individual courage, tenacity (what is 25 duck dives ...), risk and imagination are primary, as in surfing. Things can sometimes be achieved relatively rapidly in Mathematical Science; in contrast, breakthrough results in Biology and Astronomy are generally achieved by determined slogs.
- Surfing also conforms well metaphorically with the relationship between pure and applied mathematics. The point of mathematical science is not to sit outside and wait for a big wave, and then catch it, and then ride it straight into the beach, to the applause of the whitewater. You can do that, and have fun, but it is not the main point. The main point is to catch the wave, drop in, harnessing the energy of the wave, and then pull in and go sideways (left or right) at high speed: a physical metaphor for applied mathematics driven by pure mathematics.

  I have always thought that Scottish Country Dancing was appropriate for a logician. Here we stick to surfing.

3   Some Selected Surf Trips and Associated Open Problems in Parameterized Complexity

I reminisce here about a few of our surf trips and locations, associated with some of my favorite open problems.

The Great North Island Road Trip

Thanks to the miminalist (and quite sensible) rules of NSERC governing Canadian scientists in the early 1990's, described above, we were able to do this wonderful thing: I came over to New Zealand, stayed at Rod's house when we were in Wellington, rented a car (thanks to NSERC) and then we made our

---

[2] Stand-up surfers look down on body-boarders, but they shouldn't! Body-boarders can routinely catch waves that stand-up guys can't, because bodyboarders have stronger sprinting ability, and the stability to handle a lip-launch as the wave breaks "late" (which involves flying through the air a bit). One could go on about this. There are lots of ways to have fun in the waves. When I first met Rod, he was a hand-gunner, which is a rare practice (except in Queensland) closely related to plain old body-surfing.

way around the North Island on our research expedition. Besides mathematical discussions that led to the framing of the core definitions of parameterized complexity and some of the fundamental theorems, we enjoyed surfing at famous surf spots, and stops at famous wineries, and reading poetry out loud in the car. The science road-trip of a lifetime!

On that adventure we engaged the four major themes of the research community we have nurtured in parameterized complexity:

(1) The mathematical frontiers.
(2) The connection between physical adventure and intellectual adventure.
(3) Poetry, love of literature and story-telling.[3]
    But there was also another theme that we discussed on the road trip that I was obsessed about at the time:
(4) Theoretical computer science (and mathematical sciences generally) for ten-year-olds.

Rod thought this fourth cultural component was a waste of time. He had put some effort into "math education" and found it time-consuming and discouraging. Geoff Whittle of VUW followed in his tracks and put a lot of effort into math-teacher-training and I think came to the same conclusion. They gave blood at the Red Cross of mathematics education and moved on.

The whole area is a fiasco, for sure. It is comical how it marches in circles, decade after decade. But Frances Rosamond and I have established an island of joy in the wasteland. We have the advantage of focusing on the fundamental mathematics of computer science, which is mostly miles from the shopkeeper arithmetic that dominates the official math curriculum and the standardized exams. We consciously operate as anarchists and skeptics with respect to curriculum. We take our stuff to elementary schools for fun. We can (and do) go, with no prior preparation, into an event with 90 ten-year-olds, tomorrow morning, for three hours and have fun with them, engaging them with the fundamental mathematics of computing (without computers). It's too easy, and joyful. Theoretical computer science offers a lot of great material.

Our principal technical objective on that trip was to try to sort out the $W[1, t]$ dilemma, as described in [D12] (where the reader can find the definitions). The bigger theme that was coming into focus at the time had two principal components:

(1) Understanding the fundamental tower of parameterized intractability classes.
(2) Worrying about whether those definitions had sufficient traction with the obvious, well-known, naturally parameterized problems of "computing practice". That is, how "natural" were these classes?

---

[3] This is a mainstay of the community. There is a sort of informal award system for the best young researchers in the field, which is to receive a copy of what we call "Mr. Opinion" — this is the book: *The New Guide to Modern World Literature* by Martin Seymour-Smith.

We wanted to give a single homogeneous definition of the $W[t]$ classes: FPT reduction to classes of weft t circuits of bounded small-gate depth. Our initial investigations were mathematically elegant for at least for $t \geq 2$, but $t = 1$ remained frustrating: it seemed that maybe it splintered into a messy hierarchy of $W[1, t]$ classes. Not pretty. Eventually (I'm not sure we achieved this on the road trip) we proved that the nice clean perspective that worked for for $t \geq 2$ did indeed work for $t = 1$, but required special arguments.[4]

With respect to (2), we were beginning to get paranoid about the possibility that various natural problems (e.g., IRREDUNDANT SET) might belong to some $W[1.5]$ degree that we didn't know how to define. Those paranoias have receded since our primordial surf trip. The $W[t]$ degree structure seems to be extremely crisp and neat. I honestly think we have spent more time over the years sweating about (2) than about sharks!

On this surf trip we spent a lot of time discussing the basic definitions that would frame this new field. We spent a lot of time discussing possible "para-meterized analogs" of landmark theorems of classical (one-dimensional) com-plexity theory. We spent some time discussing the still-open problem of upward or downward collapse of the $W[t]$ degrees: if, for example, $W[3] = W[4]$ does this imply either: (1) $W[t] = W[3]$ for all $t \geq 3$ (upward collapse), or (2) $W[2] = W[3] = W[4]$ (downward collapse)?

My favorite open problem from this era (as it developed over the years) is:

OPEN: Is $W[t] = W^*[t]$ for $t \geq 3$?

Our original definition of the $W[t]$ classes was aimed at making membership arguments easy: t-bounded circuit weft plus constant-bounded small-gate depth. But for many natural parameterized problems, the translation of the problem into parameterized Hamming weight CIRCUIT SATISFIABILITY requires small-gate depth that increases with the parameter.

Relaxing the constant small gate depth requirement to a connection k' depen-dent only on the parameter value k gives the broader class $W^*[t]$. Two of our most important technical results (with Udayan Taylor, and Ken Regan, respec-tively) are that $W^*[1] = W[1]$ [DFT96] and $W^*[2] = W[2]$ [DFR98, DFR98b].

What happens for $t \geq 3$? Our proofs that $W^*[1] = W[1]$ and that $W^*[2] = W[2]$ are quite different. They do not seem to be provable in a unified sweep of argumentation. The fact that the cases of $t = 1$ and $t = 2$ have been settled suggests that this open problem might be approachable.

Sombrio

We have surfed this wonderful break (there are two, actually) over and over — a beautiful walk through the woods, way out on the west coast of Vancouver Island, a bit of a drive from Victoria (like 2 h). Once we even walked down to the beach through a light blanket of snow. Rod said, "Crikey, this is cold!" This is saying a lot, as the sea temperatures around Wellington are far from tropical!

---

[4] This is not unheard-of in mathematics, that smaller "dimensions" require special approaches.

Everytime we have gone to Sombrio, we have had in tow wonderful new mathematical ideas. On one expedition to Sombrio in 1992, we had along my new PhD student at the time, Michael Hallett. Rod had told me, "Mike, it is your sacred duty to teach Hallett to surf!" I had borrowed a wetsuit for Hallett from my neighbor, Don Beckner, but it was this super-thick wetsuit, the kind of thing you wear with oxygen tanks doing slow-motion gathering of abalone in the North Pacific.

At the time, Hallett and I had just proved (with Hans Bodlaender) that BANDWIDTH is hard for $W[t]$ for all t (eventually a STOC paper) [BFH94].

So that is a complexity *lower bound*. But what about an upper bound? I had tasked Hallett with showing that BANDWIDTH is in $W[P]$. But he came to my office with a very interesting response:

It cannot be in $W[P]$! $W[P]$ is basically $k \log n$ bits of nondeterminism plus P-time verification. But look! (using modern parlance) BANDWIDTH is AND-compositional. If I have one graph G_1 on n_1 vertices and BAND-WIDTH is in $W[P]$ then $k \log n_1$ bits of nondeterministic information are sufficient for a polynomial-time verification. But look! What if I am concerned with $G_1 \ldots G_m$ and I take G to be the disjoint union of the G_i? Then you are asking for a small amount of information to P-time verify that *all* of these graphs have bandwidth at most k, and that is unreasonable.

This is essentially the intuition behind the lower bounds methods for kernelization [BDFH08]. The central issue is, "too much information compression."

We got Hallett out in the water, with his massive-amount-of-rubber wetsuit. This was all new to him. Sombrio's main break involves taking off a moderate distance in front of a large rock. It is imperative to not hit the rock. A wave came along and Hallett gamely paddled into it, and Rod and I were shouting, "Don't hit the rock!" The next thing we saw was Hallett's bodyboard popping into the air quite dramatically, but not attached to Hallett! What happened? Apparently, he ended up (so floatable!) on his back, looking up at the sky, being washed into the beach on the white-water, like the ginger-bread man.

Rod and I have had several expeditions to Sombrio, and they mathematically connect. On a different expedition than the one related above, we were driving out to Sombrio with Neal Koblitz (of cryptography, elliptic curve and number theory fame). At the time, Rod and I were excited about the notion of kernelization. We had worked out the following definition:

Definition. A parameterized decision problem Π with input (x, k) with $|x| = n$ and parameter k is *kernelizable* if and only if there is a polynomial-time transformation of (x, k) to (x', k') (where k' depends only on k) such that:

(1) (x, k) is a yes-instance of Π if and only if (x', k') is a yes-instance of Π,
(2) $k' \leq k$
(3) $|(x', k')| \leq g(k)$ for some function $g(k)$.

For about 2 days Rod and I thought that the kernelizable parameterized problems might be an *interesting proper subset* of FPT. We were talking about it in the car on the way to Sombrio with Neal. About half-way there, Neal spoke up and said; "But can't you just ..." (with the trivial proof that FPT is the same as kernelizable). This was included in [DFS98] as a lemma. It is also implicit in [CCDF97]. It is often currently attributed to "folklore" but this is not accurate. It is a truly foundational observation, and sets up a whole new game, the kernelization races and the lower bounds program for kernelization. Rod and I thought about it for five seconds and said something like, "Oh, right." It should perhaps be called "Neal's Lemma".

I had another wonderful graduate student at the time: Michael Dinneen. We were working on theory and implementation of algorithms for computing the finite obstruction sets of minor ideals. So a very natural question is the following:

Question: For genus 0, there are two obstructions in the minor order: $K_{3,3}$ and K_5. How many minor-minimal obstructions characterize genus g? Could the size of the minor order obstruction set for genus g be bounded by a polynomial in g?

Theorem (Dinneen) [D97]. Not unless $coNP \subseteq NP/poly$.

Proof. Given (G, k) we must determine whether G does *not* have genus k. This is a $co - NP$ complete problem, since GRAPH GENUS is NP-complete. Our polynomial-sized advice string for the inclusion question, which needs to be polynomial-sized in the input size n of G, consists of the obstructions to genus k that have size at most n (any larger obstructions are irrelevant). Given this polynomial-sized advice, which consists of the relevant obstructions $H_1...H_m$ and access to an NP machine, to answer the question we guess how one of the H_i lives in G as a folio and this can be verified in polynomial time. *QED*

OPEN: For basically the same intuitive reasons, can we prove that BANDWIDTH is not in $W[P]$ unless $coNP \subseteq NP/poly$?

Of course, these early results informed the investigation of kernelization lower bounds.

Newcastle

After serious consideration, Rod and I agreed that the University of Newcastle, Australia, is probably the best tradeoff available on the planet for a mathematical sciences researcher who loves surfing. Any wave / wind conditions, there is always something on: Newcastle Main Break, Flatrock, Kauri Hole, Nobbies Reef, The Spot, The Spit, The Wedge, The Harbor ...

What Rod and I were obsessing about the last time he came around to Newcastle is actually an elegant and fundamental "applied" problem. I give you a linear error-correcting code over $GF[2]$ presented as a generator matrix. Is the minimum Hamming distance for decoding at most k?

This is polynomial-time equivalent to the graph problem:
EVEN SET
Instance: $G = (V, E), k$
Question: Does G have a non-empty vertex subset $V' \subseteq V$ of size at most k such that for every vertex $v \in V$, $|N[v] \cap V'|$ is even.

This is kind of a "parity variation" on DOMINATING SET, but equivalent to a really fundamental algorithmic problem in Coding Theory.

We were obsessing about this in Newcastle, and long afterwards, and have spent hundreds of hours on the problem with many seductive false starts building on our results in the paper with Geoff Whittle and Alex Vardy [DFVW99]. I don't think we work on it anymore. It remains a challenge for the younger and smarter. Our (weak) conjecture is that it is hard for $W[1]$.

4   Horizons

We're still kick'in Rod! (With flippers, on bodyboards.) I'm still looking forward to a parameterized complexity workshop at Raymond's surf-camp at G-land, Java. That is a truly amazing place to surf and think. We have been talking about the possibility for years. When are we going to do it? Happy Birthday, Rod!

References

[AEFM89] Abrahamson, K., Ellis, J., Fellows, M., Mata, M.: On the complexity of fixed-parameter problems. In: Proceedings of 13th FOCS, pp. 210–215 (1989)

[BDFH08] Bodlaender, H.L., Downey, R.G., Fellows, M.R., Hermelin, D.: On problems without polynomial kernels (extended abstract). In: Aceto, L., Damgård, I., Goldberg, L.A., Halldórsson, M.M., Ingólfsdóttir, A., Walukiewicz, I. (eds.) ICALP 2008. LNCS, vol. 5125, pp. 563–574. Springer, Heidelberg (2008). doi:10.1007/978-3-540-70575-8_46

[BFH94] Bodlaender, H., Fellows, M.R., Hallett, M.T.: Beyond NP-completeness for problems of bounded width: hardness for the W hierarchy. In: Proceedings of ACM Symposium on Theory of Computing (STOC), pp. 449–458 (1994)

[CCDF97] Cai, L., Chen, J., Downey, R., Fellows, M.: Advice classes of parameterized tractability. Ann. Pure Appl. Logic **84**, 119–138 (1997)

[D97] Dinneen, M.: Too many minor order obstructions (for parameterized lower ideals). J. Univers. Comput. Sci. **3**, 1199–1206 (1997)

[D12] Downey, R.: The birth and early years of parameterized complexity. In: Bodlaender, H.L., Downey, R., Fomin, F.V., Marx, D. (eds.) The Multivariate Algorithmic Revolution and Beyond. LNCS, vol. 7370, pp. 17–38. Springer, Heidelberg (2012). doi:10.1007/978-3-642-30891-8_2

[DFS98] Downey, R., Fellows, M., Stege, U.: Parameterized complexity: a framework for systematically confronting computational intractability. In: Graham, R., Krachovil, J., Nesetril, J., Roberts, F. (eds.) Contemporary Trends in Discrete Mathematics, DIMACS, vol. 49, pp. 49–100. American Mathematical Society, Providence (1999)

[DFR98] Downey, R.G., Fellows, M.R., Regan, K.W.: Parameterized circuit complexity and the W hierarchy. Theoret. Comput. Sci. A **191**, 91–115 (1998)

[DFR98b] Downey, R.G., Fellows, M., Regan, K.: Threshold dominating sets and an improved characterization of $W[2]$. Theoret. Comput. Sci. A **209**, 123–140 (1998)

[DFT96] Downey, R.G., Fellows, M., Taylor, U.: The parameterized complexity of relational database queries, an improved characterization of $W[1]$. In: Combinatorics, Complexity and Logic: Proceedings of DMTCS 1996, pp. 194–213. Springer, Heidelberg (1997)

[DFVW99] Downey, R., Fellows, M., Vardy, A., Whittle, G.: The parameterized complexity of some fundamental problems in coding theory. SIAM J. Comput. **29**, 545–570 (1999)

[FL87] Fellows, M., Langston, M.: Nonconstructive proofs of polynomial-time complexity. Inf. Process. Lett. **26**, 157–162 (1987/1988)

[FL88] Fellows, M., Langston, M.: Nonconstructive tools for proving polynomial-time complexity. J. Assoc. Comput. Mach. **35**, 727–739 (1988)

Prequel to the Cornell Computer Science Department

Anil Nerode[⊠]

Goldwin Smith Professor of Mathematics, Cornell University, Ithaca, NY 14850, USA
anerode1@twcny.rr.com

Preface. Rod Downey has a long career of important discoveries in the recently developed subjects of computability and computer science. These remarks were delivered by invitation at the 50th anniversary celebration of the founding of the Computer Science Department at Cornell. This was a time of transition from computers and computer programming to computer science.

Prequel. In 1963 the Vice President for Research at Cornell, Frank Long, was informed by the Sloan Foundation that it was Cornell's turn to get a large grant in some new exciting direction in the sciences. I was Acting Director of the two year old Center for Applied Mathematics. In that capacity I was asked for a suggestion. I was and am a computability theorist and also an applied mathematician. I knew the hardware and software and theory advances at Bell, IBM, Carnegie Mellon, MIT, and Stanford because I knew the people there. I was certain then that computer science would become ubiquitous in all branches of human learning (as indeed it has become). I suggested we form a computer science department, seconded by former mathematics chair Robert J. Walker who wanted machine-oriented numerical analysis research at Cornell. Long appointed a committee of three to prepare a formal proposal to Sloan, get whatever we proposed approved by the faculties of Arts and Sciences, the Graduate School, and the State of New York, and then hire a chair and a couple of faculty, all at the same time. I was chair due to my directorship; the other members were Walker and Richard Conway. I believe that no one other than me at Cornell knew the computer science community.

Here is my personal recollection of the Cornell environment into which we were to insert Computer Science. Others may have a more nuanced view. We had to deal with the universal attitude of the Cornell physicists, chemists, and engineers that computer science was Fortran or Algol or Cobol on an IBM, so why did we need a department? We needed to deal with the total preoccupation of the Dean of Engineering Andy Schultz in executing a catch-up. The engineering college was outstanding from its formation to the beginning of World War II. Apparently living in the prewar past, after the war Dean of Engineering Hollister had expected a surge of demand for undergraduate and master's students. He hired professors to do lots of undergraduate teaching and little graduate training or research, and built buildings to match. But engineering had moved on at MIT, Carnegie Mellon, and Stanford, who hired professors with high quality research credentials and introduced graduate research in newly emerging disciplines. The

© Springer International Publishing AG 2017
A. Day et al. (Eds.): Downey Festschrift, LNCS 10010, pp. 19–21, 2017.
DOI: 10.1007/978-3-319-50062-1_3

Dean of Engineering had his plate full and left it to us to design a department. The Dean of Arts and Sciences trusted Walker and possibly me.

First we created a graduate field of computer science. I knew how to do this since we had done this with applied mathematics two years earlier. We applied to New York state for approval of a Ph.D. program and asked the General Committee of the graduate school to introduce the graduate field of computer science. We introduced as founding members of that field Dave Block, Richard Conway, Chris Pottle, Syd Saltzman, Robert Walker, and myself. This established the graduate field and its initial membership before there was a computer science department, before a department was formed, before the Sloan application had been submitted, and before any external candidates had been selected or interviewed. This required effort but no funds. It created no waves.

The next step was to create an actual department. To what Dean should the new department report? I wanted a strong theory department, for which there was no Cornell academic support outside of Walker and me. I worried that if the computer science department were introduced as a department of the College of Engineering, the narrow view of computer science that the engineering faculty had at that time would mean that the department was likely to become a service department of no great distinction. To make sure this did not happen, I first proposed an all-university department reporting directly to the provost or vice president for research. This was turned down as administratively impossible, only schools report to the provost and no one reports to the vice president for research. I then proposed that the Department report to both the Dean of Engineering and the Dean of Arts and Sciences. I proposed that faculty be given appointments in both colleges. I was quite frank that this was to provide protection against parochial interests and to meet Walker's and my concerns. This was grudgingly accepted by both deans. I believe it was effective as protection for some years as the department grew. But then there was the year when Alain Seznec, a good friend and humanist, was Dean of Arts and Sciences; he complained he knew nothing about computer science, and ceased to monitor it. He also did not understand physics, chemistry, or biology in his own college, but that is neither here nor there. Around 10–15 years after its founding, the department became an engineering department. By that time, protection was not necessary; it had become a jewel beyond price.

How should we recruit a new faculty when there were no students and substantial ignorance of computer science at Cornell? We recognized that we could not hire a top flight software or hardware or computer language specialist. They were few and those wanting academic posts had been hired into the four or so departments founded before ours. We looked for theorists since they would have as complement our then four professor mathematical logic group. Our intention was to build top down from theory to practice as the years went by. Indeed this is what Juris and his successors succeeded in doing.

The semester before trying to recruit, I organized with the cooperation of Dave Block a course in topics in computer science with an audience of about thirty. This gave an appreciative audience for prospective hires, so they would not feel they were entering a computer science desert.

If hordes of undergraduates appeared soon (and they did), how could the initial three appointments handle them? I and Walker arranged that the students would have math advisors from the large pool of math professors and have a math-cs major track in mathematics until the computer science department had expanded. This went quite smoothly as courses were developed and students came, and in the course of time was terminated when the computer science faculty grew large enough to handle the load.

Who to interview as prospective chairs? I do not wish to recall the other candidates beyond Hartmanis. I suggested him as a candidate because of his work with Stearns which I knew. His lecture and his interview were incomparably better than all the others. He had vision, the others did not. I suggested that we should interview Pat Fischer, an MIT Ph.D. of my close colleague Hartley Rogers and an assistant professor at Harvard. Pat suggested we consider his colleague Gerald Salton, a Ph.D. under my friend Howard Aikin at Harvard. I read Salton's work, and agreed. All three were mathematics Ph.D.'s (Cal Tech, MIT, and Harvard.).

Hartmanis looked them over and accepted them. We made the three offers, Sloan released the funds, and Juris created the future.

9/19/2014

Some Questions in Computable Mathematics

Denis R. Hirschfeldt[✉]

Department of Mathematics, University of Chicago, Chicago, IL, USA
drh@math.uchicago.edu

To Rod Downey on his 60th Birthday.

I had the good fortune to be among Rod Downey's long and distinguished list of postdocs, in my case in 1999–2000. I recall Rod saying once that he had hoped that his young postdocs would be interested in joining him in his many athletic activities, but ended up with a bunch of drunks instead. I did learn a lot about wine from Rod, but I think I managed to squeeze some learning about mathematics as well while I was in Wellington. In any case, to the extent that I was able to hold my own with Rod at the blackboard and around the decanter, I am proud.

There is no denying that Rod is a theory-builder, parameterized complexity being a shining example, but he is also a problem-solver, problem-creator, and problem-disseminator of the first water. So in honor of his 60th birthday, I have chosen to discuss a few open problems I particularly like, and that are connected in one way or another with his work and my mathematical interactions with him. Most of these problems are well-known to experts in their areas (computable structure theory, reverse mathematics, algorithmic randomness, and asymptotic computability), but I hope there is some value in bringing them together, with some background and a bit of personal history thrown in.

I will assume familiarity with the basics of computability theory throughout, as well as those of reverse mathematics, algorithmic randomness, and model theory in places.

1 The Slaman-Wehner Theorem for Linear Orders

My dissertation was in computable structure theory. Rod's research in that area was deeply influential, as was his expository work in papers such as [14–16,30]. Russell Miller was working on his dissertation at around the same time as I, and I believe it was Rod who first told me about an exciting result by Russell that answered a couple of questions Rod had asked in [14], while leaving a third tantalizingly open.

In model theory, one identifies isomorphic structures, but in computable mathematics, structures that are isomorphic but not computably isomorphic

Partially supported by a Collaboration Grant for Mathematicians from the Simons Foundation. I thank Russell Miller, Benoit Monin, and Ludovic Patey for useful comments.

A. Day et al. (Eds.): Downey Festschrift, LNCS 10010, pp. 22–55, 2017.
DOI: 10.1007/978-3-319-50062-1_4

can be quite different from each other. Thus one of the main concerns of computable structure theory is the study of concrete copies of a countable structure (in a computable language) up to computable isomorphism.

Definition 1.1. A *presentation* of a countably infinite structure $\mathcal{M}$ is a structure $\mathcal{A} \cong \mathcal{M}$ with universe ω. A structure is *computably presentable* if it has a presentation whose atomic diagram is computable. More generally, the *degree* of a presentation $\mathcal{A}$ is the (Turing) degree of the atomic diagram of $\mathcal{A}$. The *(atomic) degree spectrum* of $\mathcal{M}$ is the set of degrees of presentations of $\mathcal{M}$.

The degree spectrum of $\mathcal{M}$ measures the computability-theoretic complexity of obtaining a concrete copy of $\mathcal{M}$. Knight [67] showed that, except in trivial situations in which the degree spectrum is a singleton, every degree spectrum is closed upwards. Thus nontrivial computably presentable structures all have the same degree spectrum.

The simplest degree spectra are those of the form $\{\mathbf{d} : \mathbf{d} \geqslant \mathbf{a}\}$, and for any degree $\mathbf{a}$, it is not difficult to find a structure $\mathcal{M}$ with this degree spectrum. In this case, it makes sense to say that $\mathbf{a}$ is the degree of (the isomorphism class of) $\mathcal{M}$, but not every degree spectrum has this form. For instance, Richter [99] showed that if a linear order is not computably presentable, then its degree spectrum has no least element.

On the other hand, not all upwards-closed sets of degrees are degree spectra of structures. For instance, if $\mathbf{a}$ and $\mathbf{b}$ are incomparable degrees, then the union of the upper cones $\{\mathbf{d} : \mathbf{d} \geqslant \mathbf{a}\}$ and $\{\mathbf{d} : \mathbf{d} \geqslant \mathbf{b}\}$ is not the degree spectrum of any structure, a fact established in unpublished work of Knight and others [personal communication] and by Soskov [110]. Thus it becomes interesting to ask whether certain natural upwards-closed classes of degrees can be degree spectra of structures. (For a broader survey of this general question, see the chapter on computable model theory by Fokina, Harizanov, and Melnikov [35] in a volume dedicated to Turing's legacy edited by Rod.)

One can think of the set of presentations of a countable structure as a mass problem (i.e., a subset of ω^ω) via some suitable encoding. One way to compare the relative complexity of two mass problems is via Muchnik reducibility, also known as weak reducibility. (Medvedev reducibility, or strong reducibility, is the uniform version of Muchnik reducibility.) For two mass problems P and Q, say that P is *Muchnik reducible* to Q if every element of Q computes some element of P. As usual, this notion leads to a degree structure on mass problems. The least Muchnik degree consists of those mass problems that have a computable member. There is also a least nontrivial Muchnik degree, namely the degree of all mass problems P such that P has no computable member, but has an X-computable member for each noncomputable X. It might seem at first that it would be difficult to find "natural" mass problems living in this degree, but that has turned out not to be the case.

Lempp (see [108,117]) asked whether there are structures whose degree spectra are in this degree (and Knight (see [108,117]) asked a closely related question about enumerations of families of sets). A positive answer was given by Slaman [108] and Wehner [117].

Theorem 1.2 (Slaman [108]; Wehner [117]). There is a structure whose degree spectrum consists of all nonzero degrees.

Whenever a structure with a particularly interesting computability-theoretic feature is found, it is natural to ask whether similar structures exist within various well-known classes of structures. For some classes $\mathcal{C}$, there are general results that show that, for certain kinds of computability-theoretic phenomena, anything that can happen in general can happen within $\mathcal{C}$. For instance, Hirschfeldt, Khoussainov, Shore, and Slinko [52] gave such results for classes such as partial orders, lattices, integral domains, commutative semigroups, and 2-step nilpotent groups, which in particular imply that the Slaman-Wehner Theorem holds in these classes. That is, each of these classes contains a structure whose degree spectrum consists of all nonzero degrees. (See also the discussion of Miller, Poonen, Schoutens, and Shlapentokh [90] in Sect. 4.)

On the other hand, there are many classes that are not "universal" in the above sense, and in particular do not contain structures realizing all the degree spectra that are possible in general. A well-known example is the class of Boolean algebras. Downey and Jockusch [23] showed that every low Boolean algebra is isomorphic to a computable one, and this result was extended to low_2 Boolean algebras by Thurber [114] and then to low_4 Boolean algebras by Knight and Stob [68]. Thus if the degree spectrum of a Boolean algebra contains any low_4 degree, then it contains all degrees. In particular, there is no Boolean algebra whose degree spectrum consists of all nonzero degrees. It is not known whether every low_n Boolean algebra is isomorphic to a computable one. This question, which goes back to Downey and Jockusch [23], remains a major one in computable structure theory.

Richter's result mentioned above shows that the class of linear orders is also not universal as far as degree spectra are concerned. On the other hand, unlike Boolean algebras, linear orders can have presentations that are close to being computable without actually being computably presentable. Jockusch and Soare [61] showed that for every nonzero c.e. degree, there is a linear order of that degree that is not isomorphic to any computable linear order. Downey and independently Seetapun (see [14]) extended this result to all nonzero Δ_2^0 degrees, and finally Knight (see [14]) extended it to all nonzero degrees.

In many ways, linear orders occupy a particularly interesting place in computable structure theory. They are neither so unstructured as to basically be the general case in disguise nor so structured as not to admit any computability-theoretic "pathologies". When I was in Wellington, Rod and I spent some time thinking about linear orders (and in particular a question about the successivity relation in computable linear orders that Rod finally solved in joint work with Lempp and Wu [29]). As I remember Rod saying several times back then, "Linear orders are hard!"

In light of the results discussed above, it was natural for Rod to ask in [14] whether there are linear orders that are not computably presentable but whose degree spectra contain all nonzero c.e. degrees, or all nonzero Δ_2^0 degrees, or even *all* nonzero degrees. The first two of these questions were the ones answered by Russell's result.

Theorem 1.3 (Miller [87,88]). There is a linear order whose degree spectrum contains every nonzero Δ^0_2 degree except **0**.

The proof consists of modifying the basic module of the Jockusch-Soare construction in [61] and combining it with Δ^0_2-permitting so that, for any noncomputable Δ^0_2 set C, the construction produces a C-computable linear order whose order type is independent of C. The resulting order type $\mathcal{L}$ is of the form $\mathcal{S}_0 + \mathcal{A}_0 + \mathcal{S}_1 + \mathcal{A}_1 + \cdots$, where each $\mathcal{A}_n$ is used to diagonalize against the possibility that the nth partial computable linear order is isomorphic to $\mathcal{L}$, and each $\mathcal{S}_n$ is a *separator* of the form $1 + \eta + i + \eta + 1$, where η is the order type of the rationals and $i \in \mathbb{N}$. The separators keep the individual diagonalization constructions apart.

Chisholm [unpublished] and Downey [unpublished] showed that the degree spectrum of $\mathcal{L}$ in fact includes all hyperimmune degrees. Barmpalias (see [36]) argued that no hyperimmune-free degree is sufficiently strong to carry out the basic module of the construction of $\mathcal{L}$, leading to the conjecture that the degree spectrum of $\mathcal{L}$ consists exactly of the hyperimmune degrees. Of course, even if this conjecture holds, it may still be possible to go beyond the hyperimmune degrees with a different order type, so Rod's third question remains open.

Open Question 1.4 (Downey [14]). *Is there a linear order whose degree spectrum consists of all nonzero degrees?*

See Frolov, Harizanov, Kalimullin, Kudinov, and Miller [36] for more on degree spectra of linear orders. In particular, they showed that for every $n \geqslant 2$, there is a linear order whose degree spectrum consists exactly of the nonlow$_n$ degrees. The $n = 1$ remains open, however. (Notice that Question 1.4 is the $n = 0$ case.)

Open Question 1.5 (Frolov, Harizanov, Kalimullin, Kudinov, and Miller [36]). *Is there a linear order whose degree spectrum consists of the nonlow degrees?*

As noted by Fokina, Harizanov, and Melnikov [35], analogs of Question 1.4 are also open for other interesting classes of structures, such as abelian groups. In that case, Khoussainov, Kalimullin, and Melnikov [65] proved the analog of Theorem 1.3 (and its extension to hyperimmune degrees), while Melnikov [78] gave a positive answer to the analog of Question 1.5.

Noah Schweber [103] has suggested an approach to giving a positive answer to Question 1.4, which goes through another set of results related to the Slaman-Wehner Theorem.

An alternative measure of the complexity of a structure can be obtained by looking at its full elementary diagram rather than just its atomic diagram.

Definition 1.6. A (presentation of a) structure is *decidable* if its elementary diagram is computable. The *elementary degree spectrum* of $\mathcal{M}$ is the set of degrees of elementary diagrams of presentations of $\mathcal{M}$.

It is easy to see that the usual Henkin proof of the completeness theorem can be effectivized to show that every complete decidable theory has a decidable model, but things are often different if one wants this model to have certain special properties. For instance, every atomic theory in a countable language has a countable atomic model, but this result does not hold effectively. (Recall that a theory T is atomic if every formula consistent with T is contained in a principal type, and a model is atomic if every type realized in it is principal.) To make this statement more precise and cast it in a form that will be more relevant to Question 1.4, consider the following definition.

A *binary tree* is a set T of finite binary strings such that if $\sigma \in T$ and $\tau \prec \sigma$ then $\tau \in T$. A string $\sigma \in T$ is a *dead end* if $\sigma 0, \sigma 1 \notin T$. A *path* on T is an infinite binary sequence P such that every finite initial segment of P is in T.

Definition 1.7. A *PAC tree* is a computable binary tree with no dead ends, each of whose paths is computable.

The motivation behind this definition is that PAC trees are essentially the trees of types of complete decidable theories all of whose types are computable. (See [43, 46] for more details.) Such a theory has only countably many types, and hence is atomic. Goncharov and Nurtazin [42] and Harrington [44] showed that a complete decidable theory T has a decidable atomic model if and only if there is a computable listing of the principal types of T. Millar [85] showed that another sufficient condition for a complete decidable theory T to have a decidable atomic model is that there be a computable listing of *all* types of T. Thus, in a sense, the simplest possible complete decidable theory with no decidable atomic model would be one such that each type is individually computable, but there is no way to uniformly compute all the types, or even all the principal types. Since isolated paths correspond to principal types, the following result has as a corollary that there exists such a theory. (For more on the computability-theoretic and proof-theoretic aspects of the existence of atomic models, see [47, Sect. 9.3] and the references mentioned there. See also Cholak and McCoy [6] in connection with Open Question 9.47 in that book.)

Theorem 1.8 (Goncharov and Nurtazin [42]; Millar [84]). There is a PAC tree whose isolated paths cannot be computably listed.

Thus the Muchnik degree of the set of listings of the isolated paths of a PAC tree is not always trivial. However, there is only one other possibility for what this degree can be.

Theorem 1.9 (Hirschfeldt [46]). Let T be a PAC tree and let $X >_T \emptyset$. Then the isolated paths on T can be X-computably listed.

Combining this result with Theorem 1.8 shows that there is a PAC tree T such that the isolated paths on T can be X-computably listed if and only if X is not computable. Restated in model-theoretic terms, Theorem 1.9 says that if T is a complete decidable theory all of whose types are computable, then the elementary degree spectrum of the atomic model of T includes all nonzero degrees. This result extends an earlier one of Csima [11,12], who showed that such a spectrum includes all nonzero Δ_2^0 degrees.

The translation of trees into theories can be done in such a way that the atomic model of the theory obtained from the PAC tree in Theorem 1.8 not only has no decidable presentation, but does not even have a computable presentation. Thus we have the following fact, which extends the Slaman-Wehner theorem to models of decidable theories.

Corollary 1.10 (Hirschfeldt [46]). There is a structure $\mathcal{M}$ whose atomic and elementary degree spectra both consist of the nonzero degrees. Furthermore, $\mathcal{M}$ can be chosen to be the atomic model of a complete decidable theory each of whose types is computable.

Let us now return to Question 1.4. For a tree $\mathcal{T}$, let $\mathcal{L}(\mathcal{T})$ be the linear order consisting of the isolated paths on $\mathcal{T}$ with the lexicographic order. Schweber [103] observed that if I is a listing of the isolated paths on $\mathcal{T}$, then $\mathcal{L}(\mathcal{T})$ has an I-computable presentation, and hence, if $\mathcal{T}$ is a PAC tree, then the degree spectrum of $\mathcal{L}(\mathcal{T})$ contains all nonzero degrees. Thus a positive answer to the following question would imply a positive answer to Question 1.4.

Open Question 1.11 (Schweber [103]). *Is there a PAC tree $\mathcal{T}$ for which $\mathcal{L}(\mathcal{T})$ has no computable presentation?*

Schweber [103] did show that there is no computable way to pass from an index for a PAC tree $\mathcal{T}$ to one for a computable presentation of $\mathcal{L}(\mathcal{T})$. Nevertheless, both he and I strongly believe that the answer to this question is negative. The linear orders $\mathcal{L}(\mathcal{T})$ arising from PAC trees $\mathcal{T}$ do not seem sufficiently complex to permit diagonalization against computable presentations. In particular, each such ordering is *scattered*, i.e., does not contain a suborder of type η, and hence cannot contain Jockusch-Soare-style separators. Indeed, it seems quite reasonable to conjecture that no scattered linear order can have degree spectrum consisting exactly of the noncomputable degrees, although this has not been shown to be the case.

Incidentally, the following question is also open.

Open Question 1.12 (Schweber [103]). *Is every computable scattered linear order isomorphic to $\mathcal{L}(\mathcal{T})$ for some PAC tree $\mathcal{T}$?*

But perhaps a little more life can be injected into this approach by considering more complicated trees. For a tree $\mathcal{T}$ and $\sigma \in \mathcal{T}$, let $\mathcal{T}_\sigma$ be the tree consisting of all τ such that $\sigma\tau \in \mathcal{T}$. Let $[\mathcal{T}]$ be the set of paths on $\mathcal{T}$.

Definition 1.13. A *quasi-PAC* tree is a computable binary tree $\mathcal{T}$ with no dead ends such that for each noncomputable path P of $\mathcal{T}$, there is a $\sigma \prec P$ for which $[\mathcal{T}_\sigma]$ is perfect (i.e., has no isolated elements).

For a quasi-PAC tree $\mathcal{T}$, let $S(\mathcal{T})$ be the set of all $\sigma \in \mathcal{T}$ such that $[\mathcal{T}_\sigma]$ is either a singleton or perfect. Let $M(\mathcal{T})$ be the set of minimal elements of $S(\mathcal{T})$

(i.e., nodes $\sigma \in S(T)$ such that if $\tau \prec \sigma$ then $\tau \notin S(T)$). Let $\mathcal{L}(T)$ be the linear order obtained by first ordering $M(T)$ lexicographically, then replacing each $\sigma \in M(T)$ such that $[T_\sigma]$ is perfect by a copy of the rationals. (If T is a PAC tree, then this definition agrees with the previous definition of $\mathcal{L}(T)$ up to isomorphism.) Notice that, unlike in the case of PAC trees, this linear order is not necessarily scattered, and indeed can include Jockusch-Soare-style separators.

Proposition 1.14. Let T be a quasi-PAC tree. Then the degree spectrum of $\mathcal{L}(T)$ contains all nonzero degrees.

Proof. Let $X >_{\mathrm{T}} \emptyset$. The idea is to first build an X-computable collection of paths on T using the same method as in the proof of Theorem 1.9 above given in [46], then use it to build an X-computable presentation of $\mathcal{L}(T)$.

Let $\sigma_0, \sigma_1, \dots$ list the nodes of T, say in length-lexicographic order. For each n, let f_n be the path on T defined as follows. Begin at σ_n, and proceed along T until there is a split in T, i.e., a $\tau \succeq \sigma_n$ such that $\tau 0$ and $\tau 1$ are both in T. (Of course, such a split might never be found.) Take the right node of this split if $0 \in X$, and take the left node if $0 \notin X$. Then continue along T until there is another split (if ever). Then take the right node of this split if $1 \in X$, and take the left node if $1 \notin X$. Continue in this way, deciding which side of splits to follow depending on successive bits of X.

Now $f_0, f_1, \dots$ are uniformly X-computable paths on T, and include all the isolated paths on T. Let $S = \{n : \forall m < n (f_n \neq f_m)\}$. Then S is c.e. Let $n_0, n_1, \dots$ be an enumeration of S and let $g_i = f_{n_i}$. Then the g_i are uniformly X-computable and list the same paths as the f_i, but without repetitions. Let $\mathcal{L}$ be the X-computable linear order with domain ω defined by letting $i <_{\mathcal{L}} j$ if g_i is to the left of g_j.

The claim now is that $\mathcal{L}$ is a presentation of $\mathcal{L}(T)$. Let $M(T)$ be as above (i.e., the minimal elements of the set of $\sigma \in T$ such that $[T_\sigma]$ is either a singleton or perfect). If g_n is not isolated then infinitely many splits are encountered in its definition. Which direction g_n takes at each split is determined by successive bits of X, so in this case X can be computed from g_n. Thus every g_n is isolated or noncomputable. So, by the definition of quasi-PAC tree, every g_n extends some element of $M(T)$. Of course, it is also the case that for every $\sigma \in M(T)$, there is a g_n extending σ, which is unique if T_σ has only one path. Thus it is enough to show that if $[T_\sigma]$ is perfect, then the set of g_n extending σ has the order type of the rationals under the lexicographic order.

Suppose that g_m and g_n both extend such a σ, for $m \neq n$. Since $g_m \neq g_n$, assume without loss of generality that there is a $\tau \succeq \sigma$ such that $\tau 0 \prec g_m$ and $\tau 1 \prec g_n$. Since g_m is not isolated, it is not computable, and hence cannot be the rightmost path on T extending $\tau 0$ (since T has no dead ends, and hence this rightmost path is computable). Thus there is a $\rho \succ \tau 0$ that is to the right of g_m. This ρ is to the left of g_n, and there must be some g_k extending ρ. Now g_k is strictly in between g_m and g_n. Similar arguments show that there cannot be a leftmost or a rightmost g_n extending σ. □

Thus, as in the case of Question 1.11, a positive answer to the following question would imply a positive answer to Question 1.4.

Open Question 1.15 (Hirschfeldt (see Schweber [103])). *Is there a quasi-PAC tree $\mathcal{T}$ for which $\mathcal{L}(\mathcal{T})$ has no computable presentation?*

Some time spent trying to give a positive answer to this question has made me lean toward believing that the answer is actually negative, but with less confidence than in the case of Question 1.11.

2 Linearizing Partial Orders

There are several other intriguing questions involving linear orders. In this section, and the next, I will briefly describe a couple of my favorite ones.

After finishing my dissertation and before going to New Zealand as Rod's postdoc, I spent a month with him visiting Steffen Lempp and Reed Solomon at Wisconsin. The four of us sat in Steffen's office for hours on end, day after day. Not exactly Rod's favorite mode of working, but productive in the event, as it yielded three papers. One of these took a reverse-mathematical look at linear extensions of partial orders.

Szpilrajn [112] showed that every partial order $(X, \leqslant_P)$ has a linear extension, that is, a linear order $(X, \leqslant_L)$ such that if $a \leqslant_P b$ then $a \leqslant_L b$. It is natural to ask which properties of a partial order can be preserved by some linear extension. For instance, if a partial order is well-founded, does it have a well-ordered linear extension? This and similar questions can be stated concisely using the following notation.

Definition 2.1. Let τ be a linear order type. Say that τ is *extendible* if every partial order with no suborder of type τ has a linear extension with no suborder of type τ. Say that τ is *weakly extendible* if every countable partial order with no suborder of type τ has a linear extension with no suborder of type τ.

Characterizations of the extendible and weakly extendible countable order types were obtained by Bonnet [4] and Jullien [62], respectively. For the purposes of reverse-mathematical and computability-theoretic analysis, weak extendibility is the natural notion to study.

Definition 2.2. Let $\mathrm{EXT}(\tau)$ be the statement that τ is weakly extendible.

$\mathrm{EXT}(\omega^*)$, for example, is the statement that every countable well-founded partial order has a well-ordered linear extension, which is indeed true. Downey, Hirschfeldt, Lempp, and Solomon [20] studied the weak extendibility of ω^*, η (which recall is the order type of the rationals), and ζ (the order type of the integers). Only in the last case did we obtain a full reverse-mathematical characterization, though. (For definitions of RCA_0, ATR_0, and other systems mentioned here, see Simpson [107].)

Theorem 2.3 (Downey, Hirschfeldt, Lempp, and Solomon [20]). *The principle* $\mathrm{EXT}(\zeta)$ *is equivalent to* ATR_0 *over* RCA_0.

For $\mathrm{EXT}(\omega^*)$ (i.e., the principle that every countable well-founded partial order has a well-ordered linearization), we were able to find the following bounds.

Theorem 2.4 (Downey, Hirschfeldt, Lempp, and Solomon [20]). *The principle* $\mathrm{EXT}(\omega^*)$ *is provable in* ACA_0, *and is strictly stronger than* WKL_0 *over* RCA_0.

The following questions remain open, however. Ramsey's Theorem for pairs (RT_2^2) and some related principles will be discussed further below.

Open Question 2.5 (Downey, Hirschfeldt, Lempp, and Solomon [20]; Hirschfeldt [47]). *Does* $\mathrm{RCA}_0 + \mathrm{EXT}(\omega^*) \vdash \mathrm{ACA}_0$? *What is the relationship between* $\mathrm{EXT}(\omega^*)$ *and* RT_2^2 *(and related principles)?*

Another way to state $\mathrm{EXT}(\eta)$ is that every scattered partial order has a scattered linear extension. Becker (see [20]) showed that $\Pi_1^1\text{-}\mathrm{CA}_0 \vdash \mathrm{EXT}(\eta)$. As part of his analysis of the reverse-mathematical strength of Julien's classification of the weakly extendible order types, Montalbán [94] improved this result by showing that $\mathrm{ATR}_0 + \mathrm{I}\Sigma_1^1 \vdash \mathrm{EXT}(\eta)$. Conversely, Joe Miller [unpublished] showed that $\mathrm{EXT}(\eta)$ implies WKL_0 over RCA_0, and implies ATR_0 over $\Sigma_1^1\text{-}\mathrm{AC}_0$. The exact strength of $\mathrm{EXT}(\eta)$ is still unknown, and in particular, the following question is open.

Open Question 2.6 (Montalbán [94]). *What is the exact relationship between* ATR_0 *and* $\mathrm{EXT}(\eta)$ *over* RCA_0?

For some further discussion of this and related questions, see [47, Sects. 10.2 and 10.3].

3   The Dushnik-Miller Theorem and Computability Theory

The paper by Downey, Lempp, and Wu [29] mentioned in Sect. 1 introduced a new method for constructing Δ_3^0 isomorphisms, which was also used by Downey, Kastermans, and Lempp [27] to give a partial answer to the longstanding Effective Dushnik-Miller Conjecture of Downey and Moses (see [15]).

A *nontrivial self-embedding* of a linear order $\mathcal{L}$ is an order preserving map from $\mathcal{L}$ into itself that is not the identity. The Dushnik-Miller Theorem [31] states that every infinite linear order has a nontrivial self-embedding. This theorem does not hold effectively, even for the simplest order type of infinite linear orders: Hay and Rosenstein (see [100]) showed that there is a computable linear order of order type ω with no computable nontrivial self-embeddings, and Downey and Lempp [28] improved this result by building a computable linear order $\mathcal{L}$ of order type ω such that any nontrivial self-embedding of $\mathcal{L}$ computes $\emptyset'$. They also showed that the latter construction can be turned into a proof that the Dushnik-Miller Theorem is equivalent to ACA_0 over RCA_0. (See Downey, Jockusch, and Miller [25] for a clarification of that proof.)

Downey, Jockusch, and Miller [25] showed that every computable infinite linear order has an $\emptyset''$-computable nontrivial self-embedding, but there is a computable infinite linear order with no $\emptyset'$-computable nontrivial self-embeddings.

Open Question 3.1 (Downey, Jockusch, and Miller [25]). *Is there a computable infinite linear order $\mathcal{L}$ such that every nontrivial self-embedding of $\mathcal{L}$ computes $\emptyset''$?*

As mentioned above, there is a computable presentation of ω with no computable nontrivial self-embeddings, and the same is true of many order types. There is one known class of computably presentable linear orders for which every computable presentation has a computable nontrivial self-embedding. A linear order $\mathcal{L}$ is η-*like* if the order type of $\mathcal{L}$ can be obtained from η by replacing each point by a nonempty block of finitely many points. A linear order $\mathcal{L}$ is *strongly η-like* if there is an n such that the order type of $\mathcal{L}$ can be obtained from η by replacing each point by a nonempty block of at most n many points. Watnick and Lerman (see [15]) noted that if a computable linear order has a strongly η-like interval, then it has a computable nontrivial self-embedding.

Since having a strongly η-like interval is a property of an order type, rather than of its presentations, if a computably presentable linear order $\mathcal{L}$ has a strongly η-like interval, then every computable presentation of $\mathcal{L}$ has a computable nontrivial self-embedding. Downey and Moses (see [15]) conjectured that this is the only situation in which this is the case, that is, that the answer to the following question is positive.

Open Question 3.2 (Downey and Moses (see [15])). *If every computable presentation of a computable linear order $\mathcal{L}$ has a computable nontrivial self-embedding, must $\mathcal{L}$ contain a strongly η-like interval?*

Downey, Kastermans, and Lempp [27] showed that this conjecture of Downey and Moses holds for all computable η-like linear orderings. In [15], Rod discussed some of the difficulties involved in proving the full conjecture.

4 Computable Dimension and Relatively Easy Isomorphisms

Another natural question to ask about a computably presentable structure is how many computable presentations it has, up to computable isomorphism. This number is known as the *computable dimension* of the structure. A structure of computable dimension 1 is said to be *computably categorical*. There are many examples of computably categorical structures, such as $(\mathbb{Q}, <)$, and of structures of computable dimension ω, such as $(\mathbb{N}, <)$. Structures of finite computable dimension greater than 1 do not seem to occur "in nature", but nevertheless exist, as shown by Goncharov [38]. Indeed, there are structures of any given finite dimension. By the kinds of general encoding results mentioned in Sect. 1, such structures also exist within various familiar classes of structures. A particularly interesting recent result in this direction by Miller, Poonen, Schoutens, and Shlapentokh [90], which resolved a longstanding open question, is that the class of fields has the same universality properties as the ones dealt with by Hirschfeldt, Khoussainov, Shore, and Slinko [52] (as discussed in Sect. 1). In particular, there are fields of any given finite dimension. (Another interesting aspect of [90] is the casting of encoding results such as the ones in [52] in terms of a new kind of computable category theory. This line of research has been further pursued by Harrison-Trainor, Melnikov, Miller, and Montalbán [45].)

There are several situations in which structures of finite computable dimension greater than 1 cannot exist, however. For instance, Goncharov and Dzgoev [41] and Remmel [98] showed that every computably presentable linear order has computable dimension 1 or ω; Goncharov [40] did the same for Boolean algebras (though the result was implicit in earlier work of Goncharov and, independently, LaRoche [70]); and Lempp, McCoy, Miller, and Solomon [71, 72] for trees (as partial orders, or under the meet function). A more computability-theoretic obstruction to the existence of structures of finite computable dimension greater than 1 is given by the following result.

Theorem 4.1 (Goncharov [39]). Let $\mathcal{A}$ and $\mathcal{B}$ be computable structures such that there is no computable isomorphism between $\mathcal{A}$ and $\mathcal{B}$, but there is a Δ_2^0 isomorphism between them. Then $\mathcal{A}$ has computable dimension ω.

Goncharov's examples in [38] of structures of finite computable dimension greater than 1 are Δ_3^0-*categorical*, i.e., for each such structure $\mathcal{M}$, there is a Δ_3^0 isomorphism between any two given presentations of $\mathcal{M}$. Thus Theorem 4.1 cannot be extended to Δ_3^0 isomorphisms. But perhaps it can be extended to some class intermediate between Δ_2^0 and full-blown Δ_3^0 isomorphisms.

One way to zero in on a potential class of this kind is to consider concrete examples. One such example is given by locally finite connected graphs, where a graph is *locally finite* if each vertex is on only finitely many edges. (It does not matter here whether the graphs are directed or undirected.) There are several examples of graphs of finite computable dimension greater than 1, and in every case they make essential use of vertices connected to infinitely many other vertices. It seems difficult to modify these constructions to produce locally finite graphs. Nevertheless, the following question, which comes from joint work with Bakh Khoussainov, remains open.

Open Question 4.2. *Is there a locally finite connected graph of finite computable dimension greater than 1?*

Another interesting example is that of algebraic fields.

Open Question 4.3 (Hirschfeldt, Kramer, Miller, and Shlapentokh [53]). *Is there an algebraic field of finite computable dimension greater than 1?*

What connects these two classes of structures is that if we take two isomorphic computable structures $\mathcal{A}$ and $\mathcal{B}$ in either of these classes, there is a computable infinite, finitely branching subtree of ω^ω each of whose paths is an isomorphism between $\mathcal{A}$ and $\mathcal{B}$. (In the sense that for each such path P, the map $n \mapsto P(n)$ is such an isomorphism.) If $\mathcal{A}$ and $\mathcal{B}$ are isomorphic computable locally finite connected graphs and we fix $a \in \mathcal{A}$ and $b \in \mathcal{B}$ such that $(\mathcal{A}, a)$ and $(\mathcal{B}, b)$ are isomorphic, it is easy to build a computable finitely branching tree whose paths are exactly the isomorphisms between $(\mathcal{A}, a)$ and $(\mathcal{B}, b)$. If $\mathcal{A}$ and $\mathcal{B}$ are isomorphic computable algebraic fields then Miller [89] showed that there is a computable finitely branching tree whose paths are exactly the isomorphisms between $\mathcal{A}$ and $\mathcal{B}$.

It is possible that Theorem 4.1 can be extended to cover this general case, giving a positive answer to the following question, which comes from discussions with Russell Miller, and hence negative answers to Questions 4.2 and 4.3.

Open Question 4.4. *Let $\mathcal{A}$ and $\mathcal{B}$ be computable structures that are not computably isomorphic. Suppose that there is a computable infinite, finitely branching subtree of ω^ω each of whose paths is an isomorphism between $\mathcal{A}$ and $\mathcal{B}$. Must the computable dimension of $\mathcal{A}$ be infinite?*

5 Ramsey's Theorem and Computability-Theoretic Reductions

Another of the papers I worked on with Rod, Steffen, and Reed at Wisconsin introduced me to a question that has continued to preoccupy me off and on since then: determining the exact relationship between Ramsey's Theorem for Pairs (RT_2^2) and its stable version SRT_2^2. (Some of this section overlaps with a recent open questions paper by Patey [97], which also contains many questions on the reverse mathematics of Ramsey-type statements not considered here.)

The computability-theoretic and reverse-mathematical analysis of versions of Ramsey's Theorem has been an important line of research since the work of Specker [111] and Jockusch [59] in the early 1970's.

Definition 5.1. For a set X, let $[X]^n$ be the collection of n-element subsets of X. A k-coloring of $[X]^n$ is a map $c : [X]^n \to k$. A set $H \subseteq X$ is *homogeneous for c* if there is an $i < k$ such that $c(s) = i$ for all $s \in [H]^n$.

Ramsey's Theorem for n-tuples and k colors RT_k^n is the statement that every k-coloring of $[\mathbb{N}]^n$ has an infinite homogeneous set. $\mathrm{RT}_{<\infty}^n$ is the statement $\forall k \, \mathrm{RT}_k^n$. RT is the statement $\forall n \, \mathrm{RT}_{<\infty}^n$.

It is not difficult to show that RT_k^n is equivalent to RT_2^n over RCA_0 for all $k \geqslant 2$, and of course RT_2^1 is provable in RCA_0. Building on computability-theoretic results of Jockusch [59], Simpson [106] showed that RT_2^n is equivalent to ACA_0 over RCA_0 for all $n \geqslant 3$. The $n = 2$ case has proved to be considerably more interesting. Building on computability-theoretic results of Jockusch [59], Hirst [56] showed that RT_2^2 is not provable in WKL_0. Seetapun (see [105]) showed that RT_2^2 does not imply ACA_0 over RCA_0. More recently, Liu [75, 76] showed that RT_2^2 does not imply WKL_0, or even $WWKL_0$ (which will be discussed further below), over RCA_0.

Unlike WKL_0, there are not many principles equivalent to RT_2^2, but there is a whole universe of principles provable from $RCA_0 + RT_2^2$. (I have told some of this story in considerably more detail in [47].) For instance, Cholak, Jockusch, and Slaman [8] found a highly productive way to split RT_2^2 into two principles, called SRT_2^2 and COH.

Definition 5.2. A coloring $c : [\mathbb{N}]^2 \to k$ is *stable* if $\lim_y c(x, y)$ exists for all x. *Stable Ramsey's Theorem for Pairs and k colors* SRT_k^2 is the statement that every stable k-coloring of $[\mathbb{N}]^2$ has an infinite homogeneous set. $SRT_{<\infty}^2$ is the statement $\forall k \, SRT_k^2$.

A set C is *cohesive* for a collection of sets $R_0, R_1, \ldots$ if C is infinite and for each i, either $C \subseteq^* R_i$ or $C \subseteq^* \overline{R_i}$ (where $X \subseteq^* Y$ means that $X \setminus Y$ is finite). The *Cohesive Set Principle* COH is the statement that every countable collection of sets has a cohesive set.

One direction of the original proof in [8] that RT_2^2 is equivalent to SRT_2^2+COH required Σ_2^0-induction, but this use of induction was removed by Mileti [83] and Jockusch and Lempp [unpublished].

Theorem 5.3 (Cholak, Jockusch, and Slaman [8]; Mileti [83]; Jockusch and Lempp [unpublished]). RT_2^2 is equivalent to $SRT_2^2 + COH$ over RCA_0.

Cholak, Jockusch, and Slaman [8] showed that COH does not imply RT_2^2 over RCA_0, but obtaining the analogous statement for SRT_2^2 in place of COH proved far more elusive. For well over a decade, many researchers, myself included, tried a variety of approaches to this problem without success.

A frustrating aspect of this problem is that, from the point of view of computability theory, stability *does* allow us to decrease the complexity of homogeneous sets in general. Jockusch [59] showed that there are computable 2-colorings of $[\mathbb{N}]^2$ with no Δ_2^0 infinite homogeneous sets. On the other hand, if the computable coloring $c : [\mathbb{N}]^2 \to 2$ is stable, then $\emptyset'$ can compute the function $x \mapsto \lim_y c(x, y)$, from which it is easy to obtain an infinite homogeneous set for c effectively. Thus c has a Δ_2^0 infinite homogeneous set. However, this fact in itself does not help to build a model of SRT_2^2 that is not a model of RT_2^2, because such a model would have to contain not only an infinite homogeneous set H for c, but one for every H-computable stable 2-coloring of pairs, at which point one might be in the realm of Δ_3^0 sets (and of course the complexity of homogeneous sets might get even higher as further iterations are considered). What *would* help would be to find a $\mathcal{C} \subset \Delta_2^0$ such that every 2-coloring of pairs $c \in \mathcal{C}$ has an infinite homogeneous set H such that $c \oplus H \in \mathcal{C}$. Cholak, Jockusch, and Slaman [8] suggested that the low sets might form such a class, but that turns out not to be the case.

Theorem 5.4 (Downey, Hirschfeldt, Lempp, and Solomon [19]). There is a computable stable 2-coloring of pairs with no low infinite homogeneous sets.

It did not occur to us (or at least to me) to ask whether this theorem holds in nonstandard models of Σ_1^0-PA (the first-order part of RCA$_0$). As it turns out, it does not, though it takes a rather intricate construction to establish this fact. Chong, Slaman, and Yang [9] built a model of RCA$_0$ + SRT$_2^2$ (in which Σ_2^0-induction fails) whose second-order part consists entirely of low sets, in the sense of the first-order part of the model. As shown by Cholak, Jockusch, and Slaman [8], BΣ_2^0 (Σ_2^0-bounding) must hold in any model of RCA$_0$ + SRT$_2^2$. Chong, Slaman, and Yang [9] also showed that Jockusch's result in [59] that there are computable 2-colorings of $[\mathbb{N}]^2$ with no Δ_2^0 infinite homogeneous sets goes through in RCA$_0$ + BΣ_2^0. Thus they were able to separate SRT$_2^2$ and RT$_2^2$ in the reverse-mathematical setting.

Theorem 5.5 (Chong, Slaman, and Yang [9]). RCA$_0$ + SRT$_2^2$ $\nvdash$ RT$_2^2$.

Remarkable as it is, this result still leaves open the question of whether any approach along more traditional lines, working in the standard first-order model, can be made to work. Such an approach would in fact establish a stronger result. Recall that an ω-model of second-order arithmetic is one with standard first-order part. Write $P \leqslant_\omega Q$ to mean that every ω-model of RCA$_0$ + Q is an ω-model of P. For example, COH and (S)RT$_2^2$ can be separated via ω-models, for instance by using a conservativity result of Hirschfeldt and Shore [55] or by considering the principle DNR, as in Hirschfeldt, Jockusch, Kjos-Hanssen, Lempp, and Slaman [49], so RT$_2^2$ $\nleqslant_\omega$ COH. The natural follow-up question to Theorem 5.5 can now be stated as follows.

Open Question 5.6 (Cholak, Jockusch, and Slaman [8]; Chong, Slaman, and Yang [9]). *Is* RT$_2^2$ $\leqslant_\omega$ SRT$_2^2$? *Equivalently, is* COH $\leqslant_\omega$ SRT$_2^2$?

In light of the methods in [9] discussed above, the following question is also of interest (and a positive answer to it would imply a positive answer to Question 5.6).

Open Question 5.7 (Chong, Slaman, and Yang [10]). *Does* RCA$_0$ + IΣ_2^0 + SRT$_2^2$ $\vdash$ RT$_2^2$?

It is possible that the approach to answering Question 5.6 ruled out in its simplest form by Theorem 5.4 could still be revived, if the answer to the following questions is positive.

Open Question 5.8 (Hirschfeldt, Jockusch, Kjos-Hanssen, Lempp, and Slaman [49]). *Does every computable stable 2-coloring of pairs have an infinite homogeneous that is both Δ_2^0 and low$_2$ (or just Δ_2^0 and low$_n$ for some n, where n could even depend on the coloring)?*

As explained in [49], a relativizable positive answer to this question would yield a negative solution to Question 5.6. On the other hand, it could be that there is a computable stable 2-coloring of pairs such that the jump of every infinite homogeneous set has PA degree relative to $\emptyset'$, which, again as explained in [49], would not only give a negative answer to Question 5.8, but (if this fact is relativizable) also a positive one to Question 5.6.

Another way to think about Question 5.6 is to study its analogs for computability-theoretic reducibilities stronger than $\leqslant_\omega$. Many interesting principles (including Ramsey's Theorem and its variants) have the form

$$\forall X\,[\Theta(X) \rightarrow \exists Y\,\Psi(X,Y)]$$

with Θ and Ψ arithmetic. Such a principle can be thought of as a *problem*. An *instance* of this problem is an X such that $\Theta(X)$ holds and a *solution* to this instance is a Y such that $\Psi(X,Y)$ holds.

For principles of this kind, the definition of $\leqslant_\omega$ can be reformulated without reference to reverse mathematics. Recall that a *Turing ideal* is a collection of sets closed under Turing reduction and finite joins. Say that a problem P *holds* in a Turing ideal $\mathcal{I}$ if every instance of P in $\mathcal{I}$ has a solution in $\mathcal{I}$. Turing ideals are exactly second-order parts of ω-models of RCA_0, so $P \leqslant_\omega Q$ if and only if P holds in every ideal in which Q holds.

Reducibilities such as the following ones allow for a finer-grained investigation of relationships between problems. All four of the notions below capture the idea of being able to solve any given instance X of a problem P by using the ability to solve an instance of another problem Q obtained computably from X. The difference between the computable and Weihrauch versions is that the latter are uniform. The difference between the normal and strong versions is that the latter do not allow the use of X itself in computing a solution to X.

Definition 5.9. Let P and Q be problems.

1. Say that P is *computably reducible* to Q, and write $P \leqslant_c Q$, if for every instance X of P, there is an X-computable instance $\widehat{X}$ of Q such that, for every solution $\widehat{Y}$ to $\widehat{X}$, there is an $X \oplus \widehat{Y}$-computable solution to X.

2. Say that P is *strongly computably reducible* to Q, and write $P \leqslant_{sc} Q$, if for every instance X of P, there is an X-computable instance $\widehat{X}$ of Q such that, for every solution $\widehat{Y}$ to $\widehat{X}$, there is a $\widehat{Y}$-computable solution to X.

3. Say that P is *Weihrauch reducible* to Q, and write $P \leqslant_w Q$, if there are Turing functionals Φ and Ψ such that, for every instance X of P, the set $\widehat{X} = \Phi^X$ is an instance of Q, and for every solution $\widehat{Y}$ to $\widehat{X}$, the set $Y = \Psi^{X \oplus \widehat{Y}}$ is a solution to X.

4. Say that P is *strongly Weihrauch reducible* to Q, and write $P \leqslant_{sw} Q$, if there are Turing functionals Φ and Ψ such that, for every instance X of P, the set $\widehat{X} = \Phi^X$ is an instance of Q, and for every solution $\widehat{Y}$ to $\widehat{X}$, the set $Y = \Psi^{\widehat{Y}}$ is a solution to X.

(Strong) Weihrauch reducibility has also been called (strong) uniform reducibility. The notion of Weihrauch reducibility is actually a broader one, introduced by Weihrauch [118,119] in the context of computable analysis and widely studied since, but the definition given above is equivalent to a special case of it. (See Dorais, Dzhafarov, Hirst, Mileti, and Shafer [13] and the papers listed in the bibliography [5].)

One approach to Question 5.6 is to seek partial answers, perhaps involving methods that can be adapted to answer the full question, by replacing $\leqslant_w$ with each of the stronger notions of reducibility above. Of course, given the computability-theoretic difference between RT_2^2 and SRT_2^2, the second of the two equivalent statements of Question 5.6 is the relevant one here. All but one of these versions of Question 5.6 have been answered by Dzhafarov [32].

Theorem 5.10 (Dzhafarov [32]). COH $\not\leqslant_{\mathrm{sc}}$ SRT_2^2 and COH $\not\leqslant_{\mathrm{w}}$ SRT_2^2 (and hence COH $\not\leqslant_{\mathrm{sw}}$ SRT_2^2).

The case of computable reducibility remains open, however, and might well be the most relevant one to a potential solution to Question 5.6.

Open Question 5.11 (Hirschfeldt and Jockusch [48]). *Is* COH $\leqslant_c$ SRT_2^2?

It should be noted that, when considering reducibilities stronger than $\leqslant_w$, the number of colors starts to matter. For instance, while it is not difficult to show that $\mathrm{RT}_k^n \leqslant_w \mathrm{RT}_j^n$ even when $2 \leqslant j \leqslant k$, Patey [96] showed that $\mathrm{RT}_k^n \not\leqslant_c \mathrm{RT}_j^n$ in this case, as long as $n \geqslant 2$. Thus the following results strengthen Theorem 5.10.

Theorem 5.12 (Dzhafarov [32]). COH $\not\leqslant_{\mathrm{w}}$ $\mathrm{SRT}_{<\infty}^2$.

Theorem 5.13 (Dzhafarov, Patey, Solomon, and Westrick [33]). COH $\not\leqslant_{\mathrm{sc}}$ $\mathrm{SRT}_{<\infty}^2$.

The analog of Question 5.11 for $\mathrm{SRT}_{<\infty}^2$ is also open.

The difference between $\leqslant_c$ and $\leqslant_w$ is that the latter covers cases in which a problem P is reducible a problem Q, but only if one is allowed to use several instances of Q to solve an instance of P. It is thus natural to seek a nonuniform version of $\leqslant_w$ that allows for multiple uses of a principle, but only if the relevant instances are produced in a uniformly computable way. Such a notion was defined in [48] using games.

Definition 5.14. For problems P and Q representing true Π_2^1 principles, the *reduction game* $G(Q \to P)$ is a two-player game that proceeds as follows.

On the first move, Player 1 plays an instance X_0 of P, and Player 2 either plays an X_0-computable solution to X_0 and declares victory, in which case the game ends, or responds with an X_0-computable instance Y_1 of Q. If Player 2 cannot move (which might happen if there is no X_0-computable instance of Q), then Player 1 wins, and the game ends.

For $n > 1$, on the nth move (if the game has not yet ended), Player 1 plays a solution X_{n-1} to the instance Y_{n-1} of Q. Then Player 2 either plays a $(\bigoplus_{i<n} X_i)$-computable solution to X_0 and declares victory, in which case again the game ends, or plays a $(\bigoplus_{i<n} X_i)$-computable instance Y_n of Q.

Player 2 wins this play of the game if it ever declares victory. Otherwise, Player 1 wins.

Reduction games can be used to give a characterization of $\leqslant_\omega$. A *strategy* for a player in a game such as the above ones is a map taking any sequence of moves by the opponent to a move by the given player. Such a strategy is *winning* if it enables the player to win no matter what the opponent does.

Theorem 5.15 (Hirschfeldt and Jockusch [48]). *If $P \leqslant_\omega Q$ then Player 2 has a winning strategy for $G(Q \to P)$. Otherwise, Player 1 has a winning strategy for $G(Q \to P)$.*

Effectivizing winning strategies yields a notion of generalized uniform reducibility between Π^1_2 principles. (See [48] for a more detailed definition.)

Definition 5.16. A *computable strategy* for Player 2 in a reduction game is a Turing functional that, given the join of Player 1's first n moves as an oracle, outputs Player 2's nth move.

Say that P is *Weihrauch (or uniformly) reducible to Q in the generalized sense*, and write $P \leqslant_{\mathrm{gW}} Q$, if Player 2 has a computable winning strategy in $G(Q \to P)$.

Assuming that, as expected, the answer to Question 5.6 is negative, the following might be an easier version of that question.

Open Question 5.17 (Hirschfeldt and Jockusch [48]). *Is $\mathrm{RT}^2_2 \leqslant_{\mathrm{gW}} \mathrm{SRT}^2_2$?*

It is also worth noting that it does not seem trivial to adapt the proof of Theorem 5.5 above given in [9] to the case of arbitrarily many colors. For one thing, Cholak, Jockusch, and Slaman [8] showed that $\mathrm{SRT}^2_{<\infty}$ implies $\mathrm{B}\Sigma^0_3$, and hence $\mathrm{I}\Sigma^0_2$, over RCA_0, so $\mathrm{SRT}^2_{<\infty}$ does not hold in the model built in that proof. Thus the following question is still open.

Open Question 5.18 (Cholak, Jockusch, and Slaman [8]). *Does $\mathrm{SRT}^2_{<\infty}$ imply $\mathrm{RT}^2_{<\infty}$ over RCA_0?*

A liability of writing an open questions paper is that some of the questions might be solved while the paper is in preparation. Indeed, while I was in the final stages of revising this paper for submission, a problem I had planned to discuss was solved by Monin and Patey [93]. I will still include this discussion here, however, as an example of ongoing work in the area, and as an opportunity to mention a couple of open questions in [93].

One of the things that looking at notions of computability-theoretic reduction does is highlight cases in which relationships between principles are less well-understood than might have been thought. Recall that Weak Weak König's Lemma (WWKL) is the statement that if T is a binary tree such that $\liminf_n \frac{|\{\sigma \in T : |\sigma| = n\}|}{2^n} > 0$, then T has a path. The system WWKL_0 obtained by adding this statement to RCA_0 has played a significant role in reverse mathematics, and there is a case for according it similar status to the area's "big five" systems (making it the John Havlicek of reverse mathematics, perhaps). This system is very closely connected with algorithmic randomness, since the "fat trees" in the statement of WWKL correspond to Π_1^0 classes of positive measure, and as shown by Kučera [69], a Π_1^0 class $\mathcal{C}$ has positive measure if and only if every 1-random set has a tail in $\mathcal{C}$.

Liu's proof in [76] that WWKL $\not\leqslant_w$ RT$_2^2$ could be seen as closing the story on the relationship between WWKL$_0$ and RT$_k^n$, since for $n \geqslant 3$ and $k \geqslant 2$, RT$_k^n$ is equivalent to ACA$_0$ over RCA$_0$, and hence considerably stronger than WWKL$_0$. Indeed, as shown by Jockusch [59], in this case, there is a k-coloring of $[\mathbb{N}]^n$ all of whose infinite homogeneous sets compute $\emptyset'$, and relativizing this result shows that WWKL $\leqslant_c$ RT$_k^n$. Jockusch's argument actually shows that WWKL $\leqslant_w$ RT$_k^n$.

But what about strong reductions? Relativizing Jockusch's theorem shows that if $n \geqslant 3$ and $k \geqslant 2$ then for any X, there is an X-computable instance of RT$_k^n$ such that $X' \leqslant_T H \oplus X$ for any solution H. However, the conclusion of this statement cannot in general be improved to $X' \leqslant_T H$. Indeed, Hirschfeldt and Jockusch [48] showed that if X is not hyperarithmetic, then there is no instance of RT (of any complexity) such that every solution computes X. In particular, RT does not allow self-encoding, where a problem P *allows self-encoding* if for every X there is an X-computable instance Z of P all of whose solutions compute X. (This notion is similar to that of *cylinder* in the theory of Weihrauch reducibility, but that notion requires the solutions of Z to compute X uniformly.) An example of a principle that *does* allow self-encoding, and indeed is a cylinder, is WKL. (Given X, consider an X-computable binary tree whose only path is X.) As noted in [48], it follows that WKL $\not\leqslant_{sc}$ RT.

WWKL, on the other hand, does not allow self-encoding. Relativizing the result of Kučera mentioned above shows that every set that is 1-random relative to a given set X computes a solution to every X-computable instance of WWKL, and it is well-known that for most X, no set that is 1-random relative to X can compute X. (The precise statement, proved by Hirschfeldt, Nies, and Stephan [54], is that this is the case unless X belongs to the countable class of K-trivial sets. Rod and researchers influenced by him have played a major role in developing the theory of these sets, which are now one of the central objects of study in algorithmic randomness.) Thus an early version of Hirschfeldt and Jockusch [48] included the following questions (which can also be asked for sW-reducibility): Let $n \geqslant 3$ and $k \geqslant 2$. Is WWKL $\leqslant_{sc}$ RT$_k^n$? Is WWKL $\leqslant_{sc}$ RT?

Another way to look at these questions is as being about the relative distribution of homogeneous and 1-random sets. For instance, the question of whether WWKL $\leqslant_{sc}$ RT can be restated as follows: Is it the case that, for every X, there is an X-computable instance of Ramsey's Theorem each of whose infinite homogeneous sets computes a set that is 1-random relative to X?

As noted above, all of these questions have now been answered by the following recent result.

Theorem 5.19 (Monin and Patey [93]). WWKL $\not\leqslant_{sc}$ RT.

A set A is *computably encodable* if every infinite set has an infinite subset that computes A. The proof of Theorem 5.19 uses a notion called Π_1^0 encodability, which is introduced in [93] as an extension of computable encodability. The definition in [93] is for subsets of ω^ω, but it is a bit simpler, and sufficient for the proof of Theorem 5.19, to consider subsets of 2^ω.

Definition 5.20. A class $\mathcal{C} \subseteq 2^\omega$ is Π_1^0 *encodable* if every infinite set has an infinite subset X such that $\mathcal{C}$ has a nonempty $\Pi_1^{0,X}$ subclass.

The key to proving Theorem 5.19 is the following result.

Theorem 5.21 (Monin and Patey [93]). A class $\mathcal{C} \subseteq 2^\omega$ is Π_1^0 encodable if and only if it has a nonempty Σ_1^1 subclass.

As explained in [93], this theorem implies Solovay's result in [109] that a set is computably encodable if and only if it is hyperarithmetic.

Theorem 5.19 follows from Theorem 5.21 by letting T be an instance of WWKL such that the class $[T]$ of paths on T has no nonempty Σ_1^1 subsets. As noted in [93], an example of such a T is an infinite tree whose paths are all 1-random relative to Kleene's $\mathcal{O}$, since every nonempty Σ_1^1 class has an element computable from $\mathcal{O}$. Now let c be an instance of RT such that any solution computes a path on T. Since every infinite set has an infinite subset that is homogeneous for c, it follows that $[T]$ is encodable, contradicting Theorem 5.21.

This argument actually proves something stronger, because there is no need for c to be computable from T. Monin and Patey [93] made the following definition.

Definition 5.22. A problem P is *strongly omnisciently computably reducible* to a problem Q, written as $P \leqslant_{\text{soc}} Q$, if for every instance X of P, there is an instance $\widehat{X}$ of Q such that, for every solution $\widehat{Y}$ to $\widehat{X}$, there is a $\widehat{Y}$-computable solution to X.

As noted in [93], several proofs that show that $P \not\leqslant_{\text{sc}} Q$ in fact show that $P \not\leqslant_{\text{soc}} Q$. As discussed above, this is in particular true of Theorem 5.19.

Theorem 5.23 (Monin and Patey [93]). WWKL $\not\leqslant_{\text{soc}}$ RT.

Monin and Patey [93] asked the following questions about soc-reducibility.

Open Question 5.24 (Monin and Patey [93]). *Let* $n, k \geqslant 2$. *Is* $\text{RT}_{k+1}^n \leqslant_{\text{soc}}$ RT_k^n? *Is* $\text{RT}_k^{n+1} \leqslant_{\text{soc}} \text{RT}_k^n$?

Before moving away from reverse mathematics, I will mention one more question, which was posed by Damir Dzhafarov and Noah Schweber (see [104]), and came from work they did in reverse mathematics.

Definition 5.25. Let f be a computable binary function such that $f(n, s + 1) \leqslant f(n, s)$ for all n and s, and let $F(n) = \lim_s f(n, s)$. A *limit-nondecreasing subsequence for f* is a set X such that if $i, j \in X$ and $i < j$ then $F(i) \leqslant F(j)$. (Such an X is called *f-good* in [104].)

It is easy to see that every f of this kind has an $\emptyset'$-computable limit-nondecreasing subsequence. Dzhafarov and Schweber (see [104]) asked whether this upper bound is tight. That is, they asked whether there is an f as above such that every limit-nondecreasing subsequence computes $\emptyset'$, and failing that, whether it is the case that a set that computes a limit-nondecreasing subsequence for every such f must compute $\emptyset'$. Patey (see [104]) has given negative answers to both of these questions, and provided further computability-theoretic information on the complexity of limit-nondecreasing subsequences. There may be more to say on this front, however.

Open Question 5.26 (Dzhafarov and Schweber (see [104])). *How complicated must a limit-nondecreasing subsequence for a function f as above be in general?*

Kolmogorov complexity functions, such as plain or prefix-free complexity, are natural examples of functions with nonincreasing approximations. Suppose for example that $f(n) = C_s(n)$, where $C_s(n)$ is the stage s approximation to the plain Kolmogorov complexity $C(n)$ of n, and let X be a limit-nondecreasing subsequence for f. Since there cannot be 2^k many numbers n with $C(n) < k$, the 2^kth element n of X must have $C(n) \geqslant k$. Thus there is an X-computable function g such that $C(g(k)) \geqslant k$ for all k. By results of Kjos-Hanssen, Merkle, and Stephan [66], X has DNC degree. (That is, there is an X-computable function h that is diagonally noncomputable, which means that $h(e) \neq \Phi_e(e)$ for all e, where Φ_e is the eth partial computable function.) Thus, as pointed out in [104], the answer to the first part of Question 5.26 is at least at the level of the DNC degrees.

Kolmogorov complexity functions are rather special, though. If A has DNC degree then, again by results in [66], A computes an increasing function g such that $C(g(k)) \geqslant k$ for all k. Let c be such that $C(n) \leqslant n + c$ for all n. Then one can A-computably find $n_0 < n_1 < \cdots$ such that $C(g(n_{i+1})) \geqslant g(n_i) + c$ for all i. The set $X = \{g(n_i) : i \in \omega\}$ is then an A-computable limit-nondecreasing subsequence for the function f in the previous paragraph. Thus in this case there is a full answer to the first part of Question 5.26, but the general case might well require more powerful oracles.

Question 5.26 can also be restated in reverse-mathematical terms. Let LNS be the following statement: If f is a binary function such that $f(n, s+1) \leqslant f(n, s)$ for all n and s then there is an infinite set X such that if $i, j \in X$ and $i < j$ then $\exists t \, \forall s > t \, (f(i, s) \leqslant f(j, s))$.

Open Question 5.27. *What is the reverse mathematical strength of* LNS*?*

Patey (see [104]) showed that LNS does not imply the principle ADS (which was studied in [55] and is strictly weaker than RT_2^2) over RCA_0.

6   Measures of Relative Randomness

Much of my time in Wellington was spent thinking about algorithmic randomness. Richard Coles brought a question of Cris Calude's down from Auckland, which Rod, André Nies, and I eventually solved [21]. In the process of working on this question, Rod and I started to get increasingly interested in the general area. This interest led to several papers, a survey article with André and Bas Terwijn [22], and a slim volume called *Algorithmic Randomness and Complexity* [17].

The Sydney Opera House was completed ten years late and almost fifteen times over budget. By those standards, Rod and I did not do too badly. Our book took about seven years longer to write and ended up being three or four times as long as we had initially projected. Some of this delay was caused by the rapidly moving target that the area became as more and more researchers— many of them brilliant young ones, and many of them mentored or influenced by Rod—began to solve its problems and unearth new ones at an alarming rate. In this section and the next, I would like to mention two old problems (by the standards of this area) that have endured despite these efforts.

The Kučera-Gács Theorem [37,69] states that every set is Turing reducible, and indeed wtt-reducible, to some 1-random set. Merkle and Mihailović [80] showed that the use of this reduction can always be taken to be of order $n+o(n)$. One of the few original results in the Downey-Hirschfeldt book [17] is that this bound cannot be improved to $n + O(1)$. Say that A is *cl-reducible* to B if there is a Turing functional Γ such that $\Gamma^B = A$ and $\gamma^B(n) \leqslant n + O(1)$, where γ is the use function of Γ. (The original name for this notion in Downey, Hirschfeldt, and LaForte [18] was "strong weak truth table (sw-) reducibility". For some reason, this adjectival salad was not popular. Lewis and Barmpalias [73,74] renamed it "computable Lipschitz (cl-) reducibility", reflecting the fact this reducibility is an effective version of the notion of Lipschitz transformation.)

Theorem 6.1 (Downey and Hirschfeldt [17]). There is a set that is not cl-reducible to any 1-random set.

Given the relationship between initial-segment complexity and randomness, if $A \leqslant_{cl} B$ then there is reason to say that A is no more random than B. (In particular, in this case $K(A \restriction n) \leqslant K(B \restriction n) + O(1)$, where K is prefix-free Kolmogorov complexity.) This is no longer the case if the bound on the use is even slightly relaxed. For instance, for any unbounded, nondecreasing computable function f and any 1-random set A, it is easy to find a non-1-random set B such that A is Turing reducible to B via a reduction with use bounded by $n + f(n)$. Other measures of relative randomness include the following.

Definition 6.2. Say that A is *K-reducible* to B if $K(A \restriction n) \leqslant K(B \restriction n) + O(1)$, and that A is *C-reducible* to B if $C(A \restriction n) \leqslant C(B \restriction n) + O(1)$ (where, as above, C is plain Kolmogorov complexity).

Say that A is *rK-reducible* to B if $K(A \restriction n \mid B \restriction n) \leqslant O(1)$. It is easy to see that this definition does not change if K is replaced by C.

The development of the theory of algorithmic randomness seems to have made these notions less significant than they once may have seemed, but the following questions, motivated by Theorem 6.1, still seem worth answering.

Open Question 6.3 (Downey, Hirschfeldt, Nies, and Terwijn [22]; Miller and Nies [86]). *Is every set K-reducible to some 1-random set? Is every set C-reducible to some 1-random set? Is every set rK-reducible to some 1-random set?*

Although these questions have not been central to the study of algorithmic randomness, I do believe they (and particularly the first one) are of intrinsic interest, given that the interplay between levels of randomness and initial-segment complexity has been a major theme in the area. Furthermore, the fact that they have remained open for so long, in the face of our greatly improved understanding of the notions involved, suggests that they may depend on aspects of the notions of 1-randomness and Kolmogorov complexity that remain underdeveloped.

7   Nonmonotonic Randomness

An even older question in algorithmic randomness is that of establishing the relationship between nonmonotonic randomness and 1-randomness. This question seems quite fundamental, since the nonmonotonic betting strategies used to define nonmonotonic randomness are natural generalizations of the usual betting strategies that can be used to define notions such as 1-randomness, computable randomness, and Schnorr randomness. Furthermore, it is the only remaining one I know of in determining implications between major notions of algorithmic randomness.

Nonmonotonic randomness (also know as Kolmogorov-Loveland randomness) was introduced by Muchnik, Semenov, and Uspensky [95]. The version of the definition below is essentially the one given by Merkle, Miller, Nies, Reimann, and Stephan [81].

In algorithmic randomness, a *martingale* is a function $d : 2^{<\omega} \to \mathbb{R}^{\geqslant 0}$ such that $d(\sigma) = \frac{d(\sigma 0) + d(\sigma 1)}{2}$, representing a strategy for betting on the successive bits of a binary sequence. The initial capital available is $d(\lambda)$, where λ is the empty string. If σ represents the bits seen so far, then the strategy is to bet $\frac{d(\sigma 0)}{2d(\sigma)}$ of the current capital on the next bit being 0, and $\frac{d(\sigma 1)}{2d(\sigma)}$ of this capital on the next bit being 1. If that strategy is followed, then for any τ, the capital available after seeing the bits of τ is $d(\tau)$. A martingale d *succeeds* on a set A if $\limsup_n d(A \restriction n) = \infty$.

A martingale is *computable* if its values are uniformly computable, and *c.e.* if its values are uniformly left-c.e. One of the several ways to define 1-randomness is to say that a set is 1-random if no c.e. martingale succeeds on it. Say that a set is *computably random* if no computable martingale succeeds on it. Schnorr [101] showed that the latter notion, which he introduced in [101,102], is strictly weaker than 1-randomness.

Schnorr [101,102] also introduced the notion of Schnorr randomness, which he believed more adequately captures the informal idea of "computable randomness" than the notion now known as computable randomness. (He saw 1-randomness itself as a notion of computably enumerable randomness.) An *order* is an unbounded, nondecreasing function from $\mathbb{N}$ to $\mathbb{N}$. A set X is *Schnorr random* if $\limsup_n \frac{d(X \upharpoonright n)}{h(n)} < \infty$ for every computable martingale d and every computable order h. Wang [115, 116] showed that Schnorr randomness is strictly weaker than computable randomness.

It is natural to ask what happens if one is allowed to bet on the bits of a sequence out of order, which leads to the idea of a nonmonotonic betting strategy. Such a strategy has two components, a scan rule and a stake function. These determine the next bit to bet on, and how much to bet on each possible value of that bit, respectively, based on the values observed at the previously selected bits. Of course, a strategy cannot be allowed to bet twice on the same bit. (In the definition in [81], the scan rules and stake functions making up nonmonotonic betting strategies are partial functions, but Merkle [79] showed that, for the purpose of defining nonmonotonic randomness, it is enough to consider total nonmonotonic betting strategies.)

Definition 7.1. A *finite assignment* is a sequence $(r_1, a_1), \ldots, (r_n, a_n)$ with $r_i \in \mathbb{N}$ and $a_i \in \{0, 1\}$, such that the r_i are pairwise distinct. The *domain* of this assignment is $\{r_1, \ldots, r_n\}$.

A *scan rule* is a function s from the set of finite assignments to $\mathbb{N}$ such that $s(x)$ is not in the domain of x for each finite assignment x.

A *stake function* is a function from the collection of finite assignments to $[-1, 1]$.

A *nonmonotonic betting strategy* is a pair consisting of a scan rule and a stake function.

The idea behind this definition of a stake function q is that, letting the current capital be d, a negative value of $q(x)$ represents a bet of $-q(x)d$ that the value of the next bit bet on is 0, while a positive value of $q(x)$ represents a bet of $q(x)d$ that the value of the next bit bet on is 1 (and hence $q(x) = 0$ represents an even bet, which is the same as not betting at all).

The nonmonotonic martingale d_b^X associated with playing a nonmonotonic strategy b on a sequence X (with starting capital 1), and the resulting notion of nonmonotonic randomness, can now be defined as follows.

Definition 7.2. Let $b = (s, q)$ be a nonmonotonic betting strategy. For a set X, let $p^X(0) = \lambda$ and

$$p^X(n+1) = p^X(n)^\frown(s(p^X(n)), X(s(p^X(n)))).$$

Then $p^X(n)$ is the finite assignment corresponding to scanning X in accordance with s. Let $c^X(0) = 1$ and

$$c^X(n+1) = \begin{cases} 1 - q(p^X(n)) & \text{if } X(s(p^X(n))) = 0 \\ 1 + q(p^X(n)) & \text{if } X(s(p^X(n))) = 1. \end{cases}$$

Let

$$d_b^X(n) = \prod_{i=0}^{n} c^X(i).$$

The strategy b *succeeds* on X if $\limsup_n d_b^X(n) = \infty$.

Say that X is *nonmonotonically random* if no computable nonmonotonic betting strategy succeeds on it.

Muchnik, Semenov, and Uspensky [95] showed that 1-randomness implies nonmonotonic randomness, which in turn clearly implies computable randomness. (Their proof also shows that the notion of randomness obtained by considering c.e. nonmonotonic betting strategies in place of computable ones is equivalent to 1-randomness.) As explained for instance in [17, Sect. 7.5], results of Muchnik (see [95]) on Kolmogorov complexity show that the latter implication is strict. The following fundamental question remains open, however.

Open Question 7.3 (Muchnik, Semenov, and Uspensky [95]). *Is there a set that is nonmonotonically random but not 1-random?*

Merkle, Miller, Nies, Reimann, and Stephan [81] obtained several interesting results related to this question. In particular, they showed that if $A \oplus B$ is nonmonotonically random, then at least one of A or B is 1-random. On the one hand, this result suggests that nonmonotonic randomness and 1-randomness are quite close (as does Muchnik's analysis of the initial-segment Kolmogorov complexity of nonmonotonically random sets in [95]). On the other hand, it is well-known that if $A \oplus B$ is random (in some sense), then one should expect the level of randomness of A and B individually to be higher than that of $A \oplus B$. (For instance, using results of Figueira, Hirschfeldt, Miller, Ng, and Nies [34], Bienvenu, Greenberg. Kučera, Nies, and Turetsky [3] showed that if $A \oplus B$ is 1-random then at least one of A or B has the stronger property of being balanced random.) Kastermans and Lempp [64] separated certain weaker versions of nonmonotonic randomness from 1-randomness.

As far as I know, the following question has not been considered so far. Say that a set X is *Schnorr nonmonotonically random* if $\limsup_n \frac{d_b^X(n)}{h(n)} < \infty$ for every computable nonmonotonic betting strategy b and every computable order h.

Open Question 7.4. *What is the strength of Schnorr nonmonotonic randomness in relation to other notions of algorithmic randomness?*

8   Asymptotic Computability

After finishing the book with Rod, I was slightly burned out on randomness. I was brought back into thinking about it by a question about coarse computability asked by Paul Schupp, which led to a paper with him, Carl Jockusch, and Rutger

Kuyper [50]. Coarse computability and other notions of asymptotic computability capture the idea of computing a set "almost everywhere". The contemporary computability-theoretic study of these notions began with a paper of Jockusch and Schupp [60], which studied the notion of generic computability introduced by Kapovich, Myasnikov, Schupp, and Shpilrain [63]. As with so many of the most interesting lines of research in computability theory, Rod got into the game early, in papers with Jockusch and Schupp [26] and Jockusch, McNicholl, and Schupp [24].

As it turns out, the idea of asymptotic computability had already occurred to Meyer [82] in the early 70's, leading him to ask a question that was answered by Lynch [77]. Much later, Terwijn [113] returned to this idea, becoming to my knowledge the first person to define coarse computability. (Meyer and Lynch were working with a different notion of asymptotic computability, defined below.)

The definitions of generic and coarse computability begin with the relevant notion of "almost everywhere".

Definition 8.1. For $S \subseteq \omega$ and $n \in \omega$, let $\rho_n(S) = \frac{|S \restriction n|}{n}$.
The *upper (asymptotic) density* $\overline{\rho}(S)$ of S is $\limsup_n \rho_n(S)$.
The *lower (asymptotic) density* $\underline{\rho}(S)$ of S is $\liminf_n \rho_n(S)$.
If $\overline{\rho}(S) = \underline{\rho}(S)$ then this number is called the *(asymptotic) density* of S.

Definition 8.2. A *partial description* of a set A is a partial function f such that $f(n) = A(n)$ whenever $f(n)$ is defined. A *generic description* of A is a partial description of A with domain of density 1. A set is *generically computable* if it has a computable generic description.

A *coarse description* of a set A is a set C such that $C(n) = A(n)$ on a set of density 1. A set is *coarsely computable* if it has a computable coarse description.

Jockusch and Schupp [61] showed that there are sets that are generically computable but not coarsely computable, and vice-versa.

These notions of asymptotic computability lead naturally to notions of asymptotic reducibility, from which degree structures are defined as usual. As with mass problems, there are both uniform and nonuniform versions.

Definition 8.3. Say that B is *nonuniformly coarsely reducible* to A if every coarse description of A computes a coarse description of B.

Say that B is *uniformly coarsely reducible* to A if there is a Turing functional Φ such that if C is a coarse description of A, then Φ^C is a coarse description of B.

Say that B is *nonuniformly generically reducible* to A if for every generic description f of A, there is an enumeration operator W such that $W^{\text{graph}(f)}$ enumerates the graph of a generic description of B.

Say that B is *uniformly generically reducible* to A if there is an enumeration operator W such that if f is a generic description of A, then $W^{\text{graph}(f)}$ is the graph of a generic description of B.

One of the basic questions one can ask about the degree structures arising from these reducibilities is whether minimal pairs exist. Recall that for a given degree structure with a minimum degree **0**, nonzero degrees **a** and **b** form a *minimal pair* if **0** is the only degree that is below both **a** and **b**. Hirschfeldt, Jockusch, Kuyper, and Schupp [50] showed that there are minimal pairs for (uniform or nonuniform) coarse reducibility, and indeed proved the stronger result that there are sets A and B that form a minimal pair for relative coarse computability; that is, A and B are not coarsely computable, but if C is coarsely computable relative both to A and to B, then C is coarsely computable. In fact, any A and B that are sufficiently mutually random form a minimal pair for relative coarse computability. (See [50] for details.)

The situation for generic reducibility is more complicated. The following question was originally asked for uniform generic reducibility, but it is open for the nonuniform version as well.

Open Question 8.4 (Jockusch and Schupp [61]; Igusa [57]). *Is there a minimal pair in the (uniform or nonuniform) generic degrees?*

One might expect that, as in the case of the coarse degrees, this question has a positive answer, which might perhaps be found by considering mutually random sets. However, if this is the case, then the proof will have to be significantly different from the one for the coarse degrees, because Igusa [57] showed that there are no minimal pairs for relative generic computability; that is, if A and B are not generically computable, then there is a C that is not generically computable but is generically computable relative both to A and to B. (A weaker form of this result, with the additional hypothesis that A and B are Δ^0_2, was proved by Downey, Jockusch, and Schupp [26].)

One approach to Question 8.4, suggested by Igusa [58], is to focus on the following question.

Open Question 8.5 (Igusa [58]). *If A is not generically computable, must there be a B that is uniformly generically reducible to A such that B is not generically computable but has density 1?*

Igusa [58] showed that answering this question in either direction would have consequences for the uniform dense degrees: a positive answer would imply that there are no minimal degrees (which is also an open question), while a negative answer would imply that there are minimal pairs. (For the nonuniform dense degrees, a positive answer to the analog of Question 8.5 would imply that there are minimal degrees, by the same argument as in the uniform case, but for minimal pairs the situation is less clear, as the argument in [58] that a negative answer to Question 8.5 implies the nonexistence of minimal pairs appears to make essential use of uniformity.)

Cholak and Igusa [7] noted that Question 8.5 can be recast in terms of the relationship between generic and coarse degrees, because, as they showed, a (uniform or nonuniform) generic degree contains a coarsely computable set if and only if it contains a set of density 1.

At the time of writing, I am working with Eric Astor and Carl Jockusch on a paper [2] that reintroduces Meyer's notion of asymptotic computability, which we call effective dense computability, and introduces another such notion, called dense computability.

Definition 8.6. A set A is *densely computable* if there is a partial computable function f such that $f(n){\downarrow} = A(n)$ on a set of density 1.

A set A is *effectively densely computable* if there is a (total) computable function $f : \omega \to \{0, 1, \square\}$ such that $\{n : f(n) = \square\}$ has density 0, and $f(n) = A(n)$ for all n outside this set.

It is easy to see that effective dense computability implies both generic and coarse computability, and that both generic and coarse computability imply dense computability. As mentioned above, generic and coarse computability are incomparable notions, so all of these implications are strict.

As with generic computability and coarse computability, one can define notions of reducibility associated with dense computability and effective dense computability, and corresponding degree structures. Eric, Carl, and I have shown that there are minimal pairs in the (uniform or nonuniform) dense degrees, but we do not know whether this is the case for the effective dense degrees. Settling this question seems likely to require methods similar to those needed to answer Question 8.4.

I had planned to finish with one more question in this area, but after the initial submission of this paper, I learned from Benoit Monin that he had recently solved it. I will still discuss it though, as I find his result quite lovely. Definitions of the classes of sets and degrees mentioned below can be found e.g. in [17].

To each notion of asymptotic computability, one can attach an asymptotic computability bound. The following two notions were introduced by Downey, Jockusch, and Schupp [26] and Hirschfeldt, Jockusch, McNicholl, and Schupp [51], respectively.

Definition 8.7. Say that A is *partially computable at density* r if there is a partial description f of A such that $\underline{\rho}(\mathrm{dom}\, f) \geqslant r$. The *partial computability bound* of A is

$$\alpha(A) = \sup\{r : A \text{ is partially computable at density } r\}.$$

Say that A is *coarsely computable at density* r if there is a computable set C such that $\underline{\rho}(\{n : C(n) = A(n)\}) \geqslant r$. The *coarse computability bound* of A is

$$\gamma(A) = \sup\{r : A \text{ is coarsely computable at density } r\}.$$

Astor, Hirschfeldt, and Jockusch [2] have shown that the analogous notions for dense and effective dense computability are equivalent to the ones for coarse and generic computability, respectively.

As shown in [51], every hyperimmune degree contains a set A such that $\gamma(A) = 0$. Andrews, Cai, Diamondstone, Jockusch, and Lempp [1] showed that the same is true of every PA degree. However, they also showed that there are degrees that do not contain any such sets. Two examples of such degrees given in that paper are the degrees of computably traceable sets, and the degrees of sets computable from a 1-random set of hyperimmune-free degree. In both cases, every set A in such degrees has $\gamma(A) \geqslant \frac{1}{2}$.

Definition 8.8. For a degree $\mathbf{a}$, let

$$\Gamma(\mathbf{a}) = \inf\{\gamma(A) : A \text{ is } \mathbf{a}\text{-computable}\}.$$

Hirschfeldt, Jockusch, McNicholl, and Schupp [51] showed that every nonzero degree contains a set A with $\gamma(A) = \frac{1}{2}$, so $\Gamma(\mathbf{a}) \leqslant \frac{1}{2}$ for all $\mathbf{a} \neq \mathbf{0}$, and the results mentioned above produce examples of degrees $\mathbf{a}$ with $\Gamma(\mathbf{a}) = 0$ and $\Gamma(\mathbf{a}) = \frac{1}{2}$. Of course, $\Gamma(\mathbf{0}) = 1$. In the original version of this paper, I wrote that "[i]t would be remarkable if these are the only possible values of $\Gamma(\mathbf{a})$" and asked the following question from Andrews, Cai, Diamondstone, Jockusch, and Lempp [1]: Is it the case that $\Gamma(\mathbf{a})$ is always 0, $\frac{1}{2}$, or 1? If not, then what are the possible values of $\Gamma(\mathbf{a})$?

Andrews, Cai, Diamondstone, Jockusch, and Lempp [1] showed that if A is truth-table reducible to a 1-random set then $\gamma(A) \geqslant \frac{1}{2}$ (from which their result on 1-random sets of hyperimmune-free degree mentioned above follows immediately). Furthermore, the proof in [51] that every Turing degree contains a set A with $\gamma(A) = \frac{1}{2}$ works for tt-degrees as well. Thus it is also interesting to consider the values of

$$\Gamma_{tt}(\mathbf{a}) = \inf\{\gamma(A) : A \text{ is tt-computable relative to } \mathbf{a}\}$$

for tt-degrees $\mathbf{a}$. In the original version of this paper, I also asked the following question: Is it the case that $\Gamma_{tt}(\mathbf{a})$ is always 0, $\frac{1}{2}$, or 1? If not, then what are the possible values of $\Gamma_{tt}(\mathbf{a})$?

Monin [91] has completely answered these questions as follows, using notions developed by Monin and Nies [92] and techniques from the theory of error-correcting codes.

Theorem 8.9 (Monin [91]). *The only possible values of $\Gamma(\mathbf{a})$ and $\Gamma_{tt}(\mathbf{a})$ are 0, $\frac{1}{2}$, and 1.*

References

1. Andrews, U., Cai, M., Diamondstone, D., Jockusch, C., Lempp, S.: Asymptotic density, computable traceability, and 1-randomness. Fundam. Math. **234**, 41–53 (2016)
2. Astor, E.P., Hirschfeldt, D.R., Jockusch Jr., C.G.: Dense computability, upper cones, and minimal pairs, in preparation
3. Bienvenu, L., Greenberg, N., Kučera, A., Nies, A., Turetsky, D.: Coherent randomness tests and computing the K-trivial sets. J. Eur. Math. Soc. **18**, 773–812 (2016)

4. Bonnet, R.: Stratifications et extension des genres de chaînes dénombrables. C. R. Acad. Sci. Ser. A-B **269**, A880–A882 (1969)
5. Brattka, V.: Maintainer, Bibliography on Weihrauch complexity, Computability and Complexity in Analysis Network. http://cca-net.de/publications/weibib.php
6. Cholak, P., McCoy, C.: Effective prime uniqueness (to appear)
7. Cholak, P., Igusa, G.: Density-1-bounding and quasiminimality in the generic degrees. J. Symbolic Logic (to appear)
8. Cholak, P.A., Jockusch Jr., C.G., Slaman, T.A.: On the strength of Ramsey's Theorem for pairs. J. Symbolic Logic **66**, 1–55 (2001). (Corrigendum in J. Symbolic Logic **74**, 1438–1439 (2009))
9. Chong, C.T., Slaman, T.A., Yang, Y.: The metamathematics of stable Ramsey's Theorem for pairs. J. Am. Math. Soc. **27**, 863–892 (2014)
10. Chong, C.T., Slaman, T.A., Yang, Y.: The inductive strength of Ramsey's Theorem for pairs (to appear)
11. Csima, B.F.: Applications of Computability Theory to Prime Models and Differential Geometry. Ph.D. Dissertation, The University of Chicago (2003)
12. Csima, B.F.: Degree spectra of prime models. J. Symbolic Logic **69**, 430–442 (2004)
13. Dorais, F.G., Dzhafarov, D.D., Hirst, J.L., Mileti, J.R., Shafer, P.: On uniform relationships between combinatorial problems. Trans. Am. Math. Soc. **368**, 1321–1359 (2016)
14. Downey, R.: On presentations of algebraic structures. In: Sorbi, A. (ed.) Complexity, Logic, and Recursion Theory, pp. 157–205. Dekker, New York (1997)
15. Downey, R.G.: Computability theory and linear orderings. In: Ershov, Y.L., Goncharov, S.S., Nerode, A., Remmel, J.B., Marek, V.W. (eds.) Handbook of Recursive Mathematics, vol. II. Studies in Logic and the Foundations of Mathematics, vol. 139, pp. 823–976 (1998)
16. Downey, R.: Computability, definability, and algebraic structures. In: Downey, R., Decheng, D., Ping, T.S., Hui, Q.Y., Yasugi, M. (eds.) Proceedings of the 7th and 8th Asian Logic Conferences, pp. 63–102. Singapore University Press and World Scientific, Singapore (2003)
17. Downey, R.G., Hirschfeldt, D.R.: Algorithmic Randomness and Complexity. Theory and Applications of Computability. Springer, New York (2010)
18. Downey, R.G., Hirschfeldt, D.R., LaForte, G.: Randomness and reducibility. J. Comput. Syst. Sci. **68**, 96–114 (2004)
19. Downey, R.G., Hirschfeldt, D.R., Lempp, S., Solomon, R.: A Δ_2^0 set with no infinite low subset in either it or its complement. J. Symbolic Logic **66**, 1371–1381 (2001)
20. Downey, R.G., Hirschfeldt, D.R., Lempp, S., Solomon, R.: Computability-theoretic and proof-theoretic aspects of partial and linear orderings. Isr. J. Math. **138**, 271–290 (2003)
21. Downey, R., Hirschfeldt, D.R., Nies, A.: Randomness, computability, and density. SIAM J. Comput. **31**, 1169–1183 (2002)
22. Downey, R.G., Hirschfeldt, D.R., Nies, A., Terwijn, S.A.: Calibrating randomness. Bull. Symbolic Logic **12**, 411–491 (2006)
23. Downey, R., Jockusch, C.G.: Every low Boolean algebra is isomorphic to a recursive one. Proc. Am. Math. Soc. **122**, 871–880 (1994)
24. Downey, R., Jockusch, C., McNicholl, T.H., Schupp, P.: Asymptotic density and the Ershov Hierarchy. Math. Logic Q. **61**, 189–195 (2015)
25. Downey, R.G., Jockusch Jr., C.G.: On self-embeddings of computable linear orderings. Ann. Pure Appl. Logic **138**, 52–76 (2006)

26. Downey, R.G., Jockusch Jr., C.G., Schupp, P.E.: Asymptotic density and computably enumerable sets. J. Math. Logic **13**, 43 (2013). 1350005
27. Downey, R.G., Kastermans, B., Lempp, S.: On computable self-embeddings of computable linear orderings. J. Symbolic Logic **74**, 1352–1366 (2009)
28. Downey, R.G., Lempp, S.: The proof-theoretic strength of the Dushnik-Miller theorem for countable linear orders. In: Arslanov, M.M., Lempp, S. (eds.) Recursion Theory and Complexity. Series in Logic and Its Applications, vol. 2, pp. 55–57. De Gruyter, Berlin (1999)
29. Downey, R., Lempp, S., Wu, G.: On the complexity of the successivity relation in computable linear orders. J. Math. Logic **10**, 83–99 (2010)
30. Downey, R., Remmel, J.B.: Questions in computable algebra and combinatorics. In: Cholak, P.A., Lempp, S., Lerman, M., Shore, R.A. (eds.) Computability Theory and its Applications (Boulder, CO, 1999). Contemporary Mathematics, vol. 257, pp. 95–125. American Mathematical Society, Providence (2000)
31. Dushnik, B., Miller, E.W.: Concerning similarity transformations of linearly ordered sets. Bull. Am. Math. Soc. **40**, 322–326 (1940)
32. Dzhafarov, D.D.: Strong reductions between combinatorial principles. J. Symbolic Logic (to appear)
33. Dzhafarov, D.D., Patey, L., Solomon, R., Westrick, L.B.: Ramsey's Theorem for singletons and strong computable reducibility. Proc. Am. Math. Soc. (to appear)
34. Figueira, S., Hirschfeldt, D.R., Miller, J.S., Ng, K.M., Nies, A.: Counting the changes of random Δ_2^0 sets. J. Logic Comput. **25**, 1073–1089 (2015)
35. Fokina, E.B., Harizanov, V., Melnikov, A.G.: Computable model theory. In: Downey, R. (ed.) Turing's Legacy: Developments from Turing's Ideas in Logic. Lecture Notes in Logic, vol. 42, pp. 124–194. Association for Symbolic Logic, La Jolla. Cambridge University Press, Cambridge (2014)
36. Frolov, A., Harizanov, V., Kalimullin, I., Kudinov, O., Miller, R.: Degree spectra of high$_n$ and nonlow$_n$ degrees. J. Logic Comput. **22**, 755–777 (2012)
37. Gács, P.: Every sequence is reducible to a random one. Inf. Control **70**, 186–192 (1986)
38. Goncharov, S.S.: Problem of the number of non-self-equivalent constructivizations. Algebra Logic **19**, 401–414 (1980)
39. Goncharov, S.S.: Limit equivalent constructivizations. In: Mathematical Logic and the Theory of Algorithms Trudy Instituta Matematiki, vol. 2, pp. 4–12. "Nauka" Sibirskoe otdelenie, Novosibirsk (1982)
40. Goncharov, S.S.: Countable Boolean Algebras and Decidability. Siberian School of Algebra and Logic. Consultants Bureau, New York (1997)
41. Goncharov, S.S., Dzgoev, V.D.: Autostability of models. Algebra Logic **19**, 28–37 (1980)
42. Goncharov, S.S., Nurtazin, A.T.: Constructive models of complete decidable theories. Algebra Logic **12**, 67–77 (1973)
43. Harizanov, V.S.: Pure computable model theory. In: Ershov, Y.L., Goncharov, S.S., Nerode, A., Remmel, J.B., Marek, V.W. (eds.) Handbook of Recursive Mathematics. Studies in Logic and the Foundations of Mathematics, vol. 138, pp. 3–114. North-Holland, Amsterdam (1998)
44. Harrington, L.: Recursively presentable prime models. J. Symbolic Logic **39**, 305–309 (1974)
45. Harrison-Trainor, M., Melnikov, A., Miller, R., Montalbán, A.: Computable functors and effective interpretability. J. Symbolic Logic (to appear)
46. Hirschfeldt, D.R.: Computable trees, prime models, and relative decidability. Proc. Am. Math. Soc. **134**, 1495–1498 (2006)

47. Hirschfeldt, D.R.: Slicing the Truth: On the Computability Theoretic and Reverse Mathematical Analysis of Combinatorial Principles. Lecture Notes Series, vol. 28. World Scientific, Singapore (2014). Institute for Mathematical Sciences, National University of Singapore
48. Hirschfeldt, D.R., Jockusch Jr., C.G.: On notions of computability theoretic reduction between Π_2^1 principles. J. Math. Logic **16**, 1650002 (2016)
49. Hirschfeldt, D.R., Jockusch Jr., C.G., Kjos-Hanssen, B., Lempp, S., Slaman, T.A.: The strength of some combinatorial principles related to Ramsey's Theorem for pairs. In: Chong, C., Feng, Q., Slaman, T.A., Woodin, W.H., Yang, Y. (eds.) Computational Prospects of Infinity, Part II: Presented Talks. Lecture Notes Series, vol. 15, pp. 143–161. World Scientific, Singapore (2008). Institute for Mathematical Sciences, National University of Singapore
50. Hirschfeldt, D.R., Jockusch Jr., C.G., Kuyper, R., Schupp, P.E.: Coarse reducibility and algorithmic randomness. J. Symbolic Logic (to appear)
51. Hirschfeldt, D.R., Jockusch Jr., C.G., McNicholl, T., Schupp, P.E.: Asymptotic density and the coarse computability bound. Computability **5**, 13–27 (2016)
52. Hirschfeldt, D.R., Khoussainov, B., Shore, R.A., Slinko, A.M.: Degree spectra and computable dimensions in algebraic structures. Ann. Pure Appl. Logic **115**, 71–113 (2002)
53. Hirschfeldt, D.R., Kramer, K., Miller, R., Shlapentokh, A.: Categoricity properties for computable algebraic fields. Trans. Am. Math. Soc. **367**, 3981–4017 (2015)
54. Hirschfeldt, D.R., Nies, A., Stephan, F.: Using random sets as oracles. J. Lond. Math. Soc. **75**, 610–622 (2007)
55. Hirschfeldt, D.R., Shore, R.A.: Combinatorial principles weaker than Ramsey's Theorem for pairs. J. Symbolic Logic **72**, 171–206 (2007)
56. Hirst, J.L.: Combinatorics in Subsystems of Second Order Arithmetic. Ph.D. Dissertation, The Pennsylvania State University (1987)
57. Igusa, G.: Nonexistence of minimal pairs for generic computation. J. Symbolic Logic **78**, 511–522 (2013)
58. Igusa, G.: The generic degrees of density-1 sets, and a characterization of the hyperarithmetic reals. J. Symbolic Logic **80**, 1290–1314 (2015)
59. Jockusch Jr., C.G.: Ramsey's Theorem and recursion theory. J. Symbolic Logic **37**, 268–280 (1972)
60. Jockusch Jr., C.G., Schupp, P.E.: Generic computability, and asymptotic density. J. Lond. Math. Soc. **85**, 472–490 (2012)
61. Jockusch, C.G., Soare, R.I.: Degrees of orderings not isomorphic to recursive linear orderings. Ann. Pure Appl. Logic **52**, 39–64 (1991)
62. Jullien, P.: Contribution à L'étude des Types D'ordre Dispersés. Ph.D. Dissertation, Université d'Aix-Marseille (1969)
63. Kapovich, I., Myasnikov, A., Schupp, P., Shpilrain, V.: Generic-case complexity, decision problems in group theory and random walks. J. Algebra **264**, 665–694 (2003)
64. Kastermans, B., Lempp, S.: Comparing notions of randomness. Theoret. Comput. Sci. **411**, 602–616 (2010)
65. Kalimullin, I., Khoussainov, B., Melnikov, A.: Limitwise monotonic sequences and degree spectra of structures. Proc. Am. Math. Soc. **141**, 3275–3289 (2013)
66. Kjos-Hanssen, B., Merkle, W., Stephan, F.: Kolmogorov complexity and the recursion theorem. Trans. Am. Math. Soc. **363**, 5465–5480 (2011)
67. Knight, J.F.: Degrees coded in jumps of orderings. J. Symbolic Logic **51**, 1034–1042 (1986)

68. Knight, J.F., Stob, M.: Computable Boolean algebras. J. Symbolic Logic **65**, 1605–1623 (2000)
69. Kučera, A.: Measure, Π_1^0-classes and complete extensions of PA. In: Ebbinghaus, H.D., Müller, G.H., Sacks, G.E. (eds.) Recursion Theory Week. Lecture Notes in Mathematics, vol. 1141, pp. 245–259. Springer-Verlag, Berlin (1985)
70. LaRoche, P.: Recursively represented Boolean algebras. Not. Am. Math. Soc. **24**, A-552 (1977). (research announcement)
71. Lempp, S., McCoy, C., Miller, R., Solomon, R.: Computable categoricity for trees of finite height. J. Symbolic Logic **70**, 151–215 (2005)
72. Lempp, S., McCoy, C., Miller, R., Solomon, R.: The computable dimension of trees of infinite height. J. Symbolic Logic **70**, 111–141 (2005)
73. Lewis, A.E.M., Barmpalias, G.: Random reals and Lipschitz continuity. Math. Struct. Comput. Sci. **16**, 737–749 (2006)
74. Lewis, A.E.M., Barmpalias, G.: Randomness and the linear degrees of computability. Ann. Pure Appl. Logic **145**, 252–257 (2007)
75. Liu, J.: RT_2^2 does not imply WKL$_0$. J. Symbolic Logic **77**, 609–620 (2012)
76. Liu, L.: Cone avoiding closed sets. Trans. Am. Math. Soc. **367**, 1609–1630 (2015)
77. Lynch, N.: Approximations to the halting problem. J. Comput. Syst. Sci. **9**, 143–150 (1974)
78. Melnikov, A.G.: Enumerations and completely decomposable torsion-free abelian groups. Theor. Comput. Syst. **45**, 897–916 (2009)
79. Merkle, W.: The Kolmogorov-Loveland stochastic sequences are not closed under selecting subsequences. J. Symbolic Logic **68**, 1362–1376 (2003)
80. Merkle, W., Mihailović, N.: On the construction of effectively random sets. J. Symbolic Logic **69**, 862–878 (2004)
81. Merkle, W., Miller, J.S., Nies, A., Reimann, J., Stephan, F.: Kolmogorov-Loveland randomness and stochasticity. Ann. Pure Appl. Logic **138**, 183–210 (2006)
82. Meyer, A.R.: An open problem on creative sets. Recursive Funct. Theor. Newsl. **4**, 15–16 (1973)
83. Mileti, J.R., Partition Theorems and Computability Theory. Ph.D. Dissertation, University of Illinois at Urbana-Champaign (2004)
84. Millar, T.S.: Foundations of recursive model theory. Ann. Math. Logic **13**, 45–72 (1978)
85. Millar, T.S.: Omitting types, type spectrums, and decidability. J. Symbolic Logic **48**, 171–181 (1983)
86. Miller, J.S., Nies, A.: Randomness and computability: open questions. Bull. Symbolic Logic **12**, 390–410 (2006)
87. Miller, R.G.: Computability, Definability, Categoricity, and Automorphisms. Ph.D. Dissertation, The University of Chicago (2000)
88. Miller, R.: The Δ_2^0-spectrum of a linear order. J. Symbolic Logic **66**, 470–486 (2001)
89. Miller, R.: d-computable categoricity for algebraic fields. J. Symbolic Logic **74**, 1325–1351 (2009)
90. Miller, R., Poonen, B., Schoutens, H., Shlapentokh, A.: A computable functor from graphs to fields (to appear)
91. Monin, B.: Asymptotic density and error-correcting codes (to appear)
92. Monin, B., Nies, A.: A unifying approach to the Gamma question. In: 30th Annual ACM/IEEE Symposium on Logic in Computer Science, LICS 2015, pp. 585–596. IEEE Computer Society (2015)
93. Monin, B., Patey, L.: Π_1^0 encodability and omniscient reductions (to appear)

94. Montalbán, A.: Equivalence between Fraïssé's Conjecture and Jullien's Theorem. Ann. Pure Appl. Logic **139**, 1–42 (2006)
95. Muchnik, A.A., Semenov, A.L., Uspensky, V.A.: Mathematical metaphysics of randomness. Theoret. Comput. Sci. **207**, 263–317 (1998)
96. Patey, L.: The weakness of being cohesive, thin or free in reverse mathematics. Isr. J. Math. (to appear)
97. Patey, L.: Open questions about Ramsey-type statements in reverse mathematics. Bull. Symbolic Logic **22**, 151–169 (2016)
98. Remmel, J.B.: Recursively categorical linear orderings. Proc. Am. Math. Soc. **83**, 387–391 (1981)
99. Richter, L.J.: Degrees of structures. J. Symbolic Logic **46**, 723–731 (1981)
100. Rosenstein, J.G.: Linear Orderings. Pure and Applied Mathematics, vol. 98. Academic Press Inc., New York-London (1982)
101. Schnorr, C.-P.: A unified approach to the definition of a random sequence. Math. Syst. Theor. **5**, 246–258 (1971)
102. Schnorr, C.-P.: Zufälligkeit und Wahrscheinlichkeit. Lecture Notes in Mathematics, vol. 218. Springer, Berlin (1971)
103. Schweber, N.: Do all linear orders in this class have computable copies? (2014). mathoverflow.net/questions/161434
104. Schweber, N.: Finding limit-nondecreasing sets for certain functions (2016). mathoverflow.net/questions/227766
105. Seetapun, D., Slaman, T.A.: On the strength of Ramsey's Theorem. Notre Dame J. Formal Logic **36**, 570–582 (1995)
106. Simpson, S.G.: Subsystems of Second Order Arithmetic. Perspectives in Mathematical Logic, 1st edn. Springer, Berlin (1999)
107. Simpson, S.G.: Subsystems of Second Order Arithmetic. Perspectives in Logic, 2nd edn. Cambridge University Press, Cambridge and Association for Symbolic Logic, Poughkeepsie (2009)
108. Slaman, T.A.: Relative to any nonrecursive set. Proc. Am. Math. Soc. **126**, 2117–2122 (1998)
109. Solovay, R.M.: Hyperarithmetically encodable sets. Trans. Am. Math. Soc. **239**, 99–122 (1978)
110. Soskov, I.N.: Degree spectra and co-spectra of structures, Annuaire de l'Université de Sofia "St. Kliment Ohrisdski", Faculté de Mathématiques et Informatique **96**, 45–68(2004)
111. Specker, E.: Ramsey's Theorem does not hold in recursive set theory. In: Gandy, R.O., Yates, C.E.M. (eds.) Logic Colloquium 1969. Studies in Logic and the Foundations of Mathematics, pp. 439–442. North-Holland, Amsterdam (1971)
112. Szpilrajn, E.: Sur l'extension de l'ordre partiel. Fundam. Math. **16**, 386–389 (1930)
113. Terwijn, S.A.: Computability and Measure. Ph.D. Dissertation, University of Amsterdam (1998)
114. Thurber, J.: Degrees of Boolean Algebras. Ph.D. Dissertation, University of Notre Dame (1994)
115. Wang, Y.: Randomness and Complexity. Ph.D. Dissertation, University of Heidelberg (1996)
116. Wang, Y.: A separation of two randomness concepts. Inf. Process. Lett. **69**, 115–118 (1999)
117. Wehner, S.: Enumerations, countable structures, and Turing degrees. Proc. Am. Math. Soc. **126**, 2131–2139 (1998)

118. Weihrauch, K.: The degrees of discontinuity of some translators between representations of the real numbers. Technical report TR-92-050. International Computer Science Institute, Berkeley (1992)
119. Weihrauch, K.: The TTE-interpretation of three hierarchies of omniscience principles. In: Informatik Berichte FernUniversität Hagen, vol. 130. Hagen (1992)

Introduction to Autoreducibility and Mitoticity

Christian Glaßer[1], Dung T. Nguyen[2], Alan L. Selman[3(✉)],
and Maximilian Witek[1]

[1] Julius-Maximilians-Universität Würzburg, Würzburg, Germany
{glasser,witek}@informatik.uni-wuerzburg.de
[2] LogicBlox Inc., Atlanta, USA
[3] University at Buffalo, Buffalo, USA
selman@buffalo.edu

Abstract. We survey results on these concepts, discover surprising similarities, and, in particular explain why autoreducibility might someday separate complexity classes.

1 Introduction

We begin with the notion of autoreducibility. This was introduced in 1970 by Trakhtenbrot [Tra70]. A set A is *T-autoreducible* (autoreducible in Trakhtenbrot's denomination) if there is an oracle Turing machine M that accepts A using A as an oracle such that for all oracles B and all x, M^B on input x never queries x. A is *m-autoreducible* if there is a total, computable function f such that for all x it holds that $f(x) \neq x$ and $(x \in A \iff f(x) \in A)$, i.e., f many-one reduces A to A.

For complexity classes such as NP and PSPACE refined measures are needed. In this spirit, Ambos-Spies [AS84] defined the notion of p-T-autoreducibility and the more restricted form of p-m-autoreducibility. A set A is *p-T-autoreducible* if it is T-autoreducible via a *polynomial-time* oracle Turing machine. A is *p-m-autoreducible* if it is m-autoreducible via a *polynomial-time* computable f. Autoreducible sets contain local redundant information. That is, if A is p-m-autoreducible, then x and $f(x)$ contain the same information about membership in A.

The question of whether complete sets for various classes are autoreducible has been studied extensively [Yao90, BF92, BFvMT00], and is currently an area of active research. Beigel and Feigenbaum [BF92] showed that polynomial-time Turing complete sets for the classes that form the polynomial-time hierarchy, Σ_i^P, Π_i^P, and Δ_i^P, are p-T-autoreducible. Thus, all polynomial-time Turing complete sets for NP are p-T-autoreducible. Buhrman et al. [BFvMT00] showed that polynomial-time Turing complete sets for EXP and Δ_i^{EXP} are p-T-autoreducible, whereas there exists a polynomial-time Turing complete set for EESPACE that

Birthday Acknowledgement: Rod Downey, whose birthday we celebrate, is the author of several papers on mitotic sets and splittings: [Dow85, Dow97, DRW87, DS89, DS93b, DS93a, DS98, DW86].
Happy Birthday, Rod!

A. Day et al. (Eds.): Downey Festschrift, LNCS 10010, pp. 56–78, 2017.
DOI: 10.1007/978-3-319-50062-1_5

is not p-T-autoreducible. Regarding NP, they showed that all polynomial-time truth-table complete sets for NP are probabilistic polynomial-time truth-table autoreducible.

We should stress that resolving some open questions about autoreducibility would lead to major class separation results. Buhrman et al. [BFvMT00] proved autoreducibility results for many different complexity classes and demonstrated strong evidence that studying structural properties of the complete sets, especially the autoreducibility property, might be an important tool to separate complexity classes (see Sect. 5.2 for details). For example, if there exists a polynomial-time Turing complete set in NEXP that is not p-T-autoreducible, then EXP is different from NEXP.

The notion of mitoticity was introduced and comprehensively studied by Ladner [Lad73]. Mitosis in biology is the process by which a cell separates its duplicated genome into two identical halves. In the same tenor, a set A is T-*mitotic* if there is a decidable set S such that A, $A \cap S$, and $A \cap \overline{S}$ are Turing equivalent. The slight difference between this and Ladner's definition [Lad73] is discussed in Sect. 4. The notion *m-mitotic* is defined similarly by requiring many-one equivalence. Note that mitoticity is a global notion of redundancy.

Ambos-Spies [AS84] formulated related notions in the polynomial time setting. A set A is *p-T-mitotic* if there is a polynomial-time decidable set S such that A, $A \cap S$, and $A \cap \overline{S}$ are polynomial-time Turing equivalent. The notion *p-m-mitotic* is defined similarly by requiring polynomial-time many-one equivalence.

A set A is *nontrivial* if it is neither finite nor cofinite. Ambos-Spies [AS84] showed that if a nontrivial set is p-m-mitotic, then it is p-m-autoreducible and he raised the question of whether the converse holds. Glaßer et al. [GPSZ08] resolved this question and showed that every nontrivial p-m-autoreducible set is p-m-mitotic. This result has an interesting consequence. First, Glaßer et al. [GOP+07] showed that nontrivial polynomial-time many-one complete sets of NP, PSPACE, and the levels of the polynomial hierarchy are p-m-autoreducible by using a left-set technique of Ogiwara and Watanabe [OW91]. Then, for example, it follows that all nontrivial polynomial-time many-one complete sets for NP are p-m-mitotic. That is, each nontrivial NP-complete set A can be split by a set S in P so that A, $A \cap S$, and $A \cap \overline{S}$ are all NP-complete.

Outline of the Paper. Section 2 introduces autoreducibility and mitoticity on the basis of easy examples and theorems from computability and complexity theory. Here we also identify fundamental questions which are raised by the theorems and which will be investigated in the subsequent sections. Section 3 studies for different reducibilities the question of whether autoreducibility implies mitoticity. Section 4 discusses an intricacy that is caused by the possibility to define the notion of mitoticity in several ways. We compare the resulting notions and show that one of them is equivalent to autoreducibility, which is a result by Ladner [Lad73]. Section 5 presents selected results on autoreducibility and mitoticity in complexity theory. We study the question of whether complete sets for typical complexity classes are autoreducible or mitotic. Section 6 extends the scope to sets that are NEXP-complete with respect to reducibility notions between polynomial-time many-one and polynomial-time Turing. This section contains some recent results by the authors.

2 Basic Results

We illustrate autoreducibility and mitoticity with a few easy examples. For the notions that we defined so far we will also use notations like $\leq_T^p$-autoreducible and $\leq_m$-mitotic instead of p-T-autoreducible and m-mitotic. We identify sets $A \subseteq \mathbb{N}$ with their characteristic functions, i.e., $A(x) = 1$ if $x \in A$, and $A(x) = 0$ otherwise. The set accepted by an oracle Turing machine M with A as oracle is defined as $L(M^A) = \{x \in \mathbb{N} \mid M^A(x) = 1\}$. Depending on the context, $M^A(x)$ denotes the computation of M with A as oracle on input x or the result of this computation. REC denotes the class of decidable sets, RE the class of computably enumerable sets (c.e. sets).[1] The Cantor pairing function is defined as $\langle x, y \rangle = y + (x + y)(x + y + 1)/2$, it is a bijection $\mathbb{N} \times \mathbb{N} \to \mathbb{N}$. Let $M_0, M_1, \ldots$ be an effective enumeration of all Turing machines, where repetitions are allowed. For a Turing machine M, let $[M] = \min\{i \mid M = M_i\}$ and observe that this function is total and computable. The halting problem is defined as $K = \{\langle x, y \rangle \mid M_x \text{ halts on input } y\}$.

Example 2.1 (K is $\leq_m$-autoreducible and $\leq_m$-mitotic). *To see autoreducibility we define the function f by $f(\langle x, y \rangle) = \langle [M'_x], y \rangle$, where M'_x is the machine obtained from M_x by adding a new state, which is not used by the machine. So f is total and computable. Since M_x and M'_x compute the same functions it holds that $\langle x, y \rangle \in K \iff f(\langle x, y \rangle) \in K$. Moreover, $[M'_x] \neq [M_x] = x$ and hence $f(\langle x, y \rangle) \neq \langle x, y \rangle$. This shows that K is $\leq_m$-autoreducible.*

Now let us argue for mitoticity. Let $S = \{\langle x, y \rangle \mid M_x$ has an even number of states$\}$, which is a decidable set. Observe that $K \cap S \leq_m K \cap \overline{S}$ and $K \cap \overline{S} \leq_m K \cap S$, both via the function f from above. Moreover, choose some $a \in \mathbb{N} - K$ and observe that $K \cap S \leq_m K$ via g and $K \leq_m K \cap S$ via h, where

$$g(\langle x, y \rangle) = \begin{cases} \langle x, y \rangle, & \text{if} \langle x, y \rangle \in S \\ a, & \text{otherwise} \end{cases}$$

$$h(\langle x, y \rangle) = \begin{cases} \langle x, y \rangle, & \text{if} \langle x, y \rangle \in S \\ f(\langle x, y \rangle), & \text{otherwise.} \end{cases}$$

This shows that K is $\leq_m$-mitotic.

Indeed *every* set A that is $\leq_m$-complete for RE is $\leq_m$-autoreducible: By Myhill's isomorphism theorem [Myh55], A and K are isomorphic, i.e., there exists a computable bijection g such that $g(A) = K$. Together with the autoreduction f we obtain

$$x \in A \iff g(x) \in K \iff f(g(x)) \in K \iff g^{-1}(f(g(x)) \in A,$$

---

[1] By now researchers favor the term "computably enumerable" over "recursively enumerable." However, it is still standard that RE denotes the class of c.e. sets.

where $g(x) \neq f(g(x))$ and hence $x \neq g^{-1}(f(g(x)))$. Therefore, A is $\leq_m$-autoreducible by the autoreduction $g^{-1}(f(g(x)))$. So we obtained the following.

Theorem 2.2. *Every $\leq_m$-complete set for RE is $\leq_m$-autoreducible.*

A similar argument shows that all A that are $\leq_m$-complete for RE are $\leq_m$-mitotic: As seen in Example 2.1, K is $\leq_m$-mitotic. Let S be the separator defined there. Recall that $K \cap S \equiv_m K \equiv_m K \cap \overline{S}$. By Myhill's isomorphism theorem, A and K are isomorphic via the computable bijection g. To show that A is $\leq_m$-mitotic, we use the separator $g^{-1}(S) \in \text{REC}$.

$$
\begin{aligned}
A \cap g^{-1}(S) &\equiv_m g(A) \cap S && \text{(via the reductions } g \text{ and } g^{-1}) \\
&= K \cap S && \text{(since } g(A) = K) \\
&\equiv_m K && \text{(via the reductions } g \text{ and } h \text{ from Example 2.1)} \\
&\equiv_m A && \text{(since } A \text{ is } \leq_m\text{-complete for RE)}
\end{aligned}
$$

Analogously we obtain $A \cap \overline{g^{-1}(S)} = A \cap g^{-1}(\overline{S}) \equiv_m A$. This shows that A is $\leq_m$-mitotic. Hence we proved the following.

Theorem 2.3. *Every $\leq_m$-complete set for RE is $\leq_m$-mitotic.*

In general, mitoticity implies autoreducibility. Let A be a nontrivial set that is $\leq_m$-mitotic. In particular, there exists a separator $S \in \text{REC}$ such that $A \cap S \leq_m A \cap \overline{S}$ via some reduction g and $A \cap \overline{S} \leq_m A \cap S$ via some reduction h. Choose two elements $a_0, a_1 \notin A$ and let

$$
f(x) = \begin{cases} g(x), & \text{if } x \in S \text{ and } g(x) \in \overline{S} \\ h(x), & \text{if } x \in \overline{S} \text{ and } h(x) \in S \\ \min(\{a_0, a_1\} - \{x\}), & \text{otherwise.} \end{cases}
$$

The definition ensures that $f(x) \neq x$ for all x. If $x \in S$ and $g(x) \in \overline{S}$, then $f(x) = g(x)$ and $x \in A \iff f(x) \in A \cap \overline{S} \iff f(x) \in A$. Similarly, if $x \in \overline{S}$ and $h(x) \in S$, then $x \in A \iff f(x) \in A$. Finally, if $f(x)$ is defined according to the third line in f's definition, then $(x \in S \wedge g(x) \in S)$ or $(x \in \overline{S} \wedge h(x) \in \overline{S})$. Each alternative implies $x \notin A$, since g and h are reductions, and hence $x \in A \iff f(x) \in A$. Thus A is $\leq_m$-autoreducible. This argumentation generalizes to further reducibilities, which results in a theorem and a question.

Theorem 2.4. *Let A be nontrivial and let $\leq$ be a reducibility from $\{\leq_m, \leq_m^p, \leq_T, \leq_T^p\}$.*

$$A \text{ is } \leq\text{-mitotic} \implies A \text{ is } \leq\text{-autoreducible}$$

Question 2.5. *Do the converse of these implications hold?*

The next sections answer these questions and we will see that the particular answer depends on the reducibility.

Now let us turn to NP, the complexity analog of RE. The satisfiability problem of Boolean formulas is denoted by $\text{SAT} = \{\varphi \mid \varphi \text{ is a satisfiable Boolean formula}\}$.

Example 2.6 (SAT is $\leq_m^P$-mitotic and $\leq_m^P$-autoreducible). *For the mitoticity we use the separator $S = \{\varphi \mid \varphi$ contains an even number of variables$\}$, which belongs to P. Moreover, $\mathrm{SAT} \cap S \leq_m^P \mathrm{SAT} \cap \overline{S}$ and $\mathrm{SAT} \cap \overline{S} \leq_m^P \mathrm{SAT} \cap S$ both via $f(\varphi) = \varphi \wedge z$, where z is a new variable. Similar to Example 2.1, this can be easily extended to obtain $\mathrm{SAT} \equiv_m^P \mathrm{SAT} \cap S \equiv_m^P \mathrm{SAT} \cap \overline{S}$, which shows that SAT is $\leq_m^P$-mitotic. By Theorem 2.4, this implies that SAT is $\leq_m^P$-autoreducible (indeed f is an autoreduction).*

This example and the analogy to computability theory raise a question.

Question 2.7. *Are all nontrivial $\leq_m^P$-complete sets for NP $\leq_m^P$-autoreducible or $\leq_m^P$-mitotic?*

Let A be an arbitrary $\leq_m^P$-complete set for NP. In contrast to the situation in computability theory, we do not know an isomorphism theorem for NP-complete sets. Moreover, the attempt to directly use SAT's autoreduction f from Example 2.6 leads to a difficulty: Assume $A \leq_m^P \mathrm{SAT}$ via g and $\mathrm{SAT} \leq_m^P A$ via h. So we have

$$x \in A \iff g(x) \in \mathrm{SAT} \iff f(g(x)) \in \mathrm{SAT} \iff h(f(g(x))) \in A.$$

We would like to define our autoreduction as $r(x) = h(f(g(x)))$. However, it is difficult to argue that r is an autoreduction, since we cannot exclude that h is a smart reduction function such that $h(f(g(x))) = x$. In Sect. 5 we will prove the $\leq_m^P$-autoreducibility of NP-complete sets with help of a left-set technique.

3 Does Autoreducibility Imply Mitoticity?

The answer to this question depends on the reduction that is used: while many-one autoreducibility implies many-one mitoticity, this implication does not hold for polynomial-time Turing reducibility [AS84]. Moreover, we do not known whether the implication holds for Turing reducibility.

3.1 Many-One Reducibility

For a total function f and $i \in \mathbb{N}$ we will denote the iterated application of f to x by

$$f^{(i)}(x) = \begin{cases} f(f^{(i-1)}(x)) & \text{if } i > 0, \\ x & \text{otherwise.} \end{cases}$$

Now suppose we are given the $\leq_m$-autoreducible set A, and we want to show that A is $\leq_m$-mitotic. Hence we are given an autoreduction function f for A, and we are looking for a decidable set S such that $A \cap S \equiv_m A \cap \overline{S} \equiv_m A$. The idea of the reduction between $A \cap S$ and $A \cap \overline{S}$ is as follows. On input x, the iterated application of f to x never changes the membership to A, because f is an autoreduction for A. We define S in such a way that after a finite number of applications of f to x there must be a change in the membership to S, which implies $x \in A \iff f^{(c)}(x) \in A$ and $x \in S \iff f^{(c)}(x) \notin S$ for some $c \in \mathbb{N}$. Mitoticity can then be shown analogously to Example 2.1.

Theorem 3.1. *For every nontrivial set $A \subseteq \mathbb{N}$, if A is $\leq_m$-autoreducible, then A is $\leq_m$-mitotic.*

Proof. Let f be an autoreduction for A. We define a function g as follows:

$$g(x) = \begin{cases} \min\{y \in \mathbb{N} \mid y < x \wedge f(y) = x\} & \text{if } y < x \text{ with } f(y) = x \text{ exists} \\ f(x) & \text{otherwise} \end{cases} \quad (1)$$

Observe that g is total and computable. Moreover, if $g(x) = y$, then either $f(x) = y$ or $f(y) = x$. In both cases we have $x \neq y$ and $x \in A \iff y \in A$, so g is also an autoreduction for A. Moreover, for all $x \in \mathbb{N}$ we claim the following:

$$x < g(x) \implies g(x) > g(g(x)) \quad (2)$$

This is seen as follows: suppose (2) does not hold and choose some $a \in \mathbb{N}$ such that $a < g(a)$ and $g(a) \leq g(g(a))$. From $a < g(a)$ it follows that $g(a) = f(a)$ and hence $a < f(a)$. So for $y = a$ it holds that $y < f(a)$ and $f(y) = f(a)$, which means that $g(g(a)) = g(f(a)) = \min\{y \in \mathbb{N} \mid y < f(a) \wedge f(y) = f(a)\} \leq a < g(a) \leq g(g(a))$, a contradiction.

Next we define the function h as follows:

$$h(x) = \begin{cases} g(x) & \text{if } x < g(x) \\ g^{(i)}(x) & \text{otherwise, where } i \in \mathbb{N} \text{ is minimal such that } g^{(i)}(x) < g^{(i+1)}(x) \end{cases} \quad (3)$$

Note that h is total and computable such that $h(x) \neq x$. By its definition and (2) we have

$$x < h(x) \iff h(x) > h(h(x)). \quad (4)$$

To show that A is $\leq_m$-mitotic, we use the separator $S = \{x \in \mathbb{N} \mid x < h(x)\}$. Since h is total and computable, we have $S \in \text{REC}$. We claim that for all x:

$$x \in A \iff h(x) \in A \quad (5)$$
$$x \in S \iff h(x) \notin S \quad (6)$$

Equivalence (5) holds, because h maps to values of g, which is an autoreduction for A, and (6) holds because of (4) and the definition of S.

Analogously to Example 2.1 we can use h to show that A is mitotic. By the equivalences (5) and (6) the function h shows that $A \cap S \leq_m A \cap \overline{S}$ and

$A \cap \overline{S} \leq_m A \cap S$. Moreover, choose some $a \in \mathbb{N} - A$ and observe that $A \cap S \leq_m A$ via h_1 and $A \leq_m A \cap S$ via h_2, where

$$h_1(x) = \begin{cases} x, & \text{if } x \in S \\ a, & \text{otherwise} \end{cases}$$

$$h_2(x) = \begin{cases} x, & \text{if } x \in S \\ h(x), & \text{otherwise.} \end{cases}$$

This shows that A is $\leq_m$-mitotic. $\square$

In the polynomial-time setting, the above implication is much harder to show, as the resources of the reduction are more restricted. Suppose for instance that we are given an $\leq_m^P$-autoreduction f for some set A and we want to show that A is $\leq_m^P$-mitotic. A straightforward adaption of the proof of Theorem 3.1 fails, because on input x we cannot efficiently check whether some $y < x$ with $f(y) = x$ exists, and so it becomes difficult to transform f into an autoreduction with some nice properties as done in the last proof.

Instead of repeating the detailed proof from [GPSZ08] we just sketch the main idea. The aim is the construction of a separator set $S \in \mathrm{P}$ with the property that for every x there exists some $i \leq p(|x|)$ such that $x \in S \iff f^{(i)}(x) \notin S$, where p is some polynomial, i.e., after a polynomial number of applications of f, the membership to S changes. For some fast growing function t, let $S_0 = \{x \in \mathbb{N} \mid$ the minimal i such that $t(i) \geq |x|$ is even$\}$ and $S_1 = \mathbb{N} - S_0$, which partitions $\mathbb{N}$ into even and odd stages. Let d be the following distance function on $\mathbb{N}$:

$$d(x, y) = \mathrm{sgn}(y - x) \cdot \lfloor \log(\mathrm{abs}(y - x)) \rfloor$$

The sets S_0, S_1 and the distance function are decidable and computable in polynomial time. They lead to the following algorithm for the separator set S. On input x:

1. $y := f(x), z := f(f(x))$
2. if $|y| > |x|$, then (if $x \in S_0$ then accept else reject)
3. if $|z| > |y|$, then (if $y \in S_1$ then accept else reject)
4. if $x = z$, then (if $x > f(x)$ then accept else reject)
5. if $d(x, y) > d(y, z)$ then reject
6. if $d(x, y) < d(y, z)$ then accept
7. if $\lfloor y/2^{\mathrm{abs}(d(x,y))+1} \rfloor$ is even, then accept else reject

One can show that for every x there is some $j < p(|x|)$ such that $f^{(j)}(x) \in S \iff f^{(j+1)}(x) \notin S$, where p is some polynomial. Given this property, we can define another autoreduction (the analog to h in the last proof) that always changes the membership to S and that is computable in polynomial time. The remainder of the proof works again analogously to Example 2.1. Hence the following holds:

Theorem 3.2 [GPSZ08]. *For every nontrivial set $A \subseteq \mathbb{N}$, if A is $\leq_m^P$-autoreducible, then A is $\leq_m^P$-mitotic.*

3.2   Turing-Reducibility

Ambos-Spies [AS84] showed that $\leq_T^p$-autoreducibility does not imply $\leq_T^p$-mitoticity.

Theorem 3.3 [AS84]. *There is a decidable A which is $\leq_T^p$-autoreducible but not $\leq_T^p$-mitotic.*

We don't know whether a similar result holds for Turing reducibility, i.e., the following question is open.

Question 3.4. *Does $\leq_T$-autoreducibility imply $\leq_T$-mitoticity?*

However, in the next section we will see that $\leq_T$-autoreducibility in fact coincides with c.e. mitoticity, a notion which is probably slightly weaker than $\leq_T$-mitoticity. So Question 3.4 is equivalent to the question of whether $\leq_T$-mitoticity and c.e. mitoticity coincide.

4   Different Notions of Mitoticity for Turing Reducibility

In this section we compare T-mitoticity with Ladner's original notion of mitoticity and show that the latter coincides with T-autoreducibility. The proof of this surprising result is more involved and more technical than the proofs in the previous sections.

Ladner [Lad73] used the following definition for mitoticity. A c.e. set A is *c.e. mitotic* (mitotic in Ladner's denomination) if there exist disjoint c.e. sets A_1, A_2 such that $A = A_1 \cup A_2$ and the sets A, A_1, A_2 are Turing equivalent.

Proposition 4.1. *For a c.e. set A the following statements are equivalent.*

1. *A is c.e. mitotic.*
2. *There exist disjoint c.e. sets A_1, A_2 such that $A = A_1 \cup A_2$, $A \leq_T A_1$, and $A \leq_T A_2$.*

Proof. By definition, 1 implies 2. For the converse, notice that $A_1 \leq_T A$ as follows: If $x \notin A$, then return 0. Otherwise, enumerate A_1 and A_2 in parallel until x is generated. If x was generated by A_1, then return 1, otherwise return 0. $A_2 \leq_T A$ is shown analogously. $\square$

Ambos-Spies [AS84] introduced a weak form of p-T-mitoticity, which translates as follows to unbounded Turing reducibility. A set A is *weakly T-mitotic* if there exist disjoint sets A_1, A_2 such that $A = A_1 \cup A_2$ and the sets A, A_1, A_2 are Turing equivalent.

Notice the slight differences in the definitions of c.e. mitotic, T-mitotic, and weakly T-mitotic: A set A is c.e. mitotic if and only if it is weakly T-mitotic such that in addition A_1 and A_2 are c.e. sets. A is T-mitotic if and only if it is c.e. mitotic such that in addition A_1 and A_2 are separable by a decidable set (i.e., $A_1 \subseteq S \subseteq \overline{A_2}$ for a decidable S). Hence for every nontrivial A,

$$A \text{ is } \leq_T\text{-mitotic} \implies A \text{ is c.e. mitotic} \implies A \text{ is weakly } \leq_T\text{-mitotic}.$$

Question 4.2. *Does the converse of these implications hold?*

We do not know the answers, but interestingly c.e. mitoticity coincides with T-autoreducibility [Lad73]. We start with the proof of the easy direction.

Theorem 4.3 [Lad73]. *If a c.e. set A is c.e. mitotic, then A is $\leq_T$-autoreducible.*

Proof. "$\Rightarrow$" Let A_1, A_2 be disjoint c.e. sets Turing equivalent to A such that $A = A_1 \cup A_2$. Let M_1 and M_2 be oracle Turing machines such that $A = L(M_1^{A_1})$ and $A = L(M_2^{A_2})$. With help of oracle A we can simulate the computation of $M_1^{A_1 - \{x\}}$ on input x by answering each query q as follows:

1. if $q = x$, then answer 'no'
2. if $q \notin A$, then answer 'no'
3. enumerate A_1 and A_2 in parallel to find out whether $q \in A_1$ or $q \in A_2$ (exactly one holds)
4. if $q \in A_1$, then answer 'yes', otherwise answer 'no'

Note that the simulation is such that we do not query x. In the same way we can simulate $M_2^{A_2 - \{x\}}$ on input x.

The autoreduction of A on input x is as follows: if both simulations return the same value, then output this value, otherwise return 1. The correctness is seen as follows: A_1 and A_2 are disjoint, which implies $x \notin A_1$ or $x \notin A_2$, and hence $L(M_1^{A_1 - \{x\}}) = L(M_1^{A_1}) = A$ or $L(M_2^{A_2 - \{x\}}) = L(M_2^{A_2}) = A$. So if the simulations agree, they provide the correct answer. If the simulations do not agree, then $L(M_1^{A_1 - \{x\}}) \neq L(M_1^{A_1})$ or $L(M_2^{A_2 - \{x\}}) \neq L(M_2^{A_2})$, hence $x \in A_1$ or $x \in A_2$, and hence $x \in A$. □

Now we turn to the difficult direction, i.e., $\leq_T$-autoreducible implies c.e. mitotic. Our proof combines Ladner's original proof [Lad73] with a proof by Downey and Stob [DS93b].

Lemma 4.4. *Let $A \subseteq \mathbb{N}$ be an infinite c.e. set and let M be a $\leq_T$-autoreduction for A. Then there exists an algorithm that on input $i \in \mathbb{N}$ computes a finite set $A_i \subseteq \mathbb{N}$ such that $A = \bigcup_i A_i$, $\emptyset = A_0 \subsetneq A_1 \subsetneq \cdots$, and for $d_j = \min(A_j - A_{j-1})$,*

$$\forall i \geq 1 \ \forall x \leq \max\{d_1, d_2, \ldots, d_i\} \ A_i(x) = M^{A_i}(x). \tag{7}$$

Proof. Let f be a total, injective, computable function with range A. The following algorithm computes finite sets A_i with the desired properties.

1. $A_0 := \emptyset$
2. for $j = 0, 1, \ldots, i - 1$
3. determine the minimal m such that $f(m) \notin A_j$
4. determine the minimal $n \geq m$ such that for $B = A_j \cup \{f(0), \ldots, f(n)\}$ it holds that $\forall x \leq \max(A_j \cup \{f(m)\}) \ [M^B(x)$ stops within n steps and $B(x) = M^B(x)]$
5. $A_{j+1} := A_j \cup \{f(0), \ldots, f(n)\}$
6. return A_i

The numbers m in step 3 exist, since A is infinite. To see that the numbers n in step 4 exist, choose $k \geq \max(A_j \cup \{f(m)\})$ large enough such that $\forall x \leq \max(A_j \cup \{f(m)\})$ it holds that $M^A(x)$ stops within k steps and all queries of this computation are $\leq k$. Now choose n large enough such that $n \geq \max(m, k)$ and $\forall r \geq n, f(r) > k$. This n satisfies the requirement in step 4, since $B \cap \{0, \ldots, k\} = A \cap \{0, \ldots, k\}$ by the choice of n, and hence by the choice of k, for all $x \leq \max(A_j \cup \{f(m)\})$ it holds that $M^B(x) = M^A(x)$ stops within $k \leq n$ steps and $B(x) = A(x) = M^A(x) = M^B(x)$. So the numbers n in step 4 exist. The lemma follows, since $\max\{d_1, d_2, \ldots, d_{j+1}\} \leq \max(A_j \cup \{f(m)\})$ and A_{j+1} equals B from step 4. $\square$

Theorem 4.5 [Lad73]. *If a c.e. set A is $\leq_T$-autoreducible, then A is c.e. mitotic.*

Proof. We may assume that A is infinite. Let M be a T-autoreduction of A and let $A_0, A_1, \ldots$ be the enumeration of A provided by Lemma 4.4. For $E \subseteq \mathbb{N} \cup \{\omega\}$ define $\sup(E) = \max E$ if E is a finite subset of $\mathbb{N}$ and $\sup(E) = \omega$ otherwise. For $D \subseteq \mathbb{N}$ let $l^D(0) = 0$ and $l^D(s+1) = \sup(\{l^D(s)+1\} \cup \{q \mid \exists x \leq l^D(s), M^D(x) \text{ queries } q\})$, where $\omega + 1 = \omega$ and $n \leq \omega$ for all $n \in \mathbb{N}$. So l^D divides $\mathbb{N}$ into stages such that if $x \leq l^D(s)$, then the queries of $M^D(x)$ are $\leq l^D(s+1)$, i.e., computations on inputs $\leq l^D(s)$ depend only on queries $\leq l^D(s+1)$.

For $i \geq 1$ let s_i be the minimal s such that $l^{A_i}(s) \geq d_i$, i.e., s_i is d_i's "stage" in A_i. We argue that s_i is defined and can be computed: By (7), $M^{A_i}(x)$ stops for all $x \leq d_i$. So we can compute $l^{A_i}(s)$ for $s = 0, 1, \ldots$ until we find the first $l^{A_i}(s) \geq d_i$. It follows that $l^{A_i}(0) < \cdots < l^{A_i}(s_i) < \omega$.

The following claim says that moving from A_{i-1} to A_i leaves the stages $< s_i$ fixed.

Claim 1: For $j < s_i$ it holds that $l^{A_i}(j) = l^{A_{i-1}}(j) < \omega$.

Follows by induction on j, since for $x \leq l^{A_i}(j-1) = l^{A_{i-1}}(j-1)$ the queries of $M^{A_i}(x)$ are $\leq l^{A_i}(j) \leq l^{A_i}(s_i - 1) < d_i = \min(A_i - A_{i-1})$. So $M^{A_{i-1}}(x)$ and $M^{A_i}(x)$ ask the same queries.

By definition, d_j is in stage s_j of A_j. The next claim tells us that if s_j is minimal in some sense, then also for the previous sets $A_{j-1}, A_{j-2}, \ldots$ it holds that d_j is in stage s_j.

Claim 2: If $i < j$ and $s_j < \min\{s_i, s_{i+1}, \ldots, s_{j-1}\}$, then $d_j \leq l^{A_{j-1}}(s_j) = \cdots = l^{A_{i-1}}(s_j) < \omega$.

By Claim 1, $l^{A_j}(s_j - 1) = l^{A_{j-1}}(s_j - 1)$ and $l^{A_{j-1}}(s_j) = \cdots = l^{A_{i-1}}(s_j) < \omega$. If there is an $x \leq l^{A_{j-1}}(s_j - 1)$ such that $M^{A_{j-1}}(x)$ queries some $q \in A_j - A_{j-1}$, then $d_j \leq q \leq l^{A_{j-1}}(s_j)$. If there is no such x, then for all $x \leq l^{A_j}(s_j - 1) = l^{A_{j-1}}(s_j - 1)$ the computations $M^{A_j}(x)$ and $M^{A_{j-1}}(x)$ ask the same queries and hence $d_j \leq l^{A_j}(s_j) = l^{A_{j-1}}(s_j)$. This proves Claim 2.

We define the splitting which proves that A is c.e. mitotic. Let $B_0 = C_0 = \emptyset$ and for $i \geq 1$:

- If s_i is even, then let $B_i = B_{i-1} \cup \{d_i\}$ and $C_i = C_{i-1} \cup ((A_i - A_{i-1}) - \{d_i\})$.
- If s_i is odd, then let $C_i = C_{i-1} \cup \{d_i\}$ and $B_i = B_{i-1} \cup ((A_i - A_{i-1}) - \{d_i\})$.

Let $B = \bigcup_i B_i$, $C = \bigcup_i C_i$, and observe that $B, C \in \mathrm{RE}$, $B \cap C = \emptyset$, and $A = B \cup C$.

We show $A \leq_T B$. The symmetric argument yields $A \leq_T C$ and by Proposition 4.1 this implies that A is c.e. mitotic. The reduction $A \leq_T B$ works as follows on input x: Compute some r such that $l^{A_r}(s_r - 2) \geq x$ and $B_r \cap \{0, \ldots, l^{A_r}(s_r - 1)\} = B \cap \{0, \ldots, l^{A_r}(s_r - 1)\}$, return $B(x) + C_r(x)$. This reduction is computable, since we have B as oracle, $l^{A_r}(s_r - 1) < \omega$, and B_i, C_i, $l^{A_r}(s_r - 2)$, and $l^{A_r}(s_r - 1)$ are computable.

Assume the reduction is not correct for some x, i.e., $A(x) \neq B(x) + C_r(x)$. From $A = B \cup C$, $B \cap C = \emptyset$, and $C_r \subseteq C$ it follows $x \in A$, $x \notin B$, and $x \in C - C_r$. So $x \in C_w - C_{w-1}$ for some $w > r$. Now choose the first $v \in \{r, \ldots, w\}$ such that s_v is minimal.

Claim 3: $s_v \leq s_r - 2$.

For $j = s_v - 1$ we have $j < s_w, s_{w-1}, \ldots, s_r$. By Claim 1, $l^{A_w}(s_v - 1) = \cdots = l^{A_r}(s_v - 1)$. If $x \leq l^{A_r}(s_v - 1) = l^{A_w}(s_v - 1)$, then $l^{A_w}(s_w - 1) < d_w \leq x \leq l^{A_w}(s_v - 1)$ and hence $s_w < s_v$, which contradicts the minimal choice of s_v. So $l^{A_r}(s_v - 1) < x \leq l^{A_r}(s_r - 2)$ and hence $s_v < s_r - 1$.

Claim 4: $d_v \leq l^{A_r}(s_r - 2)$.

By Claim 3, $r \neq v$. So from Claim 2 we obtain $d_v \leq l^{A_{v-1}}(s_v) = \cdots = l^{A_r}(s_v)$. By Claim 3, $l^{A_r}(s_v) \leq l^{A_r}(s_r - 2)$. This proves Claim 4.

If s_v is even, then $d_v \in B_v - B_{v-1} \subseteq B - B_r$, and by Claim 4, $d_v \leq l^{A_r}(s_r - 1)$, which contradicts the choice of r by the reduction. So s_v must be odd and hence $d_v \in C_v - C_{v-1}$.

Claim 5: For all u such that $r < u < v$ it holds that $s_u \geq s_v + 2$.

Assume the contrary. By the minimal choice of s_v we have $s_u \geq s_v + 1$. Hence there exists u such that $r < u < v$ and $s_u = s_v + 1$, choose the smallest such u. So s_u is even and $d_u \in B - B_r$. By Claim 2, $d_u \leq l^{A_r}(s_u) = l^{A_r}(s_v + 1)$. By Claim 3, $s_v \leq s_r - 2$ and hence $d_u \leq l^{A_r}(s_r - 1)$, which contradicts the choice of r by the reduction. This proves Claim 5.

By Claim 4, $d_v \leq l^{A_r}(s_r - 2) < d_r$. By Claim 3, $r \neq v$ and hence $r \leq v - 1$. So from (7) we obtain $M^{A_{v-1}}(x) = A_{v-1}(x)$ for all $x \leq d_r$ and hence $M^{A_{v-1}}(d_v) = A_{v-1}(d_v)$. Also by (7), $A_v(d_v) = M^{A_v}(d_v)$. Together with $d_v \in A_v - A_{v-1}$ we obtain $M^{A_{v-1}}(d_v) = A_{v-1}(d_v) \neq A_v(d_v) = M^{A_v}(d_v)$. So there exists some $e \in A_v - A_{v-1}$ that is queried by $M^{A_{v-1}}(d_v)$ and $M^{A_v}(d_v)$. Note that $e > d_v$, since $M^{A_{v-1}}(d_v)$ cannot query d_v. By Claim 2, $d_v \leq l^{A_{v-1}}(s_v)$ and hence $e \leq l^{A_{v-1}}(s_v + 1)$, since e is a query of $M^{A_{v-1}}(d_v)$. By Claim 5, for $j = s_v + 1$ it holds that $j < s_{r+1}, s_{r+2}, \ldots, s_{v-1}$. From Claim 1 we obtain $e \leq l^{A_{v-1}}(s_v + 1) = \cdots = l^{A_r}(s_v + 1)$ and by Claim 3, $e \leq l^{A_r}(s_v + 1) \leq l^{A_r}(s_r - 1)$. Moreover, $e \in B_v - B_{v-1} \subseteq B - B_r$, since s_v is odd and $e > d_v$. The existence of such e contradicts the choice of r by the reduction. $\square$

Corollary 4.6 [Lad73]. *A c.e. set A is c.e. mitotic if and only if it is $\leq_T$-autoreducible.*

In this section we gained the following knowledge on the different notions of unbounded mitoticity.

$$A \text{ is } \leq_T\text{-mitotic} \implies A \text{ is c.e. mitotic} \implies A \text{ is weakly } \leq_T\text{-mitotic}$$

$$\updownarrow$$

$$A \text{ is } \leq_T\text{-autoreducible}$$

5   Selected Results from Computational Complexity

In this section we turn our focus to the complexity classes NP, EXP, NEXP and discuss whether their complete sets are autoreducible or even mitotic.

5.1   Complete Sets for NP

Recall that for every set $A \in$ NP there exists a witness set $B \in$ P and a polynomial p such that for all x it holds that $x \in A$ if and only if there exists a witness $y \in \{0,1\}^{p(|x|)}$ such that $(x,y) \in B$. For all x and y we may assume that $(x,y) \in B$ implies $|y| = p(|x|)$ and $y \notin \{1\}^*$. We define the *left set of B, p* by

$$\mathrm{Left}(B,p) = \{(x,y) \mid \exists z \in \{0,1\}^{p(|x|)} \text{ such that } z \leq y \text{ and } (x,z) \in B\}$$
$$\cup \ \{(x,y) \mid y = 1^{p(|x|)}\},$$

where $\leq$ denotes the quasi-lexicographical order on strings over $\{0,1\}$. Observe that $\mathrm{Left}(B,p) \in$ NP. We use the left set to show that complete sets for NP are autoreducible.

Theorem 5.1 [BF92]. *Every $\leq_T^P$-complete set for NP is $\leq_T^P$-autoreducible.*

Proof. Let A be $\leq_T^P$-complete for NP, and let $B \in$ P be a witness set for A whose members have length p. Since $\mathrm{Left}(B,p) \in$ NP, there exists a polynomial-time oracle Turing machine M such that $\mathrm{Left}(B,p) = L(M^A)$. We will use M to define an autoreduction for A as follows.

On input x, let y be minimal with $(x,y) \in \mathrm{Left}(B,p)$ and observe that $x \in A$ if and only if $(x,y) \in B$. Moreover, y can be computed in polynomial time by a binary search using M with oracle A. However, since we are looking for an autoreduction, we may not query the input x, and this approach fails. Instead we use the oracle sets $A \cup \{x\}$ and $A \setminus \{x\}$ to compute the candidates y_1 and y_2 for y. Since one of the two oracles must have been correct, either y_1 or y_2 (or both) are equal to y, and it remains to check whether $(x,y_1) \in B$ or $(x,y_2) \in B$ (or both) holds: if $x \in A$, then y_1 or y_2 is correct and witnesses $x \in A$, and if $x \notin A$, then there does not exist a witness, so $(x,y_1) \notin B$ and $(x,y_2) \notin B$. $\square$

Theorem 5.2 [GOP+07, GPSZ08]. *Every $\leq_m^p$-complete set for* NP *is $\leq_m^p$-autoreducible and $\leq_m^p$-mitotic.*

Proof. Let A be $\leq_m^p$-complete for NP, and let $B \in$ P be a witness set for A whose members have length p. Since $\text{Left}(B, p) \in$ NP, there exists $f \in$ FP with $(x, y) \in \text{Left}(B, p) \iff f(x, y) \in A$. We will use f to define an autoreduction for A as follows.

On input x, we first compute $f(x, 0^{p(|x|)})$ and $f(x, 1^{p(|x|)})$:

- If $f(x, 0^{p(|x|)}) = x$, then $x \in A \iff (x, 0^{p(|x|)}) \in \text{Left}(B, p) \iff (x, 0^{p(|x|)}) \in B$, which can be tested in polynomial time.
- If $f(x, 1^{p(|x|)}) \neq x$, then $x \in A \iff (x, 1^{p(|x|)}) \in \text{Left}(B, p) \iff f(x, 1^{p(|x|)}) \in A$, and hence $f(x, 1^{p(|x|)})$ can be used as the return value of the autoreduction for x.

So at this point we can assume that $f(x, 0^{p(|x|)}) \neq x$ and $f(x, 1^{p(|x|)}) = x$. This means that by binary search we can find in polynomial time two adjacent strings $y_1, y_2 \in \{0, 1\}^{p(|x|)}$ with $y_1 < y_2$ such that $f(x, y_1) \neq x$ and $f(x, y_2) = x$. If $(x, y_2) \in B$, then $x \in A$, and we are done. So assume that $(x, y_2) \notin B$. Now we can use $f(x, y_1)$ as the return value of the autoreduction for x, because
$$x \in A \iff (x, y_2) \in \text{Left}(B, p) \iff (x, y_1) \in \text{Left}(B, p) \iff f(x, y_1) \in A.$$
Hence A is $\leq_m^p$-autoreducible. By Theorem 3.2, A is also $\leq_m^p$-mitotic.   □

The following question remains open.

Question 5.3. *Are all $\leq_T^p$-complete sets for* NP $\leq_T^p$-*mitotic?*

5.2   Complete Sets for EXP

Berman [Ber77] showed that all $\leq_m^p$-complete sets for EXP are complete with respect to length-increasing reductions. So if A is $\leq_m^p$-complete for EXP and $B \in$ EXP, then there exists a function f such that $x \in B \iff f(x) \in A$ and $|f(x)| > |x|$ for all x. In particular this means that A can be reduced to itself via a length-increasing reduction, which can be used as an autoreduction function for A. By Theorem 3.2, A is also mitotic. We obtain:

Theorem 5.4 [Kur05, GOP+07]. *Every $\leq_m^p$-complete set for* EXP *is $\leq_m^p$-autoreducible and $\leq_m^p$-mitotic.*

Regarding $\leq_T^p$-complete sets for EXP, Buhrman and Torenvliet [BT05] gave an elegant proof that shows autoreducibility. For a $\leq_T^p$-complete set A for EXP and an exponential time Turing machine M with $L(M) = A$ they consider the computation tableau of $M(x)$, i.e., the sequence of configurations C_i (consisting of the tape content including head position and machine state) for each step i of the computation of M on x. The contents of each cell of the tableau can be determined in exponential time and hence can be reduced to A via some polynomial-time oracle Turing machine. Note that either $A \cup \{x\}$ or $A - \{x\}$ can be used as the correct oracle without querying x. The key idea is that the outcome of the computation of M on input x can be determined using the one oracle, and the *consistency* of this computation can be verified using the other oracle. An *inconsistency* in a tableau is a pair (i, j) such that cell j in configuration C_i differs from cell j in the configuration that is obtained by simulating one step of M on C_{i-1}. Note that for testing whether (i, j) is an inconsistency, we just have to inspect a constant number of cells in C_{i-1}. Suppose for instance that B is the correct oracle. Then the outcome of the computation of M on input x as determined with oracle B is correct, and even using the incorrect oracle $B\triangle\{x\}$ we cannot find an inconsistency in this computation. On the other hand, if B is the incorrect oracle, then the outcome of the computation of M on input x as determined with oracle B might be incorrect, but the correct oracle $B\triangle\{x\}$ will show us the inconsistency within this computation. This argumentation leads to the following result.

Theorem 5.5 [BFvMT00, BT05]. *Every $\leq_T^P$-complete set for EXP is $\leq_T^P$-autoreducible.*

The following question remains open.

Question 5.6. *Are all $\leq_T^P$-complete sets for EXP $\leq_T^P$-mitotic?*

Buhrman and Torenvliet [BT05] proposed a Post's program for complexity theory, i.e., the search for structural properties that the complete sets of two complexity classes do not share. Thus if all complete sets of one class have the property, and at least one complete set in another class does not have the property, then the classes are different. Buhrman and Torenvliet mention autoreducibility and mitoticity as reasonable examples of such properties. Not least because Buhrman et al. [BFvMT00] showed that answering certain questions about autoreducibility results in the separation of complexity classes. For instance, consider the following question: Are all polynomial-time truth-table complete sets for EXP polynomial-time truth-table autoreducible? If the answer is "yes", then $P \neq PSPACE$. If the answer is "no", then $PH \neq EXP$.

5.3   Complete Sets for NEXP

Ganesan and Homer [GH92] showed that $\leq_m^P$-complete sets for NEXP are complete w.r.t. one-to-one reductions, which yields the autoreducibility and mitoticity of these sets.

Theorem 5.7 [GH92, GPSZ08]. *Every $\leq_m^P$-complete set for NEXP is $\leq_m^P$-autoreducible and $\leq_m^P$-mitotic.*

Proof. The theorem can be proved briefly by observing a nice result about polynomial-time many-one reductions among NEXP-complete sets. Specifically, Ganesan and Homer [GH92] showed that for any two $\leq_m^P$-complete sets for NEXP, there is a polynomial time *one-to-one* reduction that reduces one set to another. Using this result, we prove the theorem as follows. Let L be any $\leq_m^P$-complete set for NEXP. Notice that $0L \cup 1L$ is also $\leq_m^P$-complete. Then there is a polynomial time *one-to-one* reduction f that reduces $0L \cup 1L$ to L. Consequently, for any string x,

$$x \in L \iff 0x \in 0L \cup 1L \iff f(0x) \in L$$

$$x \in L \iff 1x \in 0L \cup 1L \iff f(1x) \in L$$

Note that f is one-to-one, so $f(0x) \neq f(1x)$. This implies that at least one of $f(0x)$ and $f(1x)$ must be different from the original input x. Hence a $\leq_m^P$-autoreduction for L is easily defined, and L is $\leq_m^P$-autoreducible. By Theorem 3.2, L is also $\leq_m^P$-mitotic.   □

The following related question is open.

Question 5.8. *Are all $\leq_T^P$-complete sets for* NEXP $\leq_T^P$-*autoreducible or even* $\leq_T^P$-*mitotic?*

To further approach the above question, we will consider complete sets for NEXP with respect to further reductions and also obtain some negative results in the next section.

6   Recent Results from the NEXP-Setting

Complete sets for NEXP have a special status with respect to autoreducibility and mitoticity: while all $\leq_m^P$-complete sets for NEXP are $\leq_m^P$-mitotic, we do not even know whether $\leq_T^P$-complete sets for NEXP are $\leq_T^P$-autoreducible. This raises the question if one can show autoreducibility or even mitoticity for sets that are complete for NEXP with respect to further reducibility notions. In this section we will show that for some notions we can show autoreducibility, while for others we can prove negative results.

Besides $\leq_m^P$ and $\leq_T^P$ we consider the following polynomial-time reducibility notions. For sets A and B we say that A is polynomial-time truth-table reducible to B ($A \leq_{tt}^P B$), if $A \leq_T^P B$ via a polynomial-time oracle Turing machine M whose queries are nonadaptive (i.e., independent of the oracle). For $k \geq 1$ we say that A is polynomial-time k-Turing reducible to B ($A \leq_{k\text{-}T}^P B$), if $A \leq_T^P B$ via a polynomial-time oracle Turing machine M that asks at most k queries. If additionally M's queries are nonadaptive, then A is polynomial-time k-truth-table reducible to B ($A \leq_{k\text{-}tt}^P B$). A is polynomial-time disjunctive-truth-table reducible to B ($A \leq_{dtt}^P B$), if there exists a polynomial-time-computable function f such that for all x, $f(x) = (q_1, \ldots, q_n)$ for some $n \geq 1$ and $c_A(x) = \max\{c_B(q_1), \ldots, c_B(q_n)\}$. If n is bounded by the constant k, then A is polynomial-time k-disjunctive-truth-table reducible to B ($A \leq_{k\text{-}dtt}^P B$). The polynomial-time conjunctive-truth-table reducibilities $\leq_{ctt}^P$ and $\leq_{k\text{-}ctt}^P$ are defined analogously.

We begin our investigation of autoreducibility problems for NEXP with disjunctive-truth-table and conjunctive-truth-table reductions. Although these are just slight generalizations of many-one reductions, there is some limit in the proof techniques regarding these reductions that occurs when investigating autoreducibility problems for NP. For example, the question of whether every $\leq_{ctt}^{P}$-complete set for NP is $\leq_{ctt}^{P}$-autoreducible is still open. Fortunately, the class NEXP is powerful enough to simulate all polynomial-time reductions in exponential time, and hence several variants of diagonalization techniques can be exploited to obtain autoreducibility results here. Indeed, we prove that all $\leq_{ctt}^{P}$- and $\leq_{dtt}^{P}$-complete sets for NEXP are autoreducible.

Theorem 6.1 [GNR+13]. *For every $k \geq 1$ and $\leq \; \in \{\leq_{k\text{-dtt}}^{P}, \leq_{k\text{-ctt}}^{P}, \leq_{dtt}^{P}, \leq_{ctt}^{P}\}$, every $\leq$-complete set for NEXP is $\leq$-autoreducible.*

Proof. We prove the theorem for the $\leq_{ctt}^{P}$ case, the other cases can be proved analogously. Let A be a $\leq_{ctt}^{P}$-complete set for NEXP. Let $\{M_i\}_{i \geq 1}$ be an enumeration of all conjunctive truth-table reductions and assume that on input x, M_i runs in time $|x|^i$. Let B be the subset of $\{\langle 0^i, x \rangle \mid i \geq 1, x \geq 0\}$ that is accepted by the following algorithm, where the input is $\langle 0^i, x \rangle$.

- Simulate M_i on input $\langle 0^i, x \rangle$.
- If x is not one of the queries, then accept $\langle 0^i, x \rangle$ if and only if $x \in A$.
- Otherwise, accept $\langle 0^i, x \rangle$.

This is a nondeterministic, exponential time algorithm. So $B \in$ NEXP. Because A is $\leq_{ctt}^{P}$-complete, there is some conjunctive truth-table reduction M_j that reduces B to A.

For an arbitrary input x, simulate the reduction M_j on input $\langle 0^j, x \rangle$ and suppose that the queries are $q_1, q_2 \ldots q_k$. If x is not one of the queries, then by the definition of the set B, $x \in A \iff \langle 0^j, x \rangle \in B \iff A(q_1) = \cdots = A(q_k) = 1$. Otherwise, suppose that $x = q_l$, then again by the definition B, $\langle 0^j, x \rangle \in B$. That implies $A(q_1) = \cdots = A(q_k) = 1$, and hence $x = q_l \in A$. This observation yields a $\leq_{ctt}^{P}$-autoreduction for A. □

Note that $\leq_{k\text{-dtt}}^{P}$- and $\leq_{k\text{-ctt}}^{P}$-reductions are $\leq_{m}^{P}$-reductions when $k = 1$. So the above theorem also subsumes the many-one autoreducibility of NEXP-complete sets as a special case.

The question for autoreducibility becomes more challenging when dealing with more powerful truth-table reducibilities. For example, consider the question of whether every $\leq_{2\text{-tt}}^{P}$-complete set is $\leq_{2\text{-tt}}^{P}$-autoreducible. Buhrman et al. [BFvMT00] show this for EXP, but their proof relies on the fact that EXP is closed under complement, which is not known for NEXP. We show this property for NEXP with a technique that is somewhat similar to the previous ones. However, the proof is more complicated in the sense that each case of a truth-table Boolean function needs to be handled separately in the diagonalization steps.

Theorem 6.2 [GNR+13]. *Every $\leq_{2\text{-tt}}^{P}$-complete set for NEXP is $\leq_{2\text{-tt}}^{P}$-autoreducible.*

Proof. Let A be a $\leq_{2\text{-tt}}^{P}$-complete set for NEXP. We will show that A is $\leq_{2\text{-tt}}^{P}$-autoreducible. Let $\{M_i\}_{i \geq 1}$ be an enumeration of all $\leq_{2\text{-tt}}^{P}$-reductions such that the computation of M_i on x can be simulated in time $|x|^i + i$. Assume that a $\leq^p_{2\text{-tt}}$-reduction from L_1 to L_2 is represented by two polynomial-time computable functions $f : \Sigma^* \to (\Sigma^*)^2$ and $g : \Sigma^* \times \{0,1\}^2 \to \{0,1\}$ such that for all x, $f(x) = \langle q_1, q_2 \rangle$ and $x \in L_1 \Leftrightarrow g(x, L_2(q_1), L_2(q_2)) = 1$. In this sense, let f_i and g_i be the functions that correspond to the reduction M_i. Without loss of generality we assume that if $f_i(x) = \langle q_1, q_2 \rangle$, then $q_1 \neq q_2$. Let B be the set of inputs $\langle 0^i, x \rangle$ accepted by the following nondeterministic, exponential time algorithm N:

- Compute $f_i(\langle 0^i, x \rangle) = \langle q_1, q_2 \rangle$ and let $Q = \{q_1, q_2\}$.
- If $x \notin Q$ then: accept $\langle 0^i, x \rangle \iff x \in A$.
- Otherwise $x \in Q$ and without loss of generality we assume $x = q_1$. Let g_i^x be the 2-ary Boolean function defined by $g_i^x(\alpha, \beta) \overset{df}{=} g_i(x, \alpha, \beta)$ and consider all possible cases for g_i^x:
  1. If g_i^x is constant 0 or constant 1: accept $\langle 0^i, x \rangle \iff g_i^x = 0$.
  2. If $g_i^x(\alpha, \beta) = \beta$ or $g_i^x(\alpha, \beta) = \neg\beta$: accept $\langle 0^i, x \rangle \iff x \in A$.
  3. If $g_i^x(\alpha, \beta) = \alpha$ or $g_i^x(\alpha, \beta) = \neg\alpha$: reject $\langle 0^i, x \rangle$.
  4. If $g_i^x \in \{\wedge, \nrightarrow, \nleftarrow, \neg\vee, \leftrightarrow\}$: accept $\langle 0^i, x \rangle$.
  5. If $g_i^x \in \{\neg\wedge, \rightarrow, \leftarrow, \vee, \oplus\}$: reject $\langle 0^i, x \rangle$.

This algorithm implies that $B \in \text{NEXP}$. Because A is $\leq^p_{2\text{-tt}}$-complete for NEXP, there is some $\leq^p_{2\text{-tt}}$-reduction M_j that reduces B to A. Consider the functions f_j and g_j associated with M_j, and note that j, f_j, and g_j are fixed. We describe a $\leq^p_{2\text{-tt}}$-autoreduction of A on input x:

- Compute $f_j(\langle 0^j, x \rangle) = \langle q_1, q_2 \rangle$ and let $Q = \{q_1, q_2\}$.
- If $x \notin Q$: accept $\iff g_j(x, A(q_1), A(q_2)) = 1$
- Otherwise, suppose $x = q_1$ (the case $x = q_2$ is similar).
  - If g_j^x is constant: This case cannot happen by the diagonalization!
  - If $g_j^x(\alpha, \beta) = \beta$: accept $\iff q_2 \in A$
  - If $g_j^x(\alpha, \beta) = \neg\beta$: accept $\iff q_2 \notin A$
  - If $g_j^x(\alpha, \beta) = \alpha$: reject
  - If $g_j^x(\alpha, \beta) = \neg\alpha$: accept
  - If $g_j^x \in \{\wedge, \neg\wedge, \rightarrow, \nrightarrow\}$: accept
  - If $g_j^x \in \{\vee, \neg\vee, \leftarrow, \nleftarrow\}$: reject
  - If $g_j^x \in \{\leftrightarrow, \oplus\}$: accept $\iff q_2 \in A$

Note that the above algorithm on input x never queries x. It remains to show that the algorithm reduces A to itself. For this we consider the following cases:

1. If $x \notin Q$: The algorithm is correct by the definition of B.
   For the cases below, suppose that $x = q_1$.
2. If g_j^x is constant: This case cannot happen because: $\langle 0^j, x \rangle \in B$ if and only if $g_j^x = 0$, which contradicts the fact that $B \leq^p_{2\text{-tt}} A$ via M_j. Hence, this case can not happen.
3. If $g_j^x(\alpha, \beta) = \beta$:

$$x \in A \iff \langle 0^j, x \rangle \in B \iff g_j^x(A(x), A(q_2)) = 1 \iff A(q_2) = 1 \iff q_2 \in A.$$

4. If $g_j^x(\alpha, \beta) = \neg\beta$:

$$x \in A \iff \langle 0^j, x \rangle \in B \iff g_j^x(A(x), A(q_2)) = 1 \iff \neg A(q_2) = 1 \iff q_2 \notin A.$$

5. If $g_j^x(\alpha, \beta) = \alpha$: By the definition of B, $\langle 0^j, x \rangle \notin B$. Hence $g_j^x(A(x), A(q_2)) = 0$, which implies $A(x) = 0$.
6. If $g_j^x(\alpha, \beta) = \neg\alpha$: By the definition of B, $\langle 0^j, x \rangle \notin B$. Hence $g_j^x(A(x), A(q_2)) = 0$, which implies $A(x) = 1$.
7. If $g_j^x \in \{\wedge, \nrightarrow, \nleftarrow, \neg\vee, \leftrightarrow\}$: Here $\langle 0^j, x \rangle \in B$, hence $g_j^x(A(x), A(q_2)) = 1$.
   If $g_j^x \in \{\wedge, \nrightarrow\}$, then $x \in A$.
   If $g_j^x \in \{\nleftarrow, \neg\vee\}$, then $x \notin A$.
   If $g_j^x = \ \leftrightarrow$, then $x \in A$ if and only if $q_2 \in A$.
8. If $g_j^x \in \{\neg\wedge, \rightarrow, \leftarrow, \vee, \oplus\}$: Here $\langle 0^j, x \rangle \notin B$, hence $g_j^x(A(x), A(q_2)) = 0$.
   If $g_j^x \in \{\neg\wedge, \rightarrow\}$, then $x \in A$.
   If $g_j^x \in \{\leftarrow, \vee\}$, then $x \notin A$.
   If $g_j^x = \ \oplus$, then $x \in A$ if and only if $q_2 \in A$.

Hence A is $\leq_{2\text{-tt}}^{\mathrm{P}}$-autoreducible. $\square$

Recall that we do not know whether all $\leq_{\mathrm{T}}^{\mathrm{P}}$-complete sets for NEXP are $\leq_{\mathrm{T}}^{\mathrm{P}}$-autoreducible or $\leq_{\mathrm{T}}^{\mathrm{P}}$-mitotic. The following additional questions remain open.

Question 6.3. 1. Is every $\leq_{\mathrm{tt}}^{\mathrm{P}}$-complete set for NEXP $\leq_{\mathrm{tt}}^{\mathrm{P}}$-autoreducible or $\leq_{\mathrm{tt}}^{\mathrm{P}}$-mitotic?
2. For $k \geq 3$, is every $\leq_{k\text{-tt}}^{\mathrm{P}}$-complete set for NEXP $\leq_{k\text{-tt}}^{\mathrm{P}}$-autoreducible or $\leq_{k\text{-tt}}^{\mathrm{P}}$-mitotic?

6.1 Non-autoreducible Sets for NEXP

In this section, we demonstrate some negative results for NEXP. Specifically, for some chosen reductions $\leq_r$ and $\leq_s$, we show that there is an $\leq_r$-complete set for NEXP that is not $\leq_s$-autoreducible.

To prove negative autoreducibility results, we can use well-known diagonalization techniques [BHT91,BFvMT00]. We refer to these papers and to Nguyen and Selman [NS16] for more details. Here we briefly sketch the general proof technique: Suppose we want to prove that there is an $\leq_r$-complete set for $\mathcal{C}$ that is not $\leq_s$-autoreducible. Let $\{M_i\}_{i\geq 1}$ be an enumeration of all $\leq_s$-reductions. Let K be a canonical complete set for $\mathcal{C}$. We construct the desired set L, which is $\leq_r$-complete but not $\leq_s$-autoreducible, in stages. For each stage, we select and add strings of size in the interval (y_i, y_{i+1}) so that the following conditions are satisfied:

- $K \leq_r L$.
- L is in $\mathcal{C}$.
- M_i fails on input 0^{y_i}, i.e., $0^{y_i} \in L \iff M_i^L$ rejects input 0^{y_i}.

Depending on what the $\leq_r$- and $\leq_s$-reductions are, the set L can be encoded appropriately to satisfy all conditions. Hence L is $\leq_r$-complete for $\mathcal{C}$, and because M_i is not an $\leq_s$-autoreduction for L for any i, L is not $\leq_s$-autoreducible.

Using this powerful approach, we obtain the following non-autoreducibility results for NEXP.

Theorem 6.4 [NS16]. *For any positive integer s and k such that $2^s - 1 \geq k$, there is a $\leq_{s\text{-T}}^{P}$-complete set for NEXP that is not $\leq_{k\text{-tt}}^{P}$-autoreducible.*

Proof. Let K be a canonical complete set for NEXP. We will construct a set L in stages such that L is $\leq_{s\text{-T}}^{P}$-complete for NEXP and is not $\leq_{k\text{-tt}}^{P}$-autoreducible.

Let $\{M_j\}_{j \geq 1}$ be an enumeration of all $\leq_{k\text{-tt}}^{P}$-reductions. Define the sequence $\{y_n\}_{n \geq 0}$ such that $y_0 = 1$ and $y_{n+1} = 2^{y_n^n} + 1$ for every $n \geq 1$. In each stage n, we construct $B_n = B^{\leq y_n^n}$, which is the subset of B that contains the strings of size less than or equal to y_n^n. This stage involves two typical steps:

- Encoding step: Select strings of size between $y_{n-1}^{n-1} + 1$ and y_n^n to put into B_n so that $K^{<y_n^n}$ is $\leq_{s\text{-T}}^{P}$-reducible to B_n.
- Diagonalization step: Put string 0^{y_n} into B_n if and only if $M_n^{B_n}$ rejects 0^{y_n}. This will ensure that B is not autoreducible by any $\leq_{k\text{-tt}}^{P}$-reduction.

During the construction, we make sure that the decision to put a string into B_n needs to be made deterministically or nondeterministically in no more than exponential time; hence $B \in$ NEXP.

Consider the computation of $M_n^{B_n}$ on input 0^{y_n} and let $Q = \{q_1, \ldots, q_k\}$ be a set of queries in that computation. Note that the membership of 0^{y_n} in B_n depends on the memberships of $q_1, \ldots, q_k$ in B_n. To assure that determining whether 0^{y_n} can be done deterministically in exponential time, we need to decide deterministically whether to put $q_1, \ldots, q_k$ into B_n as well. Also we want to make sure that this process is consistent to the encoding step. For example, consider some string b and a $\leq_{s\text{-T}}^{P}$-reduction R that reduces K to B. Also assume that $q_1, \ldots, q_k$ are among the queries of R on input x. So a decision to put $q_1, \ldots, q_k$ into B_n should be: (1) independent of whether b is in K, (2) such that R correctly reduces K to B, i.e., $b \in K$ if and only if R^B accepts b. The challenge is to construct the $\leq_{s\text{-T}}^{P}$-reduction such that it reduces K to B and at the same time allows the encoding and diagonalization steps go through.

Let's consider the following $\leq_{s\text{-T}}^{P}$-reduction. For every input x, we have a query tree, which is a binary tree of height s. Each node is a query, and depending on the answer, the next query is the left or right child node. The reduction accepts if and only if the last query, which is a leaf node, belongs to the oracle. Figure 1 depicts how this process works.

Using this $\leq_{s\text{-T}}^{P}$-reduction, we can construct the set B_n such that K is reducible to B_n. Now we come back to the encoding and diagonalization steps that determine whether to put $q_1, \ldots, q_k$ into B_n. Note that $2^s - 1 > k$, hence there exists a node $\mathcal{F}$ in the query tree such that $\mathcal{F}$ is not in $\{q_1, \ldots, q_k\}$. We put nodes in the query tree into B_n as follows: For every node $\mathcal{N}$ in the path from the root node to $\mathcal{F}$ (not including $\mathcal{F}$), $\mathcal{N}$ is put into B_n if and only if its

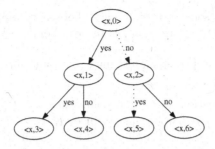

Fig. 1. Height-2 query tree of the $\leq^p_{3\text{-}T}$-reduction from K to B on input x. Start with the root node, a computation is as follows. If a current node is in the oracle B, it follows a left branch. Otherwise, the right branch is used. The dotted path (from the root node to leaf node $\langle x, 5\rangle$) shows how the reduction works in case $\langle x, 0\rangle \notin B$ and $\langle x, 2\rangle \in B$. The reduction accepts x if and only if $\langle x, 5\rangle$ is in B.

left child node is in the path. For every node $\mathcal{N}$ in the left path from $\mathcal{F}$ to a leaf node (not including $\mathcal{F}$), put $\mathcal{N}$ into B_n. Finally, $\mathcal{F}$ is put into B_n if and only if b is in K. It can be verified that K is reducible to B_n correctly by this $\leq^p_{s\text{-}T}$-reduction and the membership of $q_1, \ldots, q_k$ in B_n and hence that of 0^{y_n} in B_n can be determined deterministically in exponential time. In this way we obtain a set B in NEXP such that B is $\leq^p_{s\text{-}T}$-complete, but not $\leq^p_{k\text{-}tt}$-autoreducible. $\square$

Honest reductions are discussed in Homer [Hom87] and Downey et al. [DHGM89]. In honest reductions, the queries are at most polynomially smaller than the input length. Here we use a stronger notion of honest reductions, where for a fixed constant c, the queries of computations on inputs of length n must have a length between $n^{1/c}$ and n^c.

Definition 6.5 Honest truth-table reduction. *Given any two sets A and B and an arbitrary positive number $c \geq 1$, we define a polynomial-time honest truth-table reduction $\leq^{h\text{-}c}_{tt}$ as follows: $A \leq^{h\text{-}c}_{tt} B$ if there exists a nonadaptive, polynomial-time Turing machine M with oracle B such that M^B accepts x if and only if $x \in A$ and for any input x, all queries q made to oracle B have length satisfying $|x|^{1/c} \leq |q| \leq |x|^c$.*

Theorem 6.6 [NS16]. *For every constant $c \geq 1$, there is a $\leq^p_{2\text{-}T}$-complete set for NEXP that is not $\leq^{h\text{-}c}_{3\text{-}tt}$-autoreducible.*

Proof. Let $\{M_j\}_{j \geq 1}$ be an enumeration of $\leq^{h\text{-}c}_{3\text{-}tt}$-reductions such that all queries have length between $n^{\frac{1}{c}}$ and n^c, where n is input size. Let K be a canonical complete set for NEXP. Define the sequence $\{y_n\}_{n \geq 1}$ such that $y_1 = 1$ and $y_{n+1} = max(y_n^n, y_n^{c^2}) + 1$. We construct a set B in NEXP such that B is not autoreducible by any M_j, but $K \leq^p_{2\text{-}T} B$.

Similar to the proof of Theorem 6.4, we build the set B in stages. In each stage n, we construct a set $B_n = B^{<y_{n+1}}$, which is the subset of B containing the strings of length smaller than y_{n+1}. This stage also contains two typical steps:

- Encoding step: put some strings of length between y_n and $y_{n+1} - 1$ into B_n. This step ensures that K is reducible to B by some $\leq^p_{2\text{-}T}$-reduction that we will describe below.
- Diagonalization step: put string 0^m into B_n if and only if $M_n^{B_n}$ rejects 0^m, where $m = y_n^c$.

During this construction, we need to make sure that B_n is in NEXP.

We describe a $\leq^p_{2\text{-}T}$-reduction that reduces K to B. For every input x of length between y_n and $y_{n+1} - 1$, x is in K if and only if $\langle B(0^m), x \rangle$ is in B.

Now consider the computation of $M_n^{B_n}$ on input 0^m and suppose that the queries are $\langle 1, q_1 \rangle, \langle 0, q_2 \rangle, \langle 0, q_3 \rangle$. Other cases can be investigated analogously. Let f be a Boolean truth-table function of M_n. Now we need to decide whether to put $0^m, \langle 1, q_1 \rangle, \langle 0, q_2 \rangle, \langle 0, q_3 \rangle$ into B_n such that the following constraints are satisfied:

- Constraint 1: $B_n \in$ NEXP.
- Constraint 2: $0^m \in B_n \iff f(B_n(\langle 1, q_1 \rangle), B_n(\langle 0, q_2 \rangle), B_n(\langle 0, q_3 \rangle)) = 0$.
- Constraint 3-1: If $0^m \in B_n$: $\langle 1, q_1 \rangle \in B \iff q_1 \in K$.
- Constraint 3-2: If $0^m \notin B_n$: $\langle 0, q_2 \rangle \in B \iff q_2 \in K$ and $\langle 0, q_3 \rangle \in B \iff q_3 \in K$.

It seems hard to satisfy all constraints simultaneously, since f can be an arbitrary Boolean function of three variables. Fortunately, any 3-variable Boolean function has the following nice property.

Lemma 6.7. *For any Boolean function $f(a, b_1, b_2)$, at least one of the following statements must be true:*

- *There exist two Boolean functions $g_1(a)$ and $g_2(a)$, where $g_1(a)$ and $g_2(a)$ are one of 0, 1, a such that $f(a, g_1(a), g_2(a)) = 0$ for every a.*
- *There exists a Boolean function $h(b_1, b_2)$, where $h(b_1, b_2)$ is one of 0, 1, b_1, b_2, $b_1 \vee b_2$, $b_1 \wedge b_2$ such that $f(h(b_1, b_2), b_1, b_2) = 1$ for every b_1 and b_2.*

Now consider the Boolean truth-table function of M_n. Suppose that $f(b_1 \wedge b_2, b_1, b_2) = 1$ for every b_1 and b_2. If we set $B_n(\langle 1, q_1 \rangle) = B(\langle 0, q_2 \rangle) \wedge B(\langle 0, q_3 \rangle)$, then $f(B_n(\langle 1, q_1 \rangle), B_n(\langle 0, q_2 \rangle), B_n(\langle 0, q_3 \rangle)) = 1$ whatever the values of $B(\langle 0, q_2 \rangle)$ and $B(\langle 0, q_3 \rangle)$ are. To satisfy Constraint 2, we do not put 0^m into B. Constraint 3-2 can be satisfied easily by putting $\langle 0, q_2 \rangle$ and $\langle 0, q_3 \rangle$ into B_n if and only if q_2 and q_3 are in K, respectively. Now we verify that Constraint 1 is also satisfied. Determining the membership of 0^m in B is straightforward. Checking whether $\langle 0, q_2 \rangle$ is in B can be done by checking whether q_2 is in K. Finally, $\langle 1, q_1 \rangle \in B \iff \langle 0, q_2 \rangle \in B \wedge \langle 0, q_3 \rangle \in B \iff q_2 \in K \wedge q_3 \in K$. This can be done in nondeterministic, exponential time as well. This shows that B is $\leq^p_{2\text{-}T}$-complete for NEXP, but not autoreducible by any $\leq^{h\text{-}c}_{3\text{-}tt}$-reduction. $\square$

Definition 6.8. NOR-reduction. *Given any two sets A and B, we define a polynomial-time NOR-truth-table reduction $\leq^p_{\text{NOR-tt}}$ as follows: $A \leq^p_{\text{NOR-tt}} B$ if there exists a nonadaptive, polynomial-time Turing machine M with oracle B such that for any input x, if $q_1, \ldots, q_k$ are all queries of M^B on input x, then*

$$x \in A \iff q_1 \notin B \land \cdots \land q_k \notin B.$$

We call a truth-table reduction *typical* if its Boolean truth-table function is neither an OR nor a NOR Boolean function. The formal definition follows.

Definition 6.9. Typical-reduction. *Given any two sets A and B, we define a typical polynomial-time truth-table reduction $\leq^{P}_{tt\text{-}t}$ as follows: $A \leq^{P}_{tt\text{-}t} B$ if and only if there exist two polynomial-time computable functions f and g such that for any input x, $f(x) = \langle q_1, \ldots, q_k \rangle$, $g(x) = h(\alpha_1, \ldots \alpha_k)$ is a Boolean function with k variables $\alpha_1, \ldots, \alpha_k$ such that h is neither an OR nor a NOR Boolean function, and*

$$x \in A \iff h(B(q_1), \ldots, B(q_k)) = 1.$$

Using the similar proof technique as above, we obtain the following non-autoreducibility result when restricting the power of Boolean truth-table function.

Theorem 6.10 [NS16]

- *For any positive integer k, there is a $\leq^{P}_{k\text{-}tt}$-complete set for NEXP that is not typically $\leq^{P}_{k\text{-}tt}$-autoreducible.*
- *For any positive integer k, there is a $\leq^{P}_{k\text{-}dtt}$-complete set for NEXP that is not typically $\leq^{P}_{k\text{-}tt}$-autoreducible.*

Recall that every $\leq^{P}_{k\text{-}dtt}$-complete set for NEXP is $\leq^{P}_{k\text{-}dtt}$-autoreducible. So we ask whether every $\leq^{P}_{k\text{-}dtt}$-complete set for NEXP is $\leq^{P}_{k\text{-}NOR\text{-}tt}$-autoreducible. For EXP we know that the answer is yes, but for NEXP this is a challenging open question. Settling this question either way leads to major results about exponential time complexity classes.

Theorem 6.11 [NS16]. *For any positive integer k, every $\leq^{P}_{k\text{-}dtt}$-complete set for NEXP is $\leq^{P}_{k\text{-}NOR\text{-}tt}$-autoreducible if and only if NEXP = coNEXP.*

References

[AS84] Ambos-Spies, K.: P-mitotic sets. In: Börger, E., Hasenjaeger, G., Rödding, D. (eds.) LaM 1983. LNCS, vol. 171, pp. 1–23. Springer, Heidelberg (1984). doi:10.1007/3-540-13331-3_30

[Ber77] Berman, L.: Polynomial reducibilities and complete sets. Ph.D. thesis, Cornell University, Ithaca, NY (1977)

[BF92] Beigel, R., Feigenbaum, J.: On being incoherent without being very hard. Comput. Complex. **2**, 1–17 (1992)

[BFvMT00] Buhrman, H., Fortnow, L., van Melkebeek, D., Torenvliet, L.: Separating complexity classes using autoreducibility. SIAM J. Comput. **29**(5), 1497–1520 (2000)

[BHT91] Buhrman, H., Homer, S., Torenvliet, L.: Completeness for nondeterministic complexity classes. Math. Syst. Theory **24**(3), 179–200 (1991)

[BT05] Buhrman, H., Torenvliet, L.: A Post's program for complexity theory. Bull. EATCS **85**, 41–51 (2005)

[DHGM89] Downey, R., Homer, S., Gasarch, W., Moses, M.: On honest polynomial reductions, relativizations, and P=NP. In: IEEE Structure in Complexity Theory Conference, pp. 196–207 (1989)

[Dow85] Downey, R.G.: The degrees of r.e. sets without the universal splitting property. Trans. Am. Math. Soc. **291**(1), 337–351 (1985)

[Dow97] Downey, R.G.: On the universal splitting property. Math. Log. Q. **43**, 311–320 (1997)

[DRW87] Downey, R.G., Remmel, J.B., Welch, L.V.: Degrees of splittings and bases of recursively enumerable subspaces. Trans. Am. Math. Soc. **302**(2), 683–714 (1987)

[DS89] Downey, R.G., Slaman, T.A.: Completely mitotic r.e. degrees. Ann. Pure Appl. Logic **41**(2), 119–152 (1989)

[DS93a] Downey, R.G., Stob, M.: Friedberg splittings of recursively enumerable sets. Ann. Pure Appl. Logic **59**(3), 175–199 (1993)

[DS93b] Downey, R.G., Stob, M.: Splitting theorems in recursion theory. Ann. Pure Appl. Logic **65**(1), 1–106 (1993)

[DS98] Downey, R.G., Shore, R.A.: Splitting theorems and the jump operator. Ann. Pure Appl. Logic **94**(1–3), 45–52 (1998)

[DW86] Downey, R.G., Welch, L.V.: Splitting properties of r. e. sets, degrees. J. Symbolic Logic **51**(1), 88–109 (1986)

[GH92] Ganesan, K., Homer, S.: Complete problems and strong polynomial reducibilities. SIAM J. Comput. **21**, 733–742 (1992)

[GNR+13] Glaßer, C., Nguyen, D.T., Reitwießner, C., Selman, A.L., Witek, M.: Autoreducibility of complete sets for log-space and polynomial-time reductions. In: Fomin, F.V., Freivalds, R., Kwiatkowska, M., Peleg, D. (eds.) ICALP 2013. LNCS, vol. 7965, pp. 473–484. Springer, Heidelberg (2013). doi:10.1007/978-3-642-39206-1_40

[GOP+07] Glaßer, C., Ogihara, M., Pavan, A., Selman, A.L., Zhang, L.: Autoreducibility, mitoticity, and immunity. J. Comput. Syst. Sci. **73**(5), 735–754 (2007)

[GPSZ08] Glaßer, C., Pavan, A., Selman, A.L., Zhang, L.: Splitting NP-complete sets. SIAM J. Comput. **37**, 1517–1535 (2008)

[Hom87] Homer, S.: Minimal degrees for polynomial reducibilities. J. ACM **34**(2), 480–491 (1987)

[Kur05] Kurtz, S.: Private communication to Buhrman and Torenvliet [BT05] (2005)

[Lad73] Ladner, R.: Mitotic recursively enumerable sets. J. Symbolic Logic **38**(2), 199–211 (1973)

[Myh55] Myhill, J.: Creative sets. Math. Logic Q. **1**(2), 97–108 (1955)

[NS16] Nguyen, D., Selman, A.: Structural properties of nonautoreducible sets. ACM Trans. Comput. Theory **8**(3) (2016). Article No. 11

[OW91] Ogiwara, M., Watanabe, O.: On polynomial-time bounded truth-table reducibility of NP sets to sparse sets. SIAM J. Comput. **20**(3), 471–483 (1991)

[Tra70] Trahtenbrot, B.: On autoreducibility. Dokl. Akad. Nauk SSSR **192**, 1224–1227 (1970). (Translation in Soviet Math. Dokl. 11: 814–817 (1970))

[Yao90] Yao, A.: Coherent functions and program checkers. In: Proceedings of the 22nd Annual Symposium on Theory of Computing, pp. 89–94 (1990)

The Complexity of Complexity

Eric Allender[✉]

Department of Computer Science, Rutgers University, Piscataway, NJ, USA
allender@cs.rutgers.edu

Abstract. Given a string, what is its complexity? We survey what is known about the computational complexity of this problem, and describe several open questions.

1 Introduction

There are many different ways to define the "complexity" of a string. Indeed, if you look up "complexity" in the index of Downey and Hirschfeldt [23] you'll find that the list of entries for this single item extends over more than two full pages. And this list barely mentions any of the *computable* notions of complexity that constitute much of the focus of this article.

All of these notions of "complexity" seem to share some common characteristics. First: determining complexity is hard; even among those notions that happen to be computable, there does not seem to be any widely-studied notion of complexity where the complexity of a string is *efficiently* computable. Secondly: there is a paucity of techniques that have been developed, in order to *show* that it is a computationally hard problem to determine the complexity of a string. Indeed, since complexity theory almost always resorts to *reducibility* in order to show that certain problems are hard, we are led to the problem of studying the class of problems that are reducible to the task of computing the "complexity" of a string, for different notions of "complexity", and under different types of reducibility.

The goal of this article is to give an overview of what is known about this topic (including some new developments that have not yet been published), and to suggest several directions for future work.

2 Preliminaries and Ancient History

One advantage of writing a high-level survey, is that the author is permitted the luxury of providing only a high-level overview of various important definitions, leaving the details to be found in the cited references. So let us indulge in this luxury.

Kolmogorov complexity (in all of its many variations) provides the best tools for quantifying how much information a string x contains. In this survey, the only two versions of Kolmogorov complexity that will be considered are the plain complexity $C(x)$ and the prefix-free complexity $K(x)$. But the reader will

© Springer International Publishing AG 2017
A. Day et al. (Eds.): Downey Festschrift, LNCS 10010, pp. 79–94, 2017.
DOI: 10.1007/978-3-319-50062-1_6

be reminded at key moments that the measures C and K actually represent an infinite collection of measures C_U and K_U, indexed by the choice of "universal" Turing machine that is used in defining Kolmogorov complexity. For more formal definitions and background, see [23].

We will be especially concerned with the set of *random* strings, and with the *overgraph* of these complexity functions:

Definition 1. *Let U be a universal Turing machine.*

- $R_{K_U} = \{x : K_U(x) \geq |x|\}$. *This is the set of K-random strings.*
- $R_{C_U} = \{x : C_U(x) \geq |x|\}$. *This is the set of C-random strings.*
- $O_{K_U} = \{(x, k) : K_U(x) \leq k\}$. *This is the overgraph for K.*
- $O_{C_U} = \{(x, k) : C_U(x) \leq k\}$. *This is the overgraph for C.*

As usual, we suppress the "U" when the choice of universal machine is not important. We will be considering several other measures μ of the complexity of strings. For any such measure μ, we will refer to $R_\mu = \{x : \mu(x) \geq |x|\}$, and similarly O_μ is $\{(x, k) : \mu(x) \leq k\}$.

A binary string x of length N can also be viewed as a representation of a function $f_x : [N] \rightarrow \{0, 1\}$. When $N = 2^n$ is a power of two, then it is customary to view f_x as the truth table of an n-ary Boolean function, in which case computational complexity theory provides a different suite of definitions to describe the complexity of x, such as circuit size $\mathsf{CSize}(x)$, branching program size $\mathsf{BPSize}(x)$, and formula size $\mathsf{FSize}(x)$.[1] For more background and definitions on circuits, formulas, and branching programs, see [50].

Definition 2. *The* Minimum Circuit Size Problem MCSP *is the overgraph of the* CSize *function:* $\mathsf{MCSP} = \{(x, k) : \mathsf{CSize}(x) \leq k\}$.

The study of MCSP dates back to the 1960's ([49], see also the discussion in [15]). More recently, Kabanets and Cai focused attention on MCSP [32] in connection with its relation to the "natural proofs" framework of Razborov and Rudich [43]. In Sect. 3 we will review what is known about MCSP as a candidate "NP-intermediate" problem: a problem in NP $-$ P that is not NP-complete.

Although it was recognized in the 1960's in the Soviet mathematical community that MCSP had somewhat the same flavor as a time-bounded notion of Kolmogorov complexity [49], a formal connection was not established until the 1990's, by way of a modification of a time-bounded version of Kolmogorov complexity that had been introduced earlier by Levin.

The "standard" way to define time-bounded Kolmogorov complexity is to define a measure such as $C_U^t(x) = \min\{|d| : U(d) = x \text{ in at most } t(|x|) \text{ steps}\}$. In contrast, Levin's definition [36] of the measure Kt combines both the running time and the description length into a single number: $\mathrm{Kt}(x) = \min\{|d| + \log t : U(d) = x \text{ in at most } t \text{ steps}\}$. Levin's motivation in formulating this definition

---

[1] These measures can be extended to *all* strings by appending bits to obtain a string whose length is a power of 2 [8].

came from its utility in defining optimal search strategies for deterministic simulations of nondeterministic computations.

The connection between CSize and time-bounded Kolmogorov complexity provided by [8] is the result of replacing "$\log t$" by "t" in the definition of Kt, to obtain a new measure KT. In order to accommodate sublinear running-times as part of this definition (so that $KT(x)$ can be much less than $|x|$), the technical description of what it means for $U(d) = x$ is changed. Now U is given random access to the description d, and – given d and i – U determines the i-th bit of x. (For formal definitions, see [8].) This time-bounded measure KT is of interest because:

- $KT(x)$ is polynomially-related to $CSize(x)$ [8]. (In contrast, no such relation is known for other polynomial-time-bounded variants of Kolmogorov complexity, such as the "standard" version C^{n^2} mentioned above.)
- By providing the universal machine U with an oracle, variants of other notions of Kolmogorov complexity are obtained. For instance, with an oracle A that is complete for deterministic exponential time, $KT^A(x)$ is essentially the same as $Kt(x)$. Similarly, if A is the halting problem (or any problem complete for the c.e. sets under linear-time reductions), then $KT^A(x)$ and $C(x)$ are linearly-related [8].
- The connection between MCSP and KT carries over also to the relativized setting. One can define $CSize^A(x)$ in terms of circuits with "oracle gates", and thus obtain the language $MCSP^A$. $KT^A(x)$ is polynomially-related to $CSize^A(x)$. This has been of interest primarily in the case of $A = QBF$, the standard complete set for PSPACE; see Sect. 3.

Other resource-bounded Kolmogorov complexity measures in a similar vein were presented in [15]: KF is polynomially-related to FSize, and KB is polynomially-related to BPSize. Note in this regard the following sequence of inequalities:

$$C(x) \leq K(x) \leq Kt(x) \leq KT(x) \leq KB(x) \leq KF(x)$$

(where, as usual, all inequalities are modulo an additive $O(1)$ term). In the other direction, $K(x) \leq C(x) + O(\log C(x))$, whereas there is no computable function f such that $Kt(x) \leq f(K(x))$. It is conjectured that the other measures can differ exponentially from their "neighbors", although it is an open question even whether $KF(x) \leq Kt(x)^2$. (In fact, this question is equivalent to a question about the nonuniform circuit complexity of the deterministic exponential time class EXP [15].) Related measures have been introduced in [2,8,15] in order to connect (deterministic and nondeterministic) space complexity, nondeterministic time complexity, and "distinguishing" complexity.

We will also need to mention one other family of "complexity" measures. Consider a $2^n \times 2^n$ matrix M with rows indexed by n-bit strings i that are viewed as possible inputs to a player in a 2-person game, named Alice. The columns are similarly indexed by strings j that are the possible inputs for another player, named Bob. The (deterministic) communication complexity of M, denoted $DCC(M)$, is the minimal number of bits that Alice and Bob must communicate with each other in order for them to know $M(i, j)$. Thus if x is any string of length 2^{2n}, we will define $\mathsf{DCC}(x)$ to be equal to $\mathsf{DCC}(M)$, where we interpret x as a $2^n \times 2^n$ matrix M. There is also a nondeterministic variant of communication complexity, which we will denote NCC. The only fact that we will need regarding NCC is that, for all x, $\mathsf{NCC}(x) \leq \mathsf{DCC}(x)$. For more about communication complexity, see [34].

The reader will encounter various complexity classes in these pages, beyond the familiar P, NP, and PSPACE. There are three versions of probabilistic polynomial time ($\mathsf{ZPP} = \mathsf{RP} \cap \mathsf{coRP} \subseteq \mathsf{RP} \subseteq \mathsf{BPP}$), and the exponential-time analogs of P and NP (EXP and NEXP), as well as small subclasses of P defined in terms of classes of circuits ($\mathsf{AC}^0 \subseteq \mathsf{TC}^0$). We refer the reader to [17,50] for more background on these topics.

3 The Minimum Circuit Size Problem

There are few "natural" problems that have as strong a claim to NP-Intermediate status, as MCSP. To be sure, there are various problems that are widely used in cryptographic applications, such as factoring and the discrete logarithm problem, that are widely suspected (or hoped) not to have polynomial-time algorithms – but if it turns out they are easy, then they don't bring a large class of other problems with them into P. In contrast, if MCSP is easy, then there are large ramifications.

3.1 Reductions to MCSP

Kabanets and Cai showed that if MCSP is in P/poly, then every potential cryptographically-secure one-way function is easy to invert on a substantial portion of its range [32]. Subsequently, it was shown that every language in the complexity class SZK lies in BPP^{MCSP} [9], thereby providing strong evidence, based on the structure of complexity classes, that MCSP lies outside of P.[2] "SZK" stands for "Statistical Zero Knowledge", and it contains the graph isomorphism problem, as well as a great many problems whose presumed intractability is essential for the security of well-studied cryptosystems. For this survey, the reader will not need to know much about zero-knowledge interactive proofs, which are used in order to define SZK. It will suffice to know that every language in SZK is contained in NP/poly ∩ coNP/poly, and thus this class is in some sense "close" to NP ∩ coNP. It will also be important to know that SZK is best defined in terms of "promise problems", and that some of these promise problems are in fact *complete* for SZK.

A *promise problem* consists of two disjoint sets Y and N, called the YES-instances and the NO-instances, respectively. A *solution* to the promise problem (Y, N) is any language that contains Y and is disjoint from N. When we say that

---

[2] In spite of this "strong evidence", the only *unconditional* lower bound known for R_{KT} and MCSP is that neither is in AC^0 [8]. It is not even known whether these problems lie in $\mathsf{AC}^0[2]$ (that is, AC^0 augmented with parity gates).

MCSP is hard for SZK under BPP reductions, we mean that, for every promise problem in SZK, there is a probabilistic oracle Turing machine M that accepts with very high probability on all of the YES-instances and rejects with very high probability on all of the NO-instances, but M might have acceptance probability close to $1/2$ on other instances.

In order to establish that MCSP is hard for SZK, [9] presents a BPP reduction from the standard complete problem for SZK, called SD, for *"Statistical Difference"*. The input to SD consists of a pair of circuits (C, D) defining probability distributions. The YES instances consist of pairs (C, D) such that these probability distributions are very close, and the NO instances consist of pairs for which the probability distributions are quite far apart.

There are a few things to mention about this reduction:

- The same proof establishes that the set of KT-random strings R_{KT} is also hard for SZK, as well as the overgraph of the KT function O_{KT}. In fact, until very recently, *every* efficient reduction to $MCSP^A$ or O_{KT^A} for any oracle A carried over to the more restrictive problem R_{KT^A}. (Note that, for every measure μ, $\overline{R_\mu} \leq^P_m O_\mu$.) This is because, until very recently, all such reductions have proceeded by using derandomization techniques, using R_{KT^A} as a statistical test to foil pseudorandom generators.
- Motivated by the fact that SZK is a class of promise problems, one can define a promise problem related to MCSP, such as the problem Gap-MCSP with YES instances consisting of $\{x : KT(x) \leq \sqrt{|x|}\}$ and NO instances consisting of $\{x : KT(x) \geq |x|/2\}$. The proof in [9] establishes that every promise problem in SZK is in Promise-BPP^A for every set A that is a solution to Gap-MCSP.

For certain problems in SZK, more restrictive reductions to MCSP are known. It is shown in [8, 45] that factoring and Discrete Log are in ZPP^{MCSP}, and Graph Isomorphism lies in RP^{MCSP} [9]. As with the reductions mentioned above, these also carry over to O_{KT} and R_{KT}, and they are proved using the same suite of techniques.

Thus it is of interest that a fundamentally different type of reduction is given in [13], showing that the Graph Automorphism problem is in ZPP$^{O_{KT}}$. It is still not known whether Graph Automorphism is in ZPP^{MCSP} or ZPP$^{R_{KT}}$.

Open Question 1. *Is Graph Automorphism in ZPP^{MCSP} and/or in ZPP$^{R_{KT}}$? Until the appearance of [13], the problems O_{KT} and R_{KT} had been viewed as convenient proxies for MCSP, such that a theorem that was proved for one of the problems would hold for all of them. Similarly, the different versions of R_{KT_U} (for different universal Turing machines U) had been viewed as being more-or-less equivalent, and different versions of MCSP (say, with "size" defined in terms of number of gates instead of number of wires, or with a slightly different set of allowable gates, etc.) were viewed as being more-or-less the same set. We do not know of any theorem that can be proved for one version of R_{KT_U} and not for another, and we do not know of any theorem that holds for one version of MCSP and not for another – but we also do not know of any efficient reduction between these different versions. [13] provides the first example of a reduction* *that is known to hold for O_{KT} but does not carry over to these related problems. Is this merely a shortcoming of the proof as presented in [13], or is there really a significant difference in the complexity of these problems?*

It is shown in [15] that the problem of factoring Blum integers is in $\mathsf{ZPP}^{R_{KB}}$ and in $\mathsf{ZPP}^{R_{KF}}$, and it is shown in [35] that factoring Blum integers is also in $\mathsf{ZPP}^{R_{DCC}}$. (A number x is a Blum integer if $x = pq$ for two primes p and q such that p and q are both congruent to 3 mod 4. Factoring Blum integers is generally considered to be roughly as difficult as the general factoring problem.)

Open Question 2. *Is there a significant subclass of* SZK *that reduces to* R_{KF}*? In particular, if we consider the restriction of the standard* SZK*-complete problem* SD*, where instead of the input being a pair of circuits* (C, D)*, the input is a pair of formulae* (C, D) *does this restricted problem reduce to* R_{KF}*? And does this restricted problem capture many of the natural problems that are known to lie in* SZK*? (Related questions can be asked about* R_{KB}*.)*

Branching programs are restricted circuits, and similarly formulae are restricted branching programs. This is the best explanation that we have, for the facts that we can currently reduce fewer problems to O_{FSize} than to O_{BPSize}, and we can currently reduce fewer problems to O_{BPSize} than to MCSP. But perhaps this intuition is not valid at all. It is known that if we further restrict the formula size problem O_{FSize} to formulae in disjunctive normal form, then the resulting problem is NP-complete under $\leq_m^P$ reductions [38]. (See also [16].) Contrariwise, the problem of computing nondeterministic communication complexity (a *more* powerful model than DCC) is also NP-complete [37,41]. This might lead the reader to wonder whether MCSP and these other related problems are not *all* NP-complete. What reason is there to believe that we cannot reduce all of NP to problems such as MCSP? This is the topic that is addressed in the next section.

3.2   Complete, or Not Complete? That Is the Question

Although there is currently no strong evidence that MCSP is *not* NP-complete, there *is* a great deal of evidence that no proof of NP-completeness will be found anytime soon.

In order to present this evidence, let us briefly recall some of the more common types of reductions. The reader is probably familiar with polynomial-time and logspace many-one reducibility, denoted by $\leq_m^P$ and $\leq_m^{\log}$, respectively. With very few exceptions [1], problems that are known to be complete for NP and other complexity classes under $\leq_m^P$ and $\leq_m^{\log}$ reductions are even complete under reductions computed by AC^0 circuits. Most of the important NP-complete problems are also complete under another type of restrictive reducibility: sublinear-time reductions. For a time bound $t(n) < n$, we say that $A \leq_m^t B$ if there is a polynomial-time computable function f such that for all $x, x \in A \Leftrightarrow f(x) \in B$ where in addition, the function that maps (x, i) to the i-th bit of $f(x)$ is computed in time $t(|x|)$ by a machine that has random access to the bits of x.

Those reductions are all *more* restrictive than $\leq_m^P$ reductions. We also need to mention the *less* restrictive notion known as polynomial-time truth-table reductions $\leq_{tt}^P$, also known as *nonadaptive* reductions.

Table 1 presents information about the consequences that will follow if MCSP is NP-complete (or even if it is hard for certain subclasses of NP). The table is incomplete (since it does not mention the influential theorems of Kabanets and Cai [32] describing various consequences if MCSP were complete under a certain restricted type of $\leq_m^P$ reduction). It also fails to adequately give credit to all of the papers that have contributed to this line of work, since – for example – some of the important contributions of [40] have subsequently been slightly improved [14,30]. But one thing should jump out at the reader from Table 1: All of the conditions listed in Column 3 (with the exception of "FALSE") are widely believed to be true, although they all seem to be far beyond the reach of current proof techniques.

Table 1. Summary of what is known about the consequences of MCSP being hard for NP under different types of reducibility. If MCSP is hard for the class in Column 1 under the reducibility shown in Column 2, then the consequence in Column 3 follows.

Class $\mathcal{C}$	Reductions $\mathcal{R}$	Statement $\mathcal{S}$	Reference
TC^0	$\leq_m^{n^{1/3}}$	FALSE	[40]
TC^0	$\leq_m^{\mathsf{AC}^0}$	$\mathsf{LTH}^a \not\subseteq \mathsf{io-SIZE}[2^{\Omega(n)}]$ and $\mathsf{P} = \mathsf{BPP}$	[14,40]
TC^0	$\leq_m^{\mathsf{AC}^0}$	$\mathsf{NP} \not\subseteq \mathsf{P/poly}$	[14]
P	$\leq_m^{\log}$	$\mathsf{PSPACE} \neq \mathsf{P}$	[14]
NP	$\leq_m^{\log}$	$\mathsf{PSPACE} \neq \mathsf{ZPP}$	[40]
NP	$\leq_{tt}^P$	$\mathsf{EXP} \neq \mathsf{ZPP}$	[30]

[a]LTH is the linear-time analog of the polynomial hierarchy. Problems in LTH are accepted by alternating Turing machines that make only $O(1)$ alternations and run for linear time.

The case in favor of MCSP being an NP-intermediate problem would be stronger if there were some *unlikely* consequences that were known to follow if MCSP were NP-complete. Some indirect evidence of this sort is available, if we consider relativized versions of MCSP and KT, such as $\mathsf{MCSP}^{\mathsf{QBF}}$ and $\mathsf{KT}^{\mathsf{QBF}}$. This is explained below.

Unlike R_{KT} and MCSP, which are not known to be complete for any interesting complexity class, $\mathsf{MCSP}^{\mathsf{QBF}}$ and $R_{\mathsf{KT}^{\mathsf{QBF}}}$ are both complete for PSPACE under ZPP reductions [8]. The analogous problems MCSP^E and R_{Kt} (and also $R_{K^{2^{n^2}}}$) are complete for EXP under P/poly-Turing reductions and NP-Turing reductions [8] (where E can be any standard complete problem for $\mathsf{DTIME}(2^{O(n)})$). However, it is rather unlikely that these are complete under *deterministic* (uniform) reductions, as is highlighted in Table 2.

The main lesson from Table 2 is that even problems that appear much harder than MCSP (such as the PSPACE-complete problem $\mathsf{MCSP}^{\mathsf{QBF}}$, and the

Table 2. Summary of what is known about the consequences of MCSP^{QBF} and $R_{\text{KT}^{\text{QBF}}}$ (and related problems in EXP) being hard for various classes (under different types of reducibility). If the problem in the first column is hard for the class in Column 2 under the reducibility shown in Column 3, then the consequence in Column 4 follows.

problem	class $\mathcal{C}$	reductions $\mathcal{R}$	statement $\mathcal{S}$	Reference
MCSP^{QBF}	TC^0	$\leq_m^{\text{AC}^0}$	FALSE[a]	[14]
MCSP^{QBF}	P	$\leq_m^{\log}$	EXP = PSPACE	[14]
MCSP^{QBF}	NP	$\leq_m^{\log}$	NEXP = PSPACE	[14]
MCSP^{QBF}	PSPACE	$\leq_m^{\log}$	FALSE	[14]
$R_{\text{KT}^{\text{QBF}}}$	PSPACE	$\leq_T^{\log}$	FALSE	[8]
MCSP^E	NP	$\leq_m^{\text{P}}$	NEXP = EXP	[14]
R_{Kt}	EXP	$\leq_{tt}^{\text{P}}$	FALSE	[8]
$R_{K2^{n^2}}$	EXP	$\leq_T^{\text{P}}$	FALSE	[19]

[a]It is not explicitly stated in [14] that this consequence is FALSE, but it is stated that, under this condition, $\text{NP}^{\text{QBF}} \not\subseteq \text{P}^{\text{QBF}}/\text{poly}$, which is equivalent to $\text{PSPACE} \not\subseteq \text{PSPACE}/\text{poly}$, which is, of course, false.

EXP-complete problem MCSP^E) cannot be hard for NP unless unlikely consequences follow. However, this does still not imply that MCSP itself is unlikely to be NP-hard, since we know of no *deterministic* reduction from MCSP to the (apparently harder) problems MCSP^{QBF} and MCSP^E.

Although it might seem intuitively clear that MCSP must be no harder than MCSP^A, this intuition is suspect. Hirahara and Watanabe have shown that, if $\text{MCSP} \notin \text{P}$, then there is an oracle A such that $\text{MCSP} \notin \text{P}^{\text{MCSP}^A}$ [28]. In the same paper, they consider problems that are "oracle-independent" reducible to MCSP by *probabilistic* reductions that make only one query. (All known reductions to MCSP – other than the identity reduction of MCSP to itself – are "oracle-independent" reductions in the sense of [28].) They show that all such problems lie in the complexity class $\text{AM} \cap \text{coAM}$.

Open Question 3. *Could the* SZK *lower bound on the complexity of* MCSP *be tight? For instance, could* Gap-MCSP *lie in* SZK*? The results of [28] are intriguing here, since every promise problem in* SZK *has a solution in* $\text{AM} \cap \text{coAM} \subseteq \text{NP}/\text{poly} \cap \text{coNP}/\text{poly}$. *It would be very interesting to place* MCSP *or* $\overline{R_{\text{KT}}}$ *in any subclass of* NP, *but we seem quite far from this goal.*

Open Question 4. *Contrariwise, might* MCSP *be complete for* NP *under* P/poly *reductions (in the same way that the corresponding problems are complete for* PSPACE *and* EXP *under* P/poly *reductions)? Or might it lie in the high hierarchy of [46]? If so, then it would be "nearly"* NP-complete.

Open Question 5. *Can one show unconditionally that* MCSP *(or* $\overline{R_{\text{KT}}}$*) is not complete for* NP *under* $\leq_m^{\text{AC}^0}$*? Or can one derive some* unlikely *consequences from these sets being complete? (Some related questions are discussed in [4].)*

Open Question 6. *There are many intriguing questions concerning the complexity of the set of Levin-random strings, R_{Kt}. Although this set is complete for* EXP *(under* P/poly *and* NP *reductions), it is not known whether R_{Kt} is in* P. *Can this be resolved? Or would it imply the resolution of some long-standing open problem in complexity? Also, it is known that R_{Kt} is in* ZPP *if and only if* EXP = ZPP *[8]. This is exactly the conclusion one would obtain if R_{Kt} were complete for* EXP *under* ZPP *reductions – and yet it remains unknown whether R_{Kt} is complete under this type of reducibility.*

3.3   Relationships Among Measures

It is believed that most of the measures mentioned in this section are not polynomially-related to each other. In fact, a large table is presented in [15], showing that, for most pairs of measures $\mu 1$ and $\mu 2$ mentioned in this section, the question of whether $\mu 1$ and $\mu 2$ are polynomially-related is equivalent to some well-known open question in complexity theory.

However, there are some noteworthy relationships that should be mentioned, regarding DCC and FSize. Kushilevitz and Weinreb [35] showed that there is a polynomial-time routine that, given a bitstring x of length $N = 2^{2n}$, will produce a string M of length 2^{4n}, which can be interpreted as the matrix for a communication game, called ENE_x with the property that $\text{DCC}(ENE_x) - n - 1 \leq \text{FDepth}(f_x) \leq .886 \cdot \text{DCC}(ENE_x) + n + O(\log n)$, where "FDepth" measures the minimal formula depth of a function. (This is a consequence of the following results of [35]: Claim 3.3(iv) yields the first inequality, while the second inequality follows from Claim 3.3(iii) and Theorem 3.6.) Since $\text{FSize}(f_x) \leq 2^{\text{FDepth}(f_x)} \leq \text{FSize}(f_x)^{1.71}$ [48], we have $\text{FSize}(f_x) \leq 2^n (2^{.886 \cdot \text{DCC}(ENE_x)}) n^{O(1)} \leq 2^{1.886n} \text{FSize}(f_x)^{1.515} n^{O(1)}$. This appears to be a very poor approximation, but since f_x is a Boolean function on $2n$ variables, even a factor of $2^{1.886n}$ is not overwhelming, and this still means that, with an oracle for the overgraph of DCC, O_{DCC}, we can distinguish between those strings x where $\text{FSize}(f_x)$ is very large, and those x where $\text{FSize}(f_x)$ is very small; this is exploited in [35]. Some additional results relating DCC to problems such as MCSP have been explored by Raviv [42].

Open Question 7. *Is there a significant subclass of* SZK *that reduces to O_{DCC}? Are there more connections between* DCC *and the other complexity measures studied here?*

Shallit and Wang introduced a complexity measure on strings based on finite-state automata, called Automatic Complexity [47]. A related measure based on nondeterministic finite automata has been introduced by Hyde and Kjoos-Hanssen [31].

Open Question 8. *Are there any interesting relationships between the Automatic Complexity measure of Shallit and Wang, and any of the other measures mentioned in this section? Is there any evidence that Automatic Complexity (or the related measure of [31]) is computationally intractable?*

4   Complexity Classes and Noncomputable Complexity Measures

Up to now, this article has focused primarily on decidable problems such as MCSP and R_{KT^A} for *computable* oracles A. In this section, we survey some intriguing connections between computational complexity theory and *noncomputable* measures such as C and K.

As discussed in the previous section, there has not been much success using *deterministic* reductions to exploit the power of problems such as MCSP; P^{MCSP} is not known to contain any problems of interest, other than MCSP itself.

The situation is different for reductions to R_C and R_K. (For ease of exposition, let "R" stand for either of these sets for the time being; the following results hold, no matter which of these measures is used.) Although there are some negative results, showing that EXP and problems outside of P/poly cannot be reduced to R using restricted $\leq^P_T$ reductions (such as disjunctive truth-table reductions or reductions that make a limited number of queries) [7, 29], there are also two striking positive results:

Theorem 9. *Let R denote either R_C or R_K. Then*

- PSPACE $\subseteq$ P^R. *[8]*.
- BPP $\subseteq$ P^R_{tt}. *[18] (where P^R_{tt} is the class of problems reducible to R via $\leq^P_{tt}$ reductions.)*

Of course, since it is still an open question whether PSPACE = P, this does not unconditionally show that access to R provides a computational speed-up. But in the context of *nondeterministic* reductions to R, there is a striking speed-up:

Theorem 10. *Let R denote either R_C or R_K. Then*

- NEXP $\subseteq$ NP^R *[7]*.
- EXP^{NP} $\subseteq$ P^{NP^R} *[26]*.

The initial reaction of the reader might be to ask if there is any real content to these theorems. Perhaps *every* computable set is efficiently reducible to R? Indeed, if we consider *nonuniform* reductions to be "efficient", then this is the case:

Theorem 11. HALT $\in$ P^R/poly *[8]*.

However, if we consider only reductions computed by Turing machines, then the situation becomes more complicated. Kummer showed that HALT$\leq_{dtt}R_C$ [33], but the running time of his reduction depends on the choice of the universal Turing machine U defining C complexity. For some choices of U the running time can be as little as doubly-exponential time [7], but for other choices of U the running time can be forced to be as slow as any given computable function [7]. On the other hand, it is still open whether or not HALT $\in$ $P^{R_{C_U}}$ for some (or even every) choice of U. If that is the case, then indeed Theorems 9 and 10 are of little interest when $R = R_C$.

Open Question 12. *Is there any universal machine U with the property that* HALT $\in P^{R_{C_U}}$?

Much more is known about the case where $R = R_K$. The question of whether or not HALT$\leq_{tt} R_{K_U}$ depends on the choice of U [39]; see also [11] for an alternate proof. (Day has explored the analogous question for other Kolmogorov complexity measures [22].)

The inclusion from Theorem 9 that states PSPACE $\subseteq P^{R_K}$ is actually shorthand for PSPACE $\subseteq \bigcap_U P^{R_{K_U}}$, since the inclusion holds for every choice of universal machine U. It turns out that by explicitly considering the intersection over all U, one can obtain useful *upper* bounds:

Theorem 13. *[12,21]* $\bigcap_U P_{tt}^{R_{K_U}} \subseteq$ PSPACE *and* $\bigcap_U NP^{R_{K_U}} \subseteq$ EXPSPACE. *The techniques of [12] also allow one to show* $\bigcap_U PNP^{R_{K_U}} \subseteq$ EXPSPACE.

This theorem relies crucially on the result of [21] that there are no undecidable sets in $\bigcap_U P^{R_{K_U}}$

Resorting back to using P^{R_K} as a shorthand for $\bigcap_U P^{R_{K_U}}$, we can thus summarize our knowledge about these classes as:

$$BPP \subseteq P_{tt}^{R_K} \subseteq PSPACE \subseteq P^{R_K}$$

$$PSPACE \subseteq NEXP \subseteq NP^{R_K}$$

$$NEXP \subseteq EXP^{NP} \subseteq P^{NP^{R_K}} \subseteq EXPSPACE$$

That is, although the oracle R_{K_U} is not even decidable (for any U), the class $P_{tt}^{R_K}$ yields a complexity class between BPP and PSPACE. Similarly, the complexity of PSPACE is bounded above and below by adaptive and non-adaptive access to the oracle R_K.

4.1   Can the **PSPACE** Bound Be Improved?

In this section, let us focus on the inclusions BPP $\subseteq P_{tt}^{R_K} \subseteq$ PSPACE.

The paper [10] investigates whether the PSPACE upper bound on $P_{tt}^{R_K}$ can be improved to PSPACE $\cap$ P/poly. [10] found a connection to proof theory, and presented a collection of theorems (provable in ZF) with the property that, if the theorems were provable in certain extensions of Peano Arithmetic, then the PSPACE $\cap$ P/poly upper bound would hold. Any hopes that this might be how to show a P/poly bound were dashed by the next paper [6], which showed that the statements in question really *are* independent of the given extensions of PA.

However, on the positive side, [6] showed that the PSPACE $\cap$ P/poly upper bound *does* hold, in a related setting. [6] defined a class analogous to $P_{tt}^{R_K}$ in terms of time-bounded K-complexity (with *very* large time bounds, so that they can be considered to be reasonable approximations to R_K). More precisely, consider the class of problems that are in $P_{tt}^{R_{K^t}}$ for *all* large-enough time bounds (such as Ackermann's function, and beyond). [6] shows that this class lies between BPP and PSPACE∩P/poly. Hirahara and Kawamura present a different restriction on reductions to R_K and R_C, yielding a class that lies between BPP and NP^{NP} [27].

This is one of the main considerations that leads us to conjecture that $P_{tt}^{R_K}$ actually coincides with BPP [3]. Although it is still open whether $P_{tt}^{R_K}$ is contained in P/poly, the results of [5] show that, if containment in P/poly does not hold, then it relies on the ability of nonadaptive poly-time reductions to distinguish between R_K and (for example) Ackermann-time-bounded K-random strings.

Let us assume for the moment that $P_{tt}^{R_K} \subseteq$ P/poly. This means, in particular, that it is unlikely that $P_{tt}^{R_K}$ contains NP. Yet EXP$_{tt}^{R_K}$ contains NEXP, and thus there must be some critical time bound T when DTIME$(T(n^{O(1)}))_{tt}^{R_K}$ first contains NP.

Open Question 14. *Does this occur for subexponential T? More generally, if* BPP $= P_{tt}^{R_K}$, *and the popular conjecture that* BPP $=$ P *also holds, then R_K provides no useful power for nonadaptive poly-time reductions. On the other hand, if* EXP $\neq$ NEXP, *then nonadaptive* EXP-*reductions to R_K do provide significant power. At what point does this additional computational advantage kick in?*

4.2   Can One Characterize NEXP?

The same paper [3] that contains the conjecture BPP $= P_{tt}^{R_K}$ also contains a conjecture that NEXP $=$ NP^{R_K}. (Weak) support for this conjecture comes largely from the fact that the inclusion NP$^{R_K} \subseteq$ EXPSPACE is proved by first observing that every NP-Turing reduction can be simulated by a *nonadaptive* EXP-reduction that asks only queries of polynomial length, and then observing that the inclusion $P_{tt}^{R_K} \subseteq$ PSPACE scales up to show that EXP$_{tt}^{R_K} \subseteq$ EXPSPACE. It seems that one is throwing a *lot* away by replacing an NP-reduction by an EXP reduction, which would seem to indicate that the EXPSPACE upper bound is not tight.

Hirahara's recent result that EXP$^{NP} \subseteq (\Delta_2^p)^{R_K}$ [26] might, at first blush, seem to argue against the conjecture that NEXP $=$ NP^{R_K}. That is, giving NP^{R_K} to a poly-time oracle Turing machine yields not merely P^{NEXP}, but also all of EXP^{NP}, which seems to be significantly larger than P^{NEXP}. (For instance, P^{NEXP} is contained in NEXP/poly, whereas EXP^{NP} is in NEXP/poly iff it collapses to P^{NEXP}.) However, this heuristic argument implicitly assumes that $\bigcap_U (\Delta_2^p)^{R_{K_U}}$ is equal to P$^{\bigcap_U NP^{R_{K_U}}}$ – and it is not at all clear that this equality should hold. Thus it is conceivable that NEXP $=$ NP^{R_K} and $(\Delta_2^p)^{R_K} =$ EXP^{NP}.

Currently, the best upper bound for NP^{R_K} is EXPSPACE. However, there has been movement on a related front, as explained in the next paragraph.

The study of distinguishing random from pseudorandom distributions has led to very powerful insights and techniques. In this context, Vadhan and Gutfreund explored the class of problems that can be reduced to distinguishing random from pseudorandom in a very general sense [25]. They showed that all languages that can be reduced in a *restricted* sense to distinguishing random from pseudorandom lie in PSPACE, but the best upper bound for the *general* class of languages is EXPSPACE [12].

Hirahara has also shown [26] a better upper bound for this class: S_2^{EXP} – the exponential-time analog of S_2^p (which is a class that lies in ZPP^{NP} [20]). In particular, this class can be recognized by exponential-time alternating machines that make at most one alternation (in contrast to the EXPSPACE upper bound, which is exponential time with *no* bound on the number of alternations).

Open Question 15. *Can the* EXPSPACE *upper bound on* NP^{R_K} *be improved, to something much closer to* NEXP?

4.3 Promise Problems, Again

An alternative approach is to seek characterizations of BPP and NEXP in terms of reductions to the *promise problem* with "yes instances" consisting of those x such that $K(x) \geq |x|/2$, and "no instances" consisting of x such that $K(x) < \sqrt{|x|}$. Let us call this the Gap-K-complexity problem.

Open Question 16. *We have not succeeded in characterizing* BPP *or* NEXP *in terms of efficient reductions to* R_K. *Might one have a greater chance of success by considering efficient reductions to the Gap-K-complexity problem?*

The complexity classes that reduce to R_K also reduce to *any* solution to the Gap-K-complexity problem. Furthermore, all of the upper bounds that are proved in [12] carry over to this setting, and one obtains several other side-benefits. Note that it is no longer necessary to take the intersection over all universal machines U, since they all satisfy the promise. In a similar say, it is no longer necessary to distinguish between C and K complexity, since the gap between the YES and NO instances dwarfs the difference between the different measures. Also, there is a useful quantifier swap that applies: If a language B reduces to a promise problem (Y, N), it means that for every solution A to the promise problem (Y, N), there is a reduction from B to A. However, it is known [24,44] that if this happens then there is also a *single* reduction that reduces B to *every* solution of (Y, N).

There is a long history of promise problems being used to understand complexity classes (such as SZK among many other examples), and this might be a better way of elucidating the connection between Kolmogorov complexity and complexity classes.

5   Conclusions

This rambling account is intended to gather together some recent (and not-so-recent) developments, regarding the complexity and computational power of determining the "complexity" of a string, using various notions of complexity. The reader should be cautioned that some of the open questions that are listed occurred to the author while he was writing the paper, and some of them might be quite easy to answer. Happy hunting!

Acknowledgments. The author acknowledges the support of NSF grant CCF-1555409, and thanks Diptarka Chakraborty (for helpful comments on an earlier draft of this work), Shuichi Hirahara (for allowing mention of his recent unpublished results), and Toni Pitassi (for helpful discussions).

References

1. Agrawal, M., Allender, E., Impagliazzo, R., Pitassi, R., Rudich, S.: Reducing the complexity of reductions. Comput. Complex. **10**, 117–138 (2001)
2. Allender, E.: NL-printable sets and nondeterministic Kolmogorov complexity. Theor. Comput. Sci. **355**(2), 127–138 (2006)
3. Allender, E.: Curiouser and curiouser: the link between incompressibility and complexity. In: Cooper, S.B., Dawar, A., Löwe, B. (eds.) CiE 2012. LNCS, vol. 7318, pp. 11–16. Springer, Heidelberg (2012). doi:10.1007/978-3-642-30870-3_2
4. Allender, E.: Investigations concerning the structure of complete sets. In: Agrawal, M., Arvind, V. (eds.) Perspectives in Computational Complexity. PCSAL, vol. 26, pp. 23–35. Springer, Heidelberg (2014). doi:10.1007/978-3-319-05446-9_2
5. Allender, E., Buhrman, H., Friedman, L., Loff, B.: Reductions to the set of random strings: the resource-bounded case. In: Rovan, B., Sassone, V., Widmayer, P. (eds.) MFCS 2012. LNCS, vol. 7464, pp. 88–99. Springer, Heidelberg (2012). doi:10.1007/978-3-642-32589-2_11
6. Allender, E., Buhrman, H., Friedman, L., Loff, B.: Reductions to the set of random strings: the resource-bounded case. Logical Methods Comput. Sci. **10**(3), 1–18 (2014). CiE 2012 Special Issue
7. Allender, E., Buhrman, H., Koucký, M.: What can be efficiently reduced to the Kolmogorov-random strings? Ann. Pure Appl. Logic **138**, 2–19 (2006)
8. Allender, E., Buhrman, H., Koucký, M., van Melkebeek, D., Ronneburger, D.: Power from random strings. SIAM J. Comput. **35**, 1467–1493 (2006)
9. Allender, E., Das, B.: Zero knowledge and circuit minimization. In: Csuhaj-Varjú, E., Dietzfelbinger, M., Ésik, Z. (eds.) MFCS 2014. LNCS, vol. 8635, pp. 25–32. Springer, Heidelberg (2014). doi:10.1007/978-3-662-44465-8_3
10. Allender, E., Davie, G., Friedman, L., Hopkins, S.B., Tzameret, I.: Kolmogorov complexity, circuits, and the strength of formal theories of arithmetic. Chicago J. Theor. Comput. Sci. **5**, 1–15 (2013)
11. Allender, E., Friedman, L., Gasarch, W.: Exposition of the Muchnik-Positselsky construction of a prefix free entropy function that is not complete under truth-table reductions. Technical report TR10-138, Electronic Colloquium on Computational Complexity (ECCC) (2010)
12. Allender, E., Friedman, L., Gasarch, W.: Limits on the computational power of random strings. Inf. Comput. **222**, 80–92 (2013). ICALP 2011 Special Issue
13. Allender, E., Grochow, J., Moore, C.: Graph isomorphism and circuit size. Technical report TR15-162, Electronic Colloquium on Computational Complexity (ECCC) (2015)
14. Allender, E., Holden, D., Kabanets, V.: The minimum oracle circuit size problem. In: Symposium on Theoretical Aspects of Computer Science (STACS), LIPIcs, pp. 21–33. Schloss Dagstuhl - Leibniz-Zentrum für Informatik (2015)
15. Allender, E., Koucký, M., Ronneburger, D., Roy, S.: The pervasive reach of resource-bounded Kolmogorov complexity in computational complexity theory. J. Comput. Syst. Sci. **77**, 14–40 (2010)

16. Allender, E., Hellerstein, L., McCabe, P., Pitassi, T., Saks, M.E.: Minimizing disjunctive normal form formulas and AC^0 circuits given a truth table. SIAM J. Comput. **38**(1), 63–84 (2008)
17. Arora, S., Barak, B.: Computational Complexity: A Modern Approach, vol. 1. Cambridge University Press, Cambridge (2009)
18. Buhrman, H., Fortnow, L., Koucký, M., Loff, B.: Derandomizing from random strings. In: 25th IEEE Conference on Computational Complexity (CCC), pp. 58–63. IEEE (2010)
19. Buhrman, H., Mayordomo, E.: An excursion to the Kolmogorov random strings. J. Comput. Syst. Sci. **54**, 393–399 (1997)
20. Cai, J.: $S_2^p \subseteq ZPP^{NP}$. J. Comput. Syst. Sci. **73**(1), 25–35 (2007)
21. Cai, M., Downey, R., Epstein, R., Lempp, S., Miller, J.: Random strings and tt-degrees of Turing complete c.e. sets. Logical Methods Comput. Sci. **10**(3), 1–24 (2014)
22. Day, A.: On the computational power of random strings. Ann. Pure Appl. Logic **160**, 214–228 (2009)
23. Downey, R., Hirschfeldt, D.: Algorithmic Randomness and Complexity. Springer, New York (2010)
24. Grollmann, J., Selman, L.: Complexity measures for public-key cryptosystems. SIAM J. Comput. **17**(2), 309–335 (1988)
25. Gutfreund, D., Vadhan, S.: Limitations of hardness vs. randomness under uniform reductions. In: Goel, A., Jansen, K., Rolim, J.D.P., Rubinfeld, R. (eds.) APPROX/RANDOM -2008. LNCS, vol. 5171, pp. 469–482. Springer, Heidelberg (2008). doi:10.1007/978-3-540-85363-3_37
26. Hirahara, S.: Personal Communication (2015)
27. Hirahara, S., Kawamura, A.: On characterizations of randomized computation using plain kolmogorov complexity. In: Csuhaj-Varjú, E., Dietzfelbinger, M., Ésik, Z. (eds.) MFCS 2014. LNCS, vol. 8635, pp. 348–359. Springer, Heidelberg (2014). doi:10.1007/978-3-662-44465-8_30
28. Hirahara, S., Watanabe, O.: Limits of minimum circuit size problem as oracle. In: 31st Conference on Computational Complexity, CCC, LIPIcs, pp. 18:1–18:20. Schloss Dagstuhl - Leibniz-Zentrum fuer Informatik (2016)
29. Hitchcock, J.M.: Limitations of efficient reducibility to the kolmogorov random strings. Computability **1**(1), 39–43 (2012)
30. Hitchcock, J.M., Pavan, A.: On the NP-completeness of the minimum circuit size problem. In: Conference on Foundations of Software Technology and Theoretical Computer Science (FST&TCS). LIPIcs, vol. 45, pp. 236–245. Schloss Dagstuhl - Leibniz-Zentrum fuer Informatik (2015)
31. Hyde, K., Kjos-Hanssen, B.: Nondeterministic automatic complexity of overlap-free and almost square-free words. Electr. J. Comb. **22**(3), P3.22 (2015)
32. Kabanets, V., Cai, J.-Y.: Circuit minimization problem. In: ACM Symposium on Theory of Computing (STOC), pp. 73–79. ACM (2000)
33. Kummer, M.: On the complexity of random strings. In: Puech, C., Reischuk, R. (eds.) STACS 1996. LNCS, vol. 1046, pp. 25–36. Springer, Heidelberg (1996). doi:10.1007/3-540-60922-9_3
34. Kushilevitz, E., Nisan, N.: Communication Complexity. Cambridge University Press, Cambridge (1997)
35. Kushilevitz, E., Weinreb, E.: On the complexity of communication complexity. In: ACM Symposium on Theory of Computing (STOC), pp. 465–474 (2009)
36. Levin, L.A.: Randomness conservation inequalities; information and independence in mathematical theories. Inf. Control **61**, 15–37 (1984)

37. Lund, C., Yannakakis, M.: On the hardness of approximating minimization problems. J. ACM **41**(5), 960–981 (1994)
38. Masek, W.J.: Some NP-complete set covering problems (1979) (Unpublished manuscript)
39. Muchnik, A.A., Positselsky, S.: Kolmogorov entropy in the context of computability. Theor. Comput. Sci. **271**, 15–35 (2002)
40. Murray, C., Williams, R.: On the (non) NP-hardness of computing circuit complexity. In: 30th Conference on Computational Complexity (CCC), LIPIcs, pp. 365–380. Schloss Dagstuhl - Leibniz-Zentrum für Informatik (2015)
41. Orlin, J.: Contentment in graph theory: covering graphs with cliques. Indagationes Math. (Proceedings) **80**(5), 406–424 (1977)
42. Raviv, N.: Truth table minimization of computational models (2013)
43. Razborov, A., Rudich, S.: Natural proofs. J. Comput. Syst. Sci. **55**, 24–35 (1997)
44. Regan, K.W.: A uniform reduction theorem extending a result of J. Grollmann and A. Selman. In: Kott, L. (ed.) ICALP 1986. LNCS, vol. 226, pp. 324–333. Springer, Heidelberg (1986). doi:10.1007/3-540-16761-7_82
45. Rudow, M.: Discrete logarithm and minimum circuit size. Technical report TR16-108, Electronic Colloquium on Computational Complexity (ECCC) (2016)
46. Schöning, U.: A low and a high hierarchy within NP. J. Comput. Syst. Sci. **27**(1), 14–28 (1983)
47. Shallit, J., Wang, M.: Automatic complexity of strings. J. Autom. Lang. Comb. **6**(4), 537–554 (2001)
48. Spira, P.M.: On time-hardware complexity tradeoffs for Boolean functions. In: Proceedings of the 4th Hawaii Symposium on System Sciences, pp. 525–527 (1971)
49. Trakhtenbrot, B.A.: A survey of Russian approaches to perebor (brute-force searches) algorithms. IEEE Ann. Hist. Comput. **6**(4), 384–400 (1984)
50. Vollmer, H.: Introduction to Circuit Complexity: A Uniform Approach. Springer-Verlag New York Inc., New York (1999)

Bounded Pushdown Dimension vs Lempel Ziv Information Density

Pilar Albert[1], Elvira Mayordomo[1(✉)], and Philippe Moser[2]

[1] Departamento de Informática e Ingeniería de Sistemas,
Instituto de Investigación en Ingeniería de Aragón,
Universidad de Zaragoza, 50018 Zaragoza, Spain
elvira@unizar.es
[2] Department of Computer Science, National University of Ireland Maynooth,
Co Kildare, Ireland
pmoser@cs.nuim.ie

Abstract. In this paper we introduce a variant of pushdown dimension called bounded pushdown (BPD) dimension, that measures the density of information contained in a sequence, relative to a BPD automata, i.e. a finite state machine equipped with an extra infinite memory stack, with the additional requirement that every input symbol only allows a bounded number of stack movements. BPD automata are a natural real-time restriction of pushdown automata. We show that BPD dimension is a robust notion by giving an equivalent characterization of BPD dimension in terms of BPD compressors. We then study the relationships between BPD compression, and the standard Lempel-Ziv (LZ) compression algorithm, and show that in contrast to the finite-state compressor case, LZ is not universal for bounded pushdown compressors in a strong sense: we construct a sequence that LZ fails to compress significantly, but that is compressed by at least a factor 2 by a BPD compressor. As a corollary we obtain a strong separation between finite-state and BPD dimension.

Keywords: Information lossless compressors · Finite state (bounded pushdown) dimension · Lempel-Ziv compression algorithm

1 Introduction

I first learned of Rod Downey through his papers with Mike Fellows on Parameterized Complexity. Their idea that the computational complexity of a problem should take into account the importance of different parameters of the input affected deeply our understanding of inherent difficulty. Their 1999 book, Parameterized Complexity, is still the reference book on the subject (later improved by their 2013 book). In 2000 Rod started taking an interest in Algorithmic Randomness which quickly made him one of the main researchers in the field, he has written hundreds of papers and the main book on the topic with Denis Hirschfeldt. He is now the driving force in the Algorithmic Randomness community and his

© Springer International Publishing AG 2017
A. Day et al. (Eds.): Downey Festschrift, LNCS 10010, pp. 95–114, 2017.
DOI: 10.1007/978-3-319-50062-1_7

work encouraging students and young researchers is simply amazing. This paper is dedicated to his 60th birthday, for many years to come Rod!

Effective versions of fractal dimension have been developed since 2000 [11,12] and used for the quantitative study of complexity classes, information theory and data compression, and back in fractal geometry (see [8,13,14]). Here we are interested in information theory and data compression, where it is known that for several different bounds on the computing power, effective dimensions capture what can be considered the inherent information content of a sequence in the corresponding setting [14]. In the today realistic context of massive data streams we need to consider very low resource-bounds, such as finite memory or finite-time per input symbol.

The finite state dimension of an infinite sequence [3], is a measure of the amount of randomness contained in the sequence within a finite-memory setting. It is a robust quantity, that has been shown to admit several characterizations in terms of finite-state information lossless compressors (introduced by Huffman [3, 9]), finite-state decompressors [4,16], finite-state predictors in the logloss model [1], and block entropy rates [2]. It is an effectivization of the general notion of Hausdorff dimension at the level of finite-state machines. Informally, the finite state dimension assigns every sequence a number $s \in [0,1]$, that characterizes the randomness density in the sequence (or equivalently its compression ratio), where the larger the dimension the more randomness is contained in the sequence.

Doty and Nichols [5] investigated a variant of finite-state dimension, where the finite state machine comes equipped with an infinite memory stack and is called a pushdown automata, yielding the notion of pushdown dimension. Hence the pushdown dimension of a sequence, is a measure of the density of randomness in the sequence as viewed by a pushdown automata. Since a finite-state automata is a special case of a pushdown automata, the pushdown dimension of a sequence is a lower bound for its finite state dimension. It was shown in [5], that there are sequences for which the pushdown dimension is at most half its finite state dimension, hence yielding a strong separation between the two notions. Unfortunately the notion of pushdown dimension is not known to enjoy any of the equivalent characterizations that finite state dimension does. Moreover, the computation time per input symbol can be unbounded, which rules out this model for many real-time applications.

In this paper we introduce a variant of pushdown dimension called bounded pushdown (BPD) dimension: Whereas pushdown automata can choose not to read their input and only work with their stack for as many steps as they wish (each such step is called a lambda transition), we add the additional real-time constraint that the sequences of lambda transitions are bounded, i.e. we only allow a bounded number of stack movements per each input symbol.

We define the notion of bounded pushdown dimension as the natural effectivitation of Hausdorff dimension via Lutz's gale characterization [11]. We provide evidence that bounded pushdown dimension is a robust notion by giving a compression characterization; i.e. we introduce BPD information-lossless compressors and show that the best compression ratio achievable on a sequence

Bounded Pushdown Dimension vs Lempel Ziv Information Density

Pilar Albert[1], Elvira Mayordomo[1(✉)], and Philippe Moser[2]

[1] Departamento de Informática e Ingeniería de Sistemas,
Instituto de Investigación en Ingeniería de Aragón,
Universidad de Zaragoza, 50018 Zaragoza, Spain
elvira@unizar.es
[2] Department of Computer Science, National University of Ireland Maynooth,
Co Kildare, Ireland
pmoser@cs.nuim.ie

Abstract. In this paper we introduce a variant of pushdown dimension called bounded pushdown (BPD) dimension, that measures the density of information contained in a sequence, relative to a BPD automata, i.e. a finite state machine equipped with an extra infinite memory stack, with the additional requirement that every input symbol only allows a bounded number of stack movements. BPD automata are a natural real-time restriction of pushdown automata. We show that BPD dimension is a robust notion by giving an equivalent characterization of BPD dimension in terms of BPD compressors. We then study the relationships between BPD compression, and the standard Lempel-Ziv (LZ) compression algorithm, and show that in contrast to the finite-state compressor case, LZ is not universal for bounded pushdown compressors in a strong sense: we construct a sequence that LZ fails to compress significantly, but that is compressed by at least a factor 2 by a BPD compressor. As a corollary we obtain a strong separation between finite-state and BPD dimension.

Keywords: Information lossless compressors · Finite state (bounded pushdown) dimension · Lempel-Ziv compression algorithm

1 Introduction

I first learned of Rod Downey through his papers with Mike Fellows on Parameterized Complexity. Their idea that the computational complexity of a problem should take into account the importance of different parameters of the input affected deeply our understanding of inherent difficulty. Their 1999 book, Parameterized Complexity, is still the reference book on the subject (later improved by their 2013 book). In 2000 Rod started taking an interest in Algorithmic Randomness which quickly made him one of the main researchers in the field, he has written hundreds of papers and the main book on the topic with Denis Hirschfeldt. He is now the driving force in the Algorithmic Randomness community and his

© Springer International Publishing AG 2017
A. Day et al. (Eds.): Downey Festschrift, LNCS 10010, pp. 95–114, 2017.
DOI: 10.1007/978-3-319-50062-1_7

by BPD compressors is exactly its BPD dimension. This BPD information-lossless compressors include all that have been used for instance in XML compression [7,10].

In the context of compression, we study the relationship between BPD compression and the standard Lempel-Ziv (LZ) compression algorithm [17]. It is well known that the LZ compression ratio of any sequence is a lower bound for its finite state compressibility [17], i.e. LZ compresses every sequence at least as well as any finite-state information lossless compressor. We show that this fails dramatically in the context of BPD compressors, by constructing a sequence that LZ fails to compress significantly, but is compressed by at least a factor 2 by a BPD compressor, thus yielding a strong separation between LZ and BPD dimension. This separation improves that achieved in [15] for (unbounded) pushdown dimension versus LZ and that of [5] between finite state dimension [3] and pushdown dimension.

Section 2 contains the preliminaries, Sect. 3 presents BPD dimension and its basic properties, Sect. 4 proves the equivalence of BPD compression and dimension and Sect. 5 contains the separation of BPD compression from Lempel Ziv compression.

2 Preliminaries

We write $\mathbb{Z}$ for the set of all integers, $\mathbb{N}$ for the set of all nonnegative integers and $\mathbb{Z}^+$ for the set of all positive integers. Let Σ be a finite alphabet, with $|\Sigma| \geq 2$. Σ^* denotes the set of finite strings, and Σ^∞ the set of infinite sequences. We write $|w|$ for the length of a string w in Σ^*. The empty string is denoted λ. For $S \in \Sigma^\infty$ and $i, j \in \mathbb{N}$, we write $S[i..j]$ for the string consisting of the ith through j^{th} symbols of S, with the convention that $S[i..j] = \lambda$ if $i > j$, and $S[0]$ is the leftmost symbol of S. We write $S[i]$ for $S[i..i]$ (the ith symbol of S). For $n \geq 0$, we write $S \upharpoonright n$ for $S[0..n-1]$. We use $S \upharpoonright 0$ for the empty string. For $w \in \Sigma^*$ and $S \in \Sigma^\infty$, we write $w \sqsubseteq S$ if w is a prefix of S, i.e., if $w = S[0..|w|-1]$. All logarithms are taken in base $|\Sigma|$.

For a string x, x^{-1} denotes x written in reverse order.

3 Bounded Pushdown Dimension

In this section we first recall Lutz's characterization of Hasudorff dimension in terms of gales that can be used to effectivize dimension. Then we introduce Bounded Pushdown dimension based on the concept of BPD gamblers and give its basic properties.

Definition [11]. Let $s \in [0, \infty)$.

1. An s-gale is a function $d : \Sigma^* \to [0, \infty)$ that satisfies the condition

$$d(w) = \frac{\sum\limits_{a \in \Sigma} d(wa)}{|\Sigma|^s} \tag{1}$$

for all $w \in \Sigma^*$.

2. A *martingale* is a 1-gale.

Intuitively, an s-gale is a strategy for betting on the successive symbols of a sequence $S \in \Sigma^\infty$. For each prefix w of S, $d(w)$ is the capital (amount of money) that d has after having bet on $S \restriction |w|$. When betting on the next symbol b of a prefix wb of S, assuming symbol b is equally likely to be any value in Σ, equation (1) guarantees that the expected value of $d(wb)$ is $|\Sigma|^{-1} \sum_{a \in \Sigma} d(wa) = |\Sigma|^{s-1} d(w)$.

If $s = 1$, this expected value is exactly $d(w)$, so the payoffs are "fair".

Definition. Let d be an s-gale, where $s \in [0, \infty)$.

1. We say that d *succeeds* on a sequence $S \in \Sigma^\infty$ if

$$\limsup_{n \to \infty} d(S \restriction n) = \infty.$$

2. The *success set* of d is

$$S^\infty[d] = \{S \in \Sigma^\infty \mid d \text{ succeeds on } S\}.$$

Observation 3.1. *Let* $s, s' \in [0, \infty)$. *For every* s-gale d, *the function* $d' : \Sigma^* \to [0, \infty)$ *defined by* $d'(w) = |\Sigma|^{(s'-s)|w|} d(w)$ *is an* s'-gale. *Moreover, if* $s \le s'$, *then* $S^\infty[d] \subseteq S^\infty[d']$.

Lutz characterized Hausdorff dimension using gales as follows.

Theorem 3.2 *[11]. Given a set* $X \subseteq \Sigma^\infty$, *if* $\dim_H(X)$ *is the Haussdorf dimension of* X *[6], then*

$$\dim_H(X) = \inf\{s \mid \text{ there is an } s - \text{gale } d \text{ such that } X \subseteq S^\infty[d]\}$$

The idea for a Bounded Pushdown dimension is to consider only s-gales that are computable by a Bounded Pushdown (BPD) gambler. Bounded Pushdown gamblers are finite-state gamblers [3] with an extra memory stack, that is used both by the transition and betting functions. Additionally, BPDGs are allowed to delay reading the next character of the input –they read λ from the input– in order to alter the content of their stack, but they cannot do this more than a constant number of times per each input symbol. During such λ-transitions, the gambler's capital remains unchanged.

The betting function returns a probability measure over the input alphabet.

Definition. Let Σ be a finite alphabet. $\Delta_\mathbb{Q}(\Sigma)$ is the set of all rational-valued probability measures over Σ, i.e., all functions $\pi : \Sigma \longrightarrow [0, 1] \cap \mathbb{Q}$ such that $\sum_{a \in \Sigma} \pi(a) = 1$.

We are ready to define BPD gamblers.

Definition. A *bounded pushdown gambler (BPDG)* is an 8-tuple $G = (Q, \Sigma, \Gamma, \delta, \beta, q_0, z_0, c)$ where

- Q is a finite set of *states*,
- Σ is the finite input alphabet,
- Γ is the finite *stack alphabet*,
- $\delta : Q \times (\Sigma \cup \{\lambda\}) \times \Gamma \to Q \times \Gamma^*$ is the *transition function* (for simplicity we use the notation $\delta(q, b, a) = \perp$ when undefined; and we write $\delta(q, b, a) = (\delta_Q(q, b, a), \delta_{\Gamma^*}(q, b, a))$,
- $\beta : Q \times \Gamma \to \Delta_{\mathbb{Q}}(\Sigma)$ is the *betting function*,
- $q_0 \in Q$ is the *start state*,
- $z_0 \in \Gamma$ is the *start stack symbol*,
- $c \in \mathbb{N}$ is a constant such that the number of λ-transitions per input symbol is at most c,

with the two additional restrictions:

1. for each $q \in Q$ and $a \in \Gamma$ at least one of the following holds
   - $\delta(q, \lambda, a) = \perp$
   - $\delta(q, b, a) = \perp$ for all $b \in \Sigma$
2. for every $q \in Q$, $b \in \Sigma \cup \{\lambda\}$, either $\delta(q, b, z_0) = \perp$, or $\delta(q, b, z_0) = (q', vz_0)$, where $q' \in Q$ and $v \in \Gamma^*$.

We denote with $BPDG$ the set of all bounded pushdown gamblers.

The transition function δ outputs a new state and a string $z' \in \Gamma^*$. Informally, $\delta(q, w, a) = (q', z')$ means that in state q, reading input w, and popping symbol a from the stack, δ enters state q' and pushes z' to the stack.

Note that w can be λ (i.e., a λ-transition: the input is ignored and δ only computes with the stack) but this only happens at most c times per input symbol. Any pair (state, stack symbol) can either be a λ-transition pair or a non λ-transition pair exclusively, because the first additional restriction enforces determinism.

Moreover, since z_0 represents the bottom of the stack, we restrict δ so that z_0 cannot be removed from the bottom by the second additional restriction.

We can extend δ in the usual way to

$$\delta^* : Q \times (\Sigma \cup \{\lambda\}) \times \Gamma^+ \to Q \times \Gamma^*,$$

where for all $q \in Q$, $a \in \Gamma$, $v \in \Gamma^*$, and $b \in \Sigma \cup \{\lambda\}$

$$\delta^*(q, b, av) = \begin{cases} (\delta_Q(q, b, a), \delta_{\Gamma^*}(q, b, a)v) & \text{if } \delta(q, b, a) \neq \perp, \\ \perp & \text{otherwise.} \end{cases}$$

We denote δ^* by δ.

For each $i \geq 2$, we will use the notation

$$\delta^i(q, \lambda, v) = \delta(\delta_Q^{i-1}(q, \lambda, v), \lambda, \delta_{\Gamma^*}^{i-1}(q, \lambda, v))$$

where

$$\delta^1(q, \lambda, v) = \delta(q, \lambda, v).$$

Since δ is c-bounded we have that for any $q \in Q$, $v \in \Gamma^*$,

$$\delta^{c+1}(q, \lambda, v) = \perp$$

We also consider the extended transition function

$$\delta^{**} : Q \times \Sigma^* \times \Gamma^+ \to Q \times \Gamma^*,$$

defined for all $q \in Q$, $a \in \Gamma$, $v \in \Gamma^*$, $w \in \Sigma^*$, and $b \in \Sigma$ by

$$\delta^{**}(q, \lambda, av) = \delta^i(q, av)$$

if $\delta^i(q, \lambda, av) \neq \perp$ and $\delta^{i+1}(q, \lambda, av) - \perp$

$$\delta^{**}(q, wb, av) = \delta^i(\delta_Q(\tilde{q}, b, \widetilde{av}), \lambda, \delta_{\Gamma^*}(\tilde{q}, b, \widetilde{av}))$$

if $\delta^{**}(q, w, av) = (\tilde{q}, \widetilde{av})$, $\delta^i(\delta_Q(\tilde{q}, b, \widetilde{av}), \lambda, \delta_{\Gamma^*}(\tilde{q}, b, \widetilde{av})) \neq \perp$ and $\delta^{i+1}(\delta_Q(\tilde{q}, b, \widetilde{av}),$ $\lambda, \delta_{\Gamma^*}(\tilde{q}, b, \widetilde{av})) = \perp$, $i \leq c$.

That is, λ-transitions are inside the definition of $\delta^{**}(q, b, av)$, for $b \in \Sigma$. Notice that δ^{**} is not defined on an empty stack string, therefore av needs to be long enough in order that $\delta^{**}(q, b, av) \neq \perp$.

We denote δ^{**} by δ, and $\delta(q_0, w, z_0)$ by $\delta(w)$. We write $\delta = (\delta_Q, \delta_{\Gamma^*})$ for simplicity.

We also consider the usual extension of β

$$\beta^* : Q \times \Gamma^+ \to \Delta_{\mathbb{Q}}(\Sigma),$$

defined for all $q \in Q$, $a \in \Gamma$, and $v \in \Gamma^*$ by

$$\beta^*(q, av) = \beta(q, a),$$

and denote β^* by β.

We use BPDG to compute martingales. Intuitively, suppose a BPDG G is to bet on sequence S, has already bet on $w \sqsubset S$, with current capital $x \in \mathbb{Q}$, current state $q \in Q$ and current top stack symbol a. Then for $b \in \Sigma$, G bets the quantity $x\beta(q, a)(b)$ of its capital that the next symbol of S is b. If the bet is correct (that is, if $wb \sqsubset S$) and since payoffs are fair, G has capital $|\Sigma|x\beta(q, a)(b)$. Formally,

Definition. Let $G = (Q, \Sigma, \Gamma, \delta, \beta, q_0, z_0, c)$ be a bounded pushdown gambler. The *martingale* of G is the function

$$d_G : \Sigma^* \to [0, \infty)$$

defined by the recursion

$$d_G(\lambda) = 1$$

$$d_G(wb) = |\Sigma|d_G(w)\beta(\delta(w))(b)$$

for all $w \in \Sigma^*$ and $b \in \Sigma$.

By Observation 3.1, a BPDG G actually yields an s-gale for every $s \in [0, \infty)$. We call it the s-gale of G, and denote it by

$$d_G^s(w) = |\Sigma|^{(s-1)|w|} d_G(w).$$

A bounded pushdown s-gale is an s-gale d for which there exists a BPDG such that $d_G^s = d$.

Let us define bounded pushdown dimension. Intuitively, the BPD dimension of a sequence is the smallest s such that there is a BPD-s-gale that succeeds on the sequence.

Definition. The *bounded pushdown dimension* of a set $X \subseteq \Sigma^\infty$ is

$$\dim_{\mathsf{BPD}}(X) = \inf\{s \,|\, \text{there is a bounded pushdown } s - \text{gale } d \text{ such that } X \subseteq S^\infty[d]\}.$$

4   Dimension and Compression

In this section we characterize the bounded pushdown dimension of individual sequences in terms of bounded pushdown compressibility, therefore BPD dimension is a natural and robust definition.

Definition. A *bounded pushdown compressor (BPDC)* is an 8-tuple

$$C = (Q, \Sigma, \Gamma, \delta, \nu, q_0, z_0, c)$$

where

- Q is a finite set of states,
- Σ is the finite input and output alphabet,
- Γ is the finite stack alphabet,
- $\delta : Q \times (\Sigma \cup \{\lambda\}) \times \Gamma \to Q \times \Gamma^*$ is the transition function,
- $\nu : Q \times \Sigma \times \Gamma \to \Sigma^*$ is the output function,
- $q_0 \in Q$ is the initial state,
- $z_0 \in \Gamma$ is the start stack symbol,
- $c \in \mathbb{N}$ is a constant such that the number of λ-transitions per input symbol is at most c,

with the two additional restrictions:

1. for each $q \in Q$ and $a \in \Gamma$ at least one of the following holds
   - $\delta(q, \lambda, a) = \perp$
   - $\delta(q, b, a) = \perp$ for all $b \in \Sigma$
2. for every $q \in Q$, $b \in \Sigma \cup \{\lambda\}$, either $\delta(q, b, z_0) = \perp$, or $\delta(q, b, z_0) = (q', v z_0)$, where $q' \in Q$ and $v \in \Gamma^*$.

We extend δ to $\delta^{**} : Q \times \Sigma^* \times \Gamma^+ \to Q \times \Gamma^*$ as in Sect. 3 for the case of BPDGs, and denote δ^{**} by δ and $\delta(q_0, w, z_0)$ by $\delta(w)$.

For $q \in Q$, $w \in \Sigma^*$ and $z \in \Gamma^+$, we define the *output* from state q on input w reading z on the top of the stack to be the string $\nu(q, w, z)$ with

$$\nu(q, \lambda, z) = \lambda$$

$$\nu(q, wb, z) = \nu(q, w, z)\nu(\delta_Q(q, w, z), b, \delta_{\Gamma^*}(q, w, z))$$

for $w \in \Sigma^*$ and $b \in \Sigma$. We then define the *output* of C on input $w \in \Sigma^*$ to be the string

$$C(w) = \nu(q_0, w, z_0).$$

We are interested in *information lossless* compressors, that is, w must be recoverable from $C(w)$ and the final state.

Definition. A BPDC $C = (Q, \Sigma, \Gamma, \delta, \nu, q_0, z_0)$ is *information-lossless* (*IL*) if the function

$$\Sigma^* \to \Sigma^* \times Q$$

$$w \to (C(w), \delta_Q(w))$$

is one-to-one. An *information-lossless bounded pushdown compressor* (*ILBPDC*) is a BPDC that is IL.

Intuitively, a BPDC *compresses* a string w if $|C(w)|$ is significantly less than $|w|$. Of course, if C is *IL*, then not all strings can be compressed. Our interest here is in the degree (if any) to which the prefixes of a given sequence $S \in \Sigma^\infty$ can be compressed by an ILBPDC.

Definition. If C is a BPDC and $S \in \Sigma^\infty$, then the *compression ratio* of C on S is

$$\rho_C(S) = \liminf_{n \to \infty} \frac{|C(S[0..n-1])|}{n}.$$

The BPD compression ratio of a sequence is the best compression ratio achievable by an ILBPDC, that is

Definition. The *bounded pushdown (i.o.) compression ratio* of a sequence $S \in \Sigma^\infty$ is

$$\rho_{\mathsf{BPD}}(S) = \inf\{\rho_C(S) \mid C \text{ is a ILBPDC}\}.$$

The main result in this section states that the BPD dimension of a sequence and its ILBPD compression ratio are the same, therefore BPD dimension is the natural concept of density of information in the BPD setting.

Theorem 4.1. *For all* $S \in \Sigma^\infty$,

$$\dim_{\mathsf{BPD}}(S) = \rho_{\mathsf{BPD}}(S).$$

The rest of this section is devoted to proving Theorem 4.1.

Definition. A BPDG $G = (Q, \Sigma, \Gamma, \delta, \beta, q_0, z_0)$ is *nonvanishing* if $0 < \beta(q,z)(b) < 1$ for all $q \in Q$, $b \in \Sigma$ and $z \in \Gamma$.

Lemma 4.2. *For every BPDG G and each $\varepsilon > 0$, there is a nonvanishing BPDG G' such that for all $w \in \Sigma^*$, $d_{G'}(w) \geq |\Sigma|^{-\varepsilon|w|} d_G(w)$.*

Proof of Lemma 4.2. Let $G = (Q, \Sigma, \delta, \beta, q_0, \Gamma, z_0)$ be a BPDG, and let $\varepsilon > 0$. For each $q \in Q$, $z \in \Gamma$, $b \in \Sigma$,

$$1 - |\Sigma|^{-\varepsilon} \sum_{b \in \Sigma} \beta(q,z)(b) = 1 - |\Sigma|^{-\varepsilon} > 0,$$

so we can choose $\beta'(q,z)(b) > 0$ rational such that

$$|\Sigma|^{-\varepsilon} \beta(q,z)(b) < \beta'(q,z)(b) < 1 - |\Sigma|^{-\varepsilon} \sum_{a \in \Sigma, a \neq b} \beta(q,z)(a)$$

and

$$\sum_{b \in \Sigma} \beta'(q_0, z)(b) = 1.$$

Then, $0 < \beta'(q,z)(b) < 1$ for each $q \in Q$, $b \in \Sigma$ and $z \in \Gamma$, therefore the BPDG $G' = (Q, \Sigma, \delta, \beta', q_0, \Gamma, z_0)$ is nonvanishing.

Also, for all $q \in Q$, $b \in \Sigma$, $z \in \Gamma$,

$$\beta'(q,z)(b) \geq |\Sigma|^{-\varepsilon} \beta(q,z)(b)$$

so for all $w \in \Sigma^*$, $d_{G'}(w) \geq |\Sigma|^{-\varepsilon|w|} d_G(w)$. $\square$

Proof of Theorem 4.1. Let $S \in \Sigma^\infty$, $n \in \mathbb{N}$.

To see that $\dim_{\mathsf{BPD}}(S) \leq \rho_{\mathsf{BPD}}(S)$, let $s > s' > \rho_{\mathsf{BPD}}(S)$. It suffices to show that $\dim_{\mathsf{BPD}}(S) \leq s$. By our choice of s', there is an ILBPDC $C = (Q, \Sigma, \Gamma, \delta, \nu, q_0, z_0)$ for which the set

$$I = \{n \in \mathbb{N} \mid |C(S \restriction n)| < s'n\}$$

is infinite.

Construction 4.1. *Given a bounded pushdown compressor (BPDC) $C = (Q, \Sigma, \Gamma, \delta, \nu, q_0, z_0)$, and $k \in \mathbb{Z}^+$, we construct the bounded pushdown gambler (BPDG) $G = G(C, k) = (Q', \Sigma, \Gamma', \delta', \beta', q_0', z_0')$ as follows:*

(i) $Q' = Q \times \{0, 1, \ldots, k-1\}$
(ii) $q_0' = (q_0, 0)$
(iii) $\Gamma' = \bigcup_{i=1}^{(c+1)k} \Gamma^i$
(iv) $z_0' = z_0^{2k}$

(v) $\forall (q,i) \in Q', b \in \Sigma, a \in \Gamma'$,

$$\delta'((q,i),b,a) = \left(\left(\delta_Q(q,b,\overline{a}), (i+1) \bmod k \right), \delta_{\Gamma^*}\widehat{(q,b,\overline{a})} \right)$$

where for each $z \in (\Gamma')^+$, $\overline{z} \in \Gamma^+$ *is the* Γ*-string obtained by concatenating the symbols of* z, *and for each* $y \in \Gamma^+$, *if* $y = y_1 y_2 \cdots y_{2kl+n}$ *with* $n < 2k$, *then* $\widehat{y} \in (\Gamma')^+$ *is such that* $\widehat{y}_1 = y_1 \cdots y_{2k+n}$, $\widehat{y}_2 = y_{2k+n+1} \cdots y_{4k+n}$, $\ldots$, $\widehat{y}_l = y_{2k(l-1)+n+1} \cdots y_{2kl+n}$.

(vi) $\forall (q,i) \in Q', a \in \Gamma'$,

$$\delta'((q,i),\lambda,a) = \left(\left(\delta_Q(q,\lambda,\overline{a}), i \right), \delta_{\Gamma^*}\widehat{(q,\lambda,\overline{a})} \right).$$

(vii) $\forall (q,i) \in Q', a \in \Gamma', b \in \Sigma$

$$\beta'((q,i),a)(b) = \frac{\sigma(q, b\Sigma^{k-i-1}, a)}{\sigma(q, \Sigma^{k-i}, a)}$$

where $\sigma(q,A,a) = \sum_{x \in A} |\Sigma|^{-|\nu(q,x,\overline{a})|}$.

Notice that the fact that C is a BPDC is needed for the Construction 4.1 to be possible, since in order to define β' we need ν on inputs of length k to depend on a bounded number of stacks symbols. For a general PDC the computation of $\nu(q,x,)$ for $|x| \leq k$ could depend on an unbounded number of stack symbols.

Lemma 4.3. *In Construction 4.1, if* $|w|$ *is a multiple of* k *and* $u \in \Sigma^{\leq k}$, *then*

$$d_G(wu) = |\Sigma|^{|u|-|\nu(\delta_Q(w),u,\delta_{\Gamma^*}(w))|} \frac{\sigma(\delta_Q(wu), \Sigma^{k-|u|}, \delta_{\Gamma^*}\widehat{(wu)})}{\sigma(\delta_Q(w), \Sigma^k, \delta_{\Gamma^*}\widehat{(w)})} d_G(w).$$

Proof of Lemma 4.3. We use induction on the string u. If $u = \lambda$, the lemma is clear. Assume that it holds for u, where $u \in \Sigma^{<k}$, and let $b \in \Sigma$. Then

$$d_G(wub) = |\Sigma| \frac{\sigma(\delta_Q(wu), b\Sigma^{k-|u|-1}, \delta_{\Gamma^*}\widehat{(wu)})}{\sigma(\delta_Q(wu), \Sigma^{k-|u|}, \delta_{\Gamma^*}\widehat{(wu)})} d_G(wu)$$

$$= |\Sigma|^{1-|\nu(\delta_Q(wu),b,\delta_{\Gamma^*}(wu))|} \frac{\sigma(\delta_Q(wub), \Sigma^{k-|u|-1}, \delta_{\Gamma^*}\widehat{(wub)})}{\sigma(\delta_Q(wu), \Sigma^{k-|u|}, \delta_{\Gamma^*}\widehat{(wu)})} d_G(wu)$$

so by the induction hypothesis the lemma holds for ub. $\qquad\square$

Lemma 4.4. *In Construction 4.1, if* $w = w_0 w_1 \cdots w_{n-1}$, *where each* $w_i \in \Sigma^k$, *then*

$$d_G(w) = \frac{|\Sigma|^{|w|-|C(w)|}}{\displaystyle\prod_{i=0}^{n-1} \sigma(\delta_Q(w_0 \cdots w_{i-1}), \Sigma^k, \delta_{\Gamma^*}\widehat{(w_0 \cdots w_{i-1})})}.$$

Proof of Lemma 4.4. We use induction on n. For $n = 0$, the identity is clear. Assume that it holds for $w = w_0 w_1 \cdots w_{n-1}$, with each $w_i \in \Sigma^k$, and let $w' = w_0 w_1 \cdots w_n$. Then Lemma 4.3 with $u = w_n$ tells us that

$$d_G(w') = \frac{|\Sigma|^{k-|\nu(\delta_Q(w), w_n, \delta_{\Gamma^*}(w))|}}{\sigma(\delta_Q(w), \Sigma^k, \widehat{\delta_{\Gamma^*}}(w))} d_G(w)$$

whence the identity holds for w' by the induction hypothesis.   □

Lemma 4.5. *In Construction 4.1, if C is IL and $|w|$ is a multiple of k, then*

$$d_G(w) \geq |\Sigma|^{|w|-|C(w)|-\frac{|w|}{k}(l+\log m+\log k+1)},$$

where $l = \lceil \log |Q| \rceil$ and $m = \max\{|\nu(q, b, a)| \mid q \in Q, b \in \Sigma, a \in \Gamma^2\}$.

Proof of Lemma 4.5. We prove that for each $z \in \Sigma^*$,

$$\sigma(\delta_Q(z), \Sigma^k, \widehat{\delta_{\Gamma^*}}(z)) \leq |\Sigma|^{l+\log m+\log k+1}.$$

To see this, fix $z \in \Sigma^*$ and observe that at most $|Q|$ strings $w \in \Sigma^k$ can have the same output from state $\delta_Q(z)$ with stack content $\delta_{\Gamma^*}(z)$. Therefore, the number of $w \in \Sigma^k$ for which $|\nu(\delta_Q(z), w, \delta_{\Gamma^*}(z))| = j$ does not exceed $|Q||\Sigma|^j$. Hence

$$\sigma(\delta_Q(z), \Sigma^k, \widehat{\delta_{\Gamma^*}}(z)) = \sum_{w \in \Sigma^k} |\Sigma|^{-|\nu(\delta_Q(z), w, \delta_{\Gamma^*}(z))|} \leq \sum_{j=0}^{mk} |Q||\Sigma|^j |\Sigma|^{-j} = |Q|(mk+1)$$

$$\leq |\Sigma|^{l+\log m+\log k+1}.$$

It follows by Lemma 4.4 that

$$d_G(w) = |\Sigma|^{|w|-|C(w)|-\frac{|w|}{k}(l+\log m+\log k+1)}.$$ □

Lemma 4.6. *In Construction 4.1, if C is IL, then for all $w \in \Sigma^*$,*

$$d_G(w) \geq |\Sigma|^{|w|-|C(w)|-\frac{|w|}{k}(l+\log m+\log k+1)-(km+l+\log m+\log k+1)},$$

where $l = \lceil \log |Q| \rceil$ and $m = \max \{|\nu(q, b, a)| \mid q \in Q, b \in \Sigma, a \in \Gamma^2\}$.

Proof of Lemma 4.6. Assume the hypothesis, let l and m be as given, and let $w \in \Sigma^*$. Fix $0 \leq j < k$ such that $|w| + j$ is divisible by k. By Lemma 4.5 we have

$$d_G(w) \geq |\Sigma|^{-j} d_G(w0^j)$$

$$\geq |\Sigma|^{-j+|w0^j|-|C(w0^j)|-\frac{|w0^j|}{k}(l+\log m+\log k+1)}$$

$$= |\Sigma|^{|w|-|C(w0^j)|-\frac{|w|}{k}(l+\log m+\log k+1)-\frac{j}{k}(l+\log m+\log k+1)}$$

$$\geq |\Sigma|^{|w|-|C(w)|-\frac{|w|}{k}(l+\log m+\log k+1)-(km+l+\log m+\log k+1)}$$ □

Let $l = \lceil \log |Q| \rceil$ and $m = \max\{|\nu(q, b, a)| \mid q \in Q, b \in \Sigma, a \in \Gamma^2\}$, and fix $k \in \mathbb{Z}^+$ such that $\frac{l + \log m + \log k + 1}{k} < s - s'$. Let $G = G(C, k)$ be as in Construction 4.1. Then, by Lemma 4.6, for all $n \in I$ we have

$$d_G^{(s)}(w_n) \geq |\Sigma|^{sn - |C(w_n)| - \frac{n}{k}(l + \log m + \log k + 1) - (km + l + \log m + \log k + 1)}$$

$$\geq |\Sigma|^{(s - s' - \frac{l + \log m + \log k + 1}{k})n - (km + l + \log m + \log k + 1)}$$

Since $s - s' - \frac{l + \log m + \log k + 1}{k} > 0$, this implies that $S \in S^\infty[d_G^{(s)}]$.

Thus, $\dim_{\mathsf{BPD}}(S) \leq s$.

To see that $\rho_{\mathsf{BPD}}(S) \leq \dim_{\mathsf{BPD}}(S)$, let $s > s' > s'' > \dim_{\mathsf{BPD}}(S)$. It suffices to show that $\rho_{\mathsf{BPD}}(S) \leq s$. By our choice of s'', there is a BPDG G such that the set

$$J = \{n \in \mathbb{N} \mid d_G^{s''}(w_n) \geq 1\}$$

is infinite. By Lemma 4.2 there is a nonvanishing BPDG $\widetilde{G}$ such that $d_{\widetilde{G}}(w) \geq |\Sigma|^{(s'' - s')|w|} d_G(w)$ for all $w \in \Sigma^*$.

Construction 4.2. *Let $G = (Q, \Sigma, \Gamma, \delta, \beta, q_0, z_0)$ be a nonvanishing BPDG, and let $k \in \mathbb{Z}^+$. For each $z \in \Gamma^*$ (long enough for $d_{G_{q,z}}(w)$ to be defined for all $w \in \Sigma^k$) and $q \in Q$, let $G_{q,z} = (Q, \Sigma, \Gamma, \delta, \beta, q, z)$, and define $p_{q,z} : \Sigma^k \to [0, 1]$ by $p_{q,z}(w) = |\Sigma|^{-k} d_{G_{q,z}}(w)$. Since G is nonvanishing and each $d_{G_{q,z}}$ is a martingale with $d_{G_{q,z}}(\lambda) = 1$, each of the functions $p_{q,z}$ is a positive probability measure on Σ^k. For each $z \in \Gamma^*$, $q \in Q$, let $\Theta_{q,z} : \Sigma^k \to \Sigma^*$ be the Shannon-Fano-Elias code given by the probability measure $p_{q,z}$. Then*

$$|\Theta_{q,z}(w)| = l_{q,z}(w)$$
$$l_{q,z}(w) = 1 + \lceil \log \frac{1}{p_{q,z}(w)} \rceil$$

for all $q \in Q$ and $w \in \Sigma^k$, and each of the sets $\mathrm{range}(\Theta_{q,z})$ is an instantaneous code. We define the BPDC $C = C(G, k) = (Q', \Sigma, \Gamma', \delta', \nu', q_0', z_0')$ whose components are as follows:

(i) $Q' = Q \times \Sigma^{<k}$

(ii) $q_0' = (q_0, \lambda)$

(iii) $\Gamma' = \bigcup_{i=1}^{(c+1)k} \Gamma^i$

(iv) $z_0' = z_0^{2k}$

(v) $\forall (q, w) \in Q', b \in \Sigma, a \in \Gamma',$

$$\delta'((q, w), b, a) = \begin{cases} ((q, wb), a) & \text{if } |w| < k - 1, \\ ((\delta_Q(q, wb, \overline{a}), \lambda), \delta_{\Gamma^*} \widehat{(q, wb, \overline{a})}) & \text{if } |w| = k - 1. \end{cases}$$

(vi) $\forall (q, w) \in Q', a \in \Gamma',$

$$\delta'((q, w), \lambda, a) = ((q, w), a).$$

(vii) $\forall (q, w) \in Q', b \in \Sigma, a \in \Gamma'$,

$$\nu'((q,w), b, a) = \begin{cases} \lambda & if \ |w| < k - 1, \\ \Theta_{q,\bar{a}}(wb) & if \ |w| = k - 1. \end{cases}$$

Since each range($\Theta_{q,z}$) is an instantaneous code, it is easy to see that the BPDC $C = C(G, k)$ is IL.

Notice that the fact that G is a BPDG is needed for the construction 4.1 to be possible, since in order to define ν' we need d_G on inputs of length k to depend on a bounded number of stacks symbols. For a general PDG the computation of $d_G(q, w,)$ for $|w| = k$ could depend on an unbounded number of stack symbols.

Lemma 4.7. *In Construction 4.2, if $|w|$ is a multiple of k, then*

$$|C(w)| \le \left(1 + \frac{2}{k}\right)|w| - \log d_G(w).$$

Proof of Lemma 4.7. Let $w = w_0 w_1 \cdots w_{n-1}$, where each $w_i \in \Sigma^k$. For each $0 \le i < n$, let $q_i = \delta_Q(w_0 \cdots w_{i-1})$ and $z_i = \delta_{\Gamma^*}(w_0 \cdots w_{i-1})$. Then,

$$|C(w)| = \sum_{i=0}^{n-1} l_{q_i, z_i}(w_i)$$

$$= \sum_{i=0}^{n-1} \left(1 + \lceil \log \frac{1}{p_{q_i, z_i}(w_i)} \rceil\right) \le \sum_{i=0}^{n-1} \left(2 + \log \frac{1}{p_{q_i, z_i}(w_i)}\right)$$

$$= \sum_{i=0}^{n-1} \left(2 + \log \frac{|\Sigma|^k}{d_{G_{q_i, z_i}}(w_i)}\right) = (k+2)n - \log \prod_{i=0}^{n-1} d_{G_{q_i, z_i}}(w_i)$$

$$= (k+2)n - \log d_G(w) = (1 + \frac{2}{k})|w| - \log d_G(w) \qquad \square$$

Lemma 4.8. *In Construction 4.2, for all $w \in \Sigma^*$,*

$$|C(w)| \le \left(1 + \frac{2}{k}\right)|w| - \log d_G(w).$$

Proof of Lemma 4.8. If $|w|$ is multiple of k, then we apply the Lemma 4.7. Otherwise, let $w = w'z$, where $|w'|$ is a multiple of k and $|z| = j$, $0 < j < k$. Then, Lemma 4.7 tell us that

$$|C(w)| = |C(w')|$$

$$\le \left(1 + \frac{2}{k}\right)|w'| - \log d_G(w')$$

$$\le \left(1 + \frac{2}{k}\right)|w'| - \log(|\Sigma|^{-j} d_G(w))$$

$$= \left(1 + \frac{2}{k}\right)|w| - \log d_G(w) - \frac{2j}{k}$$

$$\le \left(1 + \frac{2}{k}\right)|w| - \log d_G(w). \qquad \square$$

Fix $k > \frac{2}{s-s'}$, and let $C = C(\widetilde{G}, k)$ be as in Construction 4.2. Then Lemma 4.8 tell us that for all $n \in J$,

$$
\begin{aligned}
\mid C(w_n) \mid &\leq \left(1 + \frac{2}{k}\right)n - \log d_{\widetilde{G}}(w_n) \\
&\leq \left(1 + \frac{2}{k} + s' - s''\right)n - \log d_G(w_n) \\
&\leq \left(\frac{2}{k} + s'\right)n - \log d_G^{s''}(w_n) \\
&\leq \left(\frac{2}{k} + s'\right)n \\
&< sn.
\end{aligned}
$$

Thus, $\rho_{\mathsf{BPD}}(S) \leq s$. $\qquad\square$

The corresponding result for strong (packing) dimension and a.e. compression ratio holds by a proof similar to that of Theorem 4.1.

Theorem 4.9. *For all* $S \in \Sigma^\infty$,

$$
Dim_{\mathsf{BPD}}(S) = R_{\mathsf{BPD}}(S).
$$

5 Separating LZ from BPD

In this section we prove that BPD compression can be much better than the compression attained with the celebrated Lempel-Ziv algorithm.

We start with a brief description of the LZ algorithm [17].

We finish relating BPD dimension (and compression) with the Lempel-Ziv algorithm. Given an input $x \in \Sigma^*$, LZ parses x in different phrases x_i, i.e., $x = x_1 x_2 \ldots x_n$ $(x_i \in \Sigma^*)$ such that every prefix $y \sqsubset x_i$, appears before x_i in the parsing (i.e. there exists $j < i$ s.t. $x_j = y$). Therefore for every i, $x_i = x_{l(i)} b_i$ for $l(i) < i$ and $b_i \in \Sigma$. We denote the *number of phrases* of x as $C(x) = n$.

LZ encodes x_i by a prefix free encoding of $l(i)$ and the symbol b_i, that is, if $x = x_1 x_2 \ldots x_n$ as before, the output of LZ on input x is

$$
LZ(x) = c_{l(1)} b_1 c_{l(2)} b_2 \ldots c_{l(n)} b_n
$$

where c_i is a prefix-free coding of i (and $x_0 = \lambda$).

LZ is usually restricted to the binary alphabet, but the description above is valid for any Σ.

For a sequence $S \in \Sigma^\infty$, the LZ compression ratio is given by

$$
\rho_{LZ}(S) = \liminf_{n \to \infty} \frac{|LZ(S \upharpoonright n)|}{n}.
$$

It is well known that LZ [17] yields a lower bound on the finite-state dimension (or finite-state compressibility) of a sequence [17], i.e., LZ is universal for finite-state compressors.

The following result shows that this is not true for BPD (hence PD) dimension, in a strong sense: we construct a sequence S that cannot be compressed by LZ, but that has BPD compression ratio less than $\frac{1}{2}$.

Theorem 5.1. *For every $m \in \mathbb{N}$, there is a sequence $S \in \{0,1\}^\infty$ such that*

$$\rho_{LZ}(S) > 1 - \frac{1}{m}$$

and

$$\dim_{\mathsf{BPD}}(S) \leq \frac{1}{2}.$$

Proof of Theorem 5.1. Let $m \in \mathbb{N}$, and let $k = k(m)$ be an integer to be determined later. For any integer n, let T_n denote the set of strings x of size n such that 1^j does not appear in x, for every $j \geq k$. Since T_n contains $\{0,1\}^{k-1} \times \{0\} \times \{0,1\}^{k-1} \times \{0\}\ldots$ (i.e. the set of strings whose every kth bit is zero), it follows that $|T_n| \geq 2^{an}$, where $a = 1 - 1/k$.

Remark 5.2. *For every string $x \in T_n$ there is a string $y \in T_{n-1}$ and a bit b such that $yb = x$.*

Let $A_n = \{a_1, \ldots a_u\}$ be the set of palindromes in T_n. Since fixing the $n/2$ first bits of a palindrome (wlog n is even) completely determines it, it follows that $|A_n| \leq 2^{\frac{n}{2}}$. Let us separate the remaining strings in $T_n - A_n$ into two sets $X_n = \{x_1, \ldots x_t\}$ and $Y_n = \{y_1, \ldots y_t\}$ with $(x_i)^{-1} = y_i$ for every $1 \leq i \leq t$. Let us choose X, Y such that x_1 and y_t start with a zero. We construct S in stages. For $n \leq k - 1$, S_n is an enumeration of all strings of size n in lexicographical order. For $n \geq k$,

$$S_n = a_1 \ldots a_u \, 1^{2n} \, x_1 \ldots x_t \, 1^{2n+1} \, y_t \ldots y_1$$

i.e. a concatenation of all strings in A_n (the A zone of S_n) followed by a flag of $2n$ ones, followed by the concatenations of all strings in X (the X-zone) and Y (the Y zone) separated by a flag of $2n + 1$ ones. Let

$$S = S_1 S_2 \ldots S_{k-1} \, 1^k \, 1^{k+1} \ldots 1^{2k-1} \, S_k S_{k+1} \ldots$$

i.e. the concatenation of the S_j's with some extra flags between S_{k-1} and S_k. We claim that the parsing of S_n ($n \geq k$) by LZ, is as follows:

$$S_n = a_1, \ldots, a_u, \, 1^{2n}, \, x_1, \ldots, x_t, \, 1^{2n+1}, \, y_t, \ldots, y_1.$$

Indeed after $S_1, \ldots S_{k-1} \, 1^k \, 1^{k+1} \ldots 1^{2k-1}$, LZ has parsed every string of size $\leq k - 1$ and the flags $1^k \, 1^{k+1} \ldots 1^{2k-1}$. Together with Remark 5.2, this guarantees that LZ parses S_n into phrases that are exactly all the strings in T_n and the two flags $1^{2n}, 1^{2n+1}$.

Let us compute the compression ratio $\rho_{LZ}(S)$. Let n, i be integers. By construction of S, LZ encodes every phrase in S_i (except the two flags), by a phrase in S_{i-1} (plus a bit). Indexing a phrase in S_{i-1} requires a code-word of length at least logarithmic in the number of phrase parsed before, i.e. $\log(C(S_1 S_2 \ldots S_{i-2}))$. Since $C(S_i) \geq |T_i| \geq 2^{ai}$, it follows

$$C(S_1 \ldots S_{i-2}) \geq \sum_{j=1}^{i-2} 2^{aj} = \frac{2^{a(i-1)} - 2^a}{2^a - 1} \geq b 2^{a(i-1)}$$

where $b = b(a)$ is arbitrarily close to 1. Letting $t_i = |T_i|$, the number of bits output by LZ on S_i is at least

$$C(S_i) \log C'(S_1 \dots S_{i-2}) \geq t_i \log b2^{a(i-1)}$$
$$\geq ct_i(i-1)$$

where $c = c(b)$ is arbitrarily close to 1. Therefore

$$|LZ(S_1 \dots S_n)| \geq \sum_{j=1}^{n} ct_j(j-1)$$

Since $|S_1 \dots S_n| \leq 2k^2 + \sum_{j=1}^{n}(jt_j + 4j)$, (the two flags plus the extra flags between S_{k-1} and S_k) the compression ratio is given by

$$\rho_{LZ}(S_1 \dots S_n) \geq c\frac{\sum_{j=1}^{n} t_j(j-1)}{2k^2 + \sum_{j=1}^{n} j(t_j+4)} \tag{2}$$

$$= c - c\frac{2k^2 + \sum_{j=1}^{n}(t_j+4j)}{2k^2 + \sum_{j=1}^{n} j(t_j+4)} \tag{3}$$

The second term in Eq. 3 can be made arbitrarily small for n large enough: Let $M \leq n$, we have

$$2k^2 + \sum_{j=1}^{n} j(t_j+4) \geq 2k^2 + \sum_{j=1}^{M} jt_j + (M+1)\sum_{j=M+1}^{n} t_j$$

$$= 2k^2 + \sum_{j=1}^{M} jt_j + M\sum_{j=M+1}^{n} t_j + \sum_{j=M+1}^{n} t_j$$

$$\geq 2k^2 + \sum_{j=1}^{M} jt_j + M\sum_{j=M+1}^{n} t_j + \sum_{j=M+1}^{n} 2^{aj}$$

$$\geq 2k^2 + \sum_{j=1}^{M} jt_j + M\sum_{j=M+1}^{n} t_j + 2^{an}$$

$$\geq M\sum_{j=M+1}^{n} t_j + M(2k^2 + 2n(n+1) + \sum_{j=1}^{M} t_j) \quad \text{for } n \text{ big enough}$$

$$= M(2k^2 + \sum_{j=1}^{n} t_j + 4\sum_{j=1}^{n} j)$$

Hence

$$\rho_{LZ}(S_1 \dots S_n) \geq c - \frac{c}{M}$$

which by definition of c, M can be made arbitrarily close to 1 by choosing k accordingly, i.e.

$$\rho_{LZ}(S_1 \dots S_n) \geq 1 - \frac{1}{m}.$$

Let us show that $\dim_{\mathsf{BPD}}(S) \leq \frac{1}{2}$. Consider the following BPD martingale d. Informally, d on S_n goes through the A_n zone until the first flag, then starts pushing the whole X zone onto its stack until it hits the second flag. It then uses the stack to bet correctly on the whole Y zone. Since the Y zone is exactly the X zone written in reverse order, d is able to double its capital on every bit of the Y zone. On the other zones, d does not bet. Before giving a detailed construction of d, let us compute the upper bound it yields on $\dim_{\mathsf{BPD}}(S)$.

$$\dim_{\mathsf{BPD}}(S) \leq 1 - \limsup_{n\to\infty} \frac{\log d(S_1 \ldots S_n)}{|S_1 \ldots S_n|}$$

$$\leq 1 - \limsup_{n\to\infty} \frac{\sum_{j=1}^{n} |Y_j|}{2k^2 + \sum_{j=1}^{n}(j|T_j| + 4j)}$$

$$\leq 1 - \limsup_{n\to\infty} \frac{\sum_{j=1}^{n} j \frac{|T_j| - |A_j|}{2}}{2k^2 + \sum_{j=1}^{n}(j|T_j| + 4j)}$$

$$\leq \frac{1}{2} + \frac{1}{2}\limsup_{n\to\infty} \frac{2k^2 + \sum_{j=1}^{n}(j|A_j| + 4j)}{2k^2 + \sum_{j=1}^{n}(j|T_j| + 4j)}.$$

Since

$$\limsup_{n\to\infty} \frac{2k^2 + \sum_{j=1}^{n}(j|A_j| + 4j)}{2k^2 + \sum_{j=1}^{n}(j|T_j| + 4j)} \leq \limsup_{n\to\infty} \frac{\sum_{j=1}^{n} j(|A_j| + 4 + 2k^2)}{\sum_{j=1}^{n} |T_j|}$$

$$\leq \limsup_{n\to\infty} \frac{\sum_{j=1}^{n} j(2^{\frac{j}{2}} + 2^{\frac{j}{4}})}{\sum_{j=1}^{n} 2^{aj}}$$

$$\leq \limsup_{n\to\infty} \frac{n2^{\frac{3n}{4}}}{2^{an}}$$

$$= 0.$$

It follows that

$$\dim_{\mathsf{BPD}}(S) \leq \frac{1}{2}.$$

Let us give a detailed description of d. Let Q be the following set of states:

- The start state q_0, and $q_1, \ldots q_v$ the "early" states that will count up to

$$v = |S_1 S_2 \ldots S_{k-1} \, 1^k \, 1^{k+1} \ldots 1^{2k-1}|.$$

- $q_0^a, \ldots, q_k^a$ the A zone states that cruise through the A zone until the first flag.
- q^{1f} the first flag state.
- $q_0^X, \ldots, q_k^X$ the X zone states that cruise through the X zone, pushing every bit on the stack, until the second flag is met.
- $q_0^r, \ldots, q_k^r$ which after the second flag is detected, pop k symbols from the stack that were erroneously pushed while reading the second flag.
- q^{2f} the second flag state.
- q^b the betting on zone Y state.

Let us describe the transition function $\delta : Q \times \{0, 1\} \times \{0, 1\} \to Q \times \{0, 1\}$. First δ counts until v i.e. for $i = 0, \ldots v - 1$

$$\delta(q_i, x, y) = (q_{i+1}, y) \quad \text{for any } x, y$$

and after reading v bits, it enters in the first A zone state, i.e. for any x, y

$$\delta(q_v, x, y) = (q_0^a, y).$$

Then δ skips through A until the string 1^k is met, i.e. for $i = 0, \ldots k - 1$ and any x, y

$$\delta(q_i^a, x, y) = \begin{cases} (q_{i+1}^a, y) & \text{if } x = 1 \\ (q_0^a, y) & \text{if } x = 0 \end{cases}$$

and

$$\delta(q_k^a, x, y) = (q^{1f}, y).$$

Once 1^k has been seen, δ knows the first flag has started, so it skips through the flag until a zero is met, i.e. for every x, y

$$\delta(q^{1f}, x, y) = \begin{cases} (q^{1f}, y) & \text{if } x = 1 \\ (q_0^X, 0y) & \text{if } x = 0 \end{cases}$$

where state q_0^X means that the first bit of the X zone (a zero bit) has been read, therefore δ pushes a zero. In the X zone, delta pushes every bit it sees until it reads a sequence of k ones, i.e. until the start of the second flag, i.e. for $i = 0, \ldots k - 1$ and any x, y

$$\delta(q_i^X, x, y) = \begin{cases} (q_{i+1}^X, xy) & \text{if } x = 1 \\ (q_0^X, xy) & \text{if } x = 0 \end{cases}$$

and

$$\delta(q_k^X, x, y) = (q_0^r, y).$$

At this point, δ has pushed all the X zone on the stack, followed by k ones. The next step is to pop k ones, i.e. for $i = 0, \ldots k - 1$ and any x, y

$$\delta(q_i^r, x, y) = (q_{i+1}^r, \lambda)$$

and

$$\delta(q_k^r, x, y) = (q_0^{2f}, y).$$

At this stage, δ is still in the second flag (the second flag is always bigger than $2k$) therefore it keeps on reading ones until a zero (the first bit of the Y zone) is met. For any x, y

$$\delta(q^{2f}, x, y) = \begin{cases} (q^{2f}, y) & \text{if } x = 1 \\ (q^b, \lambda) & \text{if } x = 0. \end{cases}$$

On the last step δ has read the first bit of the Y zone, therefore it pops it. At this stage, the stack exactly contains the Y zone (i.e. the X zone written in reverse order) except the first bit; δ thus uses its stack to bet and double its capital on every bit in the Y zone. Once the stack is empty, a new A zone begins. Thus, for any x, y

$$\delta(q^b, x, y) = (q^b, \lambda).$$

and

$$\delta(q^b, x, z_0) = \begin{cases} (q_1^a, z_0) & \text{if } x = 1 \\ (q_0^a, z_0) & \text{if } x = 0. \end{cases}$$

The betting function is equal to $1/2$ everywhere (i.e. no bet) except on state q^b, where

$$\beta(q^b, y)(z) = \begin{cases} 1 & \text{if } y = z \\ 0 & \text{if } y \neq z. \end{cases}$$

and β stops betting once start stack symbol is met, i.e.

$$\beta(q^b, z_0) = \frac{1}{2}.$$

$\square$

As a corollary we obtain a separation of finite-state dimension and bounded pushdown dimension. A similar result between finite-state dimension and pushdown dimension was proven in [5].

Corollary 5.3. *For any* $m \in \mathbb{N}$, *there exists a sequence* $S \in \{0,1\}^\infty$ *such that*

$$\dim_{\mathsf{FS}}(S) > 1 - \frac{1}{m}$$

and

$$\dim_{\mathsf{BPD}}(S) \leq \frac{1}{2}.$$

6   Conclusion

We have introduced Bounded Pushdown dimension, characterized it with compression and compared it with Lempel-Ziv compression. It is open whether BPD compression is universal for Finite-State compression, which is true for the Lempel-Ziv algorithm.

References

1. Athreya, K.B., Hitchcock, J.M., Lutz, J.H., Mayordomo, E.: Effective strong dimension in algorithmic information and computational complexity. SIAM J. Comput. **37**, 671–705 (2007)
2. Bourke, C., Hitchcock, J.M., Vinodchandran, N.V.: Entropy rates and finite-state dimension. Theor. Comput. Sci. **349**(3), 392–406 (2005)

3. Dai, J.J., Lathrop, J.I., Lutz, J.H., Mayordomo, E.: Finite-state dimension. Theor. Comput. Sci. **310**, 1–33 (2004)
4. Doty, D., Moser, P.: Finite-state dimension and lossy decompressors. Technical report, Technical report cs.CC/0609096, arXiv (2006)
5. Doty, D., Nichols, J.: Nichols.: pushdown dimension. Theor. Comput. Sci. **381**, 105–123 (2007)
6. Falconer, K.: The Geometry of Fractal Sets. Cambridge University Press, Cambridge (1985)
7. Hariharan, S., Shankar, P.: Evaluating the role of context in syntax directed compression of xml documents. In: Proceedings of the 2006 IEEE Data Compression Conference (DCC 2006), p. 453 (2006)
8. Hitchcock, J.M., Lutz, J.H.: The fractal geometry of complexity. SIGACT News Complex. Theory Column **36**, 24–38 (2005)
9. Huffman, D.A.: Canonical forms for information-lossless finite-state logical machines. Trans. Circ. Theory **CT–6**, 41–59 (1959)
10. League, C., Eng, K.: Type-based compression of xml data. In: Proceedings of the 2007 IEEE Data Compression Conference (DCC 2007), pp. 272–282 (2007)
11. Lutz, J.H.: Dimension in complexity classes. SIAM J. Comput. **32**, 1236–1259 (2003)
12. Lutz, J.H.: The dimensions of individual strings and sequences. Inf. Comput. **187**, 49–79 (2003)
13. Lutz, J.H.: Effective fractal dimensions. Math. Logic Q. **51**, 62–72 (2005)
14. Mayordomo, E.: Effective fractal dimension in algorithmic information theory. In: Cooper, S.B., Löwe, B., Sorbi, A. (eds.) New Computational Paradigms: Changing Conceptions of What is Computable, pp. 259–285. Springer, New York (2008)
15. Mayordomo, E., Moser, P., Perifel, S.: Polylog space compression, pushdown compression, and lempel-ziv are incomparable. Theory Comput. Syst. **48**, 731–766 (2011)
16. Sheinwald, D., Lempel, A., Ziv, J.: On compression with two-way head machines. In: Data Compression Conference, pp. 218–227 (1991)
17. Ziv, J., Lempel, A.: Compression of individual sequences via variable-rate coding. IEEE Trans. Inf. Theor. **24**(5), 530–536 (1978)

Complexity with Rod

Lance Fortnow[✉]

Georgia Institute of Technology, Atlanta, USA
fortnow@cc.gatech.edu

Abstract. Rod Downey and I have had a fruitful relationship though direct and indirect collaboration. I explore two research directions, the limitations of distillation and instance compression, and whether or not we can create NP-incomplete problems without punching holes in NP-complete problems.

1 Introduction

I first met Rod Downey at the first Dagstuhl seminar on "Structure and Complexity Theory" in February 1992. Even though we heralded from different communities, me as a computer scientist working on computational complexity, and Rod as a mathematician working primarily in computability, our paths have crossed many times on many continents, from Germany to Chicago, from Singapore to Honolulu. While we only have had one joint publication [1], we have profoundly affected each other's research careers.

In 2000 I made my first pilgrimage to New Zealand, to the amazingly beautiful town of Kaikoura on the South Island. Rod Downey had invited me to give three lectures [2] in the summer New Zealand Mathematics Research Institute graduate seminar on Kolmogorov complexity, the algorithmic study of information and randomness. Those lectures helped get Rod and Denis Hirschfeldt interested in Kolmogorov complexity, and their interest spread to much of the computability community. Which led to a US National Science Foundation Focused Research Group grant among several US researchers in the area including myself. What comes around goes around. In 2010 Rod Downey and Denis Hirschfeldt published an 855 page book *Algorithmic Randomness and Complexity* [3] on this line of research.

In this short paper I recount two other research directions developed with interactions with Rod Downey. In Sect. 2, I recall how an email from Rod led to a paper with Rahul Santhanam on instance compression [4], easily my most important paper of this century. In Sect. 3, I discuss my joint paper with Rod on the limitations of Ladner's theorem, that if P is different than NP, there are NP-incomplete sets and that continues to affect my current research.

2 Distillation

In the 1992 Dagstuhl workshop, Rod Downey gave a lecture entitled "A Completeness Theory for Parameterized Intractability," my first taste of fixed-parameter tractability (FPT). FPT looks at NP problems with a parameter,

© Springer International Publishing AG 2017
A. Day et al. (Eds.): Downey Festschrift, LNCS 10010, pp. 115–121, 2017.
DOI: 10.1007/978-3-319-50062-1_8

like whether a given graph G with n vertices has a vertex cover of size k. A problem is fixed-parameter tractable if there is an algorithm whose running time is of the form $f(k)n^c$ for an arbitrary f that does not depend on n. Vertex cover has such an algorithm but clique does not seem to. Downey and Michael Fellows had developed a series of complexity classes to capture these questions. I learned much more about FPT in a series of lectures of Michael Fellows at the aforementioned Kaikoura workshop in 2000.

On March 11, 2007 I travelled to Toronto to visit Rahul Santhanam, a former student and then a postdoc at the University of Toronto. On March 12th Rod Downey sent me a question by email (with a lucky typo) that came from a paper "On problems without polynomial kernels" [5] that Downey was working on with Fellows, Hans Bodlaender and Danny Hermelin. This confluence of events would lead my paper with Rahul Santhanam "Infeasibility of Instance Compression and Succinct PCPs for NP" [4]. These two papers would go on to be the co-winners of the 2014 EATCS-IPEC Nerode Prize and would eventually have over 500 combined citations.

Here is a formatted version of the email sent by Rod.

Say a language L has a distillation algorithm if there is an algorithm A which when applied to a sequence (perhaps exponentially long) $x_1, \ldots, x_n$ outputs in time polynomial in $\sum_i |x_i|$ a single string t such that
1. t is small: $|t| \leq \max\{x_i : i \leq n\}$, and
2. there exists an i, $x_i \in L$ iff $t \in L$
Clearly either all or no NP complete problems have distillation algorithms. Conjecture: No NP complete L has a distillation algorithm.
Can you think of any classical consequence of the failure of this conjecture? I had thought it implied $\mathrm{NP}^{\mathrm{NP}} \in \mathrm{NP/poly}$, but the proof was flawed.

I discussed the problem with Rahul and responded.

I believe I can show you get co-NP in NP/poly (thus PH in Σ_3^p) under this condition.
Fix a length m. Let S be the set of all strings not in L of length at most m. We will get a subset V of S with $|V|$ of size at most $\mathrm{poly}(m)$ and an $r \leq \mathrm{poly}(m)$ such that for all x in S there are $y_1, \ldots, y_r$ with
1. $x = y_i$ for some i.
2. the procedure on $y_1, \ldots, y_r$ outputs a t in V. Then we have an NP test for x in S with V as the advice.
Let $N = |S|$. There are N^r tuples $y_1, \ldots, y_r$. On each of them the procedure maps to something in S. For some z in S at least N^{r-1} tuples map to z. The number of x's covered by z is at least $N^{(r-1)/r}$ covering a $N^{-1/r}$ fraction of the elements of S. Picking $r = \log N$ (which is $\mathrm{poly}(m)$) makes this a constant fraction. Then we recurse on the remaining strings in S.

Rod Downey conferred with Michael Fellows and the next day realized he had slightly misstated the problem.

Sorry I realized that I made a mistake in the way that I defined distillation.

t is small means that $|t|$ is polynomial in $\max\{|x_i| : i \leq n\}$, not $\leq \max\{|x_i| : i \leq n\}$. I knew how to do it for $\leq |x_i|$ since then the language would be weakly p-selective (or something similar) and hence $\text{PH} = \Sigma_2^p$.

This is the problem. I cannot see how to fix your proof either, since the recursion goes awry.

Rahul and I looked it over and I responded to Rod

I worked this out with Rahul Santhanam. The same basic argument does go through if you pick r at least $|t|$.

Rod expressed surprise and when I got back to Chicago I wrote up a quick proof that would become the main lemma in my paper with Rahul [4]. A month later I cleaned up the statement and proof and present that version below (Lemma 1).

We generalized the proof for Rod's first question to get the proof for the question Rod had meant to ask. This two step approach helped us dramatically. If Rod didn't have the typo in the first question, we may never have discovered the proof. Just goes to show the role of pure luck in research.

Lemma 1. *Let L be any language. Suppose there exists a polynomial-time computable function f such that $f(\phi_1, \ldots, \phi_m) = y$ with*

1. *Each $|\phi_i| \leq n$.*
2. *$|y| \leq n^c$ with c independent of m.*
3. *y is in L if and only if there is an i such that ϕ_i is in SAT.*

then NP is in co-NP/poly.

Proof. Let $A \subseteq \overline{\text{SAT}} \cap \Sigma^{\leq n}$ with $A \neq \emptyset$. Let $B = \overline{L} \cap \Sigma^{\leq n^c}$. Let $N = |A|$ and $M = |B| \leq 2^{n^c}$. Let $m = n^c$.

Claim. There must be some y in B such that for at least half of the ψ in A, there exists $\phi_1, \ldots, \phi_m$ such that

1. For some i, $\phi = \phi_i$.
2. $f(\phi_1, \ldots, \phi_m) = y$.

Let ϕ be y-good if the above holds. Given y, we have a NP-proof that ϕ is not satisfiable for all y-good ϕ.

Now consider the N^m tuples $(\phi_1, \ldots, \phi_m)$ in A^m. The function f maps these tuples into elements of B. So for some y in B must have $\frac{N^m}{M}$ inverses in A^m.

If there are k y-good ϕ then $\frac{N^m}{M} \leq k^m$, so $k \geq \frac{N}{M^{1/m}}$. Since $m = n^c$, $M^{1/m} \leq 2$ and $k \geq \frac{N}{2}$ which proves our claim.

Now we start with $A = \overline{\text{SAT}} \cap \Sigma^{\leq n}$ and $S = \emptyset$. Applying the claim gives us a y in B. We put y in S, remove all of the y-good ϕ from A and repeat. Since $|A| \leq 2^{n+1}$ we only need to recurse at most $n + 2$ times before A becomes empty.

We then have the following NP algorithm for $\overline{\text{SAT}}$ on input ϕ using advice S:

– Guess $\phi_1, \ldots, \phi_m$.
– If $\phi = \phi_i$ for some i and $f(\phi_1, \ldots, \phi_m)$ is in S then accept.

Every language that is fixed-parameter tractable language can be mapped in polynomial time to an input whose size is a function of the parameter. Vertex cover, for example, has a stronger property that the kernel of an input is polynomial in the size of the parameter, i.e., the size of the vertex cover to check. The paper of Bodlaender, Downey, Fellows and Hermelin [5] would use Lemma 1 to show a number of fixed-parameter tractable problems do not have short kernelizations without complexity consequences.

The paper of Rahul and myself [4] had made some connections also to a paper by Danny Harnik and Moni Naor [6] on instance compression with some connections to cryptography. Later Harry Buhrman and John Hitchcock [7] would build on our lemma to show that NP can't have subexponential-sized complete sets unless the polynomial-time hierarchy collapsed. Andrew Drucker [8] generalized the lemma to problems like AND-SAT (for all i instead of there exists an i in condition 3 of Lemma 1) and to probabilistic and quantum reductions.

3   Punching Holes in SAT

In a 1944 address to the American Mathematical Society, Emil Post [9] laid out the landscape of recursive and recursively enumerable languages (now commonly called computable and computably enumerable), as well as reductions between languages.

> A primary problem in the theory of recursively enumerable sets is the problem of determining the degrees of unsolvability of the unsolvable decision problems thereof. We shall early see that for such problems there is certainly a highest degree of unsolvability. Our whole development largely centers on the single question of whether there is, among these problems, a lower degree of unsolvability than that, or whether they are all of the same degree of unsolvability.

In the paper, Post laid out his program to tackle that question but ultimately leaves it unresolved. It would take a dozen years for Friedberg and Muchnik (see [10]) to show the existence of a computably enumerable set that was not computable and not all other computably enumerable sets reduce to it.

After Steve Cook [11] and Richard Karp [12] defined the complexity classes NP and NP-complete in the early 70s, one could ask a similar question: Is there a problem in NP that is not computable in polynomial-time and not complete? Unlike in the computability world, we had several natural candidates for those classes including graph-isomorphism and factoring, where factoring as a language problem is the set of tuples (m, r) such that there is a prime factor p of m with $p \leq r$. It took just a couple of years after the introduction of NP-completeness for Richard Ladner [13] in 1975 to answer the question in the affirmative under the assumption that P differs from NP.

Ladner's proof works by "blowing holes in satisfiability". Ladner creates a language that for some input lengths is empty and other input lengths is some NP-complete problem like Boolean formula satisfiability. The lengths are chosen to diagonalize both from every polynomial-time algorithm and every reduction from satisfiability, thus ensuring that the language is not in P or NP-complete. To get the language in NP, Ladner develops a delayed diagonalization technique that doesn't move to the next requirement until it has had time to check that the previous requirement is fulfilled. We present Ladner's full proof as well as an alternate proof in our paper [1]. Both proofs leave large gaps in satisfiability.

I personally find Ladner's proof quite unsatisfying. We don't expect the natural candidates to behave like Ladner's set, as hard as satisfiability on some input lengths and computable in polynomial time on others. Rather every in every length we expect, for example, factoring to be difficult to compute but not complex enough for satisfiability to reduce to it. Is this a necessary factor to prove an intermediate set?

I visited Rod Downey in 1995 during his sabbatical year at Cornell. We discovered our shared concern about Ladner's proof. We formalized the issue by creating a definition of uniformly hard, basically that a set that is hard over polynomially long ranges of the input lengths.

Definition 2. *A language A is* uniformly hard *if for every language B computable in polynomial-time there is a k such that for every integer $n > 1$, A and B differ on some input of length between n and n^k.*

To justify uniformly hard we define an honest reduction with a slight variation to allow for giving a direct answer.

Definition 3. *An* honest reduction *from A to B is a polynomial-time computable function mapping Σ^* to $\Sigma^* \cup \{+, -\}$ such that*

1. *For some integer k, for all $n > 1$ and for all x, either $f(x) \in \{+, -\}$ or $|x| \geq |f(x)|^k$ where $|x|$ is the length of the string x.*
2. *If x is in A then $f(x) \subset B \cup \{ | \}$.*
3. *If x is not in A then $f(x) \in \overline{B} \cup \{-\}$, where $\overline{B} = \Sigma^* - B$.*

Uniformly hard sets are upwardly closed under these honest reductions.

Downey and I [1] looked at the question: If NP has uniformly-hard sets, is there an incomplete-set that is uniformly-hard under honest reductions? We conjecture such sets exist and in particular factoring should be such an example. However no proof exists that shows there are incomplete uniformly-hard sets. Why is proving such a result so difficult?

To answer that question, Downey and I look at a stronger version of Ladner's Theorem, with essentially the same proof, that there is no computable polynomial-time minimal degree.

Theorem 4. *For every computable B not in P there is a set A such that*

1. *A is not in P*
2. *A polynomial-time honestly reduces to B, in fact the reduction $f(x) \in \{x, -\}$ for all x, and*
3. *B does not even polynomial-time Turing reduce to A.*

Downey and I show that there could be a minimum uniformly-hard set if every problem computable in a polynomial amount of memory can also be computed in a polynomial amount of time.

Theorem 5. *If P = PSPACE, there is a computable minimum uniformly-hard set under polynomial-time honest reductions.*

We don't believe P = PSPACE but on the other hand complexity theorists have no approach to separating P and PSPACE. In particular that means we have no known way to avoid the large gaps given by Ladner's proof of Theorem 4.

Years later, Rahul Santhanam and I [14] generalized the uniformly hardness notion into a concept we called robustly-often which led to a new proof of the nondeterministic-time hierarchy. That work led to a paper by Rahul and myself [15] in the 2016 Computational Complexity Conference that gave new lower bounds for non-uniform classes, in particular we showed, for any a, b with $1 \le a < b$, NTIME(n^b) is not contained in NTIME(n^a) with $n^{1/b}$ bits of advice.

All of these results show that the ideas that Rod Downey helps generate have ripples that continue to push my research today.

Acknowledgments. I'd like to thank the anonymous referee for several helpful comments.

References

1. Downey, R., Fortnow, L.: Uniformly hard languages. Theor. Comput. Sci. **298**(2), 303–315 (2003)
2. Fortnow, L.: Kolmogorov complexity. In: Downey, R., Hirschfeldt, D. (eds.) Aspects of Complexity, Minicourses in Algorithmics, Complexity, and Computational Algebra, NZMRI Mathematics Summer Meeting. de Gruyter Series in Logic and Its Applications, Kaikoura, New Zealand, 7–15 January 2000, vol. 4. de Gruyter (2001)
3. Downey, R., Hirschfeldt, D.: Algorithmic Randomness and Complexity. Springer Science & Business Media, Heidelberg (2010)
4. Fortnow, L., Santhanam, R.: Infeasibility of instance compression and succinct PCPs for NP. J. Comput. Syst. Sci. **77**(1), 91–106 (2011)
5. Bodlaender, H., Downey, R., Fellows, M., Hermelin, D.: On problems without polynomial kernels. J. Comput. Syst. Sci. **75**(8), 423–434 (2009)
6. Harnik, D., Naor, M.: On the compressibility of NP instances and cryptographic applications. SIAM J. Comput. **39**(5), 1667–1713 (2010)
7. Buhrman, H., Hitchcock, J.: NP-hard sets are exponentially dense unless coNP is contained in NP/poly. In: 23rd Annual IEEE Conference on Computational Complexity, CCC 2008, pp. 1–7. IEEE, New York, June 2008

8. Drucker, A.: New limits to classical and quantum instance compression. SIAM J. Comput. **44**(5), 1443–1479 (2015)
9. Post, E.: Recursively enumerable sets of positive integers and their decision problems. Bullet. Am. Math. Soc. **50**(5), 284–316 (1944)
10. Soare, R.: Recursively Enumerable Sets and Degrees. Springer, Berlin (1987)
11. Cook, S.: The complexity of theorem-proving procedures. In: Proceedings of the 3rd ACM Symposium on the Theory of Computing, pp. 151–158. ACM, New York (1971)
12. Karp, R.: Reducibility among combinatorial problems. In: Miller, R., Thatcher, J. (eds.) Complexity of Computer Computations. The IBM Research Symposia Series, pp. 85–103. Springer, New York (1972). doi:10.1007/978-1-4684-2001-2_9
13. Ladner, R.: On the structure of polynomial time reducibility. J. ACM **22**, 155–171 (1975)
14. Fortnow, L., Santhanam, R.: Robust Simulations and Significant Separations, pp. 569–580. Springer, Heidelberg (2011)
15. Fortnow, L., Santhanam, R.: New non-uniform lower bounds for uniform classes. In: Raz, R. (ed.) 31st Conference on Computational Complexity (CCC 2016). Leibniz International Proceedings in Informatics (LIPIcs), vol. 50, pp. 19:1–19:14. Schloss Dagstuhl-Leibniz-Zentrum fuer Informatik, Dagstuhl, Germany (2016)

On Being Rod's Graduate Student

Keng Meng Ng[✉]

Division of Mathematical Sciences, School of Physical and Mathematical Sciences,
Nanyang Technological University, Singapore, Singapore
kmng@ntu.edu.sg

Rod's accomplishments as a mathematician are well-known to many in his area, and there are perhaps many others who are able to describe his numerous astonishing contributions to mathematical logic. My aim here is to say something about my experiences of being his graduate student, and to describe in some detail a non-mathematical side of the man.

In truth I had initially found it very tough to be his graduate student at VUW. I had only a limited amount of experience in mathematical research during my undergraduate years, and my then advisor, Yang Yue, suggested for me to go to Professor Downey for a PhD. And thus I was thrust onto the path of the publish-or-perish. Rod, I suspect, could sense that I was more unprepared for a creative career than he had hoped - the rigid educational system in Singapore did not help - and was thus initially rather stern with me. This had the effect of me having to work doubly hard, and after several months of frustration I was quite ready to give up. Fortunately one of the first positive qualities of Rod to rub off on me was the idea of persistence and obsession, those who know him well will attest to the fact that these two qualities are some of his more dominant traits. Thus I stuck through the steep learning curve, and was eventually able to benefit from his mentorship.

Rod was not only a great mathematician, but an amazing mathematical mentor. Perhaps it would be much easier to benefit from his guidance after one has obtained a certain amount of mathematical maturity and independence, which might be the reason why he's had far more postdocs than graduate students. Wellington has in recent years come to be affectionately known as the "rite of passage" for budding recursion/computability theorists. In any case I have learnt how to think mathematically, known about the value of persistence, and gained much valuable intuition under his tutelage. The value of working with him lies in the synergy he brings to the group, and the intuition that he shares often points the collaboration in the right direction.

One of the curious points of collaborating with Rod was his aversion to spend an extended period of time working in the same room with his collaborators. His work style was to brainstorm together for a while, and then withdraw to think alone about the problem. Being in the academia, the flexible work hours complements this very well. There are mathematicians who get things done just by pounding away persistently at problems, and there are others who obtain results through pure talent. Rod has the rare combination of both - which is why he is one of the most productive logicians around.

A. Day et al. (Eds.): Downey Festschrift, LNCS 10010, pp. 122–123, 2017.
DOI: 10.1007/978-3-319-50062-1_9

On a more personal note, Rod was never aloof or overly formal towards any of his graduate students or postdocs. Being the typically informal Australian he was quick to induct my wife and myself into his family. Over the course of my time in Wellington, we were never socially deprived. Rod made sure that we settled very nicely in the foreign land, and we've had as much social interactions as mathematical ones.

As many people know, Rod is a competent surfer, and plays squash, volleyball, tennis and table-tennis at a high level, though these activities had to be cut back due to various physical conditions. An activity which he has been actively engaged in - even till today - is Scottish Country Dancing. He initially took up Scottish Country Dancing at the suggestion of his wife, and over time has gone on to be one of the most competent dancers around, even becoming a qualified teacher. He now teaches at a club in Wellington. Unsurprisingly, part of the reason for this is due to the fact that there are definite connections between Scottish Country Dancing and abstract patterns literacy, for instance, the well-known codebreaker Hugh Foss, who had worked at Bletchley park during the war, is also a devisor of Scottish Country Dances.

Over the years Rod has stayed in active contact with most of his postdocs and students, whom he still regards as close friends. Loyalty is one of his most admirable traits. He often goes out of his way to help former postdocs with various favors, even under short notice. So, to the man who has been a mentor, collaborator and loyal friend, I wish him the best of health, and for him to find joy in whatever he does.

Computable Combinatorics, Reverse Mathematics

Herrmann's Beautiful Theorem on Computable Partial Orderings

Carl G. Jockusch Jr.$^{(\boxtimes)}$

Department of Mathematics,
University of Illinois at Urbana-Champaign, Urbana, USA
jockusch@math.uiuc.edu

Abstract. We give an exposition of Herrmann's proof that there is an infinite computable partial ordering with no infinite Σ_2^0 chains or antichains.

Keywords: Computable partial orderings · Chains · Antichains

2010 Mathematics Subject Classification: Primary 03D45

1 Introduction

Let $(P, \leqslant_P)$ be a partial ordering, so $\leqslant_P$ is a reflexive, transitive, antisymmetric relation on P. Two elements x, y of P are called *comparable* if $x \leqslant_P y$ or $y \leqslant_P x$ and otherwise are called *incomparable*. A subset S of P is called a *chain* if any two elements of S are comparable and is called an *antichain* if any two distinct elements of S are incomparable.

The *chain antichain principle* CAC asserts that every infinite partial ordering $(P, \leqslant_P)$ has either an infinite chain or an infinite antichain. Let $[A]^k$ denote the set of k-element subsets of the set A. Ramsey's Theorem for 2-colorings of pairs RT_2^2 asserts for every function $c : [\omega]^2 \to \{0, 1\}$ there is an infinite set $A \subseteq \omega$ which is *homogeneous* in the sense that c is constant on $[A]^2$. (Such a function c is called a *2-coloring* of $[\omega]^2$, and the homogeneous sets are defined by the property that all of their two-element subsets have the same color.) Note that CAC follows at once from RT_2^2. To see this, consider an infinite partial ordering $(P, \leqslant_P)$. Since every infinite set has an infinite countable subset, there is no loss of generality in assuming that $P = \omega$. Then consider the coloring $c : [\omega]^2 \to \{0, 1\}$ which maps comparable pairs to 0 and incomparable pairs to 1 and note that the homogeneous sets are exactly the chains and antichains. This argument can be

I am grateful to Eberhard Herrmann for finding the lovely proof on which this paper is based. I thank Rod Downey for his hospitality, collaboration, and support on my numerous trips to Wellington to work with him over the years. I also appreciate the opportunity to collaborate with him in many other places, including Urbana and Berkeley. Finally, I thank the Simons Foundation for supporting my travel to the meeting in New Zealand in 2017 in honor of his 60th birthday.

A. Day et al. (Eds.): Downey Festschrift, LNCS 10010, pp. 127–133, 2017.
DOI: 10.1007/978-3-319-50062-1_10

easily formalized in RCA_0, the base system for reverse mathematics, so $RCA_0 \vdash RT_2^2 \rightarrow CAC$. It is shown in [4] that CAC does not imply RT_2^2 in RCA_0, and the same paper has many further results on the strength of CAC and similar principles as well as effective analyses of these principles. Additional results on the complexity of chains and antichains in computable partial orderings may be found in [2,7], for example.

The current paper concerns the effective analysis of the principle CAC. A partial ordering $(P, \leqslant_P)$ is called *computable* if P is a computable subset of ω and $\leqslant_P$ is a computable relation on P. In particular, we want to determine the least n such that every computable partial ordering has a Π_n^0 chain or antichain, and similarly for Σ_n^0 chains and antichains. The corresponding problem for RT_2^2 was solved in [5] (Corollary 3.2 and Theorem 4.2) where it was shown that every computable 2-coloring of $[\omega]^2$ has an infinite Π_2^0 homogeneous set, but there exists a computable 2-coloring c_0 of $[\omega]^2$ with no infinite Σ_2^0 homogeneous set. It follows at once by the argument above that every infinite computable partial ordering has an infinite Π_2^0 chain or antichain. However, the 2-coloring c_0 mentioned above is not associated with a computable partial ordering. Hence the methods of [5] do not seem sufficient to show that there is an infinite computable partial ordering with no infinite Σ_2^0 chains or antichains. On the other hand it is straightforward to show by direct construction that there is an infinite computable partial ordering with no infinite Π_1^0 chains or antichains, and hence also no infinite Σ_1^0 chains or antichains.

The question raised above was solved by Eberhard Herrmann [3], who proved that there is an infinite computable partial ordering with no infinite Σ_2^0 chains or antichains, using a highly novel and ingenious argument. The purpose of the current paper is to give an exposition of Herrmann's proof.

Our terminology is standard. Let $(P, \leqslant_P)$ be a partial ordering. We write $a <_P b$ if $a \leqslant_P b$ and $a \neq b$. If $U, V \subseteq P$, we write $U <_P V$ if $(\forall u \in U)(\forall v \in V)[u <_P v]$. Similarly, we write $U \mid_P V$ if every element of U is incomparable with every element of V in $(P, \leqslant_P)$. We write (u, v) for the open interval $\{w : u <_P w <_P v\}$. Sometimes we write $\leqslant_P$ for the partial ordering $(P, \leqslant_P)$.

2   The Main Result

As explained in the introduction, the goal of this paper is to present a proof of the following theorem of Eberhard Herrmann.

Theorem 2.1 ([3], *Theorem 3.1*). *There is an infinite computable partial ordering with no infinite Σ_2^0 chains or antichains.*

Proof. We start with a certain computable partial ordering $(\omega, \leqslant_u)$ to be specified later, when we are better able to motivate its choice. The main step is to define an infinite computable set R such that for every Σ_2^0 chain or antichain S of $(\omega, \leqslant_u)$ the set $S \cap R$ is finite. It follows that the restriction of $\leqslant_u$ to R witnesses the truth of the theorem.

It remains to carry out the main step specified above and in particular to specify $\leqslant_u$. The first step is the following easy but crucial lemma.

Lemma 2.2. *Let $\leqslant_u$ be an arbitrary computable partial ordering. Then there is a uniformly Σ_2^0 sequence of sets $S_0, S_1, \ldots$ such that $S_0, S_1, \ldots$ are exactly the Σ_2^0 chains and antichains of $\leqslant_u$.*

Proof. By Post's Theorem, the Σ_2^0 sets are exactly the sets which are c.e. in K. Using an oracle for K, we list S_e uniformly in e as follows. List W_e^K (the eth set c.e. in K) and whenever a new element appears, add it to S_e provided that the resulting set is a chain or an antichain of $\leqslant_u$. Clearly, each S_e is a chain or an antichain of $\leqslant_u$, and $S_e = W_e^K$ if W_e^K is a chain or an antichain of $\leqslant_u$. Further, the sets $S_0, S_1, \ldots$ are uniformly c.e in K, and hence uniformly Σ_2^0, since Post's Theorem holds uniformly. □

We are now in a position to formulate the requirements whose satisfaction (along with making R computable) suffices to carry out the main step above. Fix sets S_e as in Lemma 2.2. The requirements are as follows:

$$P_s : (\exists x)[x \geqslant s \ \& \ x \in R]$$

$$N_e : S_e \cap R \text{ is finite}$$

Meeting any one requirement in isolation is trivial, but it is instructive to consider how to meet one negative requirement N_e and all positive requirements P_s simultaneously. We will then show how to meet all requirements.

Fix e. To be able to meet N_e and all the requirements P_s we assume that the ordering $\leqslant_u$ has three pairwise disjoint infinite computable sets B_0, B_1, B_2 called "boxes" such that $B_1 <_u B_0$ and $B_0 \mid_u B_2$. Note that every chain or antichain of $<_u$ is disjoint from some box B_i since no chain intersects both B_0 and B_2 and no antichain intersects both B_0 and B_1. To ensure the existence of the boxes B_0, B_1 and B_2 we require that the ordering $\leqslant_u$ have incomparable elements c and d whose supremum $c \sqcup d$ and infimum $c \sqcap d$ both exist and such that the three open intervals $(c, c \cup d)$, $(c \cap d, c)$, and $(d, c \cup d)$ in $\leqslant_u$ are all infinite. We can then let B_0, B_1, and B_2 be these three intervals, in the order listed.

At each stage s of the construction we enumerate some number $x_s \geqslant s$ into R, thus meeting P_s forever. To respect N_e we would like to ensure that $x_s \notin S_e$. However, our construction must be computable to make R computable, so we cannot ensure this. Instead we *try* to choose x_s so that $x_s \in B_i$ for some i such that B_i is disjoint from S_e. Again we cannot always do this in a computable construction, but we can do it for all sufficiently large s using a suitable computable approximation, as described in the next paragraph.

Let the predicate $D(e, i)$ hold if B_i is disjoint from S_e. Note that by routine expansion, the predicate $D(e, i)$ is Π_2^0, uniformly in e and an index for B_i as a c.e. set. Hence, using that every Π_2^0 set is many-one reducible to $\{k : W_k \text{ is infinite}\}$ we can uniformly obtain a ternary computable predicate $\hat{D}(e, i, t)$ such that, for all $i < 3$, $D(e, i)$ holds if and only if there are infinitely many t such that $\hat{D}(e, i, t)$ holds.

We now describe stage s of the construction. Search effectively for a pair of numbers (i, t) such that $i < 3$, $t \geqslant s$, and $\hat{D}(e, i, t)$ holds. To see that such a pair exists, recall that S_e is disjoint from B_i for some i, and for any such i, there are infinitely many t such that $\hat{D}(e, i, t)$ holds, by the previous paragraph. Let (i_s, t_s) be the first such pair found, and let x_s be the least element x of B_{i_s} with $x \geqslant s$, which exists because B_{i_s} is infinite. Enumerate x_s into R and proceed to stage $s + 1$.

We define $R = \{x_s : s \in \omega\}$ and note that R is infinite and computable.

We now check that the requirement N_e is met. Choose the number b so large so that, for all $i < 3$, if $D(e, i)$ is false, then $\hat{D}(e, i, t)$ is false for all $t \geqslant b$. It follows that if $s \geqslant b$, then B_{i_s} is disjoint from S_e. Since $x_s \in B_{i_s}$, it follows that $x_s \notin S_e$ for all $s \geqslant b$, and hence N_e is met.

We next consider how to meet all requirements simultaneously. If we do this naively by letting e vary in the above construction, there are severe problems since, for example, N_0 and N_1 may choose different values of i_s, and then we cannot require x_s to belong to B_i for more than one box B_i since the boxes are pairwise disjoint.

We overcome the difficulty above by nesting the boxes. We first explain how to meet N_0, N_1 and all the positive requirements P_s. To do this, we require that each box B_i for $i < 3$ have three infinite, computable pairwise disjoint subboxes $B_{i,j}$ for $j < 3$ such that $B_{i,1} <_u B_{i,0}$ and $B_{i,2} \mid_u B_{i,0}$. For $e = 0$ we choose i_s as in the above strategy for $e = 0$. For $e = 1$, we play the above strategy at each stage s replacing each B_j in the strategy by $B_{i_s,j}$. In other words, we play the above strategy for $e = 1$ within the box B_{i_s} chosen by the strategy for $e = 0$. Let B_{i_s,j_s} be the subbox of B_{i_s} chosen by the strategy for $e = 1$. We choose x_s to be the least element of B_{i_s,j_s} exceeding s, and enumerate x_s into R at stage s. Of course, there is now no problem if $i_s \neq j_s$, i.e. these strategies do not conflict since the relevant boxes are nested rather than disjoint. Since x_s belongs to both B_{i_s} and B_{i_s,j_s}, the verification is essentially as before.

To meet all requirements simultaneously, we simply iterate the nesting. At stage s, we meet P_s and respect the requirements $N_0, N_1, \ldots, N_s$ using $s + 1$ levels of nesting of the boxes. Each requirement N_e is respected at cofinitely many stages and hence is met as before. We will give some further details on this below, but first we consider how to choose our starting partial order $(\omega, \leqslant_u)$ so that all of these iteratively nested boxes exist.

One method to obtain $\leqslant_u$ is by a direct construction ensuring that the required nested boxes exist. This is done in [3]. Alternatively, one could choose a computable presentation $(\omega, \cup, \cap, -, 0, 1)$ of the countable atomless Boolean algebra, and let $\leqslant_u$ be the associated partial ordering given by $a \leqslant_u b$ if and only if $a \cup b = b$. We follow the latter approach. Note that if $a <_u b$ there are elements c, d such that $a <_u c, d <_u b$, $c \cup d = b$, and $c \cap d = a$. Furthermore, such c, d can be found computably from a and b by effective search. Also every open interval (a, b) with $a <_u b$ is infinite.

We now define a box B_σ for each string $\sigma \in 3^{<\omega}$. The box B_σ will be an open interval (b_σ, t_σ) in $\leqslant_u$, so B_σ will be computable. Since we will have $b_\sigma < t_\sigma$,

B_σ will be infinite. We define b_σ and t_σ by induction on the length of σ. For the empty string λ, let b_λ be the least element of ω under $\leqslant_u$, and let t_λ be the greatest element of ω under $\leqslant_u$. If b_σ and t_σ have already been defined, define $b_{\sigma \frown i}$ and $t_{\sigma \frown i}$ for $i < 3$ as follows. First, find incomparable elements c, d in the open interval (b_σ, t_σ) such that $c \cup d = t_\sigma$ and $c \cap d = b_\sigma$. Then define $t_{\sigma \frown i} = t_\sigma$ for $i \in \{0, 2\}$, $b_{\sigma \frown 0} = c$, $b_{\sigma \frown 2} = d$, $t_{\sigma \frown 1} = c$, and $b_{\sigma \frown 1} = b_\sigma$. This completes the inductive definition of b_σ and t_σ and hence of $B_\sigma = (b_\sigma, t_\sigma)$. The boxes $B_{\sigma \frown i}$ for $i = 0, 1, 2$ are similar in structure to the boxes B_i for $i = 0, 1, 2$ used in the basic strategy for meeting one negative requirement and all positive requirements, except that we now have $B_{\sigma \frown i} \subseteq B_\sigma$ for $i = 0, 1, 2$. Thus, for all σ, $B_{\sigma \frown 1} <_u B_{\sigma \frown 0}$ and $B_{\sigma \frown 0} |_u B_{\sigma \frown 2}$. The boxes B_σ are computable uniformly in σ.

Let $S_0, S_1, \ldots$ be as in Lemma 2.2. For $\sigma \in 3^{<\omega}$ let the predicate $D(e, \sigma)$ hold if $S_e \cap B_\sigma = \emptyset$. Since the predicate D is Π_2^0, there is a computable ternary predicate $\hat{D}$ such that, for all $e \in \omega$ and $\sigma \in 3^{<\omega}$, $D(e, \sigma)$ holds if and only if there are infinitely many t such that $\hat{D}(e, \sigma, t)$ holds.

We now give the construction of a computable set R satisfying all of our requirements. At stage s, we define strings $\sigma_{s,e} \in 3^{<\omega}$ for $e \leqslant s$ by recursion on e such that $\sigma_{s,0} \prec \sigma_{s,1} \prec \cdots \prec \sigma_{s,s}$ and $\sigma_{s,e}$ has length e for each $e \leqslant s$. Let $\sigma_{s,0}$ be the empty string. Now suppose that $e < s$ and $\sigma_{s,e}$ has been defined. Search effectively for a number $i < 3$ and a number $t \geqslant s$ such that $\hat{D}(e, \sigma_{s,e} \frown i, t)$ holds. Let $\sigma_{s,e+1} = \sigma_{s,e} \frown i$ for the first such i which is found. Such i and t exist because $B_{\sigma_{s,e} \frown i}$ is disjoint from S_e for some $i < 3$. Let x_s be the least $x \in B_{\sigma_{s,s}}$ with $x \geqslant s$. Enumerate x_s in R and proceed to stage $s + 1$. Clearly, $R = \{x_0, x_1, \ldots\}$ is computable and infinite.

It remains to verify that each requirement N_e is satisfied, and the argument is familiar by now. Fix e. For all sufficiently large s, $B_{\sigma_{s,e+1}}$ is disjoint from S_e since $\hat{D}(e, \sigma_{s,e+1}, t)$ holds for some $t \geqslant s$. Also, for $s > e$, $x_s \in B_{\sigma_{s,s}} \subseteq B_{\sigma_{s,e+1}}$. Hence, for all sufficiently large s we have that $x_s \notin S_e$, so $R \cap S_e$ is finite, and so the requirement N_e is met. $\square$

3   Computable Linear Extensions of Herrmann's Ordering

If $(P, \leqslant_P)$ is an infinite computable partial ordering with no infinite Σ_2^0 chains or antichains, call $\leqslant_P$ a *Herrmann ordering*. By the main result of the last section, Herrmann orderings exist. Such orderings have many special properties, as Herrmann [3] pointed out. For example, they have only finitely many minimal elements, because the set of minimal elements is a Π_1^0 antichain. As Herrmann also pointed out, there are infinitely many order types of such orderings because, for each $n > 0$ there is a Herrmann ordering with exactly n minimal elements, as can be seen by considering any Herrmann ordering and adding n new elements which are pairwise incomparable and lie below all the old elements.

On the other hand, Herrmann [3], Corollary 4.2, did prove a sort of uniqueness theorem for Herrmann orderings. A "linearization" of a partial ordering $(P, \leqslant_P)$ is any linear ordering $(P, \leqslant_L)$ such that, for all $u, v \in P$, $u \leqslant_P v$ implies

$u \leqslant_L v$. Szpilrajn [8] proved that every partial ordering has a linearization. By a well-known folklore result proved in Downey [1], Observation 6.1, this holds effectively in the sense that every computable partial ordering has a computable linearization. Herrmann's uniqueness theorem [3], Corollary 4.2, is that all computable linearizations of Herrmann orders have the same order type. However, as remarked in the author's review [6] of [3] the proof in [3] is not quite correct. First, as Joseph Mileti pointed out, in several places in the argument "minimal" should be changed to "maximal". Even after these changes are made, a gap remained, but Mileti supplied a lemma to fill this gap. This lemma is stated in [6]. Thus Herrmann's uniqueness theorem holds, and in fact any computable linearization of any Herrmann ordering has order type $\omega + (\omega^* + \omega) \cdot \eta + \omega^*$. We omit further details.

Let $\mathcal{L}$ be any computable linear ordering which is a linearization of a Herrmann ordering. It is shown in [4], proof of Proposition 2.15, that $\mathcal{L}$ is an example of an infinite computable linear ordering with no low infinite ascending or descending chains and hence no low subordering of order type ω, ω^* or $\omega + \omega^*$. No direct method is known for constructing a linear ordering with this property. See Corollary 2.16 of [4] for applications of the existence of such a linear ordering $\mathcal{L}$ to the reverse mathematics of linear orderings.

Personal Note: The reader may well ask why the current paper appears in a Festschrift for Rod Downey. Rod and I have collaborated extensively and yet our published joint work (consisting so far of twelve papers with publication dates from 1987 to 2015) does not touch on Herrmann's theorem. Our connection with Herrmann's theorem is that I travelled to Wellington to work with Rod in 1995 (before Herrmann proved his theorem), and we spent almost the entire two-week visit obsessing in vain over the problem whether every infinite computable partial ordering of ω has an infinite Σ_2^0 chain or antichain. Then Herrmann's paper [3] came along. Once I finally understood it I was deeply impressed by the beauty of his proof (which was not related to the methods that Rod and I tried). My goal in this paper is to convey that beauty and, I hope, make the proof easier to follow, clarifying the motivation and simplifying some technical details while still making heavy use of Herrmann's ideas.

References

1. Downey, R.G.: Computability theory and linear orderings. Stud. Logic Found. Math. **138**, 823–976 (1998). Handbook of Recursive Mathematics (Ershov, Goncharov, Nerode, and Remmel, editors). Elsevier, Amsterdam
2. Harizanov, V., Jockusch, C.G., Knight, J.F.: Chains and antichains in partial orderings. Archiv. Math. Logic **48**, 39–53 (2009)
3. Herrmann, E.: Infinite chains and antichains in computable partial orderings. J. Symbolic Logic **66**, 923–934 (2001)
4. Hirschfeldt, D., Shore, R.A.: Combinatorial principles weaker than Ramsey's theorem for pairs. J. Symbolic Logic **72**, 171–206 (2007)

5. Jockusch, C.G.: Ramsey's theorem and recursion theory. J. Symbolic Logic **37**, 268–280 (1972)
6. Jockusch, C.G.: Review of Herrmann's paper [3]. Math. Rev. 1833487(2003e:03082) (2003)
7. Jockusch, C., Kastermans, B., Lempp, S., Lerman, M., Solomon, D.R.: Stability and posets. J. Symbolic Logic **74**(2), 693–711 (2009)
8. Szpilrajn, E.: Sur l'extension de l'order partiel. Fundam. Math. **16**, 386–389 (1930)

Effectiveness of Hindman's Theorem
for Bounded Sums

Damir D. Dzhafarov[1,2], Carl G. Jockusch Jr.[1,2], Reed Solomon[1,2(✉)],
and Linda Brown Westrick[1,2]

[1] Department of Mathematics, University of Connecticut,
196 Auditorium Road, Storrs, CT 06269, USA
damir@math.uconn.edu, jockusch@math.uiuc.edu,
{david.solomon,linda.westrick}@uconn.edu
[2] Department of Mathematics, University of Illinois,
1409 W. Green Street, Urbana, IL 61801, USA

*This paper is dedicated to Rod Downey in
honor of his outstanding contributions to
computability theory and his leadership role
in mentoring and exposition. Two of the
authors had the pleasure of being mentored
by Rod as postdocs.*

Abstract. We consider the strength and effective content of restricted
versions of Hindman's Theorem in which the number of colors is specified
and the length of the sums has a specified finite bound. Let $\mathsf{HT}_k^{\leq n}$ denote
the assertion that for each k-coloring c of $\mathbb{N}$ there is an infinite set $X \subseteq \mathbb{N}$
such that all sums $\sum_{x \in F} x$ for $F \subseteq X$ and $0 < |F| \leq n$ have the same
color. We prove that there is a computable 2-coloring c of $\mathbb{N}$ such that
there is no infinite computable set X such that all nonempty sums of
at most 2 elements of X have the same color. It follows that $\mathsf{HT}_2^{\leq 2}$
is not provable in RCA_0 and in fact we show that it implies SRT_2^2 in
$\mathsf{RCA}_0 + \mathsf{B}\Pi_1^0$. We also show that there is a computable instance of $\mathsf{HT}_3^{\leq 3}$
with all solutions computing $0'$. The proof of this result shows that $\mathsf{HT}_3^{\leq 3}$
implies ACA_0 in RCA_0.

Keywords: Hindman's Theorem · Computable combinatorics ·
Ramsey's Theorem · Reverse mathematics

2010 AMS Subject Classification: 03D80 · 05D10 (03B30, 03F35)

Dzhafarov was partially supported by NSF grant DMS-1400267. Jockusch thanks his
coauthors for their hospitality during a visit to the University of Connecticut in May,
2015, during which the work in this paper was largely done. His visit was supported
by the University of Connecticut Department of Mathematics. The authors thank
Denis Hirschfeldt for numerous valuable insights.

A. Day et al. (Eds.): Downey Festschrift, LNCS 10010, pp. 134–142, 2017.
DOI: 10.1007/978-3-319-50062-1_11

1   Introduction

Hindman's Theorem (denoted HT) asserts that for every coloring of $\mathbb{N}$ with finitely many colors there is an infinite set $X \subseteq \mathbb{N}$ such that all nonempty finite sums of distinct elements of X have the same color. Hindman's Theorem was proved by Neil Hindman [6]. Hindman's original proof was a complicated combinatorial argument, and simpler proofs have been subsequently found. These include combinatorial proofs by Baumgartner [1] and by Towsner [12] and a proof using ultrafilters by Galvin and Glazer (see [4]).

We assume that the reader is familiar with the basic concepts of computability theory and of reverse mathematics. For information on these topics see, respectively, the books by Soare [11] and Simpson [10]. Our notation is standard. In particular, let $\mathbb{N}$ be the set of positive integers, and for $k \in \mathbb{N}$ we identify k and $\{0, 1, \ldots, k-1\}$. A k-coloring of $\mathbb{N}$ is a function $c : \mathbb{N} \to k$. A set $Z \subseteq \mathbb{N}$ is monochromatic for a coloring c if $c(x) = c(y)$ for all $x, y \in Z$.

The effective content of Hindman's Theorem and its strength as a sentence of second-order arithmetic were studied by Blass, Hirst, and Simpson [2]. They showed that every computable instance c of HT has a solution X computable from $0^{(\omega+1)}$ and, correspondingly, that HT is provable in the system ACA_0^+ obtained by adding to RCA_0 the statement $(\forall X)[X^{(\omega)}$ exists]. In the other direction, they showed that there is a computable instance c of HT such that all solutions X compute $0'$ and, correspondingly, that HT implies ACA_0 in RCA_0.

There is obviously a significant gap between the upper and lower bounds given in the previous paragraph, and closing these gaps has been a major issue in reverse mathematics. In particular it is not known whether there is an n such that every computable instance of Hindman's Theorem has a Σ_n^0 solution and, correspondingly, whether HT is provable from ACA_0 in RCA_0.

In the current paper we study the strength and effective content of Hindman's Theorem when it is restricted to sums of bounded length. One might think that such restricted versions of Hindman's Theorem are far weaker than Hindman's Theorem itself, but in fact it is unknown whether this is true. In fact it is a major open problem in combinatorics (see [7], Question 12) whether every proof of Hindman's Theorem for sums of length at most two also proves Hindman's Theorem. We now state these bounded versions formally.

Definition 1.1. For a finite nonempty set $F \subseteq \mathbb{N}$, we let $\sum F$ denote the sum of the elements of F. For $X \subseteq \mathbb{N}$ and $n \geq 1$, we define

$$\mathrm{FS}^{\leq n}(X) = \left\{ \sum F \mid F \subseteq X \text{ and } 1 \leq |F| \leq n \right\}.$$

Definition 1.2. Let $\mathsf{HT}_k^{\leq n}$ denote the statement that for every coloring $c : \mathbb{N} \to k$, there is an infinite set X such that $\mathrm{FS}^{\leq n}(X)$ is monochromatic.

We show in Sect. 2 that for every Δ_2^0 set X there is a computable instance c of $\mathsf{HT}_2^{\leq 2}$ such that every solution H to c computes an infinite subset of X or $\overline{X}$. It follows that $\mathsf{HT}_2^{\leq 2}$ has a computable instance with no computable solution and hence is not provable in RCA_0. In fact, our proof shows that $\mathsf{HT}_2^{\leq 2}$ implies SRT_2^2 (Stable Ramsey's Theorem for 2-colorings of pairs) in $\mathsf{RCA}_0 + \mathsf{B\Pi}_1^0$, where $\mathsf{B\Pi}_1^0$ is the bounding principle for Π_1^0 formulas. Next we show in Sect. 3 that there is a computable instance of $\mathsf{HT}_3^{\leq 3}$ such that every solution computes $0'$ and, correspondingly, that $\mathsf{HT}_3^{\leq 3}$ implies ACA_0 in RCA_0. Our proof uses a very ingenious trick from Blass, Hirst, and Simpson [2], combined with some new ideas.

The final section lists many open questions.

2 Hindman's Theorem for Sums of Length at Most 2

Our first theorem concerns $\mathsf{HT}_2^{\leq 2}$ and implies that it has a computable instance c with no computable solution X.

Theorem 2.1. *Let A be a Δ_2^0 set. There is a computable coloring $c : \mathbb{N} \to 2$ such that if W is an infinite set with $\mathrm{FS}^{\leq 2}(W)$ monochromatic, then there is an infinite set $Y <_T W$ such that $Y \subseteq A$ or $Y \subseteq \overline{A}$.*

Proof. Fix a Δ_2^0 set A and a computable $\{0,1\}$-valued function $f(k,s)$ such that $A(k) = \lim_s f(k,s)$. For $k \geq 0$ and $i \in \{1,2\}$, define

$$\mathcal{O}_{k,i} = \{s \in \mathbb{N} \mid s \equiv i \cdot 3^k \bmod 3^{k+1}\}.$$

If s is written as $s = i_0 \cdot 3^{k_0} + \cdots + i_m \cdot 3^{k_m}$ with $k_0 < \cdots < k_m$ and each $i_j \in \{1,2\}$, then $s \in \mathcal{O}_{k,i}$ if and only if $k = k_0$ and $i = i_0$. The sets $\mathcal{O}_{k,i}$ give a computable partition of $\mathbb{N}$ such that if $s,t \in \mathcal{O}_{k,1}$, then $s + t \in \mathcal{O}_{k,2}$ and if $s,t \in \mathcal{O}_{k,2}$, then $s+t \in \mathcal{O}_{k,1}$. Furthermore, if $s \in \mathcal{O}_{k,i}$ and $t \in \mathcal{O}_{k',i'}$ with $k < k'$ and $i' \in \{1,2\}$, then $s + t \in \mathcal{O}_{k,i}$. For any $s \in \mathbb{N}$, we let k_s, i_s be the unique numbers k, i such that $s \in \mathcal{O}_{k,i}$. We define our coloring c by

$$c(s) = \begin{cases} f(k_s, s) & \text{if } i_s = 1, \\ 1 - f(k_s, s) & \text{if } i_s = 2. \end{cases}$$

The first important property of this coloring is that for each k we have $c(s) \neq c(t)$ whenever $s \in \mathcal{O}_{k,1}$ and $t \in \mathcal{O}_{k,2}$ are both sufficiently large. This holds since for sufficiently large $s \in \mathcal{O}_{k,1}$ and $t \in \mathcal{O}_{k,2}$ we have $c(s) = f(k,s) = A(k)$ and $c(t) = 1 - f(k,t) = 1 - A(k)$. It follows that for any monochromatic set Z, either $Z \cap \mathcal{O}_{k,1}$ is finite or $Z \cap \mathcal{O}_{k,2}$ is finite.

Fix an infinite set W with $\mathrm{FS}^{\leq 2}(W)$ monochromatic. We claim that $W \cap \mathcal{O}_{k,i}$ is finite for each $k \in \mathbb{N}$ and $i \in \{1,2\}$. Suppose first that $W \cap \mathcal{O}_{k,1}$ is infinite. Let S be the set of all sums $a + b$ where a, b are distinct elements of $W \cap \mathcal{O}_{k,1}$. Then S is infinite and $S \subseteq \mathcal{O}_{k,2} \cap \mathrm{FS}^{\leq 2}(W)$. Let $Z = W \cup S$. Then Z is monochromatic since $Z \subseteq \mathrm{FS}^{\leq 2}(W)$. Furthermore, $Z \cap \mathcal{O}_{k,1}$ and $Z \cap \mathcal{O}_{k,2}$ are both infinite, contradicting the previous paragraph. This shows that $W \cap \mathcal{O}_{k,1}$ is finite, and the proof that $W \cap \mathcal{O}_{k,2}$ is finite is analogous. It follows that there are infinitely many

k such that $W \cap (\mathcal{O}_{k,1} \cup \mathcal{O}_{k,2})$ is nonempty. We call such numbers k *informative* since, as the next claim shows, W can compute $A(k)$ for all informative k.

We claim that if $s \in W \cap (\mathcal{O}_{k,1} \cup \mathcal{O}_{k,2})$ then $f(k, s) = A(k)$. To prove this claim, assume first that $s \in W \cap \mathcal{O}_{k,1}$. Note that $\mathrm{FS}^{\leq 2}(W) \cap \mathcal{O}_{k,1}$ is infinite, since it contains all sums $s + b$ with $b \in W \cap \mathcal{O}_{k',i'}$ for some $k' > k$, and $i' \in \{1, 2\}$, and there are infinitely many such b. Let t be an element of $\mathrm{FS}^{\leq 2}(W) \cap \mathcal{O}_{k,1}$ sufficiently large that $f(k, t) = A(k)$. Since $\mathrm{FS}^{\leq 2}(W)$ is monochromatic, $c(s) = c(t)$. Hence $f(k, s) = c(s) = c(t) = f(k, t) = A(k)$. The proof for $s \in W \cap \mathcal{O}_{k,2}$ is analogous. The claim is proved.

For $i \in \{0, 1\}$ let B_i be the set of numbers k such that W can compute that $A(k) = i$. More precisely, define

$$B_i = \{k \mid (\exists s)[s \in W \cap (\mathcal{O}_{k,1} \cup \mathcal{O}_{k,2}) \ \& \ f(k, s) = i]\}$$

By the above claim, $B_1 \subseteq A$ and $B_0 \subseteq \overline{A}$. Also, each set B_i is c.e. in W. Finally, if k is informative, then $k \in B_0 \cup B_1$. Since there are infinitely many informative numbers, $B_0 \cup B_1$ is infinite, and so B_0 or B_1 is infinite. Fix i such that B_i is infinite, and let Y be an infinite W-computable subset of B_i. Then Y is the desired infinite W-computable subset of A or $\overline{A}$.   □

The next corollary follows by taking A to be a bi-immune Δ_2^0 set, for example a Δ_2^0 1-generic set.

Corollary 2.2. *There is a computable coloring $c : \mathbb{N} \to 2$ such that if X is an infinite computable set, then $FS^{\leq 2}(X)$ is not monochromatic.*

The next corollary follows immediately.

Corollary 2.3. $\mathsf{HT}_2^{\leq 2}$ *is not provable in* RCA_0.

We now sharpen the previous corollary. Let SRT_2^2 be Stable Ramsey's Theorem for 2-colorings of pairs as defined in Statement 7.5 of [5].

Corollary 2.4. $\mathsf{RCA}_0 + \mathsf{B\Pi}_1^0 \vdash \mathsf{HT}_2^{\leq 2} \to \mathsf{SRT}_2^2$.

To prove the corollary, first let D_2^2 be the assertion that for every $\{0, 1\}$-valued function $f(x, s)$ such that for all x, $\lim_s f(x, s)$ exists there is an infinite set G and $j < 2$ such that $\lim_s f(x, s) = j$ for all $x \in G$. (The principle D_2^2 was defined in Statement 7.8 of [5].) Formalizing the proof of the theorem shows that $\mathsf{HT}_2^{\leq 2}$ implies the principle D_2^2 in $\mathsf{RCA}_0 + \mathsf{B\Pi}_1^0$. (We thank Denis Hirschfeldt for pointing out to us that $\mathsf{B\Pi}_1^0$ is apparently needed to show in the proof of Theorem 2.1 that there are infinitely many k such that $W \cap (\mathcal{O}_{k,1} \cup \mathcal{O}_{k,2})$ is nonempty from the facts that W is infinite and has finite intersection with each $\mathcal{O}_{k,i}$.) Then SRT_2^2 follows from $\mathsf{D}_2^2 + \mathsf{B\Pi}_1^0$ by the proof of Lemma 7.10 of [5]. (The latter proof contains a hidden use of hidden use of $\mathsf{B\Pi}_1^0$.) We do not know whether the use of $\mathsf{B\Pi}_1^0$ in this corollary is necessary, though it can be eliminated from the proof that D_2^2 implies SRT_2^2 by Theorem 1.4 of Chong, Lempp, and Yang [3].

3   Hindman's Theorem for Sums of Length at Most 3

We now strengthen the results of the previous section, at the cost of allowing longer sums and more colors. We start by considering $HT_4^{\leq 3}$ and then improve the results to $HT_3^{\leq 3}$.

Theorem 3.1. *There is a computable coloring $c : \mathbb{N} \to 4$ such that if X is infinite with $FS^{\leq 3}(X)$ monochromatic, then $0' \leq_T X$.*

Proof. Let $f : \mathbb{N} \to \mathbb{N}$ be a computable 1–1 function. We will define a computable coloring $c : \mathbb{N} \to 4$ such that if X is infinite with $FS^{\leq 3}(X)$ monochromatic, then X computes range(f).

For $n \in \mathbb{N}$, write $n = i_0 \cdot 3^{k_0} + \cdots + i_\ell \cdot 3^{k_\ell}$ with $k_0 < \cdots < k_\ell$ and each $i_j \in \{1, 2\}$. Define $\lambda(n) = k_0$, $\mu(n) = k_\ell$ and $i(n) = i_0$. We will use several properties of the functions $\lambda(n)$, $\mu(n)$ and $i(n)$. The following are all straightforward to establish.

(P1) If $\lambda(n) < \lambda(m)$, then $\lambda(n + m) = \lambda(n)$ and $i(n + m) = i(n)$.
(P2) If $\lambda(n) = \lambda(m)$ and $i(n) = i(m) = 1$, then $\lambda(n + m) = \lambda(n)$ and $i(n + m) = 2$.
(P3) If $\lambda(n) = \lambda(m)$ and $i(n) = i(m) = 2$, then $\lambda(n + m) = \lambda(n)$ and $i(n + m) = 1$.
(P4) If $\mu(n) < \lambda(m)$, then $\lambda(n + m) = \lambda(n)$ and $\mu(n + m) = \mu(m)$.

For $n = i_0 \cdot 3^{k_0} + \cdots + i_\ell \cdot 3^{k_\ell}$ with the i_j and k_j as above, we refer to the intervals (k_j, k_{j+1}) for $j < \ell$ as the *gaps* of n. A gap (a, b) of n is a *short gap* in n if there is a $y \leq a$ such that $y \in$ range(f) but there is no $x \leq b$ such that $f(x) = y$. (Note that whether a gap (a, b) in n is short does not depend on n.) A gap (a, b) of n is a *very short gap in n* if there is a $y \leq a$ for which there is an $x \leq \mu(n)$ with $f(x) = y$ but no $x \leq b$ for which $f(x) = y$. Note that we can computably determine the very short gaps in n but can only computably enumerate the short gaps in n.

For each n, we let $SG(n)$ be the number of short gaps in n and we let $VSG(n)$ be the number of very short gaps in n. As above, we can compute $VSG(n)$ but in general can only approximate $SG(n)$ in an increasing fashion as we discover the short gaps. We define our computable coloring by

$$
c(n) = \begin{cases} VSG(n) \bmod 2 & \text{if } i(n) = 1, \\ 2 + (VSG(n) \bmod 2) & \text{if } i(n) = 2. \end{cases}
$$

Let X be an infinite set such that $FS^{\leq 3}(X)$ is monochromatic. We establish the following two properties.

(P5) For all $n, m \in X$, $i(n) = i(m)$.
(P6) For $k \geq 0$, there is at most one $n \in X$ such that $\lambda(n) = k$.

(P5) holds because $i(n) = 1$ implies $c(n) \in \{0, 1\}$ and $i(m) = 2$ implies $c(m) \in \{2, 3\}$. (P6) holds since if $n \neq m \in X$ with $\lambda(n) = \lambda(m)$ (and by (P5), $i(n) = i(m)$), then by (P2) and (P3), $i(n + m) \neq i(n)$ contradicting (P5).

By (P6), we can assume without loss of generality (by computably thinning out X) that if $n, m \in X$ with $n < m$, then $\mu(n) < \lambda(m)$. The argument now proceeds almost identically to the proof of Theorem 2.2 in Blass, Hirst and Simpson with one minor change.

First, we claim that for all $n \in \mathrm{FS}^{\leq 2}(X)$, $\mathrm{SG}(n)$ is even. For this claim, it is important that n is a sum of at most two elements of X. In particular, this claim need not hold for an arbitrary element of $\mathrm{FS}^{\leq 3}(X)$.

Fix $m \in X$ such that $n < m$, $\mu(n) < \lambda(m)$ and for all $y \leq \mu(n)$, if $y \in \mathrm{range}(f)$, then there is an $x \leq \lambda(m)$ with $f(x) = y$. Since n is a sum of at most two elements of X, $n + m \in \mathrm{FS}^{\leq 3}(X)$. Because $\mu(n) < \lambda(m)$, the gaps in $n + m$ consist of the gaps in n, the gaps in m, and the gap $(\mu(n), \lambda(m))$. We want to count the number of very short gaps in $n + m$. By the choice of m, the gap $(\mu(n), \lambda(m))$ is not very short in $n + m$. By (P4), $\mu(n + m) = \mu(m)$, so each gap in m is very short in $n + m$ if and only if it is very short in m. Finally, if (a, b) is a gap in n, then $b \leq \mu(n)$ and hence by the choice of m, (a, b) is very short in $n + m$ if and only if it is short in n. Therefore, we have

$$\mathrm{VSG}(n + m) = \mathrm{SG}(n) + \mathrm{VSG}(m).$$

Since $c(m) = c(n+m)$, the parity of $\mathrm{VSG}(m)$ is equal to the parity of $\mathrm{VSG}(n+m)$ and therefore $\mathrm{SG}(n)$ is even.

The last claim we need is that if $n, m \in X$ with $n < m$, then for all $y \leq \mu(n)$, $y \in \mathrm{range}(f)$ if and only if there is an $x \leq \lambda(m)$ with $f(x) = y$. Note that this claim gives us a method to compute $\mathrm{range}(f)$ from X, completing the proof. To prove the claim, suppose for a contradiction that there is a $y \leq \mu(n)$ such that $y \in \mathrm{range}(f)$ but there is no $x \leq \lambda(m)$ with $f(x) = y$. In this case, the gap $(\mu(n), \lambda(m))$ is short in $n + m$. Therefore, because the gaps of n (respectively m) are short in $n + m$ if and only if they are short in n (respectively m), we have

$$\mathrm{SG}(n + m) = \mathrm{SG}(n) + \mathrm{SG}(m) + 1.$$

Since $n \neq m \in X$, we have $n + m \in \mathrm{FS}^{\leq 2}(X)$ and hence $\mathrm{SG}(n)$, $\mathrm{SG}(m)$ and $\mathrm{SG}(n + m)$ are all even, giving the desired contradiction. $\qquad \square$

Formalizing the proof of this theorem in RCA_0, we obtain the following corollary.

Corollary 3.2. $\mathrm{RCA}_0 \vdash \mathrm{HT}_4^{\leq 3} \to \mathrm{ACA}_0$.

We now improve the previous theorem and corollary from 4 colors to 3 colors.

Theorem 3.3. *There is a computable coloring* $c : \mathbb{N} \to 3$ *such that if* X *is infinite with* $\mathrm{FS}^{\leq 3}(X)$ *monochromatic, then* $0' \leq_T X$.

Proof. For any k and $i \in \{1,2,3,4,5,6\}$, let $\mathcal{O}_{k,i} = \{n : n \equiv i \cdot 7^k \mod 7^{k+1}\}$. Let i_n denote the first nonzero heptary bit of n, which occurs in the k_nth place, so that $n \in \mathcal{O}_{k_n,i_n}$. Color each $n \in \mathbb{N}$ red, green or blue as follows with the slash indicating a choice between two colors depending on whether $\mathrm{VSG}(n)$ is even or odd.

$$c(n) = \begin{cases} R/G & \text{if } \mathrm{VSG}(n) \text{ is even/odd and } i_n \equiv \pm 1 \mod 7, \\ G/B & \text{if } \mathrm{VSG}(n) \text{ is even/odd and } i_n \equiv \pm 2 \mod 7, \\ B/R & \text{if } \mathrm{VSG}(n) \text{ is even/odd and } i_n \equiv \pm 3 \mod 7. \end{cases}$$

Let $X \subseteq \mathbb{N}$ be an infinite set such that $\mathrm{FS}^{\leq 3}(X)$ is monochromatic. We claim that $X \cap \mathcal{O}_{k,i}$ cannot contain more than 2 elements. To prove this claim, assume that x, y, z are distinct elements of $X \cap \mathcal{O}_{k,i}$ and hence $x + y \in \mathcal{O}_{k,(2i \mod 7)} \cap \mathrm{FS}^{\leq 3}(X)$ and $x + y + z \in \mathcal{O}_{k,(3i \mod 7)} \cap \mathrm{FS}^{\leq 3}(X)$. Consider the following table of multiplication facts.

i	$2i \mod 7$	$3i \mod 7$
± 1	± 2	± 3
± 2	± 3	± 1
± 3	± 1	± 2

The table shows that $\mathrm{FS}^{\leq 3}(X)$ must contain elements from each of the sets $\mathcal{O}_{k,\pm 1 \mod 7}, \mathcal{O}_{k,\pm 2 \mod 7}$, and $\mathcal{O}_{k,\pm 3 \mod 7}$ (where $\mathcal{O}_{k,\pm 1 \mod 7} = \mathcal{O}_{k,1} \cup \mathcal{O}_{k,6}$ and similarly for the other sets). However, by the definition of the coloring c, it is not possible for a monochromatic set to intersect all three of these sets. Therefore, if $x, y, z \in X \cap \mathcal{O}_{k,i}$ are distinct, then $\mathrm{FS}^{\leq 3}(X)$ is not monochromatic, proving the claim.

By the claim, if $\mathrm{FS}^{\leq 3}(X)$ is monochromatic, then X must include elements n for which k_n is arbitrarily large. Also, we can computably thin X so that all of its elements n share the same value for i_n and thus share the same coloring convention, guaranteeing a common parity for $\mathrm{VSG}(n)$. From here, we proceed as in the proof of the previous theorem. $\qquad\square$

Corollary 3.4. $\mathrm{RCA}_0 \vdash \mathrm{HT}_3^{\leq 3} \to \mathrm{ACA}_0$.

4   Open Questions

Some of the open questions involve comparing bounded versions of Hindman's Theorem with special cases of Ramsey's Theorem. As usual, let RT_k^n denote Ramsey's Theorem for k-colorings of n-element sets. Thus, RT_k^n asserts that whenever the n-element subsets of $\mathbb{N}$ are k-colored, there is an infinite set $X \subseteq \mathbb{N}$ such that all n-element subsets of X have the same color.

We have provided some lower bounds on the strength and effective content of some versions of Hindman's Theorem for bounded sums. However, we do not know any upper bounds for the effective content and strength of $\mathrm{HT}_k^{\leq n}$ for

$n > 1, k > 1$ beyond those known from [2] for Hindman's Theorem itself. In particular, we do not know whether any of these bounded versions of Hindman's Theorem are provable in ACA_0, or whether any of them imply HT. We also do not know whether $\mathsf{HT}_2^{\leq 2}$ implies ACA_0 in RCA_0, or whether Ramsey's Theorem for 2-coloring of pairs RT_2^2 implies $\mathsf{HT}_2^{\leq 2}$ in RCA_0.

One might also consider the restriction of Hindman's Theorem to sums of length exactly n. Let $\mathsf{HT}_k^{=n}$ denote the assertion that for each k-coloring $c :$ $\mathbb{N} \to k$ there is an infinite set $X \subseteq \mathbb{N}$ such that $\{\sum F | F \subseteq X \text{ and } |F| = n\}$ is monochromatic. It is clear that RT_k^n implies $\mathsf{HT}_k^{=n}$ in RCA_0 for each $n, k \geq 1$, and indeed $\mathsf{HT}_k^{=n}$ is just the restriction of RT_k^n to colorings c of n-element sets F such that $c(F)$ depends only on $\sum F$. It follows from [8], Theorem 5.5, that each computable instance of $\mathsf{HT}_k^{=n}$ has a Π_n^0 solution. It is unknown whether this result can be improved to Σ_n^0 or better. It also remains open for each $n, k \geq 2$ whether $\mathsf{HT}_k^{=n}$ implies RT_k^n in RCA_0. We do not even know whether each computable instance of $\mathsf{HT}_2^{=2}$ has a computable solution.

Added Note (June 28, 2016): After this paper was submitted for publication, Denis Hirschfeldt pointed out to the authors that a result of Rumyantsev and Shen ([9], Corollary 2) can be used to give a quick proof that there is a computable instance of $\mathsf{HT}_2^{=2}$ with no Σ_2^0 solution. Indeed, he and Barbara Csima had used the same result from [9] to obtain a similar result with subtraction in place of addition. The details of the argument and further results will appear in a future paper.

References

1. Baumgartner, J.E.: A short proof of Hindman's theorem. J. Comb. Theor. Ser. A **17**, 384–386 (1974)
2. Blass, A.R., Hirst, J.L., Simpson, S.G.: Logical analysis of some theorems of combinatorics and topological dynamics In: Logic and Combinatorics (Arcata, California, 1985). Contemporary Mathematics, American Mathematical Society, Providence R.I., vol. 65, pp. 125 156 (1987)
3. Chong, C.T., Lempp, S., Yang, Y.: On the role of the collection principle for Σ_2 formulas in second-order reverse mathematics. Proc. Am. Math. Soc. **138**, 1093–1100 (2010)
4. Comfort, W.W.: Ultrafilters: some old and some new results. Bull. Am. Math. Soc. **83**, 417–455 (1977)
5. Cholak, P.A., Jockusch, C.G., Slaman, T.A.: On the strength of Ramsey's Theorem for pairs. J. Symbolic Logic **66**, 1–55 (2001)
6. Hindman, N.: Finite sums from sequences within cells of a partition of $\mathbb{N}$. J. Comb. Theor. Ser. A **17**, 1–11 (1974)
7. Hindman, N., Leader, I., Strauss, D.: Open problems in partition regularity. Comb. Probab. Comput. **12**, 571–583 (2003)
8. Jockusch, C.: Ramsey's theorem and recursion theory. J. Symbolic Logic **37**, 268–280 (1972)
9. Rumyantsev, A., Shen, A.: Probabilistic constructions of computable objects and a computable version of Lovász local lemma. Fund. Inform. Russ. Finnish Symp. Discrete Math. **132**, 1–14 (2014)

10. Simpson, S.: Subsystems of Second Order Arithmetic, 2nd edn. Cambridge University Press, Association for Symbolic Logic, New York (2009)
11. Soare, R.I.: Recursively Enumerable Sets and Degrees. Perspectives in Mathematical Logic. Springer, Heidelberg (1987)
12. Towsner, H.: A simple proof and some difficult examples for Hindman's Theorem. Notre Dame J. Formal Logic **53**(1), 53–65 (2012)

Reverse Mathematics of Matroids

Jeffry L. Hirst[1]([✉]) and Carl Mummert[2]

[1] Appalachian State University, Boone, NC 28608, USA
hirstjl@appstate.edu
[2] Marshall University, Huntington, WV 25755, USA
mummertc@marshall.edu
http://mathsci.appstate.edu/~jlh/, http://m6c.org/w/

Abstract. Matroids generalize the familiar notion of linear dependence from linear algebra. Following a brief discussion of founding work in computability and matroids, we use the techniques of reverse mathematics to determine the logical strength of some basis theorems for matroids and enumerated matroids. Next, using Weihrauch reducibility, we relate the basis results to combinatorial choice principles and statements about vector spaces. Finally, we formalize some of the Weihrauch reductions to extract related reverse mathematics results. In particular, we show that the existence of bases for vector spaces of bounded dimension is equivalent to the induction scheme for Σ^0_2 formulas.

Keywords: Reverse mathematics · Matroid · Induction · Graph · Connected component

MSC Subject Class (2000): 03B30 · 03F35 · 05B35

The study of computable and computably enumerable matroids links the work in this paper to the theme of this volume. The following incomplete survey establishes a framework for this connection and provides a few pointers into the substantial literature on computability and matroids.

In a seminal paper on computable and c.e. vector spaces, Metakides and Nerode [14] defined a vector space V_∞, the $\aleph_0$-dimensional vector space over a countable computable field F consisting of ω-sequences of elements of F with finite support, with point-wise operations. The lattice of c.e. subspaces of V_∞ is denoted $\mathcal{L}(V_\infty)$. A vector space V over a computable field F is *c.e. presented* if it has an effective enumeration of the vectors, partial recursive addition and scalar multiplication operations, and a c.e. congruence relation $\equiv$ such that the quotient $V/\equiv$ is a vector space. Metakides and Nerode proved that a vector space is c.e. presented if and only if it is computably isomorphic to V_∞/W for some $W \in \mathcal{L}(V_\infty)$.

Many proofs of results for $\mathcal{L}(V_\infty)$ rely on the structure of V_∞, hampering their adaptation to $\mathcal{L}(F_\infty)$, the lattice of c.e. algebraically closed subfields of a sufficiently computable algebraically closed field F_∞ with countably infinite

© Springer International Publishing AG 2017
A. Day et al. (Eds.): Downey Festschrift, LNCS 10010, pp. 143–159, 2017.
DOI: 10.1007/978-3-319-50062-1_12

transcendence degree. Matroids restrict interest to dependence properties common to both vector spaces and algebraic extensions, so proofs based on matroids can often be adapted to both vector space and field settings.

In computability theoretic papers, matroids are often described in terms of *Steinitz systems*. These are also called Steinitz *closure* systems [15] or Steinitz *exchange* systems [16]. Downey [8] defines a Steinitz system as a set U and a closure operator cl mapping subsets of U to subsets of U such that if $A, B \subset U$,

(1) $A \subset \mathrm{cl}(A)$,
(2) $A \subset B$ implies $\mathrm{cl}(A) \subset \mathrm{cl}(B)$,
(3) $\mathrm{cl}(\mathrm{cl}(A)) = \mathrm{cl}(A)$,
(4) $x \in \mathrm{cl}(A)$ implies that, for some finite $A' \subset A$, $x \in \mathrm{cl}(A')$, and
(5) (exchange) $x \in \mathrm{cl}(A \cup \{y\}) - \mathrm{cl}(A)$ implies $y \in \mathrm{cl}(A \cup \{x\})$.

As an intuitive example, we can think of U as a vector space and $\mathrm{cl}(A)$ as the linear span of the vectors in the set A. The Steinitz system (U, cl) has *computable dependence* if U is computable and there is a uniformly effective procedure that, when applied to $a, b_1, \ldots b_n \in U$, computes whether $a \in \mathrm{cl}(\{b_1, \ldots b_n\})$.

A central goal in computable matroid research is to discover algebraic properties of matroids with significant computability theoretic consequences. For example, the Steinitz system (U, cl) has the *closure intersection property* if whenever

- D is closed, that is, $\mathrm{cl}(D) = D$,
- A is independent over D, that is, for every $a \in A$, $a \notin \mathrm{cl}(D \cup A \setminus \{a\})$,
- B is independent over D, and
- $\mathrm{cl}(A \cup D) \cap \mathrm{cl}(B \cup D) = \mathrm{cl}(D)$,

then $A \cup B$ is independent over D. The system is *semiregular* (called *Downey's semiregularity* by Nerode and Remmel [16]) if no finite dimensional closed set is the union of two closed proper subsets. Downey established in his thesis [6] (abstracted in [7]) that if (U, cl) is semiregular and has the closure intersection property then the theory of $\mathcal{L}(\mathcal{U})$ is undecidable.

1   Reverse Mathematics

In his development of the theory of matroids, Whitney [18, Sect. 6] formulates matroids in terms of a ground set of elements and a specification of every set as being either dependent or independent. We define an *enumerated* matroid (*e-matroid*) to consist of a set and an enumeration of its finite dependent sets.

Definition 1. A (nontrivial) *e-matroid* is a pair (M, e) consisting of a set M and a function $e \colon \mathbb{N} \to [M]^{<\mathbb{N}}$ satisfying:

(1) The empty set is independent.

$$(\forall n)[e(n) \neq \emptyset]$$

(2) Finite supersets of dependent sets are dependent.

$$(\forall n)(\forall Y \in [M]^{<\mathbb{N}})[e(n) \subseteq Y \rightarrow \exists m(e(m) = Y)]$$

(3) If X is an independent set that is smaller than an independent set Y, then Y contains an element that is independent of X.

$$(\forall X, Y \in [M]^{<\mathbb{N}})(\text{if } |X| < |Y| \text{ and } (\forall n)[e(n) \neq X \wedge e(n) \neq Y]$$
$$\text{then } (\exists y \in Y)(\forall n)[e(n) \neq X \cup \{y\}])$$

An infinite set is independent if and only if each of its finite subsets is independent. We assume $e(0)$ is defined, so for every e-matroid, $M \neq \emptyset$ and there is at least one finite dependent set.

Although dependence in this setting is not directly related to linear combinations, it is still possible to formulate concepts of span and bases.

Definition 2. A subset B of an e-matroid (M, e) *spans* the e-matroid if adjoining any additional element to B produces a dependent set, that is,

$$(\forall x \in M)[x \notin B \rightarrow (\exists n)(e(n) \subseteq B \cup \{x\})].$$

A subset $B \subseteq M$ is a *basis* for the e-matroid if B is independent (that is, $(\forall n)[e(n) \nsubseteq B]$) and B spans M.

We can now state our first basis theorem. The analogous result showing the equivalence of ACA_0 and the existence of bases for vector spaces is included in Theorem 4.3 of Friedman, Simpson, and Smith [9].

Theorem 3. (RCA_0) *The following are equivalent:*

(1) ACA_0.
(2) *Every e-matroid has a basis.*

Proof. To show that (1) implies (2), fix an e-matroid (M, e). Let $m_0, m_1, \ldots$ be a non-repeating enumeration of M. Consider the function $g \colon \mathbb{N} \rightarrow [M]^{<\mathbb{N}}$ defined by $g(0) = \emptyset$ and for $i > 0$,

$$g(i) = \begin{cases} g(i-1) & \text{if } (\exists n)[e(n) = g(i-1) \cup \{m_{i-1}\}], \\ g(i-1) \cup \{m_{i-1}\} & \text{otherwise.} \end{cases}$$

By arithmetical comprehension, the union of the range of g exists; call this union B. Straightforward arguments verify that B is a basis for M.

To prove the converse, by Lemma III.1.3 of Simpson [17], it suffices to use (2) to prove the existence of the range of an arbitrary injection from $\mathbb{N}$ to $\mathbb{N}$. Suppose $f \colon \mathbb{N} \rightarrow \mathbb{N}$ is an injection. Let $M = \{(i, \varepsilon) : i \in \mathbb{N} \wedge \varepsilon < 2\}$ be the ground set for an e-matroid. Let $M_0, M_1, \ldots$ be an enumeration of $[M]^{<\mathbb{N}}$. Fix a bijective pairing function mapping $\mathbb{N} \times \mathbb{N}$ onto $\mathbb{N}$. Using the notation (j, k) for both the pair and its integer code, define $e((j, k)) = \{(f(j), 0), (f(j), 1)\} \cup M_k$. Because

$(f(j), 0) \in e((j, k))$, item (1) of the definition of an e-matroid holds. The inclusion of M_k in $e((j, k))$ ensures that supersets of dependent sets are dependent, satisfying item (2) of the definition. To verify item (3), suppose X and Y are finite independent sets with $|X| < |Y|$. If there is a $y \in X \cap Y$, then $X \cup \{y\} = X$ so $\forall n(e(n) \neq X \cup \{y\})$. Thus we need only consider the case where $X \cap Y = \emptyset$. We hypothesized that $|Y| > |X|$, so there must be a $y = (z, \varepsilon) \in Y$ such that for all ε', $(z, \varepsilon') \notin X$. Suppose by way of contradiction that $e(n) = X \cup \{y\}$ for some n. Then, for some j, we have $\{(f(j), 0), (f(j), 1)\} \subset X \cup \{y\}$. By the choice of y, we know $f(j) \neq z$, so $\{(f(j), 0), (f(j), 1)\} \subset X$, contradicting $(\forall n)[e(n) \neq X]$. Thus item (3) of the definition holds, and we have shown that (M, e) is an e-matroid.

Finally, we claim that if B is a basis for M, then k is in the range of f if and only if $(k, 0) \notin B$ or $(k, 1) \notin B$. First note that if $k = f(j)$ then, assuming 0 is the code for $\emptyset$, we have $e((j, 0)) = \{(k, 0), (k, 1)\}$. B is a basis, so $e((j, 0)) \not\subset B$, and thus $(k, 0) \notin B$ or $(k, 1) \notin B$. Conversely, if for example $(k, 0) \notin B$, then $(\exists n)[e(n) \subseteq B \cup \{(k, 0)\}]$. Because $e(n)$ is dependent and B is independent, both $(k, 1) \in e(n)$ and for all $f(j) \neq k$, at least one of $(f(j), 0)$ and $(f(j), 1)$ is not in $e(n)$. By the definition of e, $e(n)$ must contain both $(a, 0)$ and $(a, 1)$ for some a in the range of f, so k is in the range of f. A similar argument holds if $(k, 1) \notin B$, completing the proof of our claim. Because k is in the range of f if and only if $(k, 0) \notin B$ or $(k, 1) \notin B$, recursive comprehension suffices to prove the existence of the range of f, completing the reversal.

Our next result shows that if we add a hypothesis bounding the dimension of the matroid, the principle asserting the existence of a basis becomes weaker. The result also illustrates the interrelatedness of matroids and graph theory. We use the concept of rank to establish the dimensional bound.

Definition 4. We say the *rank* of an e-matroid (M, e) is *no more than n* if every subset of M of size n is dependent, that is, in the range of e.

Theorem 5. (RCA$_0$) *The following are equivalent:*

(1) For every n, every e-matroid of rank no more than n has a basis.

(2) For every n, if $G = (V, E)$ is a countable graph and every collection of n vertices contains at least one path connected pair, then G can be decomposed into its connected components.

(3) $\mathsf{I}\Sigma_2^0$, the induction scheme for Σ_2^0 formulas with set parameters.

Proof. Proofs that (2) implies (3) appear as Theorem 4.5 of Hirst [13] and also as Theorem 3.2 of Gura, Hirst, and Mummert [11]. Here, we will prove that (3) implies (1) and (1) implies (2).

To see that (3) implies (1), fix n and let (M, e) be an e-matroid of rank no more than n. Let $\psi(j)$ formalize the existence of an independent set of size $n - j$. If we use X_t to denote the finite subset of $\mathbb{N}$ encoded by t, then $\psi(j)$ can be written as $(\exists t)[|X_t| = n - j \wedge \forall k(e(k) \neq X_t)]$. Note that $\psi(j)$ is a Σ_2^0 formula, and the empty set witnesses $\psi(n)$. By the Σ_2^0 least element principle (which is easily deduced from the bounded Σ_2^0 comprehension, and is therefore a consequence of (3) by Exercise II.3.13 of Simpson [17]), there is a least j_0 such that $\psi(j_0)$. Let X_{t_0} witness $\psi(j_0)$. We claim that X_{t_0} is a basis. The range of e is closed under supersets, so no subset of X_{t_0} appears in the range of e. By the minimality of j_0, if $x \notin X_{t_0}$, then $X_{t_0} \cup \{x\}$ is dependent, so for some n, $e(n) = X_{t_0} \cup \{x\}$. Thus X_{t_0} spans M.

To show that (1) implies (2), let $G(V, E)$ be a graph in which every collection of n vertices contains at least one path connected pair. The independent sets of our e-matroid will consist of subsets of V with no path connected pairs. If G contains no edges, the identity function on V decomposes G into connected components. Suppose G has an edge connecting the vertices v_0 and v_1. Let $(V_i)_{i\in\mathbb{N}}$ be an enumeration of the finite subsets of V such that every subset appears infinitely often. Define $e(j)$ by $e(j) = V_j$ if there is some $t < j$ that encodes a path between two vertices of V_j, and $e(j) = \{v_0, v_1\}$ otherwise. It is easy to verify that (V, e) satisfies the first two clauses of the definition of an e-matroid. For the third clause, suppose X and Y are finite sets of vertices such that no pair in either set is path connected, and that $|X| < |Y|$. Suppose by way of contradiction that every vertex in Y is path connected to some vertex in X. RCA_0 can prove the existence of the function mapping each $y \in Y$ to some $x \in X$ to which it is path connected, and because $|X| < |Y|$, f must map two elements of Y to the same x. These two vertices of Y are path connected, yielding the desired contradiction. Thus (V, e) is a matroid. By (1), (V, e) has a basis, which is a maximal set of disconnected vertices in G. The function which is the identity on this basis and maps very other vertex of G to the element of the basis to which it is path connected is a decomposition of G into connected components. This decomposition is computable from the basis, so RCA_0 proves (1) implies (2). $\qquad\blacksquare$

2   Why e-Matroids?

We can define a matroid as a pair (M, D) where D is the set of all finite dependent subsets of M. In this case, D satisfies the set-based analogs of the three items in the definition of e-matroid. To express this definition within RCA_0, we represent each finite subset of M via its characteristic index. Using the set-based analog of the definition of basis, we can state and prove the following result.

Theorem 6. (RCA_0) *Every matroid has a basis.*

Proof. Let (M, D) be a matroid and let $m_1, m_2, \ldots$ be a non-repeating enumeration of M. Define a nested sequence of finite independent sets $\langle I_j \rangle_{j\in\mathbb{N}}$ as follows. Let $I_0 = \emptyset$. For $j > 0$, let $I_j = I_{j-1}$ if $I_{j-1} \cup \{m_j\} \in D$, and let $I_j = I_{j-1} \cup \{m_j\}$ otherwise. Define the basis B by $m_j \in B$ if and only if $m_j \in I_j$. To see that B is independent, suppose X is a finite dependent set. Let m_j be the element of largest index in X. If $X \setminus \{m_j\} \subset I_{j-1}$, then $m_j \notin I_j$, so $m_j \notin B$ and $X \not\subset B$. If $X \setminus \{m_j\} \not\subset I_{j-1}$ then $X \not\subset I_j$, so $X \not\subset B$. Summarizing, B has no finite dependent subsets, so B is independent. To see that B spans, fix $m_j \in M$. Either $m_j \in B$, or both $B \supset I_{j-1} \notin D$ and $I_{j-1} \cup \{m_j\} \in D$. In either case, m_j is in the span of B.

The preceding result can be viewed as a reverse mathematical reframing of the statement: *Every computably presented matroid has a computable basis.* This principle was stated by Crossley and Remmel [5, Sect. 5, Lemma 1], who describe it as common knowledge and implicit in the work of Metakides and Nerode [14]. The representations of the matroid by a computable dependence relationship or by a dependence algorithm for a Steinitz system with computable dependence are equivalent. The next theorem is a reverse mathematics analog of the fact that not every c.e. presented matroid is computably isomorphic to a computably presented matroid.

Theorem 7. (RCA_0) *The following are equivalent:*

(1) ACA_0.
(2) Every e-matroid is isomorphic to a matroid. That is, if (M, e) is an e-matroid, then there is a matroid (N, D) and a bijection $h \colon M \to N$ such that for all finite sets $X \subset M$, there is an n such that $e(n) = X$ if and only if $\{h(x) : x \in X\} \in D$.

Proof. To see that (1) implies (2), suppose (M, e) is an e-matroid. The range of e is arithmetically definable using e as a parameter, so ACA_0 proves the existence of the range as a set D. Then (M, D) is a matroid and the identity is the desired isomorphism.

To prove the converse, we capitalize on the construction from the proof of the reversal of Theorem 3. As in that proof, fix an injection f and construct the associated e-matroid (M, e). Apply (2) above to find a matroid (N, D) and an isomorphism $h \colon M \to N$. By the construction of (M, e), for each $k \in \mathbb{N}$, k is in the range of f if and only if $\{(k, 0), (k, 1)\}$ is in the range of e, which holds if and only if $\{h((k, 0)), h((k, 1))\} \in D$. Thus, the range of f is computable from D and h, completing the proof of the reversal.

In terms of Turing degrees, the previous theorem only shows that each c.e. presented matroid is computable from $\mathbf{0}'$. The next corollary shows that, if a c.e presented matroid is isomorphic to a computable matroid, the isomorphism may necessarily be noncomputable.

Corollary 8. *There is a c.e. presented matroid M, which is isomorphic to a computable matroid, such that if φ is any isomorphism between M and a computable matroid then $\mathbf{0}'$ is Turing computable from φ.*

Proof. Let f be any computable injection with a range that computes $\mathbf{0}'$. Use the construction of (M, e) from the proof of the reversal of Theorem 3. This is the desired c.e. presented matroid. The proof of Theorem 7 shows that any isomorphism between (M, e) and a computable matroid computes the range of f and consequently computes $\mathbf{0}'$. Since the range of f is both infinite and co-infinite, (M, e) is isomorphic to the computable matroid with ground set $\mathbb{N}$ and D consisting of all finite supersets of sets of the form $\{3k, 3k + 1\}$ where $k \in \mathbb{N}$.

A recent paper of Harrison-Trainor, Melnikov, and Montalbán [12] presents more results and applications for c.e. presented matroids. The pregeometries of their Sect. 2 are Steinitz systems.

3   Weihrauch Reducibility

In Theorem 6, we used reverse mathematics to study the problem of finding a basis for an e-matroid. In this section, we study the same problem using Weihrauch reducibility. For additional information on Weihrauch reducibility, see Brattka and Gherardi [2] and Dorais, Dzhafarov, Hirst, Mileti, and Shafer [4]. The following simplified definition of Weihrauch problems will be sufficient for our purposes.

Definition 9. A *Weihrauch problem* is a subset of $\mathbb{N}^{\mathbb{N}} \times \mathbb{N}^{\mathbb{N}}$, $\mathbb{N}^{\mathbb{N}} \times \mathbb{N}$, $\mathbb{N} \times \mathbb{N}^{\mathbb{N}}$, or $\mathbb{N} \times \mathbb{N}$. For a Weihrauch problem P, the "problem" is: given an "instance" $I \in \mathrm{dom}(P)$, produce a "solution" S with $(I, S) \in P$.

A Weihrauch problem P is *Weihrauch reducible* to a Weihrauch problem Q, written $P \leq_{\mathrm{W}} Q$, if there are computable functions or functionals Φ, Ψ such that, for all $S \in \mathrm{dom}(P)$, $\Phi(S) \in \mathrm{dom}(Q)$, and for all R such that $(\Phi(S), R) \in Q$, we have $(S, \Psi(R, S)) \in P$. If this can be done with a functional Ψ that does not depend on S, we say that P is *strongly Weihrauch reducible* to Q, written $P \leq_{\mathrm{sW}} Q$. The relations $\leq_{\mathrm{W}}$ and $\leq_{\mathrm{sW}}$ are reflexive and transitive, and thus they induce equivalence relations, which are denoted $\equiv_{\mathrm{W}}$ and $\equiv_{\mathrm{sW}}$, respectively.

The *parallelization* of a Weihrauch problem P is the problem

$$\widehat{P} = \{(f, g) : (f(n), g(n)) \in P \text{ for all } n \in \mathbb{N}\}$$

whose instances are sequences of instances of P and whose solutions are sequences of solutions corresponding to those instances.

Definition 10. We define the following Weihrauch principles. The first two are well known in the literature [1].

- $\mathsf{C}_{\mathbb{N}}$: closed choice for subsets of $\mathbb{N}$.

$$\mathsf{C}_{\mathbb{N}} = \{(f, n) : f \in \mathbb{N}^{\mathbb{N}}, n \notin \mathrm{range}(f)\}$$

- $\widehat{\mathsf{C}}_{\mathbb{N}}$: the parallelization of $\mathsf{C}_{\mathbb{N}}$.

$$\widehat{\mathsf{C}}_{\mathbb{N}} = \{(f, g) : ((f)_n, g(n)) \in \mathsf{C}_{\mathbb{N}} \text{ for all } n \in \mathbb{N}\}$$

- GAC: the graph antichain problem. For a countable graph G, an antichain is a set of vertices no two of which are connected by a path in G. Letting $\mathrm{Max}(G)$ be the set of maximal antichains of G, we have

$$\mathsf{GAC} = \{(G, A) : G \text{ is a countable graph}, A \in \mathrm{Max}(G)\}$$

- EMB: the e-matroid basis problem.

$$\mathsf{EMB} = \{(M, B) : M \text{ is a countable e-matroid}, B \text{ is a basis for } M\}$$

- VSB: the vector space basis problem, for countable vector spaces over countable fields, coded as in Definition III.4.1 of Simpson [17].

$$\mathsf{VSB} = \{(V, B) : V \text{ is a countable vector space and } B \text{ is a basis for } V\}$$

For each $n > 1$ in $\mathbb{N}$, we define the following restricted principles:

- GAC_n: the restriction to GAC to graphs with n connected components.
- EMB_n: the restriction of EMB to e-matroids with a basis of size n.
- VSB_n: the restriction of VSB to vector spaces with dimension n.

In previous work [11], we considered another well known Weihrauch problem, LPO:

$$LPO = \{(f, n) : f \in [\mathbb{N}]^{<\mathbb{N}} \text{ and } f(n) = 0 \leftrightarrow (\exists m)[f(m) = 0]\}$$

The following lemma shows that the parallelization of LPO is strict Weihrauch equivalent to the parallelization of $C_\mathbb{N}$. This equivalence is implicit in work of Brattka and Gherardi [2,3], but the reductions obtained by combining their results are very indirect. The next lemma provides a pair of direct reductions.

Lemma 11. $\widehat{C_\mathbb{N}}$ *is strongly Weihrauch equivalent to* $\widehat{LPO}$.

Proof. First, suppose we are given an instance f of $C_\mathbb{N}$. The function f enumerates the complement of some nonempty set. We form a sequence (p_n) of instances of LPO such that p_n has 0 in its range if and only if n is in the range of f. Then, given solutions to the instance $(p_n)_{n \in \mathbb{N}}$ of $\widehat{LPO}$, we can search effectively for the least n such that p_n does not have 0 in its range, which will be the least n not in the range of f. Thus, by effective dovetailing, $\widehat{C_\mathbb{N}}$ is strict Weihrauch reducible to $\widehat{LPO}$.

For the converse, we first reduce LPO to $C_\mathbb{N}$, as follows. Given an instance p of LPO, we enumerate in stages the complement of a nonempty set $A = A(p)$. If $p(0) > 0$, we enumerate 1 into the complement of A. Then if $p(1) > 0$ we enumerate 2 into the complement of A. We continue in this way. If we ever find that $p(n) = 0$ for some n, we enumerate 0 into the complement of A, after which we do not enumerate anything else into the complement, so we will have $A = \{n + 1, n + 2, \ldots\}$. On the other hand, if 0 is not in the range of p, then we continue enumerating elements into the complement of A, so that we will obtain $A = \{0\}$. Hence, if we view A as an instance of $C_\mathbb{N}$, we can determine whether $(\exists m)[p(m) = 0]$ by looking at the value of any solution. Thus LPO is strict Weihrauch reducible to $C_\mathbb{N}$, and so the parallelization of LPO is strict Weihrauch reducible to the parallelization of $C_\mathbb{N}$.

Theorem 12. *The following strong Weihrauch equivalences hold:*

$$GAC \equiv_{sW} EMB \equiv_{sW} VSB \equiv_{sW} \widehat{C_\mathbb{N}}.$$

Proof. Gura, Hirst, and Mummert [11] proved that $GAC \equiv_{sW} \widehat{C_\mathbb{N}}$. Therefore, it is sufficient to establish the following four reductions:

$$GAC \leq_{sW} EMB \leq_{sW} \widehat{C_\mathbb{N}}, \qquad \widehat{C_\mathbb{N}} \leq_{sW} VSB \leq_{sW} EMB.$$

Three of these reductions are straightforward. First, to show that $VSB \leq_{sW} EMB$, modify the construction used to prove (1) implies (2) in Theorem 5. Given a vector space with vector set V and zero vector 0_V, let $(V_i)_{i \in \mathbb{N}}$ be an enumeration of all the finite subsets of V in which each subset appears infinitely often. Define $e \colon \mathbb{N} \to [V]^{<\mathbb{N}}$ by setting $e(j) = V_j = \{v_0, \ldots, v_k\}$ if there is a sequence of field elements $\{a_0, \ldots, a_k\}$ with canonical code less than j such that $\sum_{i \le k} a_i v_i = 0$, and set $e_j = \{0_V\}$ otherwise. Because e enumerates the finite dependent subsets of V, it is easy to verify that (V, e) is a matroid and any basis for the matroid is a basis for the vector space.

Second, to show that $\mathsf{GAC} \le_{\mathrm{sW}} \mathsf{EMB}$, let $G = (V, E)$ be a graph. We wish to ensure that G has at least one edge. To this end, choose a vertex $v_1 \in V$ and add a new vertex v_0 to V and a new edge (v_0, v_1) to E, yielding a graph $G' = (V', E')$. Construct a matroid (V', e) as in the proof that (1) implies (2) in Theorem 5. (Note that in that argument, the bound on the number of components is used only to bound the rank of the matroid.) As in that proof, any basis for (V', e) is a maximal set of disconnected vertices of G'. If v_0 is in the basis, it can be replaced by v_1 to form a new basis which is a maximal set of disconnected vertices of G.

Third, to show that $\mathsf{EMB} \le_{\mathrm{sW}} \widehat{\mathsf{C}}_{\mathbb{N}}$, let (M, e) be a countable e-matroid. Construct an enumeration e' of the finite sets in $\mathrm{Range}(e) \cup \{F \mid F \not\subseteq M\}$. Then $M' = (\mathbb{N}, e')$ is an e-matroid with domain $\mathbb{N}$ which has exactly the same independent sets and exactly the same bases as M. Fix an enumeration $(F_n)_{n \in \mathbb{N}}$ of $[\mathbb{N}]^{<\mathbb{N}}$. Define an instance $(f_n)_{n \in \mathbb{N}}$ of $\widehat{\mathsf{C}}_{\mathbb{N}}$ by

$$f_n(j) = \begin{cases} j+1 & \text{if } (\forall t < j)[e_{M'}(t) \ne F_n], \\ 0 & \text{otherwise.} \end{cases}$$

Note that F_n is independent if and only if $\mathrm{Range}(f_n) = \mathbb{N} \setminus \{0\}$. Also, if F_n is dependent, then $0 \in \mathrm{Range}(f_n)$. Thus, if g is a solution to this instance of $\widehat{\mathsf{C}}_{\mathbb{N}}$, then for every $n \in \mathbb{N}$, F_n is independent if and only if $g(n) = 0$. To simplify notation, if F is finite, we can let n be the smallest value such that $F_n = F$, and write $g(F) = g(n)$. We define the basis in stages. Let $B_0 = \{0\}$ if $g(\{0\}) = 0$ and $B_0 = \emptyset$ otherwise. If B_j is defined, let $B_{j+1} = B_j \cup \{j+1\}$ if $g(B_j \cup \{j+1\}) = 0$ and $B_{j+1} = B_j$ otherwise. Then $B = \{j \mid j \in B_j\}$ is a basis for M' and thus also for M.

It remains to show that $\widehat{\mathsf{C}}_{\mathbb{N}} \le_{\mathrm{sW}} \mathsf{VSB}$. We adapt the construction presented by Simpson [17, Theorem III.4.3] showing that the principle "every countable vector space over $\mathbb{Q}$ has a basis" is equivalent to ACA_0 in the sense of reverse mathematics. The proof presented by Simpson shows, more specifically, that given an injective function $f \colon \mathbb{N} \to \mathbb{N}$ we may uniformly compute a $\mathbb{Q}$-vector space V_f such that the range of f is uniformly computable from any basis of V_f. This shows, in particular, that $\mathsf{C}_{\mathbb{N}} \le_{\mathrm{sW}} \mathsf{VSB}$.

To complete the proof, it is sufficient for us to verify that $\widehat{\mathsf{VSB}} \le_{\mathrm{sW}} \mathsf{VSB}$, because then we have $\widehat{\mathsf{C}}_{\mathbb{N}} \le_{\mathrm{sW}} \widehat{\mathsf{VSB}} \le_{\mathrm{sW}} \mathsf{VSB}$. The proof uses an effective direct sum construction. Given a sequence $(V_n)_{n \in \mathbb{N}}$ of countable vector spaces, we may assume without loss of generality that their underlying sets of vectors are pairwise disjoint. We may then form a countable vector space V whose elements are finite formal $\mathbb{Q}$-linear combinations of the form

$$a_1u_1 + \cdots + a_mu_m$$

where $a_i \in \mathbb{Q}$ and $u_i \in V_i$ for $i \le m$. The scalar multiplication on V is the obvious one, and the vector addition is so that

$$\left(\sum_{i \le m} a_iu_i\right) + \left(\sum_{i \le n} b_iv_i\right) = \sum_{i \le \max m,n} (a_iu_i + b_iv_i)$$

where each addition $a_iu_i + b_iv_i$ is carried out in V_i, and terms that did not appear in the left are treated vacuously as zero vectors. Then V is a countable vector space that is uniformly computable from the sequence $(V_n)_{n \in \mathbb{N}}$. Moreover, if B is a basis of V then $B \cap V_i$ is a basis of V_i for each $i \in \mathbb{N}$. To see this, note that on one hand $B \cap V_i$ must span V_i for each i, and on the other hand any dependency of the set $B \cap V_i$ within V_i would induce a dependency of B within V.

We next consider the restricted versions of two principles from Theorem 12.

Theorem 13. *For $n \ge 2$, the following equivalences hold:*

$$\mathsf{GAC}_n \equiv_{sW} \mathsf{EMB}_n \equiv_{sW} \mathsf{C}_{\mathbb{N}}.$$

Proof. Let $n \ge 2$ be fixed for the remainder of this proof. Gura, Hirst, and Mummert [11, Theorem 6.6] proved that $\mathsf{GAC}_n \equiv_{sW} \mathsf{C}_{\mathbb{N}}$. Therefore, it is sufficient to establish the reductions $\mathsf{GAC}_n \le_{sW} \mathsf{EMB}_n$ and $\mathsf{EMB}_n \le_{sW} \mathsf{C}_{\mathbb{N}}$.

The reduction $\mathsf{GAC}_n \le_{sW} \mathsf{EMB}_n$ follows from the proof of Theorem 12, because the construction there produces an e-matroid whose dimension is the same as the number of components of the graph.

To show that $\mathsf{EMB}_n \le_{sW} \mathsf{C}_{\mathbb{N}}$, let (M, e) be an e-matroid with a basis of size n. As in the proof of Theorem 12, construct an enumeration e' of the finite sets in $\mathrm{Range}(e) \cup \{F \mid F \not\subseteq M\}$, so that $M' = (\mathbb{N}, e')$ is an e-matroid with domain $\mathbb{N}$ and with exactly the same bases as (M, e). Let $(F_i)_{i \in \mathbb{N}}$ be an enumeration of $[\mathbb{N}]^n$ in which each set appears infinitely often. Let $(G_i)_{i \in \mathbb{N}}$ be an enumeration of $[\mathbb{N}]^n$ in which each set appears exactly once, and such that $G_0 = F_0$.

We define an instance f of $\mathsf{C}_{\mathbb{N}}$ inductively along with an auxiliary sequence $(m_j)_{j \in \mathbb{N}}$. At stage 0, let $m_0 = 0$ and $f(0) = 1$. At stage $j + 1$, suppose m_j and $f(j)$ have been defined. If $e'(j) = F_{m_j}$, set $f(j + 1) = m_j$, let k be the smallest integer such that $(\forall t \le j)[e'(t) \ne G_k]$ and set

$$m_{j+1} = (\mu s)[G_k = F_s \wedge (\forall t \le j)(s > f(t))].$$

At stage $j + 1$, if $e'(j) \ne F_{m_j}$, set $f(j + 1) = \min(\mathbb{N} \setminus (\{f(t) \mid t \le j\} \cup \{m_j\}))$ and let $m_{j+1} = m_j$.

The range of f will include all integers except one, namely some m such that $F_m = G_k$ for the least k for which G_k is a basis for M. Thus F_m will be a basis for M, as desired.

The next lemma, which is well known, extends the list of principles in Theorem 13 slightly, simplifying the proof of the next theorem.

Lemma 14. *Let $C_\mathbb{N}^u$ denote the restriction of $C_\mathbb{N}$ to functions for which the complement of the range consists of a unique natural number. Then $C_\mathbb{N}^u \equiv_{sW} C_\mathbb{N}$.*

Proof. Because $C_\mathbb{N}^u$ restricts $C_\mathbb{N}$ to a smaller class of inputs, $C_\mathbb{N}^u \leq_{sW} C_\mathbb{N}$. To prove $C_\mathbb{N} \leq_{sW} C_\mathbb{N}^u$, suppose $f \colon \mathbb{N} \to \mathbb{N}$ is not surjective. In the following construction, we will conflate the pair (i, j) with its integer code via a fixed bijection between $\mathbb{N}$ and $\mathbb{N} \times \mathbb{N}$. Define $g \colon \mathbb{N} \to \mathbb{N}$ by the following moving marker construction. Let $m_0 = (0,0)$ be the initial marker. Suppose $m_k = (m_k^0, m_k^1)$ has been defined. If $f(k) \neq m_k^0$, set $m_{k+1} = m_k$ and set $g(k)$ to the least code for a pair not included in $\{g(j) : j < k\}$. If $f(k) = m_k^0$, define a pair (y_0, t_0) so that

$$y_0 = (\mu\, y. \leq k + 1)(\forall j \leq k)[f(j) \neq y],$$
$$t_0 = (\mu t)(\forall j < k)[g(j) \neq (y_0, t)],$$

and then set $m_{k+1} = (y_0, t_0)$ and $g(k) = m_k$.

Intuitively, if y is the smallest natural number not in the range of f, then at some stage in the construction the marker is set to (y, n) for some n, and does not move after that point. The code (y, n) is not in the range of g, but every other code and consequently every other natural number is in the range of g. Thus g satisfies the input requirements for $C_\mathbb{N}^u$, and the process yields (y, n) as an output. The number y (retrievable by a projection function) is a solution to $C_\mathbb{N}$ for input f.

The following theorem adds the fixed dimension vector space basis problem to the list of equivalent problems of Theorem 13.

Theorem 15. *For $n \geq 2$, $VSB_n \equiv_{sW} C_\mathbb{N}$.*

Proof. By Theorem 13, $EMB_n \leq_{sW} C_\mathbb{N}$. In the proof of Theorem 12, the argument showing $VSB \leq_{sW} EMB$ preserves the dimension of input vector space, and so shows $VSB_n \leq_{sW} EMB_n$. By transitivity, $VSB_n \leq_{sW} C_\mathbb{N}$.

Next we will show that $C_\mathbb{N}^u \leq_{sW} VSB_2$. Our proof uses ideas and notation from the proof of Theorem III.4.2 of Simpson [17]. Fix $f \colon \mathbb{N} \to \mathbb{N}$ with the range of f including all of $\mathbb{N}$ except for one value. Let V_0 be the set of all formal sums $\sum_{i \in I} q_i x_i$ with I finite and $0 \neq q_i \in \mathbb{Q}$. We can identify formal sums with their sequence codes, yielding a well-ordering on V_0. Without loss of generality, we may assume that x_i is minimal in this ordering among all vectors with a nonzero coefficient on x_i. As in Simpson's proof, let $x_m' = x_{2f(m)} + (m + 1)x_{2f(m)+1}$ and $X' = \{x_m' : m \in \mathbb{N}\}$. Let U_0 denote the subspace consisting of the linear span of X'. Note that $\sum_{i \in I} q_i x_i \in U_0$ if and only if

$$(\forall n)\left[(q_{2n} \neq 0 \to f(q_{2n+1}/q_{2n} - 1) = n) \wedge (q_{2n} = 0 \to q_{2n+1} = 0)\right],$$

so U_0 is computable from f. Let V_1 be V_0/U_0, where a vector v is in V_1 if and only if it is the element of $\{v - u : u \in U_0\}$ which is least in the well ordering on V_0. Only finitely many sequence codes are less than the code for v, so V_1 is computable.

By our choice of ordering and the construction of U_0, for every $i \in \mathbb{N}$, $x_{2i} \in V_1$. Let U_1 be the linear span of $\{x_{2i} : i \in \mathbb{N}\}$ in V_1. Then U_1 is a vector subspace of V_1 computable from f, and we may construct the quotient space $V = V_1/U_1$, using minimal representatives as before. For any $j \in \mathbb{N}$,

$$x_0 = x_{2f(j)+1} - (-\frac{1}{j+1}x_{2f(j)} - x_0) - \frac{1}{j+1}(x_{2f(j)} + (j+1)x_{2f(j)+1}).$$

The vector $-\frac{1}{j+1}x_{2f(j)} - x_0$ is in U_1 and $\frac{1}{j+1}(x_{2f(j)} + (j+1)x_{2f(j)+1})$ is in U_0, so x_0 and $x_{2f(j)+1}$ correspond to the same vector in V. The range of f excludes only one element, so the dimension of V is 2. Let $\{v_1, v_2\}$ be a basis for V. Let P be the finite collection of odd indices in the formal sums for v_1 and v_2, and let $R = \{m : 2m + 1 \in P\}$. Exactly one m in R does not appear in the range of f. Thus, for exactly one m in R, $\{x_0, x_{2m+1}\}$ is linearly independent. Sequentially enumerate linear combinations of the form $q_0 x_0 + q_1 x_{2m+1}$, ejecting values from R corresponding to linear combinations that equal 0 in V. The last value left in R is the sole natural number that is not in the range of f. Thus $\mathsf{C}_{\mathbb{N}}^u \leq_{\mathrm{sW}} \mathsf{VSB}_2$. By Lemma 14, $\mathsf{C}_{\mathbb{N}} \leq_{\mathrm{sW}} \mathsf{VSB}_2$.

To prove $\mathsf{C}_{\mathbb{N}} \leq_{\mathrm{sW}} \mathsf{VSB}_n$ for $n > 2$, add $n - 1$ dummy vectors to the basis of V_0 in the preceding argument.

The reduction of EMB_n to $\mathsf{C}_{\mathbb{N}}$ in the proof of Theorem 13 relies heavily on knowing the precise dimensions (in the appropriate sense) of the objects being studied. This suggests a variation in which we only place an upper bound on their dimensions. We begin with definitions of bounded versions of some Weihrauch principles.

Definition 16. We define the following Weihrauch principles. In the first three principles, the output can be viewed either as a canonical code for a finite set, or equivalently as a set together with the integer corresponding to its cardinality.

– $\mathsf{EMB}_{<\omega}$: the bounded e-matroid basis problem.

$$\mathsf{EMB}_{<\omega} = \{(n, M, B) : n \in \mathbb{N},\ M \text{ is an e-matroid}, \mathrm{rank}(M) \leq n,$$
$$\text{and } B \text{ is a basis for } M\}$$

– $\mathsf{GAC}_{<\omega}$: The bounded graph antichain problem. Letting $\mathrm{Max}(G)$ be the set of maximal antichains of G, we have

$$\mathsf{GAC}_{<\omega} = \{(n, G, A) : n \in \mathbb{N},\ G \text{ is a graph},$$
$$\text{each set of } n \text{ vertices has a path connected pair},$$
$$\text{and } A \in \mathrm{Max}(G)\}$$

– $\mathsf{C}_{\mathrm{max}}^{\subseteq}$: Picking a maximal element (relative to the containment partial ordering) in the complement of an enumeration of finite nonempty sets whose range includes all sets larger than some bound.

$$C_{\text{max}}^C = \{(n, f, X) \ : \ n \in \mathbb{N}, \ f \colon \mathbb{N} \to [\mathbb{N}]_{\neq\emptyset}^{<\mathbb{N}}, \ X \in [\mathbb{N}]^{<\mathbb{N}},$$

$$\text{range}(f) \text{ includes all sets of cardinality} \geq n,$$

$$X \notin \text{range}(f), \text{ and}$$

$$(\forall Y \in [\mathbb{N}]^{<\mathbb{N}})[Y \supsetneq X \to Y \in \text{range}(f)]\}$$

- $C_{\text{max}}^\#$: Picking an element of maximal cardinality in the complement of an enumeration of finite nonempty sets whose range includes all sets larger than some bound.

$$C_{\text{max}}^\# = \{(n, f, X) \ : \ n \in \mathbb{N}, \ f \colon \mathbb{N} \to [\mathbb{N}]_{\neq\emptyset}^{<\mathbb{N}}, \ X \in [\mathbb{N}]^{<\mathbb{N}},$$

$$\text{range}(f) \text{ includes all sets of cardinality} \geq n,$$

$$X \notin \text{range}(f), \text{ and}$$

$$(\forall Y \in [\mathbb{N}]^{<\mathbb{N}})[|Y| > |X| \to Y \in \text{range}(f)]\}$$

Theorem 17. *The following strong Weihrauch equivalences hold:*

$$\mathsf{EMB}_{<\omega} \equiv_{\text{sW}} \mathsf{GAC}_{<\omega} \equiv_{\text{sW}} \mathsf{C}_{max}^C \equiv_{\text{sW}} \mathsf{C}_{max}^\#.$$

Proof. We will prove each of the following reductions, proceeding from right to left:

$$\mathsf{C}_{\text{max}}^C \leq_{\text{sW}} \mathsf{C}_{\text{max}}^\# \leq_{\text{sW}} \mathsf{GAC}_{<\omega} \leq_{\text{sW}} \mathsf{EMB}_{<\omega} \leq_{\text{sW}} \mathsf{C}_{\text{max}}^C.$$

To prove $\mathsf{EMB}_{<\omega} \leq_{\text{sW}} \mathsf{C}_{\text{max}}^C$, suppose (M, e) is an e-matroid such that every subset of M of size at least n is in the range of e. Let $\{X_j : j \in \mathbb{N}\}$ be an enumeration of $[\mathbb{N}]^{<\mathbb{N}}$ and let (i, j) denote the output of a bijective pairing function. Note that every $m \in \mathbb{N}$ has a unique representation of the form $2(i, j) + \varepsilon$ where $i, j \in \mathbb{N}$ and $\varepsilon \in \{0, 1\}$. Define $f \colon \mathbb{N} \to [\mathbb{N}]^{<\mathbb{N}}$ by

$$f(2(i, j) + \varepsilon) = \begin{cases} X_j & \text{if } \varepsilon = 0 \land i \notin M \land i \in X_j, \\ e((i, j)) & \text{otherwise.} \end{cases}$$

The range of f consists of the range of e plus all finite sets containing any elements of the complement of M. Apply $\mathsf{C}_{\text{max}}^C$ to f to obtain a finite set $B \subseteq \mathbb{N}$ in the complement of the range of f that is maximal with respect to the containment partial ordering. The range of f includes all finite sets containing elements of the complement of M, so $B \subseteq M$. Furthermore, the range of f includes the range of e, so B is independent in (M, e). By maximality, B spans (M, e), so B is a basis for (M, e).

To prove $\mathsf{GAC}_{<\omega} \leq_{\text{sW}} \mathsf{EMB}_{<\omega}$, emulate the reduction of GAC to EMB from the proof of Theorem 12. Because G has at most n connected components, every set of $n + 1$ elements in the related matroid is dependent and so appears in the range of the enumeration.

To prove $\mathsf{C}_{\text{max}}^\# \leq_{\text{sW}} \mathsf{GAC}_{<\omega}$, suppose $f \colon \mathbb{N} \to [\mathbb{N}]_{\neq\emptyset}^{<\mathbb{N}}$ and the range of f includes all finite subsets of cardinality at least n. For each b with $1 \leq b < n$, let $g_b \colon \mathbb{N} \to [\mathbb{N}]^{<\mathbb{N}}$ be an enumeration of all subsets of $\mathbb{N}$ of cardinality exactly b.

We will construct a graph G consisting of $n-1$ subgraphs each with one or two connected components. The vertices of G are $\{u_j^b, v_j^b : 1 \le b < n \wedge j \in \mathbb{N}\}$. For each b with $1 \le b < n$ and each $j \in \mathbb{N}$, add the edge (u_j^b, u_{j+1}^b) to the edge set E of G. For each b with $1 \le b < n$, define $k_0^b = 0$. Suppose k_j^b is defined. If $(\exists t \le j)[f(t) = g_b(k_j^b)]$, add (v_j^b, u_j^b) to E and set $k_{j+1}^b = k_j^b + 1$. Otherwise, if $(\forall t \le j)[f(t) \neq g_b(k_j^b)]$, add (v_j^b, v_{j+1}^b) to E and set $k_{j+1}^b = k_j^b$. Note that the graph G is uniformly computable from f.

Apply $\mathsf{GAC}_{<\omega}$ to find a maximal (finite) antichain D in G. Let b_0 be the largest number less than n such that D contains two vertices with superscript b_0. (If no such b_0 exists, $\emptyset$ is the largest set in the complement of the range of f.) At least one of these vertices must be $v_j^{b_0}$ for some j. Let j_0 be the largest value such that $v_{j_0}^{b_0} \in D$. Then $g_{b_0}(k_{j_0}^{b_0})$ is a set of maximal cardinality in the complement of the range of f.

To conclude the proof, we need only show that $\mathsf{C}_{\max}^\mathsf{C} \le_{\mathrm{sW}} \mathsf{C}_{\max}^\#$. Any f and n satisfying the hypotheses of $\mathsf{C}_{\max}^\mathsf{C}$ also satisfy those of $\mathsf{C}_{\max}^\#$. Any subset in the complement of the range of f that is maximal in cardinality is also maximal with respect to the containment partial ordering, so the identity functionals witness the desired reduction.

We close our discussion of Weihrauch reducibility with the following theorem that adds $\mathsf{VSB}_{<\omega}$ to the equivalences of Theorem 17. Here $\mathsf{VSB}_{<\omega}$ is the problem which, given an input of $n \in \mathbb{N}$ and a vector space in which every set of n vectors is linearly dependent, returns a basis for the vector space.

Theorem 18. $\mathsf{VSB}_{<\omega} \equiv_{\mathrm{sW}} \mathsf{C}_{max}^\mathsf{C}$.

Proof. By Theorem 17, $\mathsf{EMB}_{<\omega} \le_{\mathrm{sW}} \mathsf{C}_{\max}^\mathsf{C}$. The proof of $\mathsf{VSB} \le_{\mathrm{sW}} \mathsf{EMB}$ in Theorem 12 preserves dimension, so that argument also witnesses that $\mathsf{VSB}_{<\omega} \le_{\mathrm{sW}} \mathsf{EMB}_{<\omega}$. By transitivity, $\mathsf{VSB}_{<\omega} \le_{\mathrm{sW}} \mathsf{C}_{max}^\mathsf{C}$.

Next we will adapt arguments from the proofs of Lemma 14 and Theorem 15 to show that $\mathsf{C}_{\max}^\# \le_{\mathrm{sW}} \mathsf{VSB}_{<\omega}$. Fix n and $f\colon \mathbb{N} \to [\mathbb{N}]^{<\mathbb{N}}$ such that the range of f includes all sets of cardinality $\ge n$. For each $j < n$, let h_j be a bijective enumeration of $\{X : X \subset \mathbb{N} \wedge j \le |X| < n\} \times \mathbb{N}$. Emulating the moving marker construction of Lemma 14, for each $j < n$ define g_j such that either the range of f includes all sets of cardinality k for $j \le k < n$ and g_j is surjective or the unique value not in the range of g_j is some m such that $h_j(m) = (X_0, m_0)$ where $j \le |X_0| < n$ and X_0 is in the complement of the range of f. (For use in the proof of Theorem 19, note that the convergence of the moving marker construction can be formally proved using the collection principle $\mathsf{B}\Sigma_1^0$, which is provable in RCA_0.)

Now we carry out an n-fold analog of the vector space construction in the proof of Theorem 15. The goal of the construction is to form a space V as a direct sum of subspaces W_i, $i < n$, such that if j_0 is the largest size of a set omitted from the range of f, then the dimension of W_i is 1 for $i > j_0$ and the dimension is 2 for $i \le j_0$. This will ensure that the dimension of V is finite, and moreover will allow us to compute the value of j_0 if we know the exact dimension of V.

Let V_0 be the set of formal sums $\sum_{(i,k)\in I_k\times[0,n)} q_{(i,k)}x_{(i,k)}$ where, for each $k < n$, each I_k is finite and $0 \neq q_{(i,k)} \in \mathbb{Q}$. Identifying $h_j(m) = (X_0, m_0)$ with the integer code for the pair, for each $k < n$ and each m, let $x'_{(m,k)} = x_{(2h_k(m),k)} + (m+1)x_{(2h_k(m)+1,k)}$ and $X' = \{x'_{(m,k)} : m \in \mathbb{N} \wedge k < n\}$. Let U_0 be the linear span of X' and set $V_1 = V_0/U_0$. Let U_1 be the linear span in V_1 of $\{x_{(2m,k)} : m \in \mathbb{N} \wedge k < n\}$ and let $V = V_1/U_1$. Then V has a two dimensional subspace corresponding to each $j < n$ such that the range of f omits a set of cardinality k with $j \leq k < n$, and a one dimensional subspace corresponding to each $j < n$ such that f maps $\mathbb{N}$ onto the sets of cardinality k with $j \leq k < n$. Thus the dimension of V is between n and $2n$, and any set of $2n + 1$ vectors is linearly dependent.

(For use in the proof of Theorem 19, note that the claim that any collection of $2n + 1$ vectors of V is linearly dependent can be proved in RCA$_0$ as follows. Fix a set of $2n + 1$ nonzero vectors, $S = \{u_0, \ldots, u_{2n}\}$. Let B_0 be the finite set of those vectors of the form $x_{(i,k)}$ that appear in the sums representing each u_i. Because S is finite, Σ_1^0 induction suffices to find the smallest subset of B_0 that spans S. Call this set B_1. By minimality, B_1 is linearly independent. For each $k < n$, the function g_k omits at most one value, so B_1 contains at most two vectors of the form $x_{(i,k)}$. Thus $|B_1| \leq 2n$. Let $B_1 = \{v_0, \ldots, v_j\}$ where $j < 2n$. The vectors of B_1 span S, so $u_0 = \sum_{i\leq j} c_i v_i$, with some $c_{i_0} \neq 0$. Solving for v_{i_0}, we see that v_{i_0} is in the span of $B_2 = \{u_0\} \cup B_1 \setminus \{v_{i_0}\}$. Thus B_2 is a linearly independent set spanning S. Iterating this process by primitive recursion, we eventually find a $u_m \in S$ which is a linear combination of $\{u_i : i < m\}$. Thus S is linearly dependent.)

Apply VSB$_{<\omega}$ to find a basis B for V. Then $k = |B| - n - 1$ is the cardinality of the largest set omitted from the range of f. Let P be the finite collection of odd numbers m such that (m, k) appears as an index in a formal sum for an element of B. Let $R = \{m \mid 2m + 1 \in P\}$. Exactly one m in R does not appear in the range of g_k. Thus for exactly one m in R, $\{x_{(0,k)}, x_{(2m+1,k)}\}$ is linearly independent. Sequentially examine linear combinations of the form $q_0 x_{(0,k)} + q_1 x_{(2m+1,k)}$, ejecting values from R corresponding to vectors equal to 0 in V, until only one is left. Viewed as a code for a pair, the first component of that value is a code for a set of maximum cardinality in the complement of the range of f. Thus $\mathsf{C}^\#_{\max} \leq_{\mathrm{sW}} \mathsf{VSB}_{<\omega}$. By Theorem 17, $\mathsf{C}^\mathsf{C}_{\max} \leq_{\mathrm{sW}} \mathsf{VSB}_{<\omega}$.

4   Reducibility and Reverse Mathematics

We conclude by extracting a final reverse mathematics result from the proofs of Theorems 17 and 18, extending the list of equivalences in Theorem 5.

Theorem 19. (RCA$_0$) *The following are equivalent:*

(1) IΣ_2^0, *the induction scheme for Σ_2^0 formulas with set parameters.*
(2) *Let V be a countable vector space such that for some n, every subset of n vectors is linearly dependent. Then V has a basis.*

(3) *A formalized version of* $C_{max}^{\#}$. *Suppose* $f: \mathbb{N} \to [\mathbb{N}]_{\neq\emptyset}^{\leq\mathbb{N}}$ *and there is an* n *such that for all* $X \in [\mathbb{N}]^{<\mathbb{N}}$, $[|X| \geq n \to \exists t(f(t) = X)]$. *Then there is an* $X \in [\mathbb{N}]^{<\mathbb{N}}$ *such that* $(\forall t)[f(t) \neq X$ *and for all* $Y \in [\mathbb{N}]^{<\mathbb{N}}$, $[|X| < |Y| \to \exists t(f(t) = Y)]$.

(4) *A formalized version of* C_{max}^{C}. *Suppose* $f: \mathbb{N} \to [\mathbb{N}]_{\neq\emptyset}^{\leq\mathbb{N}}$ *and there is an* n *such that for all* $X \in [\mathbb{N}]^{<\mathbb{N}}$, $(|X| \geq n \to \exists t(f(t) = X))$. *Then there is an* $X \in [\mathbb{N}]^{<\mathbb{N}}$ *such that* $(\forall t)[f(t) \neq X]$ *and for all* $Y \in [\mathbb{N}]^{<\mathbb{N}}$, $[X \subsetneq Y \to \exists t(f(t) = Y)]$.

Proof. First, we use (1) to prove (2). If V is a vector space and every set of n vectors is linearly dependent, the construction from the proof of Theorem 12 can be formalized to yield an e-matroid of rank no more than n. By Theorem 5, $\mathsf{I}\Sigma_2^0$ implies that this matroid has a basis which is also a basis of V.

To show that (2) implies (3), formalize the argument form the proof of Theorem 18 showing that $C_{max}^{\#} \leq_{sW} VSB_{<\omega}$, using the parenthetical comments. As noted, the convergence of the moving marker construction is provable in RCA_0, as is the claim that every set of $2n + 1$ vectors is linearly dependent.

The proof that (3) implies (4) follows immediately from the fact that any set that is maximal in the sense of (3) is automatically maximal in the sense of (4).

The proof that $EMB_{<\omega} \leq_{sW} C_{max}^{C}$ from Theorem 17 can be formalized in RCA_0 to show that (4) implies item (1) of Theorem 5. By Theorem 5, this implies $\mathsf{I}\Sigma_2^0$, completing the proof.

References

1. Brattka, V., de Brecht, M., Pauly, A.: Closed choice and a uniform low basis theorem. Ann. Pure Appl. Logic **163**(8), 968–1008 (2012). doi:10.1016/j.apal.2011. 12.020
2. Brattka, V., Gherardi, G.: Weihrauch degrees, omniscience principles, weak computability. J. Symb. Log. **76**(1), 143–176 (2011). doi:10.2178/jsl/1294170993. MR2791341 (2012c:03186)
3. Brattka, V., Gherardi, G.: Effective choice and boundedness principles in computable analysis. Bull. Symb. Log. **1**(1), 73–117 (2011). doi:10.2178/bsl/ 1294186663
4. Dorais, F.G., Dzhafarov, D.D., Hirst, J.L., Mileti, J.R., Shafer, P.: On uniform relationships between combinatorial problems. Trans. Am. Math. Soc. **368**, 1321–1359 (2014). doi:10.1090/tran/6465
5. Crossley, J.N., Remmel, J.B.: Undecidability and recursive equivalence II. In: Börger, E., Oberschelp, W., Richter, M.M., Schinzel, B., Thomas, W. (eds.) Computation and Proof Theory. LNM, vol. 1104, pp. 79–100. Springer, Heidelberg (1984). doi:10.1007/BFb0099480. MR775710
6. Downey, R.: Abstract dependence and recursion theory and the lattice of recursively enumerable filters, Ph.D. thesis, Monash University, Victoria, Australia (1982)
7. Downey, R.: Abstract dependence, recursion theory, and the lattice of recursively enumerable filters. Bull. Aust. Math. Soc. **27**, 461–464 (1983). doi:10.1017/ S0004972700025958

8. Downey, R.: Nowhere simplicity in matroids. J. Aust. Math. Soc. Ser. A **35**(1), 28–45 (1983). doi:10.1017/S1446788700024757. MR697655
9. Friedman, H.M., Simpson, S.G., Smith, R.L.: Countable algebra and set existence axioms. Ann. Pure Appl. Logic **25**(2), 141–181 (1983). doi:10.1016/0168-0072(83)90012-X. MR725732
10. Friedman, H.M., Simpson, S.G., Smith, R.L.: Addendum to: "Countable algebra and set existence axioms". Ann. Pure Appl. Logic **28**(3), 319–320 (1985). doi:10.1016/0168-0072(85)90020-X. MR790391
11. Gura, K., Hirst, J.L., Mummert, C.: On the existence of a connected component of a graph. Computability **4**(2), 103–117 (2015). doi:10.3233/COM-150039. MR3393974
12. Harrison-Trainor, M., Melnikov, A., Montalbán, A.: Independence in computable algebra. J. Algebra **443**, 441–468 (2015). doi:10.1016/j.jalgebra.2015.06.004. MR3400410
13. Hirst, J.L.: Connected components of graphs and reverse mathematics. Arch. Math. Logic **31**(3), 183–192 (1992). doi:10.1007/BF01269946. MR1147740
14. Metakides, G., Nerode, A.: Recursively enumerable vector spaces. Ann. Math. Logic **11**(2), 147–171 (1977). doi:10.1016/0003-4843(77)90015-8. MR0446936
15. Metakides, G., Nerode, A.: Recursion theory on fields and abstract dependence. J. Algebra **65**(1), 36–59 (1980). doi:10.1016/0021-8693(80)90237-9. MR578794
16. Nerode, A., Remmel, J.: Recursion theory on matroids. Stud. Logic Found. Math. **109**, 41–65 (1982). doi:10.1016/S0049-237X(08)71356-9. Patras Logic Symposion (Patras, 1980). MR694252
17. Simpson, S.G.: Subsystems of Second Order Arithmetic. Perspectives in Logic, 2nd edn. Cambridge University Press, Association for Symbolic Logic, Cambridge, Poughkeepsie (2009). MR2517689 (2010e:03073)
18. Whitney, H.: On the abstract properties of linear dependence. Am. J. Math. **57**(3), 509–533 (1935). doi:10.2307/2371182. MR1507091

Weakly Represented Families
in Reverse Mathematics

Rupert Hölzl[1], Dilip Raghavan[2], Frank Stephan[2,3(✉)], and Jing Zhang[4]

[1] Faculty of Computer Science, Universität der Bundeswehr München,
Werner-Heisenberg-Weg 39, 85577 Neubiberg, Germany
r@hoelzl.fr
[2] Department of Mathematics, The National University of Singapore,
10 Lower Kent Ridge Road, S17, Singapore 119076, Republic of Singapore
raghavan@math.nus.edu.sg
[3] Department of Computer Science, National University of Singapore,
13 Computing Drive, COM1, Singapore 117417, Republic of Singapore
fstephan@comp.nus.edu.sg
[4] Department of Mathematical Sciences, Carnegie Mellon University,
Pittsburgh, USA
jingzhang@cmu.edu

Abstract. We study the proof strength of various second order logic
principles that make statements about families of sets and functions.
Usually, families of sets or functions are represented in a uniform way by
a single object. In order to be able to go beyond the limitations imposed
by this approach, we introduce the concept of weakly represented families
of sets and functions. This allows us to study various types of families
in the context of reverse mathematics that have been studied in set
theory before. The results obtained witness that the concept of weakly
represented families is a useful and robust tool in reverse mathematics.

1 Introduction

The study of cardinal invariants of the continuum is an important and well-
studied branch of set-theory. A cardinal invariant is a cardinal that lies
between ω_1 and the continuum $2^{\aleph_0}$. Their study has been important both for
forcing theory and for the development of techniques for constructing certain
special sets of real numbers in ZFC.

In this work we try to formulate analogues of some of these cardinal invariants
in the context of models of second order arithmetic and reverse mathematics.
Consider a model of second order arithmetic $(M, S, +, \cdot, 0, 1)$. The basic idea of
the present study is that if a suitably "nice" coding of a set of subsets of M

R. Hölzl and F. Stephan were supported in part by MOE/NUS grants R146-000-
181-112 and R146-000-184-112 (MOE2013-T2-1-062); D. Raghavan was supported in
part by MOE/NUS grant R146-000-184-112 (MOE2013-T2-1-062). Work on this article begun as J. Zhang's Undergraduate Research Opportunities Programme project
while J. Zhang was an undergraduate students of NUS and R. Hölzl worked at NUS
financed by MOE/NUS grant R146-000-184-112 (MOE2013-T2-1-062).

A. Day et al. (Eds.): Downey Festschrift, LNCS 10010, pp. 160–187, 2017.
DOI: 10.1007/978-3-319-50062-1_13

satisfying certain combinatorial properties is present in the second order part S of this model, then this corresponds to the set-theoretic statement that a certain cardinal invariant of the continuum is small. The notion of "nice coding" that we will use is that of weakly represented families, the definition of which will be made precise in Definition 4.

In the next section we give a short introduction to reverse mathematics, which will then allow us to formulate the second order arithmetical principles that we wish to study in Sect. 3. In Sect. 4 we can then discuss the connections with cardinal invariants.

We point out that connections between recursion theory and cardinal invariants have previously been studied by Rupprecht [28] as well as by Brendle, Brooke-Taylor, Ng and Nies [3]; however, their work is only loosely related to the present study.

2   Second Order Arithmetic and Its Base System

Second order arithmetic is the two-sorted strengthening of first order logic, that is, it is obtained as follows: We introduce set variables in addition to the number variables existing in first order logic. The function and relation symbols "$\cdot$", "$+$", "$=$" and "$<$" of the language of first order logic remain unchanged, and are supplemented by a new relation symbol "$\in$".

Adopting the convention of Simpson [30], we let $\mathcal{L}_2$ denote the language of second order arithmetic. In the following, without explicit mention, we will let capital letters denote set variables while lower-case letters will denote number variables.

Definition 1 (Second order arithmetic). The axioms of second order arithmetic consist of the universal closure of the following $\mathcal{L}_2$-formulas.

1. Basic Axioms:
   - $n + 1 \neq 0$
   - $m + 1 = n + 1 \rightarrow m = n$
   - $m + 0 = m$
   - $m + (n + 1) = (m + n) + 1$
   - $m \cdot 0 = 0$
   - $m \cdot (n + 1) = (m \cdot n) + m$
   - $\neg(m < 0)$
   - $m < n + 1 \rightarrow (m < n \lor m = n)$
   - $\neg(n \in m)$
   - $\neg(X \in n)$
   - $\neg(X \in Y)$
2. Induction Axiom: $(0 \in X \land \forall n\,(n \in X \rightarrow n + 1 \in X)) \rightarrow \forall n\,(n \in X)$
3. Comprehension Axioms:

$$\exists X\ \forall n\,(n \in X \leftrightarrow \varphi(n)),$$

where $\varphi(n)$ is any $\mathcal{L}_2$-formula in which X does not occur freely.

In the context of reverse mathematics, in order to investigate the strength of different axiom systems, we need to first agree on a base system, that is, on the basic logical facts that we take for granted.

Definition 2 (Induction schemes). Given a set of formulae $\mathcal{B}$, the $\mathcal{B}$-induction scheme consists of all axioms of the form

$$(\varphi(0) \wedge \forall n \, (\varphi(n) \to \varphi(n+1))) \ \to \ \forall n \, (\varphi(n))$$

for any formula $\varphi(n) \in \mathcal{B}$ in which X does not occur freely.

Definition 3 (Base system RCA_0). RCA_0 is the subsystem of second order arithmetic consisting of the Basic Axioms as in Definition 1 (1), the Σ_1^0-induction scheme as in Definition 2, and the Comprehension Axioms as in Definition 1 (3) restricted to the class of Δ_1^0-formulas.

It is reasonable to use RCA_0 as base system for the investigation of stronger axiom systems in the context of reverse mathematics as it captures the effective aspects of mathematics. Additionally, it was shown that a fair number of mathematical theories can be developed relying solely on RCA_0; for details, see Simpson [30]. In this article we will also follow this established pratice unless otherwise stated.

It is common to informally refer to different base systems as different logical *principles*, and we will employ this expression frequently in the following.

3   Some Second Order Combinatorial Principles

A model of a set of second order arithmetical principles in general takes the form $\mathcal{M} = (M, S, +, \cdot, 0, 1)$ where M is the first order part of the structure and S is the second order part. If we decide not to require that all of the axioms of Definiton 1 hold, but only a subset of them, such as RCA_0, then it is not guaranteed anymore that a model of such an axiom set has $S = \mathcal{P}(M)$; typically, S will be much smaller. If $M = \omega$, $\mathcal{M}$ is called an ω-model, and S is a Turing ideal in this case.

The major textbook of reverse mathematics, Simpson [30], describes the five major axiom systems of reverse mathematics that cover many branches of mathematics, such as algebra, analysis, etc. Another recent textbook by Hirschfeldt [12] puts a particular focus on the role of Ramsey theory for reverse mathematics.

Before we can define the principles that we will study in this article, we need the following definitions.

Definition 4 (Weakly represented partial functions). A partial function f is said to be weakly represented by a set A if, for every x and y, there exists a z with $\langle x, y, z \rangle \in A$ iff

1. $x \in \mathrm{dom}(f) \wedge f(x) = y$ and   (representation)
2. $\forall x, y, y', z, z' \, [(\langle x, y, z \rangle \in A \wedge \langle x, y', z' \rangle \in A) \ \to \ y = y']$ and   (consistency)
3. $\forall x, y, y', z, z' \, [(\langle x, y, z \rangle \in A \wedge z < z') \ \to \ \langle x, y, z' \rangle \in A]$ and   (monotonicity)
4. $\exists z \, \langle x, y, z \rangle \in A \ \to \ \forall t < x \, \exists y' \, \exists z' \, \langle t, y', z' \rangle \in A.$ (downward closure)

Definition 5 (Weakly represented families of functions). Let $A \in S$ be given and write $A_e = \{n \colon \langle e, n \rangle \in A\}$, $e \in M$, for its rows. For each e, write f_e for the (possibly partial) function weakly represented by A_e.

Then a set of total functions $\mathcal{F}$ is said to be a weakly represented family of functions represented by A if we have that $\mathcal{F}$ contains exactly those f_e, $e \in M$, that are total.

Note that all functions in a weakly represented family are by definition total. Rows A_e of A that do not represent such a function are ignored.

Definition 6 (Weakly represented families of sets). A set of sets S is said to be a weakly represented family of sets if their corresponding characteristic functions form a weakly represented family of functions.

Definition 7. $\mathcal{F}$ is said to be a uniform family of sets represented by A if

$$\mathcal{F} = \{A_e \colon e \in M\}$$

where $A_e = \{n \colon \langle e, n \rangle \in A\}$, $e \in M$.

Remark 8. It is easy to see that every uniform family of sets represented by some A is also a weakly represented family of sets represented by some B where $A =_{\mathrm{T}} B$.

One motivation for introducing weakly represented families is that the set of all partial recursive functions is a weakly represented family of functions. Similarly, it can easily be seen that in the classical setting the collection of all recursive sets is a weakly represented family of sets. This is because the class of characteristic functions of recursive sets

$$\mathcal{F} = \{\varphi_e \colon \varphi_e \text{ is total} \wedge \mathrm{range}(\varphi_e) \subseteq \{0, 1\}\}$$

can be weakly represented by a recursive set in any model of RCA_0.

These are examples of how the notion of weakly represented families enables us to talk about more and larger sets of functions; and this new ability then allows us to define new reverse mathematics principles, as we will now see.

Friedberg [10] constructed a maximal set, that is, an r.e. set A with infinite complement such that any other r.e. set B either contains almost all or almost none of the elements of the complement of A. As it turned out, the property of the complement being either almost contained in or being almost disjoint from every recursively enumerable set plays an important role in recursion theory, and thus it was given a name of its own, *cohesiveness*. This is a special case of the following more general definition.

Definition 9 (Cohesive set). For a set $A \subseteq M$, write $\bar{A}$ for $M \setminus A$. Then given a set of sets $\mathcal{F} \subseteq \{0, 1\}^M$, a set G is said to be $\mathcal{F}$-cohesive if for any $A \in \mathcal{F}$, either $G \subseteq^* A$ or $G \subseteq^* \bar{A}$. If $\mathcal{F}$ is the collection of all recursive sets, then G is called r-cohesive.

Statement 10 (Cohesion Principle COHW). *For every uniform family $\mathcal{F}$ of sets, there exists an $\mathcal{F}$-cohesive set.*

While recursion theorists were originally interested in the degree-theoretic properties of cohesiveness, it turned out that it was relevant in reverse mathematics as well: Mileti [23] showed that Ramsey's Theorem for Pairs implies COH; and Cholak, Jockusch and Slaman [4] showed that Ramsey's Theorem for Pairs is equivalent to Stable Ramsey's Theorem for Pairs together with COH. For a detailed account of the role that COH has played in reverse mathematics, see Hirschfeldt [12].

In this article we will also study COHW, a variant of COH that takes advantage of the new possibilities introduced with the notion of weakly represented families of sets.

Statement 11 (Cohesion for weakly represented families COHW). *For every weakly represented family $\mathcal{F}$ of sets, there exists an $\mathcal{F}$-cohesive set.*

By Remark 8, COHW trivially implies COH. But we will show that the other implication does not hold, not even over ω-models.

Statement 12 (Domination Principle DOM). *Given any weakly represented family of functions $\mathcal{F}$, there exists a function g such that for every $f \in \mathcal{F}$ there is some $b \in M$ such that $g(x) > f(x)$ for all $x > b$.*

In a follow-up study to the present article, Hölzl, Jain and Stephan [14] establish further properties of DOM, including the following.

– Over RCA_0, $B\Sigma_2 + DOM \vdash I\Sigma_2$;
– Over $RCA_0 + DOM$, the index set E of a weakly represented family is limit-recursive, that is, there is a binary $\{0,1\}$-valued function g such that for all $e \in M$, if $e \in E$ then $\exists s \forall t > s\,[g(e,t) = 1]$ else $\exists s \forall t > s\,[g(e,t) = 0]$.
   (Here, for a weakly represented family $\mathcal{F}$ of functions represented by A, we call the set of $e \in M$ for which f_e, as in Definition 5, is total, the *index set* of $\mathcal{F}$.)

We will show that over RCA_0 and $B\Sigma_2$, DOM implies COH and COHW.

Statement 13 (Hyperimmunity Principle HI). *Given any weakly represented family of functions $\mathcal{F}$, there exits a function g such that for each $f \in \mathcal{F}$ and each $b \in M$ we have $g(x) > f(x)$ for some $x > b$.*

Note that HI is weaker than DOM. Hirschfeldt, Shore and Slaman [13] define the principle OPT, which they show [13, Theorem 5.7] to be equivalent to the statement that for every $f \in S$ there is a $g \in S$ such that f does not compute a function majorising g; thus this principle is equivalent to HI.

For $f, g \in M^M$ we write $f <^* g$ to express that $\{n \in M : g(n) \leq f(n)\}$ is finite. The symbol "$\leq^*$" is defined accordingly. A subset $\mathcal{F} \subseteq M^M$ is called *bounded* if there exists $g \in M^M$ such that for all $f \in \mathcal{F}$ we have $f <^* g$. Otherwise $\mathcal{F}$ is said to be *unbounded*.

Statement 14 (Meeting Principle MEET). *Given any weakly represented family of functions $\mathcal{F}$, there exits a function g such that for each $f \in \mathcal{F}$ the set $\{n \in M : f(n) = g(n)\}$ is infinite.*

We will show that HI and MEET are equivalent.

Definition 15. We say that a function g *avoids* a function f if

$$\{n \in M : f(n) = g(n)\}$$

is finite.

Statement 16 (Avoidance Principle AVOID). *Given any weakly represented family of functions $\mathcal{F}$, there exits a function g avoiding all $f \in \mathcal{F}$.*

Two subsets A and B of M are said to be *almost disjoint* if $A \cap B$ is finite. A set $\mathcal{F} \subseteq \{0, 1\}^M$ is called *almost disjoint* if any two distinct elements of $\mathcal{F}$ are almost disjoint. A set $\mathcal{F} \subseteq \{0, 1\}^M$ is called *maximal almost disjoint* if it is infinite and almost disjoint and is not properly contained in any larger almost disjoint set. Formalising that a family is infinite is somewhat tricky; we use the following approach.

Definition 17. We call a weakly represented family $\mathcal{F}$ *finite* if there is a weakly represented family $\mathcal{G}$ with finite index set such that $\mathcal{F} = \mathcal{G}$. Otherwise we call $\mathcal{F}$ *infinite*.

Statement 18 (Maximal Almost Disjoint Family Principle MAD). *There exists a weakly represented family $\mathcal{F}$ of infinite sets such that the following three conditions hold:*

- *$\mathcal{F}$ is infinite;*
- *if $A, B \in \mathcal{F}$ are pairwise different, then $A \cap B$ is finite;*
- *for every infinite set $C \in S$ there is a $D \in \mathcal{F}$ such that $C \cap D$ is infinite.*

For a set $A \subseteq M$, let us temporarily write A^0 for A and A^1 for $\bar{A}$. A family $\mathcal{F} \subseteq \mathcal{P}(M)$ is said to be *independent* if for any $n \geq 1$, any collection $\{A_0, \ldots, A_{n-1}\} \subseteq \mathcal{F}$, and any string $\sigma \in 2^n$, the set $\bigcap_{i<n} A_i^{\sigma(i)}$ is infinite. A *maximal independent* family is an independent family that can not be extended to a strictly larger independent family.

Statement 19 (Maximal Independent Family Principle MIND). *There exists a weakly represented family of infinite sets that is maximal independent.*

Statement 20 (Biimmunity Principle BI). *For every weakly represented family $\mathcal{F}$ of infinite sets there is a set $B \in S$ such that there is no set $A \in \mathcal{F}$ with $A \subseteq B$ or $A \subseteq \bar{B}$.*

4   Cardinal Invariants

We now discuss the nine cardinal invariants of the continuum that are considered in this paper, the most basic being the cardinality of the continuum.

Definition 21. $\mathfrak{c} = 2^{\aleph_0} = |\mathbb{R}|$.

Recall that the Continuum Hypothesis CH is the statement that $\mathfrak{c} = \aleph_1$. The analogue of CH in a model $\mathcal{M} = (M, S, \cdot, +, 0, 1)$ is the statement that there is a weakly represented family of sets represented by $A \in S$ such that the characteristic function of every element of S appears in A. In other words, this states that there is a set in S which "encodes in a nice way" all the subsets of M that can be "seen by" $\mathcal{M}$. The simplest example of this is the case where S consists exactly of the recursive sets.

Recall the partial order $\langle M^M, <^* \rangle$ defined in the previous section. We assume that ZFC is our base theory when talking about cardinal invariants of the continuum. Therefore, we only consider the restriction of this partial order to ω^ω in this section. So for $f, g \in \omega^\omega$, $g <^* f$ means that $\{n \in \omega : g(n) \geq f(n)\}$ is finite; and "finite" here does not mean finite in the sense of some specific model of second order arithmetic, but finite as defined within ZFC. Recall that a family $\mathcal{F} \subseteq \omega^\omega$ is *unbounded* if there is no $g \in \omega^\omega$ such that $\forall f \in \mathcal{F}\ [f <^* g]$ and $\mathcal{F}$ is *dominating* if for all $g \in \omega^\omega$ there exists an $f \in \mathcal{F}$ with $g <^* f$. It is clear that every dominating set is unbounded. Based on these definitions, we define the following two cardinal invariants.

Definition 22. $\mathfrak{b} = \min\{|\mathcal{F}| : \mathcal{F} \subseteq \omega^\omega \wedge \mathcal{F} \text{ is unbounded}\}$

$\mathfrak{d} = \min\{|\mathcal{F}| : \mathcal{F} \subseteq \omega^\omega \wedge \mathcal{F} \text{ is dominating}\}$

It is easy to prove that $\aleph_1 \leq \mathrm{cf}(\mathfrak{b}) = \mathfrak{b} \leq \mathrm{cf}(\mathfrak{d}) \leq \mathfrak{d} \leq \mathfrak{c}$, where $\mathrm{cf}(\kappa)$ denotes the cofinality of the cardinal κ. It is also a classical theorem of Hechler [11] that these are the only restrictions that are provable in ZFC. In keeping with the intuition described in the introduction, in a second order model $\mathcal{M} = (M, S, \cdot, +, 0, 1)$, the statement $\mathfrak{b} = \aleph_1$ should correspond to the statement that there exists a set in S which "nicely encodes" an unbounded family of functions from the point of view of $\mathcal{M}$. In other words, $\mathfrak{b} = \aleph_1$ should correspond to the statement that there is a weakly represented family of functions $\mathcal{F}$ represented by some $A \in S$ such that no function in S dominates, in the sense of the partial order $\langle M^M, <^* \rangle$, all the elements of $\mathcal{F}$. This is the negation of the principle DOM. So DOM is the analogue of $\mathfrak{b} > \aleph_1$. Similarly HI corresponds to $\mathfrak{d} > \aleph_1$.

Another important pair of cardinals come from the notion of splitting. Recall that for a set X and a cardinal κ, $[X]^\kappa = \{A \subseteq X : |A| = \kappa\}$, in particular $[\omega]^\omega$ denotes the set of infinite subsets of ω. Let $A, B \subseteq \omega$. We say that A *splits* B if both $B \cap A$ and $B \cap \bar{A}$ are infinite. A set $\mathcal{F} \subseteq \mathcal{P}(\omega)$ is called a *splitting family* if $\forall B \in [\omega]^\omega\ \exists A \in \mathcal{F}\ [A \text{ splits } B]$. A set $A \subseteq \omega$ is said to *reap* a family $\mathcal{F} \subseteq \mathcal{P}(\omega)$ if for all $B \in \mathcal{F}$ we have that A splits B. A family $\mathcal{F} \subseteq [\omega]^\omega$ is *unreaped* if there is no $A \in \mathcal{P}(\omega)$ which reaps $\mathcal{F}$. The following cardinals correspond to the notions of splitting and reaping.

Definition 23. $\mathfrak{s} = \min\{|\mathcal{F}| : \mathcal{F} \subseteq \mathcal{P}(\omega) \wedge \mathcal{F}$ is a splitting family$\}$

 $\mathfrak{r} = \min\{|\mathcal{F}| : \mathcal{F} \subseteq [\omega]^\omega \wedge \mathcal{F}$ is an unreaped family$\}$

It is not difficut to prove that $\mathfrak{s} \leq \mathfrak{d}$, and this proof dualizes to show that $\mathfrak{b} \leq \mathfrak{r}$ (see Blass [2]). Blass and Shelah constructed a model with $\aleph_1 = \mathfrak{r} < \mathfrak{s} = \aleph_2$ (see Bartoszyński and Judah [1, Sect. 7.4.D]) and $\aleph_1 = \mathfrak{s} < \mathfrak{b} = \aleph_2$ holds in the Laver model (see Bartoszyński and Judah [1, Sect. 7.3.D]). The notion of a cohesive set in recursion theory is related to the notion of splitting. To say that G is $\mathcal{F}$-cohesive is the same as saying that G is not split by any member of $\mathcal{F}$. So the principle COHW corresponds to the statement $\mathfrak{s} > \aleph_1$ because it says that no weakly represented family $\mathcal{F} \in S$ has the property that every $A \in S$ is split by some member of $\mathcal{F}$ — in other words, S does not "nicely encode" any splitting family in the sense of $\mathcal{M}$. The principle COH is related to COHW and satisfies COHW $\vdash$ COH properly; there is no direct analogue of COH in set theory. The principle BI corresponds to the statement $\mathfrak{r} > \aleph_1$ and the reverse mathematical analogue of the ZFC theorem $\mathfrak{s} \leq \mathfrak{d}$ is the statement that COHW implies HI. However in parallel with Blass and Shelah's result that the statement $\aleph_1 = \mathfrak{b} = \mathfrak{r} < \mathfrak{s} = \aleph_2$ is consistent with ZFC, it holds that COHW does not imply DOM; the full result has no analogue as COHW implies HI and HI implies BI. Furthermore, the implication DOM $\vdash$ HI has the analogue $\mathfrak{b} \leq \mathfrak{d}$ in ZFC. In both cases, the inverse implication does not hold.

The next group of cardinals that we define stem from the context of categoricity. Recall that a set $X \subseteq \mathbb{R}$ is called *nowhere dense* if the interior of its closure is empty. A subset of $\mathbb{R}$ is *meager* if it is the union of countably many nowhere dense sets. We define the following cardinals.

Definition 24. $\mathrm{cov}(\mathcal{C}) = \min\left\{|\mathcal{F}| : \begin{array}{l} \mathcal{F} \text{ consists of meager subsets of } \mathbb{R} \\ \text{and } \bigcup \mathcal{F} = \mathbb{R} \end{array}\right\}$

 $\mathrm{non}(\mathcal{C}) = \min\{|A| : A$ is a non-meager subset of $\mathbb{R}\}$

Here $\mathcal{C}$ stands for category. These topologically defined cardinals have purely combinatorial characterizations, as the following theorem shows.

Theorem 25 (Miller [24])

1. $\mathrm{cov}(\mathcal{C})$ *is the minimal cardinal* κ *such that there exists an* $\mathcal{F} \subseteq \omega^\omega$ *with* $|\mathcal{F}| = \kappa$ *and such that for all* $g \in \omega^\omega$ *there is an* $f \in \mathcal{F}$ *such that*

$$\{n \in \omega : f(n) = g(n)\}$$

   is finite.

2. $\mathrm{non}(\mathcal{C})$ *is the minimal cardinal* κ *such that there exists an* $\mathcal{F} \subseteq \omega^\omega$ *with* $|\mathcal{F}| = \kappa$ *and such that for all* $g \in \omega^\omega$ *there is an* $f \in \mathcal{F}$ *such that*

$$\{n \in \omega : g(n) = f(n)\}$$

   is infinite.

We remind the reader that in the above theorem "finite" and "infinite" are not to be understood in the sense of $\mathcal{M}$, but as those terms as defined within ZFC.

The above theorem allows us to formulate analogues of these topological invariants in any model of second order arithmetic. For $\mathrm{cov}(\mathcal{C})$ to be "small" in a second order model $\mathcal{M} = (M, S, \cdot, +, 0, 1)$, we would like to have a weakly represented family of functions $\mathcal{F} \in S$ with the property that for any function $g \in S$, $\{n \in M : f(n) = g(n)\}$ is finite in the sense of $\mathcal{M}$. The principle MEET says that no such weakly represented family exists. Thus MEET corresponds to the statement that $\mathrm{cov}(\mathcal{C}) > \aleph_1$. Similarly, AVOID is the analogue of $\mathrm{non}(\mathcal{C}) > \aleph_1$. As it is easy to prove in ZFC that $\mathrm{cov}(\mathcal{C}) \leq \mathfrak{d}$, one would expect MEET to imply HI, and indeed this is easy to check. But somewhat unexpectedly we will prove that MEET and HI are equivalent — at least for ω-models. This contrasts with the fact that $\mathrm{cov}(\mathcal{C}) = \aleph_1 < \aleph_2 = \mathfrak{b} = \mathfrak{d}$ holds in the Laver model (see Bartoszyński and Judah [1, Sect. 7.3.D]). As a result, in the classical ZFC context, we do not even have that DOM implies MEET. Dualizing the equivalence of MEET and HI one would expect AVOID to be equivalent to DOM. Indeed, DOM implies AVOID by definition; however, we show in Theorem 41 that AVOID does not imply HI, and therefore not DOM. Also it is consistent that $\mathfrak{b} = \aleph_1 < \aleph_2 = \mathrm{cov}(\mathcal{C})$; in fact it is folklore that this holds in the Cohen model. This is reflected by the fact that MEET does not imply DOM, which follows immediately from Theorem 38. Next, regarding $\mathrm{non}(\mathcal{C})$, it is easy to see by Theorem 25 (2), that $\mathfrak{b} \leq \mathrm{non}(\mathcal{C})$ holds in ZFC, and, accordingly, DOM implies AVOID. It is also easy to prove in ZFC that $\mathfrak{s} \leq \mathrm{non}(\mathcal{C})$. This is only partially true in the reverse mathematical context. Namely, we will prove that COHW implies AVOID in ω-models. However this is not true in all non-ω-models, as we will show. Finally, in the classical ZFC context, $\mathfrak{d}$ and $\mathrm{non}(\mathcal{C})$ are independent, meaning that while it is consistent to have $\aleph_1 = \mathfrak{d} < \mathrm{non}(\mathcal{C}) = \aleph_2$ (see Bartoszyński and Judah [1, Sect. 7.3.B]) it is also consistent to have $\aleph_1 = \mathrm{non}(\mathcal{C}) < \mathfrak{d} = \aleph_2$ (see Bartoszyński and Judah [1, Sect. 7.3.E]). This is reflected by the independence of AVOID and MEET, even in ω-models.

We also considered cardinal invariants associated with almost disjointness and independence. In the ZFC context, $A, B \subseteq \omega$ are said to be *almost disjoint* if $|A \cap B| < \aleph_0$. A family $\mathscr{A} \subseteq [\omega]^\omega$ is *almost disjoint* if its members are pairwise almost disjoint. An infinite almost disjoint family $\mathscr{A}$ is said to be *maximal almost disjoint* if it is not properly contained in any larger almost disjoint family. Any infinite almost disjoint set can be extended to a maximal almost disjoint set by Zorn's Lemma.

Similarly a family $\mathcal{F} \subset [\omega]^\omega$ is called *independent* if for each $n \geq 1$, each collection $\{A_0, \ldots, A_{n-1}\} \subset \mathcal{F}$, and each string $\sigma \in 2^n$, $\left| \bigcap_{i<n} A_i^{\sigma(i)} \right| = \aleph_0$, where A_i^0 is A_i and A_i^1 is $\bar{A}_i$. A *maximal independent family* is an independent family $\mathcal{F} \subset [\omega]^\omega$ which is not properly contained in any larger independent family. Zorn's Lemma also guarantees the existence of maximal independent families.

Observe that a second order model $\mathcal{M} = (M, S, \cdot, +, 0, 1)$ need not satisfy any principle akin to Zorn's lemma. Therefore there need not be any maximal almost disjoint or maximal independent families in S, weakly represented or otherwise.

Definition 26. $\mathfrak{a} = \min\{|\mathscr{A}| : \mathscr{A} \subseteq [\omega]^{\omega}$ is a maximal almost disjoint family$\}$

$\quad\quad\quad\quad$ $\mathfrak{i} = \min\{|\mathscr{A}| : \mathscr{A} \subseteq [\omega]^{\omega}$ is a maximal independent family$\}$

The principle MAD says that there is a weakly respresented maximal almost disjoint family and so it corresponds to $\mathfrak{a} = \aleph_1$. Similarly MIND corresponds to $\mathfrak{i} = \aleph_1$.

We prove that, at least for ω-models, MAD holds iff DOM fails. Since in ZFC the inequality $\mathfrak{b} \leq \mathfrak{a}$ holds by folklore, it does not come as a surprise that DOM implies ¬MAD. However, $\aleph_1 = \mathfrak{b} < \mathfrak{a} = \aleph_2$ is consistent by a theorem of Shelah [29], so the fact that ¬DOM implies MAD is unexpected.

Similarly, we prove that, at least for ω-models, MIND holds iff BI fails. One can easily prove in ZFC that $\mathfrak{r} \leq \mathfrak{i}$ and so the direction from BI to ¬MIND is unsurprising. But once again the consistency of $\mathfrak{r} = \aleph_1 < \aleph_2 = \mathfrak{i}$ was proved by Blass and Shelah (see Bartoszyński and Judah [1, Sect. 7.4.D]), making the implication from ¬BI to MIND unexpected.

5 Cohesion Principles

In the following we will introduce definitions needed for this paper; for the recursion-theoretic background, the reader is referred to the textbooks of Downey and Hirschfeldt [8], Nies [25], Odifreddi [26,27], Simpson [30] and Soare [32].

Definition 27. Let A and B be sets. A is PA-complete with respect to B (written as $A \gg B$) if for every partial B-recursive $\{0,1\}$-valued function f, there exists an A-recursive total extension g of f. In this definition we can replace sets by degrees in the canonical way.

Definition 28. Let A and B be sets. A is hyperimmune-free with respect to B if every function recursive in $A \oplus B$ is dominated by some B-recursive function.

Theorem 29. *Over* RCA_0, *COH does not imply* COHW. *This even holds for* ω-*models.*

To proof the non-implication for ω-models, we need the following lemmata and theorem. The first lemma establishes a relationship between two 1-generic sets and their join. It is the genericity analogue of van Lambalgen's Theorem 40.

Lemma 30 (Yu [34]). *The following are equivalent for $n \geq 1$.*

1. *$A \oplus B$ is n-generic;*
2. *A is n-generic and B is n-generic relative to A;*
3. *B is n-generic and A is n-generic relative to B.*

The following theorem is a reformulation and slight variation of a result of Jockusch and Stephan [17, Theorem 2.1]; the proof is largely identical and omitted here.

Theorem 31. *Let $\mathcal{F}$ be a uniform family represented by A. If $B' \gg A'$ then there is a B-recursive $\mathcal{F}$-cohesive set.*

Lemma 32. *There is a sequence of sets $(A_i : i \in \omega)$ such that, for every $i \in \omega$, A_i is 1-generic and not high, $A_{i+1} \geq_T A_i$ and $A'_{i+1} \gg A'_i$.*

Proof. Let $B_0 = \emptyset'$. If, for some $i \in \omega$, B_i with $B'_i \leq_T \emptyset''$ has been inductively defined, then we compute relative to B_i a tree T each path of which is a complete extension of PA relative to B_i, see Odifreddi [26]. Then by Jockusch and Soare's Low Basis Theorem [16] relative to B_i we have a path $B_{i+1} \in [T]$ such that $B_{i+1} \gg B_i$, $B_{i+1} \geq_T B_i$ and such that B_{i+1} is low relative to B_i, that is, $B'_{i+1} \leq_T B'_{i+1} \oplus B_i \leq_T B'_i \leq_T \emptyset''$.

It is well-known that the sets provided by Friedberg's Jump Inversion Theorem [9] can be assumed to be 1-generic; see, for example, Stephan [33, Theorem 5.4]. By applying this result to B_0 we obtain a 1-generic set A_0 such that $A'_0 = B_0$ (that is, A_0 is low). With A_i defined for some $i \in \omega$, and using that $B_{i+1} \geq_T B_i = A'_i$, we can apply the Jump Inversion Theorem relative to A_i to the set B_{i+1} to obtain a set C_{i+1} with $C'_{i+1} = B_{i+1}$ and C_{i+1} being 1-generic relative to A_i. We let $A_{i+1} = C_{i+1} \oplus A_i$ and by Lemma 30 and using that A_i is 1-generic by induction hypothesis we again have that A_{i+1} is 1-generic.

Note that for all $i \in \omega$ we have $A''_i = B'_i \leq_T \emptyset''$ and thus $A'_i <_T \emptyset''$. It follows that A_i is not high. $\qquad \square$

Now let $S = \{A \subseteq \omega : A \leq_T A_0 \oplus \ldots \oplus A_n \text{ for some } n \in \omega\}$, where the sets A_i, $i \in \omega$, are as in Lemma 32. The following lemmata show that the ω-model $\mathcal{M} = (\omega, S, +, \cdot, 0, 1)$ satisfies COH and RCA$_0$, but not COHW.

Lemma 33. $\mathcal{M} \models \text{COH} + \text{RCA}_0$

Proof. First note that S is closed under Turing reducibility and join, thus $\mathcal{M}$ is a model of RCA$_0$.

To see that $\mathcal{M}$ is also a model of COH, let a uniform familiy $\mathcal{F}$ represented by $A \leq_T A_0 \oplus \ldots \oplus A_n =_T A_n$ for large enough n be given. Then, by construction, $A'_{n+1} \gg A'_n$, and we can apply Theorem 31 with A_{n+1} substituted for B and A_n substituted for A to see that there is an A_{n+1}-recursive $\mathcal{F}$-cohesive set. $\quad \square$

Corollary 34. $\mathcal{M} \not\models \text{COHW}$

Proof. By Jockusch and Stephan [17, Theorem 2.9] each cohesive 1-generic degree is high. But by Lemma 32, no set in S is high, therefore none of the 1-generic sets A_i, $i \in \omega$, is cohesive. Since by Jockusch [15] the cohesive degrees are upwards closed this implies that no set in S is cohesive. Now by Jockusch and Stephan [17, Corollary 2.4], the cohesive and the r-cohesive degrees coincide, so no set in S is r-cohesive. In particular, if $\mathcal{F}$ is the weakly represented family consisting of all recursive sets, there exists no $\mathcal{F}$-cohesive set in S. So COHW fails to hold. $\qquad \square$

This concludes the proof of Theorem 29, separating COH and COHW over RCA$_0$.

Theorem 35. COH *does not imply* AVOID. *This even holds for ω-models.*

We use the following well-known lemma.

Lemma 36 (Demuth and Kučera [7]). *No 1-generic set computes a diagonally non-recursive function.*

Proof (Theorem 35). We again use the above ω-model $\mathcal{M}$ with second order part S. Observe that S is a downward closure of non-high 1-generic sets. In particular, by Lemma 36, all sets in S are neither diagonally non-recursive nor high. By Kjos-Hanssen, Merkle, and Stephan [18, Theorem 5.1 ($\neg(3) \Rightarrow \neg(1)$)], this implies that no $A \in S$ computes a function avoiding all total recursive functions. As the set of all total recursive functions is a weakly represented family, this contradicts AVOID.   □

The next theorem illustrates once more the difference in reverse mathematics strength between COH and COHW.

Theorem 37. COHW $\vdash$ AVOID *for ω-models.*

 More precisely, given any r-cohesive set G, one can recursively produce a total function g such that $\{n \in \omega : g(n) = \varphi_e(n)\}$ is finite for every total recursive function φ_e.

Proof. Let $\mathcal{F}$ be the collection of all total recursive functions. We will show that there exists a function $g \in S$ such that for every $f \in \mathcal{F}$ we have that $\{n \in \omega : f(n) = g(n)\}$ is finite. Then the general case follows by relativization.

 Let $\mathcal{F}'$ be the collection of all recursive sets. COHW ensures the existence of an $\mathcal{F}'$-cohesive set, say G. If G is high, then by Martin [22], there exists a function g recursive in G that dominates every total recursive function, and we are done.

 If G is not high, then by Jockusch and Stephan [17] there exists an effectively immune set A recursive in G. Here we call A effectively immune if there is a recursive function p such that for any r.e. set W_e we have $W_e \subseteq A \to |W_e| < p(e)$. Fix this p and assume without loss of generality that it is increasing.

 Let f be the total recursive function such that

$$W_{f(e,i)} = \begin{cases} W_{\varphi_i(e)} & \text{if } \varphi_i(e){\downarrow}, \\ \emptyset & \text{otherwise.} \end{cases}$$

Let g be the total recursive function such that $W_{g(e)}$ consists of the first

$$p(\max\{f(i,e) : i \le e\}) + 1$$

elements of A.

Claim. For all $i \le e$ we have $g(e) \ne \varphi_i(e)$.

Proof. Suppose otherwise, then $g(e) = \varphi_i(e)$ for some $i \leq e$. Then $\varphi_i(e)\!\downarrow$ and $W_{f(e,i)} = W_{\varphi_i(e)} = W_{g(e)} \subseteq A$, so

$$p(f(e,i)) < p(\max\{f(e,j)\colon j \leq e\}) + 1 = |W_{g(e)}| = |W_{f(e,i)}| < p(f(e,i)),$$

which is a contradiction.   ◇

Then g is the required function.   □

Given that the previous proof was carried out in the standard model, it is natural to ask how COHW interacts with AVOID in non-standard models.

6   The Meeting and Hyperimmunity Principles

In this section we investigate the principles MEET and HI and their relations to each other as well as to other principles.

Theorem 38. *Over* RCA$_0$, MEET *and* HI *are equivalent.*

Proof. MEET $\vdash$ HI: If g is as in the statement of MEET, then HI holds with $g+1$ substituted for g.

HI $\vdash$ MEET: Let an arbitrary model $\mathcal{M} = (M, S, +, \cdot, 0, 1)$ be given. Let $\mathcal{F}$ be a weakly represented family represented by A and let A_e and f_e, for $e \in M$, be as in Definition 5. Define a function $\widetilde{f}_e$ via $x \mapsto n_x$ for $x \in M$, where n_x is the first number of the form $\langle\langle e, x\rangle, y, z\rangle$ inside A_e, if it exists; note that $y = f_e(\langle e, x\rangle)$ whenever n_x exists. Then $\widetilde{f}_e$ is total iff f_e is total.

Note that the set $\mathcal{F}' = \{\widetilde{f}_e\colon \widetilde{f}_e \text{ is total}\}$ is again a weakly represented family. By applying HI to $\mathcal{F}'$ we obtain a function $\widetilde{g}$ such that for each total $\widetilde{f}_e$ there are infinitely many x with $\widetilde{g}(x) > \widetilde{f}_e(x)$.

Then define $g(\langle e, x\rangle)$ as follows: If there is a number m of the form $\langle\langle e, x\rangle, y, z\rangle$ in A_e such that $m < \widetilde{g}(x)$, then let $g(\langle e, x\rangle) = y$, else $g(\langle e, x\rangle) = 0$. The function g is in S and is total; furthermore, whenever $\widetilde{g}(x) > \widetilde{f}_e(x)$ then we have $g(\langle e, x\rangle) = f_e(\langle e, x\rangle)$ and thus for all total f_e there are infinitely many n with $g(n) = f(n)$. This implies MEET.   □

Our next result shows that AVOID is incomparable with HI. As essential tools we employ the following two well-known results; to see the first, apply the hyperimmune-free basis theorem of Jockusch and Soare [16] to the complement of the first component of the universal Martin-Löf test.

Lemma 39. *There exists a hyperimmune-free Martin-Löf random set.*

Theorem 40 (van Lambalgen [21]). *The following are equivalent.*

1. *$A \oplus B$ is n-random.*
2. *A is n-random and B is n-random relative to A.*
3. *B is n-random and A is n-random relative to B.*

Theorem 41. AVOID *does not imply* HI. *This even holds for ω-models.*

Proof. Let A be a hyperimmune-free Martin-Löf random, as in Lemma 39. For $i \in \omega$, let $A_i = \{x \colon \langle i, x \rangle \in A\}$. Then by Theorem 40, for every $i \in \omega$, A_{i+1} is Martin-Löf random relative to $A_0 \oplus \ldots \oplus A_i$. Fix the model $\mathcal{M} = (\omega, S, +, \cdot, 0, 1)$ with second order part $S = \{B \subseteq \omega \colon B \leq_T A_0 \oplus \ldots \oplus A_i \text{ for some } i \in \omega\}$.

Let a weakly represented family $\mathcal{F}$ represented by $B \leq_T A_0 \oplus \ldots \oplus A_i$ with $i \in \omega$ large enough be given. Fix a computably bijective map $\nu \colon \{0,1\}^* \to \omega$, and let g be the function $n \mapsto \nu(A_{i+1}(0) \ldots A_{i+1}(n))$. Fix any $f \in \mathcal{F}$; trivially, $f \leq_T B$. Assume that g does not avoid f. Then there are infinitely many n such that the Kolmogorov complexity relative to B of $A_{i+1}(0) \ldots A_{i+1}(n)$ is less than $2 \log(n)$, which contradicts that A_{n+1} is Martin-Löf random relative to $A_0 \oplus \ldots \oplus A_i$. Therefore $g \leq_T A_{i+1} \in S$ is a function as required by AVOID.

On the other hand, for every $C \in S$ we have $C \leq_T A$, and since A is hyperimmune-free, C is hyperimmune-free as well. As C was arbitrary, this implies that $\mathcal{M}$ does not satisfy HI. $\qquad\square$

We now turn to the other direction.

Theorem 42. HI *does not imply* AVOID.

Proof. We again use the ω-model $\mathcal{M}$ from the proof of Theorem 35. As shown there, $\mathcal{M}$ does not satisfy AVOID.

To see that $\mathcal{M}$ satisfies HI, let a weakly represented family $\mathcal{F}$ represented by $A \leq_T A_n$ be given. As A_{n+1} is by construction 1-generic relative to A_n, it is in particular hyperimmune relative to A. Then it computes a function g that is infinitely often larger than any function $f \leq_T A$, and in particular g is for $\mathcal{F}$ as required by HI. $\qquad\square$

A further interesting result is the following, which is in line with the fact from recursion theory that the Turing degrees of cohesive sets are hyperimmune.

Theorem 43. COHW *implies* HI.

Proof. Let $\mathcal{F}$ be a weakly represented family of functions represented by A, let f_e be as in Definition 5 and let $E = \{e \colon f_e \text{ is total}\}$.

Define for each $e \in M$ inductively a function g_e such that $g_e(0) = 1$ and $g_e(x+1) = \max\{f_e(x') + 1 \colon x' \leq g_e(x) + 1\}$. Next define for each $e \in M$ the set $B_e = \{y \colon \exists x \, [g_e(2x) \leq y < g_e(2x+1)]\}$. It is easy to see that $\mathcal{F}' = \{B_e \colon e \in E\}$ is a weakly represented family of sets.

By COHW there is an $\mathcal{F}'$-cohesive set $C \in S$. Then let h be the principal function of C; that is, h is strictly monotonically increasing and $C = \{h(0), h(1), \ldots\}$. Then $h \in S$ as well. Also note that $h(n) \geq n$ holds trivially for all $n \in M$.

Fix $e \in E$. Firstly, consider the case that there is some $b \in M$ such that $C \cap \{x \colon x \geq b\} \subseteq B_e$. Then $C \cap \{y \colon g_e(2x+1) \leq y < g_e(2x+2)\}$ is empty for almost all x. We claim that $h(g_e(2x+1)) > f_e(g_e(2x+1))$ for sufficiently large numbers x. To see this observe that $h(g_e(2x+1)) \in C$ and thus in B_e, which implies $h(g_e(2x+1)) \geq g_e(2x+2)$ as the smallest element of B_e larger than $g_e(2x+1)$ is $g_e(2x+2)$. Then, by definition of g_e,

$$h(g_e(2x+1)) \geq g_e(2x+2)$$
$$= \max\{f_e(x') + 1 \colon x' \leq g_e(2x+1) + 1\}$$
$$> f_e(g_e(2x+1)).$$

Secondly, consider the case that there is some $b \in M$ with the property that $(C \cap \{x \colon x \geq b\}) \cap B_e = \emptyset$. Then $C \cap \{y \colon g_e(2x) \leq y < g_e(2x+1)\}$ is empty for almost all x. For similar reasons as in the previous case, $h(g_e(2x)) > f_e(g_e(2x))$ for almost all x.

Due to C's $\mathcal{F}'$-cohesiveness, one of the two cases must occur. As a result, the set $\{y \in M \colon h(y) > f_e(y)\}$ is guaranteed to be infinite. Since $e \in E$ was chosen arbitrarily, the requirements of HI with regards to $\mathcal{F}$ are satisfied; furthermore, since $\mathcal{F}$ was chosen arbitrarily as well, HI holds in general.   □

7   The Domination Principle

In this section we show that over $\mathsf{RCA}_0 + \mathsf{B}\Sigma_2$, the principle DOM implies COH and COHW. It is an open question whether the assumption $\mathsf{B}\Sigma_2$ is needed.

Theorem 44. *Over* $\mathsf{RCA}_0 + \mathsf{B}\Sigma_2$, *DOM implies* COH.

Proof. Hölzl, Jain and Stephan [14, Theorem 20] showed that over $\mathsf{RCA}_0 + \mathsf{B}\Sigma_2$, DOM implies $\mathsf{I}\Sigma_2$. Thus we can assume that $\mathsf{I}\Sigma_2$ holds for the purposes of this proof.

Let $\mathcal{M} = (M, S, +, \cdot, 0, 1)$ be a model of DOM and let $\mathcal{F}$ be a uniform family of sets represented by $A \in S$. For $e \in M$ let A_e be as in Definition 7 and let $\widetilde{f}_{e,x}(y)$ be the first $z > y$ with $\forall d \leq e\, [A_d(z) = A_d(x)]$ and let $\widetilde{\mathcal{F}}$ be the weakly represented family of those $\widetilde{f}_{e,x}$ which are total. By DOM there is a function $g \in S$ which dominates all members of $\widetilde{\mathcal{F}}$. Define an infinite set $G = \{x_0, x_1, \ldots\} \in S$ as follows:

- $x_0 = 0$ and
- Let $X_n = \{x_n + 1, x_n + 2, \ldots, x_n + g(x_n)\}$ and define x_{n+1} as the minimal $y \in X_n$ such that

$$A_0(y)A_1(y)\ldots A_{x_n}(y) = \max\{A_0(z)A_1(z)\ldots A_{x_n}(z) \colon z \in X_n\},$$

where the maximum is with respect to $\leq_{\mathrm{lex}}$, the lexigraphic ordering on strings.

Let $\Psi(e, x)$ be the statement

$$x \in G \wedge \forall y \geq x\, [y \in G \to A_0(y)A_1(y)\ldots A_{e-1}(y) = A_0(x)A_1(x)\ldots A_{e-1}(x)].$$

Claim. For all e, $\exists x\, (\Psi(e, x))$ holds.

Proof. As $\exists x\, \Psi(e, x)$ is a Σ_2^0-statement, using $\mathsf{I}\Sigma_2$, we can prove it by induction over $e \in M$.

The statement $\Psi(0, x)$ holds vacuously for all $x \in G$. So assume by induction that for a given $e \in M$, $\Psi(e, x')$ is true for some $x' \in G$. We distinguish two cases:

Case 1. $G \cap A_e$ is finite. Then there exists an $x'' \geq x'$ with $x'' \in G$ and $x'' > \max(A_e \cap G)$. Then for all $y \in G$ with $y \geq x''$, we have $A_e(y) = A_e(x'') = 0$ on the one hand; and by the induction hypothesis

$$A_0(y)A_1(y)\ldots A_{e-1}(y) = A_0(x'')A_1(x'')\ldots A_{e-1}(x'')$$

on the other hand. Thus $\Psi(e+1, x'')$ holds and $\exists x\, \Psi(e+1, x)$ is satisfied.

Case 2. $G \cap A_e$ is infinite. Then let x'' be any element of $G \cap A_e$ with $x'' \geq x'$. For all such x'' the function $\widetilde{f}_{e,x''}$ is the same and thus one can, without loss of generality, assume that x'' is large enough that $g(y) > \widetilde{f}_{e,x''}(y)$ for all $y \geq x''$ and $x'' > e+1$. Now let $n \in M$ be arbitrary such that $x_n \geq x''$. Then

$$A_0(x_{n+1})A_1(x_{n+1})\ldots A_{x_n}(x_{n+1}) \geq_{\text{lex}} A_0(x'')A_1(x'')\ldots A_e(x'')0^{x_n - e - 1};$$

and thus,

$$A_0(x_{n+1})A_1(x_{n+1})\ldots A_e(x_{n+1}) = A_0(x'')A_1(x'')\ldots A_e(x'').$$

As $n \in M$ was arbitrary with $x_n \geq x''$ it follows that $\Psi(e+1, x'')$ holds and that $\exists x\, \Psi(e+1, x)$ is satisfied. $\diamond$

Thus $\exists x\, \Psi(e, x)$ holds for all e, and in particular for each e there is an $x \in \dot{G}$ with $A_e(y) = A_e(x)$ for all $y \geq x$ with $y \in G$. Thus COH is satisfied. $\square$

In fact, we can obtain the following stronger result.

Corollary 45. *Over* RCA$_0$ + BΣ_2, DOM *implies* COHW.

This corollary follows immediately from Theorem 44 and the following observation.

Proposition 46. *Over* RCA$_0$ + DOM, COH *implies* COHW.

Proof. Let $\mathcal{F}$ be a weakly represented family of sets represented by A, let f_e be as in Definition 5, let $E = \{e \colon f_e \text{ is total}\}$ and for all $e \in E$ write B_e for the set whose characteristic function is f_e.

For every $e \in M$, define a function $\widetilde{f}_e$ which on input x outputs the smallest $z \in M$ such that either $\langle e, x, 0, z\rangle$ or $\langle e, x, 1, z\rangle$ is in A. It is easy to see that $\{\widetilde{f}_e \colon e \in E\}$ is a weakly represented family of functions. Then, by DOM, there is a function $g \in S$ dominating all functions $\widetilde{f}_e$, $e \in E$. Observe that then $\{C_e \colon e \in M\}$, as defined by $C_e = \{x \colon \exists z \leq g(x)\,[\langle e, x, 1, z\rangle \in A]\}$, is a uniform family of sets.

Let $e \in M$. If $e \in E$ then $C_e \subseteq B_e$ by definition, and, as g dominates all $\widetilde{f}_e$ with $e \in E$, there is a b_1 such that all $x > b_1$ satisfy $C_e(x) = B_e(x)$. If, on the other hand, $e \notin E$ then there is a b_0 such that $\langle e, x, 1, z\rangle \notin A$ for all $x > b_0$ and all $z \in M$; thus C_e is finite. Let $b = b_0$ if $e \notin E$, and let $b = b_1$ otherwise.

Now, by COH, there is an infinite set $D \in S$ such that, for every $e \in M$, there is a bound b' satisfying that for all $x, x' > b'$, if $x, x' \in D$ then $C_e(x) = C_e(x')$.

Thus for all $x, x' > \max(b, b')$, if $x, x' \in D$ then $B_e(x) = B_e(x')$. That is, D witnesses that the requirements of COHW concerning $\mathcal{F} = \{B_e \colon e \in E\}$ are satisfied. As $\mathcal{F}$ was arbitrary, COHW is satisfied in general. $\square$

Note that a similar result also holds for WKL$_0$ in place of DOM, that is, over RCA$_0$ + WKL$_0$, COH implies COHW. The reason is that WKL$_0$ proves that every weakly represented family $\mathcal{F}$ of sets is contained in a uniformly represented family $\mathcal{G}$ of sets, from which it follows that COH is equivalent to COHW.

Hirschfeldt [12, Open Question 9.18] asked if RCA$_0$ + CADS implies COH. Here CADS is the principle that whenever $\sqsubseteq \in S$ is a linear ordering on M then there is an infinite set $A \in S$ such that for every $i \in A$ there is a $k \in M$ such that either all $j \in A$ satisfy $k < j \rightarrow i \sqsubseteq j$ or all $j \in A$ satisfy $k < j \rightarrow j \sqsubseteq i$. We now show that RCA$_0$ + DOM $\vdash$ CADS, and thus an affirmative answer to Hirschfeldt's question would also prove RCA$_0$ + DOM $\vdash$ COH. Note that RCA$_0$ + DOM does not imply the closely related principle SADS (Hirschfeldt [12, Definition 9.16]); this is because SADS $\vdash$ BΣ_2 while DOM $\nvdash$ BΣ_2.

Theorem 47. *Over* RCA$_0$, DOM *implies* CADS.

Proof. Let a linear ordering $\sqsubseteq \in S$ be given and define for each $e \in M$ the function f_e via $f_e(i) = \min\{j \geq i : e \sqsubseteq j\}$. Note that f_e is total iff there are infinitely many j with $e \sqsubseteq j$. Then $\mathcal{F} = \{f_e : f_e \text{ total}\}$ forms a weakly represented family and so all functions $f \in \mathcal{F}$ are dominated by a single function $g \in S$. Let

$$h(i) = \max_{\sqsubseteq}\{j : i \leq j \leq i + g(i)\}$$

and let A be the range of h. Then $i \in A \Leftrightarrow i \in \{h(0), h(1), \ldots, h(i)\}$.

Now let e be given. If there are infinitely many j with $e \leq j$ then, for almost all i, there is a $j \in \{i, i+1, \ldots, i+g(i)\}$ with $e \sqsubseteq j$; it follows that $e \sqsubseteq h(i)$. If there are only finitely many such j then $h(i) \sqsubseteq e$ for almost all i. Thus for each e it holds that either almost all $j \in A$ satisfy $e \sqsubseteq j$ or almost all $j \in A$ satisfy $j \sqsubseteq e$.

As the choice of $\sqsubseteq$ was arbitrary, CADS holds. $\qquad\square$

8   DOM does not Imply SRT$_2^2$

We now construct an ω-model witnessing that DOM does not imply SRT$_2^2$. We require the following lemma.

Lemma 48. *Let A be Martin-Löf random relative to Ω. Then A does not compute any infinite subset of Ω or $\overline{\Omega}$.*

Proof. Without loss of generality, assume that A computes an infinite subset G of Ω; the case $\overline{\Omega}$ is symmetric. By Theorem 40 we have that Ω is Martin-Löf random relative to A. Since $G \leq_T A$, Ω is also Martin-Löf random relative to G. Let $(b_i : i \in \omega)$ be a strictly monotone listing of the elements of G. Then it is easy to see that the sequence $(U_n : n \in \omega)$ defined via

$$U_n = [\{\sigma \in \{0,1\}^{b_n+1} : \sigma(b_i) = 1 \text{ for all } i \in \{0, 1, \ldots, n\}\}]$$

is a G-Martin-Löf test covering Ω, contradiction. $\qquad\square$

Theorem 49. DOM *does not imply* SRT_2^2.

Proof. We construct an ω-model of $\mathsf{DOM} + \neg\mathsf{SRT}_2^2$. To achieve this, we will use a result of Chong, Lempp and Yang [5] and Cholak, Jockusch and Slaman [4] who proved that SRT_2^2 is equivalent to the following principle D_2^2:

For every Δ_2^0 set $G \subseteq \omega$ there exists an infinite $A \subseteq \omega$ such that $A \subseteq G$ or $A \subseteq \overline{G}$.

To ensure $\neg\mathsf{SRT}_2^2$ it is therefore enough to ensure $\neg\mathsf{D}_2^2$. To this end, for all $n \in \omega$, let $A_n = \Omega^{\emptyset'} \oplus \Omega^{\emptyset''} \oplus \ldots \oplus \Omega^{\emptyset^{(n+1)}}$, where $\emptyset^{(i)}$ is the i-th Turing jump for $i \in \omega$. Now let

$$S = \{A \subseteq \omega \colon A \leq_T A_n \text{ for some } n \in \omega\}$$

and let $\mathcal{M} = (\omega, S, +, \cdot, 0, 1)$. As $\emptyset' =_T \Omega$ we have that $\Omega^{\emptyset'}$ is Martin-Löf random relative to Ω, and by repeated application of Theorem 40 it follows that, for any $n \in \omega$, A_n is Martin-Löf random relative to Ω. By Lemma 48 we obtain that no set in S computes an infinite subset of Ω or $\overline{\Omega}$. But since Ω is Δ_2^0 this implies $\neg\mathsf{D}_2^2$.

To see that DOM is satisfied by $\mathcal{M}$ let a weakly represented family $\mathcal{F}$ represented by $A \leq_T A_i$ for $i \in \omega$ large enough be given. Note that A_{i+1} is high relative to A_i, and that it therefore computes a function g dominating all functions computable from A, in particular g dominates all $f \in \mathcal{F}$. As $\mathcal{F}$ was arbitrary, this establishes DOM. □

9 Restricted Π_2^1-conservativeness of DOM over RCA$_0$

In this section we will prove that given any restricted Π_2^1-sentence φ,

if $\mathsf{DOM} + \mathsf{RCA}_0 \vdash \varphi$, then $\mathsf{RCA}_0 \vdash \varphi$.

Here a formula φ is called a restricted Π_2^1-sentence iff it is of the form

$$\forall X\, [\alpha(X) \rightarrow \exists Y\, [\beta(X, Y)]]$$

where X, Y are quantified variables ranging over the second order part of the model in question, α is any arithmetical formula and β is a Σ_3^0-formula. We begin by introducing the following concepts.

Definition 50. Given a structure $\mathcal{M} = (M, S, +, \times, 0, 1, <)$ of second order arithmetic and $g \subseteq M$, let $\mathcal{M}_g$ be the $\mathcal{L}_2$-structure $(M, S \cup \{g\}, +, \times, 0, 1, <)$ and $\mathcal{M}[g]$ be the $\mathcal{L}_2$-structure $(M, \Delta_1^0(\mathcal{M}_g), +, \times, 0, 1, <)$ where

$$\Delta_1^0(\mathcal{M}_g) = \{X \subseteq M : X \text{ is } \Delta_1^0 \text{ definable over } \mathcal{M}_g\}.$$

Remark 51. By a result of Simpson [30, Lemma IX.1.8], for every $\mathcal{L}_2$-structure $\mathcal{M}$ and $g \subseteq M$, if $\mathcal{M}_g$ satisfies the basic axioms and $\mathsf{I}\Sigma_1$, then $\mathcal{M}[g]$ is a model of RCA_0.

Hirschfeldt [12, Proposition 7.16] proved that a statement of the form

$$\forall X\,[\vartheta(X) \to \exists Y\,[\psi(X,Y)]], \tag{$\dagger$}$$

where ϑ and ψ are arithmetic formulas, is restricted Π_2^1-conservative over RCA_0 iff one can for every countable structure $\mathcal{M} = (M, S, +, \times, 0, 1, <)$ and every $X \in S$ with $\mathcal{M} \models \vartheta(X)$ find an extension $\mathcal{N} = (M, \Delta_1^0(\mathcal{M}_g), +, \times, 0, 1, <)$ such that

- $\mathcal{N} \models \mathsf{I}\Sigma_1$,
- $\mathcal{N} \models \psi(X, g)$,
- for every Σ_3-formula $\rho(Y, Z)$ whose free variables are exactly $\{Y, Z\}$ and every $Z \in S$, if $\mathcal{N} \models \exists Y\,[\rho(Z, Y)]$ then $\mathcal{M} \models \exists Y\,[\rho(Z, Y)]$.

Note that DOM is given by a Π_2^1-formula of the form ($\dagger$) where $\vartheta(X)$ states that X represents a weakly represented family $\mathcal{F}$ of functions and $\psi(X, Y)$ states that Y dominates every function in $\mathcal{F}$. Thus we can use Hirschfeldt's criterion to prove the following theorem.

Theorem 52. DOM *is restricted* Π_2^1-*conservative over* RCA_0.

Proof. We use the coinfinite extension method of Kleene and Post [19], Lacombe [20] and Spector [31] as described by Odifreddi [26, Theorem V.4.3] to prove the result; these methods will be adjusted to work on countable models of arithmetic. The function g above will be constructed by an induction over the natural numbers for which we use a list covering the countable set of requirements listed below; furthermore, we use that there is a cofinal ascending sequence $a_0, a_1, \ldots$ of elements of M and that every ascending sequence $b_0, b_1, \ldots$ with $b_n \geq a_n$ for all n is also cofinal. The following invariant will be maintained at all stages n:

At the beginning of stage n, the function g is defined for all $\langle x, y \rangle$ with $x < b_n$ and its extension $\tilde{g}$ is in S where $\tilde{g}$ takes the value 0 on those places where g is not yet defined. Furthermore, when b_{n+1} is chosen to satisfy the requirement, it is done in such a way that $\max\{b_n, a_{n+1}\} \leq b_{n+1}$.

The ideas of this construction combine the original result of Spector as presented by Odifreddi with ideas of Hirschfeldt [12, Chaps. 6 and 7]. The requirements used are the three items below; they are stated together with a description of how they are realised at the stage n where they get attention; and as there are only countably many of these (all parameters range over S and M), there is a non-effective enumeration of these conditions by natural numbers.

- For all $X \in S$ and Turing reductions $F \in S$ and $u \in M$, if $F^{g \oplus X}$ is total and $\{0, 1\}$-valued and for all v there is an w with $F^{g \oplus X}(u, v, w) = 1$ then there is an $h \in S$ with $F^{h \oplus X}$ being total and $\{0, 1\}$-valued and for all v there is a w with $F^{h \oplus X}(u, v, w) = 1$.

This requirement is satisfied as follows: Let $c_0 = 0$ and η_0 be the everywhere undefined function; for $m = 0, 1, \ldots$ we search for a finite function η_{m+1} and a value c_{m+1} such that the following conditions hold:

- $c_{m+1} > m$;
- the domain of η_{m+1} is $\{\langle x, y\rangle : x, y < c_{m+1}\}$ and η_{m+1} can be coded using an element of M;
- all $x, y < c_m$ satisfy $\eta_{m+1}(\langle x, y\rangle) = \eta_m(\langle x, m\rangle)$;
- all $x < \min\{b_n, c_{m+1}\}$ and all $y < c_{m+1}$ satisfy $\eta_{m+1}(\langle x, y\rangle) = g(\langle x, y\rangle)$;
- $F^{\eta_{m+1} \oplus X}(u, m, w) = 1$ for some $w < c_{m+1}$ without F querying the first component of the join $\eta_{m+1} \oplus X$ outside the domain of η_{m+1};
- $F^{\eta_{m+1} \oplus X}(\tilde{u}, \tilde{v}, \tilde{w})$ terminates with a value from $\{0,1\}$ without querying outside the domain of η_{m+1} for all $\tilde{u}, \tilde{v}, \tilde{w} < c_m$.

By $\mathsf{I}\Sigma_1$ there are only two cases.

Case 1. The construction goes through for all m yealding in the limit a total extension h of the part of g constructed so far such that $F^{h \oplus X}$ is total and $\{0,1\}$-valued and the value u satisfies that for every v there is a w with $F^{h \oplus X}(u, v, w) = 1$. As the part of g constructed prior to stage n is the restriction of a function in S to a domain in S, the so constructed h is also in S. In this case the requirement is satisfied and one selects $b_{n+1} = \max\{a_{n+1}, b_n\}$ and defines for all x with $b_n \leq x < b_{n+1}$ and all y that $g(\langle x, y\rangle) = 0$.

Case 2. The construction progresses until it reaches an m for which the extension η_{m+1} cannot be found; the existence of such an m in the case that not all m are used follows from $\mathsf{I}\Sigma_1$. Now any common extension $\tilde{g}$ of the part of g built so far and of η_m found so far satisfies that either $F^{\tilde{g} \oplus X}$ is undefined or above 2 for some inputs $(\tilde{u}, \tilde{v}, \tilde{w})$ or $v = m$ satisfies that there is no w with $F^{\tilde{g} \oplus X}(u, v, w) = 1$. Now one extends g as follows: $b_{n+1} = \max\{a_{n+1}, b_n, c_m\}$ and for all $x < b_{n+1}$ and all $y \in M$ one defines if $x < b_n$ then $g(\langle x, y\rangle)$ is defined as done previously else if $x < c_m$ and $y < c_m$ then $g(\langle x, y\rangle) = \eta_m(\langle x, y\rangle)$ else $g(\langle x, y\rangle) = 0$.

Note that in both cases, one extends the function with finite case distinction between finite functions codable in S and existing functions which are restrictions of functions in S to a domain in S; then the newly extended part of g is also a restriction of a function in S to the domain $\{\langle x, y\rangle : x < b_{n+1}\}$, which is a set in S.

– For all $X \in S$ and Turing reductions $F \in S$, the range of $F^{g \oplus X}(M)$ has a minimum.

This requirement is satisfied as follows: Let $c_0 = 0$ and η_0 be the everywhere undefined function; for $m = 0, 1, \ldots$ we search for a finite function η_{m+1} and a value c_{m+1} such that the following conditions hold:

- $c_{m+1} > m$;
- the domain of η_{m+1} is $\{\langle x, y\rangle : x, y < c_{m+1}\}$ and η_{m+1} can be coded using an element of M;
- all $x, y < c_m$ satisfy $\eta_{m+1}(\langle x, y\rangle) = \eta_m(\langle x, m\rangle)$;
- all $x < \min\{b_n, c_{m+1}\}$ and all $y < c_{m+1}$ satisfy $\eta_{m+1}(\langle x, y\rangle) = g(\langle x, y\rangle)$;
- if $m = 0$ then $v = F^{\eta_{m+1} \oplus X}(w)$ is defined for some w; else there is a w such that $F^{\eta_{m+1} \oplus X}(w)$ is defined and bounded by $v - m$ with the v from

the case $m = 0$; furthermore, the computation of $F^{\eta_{m+1} \oplus X}(w)$ does not query any elements of the η_{m+1}-part of the join $\eta_{m+1} \oplus X$ except those where η_{m+1} is defined.

By $\mathsf{I}\Sigma_1$ this construction runs only up to some m; this m is at most v for the v chosen at $m = 0$. The reason for this is that afterwards the requirement would be that there is a w for which $F^{\eta_{m+1} \oplus X}(w)$ is defined and negative; however, this is not allowed as the outputs of the function are all in M. So now let m be the maximum number for which η_m is defined, this number exists by $\mathsf{I}\Sigma_1$. Then any total common extension $\tilde{g}$ of the part of g constructed so far and of η_m satisfies that the so defined function does not take values below $v - m$ while the value $v - m$ exists by the existence of η_m.

Let $b_{n+1} = \max\{a_{n+1}, b_n, c_m\}$ and define that $g(\langle x, y \rangle)$ with $b_n \leq x < b_{n+1}$ takes the value $\eta_m(\langle x, y \rangle)$ in the case that $x, y < c_m$ and takes the value 0 otherwise. The so chosen extension is again the restriction of a function in S to the domain $\{\langle x, y \rangle : x < b_{n+1}\}$ which is also a set in S. Furthermore, the function computed by F from g takes a minimum and so this necessary requirement towards satisfying $\mathsf{I}\Sigma_1$ in the model $\mathcal{M} \cup \{g\}$ is satisfied.

– For all $f \in S$ there is a $x \in M$ such that $\forall y \, [g(\langle x, y \rangle) = f(y)]$.

This is the easiest requirement to satisfy: If $f \in S$ is the function in question, then we select $b_{n+1} = \max\{a_{n+1}, b_n + 1\}$ and define $g(\langle x, y \rangle) = f(y)$ for all $y \in M$ and all x with $b_n \leq x < b_{n+1}$.

The last requirement ensures that g codes a uniform family which contains all functions contained in any weakly represented family in $\mathcal{M}$. Thus it follows that the function $h(y) = 1 + (\sum_{x=0}^{y} g(\langle x, y \rangle))$ dominates every weakly represented family in the model $\mathcal{M}$ and h is clearly a function in $\mathcal{M} \cup \{g\}$. It follows that the preconditions of Proposition 7.16 by Hirschfeldt [12] are satisfied and therefore DOM is restricted Π_2^1-conservative over RCA_0.   □

Theorem 53. DOM *is not* Π_1^1-*conservative over* $\mathsf{RCA}_0 + \mathsf{B}\Sigma_2$.

Proof. Hölzl, Jain and Stephan [14, Theorem 20] showed that over $\mathsf{RCA}_0 + \mathsf{B}\Sigma_2$, DOM implies $\mathsf{I}\Sigma_2$. So we have that DOM $+ \mathsf{RCA}_0 + \mathsf{B}\Sigma_2 \vdash \mathsf{I}\Sigma_2$, while it is well-known that $\mathsf{RCA}_0 + \mathsf{B}\Sigma_2 \nvdash \mathsf{I}\Sigma_2$. But since $\mathsf{I}\Sigma_2$ can be formalised by a Π_1^1-statement, DOM is not Π_1^1-conservative over $\mathsf{RCA}_0 + \mathsf{B}\Sigma_2$.   □

This result stands in contrast to the result of Chong, Slaman and Yang [6] that COH is Π_1^1-conservative over $\mathsf{RCA}_0 + \mathsf{B}\Sigma_2$. Furthermore, as DOM implies AVOID, MEET and BI, we obtain the following immediate consequence.

Corollary 54. *The following are restricted* Π_2^1-*conservative over* RCA_0:

(a) AVOID,
(b) MEET,
(c) BI.

Finally, the results in this section provide another proof of Theorem 49; the argument being that $\mathsf{RCA}_0 + \mathsf{DOM} + \neg\mathsf{B}\Sigma_2$ has a model; and that such a model cannot satisfy SRT_2^2, as this would contradict the result of Cholak, Jockusch and Slaman [4] that $\mathsf{RCA}_0 + \mathsf{SRT}_2^2 \vdash \mathsf{B}\Sigma_2$.

10   Almost Disjointness and Independence

In this section we prove that in ω-models MAD and MIND coincide with the negations of previously known principles.

Theorem 55. *An ω-model satisfies* MAD *iff it does not satisfy* DOM.

Proof. ($\Rightarrow$): Let $\mathcal{F}$ be a weakly represented family of sets represented by $A \in S$ that is almost disjoint. Suppose that DOM holds; we will show that this implies $\neg$MAD.

Assume without loss of generality that for the characteristic function f of every set in $\mathcal{F}$ there is a unique $e \in \omega$ such that f is weakly represented by A_e (where A_e is as in Definition 5). Indeed, this can be achieved by replacing A with a set A' derived from it, where A' and $A'_e = \{n : \langle e, n \rangle \in A'\}$ are such that whenever f'_e (the function weakly represented by A'_e) looks identical to f'_d for some $d < e$, the enumeration of elements into A'_e is suspended; this way, should there indeed be a $d < e$ with $f_e = f_d$ in the limit, then f'_e will become non-total, and A'_e will not weakly represent any function in $\mathcal{F}$.

As a consequence of the previous assumption, if, for some $d \neq e$, A_d and A_e weakly represent the characteristic functions of sets $F \in \mathcal{F}$ and $G \in \mathcal{F}$ respectively, then $F \cap G$ is finite.

Let $(\varphi_e^A : e \in \omega)$ be an enumeration of all A-recursive functions.

Claim. There is a function $g \in S$ that dominates every A-recursive function in the following sense: For every total φ_e^A and almost all n it holds that

$$g(n + 1) > \varphi_e^A(g(n)).$$

Proof. Consider the weakly represented family of all A-recursive functions, and apply DOM to obtain a function $\widehat{g}$ dominating it. Without loss of generality, assume that $\widehat{g}$ is strictly increasing, let $g(0) = 0$ and define for all n that $g(n + 1) = \widehat{g}(g(n))$. $\diamond$

Let $h(x, n)$ be the least number d such that either $\varphi_d^A(x)[g(n + 2)]\!\downarrow = 1$ or $d = n$. Let B be the set consisting of numbers $b_n \in \omega$, $n \in \omega$, with $g(n) < b_n < g(n + 1)$ and $h(b_n, n) \geq h(x, n)$ for all x with $g(n) < x < g(n + 1)$.

Informally, for an element x, the value $h(x)$ tells us that x does not seem to show up in those sets that have characteristic functions who have an A-recursive index up to $h(x)$; of course this can only be determined given an enumeration timebound, which is provided by the dominating function g here. Then B picks elements where this number is as large as possible.

More formally, note that by the choice of g, if φ_d^A coincides with the characteristic function f_e of a set in $\mathcal{F}$, then for almost all n there is an x with $g(n) < x < g(n + 1)$ such that $f_e(x)[g(n + 2)]\!\downarrow = 1$ and, due to the almost disjointness, for all $d < e$ either $f_d(x)\!\uparrow$ or $f_d(x)\!\downarrow = 0$. As by construction B consists only of numbers of this type, for almost all n it holds that $b_n \notin A_d$ for $d < e$ and therefore the set B has finite intersection with every $C \in \mathcal{F}$. Thus MAD is not satisfied.

($\Leftarrow$): Let $\mathcal{M} = (\omega, S, +, \cdot, 0, 1)$ be an ω-model of $\neg$DOM. We assume without loss of generality that S contains no high set; otherwise carry out the construction below relative to an oracle relative to which no high set in S exists.

The fact that we don't know which indeces e describe total recursive functions φ_e is a complication in the construction that follows. To circumvent this issue, we take advantage of the possibilities that the concept of weakly represented families offer, namely that partial information about functions is ignored when defining such a family; only total functions are considered a member of the family. Using this, we build a recursive numbering of partial-recursive functions such that the total functions appearing in it are all $\{0, 1\}$-valued and when interpreted as characteristic functions of sets, the collection of these sets is a maximal almost disjoint family.

Let $(\varphi_e : e \in \omega)$ be an enumeration of all recursive functions. First we build the uniformly recursive helper procedures $\psi_{c_0, c_1, \ldots, c_e}$ for all $e \in \omega$ with $c_d \in \{0, 1, \ldots, \infty\}$ for $d \leq e$. We call $(c_0, c_1, \ldots, c_e)$ *true parameters* if, for all $d \leq e$, c_d is the minimal i such that $\varphi_d(i)\uparrow$ if such an i exists, and $c_d = \infty$ if φ_d is total.

The procedure $\psi_{c_0, c_1, \ldots, c_e}$ has three states: *wait*, *success*, and *aborted*. When we define the enumeration of the characteristic function of A_e below, we will only enumerate a new function value whenever $\psi_{c_0, c_1, \ldots, c_e}$ is in state *success*. The idea is that this will only happen infinitely often, when $(c_0, c_1, \ldots, c_e)$ are the true parameters. If $(c_0, c_1, \ldots, c_e)$ are not true parameters, then $\psi_{c_0, c_1, \ldots, c_e}$ will either be stuck in state *wait* forever, or it will enter state *aborted* and stay in it forever. Then the true parameters will be the only parameters used to define A_e.

To achieve what we just described, we proceed as follows: $\psi_{c_0, \ldots, c_e}$ starts in state *wait* and runs the following $e + 1$ parallel procedures:

- For all $d \leq e$, the computations $\varphi_d(c_d)$, $d \leq e$, are run in parallel. If one of them ever terminates, then by definition $(c_0, c_1, \ldots, c_e)$ are not true parameters. Then $\psi_{c_0, \ldots, c_e}$ stops all computations, enters state *aborted*, and remains in this state permanently.
- In a single procedure, all computations $\varphi_d(c)$ with $d \leq e$ and $c < c_d$ are run *sequentially* and in order ascending with $\langle d, c \rangle$. While one of the computations runs, $\psi_{c_0, \ldots, c_e}$ is in state *wait*. Every time one of the computations $\varphi_d(c)$ terminates, $\psi_{c_0, \ldots, c_e}$ enters state *success*. If (d, c) was the last pair of parameters as above (which can only happen if all c_d, $d \leq e$, are finite) then remain in state *success* permanently. Otherwise enter state *wait* again, and continue with the next pair (d', c'), that is, with the smallest pair as above such that $\langle d', c' \rangle > \langle d, c \rangle$.

Note that this arrangement ensures that $\psi_{c_0, \ldots, c_e}$ is in state *success* at infinitely many stages if and only if $(c_0, c_1, \ldots, c_e)$ are the true parameters.

We can now describe how to produce a maximal almost disjoint family. In parallel, for all $e \in \omega$ and all possible sets of parameters $(c_0, c_1, \ldots, c_e)$ we run the following procedure.[1]

Run $\psi_{c_0,\ldots,c_e}$ step by step.
At every stage, check if $\psi_{c_0,\ldots,c_e}$ is currently in state *success*.
If so, let m be the smallest number not in $A_0 \cup A_1 \cup \ldots \cup A_{e-1}$, and check whether

$$m = n + \varphi_e(0) + \varphi_e(1) + \ldots + \varphi_e(n) \text{ for some } n. \qquad (*)$$

If not, enumerate m into A_e.

Note that if $(c_0, c_1, \ldots, c_e)$ are the true parameters, then checking $(*)$ is recursive, and the procedure never gets stuck. This finishes the construction.
We need to prove that the weakly represented family $\{A_n : n \in \omega\}$ constructed by this procedure is maximal almost disjoint. First note that for each $e \in \omega$ the complement of $A_0 \cup \ldots \cup A_e$ is infinite and contains at most n elements below $\varphi_e(n)$. Furthermore, A_e is disjoint with all A_d for $d < e$. As a consequence, $\{A_n : n \in \omega\}$ is almost disjoint.
It remains to show that $\{A_n : n \in \omega\}$ is also maximal almost disjoint. To see this let B be an infinite non-high set. Then there is a recursive function φ_e such that, for infinitely many n, there are more than $2n$ elements of B below $\varphi_e(n)$. It follows that the intersection of B with $A_0 \cup \ldots \cup A_e$ is infinite and therefore $B \cap A_d$ is infinite for some $d \leq e$. This completes the proof.   □

Theorem 56. *An ω-model satisfies* MIND *iff it does not satisfy* BI.

Proof. ($\Rightarrow$): As before, for a set $C \subseteq \omega$, let us write C^0 for C and C^1 for $\omega \setminus C$.

Let a weakly represented family $\mathcal{F}$ of sets represented by A be given. Also fix any collection $\{A_0, \ldots, A_{n-1}\} \subseteq \mathcal{F}$ and any string $\sigma \in {0,1}^n$, as well as a set B which is biimmune relative to A. Observe that the set $\widehat{A} = \bigcap_{i<n} A_i^{\sigma(i)}$ is A-recursive. Then B's biimmunity relative to A implies that both $\widehat{A} \cap B$ and $\widehat{A} \cap \overline{B}$ are infinite.

As $\{A_0, \ldots, A_{n-1}\}$ was arbitrary, it follows that $\mathcal{F} \cup \{B\}$ is still an independent family, which contradicts the assumption that $\mathcal{F}$ is maximal independent.
($\Leftarrow$): Similarly to the proof of the previous theorem, we work with lists of parameters where true parameters define sets in the maximal independent family that we need to construct. So assume that an ω-model $\mathcal{M} = (\omega, S, +, \cdot, 0, 1)$ and a set $A \in S$ are given such that no set $B \in S$ is biimmune relative to A; to simplify notation, we assume that A is recursive; otherwise carry out the construction relative to A.

---

[1] Note that to simplify notation, we do not explicitly define total characteristic functions of the sets A_e, $e \in \omega$, or the enumeration of a set that represents these functions as a weakly represented family. But since the elements of every A_e are enumerated in increasing order by the given procedure, it is easy to convert it into one defining the enumeration of such a set.

In the following, a stream is an infinite sequence of natural numbers in strictly ascending order. Each stream will be indexed with a string; the range of streams x^σ, x^τ is disjoint if σ, τ are incomparable as strings, and the range of x^σ is a superset of the range of x^τ when σ is a prefix of τ.

We begin the construction with the initial stream x^ε which is the sequence of all natural numbers, that is, $x_n^\varepsilon = n$ for all $n \in \omega$. We now describe how to define the streams x^σ, for strings σ, and then we argue that the set $\{E_e : e \in \omega\}$ with

$$E_e = \{m : \exists \sigma \in \{0,1\}^e \; \exists n \; [m = x_n^{\sigma 1}]\}$$

is a weakly represented family that is maximal independent. To define the streams x^σ, for strings σ, proceed as follows for all $e \in \omega$:

Let $R_e(n)$ be defined and let it equal $\varphi_e(n)$ iff the values $\varphi_e(0), \ldots, \varphi_e(n)$ are defined and in $\{0,1\}$; let $R_e(n)$ be undefined if there is $m \le n$ where $\varphi_e(m)$ is undefined or defined and at least 2. We define a function $\eta_e : \{0,1\}^e \to \{0,1\}$ as follows:

(a) If R_e is total and there exist both infinitely many $n \in \text{range}(x^\sigma)$ such that $R_e(n) = 0$ and infinitely many $n \in \text{range}(x^\sigma)$ such that $R_e(n) = 1$, then let, for $a = 0, 1$ and all $n \in \omega$, $x_n^{\sigma a}$ be the n-th element m of x^σ with $R_e(m) = a$. Informally this means that x^σ is split into $x^{\sigma 0}$ and $x^{\sigma 1}$ according to the values of R_e. Furthermore, let $\eta_e(\sigma) = 1$.

(b) Else let $x_n^{\sigma 0} = x_{2n}^\sigma$ and $x_n^{\sigma 1} = x_{2n+1}^\sigma$ for all $n \in \omega$. Informally this means that x^σ is split evenly into $x^{\sigma 0}$ and $x^{\sigma 1}$. Furthermore, let $\eta_e(\sigma) = 0$.

Note that η_e stores the information for which σ of length e cases (a) and (b) applied, respectively. This finishes the construction.

As an auxiliary notion needed for the verification, we define

t_d as the maximum n such that $n = 0$ or one can find a $\tau \in \{0,1\}^d$ and x_i^τ, x_j^τ with $\eta_d(\tau) = 0$, $n = \min\{x_i^\tau, x_j^\tau\}$, $R_d(x_i^\tau) = 0$ and $R_d(x_j^\tau) = 1$.

Note that for a given $\tau \in \{0,1\}^d$ the statement $\eta_d(\tau) = 0$ means that case (b) applied to τ above, and that $R_d(k)$ is the same value for all $k > t_d$ with $k \in \text{range}(x^\tau)$ such that $R_d(k)$ is defined. The same holds for all other τ of length d with $\eta_d(\tau) = 0$.

The verification consists of establishing the following two claims.

Claim. $\{E_e : e \in \omega\}$ is maximal independent.

Proof. Note that for $\sigma \in \{0,1\}^e$, $\text{range}(x^\sigma)$ is the intersection of all E_d with $d < e$ and $\sigma(d) = 1$ and all $\overline{E_d}$ with $d < e$ and $\sigma(d) = 0$. Further note that by construction $\text{range}(x^\sigma)$ has infinite cardinality. Thus $\{E_e : e \in \omega\}$ is independent.

To see that it is also maximal independent, consider any set B. As B cannot be biimmune it either has an infinite recursive subset or $\overline{B}$ has an infinite recursive subset; let e be such that R_e is the characteristic function of this set which we also denote R_e, slightly abusing notation. Now, for some $\sigma \in \{0,1\}^e$, x^σ has infinite intersection with R_e and therefore by construction almost all elements in $\text{range}(x^{\sigma 1})$ are also elements of R_e.

So for the Boolean combination of $E_0, E_1, \ldots, E_e$ that equals $\mathrm{range}(x^{\sigma 1})$ we have that it either equals an infinite subset of B or an infinite set disjoint with B. Thus $\{E_e \colon e \in \omega\} \cup \{B\}$ cannot be independent. $\diamond$

Claim. $\{E_e \colon e \in \omega\}$ is a weakly represented family.

Proof. Recall that the parameters η_e, $e \in \omega$, store for which σ of length e which of the two cases (a) and (b) was applied during the construction. The following construction is described for arbitrary parameter sets $(\widetilde{\eta}_0, \widetilde{\eta}_1, \ldots, \widetilde{\eta}_e, s)$. As in the proof of Theorem 55, for each e, the construction below will only define a set E_e if $(\widetilde{\eta}_0, \widetilde{\eta}_1, \ldots, \widetilde{\eta}_e) = (\eta_0, \eta_1, \ldots, \eta_e)$ and if s is a sufficiently large timebound. For all other parameter sets, the construction will get stuck eventually.

More formally, let $c = (\widetilde{\eta}_0, \widetilde{\eta}_1, \ldots, \widetilde{\eta}_e, s)$ be given, where $\widetilde{\eta}_d \in \{0,1\}^{2^d}$ for all $d \leq e$ and let $s \in \omega$. We describe how to inductively construct streams $\widetilde{x}^\sigma$ for each string σ based on c:

(a) If $\widetilde{\eta}_e(\sigma) = 1$, then let, for $a = 0, 1$ and all $n \in \omega$, $\widetilde{x}_n^{\sigma a}$ be the n-th element m of $\widetilde{x}^\sigma$ with $R_e(m) = a$.
(b) Else let $\widetilde{x}_n^{\sigma 0} = \widetilde{x}_{2n}^\sigma$ and $\widetilde{x}_n^{\sigma 1} = \widetilde{x}_{2n+1}^\sigma$ for all $n \in \omega$.

In other words, we try to mimic the previous construction, hoping that c is a set of true parameters. Now let

$\widetilde{t}_d$ be the maximum $n \leq s$ such that either $n = 0$ or such that one can find within time s some $\tau \in \{0,1\}^d$ and some $\widetilde{x}_i^\tau$, $\widetilde{x}_j^\tau$ with $\widetilde{\eta}_d(\tau) = 0$ and $n = \min\{\widetilde{x}_i^\tau, \widetilde{x}_j^\tau\}$ and such that $R_d(\widetilde{x}_i^\tau), R_d(\widetilde{x}_j^\tau)$ become defined within time s and $R_d(\widetilde{x}_i^\tau) \neq R_d(\widetilde{x}_j^\tau)$.

We now need to define an algorithm that uniformly from c produces a partial function F_c such that on the one hand, if for all $d \leq e$, $\widetilde{\eta}_d = \eta_d$, then for sufficiently large s, $\widetilde{t}_d = t_d$ for all $d \leq e$ and F_c is the characteristic function of E_e; and such that on the other hand, if for some $d \leq e$, $\widetilde{\eta}_d \neq \eta_d$ or $\widetilde{t}_d \neq t_d$, then F_c is only defined on finitely many inputs. The algorithm to compute $F_c(n)$ is as follows:

(1) Compute $\widetilde{x}_0^\sigma, \ldots, \widetilde{x}_n^\sigma$ for all $|\sigma| \leq e + 1$.
(2) Determine the unique $\sigma \in \{0,1\}^{e+1}$ such that there is an m with $\widetilde{x}_m^\sigma = n$.
(3) Search for n computation steps for a d and $\tau \in \{0,1\}^d$ such that $\widetilde{\eta}_d(\tau) = 0$ and for some $\widetilde{x}_i^\tau, \widetilde{x}_j^\tau > \widetilde{t}_d$ with $R_d(\widetilde{x}_i^\tau) \downarrow = 0$ and $R_d(\widetilde{x}_j^\tau) \downarrow > 0$. (If we find these, then $\widetilde{\eta}_d$ must be wrong, or $\widetilde{t}_d < t_d$.)
(4) If the computations in (1) terminate, the search in (2) is successful, and the search in (3) is *unsuccessful*, then output $F_c(n) = \sigma(e)$, else let $F_c(n)$ be undefined.

We verify that this algorithm behaves as required. First assume that all $\widetilde{\eta}_d = \eta_d$. Then all $\widetilde{x}^\sigma$ with $|\sigma| \leq e + 1$ are equal to x^σ. Furthermore, for large enough s the definition of $\widetilde{t}_d$ ensures that $\widetilde{t}_d = t_d$. Then the algorithm above produces a total function F_c and by (4) we have that F_c is the characteristic function of E_e.

If, on the other hand, there is a d such that either $\widetilde{\eta}_d \neq \eta_d$ or $\widetilde{t}_d < t_d$ then let d be the least such d. Note that then for σ with $|\sigma| \leq d$, it holds that $\widetilde{x}^\sigma = x^\sigma$. We argue that F_c is not total; there are several cases to consider.

- If $\eta_d(\tau) = 1$ and $\widetilde{\eta}_d(\tau) = 0$ for some $\tau \in \{0,1\}^d$, then there are infinitely many elements of x^τ for which R_e takes the value 1 and infinitely many for which R_e takes the value 0. However, as $\widetilde{\eta}_d(\tau) = 0$, for large enough n, some of these elements will be found in step (3) of the above algorithm, and $F_c(n)$ will be undefined.
- If $\eta_d(\tau) = 0$ and $\widetilde{\eta}_d(\tau) = 1$ for some $\tau \in \{0,1\}^d$, then by definition the streams $\widetilde{x}^{\tau 0}$ and $\widetilde{x}^{\tau 1}$ are defined from $\widetilde{x}^\tau = x^\tau$ by splitting according to the values of R_d; however, since $\eta_d(\tau) = 0$, one of $\widetilde{x}^{\tau 0}$ and $\widetilde{x}^{\tau 1}$ will then only contain finitely many elements. Then for sufficiently large n the algorithm will get stuck in step (1) when calculating $F_c(n)$.
- If $\widetilde{t}_d < t_d$, and the two previous cases do not apply, then $\widetilde{x}^{\tau a} = x^{\tau a}$ for all $\tau \in \{0,1\}^d$ and $a \in \{0,1\}$, and for sufficiently large n, the algorithm will in step (3) find the values $\widetilde{x}_i^\tau, \widetilde{x}_j^\tau > \widetilde{t}_d$ and $F_c(n)$ will be undefined.

Thus, the construction above only produces total functions F_c if

$$c = (\eta_0, \eta_1, \ldots, \eta_e, s)$$

for a sufficiently large $s \in \omega$; and in this case F_c is the characteristic function of E_e. As the construction is uniform in c, it is easy to see that $\{E_e : e \in \omega\}$ is a weakly represented family.   ◇

This completes the proof of Theorem 56.   □

Acknowledgments. The authors would like to thank C.T. Chong, Wei Li and Yue Yang for fruitful discussions and suggestions. They are also grateful to the anonymous referee for detailed and helpful comments, in particular for pointing out a simplification of the proof of Theorem 29.

References

1. Bartoszyński, T., Judah, H.: Set Theory: On the Structure of the Real Line. A K Peters Ltd., Wellesley (1995)
2. Blass, A.: Combinatorial cardinal characteristics of the continuum. In: Foreman, M., Kanamori, A. (eds.) Handbook of Set Theory, pp. 395–489. Springer, Dordrecht (2010). doi:10.1007/978-1-4020-5764-9_7. vols. 1, 2, 3
3. Brendle, J., Brooke-Taylor, A., Ng, K.M., Nies, A.: An analogy between cardinal characteristics and highness properties of oracles. In: Proceedings of the 13th Asian Logic Conference, Guangzhou, China, 16–20 September 2013, pp. 1–28. World Scientific (2013)
4. Cholak, P.A., Jockusch Jr., C.G., Slaman, T.A.: On the strength of Ramsey's theorem for pairs. J. Symbolic Logic **66**(1), 1–55 (2001)
5. Chong, C.T., Lempp, S., Yang, Y.: On the role of the collection principle for Σ_2^0 formulas in second order reverse mathematics. Proc. Am. Math. Soc. **138**, 1093–1100 (2010)
6. Chong, C.T., Slaman, T.A., Yang, Y.: Π_1^1-conservation of combinatorial principles weaker than Ramsey's theorem for pairs. Adv. Math. **230**, 1060–1077 (2012)
7. Demuth, O., Kučera, A.: Remarks on 1-genericity, semigenericity and related concepts. Commentationes Math. Univ. Carol. **028**(1), 85–94 (1987)

8. Downey, R.G., Hirschfeldt, D.R.: Algorithmic Randomness and Complexity. Springer, Heidelberg (2010)
9. Friedberg, R.: A criterion for completeness of degrees of unsolvability. J. Symbolic Logic **22**, 159–160 (1957)
10. Friedberg, R.: Three theorems on recursive enumeration. J. Symbolic Logic **23**, 309–316 (1958)
11. Hechler, S.H.: On the existence of certain cofinal subsets of $^\omega\omega$. In: Axiomatic Set Theory Proceedings of Symposia in Pure Mathematics, Part II, University of California, Los Angeles, California, 1967, vol. 13, pp. 155–173. American Mathematical Society, Providence (1974)
12. Hirschfeldt, D.R.: Slicing the Truth. World Scientific, River Edge (2015)
13. Hirschfeldt, D.R., Shore, R.A., Slaman, T.A.: The atomic model theorem and type omitting. Trans. Am. Math. Soc. **361**, 5805–5837 (2009)
14. Hölzl, R., Jain, S., Stephan, F.: Inductive inference and reverse mathematics. Ann. Pure Appl. Logic (2016). http://dx.doi.org/10.1016/j.apal.2016.06.002
15. Jockusch Jr., C.G.: Upward closure and cohesive degrees. Isr. J. Math. **15**, 332–335 (1973)
16. Jockusch, C.G., Soare, R.I.: Π_1^0 classes and degrees of theories. Trans. Am. Math. Soc. **173**, 33–56 (1972)
17. Jockusch, C.G., Stephan, F.: A cohesive set which is not high. Math. Logic Q. **39**, 515–530 (1993)
18. Kjos-Hanssen, B., Merkle, W., Stephan, F.: Kolmogorov complexity and recursion theorem. Trans. Am. Math. Soc. **263**, 5465–5480 (2011)
19. Kleene, S.C., Post, E.L.: The uppersemilattice of degrees of recursive unsolvability. Ann. Math. **59**, 379–407 (1954)
20. Lacombe, D.: Sur le semi-réseau constitué par les degrès d'indecidabilité récursive. Comptes rendus hebdomadaires des séances de l'Académie des Sciences **239**, 1108–1109 (1954)
21. van Lambalgen, M.: The axiomatization of randomness. J. Symbolic Logic **55**, 1143–1167 (1990)
22. Martin, D.A.: Classes of recursively enumerable sets and degrees of unsolvability. Zeitschrift für Mathematische Logik und Grundlagen der Mathematik **12**, 295–310 (1966)
23. Mileti, J.: Partition theorems and computability theory, Ph.D. dissertation, University of Illinois at Urbana-Champaign (2004)
24. Miller, A.W.: Some properties of measure and category. Trans. Am. Math. Soc. **266**, 93–114 (1981)
25. Nies, A.: Computability and Randomness. Oxford Science Publications, Oxford (2009)
26. Odifreddi, P.: Classical Recursion Theory. Elsevier, North Holland (1989)
27. Odifreddi, P.: Classical Recursion Theory, vol. 2. Elsevier, North Holland (1999)
28. Rupprecht, N.: Relativized Schnorr tests with universal behavior. Arch. Math. Logic **49**(5), 555–570 (2010)
29. Shelah, S.: Proper and Improper Forcing. Perspectives in Mathematical Logic, 2nd edn. Springer, Berlin (1998)
30. Simpson, S.G.: Subsystems of Second Order Arithmetic, 2nd edn. Cambridge University Press, Cambridge (2006)
31. Spector, C.: On degrees of unsolvability. Ann. Math. **64**, 581–592 (1956)
32. Soare, R.I.: Recursively Enumerable Sets and Degrees. Springer, Heidelberg (1987)
33. Stephan, F.: Recursion theory, lecture notes, school of computing, National University of Singapore, Technical report TR10/12 (2012)
34. Liang, Y.: Lowness for genericity. Arch. Math. Logic **45**, 233–238 (2006)

The Vitali Covering Theorem
in the Weihrauch Lattice

Vasco Brattka[1,2]([⊠]), Guido Gherardi[3], Rupert Hölzl[2], and Arno Pauly[4]

[1] Deptartment of Mathematics and Applied Mathematics, University of Cape Town,
Cape Town, South Africa
Vasco.Brattka@cca-net.de
[2] Faculty of Computer Science, Universität der Bundeswehr München,
Neubiberg, Germany
r@hoelzl.fr
[3] Dipartimento di Filosofia e Comunicazione, Università di Bologna, Bologna, Italy
Guido.Gherardi@unibo.it
[4] Départment d'Informatique, Université libre de Bruxelles, Brussels, Belgium
Arno.Pauly@cl.cam.ac.uk

Abstract. We study the uniform computational content of the Vitali
Covering Theorem for intervals using the tool of Weihrauch reducibility. We show that a more detailed picture emerges than what a related
study by Giusto, Brown, and Simpson has revealed in the setting of
reverse mathematics. In particular, different formulations of the Vitali
Covering Theorem turn out to have different uniform computational content. These versions are either computable or closely related to uniform
variants of Weak Weak König's Lemma.

1 Introduction

In order to analyze the uniform computational content of the Vitali Covering
Theorem in different versions it is useful to introduce some terminology that will
allow us to phrase these versions in succinct terms.

Let $\mathcal{I} = (I_n)_n$ be a sequence of open intervals $I_n \subseteq \mathbb{R}$, let $x \in \mathbb{R}$ and $A \subseteq \mathbb{R}$.
We say that $x \in \mathbb{R}$ is *captured* by $\mathcal{I}$, if for every $\varepsilon > 0$ there exists some $n \in \mathbb{N}$ with
$\mathrm{diam}(I_n) < \varepsilon$ and $x \in I_n$. We call $\mathcal{I}$ a *Vitali cover* of A, if every $x \in A$ is captured
by $\mathcal{I}$. We say that $\mathcal{I}$ is *saturated*, if $\mathcal{I}$ is a Vitali cover of $\bigcup \mathcal{I} := \bigcup_{n=0}^{\infty} I_n$. Finally,
we say that $\mathcal{I}$ *eliminates* A, if the I_n are pairwise disjoint and $\lambda(A \setminus \bigcup \mathcal{I}) = 0$,
where λ denotes the Lebesgue measure on $\mathbb{R}$.

Using this terminology we can now formulate the Vitali Covering Theorem
(see Richardson [18, Theorem 7.3.2]).

Theorem 1 (Vitali Covering Theorem). *Let $A \subseteq [0,1]$ be Lebesgue measurable and let $\mathcal{I}$ be a sequence of intervals. If $\mathcal{I}$ is a Vitali cover of A, then
there exists a subsequence $\mathcal{J}$ of $\mathcal{I}$ that eliminates A.*

This article is dedicated to Rod Downey on the occasion of his sixtieth birthday.
Vasco Brattka is supported by the National Research Foundation of South Africa.
Rupert Hölzl was partly supported by the Ministry of Education of Singapore
through grant R146-000-184-112 (MOE2013-T2-1-062).

© Springer International Publishing AG 2017
A. Day et al. (Eds.): Downey Festschrift, LNCS 10010, pp. 188–200, 2017.
DOI: 10.1007/978-3-319-50062-1_14

The Vitali Covering Theorem has been studied in reverse mathematics by Brown, Giusto, and Simpson [10] and was shown to coincide in proof strength with the well-known principle $\mathsf{WWKL_0}$ that stands for Weak Weak Kőnig's Lemma, see Simpson [19]. The following result can be found in Brown, Giusto, and Simpson [10, Theorems 3.3 and 5.5] and also in Simpson [19, Theorems X.1.9 and X.1.13]. For a related study in constructive analysis, see Diener and Hedin [11].

Theorem 2 (Brown, Giusto, and Simpson [10]). *Over* $\mathsf{RCA_0}$, *the following statements are equivalent to each other:*

1. *Weak Weak Kőnig's Lemma* $\mathsf{WWKL_0}$,
2. *The Vitali Covering Theorem (Theorem 1) for* $A = [0, 1]$,
3. *For any sequence of intervals* $\mathcal{I} = (I_n)_n$ *with* $[0, 1] \subseteq \bigcup \mathcal{I}$ *it holds that* $\sum_{n=0}^{\infty} \lambda(I_n) \geq 1$.

In a series of articles [1–3,5,6,12,13,15,16] by different authors the Weihrauch lattice was established as a uniform, resource-sensitive and hence more fine-grained version of reverse mathematics. Starting with work of Brattka and Pauly [8], Dorais, Dzhafarov, Hirst, Mileti and Shafer [12] and Brattka, Gherardi and Hölzl [4,5], probabilistic problems were studied in the Weihrauch lattice. In particular *positive choice* PC_X was considered, which is the problem of finding a point in a closed $A \subseteq X$ of positive measure, and the following relation to Weak Weak Kőnig's Lemma was established in the Weihrauch lattice [5, Proposition 8.2 and Theorem 9.3 and its proof].

Fact 3 (Weak Weak Kőnig's Lemma)

1. $\mathsf{WWKL} \equiv_{sW} \mathsf{PC}_{2^\mathbb{N}} \equiv_{sW} \mathsf{PC}_{[0,1]}$,
2. $\mathsf{WWKL} \times \mathsf{C}_\mathbb{N} \equiv_{sW} \mathsf{PC}_{\mathbb{N} \times 2^\mathbb{N}} \equiv_{sW} \mathsf{PC}_\mathbb{R}$.

Here $\equiv_{sW}$ stands for equivalence with respect to strong Weihrauch reducibility. We will provide exact definitions of the relevant terms in the following Sect. 2. In this article we are going to extend the work by Brown, Giusto and Simpson [10] using the tools of the Weihrauch lattice and we will demonstrate how the above mentioned equivalence classes and others feature in this approach.

One of our main insights is related to the observation that different logical formulations of the Vitali Covering Theorem turn out to have different uniform computational content, a phenomenon that appeared in a similar way in the study of the Baire Category Theorem by Brattka and Gherardi [2] and Brattka, Hendtlass and Kreuzer [7]. The following three propositional formulas essentially correspond to the different logical formulations of the Vital Covering Theorem that we consider:

0. $(S \wedge C) \to E$,
1. $(S \wedge \neg E) \to \neg C$,
2. $\neg E \to (\neg S \vee \neg C)$.

Here S corresponds to the statement that the input sequence is saturated, C to the statement that it is a cover and E to the statement that there is an eliminating subsequence. The stated propositional formulas are equivalent to each other when we have the full strength of classical logic at our disposal. More precisely, we are going to use the following versions of the Vitali Covering Theorem for the special case $A = [0,1]$:

0. VCT_0: For every Vitali cover $\mathcal{I}$ of $[0,1]$ there exists a subsequence $\mathcal{J}$ of $\mathcal{I}$ that eliminates $[0,1]$.
1. VCT_1: For every saturated $\mathcal{I}$ that does not admit a subsequence which eliminates $[0,1]$, there exists a point $x \in [0,1]$ that is not covered by $\mathcal{I}$.
2. VCT_2: For every sequence $\mathcal{I}$ that does not admit a subsequence which eliminates $[0,1]$, there exists a point $x \in [0,1]$ that is not captured by $\mathcal{I}$.

It is clear that 0. is equivalent to 2. since they are contrapositive forms of each other. We also obtain "0.⇒1." since every saturated cover of $[0,1]$ is a Vitali cover of $[0,1]$. Finally, we obtain "1.⇒0." since every Vitali cover $\mathcal{I}$ of $[0,1]$ can be extended to a saturated sequence $\mathcal{I}'$ by only adding intervals that do not overlap with the closed set $[0,1]$. Every subsequence $\mathcal{J}'$ of $\mathcal{I}'$ that eliminates $[0,1]$ then leads to a subsequence $\mathcal{J}$ of $\mathcal{I}$ that eliminates $[0,1]$. Our main results on the Vitali Covering Theorem can now be phrased as follows. The proofs will be presented in Sect. 3.

Theorem 4 (Vitali Covering Theorem). *We obtain that*

0. VCT_0 is computable,
1. $\mathsf{VCT}_1 \equiv_{sW} \mathsf{PC}_{[0,1]} \equiv_{sW} \mathsf{WWKL}$ and
2. $\mathsf{VCT}_2 \equiv_{sW} \mathsf{PC}_{\mathbb{R}} \equiv_{sW} \mathsf{WWKL} \times \mathsf{C}_{\mathbb{N}}$.

It can be argued that $\mathsf{C}_{\mathbb{N}}$ is the analogue of Σ_1^0–induction in the Weihrauch lattice (see Brattka and Rakotoniaina [9]) and hence the classes WWKL and $\mathsf{WWKL} \times \mathsf{C}_{\mathbb{N}}$ have no distinguishable non-uniform content in reverse mathematics, where Σ_1^0–induction is already included in RCA_0.

In this context it is also interesting to note that the equivalence classes of WWKL and $\mathsf{WWKL} \times \mathsf{C}_{\mathbb{N}}$ characterize certain natural classes of probabilistic problems. In [5, Corollary 3.4] the following was proved.

Fact 5 (Las Vegas Computability). *The following holds for any f.*

1. $f \leq_W \mathsf{PC}_{[0,1]} \iff f$ is Las Vegas computable,
2. $f \leq_W \mathsf{PC}_{\mathbb{R}} \iff f$ is Las Vegas computable with finitely many mind changes.

Since these classes of probabilistically computable maps will not play any further role in this article, we will skip the precise definitions and refer the interested reader to Brattka, Gherardi and Hölzl [4,5].

In Sect. 4 we further analyze item 3. of Theorem 2, a statement which is related to countable additivity in reverse mathematics. We show that there is a formalization ACT of this statement that we call *Additive Covering Theorem* and that turns out to be equivalent to $*$-WWKL, yet another variant of Weak Weak Kőnig's Lemma that is even weaker than WWKL from the uniform perspective. In the diagram in Fig. 2 we present a survey of our results.

2   Preliminaries

We assume that the reader is familiar with the concepts defined in the introductory part of Brattka, Gherardi, and Hölzl [5, Sect. 2]. We recall some of the most central concepts. Firstly, Weihrauch reducibility and its strong counterpart are defined for multi-valued functions $f :\subseteq X \rightrightarrows Y$ on represented spaces X, Y. Representations are surjective partial mappings from Baire space $\mathbb{N}^{\mathbb{N}}$ onto the represented spaces and they provide the necessary structures to speak about computability and other concepts. Since we are not using representations in any formal way here, we refrain from presenting further details and we point the reader to Weihrauch [20] and Pauly [17].

Definition 6 (Weihrauch reducibility). Let $f :\subseteq X \rightrightarrows Y$ and $g :\subseteq W \rightrightarrows Z$ be multi-valued functions on represented spaces.

1. f is said to be *Weihrauch reducible* to g, in symbols $f \leq_W g$, if there are computable $K :\subseteq X \rightrightarrows W$, $H :\subseteq X \times Z \rightrightarrows Y$ with $\emptyset \neq H(x, gK(x)) \subseteq f(x)$ for all $x \in \mathrm{dom}(f)$.
2. f is said to be *strongly Weihrauch reducible* to g, in symbols $f \leq_{sW} g$, if there are computable $K :\subseteq X \rightrightarrows W$, $H :\subseteq Z \rightrightarrows Y$ with $\emptyset \neq HgK(x) \subseteq f(x)$ for all $x \in \mathrm{dom}(f)$.

The corresponding equivalences are denoted by $\equiv_W$ and $\equiv_{sW}$, respectively.

In some results we are referring to products of multi-valued functions, which we define next.

Definition 7 (Products). For $f :\subseteq X \rightrightarrows Y$ and $g :\subseteq W \rightrightarrows Z$ we define $f \times g :\subseteq X \times W \rightrightarrows Y \times Z$ by $(f \times g)(x, w) := f(x) \times g(w)$ and $\mathrm{dom}(f \times g) := \mathrm{dom}(f) \times \mathrm{dom}(g)$.

Since we are going to prove that certain versions of the Vitali Covering Theorem can be characterized with the help of certain versions of positive choice, we need to define positive choice next. For this purpose we use the negative information representation of $\mathcal{A}_-(X)$, which represents closed sets $A \subseteq X$ by enumerating rational open balls $B(x_i, r_i)$ that exhaust the complement of A, that is $X \setminus A = \bigcup_{i=0}^{\infty} B(x_i, r_i)$. The x_i are taken from some canonical dense subset of X and the r_i are rational numbers. For details we refer the reader to Brattka, Gherardi and Hölzl [5].

Definition 8 (Choice and positive choice). Let X be a separable metric space with a Borel measure μ and let $\mathcal{A}_-(X)$ denote the set of closed subsets $A \subseteq X$ with respect to negative information.

1. By $\mathsf{C}_X :\subseteq \mathcal{A}_-(X) \rightrightarrows X, A \mapsto A$ we denote the *choice problem* of X with $\mathrm{dom}(\mathsf{C}_X) := \{A : A \neq \emptyset\}$.
2. By $\mathsf{PC}_X :\subseteq \mathcal{A}_-(X) \rightrightarrows X, A \mapsto A$ we denote the *positive choice problem* of X with $\mathrm{dom}(\mathsf{PC}_X) := \{A : \mu(A) > 0\}$.

We will mostly work with the real numbers $\mathbb{R}$ or the unit interval $[0,1]$, both equipped with the Lebesgue measure λ. In Sect. 4 we will also use a quantitative version $\mathsf{P}_{>\varepsilon}\mathsf{C}_{[0,1]}$ of $\mathsf{PC}_{[0,1]}$ which is the restriction of $\mathsf{PC}_{[0,1]}$ to closed sets $A \subseteq [0,1]$ with $\lambda(A) > \varepsilon$ for $\varepsilon > 0$.

3   Vitali Covering in the Weihrauch Degrees

We now translate the three logically equivalent versions of the Vitali Covering Theorem that were presented in the introduction into their corresponding multivalued functions and hence into Weihrauch degrees.

By Int we denote the set of sequences $(I_n)_n$ of open Intervals $I_n = (a,b)$ with $a,b \in \mathbb{Q}$ where we let $(a,b) = \emptyset$ if $b \leq a$. Formally we represent Int using the canonical representation of the set $(\mathbb{Q}^2)^{\mathbb{N}}$.

Definition 9 (Vitali Covering Theorem). We define the following multivalued functions.

0. $\mathsf{VCT}_0 :\subseteq \mathrm{Int} \rightrightarrows \mathrm{Int}, \mathcal{I} \mapsto \{\mathcal{J} : \mathcal{J}$ is a subsequence of $\mathcal{I}$ that eliminates $[0,1]\}$
   and $\mathrm{dom}(\mathsf{VCT}_0)$ contains all $\mathcal{I} \in \mathrm{Int}$ that are Vitali covers of $[0,1]$.
1. $\mathsf{VCT}_1 :\subseteq \mathrm{Int} \rightrightarrows [0,1], \mathcal{I} \mapsto [0,1] \setminus \bigcup \mathcal{I}$ and $\mathrm{dom}(\mathsf{VCT}_1)$ contains all $\mathcal{I} \in \mathrm{Int}$ that are saturated and that do not have a subsequence that eliminates $[0,1]$.
2. $\mathsf{VCT}_2 :\subseteq \mathrm{Int} \rightrightarrows [0,1], \mathcal{I} \mapsto \{x \in [0,1] : x$ is not captured by $\mathcal{I}\}$ and $\mathrm{dom}(\mathsf{VCT}_2)$ contains all $\mathcal{I} \in \mathrm{Int}$ that do not have a subsequence that eliminates $[0,1]$.

We note that $\mathrm{dom}(\mathsf{VCT}_1) \subseteq \mathrm{dom}(\mathsf{VCT}_2)$ and that VCT_1 is a restriction of VCT_2 (see Proposition 15). By the Vitali Covering Theorem (Theorem 1) the sequences $\mathcal{I} \in \mathrm{dom}(\mathsf{VCT}_2)$ cannot be Vitali covers of $[0,1]$.

3.1   The Computable Version

Brattka and Pauly [8] noticed that VCT_0 is computable; we will give a formal proof in this subsection. As a preparation we need the following lemma, where for $A \subseteq \mathbb{R}$ we denote by A° and ∂A the interior and the boundary of A, respectively.

Lemma 10. *Let $A \subseteq [0,1]$ be a closed set with $\lambda(A) > 0$ and $\lambda(\partial A) = 0$. If $\mathcal{I} = (I_n)_{n \in \mathbb{N}}$ is a Vitali cover of A, then the subsequence $\mathcal{I}_A$ of $\mathcal{I}$ that consists only of those I_n with $I_n \subseteq A$ is a Vitali cover of A°.*

Proof. We note that $\lambda(A) > 0$ and $\lambda(\partial A) = 0$ implies $\lambda(A^\circ) = \lambda(A \setminus \partial A) > 0$. In particular, $A^\circ \neq \emptyset$ and the sequence $\mathcal{I}_A$ is well-defined. We claim that $\mathcal{I}_A$ is saturated. Let $x \in \bigcup \mathcal{I}_A$ and $\varepsilon > 0$. Then there is an n such that $x \in I_n \subseteq A$. Since $\mathcal{I}$ is a Vitali cover of A, there is some k such that $x \in I_k \subseteq I_n$ and $\mathrm{diam}(I_k) < \varepsilon$. In particular, $I_k \subseteq A$ and hence I_k is a component of $\mathcal{I}_A$. Thus $\mathcal{I}_A$ is saturated. Similarly, it follows that $\bigcup \mathcal{I}_A = A^\circ$. Here the inclusion "$\subseteq$" follows from the definition of $\mathcal{I}_A$ and we only need to prove "$\supseteq$". For every $x \in A^\circ$ there is some $\varepsilon > 0$ with $(x - \varepsilon, x + \varepsilon) \subseteq A$ and since $\mathcal{I}$ is saturated there is some k with $x \in I_k \subseteq (x - \varepsilon, x + \varepsilon)$. Hence I_k is part of $\mathcal{I}_A$ and $x \in \bigcup \mathcal{I}_A$. This shows that $\bigcup \mathcal{I}_A = A^\circ$, and hence $\mathcal{I}_A$ is a Vitali cover of A°. $\square$

Now we are prepared to prove that VCT_0 is computable.

Theorem 11. VCT_0 *is computable.*

Proof. Given a Vitali cover $\mathcal{I}$ of $[0,1]$, we need to find a subsequence $\mathcal{J}$ of $\mathcal{I}$ that eliminates $[0,1]$. We will compute such a subsequence inductively. Initially, $\mathcal{J}$ is an empty sequence. We start with $A_0 := [0,1]$ and $\mathcal{I}_0 := \mathcal{I}$. We assume that in step n of the computation the set A_n is a non-empty finite union of closed rational intervals with $\lambda(A_n) > 0$ and that $\mathcal{I}_n$ is a Vitali cover of the interior A_n°. The fact that A_n is a non-empty finite union of rational intervals implies $\lambda(\partial A_n) = 0$. Given a Vitali cover $\mathcal{I}_n$ of A_n° there exists a subsequence $\mathcal{J}_n$ of $\mathcal{I}_n$ that eliminates A_n° by the Vitali Covering Theorem (Theorem 1). Since the Lebesgue measure λ is upper semi-computable on closed sets $A \subseteq [0,1]$, by a systematic search we can find a $k_n \in \mathbb{N}$ and a finite subsequence $(I_0, ..., I_{k_n})$ of $\mathcal{I}_n$ of pairwise disjoint intervals such that

$$0 < \lambda\left(A_n^\circ \setminus \bigcup_{i=0}^{k_n} I_i \right) < 2^{-n}.$$

We compute $A_{n+1} := A_n \setminus \bigcup_{i=0}^{k_n} I_i$ as a finite union of closed rational intervals and we add the intervals $I_0, ..., I_{k_n}$ to the set $\mathcal{J}$. Since $\lambda(\partial A_n) = 0$, we obtain that $0 < \lambda(A_{n+1}) < 2^{-n}$. We now compute $\mathcal{I}_{n+1} := (\mathcal{I}_n)_{A_{n+1}}$ (as defined in Lemma 10). Then $\mathcal{I}_{n+1}$ is a Vitali cover of A_{n+1}° by Lemma 10 and we can continue the inductive construction in step $n+1$. Altogether, this construction leads to a subsequence $\mathcal{J}$ of $\mathcal{I}$ of pairwise disjoint intervals $\mathcal{J}$ such that $[0,1] \setminus \bigcup \mathcal{J} = \bigcap_{n=0}^\infty A_n$. Since $\lambda(A_n) < 2^{-n}$, it follows that $\lambda([0,1] \setminus \bigcup \mathcal{J}) = 0$. Hence $\mathcal{J}$ eliminates $[0,1]$. □

3.2 The First Non-computable Version

In the previous subsection we observed that the most straight-forward way of formalizing the Vitali Covering Theorem in the Weihrauch degrees is computable. To obtain non-computability results, we need to look at contrapositive versions of the theorem. The idea here is that given a collection of intervals $\mathcal{I}$ that violates some of the requirements for being a Vitali cover, we want to find an $x \in [0,1]$ witnessing this violation. Again, there is more than one formalization for this idea, as there are different ways and degrees of violating the requirements.

It will turn out that these different formalizations produce mathematical tasks of different computational strengths, that is, falling into different Weihrauch degrees. The first result in this direction that we will prove is that VCT_1 is strongly equivalent to Weak Weak Kőnig's Lemma. This corresponds to Theorem 2 by Brown, Giusto, and Simpson.

To show $\mathsf{WWKL} \leq_{\mathsf{sW}} \mathsf{VCT}_1$ we will use the following lemma that shows that we can computably refine any sequence of open intervals to a saturated one.

Lemma 12 (Vitalization). *There exists a computable map $V : \mathrm{Int} \to \mathrm{Int}$ such that $\bigcup \mathcal{I} = \bigcup V(\mathcal{I})$ for all $\mathcal{I} \in \mathrm{Int}$ and $\mathrm{range}(V)$ only consists of saturated sequences of intervals.*

Proof. Given $\mathcal{I} = (I_n)_n$ we systematically add to $\mathcal{I}$ all rational intervals $I = (a, b)$ for which there is an $n \in \mathbb{N}$ with $I \subseteq I_n$. This leads in a computable way to a saturated sequence $\mathcal{J}$ with $\bigcup \mathcal{I} = \bigcup \mathcal{J}$. $\square$

Now we are prepared to prove that VCT_1 is strongly equivalent to $\mathsf{PC}_{[0,1]}$.

Theorem 13. $\mathsf{VCT}_1 \equiv_{\mathsf{sW}} \mathsf{PC}_{[0,1]}$.

Proof. Given a sequence $\mathcal{I}$ of open intervals with $A = [0,1] \setminus \bigcup \mathcal{I}$ and $\lambda(A) > 0$, by Lemma 12 we can compute a saturated sequence $V(\mathcal{I})$ with $A = [0,1] \setminus \bigcup V(\mathcal{I})$. Since $\lambda(A) > 0$, it is clear that $V(\mathcal{I})$ does not have a subsequence that eliminates $[0,1]$. Hence $V(\mathcal{I}) \in \mathrm{dom}(\mathsf{VCT}_1)$ and $\mathsf{VCT}_1(V(\mathcal{I})) = A$, which implies $\mathsf{PC}_{[0,1]} \leq_{\mathsf{sW}} \mathsf{VCT}_1$.

Now let $\mathcal{I}$ be a saturated sequence of intervals that does not have a subsequence that eliminates $[0,1]$. Clearly we can compute $A := [0,1] \setminus \bigcup \mathcal{I}$. Since $\mathcal{I}$ is a Vitali cover of $\bigcup \mathcal{I}$, there is a subsequence $\mathcal{J}$ of $\mathcal{I}$ that eliminates $\bigcup \mathcal{I}$ by the Vitali Covering Theorem (Theorem 1). If $\lambda(A) = 0$, then this subsequence $\mathcal{J}$ also eliminates $[0,1]$. This is not possible by assumption and hence $\lambda(A) > 0$. Consequently, $\mathsf{VCT}_1(\mathcal{I}) = \mathsf{PC}_{[0,1]}(A)$, which proves $\mathsf{VCT}_1 \leq_{\mathsf{sW}} \mathsf{PC}_{[0,1]}$. $\square$

Since it is known that $\mathsf{PC}_{[0,1]}$ has computable inputs that do not admit computable outputs (see for example Brattka, Gherardi and Hölzl [4, Theorem 12]), we obtain the following corollary as an immediate consequence (which also follows by Lemma 12 from a classical result of Kreisel and Lacombe [14, Théorème VI] on singular coverings, see also [20, Theorem 4.28]).

Corollary 14 (Diener and Hedin [11, Theorem 9]). *There exists a computable Vitali cover $\mathcal{J}$ of the computable points in $[0,1]$ so that every subsequence $\mathcal{I} = (I_n)_n$ consisting of pairwise disjoint intervals satisfies $\sum_{n=0}^{\infty} \lambda(I_n) < 1$.*

3.3   The Second Non-computable Version

The previous result identifies the computational strength of VCT_1 with that of the well-studied Weihrauch degree WWKL. The natural next question to ask is whether VCT_2 is of different strength and, if yes, to determine that strength precisely. Both questions will be answered in this section. We begin with the following observation.

Proposition 15. $\mathsf{VCT}_1 \leq_{\mathsf{sW}} \mathsf{VCT}_2$.

Proof. If $\mathcal{I}$ is a saturated sequence of rational open intervals that contains no subsequence that eliminates $[0,1]$, then $\mathcal{I}$ does not cover $[0,1]$ by the Vitali Covering Theorem (Theorem 1) and every point $x \in [0,1]$ which is not captured by $\mathcal{I}$ is a point that is not covered by $\mathcal{I}$, that is, $x \in [0,1] \setminus \bigcup \mathcal{I}$. Hence VCT_1 is a restriction of VCT_2 and, in particular, $\mathsf{VCT}_1 \leq_{\mathsf{sW}} \mathsf{VCT}_2$. $\square$

On the other hand, VCT_2 can be reduced to $\mathsf{PC}_\mathbb{R}$, as the next result shows. Within the proof we will use the following definition from Brown, Giusto, and Simpson [10]. A sequence $\mathcal{I} = (I_n)_n$ of intervals is an *almost Vitali cover* of a Lebesgue measurable set $A \subseteq [0,1]$ if for all $\varepsilon > 0$ and

$$U_\varepsilon := \bigcup \{I_n : n \in \mathbb{N} \text{ and } \mathrm{diam}(I_n) < \varepsilon\}$$

it holds that $\lambda(A \backslash U_\varepsilon) = 0$. In fact, Brown, Giusto, and Simpson [10, Theorem 5.6] (see Simpson [19, Theorem X.1.13]) proved the following strengthening of the Vitali Covering Theorem (Theorem 1): every almost Vitali cover $\mathcal{I}$ of $[0,1]$ admits a subsequence $\mathcal{J}$ that eliminates $[0,1]$. We use this result to obtain the following reduction.

Proposition 16. $\mathsf{VCT}_2 \leq_{\mathsf{sW}} \mathsf{PC}_\mathbb{R}$.

Proof. Let $\mathcal{I} = (I_n)_{n \in \mathbb{N}}$ be a sequence of rational open intervals that does not contain a subsequence that eliminates $[0,1]$. By Brown, Giusto, and Simpson [10, Theorem 5.6] we obtain that $\mathcal{I}$ is not even an almost Vitali cover of $[0,1]$, that is, there exists some $n \in \mathbb{N}$ such that $\lambda([0,1] \setminus U_{2^{-n}}) > 0$, with U_ε as defined above. We let $A_n := [0,1] \setminus U_{2^{-n}}$ for all n. Clearly $A_n \subseteq \mathsf{VCT}_2(\mathcal{I})$ for all n. Now we compute $A := \bigcup_{n=0}^\infty (2n + A_n)$, where $n + X := \{n + x : x \in X\}$ for all $X \subseteq \mathbb{N}$. Then $\lambda(A) > 0$ and $\mathsf{PC}_\mathbb{R}(A)$ yields a point x with $(x \bmod 2) \in \mathsf{VCT}_2(\mathcal{I})$. This proves $\mathsf{VCT}_2 \leq_{\mathsf{sW}} \mathsf{PC}_\mathbb{R}$. □

Now we prove by a direct construction that VCT_2 can compute itself concurrently with $\mathsf{C}_\mathbb{N}$.

Proposition 17. $\mathsf{C}_\mathbb{N} \times \mathsf{VCT}_2 \leq_{\mathsf{sW}} \mathsf{VCT}_2$.

Proof. For the purposes of this proof we treat sequences of intervals $\mathcal{I} = (I_n)_n$ as sets $\mathcal{I} = \{I_n : n \in \mathbb{N}\}$ of intervals. All sets of intervals that we are going to use can be enumerated in a natural way.

Let A be an instance of $\mathsf{C}_\mathbb{N}$ and $\mathcal{I}$ an instance of VCT_2, that is $\mathcal{I}$ does not have a subsequence that eliminates $[0,1]$. By $\mathcal{I}_{[a,b]}$ we denote the image of $\mathcal{I}$ under rescaling $[0,1]$ to $[a,b]$.[1] By $\mathcal{S}_{(a,b)}$ we denote some saturated and computably enumerable set of intervals with $\bigcup \mathcal{S}_{(a,b)} = (a,b)$, which exists by Lemma 12.

We use points of the form $x_n := 1 - \frac{1}{n}$ for $n > 1$ to subdivide the unit interval $[0,1]$ into countably many regions. In each of these regions with $n > 1$ we will place countably many scaled copies of $\mathcal{I}$ into certain intervals of the form

---

[1] There is a slight ambiguity here, as we need to deal with open sets ranging beyond $[0,1]$. We shall understand these to be *small enough* in the sense that we cut away everything from a certain distance ε_n on. The exact constraints that these values ε_n need to satisfy are given in the proof.

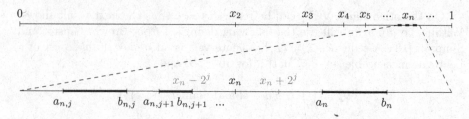

Fig. 1. Illustration of the intervals $[a_{n,j}, b_{n,j}]$ and $[a_n, b_n]$ in correct order, but oversized.

$[a_n, b_n] := [x_n + 2^{-n-1}, x_n + 2^{-n}]$ and $[a_{n,j}, b_{n,j}] := [x_n - 2^{-2j}, x_n - 2^{-2j-1}]$ for $j > n$. We construct an instance $\mathcal{J} := \mathcal{J}_P \cup \mathcal{J}_\mathcal{I} \cup \mathcal{J}_R \cup \mathcal{J}_A$ of VCT$_2$ in four parts:

$$\mathcal{J}_P := \{(x_n - 2^{-j}, x_n + 2^{-j}) : n > 1, j > n\} \cup \{(x_n, 1 + 2^{-n}) : n > 1\}$$

$$\mathcal{J}_\mathcal{I} := \bigcup_{\substack{n>1 \\ j>n}} \mathcal{I}_{[a_{n,j}, b_{n,j}]} \cup \bigcup_{n>1} \mathcal{I}_{[a_n, b_n]}$$

$$\mathcal{J}_R := \mathcal{S}_{(-2^{-1}, a_{2,3})} \cup \bigcup_{\substack{n>1 \\ j>n}} \mathcal{S}_{(b_{n,j}, a_{n,j+1})} \cup \bigcup_{n>1} (\mathcal{S}_{(x_n, a_n)} \cup \mathcal{S}_{(b_n, a_{n+1,n+2})})$$

$$\mathcal{J}_A := \bigcup_{\substack{n>1 \\ n-2 \notin A}} \mathcal{S}_{(a_n - \varepsilon_n, b_n + \varepsilon_n)} \cup \bigcup_{\substack{n>1 \\ j>n \\ j-n-1 \notin A}} \mathcal{S}_{(a_{n,j} - \varepsilon_j, b_{n,j} + \varepsilon_j)}$$

Here $(\varepsilon_n)_n$ is a computable sequence of positive rational numbers that are subject to the following constraints for all $n > 1$ and $j > n$:

$$x_n < a_n - \varepsilon_n, \quad b_n + \varepsilon_n < a_{n+1,n+2} - \varepsilon_{n+2} \text{ and } b_{n,j} + \varepsilon_j < a_{n,j+1} - \varepsilon_{j+1}.$$

In Fig. 1 the construction is visualized. Intuitively, we capture the point 1 and all points $x_n = 1 - \frac{1}{n}$ using $\mathcal{J}_P$. Using $\mathcal{J}_\mathcal{I}$ we place scaled copies of $\mathcal{I}$ into the intervals $[a_n, b_n]$ and $[a_{n,j}, b_{n,j}]$ for $n > 1$ and $j > n$. The remainder of the unit interval is captured using $\mathcal{J}_R$. Finally, those regions not corresponding to an index from A are rendered invalid responses by capturing them using $\mathcal{J}_A$, where the constraints on ε_n above guarantee that no neighbor regions are touched.

Any point not captured by $\mathcal{J}$ must lie in one of the regions designated in the definition of $\mathcal{J}_\mathcal{I}$, and, as these are separated, we can compute the parameters of the region (thus producing the answer for the instance A of C$_\mathbb{N}$), and then scale the point back up to produce the answer to the instance $\mathcal{I}$ of VCT$_2$.

It remains to prove that $\mathcal{J}$ actually is a valid input to VCT$_2$, that is, that no collection $\mathcal{S} \subseteq \mathcal{J}$ of disjoint intervals eliminates $[0,1]$. Let $\mathcal{S} \subseteq \mathcal{J}$ be a disjoint collection of intervals. We distinguish two cases:

Case 1: $(\exists n)$ $(x_n, 1 + 2^{-n}) \in \mathcal{S}$. Then no set of the form $(x_n - 2^{-j}, x_n + 2^{-j})$ can be in $\mathcal{S}$. Choose j such that $j - n - 1 \in A$. We claim that $\mathcal{S}$ cannot eliminate $[a_{n,j}, b_{n,j}]$: We have already seen that under the given conditions, we have for every $U \in \mathcal{S} \cap \mathcal{J}_P$ that $U \cap [a_{n,j}, b_{n,j}] = \emptyset$. The same is true for $U \in \mathcal{S} \cap (\mathcal{J}_R \cup \mathcal{J}_A)$

by construction and because $j - n - 1 \in A$. Thus, the only sets which could contribute to eliminating the interval $[a_{n,j}, b_{n,j}]$ come from $\mathcal{J}_\mathcal{I}$, and more specifically, $\mathcal{I}_{[a_{n,j}, b_{n,j}]}$; but if these sets would eliminate $[a_{n,j}, b_{n,j}]$, then $\mathcal{I}$ would eliminate $[0, 1]$, which is impossible.

Case 2: $(\forall n)$ $(x_n, 1 + 2^{-n}) \notin S$. Let n be such that $n - 2 \in A$. We claim that S cannot eliminate $[a_n, b_n]$. For $U \in S \cap \mathcal{J}_P$ we have that $U \cap [a_n, b_n] = \emptyset$ because $(x_n - 2^{-j}, x_n + 2^{-j}) \cap [a_n, b_n] = \emptyset$ for all $j > n$. For $U \in S \cap \mathcal{J}_R$ the same statement holds by construction; and for $U \in S \cap \mathcal{J}_A$ it holds since $n - 2 \in A$. If $\mathcal{J}_\mathcal{I}$ would eliminate $[a_n, b_n]$, then $\mathcal{I}$ would eliminate $[0, 1]$, which is impossible. $\square$

Using Fact 3, Theorem 13 and Propositions 15, 16 and 17 we obtain the following characterization of VCT_2.

Corollary 18. $\mathsf{VCT}_2 \equiv_{\mathsf{sW}} \mathsf{PC}_\mathbb{R}$.

We note that the proof of Proposition 16 shows that we can extend the domain of VCT_2 to sequences $\mathcal{I}$ of intervals that are not almost Vitali covers of $[0, 1]$ and Corollary 18 remains correct for this generalized version of VCT_2.

4   Countable Additivity

In reverse mathematics Brown, Giusto, and Simpson [10, Theorem 3.3] (see also Simpson [19, Theorem X.1.9]) have discussed countable additivity of measures and condition 3. of Theorem 2 turned out to characterize this property. In this section we would like to analyze this condition in the Weihrauch lattice and we formulate the condition and a contrapositive version of it in a slightly different way.

1. Any $\mathcal{I} = (I_n)_n$ that covers $[0, 1]$ satisfies $\sum_{n=0}^{\infty} \lambda(I_n) \geq 1$.
2. For any non-disjoint $\mathcal{I} = (I_n)_n$ that satisfies $\sum_{n=0}^{\infty} \lambda(I_n) < 1$, there exists a point $x \in [0, 1] \setminus \bigcup \mathcal{I}$.

By a *non-disjoint* $\mathcal{I} = (I_n)_n$ we mean one that satisfies $I_i \cap I_j \neq \emptyset$ for some $i \neq j$. For the correctness of the second statement non-disjointness is not relevant. However, it matters for the computational content. While the first statement has no immediate computational content (more precisely, any reasonable straightforward formalization is computable), the second one turns out to be equivalent to $*$-WWKL, which we define below. First we formalize the second statement above as a multi-valued function, which we call the *Additive Covering Theorem*.

Definition 19 (Additive Covering Theorem). The *Additive Covering Theorem* is the multi-valued function $\mathsf{ACT} :\subseteq \mathsf{Int} \rightrightarrows [0, 1], \mathcal{I} \mapsto [0, 1] \setminus \bigcup \mathcal{I}$, where $\mathrm{dom}(\mathsf{ACT})$ is the set of all non-disjoint $\mathcal{I} = (I_n)_n$ with $\sum_{n=0}^{\infty} \lambda(I_n) < 1$.

In order to define $*$-WWKL, we recall that for a sequence $f_i :\subseteq X_i \rightrightarrows Y_i$ we can define the *coproduct* $\bigsqcup_{i=0}^{\infty} f_i :\subseteq \bigsqcup_{i=0}^{\infty} X_i \rightrightarrows \bigsqcup_{i=0}^{\infty} Y_i$, where $\bigsqcup_{i=0}^{\infty} Z_i$ denotes the disjoint union of the sets Z_i. Now we define $*$-WWKL $:= \bigsqcup_{n=0}^{\infty} \mathsf{P}_{>2^{-n}} \mathsf{C}_{[0,1]}$, where $\mathsf{P}_{>\varepsilon} \mathsf{C}_{[0,1]}$ is the choice principle for closed subsets $A \subseteq [0,1]$ with $\lambda(A) > \varepsilon$, as defined in Sect. 2. Hence, intuitively, $*$-WWKL takes as input a number $n \in \mathbb{N}$ together with a closed set A of measure $\lambda(A) > 2^{-n}$ and has to produce a point $x \in A$. This could equivalently be defined using quantitative versions of WWKL, hence the name $*$-WWKL (see Brattka, Gherardi and Hölzl [5, Proposition 7.2]). Now we can formulate and prove our main result on ACT.

Theorem 20. $\mathsf{ACT} \equiv_{\mathsf{sW}} *$-WWKL.

Proof. We first prove $\mathsf{ACT} \leq_{\mathsf{sW}} *$-WWKL. Let $\mathcal{I} = (I_n)_n$ be a given non-disjoint sequence of intervals such that $\sum_{n=0}^{\infty} \lambda(I_n) < 1$. Then we can search for some numbers $i, j, k \in \mathbb{N}$ such that $\varepsilon := \lambda(I_i \cap I_j) > 2^{-k}$. In this situation we obtain by countable additivity $\lambda\left(\bigcup_{n=0}^{\infty} I_n\right) + \varepsilon \leq \sum_{n=0}^{\infty} \lambda(I_n) < 1$. Hence we obtain for the closed set $A := [0,1] \setminus \bigcup \mathcal{I}$ that

$$\lambda(A) \geq 1 - \sum_{n=0}^{\infty} \lambda(I_n) + \varepsilon > \varepsilon > 2^{-k}.$$

Therefore, we can find a point in A using $\mathsf{P}_{>2^{-k}} \mathsf{C}_{[0,1]}(A)$. This proves the desired reduction $\mathsf{ACT} \leq_{\mathsf{sW}} *$-WWKL.

We now prove $*$-WWKL $\leq_{\mathsf{sW}} \mathsf{ACT}$. Given $k \in \mathbb{N}$ and a closed set $A \subseteq [0,1]$ such that $\lambda(A) > 2^{-k}$ we need to find a point $x \in A$. The set A is given by a sequence $\mathcal{J}$ of open intervals with $A = [0,1] \setminus \bigcup \mathcal{J}$. We can now computably convert the sequence $\mathcal{J}$ into a non-disjoint sequence $\mathcal{I} = (I_n)_n$ of open intervals such that $A = [0,1] \setminus \bigcup \mathcal{I}$ and

$$\sum_{n=0}^{\infty} \lambda(I_n) \leq \lambda\left(\bigcup_{n=0}^{\infty} ([0,1] \cap I_n)\right) + 2^{-k-1}.$$

This can be achieved if for every J in $\mathcal{J}$ we select finitely many intervals $I_n \subseteq J$ such that all intervals selected so far cover J and such that the overlapping measure of I_n with the union of the previous intervals (and the exterior of $[0,1]$) is at most $2^{-k-1-n-1}$ for each $n \in \mathbb{N}$ (and non-zero for at least one n). Since $\lambda(A) > 2^{-k}$ we obtain $\lambda\left(\bigcup_{n=0}^{\infty}([0,1] \cap I_n)\right) < 1 - 2^{-k}$ and the above condition implies $\sum_{n=0}^{\infty} \lambda(I_n) < 1 - 2^{-k} + 2^{-k-1} < 1$ and hence $\mathsf{ACT}(\mathcal{I}) = A$. This yields the desired reduction. $\qquad\square$

Like WWKL $\times \mathsf{C}_\mathbb{N}$ the problem $*$-WWKL can be seen as a uniform modification of WWKL that is indistinguishable from WWKL when seen from the non-uniform perspective of reverse mathematics.

5   Conclusions

We have demonstrated that the Vitali Covering Theorem and related results split into several uniform equivalence classes when analyzed in the Weihrauch

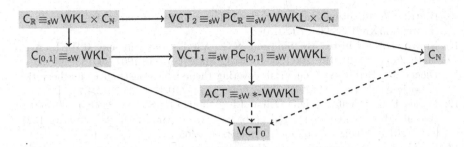

Fig. 2. The Vitali Covering Theorem in the Weihrauch lattice. Strong Weihrauch reductions $f \leq_{sW} g$ are indicated by a solid arrow $f \leftarrow g$ and similarly ordinary Weihrauch reductions are indicated by a dashed arrow $f \leftarrow\!\!- g$.

lattice. We have summarized the results in the diagram in Fig. 2. The diagram also indicates some equivalence classes in the neighborhood that are related to Weak Kőnig's Lemma WKL. These classes have not been discussed in this article and some related results can be found in Brattka, de Brecht and Pauly [1] and Brattka, Gherardi and Hölzl [5].

References

1. Brattka, V., de Brecht, M., Pauly, A.: Closed choice and a uniform low basis theorem. Ann. Pure Appl. Logic **163**, 986–1008 (2012). http://dx.doi.org/10.1016/j.apal.2011.12.020
2. Brattka, V., Gherardi, G.: Effective choice and boundedness principles in computable analysis. Bull. Symbolic Logic **17**(1), 73–117 (2011). http://dx.doi.org/10.2178/bsl/1294186663
3. Brattka, V., Gherardi, G.: Weihrauch degrees, omniscience principles and weak computability. J. Symbolic Logic **76**(1), 143–176 (2011). http://dx.doi.org/10.2178/jsl/1294170993
4. Brattka, V., Gherardi, G., Hölzl, R.: Las Vegas computability and algorithmic randomness. In: Mayr, E.W., Ollinger, N. (eds.) 32nd International Symposium on Theoretical Aspects of Computer Science (STACS 2015). Leibniz International Proceedings in Informatics (LIPIcs), vol. 30, pp. 130–142. Schloss Dagstuhl-Leibniz-Zentrum für Informatik, Dagstuhl, Germany (2015). http://drops.dagstuhl.de/opus/volltexte/2015/4909
5. Brattka, V., Gherardi, G., Hölzl, R.: Probabilistic computability and choice. Inf. Comput. **242**, 249–286 (2015). http://dx.doi.org/10.1016/j.ic.2015.03.005
6. Brattka, V., Gherardi, G., Marcone, A.: The Bolzano-Weierstrass theorem is the jump of weak Kőnig's lemma. Ann. Pure Appl. Logic **163**, 623–655 (2012). http://dx.doi.org/10.1016/j.apal.2011.10.006
7. Brattka, V., Hendtlass, M., Kreuzer, A.P.: On the uniform computational content of the Baire category theorem. Notre Dame J. Formal Logic (2016). http://arxiv.org/abs/1510.01913
8. Brattka, V., Pauly, A.: Computation with advice. In: Zheng, X., Zhong, N. (eds.) CCA 2010, Proceedings of the Seventh International Conference on Computability and Complexity in Analysis, Electronic Proceedings in Theoretical Computer Science, pp. 41–55 (2010). http://dx.doi.org/10.4204/EPTCS.24.9

9. Brattka, V., Rakotoniaina, T.: On the uniform computational content of Ramsey's theorem. arXiv:1508.00471 (2015)
10. Brown, D.K., Giusto, M., Simpson, S.G.: Vitali's theorem and WWKL. Arch. Math. Logic **41**(2), 191–206 (2002). http://dx.doi.org/10.1007/s001530100100
11. Diener, H., Hedin, A.: The Vitali covering theorem in constructive mathematics. J. Logic Anal. **4**(7), 22 (2012). http://dx.doi.org/10.4115/jla.2012.4.7
12. Dorais, F.G., Dzhafarov, D.D., Hirst, J.L., Mileti, J.R., Shafer, P.: On uniform relationships between combinatorial problems. Trans. Am. Math. Soc. **368**(2), 1321–1359 (2016). http://dx.doi.org/10.1090/tran/6465
13. Gherardi, G., Marcone, A.: How incomputable is the separable Hahn-Banach theorem? Notre Dame J. Formal Logic **50**(4), 393–425 (2009). http://dx.doi.org/10.1215/00294527-2009-018
14. Kreisel, G., Lacombe, D.: Ensembles récursivement mesurables et ensembles récursivement ouverts et fermés. Comptes Rendus Académie des Sciences Paris **245**, 1106–1109 (1957)
15. Pauly, A.: How incomputable is finding Nash equilibria? J. Univ. Comput. Sci. **16**(18), 2686–2710 (2010). http://dx.doi.org/10.3217/jucs-016-18-2686
16. Pauly, A.: On the (semi)lattices induced by continuous reducibilities. Math. Logic Q. **56**(5), 488–502 (2010). http://dx.doi.org/10.1002/malq.200910104
17. Pauly, A.: On the topological aspects of the theory of represented spaces. Computability **5**(2), 159–180 (2016). http://dx.doi.org/10.3233/COM-150049
18. Richardson, L.F.: Measure and Integration. Wiley, Hoboken (2009). http://dx.doi.org/10.1002/9780470501153. (A concise introduction to real analysis)
19. Simpson, S.G.: Subsystems of Second Order Arithmetic. Perspectives in Logic, Association for Symbolic Logic, 2nd edn. Cambridge University Press, Poughkeepsie (2009)
20. Weihrauch, K.: Computable Analysis. Springer, Berlin (2000)

Parallel and Serial Jumps
of Weak Weak König's Lemma

Laurent Bienvenu[1] and Rutger Kuyper[2(✉)]

[1] LIRMM, CNRS & Université de Montpellier, 161 Rue Ada,
34095 Montpellier Cedex 5, France
laurent.bienvenu@computability.fr
[2] Department of Mathematics, University of Wisconsin–Madison,
Madison, WI 53706, USA
mail@rutgerkuyper.com

Abstract. We study the principle of positive choice in the Weihrauch degrees. In particular, we study its behaviour under composition and jumps, and answer three questions asked by Brattka, Gherardi and Hölzl.

1 Introduction

In this paper we study the computational strength of *positive choice* for the spaces $\mathcal{X} \in \{2^\omega, \omega \times 2^\omega, \omega^\omega\}$. Here, positive choice is the principle which assigns to a tree of positive measure the collection of paths through that tree; a different name for PC_{2^ω} is *weak weak König's lemma* or WWKL. There are several different approaches to classifying the relative strength of different principles; for example, one could study the relative strength over a weak base system such as RCA_0, as is commonly done in reverse mathematics. However, in this paper we study these principles in the Weihrauch degrees, which imposes several restrictions when we are comparing two principles Φ and Ψ: for Φ to Weihrauch-reduce to Ψ, which intuitively means that Φ is 'easier' than Ψ, we should be able to solve Φ using *one instance* of Ψ in a *uniform* way.

In particular, in the Weihrauch degrees it makes sense to ask whether applying a principle twice in a row is strictly stronger than only using it once. In fact, given two principles Φ and Ψ there is a natural degree $\Phi \star \Psi$ corresponding to applying Φ after Ψ, as shown by Brattka and Pauly [6]; we call $\Phi \star \Psi$ the *compositional product* of Φ and Ψ.

One natural way of strengthening a principle Φ is by weakening the representation of its input. For example, when talking about positive choice, instead of considering the principle which takes as input a tree of positive measure and outputs a path through the tree, we could consider the principle which takes as input a sequence of trees which converges pointwise to a tree of positive measure;

The research of the second author was supported by John Templeton Foundation grant 15619: 'Mind, Mechanism and Mathematics: Turing Centenary Research Project'.

i.e., the input is only a Δ_2^0-representation of the intended tree. This can be done in general, and so, for every principle Φ there is a principle Φ', the *jump* of Φ.

It is now natural to study the interaction of these different operations. For example, Brattka, Gherardi and Marcone [4] studied the jump of weak König's lemma (the principle which assigns to an infinite binary tree the set of paths through that tree), and showed that

$$\mathrm{WKL}' \star \mathrm{WKL}' \equiv_W \mathrm{WKL}''.$$

Brattka, Gherardi and Hölzl [3] studied various properties of probabilistic choice; for example, they showed that

$$\mathrm{PC}_\mathcal{X} \star \mathrm{PC}_\mathcal{X} \equiv_W \mathrm{PC}_\mathcal{X}.$$

They concluded their paper with several questions. First of all, they asked:

Is WWKL' *closed under composition?*

Given their result, this would be a natural relativisation 'one jump up'.

On the other hand, by the result from [4], iterating WKL' brings us up to WKL''. So, another natural question is:

Or is $\mathrm{WWKL}' \star \mathrm{WWKL}' \equiv_W \mathrm{WWKL}''$?

We will show that neither of these is the case. In fact, we will show in Sect. 4 that

$$\mathrm{WWKL}' \star \mathrm{WWKL}' \equiv_W \mathrm{PC}'_{\omega \times 2^\omega},$$

and relativising a result from Brattka and Pauly [7] we show that

$$\mathrm{WWKL}' <_W \mathrm{PC}'_{\omega \times 2^\omega} <_W \mathrm{WWKL}''.$$

The third question asked in [3] is

Is $\mathrm{WWKL}' \leq_W \mathrm{PC}_{\omega^\omega}$?

We also give a negative answer to this question. In fact, we show in Sect. 5 that both $\mathrm{PC}_{\omega \times 2^\omega} \equiv_{sW} \mathrm{PC}_{\omega^\omega}$ and $\mathrm{PC}'_{\omega \times 2^\omega} \equiv_{sW} \mathrm{PC}'_{\omega^\omega}$, which we combine with the theorem from [7] that $\mathrm{PC}_{\omega \times 2^\omega} <_W \mathrm{WWKL}'$.

Finally, in Sect. 6 we study the remaining compositions $f \star g$ for $f, g \in \{\mathrm{WWKL}, \mathrm{WWKL}', \mathrm{PC}_{\omega \times 2^\omega}, \mathrm{PC}'_{\omega \times 2^\omega}\}$. The results are summarised in Sect. 7.

We assume that the reader is familiar with basic notions of computability theory and algorithmic randomness, and refer to [8,9] for a good treatment of both subjects.

Our notation is mostly standard. We use $f :\subseteq \mathcal{X} \to \mathcal{Y}$ to denote that f is a partial map, and we use $f : \mathcal{X} \rightrightarrows \mathcal{Y}$ to denote that f is a multi-valued function. Whenever we talk about a path through a tree, we mean an infinite path. When $T^{\emptyset'}$ is a Δ_2^0 tree, we denote by $T[s]$ the set of strings σ such that for no string $\tau \subseteq \sigma$ we have that $T^{\emptyset'[s]}(\tau)[s]\!\downarrow = 0$, i.e., we make sure $T[s]$ is also a tree. We fix once and for all a computable bijection $\langle .,. \rangle$ between ω^2 and ω. If $f : \omega \to \mathcal{Y}$ is a function, $f^{[i]}$ is the function defined by $f^{[i]}(n) = f(\langle i, n \rangle)$.

2 Weihrauch Degrees

In this section we will repeat the necessary definitions and background on Weihrauch reducibility. The definition of Weihrauch reducibility has gone through several generalisations, culminating in the current definition of Brattka and Gherardi [2]. This is the definition we give here.

As is known in computability theory, Baire space ω^ω can be used to represent many different kind of things, from trees to real numbers. In order to properly define Weihrauch reducibility, we need to make these representations explicit. We do this through the notion of a represented space.

Definition 2.1. A *representation* of a set $\mathcal{X}$ is a surjective partial map $\delta :\subseteq \omega^\omega \to \mathcal{X}$. We say that $(\mathcal{X}, \delta_{\mathcal{X}})$ is a *represented space*.

Now, we consider *multi-valued (partial) functions* $f :\subseteq \mathcal{X} \rightrightarrows \mathcal{Y}$, i.e., partial maps which send an $x \in \text{dom}(f)$ to a non-empty subset of Y. Henceforth, we will omit the word 'partial' and talk about multi-valued partial functions just as multi-valued functions. An easy example of a multi-valued function is the following, which will be relevant throughout this paper.

Definition 2.2. For n a positive integer, let $n-\text{Ran} : 2^\omega \rightrightarrows 2^\omega$ be the multi-valued function which sends X to the set of n-random reals relative to X, i.e., those reals which are Martin-Löf random relative to $X^{(n-1)}$.

To any multi-valued function $f :\subseteq \mathcal{X} \rightrightarrows \mathcal{Y}$ we can assign a set of partial functions from $\mathcal{X}$ to $\mathcal{Y}$ in a natural way: take the set of choice functions, i.e., the set of functions $F :\subseteq \mathcal{X} \to \mathcal{Y}$ such that $\text{dom}(F) = \text{dom}(f)$, and $F(x) \in f(x)$ for every $x \in \text{dom}(f)$.

Furthermore, given any multi-valued function $f :\subseteq \mathcal{X} \rightrightarrows \mathcal{Y}$ and representations $\delta_{\mathcal{X}}$ of $\mathcal{X}$ and $\delta_{\mathcal{Y}}$ of $\mathcal{Y}$, we can represent the function f as a function $g :\subseteq \omega^\omega \rightrightarrows \omega^\omega$, although not necessarily uniquely if $\delta_{\mathcal{Y}}$ is not injective. For example, looking at $n-\text{Ran}$, if we let δ_{2^ω} be the inclusion of 2^ω in ω^ω, we can 'pull back' this function to ω^ω by letting g be the function sending $x \in 2^\omega$ to $f(x)$, and being undefined outside of 2^ω.

In what follows, we do not just think of a multi-valued function $f :\subseteq \mathcal{X} \rightrightarrows \mathcal{Y}$ as a set-theoretic multi-valued function, but we actually think of it as a multi-valued function $f :\subseteq (X, \delta_X) \rightrightarrows (Y, \delta_Y)$ from a represented space to a represented space. In other words, we are thinking of an explicit representation of the domain and codomain. However, when the representations are clear and there is no possible confusion we will often write $f :\subseteq X \rightrightarrows Y$ anyway.

Combining these ideas, we are lead to the notion of a realiser.

Definition 2.3. Let $f :\subseteq (\mathcal{X}, \delta_{\mathcal{X}}) \rightrightarrows (\mathcal{Y}, \delta_{\mathcal{Y}})$ be a multi-valued function. We say that $F :\subseteq \omega^\omega \to \omega^\omega$ is a *realiser* of f, written as $F \vdash f$, if for every $z \in \text{dom}(f \circ \delta_{\mathcal{X}})$ we have $z \in \text{dom}(F)$, and $\delta_{\mathcal{Y}}(F(z)) \in f(\delta_{\mathcal{X}}(z))$.

As an example, a realiser of $n-\text{Ran}$ is now just a function which assigns to every $X \in 2^\omega$ an n-random real relative to X.

The notion of *Weihrauch reducibility* now defines what it means for a multi-valued function $f :\subseteq \mathcal{X} \rightrightarrows \mathcal{Y}$ to be 'easier' than a multi-valued function $g :\subseteq \mathcal{U} \rightrightarrows \mathcal{V}$, in the sense that the realisers of g uniformly compute realisers of f. This is made precise in the definition below.

Definition 2.4. Let f, g be multi-valued functions (on represented spaces). Then we say that f is *Weihrauch reducible* to g, written as $f \leq_W g$, if there exist Turing functionals $K :\subseteq \omega^\omega \to \omega^\omega$ and $H :\subseteq \omega^\omega \times \omega^\omega \to \omega^\omega$ such that for every G with $G \vdash g$ we have that $H(\mathrm{id}, G \circ K) \vdash f$.

Furthermore, we say that f is *strongly Weihrauch reducible* to g, written as $f \leq_{sW} g$, if there exist Turing functionals $K, H :\subseteq \omega^\omega \to \omega^\omega$ such that for every G with $G \vdash g$ we have that $H(G \circ K) \vdash f$.

If $f \leq_W g$ and K and H are as in the definition above, we say that K and H *witness* that $f \leq_W g$.

The difference between regular Weihrauch reducibility and strong Weihrauch reducibility is that in the first case the post-processor H has access to the original input, while this is lost in the latter case.

As for other reducibilities in computability theory, this induces a degree structure in the usual way. That is, we say that f is *Weihrauch equivalent* to g, or $f \equiv_W g$, if both $f \leq_W g$ and $g \leq_W f$. We say that the equivalence class of f under $\equiv_W$ is the *Weihrauch degree* of f. We can introduce *strong Weihrauch equivalence* $\equiv_{sW}$ and the *strong Weihrauch degrees* in the same way.

In [4], Brattka, Gherardi and Marcone have introduced a natural operation related to composition, as mentioned in the introduction. This notion is called the *compositional product*.

Definition 2.5. Let $f :\subseteq \mathcal{X} \rightrightarrows \mathcal{Y}$ and $g :\subseteq \mathcal{Y} \rightrightarrows \mathcal{Z}$. Then $g \circ f :\subseteq \mathcal{X} \rightrightarrows \mathcal{Z}$ is the multi-valued function with domain

$$\{x \in \mathcal{X} \mid x \in \mathrm{dom}(f) \wedge f(x) \subseteq \mathrm{dom}(g)\},$$

and for x in the domain of $g \circ f$ we have

$$g \circ f(x) = \{z \in \mathcal{Z} \mid \exists y \in \mathcal{Y}(z \in g(y) \wedge y \in f(x))\}.$$

Definition 2.6. Let f, g be multi-valued functions (on represented spaces). Then

$$f \star g = \max (f_0 \circ g_0 \mid f_0 \leq_W f \text{ and } g_0 \leq_W g),$$

where the maximum is taken over those f_0 and g_0 where the codomain of g_0 and the domain of f_0 coincide.

That the supremum exists and that it is even a maximum was proven in [6]. Let us give an example using randomness to illustrate how to work with these compositional products and how to formally work with Weihrauch reducibility.

Proposition 2.7

$$n{-}\mathrm{Ran} \star n{-}\mathrm{Ran} \equiv_W n{-}\mathrm{Ran}.$$

Proof. It is not hard to see that $f \leq_W f \star f$ always holds: consider the composition of f and the identity.

Conversely, let $f_0 :\subseteq (X, \delta_X) \rightrightarrows (Y, \delta_Y)$ and $g_0 :\subseteq (Z, \delta_Z) \rightrightarrows (X, \delta_X)$ with $f_0, g_0 \leq_W n-\text{Ran}$, and let this be witnessed by V_0 and U_0 for f_0, and by K_0 and H_0 for g_0. We now define K to be the identity, and we let H be the function sending $(x, y \oplus z) \in \omega^\omega \times \omega^\omega$ to $U_0(H_0(x, y), z)$. We claim: K and H witness that $f_0 \circ g_0 \leq_W n-\text{Ran}$.

Thus, let $G \vdash n-\text{Ran}$; we need to show that $H(\text{id}, G \circ K) \vdash f_0 \circ g_0$. So, let $x \in \omega^\omega$ be in the domain of $f_0 \circ g_0 \circ \delta_Z$. Then

$$H(x, G(K(x))) = U_0(H_0(x, G_0(x)), G_1(x)),$$

where $G(x) = G_0(x) \oplus G_1(x)$. Note that $G_0(x)$ is n-random relative to x, hence it is also n-random relative to $K_0(x)$. Thus, per choice of K_0 and H_0 we see that $\delta_X(H_0(x, G_0(x))) \in g_0(\delta_Z(x))$. Next, by van Lambalgen's theorem for n-randomness (see e.g. [8, Corollary 6.9.3]) we know that $G_1(x)$ is n-random relative to $x \oplus G_0(x)$, so it is also n-random relative to $V_0(H_0(x, G_0(x)))$. Therefore

$$\delta_Y(U_0(H_0(x, G_0(x)), G_1(x))) \in f_0(\delta_X(H_0(x, G_0(x)))) \subseteq (f_0 \circ g_0)(\delta_Z(x)),$$

as desired.   □

3   Probabilistic Choice

In Brattka, Gherardi and Hölzl [3], various choice principles are studied within the Weihrauch degrees. The main focus of this paper will be *probabilistic choice*, for which we will recall the definition shortly. While in [3] various spaces are studied, we will only study Cantor space 2^ω, Baire space ω^ω and the intermediate space $\omega \times 2^\omega$, which allows us to simplify the necessary definitions.

Definition 3.1. We let Tree_{2^ω} be the set of trees in $2^{<\omega}$, we let $\text{Tree}_{\omega^\omega}$ be the set of trees in $\omega^{<\omega}$ and we let $\text{Tree}_{\omega \times 2^\omega}$ be the set of trees in $\{\emptyset\} \cup (\omega \times 2^{<\omega})$, where a *tree* in Y is a subset of Y closed under taking substrings. For any tree T, we let $[T]$ be the set of infinite paths through T.

In what follows, we will assume reasonable fixed representations of Tree_{2^ω}, $\text{Tree}_{\omega^\omega}$ and $\text{Tree}_{\omega \times 2^\omega}$, where 'reasonable' means that membership of a string σ in a tree $\delta(X)$ should be uniformly decidable in X.

There are natural Borel measures on the three spaces mentioned above: on Cantor space, we have the measure μ_{2^ω} induced by

$$\mu_{2^\omega}([\![\sigma]\!]) = 2^{-|\sigma|}$$

(where $[\![\sigma]\!]$ is the set of $x \in 2^\omega$ extending σ). This corresponds to the probability measure where each bit has value 0 or 1, each with probability $1/2$, independently of other bits. On Baire space we have the measure μ_{ω^ω} induced by

$$\mu_{\omega^\omega}([\![\sigma]\!]) = \prod_{i < |\sigma|} 2^{-\sigma(i)-1},$$

that is, each value of the sequence is equal to n with probability 2^{-n-1}, independently of all other values. On $\omega \times 2^\omega$ we have the measure induced by

$$\mu_{\omega \times 2^\omega}(\llbracket \sigma \rrbracket) = 2^{-\sigma(0)-1}2^{-|\sigma|+1},$$

that is, the first value equal to n with probability 2^{-n-1} and every other value is 0 or 1 with probability $1/2$, all values being independent.

Given any tree, we define the 'measure of T' to be the measure of $[T]$. Clearly, every tree of positive measure has an infinite path. Probabilistic choice is the multi-valued function assigning to such a tree of positive measure the collection of its paths.

Definition 3.2. Given $\mathcal{X} \in \{2^\omega, \omega^\omega, \omega \times 2^\omega\}$, we let $\text{Tree}_{\mathcal{X}}^{>0} \subseteq \text{Tree}_{\mathcal{X}}$ be the set of trees of positive measure.

Definition 3.3. Given $\mathcal{X} \in \{2^\omega, \omega^\omega, \omega \times 2^\omega\}$, we let $\text{PC}_{\mathcal{X}} : \text{Tree}_{\mathcal{X}}^{>0} \rightrightarrows \mathcal{X}$ be the multi-valued function sending a tree T of positive measure to $[T]$. Alternatively, we call PC_{2^ω} *weak weak König's lemma*, or WWKL.

As a warmup, let us compare randomness and PC.

Proposition 3.4. *We have* $1-\text{Ran} \leq_{sW} \text{PC}_{2^\omega}$ *but* $\text{PC}_{2^\omega} \not\leq_W 1-\text{Ran}$.

Proof. Fix a universal oracle Martin-Löf test $\mathcal{U}_0^X, \mathcal{U}_1^X, \ldots$ and let T^X be a tree uniformly computable in X such that $[T^X]$ is the complement of $\mathcal{U}_0^X$. Now let K be the total Turing functional sending X to T^X, and let H be the identity. Then K and H witness that $1-\text{Ran} \leq_{sW} \text{PC}_{2^\omega}$.

For the converse, see Brattka, Hendtlass and Kreuzer [5]. $\qquad\square$

As informally explained in the introduction, there is a notion of a *jump* in the Weihrauch degrees, introduced in Brattka, Gherardi and Marcone [4].

Definition 3.5. Given any multi-valued function $f :\subseteq (X, \delta_X) \rightrightarrows (Y, \delta_Y)$, we obtain its jump f' by replacing the representation δ_X by $\delta_X' = \delta_X \circ \lim$, where $\lim :\subseteq \omega^\omega \to \omega^\omega$ is the partial function sending f to the pointwise limit of $f^{[0]}, f^{[1]}, \ldots$, where the domain of $\lim$ is exactly the set of f for which this limit exists.

In other words, as a set-theoretic function f' is the same as f, but its input representation is weakened by only giving a sequence converging to some z, instead of the actual intended input z. In the case of PC, this leads to the following.

Definition 3.6. Let $\mathcal{X} \in \{2^\omega, \omega^\omega, \omega \times 2^\omega\}$. We let $\text{limtree}_{\mathcal{X}}^{>0}$ be the collection of sequences $(T_i)_{i \in \omega}$ with $T_i \in \text{Tree}_{\mathcal{X}}$ such that $(T_i)_{i \in \omega}$ converges pointwise to a tree T_∞ of positive measure.

Again, we have a natural representation of $\text{limtree}_{\mathcal{X}}^{>0}$ by sending $f \in \omega^\omega$ to $\left(\delta_{\text{Tree}_{\mathcal{X}}}(f^{[i]})\right)_{i \in \omega}$.

Proposition 3.7. *Let $\mathcal{X} \in \{2^\omega, \omega^\omega, \omega \times 2^\omega\}$. Given any multi-valued function $f :\subseteq \mathrm{Tree}_{\mathcal{X}}^{\geq 0} \rightrightarrows Y$, let $\phi(f) :\subseteq \mathrm{limtree}_{\mathcal{X}}^{\geq 0} \rightrightarrows Y$ be the multi-valued function sending $(T_i)_{i \in \omega}$ to $f(T_\infty)$. Then ϕ is a bijection between $\{f \mid f :\subseteq \mathrm{Tree}_{\mathcal{X}}^{\geq 0} \rightrightarrows Y\}$ and $\{g \mid g :\subseteq \mathrm{limtree}_{\mathcal{X}}^{\geq 0} \rightrightarrows Y\}$. Furthermore, f' and $\phi(f)$ have exactly the same realisers.*

Proof. The inverse of ϕ is the function sending $g :\subseteq \mathrm{limtree}_{\mathcal{X}}^{\geq 0} \rightrightarrows Y$ to the multi-valued function sending a tree T to $g((T)_{i \in \omega})$, where $(T)_{i \in \omega}$ is the sequence that is constantly T. That f' and $\phi(f)$ have the same realisers follows directly from unfolding the definitions. $\qquad \square$

Thus, in particular we can identify $\mathrm{PC}_{\mathcal{X}}'$ with the multi-valued function sending an element $(T_i)_{i \in \omega} \in \mathrm{limtree}_{\mathcal{X}}^{\geq 0}$ to $[T_\infty]$, which we will henceforth do.

4   Iterating $\mathrm{PC}_{\mathcal{X}}'$

In this section we study what happens when we iterate $\mathrm{PC}_{\mathcal{X}}'$, i.e., we look at $\mathrm{PC}_{\mathcal{X}}' \star \mathrm{PC}_{\mathcal{X}}'$. As mentioned in the introduction, we will show that $\mathrm{WWKL}' \star \mathrm{WWKL}' \equiv_W \mathrm{PC}_{\omega \times 2^\omega}'$. However, we will first show that $\mathrm{PC}_{\omega \times 2^\omega}'$ is closed under iteration. For this, we relate $\mathrm{PC}_{\omega \times 2^\omega}'$ to 2-randomness. In what follows, we assume a fixed universal oracle Martin-Löf test $\mathcal{U}_0^X, \mathcal{U}_1^X, \ldots$.

Definition 4.1. *Let X be n-random relative to Y. Then the n-randomness deficiency of X relative to Y is the least $m \in \omega$ such that $X \notin \mathcal{U}_m^{Y^{(n-1)}}$.*

Definition 4.2. *Let $\mathrm{WWKL}_{\neq 0^\omega}'$ be the multi-valued function sending $(T_i)_{i \in \omega} \in \mathrm{limtree}_{2^\omega}^{\geq 0}$ to a non-zero element of $[T_\infty]$.*

Similarly, let $\mathrm{WWKL}_{2-\mathrm{Ran}}'$ be the multi-valued function sending $(T_i)_{i \in \omega} \in \mathrm{limtree}_{2^\omega}^{\geq 0}$ to an $X \in [T_\infty]$ which is 2-random relative to $(T_i)_{i \in \omega}$.

Finally, let $\mathrm{WWKL}_{2-\mathrm{Ran+Def}}'$ be the multi-valued function sending $(T_i)_{i \in \omega} \in \mathrm{limtree}_{2^\omega}^{\geq 0}$ to an $X \in [T_\infty]$ which is 2-random relative to $(T_i)_{i \in \omega}$ and an upper bound on its 2-randomness deficiency relative to $(T_i)_{i \in \omega}$.

Theorem 4.3

$$\mathrm{PC}_{\omega \times 2^\omega}' \equiv_{sW} \mathrm{WWKL}_{\neq 0^\omega}' \equiv_{sW} \mathrm{WWKL}_{2-\mathrm{Ran}}' \equiv_{sW} \mathrm{WWKL}_{2-\mathrm{Ran+Def}}'.$$

Proof. First, we show that $\mathrm{PC}_{\omega \times 2^\omega}' \leq_{sW} \mathrm{WWKL}_{\neq 0^\omega}'$. Let $(T_i)_{i \in \omega} \in \mathrm{limtree}_{\omega \times 2^\omega}^{\geq 0}$. Now let $(S_i)_{i \in \omega} \in \mathrm{limtree}_{2^\omega}^{\geq 0}$ be the sequence of trees where

$$S_i = \{\emptyset, 0, 00, \ldots\} \cup \bigcup_{n \in \omega} 0^n 1 T_i^n,$$

where $T_i^n = \{\sigma \in 2^{<\omega} \mid n\sigma \in T_i\}$. Then

$$[S_\infty] = \{0^\omega\} \cup \bigcup_{n \in \omega} 0^n 1 [T_\infty^n],$$

so the measure of S_∞ is the same as the measure of T_∞; in particular we see that indeed $(S_i)_{i \in \omega} \in \text{limtree}_{2^\omega}^{>0}$. Furthermore, every $X \in [S_\infty]$ different from 0^ω computes an element of $[T_\infty]$ by sending $0^n 1Y$ to nY.

Next, it is clear that every 2-random is different from 0^ω, which shows that $\text{WWKL}'_{\neq 0^\omega} \leq_{sW} \text{WWKL}'_{2-\text{Ran}}$. We also get $\text{WWKL}'_{2-\text{Ran}} \leq_{sW} \text{WWKL}'_{2-\text{Ran}+\text{Def}}$ by just forgetting the bound on the randomness deficiency.

Finally, we show that $\text{WWKL}'_{2-\text{Ran}+\text{Def}} \leq_{sW} \text{PC}'_{\omega \times 2^\omega}$. Given any $(T_i)_{i \in \omega} \in \text{limtree}_{2^\omega}^{>0}$, we can uniformly compute trees $P_i^n \in \text{Tree}_{2^\omega}^{>0}$ such that $(P_i^n)_{i \in \omega}$ converges to a tree P_∞^n with $[P_\infty^n] = 2^\omega \setminus \mathcal{U}_n^{(T_i)'_{i \in \omega}}$. Now, consider the sequence of trees $(S_i)_{i \in \omega}$ with S_i given by:

$$S_i = \bigcup_{n \in \omega} n(P_i^n \cap T_i).$$

Then $(S_i)_{i \in \omega}$ is uniformly computable in $(T_i)_{i \in \omega}$. Let $n \in \omega$ and $q > 0$ be such that $\mu_{2^\omega}(T_\infty) \geq 2^{-n} + q$, which exists because T_∞ has positive measure. Then $\mu_{2^\omega}(P_\infty^n \cap T_\infty) \geq q$, so also $\mu_{\omega \times 2^\omega}(S_\infty) \geq q > 0$. Therefore $(S_i)_{i \in \omega} \in \text{limtree}_{\omega \times 2^\omega}^{>0}$. Finally, for every element of $nX \in [S_\infty]$ we have that $X \in [T_\infty]$, that X is 2-random relative to $(T_i)_{i \in \omega}$ and that n is a bound on its 2-randomness deficiency relative to $(T_i)_{i \in \omega}$.   □

To show that $\text{PC}'_{\omega \times 2^\omega}$ is closed under composition, we use the following lemma (see [1, Proposition 2.12] for a proof with $Y = \emptyset'$, which relativizes in a straightforward way).

Lemma 4.4. *There is a single Turing functional Φ such that for every 2-random X relative to Y, if n bounds the 2-randomness deficiency of X relative to Y then $\Phi(X \oplus Y', n) = (X \oplus Y)'$.*

Theorem 4.5. *We have*

$$\text{WWKL}'_{2-\text{Ran}} \star \text{WWKL}'_{2-\text{Ran}+\text{Def}} \equiv_W \text{PC}'_{\omega \times 2^\omega},$$

and hence also by Theorem 4.3:

$$\text{PC}'_{\omega \times 2^\omega} \star \text{PC}'_{\omega \times 2^\omega} \equiv_W \text{PC}'_{\omega \times 2^\omega}.$$

Proof. The fact that $\text{WWKL}'_{2-\text{Ran}} \star \text{WWKL}'_{2-\text{Ran}+\text{Def}} \geq_W \text{PC}'_{\omega \times 2^\omega}$ is a direct consequence of Theorem 4.3.

For the converse, let f and g be multi-valued functions such that both $f \leq_W \text{WWKL}'_{2-\text{Ran}}$ and $g \leq_W \text{WWKL}'_{2-\text{Ran}+\text{Def}}$. Without loss of generality, we can assume that the domain and range of f and g are contained in ω^ω. We want to show that $f \circ g \leq_W \text{PC}'_{\omega \times 2^\omega}$. Unfolding the definition of Weihrauch reducibility, and using the assumption on f and g, we know that there exist three computable functions $T : \omega^\omega \rightarrow \text{limtree}_{2^\omega}^{>0}$, $S :\subseteq \omega^\omega \times 2^\omega \times \omega \rightarrow \text{limtree}_{2^\omega}^{>0}$ and $H :\subseteq$

$\omega^\omega \times 2^\omega \times \omega \times 2^\omega \to \omega^\omega$ such that for every X, Y, n, such that $Y \in [\lim T(X)]$ and Y is 2-random relative to X with 2-randomness deficiency at most n, we have that (X, Y, n) is in the domain of S, and for every $Z \in [\lim S(X, Y, n)]$, $H(X, Y, n, Z) \in (f \circ g)(X)$.

/ The core of the argument is to show that for all X, the set $Q(X)$ of pairs $(n, Y \oplus Z)$ such that $Y \in [\lim T(X)]$, Y is 2-random relative to X with 2-randomness deficiency at most n and $Z \in \lim[S(X, Y, n)]$ is a $\Pi_1^0(X')$ subset of $\omega \times 2^\omega$, uniformly in X. This is a consequence of Lemma 4.4. Indeed, given X the sequence $T(X)$ is computable in X, thus $\lim T(X)$ is X'-computable, uniformly in X. Thus, the set of pairs (Y, n) such that Y is a path of $\lim T(X)$ and Y is 2-random relative to X with randomness deficiency at most n is $\Pi_1^0(X')$ uniformly in X. Furthermore, the tree $\lim S(X, Y, n)$ is $(X, Y, n)'$-computable uniformly, but because of Lemma 4.4 it is in fact (X', Y, n)-computable uniformly. Thus the set of paths of $\lim S(X, Y, n)$ is $\Pi_1^0(X' \oplus Y)$. Putting all this together, we get that $Q(X)$ is indeed $\Pi_1^0(X')$ uniformly in X.

Furthermore, $Q(X)$ is a subset of $\omega \times 2^\omega$ of positive measure: given X, there is a positive probability that Y chosen at random (w.r.t. the uniform measure) is in $[\lim T(X)]$ (because $T(X)$ has positive measure!), a positive probability that an integer chosen at random bounds the 2-randomness deficiency of Y relative to X, and conditional to this, a positive probability that Z chosen at random belongs to $[S(X, Y, n)]$ (which, assuming Y and n are as above, has positive measure).

We have established that $Q(X)$ is a $\Pi_1^0(X')$ subset of $\omega \times 2^\omega$ uniformly in X, hence can be represented by an X'-computable tree over $\omega \times 2^{<\omega}$, and thus as the limit of an X-computable sequence of trees over $\omega \times 2^{<\omega}$. Now we immediately get the desired result: given X, one can compute a sequence in $\mathrm{limtree}_{\omega \times 2^\omega}^{>0}$ representing $Q(X)$, and for any path $(n, Y \oplus Z)$ of the limit tree (that is, a member of $Q(X)$), we get an element of $(f \circ g)(X)$ by simply computing $H(X, Y, n, Z)$. This shows $f \circ g \leq_W \mathrm{PC}'_{\omega \times 2^\omega}$, as wanted.   □

Thus, in particular we see that $\mathrm{WWKL}' \star \mathrm{WWKL}' \leq_W \mathrm{PC}'_{\omega \times 2^\omega}$. Perhaps surprisingly, the converse is also true, which is expressed by the next theorem.

Theorem 4.6
$$\mathrm{WWKL}'_{2-\mathrm{Ran}} \equiv_W \mathrm{WWKL}' \star \mathrm{WWKL}'.$$

Proof. We need to show that $\mathrm{WWKL}'_{2-\mathrm{Ran}} \leq_W \mathrm{WWKL}' \star \mathrm{WWKL}'$. Our idea is as follows. Given a $(T_i)_{i \in \omega} \in \mathrm{limtree}_{2^\omega}^{>0}$, we want to know the measure of T_∞, so that we can intersect it with a large enough set of 2-randoms. Using the first instance of WWKL', we will compute an X such that X' computes a lower bound on the measure of T_∞. To do this, our construction uses a partition $(I_n)_{n \in \omega}$ of ω, where the I_n should be sufficiently large. We build our first tree to which we apply WWKL' in such a way that $X \restriction I_n$ is constantly 0 for X on this tree if and only if the measure of $T_\infty \restriction n$ (i.e., the measure of $\{x \in 2^\omega \mid x \restriction n \in T_\infty\}$) drops significantly lower than the measure of $T_\infty \restriction (n-1)$. If we define 'significantly' in the right way, this will only happen finitely often; hence if our intervals I_n are large enough we do not lose too much measure by adding this restriction.

Furthermore, X' can compute how often $X \restriction I_n$ is constantly 0, and hence compute a lower bound on the measure of T_∞.

We now give the details. Let $(T_i)_{i \in \omega} \in \mathrm{limtree}_{2^\omega}^{>0}$. Define a partition of ω by $I_n = [n(n+1), (n+1)(n+2))$; hence each I_n has $2(n+1)$ elements. We define $(S_i)_{i \in \omega} \in \mathrm{limtree}_{2^\omega}^{\geq 0}$ as follows. Each S_i will be of the form $S_i^0 S_i^1 \ldots$, where each $S_i^n \subseteq \{0,1\}^{2(n+1)}$. In other words, a string σ is in S_i if and only if for each $n \in \omega$ we have $\sigma \restriction I_n \in S_i^n$.

We show how to define S_i^n by recursion on n. Simultaneously, we will define an auxiliary $k_{i,n} \in \omega$, where we initialise $k_{i,-1} = 0$. Fix $i, n \in \omega$. We consider two cases:

- $\mu(T_i \restriction n) < 2^{-k_{i,n-1}}$. Then we let $S_i^n = \{0^{2(n+1)}\}$. Let $k_{i,n}$ be least such that $\mu(T_i \restriction n) \geq 2^{-k_{i,n}}$ if $\mu(T_i \restriction n) > 0$; otherwise let $k_{i,n} = 0$.
- $\mu(T_i \restriction n) \geq 2^{-k_{i,n-1}}$. Then we let $S_i^n = \{0,1\}^{2(n+1)} \setminus \{0^{2(n+1)}\}$. Let $k_{i,n} = k_{i,n-1}$.

Then for every $n \in \omega$ we have that $(S_i^n)_{i \in \omega}$ converges to some tree S_∞^n, because if s is large enough such that T_i has settled below n for $i \geq s$, then $S_i^n = S_j^n$ for all $i, j \geq s$. For the same reason, $(k_{i,n})_{i \in \omega}$ converges to some k_n. Since T_∞ has positive measure, note that every k_n is positive. In fact, $(k_n)_{n \in \omega}$ converges to the least $k \in \omega$ with $\mu([T_\infty]) \geq 2^{-k}$.

We also claim that S_∞ has positive measure. Let m be large enough such that $k_n = k_m$ for all $n \geq m$. Then $S_\infty^n = \{0,1\}^{2(n+1)} \setminus \{0^{2(n+1)}\}$ for $n \geq m$. Fix any string $\sigma \in S_\infty^0 S_\infty^1 \ldots S_\infty^{m-1}$. Then

$$\mu(\overline{S} \mid \sigma) \leq \sum_{n \geq m} 2^{-2(n+1)} < 1,$$

hence

$$\mu(S) \geq 2^{-m(m+1)} \mu(S \mid \sigma) > 0.$$

Now, given any $X \in [S_\infty]$, note that the number of n such that $X \restriction I_n = 0^{2(n+1)}$ is exactly k. Thus, there is a Turing functional Φ, independent of $(T_i)_{i \in \omega}$, such that for every $X \in [S_\infty]$ we have that $\Phi(X, i)$ converges to k as i goes to infinity.

Next, we define $(P_i)_{i \in \omega} \in \mathrm{limtree}_{2^\omega}^{>0}$ uniformly in X and $(T_i)_{i \in \omega}$. Let $P_i = T_i \cap \overline{\mathcal{U}_{\Phi(X,i)+1}^{\emptyset'[i]}}$. Then $(P_i)_{i \in \omega}$ converges to $T_\infty \cap \overline{\mathcal{U}_{k+1}^{\emptyset'}}$ which has positive measure and only has 2-random paths, as desired. $\square$

It is known from Brattka and Pauly [7, Proposition 22] that $\mathrm{PC}_{\omega \times 2^\omega} >_W \mathrm{WWKL}$. Relativising this result we also get that $\mathrm{PC}'_{\omega \times 2^\omega} >_W \mathrm{WWKL}'$, and hence $\mathrm{WWKL}' \star \mathrm{WWKL}' >_W \mathrm{WWKL}'$.

Proposition 4.7. *We have* $\mathrm{WWKL}' \leq_{sW} \mathrm{PC}'_{\omega \times 2^\omega}$, *but* $\mathrm{PC}'_{\omega \times 2^\omega} \not\leq_W \mathrm{WWKL}'$.

Proof. Clearly, WWKL$'$ $\leq_W$ PC$'_{\omega \times 2^\omega}$. For the converse, consider the multi-valued function C$_\omega$ which assigns to a non-surjective function $f \in \omega^\omega$ an element not in the range of f; in other words, the computable instances represent finding an element of a non-empty co-c.e. set.[1] Then C$'_\omega$ $\leq_{sW}$ PC$'_{\omega \times 2^\omega}$: given $(f_i)_{i \in \omega}$, consider $(T_i)_{i \in \omega} \in \text{limtree}^{>0}_{\omega \times 2^\omega}$ given by

$$T_i = \bigcup_{j \in \omega} \bigcup_{n \notin f_i(\{0,\dots,j\})} n2^{<j}.$$

Then T_i is uniformly computable in $(T_i)_{i \in \omega}$, and it converges to

$$T_\infty = \bigcup_{j \in \omega} \bigcup_{n \notin f_\infty(\{0,\dots,j\})} n2^{<j}.$$

Furthermore, T_∞ has positive measure because f_∞ is not surjective. Thus, $(T_i)_{i \in \omega}$ is indeed an element of limtree$^{>0}_{\omega \times 2^\omega}$. Finally, every element X of $[T_\infty]$ computes an element not in the range of f, namely $X(0)$.

However, C$'_\omega$ $\not\leq_W$ WWKL$'$ (as pointed out by the referee, this also follows from [4, Corollary 12.3]; but we give a direct proof). Indeed, assume there are $K :\subseteq \omega^\omega \rightarrow \text{limtree}^{>0}_{2^\omega}$ and $H :\subseteq \omega^\omega \times 2^\omega \rightarrow \omega$ such that for every f with $(f^{[i]})_{i \in \omega}$ converging to some f_∞ with ran$(f_\infty) \neq \omega$ we have that $K(f)$ is total, and for every X on $K(f)_\infty$ we have $H(f, X) \notin \text{ran}(f_\infty)$. Given such an f the complement of its range is a non-empty set A which is co-c.e. in $\emptyset'$, and for every non-empty set A which is co-c.e. in $\emptyset'$ we can effectively find an index for such a function f from an index for A, so we will implicitly identify these two. We will therefore build a set A which is co-c.e. in $\emptyset'$ for which $H(f, X) \notin A$ for *any* $X \in K(f)_\infty$.

By the recursion theorem we may assume we know an index e for f. So, using $\emptyset'$ and e we can compute $K(\{e\})_\infty$. Now look for the least s such that for every $\sigma \in K(\{e\})_\infty$ of length s we have that $H(\{e\}, \sigma)[s]\downarrow$ if such an s exists, and enumerate the finitely many values $H(\{e\}, \sigma)$ into the complement of A. If such an s does not exist, we let $A = \omega$.

Now we know by compactness that an s as above exists, and therefore A is co-finite (hence non-empty). However, by construction we now have that $U(\{e\}, X) \notin A$ for every X on $K(\{e\})_\infty$, which is a contradiction. □

Next, we separate WWKL$'_{2-\text{Ran}}$ and hence WWKL$'\star$WWKL$'$ from WWKL$''$, using randomness and the effective 0-1-law. It is known from Brattka and Pauly [7] that PC$_{\omega \times 2^\omega}$ $<_W$ WWKL$'$. We now show that this also holds one jump higher.

Proposition 4.8. *We have* WWKL$'_{2-\text{Ran}}$ $\leq_{sW}$ WWKL$''$, *but on the other hand* WWKL$''$ $\not\leq_W$ WWKL$'_{2-\text{Ran}}$.

---

[1] Our definition is not exactly the same as the definition of C$_\omega$ in Brattka, Gherardi and Hölzl [3], but it is easily seen to be strongly Weihrauch-equivalent to it.

Proof. Given any $(T_i)_{i \in \omega} \in \mathrm{limtree}_{2^\omega}^{\geq 0}$, we can use $(T_i)_{i \in \omega}''$ to compute a lower bound 2^{-n} on the measure of T_∞, and then let S be a tree such that $[S] = [T_\infty] \setminus \mathcal{U}_{n+1}^{(T_i)_{i \in \omega}'}$. Since S is $(T_i)_{i \in \omega}''$-computable we can uniformly transform S into a $(T_i)_{i \in \omega}$-computable double sequence $((S_i^j)_{i \in \omega})_{j \in \omega}$ with $\lim_{j \to \infty} \lim_{i \to \infty} S_i^j = S$. Since every infinite path of S is a 2-random$^{(T_i)_{i \in \omega}}$ path of T_∞ and all the computations are uniform, this shows that $\mathrm{PC}_{\omega \times 2^\omega}' \leq_{sW} \mathrm{WWKL}''$.

Conversely, assume towards a contradiction that $\mathrm{WWKL}'' \leq_W \mathrm{WWKL}_{2-\mathrm{Ran}}'$. Consider any $\emptyset''$-computable tree S of positive measure without $\emptyset''$-computable paths. Then S is the limit of a computable double sequence $((S_i^j)_{i \in \omega})_{j \in \omega}$ converging to S. Therefore, from our assumption that $\mathrm{WWKL}'' \leq_W \mathrm{WWKL}_{2-\mathrm{Ran}}'$ it follows that there should be a computable $(T_i)_{i \in \omega} \in \mathrm{limtree}_{2^\omega}^{\geq 0}$ such that every 2-random path of T_∞ computes a path of S. Let X be a $\emptyset''$-computable 2-random set. Then the effective 0-1-law tells us that $Y = \sigma X$ is a path of T_∞ for some $\sigma \in 2^{<\omega}$. So, Y is a 2-random path of T_∞, but since it is $\emptyset''$-computable it clearly does not compute a path of S. $\qquad\square$

Thus, combining everything from this section we have the following.

Corollary 4.9

$$\mathrm{WWKL}' <_W \mathrm{WWKL}' \star \mathrm{WWKL}' <_W \mathrm{WWKL}''.$$

5 $\mathrm{PC}_{\omega^\omega}$

In this section we will show that $\mathrm{PC}_{\omega^\omega} \equiv_{sW} \mathrm{PC}_{\omega \times 2^\omega}$. First, let us remark that, replacing every occurrence of 2-randomness by 1-randomness in Definition 4.2 and Theorem 4.3, we also get the following result by the same proof as for Theorem 4.3.

Theorem 5.1

$$\mathrm{PC}_{\omega \times 2^\omega} \equiv_{sW} \mathrm{WWKL}_{\neq 0^\omega} \equiv_{sW} \mathrm{WWKL}_{1-\mathrm{Ran}} \equiv_{sW} \mathrm{WWKL}_{1-\mathrm{Ran}+\mathrm{Def}}.$$

The following result, due to Brattka, Gherardi and Hölzl, was previously announced by Hölzl at ARA 2014 in Gotemba, but a proof has not yet appeared in print.

Theorem 5.2

$$\mathrm{PC}_{\omega^\omega} \equiv_{sW} \mathrm{PC}_{\omega \times 2^\omega}.$$

Proof. Clearly $\mathrm{PC}_{\omega \times 2^\omega} \leq_{sW} \mathrm{PC}_{\omega^\omega}$. We show the converse. Using the previous theorem, this is equivalent to showing that $\mathrm{PC}_{\omega^\omega} \leq_{sW} \mathrm{WWKL}_{1-\mathrm{Ran}}$.

We will use the function $\alpha : 2^{<\omega} \to \omega^{<\omega}$ which maps a string σ to the string enumerating σ in increasing order, i.e., the length of $\alpha(\sigma)$ is the number of ones in σ, and $\alpha(\sigma)(n)$ is the position of the nth one.

First we define a computable function K which maps trees in ω^ω to trees in 2^ω. Let T be a tree in ω^ω. Given any string $\sigma \in 2^{<\omega}$, we put σ into $K(T)$ if and only if $\alpha(\sigma) \in T$.

Then K is clearly computable. Furthermore, if $X \in [K(T)]$ has infinitely many ones, then $\alpha(X) := \bigcup_{n \in \omega} \alpha(X \restriction n) \in \omega^\omega$ is an infinite path of T. In particular this holds for the random paths of $K(T)$, thus every random path of $K(T)$ computes a path of T. Finally, by the way we have defined our measures we know that α is measure-preserving, and hence

$$\mu_{2^\omega}\left([K(T)]\right) \geq \mu_{2^\omega}\left(\alpha^{-1}([T])\right) = \mu_{\omega^\omega}([T]) > 0. \qquad \square$$

Note that this proof directly relativises, giving us the following result as well. Alternatively, this follows from the fact that if $f \leq_{sW} g$ then $f' \leq_{sW} g'$, as proven in Brattka, Gherardi and Marcone [4].

Corollary 5.3

$$\mathrm{PC}'_{\omega^\omega} \equiv_{sW} \mathrm{PC}'_{\omega \times 2^\omega}.$$

6   Mixing Jumps and Iterations

Next, we wonder: what happens if we mix the composition of WWKL and WWKL'? First, let us show that WWKL $\star$ WWKL' $\equiv_W$ WWKL'.

Proposition 6.1

$$\mathrm{WWKL} \star \mathrm{WWKL}' \equiv_W \mathrm{WWKL}'.$$

Proof. Let $H :\subseteq \mathrm{limtree}_{2^\omega}^{>0} \times 2^\omega \to \mathrm{Tree}_{2^\omega}^{>0}$ be a Turing functional. We show how to, given $(T_i)_{i \in \omega} \in \mathrm{limtree}_{2^\omega}^{>0}$ such that $((T_i)_{i \in \omega}, X) \in \mathrm{dom}(H)$ for every $X \in [T_\infty]$, uniformly construct an $(S_i)_{i \in \omega} \in \mathrm{limtree}_{2^\omega}^{>0}$ such that every path X of S_∞ uniformly computes a path of both T_∞ and $H((T_i)_{i \in \omega}, X)$.

For this, we let S_i be the tree where a string $\sigma \oplus \tau \in S_i$ if and only if $\sigma \in S_i$ and, if $H((T_i)_{i \in \omega}, \sigma)[\sigma](\tau)\downarrow$, then $H((T_i)_{i \in \omega}, \sigma)(\tau) = 1$. Then it is not hard to verify that $(S_i)_{i \subset \omega}$ converges to some tree S_∞. Furthermore, for every $X \in [T_\infty]$ and $Y \in [H((T_i)_{i \in \omega}, X)]$ we have that $X \oplus Y \in [S_\infty]$, and vice versa. Finally, by Fubini's theorem, S_∞ has positive measure, completing the proof. $\qquad \square$

It turns out the converse also holds. In fact, we even have that WWKL' $\star$ $\mathrm{PC}_{\omega \times 2^\omega} \equiv_W$ WWKL'.

Theorem 6.2

$$\mathrm{WWKL}' \star \mathrm{PC}_{\omega \times 2^\omega} \equiv_W \mathrm{WWKL}' \star \mathrm{WWKL} \equiv_W \mathrm{WWKL}'.$$

Proof. Clearly,

$$\mathrm{WWKL}' \leq_W \mathrm{WWKL}' \star \mathrm{WWKL} \leq_W \mathrm{WWKL}' \star \mathrm{PC}_{\omega \times 2^\omega}.$$

We need to show that $\mathrm{WWKL}' \star \mathrm{PC}_{\omega \times 2^\omega} \leq_W \mathrm{WWKL}'$; or, by Theorem 5.1, that $\mathrm{WWKL}' \star \mathrm{WWKL}_{1-\mathrm{Ran}+\mathrm{Def}} \leq_W \mathrm{WWKL}'$. In fact, we show that even

$$\mathrm{WWKL}' \star \mathrm{WWKL}_{2-\mathrm{Ran}+\mathrm{Def}} \leq_W \mathrm{WWKL}',$$

where $\mathrm{WWKL}_{2-\mathrm{Ran}+\mathrm{Def}}$ is the multi-valued function assigning to a tree T of positive measure a 2-random^T path through T together with a bound on its 2-randomness^T deficiency. The proof is a variation on the proof of Theorem 4.5. Let f and g be multi-valued functions such that $f \leq_W \mathrm{WWKL}'$ and $g \leq_W \mathrm{WWKL}_{2-\mathrm{Ran}+\mathrm{Def}}$. Again, without loss of generality, we assume that the domain and range of f and g are contained in ω^ω. We want to show that $f \circ g \leq_W \mathrm{WWKL}'$. This time, we have three computable functions $T : \omega^\omega \to \mathrm{Tree}_{2^\omega}^{>0}$, $S :\subseteq \omega^\omega \times 2^\omega \times \omega \to \mathrm{limtree}_{2^\omega}^{>0}$ and $H :\subseteq \omega^\omega \times 2^\omega \times \omega \times 2^\omega \to \omega^\omega$ such that for every X, Y, n, such that $Y \in [T(X)]$ and Y is 2-random relative to X with 2-randomness deficiency at most n, we have that (X, Y, n) is in the domain of S, and for every $Z \in [\lim S(X, Y, n)]$, $H(X, Y, n, Z) \in (f \circ g)(X)$.

Given X, we can compute, uniformly relative to X', an $n = n(X)$ such that $[T(X)]$ has measure strictly greater than 2^{-n} and thus there is some member of $T[X]$ of 2-randomness deficiency at most n. Similarly to the proof of Theorem 4.5, the set $Q(X)$ of sequences $Y \oplus Z$ such that Y belongs to $T(X)$ and is 2-random relative to X with 2-randomness deficiency bounded by $n(X)$, and $Z \in [\lim S(X, Y, n)]$ is a $\Pi_1^0(X')$ subset of 2^ω, uniformly in X' (and has positive measure). Thus one can, given X, uniformly compute a sequence of trees converging to a tree whose paths are members of $Q(X)$, and for every such path $Y \oplus Z$, one gets a member of $(f \circ g)(X)$ by computing $H(X, Y, n(X), Z)$.   □

Thus, we have now studied all combinations of WWKL, WWKL', $\mathrm{PC}_{\omega \times 2^\omega}$ and $\mathrm{PC}_{\omega \times 2^\omega}'$, except for $\mathrm{PC}_{\omega \times 2^\omega} \star \mathrm{WWKL}'$. An earlier draft of this paper contained an incorrect statement about this principle, which was pointed out by Brattka and Hölzl. In fact, they pointed out the following could be proven using techniques from this paper.

Lemma 6.3. (Brattka and Hölzl [private communication]). *Let* $\mathrm{Fin} \subseteq 2^\omega$ *be the set of binary sequences with only finitely many ones. Then*

$$\mathrm{Id}_{\mathrm{Fin}} \leq_{sW} \mathrm{WWKL}.$$

Proof. As in the proof of Theorem 4.6, let $I_n = [n(n+1), (n+1)(n+2))$. Given any $X \in \mathrm{Fin}$, let T be the tree such that $Y \in [T]$ if and only if, for all $n \in \omega$ we have that $Y \restriction I_n = 0^{2(n+1)}$ if and only if $X(n) = 1$. Then T has positive measure because X only contains finitely many ones, as in the proof of Theorem 4.6. Furthermore, any path Y through T clearly computes X, as desired.   □

Lemma 6.4. (Brattka and Hölzl [private communication])

$$\mathrm{C}_\omega' \leq_W \mathrm{C}_\omega \star \mathrm{WWKL}'.$$

Proof. From the previous lemma, together with the fact that, if $f \leq_{sW} g$, then $f' \leq_{sW} g'$ (Brattka, Gherardi and Marcone [4]), we see that $\mathrm{Id}'_{\mathrm{Fin}} \leq_{sW} \mathrm{WWKL}'$. Furthermore, from [4] we also know that $\mathrm{Id}'_\omega \equiv_{sW} C_\omega$. Thus, it is enough if we show that $C'_\omega \leq_W \mathrm{Id}'_\omega \star \mathrm{Id}'_{\mathrm{Fin}}$.

Now, given any function f converging to a non-surjective function, let X be such that $X(\langle i, n, m \rangle) = 1$ if and only if the least element not in the range of $f^{[i]} \upharpoonright n+1$ is different from the least element not in the range of $f^{[i]} \upharpoonright n$, and this new least element is equal to m. Then X converges to an element $X_\infty \in \mathrm{Fin}$, so the function mapping f to X_∞ is reducible to $\mathrm{Id}'_{\mathrm{Fin}}$.

Finally, given X_∞, let m be the largest element of the finite set $\{m \mid \exists n.\langle n, m \rangle \in X_\infty \}$. Since this m can be found as the limit of an X_∞-computable sequence, we obtain that $C'_\omega \leq_W C_\omega \star \mathrm{WWKL}'$, as desired. $\square$

Proposition 6.5. (Brattka and Hölzl [private communication])

$$\mathrm{PC}_{\omega \times 2^\omega} \star \mathrm{WWKL}' \equiv_W \mathrm{PC}'_{\omega \times 2^\omega}.$$

Proof. That $\mathrm{PC}_{\omega \times 2^\omega} \star \mathrm{WWKL}' \leq_W \mathrm{PC}'_{\omega \times 2^\omega}$ follows directly from Theorem 4.5. For the converse, we use the previous lemma, together with the fact that, if $f \leq_{sW} g$, then $f' \leq_{sW} g'$, the fact that $\mathrm{PC}_{\omega \times 2^\omega} \equiv_{sW} C_\omega \times \mathrm{WWKL}$ (Brattka, Gherardi and Hölzl [3][2]) and the easy fact that $\mathrm{WWKL}' \equiv_{sW} \mathrm{WWKL}' \times \mathrm{WWKL}'$. We now have

$$
\begin{aligned}
\mathrm{PC}_{\omega \times 2^\omega} \star \mathrm{WWKL}' &\equiv_W (C_\omega \times \mathrm{WWKL}) \star \mathrm{WWKL}' \\
&\geq_W C_\omega \star \mathrm{WWKL}' \\
&\equiv_W (C_\omega \star \mathrm{WWKL}') \times \mathrm{WWKL}' \\
&\geq_W C'_\omega \times \mathrm{WWKL}' \\
&\equiv_W (C_\omega \times \mathrm{WWKL})' \\
&\equiv_W \mathrm{PC}'_{\omega \times 2^\omega}.
\end{aligned}
$$

$\square$

We would like to finish this paper with an alternative proof of the separation of $\mathrm{WWKL}' \star \mathrm{WWKL}'$ and WWKL'. The techniques for this proof were originally developed by the authors to separate $\mathrm{WWKL}' \star \mathrm{WWKL}'$ from WWKL' before they knew that $\mathrm{WWKL}' \star \mathrm{WWKL}' \equiv_W \mathrm{PC}'_{\omega \times 2^\omega}$. We hope this alternative approach, avoiding notions from algorithmic randomness, might be useful for other purposes in the future.

For this result, we will make use of another well-known Weihrauch degree, namely the degree LPO, which is the degree associated to the function (which we also denote by LPO for simplicity) sending $X \in 2^\omega$ to 0 if $X = 0^\omega$ and to 1 otherwise.

The following was proven in [3].

---

[2] They do not explicitly state the strong Weihrauch equivalence, but it follows directly from their proof.

Proposition 6.6. LPO $\leq_W$ PC$_{\omega \times 2^\omega}$.

(Indeed, given X, one can compute the tree T^X - of positive measure - such that $0\sigma \in T^X$ if and only if $X \upharpoonright |\sigma| = 0^{|\sigma|}$ and for $n > 0$, $n\sigma \in T^X$ if and only if $X \upharpoonright n$ contains a 1. Then, given a path nZ of T^X, we have $n = 0$ if and only if LPO$(X) = 0$). This immediately gives us the following corollary.

Corollary 6.7

$$\text{LPO} \star \text{WWKL}' \leq_W \text{WWKL}' \star \text{WWKL}'.$$

However, we show that one cannot do this with just one application of WWKL$'$.

Theorem 6.8

$$\text{LPO} \star \text{WWKL}' \not\leq_W \text{WWKL}'.$$

Proof. Suppose for the sake of contradiction that LPO $\star$ WWKL$' \leq_W$ WWKL$'$. This means in particular that there exist two computable functions K : limtree$_{2^\omega}^{>0} \times \omega$ $\to$ limtree$_{\omega \times 2^\omega}^{>0}$ and H : limtree$_{2^\omega}^{>0} \times 2^\omega \to 2^\omega \times \{0,1\}$ such that for every $(T_i)_{i \in \omega} \in$ limtree$_{2^\omega}^{>0}$, we have that for every path X of $\lim K((T_i)_{i \in \omega})$ that $H((T_i)_{i \in \omega}, X) = (Z, \text{LPO}(Z))$ for some path Z of T_∞. To get our contradiction, we will make use of the recursion theorem relative to $\emptyset'$ to build a suitable $(T_i)_{i \in \omega}$. What we do is build a $\emptyset'$-c.e. set of strings (elements of $2^{<\omega}$) $W_e^{\emptyset'}$ whose index we know in advance. Then we get a $\emptyset'$-computable tree T by putting $\sigma \in T$ if and only if σ has no prefix in $W_e^{\emptyset'}[|\sigma|]$. We thus know a $\emptyset'$-index for T, and therefore we also know an index for a computable sequence $(T_i)_{i \in \omega}$ of trees converging to T.

Now, we can $\emptyset'$-compute $S = \lim K((T_i)_{i \in \omega})$ (to make sure this limit exists, we will need to ensure that T has indeed positive measure, but we will see at the end of construction that it is indeed the case). Let H_0 and H_1 be the first and second projection of H, respectively. By compactness one can, relatively to $\emptyset'$, find a clopen set D such that $H_1((T_i)_{i \in \omega}, X) = 0$ for all $X \in [S_0] = [S] \cap D$ and $H_1((T_i)_{i \in \omega}, X) = 1$ for all $X \in [S_1] = [S] \cap D^c$. It is well-known that the image of an effectively compact set by a computable function which is total on this set is itself effectively compact (and an index of the image can be uniformly obtained from an index of the source). Relativizing this to $\emptyset'$, we see that the image of S_1 under $X \mapsto H_0((T_i)_{i \in \omega}, X)$ is a $\emptyset'$-effectively compact set. This image cannot contain 0^ω by definition of $[S_1]$ (because otherwise $H((T_i)_{i \in \omega}, X) = (0^\omega, 1)$ for some $X \in [S]$, contradicting the assumption on H and the definition of LPO). Therefore, we can $\emptyset'$-effectively wait until we find an l such that the image of $[S_1]$ under $X \mapsto H_0((T_i)_{i \in \omega}, X)$ is disjoint from $[0^l]$. When such an l is found, we enumerate in $W_e^{\emptyset'}$ all strings that are incompatible with $0^l 1$, so as to get $[T] = [0^l 1]$. Now, for any $X \in [S]$, either $X \in [S_0]$, in which case $H((T_i)_{i \in \omega}, X) = (Z, 0)$ for some $Z \in [T]$, which is not possible because LPO$(Z) = 1$ for all $Z \in [T]$, or $X \in [S_1]$, in which case $H((T_i)_{i \in \omega}, X) = (Z, 1)$ where 0^l is not a prefix of Z, and again this is not possible since Z is supposedly in $[T]$. Noting that the tree T does have positive measure as promised, we have obtained a contradiction. $\square$

Alternatively, the proof of the previous Theorem also follows from the fact that $\mathrm{LPO}' \leq_W \mathrm{LPO} \star \mathrm{WWKL}'$, using similar arguments as in Proposition 6.5, and the fact that $\mathrm{LPO}' \not\leq_W \mathrm{WWKL}'$ from [4].

Corollary 6.9.
$$\mathrm{WWKL}' <_W \mathrm{WWKL}' \star \mathrm{WWKL}'.$$

7   Summary

The following table summarises our results.

$\downarrow \star \rightarrow$	WWKL	$\mathrm{PC}_{\omega \times 2^\omega}$	WWKL$'$	$\mathrm{PC}'_{\omega \times 2^\omega}$
WWKL	WWKL	$\mathrm{PC}_{\omega \times 2^\omega}$	WWKL$'$	$\mathrm{PC}'_{\omega \times 2^\omega}$
$\mathrm{PC}_{\omega \times 2^\omega} \equiv_{sW} \mathrm{PC}_{\omega^\omega}$	$\mathrm{PC}_{\omega \times 2^\omega}$	$\mathrm{PC}_{\omega \times 2^\omega}$	$\mathrm{PC}'_{\omega \times 2^\omega}$	$\mathrm{PC}'_{\omega \times 2^\omega}$
WWKL$'$	WWKL$'$	WWKL$'$	$\mathrm{PC}'_{\omega \times 2^\omega}$	$\mathrm{PC}'_{\omega \times 2^\omega}$
$\mathrm{PC}'_{\omega \times 2^\omega} \equiv_{sW} \mathrm{PC}'_{\omega^\omega}$	$\mathrm{PC}'_{\omega \times 2^\omega}$	$\mathrm{PC}'_{\omega \times 2^\omega}$	$\mathrm{PC}'_{\omega \times 2^\omega}$	$\mathrm{PC}'_{\omega \times 2^\omega}$

Between these principles, we have the following relations:

$$\mathrm{WWKL} <_W \mathrm{PC}_{\omega \times 2^\omega} <_W \mathrm{WWKL}' <_W \mathrm{PC}_{\omega \times 2^\omega} \star \mathrm{WWKL}' <_W \mathrm{PC}'_{\omega \times 2^\omega}.$$

Finally, let us remark that many of the results studied in this paper also hold for more than one jump, but to avoid cluttering the notation we have not mentioned these results explicitly.

References

1. Bienvenu, L., Greenberg, N., Monin, B.: Continuous higher randomness. http://homepages.mcs.vuw.ac.nz/~greenberg/Papers
2. Brattka, V., Gherardi, G.: Weihrauch degrees, omniscience principles and weak computability. J. Symbolic Logic **76**(1), 143–176 (2011)
3. Brattka, V., Gherardi, G., Hölzl, R.: Probabilistic computability and choice. Inf. Comput. **242**, 249–286 (2015)
4. Brattka, V., Gherardi, G., Marcone, A.: The Bolzano-Weierstrass theorem is the jump of weak König's lemma. Ann. Pure Appl. Logic **163**, 623–655 (2012)
5. Brattka, V., Hendtlass, M., Kreuzer, A.P.: On the uniform computational content of computability theory. http://arxiv.org/pdf/1501.00433v3.pdf
6. Brattka, V., Pauly, A.: On the algebraic structure of Weihrauch degrees. http://arxiv.org/pdf/1604.08348v1.pdf
7. Brattka, V., Pauly, A.: Computation with advice. In: Zheng, X., Zhong, N. (eds.) CCA 2010, Proceedings of the Seventh International Conference on Computability and Complexity in Analysis, Electronic Proceedings in Theoretical Computer Science, pp. 41–55 (2010)
8. Downey, R.G., Hirschfeldt, D.R.: Algorithmic Randomness and Complexity. Theory and Applications of Computability. Springer, Heidelberg (2010)
9. Nies, A.: Computability and Randomness. Oxford Logic Guides. Oxford University Press, Oxford (2009)

Computable Model Theory, Computable Algebra

Effectively Existentially-Atomic Structures

Antonio Montalbán[(⊠)]

Department of Mathematics, University of California, Berkeley, USA
antonio@math.berkeley.edu
http://www.math.berkeley.edu/~antonio/index.html

1 Introduction

The notions we study in this paper, those of *existentially-atomic structure* and *effectively existentially-atomic structure*, are not really new. The objective of this paper is to single them out, survey their properties from a computability-theoretic viewpoint, and prove a few new results about them. These structures are the simplest ones around, and for that reason alone, it is worth analyzing them. As we will see, they are the simplest ones in terms of how complicated it is to find isomorphisms between different copies and in terms of the complexity of their descriptions. Despite their simplicity, they are very general in the following sense: every structure is existentially atomic if one takes enough jumps, the number of jumps being (essentially) the Scott rank of the structure. That balance between simplicity and generality is what makes them important.

Existentially atomic structures are nothing more than atomic structures, as in model theory, except that the generating formulas for the principal types are required to be existential. They were analyzed by Simmons in [Sim76, Sect. 2] who calls them $\exists_1$-*atomic*, or *strongly existentially closed*. Simmons referred to [Pou72] as an earlier occurrence of these structures in the literature. Here is the formal definition:

Definition 1.1. Let $\mathcal{A}$ be a structure. We define the *automorphism orbit* of a tuple $\bar{a} \in A^{<\omega}$ to be the set

$$\mathrm{orb}_{\mathcal{A}}(\bar{a}) = \{\bar{b} \in A^{|\bar{a}|} : \text{there is an automorphism of } \mathcal{A} \text{ mapping } \bar{a} \text{ to } \bar{b}\}.$$

We say that $\mathcal{A}$ is *existentially atomic* or $\exists$-*atomic* if, for every tuple $\bar{a} \in A^{<\omega}$, there is an $\exists$-formula $\varphi_{\bar{a}}(\bar{x})$ which defines the automorphism orbit of $\bar{a}$; that is, such that

$$\mathrm{orb}_{\mathcal{A}}(\bar{a}) = \{\bar{b} \in A^{|\bar{a}|} : \mathcal{A} \models \varphi_{\bar{a}}(\bar{b})\}.$$

For instance, a linear ordering is $\exists$-atomic if and only if it is either dense or finite. A field is $\exists$-atomic if and only if it is algebraic over its prime subfield. A good source of examples of $\exists$-atomic structures are the $\exists$-algebraic structures which we introduce in Sect. 3. Other than algebraic fields, other examples

The author was partially supported by the Packard Fellowship and NSF grant DMS-0901169.

© Springer International Publishing AG 2017
A. Day et al. (Eds.): Downey Festschrift, LNCS 10010, pp. 221–237, 2017.
DOI: 10.1007/978-3-319-50062-1_16

of $\exists$-algebraic structures are connected graphs of finite valence with a named root and finite-dimensional torsion-free abelian groups with a named basis (see Example 3.2).

Existentially atomic structures can be characterized in various different ways as stated in the following theorem. We will review the notions involved in the theorem later in this introduction.

Theorem 1.2. *Let $\mathcal{A}$ be a countable structure. The following are equivalent:*

(A1) $\mathcal{A}$ *is $\exists$-atomic.*
(A2) $\mathcal{A}$ *has an infinitary Π_2 Scott sentence.*
(A3) $\mathcal{A}$ *is uniformly continuously categorical.*
(A4) *Every first-order type realized in $\mathcal{A}$ is $\exists$-supported in $\mathcal{A}$.*
(A5) *Every $\forall$-type realized in $\mathcal{A}$ is $\exists$-supported in $\mathcal{A}$.*
(A6) $\mathcal{A}$ *is 1-prime.*

We prove this theorem in parts throughout the paper. The ideas for the proof are a combination of ideas from the literature which we will refer to as we use them.

This theorem is the particular case $\alpha = 1$ of [Mon, Theorem 1.1], which was for all $\alpha \in \omega_1$ and had a slightly different terminology. However, we can also view [Mon, Theorem 1.1] as a particular case of the theorem above: [Mon, Theorem 1.1] is essentially equivalent to the theorem above applied to the (relativized) $(<\alpha)$th-jump of $\mathcal{A}$, where the $(<\alpha)$*th-jump of $\mathcal{A}$* is defined to be the structure obtained by adding to $\mathcal{A}$ one relation for each computable infinitary Σ_β formula for $\beta < \alpha$. (See [Mon12, Mon13] for more on the jump of structures.) In [Mon], we defined the *Scott rank* of a structure to be the least α such that all its orbits are infinitary Σ_α-definable, and we argued that this is the best-behaved definition of Scott rank among the many in the literature. We thus have that the Scott rank of $\mathcal{A}$ is the least α such that, relative to some oracle X, the $(<\alpha)$th-jump of $\mathcal{A}$ is $\exists$-atomic. It follows that all the results we show about $\exists$-atomic structures apply to any structure so long as we take enough jumps.

Types and Scott Families. Let us now review the notions used in Theorem 1.2. Part (A4) is the definition of $\exists$-atomic structure a model theorist would give. Part (A5) states that it is enough to look at $\forall$-types instead of full first-order types. Recall that a $\forall$-*type on the variables $x_1, ..., x_n$* is a set $p(\bar{x})$ of $\forall$-formulas with free variables among $x_1, ..., x_n$ that is *satisfiable*: We say that a $\forall$-type is *realized in* a structure $\mathcal{A}$ if it is satisfied by some tuple in $\mathcal{A}$. Given $\bar{a} \in A^{<\omega}$, *the $\forall$-type of $\bar{a}$ in $\mathcal{A}$* is the set of $\forall$-formulas true of $\bar{a}$:

$$\forall\text{-}tp_{\mathcal{A}}(\bar{a}) = \{\varphi(\bar{x}) : \varphi \text{ is a } \forall\text{-formula and } \mathcal{A} \models \varphi(\bar{a})\}.$$

Note that by *type*, we do not mean *complete type*, as $\forall$-types are necessarily partial. For that same reason, instead of principal types, we have to deal with supported types:

Definition 1.3. A type $p(\bar{x})$ is $\exists$-*supported within a class of structures* $\mathbb{K}$ if there exists an $\exists$-formula $\varphi(\bar{x})$ which is realized in some structure in $\mathbb{K}$ and which implies all of $p(\bar{x})$ within $\mathbb{K}$; that is, $\mathcal{A} \models \forall \bar{x}(\varphi(\bar{x}) \rightarrow \psi(\bar{x}))$ for every $\psi(\bar{x}) \in p(\bar{x})$ and $\mathcal{A} \in \mathbb{K}$. We say that $p(\bar{x})$ is $\exists$-*supported in a structure* $\mathcal{A}$ if it is $\exists$-supported in $\mathbb{K} = \{\mathcal{A}\}$.

It is not hard to see that (A1) implies (A4) and that (A4) implies (A5). The proof that (A5) implies (A1) is given in Sect. 4.

Part (A2) states that $\exists$-atomic structures are among the simplest ones in terms of the complexity of their Scott sentences:

Definition 1.4. A sentence ψ is a *Scott sentence* for a structure $\mathcal{A}$ if $\mathcal{A}$ is the only countable structure satisfying ψ.

Scott [Sco65] proved that every countable structure has a Scott sentence in the infinitary language $L_{\omega_1,\omega}$. His proof used what we now call Scott families.

Definition 1.5. A *Scott family* for a structure $\mathcal{A}$ is a set S of formulas such that each $\bar{a} \in A^{<\omega}$ satisfies some formula $\varphi(\bar{x}) \in S$, and if $\bar{a}$ and $\bar{b}$ satisfy the same formula $\varphi(\bar{x}) \in S$, they are automorphic.

The set of all the defining formulas $\{\varphi_{\bar{a}} : \bar{a} \in A^{<\omega}\}$ from Definition 1.1 makes a Scott family. Thus, a structure is $\exists$-atomic if and only if it has a Soctt family of $\exists$-formulas. The proof that (A1) implies (A2) is essentially Scott's original construction of a Scott sentence. The proof that (A2) implies (A5) uses a variation of the type-omitting theorem which we present in Sect. 5.

Having access to a Scott family for a structure $\mathcal{A}$ allows us to recognize the different tuples in $\mathcal{A}$ up to automorphism. This is exactly what is necessary to build isomorphisms between different copies of $\mathcal{A}$, as we will see below. If we want to build a computable isomorphism, we need the Scott family to be computably enumerable.

Definition 1.6. We say that a Scott family is *c.e.* if the set of indices for its formulas is c.e. A structure $\mathcal{A}$ is *effectively* $\exists$-*atomic* if it has a c.e. Scott family of $\exists$-formulas.

A reader familiar with computable structure theory has surely heard of structures with c.e. Scott families of $\exists$-formulas before and their connection with relative computable categoricity. We will discuss this connection later.

Primality. In model theory, a *prime* model is one that elementary embeds into every model of its theory. We look at the one-quantifier version of this notion.

Definition 1.7. A structure $\mathcal{A}$ is *1-prime* if, for every countable model $\mathcal{B}$ of the $\forall_2$-theory of $\mathcal{A}$, there is an embedding from $\mathcal{A}$ to $\mathcal{B}$ which preserves $\forall$-formulas. We call such embeddings preserving $\forall$-formulas *1-embeddings*.

The proof that $\exists$-atomic implies 1-prime (i.e., (A1) $\Rightarrow$ (A6)) is quite straight-forward. The reversal needs the $\forall$-type omitting theorem. We give these proofs in Lemma 6.1.

We will also consider an effective version:

Definition 1.8. A computable structure $\mathcal{A}$ is *uniformly effectively 1-prime* if there is a computable operator Φ such that, for every computable model $\mathcal{B}$ of the $\forall_2$-theory of $\mathcal{A}$, $\Phi^{D(\mathcal{B})}$ is a 1-embedding from $\mathcal{A}$ to $\mathcal{B}$.

We will prove that the notion of uniformly effectively 1-prime is equivalent to that of effectively $\exists$-atomic. The notion of *uniformly effectively prime* for full-first order theories and elementary embeddings (instead of just one-quantifier formulas) was introduced by Cholak and McCoy in [CM]. There, they showed that it is equivalent to that of effectively atomic and that a theory can have at most one uniformly effectively prime model up to computable isomorphism. Their results follow from Theorem 1.11 below if one adds to the language relations for all first-order formulas, although the proofs are quite different.

Categoricity. Part (A3) is very different in form from the rest in the sense that it is computational in nature, rather than syntactical. For the boldface version, we use continuous rather than computable operators:

Definition 1.9. A structure $\mathcal{A}$ is *uniformly continuously categorical* if there is a continuous operator $\Phi: 2^\omega \to \omega^\omega$ that, when given as input the atomic diagram $D(\mathcal{B})$ of a copy $\mathcal{B}$ of $\mathcal{A}$, outputs an isomorphism $\Phi^{D(\mathcal{B})}$ form $\mathcal{B}$ to $\mathcal{A}$.[1]

The definition above is one of the many variations of the notion of *computable categoricity*, a notion that tries to measure the complexity of a structure in terms of how difficult it is to build isomorphisms between its different presentations. A structure is *computably categorical* if any two computable copies are computably isomorphic. Despite computable categoricity being the most natural definition for most computability theorists, the definition above is the one that has the cleanest syntactical characterization — it is equivalent to the structure being $\exists$-atomic. The connection between categoricity and atomicity was first noticed by Nurtazin [Nur74], who showed that a decidable structure is *computably categorical for decidable copies*[2] if and only if it is effectively atomic[3] over a finite set of parameters. Goncharov then improved this result and showed that a 2-decidable

---

[1] Melnikov and the author [MM] proved the equivalence between (A2) and (A3) in a much more general setting, that of Polish groups (S_∞ in this case) acting continuously on Polish spaces (the space of presentations of structures in this case). Furthermore, they showed that the equivalence is an easy corollary of a theorem of Effros from 1965 [Eff65].

[2] $\mathcal{A}$ is *computably categorical for decidable copies* if every decidable copy of $\mathcal{A}$ is computably isomorphic to $\mathcal{A}$.

[3] A structure is *effectively atomic* if it has a c.e. Scott family of elementary first-order formulas.

structure is computably categorical if and only if it is effectively $\exists$-atomic over a finite set of parameters. Ash, Knight, Manasse, Slaman [AKMS89, Theorem 4], and Chisholm [Chi90, Theorem V.10], removed the 2-decidability assumption and proved that a structure is *relatively computably categorical* if and only if it is effectively $\exists$-atomic over a finite set of parameters. Relativizing their result, we get the following theorem, which is a version of Theorem 1.2, now with parameters:

Theorem 1.10. *Let $\mathcal{A}$ be a countable structure. The following are equivalent:*

(B1) $\mathcal{A}$ *is $\exists$-atomic over a finite set of parameters.*

(B2) $\mathcal{A}$ *has an infinitary Σ_3 Scott sentence.*

(B3) $\mathcal{A}$ *is computably categorical on a cone. That is, there is a $C \in 2^\omega$ such that, for every $X \geq_T C$, every X-computable copy of $\mathcal{A}$ is X-computably-isomorphic to $\mathcal{A}$.*

The equivalence between (B1) and (B3) is just the boldface version of [AKMS89, Theorem 4] and [Chi90, Theorem V.10]. That (B1) implies (B2) easily follows from the corresponding parts in Theorem 1.2. The opposite direction is a slightly more subtle and is proved in Lemma 7.2.

The following theorem is the effective version of the equivalence between (A1), (A3) and (A6). The notion of uniform computably categorical structure was introduced by Ventsov [Ven92]. Other notions of uniform categoricity were studied by Kudinov [Kud96a, Kud96b, Kud97] and by Downey, Hirschfeldt and Khoussainov [DHK03].

Theorem 1.11. *Let $\mathcal{A}$ be a computable structure. The following are equivalent:*

(C1) $\mathcal{A}$ *is effectively $\exists$-atomic.*

(C2) $\mathcal{A}$ *is uniformly relatively computably categorical; that is, the operator Φ in Definition 1.9 is computable.*

(C3) $\mathcal{A}$ *is uniformly computably categorical; that is, the operator Φ in Definition 1.9 is computable and is only required to work when the input $\mathcal{B}$ is a computable structure, i.e., when given as oracle the atomic diagram $D(\mathcal{B})$ of a computable copy $\mathcal{B}$ of $\mathcal{A}$, Φ outputs an isomorphism $\Phi^{D(\mathcal{B})}$ form $\mathcal{B}$ to $\mathcal{A}$.*

(C4) $\mathcal{A}$ *is uniformly effectively 1-prime.*

The equivalence between (C1), (C2) and (C3) was proved by Ventsov in [Ven92]. The fact that effectively atomic structures are the same as uniformly effectively prime structures was proved by Cholak and McCoy in [CM]. We prove the equivalence between (C1) and (C4) in Lemma 6.2 using a very different proof.

Turing Degree and Enumeration Degree. The most common tool to measure the computational complexity of a structure is the degree spectrum. Prior to the introduction of the degree spectrum, Jockusch considered a much more natural notion, which unfortunately does not always apply:

Definition 1.12 (Jockusch). A structure $\mathcal{A}$ has *Turing degree* $X \in 2^\omega$ if X computes a copy of $\mathcal{A}$, and every copy of $\mathcal{A}$ computes X.

It turns out that if we consider the same definition, but on the enumeration degrees (as Knight implicitly did in [Kni98]), we obtain a better-behaved notion.

Definition 1.13. A structure $\mathcal{A}$ has *enumeration degree* $X \subseteq \omega$ if every enumeration of X computes a copy of $\mathcal{A}$, and every copy of $\mathcal{A}$ computes an enumeration of X. Recall that an *enumeration of X* is an onto function $f: \omega \to X$.

Equivalently, $\mathcal{A}$ has enumeration degree X if and only if, for every $Y \in 2^\omega$, Y computes a copy of $\mathcal{A}$ if and only if X is c.e. in Y. Notice that, for $X, Z \subseteq \omega$, if $\mathcal{A}$ has enumeration degree X, then $\mathcal{A}$ has enumeration degree Z if and only if X and Z are enumeration equivalent.

As an example, we let the reader verify that the group $\mathcal{G}_X = \bigoplus_{i \in X} \mathbb{Z}_{p_i}$, where p_i is the ith prime number, has enumeration degree X.

The enumeration degree of a structure is indeed a good way to measure its computational complexity. Unfortunately, in general, a structure need not have enumeration degree. Furthermore, there are whole classes of structures, like linear orderings for instance, where no structure has enumeration degrees unless it is already computable (this was shown by Richter [Ric81]). On the other hand, there are whole classes of structures which all have enumeration degree. For instance, Frolov, Kalimullin and Miller [FKM09] proved that all fields of finite transcendence degree over $\mathbb{Q}$ have enumeration degree. Calvert, Harizanov, Shlapentokh [CHS07] showed that every torsion-free abelian groups of finite rank always has enumeration degree. Steiner [Ste13] showed that graphs of finite valance with finitely many connected components always have enumeration degree. The following theorem (which is new) shows how all these results fit in a much more general framework. All the examples above can be easily seen to be $\exists$-algebraic over a finite tuple of parameters, and are Π^c_2-axiomatizable once that tuple of parameters is fixed.

Theorem 1.14. *Let $\mathbb{K}$ be a Π^c_2 class, all whose structures are $\exists$-atomic. Then every structure in $\mathbb{K}$ has enumeration degree, and that enumeration degree is given by its $\exists$-theory.*

2   Background and Notation

An $\exists$-*formula* is one of the form $\exists x_1 \exists x_2 ... \exists x_n \; \varphi(x_1, ..., x_n)$, where φ is finitary and quantifier free. A $\forall_2$-*formula* is one of the form $\forall y_1 \forall y_2 ... \forall y_m \; \psi(y_1, ..., y_m)$, where ψ is an $\exists$-formula. For background on infinitary formulas and computably infinitary formulas, see [AK00, Chaps. 6 and 7]. We will use Σ^{in}_α to denote the set of infinitary Σ_α-formulas, Σ^c_α for the computable infinitary formulas, and $\Sigma^{c,X}_\alpha$ for the X-computable infinitary formulas.

Given a presentation of a structure $\mathcal{B}$, we define its *atomic diagram* $D(\mathcal{B}) \in 2^\omega$ as follows: First, consider an effective enumeration of $\{\varphi^{at}_i : i \in \omega\}$ of the atomic formulas on the variables $x_0, x_1, ...$, and assume φ_i^{at} only uses variables x_j for $j < i$. Then, define $D(\mathcal{B})(i) = 1$ if and only if $\mathcal{B} \models \varphi_i^{at}[x_j \mapsto j]$, and let $D(\mathcal{B})(i) = 0$ otherwise. Recall that the domain of $\mathcal{B}$ is a subset of the natural numbers, so we are assigning to x_j the natural number j. If φ_i^{at} uses a variable x_j and j is not in the domain of $\mathcal{B}$, we let $D(\mathcal{B})(i) = 0$.

Given a tuple $\bar{b} \in B^{<\omega}$, we define $D_{\mathcal{B}}(\bar{b})$ to be the length-$|\bar{b}|$ approximation to the atomic type of $\bar{b}$: That is, $D_{\mathcal{B}}(\bar{b})$ is the string $\sigma \in 2^{|\bar{b}|}$ defined by $\sigma(i) = 1$ if and only if $\mathcal{B} \models \varphi_i^{at}(x_j \mapsto b_j)$. For each $\sigma \in 2^{<\omega}$, there is a formula $\varphi_\sigma^{at}(\bar{x})$, where $|\bar{x}| = |\sigma|$, which states that the atomic diagram of $\bar{x}$ is σ. That is:

$$\varphi_\sigma^{at}(\bar{x}) \quad \equiv \quad \left(\bigwedge_{i < |\bar{x}|, \sigma(i)=1} \varphi_i^{at}(\bar{x}) \right) \wedge \left(\bigwedge_{i < |\bar{x}|, \sigma(i)=0} \neg\varphi_i^{at}(\bar{x}) \right)$$

3   Existentially Algebraic Structures

A important source of examples of $\exists$-atomic structure are the $\exists$-algebraic structures.

Definition 3.1. An element a of a structure $\mathcal{A}$ is $\exists$-*algebraic* if there is an $\exists$-formula $\varphi(x)$ true of a such that $\{b \in A : \mathcal{A} \models \varphi(b)\}$ is finite. A structure $\mathcal{A}$ is $\exists$-*algebraic* if all its elements are.

Here are some examples.

Example 3.2. A *field that is algebraic over its prime sub-field* is $\exists$-algebraic because every element is among the finitely many roots of some polynomial with coefficients on the prime sub-field, and the elements in the prime sub-field can be defined by quantifier-free formulas.

A *connected graph of finite valance with a selected root vertex* is $\exists$-algebraic because every element is among the finitely many that are at a given distance from the root.

An *abelian, torsion-free group with a selected basis* is $\exists$-algebraic because every element can be defined by a $\mathbb{Q}$-linear combination of the basis elements.

We prove that $\exists$-algebraic structures are $\exists$-atomic in two lemmas. The core of the argument is an application of König's lemma that appears in the first one.

Lemma 3.3. *Two countable structures that are $\exists$-algebraic and have the same $\exists$-theories are isomorphic.*

Proof. Let $\mathcal{A}$ and $\mathcal{B}$ be $\exists$-algebraic structures with the same $\exists$-theories. List the elements of A as $\{a_0, a_1, ...\}$. For each n, let $\varphi_n(x_0, ..., x_{n-1})$ be an $\exists$-formula which is true of the tuple $\langle a_0, ..., a_{n-1} \rangle$, has finitely many solutions, and implies $\varphi_{n-1}(x_0, ..., x_{n-2})$. (By *solution* for a formula, we mean a tuple that makes it true.) Consider the tree

$$T = \{\bar{b} \in B^{<\omega} : D_{\mathcal{B}}(\bar{b}) = D_{\mathcal{A}}(a_0, ..., a_{|\bar{b}|-1}) \ \& \ \mathcal{B} \models \varphi_{|\bar{b}|}(\bar{b})\}.$$

T is clearly a tree in the sense that it is closed under taking initial segments of tuples. It is finitely branching because, for each n, φ_n is true for only finitely many tuples. To show that T is infinite, notice that, for each n, the tuple $(a_0, ..., a_{n-1})$ itself witnesses that

$$\mathcal{A} \models \exists x_0, ..., x_{n-1}(\varphi_\sigma^{at}(\bar{x}) \ \& \ \varphi_n(\bar{x})), \qquad \text{where } \sigma = D_\mathcal{A}(a_0, ..., a_{n-1}) \in 2^n.$$

(Here, $\varphi_\sigma^{at}(\bar{x})$ is the formula that states that "$D(\bar{x}) = \sigma$," as defined in the background section.) Since $\mathcal{A}$ and $\mathcal{B}$ have the same $\exists$-theories, $\mathcal{B}$ models this sentence too, and the witness is an n-tuple that belongs to T. König's lemma states that every infinite, finitely branching tree must have an infinite path. Thus, T must have an infinite path $P \in B^\omega$. From this path, we obtain a map $a_n \mapsto P(n) \colon A \to B$, which we claim is an isomorphism. The map is an embedding because, by the definition of T, it preserves finite atomic diagrams. But then it must be an isomorphism: If $b \in B$ is a solution of an $\exists$-formula φ with finitely many solutions, then φ must have the same number of solutions in $\mathcal{A}$ (because $\exists$-$Th(\mathcal{A}) = \exists$-$Th(\mathcal{B})$), and since $\exists$-formulas are preserved under embeddings, one of those solutions has to be mapped to b. □

Lemma 3.4. *Every $\exists$-algebraic structure is $\exists$-atomic.*

Proof. Let $\mathcal{A}$ be $\exists$-algebraic and take $\bar{a} \in A^{<\omega}$. Let $\varphi(\bar{x})$ be an $\exists$-formula true of $\bar{a}$ with the least possible number of solutions, say k solutions. We claim that every solution to φ is automorphic to $\bar{a}$, and hence that φ defines the orbit of $\bar{a}$. Suppose, towards a contradiction, that $\bar{b}$ satisfies φ but is not automorphic to $\bar{a}$. Then there must exist an $\exists$-formula $\psi(\bar{x})$ that is true of either $\bar{a}$ or $\bar{b}$, but not of both: This follows from the previous lemma, as $(\mathcal{A}, \bar{a})$ and $(\mathcal{A}, \bar{b})$ are not isomorphic and are both $\exists$-algebraic. If $\psi(\bar{x})$ is true of $\bar{a}$, then $\varphi(\bar{x}) \wedge \psi(\bar{x})$ would be true of $\bar{a}$ and have fewer solutions than φ, contradicting our choice of φ. Suppose now that $\psi(\bar{x})$ is not true of $\bar{a}$. Let i be the number of solutions of $\psi(\bar{x}) \wedge \varphi(\bar{x})$. Then the formula about $\bar{y}$ saying

"$\varphi(\bar{y})$ and there are i solutions to $\varphi \wedge \psi$ all different from $\bar{y}$"

is an $\exists$-formula true of $\bar{a}$ with $k - i$ solutions, again contradicting our choice of φ. □

The statements of the lemmas in this section are new, but the ideas behind them are not. Proofs like that of Lemma 3.3 using König's lemma have appeared in many other places before, for instance [HLZ99]. The ideas for the proof of Lemma 3.4 are similar to those one would use in a proof that algebraic structures are atomic (without the $\exists$-), except that here one has to be slightly more careful.

4   Existentially Atomicity in Terms of Types

In this short section, we prove that if every $\forall$-type is $\exists$-supported in a structure $\mathcal{A}$, the structure is $\exists$-atomic (that is, that (A5) $\Rightarrow$ (A1)). The proof is an adaptation of classical arguments with back-and-forth relations.

Definition 4.1. Given structures $\mathcal{A}$ and $\mathcal{B}$, we say that a set $I \subseteq \mathcal{A}^{<\omega} \times \mathcal{B}^{<\omega}$ has the *back-and-forth property* if, for every $\langle \bar{a}, \bar{b} \rangle \in I$,

- $D_{\mathcal{A}}(\bar{a}) = D_{\mathcal{B}}(\bar{b})$ (i.e., $|\bar{a}| = |\bar{b}|$ and $\bar{a}$ and $\bar{b}$ satisfy the same atomic formulas among the first $|\bar{a}|$ many);
- for every $c \in A$, there exists $d \in B$ such that $\langle \bar{a}c, \bar{b}d \rangle \in I$; and
- for every $d \in B$, there exists $c \in A$ such that $\langle \bar{a}c, \bar{b}d \rangle \in I$.

A standard back-and-forth argument shows that if I has the back-and-forth property and $\langle \bar{a}, \bar{b} \rangle \in I$, then there is an isomorphism from $\mathcal{A}$ to $\mathcal{B}$ mapping $\bar{a}$ to $\bar{b}$. Furthermore, if I is c.e., then there is a computable such isomorphism.

Proof of (A5) $\Rightarrow$ (A1) in Theorem 1.2. For each $\bar{a} \in A^{<\omega}$, let $\varphi_{\bar{a}}(\bar{x})$ be an $\exists$-formula supporting the $\forall$-type of $\bar{a}$. We need to show that $S = \{\varphi_{\bar{a}} : \bar{a} \in A^{<\omega}\}$ is a Scott family for $\mathcal{A}$. Consider the set

$$I_{\mathcal{A}} = \{\langle \bar{a}, \bar{b} \rangle \in A^{<\omega} \times A^{<\omega} : \mathcal{A} \models \varphi_{\bar{a}}(\bar{b})\}.$$

We will show that $\mathcal{I}_{\mathcal{A}}$ has the *back-and-forth* property.

Before we prove these three properties, we need to prove a couple of smaller facts. First, notice that, for every $\bar{a} \in A^{<\omega}$, $\mathcal{A} \models \varphi_{\bar{a}}(\bar{a})$: This is because otherwise $\neg\varphi_{\bar{a}}$ would be part of the $\forall$-type of $\bar{a}$, and hence implied by $\varphi_{\bar{a}}$, which cannot be the case, as $\varphi_{\bar{a}}$ is realizable in $\mathcal{A}$. Second, let us show that $I_{\mathcal{A}}$ is symmetric; that is, that if $\mathcal{A} \models \varphi_{\bar{a}}(\bar{b})$, then $\mathcal{A} \models \varphi_{\bar{b}}(\bar{a})$. Suppose $\mathcal{A} \models \varphi_{\bar{a}}(\bar{b})$. Then, we cannot have $\mathcal{A} \models \varphi_{\bar{a}}(\bar{x}) \to \neg\varphi_{\bar{b}}(\bar{x})$, as the negation is witnessed by $\bar{b}$. It thus follows that $\neg\varphi_{\bar{b}}(\bar{x})$ is not part of the $\forall$-type of $\bar{a}$, and hence that $\mathcal{A} \models \varphi_{\bar{b}}(\bar{a})$.

We can now prove that $I_{\mathcal{A}}$ has the back-and-forth property. Suppose $\langle \bar{a}, \bar{b} \rangle \in I_{\mathcal{A}}$. Notice that $\bar{a}$ and $\bar{b}$ must then satisfy the same $\forall$-formulas. In particular, they must satisfy the same atomic formulas, and hence have the same atomic diagrams. To show the second condition, take $c \in A$. If there was no $d \in A$ with $\langle \bar{a}c, \bar{b}d \rangle \in I_{\mathcal{A}}$, we would have that $\mathcal{A} \models \neg\exists y \varphi_{\bar{a}c}(\bar{b}, y)$. This formula would be part of the $\forall$-type of $\bar{b}$, and hence implied by $\varphi_{\bar{b}}$. But then, since $\mathcal{A} \models \varphi_{\bar{b}}(\bar{a})$, we would have $\mathcal{A} \models \neg\exists y \varphi_{\bar{a}c}(\bar{a}, y)$, which is not true as witnessed by c. The third condition of the back-and-forth property follows from the symmetry of $I_{\mathcal{A}}$.

Now that we know that $I_{\mathcal{A}}$ has the back-and-forth property, through a standard back-and-forth argument we get that if $\mathcal{A} \models \varphi_{\bar{a}}(\bar{b})$, then there exists an automorphism of $\mathcal{A}$ taking $\bar{a}$ to $\bar{b}$. In particular, we get that if $\varphi_{\bar{a}}(\bar{b})$ and $\varphi_{\bar{a}}(\bar{c})$ both hold, then $\bar{b}$ and $\bar{c}$ are automorphic. This proves that S is a Scott family for $\mathcal{A}$. $\qquad\qquad\square$

5   Building Structures and Omitting Types

Before we continue studying the properties of $\exists$-atomic structures, we need to make a stop to prove some general lemmas that will be useful in future sections. First, we prove a lemma that will allow us to find computable structures in a given class of structures. Second, using similar techniques, we prove the type-omitting lemma for $\forall$-types and its effective version.

Assume, without loss of generality, we are working with a relational vocabulary τ. Given a class of structure $\mathbb{K}$, we let $\mathbb{K}^{fin}$ be — essentially — the set of all the finite substructures of the structures in $\mathbb{K}$:

$$\mathbb{K}^{fin} = \{D_{\mathcal{A}}(\bar{a}) : \mathcal{A} \in \mathbb{K}, \bar{a} \in \mathcal{A}^{<\omega}\} \subseteq 2^{<\omega}.$$

Lemma 5.1. *Let $\mathbb{K}$ be a Π_2^c class for which $\mathbb{K}^{fin}$ is c.e. Then there is at least one computable structure in $\mathbb{K}$.*

Proof. We build a structure $\mathcal{A}$ in $\mathbb{K}$ by building a finite approximation to it. That is, we build a nested sequence of finite structures $\mathcal{A}_s$ for $s \in \omega$. Formally, that is not precisely correct: We will build the diagram $D(\mathcal{A})$ as the limit of a nested sequence $\sigma_s \in 2^{<\omega}$ for $s \in \omega$, where each σ_s is in $\mathbb{K}^{fin}$. We then think of $\mathcal{A}_s$ as the *partial* finite structure with domain $|\sigma_s|$, where only the atomic formulas φ_i^{at} for $i < |\sigma_s|$ are decided, and the rest are not decided yet. Working with the $\mathcal{A}_s$'s is closer to our intuition of what is going on, but a formal proof would only use the σ_s's.

Of course, we require that $\mathcal{A}_s \subseteq \mathcal{A}_{s+1}$ (that is, that $\sigma_s \subset \sigma_{s+1}$ as binary strings). At the end of stages, we define the structure $\mathcal{A} = \bigcup_{s \in \omega} \mathcal{A}_s$, and hence $D(\mathcal{A}) = \bigcup_s \sigma_s$.

Let $\bigwedge_{i \in I} \forall \bar{y}_i \psi_i(\bar{y}_i)$ be the Π_2^c sentence that axiomatizes $\mathbb{K}$, where each ψ_i is Σ_1^c. To get $\mathcal{A} \in \mathbb{K}$, we need to guarantee that, for each i and each $\bar{a} \in A^{|\bar{y}_i|}$, we have $\mathcal{A} \models \psi_i(\bar{a})$. For this, when we build $\mathcal{A}_{s+1}$, we will make sure that,

$$\text{for every } i < s \text{ and every } \bar{a} \in A_s^{|\bar{y}_i|}, \mathcal{A}_{s+1} \models \psi_i(\bar{a}). \tag{1}$$

Notice that, since ψ_i is Σ_1^c, $\mathcal{A}_{s+1} \models \psi_i(\bar{a})$ implies $\mathcal{A} \models \psi_i(\bar{a})$. Thus, we would end up with $\mathcal{A} \models \bigwedge_{i \in I} \forall \bar{y}_i \psi_i(\bar{y}_i)$.

Now that we know what we need, let us build the sequence of $\mathcal{A}_s$'s. Suppose we have already built $\mathcal{A}_0, ..., \mathcal{A}_s$ and we want to define $\mathcal{A}_{s+1} \supseteq \mathcal{A}_s$. All we need to do is search for a partial finite structure in $\mathbb{K}^{fin}$ satisfying (1). Notice that, given a finite diagram σ for a finite partial structure, we can check if it satisfies (1). Since $\mathbb{K}^{fin}$ is c.e., all we have to do is search for such a $\sigma_{s+1} \in \mathbb{K}^{fin}$ — well, except that we need to show that at least one such structure exists. Since $\mathcal{A}_s \in \mathbb{K}^{fin}$, there is some $\mathcal{B} \in \mathbb{K}$ which has a partial finite substructure $\mathcal{B}_s$ isomorphic to $\mathcal{A}_s$. (That is, modulo a permutation of the presentation, we can assume that σ_s is an initial segment of the atomic diagram of $\mathcal{B}$.) Since $\mathcal{B} \models \bigwedge_{i \in I} \forall \bar{y}_i \psi_i(\bar{y}_i)$, for every $i < s$ and every $\bar{b} \in \mathcal{B}_s^{|\bar{y}_i|}$, there exists a tuple in $\mathcal{B}$ witnessing that $\mathcal{B} \models \psi_i(\bar{b})$. Let $\mathcal{B}_{s+1}$ be a finite substructure of $\mathcal{B}$ containing $\mathcal{B}_s$ and all those witnessing tuples. Let σ_{s+1} be the initial segment of the atomic diagram of $\mathcal{B}$, witnessing that $\mathcal{B}_{s+1}$ satisfies (1) with respect to $\mathcal{B}_s$. □

Corollary 5.2. *Let $\mathbb{K}$ be a Π_2^c class of structures, and S be the $\exists$-theory of some structure in $\mathbb{K}$. If S is c.e. in a set X, then there is an X-computable presentation of a structure in $\mathbb{K}$ with $\exists$-theory S.*

Proof. Add to the Π_2^c axiom for $\mathbb{K}$ the $\Pi_2^{c,X}$ sentence saying that the structure must have $\exists$-theory S:

$$\left(\bigwedge_{``\exists\bar{y}\psi(\bar{y})" \in S} \exists\bar{y}\psi(\bar{y}) \right) \wedge \left( \forall\bar{x} \bigvee_{\substack{\sigma\in 2^{|\bar{x}|} \\ ``\exists\bar{y}\varphi_\sigma(\bar{y})" \in S}} \varphi_\sigma(\bar{x}) \right),$$

where $\varphi_\sigma^{at}(\bar{x})$ is the formula "$D(\bar{x}) = \sigma$" (as in the background section). Let $\mathbb{K}_S$ be the new $\Pi_2^{c,X}$ class of structures. All the models in $\mathbb{K}_S$ have $\exists$-theory S, and hence $\mathbb{K}_S^{fin}$ is enumeration reducible to S, and hence is c.e. in X too. Applying Lemma 5.1 relative to X, we get an X-computable structure in $\mathbb{K}_S$ as wanted. $\square$

Not only can we build a computable structure in such a class $\mathbb{K}$, we can build one omitting certain types.

Lemma 5.3. *Let $\mathbb{K}$ be a Π_2^{in} class of structures. Let $\{p_i(\bar{x}_i) : i \in \omega\}$ be a sequence of $\forall$-types which are not $\exists$-supported in $\mathbb{K}$. Then there is a structure $\mathcal{A} \in \mathbb{K}$ which omits all the types $p_i(\bar{x}_i)$ for $i \in \omega$.*

Furthermore, if $\mathbb{K}$ is Π_2^c, $\mathbb{K}^{fin}$ is c.e. and the list $\{p_i(\bar{x}_i) : i \in \omega\}$ is c.e., we can make $\mathcal{A}$ computable.

Proof. We construct $\mathcal{A}$ by stages as in the proof of Lemma 5.1, the difference being that now we need to omit the types p_i. So, on the even stages s, we do exactly the same thing we did in Lemma 5.1, and we use the odd stages to omit the types. At stage $s + 1 = 2\langle i, j \rangle + 1$, we ensure that the jth tuple $\bar{a}$ does not satisfy p_i as follows. Let $\bar{b} = \mathcal{A}_s \smallsetminus \bar{a}$, and let $\sigma = D_{\mathcal{A}_s}(\bar{a}, \bar{b})$. So we have that $\bar{a}$ satisfies $\exists\bar{y}\, \varphi_\sigma^{at}(\bar{a}, \bar{y})$. Since p_i is not $\exists$-supported in $\mathbb{K}$, there exists a $\forall$-formula $\psi(\bar{x}) \in p_i$ which is not implied by $\exists\bar{y}\, \varphi_\sigma^{at}(\bar{a}, \bar{y})$ within $\mathbb{K}$. That means that, for some finite $\mathcal{B} \in \mathbb{K}^{fin}$ and some $\bar{d} \in \mathcal{B}^{<\omega}$, we have $\mathcal{B} \models \exists\bar{y}\, \varphi_\sigma^{at}(\bar{b}, \bar{y}) \wedge \neg\psi(\bar{a})$. Since $\mathcal{B} \models \exists\bar{y}\, \varphi_\sigma^{at}(\bar{b}, \bar{y})$, we can assume $\mathcal{B}$ extends $\mathcal{A}_s$. Since such $\mathcal{B}$ and ψ exist, we can wait until we find them and then define $\mathcal{A}_{s+1}$ to be such $\mathcal{B}$. $\square$

6   1-Prime Structures

In this section, we prove the equivalences that have to do with the notion of 1-prime structures. The first lemma proves the equivalence between (A1) and (A6), and the second lemma the equivalence between (C1) and (C4).

Lemma 6.1. *A structure is $\exists$-atomic if and only if it is 1-prime.*

Proof. Suppose first that $\mathcal{A}$ is $\exists$-atomic. Let $\mathcal{B}$ be a model of the $\forall_2$-theory of $\mathcal{A}$. (A $\forall_2$-formula is one of the form $\forall x_1 ... \forall x_k \exists y_1 ... \exists y_\ell\, \psi$ where ψ is finitary and quantifier free.) Let $\{a_1, a_2, ...\}$ be an enumeration of A and, for each $s \in \omega$, let φ_s be an $\exists$-formula defining the orbit of $(a_1, ..., a_s)$. We define a 1-embedding

f from $\mathcal{A}$ to $\mathcal{B}$ by stages. We define $f(a_s)$ at stage s, always making sure that $\mathcal{B} \models \varphi_s(f(a_1), ..., f(a_s))$. To see we can do this, notice that the formula

$$\forall x_1, ..., x_s \; (\varphi_s(x_1, ..., x_s) \to \exists x_{s+1} \; \varphi_{s+1}(x_1, ..., x_s, x_{s+1}))$$

is true of $\mathcal{A}$, and hence part of the $\forall_2$-theory of $\mathcal{B}$ too. To see that f is a 1-embedding, notice that for every $\forall$-formula $\psi(x_1, ...x_s)$ true of $(a_1, ..., a_s)$ in $\mathcal{A}$, we have that

$$\forall x_1, ..., x_s \; (\varphi_s(x_1, ..., x_s) \to \psi(x_1, ..., x_s))$$

is part of the $\forall_2$ theory of $\mathcal{A}$, and hence of $\mathcal{B}$ too.

It is the reverse direction that uses the type-omitting theorem. Suppose $\mathcal{A}$ is not $\exists$-atomic. We have already proved that (A1) implies (A5), so we have that the $\forall$-type of some tuple $\bar{a} \in A^{<\omega}$ is not $\exists$-supported within $\mathcal{A}$. Let ψ be the conjunction of the $\forall_2$-theory of $\mathcal{A}$. Since ψ is Π_2^{in}, by Lemma 5.3, there is a model $\mathcal{B}$ of ψ which omits the $\forall$-type of $\bar{a}$. But then we cannot have a 1-embedding of $\mathcal{A}$ into $\mathcal{B}$, as 1-embeddings preserve $\forall$-types, and hence there is nowhere to map $\bar{a}$ in $\mathcal{B}$. Thus, $\mathcal{A}$ is not 1-prime. $\square$

Lemma 6.2. *A computable structure $\mathcal{A}$ is effectively $\exists$-atomic if and only if it is uniformly effectively 1-prime.*

Proof. For the left-to-right direction, notice that, given a computable model $\mathcal{B}$ of the $\forall_2$-theory of $\mathcal{A}$, we can use the c.e. Scott family of $\mathcal{A}$ to build a 1-embedding f for $\mathcal{A}$ to $\mathcal{B}$ exactly as in the proof of the lemma above. Notice also that f can be computed uniformly in $D(\mathcal{B})$.

For the right-to-left direction, let Φ be a computable operator witnessing that $\mathcal{A}$ is 1-prime. Consider $\Phi^{D(\mathcal{A})}$, which is a 1-embedding form $\mathcal{A}$ into itself. Again, let $\{a_0, a_1, ...\}$ be an enumeration of $\mathcal{A}$, and let $\bar{a}$ be an initial segment of that enumeration; we will use Φ to find an $\exists$-formula defining the orbit of $\bar{a}$, effectively uniformly in $\bar{a}$. (We are assuming the domain of $\mathcal{A}$ is ω, so actually a_i is the natural number i, but we think of a_i as a member of $\mathcal{A}$ rather than as a natural number.) Let $\tilde{a}$ be such that $\Phi^{D(\mathcal{A})}(\tilde{a}) = (0, 1, ..., |\bar{a}| - 1)$. Thus, $\Phi^{D(\mathcal{A})}$ maps $\tilde{a}$ to $\bar{a}$ in $\mathcal{A}$. Let s be such that $\Phi^{D(\mathcal{A}) \restriction s}$ is defined on $\tilde{a}$. (I.e., let s be the *use* of the computation. As a convention, when we run a computable functional on a finite oracle $\sigma \in 2^{<\omega}$, we only run it for at most $|\sigma|$-many steps.) Let $\bar{c} = (a_{|\bar{a}|}, ..., a_{s-1})$, so $\bar{a}\bar{c} = (a_0, ..., a_{s-1})$. Recall from the background section that $\varphi_{D(\mathcal{A}) \restriction s}^{at}$ is the conjunction of the first s atomic (and negation of atomic) facts about $a_0, ..., a_{s-1}$. Thus, $\mathcal{A} \models \varphi_{D(\mathcal{A}) \restriction s}^{at}(\bar{a}, \bar{c})$. Finally, define

$$\varphi_{\bar{a}}(\bar{x}) \quad \equiv \quad \exists \bar{y} \; \varphi_{D(\mathcal{A}) \restriction s}^{at}(\bar{x}, \bar{y}).$$

We claim that $\varphi_{\bar{a}}$ supports the $\forall$-type of $\bar{a}$, and hence that it defines the orbit of $\bar{a}$. Let $\bar{b}$ be another tuple in $\mathcal{A}$ satisfying $\varphi_{\bar{a}}$; we need to show that $\bar{a}$ and $\bar{b}$ satisfy the same $\forall$-types. Let $\bar{d}$ be the witnesses for $\varphi_{\bar{a}}(\bar{b})$, i.e., such that $\mathcal{A} \models \varphi_{D(\mathcal{A}) \restriction s}^{at}(\bar{b}, \bar{d})$. Consider a new presentation of $\mathcal{A}$, call it $\tilde{\mathcal{A}}$, where we permute $\bar{a}\bar{c}$ for $\bar{b}\bar{d}$ and leave the rest the same. Since the first s elements of the presentation $\tilde{\mathcal{A}}$ are $\bar{b}\bar{d}$, we have that

$$D(\tilde{\mathcal{A}}) \upharpoonright s = \mathcal{D}_{\tilde{\mathcal{A}}}(\bar{b}\bar{d}) = \mathcal{D}_{\mathcal{A}}(\bar{a}\bar{c}) = D(\mathcal{A}) \upharpoonright s.$$

It follows that $\Phi^{D(\tilde{\mathcal{A}})}(\tilde{a}) = \Phi^{D(\mathcal{A})}(\tilde{a}) = (0, 1, ..., |\bar{a}| - 1)$. But now, in the new presentation $\tilde{\mathcal{A}}$, $(0, 1, ..., |\bar{a}| - 1)$ corresponds to $\bar{b}$. Since both $\Phi^{D(\mathcal{A})}$ and $\Phi^{D(\tilde{\mathcal{A}})}$ preserve $\forall$-types, and $\Phi^{D(\mathcal{A})}(\tilde{a}) = \bar{a}$ and $\Phi^{D(\tilde{\mathcal{A}})}(\tilde{a}) = \bar{b}$, we have that

$$\forall\text{-}tp_{\mathcal{A}}(\bar{a}) = \forall\text{-}tp_{\mathcal{A}}(\tilde{a}) = \forall\text{-}tp(\bar{b}).$$

This proves that $\varphi_{\bar{a}}$ supports the $\forall$-type of $\bar{a}$ and hence defines its orbit. Since the definition of $\varphi_{\bar{a}}$ was uniform, we can build a whole c.e. Scott family for $\mathcal{A}$.   □

7   Scott Sentences of Existentially Atomic Structures

Scott [Sco65] showed that every countable structure has a Scott sentence in $\mathcal{L}_{\omega_1,\omega}$. We prove it below for $\exists$-atomic structures. The same proof would show that if a structure has a Scott family of $\Sigma_\alpha^{\mathrm{in}}$-formulas, it has a $\Pi_{\alpha+1}^{\mathrm{in}}$-Scott sentence. The key remaining step in Scott's proof is to show that every orbit in a countable structure is $\mathcal{L}_{\omega_1,\omega}$-definable by showing that if two elements satisfy the same $\mathcal{L}_{\omega_1,\omega}$-formulas, they are automorphic.

Lemma 7.1. *Every $\exists$-atomic structure has a Π_2^{in} Scott sentence. Furthermore, every effectively $\exists$-atomic computable structure has a Π_2^{c} Scott sentence.*

Proof. Let S be a Scott family of $\exists$-formulas for $\mathcal{A}$. For each $\bar{a} \in A^{<\omega}$, let $\varphi_{\bar{a}}(\bar{x})$ be the $\exists$-formula in S defining the orbit of $\mathcal{A}$. (For the empty tuple, let $\varphi_\emptyset()$ be a sentence that is always true.) For any other structure $\mathcal{B}$, consider the set

$$I_{\mathcal{B}} = \{(\bar{a}, \bar{b}) \in A^{<\omega} \times \mathcal{B}^{<\omega} : \mathcal{B} \models \varphi_{\bar{a}}(\bar{b})\}.$$

If $I_{\mathcal{B}}$ had the back-and-forth property (see Definition 4.1), we would know that $\mathcal{B}$ is isomorphic to $\mathcal{A}$. Since $I_{\mathcal{A}}$ has the back-and-forth property (see proof of Theorem 1.2), we get that $I_{\mathcal{B}}$ has the back-and-forth property if and only if $\mathcal{B}$ is isomorphic to $\mathcal{A}$. Recall from Definition 4.1 that $I_{\mathcal{B}}$ has the back-and-forth property if and only if:

$$\bigwedge_{\bar{a} \in A^{<\omega}} \forall\bar{x} \in B^{|\bar{a}|} \Bigg(\langle \bar{a}, \bar{x} \rangle \in I_{\mathcal{B}} \Rightarrow$$

$$\Bigg(\Big(D_{\mathcal{A}}(\bar{a}) = D_{\mathcal{B}}(\bar{x}) \Big) \wedge \Big(\forall y \in B \bigvee_{c \in A}(\langle \bar{a}c, \bar{x}y \rangle \in I_{\mathcal{B}}) \Big) \wedge \Big(\bigwedge_{c \in A} \exists y \in B(\langle \bar{a}c, \bar{x}y \rangle \in I_{\mathcal{B}}) \Big) \Bigg) \Bigg).$$

The Scott sentence for $\mathcal{A}$ is a sentence that is true of a structure $\mathcal{B}$ if and only if $I_\mathcal{B}$ has the back-and-forth property:

$$\bigwedge_{\bar{a} \in A^{<\omega}} \forall x_1, ..., x_{|\bar{a}|} \left(\varphi_{\bar{a}}(\bar{x}) \Rightarrow \right.$$

$$\left. \left(\left(\varphi^{at}_{D_\mathcal{A}(\bar{a})}(\bar{x}) \right) \wedge \left(\forall y \bigvee_{c \in A} \varphi_{\bar{a}c}(\bar{x}y) \right) \wedge \left(\bigwedge_{c \in A} \exists y \varphi_{\bar{a}c}(\bar{x}y) \right) \right) \right).$$

As for the effectivity claim, if $\mathcal{A}$ is a computable presentation and S is c.e., then the map $\bar{a} \mapsto \varphi_{\bar{a}}$ is computable, and the conjunctions and disjunctions in the Scott sentence above are all computable. □

To prove the other direction, we need to go through the type-omitting theorem for $\forall$-types.

Proof of (A2) $\Rightarrow$ (A1) in Theorem 1.2. Suppose ψ is a Π_2^{in} Scott sentence for $\mathcal{A}$, but that $\mathcal{A}$ is not atomic. We have already shown that (A1) implies (A5). Thus, there is a $\forall$-type realized in $\mathcal{A}$ which is not $\exists$-supported. But then, by Lemma 5.3, there exists a model of ψ which omits that type. This structure could not be isomorphic to $\mathcal{A}$, as they do not realize the same types. This contradicts that ψ is a Scott sentence for $\mathcal{A}$. □

Lemma 7.2. *Let $\mathcal{A}$ be a structure. The following are equivalent:*

(1) *$\mathcal{A}$ is $\exists$-atomic over a finite tuple of parameters.*
(2) *$\mathcal{A}$ has a Σ_3^{in}-Scott sentence.*

Proof. If $\mathcal{A}$ is $\exists$-atomic over a finite tuple of parameters $\bar{a}$, then $(\mathcal{A}, \bar{a})$ has a Π_2^{in} Scott sentence $\varphi(\bar{c})$. Then $\exists \bar{y} \varphi(\bar{y})$ is a Scott sentence for $\mathcal{A}$.

Suppose now that $\mathcal{A}$ has a Scott sentence $\bigvee_{i \in \omega} \exists \bar{y}_i \psi_i(\bar{y}_i)$. $\mathcal{A}$ must satisfy one of the disjuncts, and that disjunct must then be a Scott sentence for $\mathcal{A}$ too. So, suppose the Scott sentence for $\mathcal{A}$ is $\exists \bar{y} \, \psi(\bar{y})$, where ψ is Π_2^{in}. Let $\bar{c}$ be a new tuple of constants of the same size as $\bar{y}$. If $\varphi(\bar{c})$ were a Scott sentence for $(\mathcal{A}, \bar{a})$, we would know $\mathcal{A}$ is $\exists$-atomic over $\bar{a}$ — but this might not be the case. Suppose $(\mathcal{B}, \bar{b}) \models \varphi(\bar{c})$. Then $\mathcal{B}$ must be isomorphic to $\mathcal{A}$, as it satisfies $\exists \bar{y} \, \psi(\bar{y})$. But $(\mathcal{B}, \bar{b})$ and $(\mathcal{A}, \bar{a})$ need not be isomorphic. However, it is enough for us to show that one of the models of $\varphi(\bar{c})$ is $\exists$-atomic over $\bar{c}$, as that model is isomorphic to $\mathcal{A}$. Since there are only countably many models of $\varphi(\bar{c})$, there are countably many $\forall$-types among the models of $\varphi(\bar{c})$. Thus, using Lemma 5.3, we can omit the non-$\exists$-supported ones while satisfying $\varphi(\bar{c})$. The resulting structure would be $\exists$-atomic over $\bar{c}$ and isomorphic to $\mathcal{A}$. □

We remark that in [Mon] we mentioned this fact, but did not give a proof, as we overlooked the fact that $(\mathcal{B}, \bar{b})$ and $(\mathcal{A}, \bar{a})$ in the proof above need not be isomorphic. The extra step in the proof above seems to be necessary.

8   Turing Degree and Enumeration Degree

The proof of Theorem 1.14 needs a couple of lemmas that are interesting in their own right.

Lemma 8.1. *Let $\mathbb{K}$ be a Π_2^c class all of whose structures have different $\exists$-theories. Then every structure in $\mathbb{K}$ has enumeration degree given by its $\exists$-theory.*

Proof. Take a structure $\mathcal{A} \in \mathbb{K}$, and let S be its $\exists$-theory. By Corollary 5.2, if X can compute an enumeration of S, then it can compute a presentation of a structure $\mathcal{B} \in \mathbb{K}$ with $\exists$-theory S. Since both $\mathcal{A}$ and $\mathcal{B}$ have the same $\exists$-theory, they must be isomorphic. So, X is computing a copy of $\mathcal{A}$. Of course, every copy of $\mathcal{A}$ can enumerate S, and hence $\mathcal{A}$ has enumeration degree S. $\qquad\square$

The following lemma is a strengthening of Lemma 3.3.

Lemma 8.2. *If $\mathcal{A}$ and $\mathcal{B}$ are $\exists$-atomic and have the same $\exists$-theory, then they are isomorphic.*

Proof. We prove that $\mathcal{A}$ and $\mathcal{B}$ are isomorphic using a back-and-forth construction. Let

$$I = \{\langle \bar{a}, \bar{b} \rangle : \forall\text{-}tp_{\mathcal{A}}(a_0, ..., a_s) = \forall\text{-}tp_{\mathcal{B}}(b_0, ..., b_s)\}.$$

By assumption, $\langle \emptyset, \emptyset \rangle \in I$. We need to show that I has the back-and-forth property (Definition 4.1), as that would imply that $\mathcal{A}$ and $\mathcal{B}$ are isomorphic. Clearly, $\forall\text{-}tp_{\mathcal{A}}(a_0, ..., a_s) = \forall\text{-}tp_{\mathcal{B}}(b_0, ..., b_s)$ implies $D_{\mathcal{A}}(a_0, ..., a_s) = D_{\mathcal{B}}(b_0, ..., b_s)$. For the second condition in Definition 4.1, suppose $\langle \bar{a}, \bar{b} \rangle \in I$ and let $c \in A$. Let ψ be the principal $\exists$-formula satisfied by $\bar{a}c$. Since $\forall\text{-}tp_{\mathcal{A}}(\bar{a}) = \forall\text{-}tp_{\mathcal{B}}(\bar{b})$, there is a d in $\mathcal{B}$ satisfying the same formula over $\bar{b}$. We need to show that $\forall\text{-}tp_{\mathcal{A}}(\bar{a}c) = \forall\text{-}tp_{\mathcal{B}}(\bar{b}d)$. Let us remark that since we do not know $\mathcal{A}$ and $\mathcal{B}$ are isomorphic yet, we do not know that ψ generates a $\forall$-type in $\mathcal{B}$.

First, to show $\forall\text{-}tp_{\mathcal{A}}(\bar{a}c) \subseteq \forall\text{-}tp_{\mathcal{B}}(\bar{b}d)$, take $\theta(\bar{x}y) \in \forall\text{-}tp_{\mathcal{A}}(\bar{a}c)$. Then

$$\text{``}\forall y(\psi(\bar{x}y) \to \theta(\bar{x}y))\text{''} \ \in \ \forall\text{-}tp_{\mathcal{A}}(\bar{a}c) \ = \ \forall\text{-}tp_{\mathcal{B}}(\bar{b}d),$$

and hence $\theta \in \forall\text{-}tp_{\mathcal{B}}(\bar{b}d)$. Let us now prove the other inclusion. Let $\tilde{\psi}(\bar{x}y)$ be the $\exists$-formula generating $\forall\text{-}tp_{\mathcal{B}}(\bar{b}d)$ in $\mathcal{B}$. Then, since $\neg\tilde{\psi} \notin \forall\text{-}tp_{\mathcal{B}}(\bar{b}d)$, by our previous argument, $\neg\tilde{\psi} \notin \forall\text{-}tp_{\mathcal{A}}(\bar{a}c)$ either, and hence $\mathcal{A} \models \tilde{\psi}(\bar{a}c)$. The rest of the proof that $\forall\text{-}tp_{\mathcal{B}}(\bar{b}d) \subseteq \forall\text{-}tp_{\mathcal{A}}(\bar{a}c)$ is now symmetrical to the one of the other inclusion: For $\tilde{\theta}(\bar{x}y) \in \forall\text{-}tp_{\mathcal{A}}(\bar{b}d)$, we have that $\text{``}\forall y(\tilde{\psi}(\bar{x}y) \to \tilde{\theta}(\bar{x}y))\text{''} \in \forall\text{-}tp_{\mathcal{A}}(\bar{a}c)$, and hence $\theta \in \forall\text{-}tp_{\mathcal{B}}(\bar{a}c)$. $\qquad\square$

Proof of Theorem 1.14. The proof is immediate from Lemmas 8.1 and 8.2. $\qquad\square$

The following gives a structural property that is sufficient for a structure to have enumeration degree. The property is far from necessary though.

Corollary 8.3. *Suppose that a structure $\mathcal{A}$ has a Σ_3^c Scott sentence. Then $\mathcal{A}$ has enumeration degree.*

Proof. Let $\bigvee_{i \in \omega} \exists \bar{x}_i \ \psi_i(\bar{x}_i)$ be the Σ_3^c Scott sentence for $\mathcal{A}$, where each ψ_i is Π_2^c. $\mathcal{A}$ satisfies one of the disjuncts, say $\exists \bar{x}_i \ (\psi_i(\bar{x}_i))$, and hence this disjunct is also a Scott sentence for $\mathcal{A}$. Let $\tilde{\tau}$ be the vocabulary τ of $\mathcal{A}$, together with $|\bar{x}_i|$ many new constant symbols $\bar{c}$, and let $\tilde{\mathcal{A}}$ be the $\tilde{\tau}$-structure $(\mathcal{A}, \bar{a})$, where $\bar{a}$ is such that $\mathcal{A} \models \psi_i(\bar{a})$. Now, even if this sentence might not be a Scott sentence for $\tilde{\mathcal{A}}$, we can still work with it. We claim that $\mathcal{A}$ has enumeration degree given by $\exists\text{-}tp_{\mathcal{A}}(\bar{a})$, which is the same as $\exists\text{-theory}(\tilde{\mathcal{A}})$. Clearly, every copy of $\mathcal{A}$ can enumerate $\exists\text{-}tp_{\mathcal{A}}(\bar{a})$. On the other hand, using $\exists\text{-theory}(\tilde{\mathcal{A}})$ and the Π_2^c sentence $\psi_i(\bar{c})$, we can build a model of $\psi_i(\bar{c})$ by Corollary 5.2. Even if this model does not turn out to be isomorphic to $\tilde{\mathcal{A}}$, when we look at it as a τ-structure, it is isomorphic to $\mathcal{A}$. $\qquad\square$

References

[AK00]   Ash, C.J., Knight, J.: Computable Structures and the Hyperarithmetical Hierarchy. Elsevier Science, Amsterdam (2000)

[AKMS89] Ash, C., Knight, J., Manasse, M., Slaman, T.: Generic copies of countable structures. Ann. Pure Appl. Log. **42**(3), 195–205 (1989)

[Chi90] Chisholm, J.: Effective model theory vs. recursive model theory. J. Symb. Log. **55**(3), 1168–1191 (1990)

[CHS07] Calvert, W., Harizanov, V., Shlapentokh, A.: Turing degrees of isomorphism types of algebraic objects. J. Lond. Math. Soc. **75**(2), 273–286 (2007)

[CM]   Cholak, P., McCoy, C.: Effective prime uniqueness. Submitted for publication

[DHK03] Downey, R., Hirschfeldt, D., Khoussainov, B.: Uniformity in the theory of computable structures. Algebra Log. **42**(5), 566–593 (2003). 637

[Eff65] Effros, E.G.: Transformation groups and C^*-algebras. Ann. Math. **2**(81), 38–55 (1965)

[FKM09] Frolov, A., Kalimullin, I., Miller, R.: Spectra of algebraic fields and subfields. In: Ambos-Spies, K., Löwe, B., Merkle, W. (eds.) CiE 2009. LNCS, vol. 5635, pp. 232–241. Springer, Heidelberg (2009). doi:10.1007/978-3-642-03073-4_24

[HLZ99] Herwig, B., Lempp, S., Ziegler, M.: Constructive models of uncountably categorical theories. Proc. Am. Math. Soc. **127**(12), 3711–3719 (1999)

[Kni98] Knight, J.F.: Degrees of models. In: Handbook of Recursive Mathematics, vol. 1. Studies in Logic and the Foundations of Mathematics, vol. 138, pp. 289–309, Amsterdam (1998)

[Kud96a] Kudinov, O.V.: An autostable 1-decidable model without a computable Scott family of $\exists$-formulas. Algebra i Log. **35**(4), 458–467 (1996). 498

[Kud96b] Kudinov, O.V.: Some properties of autostable models. Algebra i Log. **35**(6), 685–698 (1996). 752

[Kud97] Kudinov, O.V.: The problem of describing autostable models. Algebra i Log. **36**(1), 26–36 (1997). 117

[MM]   Melnikov, A., Montalbán, A.: Computable Polish group actions. Submitted for publication

[Mon]   Montalbán, A.: A robuster Scott rank. In: The Proceedings of the American Mathematical Society. To appear

[Mon12] Montalbán, A.: Rice sequences of relations. Philos. Trans. R. Soc. A **370**, 3464–3487 (2012)

[Mon13] Montalbán, A.: A fixed point for the jump operator on structures. J. Symb. Log. **78**(2), 425–438 (2013)

[Nur74] Nurtazin, A.T.: Computable classes and algebraic criteria for autostability. PhD thesis, Institute of Mathematics and Mechanics, Alma-Ata (1974)

[Pou72] Pouzet, M.: Modéle uniforméement préhomogéne. C. R. Acad. Sci. Paris Sér **274**, A695–A698 (1972)

[Ric81] Richter, L.J.: Degrees of structures. J. Symb. Log. **46**(4), 723–731 (1981)

[Sco65] Scott, D.: Logic with denumerably long formulas and finite strings of quantifiers. In: Theory of Models (Proceedings of 1963 International Symposium at Berkeley), pp. 329–341 (1965)

[Sim76] Simmons, H.: Large and small existentially closed structures. J. Symb. Log. **41**(2), 379–390 (1976)

[Ste13] Steiner, R.M.: Effective algebraicity. Arch. Math. Log. **52**(1–2), 91–112 (2013)

[Ven92] Ventsov, Y.G.: The effective choice problem for relations and reducibilities inclasses of constructive and positive models. Algebra i Log. **31**(2), 101–118 (1992). 220

Irreducibles and Primes in Computable Integral Domains

Leigh Evron, Joseph R. Mileti$^{(\boxtimes)}$, and Ethan Ratliff-Crain

Department of Mathematics and Statistics, Grinnell College,
Grinnell, IA 50112, USA
miletijo@grinnell.edu

Abstract. A computable ring is a ring equipped with a mechanical procedure to add and multiply elements. In most natural computable integral domains, there is a computational procedure to determine if a given element is prime/irreducible. However, there do exist computable UFDs (in fact, polynomial rings over computable fields) where the set of prime/irreducible elements is not computable. Outside of the class of UFDs, the notions of irreducible and prime may not coincide. We demonstrate how different these concepts can be by constructing computable integral domains where the set of irreducible elements is computable while the set of prime elements is not, and vice versa. Along the way, we will generalize Kronecker's method for computing irreducibles and factorizations in $\mathbb{Z}[x]$.

1 Introduction

In an integral domain, there are two natural definitions of basic "atomic" elements: irreducibles and primes. We recall these standard algebraic definitions.

Definition 1.1. *Let A be an integral domain, i.e. a commutative ring with $1 \neq 0$ and with no zero divisors (so $ab = 0$ implies either $a = 0$ or $b = 0$).*

(1) An element $u \in A$ is a unit *if there exists $w \in A$ with $uw = 1$. We denote the set of units by $U(A)$. Notice that $U(A)$ is a multiplicative group.*

(2) Given $a, b \in A$, we say that a and b are associates *if there exists $u \in U(A)$ with $au = b$.*

(3) An element $p \in A$ is irreducible *if it nonzero, not a unit, and has the property that whenever $p = ab$, either a is a unit or b is a unit. An equivalent definition is that $p \in A$ is irreducible if it is nonzero, not a unit, and its divisors are precisely the units and the associates of p.*

(4) An element $p \in A$ is prime *if it nonzero, not a unit, and has the property that whenever $p \mid ab$, either $p \mid a$ or $p \mid b$.*

The authors thank Grinnell College for its generous support through the MAP program for research with undergraduates. They also thank the referee for providing several corrections and helpful suggestions.

ⓒ Springer International Publishing AG 2017
A. Day et al. (Eds.): Downey Festschrift, LNCS 10010, pp. 238–253, 2017.
DOI: 10.1007/978-3-319-50062-1_17

(5) A is a unique factorization domain, *or* UFD, *if it has the following two properties:*

- *For each $a \in A$ such that a is nonzero and not a unit, there exist irreducible elements $r_1, r_2, \ldots, r_n \in A$ with $a = r_1 r_2 \cdots r_n$.*
- *If $r_1, r_2, \ldots, r_n, q_1, q_2, \ldots, q_m \in A$ are all irreducible and $r_1 r_2 \cdots r_n = q_1 q_2 \cdots q_m$, then $n = m$ and there exists a permutation σ of $\{1, 2, \ldots, n\}$ such that r_i and $q_{\sigma(i)}$ are associates for all i.*

It is a simple fact that if A is an integral domain, then every prime element of A is irreducible. Although the converse is true in any UFD, it does fail for general integral domains. For example, in the integral domain $\mathbb{Z}[\sqrt{-5}]$, there are two different factorizations of 6 into irreducibles:

$$2 \cdot 3 = 6 = (1 + \sqrt{-5})(1 - \sqrt{-5}).$$

Since $U(\mathbb{Z}(\sqrt{-5})) = \{1, -1\}$, these two factorizations are indeed distinct. This example also shows that 2 is an irreducible element that is not prime because $2 \mid (1 + \sqrt{-5})(1 - \sqrt{-5})$ but $2 \nmid 1 + \sqrt{-5}$ and $2 \nmid 1 - \sqrt{-5}$. In fact, all four of the above irreducible factors are not prime.

For another example that will be particularly relevant for our purposes, let A be the subring of $\mathbb{Q}[x]$ consisting of those polynomials whose constant term and coefficient of x are both integers, i.e.

$$A = \{a_0 + a_1 x + a_2 x^2 + \cdots + a_n x^n \in \mathbb{Q}[x] : a_0 \in \mathbb{Z} \text{ and } a_1 \in \mathbb{Z}\}.$$

In this integral domain, all of the normal integer primes are still irreducible (by a simple degree argument), but none of them are prime in A because given any integer prime $p \in \mathbb{Z}$, we have that $p \mid x^2$ since $\frac{x^2}{p} \in A$, but $p \nmid x$ as $\frac{x}{p} \notin A$.

We are interested in the extent to which the irreducible and prime elements can differ in an integral domain. As mentioned above, the set of prime elements is always a subset of the set of irreducible elements, but it may be a proper subset. Can one of these sets be significantly more complicated than the other? We approach this question from the point of view of computability theory. We begin with the following fundamental definition.

Definition 1.2. *A* computable ring *is a ring whose underlying set is a computable set $A \subseteq \mathbb{N}$, with the property that $+$ and $\cdot$ are computable functions from $A \times A$ to A.*

For a general overview of results about computable rings and fields, see [9]. Computable fields together with computable factorizations in polynomial rings over those fields have received a great deal of attention [5–8] provides an excellent overview of work in this area. In particular, there exists a computable field F such that the set of primes in $F[x]$ is not computable (see [7, Lemma 3.4] or [9, Sect. 3.2] for an example). Moreover, while the set of primes in any computable integral domain is easily seen to be Π_2^0, there is a computable UFD such that the set of primes is Π_2^0-complete (see [4]).

We will prove that there exists a computable integral domain where the set of irreducible elements is computable while the set of prime elements is not, and also there exists a computable integral domain where the set of prime elements is computable while the set of irreducible elements is not. Thus, these two notions can be wildly different. Our approach will be to code an arbitrary Π_1^0 set into the set of irreducible (resp. prime) elements while maintaining control over the set of prime (resp. irreducible) elements. Moreover, our integral domains will extend $\mathbb{Z}$ and we will perform our noncomputable coding into the normal integer primes as in [4].

2   Strongly Computable Finite Factorization Domains

In Sect. 3, we will build a computable integral domain A such that the set of irreducible elements of A is computable but the set of prime elements of A is not computable. The idea is that we will turn off the primeness of a normal integer prime p_i in response to a Σ_1^0 event (such as program i halting) by introducing a new element x with $p_i \mid x^2$ but $p_i \nmid x$. In doing this, we will expand A and we will want to ensure that we can compute the irreducible elements in the resulting integral domain. Since we are adding a new element, this construction will be analogous to expanding our original A to the polynomial ring $A[x]$. However, there is a potential problem here in that even if the irreducible elements of an integral domain A are computable, it need not be the case the irreducible elements of $A[x]$ are computable. In fact, as mentioned in the introduction, there are computable fields F (where the irreducibles are trivially computable because no element is irreducible) such that the irreducibles of $F[x]$ are not computable.

To remedy this situation, we will ensure that the integral domains in our construction have a stronger property. As motivation, we first summarize Kronecker's method for finding the divisors of an element $\mathbb{Z}[x]$, and hence for determining whether an element is irreducible. Let $f(x) \in \mathbb{Z}[x]$ be nonzero, and let $n = \deg(f(x))$. We try to restrict the set of possible divisors to a finite set that we need to check. Since the degree function is additive (i.e. the degree of a product is the sum of the degrees), notice that any divisor of $f(x)$ has degree at most n. Now perform the following:

- Notice that if $g(x) \in \mathbb{Z}[x]$ and $g(x) \mid f(x)$ in $\mathbb{Z}[x]$, then $g(a) \mid f(a)$ for all $a \in \mathbb{Z}$.
- Find $n+1$ many points $a \in \mathbb{Z}$ with $f(a) \neq 0$ (which exist because $f(x)$ has at most n roots). Notice that each such $f(a)$ has only finitely many divisors in $\mathbb{Z}$.
- For each of the possible choices of the divisors of these values in $\mathbb{Z}$, find the unique interpolating polynomial (i.e. the polynomial that outputs these values at the corresponding $n+1$ points) in $\mathbb{Q}[x]$ of degree at most n.
- Check if any of these polynomials are in $\mathbb{Z}[x]$, and if so, check if they divide $f(x)$ in $\mathbb{Z}[x]$.
- Compile the resulting list of divisors.

Therefore, we can compute the finite set of divisors of any element of $\mathbb{Z}[x]$. Since we know the units of $\mathbb{Z}[x]$, it follows that we can computably determine if an element of $\mathbb{Z}[x]$ is irreducible.

The key algebraic fact that makes Kronecker's method work is that every nonzero element of $\mathbb{Z}$ has only finitely many divisors. Integral domains with this property were defined and studied in [1–3].

Definition 2.1. *Let A be an integral domain.*

- *A is a* finite factorization domain, *or FFD, if every nonzero element has only finitely many divisors up to associates.*
- *A is a* strong finite factorization domain *if every nonzero element has only finitely many divisors.*

Notice that a strong finite factorization domain is just an FFD in which $U(A)$ is finite. We now define an effective analogue of this concept. In addition to wanting our ring to be computable, we also want the stronger property that we can compute the finite set of divisors of any nonzero element. Instead of using the word "strong" twice, we adopt the following definition.

Definition 2.2. *A* strongly computable finite factorization domain, *or SCFFD, is a computable integral domain A equipped with a computable function D such that for all $a \in A\backslash\{0\}$, the set $D(a)$ is (a canonical index for) the finite set of divisors of a in A.*

Proposition 2.3. *Let A be an SCFFD equipped with divisor function D.*

(1) The set $U(A)$ is a finite set that can be uniformly computed from A.
(2) The set of irreducible elements of A is computable.

Proof. For the first claim, simply notice that $U(A) = D(1)$. For the second, given any $a \in A$, we have that a is irreducible if and only it nonzero, not a unit, and its only divisors are units and associates. Suppose then that we are given an arbitrary $a \in A$. We can check whether a is zero or a unit (by part 1), and if either is true, then a is not irreducible. Otherwise, then since $a \neq 0$, we can compute the finite set $D(a)$ of divisors of a. Since we can also compute the finite set $U(A)$, we can examine each $b \in D(a)$ in turn to determine whether $b \in U(A)$ or whether there exists $u \in U(A)$ with $b = au$. If this is true for all $b \in D(a)$, then a is irreducible in A, and otherwise it is not. $\square$

If we include an additional assumption that A is a UFD, then we have a converse to the previous result.

Proposition 2.4. *Let A be a computable integral domain with the following properties:*

- *A is a UFD.*
- *$U(A)$ is finite.*
- *The set of irreducible elements of A is computable.*

We can then equip A with a computable function D so that A becomes an SCFFD.

Proof. We first argue that we can computably factor elements of A into irreducibles. Let $a \in A$ be nonzero and not a unit. Since the set of irreducibles of A is computable, we can check whether a is irreducible. If not, we search until we find two nonzero nonunit elements of A whose product is a. We can now check if these factors are irreducible, and if not we can repeat to factor them. Notice that this process must eventually produce finitely many irreducibles whose product is a by König's Lemma together with the fact that there are no infinite descending chains of strict divisibilities in a UFD.

We now define our function D. Let $a \in A \backslash \{0\}$ be arbitrary. Check if $a \in U(A)$ (which is possible because $U(A)$ is finite), and if so, define $D(a)$ to equal $U(A)$. If $a \notin U(A)$, then we can computably factor it into irreducibles q_i so that $a = q_1 q_2 \ldots q_n$. Since $U(A)$ is finite, we can now computably check if any of the q_i are associates of each other, and if so we can find witnessing units. Thus, we can write $a = u p_1^{k_1} \cdots p_m^{k_m}$ where $u \in U(A)$, each p_i is irreducible, each $k_i \in \mathbb{N}^+$, and p_i and p_j are not associates whenever $i \neq j$. Since A is a UFD, we then have that the set of divisors of a equals the set of elements of the form $w p_1^{\ell_1} \cdots p_m^{\ell_m}$ where $w \in U(A)$ and $0 \leq \ell_i \leq k_i$ for all i. Thus, we can define $D(a)$ to be this finite set. □

In contrast, there are SCFFDs that are not UFDs, such as $\mathbb{Z}[\sqrt{-5}]$. More generally, the ring of integers in any imaginary quadratic number field is an SCFFD. To see this, let K be an imaginary quadratic number field, and fix an integral basis of $\mathcal{O}_K$. Using this integral basis, we can view $\mathcal{O}_K$ as a computable integral domain in such a way that the norm function and divisibility relation are both computable on $\mathcal{O}_K$ (see [4, Proposition 1.4]). Given any $n \in \mathbb{N}$, there are only finitely many elements of norm n, and moreover we can compute the finite set of such elements. Now given any nonzero $a \in A$, we can compute $N(a)$, examine all elements of norm dividing $N(a)$, and check which of them divide a (since the divisibility relation is computable) to compute the set of divisors of a.

Let A be a computable integral domain and let F be the field of fractions of A. Recall that elements of F are equivalence classes of pairs of elements of A. If we were to allow multiple representations of elements, we can of course work with pairs of elements of A and define addition and multiplication on these elements computably. Nonetheless, a computable ring is defined in a way that forbids such multiple representations, so it is not immediately obvious that we can view F as a computable field. However, since a computable integral domain is coded as a subset of $\mathbb{N}$, we can view pairs of elements $(a, b) \in A^2$ with $b \neq 0$ as being coded by elements of $\mathbb{N}^2$, which in turn can be coded by elements of $\mathbb{N}$. Thus, we can view the field of fractions F as a computable field by working only with pairs (a, b) such that there is no strictly smaller pair (c, d) in the usual ordering of $\mathbb{N}$ with $ad = bc$. In this way, we can still define addition and multiplication computably by searching back for the smallest equivalent representative.

Suppose that A is a computable ring, and that A is a subring of a larger computable ring B. Given a computable presentation of A, where the elements of A are coded by the elements from a coinfinite subset of $\mathbb{N}$, it might not be possible to expand the given computable presentation of A to a computable presentation of B (by only using the elements of $\mathbb{N}\backslash A$ to code elements of $B\backslash A$). For example, although every computable field K can be embedded in a computable algebraic closure, it is not always possible to do so in such a way that the image of K is a computable subset of the algebraic closure (see [8]). In such a case, we can not build an algebraic closure of K as a computable extension of K in the above sense. Similarly, for a computable integral domain A, it may not be possible to build the field of fractions as a computable extension of A, because it may not be possible to determine when an element $\frac{a}{b} \in F$ is actually an element of A. The issue is that we may not be able to determine if $b \mid a$ because the divisibility relation may not be computable. However, we have the following result.

Corollary 2.5. *If A is an SCFFD, then the field of fractions of A is a computable field, and we can computably build it as an extension of A.*

Proof. Notice that in the field of fractions of A, we have $\frac{a}{b} \in A$ if and only if $b \mid a$, which is if and only if $b \in D(a)$. Now since A is a computable integral domain, it is coded as a subset of $\mathbb{N}$. We can now add on minimal pairs (a, b) such that $b \nmid a$, and then computably define addition and multiplication on this representation of the field of fractions. $\qquad\square$

In fact, we can computably "reduce" fractions over an SCFFD to lowest terms, as we now show.

Proposition 2.6. *Let A be an SCFFD and let F be the field of fractions of A. Given an arbitrary pair of elements $a, b \in A$ with $b \neq 0$, we can computably find a pair of elements $c, d \in A$ with $d \neq 0$, with $\frac{c}{d} = \frac{a}{b}$ in F, and such that the only common divisors of c and d are the units of A.*

Proof. First notice that if $a = 0$, then we may take $c = 0$ and $d = 1$. Suppose then that $a \neq 0$. Since we also have that $b \neq 0$, we can now computably determine the finite set of divisors of each of a and b, and thus can computably build the finite set S of common divisors of a and b, i.e. $S = D(a) \cap D(b)$. For each $r \in S$, we can computably determine the number $|\{s \in S : s \mid r\}| = |D(r) \cap S|$. Fix an $r \in S$ such that $|\{s \in S : s \mid r\}|$ is as large as possible. Since r is a common divisor of a and b, we can computably search for $c, d \in A$ such that $rc = a$ and $rd = b$. Notice that $d \neq 0$ (because $b \neq 0$) and $\frac{a}{b} = \frac{c}{d}$. Suppose now that t is a common divisor of c and d. We then have that rt is a common divisor of a and b, so $rt \in S$. By definition of r, this implies that $|\{s \in S : s \mid rt\}| \leq |\{s \in S : s \mid r\}|$. Since $\{s \in S : s \mid r\} \subseteq \{s \in S : s \mid rt\}$, it follows that $\{s \in S : s \mid rt\} = \{s \in S : s \mid r\}$. In particular, we must have $rt \mid r$, so $t \in U(A)$. $\qquad\square$

Notice this reduction need not be unique, even up to units. In the SCFFD $\mathbb{Z}[\sqrt{-5}]$ we have

$$\frac{2}{1 + \sqrt{-5}} = \frac{1 - \sqrt{-5}}{3},$$

where there are no nonunit common factors for the numerator and denominator of either side.

By [1, Proposition 5.3] and [3, Theorem 5], if A is a (strong) finite factorization domain, then so is $A[x]$. We now prove an effective analogue of this result. Notice first that if A is a finite integral domain, then A is a finite field, and $A[x]$ is trivially an SCFFD because given $f(x) \in A[x] \backslash \{0\}$, every divisor $g(x)$ of $f(x)$ must satisfy $\deg(g(x)) \leq \deg(f(x))$, and so we need only check each of the finitely many possibilities (which is possible because we can computably search for quotients and remainders). We now handle the infinite case.

Theorem 2.7. *If A is an infinite SCFFD, then so is $A[x]$. Moreover, given an index for a function D witnessing that A is an SCFFD, we can computably obtain an index for a function D' extending D to witness the fact that $A[x]$ is an SCFFD.*

Before jumping into the proof, we give two lemmas.

Lemma 2.8. *Let A be an SCFFD, let $n \in \mathbb{N}^+$, let $a_0, a_1, \ldots, a_n \in A$ be distinct and let $b_0, b_1, \ldots, b_n \in A$. Let F be the field of fractions of A. There is exactly one polynomial $p(x) \in F[x]$ of degree at most n with $p(a_i) = b_i$ for all i. Furthermore, we can computably construct $p(x)$ in $F[x]$, and can computably determine if $p(x) \in A[x]$.*

Proof. Uniqueness follows from that fact that if two polynomials over a field having degree at most n agree at $n + 1$ points, then they must be the same polynomial. For existence, using Lagrange's method of interpolation for $n + 1$ distinct points of the form (a_i, b_i) will result in a polynomial of the following form:

$$p(x) = \sum_{i=0}^{n} b_i \cdot \frac{(x - a_0) \cdots (x - a_{i-1})(x - a_{i+1}) \cdots (x - a_n)}{(a_i - a_0) \cdots (a_i - a_{i-1})(a_i - a_{i+1}) \cdots (a_i - a_n)}$$

Notice that the denominator is nonzero because A is an integral domain and $a_i \neq a_j$ whenever $i \neq j$. We can computably expand $p(x)$ to write it as $p(x) = \sum_{i=0}^{n} \frac{c_i}{d_i} x^i$. We then have that $p(x) \in A[x]$ if and only if $d_i \mid c_i$ for all i, which we can verify by checking if $d_i \in D(c_i)$ for all i. $\square$

Lemma 2.9. *Suppose that A is an SCFFD. The divisibility relation on $A[x]$ is computable, i.e. given $f(x), g(x) \in A[x]$, we can computably determine if $f(x) \mid g(x)$ in $A[x]$.*

Proof. Let $f(x), g(x) \in A[x]$ be arbitrary. If $g(x) = 0$, then trivially we have $f(x) \mid g(x)$. Suppose then that both $g(x)$ is nonzero. Perform polynomial long division (or search) to find $q(x), r(x) \in F[x]$ with $f(x) = q(x)g(x) + r(x)$ and either $r(x) = 0$ or $\deg(r(x)) < \deg(g(x))$. Since quotients and remainders are unique in $F[x]$, we have $g(x) \mid f(x)$ in $A[x]$ if and only if $q(x) \in A[x]$ and $r(x) = 0$. Since we can computably determine if an element of $F[x]$ is in $A[x]$ as in Lemma 2.8, this completes the proof. $\square$

We now prove our theorem, imitating the Kronecker's argument for $\mathbb{Z}[x]$ described at the beginning of this section.

Proof of Theorem 2.7. Let $f(x) \in A[x]$ be arbitrary, and let $n = \deg(f(x))$. Suppose that $g(x) \in A[x]$ is such that $g(x) \mid f(x)$. First notice that $\deg(g(x)) \leq n$ because the degree function is additive (as A is an integral domain). Now if we fix $h(x) \in A[x]$ with $g(x)h(x) = f(x)$, we then have $g(a)h(a) = f(a)$ for all $a \in A$, so since $f(a), g(a), h(a) \in A$ for all $a \in A$, we have $g(a) \mid f(a)$ for all $a \in A$.

Search until we find $n + 1$ many distinct elements $a_0, a_1, \ldots, a_n \in A$ such that $f(a_i) \neq 0$ for all i (such a_i exist because A is infinite and $f(x)$ has at most n roots in A). Since A is an SCFFD, we have that $f(a_i)$ has only finitely many divisors for each i, and we can compute the finite sets $D(f(a_i))$. Suppose that we pick elements $b_i \in D(f(a_i))$ for each i. From Lemma 2.8, there is a unique element $p(x) \in F[x]$ with $\deg(p(x)) \leq n$ and $p(a_i) = b_i$ for all i, and we can compute this polynomial $p(x)$ and determine if $p(x) \in A[x]$. As we do this for each choice of the b_i, we obtain a finite subset of $A[x]$ of all possible divisors of $f(x)$. Now using Lemma 2.9, we can thin out this set to form the actual finite set of divisors of $f(x)$. $\square$

Proposition 2.10. *Let A be an SCFFD. Suppose that B is a subring of A, and that B is also a computable subset of A. We then have that B is an SCFFD. Moreover, given A, and an index for a function D witnessing that A is an SCFFD, we can computably build an index for a function D' witnessing that B is an SCFFD.*

Proof. Consider to arbitrary nonzero $b \in B$. First notice that if $c \in B$ is a divisor of b in the ring B, then it is trivially a divisor of b in A. Thus, to determine $D'(b)$, we first compute $D(b)$. Now for each $c \in D(b)$, we computably search until we find the unique $d \in A$ such that $cd = b$. We then have that $c \in D'(b)$ if and only if both $c \in B$ and $d \in B$. $\square$

3 Irreducibles Computable and Primes Noncomputable

We seek to build a computable integral domain where the set of irreducible elements is computable, but the set of prime elements is not. To accomplish this, the key to our construction is to work inside an SCFFD in order to maintain control over divisibility, thus allowing us to understand the irreducibles. At various points, we will want to take a given prime element q, and extend our SCFFD to a larger one, where q is no longer prime, but all of the prime elements that are not associates of q remain prime. We accomplish this with the following construction.

Theorem 3.1. *Let A be an SCFFD and let $q \in A$ be prime. Let R be the subset of $A[x]$ consisting of those polynomials of the form*

$$a_0 + a_1 x + a_2 x^2 + \cdots + a_n x^n,$$

where each $a_i \in A$ and where $q \mid a_1$ in A. We then have the following:

(1) R *is a computable subring of $A[x]$ containing A.*

(2) *For any $a \in A$, the set of divisors of a in A equals the set of divisors of a in R.*

(3) R *is an SCFFD. Moreover, given A, q, and an index for a function D witnessing that A is an SCFFD, we can computably build R as an extension of A and obtain an index for a function D' witnessing that R is an SCFFD with the property that $D'(a) = D(a)$ for all $a \in A$.*

(4) $U(R) = U(A)$.

(5) *If p is irreducible in A, then p is irreducible in R.*

(6) *If $p_1, p_2 \in A$ are irreducibles that are not associates in A, then they are not associates in R.*

(7) q *is not prime in R.*

(8) *If p is a prime of A that is not an associate of q, and $a_0 + a_1 x + a_2 x^2 + \cdots + a_n x^n \in R$, then $p \mid a_0 + a_1 x + a_2 x^2 + \cdots + a_n x^n$ in R if and only if $p \mid a_i$ in A for all i.*

(9) *If p is a prime of A that is not an associate of q, then p is prime in R.*

Proof

(1) It is straightforward to check that R is a subring of $A[x]$, and it clearly contains A. Notice that R is a computable because A is an SCFFD, and hence we can compute $D(q)$ in order to determine whether an element of $A[x]$ is in R.

(2) Let $a \in A$. Clearly, if an element of A divides a in A, then it divides a in R. For the converse, since the degree of a product of elements of $A[x]$ is the sum of the degrees, if $f(x), g(x) \in R$ are such that $a = f(x)g(x)$, then we must have $\deg(f(x)) = 0 = \deg(g(x))$, and hence $f(x), g(x) \in A$.

(3) Since R is a computable subring of $A[x]$, this follows from Theorem 2.7 and Proposition 2.10.

(4) Immediate from 2 and the fact that $U(R) = D(1)$.

(5) This follow from 2 and 4.

(6) Immediate from 4.

(7) Notice that q is nonzero and not a unit by 4. We have that $q \mid (qx)^2$ in R because $qx^2 \in R$ and $q \cdot qx^2 = (qx)^2$. However, $q \nmid qx$ in R because $x \notin R$ as q is not a unit (and this is the only possible witness for divisibility because $A[x]$ is an integral domain). Therefore, q is not prime in R.

(8) Let p be a prime of A that is not an associate of q, and let $a_0 + a_1 x + a_2 x^2 + \cdots + a_n x^n \in R$. By definition, we then have that $q \mid a_1$ in A.

Suppose first that $p \mid a_0 + a_1 x + a_2 x^2 + \cdots + a_n x^n$ in R, and fix $b_0 + b_1 x + b_2 x^2 + \cdots + b_n x^n \in R$ with $p \cdot (b_0 + b_1 x + b_2 x^2 + \cdots + b_n x^n) = a_0 + a_1 x + a_2 x^2 + \cdots + a_n x^n$. We then have $pb_i = a_i$ for all i, so $p \mid a_i$ in A for all i.

Conversely, suppose that $p \mid a_i$ in A for all i. For each i, fix $b_i \in A$ with $pb_i = a_i$. We have $q \mid a_1$ in A, so $q \mid pb_1$ in A. Since q is prime in A, it follows that either $q \mid p$ in A or $q \mid b_1$ in A. The former is impossible because p is irreducible in A, but q is neither a unit nor an associate of p in A. Therefore, we must have that $q \mid b_1$ in A. It follows that $b_0 + b_1 x + b_2 x^2 + \cdots + b_n x^n \in R$, and that $p \cdot (b_0 + b_1 x + b_2 x^2 + \cdots + b_n x^n) = a_0 + a_1 x + a_2 x^2 + \cdots + a_n x^n$. Therefore, $p \mid a_0 + a_1 x + a_2 x^2 + \cdots + a_n x^n$ in R.

(9) Let p be a prime of A that is not an associate of q. Notice that p is nonzero and not a unit of R by 4. Let $f(x), g(x) \in R$, and suppose that $p \mid f(x)g(x)$ in R. Write out

$$f(x) = a_0 + a_1 x + a_2 x^2 + \cdots + a_n x^n$$
$$g(x) = b_0 + b_1 x + b_2 x^2 + \cdots + b_n x^n$$
$$f(x)g(x) = c_0 + c_1 x + c_2 x^2 + \cdots + c_n x^n.$$

Since $p \mid f(x)g(x)$ in R, we know from 8 that $p \mid c_i$ in A for all i. Assume that $p \nmid f(x)$ and $p \nmid g(x)$ in R. Then by 8 again, there must exist i and j such that $p \nmid a_i$ in A and $p \nmid b_j$ in A. Let k and ℓ be largest possible such that $p \nmid a_k$ in A and $p \nmid b_\ell$ in A. Now element $c_{k+\ell}$ will be a sum of terms, one of which will be $a_k b_\ell$, while other terms will be divisible by p in A. Since p divides $c_{k+\ell}$, it follows that $p \mid a_k b_\ell$ in A. However, this is a contradiction because p is prime in A but divides neither a_k nor b_ℓ. □

We now show that we can code an arbitrary Π_1^0 set into the primes of an integral domain A while maintaining the computability of the irreducible elements. In fact, we perform our coding within the normal integer primes and can make the resulting integral domain an SCFFD.

Theorem 3.2. *Let S be a Σ_1^0 set, and let $p_0, p_1, p_2, \ldots$ list the usual primes from $\mathbb{N}$ in increasing order. There exists an SCFFD A such that:*

- $\mathbb{Z}$ *is a subring of A.*
- $U(A) = \{1, -1\}$.
- *Every p_i is irreducible in A.*
- p_i *is prime in A if and only if $i \notin S$.*

Proof. If $S = \emptyset$, this is trivial by letting $A = \mathbb{Z}$. Assume then that $S \neq \emptyset$. If S is finite, say $|S| = n$, then we can trivially fix a computable injective function $\alpha \colon \{1, 2, \ldots, n\} \to \mathbb{N}$ with range$(\alpha) = S$. If S is infinite, then we can fix a computable injective function $\alpha \colon \mathbb{N} \to \mathbb{N}$ with range$(\alpha) = S$.

We build our computable SCFFD A in stages, starting by letting $A_0 = \mathbb{Z}$ and letting $D_0(a)$ be the finite set of divisors of a for all $a \in \mathbb{Z} \backslash \{0\}$. Suppose that we are at a stage k and have constructed an SCFFD A_k together with witnessing function D_k. We now extend A_k to A_{k+1} by destroying the primality of $p_{\alpha(k)}$ as in the construction of Theorem 3.1 using a new indeterminate x_k. In other words, we let A_{k+1} be the subring of $A_k[x_k]$ consisting of those polynomials whose coefficient of x_k is divisible by $p_{\alpha(k)}$ in A_k. We continue this process through the construction of A_n if $|S| = n$, and infinitely often if S is infinite. Using Theorem 3.1, the following properties hold by induction on k:

- A_k is an SCFFD with witnessing function D_k extending D_i for all $i < k$.
- $U(A_k) = \{1, -1\}$.
- Every p_i is irreducible in A_k.
- p_i is prime in A_k if and only if $i \notin \{\alpha(1), \alpha(2), \ldots, \alpha(k)\}$.

Now if S is finite, say $|S| = n$, then it follows that the integral domain A_n has the required properties.

Suppose then that S is infinite, and let $A = A_\infty = \bigcup_{k=0}^\infty A_k$. Also, let $D = \bigcup_{k=1}^\infty D_k$, which makes sense because the D_i extend each other as functions. Notice that D is a computable function and that for any $a \in A_k$, the set of divisors of a in A equals the set of divisors of a in A_k, so $D(a) = D_k(a)$ is the finite set of divisors of a in A. Therefore, A is an SCFFD as witnessed by D. Since $U(A_k) = \{1, -1\}$ for all $k \in \mathbb{N}$, it follows that $U(A) = \{1, -1\}$. Since we maintain the units and divisibility at each stage, it also follows that every p_i is irreducible in A.

We now show that p_i is prime in A if and only if $i \notin S$. First notice that each p_i is nonzero and not a unit of A.

- Suppose first that $i \notin S$. We then have that $i \notin \text{range}(\alpha)$, so p_i is prime in every A_k by the last property above. Let $a, b \in A$, and suppose that $p_i \mid ab$ in A. Fix $c \in A$ with $p_i c = ab$. Go to a point k where each of p_i, a, b, c exist. We then have that $p_i \mid ab$ in A_k, so as p_i is prime in A_k, either $p_i \mid a$ in A_k or $p_i \mid b$ in A_k. Therefore, either $p_i \mid a$ in A or $p_i \mid b$ in A. It follows that p_i is prime in A.
- Suppose now that $i \in S$. Thus, we can fix $k \in \mathbb{N}$ with $\alpha(k) = i$. We then have that p_i is not prime in A_{k+1} by the last property above. Fix $a, b \in A_{k+1}$ such that $p_i \mid ab$ in A_{k+1} but $p_i \nmid a$ in A_{k+1} and $p_i \nmid b$ in A_{k+1}. Since the D_i extend each other as functions, and A is an SCFFD as witnessed by D, it follows that $p_i \mid ab$ in A but $p_i \nmid a$ in A and $p_i \nmid b$ in A. Therefore, p_i is not prime in A. $\square$

Suppose that S is an infinite Σ_1^0 set. In the above proof, the ring A is constructed in infinitely many stages, where we add a new indeterminate as each element enters S. Tracing through the construction, A can be viewed as the subring of $\mathbb{Z}[x_1, x_2, x_3, \ldots]$ consisting of those polynomials with the property that whenever x_i appears raised to the first power within a monomial, the corresponding coefficient must be divisible by $p_{\alpha(i)}$.

In the previous theorem, we coded an arbitrary Σ_1^0 set into the set of primes of a computable integral domain. In fact, if S is the given Σ_1^0 set, then $\{i : p_i$ is prime in $A\}$ is Turing equivalent to S. However, notice the set of prime elements of A might have strictly larger Turing degree that S, because A will have many prime elements other than $\{p_i : i \notin S\}$. Nonetheless, the fact that we can code arbitrary Σ_1^0 sets into the primes of A yields the following result.

Corollary 3.3. *There exists a computable integral domain A such that the set of irreducible elements of A is computable but the set of prime elements of A is not computable.*

Proof. Fix a noncomputable Σ_1^0 set S, and let A be the SCFFD given by Theorem 3.2. Since A is an SCFFD, it is a computable integral domain and the set of irreducible elements of A is computable. However, the set of prime elements of A is not computable, because if we could compute it, then we could compute S, which is a contradiction. □

4 Primes Computable and Irreducibles Noncomputable

Consider the subring $A = \mathbb{Z} + x\mathbb{Z} + x^2\mathbb{Q}[x]$ of $\mathbb{Q}[x]$. In other words, A is the set of polynomials of the form $q_0 + q_1 x + q_2 x^2 + \cdots + q_n x^n$ where $q_0 \in \mathbb{Z}$ and $q_1 \in \mathbb{Z}$. As mentioned in the introduction, each normal integer prime is irreducible in A but is not prime in A. It is also a standard fact for $p(x) \in A$, we have that $p(x)$ is prime in A if and only if $p(x)$ is irreducible in $\mathbb{Q}[x]$ and $p(0) \in \{1, -1\}$.

We will generalize this construction by replacing $\mathbb{Z}$ with an arbitrary integral domain. Suppose that R is an integral domain, and let F be its field of fractions. Consider the subring $A = R + xR + x^2 F[x]$ of $F[x]$, i.e. A is the set of polynomials of the form $q_0 + q_1 x + q_2 x^2 + \cdots + q_n x^n$ where $q_0 \in R$ and $q_1 \in R$. Such an integral domain A is particularly nice from our perspective because the irreducibles in R will remain irreducible in A (so all of the complexity of irreducibles remain), but no element of R is prime in A (so any complexity of primes is "erased"). Moreover, we can reduce the complexity of primality of elements of A to that of irreducibles in the polynomial ring over a field, about which a great deal is understood.

Lemma 4.1. *Let R be an integral domain with field of fractions F. Consider the subring $A = R + xR + x^2 F[x]$ of $F[x]$. Let $p(x) \in A$. If $p(x)$ is prime in A, then $p(x)$ is non-constant and irreducible in $F[x]$.*

Proof. We prove the contrapositive, i.e. if $p(x) \in A$ is either constant or not irreducible, then $p(x)$ is not prime in A.

Suppose first that $p(x)$ is a constant, and fix $k \in R$ with $p(x) = k$. If $k \in \{0\} \cup U(R)$, then k is either zero or a unit, so k is not prime in A by definition. Suppose then that $k \notin \{0\} \cup U(R)$. Notice that $k \mid x^2$ in A because $\frac{1}{k} \cdot x^2 \in A$, but $k \nmid x$ in A because $\frac{1}{k} \cdot x \notin A$. Therefore, $p(x) = k$ is not prime in A.

Suppose now that $p(x) \in A$ is non-constant and not irreducible in $F[x]$. Since $p(x)$ is non-constant, it is not a unit in $F[x]$. Fix $g(x), h(x) \in F[x]$ with $p(x) = g(x)h(x)$ and such that $0 < \deg(g(x)) < \deg(p(x))$ and $0 < \deg(h(x)) < \deg(p(x))$. Now since $g(x), h(x) \in F[x]$, the constant terms and coefficients of x in these polynomials need not be in R. Let b be the product of the denominators of these coefficients in $g(x)$, and let c be the product of the denominators of these coefficients in $h(x)$. We then have that $p(x) \cdot bc = (b \cdot g(x)) \cdot (c \cdot h(x))$ where both $b \cdot g(x) \in A$ and $c \cdot h(x) \in A$. Since $bc \in R \subseteq A$, we have $p(x) \mid (b \cdot g(x)) \cdot (c \cdot h(x))$ in A. However, notice that $p(x) \nmid b \cdot g(x)$ in A because $\deg(b \cdot g(x)) < \deg(p(x))$ and $p(x) \nmid c \cdot h(x)$ because $\deg(c \cdot h(x)) < \deg(p(x))$. Therefore, $p(x)$ is not prime in A. □

Lemma 4.2. *Let R be an integral domain with field of fractions F. Consider the subring $A = R + xR + x^2 F[x]$ of $F[x]$. Let $p(x) \in A$ and suppose that $p(x)$ is irreducible in $F[x]$. The following are equivalent.*

(1) $p(x)$ is prime in A.
(2) For all $f(x) \in F[x]$, if $p(x)f(x) \in A$, then $f(x) \in A$.
(3) For all $g(x) \in A$ such that $p(x) \mid g(x)$ in $F[x]$, we have $p(x) \mid g(x)$ in A.

Proof. $(1) \to (2)$: Suppose first that $p(x)$ is prime in A. We know that no constants are prime in A from above, so $p(x)$ is non-constant. Let $f(x) \in F[x]$ be such that $p(x)f(x) \in A$. We prove that $f(x) \in A$. Write $f(x) = q_0 + q_1 x + \cdots + q_n x^n$ where each $q_i \in F$. Let d be the product of the denominators of q_0 and q_1. Now $d \in R \subseteq A$ and $d \cdot f(x) \in A$, hence $p(x) \mid p(x) \cdot d \cdot f(x)$ in A, i.e. $p(x) \mid d \cdot (p(x)f(x))$ in A. Since $p(x)$ is prime in A, either $p(x) \mid d$ in A or $p(x) \mid p(x)f(x)$ in A. The former is impossible because $p(x)$ is non-constant, so we must have that $p(x) \mid p(x)f(x)$ in A. Fix $h(x) \in A$ with $p(x)h(x) = p(x)f(x)$. Since $F[x]$ is an integral domain, we conclude that $f(x) = h(x) \in A$.

$(2) \to (3)$: Immediate.

$(3) \to (1)$: Let $g(x), h(x) \in A$ and suppose that $p(x) \mid g(x)h(x)$ in A. Since A is a subring of $F[x]$, we then have that $p(x) \mid g(x)h(x)$ in $F[x]$. Now $p(x)$ is irreducible in $F[x]$, so since $F[x]$ is a UFD, we know that $p(x)$ is prime in $F[x]$. Thus, either $p(x) \mid g(x)$ in $F[x]$ or $p(x) \mid h(x)$ in $F[x]$. Using (3), we conclude that either $p(x) \mid g(x)$ in A or $p(x) \mid h(x)$ in A. Therefore, $p(x)$ is prime in A. $\square$

Proposition 4.3. *Let R be an integral domain that is not a field, and let F be its field of fractions. Consider the subring $A = R + xR + x^2 F[x]$ of $F[x]$. An element $p(x) \in A$ is prime in A if and only if $p(x)$ is irreducible in $F[x]$ and $p(0) \in U(R)$.*

Proof. We first prove that if $p(x) \in A$ does not satisfy $p(0) \in U(R)$, then $p(x)$ is not prime in A. If $p(0) = 0$, then fixing any nonzero nonunit $b \in R$ (which exists because R is not a field), we have $p(x) \cdot \frac{x}{b} \in A$ but $\frac{x}{b} \notin A$, so $p(x)$ is not prime in A by Lemma 4.2. Suppose then that $p(0) \notin \{0\} \cup U(R)$. Write $p(x) = q_n x^n + \cdots + q_2 x^2 + ax + b$ where $a, b \in R$ and $b \notin \{0\} \cup U(R)$. We have

$$p(x) \cdot \left(\frac{1}{b} \cdot x \right) = (q_n x^n + \cdots + q_2 x^2 + ax + b) \cdot \left(\frac{1}{b} \cdot x \right)$$

$$= \left(\frac{q_n}{b} \right) \cdot x^{n+1} + \cdots + \left(\frac{q_2}{b} \right) \cdot x^3 + \left(\frac{a}{b} \right) \cdot x^2 + x$$

Thus, $f(x) \cdot \frac{1}{b} \cdot x \in A$ but $\frac{1}{b} \cdot x \notin A$, so $f(x)$ is not prime in A by Lemma 4.2.

We have just shown that if $p(x) \in A$ is prime in A, then $p(0) \in U(R)$. We also know that if $p(x) \in A$ is prime in A, then $p(x)$ is irreducible in $F[x]$ by Lemma 4.1.

Suppose conversely that $p(x)$ is irreducible in $F[x]$ and that $p(0) \in U(R)$. Using Lemma 4.2, to show that $p(x)$ is prime in A it suffices to show that whenever $f(x) \in F[x]$ is such that $p(x)f(x) \in A$, then we must have $f(x) \in A$. Suppose then that $f(x) \in F[x]$ and $p(x)f(x) \in A$. Write

$$f(x) = q_0 + q_1 x + q_2 x^2 + \cdots + q_n x^n$$
$$p(x) = a_0 + a_1 x + r_2 x^2 + \cdots + r_n x^n$$

where $a_0 \in U(R)$, $a_1 \in R$, each $q_i \in F$, and each $r_i \in F$. We then have that $p(x)f(x) \in F[x]$ with

$$p(x)f(x) = q_0 a_0 + (q_0 a_1 + a_0 q_1)x + \ldots$$

As $p(x)f(x) \in A$, we know that $q_0 a_0 \in R$ and $q_0 a_1 + a_0 q_1 \in R$. Since $q_0 a_0 \in R$ and $a_0 \in U(R)$, it follows that $q_0 \in R$. Using this together with the facts that $a_1 \in R$ and $q_0 a_1 + a_0 q_1 \in R$, it follows that $a_0 q_1 \in R$. Applying again the fact that $a_0 \in U(R)$, we conclude that $q_1 \in R$. Since $q_0, q_1 \in R$, it follows that $p(x) \in A$. □

With these results in hand, we now proceed to construct an integral domain R with a complicated set of irreducible elements. We will want our R to have a "nice" field of fractions F in the sense that the irreducibles of $F[y]$ will be computable.

Lemma 4.4. *Let S be a Σ_1^0 set, and let $p_0, p_1, p_2, \ldots$ list the usual primes from $\mathbb{N}$ in increasing order. There exists a computable UFD R such that:*

- *$\mathbb{Z}$ is a subring of R, and in fact*

$$\mathbb{Z}[x_1, x_2, \ldots] \subseteq R \subseteq \mathbb{Q}(x_1, x_2, \ldots),$$

  where there are infinitely many indeterminates if S is infinite, and exactly n of them if $|S| = n$.
- *$U(R) = \{1, -1\}$.*
- *p_i is irreducible in R if and only if $i \notin S$.*

Proof. If $S = \emptyset$, this is trivial by letting $A = \mathbb{Z}$. Assume then that $S \neq \emptyset$. If S is finite, say $|S| = n$, then we can trivially fix a computable injective function $\alpha\colon \{1, , 2 \ldots, n\} \to \mathbb{N}$ with range$(\alpha) = S$. If S is infinite, then we can fix a computable injective function $\alpha\colon \mathbb{N} \to \mathbb{N}$ with range$(\alpha) = S$.

We build our computable UFD R in stages, starting by letting $R_0 = \mathbb{Z}$. Suppose that we are at a stage k and have constructed through the integral domain R_k. We now destroy the irreducibility of $p_{\alpha(k)}$ by letting $R_{k+1} = R_k[x_k, \frac{p_{\alpha(k)}}{x_k}]$ as in [4, Sect. 3]. We continue this process through the construction of R_{n+1} if $|S| = n$, and infinitely often if S is infinite. Using [4, Proposition 3.3 and Theorem 3.10], the following properties hold by induction on k:

- R_k is a Noetherian UFD.
- $\mathbb{Z}[x_1, x_2, \ldots, x_k] \subseteq R_k \subseteq \mathbb{Q}(x_1, x_2, \ldots, x_k)$.
- $U(R_k) = \{1, -1\}$.
- p_i is irreducible in R_k if and only if $i \notin \{\alpha(1), \alpha(2), \ldots, \alpha(k)\}$.

Now if S is finite, say $|S| = n$, then it follows that the integral domain R_n has the required properties.

Suppose then that S is infinite, and let $R = R_\infty = \bigcup_{k=0}^{\infty} R_k$. We then have that R has the required properties by the proofs in [4, Sect. 4] (although they are significantly easier in this case because we never change the units).   □

Theorem 4.5. *Let S be a Σ_1^0 set, and let $p_0, p_1, p_2, \ldots$ list the usual primes from $\mathbb{N}$ in increasing order. There exists a computable integral domain A such that:*

- *$\mathbb{Z}$ is a subring of A.*
- *$U(A) = \{1, -1\}$.*
- *No p_i is prime in A.*
- *The set of prime elements of A is computable.*
- *p_i is irreducible in A if and only if $i \notin S$.*

Proof. Let R be the integral domain given by Lemma 4.4. Let F be the field of fractions of R. Since

$$\mathbb{Z}[x_1, x_2, \ldots] \subseteq R \subseteq \mathbb{Q}(x_1, x_2, \ldots)$$

(where there are infinitely many indeterminates if S is infinite, and exactly n of them if $|S| = n$) and the field of fractions of $\mathbb{Z}[x_1, x_2, \ldots]$ is $\mathbb{Q}(x_1, x_2, \ldots)$, it follows that $F = \mathbb{Q}(x_1, x_2, \ldots)$. Let A be the subring $R + yR + y^2 F[y]$ of $F[y]$. Now we clearly have that $\mathbb{Z}$ is a subring of A and $U(A) = \{1, -1\}$. Also, each p_i is a constant polynomial in A, so is not prime in A by Lemma 4.1. By [5, Theorem 4.5], the set of irreducible elements of $F[y]$ is computable, so since $U(R) = \{1, -1\}$, we may use Proposition 4.3 to conclude that the set of prime elements of A is computable.

Finally, by Lemma 4.4, we have that p_i is irreducible in R if and only if $i \notin S$. Now R is the subring of A consisting of the constant polynomials, so as $U(A) = U(R)$ and divisors of the constant polynomials in A must be constants, it follows that p_i is irreducible in A if and only p_i is irreducible in R, which is if and only if $i \notin S$.   □

Corollary 4.6. *There exists a computable integral domain A such that the set of prime elements of A is computable but the set of irreducible elements of A is not computable.*

Proof. Fix a noncomputable Σ_1^0 set S, and let A be the integral domain given by Theorem 4.5. Then the set of prime elements of A is computable. However, the set of irreducible elements of A is not computable, because if we could compute it, then we could compute S, which is a contradiction.   □

References

1. Anderson, D.D., Anderson, D.F., Zafrullah, M.: Factorization in integral domains. J. Pure Appl. Algebra **69**(1), 1–19 (1990)
2. Anderson, D.D., Anderson, D.F., Zafrullah, M.: Factorization in integral domains. II. J. Algebra **152**(1), 78–93 (1992)
3. Anderson, D.D., Mullins, B.: Finite factorization domains. Proc. Am. Math. Soc. **124**(2), 389–396 (1996)
4. Dzhafarov, D., Mileti, J.: The complexity of primes in computable UFDs. Notre Dame J. Form. Log. To appear
5. Fröhlich, A., Shepherdson, J.C.: Effective procedures in field theory. Philos. Trans. R. Soc. Lond. Ser. A. **248**, 407–432 (1956)
6. Metakides, G., Nerode, A.: Effective content of field theory. Ann. Math. Log. **17**(3), 289–320 (1979)
7. Miller, R.: Computable fields and Galois theory. Not. Am. Math. Soc. **55**(7), 798–807 (2008)
8. Rabin, M.: Computable algebra, general theory and theory of computable fields. Trans. Am. Math. Soc. **95**, 341–360 (1960)
9. Stoltenberg-Hansen, V., Tucker, J.V.: Computable Rings and Fields. Handbook of Computability Theory. Elsevier, Amsterdam (1999)

Revisiting Uniform Computable Categoricity: For the Sixtieth Birthday of Prof. Rod Downey

Russell Miller[1,2(✉)]

[1] Queens College – C.U.N.Y., 65-30 Kissena Blvd., Queens, NY 11367, USA
`Russell.Miller@qc.cuny.edu`
[2] Graduate Center of C.U.N.Y., 365 Fifth Avenue, New York, NY 10016, USA
`http://qcpages.qc.cuny.edu/~rmiller`

Abstract. Inspired by recent work of Csima and Harrison-Trainor and of Montalbán in relativizing the notion of degrees of categoricity, we return to uniform computable categoricity, as described in work of Downey, Hirschfeldt and Khoussainov. Our attempt to integrate these notions together leads to certain new questions about relativizing the concept of the jump of a structure, as well as to an idea of the structural information content of a countable structure, i.e., that information which can be recovered uniformly from copies of the structure.

1 Rod

For certain mathematicians, a sixtieth-birthday conference is mainly an opportunity to reflect on the body of their work and to start to view it as a whole. This is particularly true if one believes them to have mostly completed that work. Rod Downey, on the other hand, shows no signs whatsoever of slowing down, and one can hardly think of his oeuvre as completed when he keeps on churning out one paper after another. For Rod's sixtieth birthday, therefore, it seems more appropriate to try to create a present to give him. Once again, this is no easy task. However, the recent work of Csima and Harrison-Trainor on degrees of categoricity "on a cone" suggested connections to work by Rod, joint with Denis Hirschfeldt and Bakh Khoussainov in 2003, on uniform versions of computable categoricity. This paper is an attempt to integrate those two concepts together: the goal is not necessarily to produce a fully formed result, but rather to inspire questions which can serve as a birthday present, giving Rod and others something to play with. As with any birthday present, the author felt the need to play with it a bit himself first – just to test it out, of course – and so some theorems will be stated, along with examples, but even these serve mainly to illustrate the important points and to raise further questions, rather than to resolve them.

Happy birthday, Rod!

R. Miller—The author was supported by Grant # DMS – 1362206 from the N.S.F., and by several grants from the PSC-CUNY Research Award Program.

A. Day et al. (Eds.): Downey Festschrift, LNCS 10010, pp. 254–270, 2017.
DOI: 10.1007/978-3-319-50062-1_18

2   Introduction

The notion of *computable categoricity* has become absolutely standard in computable model theory. A computable structure $\mathcal{A}$ is computably categorical if every computable structure $\mathcal{B}$ isomorphic to $\mathcal{A}$ is computably isomorphic to $\mathcal{A}$. This does not mean that all isomorphisms between $\mathcal{A}$ and $\mathcal{B}$ need be computable, of course; but it implies that, to determine whether or not $\mathcal{A}$ is isomorphic to an arbitrary computable structure $\mathcal{C}$, one need look no further than the computable functions to determine whether an isomorphism exists.

Computable model theorists have modified this definition in a number of ways. *Uniform computable categoricity* was promulgated, in two different versions, by Downey, Hirschfeldt, and Khoussainov in [5], and earlier by Kudinov [9,10] and Ventsov [17]. For this property, we require not only that the computable isomorphism between the computable isomorphic structures $\mathcal{A}$ and $\mathcal{B}$ must exist, but that there must be an effective method of finding it. The main version demands a Turing functional Γ which, given the (computable) atomic diagrams of $\mathcal{A}$ and $\mathcal{B}$ as an oracle, always computes an isomorphism from $\mathcal{A}$ onto $\mathcal{B}$; this version is equivalent (and very similar) to our Definition 1 below. A weaker version, in [5,9,10], demands a computable function f which, given any i and j such that φ_i and φ_j compute the atomic diagrams of $\mathcal{A}$ and $\mathcal{B}$, outputs the index e of a computable isomorphism φ_e from $\mathcal{A}$ onto $\mathcal{B}$. One of the surprises of [5] was that these notions turned out to be distinct: the first one always implies the second, of course, but not vice versa.

Relative computable categoricity of $\mathcal{A}$ broadens the original definition in a different way, by extending it to all structures $\mathcal{B}$ on the domain ω, whether or not they are computable. Of course, requiring a computable isomorphism to map the computable structure $\mathcal{A}$ onto a noncomputable $\mathcal{B}$ would be untenable. Rather, we say that $\mathcal{A}$ is relatively computably categorical if, for every $\mathcal{B}$ with domain ω which is isomorphic to $\mathcal{A}$, there exists a $\mathcal{B}$-computable isomorphism from $\mathcal{A}$ onto $\mathcal{B}$. (It follows that, for every $\mathcal{B}$ and $\mathcal{C}$ isomorphic to $\mathcal{A}$, there is a $(\mathcal{B} \oplus \mathcal{C})$-computable isomorphism from $\mathcal{B}$ onto $\mathcal{C}$.) This version has been shown, in [1] and independently in [2], to be equivalent to a syntactic characterization using computable infinitary formulas, the *Scott family*, which we describe below. Moreover, relative computable categoricity of $\mathcal{A}$ is equivalent to the existence of a finite tuple of elements $\boldsymbol{a}$ from $\mathcal{A}$ such that $(\mathcal{A}, \boldsymbol{a})$ is uniformly computably categorical. However, it was soon shown, for instance in [9], that a computable structure can be computably categorical without being relatively computably categorical.

Finally, for computable structures which fail these criteria, we can ask how close they come to satisfying them. For example, for a computable ordinal α, a computable structure $\mathcal{A}$ is said to be *relatively $\Delta_{1+\alpha}$ -categorical* if, for every $\mathcal{B}$ isomorphic to $\mathcal{A}$ with domain ω, there exists an isomorphism from $\mathcal{A}$ onto $\mathcal{B}$ which is computable from the α-th jump of the degree of $\mathcal{B}$. (The irritating use of $(1 + \alpha)$ is necessary to make this definition work for both finite and infinite ordinals α.) This too has a very pleasing syntactic characterization, by computable enumerability of a Scott family of $\Sigma^c_{\alpha+1}$-formulas. Plain *$\Delta_{1+\alpha}$-computable cate-*

goricity is defined by analogy, restricting the relative definition to computable structures $\mathcal{B}$ only, and under this restriction to computable structures, a further generalization is explored in [7]: *d-computable categoricity*, in which all computable copies $\mathcal{B}$ of $\mathcal{A}$ are required to have isomorphisms from $\mathcal{A}$ which are computable from the Turing degree d. When $d = \mathbf{0}^{(\alpha)}$, this is just $\Delta_{1+\alpha}$-computable categoricity, but the generalization to the relative version does not work smoothly when d is not of the form $\mathbf{0}^{(\alpha)}$.

The work [7] explored the possibility of a computable structure having a specific *degree of categoricity*, i.e., having a least degree d such that $\mathcal{A}$ is d-computably categorical. Degrees of categoricity were shown there to include all c.e. and d.c.e. degrees, as well as degrees of the form $\mathbf{0}^{(\alpha)}$ with $\alpha \leq \omega$. The results there were nicely extended in [3], to all $\alpha < \omega_1^{CK}$, but the papers which largely inspired our approach here were [4,14]. In the first of these, Csima and Harrison-Trainor showed that every computable structure has a specific level of categoricity: relative to some fixed degree d, its degree of categoricity is precisely some jump $d^{(\alpha)}$ of d. (In their language, a structure will have degree of categoricity $\mathbf{0}^{(\alpha)}$ *on the cone above* d, i.e., with all definitions relativized to d.) The results here through Sect. 3, and some of those beyond that section, are mostly implicit in their work and [14], if not explicitly stated there. Our goal, in addition to calling attention to their work, is to show how it can be integrated together with the notions of uniform computable categoricity.

With such a glut of definitions on hand, the newcomer to the subject may feel somewhat dazed. Nevertheless, each of these definitions arises out of reasonable questions. Here, to justify extending one of these definitions even further (below), we offer an example of a shortcoming in the foregoing catalogue, using two computable fields E and F.

Our E is well-known: it is simply the algebraic closure of the purely transcendental extension $\mathbb{Q}(t_0, t_1, \ldots)$ of the rational numbers. Thus, E is the unique countable algebraically closed field of characteristic 0 with infinite transcendence degree over its prime field, and this field is well-known to be computably presentable. Ershov was the first to show that E is not computably categorical (see [6]). Indeed, there are computable presentations in which the algebraic dependence set

$$\{(x_0, \ldots, x_n) \in E^{<\omega} : (x_0, \ldots, x_n) \text{ is algebraically dependent over } \mathbb{Q}\}$$

can have arbitrary computably enumerable Turing degree, whereas a computable isomorphism between computable copies of E must preserve the Turing degree of this set. E is relatively Δ_2-categorical, however, since, for an arbitrary copy K of E, one can use a $(\deg(K))'$-oracle to pick out a transcendence basis in K and another in E (since $\mathbf{0}' \leq (\deg(K))'$), and every bijection between these bases extends effectively to an isomorphism from K onto E.

Our F requires a little more description, and uses the computably enumerable set $\emptyset'$, the Halting Problem. Let $p_0 < p_1 < \cdots$ enumerate the prime numbers $2 < 3 < \cdots$. F contains two square roots (arbitrarily named $\pm\sqrt{p_n}$) of each prime p_n. We now give a simplified version of the process for one number n.

First, F also contains a square root of $+\sqrt{p_n}$. If $n \in \emptyset'$, then we adjoin a fourth root of $-\sqrt{p_n}$; in this case, both of $\pm\sqrt{p_n}$ have square roots, of course, but $+\sqrt{p_n}$ has no fourth root in F. If $n \notin \emptyset'$, then no such fourth root is ever adjoined, so $+\sqrt{p_n}$ has a square root of its own, whereas $-\sqrt{p_n}$ does not. So in both cases, the elements $\pm\sqrt{p_n}$ are in distinct orbits under automorphisms of F, but the reason for the distinction depends on whether $n \in \emptyset'$ or not.

Unfortunately, this exact procedure cannot be used for every n: once a square root of $-\sqrt{p_m}$ has been adjoined for one m, F will contain a square root of -1, and therefore any subsequent square root of any other $+\sqrt{p_n}$ would generate a square root of $-\sqrt{p_n}$ as well. However, one can follow the same plan used in [11], to give a process which accomplishes the same purpose for each single prime p_n without any interference between them. Start by adjoining $\pm\sqrt{p_n}$ to F for every n, and use the polynomials given in [11, Proposition 2.15] to "tag" them, as follows. First, picking one polynomial h_n (of a new prime degree) from that proposition, adjoin one root of $h_n(+\sqrt{p_n}, Y)$ to F. Then, if ever m enters $\emptyset'$, adjoin a root of $h_n(-\sqrt{p_n}, Y)$ to F; moreover, pick a new polynomial g_n (just like h_n, but of a new prime degree) from the proposition, and adjoin one root of $g_n(-\sqrt{p_n}, Y)$ to F. As long as all these g_n and h_n are chosen with distinct prime degrees, no extraneous roots of any of them will ever appear, as shown in Proposition 2.15 of [11], and so the procedure here will succeed. The root of $h_n(+\sqrt{p_n}, Y)$ is called the "initial tag" of $+\sqrt{p_n}$. If later n enters $\emptyset'$, the root of $h_n(-\sqrt{p_n}, Y)$ is the "balancing tag," and then the root of $g_n(-\sqrt{p_n}, Y)$ is the "secondary tag" of $-\sqrt{p_n}$.

Since $\emptyset'$ is computably enumerable, one can give a computable presentation of F in exactly this manner. However, there is another computable presentation $\widetilde{F} \cong F$ (in which we name the primes $\tilde{p}_n$, for clarity). Here again $+\sqrt{\tilde{p}_n}$ always has two square roots of its own, but if $n \in \emptyset'$, we adjoin both the initial tag and the secondary tag to $+\sqrt{\tilde{p}_n}$, with $-\sqrt{\tilde{p}_n}$ having only a balancing tag in $\widetilde{F}$. Therefore, the two fields are isomorphic, but each isomorphism f from F onto $\widetilde{F}$ must satisfy

$$f(+\sqrt{p_n}) = \begin{cases} +\sqrt{\tilde{p}_n}, & \text{if } n \notin \emptyset'; \\ -\sqrt{\tilde{p}_n}, & \text{if } n \in \emptyset'. \end{cases}$$

It follows that every such isomorphism f computes $\mathbf{0'}$.

On the other hand, this field F is relatively Δ_2-categorical. Given any field K isomorphic to F, we can use a $(\deg(K))'$-oracle to compute $\emptyset'$. Then, for each $n \in \emptyset'$, we wait until a secondary tag of one of $\pm\sqrt{p_n}$ appears in K. When we find it, we map it to the secondary tag of $-\sqrt{p_n}$ in F. For each $n \notin \emptyset'$, no secondary tags of $\pm\sqrt{p_n}$ will ever appear, and we simply find an initial tag of one of $\pm\sqrt{p_n}$ in K and map it to the initial tag of $+\sqrt{p_n}$ in F. Since these elements generate all of K, we can now extend our isomorphism effectively to all of K, proving relative Δ_2-categoricity.

None of the flavors of categoricity we have mentioned so far distinguishes E from F. Nevertheless, the proofs given here should feel different from each other: for E, the proof of relative Δ_2-categoricity made real use of the $(\deg(K))'$-oracle, whereas the proof for F only used this oracle to compute $\emptyset'$. To address this difference, in the next section, we will define yet another version of categoricity, which will distinguish these two situations. In essence it is the same definition used in [5], only allowing noncomputable structures as well as computable ones, as well as generalizing to consider $\Delta_{1+\alpha}$-categoricity for $\alpha > 0$. We believe it will strike the reader as a natural uniform version of the concept of effective categoricity.

3   Uniformly Computable Categoricity

The rationale behind the original definition of computable categoricity is standard in computable model theory, and has been used to define effective versions of many completely separate concepts as well. Roughly speaking the situation is this: we would like to investigate how difficult it is to compute isomorphisms among copies of the structure $\mathcal{A}$. Of course, the answer may be arbitrarily difficult, since (by a result of Knight in [8]) the copies of $\mathcal{A}$ themselves may be extremely difficult to compute, assuming that $\mathcal{A}$ satisfies a simple condition called automorphic non-triviality. In order to make the question about complexity of isomorphisms manageable, therefore, we restrict it: under the assumption that the copy $\mathcal{B}$ (and $\mathcal{A}$ itself) are computable structures, we ask how difficult it is to compute an isomorphism between them. This allows us to leave the structural complexity of $\mathcal{B}$ out of the question, and to focus on the difficulty of computing the isomorphisms themselves. (Requiring $\mathcal{B}$ to have domain ω is a similar restriction: it stops us from using the domain itself to encode complexity into $\mathcal{B}$. In this paper we will be able to continue to require all structures to have domain ω.)

A great deal of intriguing mathematics has arisen out of this original definition of computable categoricity, and it is certainly not our intention to disparage it. However, by reframing the question, we will be able to address the shortcoming exemplified by the example above with the fields E and F. In the definition below, we do not attempt to exclude any complexity from $\mathcal{B}$; instead, we assume that we have access to the entire atomic diagram of $\mathcal{B}$, no matter how complex it may be. The basic version of this definition was given in [5] and is shown there to be equivalent to their notion of uniform computable categoricity, and also (modulo use of parameters) to relative computable categoricity. Here we generalize first by adding an oracle X, and then (in Definition 3 below) by considering Δ_α-categoricity.

Definition 1. *In a computable language $\mathcal{L}$ with equality, a countable infinite $\mathcal{L}$-structure $\mathcal{A}$ is uniformly computably categorical if there exists a Turing functional Φ such that, for every pair of structures $\mathcal{B}$ and $\mathcal{C}$ both isomorphic to $\mathcal{A}$ (and with domains $\subseteq \omega$), the function*

$$\Phi^{\mathcal{B} \oplus \mathcal{C}} : \omega \to \omega$$

defines an isomorphism from $\mathcal{B}$ onto $\mathcal{C}$. More generally, for a subset $X \subseteq \omega$, $\mathcal{A}$ is $\deg(X)$-uniformly categorical if there is some Φ such that, in the situation above,

$$\Phi^{X \oplus \mathcal{B} \oplus \mathcal{C}} : \omega \to \omega$$

always defines an isomorphism from $\mathcal{B}$ onto $\mathcal{C}$. (Clearly this same property then holds of all sets $Y \geq_T X$.)

Finally, if there exists an $X \subseteq \omega$ for which the preceding holds, then we will call $\mathcal{A}$ continuously categorical, since the categoricity is witnessed by isomorphisms given continuously in the copies of $\mathcal{A}$.

Here the oracles $\mathcal{B}$ and $\mathcal{C}$ stand for the atomic diagrams of the structures, under some coding into ω of all atomic formulas in the language $\mathcal{L} \cup \{c_0, c_1, \ldots\}$ with a new constant c_n for each $n \in \omega$. We have momentarily allowed $\mathcal{B}$ and $\mathcal{C}$ to have domains $\subseteq \omega$, but this is immediately rectified: we have $n \in \mathrm{dom}(\mathcal{B})$ if and only if the formula $c_n = c_n$ lies in the atomic diagram of $\mathcal{B}$, and so we can decide the domain from the $\mathcal{B}$-oracle, and likewise for $\mathcal{C}$. With $\mathcal{A}$ being countably infinite, therefore, we will hereafter assume all structures to have domain ω.

Notice that this notion immediately distinguishes the fields E and F. F is $\emptyset'$-uniformly categorical, since the method given in the previous sections for computing an isomorphism onto F from an arbitrary copy $\mathcal{B}$ requires only $\emptyset'$ and $\mathcal{B}$ as oracles. On the other hand, E, the algebraically closed field of infinite transcendence degree over $\mathbb{Q}$, cannot be continuously categorical, no matter what oracle set X is used. It is not difficult to use Ershov's method, relativized to any X, to produce two X-computable copies $\mathcal{B}$ and $\mathcal{C}$ of E, one with an X-computable algebraic dependence set and the other without, and clearly no $\Phi^{X \oplus \mathcal{B} \oplus \mathcal{C}}$ could compute an isomorphism between them. Indeed, one can make the second copy have algebraic dependence set Turing equivalent to X', so X' is the degree of categoricity for X-computable copies.

One's intuition that categoricity of the field E requires precisely one jump – equivalently, one quantifier – over the atomic diagram is justified by its relative Δ_2-categoricity (along with the comments above). Indeed, relative Δ_2-categoricity without parameters will be exactly equivalent to the natural extension we now give of Definition 1. Recall first the definition of the *jump* of a structure $\mathcal{A}$, which was established by general agreement after initial work by Montalbán [12] and by Soskov and Soskova [16]. (From now on, in our notation, Σ_α^c denotes the set of computable infinitary formulas of complexity Σ_α.)

Definition 2. *For a countable structure $\mathcal{A}$ in a language $\mathcal{L}$, the jump of $\mathcal{A}$ is another structure $\mathcal{A}'$ with the same domain, functions, relations, and constants as $\mathcal{A}$, but in an expanded language $\mathcal{L}'$. This $\mathcal{L}'$ contains an additional n-ary predicate R_φ for each infinitary Σ_1^c-formula φ in the free variables $v_1, \ldots, v_n$ (for all n), and*

$$\models_{\mathcal{A}'} R_\varphi(a_1, \ldots, a_n) \iff \models_{\mathcal{A}} \varphi(a_1, \ldots, a_n).$$

This jump operation iterates through the computable ordinals. At a limit ordinal α, the result is a structure $\mathcal{A}^{(\alpha)}$ with reduct $\mathcal{A}$ in $\mathcal{L}$, but with predicates for all infinitary Σ_α^c $\mathcal{L}$-formulas (i.e., all infinitary Σ_β^c $\mathcal{L}$-formulas for all $\beta < \alpha$).

With this definition, it is now natural to extend continuous categoricity as follows. We use the ordinal $1+\alpha$ here in order to accommodate the existing system of nomenclature: Δ_2-*categorical* means that the first jump $\mathcal{A}^{(1)}$ is computably categorical, whereas Δ_ω-*categorical* means that $\mathcal{A}^{(\omega)}$ is computably categorical.

Definition 3. *A countable structure $\mathcal{A}$ is X-uniformly $\Delta_{1+\alpha}$-categorical if its α-th jump $\mathcal{A}^{(\alpha)}$ is X-uniformly categorical. If an $X \subseteq \omega$ exists for which this holds, then $\mathcal{A}$ is* continuously $\Delta_{1+\alpha}$-categorical.

It is quickly seen that the field E is uniformly (i.e., $\emptyset$-uniformly) Δ_2-categorical, using the same argument as for relative Δ_2-categoricity. Of course, F is also uniformly Δ_2-categorical; the distinction between E and F occurs with the stronger notion of $\emptyset'$-uniform categoricity, as seen earlier.

One naturally asks, given a structure $\mathcal{A}$, for the smallest ordinal α such that $\mathcal{A}$ is continuously Δ_α-categorical. This question – and also the question of the existence of such an α – is readily addressed, using the existing notion of the Scott rank of a structure.

Definition 4. *The* computable Scott rank *of a countable $\mathcal{L}$-structure $\mathcal{A}$ is the least ordinal $\alpha > 0$ such that, for every finite tuple $(a_1, \ldots, a_n)$ from $\mathcal{A}$, there exists a computable infinitary Σ_α^c $\mathcal{L}$-formula $\varphi(v_1, \ldots, v_n)$ for which, for all tuples $\boldsymbol{b} \in \mathcal{A}^n$,*

$$\models_{\mathcal{A}} \varphi(\boldsymbol{b}) \iff (\exists f \in Aut(\mathcal{A}))(\forall i \leq n) f(a_i) = b_i.$$

(That is, φ defines an orbit of n-tuples under the action of $Aut(\mathcal{A})$.) A set $\mathfrak{F}$ of Σ_β^c-formulas all satisfying this condition, such that every tuple from $\mathcal{A}^{<\omega}$ realizes at least one formula in $\mathfrak{F}$, is called a Scott family *(of rank β) for $\mathcal{A}$.*

The absolute Scott rank *of $\mathcal{A}$ is defined the same way, but with Σ_α^c replaced by Σ_α. That is, absolute Scott rank allows any $L_{\omega_1\omega}$ formula to be used, whether or not it is computable.*

We use the term *computable* Scott rank to emphasize that we only allow computable infinitary formulas. In Sect. 5, we will present some examples and questions regarding the use of arbitrary $L_{\omega_1\omega}$ formulas. It should be noted that several distinct definitions of Scott rank exist, and they do not all define the same ordinal for a single $\mathcal{A}$. Computable Scott rank is based on our needs here: it requires the individual formulas to be computable, but the family $\mathfrak{F}$ need not be given effectively. There is a connection: if $\mathcal{A}$ has an X-computably enumerable Scott family of rank α, and $X \leq_T \emptyset^{(\beta)}$, then $\mathcal{A}$ has a computably enumerable Scott family whose rank is $\max(\alpha, \beta + 1)$, built by folding the definition of X into the new Scott family.

Proposition 1. *Suppose that a computable structure $\mathcal{A}$ has computable Scott rank $\alpha + 1$, and that some Scott family of $\Sigma_{\alpha+1}^c$ formulas for $\mathcal{A}$ is X-computably enumerable. Then $\mathcal{A}$ is X-uniformly $\Delta_{1+\alpha}$-categorical.*

It is both important and difficult to get the indices correct here. First, for $\alpha = n \in \omega$, uniform Δ_n-categoricity corresponds to a Scott family of Σ_n^c formulas (since Δ_n means that we are given the $(n-1)$-st jump $\mathcal{A}^{(n-1)}$). However, in the case $\alpha = \omega$, uniform Δ_ω-categoricity means that we can compute an isomorphism from $\mathcal{A}$ onto $\mathcal{B}$, given the atomic diagrams of $\mathcal{A}^{(\omega)}$ and $\mathcal{B}^{(\omega)}$, i.e., given the Σ_n^c-diagrams of $\mathcal{A}$ and $\mathcal{B}$ uniformly for all n. Now a $\Sigma_{\omega+1}^c$-formula $\varphi(x)$ is an effective disjunction over k of formulas $\exists \boldsymbol{y}\, \psi_k(x, \boldsymbol{y})$, with each ψ_k in Π_ω. This means that, uniformly in k and $(a, \boldsymbol{a})$, we can decide whether $\models_\mathcal{A} \psi_k(a, \boldsymbol{a})$, and so, given $a \in \mathcal{A}$, we can find a formula φ with $\models_\mathcal{A} \varphi(a)$ in the X-c.e. Scott family of $\Sigma_{\omega+1}^c$ formulas. This sets up the usual argument for Scott families and categoricity, and so uniform Δ_ω-categoricity corresponds to a Scott family of $\Sigma_{\omega+1}^c$ formulas, just as stated in the Proposition with $\alpha = \omega$. This correspondence continues from $\alpha = \omega$ on up through the hyperarithmetical hierarchy: with the atomic diagram of $\mathcal{A}^{(\omega+1)}$, one can enumerate the $\Sigma_{\omega+2}^c$-statements true in $\mathcal{A}$, and thus use a Scott family of $\Sigma_{\omega+2}^c$-formulas to build an isomorphism, and so on. The Proposition states this for all α in one fell swoop, since $1 + \alpha = \alpha$ for $\alpha \geq \omega$ and $1 + \alpha = \alpha + 1$ for $\alpha < \omega$.

Proof. This is the standard use of Scott families to demonstrate categoricity. The Turing functional Φ, with oracle $X \oplus \mathcal{B}^{(\alpha)} \oplus \mathcal{C}^{(\alpha)}$, uses X to enumerate a Scott family for $\mathcal{A}$ until it finds a formula $\varphi(v_1)$ and atomic facts about $\mathcal{B}^{(\alpha)}$ (that is, $\Delta_{1+\alpha}$-facts about $\mathcal{B}$) showing that $\varphi(0)$ holds in $\mathcal{B}$. Then it searches in $\mathcal{C}^{(\alpha)}$ to find a y_0 and a tuple witnessing that $\varphi(y_0)$ holds in $\mathcal{C}$. With $\mathcal{A} \cong \mathcal{B} \cong \mathcal{C}$, the definition of Scott family shows that this search will eventually succeed, and when it does, Φ defines $\Phi^{X \oplus \mathcal{B}^{(\alpha)} \oplus \mathcal{C}^{(\alpha)}}(0) = y_0$. Next it goes backwards, finding a formula in the Scott family which holds in $\mathcal{C}$ of the tuple $(y_0, 0)$, and then finding an $x_0 \in \mathcal{B}$ such that the same formula holds of $(0, x_0)$. (If $y_0 = 0$, then $x_0 = 0$, of course.) Setting $\Phi^{X \oplus \mathcal{B}^{(\alpha)} \oplus \mathcal{C}^{(\alpha)}}(x_0) = 0$, it then proceeds to the tuple $(0, x_0, 1)$ from $\mathcal{B}$, and so on, by a back-and-forth procedure which ensures that $\Phi^{X \oplus \mathcal{B}^{(\alpha)} \oplus \mathcal{C}^{(\alpha)}}$ will be bijective and will be an isomorphism. $\qquad\square$

For $\alpha = 0$, the converse also holds.

Proposition 2. *Suppose that a computable structure $\mathcal{A}$ is X-uniformly categorical. Then $\mathcal{A}$ has an X-computably enumerable Scott family of Σ_1^0 formulas.*

Proof. This follows from the methods used in [5], relativized to the degree of X. Notice that the proof there requires $\alpha = 0$: it does not consider Δ_2-categoricity or higher. Also, we can take the formulas in the Scott family to be finitary, so there is no need to worry that an individual formula might require an X-oracle to list out its disjuncts. $\qquad\square$

One might expect this converse also to hold when $\alpha > 0$. It does, but we will first consider the example and the notions in Sect. 4. Before continuing there, we note that, in light of Propositions 1 and 2, it is more reasonable to define X-uniform categoricity for enumeration degrees, rather than for Turing degrees. All we need is an enumeration of the Scott family, and so, if we can enumerate a Scott family from an enumeration of X, then we can do the same for every set in the enumeration degree of X. A natural question now arises.

Question 1. Suppose a computable structure $\mathcal{A}$ has computable Scott rank α, as defined above: it has a Scott family of Σ^c_α-formulas, and α is least with this property. Is there a least enumeration degree $\mathbf{c}$ such that $\mathcal{A}$ is $\mathbf{c}$-continuously Δ_α-categorical? The e-degree of the Scott family itself is the obvious candidate for $\mathbf{c}$; the question really asks whether $\mathcal{A}$ could have another Scott family of Σ^c_α-formulas whose e-degree is incomparable (under $\leq_e$) with this $\mathbf{c}$.

An analogous question could be asked in the next section for structures $\mathcal{A}$ which are countable but not computably presentable.

4   Continuity for Spectra of Structures

To see how the questions and definitions above lead into questions about spectra of structures, we now introduce another countable field L. Notice first that Definition 1 applies to all countable structures, not just computable ones, and indeed our L will have no computable presentation. It is simplest to view L as a sort of reverse of F, with the roles of $\emptyset'$ and its complement $\overline{\emptyset}'$ interchanged. Like F, L contains two square roots $\pm\sqrt{p_n}$ of each prime number p_n, and also contains an initial tag of $+\sqrt{p_n}$ for every n. For those $n \notin \emptyset'$, we adjoin to L a balancing tag of $-\sqrt{p_n}$, and also a secondary tag of $-\sqrt{p_n}$. It follows that, with an oracle for an arbitrary presentation of L, we could enumerate $\overline{\emptyset}'$, simply by enumerating those n for which either of $\pm\sqrt{p_n}$ has a secondary tag, and thus could compute $\emptyset'$. Indeed, the Turing degree spectrum of L is precisely the upper cone above (and including) $\mathbf{0}'$.

The reason why L upsets our ideas about continuous categoricity is that, whereas F was only $\emptyset'$-uniformly categorical, L is uniformly computably categorical. To see this, suppose that $\widetilde{L}$ is an arbitrary copy of L with domain ω. For each n, we wait until either n enters $\emptyset'$ or a secondary tag of $\pm\sqrt{p_n}$ appears in $\widetilde{L}$. If $n \in \emptyset'$, we find an initial tag of one of $\pm\sqrt{\widetilde{p}_n}$ in $\widetilde{L}$ and map it to the corresponding initial tag in L (and map this $\pm\sqrt{\widetilde{p}_n}$ itself to $+\sqrt{p_n}$ in L). On the other hand, if we find a secondary tag $\widetilde{x}$ of $\pm\sqrt{\widetilde{p}_n}$ in $\widetilde{L}$, then we wait for a secondary tag of $-\sqrt{p_n}$ to appear in L, and map $\widetilde{x}$ to that secondary tag. In this case, the initial tag of $\pm\sqrt{\widetilde{p}_n}$ in $\widetilde{L}$ will have be balanced by a tag of $\mp\sqrt{\widetilde{p}_n}$, so we can find the balancing tag and map it to the balancing tag of $+\sqrt{p_n}$ in L. All of this can be determined from our oracles (the atomic diagrams of L and $\widetilde{L}$), and the map thus defined extends uniquely to an isomorphism from $\widetilde{L}$ onto L, since we have defined it on a generating set for $\widetilde{L}$. This proves the uniform computable categoricity of L.

This result does not contradict any previous statements, but it explains why the hypothesis of computable presentability of $\mathcal{A}$ was included in Propositions 1 and 2. Indeed, while L itself is not computably presentable, it does have a c.e. Scott family of Σ_1 formulas. For the elements $\pm\sqrt{p_n}$, this family includes all formulas saying that $\sqrt{p_n}$ has a secondary tag; it also includes, for those $n \in \emptyset'$, the formula saying that $\sqrt{p_n}$ has an initial tag. If $n \in \emptyset'$, then the formula saying that $\sqrt{p_n}$ has a secondary tag is never realized in L, and hence could have been eliminated from the Scott family, but in this case the family would no longer be c.e. On the other hand, even with these formulas included, the Scott family still allows computation of isomorphisms; unrealizable formulas clutter up the process but do not disrupt it.

We note, without going into details here, that the same process could be used with other sets in place of $\overline{\emptyset'}$. For instance, let A and B be Turing-incomparable c.e. sets. If a field J has initial tags of $+\sqrt{p_n}$ for every $n \in A \oplus \overline{B}$, and has balancing tags and secondary tags for every $n \in \overline{A} \oplus B$, then the degree spectrum of J is the upper cone above $\deg(A)$, and J has a B-c.e. Scott family of Σ_1^c formulas.

Finally, we combine two of our examples. Let $\mathcal{A}$ be the cardinal sum of the fields L (above) and F (from Sect. 2). That is, $\mathcal{A}$ is the disjoint union of these two structures, in the language of fields with one additional unary predicate R which holds of all elements of L but of no elements of F. So $\mathrm{Spec}(\mathcal{A}) = \mathrm{Spec}(F) \cap \mathrm{Spec}(L)$, which is the upper cone above $\mathbf{0}'$, and we get a Scott family $\mathfrak{F}$ of Σ_1^c formulas for $\mathcal{A}$ essentially just by taking the union of the Scott families for F and L, with obvious adjustments involving R. This $\mathfrak{F}$ is c.e. in $\emptyset'$ but not in any smaller or incomparable degree; indeed, $\mathfrak{F} \equiv_e \overline{\emptyset'}$. However, $\mathcal{A}$ is uniformly computably categorical! The process for computing isomorphisms between arbitrary copies of $\mathcal{A}$ is to use the L-side of one of the copies to enumerate $\overline{\emptyset'}$, and then to use that enumeration to enumerate $\mathfrak{F}$. Ultimately, therefore, the uniform computable categoricity of this $\mathcal{A}$ follows from the e-reduction $\mathfrak{F} \leq_e \mathrm{Th}_{\Sigma_1}(\mathcal{A})$.

Nevertheless, Proposition 3 and Theorem 1 below do *not* require an e-reduction such as $\mathfrak{F} \leq_e \mathrm{Th}_{\Sigma_1}(\mathcal{A})$; they actually show that our $\mathcal{A}$ has a c.e. Scott family of Σ_1-formulas. This family is not the $\mathfrak{F}$ described above; instead, it integrates the e-reduction $\mathfrak{F} \leq_e \mathrm{Th}_{\Sigma_1}(\mathcal{A})$ into its formulas. For each n, the new Scott family has one formula saying

if $n \in \emptyset'$, then use the appropriate $\exists$-formula on F,

and another one saying

if there exists a configuration in L showing that $n \notin \emptyset'$,
then use the other kind of $\exists$-formula on F.

So, somewhat surprisingly, we may extend Propositions 1 and 2 to all countable structures, and give the promised converse for the case $\alpha > 0$, without any use of $\mathrm{Th}_{\Sigma_\alpha^c}(\mathcal{A})$.

Proposition 3. *Fix any oracle set $X \subseteq \omega$ and any nonzero $\alpha < \omega_1^{CK}$. A countable structure $\mathcal{A}$ is X-uniformly $\Delta_{1+\alpha}$-categorical if and only if $\mathcal{A}$ has an X-c.e. Scott family of $\Sigma_{\alpha+1}^c$ formulas.*

A more precise statement is possible if we integrate e-reducibility into the notion of uniform Δ_α-categoricity. Proposition 3 follows from this version, which we now state as Theorem 1 and prove.

Definition 5. *An* enumeration *of a set* $S \subseteq \omega$ *is a set* $T \subseteq \omega$ *such that* S *is the projection of* T:

$$S = proj(T) = \{m \in \omega : (\exists n)\langle m, n \rangle \in T\}.$$

So a set S is $\boldsymbol{d}$-c.e. if and only if it has a $\boldsymbol{d}$-computable enumeration.

Theorem 1. *Fix any oracle set* $X \subseteq \omega$ *and any nonzero* $\alpha < \omega_1^{CK}$. *A countable structure* $\mathcal{A}$ *has a Scott family of* $\Sigma_{\alpha+1}^c$ *formulas which is e-reducible to* X *if and only if* $\mathcal{A}$ *satisfies:*

There exists a Turing functional Φ *such that, for all copies* $\mathcal{B} \cong \mathcal{C} \cong \mathcal{A}$ *with domain* ω *and all enumerations* Y *of* X, *the function* $\Phi^{Y \oplus \mathcal{B}^{(\alpha)} \oplus \mathcal{C}^{(\alpha)}}$ *is an isomorphism from* $\mathcal{B}$ *onto* $\mathcal{C}$.

Proof. Suppose first that $\mathcal{A}$ has such a Scott family $\mathfrak{F}$, and fix any Y as described and any $\mathcal{B} \cong \mathcal{A}$. (It is sufficient to show the second statement with $\mathcal{C} = \mathcal{A}$.) With its enumeration of X, Φ applies the given e-reduction to produce an enumeration of the Scott family $\mathfrak{F}$. From here it is standard to compute an isomorphism from $\mathcal{B}$ onto $\mathcal{A}$, going back and forth using the Scott family and the Σ_α^c-diagrams of the two structures.

Now assume Φ is a functional satisfying the second statement. We run Φ simultaneously on input 0 with each binary string $(\sigma \oplus \beta \oplus \gamma) \in 2^{<\omega}$ as oracle; moreover, whenever $\Phi^{\sigma \oplus \beta \oplus \gamma}(0)$ converges, we then run $\Phi^{\sigma \oplus \beta \oplus \gamma}(1)$ until it converges (if ever), then $\Phi^{\sigma \oplus \beta \oplus \gamma}(2)$, and so on. Thus we produce an enumeration of those tuples $(\sigma \oplus \beta \oplus \gamma, \boldsymbol{y}) \in 2^{<\omega} \times \omega^{<\omega}$ such that, for all $i < |\boldsymbol{y}|$, $\Phi^{\sigma \oplus \beta \oplus \gamma}(i) = y_i$.

Each $(\sigma \oplus \beta \oplus \gamma, \boldsymbol{y})$ in this enumeration defines a set

$$X_0 = \{a : (\exists b)\ \sigma(\langle a, b \rangle) = 1\},$$

and yields a strong index for this finite set X_0. It also includes finite initial segments β and γ of the atomic diagrams of the structures $\mathcal{B}^{(\alpha)}$ and $\mathcal{C}^{(\alpha)}$. The convergence of Φ using this oracle means that, whenever $X_0 \subset proj(Y)$ and the diagram of $\mathcal{B}$ realizes

$$\left(\bigwedge_{\beta(\varphi)=1} \varphi \right) \wedge \left(\bigwedge_{\beta(\varphi)=0} \neg\varphi \right),$$

and likewise for $\mathcal{C}$ and γ, then the map sending each $i < |\boldsymbol{y}|$ to y_i extends to an isomorphism from $\mathcal{B}$ onto $\mathcal{C}$. If m is the largest domain element of $\mathcal{B}$ mentioned in the conjunction above, and n the largest domain element of $\mathcal{C}$, then we enumerate into our e-reduction an axiom saying that if $X_0 \subseteq X$, then the following formula (with $x_0, \ldots, x_{|\boldsymbol{y}|-1}$ free) is in the Scott family:

$$\exists x_{|\boldsymbol{y}|} \cdots \exists x_{m+n+1} \left(\bigwedge_{\beta(\varphi)=1} \varphi(x_0, \ldots, x_m) \right) \wedge \left(\bigwedge_{\beta(\varphi)=0} \neg\varphi(x_0, \ldots, x_m) \right)$$

$$\wedge \left(\bigwedge_{\gamma(\psi)=1} \psi^* \right) \wedge \left(\bigwedge_{\gamma(\psi)=0} \neg\psi^* \right)$$

$$\wedge \left(\bigwedge_{|\boldsymbol{y}| \leq i < j \leq n} x_i \neq x_j \right) \wedge \left(\bigwedge_{m < i < j \leq m+1+n} x_i \neq x_j \right)$$

Here, if ψ is a sentence mentioning the domain elements $0, \ldots, n$ of $\mathcal{C}$, ψ^* is the same sentence, but with each domain element y_i replaced by the variable x_i, and with each domain element $j \notin \boldsymbol{y}$ replaced by the variable x_{m+1+j}. For φ, the replacements were simpler: each domain element i was replaced by the variable x_i. Thus, the formula defined here is $\Sigma_{\alpha+1}$ and says that $(x_0, \ldots, x_{|\boldsymbol{y}|-1})$ satisfies all the existential conditions given by β on $\mathcal{B}$ and all those given by γ on $\mathcal{C}$. (It is likely but not assumed that these conditions repeat each other; there is no danger in including conditions from β even if they are not in γ, or vice versa, but there would be a danger in excluding any of them, since these are the conditions which Φ actually checks before defining its isomorphism.) It is now clear from this definition and the conditions of the theorem that, given any enumeration of X, our e-reduction will enumerate a Scott family of $\Sigma_{\alpha+1}^c$-formulas for $\mathcal{A}$. □

In a certain sense, the reason why the field L can be uniformly computably categorical is that $\emptyset'$ is computable in every copy of L, and moreover, that this computation of $\emptyset'$ is uniform across all copies of L. This property was studied in [8], and we give it a name here, which will only be used until we can demonstrate (in Proposition 4 below) its equivalence to a known condition. The reader may wish to try to identify this known condition right now, without skipping ahead to the proposition to peek.

Definition 6. A set $S \subseteq \omega$ is uniformly intrinsically computable from a countable infinite structure $\mathcal{A}$ if there exists a Turing functional Γ such that, for every $\mathcal{B} \cong \mathcal{A}$ with domain ω, $\Gamma^{\mathcal{B}}$ computes the characteristic function of S.

Likewise, S is uniformly intrinsically computably enumerable in $\mathcal{A}$ if there exists a Turing functional Θ such that, for every $\mathcal{B} \cong \mathcal{A}$ with domain ω, the function $\Theta^{\mathcal{B}}$ has domain S.

Clearly S is uniformly intrinsically computable from $\mathcal{A}$ if and only if both S and $\overline{S}$ are uniformly intrinsically c.e. in $\mathcal{A}$. The latter of these two properties will in fact be more natural and relevant; it is the property well-known to the Bulgarian school of computable model theory, where the collection of sets uniformly intrinsically c.e. in $\mathcal{A}$ is called the co-spectrum of $\mathcal{A}$. The following result was proven by Knight in [8].

Proposition 4. A set $S \subseteq \omega$ is uniformly intrinsically c.e. in a countable structure $\mathcal{A}$ if and only if S is e-reducible to the existential theory $Th_{\Sigma_1}(\mathcal{A})$ of $\mathcal{A}$.

Consequently, S is uniformly intrinsically computable in $\mathcal{A}$ if and only if both S and $\overline{S}$ are e-reducible to $\mathrm{Th}_{\Sigma_1}(\mathcal{A})$.

Proof. The backwards direction is immediate. With an oracle for any copy of $\mathcal{A}$, we can (uniformly) enumerate $\mathrm{Th}_{\Sigma_1}(\mathcal{A})$, and therefore can enumerate S, uniformly, using its e-reduction to $\mathrm{Th}_{\Sigma_1}(\mathcal{A})$.

For the forwards direction, in order to show that $S \leq_e \mathrm{Th}_{\Sigma_1}(\mathcal{A})$, we enumerate a set Ψ of axioms $(n, \exists \boldsymbol{x} \beta(\boldsymbol{x}))$ for an e-reduction. Such an axiom represents the instruction "if $\models_{\mathcal{A}} \exists \boldsymbol{x} \beta(\boldsymbol{x})$, then enumerate n." The nature of the (finitary) Σ_1-theory is such that each axiom need contain only one formula, although e-reductions in general allow us to use a finite conjunction. (One can call our Ψ an *e-reduction of norm* 1.)

Recall the basics. We have a Gödel coding $\gamma \mapsto \ulcorner \gamma \urcorner$ of atomic sentences in the language $\mathcal{L}'$, which is the language of $\mathcal{L}$ extended by new constants $c_0, c_1, \ldots$ representing elements of the domain ω. A $\mathcal{B}$-oracle is simply the subset $\{\ulcorner \beta(c_{i_0}, \ldots, c_{i_n}) \urcorner : \models_{\mathcal{B}} \beta(i_0, \ldots, i_n)\}$ of ω, and we know that whenever $\mathcal{B} \cong \mathcal{A}$, $\mathrm{dom}(\Phi^{\mathcal{B}}) = S$.

To build Ψ, we simply run $\Phi_s^\sigma(n)$ for all $n, s \in \omega$ and all $\sigma \subset 2^{<\omega}$. If this computation halts within the allotted s steps, we enumerate into Ψ the axiom

$$(n, \exists \boldsymbol{x} \beta) = \left(n, \exists \boldsymbol{x} \left(\left(\bigwedge_{\sigma(\ulcorner \gamma \urcorner)=1} \gamma_{\boldsymbol{x}}^c \right) \wedge \left(\bigwedge_{\sigma(\ulcorner \gamma \urcorner)=0} \neg \gamma_{\boldsymbol{x}}^c \right) \right) \right),$$

where $\gamma_{\boldsymbol{x}}^c$ has x_i substituted for each c_i in γ. This is less complicated than it appears: β is simply the configuration described by σ, where σ is seen as a (partial) characteristic function deeming certain atomic facts to be true and certain others to be false.

Now if $(n, \exists \boldsymbol{x} \beta) \in \Psi$, say with $\boldsymbol{x} = (x_0, \ldots, x_m)$, and if $\models_{\mathcal{A}} \exists \boldsymbol{x} \beta$, then we can easily build a structure $\mathcal{B} \cong \mathcal{A}$ whose elements $0, \ldots, m$ realize β: let $\mathcal{B}$ be the image of $\mathcal{A}$ under an isomorphism which permutes a finite subset of ω to make this happen. It follows that $n \in S$, since now $n \in \mathrm{dom}(\Phi^{\mathcal{B}})$ for this $\mathcal{B}$, and so our e-reduction Ψ only ever enumerates elements of S when we run it using an arbitrary enumeration of $\mathrm{Th}_{\Sigma_1}(\mathcal{A})$. Of course, it may happen that $(n, \exists \boldsymbol{x} \beta) \in \Psi$ yet $\not\models_{\mathcal{A}} \exists \boldsymbol{x} \beta$; but in this case the instruction $(n, \exists \boldsymbol{x} \beta) \in \Psi$ will have no effect when we use $\Psi^{\mathrm{Th}_{\Sigma_1}(\mathcal{A})}$ to enumerate S.

On the other hand, if $n \in S$, then $\Phi^{\mathcal{A}}(n)$ itself halts, after examining only a finite initial segment σ of its oracle $\mathcal{A}$, i.e., of the atomic diagram of the structure $\mathcal{A}$. Our construction of Ψ will have found this σ and will have enumerated a corresponding axiom $(n, \exists \boldsymbol{x} \beta)$ into Ψ. Since $\models_{\mathcal{A}} \exists \boldsymbol{x} \beta$, we certainly have $(\exists \boldsymbol{x} \beta) \in \mathrm{Th}_{\Sigma_1}(\mathcal{A})$, and so, when Ψ runs using any enumeration of $\mathrm{Th}_{\Sigma_1}(\mathcal{A})$, it will enumerate n. This completes the proof. $\qquad \square$

As with categoricity, this proposition reflects various concepts and facts already known about countable structures, such as relative intrinsic computable enumerability (see e.g. [13]). The obvious distinction is that here we consider information content (that is, arbitrary subsets of ω given uniformly in copies of

$\mathcal{A}$) rather than definable subsets of the structure $\mathcal{A}$ itself. One naturally asks whether this distinction is significant, but we leave that question for future study.

By way of piquing interest in uniform intrinsic computability, we recall a theorem of Richter from [15]. This theorem is usually quoted as saying that for countable infinite linear orders and for countable infinite trees (as partial orders) $\mathcal{A}$, the only possible least degree in the Turing degree spectrum of $\mathcal{A}$ is the degree $\mathbf{0}$. (Richter did not mention Boolean algebras, but her proof is quickly seen to apply to them as well.) In fact, Richter proved slightly more: that every such structure $\mathcal{A}$ has spectrum containing a minimal pair of Turing degrees, and thus the spectrum cannot be contained within any nontrivial upper cone. One might say that the structure $\mathcal{A}$ cannot *intrinsically compute* any noncomputable set.

Since uniform intrinsic computation is a form of intrinsic computation, Richter's result immediately implies the following special case as a corollary. However, our notions yield a far more direct proof.

Corollary 1. *For any countable infinite linear order, tree (viewed as a partial order), or Boolean algebra $\mathcal{A}$, only the computably enumerable sets are uniformly intrinsically computably enumerable in $\mathcal{A}$.*

Proof. Apply Proposition 4, since the existential theory of any such $\mathcal{A}$ is decidable. (For the trees, this decidability requires an application of Kruskal's Theorem – as did Richter's original result.) ☐

Uniform categoricity is in much the same spirit as past investigations into intrinsic computability, as in [1,13], for example. If the images $f(R)$ of a subset R of the domain of a countable structure $\mathcal{A}$ are computably enumerable relative to $\mathcal{B}$ under every isomorphism f from $\mathcal{A}$ onto any copy $\mathcal{B}$ with domain ω, then R must be defined in $\mathcal{A}$ by a Σ_1^c formula, possibly using finitely many parameters $\boldsymbol{a}$ from $\mathcal{A}$. The parameters create a nonuniformity, but in the structure $(\mathcal{A}, \boldsymbol{a})$, this definition yields a Turing functional Γ such that $f(R) = \mathrm{dom}(\Gamma^{(\mathcal{B}, f(\boldsymbol{a}))})$ under every isomorphism f from $\mathcal{A}$ onto any $\mathcal{B}$. That is, relative intrinsic computable enumerability is equivalent (up to those parameters) to the uniform version, i.e., to the existence of such a Γ, since the latter clearly implies the former. Proposition 4 is a natural extension of these results.

5 Noncomputable Infinitary Formulas

So far, the only infinitary formulas we have used have been computable ones, in the classes Σ_α^c (for various $\alpha < \omega_1^{CK}$). The main point of this section is to suggest that these formulas are not sufficient: we give examples of structures which would have lower levels of categoricity if one allowed certain noncomputable infinitary formulas. In the general setting of [4], working on a cone, one simply chooses the base degree of the cone to include sufficient information to be able to compute the necessary formulas. This is improved a bit further in [14]. However, it appears that our Definition 3 could be improved by adding a parameter Y, representing a Turing degree, and allowing Y-jumps (that is, jumps of structures defined by adding predicates for all Y-computable infinitary Σ_1-formulas). This section is mostly conjectural; we would welcome proofs of precise results about the examples described here.

Fix an arbitrary set $Y \subseteq \omega$, and define the following (symmetric, irreflexive) graph $\mathcal{A}_Y$. We start with a single node u, with countably many nodes z_{n0} (for all $n \in \omega$) adjacent to u. Each z_{n0} is then adjacent to z_{n1}, which is adjacent to z_{n2} and so on, so that countably many "ω-chains" are attached to u. For identification purposes, we also attach to u a single loop of length 3 (that is, we make u adjacent to one of the three nodes in this loop), and attach a loop of length $2i + 5$ to each node z_{ni}, (that is, one unique loop for each pair $\langle n, i \rangle$).

We now use loops of even length to add the desired complexity to $\mathcal{A}_Y$. Write $Y^{[n]} = \{j : \langle n, j \rangle \in Y\}$ for the n-th column of Y. To each node z_{ni}, we attach one loop of length $2j+4$ for every $j \notin Y^{[n]}$. Finally, writing $Y^{[n]} = \{k_{n0} < k_{n1} < \cdots\}$, we attach to z_{ni} a single loop of length $2k_{ni}+4$. Hereafter this one will be known as the special loop for z_{ni}.

Now $\mathcal{A}_Y$ has a Scott family of infinitary Y-computable Σ_2 formulas (in fact, Π_1 formulas). The principal difficulty is to distinguish the nodes z_{n0} for different n; everything else is well labeled by loops. (To compute an isomorphism between copies, clearly it would suffice to map the z_{n0}'s to correct images.) A Y-oracle allows us to specify exactly what loops should be attached to each z_{ni} in the n-th ω-chain. Specifically, each $z = z_{n0}$ satisfies a formula saying,

$$(\forall i)(\forall \text{ loops } L \text{ attached to } z_i \text{ in } z's \text{ chain}) \, [2 \cdot |L| + 4 = k_{ni} \text{ or } \notin Y^{[n]}],$$

along with the statements specifying that each of these z_i is connected to z by a chain of length i and is attached to a loop of length $2i+5$, and that z is adjacent to some u adjacent to a loop of length 3.

One might therefore expect $\mathcal{A}_Y$ to be Y-uniformly Δ_3-categorical. In fact, though, this can fail for certain Y. (Thanks are due to an anonymous referee for the following proof of this fact!) Let $\varphi_0, \varphi_1, \ldots$ be a list of all computable infinitary formulas. Set $\mathcal{Y}_0 = 2^\omega$. For each n, if only countably many $Z \in \mathcal{Y}_n$ have the property that $\models_{\mathcal{B}_Z} \varphi_n$, then let $\mathcal{Y}_{n+1}$ be $\mathcal{Y}_n$ with these countably many Z deleted. Otherwise, let $\mathcal{Y}_{n+1} = \mathcal{Y}_n$. By induction, every $\mathcal{Y}_n$ is co-countable, so there exists some $Z \in \cap_n \mathcal{Y}_n$, and this Z has the property that, for every n, either $\not\models_{\mathcal{B}_Z} \varphi_n$ or else uncountably many U have $\models_{\mathcal{B}_U} \varphi_n$.

Now, for each n, choose some $Z_n \neq Z$ for which

$$\models_{\mathcal{B}_{Z_n}} \varphi_n \iff \models_{\mathcal{B}_Z} \varphi_n.$$

Now set $Y = Z \oplus (\oplus_{(j,k) \in \omega^2} Z_j)$ to be the set with Z and infinitely many copies of each Z_n as its columns. We claim that $\varphi_n(x)$ cannot identify the node z_{00} at the top of the chain for Z. Indeed, if $\varphi_n(z_{00})$ holds in $\mathcal{A}_Y$, then $\varphi_n(z_{\langle n,0 \rangle+1,0})$ (the node at the top of one of the Z_n-chains) must also hold there, since $\models_{\mathcal{B}_{Z_n}} \varphi_n$ if and only if $\models_{\mathcal{B}_Z} \varphi_n$. Since $Z \neq Z_n$, this means that φ_n cannot be part of a Scott family for $\mathcal{A}_Y$: it holds of two nodes not in the same orbit under automorphisms. So in fact this $\mathcal{A}_Y$ has no Scott family of computable formulas at all, and thus cannot be continuously Δ_α-categorical for any α.

The problem with the Scott family of infinitary Y-computable Σ_2 formulas is that those formulas are not computable infinitary; they are only Y-computable infinitary. Consequently, the second jump $(\mathcal{A}_Y)''$ generally does not give information about whether a specific node satisfies such a formula or not: the predicates in the second jump of a structure only describe satisfaction of computable infinitary Σ_2-formulas.

For certain structures, one can convert a Scott family of Y-computable infinitary formulas into an Y-computably enumerable family of computable infinitary formulas, possibly of higher rank. However, the $\mathcal{A}_Y$ here in general is sufficiently complex to preclude this possibility, with the use of the Y-oracle hidden within the Π_1 part of the Σ_2 formulas. So we have here structures $\mathcal{A}_Y$ which do have absolute Scott rank 2, yet do not appear to satisfy any of our versions of continuous Δ_3-categoricity.

On the bright side, there does exist a single Turing functional Γ such that, for every $Y \subseteq \omega$ and every $\mathcal{B} \cong \mathcal{A}_Y$, the function

$$\Gamma^{(\mathcal{A}_Y \oplus \mathcal{B})''}$$

is an isomorphism from $\mathcal{A}_Y$ onto $\mathcal{B}$. With this oracle, Γ searches for some $z \in \mathcal{B}$ adjacent to the u in $\mathcal{B}$ such that, for every loop attached to every z_i below z, there is a loop of the same length attached to the same z_{ni} below z_{n0} (in $\mathcal{A}_Y$), and likewise with the roles of $\mathcal{A}_Y$ and $\mathcal{B}$ reversed. So we conjecture that using the jump(s) of the join of (the atomic diagrams of) $\mathcal{A}_Y$ and $\mathcal{B}$, rather than the join of their jump(s), may allow us to extend uniform notions of categoricity to other computably non-presentable structures.

Finally, we note that Y is not in general e-reducible to the existential theory of the structure $\mathcal{A}_Y$. Indeed, given any Y, every set U with the same columns as Y (and having each column occur with the same multiplicity) will yield an $\mathcal{A}_U$ with the same existential theory, indeed with $\mathcal{A}_U \cong \mathcal{A}_Y$. However, unless cofinitely many columns of U are all equal to each other, there will be uncountably many distinct sets U with the same columns as Y, and all but countably many of those U will have no e-reduction to the existential theory of their structures $\mathcal{A}_U$.

It is the case that each individual column $Y^{[n]}$ is decidable from the jump $(\mathcal{A}_Y)'$: the i-th element $k_{ni} \in Y^{[n]}$ has the property that there is a loop of length $2k_{ni} + 4$ attached to z_{ni} but no such loop attached to $z_{n(i+1)}$. However, this procedure is not uniform: starting this process in the jump of a copy $\mathcal{B} \cong \mathcal{A}_Y$ requires knowing the image of z_{n0} in $\mathcal{B}$, i.e., knowing which ω-chain in $\mathcal{B}$ to use.

To sum up all the loose ends in this section is a challenge, but in general they suggest that it would be useful to define a relative notion of the jump of a structure, using Y-computable infinitary Σ_1-formulas in place of computable ones. Definition 3 could likely be sharpened by using such jumps, and/or by allowing the Turing functional to use a jump of the join $(X \oplus \mathcal{A})^{(\alpha)}$ in place of the join $(X' \oplus \mathcal{A}^{(\alpha+1)})$ of their jumps. The conjectures and examples in this section make it appear that under Definition 3, there exist countable structures $\mathcal{B}$ which cannot be continuously Δ_α-categorical, no matter what α one chooses: if $\mathcal{B}$ has noncomputable Scott rank β, then the jump $\mathcal{B}^{(\beta)}$ is not even defined;

and even for structures such as many of the $\mathcal{A}_Y$ constructed here, it seems likely that no jump $(\mathcal{A}_Y)^{(\alpha)}$ with $\alpha < \omega_1^{CK}$ is continuously categorical. This contradicts one's intuition, based on the original results of Scott, that categoricity should be continuous for every countable structure, although at arbitrarily high countable levels β. So it would be natural to develop a relativized notion of the jump of a structure – presumably using X-computable infinitary Σ_1 formulas to relativize to X – and to extend our notion of uniform Δ_α-categoricity to a uniform Δ_α^X-categoricity which includes this relativization.

References

1. Ash, C.J., Knight, J.F., Manasse, M.S., Slaman, T.A.: Generic copies of countable structures. Ann. Pure Appl. Log. **42**, 195–205 (1989)
2. Chisholm, J.: On intrisically 1-computable trees. Unpublished MS
3. Csima, B.F., Franklin, J.N.Y., Shore, R.A.: Degrees of categoricity and the hyperarithmetic hierarchy. Notre Dame J. Form. Log. **54**(2), 215–231 (2013)
4. Csima, B.F., Harrison-Trainor, M.: Degrees of categoricity on a cone. Submitted for publication
5. Downey, R.G., Hirschfeldt, D.R., Khoussainov, B.: Uniformity in computable structure theory. Algebra Log. **42**, 318–332 (2003)
6. Ershov, Y.L.: Theorie der Numerierungen. Zeits. Math. Logik Grund. Math. **23**, 289–371 (1977)
7. Fokina, E., Kalimullin, I., Miller, R.G.: Degrees of categoricity of computable structures. Arch. Math. Log. **49**(1), 51–67 (2010)
8. Knight, J.F.: Degrees coded in jumps of orderings. J. Symb. Log. **51**, 1034–1042 (1986)
9. Kudinov, O.V.: An autostable 1-decidable model without a computable Scott family of ∃-formulas. Algebra Log. **35**, 255–260 (1996)
10. Kudinov, O.V.: A description of autostable models. Algebra Log. **36**, 16–22 (1997)
11. Miller, R.G.: d-Computable categoricity for algebraic fields. J. Symb. Log. **74**(4), 1325–1351 (2009)
12. Montalbán, A.: Notes on the Jump of a Structure. In: Ambos-Spies, K., Löwe, B., Merkle, W. (eds.) CiE 2009. LNCS, vol. 5635, pp. 372–378. Springer Berlin Heidelberg, Berlin, Heidelberg (2009). doi:10.1007/978-3-642-03073-4_38
13. Montalbán, A.: Rice sequences of relations. Philos. Trans. R. Soc. Lond. Ser. A Math. Phys Eng. Sci. **370**, 3464–3487 (2012)
14. Montalbán, A.: A robuster Scott rank. Proc. Am. Math. Soc. **143**, 5427–5436 (2015)
15. Richter, L.J.: Degrees of structures. J. Symb. Log. **46**, 723–731 (1981)
16. Soskova, A.A., Soskov, I.N.: A jump inversion theorem for the degree spectra. J. Log. Comput. **19**(1), 199–215 (2009)
17. Ventsov, Y.G.: Effective choice for relations and reducibilities in classes of constructive and positive models. Algebra Log. **31**, 63–73 (1992)

Enumeration Reducibility and Computable Structure Theory

Alexandra A. Soskova and Mariya I. Soskova[✉]

Faculty of Mathematics and Computer Science, Sofia University,
5 James Bourchier Blvd, 1164 Sofia, Bulgaria
{asoskova,msoskova}@fmi.uni-sofia.bg

Keywords: Enumeration reducibility · Enumeration jump · Degree spectra

2010 Mathematics Subject Classification: 03D30 · 03D45 · 03C57

1 Introduction

In classical computability theory the main underlying structure is that of the natural numbers or equivalently a structure consisting of some constructive objects, such as words in a finite alphabet. In the 1960's computability theorists saw it as a challenge to extend the notion of *computable* to arbitrary structure. The resulting subfield of computability theory is commonly referred to as *computability on abstract structures*. One approach towards this is the theory of computability in admissible sets of the hereditarily finite superstructure $\mathbb{HF}(\mathfrak{A})$ over a structure $\mathfrak{A}$. The development of computability on ordinals was initiated by Kreisel and Sacks [42,43], who investigated computability notions on the first incomputable ordinal, and then further developed by Kripke and Platek [44,58] on arbitrary admissible ordinals and by Barwise [6], who considered admissible sets with urelements. The notion of Σ-definability on $\mathbb{HF}(\mathfrak{A})$, introduced and studied by Ershov [16,17] and his students Goncharov, Morozov, Puzarenko, Stukachev, Korovina, etc., is a model of nondeterministic computability on $\mathfrak{A}$. A survey of results on $\mathbb{HF}$-computability and on abstract computability based on the notion of Σ-definability can be found in [18,95]. Montague [53] took a model theoretic approach to generalized computability theory, considering computability as definability in higher order logics.

The approach towards abstract computability that ultimately lead to the results discussed in this article starts with searching for ways in which one can identify abstract computability on a structure internally. Let $\mathfrak{A}$ be an arbitrary abstract structure. There are many different internal ways to define a class of

---

This research was supported by Sofia university Science Fund, contract 54/12.04.2016. The second author was also supported by the L'Oréal-UNESCO program "For women in science".

A. Day et al. (Eds.): Downey Festschrift, LNCS 10010, pp. 271–301, 2017.
DOI: 10.1007/978-3-319-50062-1_19

functions that can be considered as the analog of classical computable functions. Different models of computation on $\mathfrak{A}$ give rise to different classes of computable functions: $PC(\mathfrak{A})$ denotes the functions that are *prime computable* in $\mathfrak{A}$, introduced by Moschovakis [54]. $REDS(\mathfrak{A})$ is the set of functions computable by means of *recursively enumerable definitional schemes*, introduced by Friedman and Shepherdson [21,65]. Finally, we have the *search computable* functions, denoted by $SC(\mathfrak{A})$, and also introduced by Moschovakis [54]. Gordon [34] proved the equivalence of search computability with Montague's approach and with computability in admissible sets. Prime computability has a deterministic (sequential) character. REDS is nondeterministic (parallel) and allows searches on the set of natural numbers. Search computability is also nondeterministic, however here one is allowed to perform a search among arbitrary elements of the domain of the structure. For every structure $\mathfrak{A}$ we have $PC(\mathfrak{A}) \subseteq REDS(\mathfrak{A}) \subseteq SC(\mathfrak{A})$. In general these inclusions do not reverse.

Another natural way to study computability on a countable first-order structure is to consider an external approach. Every enumeration of the domain of a structure gives rise to an isomorphic structure on the natural numbers, called its *representation*. Fraisse [19] and Lacombe [45] suggest the notion of $\forall$-recursiveness: a function falls in this class if every enumeration of the domain of the given structure transforms this function into a function on the natural numbers that is recursive in the diagram of the corresponding representation. The equivalence between $\forall$-recursiveness and search computability on countable (total) structures with equality is proved by Moschovakis [56].

In the 1970s Skordev initiated the development of algebraic recursion theory, presented in his monograph [66]. The main goal of this program is to further clarify the connections between the two basic approaches to abstract computability: the internal approach, based on specific models of computation, and the external approach, which defines the computable functions through invariance relative to all enumerations of a structure, in the more general setting of partial structures, structures whose functions and relations can be partial. To find natural external analogues for partial structures we must extend classical relative computability to partial functions. Here as well, there are two different approaches: one corresponding to deterministic computational procedures and one corresponding to arbitrary effective ones. The first one can be mathematically described as relative μ-*recursiveness*: a partial function φ is μ-recursive relative to partial functions $\varphi_1, \ldots, \varphi_n$ if φ can be obtained from $\varphi_1, \ldots, \varphi_n$, the constant 0, the successor function S, and the projection functions, using superposition, primitive recursion and the minimization operation μ. The other notion is called *relative partial recursiveness* and it can be described via enumeration reducibility: the graph of φ is enumeration reducible to the graphs of $\varphi_1, \ldots, \varphi_n$. If we restrict these notions to total functions then they coincide. However there are easy examples of partial objects for which they do not. Let φ be the characteristic function of the complement of the halting set $\overline{K}$ and ψ be the partial function that equals zero when the argument is in $\overline{K}$ and is not defined otherwise. Then φ is partial recursive relative to ψ but not μ-recursive relative ψ.

In 1977 Skordev conjectures that the partial functions which are invariantly computable in all computable presentations of a countable partial structure $\mathfrak{A}$ on the natural numbers are exactly the ones that are search computable on $\mathfrak{A}$. Soskov [69–71] modifies and extends this hypothesis to give a full classification. He proves that the invariantly partial computable functions in all total representations of $\mathfrak{A}$ are exactly $SC(\mathfrak{A})$, the invariantly partial computable functions in all partial representations of $\mathfrak{A}$ are exactly $REDS(\mathfrak{A})$ and the invariantly μ-recursive functions in all partial representations of $\mathfrak{A}$ are exactly $PC(\mathfrak{A})$.

The next theme investigated in this context is a reducibility between a certain class of abstract structures, considered natural for the purposes of abstract computability. These are partial two-sorted relational structures, with an abstract sort and the sort of the natural numbers. Partial functions can be represented through their graphs, provided that we have included equality and non-equality among the basic predicates. The reducibility is defined between structures with the same abstract sort: a structure $\mathfrak{A}$ is s-reducible to a structure $\mathfrak{B}$ if all the basic predicates of $\mathfrak{A}$ are search computable in $\mathfrak{B}$. The properties of this reducibility are very similar to the properties of enumeration reducibility. The obtained results [5,7,36] about the structure of the s-degrees have natural analogs in the enumeration degrees. On the other hand many of the techniques used in this area, could be adapted to study the enumeration degrees. This leads Soskov to transfer his focus towards degree theory, where he explored the ideological connections between one of the models of abstract computability, search computability, and enumeration reducibility. Soskov and his students [72–74,77] develop the theory of regular enumerations and apply it to the enumeration degrees, obtaining a series of new results, mainly in relation to the enumeration jump.

The relationship between enumeration degrees and abstract models of computability inspires a new direction in the field of computable structure theory. Computable structure theory uses the notions and methods of computability theory in order to find the effective contents of some mathematical problems and constructions. One of the fundamental problems is to characterize the abstract structures from the point of view of their computability theoretic complexity and definability strength. A well studied measure of the computability theoretic complexity of a given structure is the notion of Turing degree spectrum. The Turing degree spectrum, introduced by Jockusch and Richter [60,61], is the set of all Turing degrees of the diagrams of the representations (the isomorphic copies) of the structure. In recent years the Sofia school in computability lead by Soskov has been exploring computable structure theory in the more general setting obtained by considering partial structures with the underlying computation model given by enumeration reducibility and measure of complexity given by their enumeration degree spectra. In this article we will outline this line of research.

2   Enumeration Reducibility

Enumeration reducibility gives a general way to compare the positive information in two sets of natural numbers. It is introduced by Friedberg and Rogers

[20] in 1959. Enumeration reducibility relates to relative partial recursiveness in the same way that Turing reducibility relates to relative μ-recursiveness, the reducibility that captures both positive and negative information between two sets.

A set A is enumeration reducible to a set B if there is an effective uniform way, given by an *enumeration operator*, to obtain an enumeration of A given any enumeration of B. The enumeration operators are interesting in themselves, as they give the semantics of the type free λ-calculus in graph models, suggested by Plotkin [59] in 1972. The interest in enumeration reducibility is also supported by the fact that the structure of the enumeration degrees contains the structure of the Turing degrees without being elementary equivalent to it. Contemporary definability results [8,29,30,92] in the theory of the enumeration degrees show that the structure is useful for the study of the structure of Turing degrees.

Definition 1. *Let A and B be sets of natural numbers. The set A is enumeration reducible to the set B, written $A \leq_e B$, if there is a c.e. set W, such that:*

$$A - W(B) - \{x \mid (\exists D)[\langle x, D \rangle \subset W \ \& \ D \subseteq B]\},$$

where D is a finite set coded in the standard way.

The definition above associates an effective operator on sets to every c.e. set, the aforementioned enumeration operator. The set $A \oplus B = \{2n \mid n \in A\} \cup \{2n + 1 \mid n \in B\}$ is a least upper bound of A and B with respect to $\leq_e$. Two sets A and B are *enumeration equivalent* ($A \equiv_e B$) if $A \leq_e B$ and $B \leq_e A$. The equivalence class of a set A under this relation is its *enumeration degree* $d_e(A)$. The set $\mathcal{D}_e$ consisting of all enumeration degrees, together with the naturally induced partial order and least upper bound operation is the *upper semi-lattice of the enumeration degrees*. It has a least element $\mathbf{0}_e$ consisting of all computably enumerable sets.

Let $A^+ = A \oplus \overline{A}$. The set A^+ codes in a positive way the positive and negative information about a set A. This suggests a relationship between Turing reducibility, enumeration reducibility and the relation "c.e. in", formally expressed as follows.

Proposition 1. *Let A and B be sets of natural numbers.*

(1) $A \leq_T B$ *if and only if* $A^+ \leq_e B^+$.
(2) A *is c.e. in* B *if and only if* $A \leq_e B^+$.

A set A is called *total* if and only if $A \equiv_e A^+$. Examples of total sets are the graphs of total functions. Proposition 1 gives rise to a natural embedding of the Turing degrees into the enumeration degrees $\iota : \mathcal{D}_T \to \mathcal{D}_e$, defined by $\iota(d_T(A)) = d_e(A^+)$ [49,57]. An enumeration degree is *total* if it contains a total set. The enumeration degrees in the range of ι coincide with the total enumeration degrees.

The pioneering work on the enumeration degrees dates back to Case [9] and Medvedev [49]. In particular, Case shows that $\mathcal{D}_e$ is not a lattice as a consequence of the exact pair theorem and Medvedev proves the existence of quasi-minimal degrees: a degree is *quasi-minimal* if it bounds no nonzero total enumeration degree. The following theorem by Selman shows that the total enumeration degrees play an important role in the structure: an enumeration degree can be characterized by the set of total degrees above it.

Theorem 1 [64]. *For any $A, B \subseteq \mathbb{N}$ the following are equivalent:*

(1) $A \leq_e B$;
(2) $\{X \mid B$ is c.e. in $X\} \subseteq \{X \mid A$ is c.e. in $X\}$;
(3) $\{\mathbf{x} \in \mathcal{D}_e \mid \mathbf{x}$ is total $\&\ d_e(B) \leq \mathbf{x}\} \subseteq \{\mathbf{x} \in \mathcal{D}_e \mid \mathbf{x}$ is total $\&\ d_e(A) \leq \mathbf{x}\}$.

Finally, we give the definition of a jump operator for the enumeration degrees, originally due to Cooper and studied by McEvoy [12,48]. Let $E_A = \{\langle e, x \rangle \mid x \in W_e(A)\}$. The set $A' = E_A^+$ is called the enumeration jump of A and $d_e(A)' = d_e(A')$. The enumeration jump is monotone and agrees with the Turing jump J_T in the following sense: $J_T(A)^+ \equiv_e (A^+)'$.

We will use Soskov's jump inversion theorem for the enumeration jump:

Theorem 2 [73]. *For every enumeration degree $\mathbf{a}$ there exists a total enumeration degree $\mathbf{b}$, such that $\mathbf{a} \leq \mathbf{b}$ and $\mathbf{a}' = \mathbf{b}'$.*

We can iterate the enumeration jump along all computable ordinals. We will identify every ordinal with its notation. In particular we will write $\alpha < \beta$ instead of $\alpha <_o \beta$. If α is a limit ordinal then by $\{\alpha(p)\}_{p\in\mathbb{N}}$ we will denote the unique strongly increasing sequence of ordinals with limit α, determined by the notation of α, and write $\alpha = \lim \alpha(p)$. For every computable ordinal α the α-th iteration of the enumeration jump $\mathbf{a}^{(\alpha)}$ is defined in a way similar to that one used in the definition the α-th iteration of the Turing jump, see [74]. Let $A^{(\alpha+1)} = (A^{(\alpha)})'$, and if $\alpha = \lim \alpha(p)$ is a limit ordinal then $A^{(\alpha)} = \{\langle p, x \rangle \mid x \in A^{(\alpha(p))}\}$. Again it turns out that both definitions are consistent on the total enumeration degrees. Using the technique of regular enumerations Soskov and Baleva extend Theorem 2 for the computable ordinals. Here is a simple version of their result.

Theorem 3 [74]. *Let B be a set of natural numbers and let Q be a total set, such that $Q \geq_e B^{(\alpha)}$. Let A be such that $A^+ \leq_e Q$ and $A \not\leq_e B^{(\beta)}$ for some $\beta < \alpha$. There exists a total set F such that:*

(1) $B \leq_e F$,
(2) $A \not\leq_e F^{(\beta)}$, and
(3) $F^{(\alpha)} \equiv_e Q$.

3 Enumeration Degree Spectra

The enumeration degree spectrum $DS(\mathfrak{A})$ of a countable structure $\mathfrak{A}$ is introduced by Soskov [75] as the set of all enumeration degrees generated by the presentations (homomorphic copies) of $\mathfrak{A}$ on the set of the natural numbers. Let $\mathfrak{A} = (\mathbb{N}; R_1, \ldots, R_k)$ be a countable relational structure. Here we consider the relations as sets instead of zero-one-valued functions. In the context of enumeration reducibility this corresponds to partial functions, i.e. the relations are true on certain elements and not defined on others. As $\mathfrak{A}$ is countable we may assume that the domain of $\mathfrak{A}$ is $\mathbb{N}$. An enumeration of $\mathfrak{A}$ is a total surjective mapping of $\mathbb{N}$ onto $\mathbb{N}$. Given an enumeration f of $\mathfrak{A}$ and a subset of A of $\mathbb{N}^a$, let

$$f^{-1}(A) = \{\langle x_1, \ldots, x_a \rangle \mid (f(x_1), \ldots, f(x_a)) \in A\}.$$

Denote by $f^{-1}(\mathfrak{A}) = f^{-1}(R_1) \oplus \cdots \oplus f^{-1}(R_k) \oplus f^{-1}(=) \oplus f^{-1}(\neq)$. If f is the identity then we refer to $f^{-1}(\mathfrak{A})$ as $D(\mathfrak{A})$—the positive atomic diagram of $\mathfrak{A}$.

Definition 2 [75]. *The enumeration degree spectrum of $\mathfrak{A}$ is the set*

$$DS(\mathfrak{A}) = \{d_e(f^{-1}(\mathfrak{A})) \mid f \text{ is an enumeration of } \mathfrak{A}\}.$$

If $\mathbf{a}$ is the least element of $DS(\mathfrak{A})$, then $\mathbf{a}$ is called the enumeration degree (e-degree) of $\mathfrak{A}$.

One noticeable difference with the standard definition of Turing degree spectra is that in the definition of the enumeration spectra we use the surjective enumerations, instead of bijective enumerations. Consider the structure $\mathfrak{A} = (\mathbb{N}; =, \neq)$ if we define the degree spectrum of $\mathfrak{A}$ by taking into account only the bijective enumerations, then it will be equal to $\{\mathbf{0}_e\}$, while if we take all surjective enumerations, then $DS(\mathfrak{A})$ will consist of all total enumeration degrees. Fortunately, this difference does not affect the notion of e-degree of a structure since for every enumeration f of $\mathfrak{A}$ there exists a bijective enumeration g of $\mathfrak{A}$ such that $g^{-1}(\mathfrak{A}) \leq_e f^{-1}(\mathfrak{A})$. On the other hand it allows us to show that the enumeration degree spectrum is always *closed upwards with respect to total degrees*, i.e. if $\mathbf{a} \in DS(\mathfrak{A})$, $\mathbf{b}$ is a total e-degree and $\mathbf{a} \leq \mathbf{b}$ then $\mathbf{b} \in DS(\mathfrak{A})$. This can be seen as follows: if g is an enumeration of $\mathfrak{A}$ and F is a total set such that $g^{-1}(\mathfrak{A}) \leq_e F$ then we can define a new enumeration f of $\mathfrak{A}$, which mimics g on the even numbers: $f(n/2) = g(n)$ and codes F on the odd numbers, by mapping all of them to one of two distinct members of $\mathfrak{A}$ depending on membership in F. In general, however, the enumeration degree spectra are not closed upwards as we shall see next.

Just like Turing reducibility can be expressed in terms of enumeration reducibility, the Turing degree spectrum [40,61] of a structure $\mathfrak{A}$ corresponds to the enumeration degree spectrum of a structure, denoted by $\mathfrak{A}^+$, which codes in a positive way both the positive and negative facts about the predicates in $\mathfrak{A}$. If $\mathfrak{A} = (\mathbb{N}, R_1, \ldots, R_k)$ then let $\mathfrak{A}^+ = (\mathbb{N}, R_1, \ldots, R_k, \overline{R}_1, \ldots, \overline{R}_k)$. The image of the Turing degree spectrum of $\mathfrak{A}$ is exactly $DS(\mathfrak{A}^+)$.

Note, that $DS(\mathfrak{A}^+)$ consists only of total enumeration degrees. A structure $\mathfrak{A}$ is called *total* if for every enumeration f of $\mathfrak{A}$ the set $f^{-1}(\mathfrak{A})$ is total. In general, if $\mathfrak{A}$ is a total structure then $DS(\mathfrak{A}) = \iota(DS_T(\mathfrak{A}))$, so if $\mathfrak{A}$ is a total structure then $\mathfrak{A}$ and $\mathfrak{A}^+$ have the same enumeration degree spectrum. Note that, however, not all structures whose degree spectrum consist only of total enumeration degrees are total. Consider for example, the structure $\mathfrak{A} = (\mathbb{N}; G_S, K)$, where G_S is the graph of the successor function and K is the halting set. Then $DS(\mathfrak{A})$ consists of all total degrees. On the other hand if $f = \lambda x.x$, then $f^{-1}(\mathfrak{A})$ is a c.e. set. Hence $\overline{K} \not\leq_e f^{-1}(\mathfrak{A})$. Clearly $\overline{K} \leq_e (f^{-1}(\mathfrak{A}))^+$, so $f^{-1}(\mathfrak{A})$ is not a total set.

A natural question arises here: if $DS(\mathfrak{A})$ consists of total degrees does there exist a total structure $\mathfrak{B}$ such that $DS(\mathfrak{A}) = DS(\mathfrak{B})$? In his last paper [81] Soskov proves the following general result, giving a much stronger relationship between Turing degree spectra and enumeration degree spectra:

Theorem 4 [81]. *For every structure $\mathfrak{A}$ there exists a total structure $\mathfrak{M}$ such that $DS(\mathfrak{M}) = \{\mathbf{a} \mid \mathbf{a}$ is total $\wedge (\exists \mathbf{x} \in DS(\mathfrak{A}))(\mathbf{x} \leq \mathbf{a})\}$.*

We will return to explain the methods developed for the proof of this result in the last section of this paper. Here we turn to some important examples of degree spectra.

Slaman [67] and independently Wehner [101] give an example of a structure whose Turing degree spectrum consists of all nonzero Turing degrees. Translated into our terms this gives a structure $\mathfrak{A}$ such that $DS(\mathfrak{A}) = \{\mathbf{a} \mid \mathbf{a}$ is total and $\mathbf{0}_e < \mathbf{a}\}$. Kalimullin [39], building on Wehner's result, transfers these ideas to enumeration degree spectra.

Theorem 5 [39]. *There is a structure $\mathfrak{A}$ such that $DS(\mathfrak{A}) = \{\mathbf{a} \mid \mathbf{a} \in \mathcal{D}_e$ & $\mathbf{a} > \mathbf{0}_e\}$.*

Kalimullin has a different definition of enumeration degree spectra: for a countable structure $\mathfrak{A}$ he considers the set of the enumeration degrees of full diagrams of isomorphic copies of $\mathfrak{A}$ with domain a *subset of* $\mathbb{N}$. He denotes this set by $e\text{-}SP(\mathfrak{A})$ and shows that for every structure $\mathfrak{A}$ there is a structure $P(\mathfrak{A})$ such that $DS(P(\mathfrak{A}))$ is the upwards closure of $e\text{-}SP(\mathfrak{A})$ in the enumeration degrees.

Following Knight [40] we define the α-th jump spectrum and α-th jump degree of a structure for computable ordinals α:

Definition 3. *Let $\alpha < \omega_1^{CK}$. Then the α-th jump spectrum of $\mathfrak{A}$ is the set*

$$DS_\alpha(\mathfrak{A}) = \{d_e(f^{-1}(\mathfrak{A})^{(\alpha)}) \mid f \text{ is an enumeration of } \mathfrak{A}\}.$$

If $\mathbf{a}$ is the least element of $DS_\alpha(\mathfrak{A})$, then $\mathbf{a}$ is called the α-th jump degree of $\mathfrak{A}$.

We will leave examples of structures with or without α-th jump degree for Sect. 6, where we also investigate the possibilities of defining the jump of a structure. Next we consider the co-spectrum of a structure, a characteristic that plays especially well with enumeration degrees.

4 Co-spectra

Let $\mathcal{A}$ be a nonempty set of enumeration degrees the *co-set of* $\mathcal{A}$ is the set $co(\mathcal{A})$ of all lower bounds of $\mathcal{A}$. Namely

$$co(\mathcal{A}) = \{\mathbf{b} \mid \mathbf{b} \in \mathcal{D}_e \ \& \ (\forall \mathbf{a} \in \mathcal{A})(\mathbf{b} \leq \mathbf{a})\}.$$

For every $\mathcal{A} \subseteq \mathcal{D}_e$ the set $co(\mathcal{A})$ is a countable ideal. We will see that every countable ideal can be represented as co-set of the spectrum of some structure $\mathfrak{A}$.

Definition 4. *Let $\mathfrak{A}$ be a countable relational structure.*

(1) *The* co-spectrum $CS(\mathfrak{A})$ *of a structure $\mathfrak{A}$ is the co-set of $DS(\mathfrak{A})$, i.e. the set of all lower bounds of the enumeration degree spectrum of the structure $\mathfrak{A}$. If $CS(\mathfrak{A})$ has a greatest element, then it is the* co-degree *of $\mathfrak{A}$.*

(2) *For every $\alpha < \omega_1^{CK}$ the co-set of $DS_\alpha(\mathfrak{A})$ is $CS_\alpha(\mathfrak{A})$, the α-th jump co-spectrum of $\mathfrak{A}$. If $CS_\alpha(\mathfrak{A})$ has a greatest element, then it is the α-th jump co-degree of $\mathfrak{A}$.*

4.1 Examples

If a structure $\mathfrak{A}$ has a degree $\mathbf{a}$ then $\mathbf{a}$ is also its co-degree. The reverse is not always true. We have already seen one such example: Kalimullin's structure $\mathfrak{A}$ with degree spectrum $DS(\mathfrak{A})$ consisting of all nonzero enumeration degrees clearly has no enumeration degree, but has co-degree $\mathbf{0}_e$. As a second example, consider Richter's [61] result on linear orderings: the Turing degree spectrum $DS_T(\mathfrak{A})$ always contains a minimal pair. Thus the co-degree of $DS(\mathfrak{A}^+)$ is always $\mathbf{0}_T$, and non computable linear orderings have co-degree but no degree. (In fact, Richter gives conditions in terms of enumeration reducibility for when a first order theory has a model with no degree). Knight [40] extends Richter's result to show that the only possible first jump Turing degree of a linear ordering is $\mathbf{0}_T'$. An analysis of her proof shows that the first jump co-spectrum of a linear ordering consists of all Σ_2^0 enumeration degrees, and so the first jump co-degree is always $\mathbf{0}_e'$, even though not every linear ordering has a first jump degree.

There are also structures with no co-degree. For example, consider $\mathfrak{A} = (\mathbb{N}; G_\Psi, P)$, where Ψ is a function such that $\Psi(\langle n, x\rangle) = \langle n, x+1\rangle$ and the relation $P(x)$ is defined and true if $(\exists t)(x = \langle 0, t\rangle)$ or $(\exists n)(\exists t)(x = \langle n+1, t\rangle \,\&\, t \in \emptyset^{(n+1)})$. For every $X \subseteq \mathbb{N}$ we have that $d_e(X) \in CS(\mathfrak{A})$ iff $(\exists n)(X \leq_e \emptyset^{(n)})$. It follows that $CS(\mathfrak{A})$ consists of all arithmetical degrees and hence has no greatest element, i.e. $\mathfrak{A}$ has no co-degree.

The co-degree of a structure is closely related to what Knight [41] and Montalbán [52] call the "enumeration degree of a structure". A set $X \subseteq \mathbb{N}$ is the "enumeration degree" of a structure $\mathfrak{A}$ if every enumeration of X computes a copy of $\mathfrak{A}$, and every copy of $\mathfrak{A}$ computes an enumeration of X. Thus by Selman's theorem the enumeration degree of X is the co-degree of the structure $\mathfrak{A}^+$. This co-degree, however has an additional property: $DS(\mathfrak{A}^+)$ is exactly the set of total enumeration degrees above $d_e(X)$. Thus, examples of structures with "enumeration degree" translate to examples of structures with co-degree and there are many of those: Given $X \subseteq \mathbb{N}$, consider the group $G_X = \bigoplus_{i \in X} \mathbb{Z}_{p_i}$, where p_i is the i-th prime number. Then G_X has "enumeration degree" X, as we can easily build G_X given any enumeration of X, and for the reverse direction, we have that $n \in X$ if and only if there is an elements $g \in G_X$ of order p_n. Montalbán [52] proves that if a class K of structures is axiomatized by some computable infinitary Π_2^c sentence and every structure $\mathfrak{A}$ in K is existentially atomic, i.e. an atomic structure with all types generated by existential formulas, then every structure in K has "enumeration degree" given by its $\exists$-theory.

A further example of this sort is given, when one considers torsion free abelian groups of rank 1, i.e. subgroups of $(\mathbb{Q}, +, =)$. Downey and Jockusch [13] analyze the computability theoretic properties of such groups. Using results that go back to Baer, they discover a way to associate a set $S(G)$, called the characteristic of G, to every torsion free abelian group G of rank 1, so that the Turing degree spectrum of G is precisely $\{d_T(Y) \mid S(G) \text{ is c.e. in } Y\}$. On the other hand, they show that for every set of natural numbers S there is a torsion free abelian group G of rank 1, such that $S(G) \equiv_1 S$. They knew from Richter [61] that this meant that not all such groups have a degree. Coles, Downey and Slaman [11] use a forcing construction to show that, however, every such group has first jump degree.

Soskov [75] considers the problem from the point of view of enumeration reducibility. Any subgroup of the rationals can be seen as a total structure, as the only relation involved is the graph of addition, which is a total function. Let G be such a group and let $\mathbf{s}_b = d_e(S(G))$. It follows that

$$DS(G) = \{\mathbf{b} \mid \mathbf{b} \text{ is total and } \mathbf{s}_b \leq_e \mathbf{b}\}.$$

It is an easy consequence of Selman's theorem that $\mathbf{s}_b$ is the co-degree of G. Furthermore, G has degree if and only if $\mathbf{s}_b$ is total. The result of Coles, Downey and Slaman now follows from Theorem 2. There is a total enumeration degree $\mathbf{f} \geq \mathbf{s}_b$ with $\mathbf{f}' = \mathbf{s}'_b$ and so the first jump spectrum of G consists of all total enumeration degrees greater than or equal to $\mathbf{s}'_b$, in particular $\mathbf{s}'_b$ is the first jump degree of G.

Another consequence of this example is that every principal ideal of enumeration degrees is a co-spectrum of a structure, namely the co-spectrum of some torsion free abelian group of rank one. To generalize this result to arbitrary countable ideals we need a characterization of the co-spectrum of a structure.

4.2   Normal Forms

Soskov [75] gives two characterizations of $CS_\alpha(\mathfrak{A})$ in terms of the structure $\mathfrak{A}$, one in terms of forcing and one in terms of definability. The first characterization is inspired by the fact that the members of $CS(\mathfrak{A})$ are exactly the degrees of the domains of the search computable functions ranging over the natural numbers and by the well known results by Ash, Knight, Manasse and Slaman [4] and by Chisholm [10]. We note that independently Ash and Knight [3] also characterize the elements of the co-spectrum for certain structures: they showed that for a computable structure $\mathfrak{A}$ a set $A \subseteq \mathbb{N}$ is c.e. relative to $f^{-1}(\mathfrak{A})$ for every bijective enumeration f of $\mathfrak{A}$ if and only if for some tuple $\bar{a}$ in $\mathfrak{A}$, the set A is enumeration reducible to the existential type of $\bar{a}$.

The natural forcing partial order associated with enumerations of a given structure $\mathfrak{A}$ with domain $\mathbb{N}$ consists of finite functions from $\mathbb{N}$ to $\mathbb{N}$ ordered by extension, called *finite parts*. An enumeration f of $\mathfrak{A}$ is α-*generic* for a computable ordinal α if for every computable ordinal $\beta < \alpha$ and for every set S of finite parts such that $S \leq_e D(\mathfrak{A})^{(\beta)}$ the enumeration f meets or avoids

S. By transfinite induction Soskov then defines the relations $\tau \Vdash_\alpha F_e(x)$ and $\tau \Vdash_\alpha \neg F_e(x)$ for every computable ordinal α, so that if f is α-generic then $x \in (f^{-1}(\mathfrak{A}))^{(\alpha)}$ if and only if there is a finite function $\tau \preceq f$, such that $\tau \Vdash_\alpha F_e(x)$ and if f is $(\alpha+1)$-generic then $x \notin (f^{-1}(\mathfrak{A}))^{(\alpha)}$ if and only if there is a finite function $\tau \preceq f$, such that $\tau \Vdash_\alpha \neg F_e(x)$.

Definition 5. *A set $A \subseteq \mathbb{N}$ is* forcing α-definable *in the structure $\mathfrak{A}$ if there exist finite part δ and a natural number e s.t.*

$$A = \{x \mid (\exists \tau \supseteq \delta)(\tau \Vdash_\alpha F_e(x))\}.$$

Soskov shows that $CS_\alpha(\mathfrak{A})$ consists of the enumeration degrees of the forcing α-definable sets.

Theorem 6 [75]. *A set $A \subseteq \mathbb{N}$ is forcing α-definable in $\mathfrak{A}$ if and only if $A \leq_e f^{-1}(\mathfrak{A})^{(\alpha)}$ for every enumeration f of $\mathfrak{A}$.*

The second characterization uses positive computable infinitary Σ_α formulas, denoted by Σ_α^+, whose structure follows that of the forcing relation. These formulas can be considered as a modification of the ones introduced by Ash [2]. Let $\mathcal{L}$ be the first order language of the structure $\mathfrak{A}$. A Σ_α^+ formula with free variables among $X_1, \ldots, X_l$ is a c.e. infinitary disjunction of *elementary* Σ_α^+ formulas with free variables among $X_1, \ldots, X_l$ which are defined by transfinite induction on α as follows. The elementary Σ_0^+ formulas are those of the form $\exists Y_1 \ldots \exists Y_m \theta(X_1, \ldots, X_l, Y_1, \ldots, Y_m)$ where θ is a finite conjunction of atomic predicates of $\mathcal{L}$. For $\alpha = \beta + 1$ an elementary Σ_α^+ formula is of the form $\exists Y_1 \ldots \exists Y_m \Psi(X_1, \ldots, X_l, Y_1, \ldots, Y_m)$, where Ψ is a finite conjunction of Σ_β^+ formulas and negations of Σ_β^+ formulas with free variables among $X_1, \ldots, X_l$, $Y_1, \ldots, Y_m$.

For $\alpha = \lim \alpha(p)$ a limit ordinal the elementary Σ_α^+ formulas are of the form $\exists Y_1 \ldots \exists Y_m \Psi(X_1, \ldots, X_l, Y_1, \ldots, Y_m)$, where Ψ is a finite conjunction of $\Sigma_{\alpha(p)}^+$ formulas with free variables among $X_1, \ldots, X_l, Y_1, \ldots, Y_m$.[1]

Definition 6. *A set $A \subseteq \mathbb{N}$ is* formally α-definable *in a structure $\mathfrak{A}$ if there exists a computable function $f(x)$ with values indices of Σ_α^+ formulas $\Phi_{f(x)}$ with free variables among $W_1, \ldots, W_r$ and parameters $t_1, \ldots, t_r \in |\mathfrak{A}|$ such that for every natural number x the following equivalence holds:*

$$x \in A \iff \mathfrak{A} \models \Phi_{f(x)}(W_1/t_1, \ldots, W_r/t_r).$$

Theorem 7 [75]. *A set $A \subseteq \mathbb{N}$ is forcing α-definable in a structure $\mathfrak{A}$ iff it is formally α-definable in $\mathfrak{A}$.*

---

[1] Note, that this indexing does not quite match the usual definition of computable infinitary formulas, namely level zero in this definition corresponds to level one in the usual definition.

Using these normal forms, as promised, we can represent every countable ideal of enumeration degrees I as the co-spectra of a structure. Fix such an ideal I, and let $\mathbf{b}_0 \leq \mathbf{b}_1 \leq \cdots \leq \mathbf{b}_k \ldots$ be a countable sequence, generating I. Fix $B_k \in \mathbf{b}_k$, for each k. Consider the structure $\mathfrak{A} = (\mathbb{N}; G_f, \sigma, =, \neq)$, where
$$f(\langle i, n \rangle) = \langle i+1, n \rangle \text{ and } \sigma = \{\langle i, n \rangle \mid n = 2k+1 \vee n = 2k \ \& \ i \in B_k\}.$$

To show that $I \subseteq CS(\mathfrak{A})$ it is sufficient to see that $B_k \leq_e g^{-1}(\mathfrak{A})$ for every enumeration g of $\mathfrak{A}$ and each k. For every x using the pre-image of G_f we can find the pre-image of the natural number $\langle x, 2k \rangle$ and enumerate x in B_k if the pre-image of $\langle x, 2k \rangle$ is in the pre-image of σ. The reverse direction requires quite a bit more work, and relies on an analysis of the formally 0-definable in $\mathfrak{A}$ sets.

Theorem 8 [75]. *Every countable ideal I of enumeration degrees is a co-spectrum of a structure.*

4.3 Structural Properties of Spectra and Co-spectra

Now that we know that every countable ideal of enumeration degrees is the co-spectrum of a structure, we might wonder if we can characterize spectra in a similar way: is every set of degrees that is upwards closed with respect to total elements the enumeration spectrum of a structure? The answer is, of course, 'No'. One way to see this is via the notion of a *base* and its relationship to the existence of a degree. A subset $\mathcal{B} \subseteq \mathcal{A}$ of a set of enumeration degrees $\mathcal{A}$ is *a base of $\mathcal{A}$* if $(\forall \mathbf{a} \in \mathcal{A})(\exists \mathbf{b} \in \mathcal{B})(\mathbf{b} \leq \mathbf{a})$. Using generic enumerations and an argument much like that used in Selman's theorem we can show the following theorem.

Theorem 9 [75]. *A structure $\mathfrak{A}$ has an e-degree if and only if $DS(\mathfrak{A})$ has a countable base.*

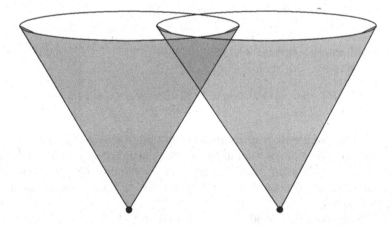

Fig. 1. An upwards closed set with respect to total degrees which is not a degree spectra of a structure

In particular the union of two cones above (Fig. 1) incomparable degrees cannot be the enumeration degree spectrum of a structure (just like it cannot be the Turing degree spectrum of a structure). Nevertheless, degree spectra play well with co-spectra and behave structurally with respect to their elements just like the cone of total degrees above a fixed enumeration degree. This is not too surprising, as a further easy application of Selman's theorem shows that the co-spectrum of $\mathfrak{A}$ depends only on the total elements of the spectrum of $\mathfrak{A}$, i.e. $CS(\mathfrak{A}) = co(DS(\mathfrak{A})_t)$, where $DS(\mathfrak{A})_t = \{\mathbf{a} \mid \mathbf{a} \text{ is total } \& \mathbf{a} \in DS(\mathfrak{A})\}$.

Our first more elaborate example of this phenomenon is an analogue, and in fact a generalization, of a result of Rozinas [62], stating that for every $\mathbf{a} \in \mathcal{D}_e$ there exist total $\mathbf{f_1}, \mathbf{f_2}$ below $\mathbf{a}''$ which are a minimal pair above $\mathbf{a}$.

Theorem 10 [75]. *Let $\alpha < \omega_1^{CK}$ and let $\mathbf{b} \in DS_\alpha(\mathfrak{A})$. There exist total elements $\mathbf{f_0}$ and $\mathbf{f_1}$ of $DS(\mathfrak{A})$ such that:*

(1) $\mathbf{f_0}^{(\alpha)} \leq \mathbf{b}$ and $\mathbf{f_1}^{(\alpha)} \leq \mathbf{b}$;
(2) $\mathbf{f_0}^{(\beta)}$ and $\mathbf{f_1}^{(\beta)}$ do not belong to $CS_\beta(\mathfrak{A})$ for $\beta < \alpha$;
(3) $co(\{\mathbf{f_0}^{(\beta)}, \mathbf{f_1}^{(\beta)}\}) = CS_\beta(\mathfrak{A})$ for every $\beta + 1 < \alpha$.

This property does not hold for arbitrary sets that are upwards closed with respect to total degrees. Consider the finite lattice L consisting of the elements $\mathbf{a}, \mathbf{b}, \mathbf{c}, \mathbf{a} \wedge \mathbf{b}, \mathbf{a} \wedge \mathbf{c}, \mathbf{b} \wedge \mathbf{c}, \top, \bot$ such that $\top$ and $\bot$ are the greatest and the least element of L, respectively, $\mathbf{a} > \mathbf{a} \wedge \mathbf{b}, \mathbf{a} > \mathbf{a} \wedge \mathbf{c}, \mathbf{b} > \mathbf{a} \wedge \mathbf{b}, \mathbf{b} > \mathbf{b} \wedge \mathbf{c}$, $\mathbf{c} > \mathbf{a} \wedge \mathbf{c}$ and $\mathbf{c} > \mathbf{b} \wedge \mathbf{c}$. The lattice L can be embedded in the enumeration degrees (see for example [46]). Then $\mathcal{A} = \{\mathbf{d} \in \mathcal{D}_e \mid \mathbf{d} \geq \mathbf{a} \vee \mathbf{d} \geq \mathbf{b} \vee \mathbf{d} \geq \mathbf{c}\}$ is a set that does not satisfy the minimal pair property, because $co(\mathcal{A}) = \{\bot\}$, but no pair of elements in $\mathcal{A}$ has greatest lower bound $\bot$ (Fig. 2).

The next property is analogue of the existence of a quasi-minimal enumeration degree proved by Medvedev [49]. Let $\mathcal{A}$ be a set of enumeration degrees. The degree $\mathbf{q}$ is *quasi-minimal with respect to* $\mathcal{A}$ if:

- $\mathbf{q} \notin co(\mathcal{A})$.
- If $\mathbf{a}$ is total and $\mathbf{a} \geq \mathbf{q}$, then $\mathbf{a} \in \mathcal{A}$.
- If $\mathbf{a}$ is total and $\mathbf{a} \leq \mathbf{q}$, then $\mathbf{a} \in co(\mathcal{A})$.

Theorem 11 [75]. *For every structure $\mathfrak{A}$ there exists a quasi-minimal with respect to $DS(\mathfrak{A})$ degree.*

To prove this theorem Soskov introduces the notion of a *partial generic enumeration* φ of $\mathfrak{A}$, generic enumeration in the forcing partial order consisting of finite functions from $\mathbb{N}$ to $\mathbb{N} \cup \{\bot\}$, where $\bot$ represents partiality. He then shows that if φ is a partial generic enumeration of $\mathfrak{A}$ then $d_e(\varphi^{-1}(\mathfrak{A}))$ is quasi-minimal with respect to $DS(\mathfrak{A})$.

Since every countable ideal of enumeration degrees is a co-spectrum of a structure as a corollary we receive a result of Slaman and Sorbi:

Corollary 1 [68]. *Let I be a countable ideal of enumeration degrees. There exists an enumeration degree $\mathbf{q}$ such that*

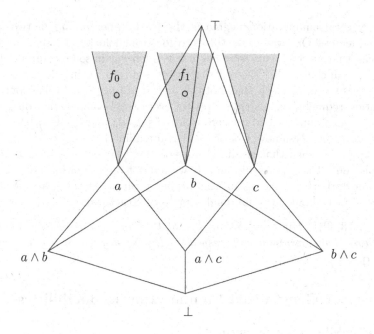

Fig. 2. An upwards closed set with no minimal pair

(1) *If* $\mathbf{a} \in I$ *then* $\mathbf{a} <_e \mathbf{q}$.
(2) *If* $\mathbf{a}$ *is total and* $\mathbf{a} \leq_e \mathbf{q}$ *then* $\mathbf{a} \in I$.

The technique of partial generic enumerations is further developed by Ganchev, Soskov and A. Soskova in [22, 24, 84]. Soskov and A. Soskova also investigate further properties of the notion of a quasi-minimal degree in [91]. They show that for every countable structure $\mathfrak{A}$ there are uncountably many quasi-minimal degrees with respect to $DS(\mathfrak{A})$. The proof relies on a diagonalization: for every countable sequence $\{X_i\}$ of sets that are not forcing 0-definable, (such as the members of a quasi-minimal degree), there is a partial generic enumeration of the structure omitting all X_i. Their main find is however a characterization of the first jump spectra in terms of the jumps of quasi-minimal degrees:

Theorem 12 [91]. *The first jump spectrum of every structure* $\mathfrak{A}$ *consists exactly of the enumeration jumps of the quasi-minimal with respect to* $DS(\mathfrak{A})$ *degrees.*

When one applies the theorem above to any computable structure, one obtains directly McEvoy's jump inversion theorem:

Corollary 2 [48]. *For every total e-degree* $\mathbf{a} \geq_e \mathbf{0}'_e$ *there is a quasi-minimal degree* $\mathbf{q}$ *with* $\mathbf{q}' = \mathbf{a}$.

The final property of quasi-minimal degrees that we will mention here, is inspired by the well-known fact from enumeration degree theory, which states that every total enumeration degree is the least upper bound of two quasi-minimal e-degrees. One way to see this is to go through Jockusch's semi-recursive sets. Recall that a set is *semi-recursive* if it is a left cut in some computable linear ordering. Jockusch [37] showed that every nonzero Turing degree contains a semi-recursive set A, such that both A and $\overline{A}$ are not c.e. In the context of enumeration reducibility this translates to: every total enumeration degree $\mathbf{a}$ is the least upper bound of two nonzero e-degrees $d_e(A)$ and $d_e(\overline{A})$, where A is a semi-recursive set. Arslanov, Cooper and Kalimullin [1] showed that if A is a semi-recursive set such that A and $\overline{A}$ are not c.e., then the e-degrees of A and its complement $\overline{A}$ are quasi-minimal. If we restrict our attention only to total degrees above $\mathbf{0}'_e$ then once again, this property turns out to be a special case of a general fact about quasi-minimal degrees of structures:

Theorem 13 [91]. *For every element* $\mathbf{a}$ *of the jump spectrum of a structure* $\mathfrak{A}$ *there exist quasi-minimal with respect to* $DS(\mathfrak{A})$ *degrees* $\mathbf{p}$ *and* $\mathbf{q}$ *such that* $\mathbf{a} = \mathbf{p} \vee \mathbf{q}$.

5 Abstract Generalized Enumeration Reducibilities

5.1 Definability on a Structure

Another way to characterize the complexity of a structure $\mathfrak{A}$ is to analyze the definable sets in $\mathfrak{A}$. This gives a finer measure as it may happen that two structures have the same degree spectra but greatly differ in their definability power and model theoretic properties. Let α be a computable ordinal and $A = |\mathfrak{A}|$. A set $B \subseteq A^a$ is $\Sigma^c_{\alpha+1}$ definable on a structure $\mathfrak{A}$ if there is a computable infinitary $\Sigma^c_{\alpha+1}$ formula $\varphi(\bar{X}, \bar{Z})$ and parameters $\bar{t} \in A$ such that $B = \{\bar{s} \mid \mathfrak{A} \models \varphi(\bar{s}, \bar{t})\}$. A set $B \subseteq A^a$ is relatively intrinsically $\Sigma^0_{\alpha+1}$ in a structure $\mathfrak{A}$ if for each $(\mathfrak{B}, X) \simeq (\mathfrak{A}, B)$ the set X is $\Sigma^0_{\alpha+1}$ in the atomic diagram $D(\mathfrak{B})$, which in our terms means that $f^{-1}(B) \leq_e f^{-1}(\mathfrak{A}^+)^{(\alpha)}$ for every enumeration f of $\mathfrak{A}$. Ash, Knight, Manasse and Slaman [4] and independently Chisholm [10] prove that these two notions coincide. Soskov and Baleva [76] give an analogue of the relatively intrinsically Σ^0_α sets on a structure $\mathfrak{A}$ from the point of view of enumeration reducibility: For every computable ordinal α a set $B \subseteq A^a$, is *relatively α-intrinsic* on the structure $\mathfrak{A}$ if for every enumeration f of $\mathfrak{A}$ the set $f^{-1}(B)$ is enumeration reducible to $(f^{-1}(\mathfrak{A}))^{(\alpha)}$. Soskov and Baleva show that the α-intrinsic sets are exactly the ones definable by computable infinitary $\Sigma^+_{\alpha+1}$ formulas with parameters.

Having moved to this setting, they go one step further and consider the following generalization in the spirit of Ash [2]. For two subsets B and C of A and two computable ordinals α and β Ash defines that B is relatively α, β-intrinsic on $\mathfrak{A}$ with respect to C if for all enumerations f such that $f^{-1}(C)$ is enumeration reducible to $f^{-1}(\mathfrak{A})^{(\beta)}$, $f^{-1}(B)$ is enumeration reducible to $f^{-1}(\mathfrak{A})^{(\alpha)}$. In other words, consider not all enumerations of $\mathfrak{A}$ but only those enumerations which "assume" that B is relatively β-intrinsic. Soskov and Baleva generalized this notion with respect to a sequence of sets $\{B_\gamma\}_{\gamma \leq \varsigma}$ of subsets of A.

Definition 7. *A subset B of A^a is relatively α-intrinsic on $\mathfrak{A}$ with respect to the sequence $\mathcal{B} = \{B_\gamma\}_{\gamma \leq \zeta}$ if for every enumeration f of $\mathfrak{A}$ such that*
$$(\forall \gamma \leq \zeta)(f^{-1}(B_\gamma) \leq_e (f^{-1}(\mathfrak{A}))^{(\gamma)}) \text{ uniformly in } \gamma, \text{ the set } f^{-1}(B) \text{ is enumeration reducible to } (f^{-1}(\mathfrak{A}))^{(\alpha)}.$$

The authors give a normal form of these sets first in terms of a forcing construction. To give a syntactic characterization, they redefine the infinitary computable $\Sigma^+_{\alpha+1}$ formulas, taking into account the sequence $\mathcal{B}$. For every γ they add a new unary predicate P_γ for the set B_γ. This predicate is included positively at level γ of the hierarchy. For example for $\alpha = \beta+1$ an elementary Σ^+_α formula is in the form $\exists Y_1 \ldots \exists Y_m \Psi(X_1, \ldots, X_l, Y_1, \ldots, Y_m)$, where Ψ is a finite conjunction of $P_\alpha(X_i)$ or $P_\alpha(Y_j)$ or Σ^+_β formulas and negations of Σ^+_β formulas with free variables among $X_1, \ldots, X_l, Y_1, \ldots, Y_m$.

Theorem 14 [76]. *A subset B of A^a is relatively α-intrinsic on $\mathfrak{A}$ with respect to the sequence $\mathcal{B} = \{B_\gamma\}_{\gamma \leq \zeta}$ if and only if B is definable in $\mathfrak{A}$ by a computable infinitary Σ^+_α-formula with parameters, constructed with respect to the sequence $\mathcal{B}$.*

The authors also give an abstract version of the Theorem 3. To formulate it we need the following definition:

Definition 8. *For any computable ordinal $\alpha \leq \zeta$ the jump sequence $\mathcal{P}(\mathcal{B}) = \{\mathcal{P}_\alpha\}_{\alpha < \zeta}$ of the sequence $\mathcal{B}$ is defined inductively as follows:*

- *$\mathcal{P}_0 = B_0$, for $\alpha = 0$;*
- *$\mathcal{P}_\alpha = (\mathcal{P}_\beta)' \oplus B_\alpha$, for $\alpha = \beta + 1$;*
- *For $\alpha = \lim \alpha(p)$, denote by $\mathcal{P}_{<\alpha} = \{\langle p, x \rangle : x \in \mathcal{P}_{\alpha(p)}\}$ and let $\mathcal{P}_\alpha = \mathcal{P}_{<\alpha} \oplus B_\alpha$.*

The abstract jump inversion says that for every $B \subseteq A$ which is not Σ^+_α-definable on $\mathfrak{A}$ and each total set $Q \geq_e A^+ \oplus \mathcal{P}_\xi$, where $\xi = \max(\alpha + 1, \zeta)$ there exists an enumeration f of $\mathfrak{A}$ satisfying the following conditions: $f \leq_e Q$, the enumeration degree of $f^{-1}(\mathfrak{A})$ is total, for all $\gamma \leq \zeta$, $f^{-1}(B_\gamma) \leq_e (f^{-1}(\mathfrak{A}))^{(\gamma)}$ uniformly in γ, $(f^{-1}(\mathfrak{A}))^{(\xi)} \equiv_e Q$ and $f^{-1}(B) \not\leq_e (f^{-1}(\mathfrak{A}))^{(\alpha)}$.

5.2   Joint Spectra and Relative Spectra

The results described so far lead Soskov and A. Soskova to the goal of generalizing the notion of degree spectrum of a structure to the degree spectrum of sequences of structures. Initially, they consider the case when the sequence is finite and introduce two generalizations: the joint spectrum [82–84] and the relative spectrum [85,86].

Fix countable structures $\mathfrak{A}_0, \ldots, \mathfrak{A}_n$.

Definition 9. *The* joint spectrum *of $\mathfrak{A}_0, \ldots, \mathfrak{A}_n$ is the set $DS(\mathfrak{A}_0, \mathfrak{A}_1, \ldots, \mathfrak{A}_n) = \{\mathbf{a} \mid \mathbf{a} \in DS(\mathfrak{A}_0), \mathbf{a}' \in DS(\mathfrak{A}_1), \ldots, \mathbf{a}^{(n)} \in DS(\mathfrak{A}_n)\}$.*

So, the joint spectrum is the set of all enumeration degrees of the $DS(\mathfrak{A}_0)$, such that for all $i \leq n$ their ith enumeration jump is in $DS(\mathfrak{A}_i)$. The k-th jump joint spectrum $DS_k(\mathfrak{A}_0, \ldots, \mathfrak{A}_n)$ and the k-th co-spectrum are defined similarly to $DS_k(\mathfrak{A})$ and $CS_k(\mathfrak{A})$. In this case as well $DS_k(\mathfrak{A}_0, \ldots, \mathfrak{A}_n)$ is closed upwards with respect to total degrees. The k-th co-spectrum of the sequence $\mathfrak{A}_0, \ldots, \mathfrak{A}_n$ depends only on the first k members.

Theorem 15. *For every $k \leq n$ we have that $CS_k(\mathfrak{A}_0, \ldots, \mathfrak{A}_k) = CS_k(\mathfrak{A}_0, \ldots, \mathfrak{A}_n)$. Moreover for every set B of natural numbers $d_e(B) \in CS_k(\mathfrak{A}_0, \ldots, \mathfrak{A}_k)$ if and only if for every $k+1$ enumerations $f_0, \ldots, f_k$, of $\mathfrak{A}_0, \ldots, \mathfrak{A}_k$ respectively, the set $B \leq_e \mathcal{P}(f_0^{-1}(\mathfrak{A}_0), \ldots, f_k^{-1}(\mathfrak{A}_k))$.*

Here $\mathcal{P}(f_0^{-1}(\mathfrak{A}_0), \ldots, f_k^{-1}(\mathfrak{A}_k))$ is the kth jump sequence of the given sequence.

Soskov and A. Soskova [82] give a syntactical normal form for the members of the degrees in the set $CS_k(\mathfrak{A}_0, \ldots, \mathfrak{A}_n)$. This time they use many-sorted Σ_k^+ infinitary computable formulas with different sorts for every structure $\mathfrak{A}_i$. A. Soskova [83,84] shows that the structural properties of co-spectra are preserved. The analog of the minimal pair theorem holds here as well: for any sequence of structures $\mathfrak{A}_0, \ldots, \mathfrak{A}_n$, there exist enumeration degrees $\mathbf{f}$ and $\mathbf{g}$ in $DS(\mathfrak{A}_0, \ldots, \mathfrak{A}_n)$, such that for any enumeration degree $\mathbf{a}$ and $k \leq n$:

$$\mathbf{a} \leq \mathbf{f}^{(k)} \;\&\; \mathbf{a} \leq \mathbf{g}^{(k)} \Rightarrow \mathbf{a} \in CS_k(\mathfrak{A}_0, \ldots, \mathfrak{A}_n).$$

Furthermore, A. Soskova proves the existence of quasi-minimal degree $\mathbf{q}$ with respect to $DS(\mathfrak{A}_0, \mathfrak{A}_1, \ldots, \mathfrak{A}_n)$. The proof techniques are based on regular enumerations introduced in [73] and partial generic enumerations used in [75].

The second generalization defines the relative spectrum of a structure with respect to finitely many structures. Consider a structure $\mathfrak{A}$ and finitely many structures $\mathfrak{A}_1, \ldots, \mathfrak{A}_n$. We will restrict the class of enumerations of $\mathfrak{A}$ to these enumerations of $\mathfrak{A}$ which "assume" that each $\mathfrak{A}_i$ is relatively intrinsically Σ_{i+1}^0 in $\mathfrak{A}$: An enumeration f of $\mathfrak{A}$ is n-*acceptable* with respect to the structures $\mathfrak{A}_1, \ldots, \mathfrak{A}_n$ if $f^{-1}(\mathfrak{A}_i)$ is enumeration reducible to $f^{-1}(\mathfrak{A})^{(i)}$ for each $i \leq n$.

Definition 10. *The* relative spectrum of the structure $\mathfrak{A}$ with respect to $\mathfrak{A}_1, \ldots, \mathfrak{A}_n$ is the set
$$RS(\mathfrak{A}, \mathfrak{A}_1, \ldots, \mathfrak{A}_n) = \{d_e(f^{-1}(\mathfrak{A})) \mid f \text{ is an } n\text{-acceptable enumeration of } \mathfrak{A}\}.$$

The elements of the co-spectrum of the k-th relative spectrum are the enumeration degrees which contain a set which is enumeration reducible to the k-th jump sequence $\mathcal{P}_k^f$ of the sequence $f^{-1}(\mathfrak{A}), f^{-1}(\mathfrak{A}_1), \ldots, f^{-1}(\mathfrak{A}_k)$, for every k-acceptable enumeration of $\mathfrak{A}$ with respect to the structures $\mathfrak{A}_1, \ldots, \mathfrak{A}_k$. The normal form of these sets is given [85,86] using a forcing construction. In this case as well there is an analog of the minimal pair theorem and the existence of quasi-minimal degree. The co-spectra of the joint spectra and the relative spectra coincide, but there are examples of sequence of structures for which the k-th co-spectra for $k > 0$ differ.

As we have seen the structural properties of the degree spectra and co-spectra obtained remain true when one relativizes to consider finite sequences of structures. The main question here is whether these generalizations give rise to new sets of degrees, or is it the case that for every finite sequence of countable structures there exists one structure whose degree spectrum is exactly the relative spectrum or the joint spectrum of the given sequence. An answer to this question will be given in the last section of this paper.

5.3   Omega-Enumeration Reducibility

In 2006 Soskov initiates the study of uniform reducibility between sequences of sets and the induced structure of the ω-degrees. Soskov, Ganchev and M. Soskova obtain many results, providing substantial proof that the structure of the ω-degrees is a natural extension of the structure of the enumeration degrees, with a jump operation that has interesting properties and with natural new members, which turn out to be extremely useful for the characterization of certain classes of enumeration degrees. These investigations appear in [23–28, 77–79, 92].

The jump class of the sequence $\mathcal{X} = \{X_n\}_{n<\omega}$ of sets of natural numbers is the set $J_{\mathcal{X}} = \{d_T(B) \mid (\forall n)(X_n$ is c.e. in $B^{(n)}$ uniformly in $n)\}$. The definition of ω-enumeration reducibility between sequences of sets is an analogue of Selman's characterization Theorem 1 of enumeration reducibility.

Definition 11. *The sequence* $\mathcal{X}$ *is* ω*-enumeration reducible to the sequence* $\mathcal{Y}$ *($\mathcal{X} \leq_\omega \mathcal{Y}$) if* $J_{\mathcal{Y}} \subseteq J_{\mathcal{X}}$.

Let $\mathcal{X} = \{X_n\}_{n<\omega}$ and $\mathcal{Y} = \{Y_n\}_{n<\omega}$ be sequences of sets of natural numbers. $\mathcal{X} \leq_e \mathcal{Y}$ if for all n, $X_n \leq_e Y_n$ uniformly in n. This reducibility is useful in many considerations, however it does not quite characterize ω-enumeration reducibility. The true characterization was given by Soskov and Kovachev:

Theorem 16 [77]. $\mathcal{X} \leq_\omega \mathcal{Y} \iff \mathcal{X} \leq_e \mathcal{P}(\mathcal{Y})$.

Clearly "$\leq_\omega$" is a reflexive and transitive relation on the set $\mathcal{S}$ of all sequences of sets of natural numbers and induces the equivalence relation $\equiv_\omega$. For every sequence $\mathcal{X}$ the set $d_\omega(\mathcal{X}) = \{\mathcal{Y} \mid \mathcal{Y} \equiv_\omega \mathcal{X}\}$ is the ω-enumeration degree of the sequence $\mathcal{X}$ and $\mathcal{D}_\omega = \{d_\omega(\mathcal{X}) \mid \mathcal{X} \in \mathcal{S}\}$ is the *structure of the ω-enumeration degrees*. The relation $\leq_\omega$ induces a partial ordering of $\mathcal{D}_\omega$ with least element $\mathbf{0}_\omega = d_\omega(\emptyset_\omega)$, where $\emptyset_\omega$ is the sequence with all members equal to $\emptyset$. $\mathcal{D}_\omega$ is further an upper semi-lattice, with least upper bound induced by $\mathcal{X} \oplus \mathcal{Y} = \{X_n \oplus Y_n\}_{n<\omega}$. There is a natural embedding of the enumeration degrees into the ω-enumeration degrees. Given a set A of natural numbers denote by $A \uparrow \omega$ the sequence $\{A_k\}_{k<\omega}$, where $A_0 = A$ and for all $k \geq 1$, $A_k = \emptyset$. The embedding is $\kappa : \mathcal{D}_e \to \mathcal{D}_\omega$ by $\kappa(d_e(A)) = d_\omega(A \uparrow \omega)$.

For every $\mathcal{X} \in \mathcal{S}$ the *ω-enumeration jump of $\mathcal{X}$* is $\mathcal{X}' = \{\mathcal{P}_{n+1}(\mathcal{X})\}_{n<\omega}$. We have that $J'_{\mathcal{X}} = \{\mathbf{a}' \mid \mathbf{a} \in J_{\mathcal{X}}\}$. The jump operator is monotone and it induces a jump operation on the ω-enumeration degrees. It agrees with the jump operation on $\mathcal{D}_e$ and the embedding κ. It turns out that the ω-enumeration degrees behave in an unusual way with respect to the considered jump operation. In [27] Soskov and Ganchev prove the following strong jump inversion theorem: for every $n \in \mathbb{N}$ and for $\mathbf{a}^{(n)} \leq \mathbf{b}$ there exists a **least** ω-enumeration degree $\mathbf{x} \geq \mathbf{a}$ such that $\mathbf{x}^{(n)} = \mathbf{b}$. So we can define an operation $I_{\mathbf{a}}^n$ on the upper cone with a least element $\mathbf{a}^{(n)}$ such that $I_{\mathbf{a}}^n(\mathbf{b})$ is the least solution $\mathbf{x}$ of this system: $\mathbf{x} \geq \mathbf{a}$ such that $\mathbf{x}^{(n)} = \mathbf{b}$. Let $\mathbf{o}_n = I_{\mathbf{0}_\omega}^n(\mathbf{0}_\omega^{(n+1)})$, i.e. $\mathbf{o}_n$ denotes the least ω-enumeration degree, such that $\mathbf{o}_n^{(n)} = \mathbf{0}_\omega^{(n+1)}$. We have $\mathbf{0}'_\omega = \mathbf{o}_0 \geq \mathbf{o}_1 \geq \cdots \geq \mathbf{o}_n \geq \cdots$. The sequence is strictly decreasing but it does not converge to the least degree $\mathbf{0}_\omega$. The authors proved the existence of almost zero nontrivial degrees which are nonzero and below all $\mathbf{o}_n$. A nontrivial almost zero ω-enumeration degree contains a sequence $\mathcal{R}$ such that $(\forall n)(\mathcal{P}_n(\mathcal{R}) \equiv_e \emptyset^{(n)})$, but non-uniformly.

A. Soskova [89] generalizes the enumeration degree spectrum with respect to an infinite sequences of sets using ω-enumeration reducibility. Let $\mathcal{B} = \{B_n\}_{n<\omega}$ be a sequence of sets of natural numbers and $\mathfrak{A}$ be a countable structure on the natural numbers.

Definition 12. The ω-degree spectrum *of the structure* $\mathfrak{A}$ *with respect to the sequence* $\mathcal{B}$ *is the set*

$$DS(\mathfrak{A}, \mathcal{B}) = \{d_e(f^{-1}(\mathfrak{A})) \mid f \text{ - enumeration of } \mathfrak{A} \text{ s.t. } \{f^{-1}(B_n)\} \leq_\omega \{f^{-1}(\mathfrak{A})^{(n)}\}\}.$$

The ω-*co-spectrum of* $DS(\mathfrak{A}, \mathcal{B})$ *is the set* $Ocsp(\mathfrak{A}, \mathcal{B})$ *of* ω-*enumeration degrees, which are lower bounds of the* ω-*spectrum.*

Note that if $\mathcal{B}$ is the sequence of empty sets then $DS(\mathfrak{A}, \mathcal{B}) = DS(\mathfrak{A})$. The set $Ocsp(\mathfrak{A}, \mathcal{B})$ is in this case a new meaningful notion and we will denote it by $Ocsp(\mathfrak{A})$.

Most properties of co-spectra, such as the existence of minimal pairs and quasi-minimal degrees, hold for the ω-co-spectra, but not all. For every structure $\mathfrak{A}$ and $n > 0$ if $\mathbf{c} \in DS_n(\mathfrak{A})$ then $CS_n(\mathfrak{A})$ is the co-set of $\mathcal{A} = \{\mathbf{a} \mid \mathbf{a} \in DS(\mathfrak{A}) \ \& \ \mathbf{a}^{(n)} = \mathbf{c}\}$. Vatev [96] shows that there is a structure $\mathfrak{A}$, a sequence $\mathcal{B}$ and $\mathbf{c} \in DS_n(\mathfrak{A}, \mathcal{B})$ such that if $\mathcal{A} = \{\mathbf{a} \in DS(\mathfrak{A}, \mathcal{B}) \mid \mathbf{a}^{(n)} = \mathbf{c}\}$ then $CS(\mathfrak{A}, \mathcal{B}) \neq co(\mathcal{A})$.

A. Soskova gives a characterization of the k-th ω-co-spectrum of a structure (the co-set of the k-th jump ω-spectrum) in terms of definability via computable sequence $\{\Phi^{\gamma(n,x)}\}_{n,x<\omega}$ of formulas such that for every n, $\Phi^{\gamma(n,x)}$ is a Σ_{n+k}^+ infinitary computable formula with parameters. This set is also characterized as the least ideal of ω-enumeration degrees containing the k-th jumps of elements of the ω-co-spectrum. The set $I = CS(\mathfrak{A}, \mathcal{B})$ is a countable ideal. By the minimal pair theorem there exist total enumeration degrees $\mathbf{f}, \mathbf{g}$ in $DS(\mathfrak{A}, \mathcal{B})$, such that $CS(\mathfrak{A}, \mathcal{B}) = I(\mathbf{f}_\omega) \cap I(\mathbf{g}_\omega)$ where $I(\mathbf{f}_\omega)$ and $I(\mathbf{g}_\omega)$ are the principal ideals of ω-enumeration degrees with greatest elements $\mathbf{f}_\omega = \kappa(\mathbf{f})$ and $\mathbf{g}_\omega = \kappa(\mathbf{g})$, the images of $\mathbf{f}$ and $\mathbf{g}$ under the embedding κ of $\mathcal{D}_e$ in $\mathcal{D}_\omega$. Denote by $I^{(k)}$ the least ideal, containing all k-th ω-jumps of the elements of I. Ganchev [23] proves that if $I = I(\mathbf{f}_\omega) \cap I(\mathbf{g}_\omega)$ then $I^{(k)} = I(\mathbf{f}_\omega^{(k)}) \cap I(\mathbf{g}_\omega^{(k)})$ for every k. But $I(\mathbf{f}_\omega^{(k)}) \cap I(\mathbf{g}_\omega^{(k)}) = CS_k(\mathfrak{A}, \mathcal{B})$ for each k. Thus $I^{(k)} = CS_k(\mathfrak{A}, \mathcal{B})$, i.e. the k-th omega co-spectrum is a minimal ideal containing the k-th jumps of elements of the ω-co-spectrum.

Using this result Ganchev, A. Soskova and Vatev show another difference between co-spectra and ω-co-spectra: There is a countable ideal I of ω-enumeration degrees for which there is no structure $\mathfrak{A}$ and sequence $\mathcal{B}$ such that $I = CS(\mathfrak{A}, \mathcal{B})$. Let
$$\mathcal{A} = \{\mathbf{0}_\omega, \mathbf{0}'_\omega, \mathbf{0}''_\omega, \ldots, \mathbf{0}^{(n)}_\omega, \ldots\}.$$
and consider the countable ideal I generated by $\mathcal{A}$. Assume now that there is a structure $\mathfrak{A}$ and a sequence $\mathcal{B}$ such that $I = CS(\mathfrak{A}, \mathcal{B})$ and let $\mathbf{f}$ and $\mathbf{g}$ be a minimal pair of total enumeration degrees for $DS(\mathfrak{A}, \mathcal{B})$. It follows that $I^{(n)} = I(\mathbf{f}^{(n)}_\omega) \cap I(\mathbf{g}^{(n)}_\omega)$ for each n. But $\mathbf{f}_\omega \geq \mathbf{0}^{(n)}_\omega$ and $\mathbf{g}_\omega \geq \mathbf{0}^{(n)}_\omega$ for each n. If $F \in \mathbf{f}$ and $G \in \mathbf{g}$ are total, then $F \geq_T \emptyset^{(n)}$ and $G \geq_T \emptyset^{(n)}$ for each n. By Enderton and Putnam (1970) [15], Sacks (1971) [63] : $F'' \geq_T \emptyset^{(\omega)}$ and $G'' \geq_T \emptyset^{(\omega)}$ and hence $\mathbf{f}'' \geq_T \mathbf{0}^{(\omega)}_T$ and $\mathbf{g}'' \geq_T \mathbf{0}^{(\omega)}_T$. Then $\kappa(\iota(\mathbf{0}^{(\omega)}_T)) \in I(\mathbf{f}''_\omega) \cap I(\mathbf{g}''_\omega)$, but $\kappa(\iota(\mathbf{0}^{(\omega)}_T)) \notin I''$ since all elements of I'' are bounded by $\mathbf{0}^{(k+2)}_\omega$ for some k. Hence $I'' \neq I(\mathbf{f}''_\omega) \cap I(\mathbf{g}''_\omega)$, a contradiction.

Inspired by this Vatev [96] investigates the principal ideal case. He shows that for every principal ideal of ω-enumeration degrees I there is sequence $\mathcal{B}$ and a structure $\mathfrak{A}$ such that $I = CS(\mathfrak{A}, \mathcal{B})$.

6   Jump of a Structure

The idea of the *jump* of a structure is first considered by Soskov and his student Baleva [5] in the context of s-reducibility between structures, a reducibility based on relative search computability. Given a structure $\mathfrak{A}$ the goal is to define a structure $\mathfrak{A}'$ so that $\mathfrak{A}'$ knows the sets definable by computable infinitary Σ^c_1 formulas in $\mathfrak{A}$. The idea to define such a structure resurfaced in computable structure theory in the period 2002–2010 independently in the work of Soskov and Soskova [90], Montalbán [50] and Stukachev [93, 94]. Soskov and A. Soskova [90] define the jump $\mathfrak{A}'$ of the structure $\mathfrak{A}$ by considering the Moschovakis' extension of $\mathfrak{A}$ together with a predicate, an analogue of the halting set, which codes all sets definable by computable infinitary Σ^c_1 formulas with parameters. This changes the domain of the structure, but keeps the language finite. Montalbán's approach was to keep the domain of the structure the same and to add a complete set of relations definable by computable infinitary Π^c_1 formulas. This can possibly make the language infinite, however Montalbán [35, 50, 51] gives some examples of structures, such as linear orderings and Boolean algebras, where the complete set of relations is finite and natural. Stukachev's approach is in terms of Σ-definability in hereditarily finite extension of the structure. We will focus on the approach taken by Soskov and Soskova.

Let $\mathfrak{A} = (A; R_1, \ldots, R_n)$ be a countable structure and let equality be among the predicates $R_1, \ldots, R_s$. Following Moschovakis [55] we define an extension of $\mathfrak{A}$ as follows. Let $\bar{0}$ be a new element, such that $\bar{0} \notin A$ and let $A_0 = A \cup \{\bar{0}\}$. Let $\langle .,. \rangle$ be a pairing function such that none of the elements of A_0 is a pair. The set A^* is the closure of A_0 under $\langle .,. \rangle$ and functions $L(\langle s, t \rangle) = s$ and $R(\langle s, t \rangle) = t$ are decoding functions. We next represent the basic relations in $\mathfrak{A}^*$ by unary relations in $\mathfrak{A}^*$ as follows: $R^*_i(\langle s_1, \ldots, s_{k_i} \rangle) = R_i(s_1, \ldots, s_{k_i})$.

Definition 13. Moschovakis' extension [55] *of* $\mathfrak{A}$ *is the structure*

$$\mathfrak{A}^* = (A^*, R_1^*, \ldots, R_n^*, A_0, G_{\langle .,. \rangle}, G_L, G_R).$$

It is straightforward to check that for any countable structure $\mathfrak{A}$ the structure $\mathfrak{A}^*$ has the same complexity as $\mathfrak{A}$, namely $DS(\mathfrak{A}) = DS(\mathfrak{A}^*)$. The advantage to considering $\mathfrak{A}^*$ is that in it we can code a copy of the natural numbers $\mathbb{N}^*$ in A^* by induction: $\bar{0}^* = \bar{0}$ and $\overline{(n+1)}^* = \langle \bar{0}, \bar{n}^* \rangle$. Using $\mathbb{N}^*$ we can now represent the graph of every finite part $\tau : \mathbb{N} \to A$ as an element τ^* of $\mathfrak{A}^*$. Let

$$K_{\mathfrak{A}} = \{\langle \delta^*, \bar{e}^*, \bar{x}^* \rangle : (\exists \tau \supseteq \delta)(\tau \Vdash_0 F_e(x))\}.$$

Soskov and A. Soskova define the jump only for total structure $\mathfrak{A}^+$. In light of Theorem 4 there is a natural way to extend this definition to non-total structures.

Definition 14. *The jump of the structure* $\mathfrak{A}^+$ *is the structure*

$$\mathfrak{A}' = ((\mathfrak{A}^+)^*, K_{\mathfrak{A}}, A^* \setminus K_{\mathfrak{A}}).$$

Note, that the structure $\mathfrak{A}'$ is also total. The next property can be viewed as a correctness statement: it reaffirms that this definition of the jump of a structure is truly an analog of the jump operator on sets of natural numbers. The main technique used in its proof is once again that of generic enumerations.

Theorem 17 [88,90][2]. *For every countable structure* $\mathfrak{A}$, $DS_1(\mathfrak{A}^+) = DS(\mathfrak{A}')$.

Another proof of this theorem was published independently by Montalbán [50]. Montalbán called it in [51] the second jump inversion theorem. Both proofs are essentially the same, even though the great differences in the underlying setting make them look quite different.

Vatev [98,99] extends the jump of a structure to the α-th jump of a structure for arbitrary computable ordinal α. Vatev's approach [97] relies on the notion of *conservative extension*. This notion provides a finer way to compare the relative definability between two structures at arbitrary levels of the Σ_α^c-hierarchy. Given two countable structures $\mathfrak{A}$ and $\mathfrak{B}$ with $|\mathfrak{A}| \subseteq |\mathfrak{B}|$ and α, β computable ordinals the structure $\mathfrak{B}$ is an (α, β) conservative extension of $\mathfrak{A}$ if for every enumeration g of $\mathfrak{B}$ there is an enumeration f of $\mathfrak{A}$ such that $\{\langle x, y \rangle \mid f(x) = g(y)\}$ is Σ_β^0 in $g^{-1}(\mathfrak{B})$ and $f^{-1}(\mathfrak{A})^{(\alpha)} \leq_T g^{-1}(\mathfrak{B})^{(\beta)}$, and the opposite, for every enumeration f of $\mathfrak{A}$ there is an enumeration g of $\mathfrak{B}$ such that $\{\langle x, y \rangle \mid f(x) = g(y)\}$ is Σ_α^0 in $f^{-1}(\mathfrak{A})$ and $g^{-1}(\mathfrak{B})^{(\beta)} \leq_T f^{-1}(\mathfrak{A})^{(\alpha)}$. He proved that if $\mathfrak{B}$ is an (α, β) conservative extension of $\mathfrak{A}$ then $(\forall X \subseteq |\mathfrak{A}|)(X \in \Sigma_\alpha^c(\mathfrak{A}) \iff X \in \Sigma_\beta^c(\mathfrak{B}))$. He showed furthermore that $\mathfrak{A}^{(\alpha+1)}$ is $(\beta+1, \beta)$ conservative extension of $\mathfrak{A}^{(\alpha)}$ and from here it follows that the $\Sigma_{\alpha+1}^c$ definable in $\mathfrak{A}^*$ subsets of A^* are exactly the Σ_α^c definable sets in $\mathfrak{A}'$. More generally, he shows that for any computable

---

[2] Theorem 17 was first announced by Soskov during his LC talk in Münster in 2002.

ordinals α, β the $\Sigma^c_{\beta+1}$ definable sets in $\mathfrak{A}^{(\alpha)}$ are exactly the Σ^c_β definable sets in $\mathfrak{A}^{(\alpha+1)}$.

Naturally, once we have a jump of a structure, the question of jump inversion arises: Given a structure $\mathfrak{B}$ with $DS(\mathfrak{B})$ consisting of total degree above $\mathbf{0}'_e$, is there a structure $\mathfrak{C}$ such that $DS_1(\mathfrak{C}) = DS(\mathfrak{B})$. Soskova and Soskov prove an even more general statement. For every structure $\mathfrak{B}$, denote by $DS_t(\mathfrak{B})$ the set of total elements in $DS(\mathfrak{B})$. (In particular, if $\mathfrak{B}$ is total then $DS(\mathfrak{B}) = DS_t(\mathfrak{B})$.)

Theorem 18 [87,88,90]. *Let $\mathfrak{A}$ and $\mathfrak{B}$ be structures such that $DS(\mathfrak{B})_t \subseteq DS_n(\mathfrak{A})$. Then there exists a structure $\mathfrak{C}$ such that $DS(\mathfrak{C}) \subseteq DS(\mathfrak{A})$ and $DS_n(\mathfrak{C}) = DS(\mathfrak{B})_t$.*

The proof of Theorem 18 uses the method of Marker extensions, which will be discussed in detail in Sect. 7. This method is also used by Stukachev [93,94] for similar jump inversion theorem for his notion of the jump of a structure based on Σ-definability. Downey and Knight [14] prove, using a fairly complicated construction, that for every computable ordinal α there exists a structure $\mathfrak{A}$ (a linear ordering, in fact) such that $\mathfrak{A}$ has α-th jump degree equal to $\mathbf{0}^{(\alpha)}$, but no β-th jump degree for any $\beta < \alpha$. Now we can obtain this theorem for the finite ordinals as an application of Theorem 18. Consider a structure $\mathfrak{B}$ such that $DS(\mathfrak{B})$ consists of total elements above $\mathbf{0}^{(n)}_e$ and has no least element, and such that $\mathbf{0}^{(n+1)}_e$ is the least element of $DS_1(\mathfrak{B})$. Let $\mathfrak{A}$ be any total computable structure. Clearly $DS(\mathfrak{B}) \subseteq DS_n(\mathfrak{A})$. By Theorem 18 there exists a structure $\mathfrak{C}$ such that $DS_n(\mathfrak{C}) = DS(\mathfrak{B})$. Therefore $\mathfrak{C}$ does not have a n-th jump degree and so no k-th jump degree for $k \leq n$. On the other hand $DS_{n+1}(\mathfrak{C}) = DS_1(\mathfrak{B})$ and hence the $(n+1)$-th jump degree of $\mathfrak{C}$ is $\mathbf{0}^{(n+1)}_e$. Why does such a structure $\mathfrak{B}$ exist? Consider a degree $\mathbf{q}$ that is quasi-minimal relative to $\mathbf{0}^{(n)}_e$ and with $\mathbf{q}' = \mathbf{0}^{(n+1)}_e$. Let $\mathfrak{B} = G$ be the torsion free abelian group of rank 1 such that $s_G = \mathbf{q}$. Recall that $DS(G) = \{\mathbf{a} \mid s_G \leq_e \mathbf{a} \text{ and } \mathbf{a} \text{ is total}\}$ and the first jump degree of G is s'_G.

The next natural questions is if one can extend the jump inversion theorem to every constructive ordinal α. Goncharov, Harizanov, Knight, McCoy, Miller and Solomon [32] show that this is true if α is a computable successor ordinal, even though they do not state their result in terms of the jump of a structure. This result was useful later on, for instance Greenberg, Montalbán and Slaman [33] use it to build a structure whose spectrum consists of the non-hyperarithmetic degrees. Vatev [98–100] proves the α-jump inversion theorem for a computable successor ordinal α based on the construction in [32].

The problem of jump inversion for $\alpha = \omega$, or, in general, any computable limit ordinal remains open for longer. In one of his last papers Soskov [80] finally proves that there is a good reason for that.

Theorem 19 [80]. *There is a total structure $\mathfrak{A}$ with $DS(\mathfrak{A}) \subseteq \{\mathbf{b} \mid \mathbf{0}^{(\omega)}_e \leq \mathbf{b}\}$ for which there is no structure $\mathfrak{M}$ with $DS_\omega(\mathfrak{M}) = DS(\mathfrak{A})$.*

The proof relies on an analysis of the ω-jump co-spectrum of a structure. Soskov shows that every member of $\mathbf{a} \in CS_\omega(\mathfrak{M})$ is bounded by a total $\mathbf{b}$, which is also a member of $CS_\omega(\mathfrak{M})$. To see this, let $R \in \mathbf{a}$ and $\mathbf{a} \in CS_\omega(\mathfrak{M})$. It follows from Theorem 7 that the set R is Σ_ω^c definable in $\mathfrak{M}$ and hence there is a computable function γ and parameters $t_1, \ldots, t_m$ of $|\mathfrak{M}|$ such that $x \in R \iff \mathfrak{M} \models F_{\gamma(x)}(t_1, \ldots, t_m)$. Each $F_{\gamma(x)}$ is a computable Σ_ω^c formula, i.e. a c.e. disjunction of computable Σ_{n+1}^c formulas, where $n < \omega$, and so there is a computable function $\delta(n, x)$ such that for all n and x, $\delta(n, x)$ yields a code of some computable Σ_{n+1}^c formula $F_{\delta(n,x)}$ and $x \in R \iff (\exists n)(\mathfrak{M} \models F_{\delta(n,x)}(t_1, \ldots, t_m))$.

Let $R_n = \{x \mid x \in \mathbb{N} \wedge \mathfrak{M} \models F_{\delta(n,x)}(t_1, \ldots, t_n)\}$ and let $\mathbf{b} = d_e(\mathcal{P}_{<\omega}(\{R_n\}))$. Note that $\mathbf{b}$ is a total enumeration degree. It is easy to see that for every enumeration f of $\mathfrak{M}$ we have that $\{R_n\} \leq_e \{f^{-1}(\mathfrak{M})^{(n)}\}$ uniformly in n. It follows that $\mathcal{P}(\{R_n\}) \leq_e \{f^{-1}(\mathfrak{M})^{(n)}\}$ and so $\mathcal{P}_{<\omega}(\{R_n\}) \leq_e f^{-1}(\mathfrak{M})^{(\omega)}$, i.e. $\mathbf{b} \in CS_\omega(\mathfrak{M})$. On the other hand $x \in R \iff (\exists n)(x \in R_n)$ and so $R \leq_e \bigoplus_n R_n \leq_e \mathcal{P}_{<\omega}(\{R_n\})$. Thus $\mathbf{a} \leq_e \mathbf{b}$.

To complete the proof of Theorem 19, let $\mathfrak{A}$ be a total structure with co-spectrum $CS(\mathfrak{A}) = \{\mathbf{a} \mid \mathbf{a} \leq_e \mathbf{y}\}$, where $\mathbf{y}$ is some quasi-minimal above $\mathbf{0}_e^{(\omega)}$ degree. We have already seen that such an $\mathfrak{A}$ exists, as every principal ideal is the co-spectrum of a total structure. Then $DS(\mathfrak{A}) \subseteq \{\mathbf{a} \mid \mathbf{0}_e^{(\omega)} \leq_e \mathbf{a}\}$, but $DS(\mathfrak{A})$ cannot be the ω-jump spectrum of any structure $\mathfrak{M}$. If we assume otherwise then $CS_\omega(\mathfrak{M}) = CS(\mathfrak{A})$ and so $\mathbf{y}$ must be bounded by a total enumeration degree $\mathbf{b} \in CS(\mathfrak{A})$. Since $\mathbf{y}$ is the greatest element of $CS(\mathfrak{A})$, $\mathbf{b} = \mathbf{y}$ contradicting the choice of $\mathbf{y}$.

7 Generalized Marker Extensions for Sequences of Structures

The last paper by Soskov [81] settles a series of questions relating to the connections between Turing degree spectra, enumeration degree spectra and spectra of sequences of structures. The main technique is that of Marker extensions. Marker's method [47] is originally used in model theory. The computable content of this construction is established in the work of Goncharov and Khoussainov [31]. Soskov gives a more general version of this approach.

We introduce Soskov's ideas with a simple example. Consider a countable structure $\mathfrak{A}$. A set $Y \subseteq |\mathfrak{A}|$ is relatively intrinsically c.e. in $\mathfrak{A}$ if for every enumeration f of $\mathfrak{A}$ we have that $f^{-1}(Y)$ is c.e. in $f^{-1}(\mathfrak{A})$, or equivalently if Y is definable by some computable infinitary Σ_1^c formula. In this definition $\mathfrak{A}$ is treated as a total object, in particular $f^{-1}(\mathfrak{A})$ is treated as a total oracle. Alternatively, we can consider sets $Y \subseteq |\mathfrak{A}|$, such that Y is (relatively intrinsically) enumeration reducible to $\mathfrak{A}$, i.e. for every enumeration f of $\mathfrak{A}$ we have that $f^{-1}(Y) \leq_e f^{-1}(\mathfrak{A})$, or equivalently if Y is definable by some positive computable infinitary Σ_1^+ formula. In the second case $f^{-1}(\mathfrak{A})$ is treated as a partial oracle. These two notions are in general different, but to what extent? Are there classes of sets that can be characterized as the ones that are enumeration reducible to a fixed structure, but cannot be characterized as the sets that are relatively intrinsically c.e. in any structure. If we move away from computable structure theory and view the analogous question simply in terms of the relations $\leq_e$ and "c.e. in" the question becomes: is it true that for every set A there is a set M

such that $\{Y \mid Y \leq_e A\} = \{Y \mid Y$ is c.e in M$\}$? The answer to this last question is clearly "no", as there are sets A that are not enumeration equivalent to any total set. So are there truly partial structures in this same sense? Soskov [81] reveals that surprisingly computable structure theory differs from degree theory in this respect: for every structure $\mathfrak{A}$, there is a structure $\mathfrak{M}$, such that for every $Y \subseteq |\mathfrak{A}|$, $Y \leq_e \mathfrak{A}$ if and only if Y is c.e. in $\mathfrak{M}$.

For simplicity let $\mathfrak{A} = (A; R)$ and $R \subseteq A$ is infinite. The 0-th Marker extension $\mathfrak{M}$ of $\mathfrak{A}$ is constructed as follows. Consider an infinite countable set X disjoint from A and a bijection $h : R \rightarrow X$. Let $M(a,x)$ be true if and only if $h(a) = x$. Let $\mathfrak{M} = (A \cup X; A, X, M)$, where A and X are unary predicates. Note that R is Σ_1^0 definable in $\mathfrak{M}$ since $R(a) \Leftrightarrow (\exists x \in X)M(a, x)$. Now consider any set $Y \subseteq A$ such that $Y \leq_e \mathfrak{A}$. It is straightforward to check that for every enumeration f of $\mathfrak{M}$ $f^{-1}(Y)$ is c.e. in $f^{-1}(\mathfrak{M})$: Indeed, using a computable in $f^{-1}(\mathfrak{M})$ bijection from $\mathbb{N}$ to $f^{-1}(A)$ we can transform f into an enumeration g of the structure $\mathfrak{A}$. Now we have that $g^{-1}(Y) \leq_e g^{-1}(\mathfrak{A}) = g^{-1}(R)$, and $f^{-1}(R)$ is c.e. in $f^{-1}(\mathfrak{M})$. Since we can pass between f and g using oracle $f^{-1}(\mathfrak{M})$ it follows that $f^{-1}(Y)$ is c.e. in $f^{-1}(\mathfrak{M})$.

For the reverse direction, we show that if $Y \not\leq_e \mathfrak{A}$ then Y is not relatively intrinsically c.e. in $\mathfrak{M}$, i.e. that there is an enumeration f of $\mathfrak{M}$ such that $f^{-1}(Y)$ is not c.e. in $f^{-1}(\mathfrak{M})$. Let g be an enumeration of $\mathfrak{A}$ such that $g^{-1}(Y) \not\leq_e g^{-1}(\mathfrak{A})$. We construct f so that $f(2n) = g(n)$. To fill in $f(2\mathbb{N})$ we construct a bijection $k : f^{-1}(R) \rightarrow 2\mathbb{N} + 1$ and complete f by $f(2n+1) = h(f(k^{-1}(2n+1)))$. Note that then we will have $f^{-1}(\mathfrak{M}) \equiv_e f^{-1}(M) \equiv_e G_k$ and $f^{-1}(Y) \equiv_e g^{-1}(Y)$. We construct k using forcing so that statements of the form $x \in \Gamma_e(G_k^+)$ are decided at finite stages. For $\sigma : f^{-1}(R) \rightarrow 2\mathbb{N} + 1$ we say that $\sigma \Vdash x \in \Gamma_e(G_k^+)$ if there exists v, such that $\langle x, v \rangle \in \Gamma_e$ and for every $u \in D_v$ we have $u = 2\langle a, x \rangle$ and $\sigma(a) = x$ or $u = 2\langle a, x \rangle + 1$ and $\sigma(b) = x$ for some $b \neq a$. Then the set $\{x \mid \exists \sigma \supseteq \tau(\sigma \Vdash x \in \Gamma_e(G_k^+))\}$ is enumeration reducible to $g^{-1}(\mathfrak{A})$. We use this to ensure that $g^{-1}(Y) \neq \Gamma_e(G_k^+)$ and thus Y is not c.e. in $\mathfrak{M}$.

Let $\vec{\mathfrak{A}} = \{\mathfrak{A}_n\}_{n<\omega}$ be a sequence of structures, where $\mathfrak{A}_n = (A_n; R_1^n, R_2^n, \ldots R_{m_n}^n)$. An enumeration f of $\vec{\mathfrak{A}}$ is a bijection from $\mathbb{N} \rightarrow A = \bigcup_n A_n$. For every n let $f^{-1}(\mathfrak{A}_n) = f^{-1}(A_n) \oplus f^{-1}(R_1^n) \cdots \oplus f^{-1}(R_{m_n}^n)$ and let $f^{-1}(\vec{\mathfrak{A}})$ be the sequence $\{f^{-1}(\mathfrak{A}_n)\}_{n<\omega}$.

In this setting we can talk about a sequence of sets that is *relatively intrinsically ω -enumeration* reducible to $\vec{\mathfrak{A}}$: a sequence $\{Y_n\}_{n<\omega}$ of subsets of A, such that for every enumeration f of $\vec{\mathfrak{A}}$, $\{f^{-1}(Y_n)\} \leq_\omega f^{-1}(\vec{\mathfrak{A}})$. Soskov and Baleva [76] and A. Soskova [89] show that sequence of this kind also have a syntactic characterization: Y_n is uniformly in n definable by a positive computable infinitary Σ_{n+1}^+ formula with predicates only from the first n structures, such that the predicates for the n-th appear for the first time at level $n+1$ positively. We can compare this notion to the following: say that a sequence $\{Y_n\}_{n<\omega}$ of subsets of A is relatively intrinsically c.e. in a structure $\mathfrak{M}$ with $A \subseteq |\mathfrak{M}|$ if for every enumeration f of $\mathfrak{M}$ the set $f^{-1}(Y_n)$ is $\Sigma_{n+1}^0(f^{-1}(\mathfrak{M}))$ uniformly in n.

The key idea is to generalize Marker extensions to the sequence $\vec{\mathfrak{A}}$. First we must define the n-th Marker extension of a predicate. Let $\mathfrak{A} = (A; R_1, \ldots, R_k)$

and $R \subseteq A^m$. The n-th *Marker extension* of R is a structure $\mathfrak{M}_n(R)$ defined as follows. Consider new infinite disjoint countable sets $X_0, X_1, \ldots X_n$ called *companions*.

Fix bijections: $h_0 : R \to X_0$

$\quad h_1 : (A^m \times X_0) \setminus G_{h_0} \to X_1$

$\quad \ldots$

$\quad h_n : (A^m \times X_0 \times X_1 \cdots \times X_{n-1}) \setminus G_{h_{n-1}} \to X_n.$

Let $M_n = G_{h_n}$ and $\mathfrak{M}_n(R) = (A \cup X_0 \cup \cdots \cup X_n; X_0, X_1, \ldots X_n, M_n)$. Notice, that R is Σ^0_{n+1} definable in $\mathfrak{M}_n(R)$ since for $\bar{a} \in A^m$ we have

$$R(\bar{a}) \iff (\exists x_0 \in X_0)(G_{h_0}(\bar{a}, x_0))$$

and for all $k < n, x_0 \in X_0, \ldots, x_k \in X_k$ we have

$$G_{h_k}(\bar{a}, x_0, \ldots, x_k) \iff (\forall x_{k+1} \in X_{k+1}) \neg G_{h_{k+1}}(\bar{a}, x_0, \ldots, x_k, x_{k+1}).$$

Next we define $\mathfrak{M}(\vec{\mathfrak{A}})$ for the sequence of structures $\vec{\mathfrak{A}} = \{\mathfrak{A}_n\}_{n<\omega}$.

(1) For every n construct the n-th Marker extensions of $A_n, R^n_1, \ldots R^n_{m_n}$ with disjoint companions.
(2) For every n let $\mathfrak{M}_n(\mathfrak{A}_n) = \mathfrak{M}_n(A_n) \cup \mathfrak{M}_n(R^n_1) \cup \cdots \cup \mathfrak{M}_n(R^n_{m_n})$.
(3) Set $\mathfrak{M}(\vec{\mathfrak{A}})$ to be $(\bigcup_n \mathfrak{M}_n(\mathfrak{A}_n))^+$ with additional predicate for $A = \bigcup_n A_n$ and $\overline{A}$.

Note that $\mathfrak{M}(\vec{\mathfrak{A}})$ is a total structure.

Soskov [81] describes the relationship between the enumerations of $\vec{\mathfrak{A}}$ and $\mathfrak{M} = \mathfrak{M}(\vec{\mathfrak{A}})$: It is not too difficult to see that for every enumeration f of $\mathfrak{M}(\vec{\mathfrak{A}})$ there is an enumeration g of $\vec{\mathfrak{A}}$ such that:

(1) the set $\{\langle x, y \rangle \mid f(i) = g(j)\}$ is computable in $f^{-1}(\mathfrak{M})$.
(2) $\mathcal{P}_n(g^{-1}(\vec{\mathfrak{A}})) \leq_e (f^{-1}(\mathfrak{M}))^{(n)}$ uniformly in n.
(3) $\mathcal{P}_{<\omega}(g^{-1}(\vec{\mathfrak{A}})) \leq_T (f^{-1}(\mathfrak{M}))^{(\omega)}$.

The reverse relationship requires an elaborate forcing construction: For every enumeration g of $\vec{\mathfrak{A}}$ and $\mathcal{Y} \not\leq_\omega g^{-1}(\vec{\mathfrak{A}})$ there is an enumeration f of $\mathfrak{M}$:

(1) the set $\{\langle x, y \rangle \mid f(i) = g(j)\}$ is computable.
(2) $\mathcal{P}_{<\omega}(g^{-1}(\vec{\mathfrak{A}})) \equiv_e (f^{-1}(\mathfrak{M}))^{(\omega)}$.
(3) $\mathcal{Y}$ is not c.e. in $f^{-1}(\mathfrak{M})$.

Our simple example is transformed to the following general theorem:

Theorem 20 [81]. *A sequence $\mathcal{Y}$ of subsets of A is relatively intrinsically ω-enumeration reducible to $\vec{\mathfrak{A}}$ if and only if $\mathcal{Y}$ is relatively intrinsically c.e. in $\mathfrak{M}(\vec{\mathfrak{A}})$.*

The structure $\mathfrak{M} = \mathfrak{M}(\vec{\mathfrak{A}})$ has very interesting properties. The first one considered in [81] is a characterization its co-spectrum.

Theorem 21

(1) *The n-th co-spectrum of $\mathfrak{M}$ is*

$$CS_n(\mathfrak{M}) = \{d_e(Y) \mid \text{ for every enumeration } g \text{ of } A,\ Y \leq_e \mathcal{P}_n(g^{-1}(\vec{\mathfrak{A}}))\}.$$

(2) *The ω-co-spectrum of $\mathfrak{M}$ is*

$$Ocsp(\mathfrak{M}) = \{d_\omega(\mathcal{Y}) \mid \text{ for every enumeration } g \text{ of } A,\ \mathcal{Y} \leq_\omega g^{-1}(\vec{\mathfrak{A}})\}.$$

Theorem 21 allows us to construct examples of structures with interesting properties in an easy way. Let $\mathcal{R} = \{R_n\}$ be a sequence of sets. Consider the sequence $\vec{\mathfrak{A}}_\mathcal{R}$, where $\mathfrak{A}_0 = (\mathbb{N}; G_S, R_0)$, here G_S is the graph of the successor function, and for all $n \geq 1$, $\mathfrak{A}_n = (\mathbb{N}; R_n)$. Then it is not too hard to see that for every n we have that $CS_n(\mathfrak{M}(\vec{\mathfrak{A}}_\mathcal{R})) = \{d_e(Y) \mid Y \leq_e \mathcal{P}_n(\mathcal{R})\}$ and for each enumeration g of $\mathbb{N}$ $\mathcal{R} \leq_\omega g^{-1}(\vec{\mathfrak{A}}_\mathcal{R})$.

When one takes $\mathcal{R}$ to be an almost zero sequence, we obtain a structure $\mathfrak{M}(\vec{\mathfrak{A}}_\mathcal{R})$ with n-th co-degree $\mathbf{0}_e^{(n)}$, but no n-th jump degree for any n. Indeed, recall that an almost zero sequence $\mathcal{R}$ is one that is not ω-enumeration reducible to $\mathbf{0}_\omega$, but has the property that $\mathcal{P}_n(\mathcal{R}) \equiv_e \emptyset^{(n)}$ for every n. If we assume that the n-th jump degree of $\mathfrak{M} = \mathfrak{M}(\vec{\mathfrak{A}}_\mathcal{R})$ exists, then it must be $\mathbf{0}_e^{(n)}$, so there is an enumeration f of $\mathfrak{M}$ such that $(f^{-1}(\mathfrak{M}))^{(n)} \equiv_e \emptyset^{(n)}$. However this would mean that there is an enumeration g of $\mathbb{N}$ such that for all $k \geq n$, $\mathcal{P}_k(\mathcal{R}) \leq_e \mathcal{P}_k(g^{-1}(\vec{\mathfrak{A}}_\mathcal{R})) \leq_e (f^{-1}(\mathfrak{M}))^{(k)}$ uniformly in k, and for $k \leq n$, $\mathcal{P}_k(\mathcal{R}) \leq_e \emptyset^{(n)}$, contradicting the fact that $d_\omega(\mathcal{R}) \not\leq_\omega \mathbf{0}_\omega$.

Next Soskov [81] turns to investigate the properties of the spectra of Marker extensions. There are two ways in which one can define the spectrum of a sequence of structures. The first one is to treat $\vec{\mathfrak{A}}$ within an underlying structure with domain $\bigcup_n A_n$ and consider enumerations f of A and the sequences $\{f^{-1}(\mathfrak{A}_n)\}_{n<\omega}$. The other possibility is to consider different enumerations: f_n an enumeration of $\mathfrak{A}_n$ for every n, giving rise to a sequence $\{f_n^{-1}(\mathfrak{A}_n)\}$. We can then collect all ω-enumeration degrees of such sequence as a measure of complexity, or better yet, collect all Turing degrees (or total enumeration degrees) in the jump class of one such sequence. For a set C let E_C denote all enumerations of the set C. The *relative spectrum* of a sequence $\vec{\mathfrak{A}}$ is the set

$$RS(\vec{\mathfrak{A}}) = \{d_T(B) \mid \exists g \in E_A(g^{-1}(\mathfrak{A}_n) \in \Sigma_{n+1}^0(B) \text{ uniformly in } n)\}.$$

The *joint spectrum* of the sequence $\vec{\mathfrak{A}}$ is the set

$$JS(\vec{\mathfrak{A}}) = \{d_T(B) \mid \exists \{g_n\}_{n<\omega}(g_n \in E_{A_n} \ \& \ g_n^{-1}(\mathfrak{A}_n) \in \Sigma_{n+1}^0(B) \text{ uniformly in } n)\}.$$

Note that in general $RS(\vec{\mathfrak{A}}) \neq JS(\vec{\mathfrak{A}})$. For example, for the sequence of structures $\vec{\mathfrak{A}}_\mathcal{R}$ obtained from an almost zero sequence $\mathcal{R}$ where $\mathfrak{A}_0 = (\mathbb{N}; G_S, R_0)$ and for all $n \geq 1$, $\mathfrak{A}_n = (\mathbb{N}; R_n)$ we have that $\mathbf{0}_T \in JS(\vec{\mathfrak{A}}_\mathcal{R}) \setminus RS(\vec{\mathfrak{A}}_\mathcal{R})$. However, if the structures in the sequence $\vec{\mathfrak{A}}$ have disjoint domains then the notions coincide.

These two notions can be seen as generalizations of ω-spectra and of joint spectra and relative spectra for finitely many structures. Recall that when these notions were investigated the main unanswered question was wether or not they give rise to new sets of degrees, or if the basic notion of degree spectrum already captures these sets. The next theorem unravels this mystery.

Theorem 22 (Soskov [81]). *Let $\mathfrak{A} = \{\mathfrak{A}_n\}_{n<\omega}$ be a sequence of structures.*

(1) *There exists a structure $\mathfrak{M}$ such that $DS_T(\mathfrak{M}) = RS(\vec{\mathfrak{A}})$.*
(2) *There exists a structure $\mathfrak{M}$ such that $DS_T(\mathfrak{M}) = JS(\vec{\mathfrak{A}})$.*

The proof of this theorem relies on a generalization of a result by Goncharov and Khoussainov [31].

Lemma 1. *Let R be a Σ^0_{n+1} set of natural numbers possessing an infinite computable subset S. Then there exist functions $\kappa_0, \dots, \kappa_n$ such that the graph of κ_n is computable and κ_0 is a bijection of R onto $\mathbb{N}$; κ_1 is a bijection of $\mathbb{N}^2 \setminus G_{\kappa_0}$ onto $\mathbb{N}$; $\dots \kappa_n$ is a bijection of $\mathbb{N}^{n+1} \setminus G_{\kappa_{n-1}}$ onto $\mathbb{N}$.*

Theorem 4 is a special case of Theorem 22 applied to the sequence $\mathfrak{A}$ where $\mathfrak{A}_0 = \mathfrak{A}$ and for every $n > 0$ we have the trivial structure $\mathfrak{A}_n = (A; =)$. To illustrate the main idea consider once again the example that we gave at the beginning of this section. We had a structure $\mathfrak{A} = (A; R)$ for which we built the Marker extension $\mathfrak{M} = (A \cup X; X, A, M)$. Assume that R is infinite, (if not we can instead use the Marker extension of the structure obtained by adding one more element $\perp$ to the domain of A and replace R by a $R_\perp = \{(m, n) \mid R(m) \vee n = \perp\}$). We showed that if f is any enumeration of $\mathfrak{M}$ then we can build an enumeration g of $\mathfrak{A}$, such that $g^{-1}(\mathfrak{A}) \leq_e f^{-1}(\mathfrak{M})^+$. Fix an enumeration g of $\mathfrak{A}$ and a total set Y such that $g^{-1}(\mathfrak{A}) \leq_e Y$. We can use the same trick as before: We construct f so that $f(2n) = g(n)$. To fill in $f(2\mathbb{N})$ we construct a bijection $k : f^{-1}(R) \to 2\mathbb{N} + 1$ and complete f by $f(2n + 1) = h(f(k^{-1}(2n + 1)))$. Then $f^{-1}(\mathfrak{M}) \equiv_e f^{-1}(M) \equiv_e G_k$. To construct G_k we use Lemma 1 relativized to Y. It follows that $DS(\mathfrak{M}^+) = \{\mathbf{y} \mid \mathbf{y}$ is total and $\mathbf{y} \geq \mathbf{x}$ for some $\mathbf{x} \in DS(\mathfrak{A})\}$.

Soskov gives several further applications of Theorem 22. He shows that the ω-enumeration degrees can be embedded into the Muchnick degrees generated by spectra of structures. To see this consider again the sequence $\vec{\mathfrak{A}}_{\mathcal{R}}$ obtained from a sequence of sets $\mathcal{R}$. Recall that for every enumeration g of $\vec{\mathfrak{A}}_{\mathcal{R}}$, we have that $\mathcal{R} \leq_\omega g^{-1}(\vec{\mathfrak{A}}_{\mathcal{R}})$. It follows that $RS(\vec{\mathfrak{A}}_{\mathcal{R}})$ is exactly the jump class of the sequence $\mathcal{R}$ and hence $DS_T(\mathfrak{M}(\vec{\mathfrak{A}}_{\mathcal{R}})) = \{d_T(B) \mid \mathcal{R}$ is c.e. in $B\}$. This induces the desired embedding as by definition we have that $\mathcal{R} \leq_\omega \mathcal{Q}$ if and only if $\{d_T(B) \mid \mathcal{R}$ is c.e. in $B\} \supseteq \{d_T(B) \mid \mathcal{Q}$ is c.e. in $B\}$ and this is true if and only if $DS_T(\mathfrak{M}(\vec{\mathfrak{A}}_{\mathcal{R}})) \supseteq DS_T(\mathfrak{M}(\vec{\mathfrak{A}}_{\mathcal{Q}}))$.

As a final application of these results we show how to build a structure $\mathfrak{M}$ whose spectrum consists of all Turing degrees, which are non-low$_n$ for every n. The previously known related examples are given by Kalimullin [38], who constructs for each low degree $\mathbf{b}$ a structure $\mathfrak{A}$ with $DS_T(\mathfrak{A}) = \{\mathbf{x} \mid \mathbf{x} \not\leq_T \mathbf{b}\}$ and by Goncharov et al., [32] who construct for every n a structure with spectrum consisting of all non-low$_n$ Turing degrees.

The construction relies on Wehner's [101] technique. Let $\mathcal{F}$ be a countable family of sets of natural numbers. An enumeration of $\mathcal{F}$ is a set $U \subseteq \mathbb{N}^2$ such that:

(1) For every a, the set $\{n \mid (a,n) \in U\} \in \mathcal{F}$.
(2) For every $F \in \mathcal{F}$ there is an a such that $\{n \mid (a,n) \in U\} = F$.

Let $\mathfrak{A}_{\mathcal{F}} = (A; S, Z, I)$ where $A = \mathcal{F} \times \mathbb{N}^2$; $Z = \{(F, x, 0) \mid F \in \mathcal{F}, x \in \mathbb{N}\}$, $S = \{((F, x, n), (F, x, n+1)) \mid F \in \mathcal{F}, x, n \in \mathbb{N}\}$ and $I = \{(F, x, n) \mid n \in F\}$. Wehner shows that there is a uniform way to compute an enumeration of $\mathcal{F}$ in any isomorphic copy $\mathcal{B}$ of $\mathfrak{A}_{\mathcal{F}}$ and that there is a uniform way to compute an isomorphic copy $\mathcal{B}$ of $\mathfrak{A}_{\mathcal{F}}$ in any enumeration of $\mathcal{F}$. Consider the relativized version of Wehner's family: $\mathcal{F}^X = \{\{n\} \oplus F \mid F \text{ is finite and } F \neq W_n^X\}$ for $X \subseteq \mathbb{N}$. No enumeration of $\mathcal{F}^X$ is c.e. in X. Furthermore, if $B \not\leq_T X$ then one can compute uniformly in B and X an enumeration of $\mathcal{F}^X$.

Finally, let $\vec{\mathfrak{A}}$ be the sequence of structures where $\mathfrak{A}_n = \mathfrak{A}_{\mathcal{F}^{\emptyset^{(n)}}}$. Let $\mathfrak{M}$ be such that $DS_T(\mathfrak{M}) = JS(\vec{\mathfrak{A}})$. If $d_T(B) \in DS_T(\mathfrak{M})$ then $B^{(n)}$ computes an enumeration of $\mathcal{F}^{\emptyset^{(n)}}$ and hence $B^{(n)} \not\leq_T \emptyset^{(n)}$. If $B^{(n)} \not\leq_T \emptyset^{(n)}$ for every n then as $\emptyset^{(n)} \leq_T B^{(n)}$ uniformly in n, it follows that $B^{(n)}$ computes an enumeration of $\mathcal{F}^{\emptyset^{(n)}}$.

Theorem 23 (Soskov [81]). *There is a structure $\mathfrak{M}$ with*

$$DS_T(\mathfrak{M}) = \{\mathbf{b} \mid \forall n(\mathbf{b}^{(n)} \not\leq \mathbf{0}_T^{(n)})\}.$$

The untimely death of Ivan Soskov left this area not fully explored. We hope that with this exposition, we will attract the interest of researchers who will join us in developing this line of investigation further.

References

1. Arslanov, M.M., Cooper, S.B., Kalimullin, I.S.: Splitting properties of total enumeration degrees. Algebra Logic **42**, 1–13 (2003)
2. Ash, C.J.: Generalizations of enumeration reducibility using recursive infinitary propositional senetences. Ann. Pure Appl. Logic **58**, 173–184 (1992)
3. Ash, C.J., Knight, J.: Computable Structures and the Hyperarithmetical Hierarchy, Volume 144. Studies in Logic and the Foundations of Mathematics. North-Holland Publishing Co., Amsterdam (2000)
4. Ash, C.J., Knight, J.F., Manasse, M., Slaman, T.: Generic copies of countable structures. Ann. Pure Appl. Logic **42**, 195–205 (1989)
5. Baleva, V.: The jump operation for structure degrees. Arch. Math. Logic **45**, 249–265 (2006)
6. Barwise, J.: Admissible Sets and Structures. Springer, Berlin (1975)
7. Boutchkova, V.: Genericity in abstract structure degrees. Ann. Sofia Univ. Fac. Math. Inf. **95**, 35–44 (2001)
8. Cai, M., Ganchev, H.A., Lempp, S., Miller, J.S., Soskova, M.I.: Defining totality in the enumeration degrees. J. Am. Math. Soc. http://dx.doi.org/10.1090/jams/848

9. Case, J.: Enumeration reducibility and partial degrees. Ann. Math. Logic **2**, 419–439 (1971)

10. Chisholm, J.: Effective model theory vs. recursive model theory. J. Symbol. Logic **55**, 1168–1191 (1990)

11. Coles, R., Downey, R., Slaman, T.: Every set has a least jump enumeration. Bull. Lond. Math. Soc. **62**, 641–649 (2000)

12. Cooper, S.B.: Partial degrees and the density problem. Part 2: the enumeration degrees of the Σ_2 sets are dense. J. Symbol. Logic **49**, 503–513 (1984)

13. Downey, R.G.: On presentations of algebraic structures. In: Sorbi, A. (ed.) Complexity, Logic and Recursion Theory. Lecture Notes in Pure and Applied Mathematics, vol. 187, pp. 157–205 (1997)

14. Downey, R.G., Knight, J.F.: Orderings with α-th jump degree $\mathbf{0}^{(\alpha)}$. Proc. Am. Math. Soc. **114**, 545–552 (1992)

15. Enderton, H.B., Putnam, H.: A note on the hyperarithmetical hierarchy. J. Symbol. Logic **35**, 429–430 (1970)

16. Ershov, Y.L.: Σ-denability in admissible sets. Sov. Math. Dokl. **3**, 767–770 (1985)

17. Ershov, Y.L.: Definability and Computability. Consultants Bureau, New York-London-Moscow (1996)

18. Ershov, Y., Puzarenko, V.G., Stukachev, A.I.: HF-computability. In: Cooper, B.S., Sorbi, A. (eds.) Computability in Context Computation and Logic in the Real World, pp. 169–242. Imperial College Press (2011)

19. Fraisse, R.: Une notion de recursivite relative. In: Proceedings of the Symposium of Mathematics 1959, Infinistic Methods, pp. 323–328. Pergamon Press, Warsaw (1961)

20. Friedberg, R.M., Rogers Jr., H.: Reducibility and completeness for sets of integers. Z. Math. Logik Grundlag. Math. **5**, 117–125 (1959)

21. Friedman, H.: Algorithmic procedures, generalized turing algorithms and elementary recursion theory. In: Gandy, R.O., Yates, C.E.M. (eds.) Logic Colloquium-69, pp. 361–389. North-Holland, Amsterdam (1971)

22. Ganchev, H.: A total degree splitting theorem and a jump inversion splitting theorem. In: Proceedings of the 5th Panhellenic Logic Symposium, Athens, Greece, pp. 79–81 (2005)

23. Ganchev, H.: Exact pair theorem for the ω-enumeration degrees. In: Cooper, S.B., Löwe, B., Sorbi, A. (eds.) CiE 2007. LNCS, vol. 4497, pp. 316–324. Springer, Heidelberg (2007). doi:10.1007/978-3-540-73001-9_33

24. Ganchev, H.: A jump inversion theorem for the infinite enumeration jump. Ann. Univ. Sofia Univ. **98**, 61–85 (2008)

25. Ganchev, H.: Definability in the local theory of the ω-enumeration degrees. In: Ambos-Spies, K., Löwe, B., Merkle, W. (eds.) CiE 2009. LNCS, vol. 5635, pp. 242–249. Springer, Heidelberg (2009). doi:10.1007/978-3-642-03073-4_25

26. Ganchev, H., Soskov, I.N.: The groups $\mathrm{Aut}(\mathcal{D}'_\omega)$ and $\mathrm{Aut}(\mathcal{D}_e)$ are isomorphic. In: Proceedings of the 6th Panhellenic Logic Symposium, Volos, Greece, pp. 53–57 (2007)

27. Ganchev, H., Soskov, I.N.: The jump operator on the ω-enumeration degrees. Ann. Pure Appl. Logic **160**, 289–301 (2009)

28. Ganchev, H., Soskova, M.: The high/low hierarchy in the local structure of the ω-enumeration degrees. Ann. Pure Appl. Logic **163**(5), 547–566 (2012)

29. Ganchev, H.A., Soskova, M.I.: Interpreting true arithmetic in the local structure of the enumeration degrees. J. Symbol. Log. **77**(4), 1184–1194 (2012)

30. Ganchev, H.A., Soskova, M.I.: Definability via Kalimullin pairs in the structure of the enumeration degrees. Trans. AMS **367**, 4873–4893 (2015)

31. Goncharov, S., Khoussainov, B.: Complexity of categorical theories with computable models. Algebra Logic **43**(6), 365–373 (2004)
32. Goncharov, S., Harizanov, V., Knight, J., McCoy, C., Miller, R., Solomon, R.: Enumerations in computable structure theory. Ann. Pure Appl. Logic **136**(3), 219–246 (2005)
33. Greenberg, N., Montalbán, A., Slaman, T.A.: Relative to any non-hyperarithmetic set. J. Math. Logic **13**(01) (2013). doi:10.1142/S0219061312500079
34. Gordon, C.: Comparisons between some generalizations of recursion theory. Compos. Math. **22**, 333–346 (1970)
35. Harris, K., Montabán, A.: On the n-back-and-forth types of Boolean algebras. Trans. AMS **364**, 827–866 (2012)
36. Jojgov, G.: Minimal pairs of structure degrees. Master's thesis, Sofia University (1997)
37. Jockusch Jr., C.G.: Semirecursive sets and positive reducibility. Trans. Am. Math. Soc. **131**, 420–436 (1968)
38. Kalimullin, I.: Some notes on degree spectra of the structures. In: Cooper, S.B., Löwe, B., Sorbi, A. (eds.) CiE 2007. LNCS, vol. 4497, pp. 389–397. Springer, Heidelberg (2007). doi:10.1007/978-3-540-73001-9_40
39. Kalimullin, I.S.: Enumeration degrees and enumerability of families. J. Logic Comput. **19**(1), 151–158 (2009)
40. Knight, J.F.: Degrees coded in jumps of orderings. J. Symbol. Logic **51**, 1034–1042 (1986)
41. Knight, J.F.: Degrees of models. In: Handbook of Recursive Mathematics, Volume 1, Studies Logic Foundations of Mathematics, vol. 138, pp. 289–309. North-Holland, Amsterdam (1998)
42. Kreisel, G.: Some reasons for generalizing recursion theory. In: Gandy, R.O., Yates, C.E.M. (eds.) Logic Colloquium 69, pp. 139–198. North Holland (1971)
43. Kreisel, G., Sacks, G.E.: Metarecursive sets. J. Symbol. Logic **30**, 318–338 (1965)
44. Kripke, S.: Transfinite recursion on admissible ordinals, I, II (abstracts). J. Symbol. Logic **29**, 161–162 (1964)
45. Lacombe, D.: Deux generalizations de la notion de recursivite relative. C. R. de l-Academie des Sciences de Paris **258**, 3410–3413 (1964)
46. Lempp, S., Slaman, T.A., Sorbi, A.: On extensions of embeddings into the enumeration degrees of the Σ_2^0-sets. J. Math. Log. **5**(2), 247–298 (2005)
47. Marker, D.: Non Σ_n-axiomatizable almost strongly minimal theories. J. Symbol. Logic **54**(3), 921–927 (1989)
48. McEvoy, K.: Jumps of quasi-minimal enumeration degrees. J. Symbol. Logic **50**, 839–848 (1985)
49. Medvedev, I.T.: Degrees of difficulty of the mass problem. Dokl. Nauk. SSSR **104**, 501–504 (1955)
50. Montalbán, A.: Notes on the jump of a structure. In: Ambos-Spies, K., Löwe, B., Merkle, W. (eds.) CiE 2009. LNCS, vol. 5635, pp. 372–378. Springer, Heidelberg (2009). doi:10.1007/978-3-642-03073-4_38
51. Montalbán, A.: Rice sequences of relations. Philos. Trans. R. Soc. A **370**, 3464–3487 (2012)
52. Montalbán, A.: Computable Structure Theory, draft
53. Montague, R.: Recursion theory as a branch of model theory. In: Proceedings of the Third International Congress for Logic Methodology and Philosophy of Science, pp. 63–86. North-Holland, Amsterdam-London (1967)
54. Moschovakis, Y.N.: Abstract first order computability I. Trans. Am. Math. Soc. **138**, 427–464 (1969)

55. Moschovakis, Y.N.: Elementary Induction of Abstract Structures. North-Holland, Amsterdam (1974)
56. Moschovakis, Y.N.: Abstract computability and invariant definability. J. Symbol. Logic **34**, 605–633 (1969)
57. Myhill, J.: A note on the degrees of partial functions. Proc. Am. Math. Soc. **12**, 519–521 (1961)
58. Platek, R.: Foundations of recursion theory. Ph.D. thesis, Stanford University (1966)
59. Plotkin, G.D.: A set-theoretical definition of application Memorandum MIP-R-95. University of Edinburgh, School of Artificial Intelligence (1972)
60. Richter, L.J.: Degrees of unsolvability of models. Ph.D. dissertation, University of Illinois, Urbana-Champaign (1977)
61. Richter, L.J.: Degrees of structures. J. Symbol. Logic **46**, 723–731 (1981)
62. Rozinas, M.: The semi-lattice of e-degrees. In: Recursive functions (Ivanovo), pp. 71–84. Ivano. Gos. Univ. (1978). (Russian)
63. Sacks, G.E.: Forcing with perfect closed sets. Proc. Symp. Pure Math. **17**, 331–355 (1971)
64. Selman, A.L.: Arithmetical reducibilities I. Z. Math. Logik Grundlag. Math. **17**, 335–350 (1971)
65. Shepherdson, J.C.: Computation over abstract structures. In: Rose, H.E., Shepherdson, J.C. (eds.) Logic Colloquium-73, pp. 445–513. North-Holland, Amsterdam (1975)
66. Skordev, D.G.: Computability in Combinatory Spaces. Kluwer Academic Publishers, Dordrecht-Boston-London (1992)
67. Slaman, T.A.: Relative to any nonrecursive set. Proc. Am. Math. Soc. **126**, 2117–2122 (1998)
68. Slaman, T.A., Sorbi, A.: Quasi-minimal enumeration degrees and minimal Turing degrees. Annali di Matematica **174**(4), 79–88 (1998)
69. Soskov, I.N.: Definability via enumerations. J. Symbol. Logic **54**, 428–440 (1989)
70. Soskov, I.N.: An external characterization of the Prime computability. Ann. Univ. Sofia Fac. Math. Inf. **83**, 89–111 (1989)
71. Soskov, I.N.: Computability by means of effectively definable schemes and definability via enumerations. Arch. Math. Logic **29**, 187–200 (1990)
72. Soskov, I.N.: Constructing minimal pairs of degrees. Ann. Univ. Sofia **91**, 101–112 (1997)
73. Soskov, I.N.: A jump inversion theorem for the enumeration jump. Arch. Math. Logic **39**, 417–437 (2000)
74. Soskov, I.N., Baleva, V.: Regular enumerations. J. Symbol. Logic **67**, 1323–1343 (2002)
75. Soskov, I.N.: Degree spectra and co-spectra of structures. Ann. Univ. Sofia **96**, 45–68 (2004)
76. Soskov, I.N., Baleva, V.: Ash's theorem for abstract structures. In: Chatzidakis, Z., Koepke, P., Pohlers, W. (eds.) Logic Colloquium 2002, Muenster, Germany, pp. 327–341. Association for Symbolic Logic (2006)
77. Soskov, I.N., Kovachev, B.: Uniform regular enumerations. Math. Struct. Comput. Sci. **16**(5), 901–924 (2006)
78. Soskov, I.N.: The ω-enumeration degrees. J. Logic Comput. **17**, 1193–1217 (2007)
79. Soskov, I., Soskova, M.: Kalimullin pairs of Σ^0_2 omega-enumeration degrees. Int. J. Softw. Inf. **5**(4), 637–658 (2011)
80. Soskov, I.N.: A note on ω-jump inversion of degree spectra of structures. In: Bonizzoni, P., Brattka, V., Löwe, B. (eds.) CiE 2013. LNCS, vol. 7921, pp. 365–370. Springer, Heidelberg (2013). doi:10.1007/978-3-642-39053-1_43

81. Soskov, I.N.: Effective properties of Marker's extensions. J. Logic Comput. **23**(6), 1335–1367 (2013). doi:10.1093/logcom/ext041
82. Soskova, A.A., Soskov, I.N.: Co-spectra of joint spectra of structures. Ann. Sofia Univ. Fac. Math. Inf. **96**, 35–44 (2004)
83. Soskova, A.A.: Minimal pairs and quasi-minimal degrees for the joint spectra of structures. In: Cooper, S.B., Löwe, B., Torenvliet, L. (eds.) CiE 2005. LNCS, vol. 3526, pp. 451–460. Springer, Heidelberg (2005). doi:10.1007/11494645_56
84. Soskova, A.A.: Properties of co-spectra of joint spectra of structures. Ann. Sofia Univ. Fac. Math. Inf. **97**, 15–32 (2005)
85. Soskova, A.A.: Relativized degree spectra. In: Beckmann, A., Berger, U., Löwe, B., Tucker, J.V. (eds.) CiE 2006. LNCS, vol. 3988, pp. 546–555. Springer, Heidelberg (2006). doi:10.1007/11780342_56
86. Soskova, A.A.: Relativized degree spectra. J. Logic Comput. **17**, 1215–1234 (2007)
87. Soskova, A.A.: A jump inversion theorem for the degree spectra. In: Cooper, S.B., Löwe, B., Sorbi, A. (eds.) CiE 2007. LNCS, vol. 4497, pp. 716–726. Springer, Heidelberg (2007). doi:10.1007/978-3-540-73001-9_76
88. Soskova, A.A., Soskov, I.N.: Jump spectra of abstract structures. In: Proceeding of the 6th Panhellenic Logic Symposium, Volos, Greece, pp. 114–117 (2007)
89. Soskova, A.A.: ω-degree spectra. In: Beckmann, A., Dimitracopoulos, C., Löwe, B. (eds.) CiE 2008. LNCS, vol. 5028, pp. 544–553. Springer, Heidelberg (2008). doi:10.1007/978-3-540-69407-6_58
90. Soskova, A.A., Soskov, I.N.: A jump inversion theorem for the degree spectra. J. Logic Comput. **19**, 199–215 (2009)
91. Soskova, A., Soskov, I.N.: Quasi-minimal degrees for degree spectra. J. Logic Comput. **23**(2), 1319–1334 (2013). doi:10.1093/logcom/ext045
92. Soskova, M., Soskov, I.: Embedding countable partial orderings in the enumeration degrees and the omega enumeration degrees. J. Logic Comput. **22**(4), 927–952 (2012). doi:10.1093/logcom/exq051. First published online October 23, 2010
93. Stukachev, A.I.: A jump inversion theorem for semilattices of Σ-degrees. Sib. Élektron. Mat. Izv. **6**, 182–190 (2009)
94. Stukachev, A.I.: A jump inversion theorem for the semilattices of Sigma-degrees. Siberian Adv. Math. **20**(1), 68–74 (2010)
95. Stukachev, A.I.: Effective model theory: an approach via Σ-definability. In: Greenberg, N., Hamkins, J.D., Hirschfeldt, D., Miller, R. (eds.) Effective Mathematics of the Uncountable. Lecture Notes in Logic, vol. 41, pp. 164–197 (2013)
96. Vatev, S.: Omega spectra and co-spectra of structures. Master thesis, Sofia University (2008)
97. Vatev, S.: Conservative extensions of abstract structures. In: Löwe, B., Normann, D., Soskov, I., Soskova, A. (eds.) CiE 2011. LNCS, vol. 6735, pp. 300–309. Springer, Heidelberg (2011). doi:10.1007/978-3-642-21875-0_32
98. Vatev, S.: Another jump inversion theorem for structures. In: Bonizzoni, P., Brattka, V., Löwe, B. (eds.) CiE 2013. LNCS, vol. 7921, pp. 414–423. Springer, Heidelberg (2013). doi:10.1007/978-3-642-39053-1_49
99. Vatev, S.: Effective properties of structures in the hyperarithmetical hierarchy. Ph.D. thesis, Sofia University (2014)
100. Vatev, S.: On the notion of jump structure. Ann. Sofia Univ. Fac. Math. Inf. **102**, 171–206 (2015)
101. Wehner, St.: Enumerations, countable structures and turing degrees. Proc. Am. Math. Soc. **126**, 2131–2139 (1998)

Strength and Weakness in Computable Structure Theory

Johanna N.Y. Franklin[✉]

Department of Mathematics, Hofstra University, Room 306, Roosevelt Hall,
Hempstead, NY 11549-0114, USA
johanna.n.franklin@hofstra.edu

Abstract. We survey the current results about degrees of categoricity and the degrees that are low for isomorphism as well as the proof techniques used in the constructions of elements of each of these classes. We conclude with an analysis of these classes, what we may deduce about them given the sorts of proof techniques used in each case, and a discussion of future lines of inquiry.

1 Introduction

The question of whether a computable isomorphism between two computable structures exists was first discussed in computable model theory sixty years ago [17]. Later, this question was generalized to the question of whether an isomorphism of a given Turing degree exists between two computable structures. There has been a great deal of recent work on Turing degrees that have been shown to be very strong with respect to computing isomorphisms between structures and those that have been shown to be very weak. The first such class of degrees is called the *degrees of categoricity*; degrees in the second such class are called *low for isomorphism*. Both of these classes of degrees have proven to be difficult to characterize completely; in fact, no full characterization exists for either class. In this paper, we will synthesize the work on these topics and the proof techniques involved, present some open questions, and discuss possible approaches to the subject.

1.1 Terminology

We begin with a discussion of the most relevant definitions; other terms will be defined as necessary throughout the paper. We assume the reader is familiar with computability theory in general and computable structure theory in particular; [18,29,30,34] are useful references for these subjects, respectively. We will use the notation from Ash and Knight [2] when we discuss the hyperarithmetic hierarchy (as do the authors of all the papers concerning this hierarchy that we survey), and we suggest [31] as a general reference.

The author would like to thank the editors of this Festschrift for the invitation to contribute.

A. Day et al. (Eds.): Downey Festschrift, LNCS 10010, pp. 302–323, 2017.
DOI: 10.1007/978-3-319-50062-1_20

The most fundamental concept in this paper is that of an isomorphism relative to a particular Turing degree **d**.

Definition 1.1. Given a Turing degree **d** and computable structures $\mathcal{A}$ and $\mathcal{B}$, we say that $\mathcal{A}$ is **d**-*computably isomorphic to* $\mathcal{B}$ (which we will write $\mathcal{A} \cong_{\mathbf{d}} \mathcal{B}$) if there is an isomorphism between $\mathcal{A}$ and $\mathcal{B}$ that is computable from **d**. If $\mathbf{d} = \mathbf{0}$, we say that $\mathcal{A}$ is *computably isomorphic to* $\mathcal{B}$ and write $\mathcal{A} \cong_{\Delta_1^0} \mathcal{B}$.

This idea is then used to define the concept of computable categoricity relative to a given Turing degree **d**.

Definition 1.2. A computable structure $\mathcal{A}$ is **d**-*computably categorical* if, for every computable structure $\mathcal{B}$ that is classically isomorphic to $\mathcal{A}$, we have $\mathcal{A} \cong_{\mathbf{d}} \mathcal{B}$.

Now we can define the central concepts in this paper: degrees of categoricity and lowness for isomorphism.

Definition 1.3 [11]. A Turing degree **d** is a *degree of categoricity* if there is a computable structure $\mathcal{A}$ such that $\mathcal{A}$ is **c**-computably categorical if and only if $\mathbf{c} \geq_T \mathbf{d}$. This degree **d** is furthermore a *strong degree of categoricity* if there is a computable structure $\mathcal{A}$ with computable copies $\mathcal{A}_1$ and $\mathcal{A}_2$ such that $\mathcal{A}$ has degree of categoricity **d** and every isomorphism from $\mathcal{A}_1$ to $\mathcal{A}_2$ computes **d**.

In short, a degree is a degree of categoricity if it is the least degree that, for some computable structure, can compute an isomorphism from that structure to any classically isomorphic computable copy of itself. This means that we can think of it as calibrating the complexity of that computable structure in some way. Furthermore, a degree is a strong degree of categoricity if it not only computes such isomorphisms but can be computed by any isomorphism from one copy of a particular computable structure to another. We can thus say that a (strong) degree of categoricity is, in some way, a very strong degree: it is guaranteed to have a certain level of computational power for some computable structure.

On the other hand, a degree that is low for isomorphism is a degree that is very weak indeed for any pair of computable structures:

Definition 1.4 [14]. A Turing degree **d** is *low for isomorphism* if for every pair of computable structures $\mathcal{A}$ and $\mathcal{B}$, $\mathcal{A} \cong_{\mathbf{d}} \mathcal{B}$ if and only if $\mathcal{A} \cong_{\Delta_1^0} \mathcal{B}$.

The word *low* is used in this definition as it has been used in computability theory since the 1970s: a degree **d** is generally called low for a relativizable class $\mathcal{C}$ if, when it is used as an oracle, the new, relativized class is no different than the original, unrelativized one (that is, when $\mathcal{C}^D = \mathcal{C}$ for $D \in \mathbf{d}$). This notion, first used in computability theory by Soare in [33], has appeared in almost every context in computability theory: degree theory [33], learning theory [32], and, more recently, algorithmic randomness [7,13,27]. Franklin and Solomon's paper introduced this concept into computable structure theory for the first time [14].

These notions appear to be entirely incompatible. Nontrivial degrees of categoricity possess some additional information required to compute an isomorphism for some structure, while degrees that are low for isomorphism have none. Clearly, the only degree that satisfies both of these conditions is $\mathbf{0}$.

At this point, there are several natural questions to ask. What sorts of closure do these classes possess? It is clear from the definition that the degrees that are low for isomorphism are closed downwards, but do they form an ideal? Are these degrees compatible or incompatible with natural classes of degrees, such as the hyperimmune-free degrees, minimal degrees, or low degrees?

Examples of degrees of categoricity and degrees that are low for isomorphism have been found, but a full characterization has been elusive for each. In this paper, we hope to present some of the constructions of these degrees and to analyze these constructions as well as to present some more general metainformation about both kinds and consider reasons that each type of degree is so difficult to characterize. We discuss degrees of categoricity in Sect. 2 and degrees that are low for isomorphism in Sect. 3, and we include an analysis of these classes in Sect. 4.

2 Degrees of Categoricity

As mentioned, the concept of a degree of categoricity was first defined by Fokina, Kalimullin, and R. Miller in [11]. In this paper, they demonstrated that certain degrees were degrees of categoricity, showed that there were only countably many strong degrees of categoricity, and considered the question of degrees of categoricity for particular classes of structures. Csima, Franklin, and Shore then extended their results through the hyperarithmetic hierarchy and proved that there were only countably many degrees of categoricity [4] and, more recently, Csima and Harrison-Trainor showed that the degrees of categoricity of "natural" structures are very limited indeed [5].

2.1 Examples of Degrees of Categoricity

All of the results in this section are centered around the Ershov hierarchy [8–10]. Fokina, Kalimullin, and R. Miller's primary results can be stated as the following theorem:

Theorem 2.1 [11]. *If* $\mathbf{d}$ *is a Turing degree that is c.e. or d.c.e. in* $\mathbf{0}^{(m)}$ *and* $\mathbf{0}^{(m)} \leq_T \mathbf{d}$ *for some* $m \in \omega$, *then* $\mathbf{d}$ *is a strong degree of categoricity. Furthermore,* $\mathbf{0}^{(\omega)}$ *is a strong degree of categoricity.*

We outline their constructions in increasing order of complexity. They begin by simply showing that a c.e. degree $\mathbf{d}$ is a degree of categoricity. To do so, they fix a c.e. Turing degree $\mathbf{d}$, a c.e. set W_e inside it, and a computable injective function h with range W_e. From this, they construct a structure $\mathcal{B}$ witnessing that $\mathbf{d}$ is a degree of categoricity.

$\mathcal{B}$ is a directed graph with two constant elements, c and d, and is constructed as follows. Four elements, α, β, γ, and δ, are dedicated to be the "origin" nodes, and the sequences $x_0, x_1, \ldots, x_i, \ldots$ and $y_0, y_1, \ldots, y_i, \ldots$ are the "target" nodes. (There is also a set of "witness" nodes $\{u_i\}_{i\in\omega}$, but those do not appear as elements of the graph and we will ignore them in this sketch.) We declare that $c^{\mathcal{B}} = \gamma$ and $d^{\mathcal{B}} = \delta$. At stage 0, $\mathcal{B}$ only has edges from β to every y_i and from δ to every x_i.

When a number i enters W_e, we add the following additional edges to our graph:

- an edge from α to x_i,
- an edge from β to x_i,
- an edge from γ to y_i, and
- an edge from δ to y_i.

At the end of our construction, we see that if $i \in W_e$, then there are edges from α, β, and δ to x_i and edges from γ, δ, and β to y_i. Therefore, any automorphism of $\mathcal{B}$ that swaps x_i and y_i must swap α and γ, and it may either fix β and δ or swap them. Furthermore, if $i \notin W_e$, then the only edge to x_i comes from δ and the only edge to y_i comes from β. This means that if an automorphism of $\mathcal{B}$ swaps x_i and y_i, then it must swap β and δ as well but its behavior on α and γ does not matter.

We first argue that $\mathbf{d}$ is a degree of categoricity for $\mathcal{B}$. Suppose we have another computable structure $\mathcal{A}$ that is classically isomorphic to $\mathcal{B}$ and we wish to build an isomorphism g from $\mathcal{B}$ to $\mathcal{A}$ computable in $\mathbf{d}$. We first note that we can identify $g(\gamma)$ as $c^{\mathcal{A}}$ and $g(\delta)$ as $d^{\mathcal{A}}$. Since we defined the x_is and y_is as sequences, we can identify the pair $g(x_i)$ and $g(y_i)$ for each i. Now we use W_e as our oracle to determine which is which: if $i \in W_e$, then the element that is connected to $c^{\mathcal{A}}$ is $g(y_i)$; if $i \notin W_e$, then the element that is connected to $d^{\mathcal{A}}$ is $g(x_i)$.

Furthermore, we can prove that $\mathbf{d}$ is a strong degree of categoricity for $\mathcal{B}$ by exhibiting a structure $\mathcal{A}$ such that any isomorphism between $\mathcal{B}$ and $\mathcal{A}$ can compute W_e. We define $\mathcal{A}$ in this case to be identical to $\mathcal{B}$ save for the choice of constants: we set $c^{\mathcal{A}} = \widehat{\alpha}$ and $d^{\mathcal{A}} = \widehat{\delta}$ (where $\widehat{\alpha}$ and $\widehat{\delta}$ are the $\mathcal{A}$-equivalents of α and δ). There is exactly one isomorphism from $\mathcal{B}$ to $\mathcal{A}$: the isomorphism that swaps the x_is that are connected to α with the corresponding y_is that are connected to γ and preserves the rest. Knowledge of this isomorphism clearly allows one to determine W_e.

The proof that every d.c.e. degree is a degree of categoricity is slightly more complicated. Again, we fix a d.c.e. degree $\mathbf{d}$ and a d.c.e. set $A - B$ in $\mathbf{d}$, where A and B are c.e. sets such that $B \subseteq A$. We construct a directed graph once more. However, this time we will have a single sequence $x_0, x_1, \ldots, x_i, \ldots$ where the ith element is connected to the four points a_i, b_i, c_i, and d_i and these four points form a square at stage 0: there are arrows from a_i to b_i, b_i to c_i, c_i to d_i, and d_i to a_i.

We now choose our witnessing computable structure $\mathcal{B}$ to be the substructure of the directed graph above with all the x_is, c_is, and d_is, but only the a_is for

$i \in A$ and only the b_is for $i \in B$. Since $\mathcal{B}$'s universe is c.e., we can proceed as though $\mathcal{B}$ is computable.

Now suppose that $\mathcal{A}$ is a computable structure that is isomorphic to $\mathcal{B}$. To compute an isomorphism g from $\mathcal{B}$ to $\mathcal{A}$, we first identify $g(x_i)$ for each i. If $i \in D$, then there is no b_i, and we can uniquely define $g(a_i)$, $g(b_i)$, and $g(c_i)$. If $i \notin D$, then either we have to identify $g(a_i)$, $g(b_i)$, $g(c_i)$, and $g(d_i)$ (if i never entered D) or only $g(c_i)$, and $g(d_i)$ (if i entered and then exited D). In any case, we define $g(c_i)$, and $g(d_i)$ as soon as we find two elements that are candidates for them; if we later determine that $i \in A$ and therefore $i \in B$ as well, we can extend the isomorphism appropriately.

To prove that $\mathbf{d}$ is actually a strong degree of categoricity, we show that an isomorphism exists between the structure previously described and the structure $\mathcal{A}$, where $\mathcal{A}$ is the substructure of the original directed graph with all the x_is, c_is, and d_is, but only the a_is for $i \in B$ and only the b_is for $i \in A$. Suppose we have such an isomorphism. Then, for each $i \in \omega$, we can see that $i \in D$ if and only if $i \notin B$ and $f(c_i) \neq c_i$.

Both of the results above can be seen to relativize to degrees c.e. and d.c.e. in and above $\mathbf{0}^{(m)}$ for any $m \in \omega$ using Marker's construction from [24], so to fully prove Theorem 2.1, we only need show that there is a computable structure whose degree of categoricity is $\mathbf{0}^{(\omega)}$. As is logical for a limit case, this structure is simply the cardinal sum of the computable structures constructed to show that the degrees $\mathbf{0}^{(n)}$ are degrees of categoricity for all $n \in \omega$.

Fokina, Kalimullin, and R. Miller's paper did not treat the 3-c.e. case, which remains unsolved to this writing. Let us discuss briefly why it is far more difficult than the d.c.e. case. In the d.c.e. case, there are three possible scenarios for each $i \in \omega$: i is in D, which means that all of a_i, c_i, and d_i are in the first structure; i never entered D, which means that a_i, b_i, c_i, and d_i are in the structure, or i entered and then exited D, which means that only c_i and d_i are in the structure. Suppose that, in addition to these scenarios, we also had to deal with the case in which i had entered, exited, and then entered D again. This would mean that there would have to be two subcases for each of $i \in D$ and $i \notin D$. So far, no way has been found to code information into a structure in such a way that the first and third "versions" can be made isomorphic (the cases where $i \notin D$), the second and fourth "versions" can also be made isomorphic (the cases where $i \in D$), and we can transition from each version to the next in a computable way.

One of the questions asked in [11] was whether or not their construction could be extended to higher hyperarithmetic degrees. This was answered by Csima, Franklin, and Shore in [4], where they proved the following result.

Theorem 2.2. *If α is a computable ordinal, then $\mathbf{0}^{(\alpha)}$ is a strong degree of categoricity. If, in addition, α is a successor ordinal, then every degree that is c.e. or d.c.e. in and above $\mathbf{0}^{(\alpha)}$ is a strong degree of categoricity.*

Once again, these constructions use directed graphs. The authors use Hirschfeldt and White's "back-and-forth trees" in their construction [22], which are computable subtrees of $\omega^{<\omega}$ with no infinite paths. We outline their construction below.

They fix a system of notation for ordinals as follows: 1 denotes the ordinal 0, 2^a denotes the ordinal $\alpha + 1$ when a denotes α, and $3 \cdot 5^e$ denotes a limit ordinal λ under certain technical conditions, including the totality of φ_e. This makes it possible to define two structures, $\mathcal{A}_a$ and $\mathcal{E}_a$, for each notation a using transfinite recursion. $\mathcal{A}_1$ is a single node, and $\mathcal{E}_1$ is a single root node that has infinitely many children, all of which are childless. If a represents the successor of a successor ordinal represented by b (so $a = 2^b$), then $\mathcal{A}_a$ consists of a single root node with infinitely many copies of $\mathcal{E}_b$ attached, and $\mathcal{E}_a$ consists of a single root node with infinitely many copies of both $\mathcal{A}_b$ and $\mathcal{E}_b$ attached. Finally, if a is the successor of a limit ordinal coded by e (so $a = 2^{3 \cdot 5^e}$), we must first define auxiliary trees $\mathcal{L}_{e,k}$ for every $k \in \omega$ as well as a structure $\mathcal{L}_{e,\infty}$ as follows:

- $\mathcal{L}_{e,k}$ consists of exactly one copy of $\mathcal{A}_{\varphi_e(n)}$ for all $n \leq k$ and exactly one copy of $\mathcal{E}_{\varphi_e(n)}$ for all $n > k$, and
- $\mathcal{L}_{e,\infty}$ consists of exactly one copy of $\mathcal{A}_{\varphi_e(n)}$ for every $n \in \omega$.

Now we can define $\mathcal{A}_a$ to consist of a root node with infinitely many copies of $\mathcal{L}_{e,k}$ for every $k \in \omega$ and $\mathcal{E}_a$ to consist of a root node with infinitely many copies of $\mathcal{L}_{e,k}$ for each $k \in \omega$ and infinitely many copies of $\mathcal{L}_{e,\infty}$. These procedures will always give us a computable tree.

We note that $\mathcal{A}_a$ can always be converted to $\mathcal{E}_a$ just by adding infinitely copies of either the appropriate $\mathcal{E}_b$ or the appropriate $\mathcal{L}_{e,\infty}$. This will be essential for our construction.

Now, in preparation for building the structures that witness the existence of the degrees of categoricity previously mentioned, we make note of several technical facts about these structures. A lemma from [22] allows us to see that, given an ordinal α and a Σ_α predicate P, for every notation a for α, there is a sequence of trees $\mathcal{T}_n$ that is uniformly computable from a and a Σ_α index for P such that for all n, $\mathcal{T}_n$ is isomorphic to one of $\mathcal{E}_a$, $\mathcal{A}_a$, $\mathcal{L}_{e,k}$, or $\mathcal{L}_{e,\infty}$ depending on whether $P(n)$ holds and whether α is a successor or limit ordinal. We can also define the rank of a back-and-forth limb of a tree ($\mathcal{S}$ is a limb of $\mathcal{T}$ if $\mathcal{S} \subseteq \mathcal{T}$ and is closed under the "child" relation within $\mathcal{T}$, and $\mathcal{S}$ is a back-and-forth limb if it is isomorphic to one of our back-and-forth trees) and then use this rank to associate a natural complexity with a back-and-forth tree based on its isomorphism type. This will let us prove that $\mathbf{0}^{(\alpha)}$ can compute an isomorphism between two back-and-forth limbs of different computable trees as long as both limbs have rank less than a (the notation for α) and are classically isomorphic, and this computation is uniform in the roots for the limbs.

Now we can prove that for every α, there is a computable structure $\mathcal{S}_a$ with strong degree of categoricity $\mathbf{0}^{(\alpha)}$. For each notation, we construct a "standard" copy of $\mathcal{S}_a$ and a "hard" copy $\widehat{\mathcal{S}}_a$. All of these structures consist of infinitely many disjoint copies of these back-and-forth trees. We present the case for $a = 2$ (the notation for $\mathbf{0}'$) here and then describe the other cases briefly.

The "standard" copy, $\mathcal{S}_2$, will consist of infinitely many disjoint copies of $\mathcal{A}_1$ and $\mathcal{E}_1$. The set of edges in this copy is $\{(\langle 2n, 0 \rangle, \langle 2n, k \rangle) \mid k > 0\}$. The elements in the odd columns are not connected to any other elements and are thus each isomorphic to $\mathcal{A}_1$; each even column is isomorphic to $\mathcal{E}_1$ with $\langle 2n, 0 \rangle$ as the root node.

Now we use an approximation $\{K_s\}_{s\in\omega}$ to $0'$ to build the "hard" copy $\widehat{\mathcal{S}}_2$. The set of edges of this copy is defined to be $\{(\langle 2n, 0\rangle, \langle 2n, t\rangle) \mid n \in K_t\}$. In this case, if $n \in 0'$, a subset of the nth even column will be isomorphic to $\mathcal{E}_1$, and all of the other elements will form substructures isomorphic to $\mathcal{A}_1$. Clearly, if one is given an isomorphism between the "standard" and "hard" copies, $0'$ can be computed by determining which of the odd columns contain copies of $\mathcal{A}_1$, and if one is given any two computable copies of $\mathcal{S}_2$, then the only questions that need to be answered to compute an isomorphism between them are Σ_1^0 and Π_1^0, which $0'$ can answer.

For an ordinal β that is the successor of a successor ordinal α with notation a, our structure will consist of infinitely many disjoint copies of $\mathcal{A}_a$ and $\mathcal{E}_a$. The "standard" copy will code the $\mathcal{E}_a$s in the even columns and the $\mathcal{A}_a$s in the odd columns; the "hard" copy will code the $\mathcal{E}_a$s in the columns corresponding to those n in the jump of $0^{(\alpha)}$ and the $\mathcal{A}_a$s in the other columns. The basic argument is the one given above, though with more bookkeeping: we must show that the root nodes of all the connected components of each structure and the back-and-forth indices of their limbs can be computed in this jump. This allows us to define a bijection between the root nodes in each structure that preserves back-and-forth indices, which is all we need to compute an isomorphism between the standard copy and an arbitrary computable copy.

For an ordinal α that is a limit ordinal coded by e, we construct the "standard" copy by coding a copy of $\mathcal{E}_{\varphi_e(n)}$ in the $\langle k, n\rangle^{th}$ column if k is even and a copy of $\mathcal{A}_{\varphi_e(n)}$ in the $\langle k, n\rangle^{th}$ column if k is odd. In the "hard" copy, we determine where to code copies of $\mathcal{E}_{\varphi_e(n)}$ and $\mathcal{A}_{\varphi_e(n)}$ depending on whether n is in $0^{(\alpha)}$.

Finally, for an ordinal that is the successor of a limit ordinal coded by e, our structure will consist of infinitely many disjoint copies of $\mathcal{L}_{e,\infty}$ and $\mathcal{L}_{e,k}$ for all $k \in \omega$. We construct the "standard" copy by coding a copy of $\mathcal{L}_{e,k}$ in the $\langle n, k, 0\rangle^{th}$ column if n is even and a copy of $\mathcal{L}_{e,\infty}$ in it otherwise. The fact that we can compute a sequence of trees of the form $\mathcal{E}_a$, $\mathcal{A}_a$, $\mathcal{L}_{e,k}$, and $\mathcal{L}_{e,\infty}$ as previously described lets us construct a "hard" copy that codes information about our ordinal. Since all the connected components are back-and-forth trees with rank below the ordinal we are considering, the corresponding Turing degree is enough to compute an isomorphism between these copies, and we can argue as before that it is enough to compute an isomorphism between any two computable copies of this structure.

We now move from the c.e. case to the d.c.e. case and argue that any degree $\mathbf{d}$ d.c.e. in and above $0^{(\alpha)}$ for a computable successor ordinal α must be a strong degree of categoricity. Once again, two different structures, $\mathcal{G}$ and $\widehat{\mathcal{G}}$ are constructed, the former the "standard" copy and the latter the "hard" copy. Let $D \in \mathbf{d}$ witness that $\mathbf{d}$ is d.c.e. in and above $0^{(\alpha)}$. We will use the same general technique as in [11]: information is coded into 4-cycles containing the nodes a, b, c, and d based on whether n enters D and then leaves it, enters and never leaves, or never enters it at all. This information is coded by attaching either $\mathcal{A}_a$s or $\mathcal{E}_a$s to each of the nodes in the 4-cycle. The nodes a and c are treated identically in $\mathcal{G}$ and $\widehat{\mathcal{G}}$, but the roles of b and d are swapped, and the choices of $\mathcal{A}_\alpha$ and $\mathcal{E}_\alpha$ are made in such a way to ensure that if n is in our set, we can create an isomorphism regardless of the way in which it entered. Furthermore, to ensure that any isomorphism between these structures can compute $\mathbf{d}$, we add a 3-cycle to each of $\mathcal{G}$ and $\widehat{\mathcal{G}}$. In $\mathcal{G}$, each node in this 3-cycle will have a copy of the "standard" structure we built previously attached to it, and $\widehat{\mathcal{G}}$ will have a copy of the "hard" structure we built previously attached to it.

The other primary example of degrees of categoricity to date comes from Csima and Harrison-Trainor [5]. In this paper, they consider computable structures on cones in the Turing degrees. They begin by defining a relativized version of degrees of categoricity:

Definition 2.3 [5]**.** A structure $\mathcal{A}$ has *degree of categoricity* $\mathbf{d}$ *relative to* $\mathbf{c}$ if $\mathbf{d}$ is the least degree that can compute an isomorphism between any two $\mathbf{c}$-computable copies of $\mathcal{A}$. If there are also two $\mathbf{c}$-computable copies of $\mathcal{A}$ such that for every isomorphism f between them, $f \oplus \mathbf{c} \geq_T \mathbf{d}$, then $\mathcal{A}$ has *strong degree of categoricity* $\mathbf{d}$ *relative to* $\mathbf{c}$.

Definition 2.4. A structure $\mathcal{A}$ has a *(strong) degree of categoricity on a cone* if there is some $\mathbf{d}$ such that for every $\mathbf{c} \geq_T \mathbf{d}$, $\mathcal{A}$ has a (strong) degree of categoricity relative to $\mathbf{c}$. Furthermore, we say that a structure $\mathcal{A}$ has a *(strong) degree of categoricity* $\mathbf{0}^{(\alpha)}$ *on a cone* if there is some $\mathbf{d}$ such that for every $\mathbf{c} \geq_T \mathbf{d}$, $\mathcal{A}$ has a (strong) degree of categoricity $\mathbf{c}^{(\alpha)}$ relative to $\mathbf{c}$.

Their main theorem is as follows:

Theorem 2.5 [5]**.** *Suppose that $\mathcal{A}$ is a computable structure. Then on a cone, $\mathcal{A}$ has a strong degree of categoricity, and this degree is $\mathbf{0}^{(\alpha)}$, where α is the least computable ordinal such that $\mathcal{A}$ is $\mathbf{0}^{(\alpha)}$-computably categorical on a cone.*

The general proof of this theorem involves a version of Ash's metatheorem [2]: Montalbán recently developed a variant on it for successor ordinals [25], and Csima and Harrison-Trainor expanded his variant to include limit ordinals. Here we will only sketch their proof for structures that are $\mathbf{0}'$-computably categorical on a cone due to the complexity of the general proof.

We begin by supposing that $\mathcal{A}$ is not computably categorical on any cone and choose a degree $\mathbf{e}$ that can compute $\mathcal{A}$ and a Scott family for $\mathcal{A}$ with certain properties and that satisfies some technical conditions. We let $\mathbf{d} \geq_T \mathbf{e}$, and then we choose $\mathbf{c}$ to be c.e. in and above $\mathbf{d}$ and choose a $C \in \mathbf{c}$ and take a $\mathbf{d}$-computable approximation to it. This allows us to build our $\mathcal{B}$ and a sequence $\langle f_s \rangle$ of partial isomorphisms computably in $\mathbf{d}$. Thus, the limit of the partial isomorphisms, f, will be a C-computable isomorphism between $\mathcal{B}$ and $\mathcal{A}$. This means that $\mathbf{c}$ will compute an isomorphism between $\mathcal{B}$ and $\mathcal{A}$, and we can further use $g \oplus d$ to compute $\mathbf{c}$ for every isomorphism f between $\mathcal{A}$ and $\mathcal{B}$.

We then use Knight's theorem on the upwards closure of degree spectra from [23] to show that every isomorphism between $\mathcal{B}$ and $\mathcal{A}$ computes $\mathbf{c}$ instead of simply that $g \oplus d$ computes $\mathbf{c}$ for every isomorphism f between $\mathcal{A}$ and $\mathcal{B}$. This lets us see that a structure cannot have a degree of categoricity properly between $\mathbf{0}$ and $\mathbf{0}'$ on a cone.

Csima and Harrison-Trainor also prove the following:

Theorem 2.6 [5]. *Suppose $\mathcal{A}$ is a countable structure. Then, on a cone, if $\mathcal{A}$ is Δ^0_α-categorical, then for every copy $\mathcal{B}$ of $\mathcal{A}$, there is a degree $\mathbf{d}$ that is $\Sigma^0_{\alpha-1}$ in $\mathcal{B}$ if α is a successor ordinal and Δ^0_α in $\mathcal{B}$ if α is a limit ordinal such that $\mathbf{d}$ computes an isomorphism between $\mathcal{A}$ and $\mathcal{B}$ and all isomorphisms between $\mathcal{A}$ and $\mathcal{B}$ compute $\mathbf{d}$.*

This theorem is proved using a more technical result. We begin by considering a structure $\mathcal{A}$ and a degree $\mathbf{c}$ such that $\mathcal{A}$ is $\mathbf{c}$-computable and Δ^0_α-categorical on the cone above $\mathbf{c}$. We can assume that $\mathcal{A}$ has a c.e. Scott family S of computable Σ_α formulas relative to $\mathbf{c}$ with a certain collection of properties. Then, given a copy $\mathcal{B}$ of $\mathcal{A}$, we can consider the set $S(\mathcal{B})$ of pairs $(\bar{b}, \varphi)$ such that $\varphi(\bar{b})$ is true in $\mathcal{B}$ and $\varphi \in S$. Our degree $\mathbf{d}$ will be the degree of $S(\mathcal{B}) \oplus \mathcal{B} \oplus \mathbf{c}$. We then show that there is an isomorphism $f : \mathcal{A} \cong \mathcal{B}$ such that $f \oplus \mathbf{c} \equiv_T \mathbf{d}$ and then, using a set closely related to $S(\mathcal{B})$, use the properties associated with this particular Scott family to show that $\mathbf{d}$ is the desired degree.

Csima and Harrison-Trainor then proceed to argue that this means that the only natural degrees of categoricity are these degrees: arguments concerning structures found naturally in mathematics tend to relativize, and therefore any natural structure has a given property exactly if it has that property on a cone.

We can see that the key to all of these constructions is the ability to approximate a set in the degree in question well enough to construct a computable structure that encodes it.

2.2   Bounding This Class from Above

In [1], Anderson and Csima turned their attention to classes of degrees that are incompatible with the degrees of categoricity. Their first result may be summarized as follows.

Theorem 2.7 [1]. *There is a degree below $\mathbf{0}''$ that is not a degree of categoricity; in fact, there is a Σ^0_2 degree that is not a degree of categoricity.*

The proof that $\mathbf{0}''$ computes a degree that is not a degree of categoricity actually shows that $\mathbf{0}''$ computes a degree that is low for isomorphism. It is quite straightforward: we simply build a set X by finite extensions using a $\mathbf{0}''$ oracle. At stage $\langle \ell, m, k \rangle + 1$, we first extend our finite approximation to ensure that our set is not computable by $\varphi_{\langle \ell, m, k \rangle}$ using $\mathbf{0}'$. We then use $\mathbf{0}'$ to determine whether our approximation can be extended to a string σ such that Φ^σ_ℓ is not a partial isomorphism from $\mathcal{A}_m$ to $\mathcal{A}_k$; if so, that extension is our new approximation. If not, we use $\mathbf{0}''$ to check to see if we can extend our approximation to a string σ such that Φ^σ_ℓ is either not total or not surjective; if so, that extension is our new approximation. Otherwise, we know that any extension of our approximation can be extended to an isomorphism from $\mathcal{A}_m$ to $\mathcal{A}_k$, so we can find a computable isomorphism from $\mathcal{A}_m$ to $\mathcal{A}_k$ and take our new approximation to be the current one.

To compute such a degree that is Σ_2^0, we simply build our set D to be left-c.e. in $0'$. For each tuple $\langle e, i, j \rangle$, we satisfy the requirement that if Φ_e^D is an isomorphism from $\mathcal{A}_i$ to $\mathcal{A}_j$, then there is a computable isomorphism between these structures as well (so, once again, we compute a degree that is low for isomorphism). At each stage, we consider the highest priority requirement (suppose it is the requirement for the tuple $\langle e, i, j \rangle$) and ask if there is a string $\sigma \succeq 1$ such that Φ_e^σ is not a partial isomorphism from $\mathcal{A}_i$ to $\mathcal{A}_j$; if so, we do it and satisfy our requirement. If not, we ask at successive stages whether we can find a string $\sigma \succeq 1$ that can always be extended to a longer partial map from ω to ω. If the answer is always yes, then we can use that functional Φ_e and get a computable isomorphism from $\mathcal{A}_i$ to $\mathcal{A}_j$; if the answer is ever no, we choose a new approximation witnessing this. This is left-c.e. in $0'$ and may injure lower-priority requirements.

Anderson and Csima also demonstrated that the degrees of categoricity are disjoint from the hyperimmune-free degrees:

Theorem 2.8 [1]. *No noncomputable hyperimmune-free degree is a degree of categoricity.*

This proof proceeds by contradiction. We assume that a structure $\mathcal{A}$ witnesses that $\mathbf{d}$ is a hyperimmune-free degree of categoricity and that $\mathbf{d}$ computes an f witnessing that $\mathcal{A} \cong \mathcal{B}$. Since $\mathbf{d}$ is hyperimmune free, there must be a computable function h that dominates both f and f^{-1}. This function h is then used to build an infinite computably bounded tree $T \subseteq \omega^{<\omega}$ whose infinite paths code isomorphisms between $\mathcal{A}$ and $\mathcal{B}$. One of these paths is guaranteed to be computable from $\mathbf{0}'$, so there is $g \leq_T 0'$ witnessing that $\mathcal{A} \cong \mathcal{B}$. This means that $\mathcal{A}$ is $\mathbf{0}'$-computably categorical and thus that $\mathbf{d} \leq_T \mathbf{0}'$, which is impossible since $\mathbf{d}$ is hyperimmune free.

Anderson and Csima also proved that if A is a set and G is Cohen 2-generic in A or if G is Cohen 2-generic relative to a perfect tree, then the degree of $G \oplus A$ is not a degree of categoricity. We note that in fact they proved here that all such degrees are low for isomorphism and that this proof is very similar to the proof of Theorem 2.7, so we reserve a comparable proof until Sect. 3.

We can further restrict the Turing degrees that may be degrees of categoricity as follows. Fokina, Kalimullin, and R. Miller proved in [11] that every strong degree of categoricity is hyperarithmetic using the Effective Perfect Set Theorem [26]. Csima, Franklin, and Shore proved later in [4] that every degree of categoricity, strong or not, is hyperarithmetic. Their proof requires Kreisel's Basis Theorem [31]. To prove this, we begin by taking an arbitrary degree $\mathbf{d}$ that is not hyperarithmetic and an arbitrary computable structure $\mathcal{A}$ and listing all the computable copies of $\mathcal{A}$: $\mathcal{A}_0, \mathcal{A}_1, \ldots$. The class of isomorphisms between $\mathcal{A}_0$ and $\mathcal{A}_1$ is Π_2^0 and thus Σ_1^1, and by Kreisel's Basis Theorem, there is an isomorphism f_1 such that $\mathbf{d} \not\leq_h f_1$. In fact, we can relativize Kreisel's Basis Theorem to find a sequence of isomorphisms $f_0, f_1, \ldots$ such that f_i is an isomorphism between $\mathcal{A}_0$ and $\mathcal{A}_i$ and $\mathbf{d} \not\leq_h f_1 \oplus \ldots \oplus f_i$ for each i. We take an exact pair $\mathbf{a}$ and $\mathbf{b}$ for this sequence and note that both of these degrees can compute an isomorphism between any two copies of $\mathcal{A}$. This means that any degree of categoricity for $\mathcal{A}$ must be below both **a** and **b**. If **d** is such a degree, then it must therefore be computable from $f_1 \oplus \ldots \oplus f_n$ for some n, which would lead to a contradiction.

2.3   Open Questions

We can see that there can only be countably many degrees of categoricity since they are all hyperarithmetic. However, the examples produced are all of the same type and come nowhere near the upper bounds we have established for this class: all known examples are d.c.e. in and above some degree of the form $\mathbf{0}^{(\alpha)}$. All efforts to extend these constructions to even 3-c.e. degrees have failed to date, and indeed Csima and Harrison-Trainor's work shows that no natural structure can even have properly d.c.e. degree. This leads to a first obvious question:

Question 2.9. Is there a degree that is n-c.e. in and above $\mathbf{0}^{(\alpha)}$ for some computable ordinal α and some $n > 2$ that is not a degree of categoricity?

We may also ask a weaker version of this question inspired by the observation that the known degrees of categoricity all have very simple approximations in the intervals $[\mathbf{0}^{(\alpha)}, \mathbf{0}^{(\alpha+1)}]$. Must this always be true?

Question 2.10. Is there a degree of categoricity that is not contained in an interval of the form $[\mathbf{0}^{(\alpha)}, \mathbf{0}^{(\alpha+1)}]$ for some computable ordinal α?

On the more technical side, we note that there is a case that Csima, Franklin, and Shore did not consider in [4]:

Question 2.11. If α is a computable limit ordinal, is every degree that is c.e. or d.c.e. in and above $\mathbf{0}^{(\alpha)}$ a (strong) degree of categoricity?

We also note that the degrees of categoricity for one particular class of structures have been studied: in [11], it is shown that any c.e. degree is the degree of categoricity of some computable algebraic field. It may be illuminating to consider the degrees of categoricity for other nonuniversal structures:

Question 2.12. Which Turing degrees may be degrees of categoricity for a particular class $\mathcal{C}$ of structures?

We now go on to the two most fundamental questions in the area. First of all, all the known degrees of categoricity are strong degrees of categoricity, which leads to the following question:

Question 2.13. Is every degree of categoricity a strong degree of categoricity?

Secondly, we can ask for a full characterization.

Question 2.14. Characterize the Turing degrees that are degrees of categoricity.

3   Degrees that Are Low for Isomorphism

We now turn our attention to Turing degrees that are very far from being degrees of categoricity: those that are low for isomorphism, introduced by Franklin and Solomon in [14]. They use directed graphs to study this concept as the authors considering degrees of categoricity have done, but here these graphs are used because of the need to quantify over all structures in all computable languages. This decision is based on work by Hirschfeldt, Khoussainov, Shore, and Slinko, who proved in [19] that directed graphs are universal in the following sense: arbitrary countable structures $\mathcal{A}$ and $\mathcal{B}$ in a computable language can be coded into countable directed graphs $G(\mathcal{A})$ and $G(\mathcal{B})$ such that

- $\mathcal{A} \cong \mathcal{B}$ if and only if $G(\mathcal{A}) \cong G(\mathcal{B})$,
- $\mathcal{A}$ is computable exactly when $G(\mathcal{A})$ is computable, and
- if $\mathcal{A}$ and $\mathcal{B}$ are computable, then for any Turing degree $\mathbf{d}$, $\mathcal{A} \cong_{\mathbf{d}} \mathcal{B}$ if and only if $G(\mathcal{A}) \cong_{\mathbf{d}} G(\mathcal{B})$.

3.1   Examples of Degrees that Are Low for Isomorphism

The most common theme in these proofs is that of forcing. In fact, any reasonable sort of computability-theoretic forcing at the right level will allow us to produce a degree that is low for isomorphism.

The first type of forcing considered in [14] is forcing with generic reals; specifically, with Cohen and Matthias generic reals. The following theorem is obtained:

Theorem 3.1 [14]. *Every Cohen 2-generic degree and every Matthias 3-generic degree is low for isomorphism.*

The proofs in this paper rely heavily on machinery from reverse mathematics [20, 21]; here we present a direct proof for the Cohen generic case.

Let G be a Cohen 2-generic real. (In the future, when we write "generic" without further qualification, we will mean "Cohen generic.") We must show that for any $\mathcal{A}$ and $\mathcal{B}$ such that $\mathcal{A} \cong_G \mathcal{B}$, $\mathcal{A} \cong_{\Delta_1^0} \mathcal{B}$. We begin by considering the following statements:

- Φ_e^X maps an element of $\mathcal{A}$ to an element of $\mathcal{B}$ that witnesses that Φ_e^X is not an isomorphism from $\mathcal{A}$ to $\mathcal{B}$.
- Φ_e^X is total.
- Φ_e^X is surjective.

We note that the first of these statements is $\Sigma_1^{0,X}$, since it states that at some stage, Φ_e^X maps an element of $\mathcal{A}$ to an element of $\mathcal{B}$ that do not satisfy the same formulas in the atomic diagram. It is clear that the latter two statements are $\Pi_2^{0,X}$. Now we fix an $\mathcal{A}$ and $\mathcal{B}$ such that $\mathcal{A} \cong_G \mathcal{B}$. Since G is 2-generic, it must force the truth or falsity of each of the above statements, and since G does compute an isomorphism between $\mathcal{A}$ and $\mathcal{B}$, we know that G must force the first statement to be false and the others to be true. Let ρ be the initial segment of G

that forces all these things. We will construct a computable sequence $\rho = \sigma_0 \preceq \sigma_1 \preceq \sigma_2 \preceq \ldots$ in stages, defining σ_i at stage i, so that each new term σ_i lets us define a longer partial isomorphism between $\mathcal{A}$ and $\mathcal{B}$.

To define σ_{i+1} for an even i, we consider the partial isomorphism found through σ_i. There is a least element n_{i+1} of $\mathcal{A}$ whose image is undefined by this partial isomorphism, so we search above σ_i for an extension σ_{i+1} that, when used as an oracle on Φ_e, will place n_{i+1} in our domain. Such an extension must exist because ρ has already forced totality, and the mapping it finds must be extendible to an isomorphism between $\mathcal{A}$ and $\mathcal{B}$ because ρ has forced the first statement to be false. Now we have extended the initial segment of the domain of our partial isomorphism.

To define σ_{i+1} for an odd i, we do the same thing, but in reverse: there is a least element m_{i+1} of $\mathcal{B}$ that is not yet mapped to by the partial isomorphism defined at the end of the previous step using σ_i as an oracle. Now, we search above σ_i for an extension σ_{i+1} that, when used as an oracle with Φ_e, will place m_{i+1} in our range. In this case, such an extension must exist because ρ has forced surjectivity, and we have preserved our ability to extend to an isomorphism between $\mathcal{A}$ and $\mathcal{B}$ as before.

Franklin and Solomon also use Sacks forcing with computable perfect trees to produce a degree that is low for isomorphism that are minimal and hyperimmune free as well (see Chapter V.5 in [29] for a discussion of this sort of forcing). Using a noneffective enumeration of all pairs $(\mathcal{A}_i, \mathcal{B}_i)$ of all infinite computable directed graphs, we build a sequence of computable perfect trees $T_0 \supseteq T_1 \supseteq \ldots$ such that T_0 is the identity tree and $T_i(\lambda) \subseteq T_{i+1}(\lambda)$ for each i. The resulting set D is the set such that $T_i(\lambda) \preceq D$ for every i. Four kinds of requirements must be satisfied in this proof:

- Noncomputability: For every e, $D \neq \Phi^e$.
- Hyperimmune-freeness: For every e, either Φ_e^D is not total or Φ_e^D is majorized by a computable function.
- Minimality: For every e, if Φ_e^D is total, then either Φ_e^D is computable or $D \leq_T \Phi_e^D$.
- Lowness for isomorphism: For every e and i, if Φ_e^D is an isomorphism from $\mathcal{A}_i$ to $\mathcal{B}_i$, then $\mathcal{A}_i \cong_{\Delta_1^0} \mathcal{B}_i$.

The construction once again proceeds by stages, and one of these requirements is satisfied at each stage. The first three requirements are satisfied in the usual way (see [29] for details). We will discuss the lowness for isomorphism requirements here.

Suppose we want to ensure that the lowness for isomorphism requirement is satisfied for Φ_e and the pair $(\mathcal{A}_i, \mathcal{B}_i)$. Without loss of generality, we can assume we have already satisfied the hyperimmune-freeness requirement for e and that we are working at stage $s + 1$. We now proceed by cases.

If we satisfied the hyperimmune-freeness requirement by guaranteeing that Φ_e^D will not be total, then our lowness for isomorphism requirement is satisfied trivially and we simply choose the root of our new tree T_{s+1} to be any nonroot element of T_s.

If we satisfied the hyperimmune-freeness requirement by guaranteeing that Φ_e^D will be total and majorized by a computable function, we know that Φ_e^A is total for every branch A of T_s. We now check to see whether there is a string σ and a number n such that $\Phi_e^{T_s(\sigma)} \restriction n$ halts and $\Phi_e^{T_s(\sigma)} \restriction n$ is not a partial isomorphism from $\mathcal{A}_i$ to $\mathcal{B}_i$. If there is such a string σ, we take T_{s+1} to be the full subtree of T_s above σ. Otherwise, we know that any branch in T_s will give us an isomorphism between $\mathcal{A}_i$ and $\mathcal{B}_i$, and we can define a new subtree inside T_s computably so a computable isomorphism can actually be found.

Franklin and Solomon also asked in [14] if one could "cap" the level of Cohen genericity associated with lowness for isomorphism at 2-genericity: in other words, if it is possible for a 1-generic that is not computed by a 2-generic to be low for isomorphism. In [16], Franklin and Turetsky answered this question in the negative by constructing a 1-generic G that satisfies the following requirements:

(**One$_e$**): G either meets or avoids the Σ_1^0 set W_e.
(**Two$_i$**): There is a Σ_2^0 set X_i such that if $\Phi_i^Y = G$, then Y neither meets nor avoids X_i.
(**IM$_{\langle i,j_1,j_2 \rangle}$**): if Φ_i^G is an isomorphism between $\mathcal{A}_{j_1}$ and $\mathcal{A}_{j_2}$, then $\mathcal{A}_{j_1} \cong_{\Delta_1^0} \mathcal{A}_{j_2}$.

The first requirement can be satisfied through a standard finite injury approach: if we find at some stage that we can extend our finite approximation to G to meet W_e, we do so, and it is satisfied automatically otherwise.

Now, to satisfy (Two$_i$), we use infinitely many subrequirements:

(**Two$_{\langle i,\tau \rangle}$**): If there is a $Y \succ \tau$ such that $\Phi_i^Y = G$, then Y does not meet X_i and there is some string $\rho \succ \tau$ such that $\rho \in X_i$.

In meeting each of these subrequirements, we construct our X_i. Suppose we have a finite approximation g to G and we are trying to satisfy (Two$_{\langle i,\tau \rangle}$). We reserve the next bit b at position $|g|$ for our use and initially require that $G(b) = 0$. Now we try to find a string $\rho \succ \tau$ such that there is no Y extending ρ with $\Phi_i^Y = G$. If at some point we see a ρ extending τ where $\Phi_i^\rho \succeq g \char94 0$, we put this ρ in X_i and change $G(b)$ to 1. Since the construction is $\mathbf{0}''$, the set of all these ρs over all τ will be Σ_2^0.

We argue briefly that this X_i serves its intended purpose: that for any i and Y such that $\Phi_i^Y = G$, Y cannot meet or avoid X_i. If Y avoids X_i, then we fix an initial segment τ of Y where this happens and consider the appropriate node on the true path. By our definition of X_i, there is no $\rho \succeq \tau$ with $\Phi_i^\rho \preceq g \char94 0$, so $\Phi_i^Y(|g|) \neq 0$. However, if there is no such ρ, $g \char94 0$ will be an initial segment of G, so Φ_i^Y and G must differ at position $|g|$.

If Y does meet X_i, then we fix an initial segment ρ where this happens and consider the (Two$_{\langle i,\tau \rangle}$)-strategy that caused us to add this ρ to X_i and the node on the priority tree that witnesses this. By definition, we know that we have a potential initial segment g of G associated with this node and that $\Phi_i^\rho \succeq g \char94 0$. There are two possible scenarios. In the first, the node in question is to the left of the true path, and the string g is not actually an initial segment of G. Therefore, we cannot have $\Phi_i^Y = G$. In the second, the node in question is actually on the true path. In this case, $g^\frown 1$ will be an initial segment of G, and Φ_i^Y and G must disagree at position $|g|$.

To satisfy $(\mathrm{IM}_{\langle i, j_1, j_2 \rangle})$, we use a standard infinitary construction. We establish a length of agreement function for the appropriate node on the priority tree. If at some stage we can find a string extending our current approximation that defines a longer isomorphism, we choose it as our new approximation and take the infinite outcome at our node on the priority tree; otherwise we choose a finite outcome.

3.2   Bounding This Class from Above

Franklin and Solomon also identify significant classes of degrees that cannot be low for isomorphism. The first such class is the nontrivial Δ_2^0 degrees:

Theorem 3.2. *No nontrivial Δ_2^0 degree is low for isomorphism and thus no degree that computes a nontrivial Δ_2^0 degree is low for isomorphism.*

The proof is quite straightforward. We take a representative D of a noncomputable Δ_2^0 degree $\mathbf{d}$ and fix a Δ_2^0 approximation $\langle D_s \rangle$ to it. We then use this approximation to construct two computable directed graphs, G and H, so that the unique isomorphism between them is Turing equivalent to D.

We begin by placing a $(n+2)$-cycle in each of G and H for every $n \in \omega$. The $(n+2)$-cycle component will code n's membership in D. Then, for each $(n+2)$-cycle in G, we add an arrow from some element x_n to a new element a_n, and for each $(n+2)$-cycle in H, we add an arrow from some element y_n to a new element b_n. At this point, $n \notin D$, G and H are isomorphic, and this isomorphism must map a_n to b_n.

If, at stage s, n enters D, we add a new element a' to G and a new element b' to H so that there are edges from a_n to a' and from x_n to a' and edges from b' to b_n and y_n to b'. We still have an isomorphism between G and H, but now the isomorphism must map a_n to b' and a' to b_n.

If n exits D at a later stage, we add new elements a'' and b'' to G and H respectively. This time, we add edges from a'' to a_n and from x_n to a'' and edges from b_n to b'' and from y_n to b''. We can see that G and H are still isomorphic, but the isomorphism maps a_n to b_n once more.

We can repeat this pattern and see that since after some point our approximation to D will be constant on n, the $(n+2)$-cycles in G and H will stabilize, and the isomorphism between G and H will map a_n to b_n if and only if $n \notin D$. This is enough to see that $G \cong_c H$ if and only if $\mathbf{d} \leq_T \mathbf{c}$.

This lets us see that no degree above $\mathbf{0}'$ is low for isomorphism either, since the degrees that are low for isomorphism are closed downward.

They also show using a similar proof that if a degree can compute a separating set for a pair of computably inseparable c.e. sets, that degree cannot be low for isomorphism.

Franklin and Solomon then turn their attention to measure and prove the following theorem:

Theorem 3.3. *No Martin-Löf random degree is low for isomorphism.*

Here, we sketch a proof that a set of degrees of measure one is not low for isomorphism and then discuss briefly how it can be modified to prove the theorem above.

We begin by observing that we can produce a class of degrees that are not low for isomorphism with some positive measure and conclude using Kolmogorov's 0–1 law that it must actually have measure 1. We construct two isomorphic computable directed graphs G and H and a Π_1^0 class $\mathcal{C}$ so that

(P1): $G \not\equiv_{\Delta_1^0} H$,
(P2): $\mu(\mathcal{C}) \geq \frac{1}{2}$, and
(P3): if $X \in \mathcal{C}$, then X can compute an isomorphism from G to H.

(P1) and (P3) clearly combine to guarantee that no element of $\mathcal{C}$ can be low for isomorphism and will not need to be modified when we require Martin-Löf randomness instead of simply positive measure; the only adaptation we will need to make to (P2) is to construct a sequence of trees whose measure increases in a very controlled way and forms the complement of a Martin-Löf test.

To satisfy (P1), we meet the following requirement:

$$R_e : \Phi_e \text{ is not an isomorphism from } G \text{ to } H.$$

As we satisfy this, we ensure that our diagonalization strategy for R_e does not remove too much measure from $\mathcal{C}$, thus satisfying (P2) at the same time. Finally, to satisfy (P3), we construct a Turing functional Γ so that for any X in $\mathcal{C}$, Γ^X is an isomorphism from G to H.

Our graphs G and H initially begin as infinitely many e-components for each $e \in \omega$, where an e-component is an $(e + 3)$-cycle with a coding node u distinguished by a loop. In the course of our construction, we will add "tails" to coinfinitely infinitely many of the e-components when we actively diagonalize to satisfy R_e: a "tail" consists of two nodes x_0 and x_1 with arrows from u to x_0 to x_1 to x_0. This guarantees that a set X can compute an isomorphism between G and H if and only if it can compute a bijection between the coding nodes in G and H and, furthermore, successfully match up the tailed and untailed coding nodes.

First we discuss how we will meet a single requirement R_e. To do this, we fix an e-component in G and diagonalize against its coding node a_e. If there is no stage s where a_e is mapped to a coding node b of an e-component in H, the requirement is satisfied trivially. Otherwise, we actively diagonalize by adding tails to an infinite coinfinite set of coding nodes of e-components in H, including b. We also add tails to an infinite coinfinite set of coding nodes of e-components in G but ensure that a_e is not among them to make sure that no isomorphism between G and H can map a_e to b as Φ_e does.

These infinite coinfinite sets are also used to define the Turing functional Γ and to ensure that enough reals can compute an isomorphism between G_0 and H_0. At stage 0, we define Γ so that for each $e \in \omega$ and each string σ of length

$e + 2$, we define Γ^σ so it maps the coding nodes for e-components in G_0 to e-components in H_0. Furthermore, we make sure that different strings of the same length do not produce the same mapping. Now observe that these mappings will continue to extend to isomorphisms at later stages as long as they map untailed components to untailed components and tailed components to tailed components (when tails are added to our structures at later stages). If, however, we satisfy R_e by adding tails to some components, we must make sure that we remove any branch X from our tree such that Γ^X maps an untailed component to a newly tailed component or vice versa.

Now we describe an abbreviated version of R_0's strategy to give an idea of how to balance these conflicting requirements. To ensure that (P2) holds, we ensure that we do not remove more than $\frac{1}{4}$ of the measure from our tree. To do this, we choose the infinite coinfinite sets that we will use to diagonalize against a_0 carefully and define Γ in such a way that, no matter what coding node Φ_0 may map a_0 to, we can diagonalize in such a way that we can remove no more than one string of measure $\frac{1}{4}$ and still have the oracles remaining in the class correctly map untailed components to untailed components and tailed components to tailed components.

The class $\mathcal{C}$ will be defined as the set of branches in the intersection of our computable sequence of trees $2^{<\omega} = T_0 \supseteq T_1 \supseteq \ldots$, and this class is obtained by removing the strings that are no longer appropriate oracles for Γ given the changes in G and H that have taken place. Since we code information about where the e-components map at a string of length $e + 2$ and we have arranged the coding nodes so no more than one of the strings of that length will fail to code an isomorphism, we remove at most $\frac{1}{4} + \frac{1}{8} + \ldots = \frac{1}{2}$ from our tree overall, and we have a tree of positive measure.

Now we explain how this proof can be modified to show that no Martin-Löf random real is low for isomorphism. We begin by recalling the definition of Martin-Löf randomness; for a more thorough discussion of algorithmic randomness, see [7].

Definition 3.4. A *Martin-Löf* test is an effectively c.e. sequence $\langle V_i \rangle$ of subsets of $2^{<\omega}$ such that $\mu([V_i]) \leq 2^{-i}$ for all i, and a real X is *Martin-Löf random* if $X \notin \cap_i [V_i]$ for every Martin-Löf test $\langle V_i \rangle$.

Note that the class $\mathcal{C}$ we built has measure at least $\frac{1}{2}$; its complement is therefore a Σ^0_1 class of measure no more than $\frac{1}{2}$. Its complement could therefore be the first component of a Martin-Löf test. We construct an entire Martin-Löf test by constructing not just one Π^0_1 class $\mathcal{C}$ but an effective sequence of nested Π^0_1 classes $\mathcal{C}_0 \subseteq \mathcal{C}_1 \subseteq \ldots$ where the ith class has measure at least $1 - \frac{1}{2^{i+1}}$. Their complements will therefore form a Martin-Löf test, and any Martin-Löf random real X will be in $\mathcal{C}_i$ for some i and will thus not be low for isomorphism.

To construct this sequence of classes, we repeat the construction we just described as follows. $\mathcal{C}_0$ will be generated as above. We arrange for each class $\mathcal{C}_{i+1}$ to be larger than the previous one as follows. When we remove a string σ from $\mathcal{C}_i$ (or, indeed, any previous class), we do not remove that string from $\mathcal{C}_{i+1}$.

Instead, we start a new version of the construction inside this string. If a new diagonalization process within these constructions requires that we remove a string from C_{i+1}, it will be longer and thus we will remove less measure from C_{i+1} than we did from C_i to satisfy any given diagonalization requirement; with some planning, we can require that $\mu(C_i) \geq 1 - \frac{1}{2^i}$ and thus that the complement of C_i can be the ith component of our Martin-Löf test.

We also observe that this is the strongest result that can be obtained concerning lowness for isomorphism and randomness: the computably random degrees and those that are low for isomorphism are not disjoint, since every high degree contains a computably random real [28], and there is a high 2-generic.

3.3   Open Questions

We first observe that Franklin and Turetsky's result still leaves a gap in the genericity hierarchy:

Question 3.5. Is there a properly 1-generic degree that is low for isomorphism and not computable from a weakly 2-generic?

While there seems to be no easy way to adapt their construction to answer this question, it may be possible to construct such a degree in some other way.

Csima has also defined a similar notion, *lowness for categoricity*. She has defined a degree **d** to be low for categoricity if every computable structure that is **d**-computably categorical is already computably categorical [3]. Lowness for isomorphism clearly implies lowness for categoricity, but whether the converse holds is uncertain.

Question 3.6. Is every degree that is low for categoricity also low for isomorphism?

As with degrees of categoricity, we may also consider the degrees that are low for isomorphism for a particular class of structures. Suggs has studied several cases, including linear orders [35]; some of this work appears in [14].

Question 3.7. Describe the degrees that are low for isomorphism for a particular class of structures C.

We end with two questions that are rather hard and closely related:

Question 3.8. Are there other natural classes of degrees that are either subsets of or disjoint to the degrees that are low for isomorphism?

Some candidates for such classes include the computably traceable degrees (a subset of the hyperimmune-free degrees) and the c.e. traceable degrees.

We end with, once more, the obvious question.

Question 3.9. Characterize the Turing degrees that are low for isomorphism.

4   Discussion and Musings

While the degrees of categoricity and those that are low for isomorphism both lack a full characterization, they lack this characterization in very different ways. It is easy to describe all known degrees of categoricity: they all belong to intervals of the form $[\mathbf{0}^{(\alpha)}, \mathbf{0}^{(\alpha+1)}]$ for some computable ordinal α; in fact, they are all even d.c.e. in and above a degree of the form $\mathbf{0}^{(\alpha)}$ for some such α. All of the known constructions are very similar—one codes information about an appropriate set in the degree in question into two copies of the same structure, using an approximation of this set—and none of them extend to the 3-c.e. case. It is known that the degrees of categoricity are all hyperarithmetic and thus that this class is countable, so it must be null. It is also small with respect to category, since no such degree can be 2-generic. Furthermore, no degree of categoricity can be hyperimmune free.

On the other hand, there is no convenient way to describe the class of degrees that are low for isomorphism. While the Cohen 2-generics and Matthias 3-generics are known to be subsets of this class and no Martin-Löf degree can belong to it, most of the results in this area consist of showing that some degree of a certain kind is low for isomorphism and another degree of the same kind is not; a summary appears in Table 1. For any category that does contain degrees that are low for isomorphism, the proof method is indicated; for any category that does not, one type of counterexample is indicated. It is clear that lowness for isomorphism is not closely related to any natural class except the Δ_2^0 degrees.

Most of the results on lowness for isomorphism were obtained by forcing. In general, any type of forcing that will allow us to force a functional to be total and surjective and never to fail to be a partial isomorphism will permit us to construct a set that is low for isomorphism.[1] However, lowness for isomorphism is not a property strictly determined by the ability to force: Anderson and Csima constructed a Σ_2^0 example of a degree that is low for isomorphism using a standard injury argument.

Table 1. Lowness for isomorphism: Categories of degrees

	Low for isomorphism	Not low for isomorphism
Nontrivial Δ_2^0	no	yes: Δ_2^0
Nontrival Δ_3^0	yes: Cohen forcing	yes: Martin-Löf random
Minimal	yes: perfect trees	yes: Δ_2^0
Not minimal	yes: Cohen forcing	yes: Δ_2^0
Hyperimmune	yes: Cohen forcing	yes: Δ_2^0
Hyperimmune-free	yes: perfect trees	yes: Martin-Löf random

---

[1] We note with some amusement that the isomorphism condition is actually lower in the arithmetic hierarchy than the others and is therefore not the condition that determines the degree of genericity necessary to force lowness for isomorphism.

We can also argue that there are very few degrees that are low for isomorphism: they have measure 0 since no Martin-Löf degree is low for isomorphism. However, unlike the degrees of categoricity, they are large with respect to category since every 2-generic degree is low for isomorphism.

In short, the degrees of categoricity and the degrees that are low for isomorphism appear to be diametrically opposed, bros. All known degrees of categoricity have a simple approximation: one that is no more than d.c.e. in some jump of 0. In some way, they form the "backbone" of the Turing degrees. The degrees that are low for isomorphism, on the other hand, are in general, those that cannot be effectively approximated (Anderson and Csima's Σ_2^0 example is, once again, a delightfully puzzling exception). Unsurprisingly, they are bounded away from each other: no degree that is comparable to $0'$ is low for isomorphism, which is where all the all the known degrees of categoricity reside.

Since both of these classes resist characterization and they are bounded away from each other, it may be of interest to consider the class of degrees that fall between them. What kinds of degrees are neither low for isomorphism nor degrees of categoricity? They must resist approximation, but not too much. Furthermore, this class of degrees is large with respect to measure and small with respect to category. There are certainly natural classes of degrees with this property, and if one of them proved to be disjoint from both of the classes we have considered in this paper, it might illuminate the features inherent in each of them.

Question 4.1. Is there a natural class of degrees that is disjoint from both the degrees of categoricity and the degrees that are low for isomorphism?

We also notice that the degrees that are low for isomorphism do not form an ideal in the Turing degrees: while they are downward closed, they are not closed under join because there is a pair of 2-generics whose join computes $0'$. However, we may ask a follow-up question: are we, in fact, considering the most appropriate degree structure? In algorithmic randomness, the Schnorr trivial reals have unexpected properties in the Turing degrees [6,12], but they behave as one expects trivial reals to behave in the truth-table degrees [15]. It may be that these notions are better understood in another degree structure:

Question 4.2. Describe the behavior of the degrees of categoricity and the degrees that are low for isomorphism in an alternate degree structure such as the weak truth-table or truth-table degrees.

References

1. Anderson, B.A., Csima, B.F.: Degrees that are not degrees of categoricity. Notre Dame J. Formal Logic (to appear)
2. Ash, C.J., Knight, J.: Computable structures and the hyperarithmetical hierarchy. In: Studies in Logic and the Foundations of Mathematics, vol. 144. North-Holland (2000)
3. Csima, B.: Degrees of categoricity and related notions. In: BIRS Computable Model Theory Workshop, November 2013

4. Csima, F.B., Franklin, J.N.Y., Shore, R.A.: Degrees of categoricity and the hyper-arithmetic hierarchy. Notre Dame J. Formal Logic **54**(2), 215–231 (2013)
5. Csima, B.F., Harrison-Trainor, M.: Degrees of categoricity on a cone. J. Symbolic Logic (to appear)
6. Downey, R., Griffiths, E., LaForte, G.: On Schnorr and computable randomness, martingales, and machines. Math. Log. Q. **50**(6), 613–627 (2004)
7. Downey, R.G., Hirschfeldt, D.R.: Algorithmic Randomness and Complexity. Springer, New York (2010)
8. Ershov, Y.: A hierarchy of sets, I. Algebra Logic **7**(1), 212–232 (1968). Originally appearing in Algebra i Logika **7**(1), 47–74 (1968) (Russian)
9. Ershov, Y.: On a hierarchy of sets, II. Algebra Logic **7**(4), 25–43 (1968). Originally appearing in Algebra i Logika **7**(4), 15–47 (1968) (Russian)
10. Ershov, Y.: On a hierarchy of sets, III. Algebra Logic **9**(1), 20–31 (1970). Originally appearing in Algebra i Logika **9**(1), 34–51 (1970) (Russian)
11. Fokina, E.B., Kalimullin, I., Miller, R.: Degrees of categoricity of computable structures. Arch. Math. Logic **49**(1), 51–67 (2010)
12. Franklin, J.N.Y.: Schnorr trivial reals: a construction. Arch. Math. Logic **46**(7–8), 665–678 (2008)
13. Franklin, J.N.Y.: Lowness and highness properties for randomness notions. In: Arai, T. et al. (ed.) Proceedings of the 10th Asian Logic Conference, pp. 124–151. World Scientific (2010)
14. Franklin, J.N.Y., Solomon, R.: Degrees that are low for isomorphism. Computability **3**(2), 73–89 (2014)
15. Franklin, J.N.Y., Stephan, F.: Schnorr trivial sets and truth-table reducibility. J. Symbol. Logic **75**(2), 501–521 (2010)
16. Franklin, J.N.Y., Turetsky, D.: Genericity and lowness for isomorphism (submitted)
17. Fröhlich, A., Shepherdson, J.C.: Effective procedures in field theory. Philos. Trans. R. Soc. Lond. Ser. A. **248**, 407–432 (1956)
18. Harizanov, V.S.,: Pure computable model theory. In: Handbook of Recursive Mathematics, Volume 1: Recursive Model Theory, Studies in Logic and the Foundations of Mathematics, vol. 138, pp. 3–114 (1998)
19. Hirschfeldt, D., Khoussainov, B., Shore, R., Slinko, A.: Degree spectra and computable dimensions in algebraic structures. Ann. Pure Appl. Logic **115**(1–3), 71–113 (2002)
20. Hirschfeldt, D.R., Shore, R.A.: Combinatorial principles weaker than Ramsey's theorem for pairs. J. Symbol. Logic **72**(1), 171–206 (2007)
21. Hirschfeldt, D.R., Shore, R.A., Slaman, T.A.: The atomic model theorem and type omitting. Trans. Am. Math. Soc. **361**(11), 5805–5837 (2009)
22. Hirschfeldt, D.R., White, W.M.: Realizing levels of the hyperarithmetic hierarchy as degree spectra of relations on computable structures. Notre Dame J. Formal Logic **43**(1), 51–64 (2003). 2002
23. Knight, J.F.: Degrees coded in jumps of orderings. J. Symbol. Logic **51**(4), 1034–1042 (1986)
24. Marker, D.: Non Σ_n axiomatizable almost strongly minimal theories. J. Symbol. Logic **54**(3), 921–927 (1989)
25. Montalbán, A.: Priority arguments via true stages. J. Symbol. Log. **79**(4), 1315–1335 (2014)
26. Moschovakis, Y.N.: Descriptive Set Theory. Studies in Logic and the Foundations of Mathematics, vol. 100. North-Holland (1980)
27. Nies, A.: Computability and Randomness. Clarendon Press, Oxford (2009)

28. Nies, A., Stephan, F., Terwijn, S.A.: Randomness, relativization and Turing degrees. J. Symbol. Logic **70**(2), 515–535 (2005)
29. Odifreddi, P.: Classical Recursion Theory. Studies in Logic and the Foundations of Mathematics, vol. 125. North-Holland (1989)
30. Odifreddi, P.: Classical Recursion Theory, Volume II. Studies in Logic and the Foundations of Mathematics, vol. 143. North-Holland (1999)
31. Sacks, G.E.: Higher Recursion Theory. Springer, New York (1990)
32. Slaman, T.A., Solovay, R.M.: When oracles do not help. In: Fourth Annual Workshop on Computational Learning Theory, pp. 379–383. Morgan Kaufman, Los Altos (1991)
33. Soare, R.: The Friedberg-Muchnik theorem re-examined. Can. J. Math. **24**, 1070–1078 (1972)
34. Soare, R.I.: Recursively Enumerable Sets and Degrees. Perspectives in Mathematical Logic. Springer, New York (1987)
35. Suggs, J.: Degrees that are low for $\mathcal{C}$ isomorphism. Ph.D. thesis, University of Connecticut (2015)

On Constructive Nilpotent Groups

Nazif G. Khisamiev[⊠] and Ivan V. Latkin

East Kazakhstan State Technical University, Ust-Kamenogorsk, Kazakhstan
hisamiev@mail.ru, lativan@yandex.ru

Brief Historical Introduction

Algorithmic problems in the areas of algebra and theory of numbers arose in an explicit form prior to the appearance of a precise concept of an *algorithm*. More than 100 years ago, intuitive notions of algorithmic issues in group theory were known as the problems of word, conjugacy, isomorphism, and occurrence, etc. [3,64,66,73,88]. A far-famed algorithmic task was Hilbert's 10th problem: find a procedure which in a finite number of steps, determines whether or not a Diophantine equation has an integer solution [36]. However, research of an algorithmic nature had been carried out less explicitly for a long time; e.g., the Euclidean algorithm, the formulae of for the solution of some algebraic equations, and many others.

Let us notice that the above-mentioned algorithmic problems have been posed in group theory for the finitely presented groups. The elements of such groups are the *words*, i.e., the finite ordered sequences of generators. All other "ancient" algorithms also executed the operations immediately with the finite ordered sets of some symbols.

The appearance of a precise concept of a computable function allowed finer techniques for in-depth scrutiny of manifold mathematical challenges. Gödel successfully used *enumerations* for the first time in his classical work on incompleteness [33]. Kolmogorov pointed to the importance of studying the enumerations of arbitrary objects in the middle 1950's. His student, Uspenskiĭ was engaged in implementation of these ideas for computable enumerations [86]. Many other investigators also studied such enumerations at that time and in the 1960's, e.g., Dekker, Ershov, Friedberg, Lachlan, Lacombe, Myhill, Pour-El, Rogers Jr., Rice, et al. Furthermore, there appeared the interesting works of Fröhlich, Shepherdson, and Rabin on the enumerated fields in the 1950's [25,78]. Fröhlich and Shepherdson formalized analysis of van der B. L. Waerden, who considered effective procedures in field theory, but without the language of computability theory (see [87]).

Mal'tsev combined the two approaches of symbol-combinatoric and numeric, and laid the foundations for the general theory of constructive algebraic systems in [69] (see definition below). The most significant results in this theory were

(International Conference Dedicated to the 60th Anniversary of the Birthday of Rodney G. Downey)

© Springer International Publishing AG 2017
A. Day et al. (Eds.): Downey Festschrift, LNCS 10010, pp. 324–353, 2017.
DOI: 10.1007/978-3-319-50062-1_21

obtained at the world famous Siberian school of Algebra and Logic in the former Soviet Union, founded by Mal'tsev. Remarkable achievements were reached by Ershov, Goncharov, and by their numerous students and their followers.

Independently, at Cornell University in the United States in 1972, A. Nerode began his programme to determine the effective content of mathematical constructions and started to develop a systematic theory of recursive structures (see definition below). The best known results on the recursive structures were obtained by Nerode, Remmel, Metakides, Millar, Kalantary, Retzlaff, Lin, and also by Crossley, Ash, Downey, Moses, Knight, Hird, Harizanov and others.

The purpose of this paper to review the results on the constructive nilpotent groups, not claiming to be complete. We consider both the fundamental questions, which are general in the computable algebra, such as the problems of existence, uniqueness, and extension of constructivizations (computable copies), and the questions that arise from the study of the effectiveness of the basic theorems, constructions and structural properties within the class of nilpotent groups.

These main problems of constructive nilpotent groups are studied slightly, namely, there are not enough algebraic structural conditions that are necessary or sufficient for a nilpotent group to have the desired algorithmic properties.

On the other hand, the computability-theoretic behavior of nilpotent groups is relatively well understood, if we consider the various computable model theoretic notions such as computable dimension, degree spectra of structures, degree spectra of relations, etc., and ask how these notions behave within the class of nilpotent groups. The several results about the commonly considered computable model theoretic notions (such as those mentioned above) remain true when we restrict our attention to the class two nilpotent groups, provided that these results are true for some model [37]. Therefore we do not give wide coverage to these questions for the nilpotent groups.

At various times, the investigations of the constructive abelian and nilpotent groups were performed by Mal'tsev, Ershov, Goncharov, Downey, Knight, Lempp, Romanovskiĭ, Molokov, Dobritsa, Nurtazin, Khisamiev, Roman'kov, Latkin, Khoussainov, Hirschfeldt, Shore, Slin'ko, Csima, Solomon, Mel'nikov, and other authors.

The techniques of computability theory have numerous applications both for the solution of purely mathematical problems and in computer science. Downey is the striking instance of such a polymath scientist. Downey is co-author of the Downey-Fellows monograph, which is the first systematic work on parameterized complexity [10]. Parameterized complexity is a branch of computational complexity theory that focuses on classifying computational problems according to their inherent difficulty with respect to multiple parameters of the input or output. The complexity of a problem is then measured as a function in those parameters. This allows the classification of NP-hard problems on a finer scale than in the classical setting, where the complexity of a problem is only measured by the number of bits in the input.

1   The Primary Definitions and Designations

In this section, we remind the reader of some basic definitions and designations on groups; constructive (computable numbered) algebras, in particular strongly constructive algebras; and recursive (computable) structure, in particular decidable structure. Other notions of these areas will be described as needed; for background on computability theory, we refer the reader to [17,18,23,32,71,79,83].

1.1   Group Theory

One can find all other terms of group theory and their definitions which are not explained here or in what follows in the books [26,34,35,39,64,66,74,88].

Let G be a group written multiplicatively. For $x, y \in G$ (the group and its universe are designated by uniform sign), the *commutator of x and y is* $[x, y] = x^{-1}y^{-1}xy$. If H and K are subgroups of G, then $[H, K]$ is the subgroup generated by the commutators $[h, k]$ with $h \in H$ and $k \in K$, i.e., $[H, K] = gr(\{[h, k] \mid h \in H, \ k \in K\})$. As usual, the entry $H \trianglelefteq G$ denotes that the H is a normal subgroup in G.

Definition 1. *The lower central series of a group G is $G = \gamma_1 G \trianglerighteq \gamma_2 G \trianglerighteq \gamma_3 G \trianglerighteq \ldots$ defined inductively by $\gamma_1 G = G$ and $\gamma_{i+1} G = [\gamma_i G, G]$. A group G is a class r nilpotent, if $\gamma_{r+1}G = 1$, and $\gamma_r G \neq 1$.*

The second term $\gamma_2 G = [G, G]$ is also named as the (first) commutant of the group G, and it is denoted by G'; the i-th term $\gamma_i G$ is sometimes termed the i-th central of the group G.

One can extend the definition of commutators inductively by $[x_1, x_2, \ldots, x_{n+1}] = [[x_1, x_2, \ldots, x_n], x_{n+1}]$. A group G is nilpotent if and only if there is a $r \geqslant 1$ such that $[x_1, x_2, \ldots, x_{r+1}] = 1$ for all $x_1, \ldots, x_{r+1} \in G$. For the least such r, G is class r nilpotent. Thus all groups that are the class no more than r nilpotent constitute *a variety $\mathfrak{N}_r$ of groups* given by the identity $[x_1, x_2, \ldots, x_{r+1}] = 1$.

Nilpotent groups can also be defined by the upper central series. For any normal subgroup H of a group G, there is a natural projection $\pi_H : G \to G/H$ given by $\pi_H(g) = gH$. The *center of G*, denoted $C(G)$, is defined by $g \in C(G)$ if and only if $gh = hg$ for all $h \in G$. Since $C(G)$ is a normal subgroup; therefore taking the center of $G/C(G)$ and pulling back to G by $\pi_{C(G)}^{-1}$, one gets another normal subgroup of G. Continuing in this spirit yields the upper central series of G.

Definition 2. *The upper central series of a group G is $1 = \zeta_0 G \trianglelefteq \zeta_1 G \trianglelefteq \zeta_2 G \trianglelefteq \ldots$ defined inductively by $\zeta_0 G = 1$ and $\zeta_{i+1} G = \pi^{-1}(C(G/(\zeta_i G))$ for $\pi : G \to G/\zeta_i G$. A group G is nilpotent if there is an r such that $\zeta_r G = G$. More specifically, G is a class r nilpotent group if r is the least such that $\zeta_r G = G$.*

The i-th term $\zeta_i G$ is sometimes termed the i-th hypercenter of the group G.

These two definitions are equivalent in the sense that a group G is class r nilpotent under the lower central series definition if and only if it is class r nilpotent under the upper central series definition.

Most generally, if H and K are the normal subgroups of group G, and $H \leq \pi_K^{-1}(C(G/K))$, then the section H/K of the series $K \trianglelefteq H \trianglelefteq G$ is called *central*. A group G is class r nilpotent if and only if there exist a series $1 = G_0 \trianglelefteq G_1 \trianglelefteq G_2 \trianglelefteq \ldots \trianglelefteq G_r = G$ whose every section is central; such a series is named also *central*.

Thus, if $1 = G_0 \trianglelefteq G_1 \trianglelefteq G_2 \trianglelefteq \ldots \trianglelefteq G_r = G$ is a central series of a class r nilpotent group G, then we have the following scheme of inclusions:

$$
\begin{array}{ccccccccc}
\zeta_0 G & \lhd & \zeta_1 G & \lhd \ldots \lhd & \zeta_{r-1} G & \lhd & \zeta_r G & & \\
\| & & \nabla | & & \nabla | & & \| & & \\
1 = & G_0 & \lhd & G_1 & \lhd \ldots \lhd & G_{r-1} & \lhd & G_r & = G \\
\| & & \nabla | & & \nabla | & & \| & & \\
\gamma_{r+1} G & \lhd & \gamma_r G & \lhd \ldots \lhd & \gamma_2 G & \lhd & \gamma_1 G & &
\end{array}
\tag{1}
$$

The class 1 nilpotent groups are exactly the abelian groups, so the nilpotent class can be thought of as giving a measure of closeness to being abelian.

Definition 3. *A periodic part τG of the group G is a set that consists of the elements $g \in G$ for which there is an integer $n > 0$ such that $g^n = 1$; the least such n is called an order of element g. When the periodic part τG is trivial, then the group G is named torsion-free.*

It is well known that the periodic part of the nilpotent group is a normal subgroup [34, 35, 39]. Moreover, if the nilpotent group G is torsion-free, then it is *a group with unique roots* (or *R-group*), i.e., for any $x, y \in G$ and $n \in \mathbb{N}$, $n > 0$, it follows from $x^n = y^n$ that $x = y$.

1.2 Constructive Algebras

We recall the notion of a (strongly) constructive group. The general concept of a constructive algebraic structure is defined similarly; see, e.g., [16, 19–22, 69]. The notion of a strongly constructive algebraic structure was introduced by Ershov [19]. This notion is a natural evolution and a logical synthesis of concepts of the constructive model and the decidable theory. The raising of problem on decidable theories belongs to Tarskiĭ.

Let $G = \langle G, \cdot, ^{-1}, 1 \rangle$ be a countable group. A mapping $\nu : D \to G$ of the computable subset D of all natural numbers $\mathbb{N}$ onto universe G is called *a numbering of G*, and a pair (G, ν) is called an *enumerated group*.

Definition 4. *Let X be either a subset of $\mathbb{N}$ or else a family of subsets of $\mathbb{N}$.*

We will say that the pair (G, ν) is a X-constructive (or X-computable numbered) group, if there are X-computable functions f, g such that for all $n, m \in D$, the equalities $\nu(n) \cdot \nu(m) = \nu f(n, m)$ and $(\nu(n))^{-1} = g(n)$ hold, provided that a set $\nu^{-1}(1)$ is X-recursive (X-computable). If the set $\nu^{-1}(1)$ is X-recursively (X-computably) enumerable, then the numbering ν will be X-positive; and when this set is a complement to the X-computably enumerable set, then the numbering ν will be X-negative.

A numbering ν of group G such that (G, ν) is X-constructive, is called a X-constructivization (or X-computable numbering) of the group G. We will name the group, which has a X-constructivization, as X-computable.

If X is computable set, in particular $\mathbb{N}$, then we will omit the prefix "$X-$" by the writing the terms defined above.

The condition, "a set $\nu^{-1}(1)$ is X-computable (or X-computably enumerable, or complement to the X-computably enumerable set)" is equally matched to the numeration equivalence under ν in G

$$\eta_\nu = \{(n, m) | \nu n = \nu m\}$$

is respectively the same. Thus an enumerated group (G, ν) is X-constructive if the set of all atomic formulae satisfied in G is X-decidable. More specifically, an enumerated group (G, ν) is X-constructive if and only if there is an X-algorithm, that for every atomic formula $\varphi(x_1, \dots, x_n)$ and for any natural numbers $m_1, \dots, m_n$ determines whether the formula $\varphi(\nu m_1, \dots, \nu m_n)$ is true on G; i.e., a set

$$D_\nu(G) \rightleftharpoons \{\langle m, n_1, \dots, n_s\rangle \, | G \models \gamma^{-1}(m)(\nu n_1, \dots, \nu n_s), \text{ and } \gamma^{-1}(m) \text{ is atomic}$$
$$\text{formula with the free variables } v_1, \dots, v_s\}$$

is X-computable, where the γ is a fixed Gödel numbering of all formulae and terms of the signature $\langle \cdot, ^{-1}, 1\rangle$ — see definition below or in [21–23].

Definition 5. *An enumerated group (G, ν) is called strongly X-constructive if there is an X-algorithm, that for every formula $\varphi(x_1, \dots, x_n)$ and for any natural numbers $m_1, \dots, m_n$ determines whether the formula $\varphi(\nu m_1, \dots, \nu m_n)$ is true on G; i.e., a set*

$$D_\nu^*(G) \rightleftharpoons \{\langle m, n_1, \dots, n_s\rangle \, | G \models \gamma^{-1}(m)(\nu n_1, \dots, \nu n_s), \text{ and } \gamma^{-1}(m) \text{ is}$$
$$\text{a formula with the free variables } v_1, \dots, v_s\}$$

is X-computable.

A group G is X-decidable if there exists a numbering ν of G such that (G, ν) is strongly X-constructive; such a numbering ν is called a strong X-constructivization (or X-decidable numbering) of G.

1.3   Computable Models

We recall the notions of decidable and computable groups only. The general concept of these notions for arbitrary countable models is introduced similarly; for background on effective algebra, we refer the reader to [1, 21, 22, 32].

Definition 6. *The countable group $G = \langle G, \cdot, ^{-1}, 1\rangle$ is named X-recursive (or X-computable) if G is an X-recursive (X-computable) subset of the natural numbers $\mathbb{N}$; there are X-computable functions f, g such that f restricted to G^2 computes the multiplication, and g restricted to G computes the multiplicative inverse, i.e., for all $n, m \in G$, $f(m, n) = m \cdot n$ and $g(m) = m^{-1}$.*

In other words, a group G is X-computable if its universe is a X-computable subset of $\mathbb{N}$ and there exists a X-computable enumeration $(g_i)_{i \in \mathbb{N}}$ of the universe G such that the atomic diagram of the group G is X-decidable.

Definition 7. *A group G is X-decidable if its universe is computable subset of $\mathbb{N}$ and there exists a X-computable enumeration $(g_i)_{i \in \mathbb{N}}$ of the universe G such that the complete diagram of the group G (that is $Th((G, g_i)_{i \in \mathbb{N}})$ is X-decidable.*

It is evident that a (decidable) recursive group which is isomorphic to a group G can be viewed as a (strong) constructivization of G. So (decidable) recursive groups (or, most generally, structures) and (strong) constructivizations are essentially interchangeable (see [21, 22] for further details). Therefore we will apply the terms "computable" and "decidable" without specifying which definition is used, since one can always understand what kind of universe has the structure under study.

Remark 1. At the present time, the terms "recursive" and "recursively" are interchangeable with "computable" and "computably", when they are used in the sense of the definitions of this section. Moreover, a group that is computable under definition 4 was earlier called *constructivizable*. We will mainly apply the new terms. A (decidable) computable group, which is isomorphic to a group G, is called a *(decidable) computable presentation* (or, sometimes, *(decidable) computable copy*) of G.

Remark 2. We simply write "computable", "constructive", "positive" and etc., although the results that are mentioned below will remain correct for the most part, if we substitute "X-computable", "X-constructive", "X-positive" and etc. for these words. However, we will adduce the exact wordings when it is essential to apply these facts hereinafter.

1.4   The Symbol-Combinatoric and Numeric Approaches to Algorithmic Problems

Let us notice that the combinatorial approach continued to successfully be employed for the investigation of algorithmic problems in group theory along with the numeric methods. Many decision problems for groups were solved by the application of symbol-combinatoric technique at the latter half of the twentieth century [68, 75]. There appeared the proofs of the algorithmic solvability of the word problem for some classes of groups, e.g., for groups with a single defining relation, for the finitely generated nilpotent and metabelian groups etc. It turned out that this problem is generally undecidable for finitely presented groups even if these groups are solvable; it follows from this that other decision problems (conjugacy problem, isomorphic problem etc.) are undecidable also for these varieties of groups [2, 3, 40, 64, 66, 73, 75].

We now establish a connection between the symbol-combinatoric and numeric approaches to the study of algorithmic problems for the abstractly defined groups.

Let M be a set of words over some alphabet A. A *Gödel numbering* of the M is a biunivocal mapping γ of the set M into the set of natural numbers $\mathbb{N}$ such that one can effectively recognize if a given natural integer is the number of an element of M (i.e., a set γM is computable), and from a number, one can recover the structure of the corresponding word.

For instance, the concrete Gödel numbering is described for all formulae and terms of the signature $\langle <, +, \cdot, s, 0 \rangle$ in variables of the set $\{v_0, v_1, \ldots\}$ in [23].

Let $Gen(G) = \{g_0, g_1, \ldots\}$ be a set of the generators of the group G; and $Rel(G) = \{R_0(\bar{g}), R_1(\bar{g}), \ldots\}$ be a set of its defining relations; so the group G has presentation

$$G = \langle Gen(G)|Rel(G) \rangle = \langle g_0, g_1, \ldots | R_0(\bar{g}), R_1(\bar{g}), \ldots \rangle,$$

— see [64, 66] for further details. We can define the Gödel numbering of the set of all words in an alphabet $\{g_0, g_0^{-1}, g_1, g_1^{-1}, \ldots\}$ just as is done in [23], replacing the variable v_i with the generator g_i. Consider an inverse mapping $\nu(\gamma) = \gamma^{-1}$; it already is the numbering of the group G. Such a numbering is named *natural, constructed by Gödel numbering of the words in the given presentation*; or briefly, *natural numbering of given presentation*.

It is obvious that when the set $Gen(G)$ is computable, in particular finite, and the set $Rel(G)$ is computably enumerable, in particular finite, then the numbering $\nu(\gamma)$ will be positive. The word problem is soluble for given group presentation $\langle Gen(G)|Rel(G) \rangle$ if and only if its natural numbering $\nu(\gamma)$ is constructive. Therefore, it is sometimes convenient to describe a group by its sets of generators and defining relations in order to obtain a numbering of group with the requisite algorithmic properties [4, 31, 58].

1.5   Operations over Constructive Groups

A sequence of (strongly) constructive groups $\{(G_i, \nu_i) \mid i \in \mathbb{N}\}$ is called *computable* if the sequence of sets $(\{D_{\nu_i}^*(G_i)\}_{i \in \mathbb{N}})$ $\{D_{\nu_i}(G_i)\}_{i \in \mathbb{N}}$ is computable [8, 19–22].

Let $\langle x_0, \ldots, x_k \rangle$ be a Gödel number of the finite sequence $(x_0, \ldots, x_k)$ of natural numbers; and $G = \prod_{i \in \mathbb{N}} G_i$ be direct product of groups G_i. A numbering μ of G will be named *canonical* (or *natural*) *numbering of direct product* $\prod_{i \in \mathbb{N}} (G_i, \nu_i)$, if it is defined by $\mu x = (\nu_0 x_0, \ldots, \nu_k x_k)$, where $x = \langle x_0, \ldots, x_k \rangle$.

It is well known that when the sequence of (strongly) constructive groups is computable, then the canonical numbering of the direct product of these groups will also be (strongly) constructive [19–22, 49]. However there exists an abelian torsion-free group A such that the direct product $A \times A$ is computable, but the A is non-computable [14].

Let (G, ν) be an enumerated group, M be a subset of universe G. We will say that this subset is *computable* (or *computably enumerable*) if its numeral set $\nu^{-1}(M)$ is the same.

Given the computably enumerable subgroup H of the enumerated group (G, ν), one can define a numbering μ of H with the help of computable function f that enumerates the numeral set $\nu^{-1}(H)$: $\mu(x) = \nu f(x)$. If the numbering ν is constructive, then so will be μ [16,20,69]. Since the periodic part and all terms of lower central series are computably enumerable in every constructive group according to their definition, therefore these subgroups of the computable group are computable. This assertion is not true for decidable groups (see Proposition 1 in the next section and [42,46,49]).

If H is a normal subgroup of group G, then for every numbering ν of the latter, one can define a *canonical* (or *natural*) *numbering* ν/H *of the quotient group* G/H by $(\nu/H)(m) = \nu(m)/H$ for any $m \in \mathbb{N}$. It is clear that the numbering ν/H will be positive (constructive), when H is computably enumerable (computable) in G under ν [16,20,69]. This statement is not true for strong constructivization (see Proposition 1 in the next section and [42,46,49]).

2   Abelian Groups

The study of constructive groups was started by Mal'tsev in 1962 [70], where he posed the general question of finding for given abstractly defined groups, what kind of constructive numberings they admit. He obtained the first results on the properties of the constructive abelian groups. Further results in this field are contained in works of Goncharov, Downey, Dzgoev, Dobritsa, Khisamiev, Lin, Molokov, Nurtazin, Richman, Smith, and many others.

Constructive abelian groups are researched rather well at the present time (see e.g. [49]). There exist a great number of papers in which the basic problems (namely, the issues of existence, uniqueness, and extensions of constructivizations) are scrutinized for almost all subclasses of abelian groups. Moreover, the theory of constructive abelian groups has extremely interesting applications. For instance, Khisamiev provided a negative answer to Macintyre's question by using methods that were devised for the study of abelian torsion-free groups. Namely, he proved that every ordered field of real numbers containing a field of primitive recursive numbers is essentially non-computable, i.e., it is not contained in any computable ordered field of real numbers; therefore the field of all primitive recursive numbers is non-computable [47,49].

Nevertheless, many interesting questions are still open. For example, there are no simple invariants defining torsion-free abelian groups of countable rank up to isomorphism, as a consequence of work by Downey and Montalbán [13] and Hjorth [38]. Therefore the problem of existence of (strong) constructivization for such groups is not solved.

We recite a brief catalog of the fundamental results on computable abelian groups below. Some recent results are not contained in the survey [49].

The abelian groups will be written additively. It is natural to use "direct sum" in place of the words "direct product".

2.1   Existence

The existence problem of constructivizations for the class of abelian groups is reduced to the same problem for the classes of periodic groups and torsion-free groups [49].

2.1.1   Torsion-Free Groups

A maximal system of linearly independent elements in a torsion-free abelian group is called its *basis,* and a *dimension of the abelian group* (or *Prüfer rank*) is the cardinality of a basis.

Theorem 1 (Khisamiev [46]**).** *Suppose that* (A, ν) *is a (strongly) X-constructive abelian group and B is a X-computably enumerable subgroup of A such that the quotient A/B is a torsion-free abelian group. Then there exists a numbering μ of A possessing the following properties:*

 (i)   the group (A, μ) is a (strongly) X-constructive;
 (ii) the subgroup B is X-computable in (A, μ);
 (iii) there exists a X-computably enumerable system $\{c_i \mid i \in I\}$ of elements of (A, μ) such that the cosets $\{c_i + B\}$ form a basis of the quotient A/B.

This theorem strengthens the results of Dobritsa [7] and Nurtazin [77].

Corollary 1 (Khisamiev [46]). *Every X-computably enumerably definable (X-positively enumerable) torsion-free abelian group is X-computable.*

From this result, we obtain an affirmative answer to the question of G. Baumslag raised in [2].

Let us notice that each of the properties (ii) and (iii) (in Theorem 1) together with (i) results in the computability of the quotient-group A/B. However it can be undecidable even if the (A, μ) is a strongly constructive.

Proposition 1 (Khisamiev [42,46]). *There are decidable abelian groups A and B such that the periodic part τA and the quotient group $B/\tau B$ are not decidable.*

2.1.2   Direct Sums of Cyclic and Quasicyclic Groups

The lowercase letter "p" denotes a prime number in what follows.
    Let $A = \bigoplus_{i \in \mathbb{N}} \mathbb{Z}_{p^{n_i}}$. The set

$$\chi(A) = \{(m,k) \mid \exists i_1, \ldots, i_k \big(\bigwedge_{1 \leqslant r < s \leqslant k} i_r \neq i_s \wedge n_{i_1} = \ldots = n_{i_k} = m \big)\}$$

is called the *characteristic of A.*

Theorem 2 (Khisamiev [49]). *Let $G = A \oplus \mathbb{Z}_{p^\infty}^\alpha$, where $\alpha \in \omega + 1$ and A is a direct sum of cyclic p-groups. The G is strongly computable if and only if the characteristic $\chi(A)$ is computable.*

An *s-function* is a function $f(j, x)$ for which the following conditions hold.

 (i) The function $\lambda x f(j, x)$ is non-decreasing for each i.
 (ii) For every j, $\lim_x f(j, x) = m_j$ exists.

If, in addition, $m_0 < m_1 < m_2 < \dots$, then $f(j,x)$ is called an s_1-*function*. Let

$$\bar{\rho}f = \{(m,k) | \exists i_1, \dots, i_k (\bigwedge_{1 \leqslant r < s \leqslant k} i_r \neq i_s \wedge m_{i_1} = \dots = m_{i_k} = m)\}$$

Theorem 3 (Khisamiev [43]). *Let A be a direct sum of cyclic p-groups whose orders are unbounded. The A is X-computable if and only if the following conditions hold.*

(i) *The characteristic $\chi(A) \in \Sigma_2^X$.*
(ii) *There exists an X-computable s_1-function $f(j,x)$ such that $\bar{\rho}f \subseteq \chi(A)$.*

2.2 Extensions and Orderability

The extension problem lies in finding the conditions that afford extending a computable numbering of subgroups on the whole group. This challenge is explored in the very few cases.

A group D (it may be non-abelian) is called *divisible* if for every integer $n > 0$ and any $g \in D$, the equation $x^n = g$ (or $nx = g$, when D is abelian) has at least one solution in D. Such a group is sometimes said to be *complete* also [39]. The pair (D, φ) is a *divisible closure of a group G* (or its *completion*) if D is a divisible group; $\varphi : G \to D$ is embedding; and for any $d \in D$, there exist $m \in \mathbb{N}$, and $g \in G$ such that $d^m = \varphi(g)$ (or $md = \varphi(g)$ when G is abelian). A constructive group (D, α) is called a *divisible closure of constructive group (G, ν)* if there is a computable embedding $\varphi : (G, \nu) \to (D, \alpha)$ and (D, φ) is a divisible closure of G, i.e., for some computable function f, the equality $\varphi \nu(n) = \alpha f(n)$ holds for each $n \in \mathbb{N}$.

Theorem 4 (Smith [84], Khisamiev [46]). *Every (strongly) constructive abelian group has a (strongly) constructive divisible closure.*

Definition 8. *An ordered abelian group is a pair $(G, \leqslant)$, where G is an abelian group and $\leqslant$ is a linear order on G such that if $a \leqslant b$, then $a + g \leqslant b + g$ for all $g \in G$.*

It is well known that an abelian group is orderable if and only if it is torsion free.

Theorem 5 (Solomon [85], Roman'kov and Khisamiev [81]). *A computable torsion-free abelian group is orderly computable.*

However, we must not believe that each computable numbering of such a group admits computable order. R.G. Downey and S.A. Kurtz constructed the series of difficult, but at the same time elegant, examples of computable copies of abelian groups with no computable orders:

Theorem 6 (Downey and Kurtz [12]).

(i) There is a constructivization of free abelian group with no computable orderings.

(ii) There exists a constructive divisible abelian group with no computable ordering.

(iii) There exists a torsion free constructive abelian group with no computable ordering and whose only computably orderable subgroups are finitely generated.

2.3 Uniqueness

The numberings ν and μ of structure A are called *autoequivalent* if there exists an automorphism φ of A and a computable functon f such that $\varphi\nu(n) = \mu f(n)$ for each $n \in \mathbb{N}$; in this case, such an automorphism is named *computable*. A structure is *autostable* or *computably categorical* if every two of its constructivizations are autoequivalent. The *computable dimension* of A is the number of constructivizations of A up to computable isomorphism.

The problem of finding algebraic criteria for the autostability of abelian groups is open. Nevertheless this question is rather well investigated.

The following two theorems give a description of the autostable abelian p-groups and torsion-free groups.

Theorem 7 (Goncharov [27], Smith [84]). *An abelian p-group A is autostable if and only if either*

$$A \simeq \mathbb{Z}_{p^\infty}^\omega \oplus F \qquad or \qquad A \simeq \mathbb{Z}_{p^\infty}^\omega \oplus \mathbb{Z}_{p^n}^\omega \oplus F,$$

where $\mathbb{Z}_{p^\infty}^\omega$ is the direct sum of ω copies of the quasi cyclic p-group $\mathbb{Z}_{p^\infty}$; $\mathbb{Z}_{p^n}$ is the cyclic group of order p^n; F is a finite p-group, and $n \in \mathbb{N}$.

Theorem 8 (Nurtazin [76], Downey and Kurtz [12]). *A torsion-free abelian group is autostable if and only if it has finite rank.*

Dobritsa [7] built a non-autostable group A such that the periodic part τA of A is autostable; and the rank of $A/\tau A$ is equal to one.

Theorem 9 (Goncharov [27,28]). *Every abelian group has computable dimension 1 or ω.*

Theorem 10 (Goncharov, Lempp, Solomon [30]). *Every computable ordered abelian group has computable dimension 1 or ω. Furthermore, such a group is computably categorical if and only if it has finite rank.*

2.4 Completely Decomposable Groups

Since there exist no simple invariants defining torsion-free abelian groups of countable rank up to isomorphism in general [13], it makes sense to describe the computable groups of the particular subclasses of such groups.

Definition 9. *A torsion-free abelian group A is completely decomposable if there is a collection of groups $(A_i)_{i \in I}$ with*

$$A = \bigoplus_{i \in I} A_i \tag{2}$$

and $A_i \leq (\mathbb{Q}, +)$ for each $i \in I$.

Let ν be (decidable) computable numbering of a completely decomposed group A. If there exists a recursively enumerable system is non-trivial elements $b_i \in A_i$, i.e., the predicates $P_i(x) \rightleftharpoons x \in A_i$ are uniformly computable, then the pair (A, ν) is called the (decidable) constructive completely decomposed group; the group A is named (strongly) effectively completely decomposable; and such a elements b_i is termed s-generative for the G_i.

If A is a group of the kind (2) and $g \in A$, then the set $\chi_s(g) = \{\langle p, r \rangle : p^r | g\}$ is a *s-characteristic of this element*. For every non-zero elements g_1 and g_2 of group A that has rank one, the s-characteristics of these elements are *almost equal*, i.e., a set $(\chi_s(g_1) \setminus \chi_s(g_2)) \cup (\chi_s(g_2) \setminus \chi_s(g_1))$ is finite. It is known that a group of rank one is uniquely determined by s-characteristic of any its non-zero element, while a completely decomposable group A of the kind (2) can be uniquely regenerated by the sequence of sets $\chi_s(b_i)$, where b_i is s-generative for the G_i. Indeed, the generators of group G_i will be the elements in the form of b_i / p^r, where $\langle p, r \rangle \in \chi_s(b_i)$. Such a sequence of the s-characteristics is called *s-characteristic of group A*, it is designated by $\chi_s(A)$.

We recall that a sequence of sets is computable if the sequence of their partial characteristic functions is computable. The sequence of the computable sets $R_0, R_1, \ldots$ is termed *well-computable* if the sequence of the total characteristic functions of these sets is computable.

The sequence $\chi_s(A)$ is called almost computable (well-computable), if there is a computable (well-computable) sequence of sets R_i ($i \in \mathbb{N}$), such that each R_i consists of the pair of natural numbers and almost equal to $\chi_s(A)$, and also the condition $(\langle p, k \rangle \in R_i \wedge s < k) \rightarrow \langle p, s \rangle \in R_i$ holds for every $i, p, k, s \in \mathbb{N}$.

Proposition 2 (Khisamiev [51]). *An abelian group of the kind (2) is strongly decomposable if and only if it have solvable theory and its s-characteristic is almost well-computable.*

Proposition 3 (Khisamiev [51]). *An abelian group is effectively completely decomposable if and only if its s-characteristic is almost computable.*

Let $p_0, p_1, \ldots, p_i, \ldots$ be some sequence of prime numbers, possibly with repetition; $\mathbb{Q}_{p_i}$ be the additive group of the rational numbers, whose denominators are the powers of p_i; and

$$B = \bigoplus_{i \in \mathbb{N}} \mathbb{Q}_{p_i}. \tag{3}$$

Then *p-characteristic* of group B is the set $\chi_p(B)$ of all pairs of integers $\langle p, k \rangle$ such that for some indices $i_1, \ldots, i_k$, the equalities $p_{i_1} = \cdots = p_{i_k} = p$ are true. A group of the form (3) is determined by its *p*-characteristic uniquely up to isomorphism.

Theorem 11 (Khisamiev, Krykpaeva [53]). *A group B of the kind (3) is effectively decomposable if and only if its p-characteristic $\chi_p(B)$ belongs to the class Σ_2^0 of the arithmetical hierarchy.*

Definition 10. *We name a set $S \subseteq \mathbb{N}$ quasi-hyperhyperimmune if there is not the well-computable sequence of the computable sets $R_0, R_1, \ldots$ such that for any $i \in \mathbb{N}$ the following conditions are true:*

(i) *the elements of R_i are the pairs of numbers;*
(ii) *if $\langle x, y \rangle \in R_i$ and $z < y$, then $\langle x, z \rangle \in R_i$;*
(iii) *the set $\{x \mid \exists y (\langle x, y \rangle \in R_i)\}$ is finite;*
(iv) *there exists a unique number r_i such that the inclusion $\langle r_i, y \rangle \in R_i$ is true for every $y \in \mathbb{N}$; such a number r_i is called main for R_i;*
(v) *if $i \neq j$, then the main numbers r_i and r_j of corresponding sets R_i and R_j are different;*
(vi) *the main number r_i belongs to the set S.*

Proposition 4 (Khisamiev [51]). *Each hyperhyperimmune set is quasi-hyperhyperimmune.*

Proposition 5 (Khisamiev [51]). *There is a quasi-hyperhyperimmune set that is not hyperhyperimmune.*

Theorem 12 (Khisamiev [51]). *A group B of the kind (3) is strongly decomposable if and only if its p-characteristic $\chi_p(B)$ belongs to the class Σ_2^0 of the arithmetical hierarchy and the set $P(B) = \{p \mid \langle p, 1 \rangle \in \chi_p(B)\}$ is not quasi-hyperhyperimmune.*

Theorem 13 (Downey et al. [11]). *A group B of the kind (3) is decidable if and only if its p-characteristic $\chi_p(B)$ belongs to the class Σ_2^0 of the arithmetical hierarchy. This group is computable if and only if $\chi_p(B) \in \Sigma_3^0$.*

2.5 Computable Classes of Constructive Groups

We recall that dimension of the factor group $A/\tau A$ of abelian group A by its periodic part τA is termed a *Prüfer rank* or *torsion-free rank* of abelian group A. A computable numbering γ (or *indexing* [8,17,18,20]) of the class of constructive groups is *principal* if any other computable numbering η of the same class is reducible to the γ, i.e., there is a total computable function f such that $\eta(n) = \gamma(f(n))$ for every $n \in \mathbb{N}$.

The following facts on the computable classes of abelian groups are known.

Theorem 14 (Khisamiev [41]). *The class of periodic abelian groups has principal computable numbering.*

Theorem 15 (Dobritsa [5]). *For every $n \in \mathbb{N}$, the class of abelian groups, whose torsion-free rank equals to n, has a principal computable numbering.*

Theorem 16 (Dobritsa [5]). *The following classes of abelian groups are computable but they do not have principal numbering:*

(i) the class of groups whose torsion-free rank is not more than n for every $n \in \mathbb{N}$;

(ii) the class of groups whose torsion-free rank is finite.

Suppose, we are given constructive abelian periodic group (T, μ), computable function $g(s, t)$, and partially recursive function $f(k, s, n)$. Using these functions, one can build a group $A[(T, \mu), f, g]$ such that its periodic part is T, and its torsion-free rank r is calculated in the following way [50]. Let s_0 be the smallest number such that either there is not $\lim_t g(s_0, t)$ or $\lim_t g(s_0, t) = 0$. If there is not such s_0, then one defines $s_0 = \omega$. Finally, $r \rightleftharpoons \max\{s_0, 1\}$. N. G. Khisamiev applied the group $A[(T, \mu), f, g]$ to the obtaining of criterion of computability of abelian groups in [50]. The following theorem is based on this criterion and gives an answer to the question of V. P. Dobritsa — (see [63] Question No 72 or [50]).

Theorem 17 (Khisamiev [50]). *For any natural $r \geqslant 1$, there exists a principal computable numbering of class of the constructive abelian groups, whose torsion-free rank is non-zero and not more than r.*

3 Nilpotent Torsion-Free Groups and Factor Groups by Periodic Part

The several assertions about algorithmic properties of abelian groups remain true for the nilpotent groups too. For instance, each finitely generated nilpotent group is computable [2,3,40,64,66,68,73]. Another example of such a property is given by the following theorem.

Theorem 18 (Ershov [19]). *Every constructive locally nilpotent torsion-free group has a unique constructive completion.*

Let us notice that the proof of this theorem is very intricate in spite of the simplicity of its formulation and the superficial resemblance to the statement of Theorem 4. This proof is based on a common approach that was developed in [15–17,19].

Nevertheless, the computability-theoretic behavior of nilpotent groups can drastically differ from that of abelian groups. These differences are pronounced in the following results.

Theorem 19 (Goncharov, Molokov, and Romanovskiĭ [31]). *For any natural number $n \geqslant 1$, there is a class two nilpotent group whose computable dimension is n.*

Furthermore, Hirschfeldt et al. [37] proved that the theory of the class two nilpotent groups is computably complete with respect to degree spectra of non-trivial structures; computable dimensions; expansion by constants; and degree spectra of relations in the sense that, if there are any examples of structures with such properties then there are such examples that are the class two nilpotent groups — see for further details [37].

Therefore, an absence of an analogue of Corollary 1 does not look too much strange.

Theorem 20 (Latkin [59]). *There exists a X-computable class two nilpotent group, in which the quotient group by its periodic part is not X-computable.*

Corollary 2 (Latkin [59]). *There is a X-positive torsion-free class two nilpotent group that is not X-computable.*

However, there exist certain analogues of Theorem 1 and its corollaries for nilpotent groups all the same. These results are given below.

3.1   Computably Enumerable Basis of Quotient Group

We recall that a group G is called *R-group* if for any $x, y \in G$ and $n \in \omega$, $n > 0$, it follows from $x^n = y^n$ that $x = y$ (see the end of Subsect. 1.1).

Theorem 21 (Khisamiev [52]). *Let (G, ν) be a positive (constructive) R-group and B be its computably enumerable subgroups such that G/B is a torsion-free abelian group, and the dimension of the quotient group $\bar{G} = C(G)/(C(G) \cap B)$ of the center $C(G)$ of group G by the $C(G) \cap B$ is infinite. Then there exists a positive (constructive) numbering μ of the G possessing the following properties:*

(1) the subgroup B is computable in (G, μ);
(2) there exists a computably enumerable system of elements $\{g_i \,|\, i \in I\}$ in (G, μ) such that the cosets $\{g_i B\}$ constitute a basis for G/B.

The set of all elements $x \in G$ whose some powers belong to the commutant G' is called the *isolator of commutant* and denoted by $I(G')$.

Corollary 3 (Khisamiev [52]). *Suppose that (G, ν) is positive (constructive) R-group and the dimension of the quotient group $\bar{G} = C(G)/C(G) \cap I(G')$ of the center $C(G)$ by the isolator of the commutant is finite. Then there exists a positive (constructive) numbering α of the G possessing the following properties:*

(1) the subgroup $I(G')$ is computable in (G, α);
(2) there exists a computably enumerable system of elements $\{g_i \,|\, i \in J\}$ in (G, α) such that the cosets $\{g_i I(G')\}$ form a basis of $G/I(G')$.

Corollary 4 (Khisamiev [52]). *For any positive (constructive) R-group (G, ν), there exists a positive (constructive) group (H, β) and a computable isomorphism $\varphi : (G, \nu) \to (H, \beta)$, which have following properties:*

(1) $\varphi(G') = H'$;
(2) the subgroup $I(H') = I(\varphi(G'))$ is computable in (H, β);
(3) there exists a computably enumerable set of elements $\{h_i \,|\, i \in J\}$ in (H, β) such that the cosets $\{h_i I(G')\}$ constitute a basis for $H/I(H')$.

Remark 3. If the group G from Corollary 4 is divisible or a class r nilpotent, then H is the same.

3.2   Extensions by System of Factors and Sequence of Automorphisms

We say that the *dimension of a torsion-free nilpotent group is finite* if there exists a central series whose every section has finite dimension.

Theorem 22 (Khisamiev [52]). *A positively enumerable R-group whose commutant has a finite dimension is computable.*

Corollary 5 *(Khisamiev [52]). A positively enumerable torsion-free nilpotent group whose commutant has a finite dimension is computable.*

This result generalizes Corollary 1, namely every computably enumerable defined torsion-free abelian groups is computable.

Let N and H be groups, and $N \cap H = \{1\}$. Assume that there exists a two-place function $f : H \times H \to N$, and also for every $u \in H$, there is an automorphism $\varphi_u : N \to N$ such that for any $a \in N$ and $u, v, w \in H$, the following identities hold:

(i) $\varphi_v(\varphi_u(a)) = f^{-1}(u, v)\varphi_{uv}(a)f(u, v)$;
(ii) $f(uv, w) \cdot \varphi_w(f(u, v)) = f(u, vw)f(u, w)$;
(iii) $f(1, 1) = 1$.

We define a set G and a binary operation on it in the following way:

$$G = \{ua | u \in H, a \in N\}; \qquad (ua) \cdot (vb) = uv \cdot f(u, v)\varphi_v(a)b.$$

One can simply enough make sure that this algebraic structure is a group. The group G is called *an extension of the group N by the group H, the system of factors f, and the sequence Φ of automorphisms* $\{\varphi_u | u \in H\}$. Such a construct is designated by $G = E(N, H, f, \Phi)$.

Corollary 6 (Khisamiev [52]). *Let G be a torsion-free class s nilpotent group whose commutant has finite dimension. Then the G is computable, if and only if there exist constructive torsion-free class less than s nilpotent groups (N, α) and (H, β), one of which is abelian, some computable system f of factors from (H, β) in (N, α), and a computable sequence of automorphisms Φ such that the G is isomorphic to a computable extension $E((N, \alpha), (H, \beta), f, \Phi)$.*

4   Computability and Matrix Groups

Let K be a commutative associative ring with unity. As usual, $GL_n(K)$ is a group of all invertible $(n \times n)$-matrices over K; $SL_n(K)$ and $UT_n(K)$ are groups of special and unitriangular n-matrices over K, respectively [39]. To be specific, we also assume that $UT_n(K)$ consists of upper unitriangular matrices. If ν is a numbering of the ring K, then one can construct the natural numbering $\nu(\gamma)$ by the Gödel numbering of $GL_n(K)$ (see Subsect. 1.4). It is easy see that $\nu(\gamma)$ is computable, provided that ν is the same. Moreover the subgroup $SL_n(K)$ and $UT_n(K)$ are computable under $\nu(\gamma)$ in this case.

4.1 Computability of Matrix Groups in General Case

It is well known (and easily proved) that the i-th hypercenter $\zeta_i UT_n(K)$ of $UT_n(K)$ consists of matrices having exactly $n-i-1$ $(i = 1, 2, ..., n-1)$ zero secondary diagonals, originating in the principal one (see, e.g., [39]). Hence, $UT_n(K)$ is a class $n-1$ nilpotent group. Therefore we have:

Proposition 6 (Romankov and Khisamiev [81]). *If the group $UT_n(K)$ is constructive then any one of its hypercenters is computable.*

Notice that the group $UT_2(K)$ is isomorphic to an additive group of K, and so if the former group is computable then the latter is likewise. In the general case, computability of $UT_2(K)$ does not imply computability of K. In fact, in [47] it was proved that the field of primitive recursive reals is not computable. The additive group of this field is a divisible torsion-free abelian group. Therefore it is computable.

Theorem 23 (Roman'kov and Khisamiev [81]). *The group $UT_n(K)$ of all unitriangular matrices of degree $n \geqslant 3$ over a commutative associative ring K with unity is computable if and only if K is computable.*

Theorem 24 (Roman'kov and Khisamiev [81]). *The group $GL_n(K)$ $(SL_n(K))$ of all matrices (of determinant 1) of degree $n \geqslant 3$ over a commutative associative ring K with unity is computable if and only if K is computable.*

Theorem 25 (Roman'kov and Khisamiev [81]). *Let K be an orderly constructible associative commutative ring with 1. Then the group $UT_n(K)$ of all unitriangular n-matrices over K is orderly computable for all non-zero $n \in \mathbb{N}$.*

Proposition 7 (Roman'kov and Khisamiev [82]). *If $GL_2(K)$ over an arbitrary commutative associative ring K with 1 is computable then the additive group of K, too, is computable.*

Theorem 26 (Roman'kov and Khisamiev [82]). *Let K be a commutative associative ring with 1 such that for some invertible element $\xi \neq 1$, the difference $\xi - 1$ is not a zero divisor. If every element of K is representable as a sum of invertible elements and the group $GL_2(K)$ is computable, then K is computable.*

Corollary 7 (Roman'kov and Khisamiev [82]). *If $GL_2(P)$ over a field P is computable then so is P.*

Corollary 8 (Roman'kov and Khisamiev [82]). *Let K be a group algebra of an abelian group A over a field of order not 2. Then K is computable if so is $GL_2(K)$.*

Theorem 27 (Roman'kov and Khisamiev [82]). *Let G be a matrix group over a field P. If G is computable then its factor group w.r.t. the center is also computable.*

Corollary 9 (Roman'kov and Khisamiev [82]). *Let G be a computable matrix group over a field P. Then the factor group $G/\zeta_i(G)$ is computable for any i.*

Proposition 8 (Roman'kov and Khisamiev [82]). *There exists computable group $GL_2(K)$ over a non-computable commutative associative ring K with 1.*

4.2 Nilpotent Group of Finite Dimension

Let γ_n be a natural numbering of the group $UT_n(\mathbb{Q})$, constructed by the Gödel enumeration of matrices (see Subsect. 1.4).

Proposition 9 (Khisamiev, Nurizinov, and Tyulyubergenev [55]). *A torsion-free nilpotent group has finite dimension if and only if it is isomorphic to a subgroup of $UT_n(\mathbb{Q})$ for some n.*

Let G be a nilpotent group of finite dimension, and

$$1 = G_0 < G_1 < \cdots < G_{k-1} = G \tag{4}$$

be its central series whose sections $\bar{G}_i = G_i/G_{i-1}$ are torsion-free for $i < k$. We will name such a series *central torsion-free*.

Theorem 28 (Khisamiev, Nurizinov, and Tyulyubergenev [55]). *Let (G, ν) be a constructive nilpotent group of finite dimension. Suppose that the series of the form (4) is central torsion-free. Then for each i, the subgroup G_i is computable in (G, ν).*

Definition 11. *Let G be a torsion-free nilpotent group of finite dimension. Assume that (4) is its central torsion-free series, and $\{g_{i,0}, g_{i,1}, \ldots, g_{i,n_i-1}\}$ is a basis of the section $\bar{G}_i$. A set $\chi(\bar{G}_i)$ of the collections of integers is called a characteristic of the section $\bar{G}_i$ if*

$$\chi(\bar{G}_i) = \{(m, s_{i,0}, \ldots, s_{i,n_i-1}) \mid \exists g \in G_i \ \exists h \in G_{i-1}$$
$$g_{i_0}^{s_{i_0}} \cdot \ldots \cdot g_{i,n_i-1}^{s_{i,n_i-1}} \cdot h = g^m, \ \sum_{j<n_i} |s_{i,j}| \neq 0\}.$$

Corollary 10 (Khisamiev, Nurizinov, and Tyulyubergenev [55]). *If G is a computable torsion-free nilpotent group of finite dimension, and (4) is central torsion-free series of G, then the characteristic $\chi(\bar{G}_i)$ is a computably enumerable set for each $i < k$.*

Theorem 29 (Khisamiev, Nurizinov, and Tyulyubergenev [55]). *Suppose that a group G is the subgroup of $UT_n(\mathbb{Q})$; ν is the computable numbering of G; and (4) is central torsion-free series of G. Then each subgroup G_i is computably enumerable in $(UT_n(\mathbb{Q}, \gamma_n)$, where γ_n is the natural numbering, constructed by the Gödel enumeration of matrices; moreover $\nu \leqslant_m \gamma_n$.*

Corollary 11 (Khisamiev, Nurizinov, and Tyulyubergenev [55]). *A subgroup $G < UT_n(\mathbb{Q})$ is computable if and only if G is a computably enumerable subgroup in $(UT_n(\mathbb{Q}), \gamma_n)$.*

Corollary 12 (Khisamiev, Nurizinov, and Tyulyubergenev [55]). *A torsion-free nilpotent group of finite dimension is computable if and only if it is isomorphic to a computable subgroup in $(UT_n(\mathbb{Q}, \gamma_n)$ for some n.*

Theorem 30 (Khisamiev, Nurizinov, and Tyulyubergenev [55]). *There exists a principal computable enumeration of the family of all computable torsion-free nilpotent groups of finite dimension.*

Theorem 31 (Khisamiev, Nurizinov, and Tyulyubergenev [55]). *Suppose that G is a subgroup in $(UT_n(\mathbb{Q}, \gamma_n)$, (4) is its central torsion-free series, and $\{g_{i,0}, g_{i,1}, \ldots, g_{i,n_i-1}\}$ is a basis of the section $\bar{G}_i$. Then G is computable if and only if the conditions are fulfilled:*

(1) $\chi(\bar{G}_i)$ is computably enumerable;
(2) there exists a partially computable choice function $ch_i^* : \chi(\bar{G}_i) \to \gamma_n^{-1} G_{i-1}$ such that, for every sequence $x = (m, s_{i,0}, \ldots, s_{i,n_i-1}) \in \chi(\bar{G}_i)$, there is $g \in G_i$ for which $g_{i_0}^{s_{i_0}} \cdot \ldots \cdot g_{i,n_i-1}^{s_{i,n_i-1}} \cdot \gamma_n ch_i^*(x) = g^m$, where $ch_1^* = 1$.

Remark 4. Condition 2 in this theorem does not depend on Condition (1).

Remark 5. All results of this subsection, starting with Theorem 28 and ending Theorem 31 remain valid if the condition "computable" is replaced by "computably enumerably defined" ("positively enumerable").

Corollary 13 (Khisamiev, Nurizinov, and Tyulyubergenev [55]). *Let G be a computably enumerably defined nilpotent group whose quotient $G/\tau G$ modulo the periodic part τG has finite dimension. Then $G/\tau G$ is computable.*

There exists an example of a non-computable torsion-free nilpotent group whose sections of every central series are computable. It is given by the following theorem.

Theorem 32 (Khisamiev, Nurizinov, and Tyulyubergenev [55]). *There exists a non-computable subgroup G in $UT_3(\mathbb{Q})$ such that all sections in its every central series are computable.*

4.3   Computability of Matrix Groups in $UT_3(\mathbb{Z}[x])$

Theorem 33 (Khisamiev, Nurizinov, and Tyulyubergenev [56]). *Let $G \leq UT_3(\mathbb{Z}[x])$ and suppose that there exists a number m, such that for every matrix $g \in G$, $\deg g_{12}(x) \leqslant m$ $(\deg g_{23}(x) \leqslant m)$. Then the group G is computable.*

Theorem 34 (Khisamiev, Nurizinov, and Tyulyubergenev [56]). *Any abelian subgroup of the group $UT_3(\mathbb{Z}[x])$ is isomorphic to a direct sum of the infinite cyclic groups.*

Corollary 14 (Khisamiev, Nurizinov, and Tyulyubergenev [56]). *Any Abelian subgroup of the group $UT_3(\mathbb{Z}[x])$ is computable.*

Corollary 15 (Khisamiev, Nurizinov, and Tyulyubergenev [56]). *Let G be a non-abelian subgroup of the group $UT_3(\mathbb{Z}[x])$ and let ν be some of its computable numberings. Then any maximal abelian subgroup A is computable in $(G; \nu)$, and therefore the subgroup A and the quotient G/A are computable.*

4.4 A Condition for Non-computability of a Periodic Nilpotent Group

Theorem 35 (Khisamiev, Nurizinov, and Tyulyubergenev [56]). *If $(G; \nu)$ is a computably numbered periodic nilpotent group of class 2, then the set $G_p = \{g \in G \mid \exists n(g^{p^n} = 1\} $ is a computable subgroup in $(G; \nu)$, for every prime number p.*

Corollary 16 (Khisamiev, Nurizinov, and Tyulyubergenev [56]). *If G is a computable periodic nilpotent group of class 2, then for any prime number p, the p-primary component G_p, and the quotient G/G_p are computable.*

Corollary 17 (Khisamiev, Nurizinov, and Tyulyubergenev [56]). *If G is a periodic nilpotent group of class 2, and one can find a prime number p such that either G_p, or the quotient G/G_p is not computable, then the group G is not computable.*

Corollary 18 (Khisamiev, Nurizinov, and Tyulyubergenev [56]). *If G is a periodic nilpotent group of class 2 and for some primary component G_p, the subgroup $G_p \cap G'$ is not computable, where G' is commutant of the group G, then the group G is not computable.*

5 An Algorithm for Root Extraction and Complexity of Computing the Centrals and Hypercenters

5.1 On an Algorithm for Root Extraction for a Nilpotent Torsion-Free Group

If $(G; \nu)$ is a numbered group and from any natural numbers k and n one can effectively determine whether there is an element x such that $x^k = \nu n$, then we say that in $(G; \nu)$ there *exists an algorithm of root extraction.*

Theorem 36 (Khisamiev, Nurizinov, and Tyulyubergenev [56]). *There exists an algorithm for root extraction in every finitely generated nilpotent group.*

Theorem 37 (Khisamiev, Nurizinov, and Tyulyubergenev [56]). *Let $(G; \nu)$ be a computable torsion-free nilpotent group, with central series given by $1 = G_0 < G_1 < G_2 = G$, where G_1 is the center of the group G, and $\nu^{-1} G_1$ is a computable set. Then the following is true: if in the factors $\hat{G}_i = G_i/G_{i-1}$, $i = 1, 2$ there exists an algorithm of root extraction, then such an algorithm exists in $(G; \nu)$ too.*

5.2 Algorithmic Complexity of the Centrals and Hypercenters in Computable Nilpotent Groups

In this section, we will be concerned with the complexity of computing the terms in the upper and lower central series of a computable group; more precisely, we will consider the problem of occurrence in commutants and terms in these central series of constructive groups. We also will consider the computability of quotient groups by these terms.

This question is interesting because many algebraic properties of nilpotent groups are proved by induction on class nilpotent; the terms of upper or lower series and factor groups by them are often considered for this purpose. It is quite natural to expect that this method is applicable for the computable groups, too. These hopes are partly reasonable — see Subsects 3.2–4.2. Furthermore, the scheme of inclusions (1) creates the illusion that the ability to solve the problem of occurrence for any of the terms of these central series will enable us easily to solve such issues for the others.

It is true for finitely generated groups. Really, let (G, ν) be such a positive enumerated group. Then $\nu/\gamma_n G$ is a positive numbering of the finitely generated nilpotent group $G/\gamma_n G$ (see Subsect. 1.5). Because the word problem for positive enumerated finitely generated nilpotent groups is solvable, the group $(G/\gamma_n G, \nu/\gamma_n G)$ is constructive. Moreover, an element g of G belongs to center if and only if the equality $[g, x] = 1$ holds for each generator x of G. Therefore, one can easily prove by induction on parameter i that every term $\zeta_i G$ in the upper central series can be effectively calculated in the computable finitely generated group G.

Thus, we focus our attention on infinitely generated nilpotent computable groups. If H and K are computably enumerable subgroups of a constructive group (G, ν), then the commutator subgroup $[H, K]$ is easily seen to be computably enumerable — see Subsect. 1.1. It follows by induction that the terms of the lower central series (or centrals) of a constructive group must be computably enumerable. It is easy to see that the terms in the upper central series (or hypercenters) are Π_1^0-sets:

$$g \in \zeta_1 G \iff \forall h(gh = hg)$$
$$g \in \zeta_{i+1} G \iff \forall h(gh = hg \mod \zeta_i G) \iff \forall h([g, h] \in \zeta_i G).$$

Therefore, the terms in the upper and lower central series of a computable group have c.e. Turing degree.

Definition 12. *Let $\langle G, \nu \rangle$ be a numbered group and $H \subseteq G$. The Turing degree of unsolvability of the set $\nu^{-1}(H)$ is called the algorithmic complexity of the problem of occurrence in the set H in the group $\langle G, \nu \rangle$ and is denoted by $T(G, H, \nu)$.*

Everywhere in the sequel, by *degree* we mean the Turing degree of unsolvability [79, 83]. If the upper central series of the computable group G coincides with its lower central series, then $T(G, \gamma_n G, \nu) = T(G, \zeta_n G, \nu) = 0$ for all n. In particular, this is true for nilpotent free groups.

But the complexity of the problem of occurrence in the centrals and/or hypercenters is enough intricate in the general case.

Theorem 38 (Latkin [58]). *For each natural number $n \geqslant 2$ there exists a torsion-free nilpotent group $G(n)$ of class n such that for each set of c.e. degrees*

$\hat{a} = \langle a_2, \ldots, a_n \rangle$ there exists a constructivization $\nu(\hat{a})$ of this group for wich $T(G(n), \gamma_2 G(n), \nu(\hat{a})) = a_2, \ldots, T(G(n), \gamma_n G(n), \nu(\hat{a})) = a_n$.

Theorem 39 (Latkin [58]). *For each natural number $n \geqslant 2$ and an arbitrary set $\hat{a} = \langle a_2, \ldots, a_n \rangle$ of c.e. degrees there exists a nilpotent group $G(\hat{a})$, of class n, possessing the following properties:*
(i) $T(G(\hat{a}), \gamma_2 G(\hat{a}), \nu) = a_2, \ldots, T(G(\hat{a}), \gamma_n G(\hat{a}), \nu) = a_n$ for each constructivization ν of $G(\hat{a})$;
(ii) the quotient-group $G(\hat{a})/\gamma_i G(\hat{a})$ is computable if and only if $a_i = 0$;
(iii) for each c.e. degree b there exists a constructivization μ of the group $G(\hat{a})$ such that $T(G(\hat{a}), \tau G(\hat{a}), \mu) = b$, where $\tau G(\hat{a})$ is the periodic part of $G(\hat{a})$. In particular, this group is non-autostable.

Theorem 40 (Csima and Solomon [4]). *Fix $n \geqslant 2$ and c.e. Turing degrees $d_1, \ldots, d_{n-1}$ and $e_2, \ldots, e_n$. There is a constructive torsion-free group (G, α) which is class n nilpotent with $T(G, \zeta_i G, \alpha) = d_i$ for $1 \leqslant i \leqslant n-1$ and $T(G, \gamma_i G, \alpha) = e_i$ for $2 \leqslant i \leqslant n$. Furthermore, (G, α) admits a computable order so this computational independence property holds for computable ordered nilpotent groups as well.*

The observation in [4]: "The construction of $G(\hat{a})$ in Theorem 39 uses torsion elements and hence $G(\hat{a})$ does not admit an order (computable or otherwise)," raises the question of whether one can obtain a similar result using a torsion-free nilpotent group, and if so, whether such a group could admit a computable order (in some or possibly all computable copies). Theorem 39 also raises the natural question of whether one can obtain a similar result for the terms in the upper central series, and if so, whether one can do it with a torsion-free (or possibly computably orderable) nilpotent computable group.

Let us notice that a partial answer to the second question was given by the following proposition.

Proposition 10 (Latkin [62]). *There exists a constructive nilpotent group H of class two with a non-computable center.*

The group H has an uncomplicated design, therefore it is easy see that $T(H, \zeta_1 H, \alpha) = d > 0$ for its computable numberings α, but this group has the torsion elements.

6 Arithmetical Hierarchies of Nilpotent Groups and Nilpotent Product

6.1 Arithmetical Hierarchies of Numbered Nilpotent Groups

Let Y be a subset of the natural numbers. In the theory of recursive functions, the classes Σ_n^Y, Π_n^Y, and Δ_n^Y, for $n \in \mathbb{N}$, of the arithmetic hierarchy of sets play an important role [79]. If $\mathcal{K}$ is class of algebras, and X is one of the symbols $\Delta_n^0, \Pi_n^0, \Sigma_n^0$, then $X\mathcal{K}$ denotes the subclass consisting of the algebras, of $\mathcal{K}$, that are X-computable. For any class $\mathcal{K}$, there appears a given below scheme of inclusions in accordance with definitions:

$$
\begin{array}{ccccccc}
& \Sigma_1^0\mathcal{K} & & & \Sigma_2^0\mathcal{K} & & \cdots \\
& \rotatebox{45}{$\subseteq$} & \rotatebox{-45}{$\subseteq$} & \rotatebox{45}{$\subseteq$} & & \rotatebox{-45}{$\subseteq$} & \rotatebox{45}{$\subseteq$} \\
\Delta_1^0\mathcal{K} & & & \Delta_2^0\mathcal{K} & & & \Delta_3^0\mathcal{K} \\
& \rotatebox{-45}{$\subseteq$} & \rotatebox{45}{$\subseteq$} & & \rotatebox{-45}{$\subseteq$} & \rotatebox{45}{$\subseteq$} & \\
& & \Pi_1^0\mathcal{K} & & & \Pi_2^0\mathcal{K} & \cdots
\end{array}
\tag{5}
$$

In addition, there exist equations $\Delta_0^0\mathcal{K} = \Pi_0^0\mathcal{K} = \Sigma_0^0\mathcal{K} = \Delta_1^0\mathcal{K}$, and also $\Delta_n^0\mathcal{K} = \Sigma_n^0\mathcal{K} \cap \Pi_n^0\mathcal{K}$ for every class $\mathcal{K}$ and any $n \in \mathbb{N}$.

This raises the question: which inclusions from (5) are strict?

In this section, the question is answered for the following classes:

(1) torsion-free abelian groups;
(2) direct sums of cyclic groups;
(3) direct sums of cyclic and quasicyclic p-groups;
(4) torsion-free class two nilpotent groups.

Let $\mathfrak{A}$ be the variety of abelian groups; $\mathfrak{N}_c$ be the variety groups, whose class nilpotent not higher than c. We denote their subquasivarieties that consisting of a torsion-free groups by $\mathfrak{A}_0$ and $\mathfrak{N}_{c,0}$, respectively.

From Corollary 1 we obtain that $\Sigma_n^0\mathfrak{A}_0 = \Delta_n^0\mathfrak{A}_0$ for every n [46,49]. Furthermore, the inclusion $\Sigma_n^0\mathfrak{A}_0 \subseteq \Pi_n^0\mathfrak{A}_0$ holds [48,49]; and the difference $\Pi_{n+1}^0\mathfrak{A}_0 \setminus \Pi_n^0\mathfrak{A}_0$ is not empty at the same time because there is a strongly constructive torsion-free abelian group whose reduced part is not computable, and the reduced part of a strongly Δ_n^0-constructive torsion-free abelian group is Π_n^0 (see [44,48,49] for further details).

It follows from this and Corollary 2 that all the inclusions in scheme (5) are strict, when the class $\mathcal{K}$ is $\mathfrak{N}_{2,0}$.

Using Theorem 3, one can prove that the following assertions are valid (see [43,48,49] for further details).

(1) If $\mathcal{K}$ is the class of direct sums of cyclic groups, then all the inclusions in scheme (5) are strict.

(2) Let $\mathcal{L}_p$ be the class of groups which are direct sums of cyclic and quasicyclic groups. Then the following relations hold: (i) $\Pi_n^0\mathcal{L}_p = \Delta_{n+1}^0\mathcal{L}_p$; (ii) $\Sigma_n^0\mathcal{L}_p \subsetneqq \Pi_n^0\mathcal{L}_p$.

Hence, the all inclusions in (5) are strict provided that the class $\mathcal{K}$ is $\mathfrak{A}$. It is all the more true in relation to the whole variety $\mathfrak{N}_c$.

6.2   Computability of Nilpotent Product

We above (see Subsect. 1.5) pointed out that the natural numbering μ of the direct product of the computable sequence of constructive groups $\{(G_i, \nu_i) \mid i \in \mathbb{N}\}$ is computable, and the numeral set $\mu^{-1}(G_i)$ of each multiplier G_i is computable. We recall that the direct product is free for variety of abelian groups.

One can reckon without losing generality, that the domains D_i of numberings ν_i such that $D_i \cap D_j = \varnothing$ for $i \neq j$. Let $\langle x_0, \ldots, x_k \rangle$ be a Gödel number of the finite sequence $(x_0, \ldots, x_k)$ of natural numbers; and $G = *_{i \in \mathbb{N}} G_i$ be a free product of groups G_i. We define a canonical (or natural) numbering σ of free product by

$$\sigma(x) = \nu_{i_0}(y_{i_0}) \cdot \nu_{i_1}(y_{i_1}) \cdot \ldots \cdot \nu_{i_t}(y_{i_t}),$$

where $x = \langle y_{i_0}, y_{i_1}, \ldots, y_{i_t} \rangle$, and $\nu_{i_0}(y_{i_0}) \cdot \nu_{i_1}(y_{i_1}) \cdot \ldots \cdot \nu_{i_t}(y_{i_t})$ is finite generalized irreducible sequences of elements of groups G_i. It is obvious that all numeral sets $\sigma^{-1}(G_i)$ are computable. Since the word problem for the free product of two groups is solvable if it is the same in each multiplier [66], therefore the numbering σ will be computable when every ν_i is a constructivization.

So the free products in the varieties of abelian groups and all groups (as this is understood in Universal algebra [72]) are computable if their multipliers constitute the computable sequence of computable groups.

The *nilpotent product* $A \circledast_k B$ *of class* k of groups A and B is the factor group of their free product $A * B$ by its normal subgroup $[A, B] \cap \gamma_{k+1}(A * B)$. This is one of the equivalent definitions of nilpotent product — see [66] for the further details. The nilpotent product of class k is the free product for the variety of groups whose class nilpotent is not more than k [66, 72].

The subgroup $[A, B] \cap \gamma_{k+1}(A * B)$ is recursively enumerable if A and B are positive numbered. Thus, the canonical factor-numbering μ (see Subsect. 1.5) of a nilpotent product obtained by the canonical numbering free product turns positive. This numbering will be called a *canonical* (or natural) numbering of nilpotent product. Note that both groups A and B have the computable sets of μ-numbers.

Theorem 41 (Faermark [24]). *If the A and B are finitely generated group with solvable word problem, then $A \circledast_k B$ also has a solvable word problem.*

This assertion is not valid for infinitely generated groups.

Proposition 11 (Latkin [57]). *There is a constructive class two nilpotent group H such that its nilpotent product of class two with infinite cyclic group is not computable.*

This group H is the group $G(a_2)$ constructed in the proof of Theorem 39 for every $a_2 > 0$. Most generally, it is rather easy to see that the computability of the members of the lower central series is the necessary condition for the computability of nilpotent product.

Proposition 12 (Latkin [60]). *Let (G, ν), (H, μ) be constructive groups; and H_1 be a normal subgroup of H such that $H/H_1 \cong \mathbb{Z}$. If the canonical numbering δ of their k-nilpotent product is computable, then $T(G, \gamma_i G, \delta) = 0$ for all $2 \leqslant i \leqslant k$.*

However, we can guarantee the computability of nilpotent product in some simple cases.

Theorem 42 (Khisamiev and Latkin [54]). *Nilpotent product $A \circledast_2 B$ of computable torsion-free abelian group A and B is computable.*

The condition "to be torsion-free" is essential. Indeed, the mutual commutant of multipliers in class two nilpotent product of abelian groups is isomorphic to their tensor product as the modules over the ring of integers [65]. In the [61], I.V. Latkin built computable abelian groups A and B, the tensor product of which is not computable. If nilpotent product C of groups A and B would be computable, then one could calculate the commutant of group C (see Subsect. 1.5), but C' coincides with the mutual commutant $[A, B]$ of A and B.

Acknowledgments. The authors acknowledge support from the Committee on Science of the Ministry of Education and Science of the Republic of Kazakhstan (project No 3953/GF4).

References

1. Ash, C.J., Knight, J.: Computable Structures and the Hyperarithmetical Hierarchy. Studies in Logic and the Foundations of Mathematics, vol. 144. North-Holland Publishing Co., Amsterdam (2000)
2. Baumslag, G., Dyer, E., Miller, C.: On the integral homology of finitely presented groups. Topology **22**, 27–46 (1983)
3. Chandler, B., Magnus, W.: The History of Combinatorial Group Theory: A Case Study in the History of Ideas. Springer, Berlin, Heidelberg, New York (1982)
4. Csima, B.F., Solomon, R.: The complexity of central series. Ann. Pure Appl. Logic **162**, 667–678 (2011)
5. Dobritsa, V.P.: Computability of certain classes of construktive algebras (Russian). Sibirsk. Math. Zh. **18**, 570–579 (1977). [translated in: Siberian Math. J. 18, 406–413 (1977)]
6. Dobritsa, V.P.: On construktivizable abelian groups (Russian). Sibirsk. Math. Zh. **22**, 208–213, 239 (1981)
7. Dobritsa, V.P.: Some construktivizations of abelian groups (Russian). Sibirsk. Math. Zh. **24**, 18–25 (1983). [translated in: Siberian Math. J. 24, 167–173 (1983)]
8. Dobritsa, V.P.: Computable classes of constructive models. In: Ershov Yu, L., Goncharov, S.S., Nerode, A., Remmel, J.B. (eds.) Handbook of Recursive Mathematics, vol. 1. Studies in Logic and the Foundations of Mathematics, vol. 138, pp. 183–233. Elsevier, New York (1998)
9. Dobritsa, V.P., Khisamiev, N.G., Nurtazin, A.T.: Construktive periodic abelian groups (Russian). Sibirsk. Math. Zh. **19**, 1260–1265 (1978). [translated in: Siberian Math. J. 19, 886-890 (1978)]
10. Downey, R.G., Fellows, M.R.: Parameterized Complexity. Monographs in Computer Science. Springer, New York (1999)
11. Downey, R.G., Goncharov, S.S., Kach, A.M., Knight, J.F., Kudinov, O.V., Melnikov, A.G., Turetsky, D.: Decidability and computability of certain torsion-free abelian groups. Notre Dame J. Formal Logic **51**(1), 85–96 (2010)
12. Downey, R.G., Kurtz, S.A.: Recursion theory and ordered groups. Ann. Pure Appl. Logic **32**, 137–151 (1986). North-Holland
13. Downey, R.G., Montalbàn, A.: The isomorphism problem for torsion-free abelian groups is analytic complete. J. Algebra **320**, 2291–2300 (2008)

14. Dzgoev, V.D.: Constructivizations of direct products of algebraic systems (Russian). Algebra i Logika **21**, 138–148 (1982). [translated in: Algebra and Logic, 21, 88–96 (1982)]
15. Ershov, Y.: Existence of constructivizations (Russian). Dokl. Akad. Nauk SSSR **204**, 1041–1044 (1972). [translated in: Soviet Math.-Dokl. 13, 779–783 (1972)]
16. Ershov, Y.L.: Constructive models (Russian). Izbr. Vopr. Alg. i Log. [Selected Questions in Algebra and Logic], Sbornik posvjascen. pamjati A. I. Malceva, [A collection dedicated to the memory of A. I. Malceva], Izdat. Nauk Sibirsk. Otdel., Novosibirsk, pp. 111–130 (1973)
17. Ershov, Y.L.: Theory of Numerations (Russian). Monographs in Mathematics Logic and Foundations of Mathematics. Nauka, Moscow (1977)
18. Ershov, Y.L.: Theorie der Numerierungen. III. Z. Math. Logik Grundlag. Math. **23**, 289–371 (1977). [translated from Russian and edited by G. Asser and H. D. Hecker]
19. Ershov, Y.L.: Decision Problems and Constructivizable Models (Russian). Mathematical Logic and Foundations of Matimatics. Nauka, Moskva (1980)
20. Ershov, Y.L., Goncharov, S.S.: Constructive Models, Siberian School of Algebra and Logic.(Russian). Nauch. Kniga, Novosibirsk (1996). [translated in: Consultants Bureau, New York, 2000]
21. Ershov, Y.L., Goncharov, S.S., Nerode, A., Remmel, J.B., Marek, V.W. (eds.): Handbook of Recursive Mathematics, vol. 1 Recursive Model Theory. Studies in Logic and the Foundations of Mathematics, vol. 139. North-Holland, Amsterdam (1998)
22. Ershov, Y.L., Goncharov, S.S., Nerode, A., Remmel, J.B., Marek, V.W. (eds.): Handbook of Recursive Mathematics, vol. 2, Recursive Algebra, Analysis and Combinatorics. Studies in Logic and the Foundations of Mathematics, vol. 139. North Holland, Amsterdam (1998)
23. Ershov, Y.L., Palutin, E.A.: Mathematical Logic (Russian), 2nd edn. Nauka, Moscow (1987). [English translation of 1st edn. translated by V. Shokurov (MIR Publishers, Moscow 1984)]
24. Faermark, D.S.: Algorithm to determine the identity of words in the nilpotent products of groups defined by a finite number of generators, defining relations (Russian). Dokl. Akad. Nauk SSSR **137**, 291–294 (1961). [translated in: Soviet Math. Dokl. 3 (1961)]
25. Fröllich, A., Shepherdson, J.C.: Effective procedures in field theory. Philos Trans R. Soc. London Ser. A **248**, 407–432 (1956)
26. Fuchs, L.: Innite Abelian Groups, vol. I, II. Academic Press, New York (1973). Pure and Applied Mathematics, vol. 36
27. Goncharov, S.S.: Autostability of models, abelian groups (Russian). Algebra i Logika **19**, 23–44 (1980). [translated in: Algebra and Logic, 19, 13-27 (1980)]
28. Goncharov, S.S.: Groups with a finite number of constructivizations (Russian). Dokl. Akad. Nauk SSSR **256**, 269–272 (1981). [translated in: Soviet Math. Dokl. 23, 58–61 (1981)]
29. Goncharov, S.S., Drobotun, B.N.: On the algorithmic dimention of nilpotent groups (Russian). Sibirsk. Math. Zh. **30**, 52–60, 225 (1989). [translated in: Siberian Math. J. 30, 210–217 (1989)]
30. Goncharov, S.S., Lempp, S., Solomon, R.: The computable dimension of ordered abelian groups. Adv. Math. **175**(1), 102–143 (2003)
31. Goncharov, S.S., Molokov, A.V., Romanovskiĭ, N.S.: Nilpotent groups of finite algorithmic dimention. Sibirsk. Math. Zh. **30**, 82–88 (1989). [translated in: Siberian Math. J. 30, 63–68 (1989)]

32. Griffor, E.R. (ed.): Handbook of Computability Theory. Studies in Logic and the Foundations of Mathematics, vol. 140. North- Holland Publishing Co., Amsterdam (1999)

33. Gödel, K.: Über formal unentscheibare Sätze der Principia mathematica und verwandter Systeme I. Monatsh. Math. Phys. **38**, 173–198 (1931). [translated as: On formally undecidable propositions of Principia Mathematica and related systems I, (translated by B. Meltzer, with an introduction by R. B. Braithwaite), (Basic Books, NY, 1963, reprinted Dover, NY, 1992); also in: From Frege to Gödel: A Source Book in logic, 1879–1931, J. van Heijeroot, (ed.), (Harvard Univ. Press, Cambridge, Mass. And Oxford Univ. Press, London, 1967, pp. 592–617]

34. Hall, M.: The Theory of Groups. Macmillan, New York (1959)

35. Hall, P.: Nilpotent Groups, Queen Mary College Math. Notes, Queen Mary Coll. (Univ. London), London (1969)

36. Hilbert, D.: Mathematische probleme. Archiv. f. Math. und Phys. Ser. **3**, **1**, 44–63, 213–237 (1901)

37. Hirschfeldt, D.R., Khoussainov, B., Shore, R.A., Slinko, A.M.: Degree spectra and computable dimensions in algebraic structures. Ann. Pure Appl. Logic **115**, 71–113 (2002)

38. Hjorth, G.: The isomorphism relation on countable torsion free abelian groups. Fundamenta Math. **175**, 241–257 (2002)

39. Kargapolov, M.I., Merzljakov, Y.I.: Fundamentals of the Theory of Groups, Translations of Osnovy Theorii Grupp, 2nd edn. Nauka, Moscow (1977). [translated by Burns, R.G., Springer, Grad. Texts in Math. 62 (1979), 4th. Edn., Fismatlit. Nauka, Moscow (1996)]

40. Kharlampovich, O.: The word problem for solvable groups and lie algebras. In: Baumslag, G., Miller III, C.F. (eds.) Algorithms and Classification in Combinatorial Group Theory, vol. 23, pp. 61–67. Mathematical Sciences Research Institute Publications, Hampton (1992)

41. Khisamiev, N.G.: Construktive periodic abelian groups (Russian). In: Proceedings of 5th Kazakhstan Conference on Mathematics and Mechanics, Alma-Ata, Extended Abstracts, part 2, vol. 253 (1974)

42. Khisamiev, N.G.: The periodic part of a strongly construktivizable abelian group (Russian). Teor. I Priklad. Zad. Mat. I Mech. Alma-Ata, Nauka Kazakh. SSR **318**, 299–303 (1977)

43. Khisamiev, N.G.: Criterion for construktivizable of a direct sum of cyclic p-groups (Russian). Izv. Akad. Nauk. Kazakh. SSR, Ser. Fiz.-Mat. **86**, 51–55 (1981)

44. Khisamiev, Z.G., Khisamiev, N.G.: Nonconstruktivizability of the reduced part of a strongly constructive torsion-free abelian group (Russian). Algebra I Logika **24**, 108–118, 123 (1985)

45. Khisamiev, N.G.: Theory of abelian groups with constructive models (Russian). Sibirsk. Math. Zh. **27**, 128–143 (1986). [translated in: Siberian Math. J. 27, 572–585 (1986)]

46. Khisamiev, N.G.: Hierarchies of torsion-free Abelian groups (Russian). Algebra i Logika **25**(2), 205–226, 244 (1986). [translated in: Algebra and Logic, 25,2, 128–142 (1986)]

47. Khisamiev, N.G.: Non construktivizability of certain ordered fields of real numbers' (Russian). Sibirsk. Math. Zh. **28**, 193–195 (1987). [translated in: Siberian Math. J. 28, 845–847 (1987)]

48. Khisamiev, N.G.: The arithmetic hierarchy of Abelian Groups (Russian). Sibirsk. Math. Zh. **29**, 144–159 (1988). [translated in: Siberian Math. J. 29, 987–999 (1988)]

49. Khisamiev, N.G.: Constructive abelian groups. In: Ershov, Y.L., Goncharov, S.S., Nerode, A., Remmel, J.B., Marek, V.M. (eds.) Handbook of Recursive Mathematics, vol. 2. Studies in Logic and the Foundations of Mathematics, vol. 139, pp. 1177–1231. Elsevier, New York (1998)

50. Khisamiev, N.G.: Constructivizability criterion for an abelian group (russian). Algebra i Logika **38**, 743–760 (1999). [translated in: Algebra and Logic, 38, 410–419 (1999)]

51. Khisamiev, N.G.: On a class of strongly decomposable abelian groups. Algebra i Logika **41**(4), 493–509, 511, 512 (2002). [translated in: Algebra and Logic, 41(4), 274–283 (2002)]

52. Khisamiev, N.G.: On positive, constructive groups (Russian). Sibirsk. Mat. Zh. **53**(5), 1133–1146 (2012). [translated in: Siberian Math. J. 53(5), 906–917 (2012)]

53. Khisamiev, N.G., Krykpaeva, A.A.: Effectively completely decomposable abelian groups. Sibersk. Mat. Zh. **38**(6), 1413–1426 (1997). [translated in: Siberian Math. J. 38(6), 1227–1229 (1997)]

54. Khisamiev, N.G., Latkin, I.V.: Computability of nilpotent product of computable torsion free Abelian groups (Russian), Abstracts of the International Conference "Actual problems of mathematics and mathematical modeling", dedicated to the 50th anniversary of the Institute of Mathematics and Mechanics, Almaty, 1–5 June, pp. 185–186 (2015)

55. Khisamiev, N.G., Nurizinov, M.K., Tyulyubergenev, R.K.: Computable torsion-free nilpotent groups of finite dimension. Sibersk. Mat. Zh. **55**(3), 580–591 (2014). [translated in: Siberian Math. J. 55(3), 471–481 (2014)]

56. Khisamiev, N.G., Nurizinov, M.K., Tyulyubergenev, R.K.: On constructive nilpotent groups. KazNU Bull. Math. Mech. Comput. Sci. Ser. **2**(82), 94–106 (2015)

57. Latkin, I.V.: Constructivizable group, nilpotent product which does not constructivizable (Russian). In: 8-Union Conference on Mathematical Logic - Moscou, p. 101 (1986)

58. Latkin, I.V.: Algorithmic complexity of the problem of occurrence in commutants, the members of the lower central series (Russian). Sibirsk. Math. Zh. **28**(5), 102–110 (1987). [translated in: Siberian J. Math. 5, 772–779 (1987)]

59. Latkin, I.V.: Arithmetic hierarchy of torsion-free nilpotent groups. Algebra i Logika **35**(3), 308–313 (1996). [translated in: Algebra and Logic, 35(3), 172–175 (1996)]

60. Latkin, I.V.: Thesis for the degree of Candidate of Physico-Mathematical Sciences. Novosibirsk State University, Russian (2001)

61. Latkin, I.V.: On constructivizability of the tensor product of modules (Russian). Siberian Math. J. **43**(2), 414–418 (2002). [translated in: Siberian Math. J. 43, 330–333 (2002)]

62. Latkin, I.V.: Constructive nilpotent groups with a non-constructivizable center (Russian). Bull. East Kazak. State Techn. Univ. **23**(1), 82–84 (2004)

63. Logic notebook (Russian), Novosibirsk, Inst. Math. SB USSR Acad. Sci. (1986)

64. Lyndon, R.C., Schupp, P.E.: Combinatorial Group Theory. Springer, Berlin, Heidelberg, New York (1977)

65. MacHenry, T.: The tensor product and the 2-nd nilpotent product of groups. Math. J. **73**(1), 134–145 (1960)

66. Magnus, W., Karrass, A., Solitar, D.: Combinatorial Group Theory: Presentations of Groups in Terms of Generators and Relations. Interscience Publishers, A division of Wiley & Sons Inc, Interscience, New York, London, Sydney (1966)

67. Maltsev, A.I.: Torsion-free abelian groups of finite rank (Russian). Math. Sb. (N.S.) **4**, 67–68 (1938)

68. Maltsev, A.I.: On homomorphisms onto finite groups (Russian). Uchen. Zapiski Ivanovsk Ped. instituta **18**(5), 49–60 (1958). [translated in: Transl., II Ser., Am. Math. Soc., vol. 119, pp. 67–79 (1983)]
69. Maltsev, A.I.: Constructive algebras I (Russian). Uspekhi Mat. Nauk **16**, 3–60 (1961). [translated in: Constructive algebras I, Russian Math. Surveys, 16: 3 77–129; also in: Metamathematics of Algebraic Systems, Collected Papers: 1936–1967, translated and edited by B. F. Wells III, Stud. Logic Found. Math., 66 (1971), Ch. 18, 148–200]
70. Maltsev, A.I.: On recursive abelian groups (Russian). Dokl. Akad. Nauk SSSR **146**, 1009–1012 (1962). [translated in: Soviet Math. Dokl., 3 (1962) 1431-1434; also in: Metamathematics of Algebraic Systems Collected Papers: 1936-1967, translated and edited by B. F. Wells III, Stud. Logic Found. Math., 66 (1971), Ch. 24, 282-286]
71. Maltsev, A.I.: Algorithms and Recursive Functions (Russian) Izdat. Nauka, Moscow (1965). [translated by Boron, L.F., Sanchis, L.E., Stilwell, J., Iseki, K. (Wolters NoordHoff Publishing, Groningen, 1970)]
72. Maltsev, A.I.: Algebraic Systems (Russian). In: Smirnov, V.D., Taiclin, M. (eds.) Izdat. Nauka, Moscow (1970). [translated by Seckler, B.D., Doohovskoy, A.P., Grundlehren Math. Wiss. 192 (1973)]
73. Miller III, C.F.: Decision problems for groups survey and reflections. In: Baumslag, G., Miller III, C.F. (eds.) Algorithms and Classification in Combinatorial Group Theory. Mathematics Sciences Research Institute Publications, vol. 23, pp. 1–59. Springer, Heidelberg (1992)
74. Neumann, H.: Varieties of Groups. Springer, Berlin (1967)
75. Novikov, P.S.: On the algorithmic unsolvability of the word problem in group theory (Russian). Trudy Math. Inst. Steklov. **44**, 3–143 (1955)
76. Nurtazin, A.T.: Computable classes and algebraic criteria of autostability, summary of scientific thesis (Russian), Math. Inst. Siberian Branch of SSSR Acad. Sci. Novosibirsk (1974)
77. Nurtazin, A.T.: On constructive groups (Russian). In: Proceedings of 4th All-Union Conference on Mathematical Logic (Kishinev), vol. 106 (1976)
78. Rabin, M.O.: Computable algebra, general theory and theory of computable fields. Trans. Amer. Math. Soc. **95**(2), 341–360 (1960)
79. Rogers Jr., H.: Theory of Recursive Functions and Effective Computability, Ist. edn. McGraw-Hill, New York, Toronto, Ontario, London (1967), 2nd edn. MIT Press, Cambridge, London (1987)
80. Roman'kov, V., Khisamiev, N.: On constructible matrix or ordered groups. In: Proceedings of Workshop Computability and Models, Kazakhstan University, Almaty, pp. 44–47(2002)
81. Roman'kov, V.A., Khisamiev, N.G.: Constructive matrix, orderable groups. Algebra i Logika **43**(3), 353–363 (2004). [translated in: Algebra and Logic, 43(3), 261–290 (2004)]
82. Roman'kov, V.A., Khisamiev, N.G.: Constructible matrix groups. Algebra i Logika **43**(5), 603–613 (2004). [translated in: Algebra and Logic, 43(5), 339–345 (2004)]
83. Soare, R.I.: Recursively Enumerable Sets and Degrees. Springer, Heidelberg (1987)
84. Smith, R.L.: Two theorems on autostability in p-groups, in Logic Year 1979–1980. In: Lerman, M., Soar, R.I. (eds.) Proceedings of Seminars and Conference on Mathematics and Logic University Connecticut, Storrs, CT, 1979/80. Lecture Notes in Mathematics, vol. 859, pp. 302–311. Springer, Heidelberg (1981)
85. Solomon, V.A.: Pi_1^0 classes and orderable groups. Ann. Pure Appl. Logic **115**, 279–302 (2002)

86. Uspenskiĭ, V.A.: The systems of enumeramerable sets and their enumerations (Russian). Docl. Akad. Nauk SSSR. **105**(6), 1155–1158 (1955)
87. van der Waerden, B.L.: Eine Bemerkung ber die Unzerlegbarkeit von Polynomen. Mathematische Annalen **102**, 738–739 (1930)
88. Wussing, H.: Die Genesis des Abstrakten Gruppenbegriff. VEB Deutcher Verlag Wiss, Berlin (1969)

Computable Model Theory over the Reals

Andrey S. Morozov[1,2](✉)

[1] Sobolev Institute of Mathematics SB RAS, Koptyug Avenue 4, Novosibirsk, Russia
morozov@math.nsc.ru
[2] The Novosibirsk State University, Pirogova Street 2, Novosibirsk 630090, Russia

Abstract. This paper is a survey of results together with a list of open questions on Σ–definability of structures over $\mathbb{HF}(\mathbb{R})$, the hereditarily finite superstructure over the ordered field of the real numbers.

1 Introduction

There are at least two groups of approaches to the study of computability over uncountable structures. In the first group, one uses approximations of the data by means of finite constructive objects, like, for example, the rational numbers. More specifically, this approach uses approximations of arguments to classically compute approximations of the limit objects, such as real numbers, matrices, etc. The bibliography of papers using this approach is rather rich and we mention here only the monograph [15] and the paper [12]. Another approach, which we are going to consider in this paper, is applicable to any structure of finite signature. This approach is based on the concept of Σ–definability over admissible sets (see [2,3]). Ershov, in the introduction to his monograph [3], notes that the specific admissible set $\mathbb{HF}(\mathbb{R})$ — the hereditarily finite superstructure over the ordered field of the real numbers — is one of the most interesting objects for further study. Informally, this approach can be viewed as a hypothetical situation when we have an advanced programming language supporting abstract data types (for example, Pascal or C) in which the ordered field of real numbers is implemented as one of the basic data types. Modern programming languages already have types for "real numbers", except that they must implement them with only finite precision. Here, we are just asking that these data types have the full precision of real numbers, and that, just like modern computers can compute square roots as an elementary operation, we can do the same with algebraic equations, i.e., we can find solutions of algebraic equations with rational coefficients and use them in further computations. It is important that here we consider actual real numbers, not their approximations. This approach can be considered as a possible mathematical model of an analog computer or, maybe, of computer systems working with algebraic reals.

The most essential difference between these two groups of approaches is that in the first one we generally cannot compare two equal reals (if they are unequal, we can eventually discover this using their approximations, but if they are equal, we will never stop our check) while in the second one it is always possible to do this.

© Springer International Publishing AG 2017
A. Day et al. (Eds.): Downey Festschrift, LNCS 10010, pp. 354–365, 2017.
DOI: 10.1007/978-3-319-50062-1_22

Inasmuch as many algorithmic languages assume the essential use of abstract data types, which are the structures we define and use for our algorithms, and in the classical case all such abstract data types are actually computable structures, in the second group of approaches it is interesting to study the analog of computable model theory in which the computability is understood in an appropriate way. In this paper, we are going to present some results on the analog of computable model theory over $\mathbb{HF}(\mathbb{R})$, i.e., the computable model theory in which the classical concept of computability is replaced with Σ–definability over $\mathbb{HF}(\mathbb{R})$, and formulate some questions the author finds to be interesting and important.

2 Basic Facts and Definitions

Generally, admissible sets are transitive parts of the set–theoretic universes over structures of finite signatures satisfying the axioms of Kripke–Platek set theory with urelements (KPU) [2]. Here we do, not need the general definitions of admissible sets and we restrict ourselves to the case of hereditarily finite superstructures over structures. Nevertheless, a basic knowledge of admissible sets will facilitate the reader's understanding of this paper. We consider the ordered field of the real numbers $\mathbb{R}$ in the signature $\langle +, \times, < \rangle$, whose elements are all predicate symbols, i.e., $+$ and $\times$ denote the graphs of the corresponding operations of addition and multiplication. In spite of this, we will use these symbols as operations and we will also use the constants 0, 1 as well as all the rational numbers. Since every quantifier–free formula with these operations and constants could be rewritten in an equivalent $\exists$– and $\forall$–form, this will not affect Σ– and Δ–formulas, which we are going to work with (definitions follow later).

Assume that $\mathfrak{M}$ is a structure of finite predicate signature (in most cases $\mathfrak{M}$ will be equal to $\mathbb{R}$). Roughly speaking, the basic set of the structure $\mathbb{HF}(\mathfrak{M})$ we are going to define consists of the elements of $\mathfrak{M}$ and all the sets which could be written down using the real numbers, $\{;\}$, and the comma symbol. Here is the formal definition of it:

$$\mathbb{HF}(\mathbb{R}) = \bigcup_{n<\omega} \mathbb{HF}_n(\mathbb{R}),$$

where the sets $\mathbb{HF}_n(\mathfrak{M})$, $n < \omega$ are defined as follows (here $S_{<\omega}(A)$ is the set of all finite subsets of A):

$$\mathbb{HF}_0(\mathfrak{M}) = \mathfrak{M},$$
$$\mathbb{HF}_{n+1}(\mathfrak{M}) = \mathbb{HF}_n(\mathfrak{M}) \cup S_{<\omega}(\mathbb{HF}_n(\mathfrak{M})).$$

The typical examples of elements of $\mathbb{HF}(\mathbb{R})$ are $\varnothing$, 1, $\{\sqrt{2}, 1\}$, $\{7, \{18, 0\}\}$. It is important to note that we assume the elements of $\mathbb{R}$ to have no set–theoretical structure: they cannot contain anything as elements but are different from the empty set $\varnothing$, i.e., they are *urelements*. For the discussion of the role of urelements and the correctness of such concept see [2].

The universe $\mathbb{HF}(\mathfrak{M})$ actually contains all standard constructions used in high level programming languages (more exactly, it contains their reflections understood in a similar way as is done in set theory), for instance the ordered pair of a and b, which is usually defined as $\{\{a\}, \{a, b\}\}$, is always in $\mathbb{HF}(\mathfrak{M})$ provided that $a, b \in \mathbb{HF}(\mathfrak{M})$. Similarly one can define such constructions as lists, matrices, texts, structures with references, etc. The set of all urelements in the transitive closure of an element $a \in \mathbb{HF}(\mathfrak{M})$ is called the support of a and is denoted by $\mathsf{sp}(a)$.

The structure $\mathbb{HF}(\mathfrak{M})$ is usually considered in the signature $\sigma = \langle U, \in, \varnothing, \sigma_0 \rangle$, where the unary predicate U distinguishes the urelements, i.e., the elements of $\mathfrak{M}$, $\in$ is the restriction of the usual membership relation to $(\mathbb{HF}(\mathfrak{M}) \setminus \mathfrak{M}) \times \mathbb{HF}(\mathfrak{M})$, $\varnothing$ is the constant whose value is the empty set, and σ_0 is the signature of $\mathfrak{M}$. The predicates contained in σ_0 have the same values as they had in $\mathfrak{M}$, in particular, if $P \in \sigma_0$ and $\mathbb{HF}(\mathfrak{M}) \models P(\bar{a})$ then all the elements of $\bar{a}$ are in $\mathfrak{M}$. Note that the structure $\mathbb{HF}(\mathfrak{M})$ is a model of KPU.

Now we are going to define the class of Σ–formulas, which express 'computable enumerability' on $\mathbb{HF}(\mathfrak{M})$. First we define the class of Δ_0–formulas as the smallest class of formulas which contains all atomic formulas and is closed under all propositional connectives ($\wedge$, $\vee$, $\rightarrow$, $\neg$) and the restricted quantifications of kind $\exists x \in y$ and $\forall x \in y$.

The class of Σ–formulas is defined as the smallest class of formulas containing all Δ_0–formulas which is closed under positive propositional connectives $\wedge$, $\vee$, restricted existential and universal quantifications and unrestricted existential quantification $\exists x$.

If we assume that we can 'effectively' check the validity of all the atomic formulas of the signature σ_0 then, of course, we can check the validity of all Δ_0–formulas as well. The following result will be useful to understand the analogy between computably enumerable and Σ–definable relations:

Theorem 1 (Σ–reflection principle, [2]). *For any Σ–formula $\varphi(\bar{x})$ there exists a Δ_0–formula $\varphi^*(x, \bar{x})$ such that*

$$\text{KPU} \vdash \varphi(\bar{x}) \leftrightarrow \exists y \varphi^*(y, \bar{x}).$$

This result shows that, to check whether a Σ–formula $\varphi(\bar{x})$ is true on $\bar{a}$, one can just perform the exhaustive search over all elements $u \in \mathbb{HF}(\mathfrak{M})$ and for each u check the truth of the Δ_0–property $\varphi^*(y, \bar{x})$ on $u, \bar{a}$. This consideration is an argument in favor of viewing Σ–formulas as those defining 'computably enumerable' ('c.e.') relations, since the c.e. relations are exactly those definable by conditions of kind $\exists y \varphi(y, x)$, where $\varphi(y, x)$ is a computable condition.

Knowing the notion of 'computably enumerable', we actually know everything about the concrete generalization of computability, of course, if we want the analogs of the graph theorem and the Post's theorem to be valid. Indeed, in this case 'computable' sets are exactly those 'c.e.' sets whose complements are 'c.e.' as well (Δ–definable sets), and computable functions are exactly functions whose graphs are 'c.e.' (Σ–definable).

The following result is important in the understanding of computability over $\mathbb{HF}(\mathbb{R})$:

Theorem 2. *For every Σ-formula $\varphi(\bar{x})$, one can uniformly find an index to enumerate the computable family $\varphi_i(\bar{x})$, $i < \omega$ of quantifier–free formulas of signature $\langle +, \times, < \rangle$ (where $+$ and $\times$ are understood as operations) such that for all $\bar{a} \in \mathbb{R}$ holds*

$$\mathbb{HF}(\mathbb{R}) \models \varphi(\bar{a}) \leftrightarrow \mathbb{R} \models \bigvee_{i < \omega} \varphi_i(\bar{a}).$$

This theorem is easily obtained from a well–known result on decomposition of Σ-formulas into computable infinite disjunctions of $\exists$-formulas; this result is known as folklore and it can be found in [16] together with the well–known fact that $\mathbb{R}$ is decidable and admits effective quantifier elimination in the functional signature [14]. Korovina was the first to use a result like this to note that the function $\sin(x)$ on $\mathbb{R}$ is not Σ-definable on $\mathbb{HF}(\mathbb{R})$ [5]. Actually, she worked with hereditarily finite list extensions, but the result can be easily rewritten for $\mathbb{HF}(\mathbb{R})$.

Conjecture (Church's thesis for $\mathbb{R}$). *The class of intuitively computable functions on $\mathbb{HF}(\mathbb{R})$ coincides with the class of functions whose graphs are definable by Σ-formulas.*

Recall that here we consider the checking of quantifier–free formulas and finding roots of polynomials with integer coefficients to be effective.

Also a useful tool in the study of computability over $\mathbb{HF}(\mathbb{R})$ is the so–called *algebraic generalization principle (AGP)*. To formulate it, we need to recall some classical definition first.

Definition 1. *We say that a tuple $\bar{\alpha} \in \mathbb{R}^{<\omega}$ is algebraically independent over a tuple $\bar{p} \in \mathbb{R}^{<\omega}$, if for any polynomial $f(\bar{x}, \bar{y}) \in \mathbb{Q}[\bar{x}, \bar{y}]$, the property $f(\bar{\alpha}, \bar{p}) = 0$ implies that $f(\bar{x}, \bar{p})$ is the zero polynomial.*

Theorem 3 (AGP, [6]). *Let $\varphi(\bar{x}, \bar{y})$ be either a first order formula in the language of ordered fields or a Σ-formula such that $\varphi(\bar{\alpha}, \bar{p})$, where $\bar{\alpha}$ is algebraically independent over $\bar{p}$. Then this formula is also true in some open neighborhood of $\bar{\alpha}$.*

Under some extra assumptions, this property can be also formulated for some more general classes of structures, for instance, for o–minimal structures.

Now we recall the notions of Σ-definable and Σ-presentable structure over $\mathbb{HF}(\mathbb{R})$. Below is an equivalent reformulation of the original definition given by Ershov [4] for the case of hereditarily finite superstructures, with a slight modification; namely we will distinguish Σ-definable structures and Σ-presentable structures, which are exactly those isomorphic to one which is Σ-definable:

Definition 2. *A structure of computable signature is called Σ-definable over $\mathbb{HF}(\mathfrak{M})$ if each of its element is coded by some elements of $\mathbb{HF}(\mathfrak{M})$ in such a way that its diagram is a Σ-subset of $\mathbb{HF}(\mathfrak{M})$. If every element of the structure* *is coded by a unique element of* $\mathbb{HF}(\mathfrak{M})$ *then we say it is* simply Σ*–definable. Structures isomorphic to those* Σ*–definable over* $\mathbb{HF}(\mathfrak{M})$ *are called* Σ*–presentable over* $\mathbb{HF}(\mathbb{R})$.

Note that since in the definition above the equality and inequality of elements are Σ–definable on their codes, the equality and inequality of the elements of the structure are Δ–definable, i.e., they are 'computable'.

One can find some results on presentability and non–presentability of structures in [3,4,16].

3   Model Existence Theorem

Of course, one of the first questions on our way is the problem of existence of a Σ–presentable model for any countable compatible theory. If we do not restrict ourselves with cardinality of this model then the answer is obtained very easily, because as it was noted in [1], any at most countable structure has a simple Σ–presentation over $\mathbb{HF}(\mathbb{R})$, with at most one parameter. The answer for the cardinality 2^ω is given by the following

Theorem 4 ([11]). *Every consistent theory having an infinite model has a model of cardinality* 2^ω Σ*–presentable over* $\mathbb{HF}(\mathbb{R})$.

Unfortunately, the proof essentially uses a real parameter which can be very complicated, even if the theory is decidable. Thus, the following questions are very interesting to answer:

Question 1. *Does every decidable theory of computable signature with infinite models have a model of cardinality* 2^ω Σ*–definable over* $\mathbb{HF}(\mathbb{R})$ *without parameters?*

Question 2. *Does every decidable theory of computable signature with infinite models have a model of cardinality* 2^ω *simply* Σ*–definable over* $\mathbb{HF}(\mathbb{R})$?

4   Countable Σ–Definable Structures

As it was already noted, every at most countable structure of computable signature has a simple Σ–presentation over $\mathbb{HF}(\mathbb{R})$.

In [1] a characterization of at most countable structures Σ–definable over $\mathbb{HF}(\mathbb{R})$ without parameters is given. The results could be summed up in the following:

Theorem 5

1. *An at most countable structure has a* Σ*–presentation over* $\mathbb{HF}(\mathbb{R})$ *without parameters in which each element has at most a countable number of codes if and only if it admits a computable presentation.*
2. *If an at most countable structure has a* Σ*–presentation over* $\mathbb{HF}(\mathbb{R})$ *without parameters then it has a hyperarithmetical isomorphic copy.*

3. *For every hyperarithmetical degree of type* $\mathbf{0}^{(\alpha)}$, $\alpha < \omega_1^{CK}$, *there exists a countable structure* $\mathfrak{M}$ Σ-*presentable over* $\mathbb{HF}(\mathbb{R})$ *without parameters such that* $\mathbf{0}^{(\alpha)}$ *is the smallest element in the set of Turing degrees*

$$\{\mathbf{d} \mid \mathfrak{M} \text{ has a } \mathbf{d} - computable \ isomorphic \ copy\},$$

i.e., $\deg \mathfrak{M} = \mathbf{0}^{(\alpha)}$.

4. *An arbitrary at most countable structure has a* Σ-*presentation over* $\mathbb{HF}(\mathbb{C})$ *with parameters, where* $\mathbb{C}$ *is the field of complex numbers, if and only if it has a computable copy.*

It was also shown in [1] that the definability over the quaternions is actually the same as over $\mathbb{R}$.

Question 3. *Is it true that any hyperarithmetical structure has a* Σ-*presentation without parameters over* $\mathbb{HF}(\mathbb{R})$?

5   Presentations of Finite Dimensions

When we deal with the computability over the natural numbers, all of them have similar properties from the point of view of computability, i.e., for instance, any two infinite computable sets with computable complements are computably isomorphic, or, if we construct an infinite computable model whose basic set is the set of all even numbers then we can easily construct its isomorphic copy whose basic set is the set of all natural numbers. The situation for $\mathbb{HF}(\mathbb{R})$ differs very much. For instance, one can prove that the sets $\mathbb{R}$ and $\mathbb{R}^2$ are not Σ-isomorphic, i.e., they cannot be identified via a Σ-definable bijection. This means that the properties of models puts some restrictions on the subsets of $\mathbb{HF}(\mathbb{R})$ we use to code their elements in Σ-presentations.

In [8], the notion of dimension of elements of $\mathbb{HF}(\mathbb{R})$ over a tuple of parameters $\bar{p}$ was introduced: if $\mathsf{sp}(a) = \{a_0, \ldots, a_{k-1}\}$ then the dimension of a over $\bar{p}$ is the cardinality of any its maximal subset algebraically independent over $\bar{p}$. It is denoted by $\dim_{\bar{p}}(a)$. We denote

$$D_m(\bar{p}) = \{x \in \mathbb{HF}(\mathbb{R}) \mid \dim_{\bar{p}}(x) \leqslant m\},$$

which is the set of all elements whose dimension over $\bar{p}$ does not exceed m.

It is known that the sets $D_m(\bar{p})$ are definable by Σ-formulas with parameters $\bar{p}$ and neither they complements nor the differences $D_{m+1}(\bar{p}) \setminus D_m(\bar{p})$ are Σ-definable with parameters $\bar{p}$ (see [8]). Examining the proof, we can see that these sets cannot be defined by Σ-formulas with any parameters, not only with $\bar{p}$.

In what follows, $\Sigma_{\bar{p}}$-definability will mean Σ-definability with parameters $\bar{p}$.

Definition 3. *Let* $\bar{p} \in \mathbb{R}$ *and* $m \in \omega$. *A simply* $\Sigma_{\bar{p}}$*-definable structure is called* *m-dimensional over* $\bar{p}$, *if its universe is a subset of* $D_m(\bar{p})$.

In [8] it was proven that a structure has an m–dimensional $\Sigma_{\bar{p}}$–presentation if and only if it has a $\Sigma_{\bar{p}}$–presentation whose universe is a subset of $\mathbb{R}^m$.

It was also proven in [8] that any free system of uncountable rank whose signature contains at least one at least binary operation has no 1–dimensional Σ–presentations but for any finite signature, the free algebra of rank 2^ω of this signature has a simple Σ–presentation.

Question 4. *Classify* Σ*-definable structures according to minimal possible* m *for which it has* m*-dimensional presentations.*

Question 5. *Is it true that for any* $m < \omega$ *there exists a structure with* $m + 1$*-dimensional* Σ*-presentation but without* m*-dimensional* Σ *presentations?*

Question 6. *Does there exists a structure with simple* Σ*-presentation without* m*-dimensional* Σ*-presentations for any* $m < \omega$?

6   One-Dimensional Σ–Categoricity for $\mathbb{R}$

Having the 'real reals' at our disposal, we still cannot compute some transcendental functions like $\sin(x)$, $\cos(x)$, e^x, etc. without approximations (see for instance [5]). But maybe we can use our reals and the power of data type constructions to create an isomorphic 'computable' copy of the reals so that these functions will be computable in this presentation? The following theorem shows it to be impossible at least if we are going to use a one–dimensional presentation of the 'new reals'. As a result, we will always obtain a Σ–isomorphic copy of $\mathbb{R}$.

Theorem 6 ([6]). *The ordered field of the reals* $\mathbb{R}$ *has only one one–dimensional* Σ*-presentation, up to* Σ*-isomorphism.*

Note that the structure $\langle \mathbb{R}, \exp, < \rangle$, where $+$ and $\times$ are not in the signature, has a Σ–presentation over $\mathbb{HF}(\mathbb{R})$ (see [6]).

Question 7. *Does* $\mathbb{R}$ *have unique* Σ*-presentation (or at least unique simple* Σ*-presentation) up to* Σ*-isomorphism?*

7   Presentations of the Real Ordering

In computable model theory it can happen that two computable structures are isomorphic but not computably isomorphic. It is known that computable structures can have any number m of computable presentations, where $0 \leqslant m \leqslant \omega$ [13]. So far we know examples of structures with 0 or 2^ω non–Σ–isomorphic simple presentations only. In particular, there is the natural ordering on the reals:

Theorem 7 ([9]). *The ordering on the reals has 2^ω simple non–Σ–isomorphic presentations.*

There are at least ω Σ–presentations of the real ordering which have no Σ–definable nontrivial self–embeddings [7]. Moreover, the class of all Σ–presentations of this ordering is somehow unobservable.

Question 8. *How many non–Σ–isomorphic presentations a structure can have? Can it be an arbitrary finite natural number?*

It should be noted that in classical computability the algorithms defined with the use of parameters (natural numbers), are actually definable without them, but in the case of computability over $\mathbb{HF}(\mathbb{R})$, we generally cannot eliminate parameters because there are 2^ω possible tuples of parameters but only ω possible Σ–formulas. For instance, if we are going to construct non–Σ–isomorphic structures without parameters, we usually have to provide that any of ω possible Σ–formulas fail to establish a Σ–isomorphism between the structures we define. Here some step–by–step constructions can be used. If we are allowed to use any parameters, we need to provide that 2^ω possible mappings intending to establish such an isomorphism fail to do so. Here step–by–step constructions may not work and some topological ideas could be used as well. The case when we fix a finite number of possible parameters actually does not differ much from the parameter–free case.

8   Metatheorems and Non–presentability

It is not difficult to understand that the structures like the field $\mathbb{C}$ of complex numbers, the body of quaternions $\mathbb{H}$, ring of polynomials over $\mathbb{R}$ or $\mathbb{C}$, rings of matrices over them, etc., have simple presentations over $\mathbb{HF}(\mathbb{R})$.

Now we outline some results on non–Σ–presentability of structures. It appears, many natural structures of cardinality 2^ω have no such presentations.

It was possible to prove some metatheorems on non–Σ–presentability over $\mathbb{HF}(\mathfrak{M})$ not only for $\mathfrak{M} = \mathbb{R}$ but also for a wider class of structures which is called $\exists$–*Steinitz* or *existentially Steinitz* structures. To define this class of structures, first we need a definition of $\exists$–algebraic elements. Actually, this definition is the restriction of the usual model–theoretic definition of algebraic elements to $\exists$–formulas. More exactly,

Definition 4. *Let $\mathfrak{M}$ be a structure and A be some its subset. An element $a \in \mathfrak{M}$ is called $\exists$–algebraic over A ([10]) if there exists an $\exists$–$\varphi(x, \bar{y})$ and parameters $\bar{b} \in A$ such that the set $\{x \mid \mathfrak{M} \models \varphi(x, \bar{b})\}$ is finite and contains a.*

The set of all elements of $\mathfrak{M}$ $\exists$–algebraic over A is denoted by $\mathbf{C}_\exists^{\mathfrak{M}}(A)$. The operator $\mathbf{C}_\exists^{\mathfrak{M}}$ has the usual properties of closure operators. If, in addition it has the classical *exchange property*

if $a \in \mathbf{C}_\exists^{\mathfrak{M}}(A \cup \{b\}) \setminus \mathbf{C}_\exists^{\mathfrak{M}}(A)$ then $b \in \mathbf{C}_\exists^{\mathfrak{M}}(A \cup \{a\})$, then the structure $\mathfrak{M}$ is called $\exists$-*Steinitz*.

As usual, a set $X \subseteq \mathfrak{M}$ is called *independent* if for any $x \in X$ holds $x \notin \mathbf{C}_{\exists}^{\mathfrak{M}}(X \setminus \{x\})$.

The main reason for this definition is that in $\exists$-Steinitz structures for each subset S one can correctly define the dimension of S as the cardinality of any its maximal independent subsets. This notion works in the proofs of the results below.

Some examples of $\exists$-Steinitz structures are: the ordered field of real numbers $\mathbb{R}$, the field of complex numbers $\mathbb{C}$, any model of a strongly minimal model complete theory, model complete fields, and model complete ordered fields. Since both $\mathbb{R}$ and $\mathbb{C}$ admit quantifier elimination, in these fields, $\exists$-independence coincides with algebraic independence.

Now we can formulate the first metatheorem.

Theorem 8 ([10]). *Assume that $\mathfrak{M}$ is an $\exists$-Steinitz structure of finite signature and that $\mathfrak{A}$ is an arbitrary structure of finite signature for which there exists a family $(F_i)_{i<\omega}$ of unary termal operations definable in the language of $\mathfrak{A}$ with parameters (the set of parameters used in these formulas can be infinite) and a family $(A_i)_{i<\omega}$ of subsets of $\mathfrak{A}$ such that*

1. *all the sets $F_i[A_i]$ are uncountable*
2. *for each family $(a_i)_{i<\omega} \in \prod_{i<\omega} A_i$ there exists an element $b \in \mathfrak{A}$ such that for all $i < \omega$ holds $F_i(b) = F_i(a_i)$.*

Then $\mathfrak{A}$ has no embeddings into a structure possessing a simple Σ-presentation over $\mathbb{HF}(\mathfrak{M})$ with parameters.

The proof uses the idea that under the conditions of the theorem, the code of a single element a can produce an infinite set of elements algebraically dependent on it whose dimension is infinite, which is a contradiction.

It follows from this theorem that

Theorem 9 ([10]). *The following structures have no isomorphic embeddings into structures having simple Σ-presentations over an $\exists$-Steinitz structure of finite signature:*

1. *the Boolean algebra $\mathcal{P}(\omega)$ of all subsets of ω.*
2. *$\mathcal{P}(\omega)^*$ the factor of the Boolean algebra $\mathcal{P}(\omega)$ of all subsets of ω modulo the ideal of all finite sets.*
3. *the lattice of all open (closed) subsets of $\mathbb{R}^m$, $m > 0$.*
4. *$\mathrm{Sym}(\omega)$, the symmetric groups of ω (even in the signature without the inversion operation);*
5. *$\mathrm{Sym}(\omega)^*$, the quotient of the symmetric groups of ω modulo the subgroup of finitary permutations (even in the signature without the inversion operation).*
6. *ω^ω, the semigroup of all mappings from ω to ω.*
7. *the group of all permutations of $\mathbb{R}$ which are Σ-definable over $\mathbb{HF}(\mathbb{R})$ (even in the signature without inversion operation).*
8. *the semigroup of all mappings from $\mathbb{R}$ to $\mathbb{R}$ which are Σ-definable over $\mathbb{HF}(\mathbb{R})$*

9. *the automorphism group* Aut $\langle \mathbb{Q}, < \rangle$ *of the ordering on the rational numbers (even in the signature without the inversion operation).*
10. *the automorphism group* Aut $\langle \mathbb{R}, < \rangle$ *of the ordering on the reals (even in the signature without the inversion operation).*
11. *the semigroup* $\mathrm{C}(\mathbb{R}^n)$ *of all continuous mappings from* $\mathbb{R}^n$ *to* $\mathbb{R}^n$, *for each* $n > 0$.
12. *the semigroup* $\mathrm{C}^1(\mathbb{R}^n)$ *of all continuously differentiable mappings from* $\mathbb{R}^n$ *to* $\mathbb{R}^n$, *for each* $n > 0$.

The following metatheorem differs from the previous one in two aspects: here we consider Σ–definable operations, not only termal ones, but the payment for this is that we do not prove the absence of embeddings into simply Σ–presentable structures; we prove the absence of such presentations only.

Theorem 10 ([11]). *Suppose that* $\mathfrak{M}$ *is an* $\exists$–*Steinitz structure of finite signature. Let* $\mathfrak{A}$ *be an arbitrary structure for which there exists a family* $(F_i)_{i<\omega}$ *of unary partial functions definable by* Σ–*formulas with parameters over* $\mathbb{HF}(\mathfrak{A})$ *and the family* $(A_i)_{i<\omega}$ *of subsets of* $\mathfrak{A}$ *with the following properties:*

1. *for each* $i < \omega$ *holds* $A_i \subseteq \mathrm{dom}(F_i)$ *and* $F_i[A_i]$ *is uncountable*
2. *for each family* $(a_i)_{i<\omega} \in \prod_{i<\omega} A_i$, *there is a* $b \in \mathfrak{A}$ *such that for all* $i < \omega$ *holds* $F_i(b) = F_i(a_i)$.

Then $\mathfrak{A}$ *has no simple* Σ–*presentations over* $\mathbb{HF}(\mathfrak{M})$ *with parameters.*

Using this metatheorem, we can prove that there are no simple Σ–presentations for some structures related to nonstandard analysis:

Theorem 11 ([11])

1. *Let* D *be an arbitrary nonprincipal filter over* ω *and* $\mathfrak{M}$ *be an* $\exists$–*Steinitz structure of finite signature. Then the filtered power*

$$\langle \mathbb{R}^\omega / D, \mathrm{St}, \mathrm{Inf} \rangle$$

expanded with unary predicates St *and* Inf *, where* St *distinguishes standard elements, i.e., elements defined by constant functions from* $\mathbb{R}^\omega$, *and* Inf *distinguishes infinitesimal elements, i.e., elements situated in* $\mathbb{R}^\omega / D$ *between all negative and all positive standard elements, has no simple* Σ–*presentations over* $\mathbb{HF}(\mathfrak{M})$.
2. *Let* $\mathfrak{M}$ *be an* $\exists$–*Steinitz structure of finite signature and let* $^*\mathbb{R}$ *be an elementary extension of the ordered field of real numbers* $\mathbb{R}$ *containing infinitesimal elements, and assume that for each function* $f : \mathbb{R}^2 \to \mathbb{R}$ *there exists a function* $^*f : (^*\mathbb{R})^2 \to {}^*\mathbb{R}$ *such that* $\langle \mathbb{R}, f \rangle_{f \in \mathcal{F}}$ *is an elementary submodel of* $\langle {}^*\mathbb{R}, {}^*f \rangle_{f \in \mathcal{F}}$, *where* $\mathcal{F}$ *is the family of all binary functions on* $\mathbb{R}$. *Then the structure* $\langle {}^*\mathbb{R}, \mathrm{St}, \mathrm{Inf} \rangle$ *in which the unary predicate* St *distinguishes standard elements of* $^*\mathbb{R}$ *(i.e.,* $\mathbb{R}$*) and* Inf *distinguishes infinitesimal elements, (i.e.,* $\mathrm{Inf} = \bigcap_{i<\omega} [-\frac{1}{n}, \frac{1}{n}]$*), has no simple* Σ–*presentations over* $\mathbb{HF}(\mathfrak{M})$.

It follows that there is not much hope for the adequate use of nonstandard reals as abstract data types even in programming with the 'real' reals.

Nevertheless, there is one more unsolved question here:

Question 9. *Does there exist a nonstandard ordered field of cardinality 2^ω elementary equivalent to $\mathbb{R}$ with Σ-presentations over $\mathbb{HF}(\mathbb{R})$?*

The same question can be formulated about nonstandard elementary extensions of $\mathbb{R}$.

Question 10. *Do the fields of formal power series $\mathbb{R}[[t]]$ and $\mathbb{C}[[t]]$ have (simple) Σ-presentations over $\mathbb{HF}(\mathbb{R})$?*

The other interesting problems are:

Question 11. *Does the field of p-adic numbers $\mathbb{Q}_p$ have a (simple) Σ-presentations over $\mathbb{HF}(\mathbb{R})$?*

Question 12. *Does $\mathbb{R}$ have a (simple) Σ-presentation over $\mathbb{HF}(\mathbb{Q}_p)$?*

Generally, it would be interesting to understand which of the classical structures have Σ-presentations over the other structures.

References

1. Morozov, A.S., Korovina, M.V.: On Σ-definability of countable structures over real numbers, complex numbers, and quaternions. Algebra Logic **47**(3), 193–209 (2008)
2. Barwise, J.: Admissible Sets and Structures. Springer, Heidelberg (1975)
3. Ershov, Y.L.: Definability and Computability. Plenum Publ. Co., New York (1996)
4. Ershov, Y.L.: Σ-definability of algebraic structures. In: Studies in Logic and Foundations of Mathematics, vol. 1, pp. 235–260. Elsevier, Amsterdam (1998)
5. Korovina, M.V.: Generalized computability of functions on the reals. Vychislitel'nye Systemi (Computing Systems) **133**, 38–68 (1990). In Russian
6. Morozov, A.S.: On some presentations of the real number field. Algebra Logic **51**(1), 66–88 (2012)
7. Morozov, A.S.: On Σ-rigid presentations of the real order. Siberian Math. J. **55**(3), 562–572 (2014)
8. Morozov, A.S.: One-dimensional Σ-presentations of structures over $\mathbb{HFR}$. In: Geschke, P.S.S., Löwe, B. (eds.) Infinity, Computability, and Metamathematics, Festschrift celebrating the 60th birthdays of Peter Koepke and Philip Welch. Tributes Series, vol. 23, pp. 285–298. College Publications, London (2014)
9. Morozov, A.S.: Σ-presentations of the ordering on the reals. Algebra Logic **53**(3), 217–237 (2014)
10. Morozov, A.S.: A sufficient condition for nonpresentability of structures in hereditarily finite superstructures. Algebra Logic **55**, 242–251 (2016)
11. Morozov, A.S.: Nonpresentability of some structures of analysis in hereditarily finite superstructures. Algebra Logic (2017, to appear)
12. Korovina, M.V., Kudinov, O.V.: Positive predicate structures for continuous data. Math. Struct. Comput. Sci. **25**(Special Issue 08), 1669–1684 (2015)

13. Ershov, Y.L., Goncharov, S.S.: Constructive Models. Siberian School of Algebra and Logic. Kluwer Academic/Plenum, New York (2000)
14. Tarski, A.: A Decision Method for Elementary Algebra and Geometry. The Rand Corporation, Santa Monica (1957)
15. Weihrauch, K.: Computable Analysis. An Introduction, p. 285. Springer, Heidelberg (2000)
16. Ershov, Y.L., Puzarenko, V.G., Stukachev, A.I.: $\mathbb{HF}$-Computability. In: Cooper, S.B., Andrea, S. (eds.) Computability in Context. Computation and Logic in the Real World, pp. 169–242. Imperial College Press, London (2011)

The Lattice of Computably Enumerable Vector Spaces

Rumen D. Dimitrov[1] and Valentina Harizanov[2(✉)]

[1] Department of Mathematics, Western Illinois University,
Macomb, IL 61455, USA
rd-dimitrov@wiu.edu
[2] Department of Mathematics, George Washington University,
Washington, DC 20052, USA
harizanv@gwu.edu

Abstract. We survey fundamental notions and results in the study of the lattice of computably enumerable vector spaces and its quotient lattice modulo finite dimension. These lattices were introduced and first studied by Metakides and Nerode in the late 1970s and later extensively investigated by Downey, Remmel and others. First, we focus on the role of the dependence algorithm, the effectiveness of the bases, and the Turing degree-theoretic complexity of the dependence relations. We present a result on the undecidability of the theories of the above lattices. We show the development of various notions of maximality for vector spaces, and role they play in the study of lattice automorphisms and automorphism bases. We establish a new result about the role of supermaximal spaces in the quotient lattice automorphism bases. Finally, we discuss the problem of finding orbits of maximal spaces and the recent progress on this topic.

1 Computable and Computably Enumerable Vector Spaces

Computable model theory uses the tools of computability theory to investigate algorithmic content (effectiveness) of notions, theorems, and constructions in classical mathematics (see [28]). Computably enumerable vector spaces and computability-theoretic complexity of their bases were first considered by Mal'tsev in [40] and Dekker in [4]. Modern study of these spaces including the use of the priority method has been introduced by Metakides and Nerode in [43]. Computably enumerable vector spaces have been further investigated in computable model theory (see Downey and Remmel [26] and Nerode and Remmel [50]). For more recent developments in the study of effective vector spaces, see [9,11]. Many of the results about vector spaces can be generalized to certain

Harizanov was partially supported by the NSF grant 1202328, by the CCFF award and CCAS Dean's research chair award of the George Washington University, as well as by the Simons Foundation Collaboration Grants for Mathematicians.

© Springer International Publishing AG 2017
A. Day et al. (Eds.): Downey Festschrift, LNCS 10010, pp. 366–393, 2017.
DOI: 10.1007/978-3-319-50062-1_23

effective closure systems (see [26]). More recently, effective vector spaces have been also studied in the context of reverse mathematics.

We will now introduce some definitions and state basic facts about computable and c.e. vector spaces. As customary in model theory, for a structure $\mathcal{A}$ we often use A to denote both the structure and its domain.

Definition 1. *Let $(F, +, \cdot)$ be a computable field and $(V, +, \cdot, \equiv)$ a structure, $V \subseteq \omega$, with a partial computable binary operation $+$ defined on $V \times V$ and a partial computable binary operation $\cdot$ defined on $F \times V$, and a congruence relation $\equiv \subseteq V \times V$ such that the quotient structure $\frac{V}{\equiv}$ is a vector space over F with vector addition induced by $+$ and scalar multiplication induced by $\cdot$.*

  (i) *The structure $\frac{V}{\equiv}$ is a c.e. vector space given by $(V, +, \cdot, \equiv)$ if V is a c.e. set and $\equiv$ is a c.e. relation.*

 (ii) *The structure $\frac{V}{\equiv}$ is a computable vector space given by $(V, +, \cdot, \equiv)$ if V is a c.e. set, $\equiv$ is a c.e. relation, and the relation $(V \times V) - \equiv$ is also c.e.*

(iii) *The structure $\frac{V}{\equiv}$ is a normal vector space given by $(V, +, \cdot, \equiv)$ if V is a c.e. set and the relation $\equiv$ is the equality, $=$.*

We usually do not write the equality explicitly. Every vector space can be thought of as a quotient space with the congruence relation being the equality. A normal vector space $(V, +, \cdot)$ has a c.e. set of vectors V, a partial computable vector addition $+$, and a partial computable scalar multiplication $\cdot$. Furthermore, since the equality is a computable binary relation on ω, both the equality on V and the inequality on V are c.e. relations. Hence every normal vector space is computable.

Example 2. *Let F be a computable field. Define*

$$V_\infty = \{u \in F^\omega : (\exists n_s)(\forall n \geq n_s)[u(n) = 0]\}.$$

Then V_∞ is a (normal) vector space with domain V_∞ and pointwise operations of vector addition and scalar multiplication of vectors. The set of vectors $E = \{\varepsilon_i \in F^\omega : i \in \omega\}$, where

$$\varepsilon_i(n) = \begin{cases} 1 & if\ n = i, \\ 0 & if\ n \neq i, \end{cases}$$

forms a computable basis for V_∞. We will call this basis a standard basis.

Thus, V_∞ is an $\aleph_0$-dimensional computable vector space. Its computable first-order language is $\{+, \{\cdot_f\}_{f \in F}\}$. It has a computable basis and hence a dependence algorithm. Intuitively, a *dependence algorithm* is an effective procedure for deciding whether a finite tuple of vectors is linearly dependent.

Lemma 3. *Every c.e. basis of V_∞ is computable.*

Proof. Assume that B is a c.e. basis of V_∞. Let $b_0, b_1, b_2, \ldots$ be a computable enumeration of B. Let $v \in V_\infty$. We effectively find $\lambda_{i_0}, \ldots, \lambda_{i_{n-1}} \in F - \{0\}$ such that

$$v = \lambda_{i_0} b_{i_0} + \cdots + \lambda_{i_{n-1}} b_{i_{n-1}}.$$

Then we have

$$v \in B \Leftrightarrow (n = 1 \wedge v = b_{i_0}).$$

∎

For any set $I \subseteq V_\infty$, by $cl(I)$ we denote the smallest (with respect to inclusion) subspace of V_∞ containing I; that is, $cl(I)$ is the linear span of I. A subspace V of V_∞ is c.e. if its domain V is a c.e. subset of V_∞. The set of all c.e. subspaces of V_∞ is denoted by $\mathcal{L}(V_\infty)$.

Example 4. *Let $W \in \mathcal{L}(V_\infty)$. Let the congruence relation $\equiv_W$ on V_∞ be defined by*

$$x \equiv_W y \Leftrightarrow x - y \in W.$$

Clearly, $\equiv_W$ is a c.e. relation because W is a c.e. set. Hence the quotient space $\frac{V_\infty}{W}$ is a c.e. vector space. If W is computable, then $\frac{V_\infty}{W}$ is a computable vector space.

Let $\underset{=}{V}$ and $\underset{=}{V'}$ be c.e. vector spaces, and let $f : \underset{=}{V} \to \underset{=}{V'}$ be a vector space isomorphism. Then we say that f is computable if the relation

$$\{(u, v) \in V \times V' : f([u]_\equiv) = [v]_{\equiv'}\}$$

is c.e.

Proposition 5. *Every c.e. vector space $\underset{=}{V}$ is computably isomorphic to $\frac{V_\infty}{W}$ for some $W \in \mathcal{L}(V_\infty)$. If $\underset{=}{V}$ is a computable vector space, then W is computable.*

Proof. Let $v_0, v_1, \ldots$ be a computable enumeration of V. Define $f : V_\infty \to \underset{=}{V}$ by $(\forall i)[f(\varepsilon_i) = [v_i]_\equiv]$ so that f is a linear function from V_∞ to $\underset{=}{V}$. Clearly, f is onto. Let $W =_{def} \ker(f) = \{v \in V_\infty : f(v) = [0]_\equiv\}$. Then W is a c.e. subspace of V_∞. If $\underset{=}{V}$ is computable, then W is also computable. Let an isomorphism $g : \frac{V_\infty}{W} \to \underset{=}{V}$ be defined by

$$g(v + W) = [f(v)]_\equiv.$$

Clearly, g is a computable isomorphism. ∎

Lemma 6. *Every computable vector space $\underset{=}{V}$ is computably isomorphic to a normal vector space.*

Proof. Let $\underset{=}{V}$ be a computable vector space given by $(V, +, \cdot, \equiv)$. Assume that $v_0, v_1, v_2, \ldots$ is a computable enumeration of V. Define $W = \{v_i : (\forall j < i) \neg[v_i \equiv v_j]\}$. The set W is a computable subset of V. Clearly, $(W, +, \cdot, \equiv)$ is a normal vector space. Let $f : \underset{=}{V} \to W$ be a linear function given by $f([v_n]_\equiv) = v_i$, where v_i is the unique element in W such that $v_n \equiv v_i$. Then f is a computable isomorphism. ∎

We will now discuss the structure on $\mathcal{L}(V_\infty)$. A lattice is a structure L in the language $\{\leq, \vee, \wedge\}$ such that $\leq$ is a partial order, and $\vee$ and $\wedge$ are supremum and infimum, respectively. If a lattice has the greatest element and the least element, then they are denoted by 1 and 0, respectively. If L is a lattice with 1, then $a \in L$ is called a *co-atom* (dual atom) if

$$a < 1 \wedge (\forall b \in L)\,[a < b \Rightarrow b = 1].$$

As usual, by $\mathcal{E}$ we denote the lattice of all c.e. subsets of ω.

Let $U, V \in \mathcal{L}(V_\infty)$. Then $U \cap V$ is the subspace with domain $U \cap V$, and $U + V$ is the subspace with domain

$$U + V = \{u + v : u \in U \wedge v \in V\}.$$

By $Y = U \oplus V$ we denote that $Y = U + V$ and $U \cap V = \{0\}$. We write $U \subseteq V$ if U is a subspace of V. Consider the lattice $(\mathcal{L}(V_\infty), \subseteq, \cap, +, \{0\}, V_\infty)$. The lattice $\mathcal{L}(V_\infty)$ modulo finite dimension is denoted by $\mathcal{L}^*(V_\infty)$.

For $A, B \in \mathcal{E}$ we will use $A =^* B$ to denote that the symmetric difference $A \triangle B$ is a finite set. Similarly, for $U, V \in \mathcal{L}(V_\infty)$ we write $U =^* V$ if there is a finite-dimensional subspace W such that $U + W = V + W$. This means that $cl(U \cup P) = cl(V \cup Q)$ for some finite sets of vectors P and Q. Hence $\mathcal{E}^* = (\mathcal{E}/_{=^*})$ and $\mathcal{L}^*(V_\infty) = (\mathcal{L}(V_\infty)/_{=^*})$. Clearly, each of the lattices $\mathcal{E}$, $\mathcal{E}^*$, $\mathcal{L}(V_\infty)$, and $\mathcal{L}^*(V_\infty)$ has both 1 and 0.

The structure and automorphisms of $\mathcal{L}(V_\infty)$ and $\mathcal{L}^*(V_\infty)$ have been studied extensively. The approach, in general, has been modelled upon the study of the distributive lattices $\mathcal{E}$ and $\mathcal{E}^*$ in computability theory. However, the study of $\mathcal{L}(V_\infty)$ and $\mathcal{L}^*(V_\infty)$ follows a more geometric approach because these lattices are modular and nondistributive. For more on lattice theory see [1].

Proposition 7. *The structure $\mathcal{L}(V_\infty)$ is a modular nondistributive lattice.*

Proof. To prove that the lattice $\mathcal{L}(V_\infty)$ is modular, we will show that

$$U \subseteq V \Rightarrow [(W + U) \cap V = (W \cap V) + U].$$

Let $U, V, W \in \mathcal{L}(V_\infty)$, where $U \subseteq V$. It is easy to see that then $(W \cap V) + U \subseteq (W + U) \cap V$. Now, let $v \in (W + U) \cap V$. Then $v = w + u$ for some $w \in W$ and $u \in U$. Hence, $w = v - u$, so, since $U \subseteq V$, $w \in V$. Thus, $w + u \in (W \cap V) + U$, i.e., $v \in (W \cap V) + U$.

To show that $\mathcal{L}(V_\infty)$ is not distributive, choose two (nonzero) independent vectors, u and v. Consider the following three subspaces: $cl(\{u\})$, $cl(\{v\})$ and $cl(\{u + v\})$. Then

$$(cl(\{u\}) + cl(\{v\})) \cap cl(\{u + v\}) = cl(\{u + v\}),$$

but

$$(cl(\{u\}) \cap cl(\{u + v\})) + (cl(\{v\}) \cap cl(\{u + v\})) = \{0\}.$$

■

Let $I_0, I_1, I_2, \ldots$ be a fixed effective enumeration of all *c.e. independent* subsets of V_∞. For $e \in \omega$, let

$$V_e =_{def} cl(I_e).$$

Hence, $V_0, V_1, V_2, \ldots$ is a fixed effective enumeration of all c.e. subspaces of V_∞. For $s \in \omega$, let $V_{e,s} =_{def} cl(I_{e,s})$. Hence $V_e = \bigcup_{s \in \omega} V_{e,s}$.

Proposition 8. *Let V be a c.e. vector space. If V has a c.e. basis, then V has a dependence algorithm.*

Proof. Assume that V has a c.e. basis $b_0, b_1, \ldots$ Let $u_0, \ldots, u_{n-1} \in V$. Effectively find the least $k \in \omega$ and $\alpha_{ij} \in F$, for $i \in \{0, \ldots, n-1\}$ and $j \in \{0, \ldots, k-1\}$, such that $u_i = \sum_{j=0}^{k-1} \alpha_{ij} b_j$. Form a matrix $M = [\alpha_{ij}]_{n \times k}$, and algorithmically find the rank of M. Then $u_0, \ldots, u_{n-1}$ are linearly dependent iff $rank(M) < n$. ∎

Theorem 9. *Let V be a c.e. vector space. If V has a dependence algorithm, then V has a computable basis.*

Proof. If V is finite-dimensional, then every basis of V is computable. Therefore, we assume that V is infinite-dimensional. Let $b_0, b_1, b_2, \ldots$ be an effective enumeration of a c.e. basis of V. We will enumerate a computable basis $a_0, a_1, a_2, \ldots$ of V. As usual, assume that $V \subseteq \omega$ with the usual ordering $<$. Inductively, let $a_0, \ldots, a_{2n}$ be defined such that

$a_0, \ldots, a_{2n}$ are linearly independent,
$b_{n-1} \in cl(\{a_0, \ldots, a_{2n}\})$, and
$a_0 < \cdots < a_{2n}$.

We will now extend the sequence $a_0, \ldots, a_{2n}$ by defining a_{2n+1} and a_{2n+2}. We first effectively check whether $b_n \in cl(\{a_0, \ldots, a_{2n}\})$.

If $b_n \in cl(\{a_0, \ldots, a_{2n}\})$, then we choose the least two vectors $b, d \in V$ such that $a_{2n} < b < d$, and $a_0, \ldots, a_{2n}, b, d$ are linearly independent. Let $a_{2n+1} =_{def} b$ and $a_{2n+2} =_{def} d$.

Assume that $b_n \notin cl(\{a_0, \ldots, a_{2n}\})$. Choose the least vector $x \in V$ such that $x > \max\{a_{2n}, a_{2n} - b_n\}$ and $a_0, \ldots, a_{2n}, b_n, x$ are linearly independent. Such x exists because V is infinite-dimensional. Hence, $b_n + x > a_{2n}$ and $a_0, \ldots, a_{2n}, x, b_n + x$ are linearly independent. We define a_{2n+1} and a_{2n+2} such that $\{a_{2n+1}, a_{2n+2}\} = \{x, b_n + x\}$.

If the underlying field for V_∞ is infinite, then there is an easier way to obtain a computable basis for V. Namely, we can choose $k_1, k_2, \ldots \in F$ such that $b_0 < k_1 b_1 < k_2 b_2 < \cdots$. Then $\{b_0, k_1 b_1, k_2 b_2, \ldots\}$ is a computable basis for V. ∎

Hence, if $V \in \mathcal{L}(V_\infty)$, then V has a computable basis, as first established by Dekker [4]. Metakides and Nerode further showed that V has a c.e. basis B such that $V \equiv_T B$. As usual, we use $\leq_T$ for Turing reducibility and $\equiv_T$ for Turing equivalence of sets. The Turing degree of X is denoted by $\deg(X) = \mathbf{x}$, the nth Turing jump of X by $X^{(n)}$, and $\mathbf{x}^{(n)} = \deg(X^{(n)})$. In particular, $\mathbf{0}'$ denotes the

Turing degree of the halting set $\emptyset'$. The Turing degrees form an upper semilattice. For more on computability theory see [56].

The result of classical mathematics that every independent set of vectors can be extended to a basis of the whole vector space does not effectivize. That is, some independent sets cannot be extended to c.e. independent sets by adding infinitely many vectors.

Let $J \subseteq V_\infty$ be an independent set. The set J is called *nonextendible* if $\dim \frac{V_\infty}{cl(J)} = \infty$ and for every $e \in \omega$:

$$J \subseteq I_e \Rightarrow |I_e - J| < \infty.$$

Otherwise, the independent set J is called *extendible*. Metakides and Nerode [43] showed that there is a c.e. nonextendible independent subset J of V_∞. We say that a c.e. subspace V has a *(fully) extendible* basis if some c.e. basis of V can be extended to a c.e. basis of V_∞.

Theorem 10 *(Metakides and Nerode [43]). Let V_∞ be over any computable field. Then there is a c.e. subspace space V of V_∞ such that no basis of V is fully extendible.*

2 Dependence Relation and k-Dependence Relations

We have already considered a dependence algorithm. Now, we formally introduce dependence relations. Let $V \subseteq V_\infty$. The *dependence relation over V*, in symbols $D(V)$, is defined by

$$D(V) = \{(u_0, \ldots, u_{k-1}) : k \in \omega \wedge u_0, \ldots, u_{k-1} \in V_\infty \wedge$$
$$(u_0, \ldots, u_{k-1} \text{ are linearly dependent over } V)\}.$$

Since for $v \in V_\infty$, we have $v \notin V$ iff $v \in D(V)$, it follows that

$$V \leq_T D(V)$$

Hence, if $D(V)$ is computable, then V is computable. The *dependence degree* of V is the Turing degree of $D(V)$, $deg(D(V))$. A space V is called *decidable* if its dependence degree is $\mathbf{0}$, that is, $D(V)$ is a computable set. Equivalently, V is decidable if $\frac{V_\infty}{V}$ has a dependence algorithm.

Proposition 11. *Let V_∞ be a vector space over a finite computable field F. Then, for $V \in \mathcal{L}(V_\infty)$, we have*

$$V \equiv_T D(V).$$

Proof. It is enough to show that $D(V) \leq_T V$. Let $|F| = n$. For any given $v_0, \ldots, v_{k-1} \in V$, there are $(n^k - 1)$ nontrivial linear combinations. To determine whether $v_0, \ldots, v_{k-1}$ are linearly dependent, list all nontrivial linear combinations, and use oracle V to test whether any of them belongs to V. ∎

Proposition 12. *Let* V, W *be vector subspaces of* V_∞ *such that* $V \subseteq W$ *and* $\dim \frac{W}{V} < \infty$.

(i) *Then*
$$D(W) \leq_T D(V).$$

(ii) *If, in addition,* $V, W \in \mathcal{L}(V_\infty)$, *then*
$$D(V) \leq_T D(W).$$

Proof. (i) Assume that $\dim \frac{W}{V} = k$ and let $w_0 + V, \ldots, w_{k-1} + V$ be a basis for $\frac{W}{V}$. Let $u_0, \ldots, u_{n-1} \in V_\infty$.

We have
$$(u_0, \ldots, u_{n-1}) \in D(W) \text{ iff}$$

$$(\exists \alpha_0, \ldots, \alpha_{n-1} \in F)(\exists w \in W)[\alpha_0 u_0 + \cdots + \alpha_{n-1} u_{n-1} = w] \text{ iff}$$

$$(\exists \alpha_0, \ldots, \alpha_{n-1} \in F)(\exists \beta_0, \ldots, \beta_{k-1} \in F)(\exists v \in V)[\alpha_0 u_0 + \cdots + \alpha_{n-1} u_{n-1} = \beta_0 w_0 + \cdots + \beta_{k-1} w_{k-1} + v] \text{ iff}$$

$$(u_0, \ldots, u_{n-1}, w_0, \ldots, w_{k-1}) \in D(V).$$

Hence $D(W) \leq_T D(V)$. ∎

Metakides and Nerode proved that if the (computable) field F for V_∞ is infinite then for an arbitrary c.e. Turing degree $\mathbf{c}$, there is a computable vector subspace V of V_∞ such that
$$\deg(D(V)) = \mathbf{c}.$$

Proposition 8 can be easily generalized to quotient c.e. vector spaces. It can also be relativized. Namely, we have the following proposition.

Proposition 13. *Let* $V \in \mathcal{L}(V_\infty)$.

(i) *Then* $\frac{V_\infty}{V}$ *has a dependence algorithm iff* $\frac{V_\infty}{V}$ *has a c.e. basis.*

(ii) *Let* $C \subseteq \omega$. *Then* $D(V) \leq_T C$ *iff* $\frac{V_\infty}{V}$ *has a basis that is computable in* C.

Let $V \in \mathcal{L}(V_\infty)$. Then we say that V is a *complemented* element of $\mathcal{L}(V_\infty)$ if there exists $W \in L(V_\infty)$ such that $V \oplus W = V_\infty$.

Theorem 14 *(Metakides and Nerode [43]). Let* $V \in \mathcal{L}(V_\infty)$. *Then the following conditions are equivalent.*

(i) *The space* V *is decidable.*

(ii) *Every c.e. basis of* V *is extendible to a computable basis of* V_∞.

(iii) *The space* V *has a computable basis that is extendible to a computable basis of* V_∞.

(iv) *The space* V *is a complemented element in* $\mathcal{L}(V_\infty)$.

Proof. (i)$\Rightarrow$(ii) Let A be a c.e. basis for V. Assume that V is decidable. Thus $\frac{V_\infty}{V}$ has a dependence algorithm, and hence a c.e. basis. Let $b_0 + V, b_1 + V, b_2 + V, \ldots$ be a computable enumeration of a basis for $\frac{V_\infty}{V}$. Let $B = \{b_0, b_1, b_2, \ldots\}$. Then $A \cup B$ is a c.e. basis, and hence a computable basis of V_∞.

(ii)$\Rightarrow$(iii) Since $V \in \mathcal{L}(V_\infty)$, V has a computable basis. Let B be a computable basis for V. Extend B to a computable basis for V_∞.

(iii)$\Rightarrow$(iv) Assume that V has a computable basis B that is extendible to a computable basis A for V_∞. Let $W = cl(A - B)$. Then $W \in \mathcal{L}(V_\infty)$, $V \cup W = V_\infty$ and $V \cap W = \{0\}$.

(iv)$\Rightarrow$(i) Assume that $V, W \in \mathcal{L}(V_\infty)$, where $V \oplus W = V_\infty$. Since $W \in \mathcal{L}(V_\infty)$, W has a c.e. basis B. Then $\{b + V : b \in B\}$ is a c.e. basis for $\frac{V_\infty}{V}$. Hence, $\frac{V_\infty}{V}$ has a dependence algorithm. ∎

The set of all decidable subspaces of V_∞ is denoted by $\mathcal{S}(V_\infty)$. In the next proposition we will establish that the structure $(\mathcal{S}(V_\infty), \subseteq, \cap, +, \{0\}, V_\infty)$ is a lower semilattice.

Proposition 15. *Let $V_0, V_1 \in \mathcal{S}(V_\infty)$. Then $V_0 \cap V_1 \in \mathcal{S}(V_\infty)$.*

Proof. Let $\vec{v} = (v_0, \ldots, v_{n-1}) \in (V_\infty)^n$ for some $n \in \omega$. We will present an algorithm that decides whether $\vec{v}$ is dependent over $V_0 \cap V_1$, equivalently, whether $cl(\vec{v}) \cap (V_0 \cap V_1) \neq \{0\}$ (where $cl(\vec{v}) =_{def} cl(rng(\vec{v}))$). If $cl(\vec{v}) \cap V_0 = \{0\}$ (that is, $\vec{v}$ is independent over V_0), then $\vec{v}$ is independent over $V_0 \cap V_1$. Assume that $cl(\vec{v}) \cap V_0 \neq \{0\}$. Now, we effectively compute a basis B of $cl(\vec{v}) \cap V_0$ in the following way. We find the least $z_0 \in V_0 - \{0\}$ such that $z_0 \in cl(\vec{v})$. Exchange z_0 with the first appropriate v_i. Now check whether $(v_0, \ldots, v_{i-1}, z_0, v_{i+1}, \ldots, v_{n-1})$ is independent over V_0. If it is, we stop. Otherwise, we look for the least $z_1 \in V_0 \cap cl(\vec{v})$ such that z_1 is independent of z_0 over V_0. We continue until we find the basis $B = \{z_0, \ldots, z_{m-1}\}$. Now, $\vec{v}$ is dependent over $V_0 \cap V_1$ iff B is dependent over V_1. ∎

Theorem 16 *(Ash and Downey [3]). Let $U, V, W \in \mathcal{L}(V_\infty)$ be such that $\dim(U) = \infty$ and $U \oplus V = W$. Then there exists $D \in \mathcal{S}(V_\infty)$ such that $U \oplus D = W$.*

As a corollary we obtain that if $U \in \mathcal{S}(V_\infty)$ and $W \in \mathcal{L}(V_\infty)$ are such that $\dim(U) = \infty$ and $U \subseteq W$, then there exists $D \in \mathcal{S}(V_\infty)$ such that $U \oplus D = W$. Furthermore, we have the following result.

Theorem 17 *(Ash and Downey [3]). For every $W \in \mathcal{L}(V_\infty)$, there are $D_0, D_1 \in \mathcal{S}(V_\infty)$ such that $D_0 \oplus D_1 = W$.*

Let $A, B \in \mathcal{L}(V_\infty)$ be such that $B \subseteq A$ and $\dim \frac{A}{B} = \infty$. Kalantari defined the space B to be a *major subspace* of A if for every $e \in \omega$:

$$(V_e + A = V_\infty) \Rightarrow (V_e + B =^* V_\infty).$$

Guichard defined the space B to a *supermajor subspace* of A if for every $e \in \omega$:

$$(V_e + A = V_\infty) \Rightarrow (V_e + B = V_\infty).$$

Theorem 18 *(Guichard [31]). Let A be a nondecidable c.e. subspace of V_∞. Then there is a supermajor subspace of A.*

For any $V \subseteq V_\infty$ and $k \geq 1$, let

$$D_k(V) =_{def} \{(x_0, \ldots, x_{k-1}) : x_0, \ldots, x_{k-1} \text{ are linearly dependent over } V\}.$$

The $k-th$ *dependence degree* of V is the Turing degree of $D_k(V)$. Therefore, $D(V) =_{def} \bigcup_{k \geq 1} D_k(V)$. We can easily establish the following facts.

(i) Uniformly in k, $D_k(V) \leq_T D(V)$.
(ii) Assume that $\dim(\frac{V_\infty}{V}) = \infty$. Then $D_k(V) \leq_T D_{k+1}(V)$.
(iii) If $V \in \mathcal{L}(V_\infty)$, then $D_k(V)$ is a c.e. set.

The next lemma will be used to establish the theorem that follows it.

Lemma 19 *(Shore [54]). Assume that V is a finite-dimensional subspace of V_∞. Let $k \in \omega$, and let the vectors $v_0, \ldots, v_k$ be linearly independent over V. Assume that X is a finite set of tuples of vectors of length $\leq k$ such that every tuple from X is independent over V. Then there are scalars $\lambda_0, \ldots, \lambda_k$ such that every tuple from X is still independent over $cl(V \cup \{\lambda_0 v_0 + \cdots + \lambda_k v_k\})$.*

Theorem 20 *(Shore [54]). Let the space V_∞ be over an infinite (computable) field. Assume that $E_1, E_2, E_3, \ldots, E_0$ is a c.e. sequence of c.e. sets such that $E_k \leq_T E_{k+1}$ and $E_k \leq_T E_0$, uniformly in k. Then there is a c.e. subspace V such that for every $k \geq 1$,*

$$D_k(V) \equiv_T E_k \wedge D(V) \equiv_T E_0.$$

Let V be a computable vector space. Its *computable automorphism group*, $Aut_0(V)$, consists of all computable automorphisms of V. An automorphism f of a vector space V is *trivial* if it maps every 1-dimensional subspace of V into itself. That is, $f = f_\alpha$ for some $\alpha \in F - \{0\}$ where

$$(\forall v \in V)[f_\alpha(v) = \alpha v].$$

Hence f also maps every subspace of V into itself. A computable vector space is called *computably rigid* if its computable automorphism group is trivial. Morozov [44] constructed a computable vector space V such that $\frac{V_\infty}{V}$ is computably rigid. We will now assume that the computable field F is infinite. In [44], Morozov asked whether it is possible to obtain for every $k \geq 2$, a computable vector space V such that $\frac{V_\infty}{V}$ is computably rigid, has the k-dependence algorithm *mod V*, does not have the $(k+1)$-dependence algorithm *mod V*, and its dependence algorithm *mod V* has an arbitrary nonzero c.e. Turing degree. Clearly, if $deg(D(V)) = \mathbf{0}$, then $\frac{V_\infty}{V}$ has a computable basis, and hence the computable automorphism group of $\frac{V_\infty}{V}$ is nontrivial. We have the following lemma for the nontrivial automorphisms of vector spaces.

Lemma 21 *(Dimitrov, Harizanov and Morozov [10]). Let ψ be a total function such that $\psi : V_\infty \to V_\infty$. If ψ does not induce a trivial automorphism of $\frac{V_\infty}{V}$, then one of the following conditions hold:*

(1) There exist $u, v \in V_\infty$ and $\alpha, \beta \in F$ such that

$$\psi(\alpha u + \beta v) \neq_{mod \ V} \alpha\psi(u) + \beta\psi(v),$$

(2) There exists $w \in V_\infty - V$ such that $\psi(w) \in V$,

(3) There exists $w \in V_\infty - V$ such that the set $\{w, \psi(w)\}$ is independent mod V.

In [10], Morozov's question was answered positively by establishing a more general result.

Theorem 22 *(Dimitrov, Harizanov and Morozov [10]). Let E_0 be a noncomputable c.e. set, and let $E_1, E_2, E_3, \ldots$ be a c.e. sequence of c.e. sets such that E_1 is computable, and*

$$E_1 \leq_T \cdots \leq_T E_k \leq_T E_{k+1} \leq_T \cdots \leq_T E_0,$$

uniformly in k. Then there is a computable subspace V of V_∞ such that $\frac{V_\infty}{V}$ is computably rigid, and for $k \geq 1$,

$$D_k(V) \equiv_T E_k \wedge D(V) \equiv_T E_0.$$

3   Maximal Vector Spaces

We now introduce the notion of a maximal vector space, which is analogous to the notion of a maximal set in classical computability theory. Maximal sets have been extensively studied within the lattice $\mathcal{E}$ of c.e. sets. Recall that an infinite set $C \subseteq \omega$ is *cohesive* if for every c.e. set W, either $W \cap C$ or $\overline{W} \cap C$ is finite. A set $M \subseteq \omega$ is *maximal* if M is c.e. and $\overline{M}$ is cohesive. Equivalently, a set $M \in \mathcal{E}$ is *maximal* if $\overline{M}$ is infinite and

$$(\forall E \in \mathcal{E})\,[(M \subseteq E \wedge |E - M| = \infty) \Rightarrow (\overline{E} =^* \omega)].$$

For $X \in \mathcal{E}$ as well as for $X \in \mathcal{L}(V_\infty)$ we will use $[X]$ to denote the equivalence class of X modulo the corresponding equivalence relation $=^*$. Hence $[M]$ is a coatom in $\mathcal{E}^*$. A maximal set was first constructed by Friedberg. Soare established that for any two maximal sets M_1 and M_2, there is an automorphism Φ of $\mathcal{E}$ such that $\Phi(M_1) = M_2$ (see [56]). A set $B \subseteq \omega$ is *quasimaximal* if it is the intersection of finitely many maximal sets, $B = \bigcap_{i=1}^{n} M_i$ where M_i's are maximal. The number n is called the *rank* of B.

Definition 23. *Let $V \in \mathcal{L}(V_\infty)$. The subspace V is maximal if $\dim(\frac{V_\infty}{V}) = \infty$ and for every c.e. space W such that $V \subseteq W$, we have that*

$$\dim(\frac{V_\infty}{W}) < \infty \vee \dim(\frac{W}{V}) < \infty.$$

Hence, a subspace $V \in \mathcal{L}(V_\infty)$ is maximal if its equivalence class $[V]$ is a co-atom in $\mathcal{L}^*(V_\infty)$. Metakides and Nerode [43] showed that a maximal space can be constructed by modifying the e-state construction of a maximal set. For $v \in V_\infty$ and $e \in \omega$, the e-state of v is the following string in $\{0,1\}^{e+1}$: $(V_0(v), \ldots, V_e(v))$. If a computable basis of V_∞ is identified with the set ω, then maximal sets generate maximal spaces.

Theorem 24 *(Shore, see [43]). Let M be a maximal subset of a computable basis B of V_∞. Then M^* is a maximal subspace of V_∞.*

There are stronger notions of maximality for vector spaces.

Definition 25. *Let $V \in \mathcal{L}(V_\infty)$.*

(i) *The subspace V is* supermaximal *if* $\dim(\frac{V_\infty}{V}) = \infty$ *and for every c.e. space W such that $V \subseteq W$, we have that*

$$V_\infty = W \vee \dim(\frac{W}{V}) < \infty.$$

(ii) *The subspace V is* strongly supermaximal *if* $\dim(\frac{V_\infty}{V}) = \infty$ *and for every c.e. set X contained in $V_\infty - V$, there are $a_0, \ldots, a_{n-1} \in V_\infty$ such that*

$$X \subseteq cl(V \cup \{a_0, \ldots, a_{n-1}\}).$$

Clearly, every supermaximal space is maximal. The existence of a supermaximal space was first established by Kalantari and Retzlaff [36].

Theorem 26 *(Kalantari and Retzlaff [36]). There is a maximal space that is not supermaximal.*

Theorem 27 *(Nerode and Remmel [49]). Let the space V_∞ be over an infinite field. Let $k \geq 1$. Assume that $E_1, E_2, E_3, \ldots, E_0$ is a c.e. sequence of c.e. sets such that E_0 is non-computable, $E_k \leq_T E_{k+1}$ and $E_k \leq_T E_0$. Then there are supermaximal non-automorphic subspaces V and W such that*

$$D(V) \equiv_T D(W) \equiv_T E_0 \text{ and}$$

$$D_k(V) \equiv_T D_k(W) \equiv_T E_k.$$

Let V be a vector space with a basis J. Let $v \in V$. The support of v with respect to J, in symbols $supp_J(v)$, is the set of all vectors appearing in the linear combination of vectors in J, which equals v.

Theorem 28 *(Downey and Hird [19]). There is a strongly supermaximal vector space.*

Proof. Let $\varepsilon_0, \varepsilon_1, \varepsilon_2, \ldots$ be an effective enumeration of a computable basis for V_∞. At every stage $s \geq 0$, we will have a finite set J_s of linearly independent vectors and an effective enumeration $b_0^s, b_1^s, b_2^s, \ldots$ of a computable set of linearly independent vectors such that $J_s \cup \{b_0^s, b_1^s, b_2^s, \ldots\}$ is a basis for V_∞. At the end of the construction we will define $J = \bigcup_{s \geq 0} J_s$ and show that J is a basis of a strongly supermaximal vector space V. That is, $V =_{def} cl(J)$. We will satisfy the following requirements for every $e \in \omega$,

$$P_e : (W_e \cap cl(J) = \emptyset) \Rightarrow (W_e \subseteq cl(J \cup \{b_0, \ldots, b_{e-1}\})),$$

$$N_e : b_e = \lim_{s \to \infty} b_e^s \text{ exists.}$$

The positive requirements P_e, $e \in \omega$, ensure that the space V is supermaximal. The negative requirements N_e, $e \in \omega$, ensure that $\dim(\frac{V_\infty}{V})$ is infinite. The priority ordering of the requirements is

$$P_0, N_0, P_1, N_1, \ldots$$

We say that P_e *requires attention* at stage $s + 1$ if

$$W_{e,s+1} \cap J_s = \emptyset, \text{ and}$$

$$W_{e,s+1} - cl(J_s \cup \{b_0^s, \ldots, b_{e-1}^s\}) \neq \emptyset.$$

Construction of J.

Stage 0. Let $J_0 = \emptyset$, and $b_i^0 = \varepsilon_i$ for $i \in \omega$.

Stage $s + 1$. If no positive requirement requires attention at stage $s + 1$, define $J_{s+1} = J_s$ and $b_i^{s+1} = b_i^s$.

Now assume that P_e is the first requirement that requires attention at $s + 1$. Let v be the least element such that $v \in W_{e,s+1}$ and $v \notin cl(J_s \cup \{b_0^s, \ldots, b_{e-1}^s\})$. Let

$$J_{s+1} =_{def} J_s \cup \{v\}.$$

Let j be the least number such that $j \geq e$ and

$$b_j^s \in supp_{J_s \cup \{b_0^s, b_1^s, \ldots\}}(v).$$

That is,

$$v = a + k_0^s b_0^s + \cdots + k_{e-1}^s b_{e-1}^s + k_j^s b_j^s + k_{j+1}^s b_{j+1}^s + \cdots,$$

where $a \in J_s$ and $k_j^s \neq 0$. Define

$$b_n^{s+1} = \begin{cases} b_n^s & \text{if } n < j; \\ b_{n+1}^s & \text{if } n \geq j. \end{cases}$$

End of construction.   ■

Proposition 29 *(Downey and Hird [19]). Every strongly supermaximal vector space is supermaximal.*

Proof. Assume that V is a strongly supermaximal space, which is not supermaximal. Let a c.e. space W be such that $V \subseteq W$, $V_\infty \neq W$ and $\dim(\frac{W}{V})$ is infinite. Choose $u \in V_\infty - W$, and let $w_0, w_1, w_2, \ldots$ be an effective enumeration of W. For every $i \in \omega$, we have $u + w_i \notin W$, since $u = (u + w_i) - w_i$, and $w_i \in W$ and $u \notin W$. Let $X =_{def} \{u, u + w_0, u + w_1, u + w_2, \ldots\}$. Thus,

$$X \subseteq V_\infty - W \subseteq V_\infty - V.$$

However,

$$W \subseteq cl(X).$$

Note that since X is a c.e. set and V is a strongly supermaximal space, there are $a_0, \ldots, a_{n-1} \in V_\infty$ such that

$$X \subseteq cl(V \cup \{a_0, \ldots, a_{n-1}\}).$$

Hence

$$W \subseteq cl(V \cup \{a_0, \ldots, a_{n-1}\}).$$

Clearly, this implies that

$$\dim(\frac{W}{V}) \leq \dim(\frac{cl(V \cup \{a_0, \ldots, a_{n-1}\})}{V}) \leq n,$$

which contradicts the fact that $\dim(\frac{W}{V})$ is infinite. ∎

Theorem 30 *(Hird [33]). There is a supermaximal space that is not strongly supermaximal.*

Hird [32] further introduced a computable model-theoretic notion of a quasi-simple subset of a model. See [2,33] for the appropriate definition. This model-theoretic quasi-simplicity translates as computability-theoretic simplicity in the structure $(\omega, =)$. However, it turns out that a vector subspace of V_∞ is quasi-simple iff it is strongly supermaximal.

The following definition generalizes the notion of a supermaximal space within the class of maximal subspaces of V_∞.

Definition 31 *(Kalantari and Retzlaff [36]). Let $V \in \mathcal{L}(V_\infty)$.*

(i) The subspace V is called 0-thin if it is supermaximal.
(ii) Let $k \in \omega - \{0\}$. The subspace V is called k-thin if $\dim(\frac{V_\infty}{V}) = \infty$, there is a c.e. space U such that

$$\dim(\frac{V_\infty}{U}) = k,$$

and for every c.e. space W such that $V \subseteq W$, we have that

$$\dim(\frac{V_\infty}{W}) \leq k \vee \dim(\frac{W}{V}) < \infty.$$

Kalantari and Retzlaff [36] showed that k-thin spaces exist for all k.

4   Undecidability of the First-Order Theories of $\mathcal{L}(V_\infty)$ and $\mathcal{L}^*(V_\infty)$

The structure of $\mathcal{L}^*(V_\infty)$ is not as well-understood as that of $\mathcal{E}^*$. Both $\mathcal{L}(V_\infty)$ and $\mathcal{L}^*(V_\infty)$ are modular nondistributive lattices. This means that the "diamond" lattice M_5 can be embedded in $\mathcal{L}(V_\infty)$ and $\mathcal{L}^*(V_\infty)$, while the "pentagon" lattice N_5 cannot. The lattice $\mathcal{L}(V_\infty)$ has both atoms and co-atoms. More generally, if V is a finite k-dimensional subspace of V_∞, then the lattice of subspaces of V is an initial segment of the lattice $\mathcal{L}(V_\infty)$ and so it has the structure of the lattice $L(k, F)$ of all subspaces of any k-dimensional vector space over the field F. Also, if $V \in \mathcal{L}(V_\infty)$ is such that $\dim(\frac{V_\infty}{V}) = k$, then the principal filter $\mathcal{L}(V, \uparrow)$ of V in $\mathcal{L}(V_\infty)$ is also isomorphic to $L(k, F)$. These finite-rank initial and final segments collapse to the least and the greatest elements in $\mathcal{L}^*(V_\infty)$, respectively. We know that the lattice $\mathcal{L}^*(V_\infty)$ has co-atoms but does not have atoms. Remmel [52] and Downey [21] showed that every Σ_0^3 Boolean algebra is isomorphic to $\mathcal{L}^*(V, \uparrow)$ for some $V \in \mathcal{L}(V_\infty)$. Downey conjectured that every bounded Σ_0^3 modular lattice is a filter in $\mathcal{L}^*(V_\infty)$. Nerode and Smith established the following key structural result about $\mathcal{L}^*(V_\infty)$.

Theorem 32 *(Nerode and Smith [51]). Every finite distributive lattice is a filter in $\mathcal{L}^*(V_\infty)$.*

The proof is based on an interesting combinatorial construction, which uses Birkhoff's characterization of finite distributive lattices. The construction has requirements similar to those used in the construction of a supermaximal space. The following undecidability results are the main corollaries of the theorem.

Theorem 33 *(Nerode and Smith [51]).*

  (i) *The first-order theory of $\mathcal{L}^*(V_\infty)$ is undecidable.*
  (ii) *The first-order theory of $\mathcal{L}(V_\infty)$ is undecidable.*

The first result (i) is a corollary of Theorem 32, and an earlier result by Ershov and Taitslin, which establishes that the theory of distributive lattices is computably inseparable from the set of sentences refutable in some finite distributive lattices. Note that $V \in \mathcal{L}(V_\infty)$ is finite-dimensional if and only if every $W \subseteq V$ is complemented in $\mathcal{L}(V_\infty)$. The second result (ii) then follows from (i) using the definability of $\subseteq^*$ in $\mathcal{L}(V_\infty)$. Later, Galminas and Rosenthal [29] established that the theory of $\mathcal{L}(V_\infty)$ has the same logical complexity as the first-order number theory. The question whether $\forall\exists$-theory of $\mathcal{L}^*(V_\infty)$ is decidable is still open.

In [21], Downey introduced the following important notion.

Definition 34 *(Downey [21]). A c.e. set A has the lifting property if A is coinfinite and for every c.e. strong array $\{D_{g(x)} : x \in \omega\}$, for almost all x, $\left| D_{g(x)} - A \right| \le 1$.*

Downey used the lifting property to obtain undecidability results for a large class of lattices of c.e. structures, including $\mathcal{L}^*(V_\infty)$. The lifting property guarantees the "lifting" of principal filters under the closure operation. We will state these results of Downey only for $\mathcal{L}^*(V_\infty)$. In particular, let B is a computable basis of V_∞ and let $A \subseteq B$ have the lifting property. If we identify B with ω, then $\mathcal{E}^*(A, \uparrow) \cong \mathcal{L}^*(cl(A), \uparrow)$. Recall that a set $A \subseteq \omega$ is *semi-low* if $\{e : W_e \cap A \neq \emptyset\} \leq_T \emptyset'$.

Theorem 35 *(Downey [24]). There exists a c.e. set A with the lifting property such that $\overline{A}$ is semi-low.*

The undecidabilty results in [21,24] are then obtained using an earlier result by Soare that for such A we have that $\mathcal{E}^*(A, \uparrow)$ is effectively isomorphic to $\mathcal{E}^*$. Therefore, it follows that the first-order theory of $\mathcal{L}^*(V_\infty)$ is undecidable.

In [21], Downey also established that every Σ_0^3 Boolean algebra is isomorphic to a principal filter for a large class of lattices of c.e. structures. This result stated only for $\mathcal{L}^*(V_\infty)$ is the following.

Theorem 36 *(Downey [21]). Let $\mathfrak{B}$ be a Σ_0^3 Boolean algebra. Then exists a c.e. set A with the lifting property such that $\mathcal{E}^*(A, \uparrow) \cong \mathfrak{B}$.*

Corollary 37 *(Downey [21]). Every Σ_0^3 Boolean algebra is a filter in $\mathcal{L}^*(V_\infty)$.*

5 The Co-atoms Form an Automorphism Basis for $\mathcal{L}^*(V_\infty)$

Recall that for $X \in \mathcal{E}$ (or $X \in \mathcal{L}(V_\infty)$), we use $[X]$ to denote the equivalence class of X modulo the corresponding equivalence relation $=^*$. If S and T are arbitrary sets of vectors, then

$$\dim(S \bmod T) =_{def} \dim(\frac{cl(S \cup T)}{cl(T)}).$$

By $\mathfrak{M}^*$ and $\mathfrak{R}^*$ we denote the classes of maximal and computable sets modulo $=^*$, respectively. Clearly, the computable, as well as the maximal sets are closed under $=^*$. Note that $\mathfrak{M}^*$ can also be described as the set of the co-atoms in $\mathcal{E}^*$, while $\mathfrak{R}^*$ is the set of the complemented elements of $\mathcal{E}^*$. Nerode asked the following questions.

(1) Is every automorphism of $\mathcal{E}^*$ uniquely determined by its action on $\mathfrak{R}^*$?
(2) Does every automorphism of $\mathfrak{R}^*$ extend to an automorphism of $\mathcal{E}^*$?

In [54], Shore answered the first question positively and the second question negatively. In particular, he established the following results.

Proposition 38 *(Shore [54]). Assume that Φ_1 and Φ_2 are automorphisms of $\mathcal{E}^*$.*

(i) If Φ_1 and Φ_2 agree on the low sets, then $\Phi_1 = \Phi_2$.

(ii) If Φ_1 and Φ_2 agree on $\mathfrak{M}^*$, then $\Phi_1 = \Phi_2$.

(iii) If Φ_1 and Φ_2 agree on $\mathfrak{R}^*$, then $\Phi_1 = \Phi_2$.

For (i) Shore used Sacks splitting theorem that every c.e. set is the union of two disjoint low sets (see Theorem 3.2 in [56]). Then the proof of (ii) uses (i) and results from Lachlan [38], while the proof of (iii) uses (ii).

Theorem 39 *(Shore [54]). Let $\mathfrak{C}^*$ be any nontrivial class of c.e. sets (i.e., none of $\emptyset$, $\{0\}$, $\{N\}$), modulo finite sets, closed under computable isomorphism. If Φ_1 and Φ_2 agree on $\mathfrak{C}^*$, then $\Phi_1 = \Phi_2$.*

The proof of Theorem 39 uses Proposition 38 (iii). In a later paper, Shore proved that nowhere simple sets generate $\mathcal{E}$, thus improving Theorem 39.

It is natural to ask which natural classes of c.e. vector spaces form automorphism bases in the lattices $\mathcal{L}(V_\infty)$ and $\mathcal{L}^*(V_\infty)$. Currently, we do not know of any analogue of Proposition 38 (i) for the lattices $\mathcal{L}(V_\infty)$ or $\mathcal{L}^*(V_\infty)$. Ash and Downey established an analogue of Proposition 38 (iii) for the lattice $\mathcal{L}(V_\infty)$ (see Corollary 40 below). The result easily extends to $\mathcal{L}^*(V_\infty)$ and we will later give a short proof of this fact. We will also give a direct proof of an analogue of Proposition 38 (ii) for $\mathcal{L}^*(V_\infty)$ (see Theorem 44 below). An analogue of Theorem 39 for $\mathcal{L}(V_\infty)$ has been given by Nerode and Remmel in [48]. An analogue of Theorem 39 for $\mathcal{L}^*(V_\infty)$ has been given by Downey and Remmel in [27]. The following result follows immediately from Theorem 17.

Corollary 40.   (i) *The lattice $\mathcal{L}(V_\infty)$ is generated, under $\oplus$, by the decidable subspaces of V_∞.*

(ii) *Each automorphism of $\mathcal{L}(V_\infty)$ is uniquely determined by its action on the decidable subspaces.*

It is known that this result of Ash and Downey extends to $\mathcal{L}^*(V_\infty)$ as follows.

(a) The lattice $\mathcal{L}^*(V_\infty)$ is generated, under $\vee$, by the equivalence classes of the decidable subspaces of V_∞.

(b) Every automorphism of $\mathcal{L}^*(V_\infty)$ is uniquely determined by its action on the complemented elements of $\mathcal{L}^*(V_\infty)$.

Before we give proofs for these statements we will establish the following result.

Proposition 41. *If $V, W \in \mathcal{L}(V_\infty)$ are such that $[V] = [W]$, then*

$$D(V) \equiv_T D(W).$$

Proof. Suppose that $A = \{a_1, \ldots, a_p\}$ and $B = \{b_1, \ldots, b_q\}$ are sets of vectors that are independent modulo V and W, respectively, such that $cl(V \cup A) = cl(W \cup B)$. We claim that

$$D(V) \equiv_T D(cl(V \cup A)) = D(cl(W \cup B)) \equiv_T D(W).$$

We will only prove $D(V) \equiv_T D(cl(V \cup A))$. (The proof that $D(cl(W \cup B)) \equiv_T D(W)$ is identical.)

To prove that $D(V) \leq_T D(cl(V \cup A))$, fix arbitrary $x_1, \ldots, x_n \in V_\infty$ and use oracle $D(cl(V \cup A))$ to decide whether $(x_1, \ldots, x_n) \in D(cl(V \cup A))$.

Case (1). Let $(x_1, \ldots, x_n) \notin D(cl(V \cup A))$. Clearly, $(x_1, \ldots, x_n) \notin D(V)$.

Case (2). Let $(x_1, \ldots, x_n) \in D(cl(V \cup A))$. Suppose that I_1 is a computable basis of V. (Recall that such a basis exists.) Using oracle $D(cl(V \cup A))$, we construct a $D(cl(V \cup A))$-computable basis I_2 of $(V_\infty \bmod cl(V \cup A))$. Then $I_1 \cup A \cup I_2$ is a $D(cl(V \cup A))$-computable basis of V_∞. Representing each element in the sequence $x_1, \ldots, x_n$ as a linear combination in the basis $I_1 \cup A \cup I_2$ and using standard linear algebra we can decide whether the set $\{x_1, \ldots, x_n\} \cup I_1$ is dependent. Therefore, $D(V) \leq_T D(cl(V \cup A))$.

To prove that $D(cl(V \cup A)) \leq_T D(V)$, we will use oracle $D(V)$ to decide whether $(x_1, \ldots, x_n) \in D(cl(V \cup A))$. We check whether $(x_1, \ldots, x_n, a_1, \ldots, a_p) \in D(V)$. If the answer is positive, then $(x_1, \ldots, x_n) \in D(cl(V \cup A))$. Otherwise, $(x_1, \ldots, x_n) \notin D(cl(V \cup A))$. Therefore, $D(cl(V \cup A)) \leq_T D(V)$.   ∎

We will use the following notation for the co-atoms and the complemented elements in $\mathcal{L}^*(V_\infty)$.

$\mathcal{M}^* = \{[M] : M \text{ is a maximal subspace of } V_\infty\}$

$\mathcal{S}^*(V_\infty) = \{[D] : D \text{ is a decidable subspace of } V_\infty\}$

Note that $\mathcal{S}^*(V_\infty)$ is well-defined by Proposition 41. It is immediate that if M_1 is a maximal subspace of V_∞ and $M_1 =^* M_2$, then the space M_2 is also maximal. Therefore, $\mathcal{M}^*$ is also well-defined.

Corollary 42

(i) $\mathcal{L}^*(V_\infty)$ is generated, under $\vee$, by $\mathcal{S}^*(V_\infty)$.

(ii) Each automorphism of $\mathcal{L}^*(V_\infty)$ is uniquely determined by its action on $\mathcal{S}^*(V_\infty)$.

Proof. (i) Let $[V] \in \mathcal{L}^*(V_\infty)$. By Corollary 17, there are decidable spaces $D_1, D_2 \in \mathcal{L}(V_\infty)$ such that $V = D_1 \oplus D_2$. Then $[V] = [D_1] \vee [D_2]$.   ∎

An analogue of Theorem 39 has been given by Nerode and Remmel in [48] and by Downey and Remmel in [27]. The result by Downey and Remmel for the lattice $\mathcal{L}^*(V_\infty)$ is as follows.

Theorem 43 (*Downey and Remmel [27]*). *Let $\mathfrak{C}^*$ be any nontrivial class of elements of $\mathcal{L}^*(V_\infty)$ (i.e., none of $\emptyset$, $\{[0]\}$, $\{[V_\infty]\}$, $\{[0], [V_\infty]\}$), which is closed under automorphisms of $\mathcal{L}^*(V_\infty)$ that are generated by invertible computable linear transformations. Then, if Φ is an automorphism of $\mathcal{L}^*(V_\infty)$ such that $\Phi \restriction_{\mathfrak{C}^*} = id \restriction_{\mathfrak{C}^*}$, then $\Phi \restriction_{\mathcal{L}^*(V_\infty)} = id$.*

Proof. Suppose that $\Phi \restriction_{\mathcal{L}^*(V_\infty)} \neq id$, and let $[D] \in \mathcal{S}^*(V_\infty)$ be such that $\Phi([D]) \neq [D]$. Since $\Phi([D])$ is complemented, without loss of generality, assume that $D_1 \in \Phi([D])$ and $dim(D_1 \bmod D) = \infty$.

Let A be a computable basis of D. Extend A to a computable basis $A \cup B \cup C$ of V_∞ such that $B \subseteq D_1$ is an infinite independent set modulo D, and C is a c.e. set. Let $[V] \in \mathfrak{C}^*$ be such that $[V] \neq [0]$ and $[V] \neq [V_\infty]$. Then V has an infinite-dimensional subspace R such that $[R] \in \mathcal{S}^*(V_\infty)$. Let S_1 be a computable basis of R, and let S_2 be a computable independent set such that $S_1 \cup S_2$ is a basis of V_∞. Let f be the computable invertible linear transformation such that $f(S_1) = A \cup C$ and $f(S_2) = B$. Let $[W] = [f(V)]$ and note that $[W] \in \mathfrak{C}^*$, so $\Phi([W]) = [W]$.

Then $S_1 \subseteq V$ and hence $[cl(f(S_1))] = [cl(A \cup C)] \leq [W]$. Thus,

$$[V_\infty] = [cl(A \cup C)] \vee [cl(B)] \leq [W] \vee [cl(B)],$$

and so

$$\Phi^{-1}([W]) \vee \Phi^{-1}([cl(B)]) = [V_\infty].$$

However, $\Phi^{-1}([cl(B)]) \leq \Phi^{-1}([D_1]) = [D] = [cl(A)] \leq [cl(A \cup C)] \leq [W]$, and so

$$[W] \vee \Phi^{-1}([cl(B)]) = [W].$$

This implies that $[W] \neq \Phi^{-1}([W])$, which is a contradiction.   ∎

The analogue of Proposition 38 (ii) for $\mathcal{L}^*(V_\infty)$ follows from Downey and Remmel's result. It will also follow from the following theorem, where we construct a certain supermaximal space.

Theorem 44. *Let Φ_1 and Φ_2 be automorphisms of the lattice $\mathcal{L}^*(V_\infty)$ such that for some $[W] \in \mathcal{L}^*(V_\infty)$ we have*

$$\Phi_1([W]) \neq \Phi_2([W]).$$

Then there is a supermaximal space M such that $\Phi_1^{-1}([M]) \neq \Phi_2^{-1}([M])$.

Proof. By Corollary 42 (ii), there is a decidable space D such that $\Phi_1([D]) \neq \Phi_2([D])$. Note that $\Phi_1([V_\infty]) = [V_\infty] = \Phi_2([V_\infty])$ since every automorphism of $\mathcal{L}^*(V_\infty)$ fixes its greatest element. Therefore, $[D] \neq [V_\infty]$. Suppose that $U, V \in \mathcal{L}(V_\infty)$ are such that

$$[U] = \Phi_1([D]) \neq \Phi_2([D]) = [V].$$

Assume also that $dim(V \bmod U) = \infty$. We will construct a supermaximal space M such that $\Phi_1^{-1}([M]) \neq \Phi_2^{-1}([M])$. The space M will be such that $U \subseteq M$, $dim(M \bmod U) = \infty$, and $dim(V \bmod M) = \infty$ (see Fig. 1).   ∎

In the language of lattices $\{\leq, \vee, \wedge\}$ these conditions are:
$[U] \lneq [M]$ ($U \subseteq M$, and $[U] \neq [M]$ since $dim(M \bmod U) = \infty$), and
$[V] \not\leq [M]$ (because $dim(V \bmod M) = \infty$).
Before we proceed with the construction of M we will prove that these requirements guarantee that

$$\Phi_1^{-1}([M]) \neq \Phi_2^{-1}([M]).$$

To see this, note that in the lattice $\mathcal{L}^*(V_\infty)$ we have:

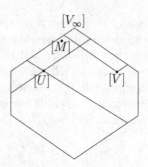

Fig. 1. Assume $[V] = \Phi_2([D])$ is not in the lower cone of $[U] = \Phi_1([D])$ in $\mathcal{L}^*(V_\infty)$. We construct a maximal space M such that $[M]$ is in the upper cone of $[U]$ while avoiding the upper cone of $[V]$. Note that we do not require that $[V]$ avoids the upper cone of $[U]$ despite our choice to draw it this way in the diagram.

(i) $[M] \vee [V] = [V_\infty]$ since $[M]$ is a co-atom in $\mathcal{L}^*(V_\infty)$ and $[V] \not\leq [M]$,

(ii) $\Phi_2^{-1}([M]) \vee \Phi_2^{-1}([V]) = \Phi_2^{-1}([M] \vee [V]) = \Phi_2^{-1}([V_\infty]) = [V_\infty]$,

(iii) $\Phi_1^{-1}([M]) \vee \Phi_1^{-1}([U]) = \Phi_1^{-1}([M] \vee [U]) = \Phi_1^{-1}([M])$ since $[U] \lesssim [M]$,

(iv) $\Phi_2^{-1}([M]) \vee \Phi_2^{-1}([V]) \neq \Phi_1^{-1}([M]) \vee \Phi_1^{-1}([U])$ by (ii) and (iii).

By substituting $\Phi_2^{-1}([V]) = [D]$ and $\Phi_1^{-1}([U]) = [D]$ in (iv) we obtain:

(v) $\Phi_2^{-1}([M]) \vee [D] \neq \Phi_1^{-1}([M]) \vee [D]$, and therefore,

(vi) $\Phi_1^{-1}([M]) \neq \Phi_2^{-1}([M])$.

We will now construct a supermaximal space the M. Note that both $[U]$ and $[V]$ are complemented in $\mathcal{L}^*(V_\infty)$ because they are images of the complemented $[D]$ under the automorphisms Φ_1, Φ_2, respectively. Therefore, U and V are decidable spaces. We can find computable bases $A, B,$ and C of $V, U,$ and $(V_\infty \bmod U)$, respectively. Let $A = \{a_0, a_1, \ldots\}, B = \{b_0, b_1, \ldots\},$ and $C = \{c_0, c_1, \ldots\}$ be fixed computable enumerations of these bases. We can regard C as a computable subset of V_∞. Thus, $B \cup C$ is a computable basis of V_∞, which extends the basis B of U. A space M will be constructed in stages. By M^s we will denote the approximation of M at the end of stage s.

At every stage s, the set B^s will be a computable basis for M^s. At stage 0, we will let $B^0 = B$ (and, therefore, $M^0 = U$). At stage $s > 0$, we will enumerate at most one vector $v \notin M^{s-1}$ into B^s, and then let $M^s = cl(B^s)$. Hence $dim(M^s \bmod M^0) < \infty$ and, therefore, M^s will be a decidable space, uniformly in s, for every $s \geq 0$.

Recall that V_e is the e-th c.e. subspace of V_∞. In the construction of M we will satisfy the following requirements for every $e \geq 0$:

$$R_e : \text{If } dim((V_e \vee M) \bmod M) = \infty, \text{ then } V_e \vee M = V_\infty.$$

Every R_e will be satisfied by satisfying the following sub-requirements for every $k \geq 0$:

$$R_{\langle e,k \rangle} : \text{If } dim((V_e \vee M) \bmod M) = \infty, \text{ then } c_k \in V_e \vee M.$$

We will also satisfy the following negative requirements for every $e \geq 0$:

$$N_e : dim(V \ mod \ M) > e.$$

Note that the satisfaction of $R_{\langle e,k \rangle}$ and N_e for each $e, k \geq 0$ will guarantee that M is a supermaximal subspace of V_∞ with the desired properties. To see this, note that if $M \subseteq V_{e_1}$ and $dim(V_{e_1} \ mod \ M) = \infty$ for some $e_1 \in \omega$, then $V_{e_1} = V_e \vee M$ for some $e \in \omega$. By construction, $B \subseteq U \subseteq M \subseteq V_{e_1}$. The satisfaction of the requirements $R_{\langle e,k \rangle}$ for all $e, k \geq 0$ will guarantee that $C \subseteq V_{e_1}$. Since $cl(B \cup C) = V_\infty$, we conclude that $V_{e_1} = V_\infty$.

At stage s, each requirement N_e will place a marker Γ_e on the first element $a_n \in A$ such that

$$dim(\{a_0, \ldots, a_n\} \ mod \ M^s) = e + 1.$$

For all $e, k \geq 0$ the requirements N_m for $m \leq \langle e, k \rangle$ will have higher priority than the requirement $R_{\langle e,k \rangle}$. The requirement $R_{\langle e,k \rangle}$ will respect the higher priority requirements N_m by not allowing markers $\Gamma_0, \ldots, \Gamma_m$ to be moved.

The requirement $R_{\langle e,k \rangle}$ *requires attention at stage* $s + 1$ if:

(1) $R_{\langle e,k \rangle}$ has not been satisfied, and

(2) there is $y \in V_e^s$ with $y \leq s$ such that the following conditions are satisfied:

   (i) $y + c_k \notin M^s$,
   (ii) if a_{n_j} is the element of A marked by the marker Γ_j at stage s, then

$$dim(\{a_{n_0}, \ldots, a_{n_{\langle e,k \rangle}}\} \ mod \ M^s) =$$

$$dim(\{a_{n_0}, \ldots, a_{n_{\langle e,k \rangle}}\} \ mod \ cl(M^s \cup \{y + c_k\})).$$

If such y exists, then we say that $R_{\langle e,k \rangle}$ *requires attention via* y at stage $s + 1$.

Construction

Stage 0. Let $B^0 = B$ and $M^0 = cl(B^0)$. For each $i \geq 0$, place the marker Γ_i on the first element $a_n \in A$ such that

$$dim(\{a_0, \ldots, a_n\} \ mod \ M^0) = i + 1.$$

Stage $s + 1$. Check if some requirement $R_{\langle e_1,k_1 \rangle}$, where $\langle e_1, k_1 \rangle \leq s + 1$, requires attention at stage $s + 1$. If there is no such requirement, let $B^{s+1} = B^s$, $M^{s+1} = cl(B^{s+1})$, and go to the next stage. Otherwise, let $\langle e, k \rangle$ be the least such that $R_{\langle e,k \rangle}$ requires attention, and let y be the least such that $R_{\langle e,k \rangle}$ requires attention via y at stage $s + 1$. Let $x =_{def} y + c_k$. Then
   (a) let $M^{s+1} = cl(B^{s+1})$,
   (b) for every $i \geq 0$ place the marker Γ_i on the first element $a_n \in A$ such that

$$dim(\{a_0, \ldots, a_n\} \ mod \ M^{s+1}) = i + 1.$$

We say that $R_{\langle e,k \rangle}$ received attention. Note that the condition above can be checked effectively since M^{s+1} is a decidable space. Note also that, because of the condition (2)(ii), only the markers $\Gamma_{\langle e,k \rangle +1}, \Gamma_{\langle e,k \rangle +2}, \ldots$ are moved from the elements they marked at the previous stage.

   End of Construction

In the following lemmas we will prove that the space M is supermaximal. Lemma 46 will imply that $dim(V \ mod \ M) = \infty$. Hence $[M]$ avoids the upper cone of $[V]$ and, therefore, $dim(V_\infty \ mod \ M) = \infty$. Lemma 47 will imply that if $dim((V_e \vee M) \ mod \ M) = \infty$, then $V_e \vee M = V_\infty$.

Lemma 45. *Each requirement $R_{\langle e,k \rangle}$ receives attention at most once.*

Proof. If $R_{\langle e,k \rangle}$ receives attention at stage $s+1$ via $y \in V_e^s$, then $x = y + c_k$ is enumerated into M^{s+1}. Then $c_k = (y + c_k) - y \in M^{s+1} \vee V_e^{s+1}$, and, therefore, $R_{\langle e,k \rangle}$ will be satisfied at stage $s+1$ and will not require attention at any later stage. ∎

Lemma 46. *Each marker Γ_m moves finitely often.*

Proof. Let s be a stage such that no $R_{\langle e,k \rangle}$ for $\langle e, k \rangle \leq m$ requires attention after stage s. Then the construction guarantees that Γ_m will not be moved after s. ∎

Lemma 47. *Each requirement $R_{\langle e,k \rangle}$ is satisfied.*

Proof. Suppose that $\langle e, k \rangle$ is the least number such that $R_{\langle e,k \rangle}$ is not satisfied. That means that $dim((V_e \vee M) \ mod \ M) = \infty$, but $c_k \notin M \vee V_e$. Suppose that s is the least stage such that no $R_{\langle e_1,k_1 \rangle}$ for $\langle e_1, k_1 \rangle < \langle e, k \rangle$ requires attention after s. This means that no marker Γ_j for $j \leq \langle e, k \rangle$ is moved after stage s. Suppose that a_{n_j} is the element marked by the marker Γ_j for $j = 0, \ldots, \langle e, k \rangle$. Since $dim((V_e \vee M) \ mod \ M) = \infty$, we also have

$$dim((V_e \vee M) \ mod \ cl(M \cup \{a_{n_0}, \ldots, a_{n_{\langle e,k \rangle}}, c_k\})) = \infty.$$

Therefore, there are a stage $s_1 > s$ and $y \in V_e^{s_1}$ such that

$$y \notin cl(M^{s_1} \cup \{a_{n_0}, \ldots, a_{n_{\langle e,k \rangle}}, c_k\}).$$

Then $y + c_k \notin cl(M^{s_1} \cup \{a_{n_0}, \ldots, a_{n_{\langle e,k \rangle}}\})$. The requirement $R_{\langle e,k \rangle}$ will receive attention via y at stage s_1, and will then remain satisfied. ∎

6   Automorphisms of the Lattices of Vector Spaces

The study of automorphisms of structures of importance in computable model theory connects computability theory with classical group theory. Let $\mathbf{d}$ be a Turing degree. For an infinite computable structure A, we define $Aut_{\mathbf{d}}(A)$ to be the set of all automorphisms of A, which are computable in $\mathbf{d}$. The set $Aut_{\mathbf{d}}(A)$ forms a group under composition and it is a subgroup of the group $Aut(A)$ of all automorphisms of A. It is natural to ask questions about computability-theoretic properties of this group and its subgroups. When the structure A is

ω with equality, then its automorphism group $Aut(A)$ is usually denoted by $Sym(\omega)$, the symmetric group of ω. Hence we have

$$Sym_{\mathbf{d}}(\omega) = \{f \in Sym(\omega) : \deg(f) \leq \mathbf{d}\}.$$

Lachlan showed that there are $2^{\aleph_0}$ automorphisms of $\mathcal{E}^*$. Every automorphism of $\mathcal{E}$ induces an automorphism of $\mathcal{E}^*$. Every computable permutation of ω induces an automorphism of $\mathcal{E}$, and hence of $\mathcal{E}^*$. Every automorphism of $\mathcal{E}^*$ is induced by some permutation of ω, which is not necessarily computable. Hence, since every automorphism of $\mathcal{E}^*$ is induced by some automorphism of $\mathcal{E}$, there are $2^{\aleph_0}$ automorphisms of $\mathcal{E}$.

By $\mathcal{L}$ we denote the lattice of all subspaces of V_∞. For a Turing degree $\mathbf{d}$, by $\mathcal{L}_{\mathbf{d}}(V_\infty)$ we denote the following sublattice of $\mathcal{L}$:

$$\mathcal{L}_{\mathbf{d}}(V_\infty) = \{V \in \mathcal{L} : V \text{ is } \mathbf{d}\text{-computably enumerable}\}.$$

Note that $\mathcal{L}_{\mathbf{0}}(V_\infty)$ is the same as $\mathcal{L}(V_\infty)$. The problem of finding the number of automorphisms of $\mathcal{L}^*(V_\infty)$ is still open. However, Guichard [30] established that there are countably many automorphisms of $\mathcal{L}(V_\infty)$ by showing that each computable automorphism is generated by a $1-1$ and onto computable semilinear transformation of V_∞.

Recall that a pair (μ, σ) is a *semilinear* transformation of V_∞ if $\mu : V_\infty \to V_\infty$ and σ is an automorphism of F such that

$$\mu(\alpha u + \beta v) = \sigma(\alpha)\mu(u) + \sigma(\beta)\mu(v)$$

for every $u, v \in V_\infty$ and every $\alpha, \beta \in F$. By $GSL_{\mathbf{d}}$ we will denote the group of $1-1$ and onto semilinear transformations (μ, σ) such that $deg(\mu) \leq \mathbf{d}$ and $deg(\sigma) \leq \mathbf{d}$. Thus, Guichard proved that every element of $Aut(\mathcal{L}_{\mathbf{0}}(V_\infty))$ is generated by an element of $GSL_{\mathbf{0}}$. It is easy to show that this result can be relativized to an arbitrary Turing degree $\mathbf{d}$.

Theorem 48. *Every $\Phi \subset Aut(\mathcal{L}_{\mathbf{d}}(V_\infty))$ is generated by some $(\mu, \sigma) \in GSL_{\mathbf{d}}$. Moreover, if Φ is also generated by some other $(\mu_1, \sigma_1) \in GSL_{\mathbf{d}}$, then there is $\gamma \in F$ such that*

$$(\forall v \in V_\infty)\,[\mu(v) = \gamma\mu_1(v)].$$

Proof. Note that each automorphism Φ of $\mathcal{L}_{\mathbf{d}}(V_\infty)$ acts on the one-dimensional subspaces of V_∞ and hence generates a unique automorphism $\overline{\Phi}$ of $\mathcal{L}$. By the fundamental theorem of projective geometry applied to the lattice $\mathcal{L}$, since $\overline{\Phi}$ is in $Aut(\mathcal{L})$, it follows that it is generated by a semilinear transformation (μ, σ). Note that (μ, σ) also generates Φ. We will now show that $deg(\mu) \leq \mathbf{d}$ and $deg(\sigma) \leq \mathbf{d}$.

Let $\alpha_0, \alpha_1, \alpha_2, \ldots$ be a fixed computable enumeration of the elements of the field F. Assume that $v_0, v_1, v_2, \ldots$ is a computable enumeration of a computable basis of V_∞. Define the following computable subspaces of V_∞:

$U_1 = cl(\{v_0, v_2, v_4, \ldots\})$,
$U_2 = cl(\{v_1, v_3, v_5, \ldots\})$,
$U_3 = cl(\{v_0 + v_1, v_2 + v_3, v_4 + v_5, \ldots\})$,

$$U_4 = cl(\{v_1 + v_2, v_3 + v_4, v_5 + v_6, \dots\}),$$
$$U_5 = cl(\{v_0 + \alpha_0 v_1, v_2 + \alpha_1 v_3, v_4 + \alpha_2 v_5, \dots\}).$$

Suppose that $\Phi(U_i) = Y_i$ for $i = 1, \dots, 5$, and note that $Y_i \in \mathcal{L}_{\mathbf{d}}(V_\infty)$ since $U_i \in \mathcal{L}_{\mathbf{d}}(V_\infty)$.

To prove that $deg(\mu) \leq \mathbf{d}$, suppose that $\mu(v_0) = w_0$ for some fixed w_0. Assume inductively that $\mu(v_{2i}) = w_{2i}$ has been found $\mathbf{d}$-computably. To find $\mathbf{d}$-computably $\mu(v_{2i+1})$, we let w_{2i+1} be the least $y \in Y_2$ such that $w_{2i} + y \in Y_3$. Then we have $\mu(v_{2i+1}) = w_{2i+1}$. Next, to find $\mathbf{d}$-computably $\mu(v_{2i+2})$, we let w_{2i+2} be the least $y \in Y_1$ such that $w_{2i+1} + y \in Y_4$. Then we have $\mu(v_{2i+2}) = w_{2i+2}$.

Finally, to find $\mathbf{d}$-computably $\sigma(\alpha_i)$, we look for the least $w \in Y_5$ and $\beta \in F$ such that $w = w_{2i} + \beta w_{2i+1}$ and note that $\sigma(\alpha_i) = \beta$. It is not difficult to prove that if our choice for $\mu(v_0)$ is a scalar multiple of the original w_0, namely, $\mu(v_0) = \gamma w_0$, then $\mu(v_i) = \gamma w_i$ for every $i \geq 1$.   ∎

The *Turing degree spectrum* of a countable structure A is

$$DgSp(A) = \{\deg(B) : B \cong A\},$$

where $\deg(B)$ is the Turing degree of the atomic diagram of B. Knight [37] proved that the degree spectrum of any structure is either a singleton or is upward closed. Jockusch and Richter (see [53]) defined the *degree of the isomorphism type* of a structure, if it exists, to be the least Turing degree in its Turing degree spectrum. Morozov [47] established that the degree of the isomorphism type of the group $Sym_{\mathbf{d}}(\omega)$ is $\mathbf{d}''$.

Theorem 49 *(Dimitrov, Harizanov and Morozov [12]). The degree of the isomorphisms type of the group $GSL_{\mathbf{d}}$ is $\mathbf{d}''$.*

In 1998, Downey and Remmel [26] raised the question of finding meaningful orbits in $\mathcal{L}^*(V_\infty)$. Recently, Dimitrov and Harizanov [9] gave a necessary and sufficient condition for quasimaximal vector spaces with extendible bases to be in the same orbit of $\mathcal{L}^*(V_\infty)$. The condition is stated in terms of m-degrees.

Unlike for the principal filters in $\mathcal{E}^*$ determined by quasimaximal sets of a fixed rank, there are several possibilities for the principal filters in $\mathcal{L}^*(V_\infty)$ determined by the closures of quasimaximal subsets of a computable basis. More precisely, Dimitrov [5,6] gave a description of all possible isomorphism types of $\mathcal{L}^*(cl(B), \uparrow)$ when B is a quasimaximal subset of rank n of any computable basis of V_∞. He proved that $\mathcal{L}^*(cl(B), \uparrow)$ is isomorphic to either:

(1) Boolean algebra $\mathbf{B_n}$ (which has 2^n elements),

(2) the lattice $L(n, \prod_C F)$ of all subspaces of an n-dimensional vector space over a certain extension $\prod_C F$ of the underlying field F, or

(3) a finite product of structures from the previous two cases.

Note that the Boolean algebra $\mathbf{B_n}$ in (1) can also be viewed as a product of n copies of the Boolean algebra $\mathbf{B_1}$. The extensions $\prod_C F$ of F mentioned in (2) are cohesive powers (see the definition below) of the field F over various cohesive sets C. Using the results in [11] it follows that these principal filters fall into infinitely many non-isomorphic classes, even when the filters are isomorphic to the lattices of subspaces of the vector spaces of the same dimension. Cohesive power is related to the versions of effective ultraproducts previously used by Hirschfeld, Wheeler, and McLaughlin [34, 35, 41, 42] in their study of models of various fragments of arithmetic. As usual, we will denote the equality of partial functions by $\simeq$.

Definition 50. *Let $\mathcal{A}$ be a computable structure with domain A in a computable language S, and let $C \subseteq \omega$ be a cohesive set. The cohesive power of $\mathcal{A}$ over C, denoted by $\prod_C \mathcal{A}$, is a structure $\mathcal{B}$ for S with domain B defined as follows.*

(1) *The set B is $D/(=_C)$, where $D = \{\varphi \mid \varphi : \omega \to A$ is a partial computable function, and $C \subseteq^* dom(\varphi)\}$.
For $\varphi_1, \varphi_2 \in D$, we have*

$$\varphi_1 =_C \varphi_2 \quad iff \quad C \subseteq^* \{x : \varphi_1(x) \downarrow = \varphi_2(x) \downarrow\}.$$

The equivalence class of φ with respect to $=_C$ will be denoted by $[\varphi]_C$, or simply by $[\varphi]$ (when the reference to C is clear from the context).

(2) *If $f \in S$ is an n-ary function symbol, then $f^{\mathcal{B}}$ is an n-ary function on B such that for every $[\varphi_1], \ldots, [\varphi_n] \in B$, we have $f^{\mathcal{B}}([\varphi_1], \ldots, [\varphi_n]) = [\varphi]$, where for every $x \in \omega$,*

$$\varphi(x) \simeq f^{\mathcal{A}}(\varphi_1(x), \ldots, \varphi_n(x)).$$

If $P \in S$ is an m-ary predicate symbol, then $P^{\mathcal{B}}$ is an m-ary relation on B such that for every $[\varphi_1], \ldots, [\varphi_m] \in B$,

$$P^{\mathcal{B}}([\varphi_1], \ldots, [\varphi_m]) \quad \text{iff} \quad C \subseteq^* \{x \in \omega \mid P^{\mathcal{A}}(\varphi_1(x), \ldots, \varphi_m(x))\}.$$

If $c \in S$ is a constant symbol, then $c^{\mathcal{B}}$ is the equivalence class of the (total) computable function on A with constant value $c^{\mathcal{A}}$.

In the context of c.e. vector spaces, the most interesting cases occur when F is finite or $F = \mathbb{Q}$. For finite F, we have $\prod_C F \cong F$. Various results about the cohesive powers of $\mathbb{Q}$ have been established in [7, 11]. These results, together with the above classification of the possible isomorphism types of $\mathcal{L}^*(cl(B), \uparrow)$, were used in the proof of the result discussed in the next paragraph.

To state the theorem, we introduced the notion of an m-*degree type* of a quasimaximal set $E = \bigcap_{i=1}^{n} M_i$ of rank n, denoted by $type(E)$. This notion captures the number and the m-degrees of the maximal sets M_i's. For $i = 1, 2$, let $E_i \subseteq A_i$

be a quasimaximal subset of a computable basis A_i. Dimitrov and Harizanov [9] proved that, assuming that the field is $\mathbb{Q}$, there is an automorphism Φ of $\mathcal{L}^*(V_\infty)$ such that $\Phi([E_1]) = [E_2]$ if and only if $type_{A_1}(E_1) = type_{A_2}(E_2)$. Since maximal sets are also quasimaximal, we have the following corollary.

Theorem 51 *(Dimitrov and Harizanov [9]). Assume that the underlying field is $\mathbb{Q}$. Let M_1 and M_2 be maximal subsets of computable bases B_1 and B_2 of V_∞, respectively. Then the following are equivalent:*

(1) There is an automorphism Φ of $\mathcal{L}^*(V_\infty)$ such that

$$\Phi([M_1]) = [M_2],$$

(2) $deg_m(M_1) = deg_m(M_2)$.

In some cases, it is also possible to connect the embeddability of the subgroups with Turing degree complexity. Morozov showed that the correspondence $\mathbf{d} \to Sym_\mathbf{d}(\omega)$ can be used to substitute Turing reducibility with group-theoretic embedding. More precisely, Morozov [45] established that for every pair $\mathbf{d}, \mathbf{s}$ of Turing degrees, we have

$$(Sym_\mathbf{d}(\omega) \hookrightarrow Sym_\mathbf{s}(\omega)) \Leftrightarrow (\mathbf{d} \leq \mathbf{s}).$$

It follows from this result that $\mathbf{d} = \mathbf{s}$ if and only if $Sym_\mathbf{d}(\omega) \cong Sym_\mathbf{s}(\omega)$. In [12], we established a similar result for the subgroups of the group of automorphisms of the lattice of the subspaces of V_∞.

Theorem 52 *(Dimitrov, Harizanov and Morozov [12]). For any pair of Turing degrees $\mathbf{d}, \mathbf{s}$ we have*

$$(Aut(\mathcal{L}_\mathbf{d}(V_\infty)) \hookrightarrow Aut(\mathcal{L}_\mathbf{s}(V_\infty))) \Leftrightarrow \mathbf{d} \leq \mathbf{s}.$$

References

1. Aigner, M.: Combinatorial Theory. Springer, Berlin (1997)
2. Ash, C.J., Knight, J.F., Remmel, J.B.: Quasi-simple relations in copies of a given recursive structure. Ann. Pure Appl. Logic **86**, 203–218 (1997)
3. Ash, C.J., Downey, R.G.: Decidable subspaces and recursively enumerable subspaces. J. Symbolic Logic **49**, 1137–1145 (1984)
4. Dekker, J.C.E.: Countable vector spaces with recursive operations, part I. J. Symbolic Logic **34**, 363–387 (1969)
5. Dimitrov, R.: Quasimaximality and principal filters isomorphism between $\mathcal{E}^*$ and $\mathcal{L}^*(V_\infty)$. Arch. Math. Logic **43**, 415–424 (2004)
6. Dimitrov, R.: A class of Σ_3^0 modular lattices embeddable as principal filters in $\mathcal{L}^*(V_\infty)$. Arch. Math. Logic **47**, 111–132 (2008)
7. Dimitrov, R.: Cohesive powers of computable structures. Annuaire de l'Université de Sofia "St. Kliment Ohridski", Faculté de Mathématiques et Informatique, vol. 99, pp. 193–201 (2009)

8. Dimitrov, R.: Extensions of certain partial automorphisms of $\mathcal{L}^*(V_\infty)$. Annuaire de l'Université de Sofia "St. Kliment Ohridski", Faculté de Mathématiques et Informatique, vol. 99, pp. 183–191 (2009)

9. Dimitrov, R.D., Harizanov, V.: Orbits of maximal vector spaces. Algebra and Logic **54**, 680–732 (2015) (in Russian). 440–477 (in English) (2016)

10. Dimitrov, R.D., Harizanov, V.S., Morozov, A.S.: Dependence relations in computably rigid computable vector spaces. Ann. Pure Appl. Logic **132**, 97–108 (2005)

11. Dimitrov, R., Harizanov, V., Miller, R., Mourad, K.J.: Isomorphisms of nonstandard fields and Ash's conjecture. In: Beckmann, A., Csuhaj-Varjú, E., Meer, K. (eds.) CiE 2014. LNCS, vol. 8493, pp. 143–152. Springer, Heidelberg (2014). doi:10.1007/978-3-319-08019-2_15

12. Dimitrov, R., Harizanov, V., Morozov, A.: Automorphism groups of substructure lattices of vector spaces in computable algebra. In: Beckmann, A., Bienvenu, L., Jonoska, N. (eds.) CiE 2016. LNCS, vol. 9709, pp. 251–260. Springer, Heidelberg (2016). doi:10.1007/978-3-319-40189-8_26

13. Downey, R.: On a question of A. Retzlaff. Zeitschrift für Mathematische Logik und Grundlagen der Mathematik **29**, 379–384 (1983)

14. Downey, R.G.: Nowhere simplicity in matroids. J. Aust. Math. Soc. Ser. A. Pure Math. Stat. **35**, 28–45 (1983)

15. Downey, R.: Bases of supermaximal subspaces and Steinitz systems, part I. J. Symbolic Logic **49**, 1146–1159 (1984)

16. Downey, R.G.: A note on decompositions of recursively enumerable subspaces. Zeitschrift für Mathematische Logik und Grundlagen der Mathematik **30**, 465–470 (1984)

17. Downey, R.: Co-immune subspaces and complementation in V_∞. J. Symbolic Logic **49**, 528–538 (1984)

18. Downey, R.G.: The degrees of r.e. sets without the universal splitting property. Trans. Am. Math. Soc. **291**, 337–351 (1985)

19. Downey, R.G., Hird, G.R.: Automorphisms of supermaximal subspaces. J. Symbolic Logic **50**, 1–9 (1985)

20. Downey, R.G.: Bases of supermaximal subspaces and Steinitz systems, part II. Zeitschrift für Mathematische Logik und Grundlagen der Mathematik **32**, 203–210 (1986)

21. Downey, R.G.: Undecidability of $L(F_\infty)$ and other lattices of r.e. substructures. Ann. Pure Appl. Logic **32**, 17–26 (1986)

22. Downey, R.G.: Maximal theories. Ann. Pure Appl. Logic **33**, 245–282 (1987)

23. Downey, R.G.: Orbits of creative subspaces. Proc. Am. Math. Soc. **99**, 163–170 (1987)

24. Downey, R.: Correction to "Undecidability of $L(F_\infty)$ and other lattices of r.e. substructures". Ann. Pure Appl. Logic **48**, 299–301 (1990)

25. Downey, R.: On the universal splitting property. Math. Logic Q. **43**, 311–320 (1997)

26. Downey, R.G., Remmel, J.B.: Computable algebras, closure systems: coding properties. In: Ershov, Y.L., Goncharov, S.S., Nerode, A., Remmel, J.B., (ed.) Handbook of Recursive Mathematics. Studies in Logic and the Foundations of Mathematics 139, vol. 2, pp. 997–1039 (1998)

27. Downey, R.G., Remmel, J.B.: Automorphisms and recursive structures. Zeitschrift für Mathematische Logik und Grundlagen der Mathematik **33**, 339–345 (1987)

28. Fokina, E., Harizanov, V., Melnikov, A.: Computable model theory. In: Downey, R. (ed.) Turing's Legacy: Developments from Turing Ideas in Logic. Lecture Notes in Logic, vol. 42, pp. 124–194. Cambridge University Press/ASL, Cambridge (2014)

29. Galminas, L.R., Rosenthal, J.W.: More undecidable lattices of Steinitz exchange systems. J. Symbolic Logic **67**, 859–878 (2002)
30. Guichard, D.R.: Automorphisms of substructure lattices in recursive algebra. Ann. Pure Appl. Logic **25**, 47–58 (1983)
31. Guichard, D.R.: A note on r-maximal subspaces of V_∞. Ann. Pure Appl. Logic **26**, 1–9 (1984)
32. Hird, G.R.: Recursive properties of relations on models. Ann. Pure Appl. Logic **63**, 241–269 (1993)
33. Hird, G.R.: Recursive properties of relations on models. Ph.D. dissertation, Monash University, Melbourne, Australia (1984)
34. Hirschfeld, J.: Models of arithmetic and recursive functions. Isr. J. Math. **20**, 111–126 (1975)
35. Hirschfeld, J., Wheeler, W.H.: Forcing, Arithmetic, Division Rings. Lecture Notes in Mathematics, vol. 454. Springer, Heidelberg (1975)
36. Kalantari, I., Retzlaff, A.: Maximal vector spaces under automorphisms of the lattice of recursively enumerable spaces. J. Symbolic Logic **42**, 481–491 (1977)
37. Knight, J.F.: Degrees coded in jumps of orderings. J. Symbolic Logic **51**, 1034–1042 (1986)
38. Lachlan, A.H.: Degrees of recursively enumerable sets which have no maximal supersets. J. Symbolic Logic **33**, 431–443 (1968)
39. Lian, W.S.: Automorphisms of the lattice of recursively enumerable vector spaces and embeddings of $\mathcal{L}^*(V_\infty)$. Senior thesis, National University of Singapore (1985)
40. Mal'tsev, A.I.: On recursive Abelian groups. Doklady Akademii Nauk SSSR **146**, 1009–1012 (1962) (in Russian). 1431–1434 (in English)
41. McLaughlin, T.G.: Sub-arithmetical ultrapowers: a survey. Ann. Pure Appl. Logic **49**, 143–191 (1990)
42. McLaughlin, T.G.: Δ_1 ultrapowers are totally rigid. Arch. Math. Logic **46**, 379–384 (2007)
43. Metakides, G., Nerode, A.: Recursively enumerable vector spaces. Ann. Math. Logic **11**, 147–171 (1977)
44. Morozov, A.S.: Rigid constructive modules. Algebra and Logic **28**, 570–583 (1989) (in Russian). 379–387 (1990) (in English)
45. Morozov, A.S.: Turing reducibility as algebraic embeddability. Siberian Math. J. **38**, 362–364 (1997) (in Russian). 312–313 (in English)
46. Morozov, A.S.: Groups of computable automorphisms. In: Ershov, Y.L., Goncharov, S.S., Nerode, A., Remmel, J.B., (eds.) Handbook of Recursive Mathematics, Studies in Logic and the Foundations of Mathematics 139, vol. 1, pp. 311–345 (1998)
47. Morozov, A.S.: Permutations and implicit definability. Algebra and Logic **27**, 19–36 (1988) (in Russian). 12–24 (in English)
48. Nerode, A., Remmel, J.: Recursion theory on matroids. In: Metakides, G., (ed.) Patras Logic Symposium (Patras, 1980), Studies in Logic and the Foundations of Mathematics, vol. 109, pp. 41–65 (1982)
49. Nerode, A., Remmel, J.B.: Recursion theory on matroids, part II. In: Chong, C.-T., Wicks, M.J., (eds.) Southeast Asian Conference on Logic (Singapore, 1981), Studies in Logic and the Foundations of Mathematics, vol. 111, pp. 133–184 (1983)
50. Nerode, A., Remmel, J.: A survey of lattices of r.e. substructures. In: Nerode, A., Shore, R. (eds.) Recursion Theory, Proceedings of Symposia in Pure Mathematics, vol. 42, pp. 323–375. American Mathematical Society, Providence (1985)

51. Nerode, A., Smith, R.L.: The undecidability of the lattice of recursively enumerable subspaces. In: Arruda, A.I., da Costa, N.C.A., Settepp, A.M. (eds.) Proceedings of the Third Brazilian Conference on Mathematical Logic (Federal University of Pernambuco, Recife, 1979), pp. 245–252. Sociedade Brasileira de Lógica, São Paulo (1980)
52. Remmel, J.B.: On the lattice of r.e. superspaces of an r.e. vector space. Notices of the American Mathematical Society **24**, abstract #77T–E26 (1977)
53. Richter, L.J.: Degrees of unsolvability of models. Ph.D. dissertation, University of Illinois at Urbana-Champaign (1977)
54. Shore, R.A.: Determining automorphisms of the recursively enumerable sets. Proc. Am. Math. Soc. **65**, 318–325 (1977)
55. Soare, R.I.: Automorphisms of the lattice of recursively enumerable sets, part I: maximal sets. Ann. Math. **100**, 80–120 (1974)
56. Soare, R.I.: Recursively Enumerable Sets and Degrees. A Study of Computable Functions and Computably Generated Sets. Springer, Berlin (1987)

Injection Structures Specified by Finite State Transducers

Sam Buss[1], Douglas Cenzer[2(✉)], Mia Minnes[3], and Jeffrey B. Remmel[1]

[1] Department of Mathematics, University of California-San Diego,
La Jolla, CA 92093-0112, USA
{sbuss,jremmel}@ucsd.edu
[2] Department of Mathematics, University of Florida, Gainesville, FL 32611, USA
cenzer@math.ufl.edu
[3] Department of Computer Science and Engineering,
University of California-San Diego, La Jolla, CA 92093-0404, USA
minnes@eng.ucsd.edu

Abstract. An injection structure $\mathcal{A} = (A, f)$ is a set A together with a one-place one-to-one function f. $\mathcal{A}$ is an FST injection structure if A is a regular set, that is, the set of words accepted by some finite automaton, and f is realized by a finite-state transducer. We initiate the study of FST injection structures. We show that the model checking problem for FST injection structures is undecidable which contrasts with the fact that the model checking problem for automatic relational structures is decidable. We also explore which isomorphisms types of injection structures can be realized by FST injections. For example, we completely characterize the isomorphism types that can be realized by FST injection structures over a unary alphabet. We show that any FST injection structure is isomorphic to an FST injection structure over a binary alphabet. We also prove a number of positive and negative results about the possible isomorphism types of FST injection structures over an arbitrary alphabet.

Keywords: Computability theory · Injection structures · Automatic structures · Finite state automata · Finite state transducers

1 Introduction and Preliminaries

The main goal of this paper is to initiate the study of FST injection structures. Throughout this paper, we will restrict our attention to countable structures for computable languages. There has been considerable work on automatic structures for languages that contain only relation symbols. A structure, $\mathcal{A} = (A; R_0, \ldots, R_m)$, is **automatic** if its domain A and all its basic relations $R_0, \ldots, R_m$ are recognized by finite automata. Independently, Hodgson [8] and

Buss was partially supported by NSF grants DMS-1101228 and CCF-1213151. Cenzer was partially supported by the NSF grant DMS-1101123. Minnes was partially supported by the NSF grant DMS-1060351.

© Springer International Publishing AG 2017
A. Day et al. (Eds.): Downey Festschrift, LNCS 10010, pp. 394–417, 2017.
DOI: 10.1007/978-3-319-50062-1_24

later Khoussainov and Nerode [9] proved that for any given automatic structure there is an algorithm that solves the model checking problem for the first-order logic in the language of the structure. In particular, the first-order theory of the structure is decidable. In fact an even stronger result is true. We denote by $\exists^\infty$ the quantifier "there exists infinitely many" and $\exists^{(n,m)}$ the quantifier "there are m many mod n." Then we let $FO + \exists^\infty + \exists^{(n,m)}$ denote first order logic extended with these quantifiers. Then Khoussainov, Rubin, and Stephan [13] proved the following.

Theorem 1. *There is an algorithm that given an automatic structure $\mathcal{A}$ and a $(FO + \exists^\infty + \exists^{(n,m)})$-formula $\phi(x_1, \ldots, x_n)$ with parameters from $\mathcal{A}$ produces an automaton recognizing those tuples $\langle a_1, \ldots, a_n \rangle$ that make the formula true in $\mathcal{A}$.*

It follows that the $(FO + \exists^\infty + \exists^{(n,m)})$-theory of any automatic structure is decidable.

Blumensath and Grädel proved a logical characterization theorem stating that automatic structures are exactly those definable in the fragment of arithmetic $(\omega; +, |_2, \leq, 0)$, where $+$ and $\leq$ have their usual meanings and $|_2$ is a weak divisibility predicate for which $x|_2 y$ if and only if x is a power of 2 and divides y (see [4]). In addition, for some classes of automatic structures, there are characterization theorems that have direct algorithmic implications. For example, in [7], Delhommé proved that automatic well-ordered sets all have order type strictly less than ω^ω. Using this characterization, [11] gives an algorithm which decides the isomorphism problem for automatic well-ordered sets. Another characterization theorem of this ilk is that automatic Boolean algebras are exactly those that are finite products of the Boolean algebra of finite and co-finite subsets of ω [12]. Again, this result can be used to show that the isomorphism problem for automatic Boolean algebras is decidable.

Another body of work is devoted to the interaction between the representation of an automatic structure and the complexity of the model checking problem. In particular, every automatic structure has a presentation over a binary alphabet but there are automatic structures which do not have presentations over a unary (one letter) alphabet. Moreover, for automatic structures with unary presentations, the monadic second-order theory (not just the first-order theory) is decidable. There are also feasible time bounds on deciding the first-order theories of automatic structures over the unary alphabet ([3,14]).

In this paper, we restrict our attention to *injection structures*. We begin by fixing notation and terminology. We let $\mathbb{N} = \{0, 1, 2, \ldots\}$ denote the natural numbers and $\mathbb{Z} = \{0, \pm 1, \pm 2, \ldots\}$ denote the integers. We let ω denote the order type of $\mathbb{N}$ under the usual ordering and Z denote the order type of $\mathbb{Z}$ under the usual ordering. For any finite nonempty set Σ, we let Σ^* denote the set of all words over the alphabet Σ. We let ϵ denote the empty word and for any word $w = w_1 \ldots w_n$, we let $|w| = n$ denote the length of w. We let $\Sigma^+ = \Sigma^* - \{\epsilon\}$ and for $n \in \mathbb{N}$, $\Sigma^{\leq n} = \{w \in \Sigma^* : |w| \leq n\}$. An injection is a one-place one-to-one function and an injection structure $\mathcal{A} = (A, f)$ consists of a set A and an injection $f : A \to A$. A is a *permutation structure* if f is a permutation of A. Given $a \in A$, the orbit $\mathcal{O}_f(a)$ of a under f is

$$\mathcal{O}_f(a) = \{b \in A : (\exists n \in \mathbb{N})(f^n(a) = b \ \lor \ f^n(b) = a)\}.$$

The order $|a|_f$ of a under f is $card(\mathcal{O}_f(a))$. We define the *character* $\chi(A)$ of an injection structure $A = (A, f)$ by

$$\chi(A) = \{(n, k) : A \text{ has at least } n \text{ orbits of size } k\}$$

and its *type* $T(A)$ by

$$T(A) = \{(n, k) : k \in \mathbb{N} - \{0\} \ \& \ A \text{ has exactly } n \text{ orbits of order type } k\}.$$

Infinite orbits of injection structures (A, f) are either of type Z, in which every element is in the range of f, or of type ω, which have the form $\mathcal{O}_f(a) = \{f^n(a) : n \in \mathbb{N}\}$ for some $a \notin ran(f)$. It is easy to see that the type of an injection structure plus the finite information about the number of Z-orbits and ω-orbits completely characterizes its isomorphism type.

The algorithmic properties of injection structures were first studied in [5,6]. Recall that if A is a structure with universe A for a language $\mathcal{L}$, then $\mathcal{L}^A$ is the language obtained by expanding $\mathcal{L}$ by constants for all elements of A. The *atomic diagram* of A is the set of all quantifier-free sentences of $\mathcal{L}^A$ true in A. A structure A is *computable* if its atomic diagram is computable. We call two structures *computably isomorphic* if there is a computable function that is an isomorphism between them. A computable structure A is *relatively computably isomorphic* to a possibly noncomputable structure B if there is an isomorphism between them that is computable in the atomic diagram of B. A computable structure A is *computably categorical* if every computable structure that is isomorphic to A is computably isomorphic to A. A computable structure A is *relatively computably categorical* if every structure that is isomorphic to A is relatively computably isomorphic to A.

In [6], Cenzer, Harizanov, and Remmel characterized computably categorical injection structures, and showed that they are all relatively computably categorical. Among other things, they proved that a computable injection structure A is computably categorical if and only if it has finitely many infinite orbits. They also characterized Δ_2^0-categorical injection structures as those with finitely many orbits of type ω, or with finitely many orbits of type $\mathbb{Z}$. They showed that they coincide with the relatively Δ_2^0-categorical structures. Finally, they proved that every computable injection structure is relatively Δ_3^0-categorical. They also showed that the character of any computable injection structure is a c.e. set and that any c.e. character may be realized by a computable injection structure.

A deterministic finite automaton (DFA) is specified by the tuple $(Q, \iota, \Sigma, \delta, F)$ where Q is the finite set of states, ι is the initial state, Σ is the input alphabet, $\delta : Q \times \Sigma \to Q$ is the (possibly partial) transition function, and $F \subseteq Q$ is the set of final, or accepting states. A DFA M *accepts* a string w if the last input of w causes M to halt in one of the accepting. The set $L(M) \subseteq \Sigma^*$ of strings accepted by M is the language *recognized* by M. A language $L \subseteq \Sigma^*$ is said to be *regular* if it is accepted by some DFA. To recognize a relation $R \subseteq \Sigma^* \times \Sigma^*$, there are two possibilities. We can have a two-tape synchronous DFA, where the transition function $\delta : Q \times \Sigma \cup \{\square\} \times \Sigma \cup \{\square\} \to Q$ and $\square$ denotes a blank square. The blank square is needed in the case that one input is longer than the other. Then M halts after reaching the end of the longer word. The other approach is to simulate this with a standard one tape DFA over the language $\Sigma \cup \{\square\} \times \Sigma \cup \{\square\}$. Automatic relations and structures have been studied by Khoussainov, Minnes, Nies, Rubin, Stephan and others [9,11,12,14].

A deterministic finite-state transducer (DFST) is specified by the tuple $(Q, \iota, \Sigma, \Gamma, \delta, \tau)$ where Q is the finite set of states, ι is the initial state, Σ is the input alphabet, Γ is the output alphabet, $\delta : Q \times \Sigma \to Q$ is the (possibly partial) transition function, and $\tau : Q \times \Sigma \to \Gamma^*$ is the (possibly partial) output function. A DFST M naturally defines a (possibly partial) function, $f_M : \Sigma^* \to \Gamma^*$. We say that the DFST M *realizes* a function f on a set $U \subseteq \Sigma^*$ if $f_M \upharpoonright U = f$.

Definition 2. *An injection structure $\mathcal{A} = (A, f)$ consists of a set A together with a one-to-one mapping $f : A \to A$. $\mathcal{A}$ is an FST injection structure if A is a regular set of words in Σ^* (for some alphabet Σ), that is, the set of words accepted by some finite automaton, and f is realized by a DFST. By convention, we will assume that if $\epsilon \in A$, then $f(\epsilon) = \epsilon$.*

Next we shall use the notion of accepting deterministic finite-state transducers to give an equivalent definition of FST injection structures.

Definition 3. *An* accepting deterministic finite-state transducer *(AFST) is specified by the tuple $(Q, \iota, \Sigma, \Gamma, \delta, \tau, F)$ where F is a subset of states designated as accepting states. The function defined by this AFST has as its domain the set of words in Σ^* such that, when processing these words, the AFST ends in a state in F.*

Since the isomorphism class of an injection structure is determined by its type, and the number of orbits of type ω and $\mathbb{Z}$, we seek to characterize those types which have presentations using transducers. We prove that the two kinds of transducers defined above realize the same isomorphism types of injection structures.

Theorem 4. *An isomorphism type of an injection structure is realized by an FST injection structure if and only if that structure can be realized by an AFST.*

Proof. An AFST $(Q, \iota, \Sigma, \Gamma, \delta, \tau, F)$ is naturally associated with a DFST M (by omitting F). The automaton for the domain of the function is then the DFA $(Q, \iota, \Sigma, \delta, F)$, recognizing the set U. Then (U, f_M) is the original FST injection structure.

Conversely, suppose the DFST $(Q, \iota, \Sigma, \delta, F)$ and the DFA $(Q', \iota', \Sigma, \delta', F')$ are given. Form the product automaton of the two (analogously to the classical automata constructions). The resulting DFST can be augmented with accepting states which are all those whose second component is an accepting state in F.

Let $\mathcal{A} = (A, f)$ and $\mathcal{B} = (B, g)$ be injection structures. We let $\mathcal{A} \otimes \mathcal{B} = (A \times B, f \times g)$ where for any $a \in A$ and $b \in B$, $f \times g(a, b) = (f(a), g(b))$. We let $\mathcal{A} \oplus \mathcal{B} = (\{x_1 a : a \in A\} \cup \{x_2 b : b \in b\}, f \oplus g)$ where x_1 and x_2 are new letters outside of $A \cup B$ and $f \oplus g(x_1 a) = x_1 f(a)$ and $f \oplus g(x_2 b) = x_2 g(b)$ for all $a \in A$ and $b \in B$. It is easy to see that if $\mathcal{A} = (A, f)$ and $\mathcal{B} = (B, g)$ are FST injection structures, then $\mathcal{A} \otimes \mathcal{B}$ and $\mathcal{A} \oplus \mathcal{B}$ are FST injection structures.

The main goal of this paper is explore the difference between computable injection structures and FST injection structures. Our first result is that the model checking problem for FST injection structures is not decidable. This contrasts with the fact that the model checking problem for automatic relational structures is decidable. We shall also study the possible types of FST injection structures. Whether a given type may be realized by an FST injection structure (A, f) depends on a number of factors such as (i) the underlying alphabet Σ, (ii) the number of states in of the underlying transducer $T = (Q, \iota, \Sigma, \Gamma, \delta, \tau)$ for f, and (iii) the nature of the output function τ of the transducer T. For example, we say that a transducer $T = (Q, \iota, \Sigma, \Gamma, \delta, \tau)$ *has ϵ-outputs* if there is state $q \in Q$ and $a \in \Sigma$ such that $\tau(q, a) = \epsilon$. We say T is *length preserving* if $\tau(q, a) \in \Gamma$ for all $q \in Q$ and $a \in \Sigma$. Thus when a length preserving transducer reads a symbol $a \in \Sigma$ in any state q, it outputs a single letter in Γ. We say that (A, f) has full domain if $\Sigma^+ \subseteq A$.

The outline of this paper is as follows. In Sect. 2, we will show that the model checking problem for FST injection structures is undecidable; this is done by coding computations of reversible Turing machines into FST injection structures. In Sect. 3, we will show that a large class of types can be realized by FST injection structures. For example, we will construct FST injection structures as summarized in the following table. Note that the columns describing the FST realization list the parameters for the examples we build; they may not be optimal.

Type	Alphabet	ϵ-outputs?	States	Full domain?	Lemma
1_m	$\{1\}$	N	1	N	1
1_∞	$\{1\}$	N	1	Y	1
$1_m, \omega_n$	$\{1\}$	N	$m + 1$	Y	2
$1_m, \omega_\infty$	$\{1\}$	N	$m + 1$	Y	3
k_1	$\{0, 1\}$	Y	$2^{k+1} - 1$	N	4
k_∞	$\{0, 1\}$	Y	$2^{k+1} - 1$	N	4
ζ_1	$\{0, 1\}$	Y	3	N	5
ζ_∞	$\{0, 1\}$	Y	3	N	5
ℓ_∞	$\{0, 1, \ldots, \ell - 1\}$	N	1	N	6
ℓ_1^i	$\{0, 1, \ldots, \ell - 1\}$	N	2	Y	7
ℓ_∞^i	$\{0, 1, \ldots, \ell - 1\}$	N	2	Y	7

Here for any positive integers k and m, we write k_m to denote the character which has m orbits of size k and we write k_∞ to denote the character with infinitely orbits of size k. Similarly, we write ω_m to denote the character that has m ω orbits, ω_∞ to denote the character with infinitely many ω orbits, ζ_m to denote the character which has m Z orbits, and ζ_∞ to denote the character with infinitely many Z orbits. Then, for example, $1_m, \omega_n$ in line 3 of the table means an injection structure with m orbits of size 1 and n orbits of type ω, ζ_1 on line 7 means an injection structure with exactly one orbit which is of type Z, and ζ_∞ on line 8 means an injection structure with infinitely many orbits of type Z. We write ℓ_1^i for the type of injection structures with exactly one orbit of size ℓ^i for each $i \geq 1$. Similarly, ℓ_∞^i is the type with infinitely many orbits of size ℓ^i for each $i \geq 1$. Of course, the fact that FST injection structures are closed under $\otimes$ and $\oplus$ means that we can use these simple types to obtain many other types which are realized by FST injection structures.

In Sect. 4, we shall characterize all types realized by FST injection structures over a unary alphabet as the ones already listed in the table above. In particular, it will follow that FST injection structures over a unary alphabet are not closed under disjoint union. However, we will show that any type realized by an FST injection structure can be realized by an FST injection structure over a two-letter alphabet. In Sect. 5, we will prove a restriction on the types that can be realized by FST injection structures which are length preserving and have full domain. In Sect. 6, we will address the question of when FST injection structures have automatic graphs. Finally, in Sect. 7, we will state some open problems.

2 Undecidability

In this section, we prove that the model checking problem for injection structures is undecidable. We fix the first-order language L_f to have a single unary function symbol f. The model checking problem for L_f is the problem of, given a presentation of an FST structure $\mathcal{A}$ over L_f and given a first-order formula φ, deciding whether $\mathcal{A} \vDash \varphi$.

Theorem 5. *The model checking problem for FST injection structures is undecidable for the formula* $\exists x(f(x) = x)$.

Proof. We use Bennett's reversible Turing machines. Theorem 1 of [1] shows that any Turing machine can be simulated by a 3-tape Turing machine N which is reversible; namely, every configuration of N has at most one predecessor configuration. As a consequence, the halting problem for 3-tape reversible Turing machines is undecidable. Such a machine N can be simulated by a single tape machine M by letting the tape contents of M encode a configuration of N. For this, M's tape contains three "tracks", one for each tape of N. Each track holds a symbol from one of N's tapes plus a flag indicating whether the corresponding tape head is positioned over that symbol. The state of M incorporates the information of the current state of N and the current symbols under N's three tape heads. This allows M to simulate N by scanning its entire tape repeatedly from left-to-right. As M scans from left-to-right, it updates the contents its tapes to reflect the change in the configuration caused by a single step of N. The scan from right-to-left merely returns to the left end and prepares for simulating the next step of N.

A configuration of M can be encoded as a string C over the alphabet L_M which contains the tape symbols of M and symbols for each state of M. In C, the symbol for M's state immediately follows the symbol under the tape head. Such a string C is called a *valid configuration* provided that the tape contents as encoded by C satisfy the following four conditions:

(a) for each of N's tapes, there is exactly one symbol in C indicating the presence of N's tape head on that tape,
(b) there is one symbol in C encoding M's tape head position,
(c) the state of M as encoded in C correctly agrees with the relative positions of N's tape heads and the symbols currently being read by N, and
(d) C does not have leading or trailing blank symbols.

We omit the straightforward details, but the valid configurations satisfy the following four properties.

(**A**) Every valid configuration C has a valid successor configuration, $Nxt(C)$, representing the next (also valid) configuration of M. $Nxt(C)$ is defined even for halting configurations of M (i.e., M continues running after reaching a halting configuration). By reversibility, the Nxt function is injective.
(**B**) Every valid configuration of M has at most one valid predecessor configuration.
(**C**) The set of valid configurations is regular.
(**D**) There is a DFST which, when given valid configuration C as input, outputs the next configuration $Nxt(C)$.

Without loss of generality, when (and if) N halts it first erases its output and work tape — leaving the input tape with its original contents. We are interested only in running N on the empty word as input. We let C_{init} represent the initial configuration for N. We likewise let C_{final} represent the accepting configuration of M which results if N halts after having the empty string as input. By the convention on how N halts, C_{final} is a known, fixed configuration.

We give a many-one reduction from the halting problem for a 3-tape reversible machine N to the question of whether an FST injection structure $\mathcal{A}$ satisfies the formula $\exists x(f(x) = x)$. The structure $\mathcal{A}$ uses the alphabet $L_M \cup \{|, \$\}$ where "$|$" and "$\$$" are new symbols. The domain of $\mathcal{A}$ is the set of words of the form

$$C_1|C_2|C_3|\cdots|C_k|\$^i||, \tag{1}$$

where each C_i is a valid configuration, $k \geq 1$, and $\$^i$ denotes i many $\$$ symbols with $i \geq 0$. The domain of $\mathcal{A}$ is regular since the set of valid configurations is regular. We define the injective function f as follows. For any string v of the form specified by (1), $f(v)$ is defined according the following four cases.

(1) If $i > 0$, then $f(v)$ is equal to $C_{\text{init}}|Nxt(C_1)|Nxt(C_2)|\cdots|Nxt(C_k)|\$^{i+1}||$. Note the additional \$ symbol.

(2) If $i = 0$ and $k > 0$ and both $Nxt(C_{k-1})$ and C_k are equal to C_{final}, then $f(v)$ is equal to $C_{\text{init}}|Nxt(C_1)|Nxt(C_2)|\cdots|Nxt(C_{k-1})|||$.

(3) If $i = 0$ and $Nxt(C_k)$ is equal to C_{final}, then $f(v)$ is equal to $C_{\text{init}}|Nxt(C_1)|Nxt(C_2)|\cdots|Nxt(C_k)|\$||$. Note the added \$ symbol.

(4) Otherwise, $f(v)$ is equal to $C_{\text{init}}|Nxt(C_1)|Nxt(C_2)|\cdots|Nxt(C_k)|||$. Note there is no \$-symbol. In this case, $Nxt(C_k)$ is not equal to C_{final} since otherwise case **(3)** would apply.

By inspection, if v is in the domain of $\mathcal{A}$, then so is $f(v)$. It is straightforward to see that f can be computed by a deterministic finite state transducer. The DFST starts out by writing "$C_{\text{init}}|$"; it then repeatedly reads a string a "$C_i|$" while writing "$C_{Nxt(C_i)}|$". The string "C_{final}" is a fixed string, so the DFST can detect when it has reached the end of v and which of the cases **(1)**-**(4)** applies by looking forward only a constant number of symbols. (The purpose of the extra "$||$" at the end of v to let the DFST detect when it has reached C_k with C_k equal to C_{final} without having to read the last symbol of v.)

To see that f is injective, fix a string w in $\mathcal{A}$ of the form (1). We must show that $f(v) = w$ can hold for at most one v. If w has the form (1) with $i > 0$, then any $f(v) = w$ must be defined according to case **(1)** or **(3)** depending on whether $i = 1$ or $i > 1$. In either case, the injectivity of the Nxt function ensures that v is unique. Otherwise $i = 0$. If the C_k in w is not equal to C_{final}, then case **(2)** does not apply and so case **(4)** was used to define $f(v)$. Thus the uniqueness of v again follows by the injectivity of the Nxt. Finally, if $i = 0$ and the C_k in w does equal C_{final}, then $f(v) = w$ is defined by case **(2)**, and once again, injectivity of Nxt implies the uniqueness of v.

To finish the proof of Theorem 5, we show that $\mathcal{A} \models \exists x(f(x) = x)$ holds if and only if N halts when started on the empty string. When calculating $f(v)$, the only way to have $f(v) = v$ is for case **(2)** to apply, since the other three cases add a symbol "$|$", and maybe a "\$". If case **(2)** gives $f(v) = v$, then we have $C_1 = C_{\text{init}}$, and $C_j = Nxt(C_{j-1})$ for all j, and $C_k = C_{\text{final}}$. Thus the configurations in v are a halting computation of M. Conversely, a halting computation of M gives v such that $f(v) = v$. $\qquad\square$

It remains open whether there is a FST injection structure with undecidable first-order theory. However, we can build FST injection structures with some level of undecidability. Let $Fin(M)$ denote $\{x : x$ belongs to a finite orbit of $M\}$.

Theorem 6. *There is a FST N such that $Fin(N)$ is c.e. complete.*

Proof. Let W be a complete c.e. set and let A be the set $\{0^n : n \in W\}$. Then there is a reversible Turing machine M such that for any string $\sigma \in \{0,1\}^*$, M converges on σ if and only $\sigma \in A$. We assume that M is realized by a single tape Turing machine as described in Theorem 5. We can modify M as follows to produce a Turing machine T. First, since M is reversible, it is easy to see that there is Turing machine $\overline{M}$ that operates as follows. For any state q of M,

$\overline{M}$ has a state $\overline{q}$ where we assume that states of M and $\overline{M}$ are disjoint. We assume that M has a single start state s and single final state f. Then the start state of $\overline{M}$ is $\overline{f}$ and the final state of $\overline{M}$ is $\overline{s}$. For any configuration C, let $\overline{C}$ denote the configuration obtained from C by replacing the state q of the configuration by $\overline{q}$. For any input $\sigma \in \{0, 1\}$, suppose that M starts on an initial configuration I_σ for input σ and proceeds through a sequence of configurations $C_1, \ldots, C_r$ and then ends with a configuration F whose state is the final state f. Then the transition table of $\overline{M}$ is defined so that if $\overline{M}$ starts on configuration $\overline{F}$, it will proceed through the sequence of configurations $\overline{C_r}, \ldots, \overline{C_1}$ and then end in configuration $\overline{I_\sigma}$. Our Turing machine T will operate as follows. It has a new start state s_0 and its final state is $\overline{s}$. Given an initial configuration I_σ for input σ, it first goes to state s and does not move. Then it uses the states and transition function of M to proceed through the sequence of states $I_\sigma, C_1, \ldots, C_r, F$. On seeing the final state f of M, T goes to state $\overline{f}$ and does not move. This will result in configuration $\overline{F}$. Then T uses the states and the transition function of $\overline{M}$ to go through the sequence of configurations $\overline{F}, \overline{C_r}, \ldots, \overline{C_1}, \overline{I_\sigma}$. Finally on seeing state $\overline{s}$, it goes to state s_0 and does not move so that we will end the initial configuration for input σ. Thus on a computation of M with input σ that converges, T will produce a cycle of configurations.

It is easy to see that configurations of T will be a regular language L and we can define an FST G so that on any configuration of T, G outputs the result of applying T to configuration C. It follows that the only cycles of (L, G) correspond to converging computations of M. Thus if we could decide whether a configuration is part of cycle, then we can decide whether M started on input 0^n converges. It follows that $Fin(L, G)$ is c.e. complete.

3   Realizing Injection Structures with Simple Types

Lemma 1. *Let $m \in \mathbb{N}$. There is a FST injection structure (over a unary alphabet) realizing each of the types $\{1_m\}$ and $\{1_\infty\}$.*

Proof. The identity function on $\{1^0, \ldots, 1^{m-1}\}$ and 1^* (respectively) has this type. Each of these domains is regular and the function can be realized by the one-state DFST $(\{\iota\}, \iota, \{1\}, \{1\}, \{(\iota, 1, \iota)\}, \{(\iota, 1, 1)\})$.

Lemma 2. *Let $m, n \in \mathbb{N}$. There is an FST injection structure (over a unary alphabet) realizing the type $\{1_m, \omega_n\}$.*

Proof. The function

$$1^i \mapsto \begin{cases} 1^i & \text{if } 0 \le i < m \\ 1^{i+n} & \text{if } i \ge m \end{cases}$$

has this type. In particular, the m-many 1-cycles have elements in the set $\{\epsilon, 1, \ldots, 1^{m-1}\}$; the n-many ω-cycles can be described using arithmetic progressions with bases $m, \ldots, m+n-1$ and successive difference n. This function can be realized by the $(m+1)$-state DFST $(\{\iota, q_1, \ldots, q_{m-1}, q'\}, \iota, \{1\}, \{1\}, \delta, \tau)$ where $\delta(\iota, 1) = q_1$, $\delta(q_i, 1) = q_{i+1}$ for $1 \leq i < m-1$, $\delta(q_{m-1}, 1) = q'$, and $\delta(q', 1) = q'$. Also, $\tau(q, 1) = 1$ for $q = \iota$ or $q = q'$ or $q = q_i$ with $1 \leq i < m-1$, and $\tau(q_{m-1}, 1) = 1^{n+1}$.

Lemma 3. *Let $m \in \mathbb{N}$. There is a FST injection structure (over a unary alphabet) realizing the type $\{1_m, \omega_\infty\}$.*

Proof. The function

$$1^i \mapsto \begin{cases} 1^i & \text{if } 0 \leq i < m \\ 1^{2i} & \text{if } i \geq m. \end{cases}$$

has this type. In particular, the m-many 1-cycles are on the domain $\{\epsilon, 1, \ldots, 1^{m-1}\}$; the infinitely-many ω-cycles can be described using arithmetic progressions each of whose base is equal to its successive difference. This function can be realized by the $(m+1)$-state DFST $(\{\iota, q_1, \ldots, q_{m-1}, q'\}, \iota, \{1\}, \{1\}, \delta, \tau)$ where $\delta(\iota, 1) = q_1$, $\delta(q_i, 1) = q_{i+1}$ for $1 \leq i < m-1$, $\delta(q_{m-1}, 1) = q'$, and $\delta(q', 1) = q'$. Also, $\tau(q, 1) = 1$ for $q = \iota$ or $q = q_i$ with $1 \leq i < m-1$, $\tau(q', 1) = 11$, and $\tau(q_{m-1}, 1) = 1^{m+1}$.

Lemma 4. *Let $k \in \mathbb{N}$ be greater than or equal to 2. The types $\{k_1\}$ and $\{k_\infty\}$ are realized by FST injection structures (with ϵ-outputs) over the alphabet $\{0, 1\}$.*

Proof. Fix k and let $\sigma^{(1)}, \ldots, \sigma^{(k)}$ be the first k binary strings of length k in lexicographic (dictionary) order. For example, if $k = 3$, then $\sigma^{(1)} = 000$, $\sigma^{(2)} = 001$ and $\sigma^{(3)} = 010$. Consider a transducer with ϵ-outputs on $\{0, 1\}$ whose set of states is the set of all $\tau \in \{0, 1\}^{\leq k}$. For each $q \in \{0, 1\}^{\leq k}$, each $i \in \{0, 1\}$ the transition function and output functions are given by

$$\delta(q, i) = \begin{cases} qi & \text{if } |q| < k \\ q & \text{if } |q| = k. \end{cases} \qquad \tau(q, i) = \begin{cases} \epsilon & \text{if } |q| < k \text{ and } qi \neq \sigma^{(s)} \text{ for any } s \\ \sigma^{(s+1)} & \text{if } |q| < k \text{ and } qi = \sigma^{(s)} \text{ and } s < k \\ \sigma^{(1)} & \text{if } |q| < k \text{ and } qi = \sigma^{(k)} \\ i & \text{if } |q| = k. \end{cases}$$

For example, the transducer for $k = 3$ is given below.

It is then easy to see that if we restrict A to be $\{\sigma^{(1)}, \ldots, \sigma^{(k)}\}$, then (A, f) will consist of a single k cycle and if we restrict A to be all strings that extend one of $\sigma^{(1)}, \ldots, \sigma^{(k)}$ (a regular set), then (A, f) will have infinitely many k cycles (Fig. 1).

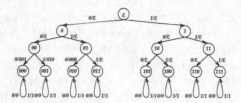

Fig. 1. The transducer constructed in Lemma 4 when $k = 3$.

Lemma 5. *There are FST injection structures (with ϵ-outputs) realizing the types $\{\zeta_1\}$ and $\{\zeta_\infty\}$ on domains which are regular subsets of $\{0,1\}^*$.*

Proof. Consider the function $1^i 01^j \mapsto 1^{i-1} 01^j$ for $i \geq 1$ and $0^i 1^j \mapsto 0^{i+1} 1^j$ for all i and all $j \in \mathbb{N}$. For each fixed $m \in \mathbb{N}$, we get the chain

$$\cdots 1^i 01^m \mapsto 1^{i-1} 01^m \mapsto \cdots \mapsto 101^m \mapsto 01^m \mapsto 0^2 1^m \cdots$$

The regular set $1^* 0 \cup 0^*$ corresponds to $m = 0$. The set $1^* 01^* \cup 0^* 1^*$ is the union of infinitely many ζ-chains, one for each $m \in \mathbb{N}$.

Regardless of the domain, the function can be realized by the following 3-state DFST with ϵ-outputs, $(\{\iota, q, q'\}, \iota, \{0,1\}, \{0,1\}, \delta, \tau)$ where $\delta(\iota, 1) = q'$, $\delta(\iota, 0) = q$, and $\delta(s, b) = s$ for $s \in \{q, q'\}$ and $b \in \{0,1\}$. Also, $\tau(\iota, 1) = \epsilon$, $\tau(\iota, 0) = 00$, and $\tau(s, b) = b$ for $s \in \{q, q'\}$ and $b \in \{0,1\}$.

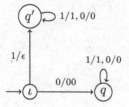

Lemma 6. *Fix $\ell \in \mathbb{N}$, $\ell > 0$. There is an FST injection structure realizing the type $\{\ell_\infty\}$ over the domain $\{0, 1, \ldots, \ell-1\}^+$.*

Proof. Consider the function which cycles each input letter through the letters in the alphabet. Then length is preserved and each nonempty word belongs to a cycle of length ℓ. This function can be realized by the one-state DFST $(\{\iota\}, \iota, \{1\}, \{1\}, \{(\iota, i, \iota) : 0 \leq i < \ell\}, \{(\iota, \ell-1, 0), (\iota, i, i+1) : 0 \leq i < \ell-1\})$.

$$0/1, 1/2, \ldots, \ell-1/0$$

$$\rightarrow \overset{\curvearrowright}{\underset{\iota}{\bigcirc}}$$

Lemma 7. *Fix $\ell \in \mathbb{N}$, $\ell > 1$. There is a two-state DFST realizing the type $\{\ell_1^i : i \in \mathbb{N}\}$ over the domain $\{0, 1, \ldots, \ell-1\}^*$. Namely, for each length k, all words of length k are in a single cycle and so there is one cycle of length each power of ℓ.*

Proof. Consider the function which treats strings over an ℓ-letter alphabet as representations of integers in base-ℓ, least significant bit first, and subtracts one. More explicitly, we associate a value $v(x_0 \cdots x_{n-1})$ with each string in $\{0, \ldots, \ell - 1\}^*$ by defining.

$$v(x_0 \cdots x_{n-1}) = \sum_i x_i \ell^i.$$

Then

$$f(x_0 \cdots x_{n-1}) = \begin{cases} \underbrace{(\ell - 1) \cdots (\ell - 1)}_{n \text{ times}} & \text{if } x_i = 0 \text{ for all } i \\ y_0 \cdots y_{n-1} & \text{such that } v(y_0 \cdots y_{n-1}) = v(x_0 \cdots x_{n-1}) - 1. \end{cases}$$

The function f is realized by the two-state DFST

$$(\{\iota, c\}, \iota, \{0, 1, \ldots, \ell - 1\}, \{0, 1 \ldots, \ell - 1\}, \{(\iota, 0, \iota), (\iota, i, c), (c, j, c) : 1 \leq i \leq \ell - 1, 0 \leq j \leq \ell - 1\},$$
$$\{(\iota, 0, \ell - 1), (\iota, i, i - 1), (c, j, j) : 1 \leq i < \ell - 1, 0 \leq j \leq \ell - 1\}).$$

Of course, the fact that FST injection structures are closed under $\oplus$ and $\otimes$ means that we can use these structures to construct more complicated FST injection structures. For example if $\mathcal{A} = (A, f)$ and $\mathcal{B} = (B, g)$ are FST injection structures realizing 2^i_∞ and 3^i_∞, respectively, then it is easy to see that if $w \in A$ has orbit size 2^s and $v \in B$ has orbit size 3^t, then (w, v) must have orbit size $2^s 3^t$ in $\mathcal{A} \otimes \mathcal{B}$ so that will have infinitely many orbits of size $2^s 3^t$ for all $s, t \geq 1$. Similarly, if $\mathcal{C} = (C, f)$ and $\mathcal{D} = (D, g)$ are FST injection structures realizing the types $\{\omega_\infty\}$ and $\{\zeta_\infty\}$, respectively, then $\mathcal{C} \oplus \mathcal{D}$ realizes the type $\{\omega_\infty, \zeta_\infty\}$. Note in this situation, $\mathcal{C}$ and $\mathcal{D}$ are Δ^0_2-categorical but not computably categorical and $\mathcal{C} \oplus \mathcal{D}$ is Δ^0_3-categorical but not Δ^0_2-categorical.

The type of an injection structure (A, f) is said to be *bounded* if there is a finite n such that the size of any finite cycle in (A, f) is less than n. Our results show the following.

Theorem 7. *Any bounded type can be realized by an FST injection structure.*

Theorem 8. *There is an FST injection structure such that (A, f) has exactly one k-cycle for every $k \geq 1$ and no infinite chains.*

Proof. First we shall build an FST injection structure (E, g) such that (E, g) has an exactly one $2n + 3$ cycle for every $n \geq 0$ and no other cycles. The underlying alphabet A will be $\{*, R, L, 1\}$ Each $2n + 3$-cycle of (E, g) will contain the string $*1^n*$. The next string in this cycle is the result of replacing the first $*$ with R

which we think of as move right symbol. This is followed by the sequence of strings that result by moving the symbol R one space to the right until it is next to the second $*$ at which point the next string in the cycle is the result of replacing R by L. We think of L as a move left symbol. Then the next set of strings in the cycle is the sequence of strings that result by moving the L one symbol to the left until it reaches the start. At that point, we replace L by $*$ in which case we are back where we started. For example if $\alpha = *111*$, then the cycle that contains α will be

$$\alpha = *111*$$
$$g(\alpha) = R111*$$
$$g^2(\alpha) = 1R11*$$
$$g^3(\alpha) = 11R1*$$
$$g^4(\alpha) = 111R*$$
$$g^5(\alpha) = 111L*$$
$$g^6(\alpha) = 11L1*$$
$$g^7(\alpha) = 1L11*$$
$$g^8(\alpha) = L111*$$
$$g^9(\alpha) = *111*$$

Let $E = *\{1\}^* * \cup \{1\}^* R\{1\}^* * \cup \{1\}^* L\{1\}^* *$ which is clearly a regular set. We shall let C stand for a copy state. That is, in state C, we read a letter a, we output a and stay in state C. Let S_0 be the start state. Then in state S_0 do the following,

1. If we read $*$, we output R and go to the copy state C.
2. If we read L, we output $*$ and go to the copy state C.
3. If we read 1, we output ϵ and go to state S_1.
4. If we read R, we output ϵ and go to state S_R.

The idea is that in state S_0 if we read either 1 or R, we output nothing but we remember where we came from. In state S_1, we do the following.

1. If we read 1, we output 1 and go to state S_1.
2. If we read L, we output $L1$ and go to the copy state C.
3. If we read R, we output $1R$ and go to the copy state C.
4. We will never read $*$ in state S_1 since we must see either R or L first, but for completeness if we read $*$, then we print $*$ and go to the copy state C.

Finally in state S_R, we do the following.

1. If we read 1, we output $1R$ and go to the copy state C.
2. If we read $*$, we output $L*$ and go to the copy state C.
3. We will never read R or L in state S_R since we must see $*$ first, but for completeness if we read R or L, then we print $*$ and go to the copy state.

It is easy to check that this defines a FST function g such that every cycle contains $*1^n*$ for some n and this cycle will be of length $2n+3$. That is, we have one step to replace $*$ by R, n steps to move R past the 1s, one step to replace R by L, n steps to move L past the 1s, and one step to replace L by $*$. Thus (E, g) has the desired behavior.

Next we modify the construction of (E, g) to produce an FST injection structure such that (F, h) has an exactly one $2n + 4$ cycle for every $n \geq 0$ and no other cycles. The underlying alphabet A will be $\{*, D, R, L, 1\}$. We will think of the D symbol as delay symbol.

Each $2n+4$-cycle of (F, h) will contain the string $*1^n*$. Then our first step is to replace $*$ by D and our second step is to replace D by R and then we proceed as in (E, g). This will essentially add an extra string to every cycle. For example, For example if $\alpha = *111*$, then the (F, h) cycle of α will be

$$\alpha = *111*$$
$$h(\alpha) = D111*$$
$$h^2(\alpha) = R111*$$
$$h^3(\alpha) = 1R11*$$
$$h^4(\alpha) = 11R1*$$
$$h^5(\alpha) = 111R*$$
$$h^6(\alpha) = 111L*$$
$$h^7(\alpha) = 11L1*$$
$$h^8(\alpha) = 1L11*$$
$$h^9(\alpha) = L111*$$
$$h^{10}(\alpha) = *111*$$

Let $F = *\{1\}^* * \cup \{1\}^* R \{1\}^* * \cup \{1\}^* L \{1\}^* * \cup D\{1\}^* *$ which is clearly a regular set. Again we let C be the copy state for this alphabet. Let S_0 be the start state. Then we only have to do the following modification in state S_0 do the following,

1. If we read $*$, we output D and go to the copy state C.
2. If we read D, we output R and go to the copy state.
3. If we read L, we output $*$ and go to the copy state C.
4. If we read 1, we output ϵ and tò state S_1.
5. If we read R, we output ϵ and go to state S_R.

In states S_1 or S_R, we will never read D, but for completeness we say that if we read D in either state S_1 or S_R, then we print $*$ and go to the copy state C.

It is easy to check that this defines a FST injection structure (F, h) which has the described behavior.

Finally we let (G, k) be a FST injection structure with one cycle of length 1 and one cycle of length 2. Then the FST injection structure $(E, g) \oplus (F, h) \oplus (G, k)$ has exactly one cycle of length k for every $k \geq 1$.

We note that it easy to modify the FST injection structure (A, f) to produce a FST injection structure (A_k, f_k) for any $m \geq 2$ such that (A_m, f_m) has exactly m cycles of length n for every $n \geq 1$. That is, let q be a symbol which is not in the underlying alphabet of A. Then the strings of (A_k, f_k) will be of the form $q^j \alpha$ where $1 \leq j \leq k$ and $\alpha \in B$. In the start state, when we read a q we output q and go back to the start state. Similarly, we can produce a FST injection structure (A_∞, f_∞) such that (A_∞, f_∞) has infinitely many cycles of length n for every $n \geq 1$. That is, let q be a symbol which is not in the underlying alphabet of A. Then the strings of (A_∞, f_∞) will be $\{q\}^* A$ where again in the state, when we read a q we output q and go back to the start state.

4   Types over Unary Alphabets

Theorem 9. *The only types that can be realized over a unary alphabet are $\{1_\infty\}$, $\{1_m, \omega_n\}$, or $\{1_m, \omega_\infty\}$.*

Proof. Our results in Sect. 3 show that the types mentioned above can each be realized by FST injection structures over the alphabet $\{1\}$.

The characterization will be complete once we prove that these are the only possible types. We do so in the next two lemmas. For simplicity, we will identify 1^i with i for this discussion. Thus the natural order on the integers i, j becomes order by length on strings $1^i, 1^j$.

Lemma 8. *If $\mathcal{A} = (A, f)$ is an FST injection structure over $\{1\}^*$, then f is monotonic, that is, $i < j$ implies $f(i) < f(j)$.*

Proof. For any $u, v \in A$, since f is an injection, if $u < v$, then $f(u) \neq f(v)$. Moreover, by definition of FSTs, extensions can only add to the output and therefore, if $u < v$ then $f(u) < f(v)$.

Lemma 9. *If $\mathcal{A}$ is an FST injection structure over $\{1\}^*$, then*

(1) $\mathcal{A}$ has no finite cycles of length greater than 1,
(2) $\mathcal{A}$ has no Z-chains, and
(3) if $\mathcal{A}$ has an ω-chain, then $\mathcal{A}$ has only finitely many 1-chains.

Proof. (1) Let $k > 1$ and suppose towards a contradiction that $w_1, w_2 = f(w_1), \ldots, w_k = f(w_{k-1})$ is a k-cycle. It follows that $w_1 \neq w_2$. Assume without loss of generality that $w_1 < w_2$. Then by Lemma 8, $w_2 < w_3 < \cdots < w_{k-1} < w_k$ and finally $w_k < w_1$, so that $w_1 < w_1$ by transitivity.

(2) Next suppose that $\ldots, w_{-2}, w_{-1}, w_0, w_1, \ldots$ is a Z-chain. Then each w_i is distinct and, since there can be no infinite decreasing sequence, there must be some i such that $w_{i+1} = f(w_i) > w_i$. It now follows from Lemma 8, by induction, that $w_{j+1} > w_j$ for all j. But then the sequence $w_0, w_{-1}, w_{-2}, \ldots$ is an infinite decreasing sequence.

(3) Finally, suppose that $w_0, w_1, \ldots$ is an ω-chain. Then as in (2), there must be some i such that $w_{i+1} = f(w_i) > w_i$. But it then follows from Lemma 8 that

$w_{n+1} > f(w_n)$ for all $n \geq i$. Now if $j > w_i$ and $f(j) = j$, then, for some $n \geq i$, $w_n < j < w_{n+1}$. It follows that $f(j) > f(w_n) = w_{n+1} > j$, so that $\{j\}$ is not a 1-cycle.

Corollary 1. *Unary FST realizable structures are not closed under disjoint unions.*

Proof. Otherwise, one could form the disjoint union of a structure with infinitely many one-cycles and a structure with one ω chain and get a type not in this list.

Corollary 2. *The FO-theory of unary FST realizable injections structures is decidable.*

Proof. In [6], it was shown that for any injection structure $\mathcal{A}$, $Th(\mathcal{A})$ is decidable if and only if $\chi(\mathcal{A})$ is computable. The result then immediately follows from Theorem 9.

In contrast to Theorem 9, we can show that are no such restrictions on types realized by FST injection structures over a two letter alphabet.

Theorem 10. *Any type realizable with a finite alphabet can be realized (with ϵ-outputs) using an alphabet with two symbols.*

Proof. Given an FST injection structure $\mathcal{A} = (A, f)$ over a finite alphabet, we may assume without loss of generality that $card(\Sigma) = card(\Gamma) = 2^n$ for some n and hence we may assume that $\Sigma = \Gamma = \{0, 1\}^n$. Now we can simulate the DFST using alphabet $\{0, 1\}$ as follows. For each state q, we will have states q^σ for every $\sigma \in \{0, 1\}^{<n}$. If $|\sigma| < n - 1$, then the transition is $\delta'(q^\sigma, a) = q^{\sigma a}$ and the output is $\tau'(q^\sigma, a) = \epsilon$. If $|\sigma| = n - 1$, then $\delta'(q^\sigma, a) = \delta(q, \sigma a)^\epsilon$ and the output $\tau'(q^\sigma, a) = \tau(q, \sigma a)$. Thus, from state q, if the next n bits of the input tape are given by σ, then after n steps in the simulation, we will read σ and output $\tau(q, \sigma)$.

5   Restrictions on Types Realized by FST Injection Structures

The main result of this section is a kind of analogue of the Pumping Lemma for length preserving FST injection structures over a finite alphabet Σ and whose domain is Σ^* or Σ^+.

Theorem 11. *Suppose that $\mathcal{A} = (\Sigma^*, f)$ is a length preserving FST injection structure where f is defined by a transducer $T = (Q, \iota, \Sigma, \Gamma, \delta, \tau)$. Then there is a constant $c_{|Q|, |\Sigma|, k}$ depending on $|Q|$, $|\Sigma|$ and k such that if $\mathcal{A}$ has more than $c_{|Q|, |\Sigma|, k}$ orbits of size k, then $\mathcal{A} = (\Sigma^*, f)$ must have infinitely many orbits of size k.*

Proof. Suppose that $\sigma^{(0)}, \ldots, \sigma^{(k-1)}$ is a k-cycle of f, i.e., $f(\sigma^{(i)}) = \sigma^{(i+1)}$ for $0 \leq i < k - 1$ and $f(\sigma^{(k-1)}) = \sigma^{(0)}$. Since $\mathcal{A} = (\Sigma^*, f)$ is length preserving, we must have $|\sigma^{(0)}| = |\sigma^{(1)}| = \cdots = |\sigma^{(k-1)}|$. Now suppose that $|\sigma^{(0)}| = n > |Q|^k |\Sigma|^k + 1$. Then for each $0 \leq i \leq k-1$, let $\sigma^{(i)} = \sigma_1^{(i)} \ldots \sigma_n^{(i)}$ and let $a_j^{(i)}$ denote the state of T after reading letters $\sigma_1^{(i)} \ldots \sigma_{j-1}^{(i)}$. Thus as our FST processes the string $\sigma_1^{(i)} \ldots \sigma_n^{(i)}$, the FST is in state $a_j^{(i)}$ when it reads the letter $\sigma_j^{(i)}$. It follows that there must be some $1 \leq s < t \leq n$ such that $(\sigma_s^{(i)}, a_s^{(i)}) = (\sigma_t^{(i)}, a_t^{(i)})$ for all $i = 1, \ldots, k$. That is, there are only $|Q|^k |\Sigma|^k$ possible choices for the sequence $((\sigma_s^{(1)}, a_s^{(1)}), \ldots, (\sigma_s^{(k)}, a_s^{(k)}))$ as s varies from 1 to n. Hence there must exist such an s and t since $n > |Q|^k |\Sigma|^k + 1$. But this means that for each $1 \leq i \leq k$, as T processes $\sigma_s^{(i)} \ldots \sigma_{t-1}^{(i)}$ starting in state $a_s^{(i)}$, it will end up in state $a_s^{(i)}$ and output $\sigma_s^{(i+1)} \ldots \sigma_{t-1}^{(i+1)}$ where for $i + 1$, we take the addition modulo k. It then follows that for all $j \geq 1$ as T processes $(\sigma_s^{(i)} \ldots \sigma_{t-1}^{(i)})^j$ starting in state $a_s^{(i)}$, it will end up in state $a_s^{(i)} = a_t^{(i)}$ and output $(\sigma_s^{(i+1)} \ldots \sigma_{t-1}^{(i+1)})^j$. Hence for all $j \geq 1$, the sequences

$$\sigma_1^{(i)} \ldots \sigma_{s-1}^{(i)} (\sigma_s^{(i)} \ldots \sigma_{t-1}^{(i)})^j \sigma_t^{(i)} \ldots \sigma_n^{(i)}$$

for $i = 1, \ldots, k$, will also be in a k-cycle in $\mathcal{A}$. Since we are assuming the domain of $\mathcal{A}$ is all of Σ^*, all these cycles are in $\mathcal{A}$ so that $\mathcal{A}$ would have infinitely many k-cycles.

Our argument allows us to give a simple bound for $c_{|Q|, |\Sigma|, k}$. That is there are

$$\sum_{i=0}^{|Q|^k |\Sigma|^k + 1} |\Sigma|^i = \frac{|\Sigma|^{|Q|^k |\Sigma|^k + 2} - 1}{|\Sigma| - 1}$$

strings in Σ^* of length $\leq |Q|^k |\Sigma|^k + 1$. If each of these strings were involved in a k-cycle, then we would have $\lfloor \frac{|\Sigma|^{|Q|^k |\Sigma|^k + 2} - 1}{k(|\Sigma| - 1)} \rfloor$ k cycles. Thus $c_{|Q|, |\Sigma|, k} = \lfloor \frac{|\Sigma|^{|Q|^k |\Sigma|^k + 2} - 1}{k(|\Sigma| - 1)} \rfloor$ and so if $\mathcal{A}$ has more than $c_{|Q|, |\Sigma|, k}$, then we will have at least one k-cycle of type described above so that $\mathcal{A}$ would have infinitely many k-cycles.

We note that the proof of Theorem 11 used the fact that the domain of $\mathcal{A}$ was all of Σ^* to ensure that all the strings $\sigma_1^{(i)} \ldots \sigma_{s-1}^{(i)} (\sigma_s^{(i)} \ldots \sigma_{t-1}^{(i)})^j \sigma_t^{(i)} \ldots \sigma_n^{(i)}$ for $j \geq 1$ are in the domain of $\mathcal{A}$. However, we did not need that T is length preserving, but only that certain types of cycles in T are length preserving. That is, a state-cycle C in T consists of state $s \in S$ and string of symbols $\sigma_1 \ldots \sigma_n \in \Sigma$ such that when T processes $\sigma_1 \ldots \sigma_i$ starting state s, it ends in state s_i for $1 \leq i \leq n$, $s_1, \ldots, s_n$ are pairwise distinct, and $s_n = s$. We then say that a state-cycle C is *state-cycle length preserving* if when T processes $\sigma_1 \ldots \sigma_n$ starting in state s, the corresponding output is of length n. We say that T has *state-cycle length preserving cycles* if all state-cycles of C are state-cycle length preserving. This condition would ensure that for each string $\sigma_s^{(i)} \ldots \sigma_{t-1}^{(i)}$ in the argument above, the output string $\sigma_s^{(i+1)} \ldots \sigma_{t-1}^{(i+1)}$ will have length $t - s$.

Note that the individual $\sigma_j^{(i)}$'s might be strings rather than symbols, when T is state-cycle length preserving.

Thus we obtain the following corollary from our proof of Theorem 11.

Corollary 3. *Suppose that $\mathcal{A} = (\Sigma^*, f)$ is an FST injection structure such that any cycle of f consists of strings of the same length and where f is defined by a transducer $T = (Q, \iota, \Sigma, \Gamma, \delta, \tau)$ that is state-cycle length preserving. Then there is a constant $c_{|Q|,|\Sigma|,k}$ depending on $|Q|$, $|\Sigma|$ and k such that if $\mathcal{A}$ has more than $c_{|Q|,|\Sigma|,k}$ orbits of size k, then $\mathcal{A}$ must have infinitely many orbits of size k.*

It is easy to construct a transducer T which satisfies the hypothesis of Corollary 3 but is not length preserving. For example, the following transducer has this property.

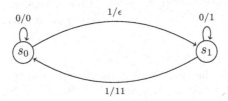

Here is another corollary to the proof of Theorem 11.

Corollary 4. *There is a computable type χ which is not recognized by any FST injection structure satisfying the conditions of Corollary 3.*

Proof. Consider the computable function $c(q, s, k) = c_{q,s,k}$, assumed to be monotonic, and, for each k, let $n(k)$ be larger than $c(q, s, k)$ for each $q, s \leq k$.

Suppose, towards a contradiction, that there is an FST injection structure realizing the type which has $n(k)$ many k-cycles for each finite k. Let q be the number of states in the DFST realizing this structure and let s be the number of letters in the alphabet. Then for $k = max\{q, s\}$, if the structure has $n(k)$ k-cycles, it must have infinitely many k-cycles, contradicting the assumption that this transducer realizes the given type.

A stronger result can be proved if we require the FST to be length preserving.

Theorem 12. *There is a computable type χ with at most one k-cycle for every finite k and which is not realized by any length preserving FST injection structure.*

Proof. It is easy to see that we can effectively determine whether a transducer $T = (Q, \iota, \Sigma, \Gamma, \delta, \tau)$ defines a length preserving injection structure by looking at the definition of τ. It follows that we can effectively list all length preserving FST injection structures by listing their corresponding FSTs $T_0, T_1, \ldots$.

Now suppose that $T_i = (Q_i, \iota_i, \Sigma_i, \Gamma_i, \delta_i, \tau_i)$ and $(\mathcal{A}_i, f_i)$ is the injection structure determined by T_i. Note that $S_i = (Q_i, \iota_i, \Sigma_i, \delta_i)$ is a deterministic finite automaton (DFA) so that it follows from the standard pumping for $DFAs$ that we can effectively decide if the language accepted by S_i, namely A_i, is infinite. It follows from results of Cenzer, Harizanov, and Remmel [5] that every c.e. character is the character of some computable injection structure so that we need only produce a character $\chi = \{(k,1) : k \in B\}$ for some c.e. set B that is not the character of (A_i, f_i) for any i. We enumerate a computable B with this property in stages. We let B_s denote the set of elements of enumerated into B by the end of stage s. Our construction will ensure that $|B_s| = s + 1$ so that a computable injection with character χ will be infinite.

Stage 0. If A_0 is finite, then let $B_0 = \{1\}$. If A_0 is infinite, then we list the strings in A_0 first by length and within strings of the same length lexicographically as $a_0^{(0)}, a_1^{(0)}, \ldots$. Since T_0 is length preserving, we can effectively compute the orbit $\mathcal{O}_{f_0}(a_s^{(0)})$ for any $a_s^{(0)}$ for any s. Then we let $k_0 = |\mathcal{O}_{f_0}(a_0^{(0)})|$ and let $B_0 = \{k_0 + 1\}$. Our construction will ensure that no element smaller that $k_0 + 1$ will be in B so that (A_0, f_0) will not be isomorphic to any injection structure with character χ.

Stage s+1. Assume that $B_s = \{b_0 < b_1 \cdots < b_s\}$ and for all $i \leq s$, either (i) A_i is finite, (ii) (A_i, f_i) has two distinct orbits of the same size, or (iii) (A_i, f_i) has an orbit of size d_i where $d_i < b_s$ and $d_i \notin B_s$. If A_{s+1} is finite, then we let $b_{s+1} = b_s + 1$. If A_{s+1} is infinite, then we list the strings in A_{s+1} first by length and within strings of the same length lexicographically as $a_0^{(s+1)}, a_1^{(s+1)}, \ldots$. Since T_{s+1} is length preserving, we can effectively compute the orbit $\mathcal{O}_{f_{s+1}}(a_t^{(s+1)})$ for any $a_t^{(s+1)}$ for any t. Then we compute the orbits $\mathcal{O}_{f_{s+1}}(a_0^{(s+1)}), \mathcal{O}_{f_{s+1}}(a_1^{(s+1)}), \ldots$ until either we find $i = j$ such that $|\mathcal{O}_{f_{s+1}}(a_i^{(s+1)})| = |\mathcal{O}_{f_{s+1}}(a_j^{(s+1)})|$ in which case we set $b_{s+1} = b_s + 1$ or we find an i such that $|\mathcal{O}_{f_{s+1}}(a_i^{(s+1)})| > b_s$ in which case we set $b_{s+1} = |\mathcal{O}_{f_{s+1}}(a_i^{(s+1)})| + 1$. Then we let $B_{s+1} = \{b_0 < b_1 \cdots < b_s < b_{s+1}\}$.

It is easy to see that B is an infinite computable set and that for all i either A_i is finite, (A_i, f_i) has two orbits of the same size, or (A_i, f_i) has an orbit of size k which is not in B. It then follows that if (A, f) is a computable injection structure with character χ, then (A, f) is not isomorphic of (A_i, f_i) for any i.

6 Graphs of FST Injection Structures

It is natural to ask whether there is a difference between FST injection structures and automatic structures $\mathcal{A} = (A, gr(f))$ where $gr(f)$ is the graph of the function f, i.e., the set of all pairs $(a, f(a))$ for $a \in f$. Since an automatic structure $\mathcal{A} = (A, gr(f))$ has a decidable theory, it would follow that if $gr(f)$ is recognizable by a DFA, then (A, f) would also have a decidable theory. However, it is easy to construct a DFST with no ϵ-outputs whose graph is not recognizable by a 2-tape synchronous automaton.

Theorem 13. *There is an FST realizable injection structure (A, f) over a unary alphabet such that the graph of f is not recognizable by a DFA.*

Proof. For example, consider the function $1^i \mapsto 1^{2i}$. The graph of this function does not satisfy the Constant Growth Lemma [15] and hence is not automatic.

On the other hand, it is easy to see that if $T = (Q, \iota, \Sigma, \Gamma, \delta, \tau)$ is length preserving, i.e. $\tau(q, a) \in \Gamma$ for all $q \in Q$ and $a \in \Sigma$, then we can use the transition diagram for δ and τ to construct of DFA which accepts the graph $(w, \tau(w))$ for all w accepted by the DFA $(Q, \iota, \Sigma, \delta)$. Thus we have the following theorem.

Theorem 14. *If a relation (not necessarily an injection, or even a function) is realized by a DFST all of whose moves are length preserving (one output symbol for each input symbol), then the graph of the relation is recognizable by a 2-tape synchronous automaton.*

In fact, we can prove a stronger theorem

Theorem 15. *If (A, f) is an FST injection structure where f is realized by a DFST which has length preserving cycles, then the graph of the relation is recognizable by a 2-tape synchronous automaton and, hence, (A, f) has a decidable theory.*

7   Open Questions and Further Research

There are two major questions that we have not yet been able to answer.

1. Does every FST injection structure (A, f) have a decidable theory?
2. Can one classify the types that can be realized by FST injection structures?

For question (1), we have a partial result, Theorem 15. Namely, that if f is realized by a transducer with length preserving cycles, the answer is yes.

Question (2) seems to be a much harder question. We should note that determining the types of injection structures realized by very simple FSTs can be challenging. For example, consider the FST injection structure $\mathcal{D}_2 = (\{0, 1\}^+, f)$ where f is specified by the following transducer T_2 where s_0 is the start state.

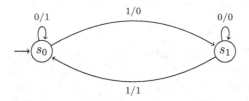

Computational evidence suggests that for all $n \geq 1$, all strings of σ such that $2^n \leq |\sigma| < 2^{n+1}$ are contained in 2^{n+1} cycles. However we have not been able to prove this even though $\mathcal{D}_2$ has a decidable theory. We have the following partial result.

Theorem 16. *All strings of the form 0^k where $2^n \leq k < 2^{n+1}$ are contained in 2^{n+1} cycles in $\mathcal{D}$. Moreover, all elements of A lie in cycles whose length is a power of 2.*

Proof. For any k, let $0^k, f(0^k), f^2(0^k), \ldots f^{n_k}(0^k)$ be the orbit of 0^k in $\mathcal{D}$ and let $a_{k,p}$ be state of the transducer after processing $f^p(0^k)$. It will be useful to picture this information in an array as follows

$$
\begin{aligned}
0^k & : a_{k,0} \\
f(0^k) & : a_{k,1} \\
f^2(0^k) & : a_{k,2} \\
\vdots & \quad \vdots \\
f^{n_k}(0^k) & : a_{k,p}.
\end{aligned}
$$

For example, the first few such arrays are as follows.

$$
\begin{array}{llll}
& & & 0000 : s_0 \\
& & & 1111 : s_0 \\
& 00 : s_0 & 000 : s_0 & 0101 : s_0 \\
0 : s_0 & 11 : s_0 & 111 : s_1 & 1001 : s_0 \\
1 : s_1 & 01 : s_1 & 010 : s_1 & 0001 : s_1 \\
& 10 : s_1 & 100 : s_1 & 1110 : s_1 \\
& & & 0100 : s_1 \\
& & & 1000 : s_1.
\end{array}
$$

We shall prove by induction that all elements lie in a cycle whose length is a power of 2. Clearly this is true for words of length 1. Now suppose that $|\sigma| \geq 1$ and σ is in a 2^r-cycle specified by

$$
\begin{aligned}
\sigma & : b_0 \\
f(\sigma) & : b_1 \\
\vdots & \quad \vdots \\
f^{2^r-1}(\sigma) & : b_{2^r-1}.
\end{aligned}
$$

Since processing 0 keeps the transducer in the same state and processing 1 always changes the state, it follows that $b_i = s_0$ if and only if $f^i(\sigma)$ has an even number of 1s. Similarly if $i \in \{0, 1\}$, then processing i in state s_0 will output $1 - i$ and processing i in state s_1 will output i. Thus if we consider the cycle of σi where $i \in \{0, 1\}$, then $f^{2^r}(\sigma i)$ will equal σi if $b_0, \ldots, b_{2^r-1}$ contains an even number of s_0s and will equal $\sigma(1 - i)$ if $b_0, \ldots, b_{2^r-1}$ contains an odd number of s_0s. It follows that σi is part a 2^r cycle if $b_0, \ldots, b_{2^r-1}$ contains an even number of s_0s and is part of 2^{r+1} cycle if $b_0, \ldots, b_{2^r-1}$ contains an odd number of s_0s.

We claim that for all $n \geq 1$, 0^{2^n} is in a 2^{n+1} cycle whose final states are $s_0^{2^n} s_1^{2^n}$ and that $0^{2^n+2^n-1}$ is in a 2^{n+1} cycle whose final states are $s_0 s_1^{2^{n+1}-1}$. We proceed by induction on n. We have verified this for $n = 1$.

Now consider the cycle of $0^{2^n+2^n-1}0 = 0^{2^{n+1}}$. Since the final states for the first 2^{n+1} steps of applying f to $0^{2^{n+2}-1}$ are $s_0 s_1^{2^{n+1}-1}$, it is easy to see that final elements of applying f to $0^{2^{n+1}}$ will be $01^{2^{n+1}}$ so that $f^{2^{n+1}}(0^{2^{n+2}}) = 0^{2^{n+1}-1}1$. Moreover, it follows that $f^i(0^{2^{n+1}})$ will have an even number of 1s for all $0 \le i < 2^{n+1}$ so that final states for those strings will be $s_0^{2^{n+1}}$ since they will contain an even number of 1s. Similarly, the final elements of the result of the first 2^{n+1} steps of applying f to $0^{2^{n+1}-1}1$ will be $10^{2^{n+1}-1}$ so that $f^{2^{n+1}}(0^{2^{n+2}-1}1) = 0^{2^{n+1}-1}0$. Moreover the final states of these strings will be $s_1^{2^{n+1}}$ since they will all contain an odd number of strings. Hence the cycle of $0^{2^{n+1}}$ is a cycle of length 2^{n+2} whose final states are $s_0^{2^{n+1}}s_0^{2^{n+1}}$.

Next consider the first 2^{n+1} steps of applying f to $0^{2^{n+1}+2^{n+1}-1} = 0^{2^{n+2}-1}$. Since the final states for first 2^{n+1} steps of applying f are $s_0^{2^{n+1}}$, the resulting final states and strings will be

$$
\begin{array}{ll}
0^{2^{n+1}}0^{2^{n+1}-1} & : s_0 \\
f(0^{2^{n+1}})f(0^{2^{n+1}-1}) & : s_1 \\
f^2(0^{2^{n+1}})f^2(0^{2^{n+1}-1}) & : s_1. \\
\quad \vdots & \quad \vdots \\
f^{2^{n+1}-1}(0^{2^{n+1}})f^{2^{n+1}-1}(0^{2^{n+1}-1}) & : s_1.
\end{array}
$$

Since $f^{2^{n+1}}(0^{2^{n+1}-1}) = 0^{2^{n+1}-1}$, when we apply f again, we first get $f^{2^{n+1}}(0^{2^{n+1}})$ which will end in state s_1 and then we will see a string of 0s which will just be copied in state s_1. It follows for the next 2^{n+1} steps of applying f, we will get

$$
\begin{array}{ll}
f^{2^{n+1}}(0^{2^{n+1}})0^{2^{n+1}-1} & : s_1 \\
f^{2^{n+1}+1}(0^{2^{n+1}})0^{2^{n+1}-1} & : s_1 \\
\quad \vdots & \quad \vdots \\
f^{2^{n+2}-1}(0^{2^{n+1}})0^{2^{n+1}-1} & : s_1.
\end{array}
$$

As $f^{2^{n+2}}(0^{2^{n+1}}) = 0^{2^{n+1}}$, it follows that the cycle of $0^{2^{n+2}-1}$ is of length 2^{n+2} whose final states are $s_0 s_1^{2^{n+2}-1}$.

Let $\overline{\mathcal{D}}_2 = (\{0,1\}^+, g)$ where g is the function induced by the same transducer T_2 except that the start state is s_1. In this case for any $\sigma \in \{0,1\}^+$, the cycle of $0^n 1\sigma$ will just be $0^n 1\sigma, 0^n 1 f(\sigma), 0^n 1 f^2(\sigma), \ldots$. As before we can show that all cycles are powers of 2 so that since we know $\mathcal{D}$ has 2^n-cycles for all $n \ge 1$, $\overline{\mathcal{D}}_2 = (\{0,1\}^+, g)$ will have infinitely many 2^n-cycles for all $n \ge 1$.

Generalizing T_2 gives a family of two-state DFSTs T_p over the alphabet $\{0, \ldots, p-1\}$ pictured below.

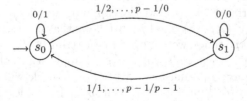

We can prove the following theorem about such transducers.

Theorem 17. *Let $\mathcal{D}_p = (\{0,\ldots,p-1\}^+, f_p)$ be the FST injection structure where f_p is induced by T_p. If $p \geq 3$ is prime, then for all $n \geq 1$, all strings of length n form a p^n-cycle in $\mathcal{D}_p$. Thus, the type p_1^n is realized by this FST injection structure.*

Proof. We shall show by induction on n that all strings of length n are part of p^n cycle. This is clear for $n = 1$. That is, since the start state is s_0, it is easy to see that $f_p(i) = i+1$ for $0 \leq i \leq p-2$ and $f_p(p-1) = 0$. Thus $0,1,\ldots,p-1,0$ is p-cycle in $\mathcal{D}_p$. Next suppose that strings of length n form an p^n-cycle in $\mathcal{D}_p$. Let $\sigma^{0,n},\ldots,\sigma^{p^n-1,n},\sigma^{0,n}$ be this p^n-cycle where $\sigma^{0,n} = 0^n$. Let $s^{j,n}$ denote the final state of the FST when we process $\sigma^{j,n}$ starting in state s_0. For example, $\sigma^{0,n} = 0^n$ so that $\sigma^{1,n} = 1^n$ and $s^{0,n} = 0$.

For any string $\sigma \in \{0,\ldots,p-1\}^*$ and any $0 \leq i \leq p-1$, let $|\sigma|_i$ denote the number of i's which occur in σ and

$$|\sigma|_{\neq 0} = |\sigma|_1 + |\sigma|_2 + \cdots + |\sigma|_{p-1}.$$

Then for any j, when we process $\sigma^{j,n}$ starting in state s_0, we will end in state s_0 if $|\sigma^{j,n}|_{\neq 0}$ is even and we will end in state s_1 if $|\sigma^{j,n}|_{\neq 0}$ is odd. Note that if $n = 2k$, then the number of of strings of length n such that $|\sigma|_{\neq 0}$ is odd equals

$$\binom{2k}{2k-1}(p-1)^{2k-1} + \binom{2k}{2k-3}(p-1)^{2k-3} + \cdots + \binom{2k}{1}(p-1)$$

which clearly is an even number. Similarly, if $n = 2k+1$, then the number of of strings of length n such that $|\sigma|_{\neq 0}$ is odd equals

$$\binom{2k+1}{2k+1}(p-1)^{2k+1} + \binom{2k+1}{2k-1}(p-1)^{2k-1} + \cdots + \binom{2k+1}{1}(p-1)$$

which is also an even number. It follows that in the sequence of state $s^{0,n}, s^{1,n}, \ldots, s^{p^n-1,n}$ the number of s_1s is equivalent to $i \mod p$ for some $0 < i \leq p-1$ which means that the number s_0s in the sequence is equivalent to $p-i \mod p$ for some $0 < i \leq p-1$. Now suppose that we fix an t such that $0 \leq t \leq p-1$ and we apply f_p p^n times starting with the string $0^n t$. Then we will get a sequence

$$\sigma^{0,n}a^{0,t}, \sigma^{1,n}a^{1,t}, \ldots, \sigma^{p^n-1}a^{p^n-1,t}$$

where $a^{0,t} = t$ since after processing 0^n we end in state s_0. For $b \geq 0$, $a^{b+1,t}$ equals $a^{b,t}$ is $s^{b,n} = s_1$ and $a^{b+1,t}$ equals $a^{b,t} + 1 \mod p$ is $s^{t,n} = s_0$. This implies that if we apply f_p p^n times starting with the string $0^n t$, we will end up with the string $0^n(t+p-i \mod p)$. Thus if we start with the string 0^{n+1} and apply f_p p^n times will produce the string $0^n(p-i)$. If we then apply f_p another p^n times, we will end up with the string $0^n(2(p-i) \mod p)$. It follows that if we start with the string 0^{n+1} and apply f_p kp^n times, we end up with the string $0^n(k(p-i) \mod p)$. Since $p \geq 3$ and p is prime, then we it will take p^{n+1} applications of f_p before we can return to 0^{n+1}. Hence, 0^{n+1} must be part of a p^{n+1}-cycle which means that the cycle determined by 0^{n+1} must consist of all the strings of length $n+1$.

Finally we should observe that our proof shows that if we choose s_1 as the start state for the FST in Theorem 17, then for any $k \geq 0$, the strings $0^k 10^n$ will generate a p^n-cycle where each string in the cycle starts with $0^k 1$. It follows that if we restrict the set of strings of the form $0^k 1\sigma$ where $k > 0$ and $\sigma \in \{0, 1, \ldots, p-1\}^+$, then we will obtain an FST injection structure such that there are infinitely many p^n-cycles for each $n \geq 1$.

References

1. Bennett, C.H.: Logical reversibility of computation. IBM J. Res. Dev. **17**, 525–532 (1973)
2. Ash, C., Knight, J.: Computable Structures and the Hyperarithmetical Hierarchy. Elsevier, Amsterdam (2000)
3. Blumensath, A.: Automatic structures. Diploma Thesis, RWTH Aachen (1999)
4. Blumensath, A., Grädel, E.: Automatic structures. In: Proceedings of the 15th LICS, pp. 51–62 (2000)
5. Cenzer, D., Harizanov, V., Remmel, J.B.: Effective categoricity of injection structures. In: Löwe, B., Normann, D., Soskov, I., Soskova, A. (eds.) CiE 2011. LNCS, vol. 6735, pp. 51–60. Springer, Heidelberg (2011). doi:10.1007/978-3-642-21875-0_6
6. Cenzer, D., Harizanov, V., Remmel, J.B.: Computability-theoretic properties of injection structures. In: Algebra and Logic (to appear)
7. Delhommé, C.: Automaticité des ordinaux et des graphes homogènes. C.R. Acadèmie des sciences Paris, Ser. I **339**, 5–10 (2004)
8. Hodgson, B.R.: On direct products of automaton decidable theories. Theoret. Comput. Sci. **19**, 331–335 (1982). North-Holland
9. Khoussainov, B., Nerode, A.: Automatic presentations of structures. In: Leivant, D. (ed.) LCC 1994. LNCS, vol. 960, pp. 367–392. Springer, Heidelberg (1995). doi:10.1007/3-540-60178-3_93
10. Khoussainov, B., Shore, R.A.: Effective model theory: the number of models and their complexity. In: Cooper, S.B., Truss, J.K. (eds.) Models and Computability, Invited Papers from LC 1997, LMSLNS vol. 259, pp. 193–240, Cambridge University Press, Cambridge (1999)
11. Khoussainov, B., Rubin, S., Stephan, F.: On automatic partial orders. In: Proceedings of the 18th LICS, pp. 168–177 (2003)
12. Khoussainov, B., Nies, A., Rubin, S., Stephan, F.: Automatic structures: richness and limitations. In: Proceedings of the 19th LICS, pp. 44–53 (2004)
13. Khoussainov, B., Rubin, S., Stephan, F.: Automatic linear orders and trees. ACM Trans. Comput. Logic **6**(4), 675–700 (2005)
14. Khoussainov, B., Liu, J., Minnes, M.: Unary automatic graphs: an algorithmic perspective. In: Agrawal, M., Du, D., Duan, Z., Li, A. (eds.) TAMC 2008. LNCS, vol. 4978, pp. 542–553. Springer, Heidelberg (2008). doi:10.1007/978-3-540-79228-4_47
15. Rubin, S.: Automatic structures. Ph.D. Thesis, University of Auckland (2004)
16. Soare, R.I.: Recursively Enumerable Sets and Degrees. Springer, Berlin (1987)

A Survey on Universal Computably Enumerable Equivalence Relations

Uri Andrews[1], Serikzhan Badaev[2], and Andrea Sorbi[3(✉)]

[1] Department of Mathematics, University of Wisconsin,
Madison, WI 53706-1388, USA
andrews@math.wisc.edu
[2] Department of Fundamental Mathematics, Al-Farabi Kazakh National University,
Almaty 050040, Kazakhstan
serikzhan.badaev@kaznu.kz
[3] Dipartimento di Ingegneria Informatica e Scienze Matematiche,
Università Degli Studi di Siena, 53100 Siena, Italy
andrea.sorbi@unisi.it
http://www.math.wisc.edu/~andrews/, http://www3.diism.unisi.it/~sorbi/

Abstract. We review the literature on universal computably enumerable equivalence relations, i.e. the computably enumerable equivalence relations (ceers) which are Σ_1^0-complete with respect to computable reducibility on equivalence relations. Special attention will be given to the so-called uniformly effectively inseparable (u.e.i.) ceers, i.e. the nontrivial ceers yielding partitions of the natural numbers in which each pair of distinct equivalence classes is effectively inseparable (uniformly in their representatives). The u.e.i. ceers comprise infinitely many isomorphism types. The relation of provable equivalence in Peano Arithmetic plays an important role in the study and classification of the u.e.i. ceers.

Keywords: Computably enumerable equivalence relation · Computable reducibility on equivalence relations

2010 Mathematics Subject Classification: 03D25

1 Introduction

Recently there has been a growing interest in studying and classifying equivalence relations on the set ω of natural numbers, by mean of the so-called *computable reducibility*, where, given equivalence relations R and S on ω, we say

Andrews was partially supported by NSF grant DMS-1201338.

Badaev was partially supported by Grant 3952/GF4 of the Science Committee of the Republic of Kazakhstan.

Sorbi is a member of INDAM-GNSAGA; he was partially supported by Grant 3952/GF4 of the Science Committee of the Republic of Kazakhstan, and by PRIN 2012 "Logica Modelli e Insiemi".

The authors thank Keng Meng Ng for his helpful comments and suggestions. The wish also to thank an anonymous referee for having pointed out several inaccuracies in a previous version of the paper.

A. Day et al. (Eds.): Downey Festschrift, LNCS 10010, pp. 418–451, 2017.
DOI: 10.1007/978-3-319-50062-1_25

that R is *computably reducible* (or simply, *reducible*) to S (in symbols: $R \leqslant S$), if there exists a computable function f such that

$$(\forall x, y)[x \, R \, y \Leftrightarrow f(x) \, S \, f(y)].$$

The first systematic study of this reducibility goes back perhaps to Ershov [13, 14], as an alternative way of looking at monomorphisms in the category of numbered sets. An obvious related notion is that of completeness: if $\mathcal{A}$ is a class of equivalence relations on ω, one says that R is $\mathcal{A}$complete, if $R \in \mathcal{A}$, and $S \leqslant R$, for every $S \in \mathcal{A}$. This reducibility, and its related notion of completeness, have been successfully applied to measure the complexity of equivalence relations naturally arising in mathematics, and in particular in computable model theory and in computability theory. For instance, the isomorphism relations for various familiar classes of computable structures (identified with numbers via suitable numberings) are Σ_1^1 complete: this includes computable groups, computable torsion abelian groups, computable torsion-free abelian groups, abelian p-groups, see [17]. Other interesting mathematical applications of reducibility $\leqslant$ appear in [11,15,16,20,21].

This paper is a survey (far from being exhaustive) on Σ_1^0-universal equivalence relations, henceforth called *universal ceers*: we shall use the acronym "ceer" for "computably enumerable equivalence relation"; ceers are called *positive* equivalence relations in the Russian literature. In Sect. 2 we focus our attention on some classes of universal ceers of particular importance in logic and computability theory. It is interesting to notice that the first example of a nontrivial and mathematically interesting universal ceer appears in the book [24], where Miller III builds a finitely presented group G for which the ceer provided by equality $=_G$ in G, is universal. If this example was most likely not motivated by any specific interest in ceers and computable reducibility, Ershov [12] on the contrary, in this case clearly motivated by studying ceers under $\leqslant$, pointed out another mathematically interesting universal ceer, see Theorem 2.7. Another universal ceer of special interest, first pointed out in [8], is the relation of provable equivalence in Peano Arithmetic, denoted by $\sim_{PA}$, which relates two numbers if the two sentences coded by these numbers are provably equivalent in PA. The class of nontrivial ceers which are quotients of $\sim_{PA}$ (i.e. computably isomorphic to nontrivial ceers extending $\sim_{PA}$) form the class of the so-called uniformly finitely precomplete (u.f.p.) ceers, which are all universal. Inside this class we find two special isomorphism types: the so-called e-complete ceers (which turn out to be computably isomorphic to $\sim_{PA}$), and the precomplete ceers (which turn out to be computably isomorphic to the restriction of $\sim_{PA}$ to the Σ_n^0-sentences, for any fixed n).

As in the case of universality with respect to m-, or 1-reducibility, for c.e. sets, or pairs of disjoint c.e. sets (where the universal sets coincide with the creative sets, and the universal pairs of disjoint c.e. sets coincide with the effectively inseparable pairs), the notions of creativeness and effective inseparability play an important role in the investigation of universal ceers. Not only can one show that a u.f.p. ceer R yields a partition of ω such that any disjoint pair $([a]_R, [b]_R)$

of equivalence classes are effectively inseparable uniformly in a, b, but it turns out that this latter notion by itself suffices to give universality: every uniformly effectively inseparable (u.e.i.) ceer R (i.e. a nontrivial ceer yielding a uniformly effectively inseparable partition of ω) is universal.

Unlike classical isomorphism theorems (in particular, Myhill's theorem on computable isomorphisms of creative sets, and Smullyan's theorem on computable isomorphisms of e.i. pairs), uniform effective inseparability for ceers does not imply computable isomorphism. Infinitely many distinct computable isomorphism types for u.e.i. ceers appear already at the level of u.f.p. ceers. Moreover, a recent result in [3] shows that there are u.e.i. ceers that are not u.f.p.

The class of u.f.p. ceers is however partitioned into infinitely many computable isomorphism types.

In Sect. 5 we review a characterization (see [1]) of universal ceers in terms of a jump operation on ceers, due to [19]: namely, a ceer is universal if and only if its jump is reducible to it.

In more than one occasion, we give new and simplified proofs of classical results, including for instance universality of u.f.p. ceers, and isomorphism of e-complete ceers.

1.1   Terminology and Notations

We use standard computability theoretic terminology and notation, which can be found in the textbooks [10,28,31]. We often identify finite sets with their canonical indices: so when for a function f we write $f(D)$ where D is a finite set, then we in fact mean $f(u)$, with $F = D_u$.

Given any set X and any equivalence relation R, we write $[X]_R = \{y : (\exists x)[y \, R \, x]\}$; and $[x]_R = [\{x\}]_R$ denotes the R-equivalence class of R.

The following category theoretic terminology is adapted from [13,14], which study the category of numberings.

Definition 1.1. Given equivalence relations R, S on ω, a *morphism* $\mu : R \longrightarrow S$ is a function from $\omega_{/R}$ to $\omega_{/S}$ (i.e. between the quotient sets), for which there exists a computable function $f : \omega \to \omega$ such that $\mu([x]_R) = [f(x)]_S$, for all x; we say in this case that f *induces* μ; a *monomorphism* is a 1-1 morphism, an *isomorphism* is an onto monomorphism. An *endomorphism* for R is a morphism $\mu : R \longrightarrow R$.

Remark 1.2. We observe that if $\mu : R \longrightarrow S$ is an isomorphism, and all R- and S-equivalence classes are infinite, then by a standard back and forth argument, it is easy to see that there is a computable permutation of ω that induces μ, or equivalently there is a computable permutation of ω that reduces R to S.

Lemma 1.3. *If R, S are ceers then $R \leqslant S$ if and only if there is a monomorphism $\mu : R \longrightarrow S$.*

Proof. Easy.   □

In the same vein, we can define a *partial morphism* from R to S to be a partial function μ from $\omega_{/R}$ to $\omega_{/S}$ for which there is a partial computable function φ such that: (1) if $\mu([x]_R)$ is defined, then there is x' such that $\varphi(x')$ is defined, $x\ R\ x'$ and $\mu([x]_R) = [\varphi(x')]_S$; (2) $[\mathrm{domain}(\varphi)]_R = \{x : [x]_R \in \mathrm{domain}(\mu)\}$. Notice that if R and S are ceers and φ is a partial computable function inducing a partial morphism from R to S then we can extend φ to a partial computable function ψ such that ψ still induces the partial morphism and the domain of ψ is R-closed, i.e. if $\psi(x)$ converges and $x\ R\ x'$ then $\psi(x')$ converges as well.

1.2 Indexing

Throughout the paper, we refer to the indexing $\{R_e : e \in \omega\}$ of all ceers, where R_e is the equivalence relation generated by W_e (viewed as a set of pairs).

We say that a sequence $\{R^s : s \in \omega\}$ of equivalence relations on ω is a *computable approximation to* a ceer R, if

(1) the set $\{\langle x, y, s\rangle : x\ R^s\ y\}$ is computable;
(2) $R^0 = \mathrm{Id}$;
(3) for all s, $R^s \subseteq R^{s+1}$; the equivalence classes of R^s are finite; there exists at most one pair $[x]_{R^s}, [y]_{R^s}$ of equivalence classes, such that $[x]_{R^s} \cap [y]_{R^s} = \varnothing$, but $[x]_{R^{s+1}} = [y]_{R^{s+1}}$ (we say in this case that the equivalence relation R-collapses x and y at stage $s + 1$);
(4) $R = \bigcup_t R^t$.

Lemma 1.4. *There exists a sequence $\{R_e^s : e, s \in \omega\}$ of equivalence relations such that the set $\{\langle e, x, y, s\rangle : x\ R_e^s\ y\}$ is computable (in fact, we may even assume that one can effectively find the canonical index of $[x]_{R^s}$, and we can decide, given e, s whether $R_e^s = R_e^{s+1}$), and the sequence $\{R_e^s : s \in \omega\}$ is a computable approximation to R_e. Therefore an equivalence relation R is a ceer if and only if R can be computably approximated. Moreover if R is a ceer and $R \setminus \{\langle x, x\rangle : x \in \omega\}$ is infinite, then one can find an approximating sequence $\{R^s : s \in \omega\}$ to R satisfying that for every s, the relation R^{s+1} is obtained from R^s by the R-collapse of exactly one pair of equivalence classes of R^s.*

Proof. Straightforward.   □

One could alternatively consider the following numbering, suggested by Ershov [12]: let
$$x\ S_e\ y \Leftrightarrow (\exists m, n)[\varphi_e^m(x)\downarrow = \varphi_e^n(y)\downarrow],$$
where, given a partial function ψ, $\psi^n(x)$ denotes the n-th iterate of ψ on x, where $\psi^0(x) = x$, and of course $\psi^n(x)$ converges if and only if both $\psi^{n-1}(x)$ and $\psi(\psi^{n-1}(x))$ converge. We may also write S_{φ_e} for S_e. Indeed, if R is a ceer, then $R = S_\varphi$ where φ is the partial computable function $\varphi(x) = (\mu(\langle y, s\rangle). [x\ R_s\ y\ \&\ y < x])_0$, where we refer to some computable approximation $\{R_s\}$ to R.

1.3 Some Special Classes of Ceers

We now introduce some important classes of ceers, which will be shown to be universal in next section.

Definition 1.5. Let R be an equivalence relation on ω.

(1) [23] R is *precomplete* if there exists a computable function $f(e, x)$ (called a *totalizer of* R) such that, for all e, x,

$$\varphi_e(x) \downarrow \Rightarrow \varphi_e(x) \; R \; f(e, x).$$

Moreover, $f(e, \_)$ is called an R-*totalizer of* φ_e, or alternatively we say that $f(e, \_)$ *makes* φ_e *total modulo* R.

(2) [25] R is *uniformly finitely precomplete* (or *u.f.p.* for short) if there exists a computable function $f(D, e, x)$ such that for every finite set D and every e, x,

$$\varphi_e(x) \downarrow \in [D]_R \; \Rightarrow \; \varphi_e(x) \; R \; f(D, e, x).$$

Moreover, $f(\_, e, \_)$ is called an R-*totalizer of* φ_e, or alternatively we say that $f(\_, e, \_)$ *makes* φ_e *total modulo* R.

(3) [1,6] We say that R is *uniformly effectively inseparable* (or *u.e.i.* for short) if there is a *uniform productive function*, i.e., a partial computable function $p(a, b, u, v)$ such that if $[a]_R \cap [b]_R = \varnothing$ then

$$(\forall u, v)[[a]_R \subseteq W_u \; \& \; [b]_R \subseteq W_v \; \& \; W_u \cap W_v = \varnothing \Rightarrow p(a, b, u, v) \downarrow \notin W_u \cup W_v].$$

Remark 1.6. We note that, as in the case of effective inseparability for pairs of c.e. sets, if R is a u.e.i. ceer then we can in fact assume that $p(a, b, u, v)$ be total. Indeed, if p is partial computable, we can always assume that if $a \not\mathrel{R} b$ then the function $p(a, b, \_, \_)$ is total, as from any partial productive function for a pair of disjoint c.e. sets, one can uniformly find a total productive function for that pair: this is similar to showing that from any productive function for a productive set, one can uniformly find a total productive function for that set, see [28]. Having such a function p, define a total productive function q for R as follows:

$$q(a, b, u, v) = \begin{cases} 0, & \text{if } [a]_R \cap [b]_R \neq \varnothing \leq p(a, b, u, v) \downarrow \\ p(a, b, u, v), & \text{otherwise,} \end{cases}$$

where given two c.e. relations $U := (\exists x)A(x)$ and $V := (\exists x)B(x)$ in Σ_1-normal form, with A, B decidable, we write as usual $U \leq V := (\exists x)A(x) \; \& \; (\forall y \leq x)\neg B(y)$.

Lemma 1.7. *The classes of Definition 1.5 are closed under isomorphisms, and are upwards $\subseteq$-closed.*

Proof. Straightforward. □

Remark 1.8. Throughout the paper, when we refer to an equivalence relation R as lying in any of the three classes of Definition 1.5, we will also always assume that R is not trivial (i.e. there are two numbers which are non-R-equivalent).

2   Precomplete and Uniformly Finitely Precomplete Ceers

As promised, in this section we show that the special ceers introduced in Sect. 1.3 are all universal.

2.1   Precomplete Ceers

Let us begin our trip through the land of universal ceers by looking at precomplete ceers. First let us recall some important properties of precomplete equivalence relations. The following theorem is in fact a characterization of all precomplete equivalence relations (including in this case the trivial one), not only the computably enumerable ones.

Theorem 2.1 (Ershov's Fixed Point Theorem). *An equivalence relation R is precomplete if and only if there is a computable function* fix *such that, for every n,*

$$\varphi_n(\mathrm{fix}(n)) \downarrow \Rightarrow \varphi_n(\mathrm{fix}(n)) \; R \; \mathrm{fix}(n).$$

Proof. $\Rightarrow$. If R is precomplete then let $\hat{u}(x)$ be a computable function that makes $\varphi_x(x)$ total modulo R. Let $\varphi_{s(n)} = \varphi_n \circ \hat{u}$, and define fix $= \hat{u} \circ s$. Then if $\varphi_n(\mathrm{fix}(n)) \downarrow$, then $\varphi_n(\mathrm{fix}(n)) = \varphi_n(\hat{u} \circ s(n)) = \varphi_n \circ \hat{u} \circ s(n) = \varphi_{s(n)}(s(n)) \; R \; \hat{u} \circ s(n) = \mathrm{fix}(n)$.

$\Leftarrow$. Given fix and a partial computable φ, let $\varphi_{f(x)}(y) = \varphi(x)$. Then we claim that $g = \mathrm{fix} \circ f$ makes φ total modulo R. If $\varphi(x) \downarrow$, then $\varphi(x) = \varphi_{f(x)}(\mathrm{fix} \circ f(x)) \; R \; \mathrm{fix} \circ f(x) = g(x)$. $\qquad\square$

Definition 2.2. A *diagonal function* for an equivalence relation R is a computable function d such that $x \not\!R d(x)$ for every x.

Corollary 2.3. *If R is precomplete then there is no diagonal function for R.*

Proof. This is an immediate consequence of the Ershov Fixed Point Theorem. $\square$

Another important property of precomplete equivalence relations is the Padding Lemma.

Theorem 2.4 (Padding Lemma). *For every precomplete R there exists a 1-1 total computable $p(x,y)$ such that, for all x, m, $p(x, m) \; R \; x$. Hence, all R-equivalence classes contain infinite c.e. sets, and R has an injective totalizer.*

Proof. Let R be a precomplete equivalence relation. We show that there is a computable p with the desired properties which is injective in the second argument; we leave it as an exercise to show that one can get an injective totalizer. We need to show that from any finite set $F = \{n_1, \ldots, n_k\}$ of numbers such that $n_1 \; R \cdots R \; n_k$ we can uniformly find $n \notin F$ such that $n \; R \; n_1$. Let $G(e, x)$ be a totalizer for R. Then by the Recursion Theorem, let e be such that

$$\varphi_e(x) = \begin{cases} n_1 & \text{if } G(e, 0) \notin F, \\ \max F + 1 & \text{otherwise.} \end{cases}$$

Then the number

$$n = \begin{cases} G(e,0) & \text{if } G(e,0) \notin F, \\ \max F + 1 & \text{otherwise.} \end{cases}$$

is the desired number. Indeed, $n \notin F$ since either $n = \max F + 1$ or $n = G(e,0)$ if $G(e,0) \notin F$. In the former case, $n = \max F + 1 = \varphi_e(0) \ R \ G(e,0) \in F$. So, n is R-equivalent to an element of F, so to n_1. In the latter case, $n = G(e,0) \ R$ $\varphi_e(0) = n_1$. □

Notice that the usual Padding Lemma for the standard numbering $\{\varphi_e\}$ of the partial computable functions is a corollary of the previous result, as the equivalence relation, in x, y, $\varphi_x = \varphi_y$ is easily seen to be precomplete, see [23].

2.2 Examples of Precomplete Ceers

Recall that a partial computable function u is called *universal*, if there exists a computable function $f(e,x)$ such that $\varphi_e(x) = u(f(e,x))$. By the Padding Lemma for the numbering $\{\varphi_e\}$, we can also assume that f is 1-1.

The following result is attributed in [12] to Lachlan.

Lemma 2.5. *If u is a universal unary partial computable function then S_u is precomplete.*

Proof. If f witnesses that u is universal, and $\varphi_e(x) \downarrow$, then $u^1(\varphi_e(x)) = u(f(e,s))$, hence $\varphi_e(x) \ S_u \ f(e,x)$, which shows that $f(e, \_)$ is a totalizer for φ_e. □

Assume that first order Peano Arithmetic PA is Σ_1-sound, and for every $n \geq 1$ let $T_n(v)$ be a Σ_n-truth predicate, i.e., for all Σ_n-sentences σ,

$$PA \vdash \sigma \leftrightarrow T_n(\ulcorner \sigma \urcorner)$$

where $\ulcorner \_ \urcorner$ is a suitable Gödel numbering for all sentences in the language of PA, and $\overline{m}$ denotes the numeral term for the number m.

For every number x there is a Σ_1-formula $F_x(u,v)$ (in fact, $F_x(u,v) := F(\overline{x}, u, v)$ for some Σ_1-formula F) *representing* φ_x in PA, i.e. such that

$$\varphi_x(n) = m \Leftrightarrow PA \vdash F_x(\overline{n}, \overline{m}).$$

We may assume that for every number m, $PA \vdash F_x(\overline{m}, v) \wedge F_x(\overline{m}, v') \rightarrow v = v'$.)

Define $\sim_n$ on ω by

$$\ulcorner \sigma \urcorner_n \sim_n \ulcorner \tau \urcorner_n \Leftrightarrow T \vdash \sigma \leftrightarrow \tau$$

where $\ulcorner \_ \urcorner_n$ is a suitable Gödel numbering identifying Σ_n sentences (which form an infinite c.e. set, and therefore is a set computably isomorphic to ω) with numbers: notice that we use here $\ulcorner \_ \urcorner_n$ instead of $\ulcorner \_ \urcorner$, as otherwise the domain of $\sim_n$ would be a proper subset of ω. Then $\sim_n$ is a precomplete ceer. Given the relevance of this example, we sketch the proof of why $\sim_n$ is precomplete.

Theorem 2.6. $\sim_n$ *is a precomplete ceer.*

Proof. We limit ourselves to the case $n = 1$. Given a partial computable function φ, let F be a representing Σ_1 formula for the partial computable function ψ, where

$$\psi(\ulcorner\sigma\urcorner_1) = \begin{cases} \ulcorner\tau\urcorner, & \text{if } \varphi(\ulcorner\sigma\urcorner_1)\!\downarrow = \ulcorner\tau\urcorner_1, \\ \uparrow, & \text{if } \varphi(\ulcorner\sigma\urcorner_1)\!\uparrow. \end{cases}$$

Define

$$f(m) = \ulcorner(\exists v)[F(\overline{m}, v) \wedge T_1(v)]\urcorner_1.$$

(Notice that the formula $(\exists v)[F(\overline{m}, v) \wedge T_1(v)]$ is Σ_1.) Assume now that $\varphi(\ulcorner\sigma\urcorner_1)\!\downarrow = \ulcorner\tau\urcorner_1$, where σ and τ are Σ_1-sentences. Then

$$PA \vdash (\exists v)[F(\overline{\ulcorner\sigma\urcorner_1}, v) \wedge T_1(v)] \leftrightarrow F(\overline{\ulcorner\sigma\urcorner_1}, \overline{\ulcorner\tau\urcorner}) \wedge T_1(\overline{\ulcorner\tau\urcorner}).$$

But $PA \vdash F(\overline{\ulcorner\sigma\urcorner_1}, \overline{\ulcorner\tau\urcorner}) \wedge T_1(\overline{\ulcorner\tau\urcorner}) \leftrightarrow T_1(\overline{\ulcorner\tau\urcorner})$, and $PA \vdash T_1(\overline{\ulcorner\tau\urcorner}) \leftrightarrow \tau$, which implies that $\varphi(\ulcorner\sigma\urcorner_1) \sim_1 f(\ulcorner\sigma\urcorner_1)$. Thus, f is the desired computable function that makes φ total modulo $\sim_1$. □

Other examples of precomplete ceers can be found in [32].

2.3 The First Universality Result

As already remarked in the introduction, one of the earliest nontrivial universality results for ceers was pointed out by Ershov [12].

Theorem 2.7. *If u is a universal unary partial computable function, then S_u is universal.*

Proof. Let u be a universal function and let φ be a partial computable function. As we have observed, we may suppose that there exists a 1-1 computable function g such that $\varphi_e(x) = g(\langle e, x\rangle)$. Thus it is easy to see that there is a computable sequence f_n of computable 1-1 functions such that $\varphi_n = u \circ f_n$. So, by the Recursion Theorem, let e be such that $u \circ f_e = f_e \circ \varphi$ (take a fixed point of a computable h, such that $\varphi_{h(e)} = f_e \circ \varphi$). Let $f = f_e$: then $f \circ \varphi = u \circ f$. Next, by induction on n it is easy to see that for every n, $f \circ \varphi^n = u^n \circ f$. It follows that for every m, n, if $\varphi^m(x) \!\downarrow = \varphi^n(y) \!\downarrow$ then $f(\varphi^m(x)) \!\downarrow = f(\varphi^n(y)) \!\downarrow$, thus $u^m(f(x)) \!\downarrow = u^n(f(y)) \!\downarrow$. On the other hand, if $u^m(f(x)) \!\downarrow = u^n(f(y)) \!\downarrow$ then $f(\varphi^m(x)) \!\downarrow = f(\varphi^n(y)) \!\downarrow$, and by injectivity, $\varphi^m(x) \!\downarrow = \varphi^n(y) \!\downarrow$. This shows that f reduces S_φ to S_u. Since for every ceer R, there is a partial computable φ such that $R = S_\varphi$, we have proved that S_u is universal. □

2.4 All Precomplete Ceers are Isomorphic

The precomplete ceers form a single isomorphism type, as shown by Lachlan [22].

Theorem 2.8 [22]. *If R, S are precomplete ceers then R is isomorphic to S, i.e., there exists a permutation h of ω which reduces R to S.*

Proof. We can assume that every ceer R has approximations $\{R_s\}$ and $\{\hat{R}_s\}$ satisfying Lemma 1.4 and in addition:

$$R_{s+1} - R_s \neq \varnothing \Rightarrow s + 1 \text{ odd}$$
$$S_{s+1} - S_s \neq \varnothing \Rightarrow s + 1 \text{ even.}$$

Let R, S be precomplete ceers, with corresponding computable approximations $\{R_s\}$ and $\{S_s\}$, as above: R may change only at odd stages, and S may change only at even stages. (Although not necessary, these additional properties of the approximations simplify the construction, since they make sure that changes for R (respectively, S) may appear only at stages when we really deal with R (respectively, S). In fact since all R- and S-equivalence classes are infinite, by Lemma 1.4 we could even assume in this case that at each stage exactly one change happens when we deal with the corresponding ceer.) Let F and G be injective totalizers for R and S respectively.

We will define two computable sequences $a_0, a_1, \ldots, a_s, \ldots$ and $b_0, b_1, \ldots, b_s, \ldots$, such that the assignment $a_s \mapsto b_s$ (we say in this case that a_s and b_s *match*) satisfies, for all i, j, indices

$$a_i \, R \, a_j \Leftrightarrow b_i \, S \, b_j,$$

and $\omega = \{a_s : s \in \omega\} = \{b_s : s \in \omega\}$. We start up with four numbers c_0, c_1, d_0, d_1 such that $c_0 \, \cancel{R} \, c_1$ and $d_0 \, \cancel{S} \, d_1$.

By the Double Recursion Theorem, we will assume that we control indices e, z of partial computable functions φ_e and φ_z. At the beginning of each stage $s + 1$, we assume that, for all $i, j < s$,

$$a_i \, R_s \, a_j \Leftrightarrow b_i \, S_s \, b_j.$$

We use in the following the symbols e', z', e'', z'' to represent suitable new indices of φ_e and φ_z, by the Padding Lemma. At stage $s + 1$ we say, for $i < s$, that $[a_i]_{R_s}$ is *right available* if there is $a \in [a_i]_{R_s}$ such that $\varphi_{e,s}(a)$ is undefined, and a already matches with a number chosen as $b = G(e', a) \in [b_i]_{S_s}$, with $\varphi_e = \varphi_{e'}$; similarly, we say that $[b_i]_{R_s}$ is *left available* if there is $b \in [b_i]_{R_s}$ such that $\varphi_{z,s}(b)$ is undefined, and b already matches with some number chosen as $a = F(z', b) \in [a_i]_{R_s}$, with $\varphi_z = \varphi_{z'}$. At the end of the stage, we define a new pair (a_s, b_s).

If a_i and b_i match, we assume by induction that either $[a_i]_{R_s}$ is right available or $[b_i]_{R_s}$ is left available.

Step 0. $\varphi_{e,0}(i)$ and $\varphi_{z,0}(i)$ are undefined for all i.

Step $s + 1$. Distinguish whether $s + 1$ is odd or even:

$s + 1$ *odd*. Perform in the order the following actions:

(1) Suppose there are $i < j$ such that a_i and a_j are R-collapsed at $s+1$. There are two subcases:

(a) at least one among $[a_i]_{R_s}$ and $[a_j]_{R_s}$ is right available, say $a \in [a_i]_{R_s}$ is such that $\varphi_{e,s}(a)$ is undefined, and matches with $b \in [b_i]_{S_s}$, of the form $b = G(e', a)$: then define $\varphi_e(a) = b_j$. This has the effect that

$$b_i \ S \ b = G(e', a) \ S \ \varphi_{e'}(a) = \varphi_e(a) = b_j;$$

(b) neither $[a_i]_{R_s}$ nor $[a_j]_{R_s}$ is right available: then $[b_i]_{S_s}$ and $[b_j]_{S_s}$ are both left available. Say $b \in [b_i]_{S_s}$, $b' \in [b_j]_{S_s}$ are such that $\varphi_{z,s}(b)$ and $\varphi_{z,s}(b')$ are still undefined and match with $a = F(z', b) \in [a_i]_{R_s}$ and $a' = F(z'', b') \in [a_j]_{R_s}$, respectively: then define $\varphi_z(b) = c_0$, and $\varphi_z(b') = c_1$. Using the fact that $\varphi_z = \varphi_{z'} = \varphi_{z''}$, this has the effect that

$$c_0 = \varphi_z(b) \ R \ F(z, b) = a \ R \ a_i$$
$$c_1 = \varphi_z(b') \ R \ F(z, b') = a' \ R \ a_j,$$

giving $c_0 \ R \ c_1$: this case *cannot* happen.

(2) Finally we define (a_s, b_s). Let a_s be the least number not in $\{a_i : i < s\}$. Let e' be an index of φ_e chosen by the Padding Lemma and the injectivity of G to be such that

$$G(e', a_s) \notin \bigcup_{i<s} [b_i]_{S_s};$$

and define $b_s = G(e', a_s)$. Now we check that the inductive assumption on availability still holds: suppose we see that a_i and a_j are R-equivalent, and b_i and b_j need to be made S-equivalent, thus we act by making $\varphi_e(a) = b_j$ (where $a \in [a_i]_{R_s}$ which is right available). If the class $[a_i]_{R_{s+1}} \cup [a_i]_{R_{s+1}}$ fails to be right available, then $[a_j]_{R_s}$ was not right available, so $[b_j]_{S_s}$ was left available by the inductive hypothesis. Therefore, $[b_j]_{S_{s+1}}$ is still left available.

Lastly, we check the inductive assumption for the new pair a_s, b_s. Since we only define φ_e in the operation above, since a_s is not in $\{a_i : i < s\}$, we have $\varphi_{e,s+1}(a_s) \uparrow$. We chose b_s to make a_s right available.

$s + 1$ *even.* Perform the same steps, inverting the roles between the a's and the b's, and between F and G.

It is easy to see that for every pair of numbers i, j,

$$a_i \ R \ a_j \Leftrightarrow b_i \ S \ b_j$$

and $\omega = \{a_i : i \in \omega\} = \{b_i : i \in \omega\}$.

Finally, note that we always maintain injectivity when we add a new pair a_s, b_s, and since at odd stages, we enter the least missing number into the domain of the reduction, and at even stages we enter the least missing number into the range of the reduction that this reduction is a permutation of ω.   □

Corollary 2.9 [8]. *Every precomplete ceer is universal.*

Proof. By Lemma 2.5, Theorems 2.7 and 2.8 and the fact that for ceers the property of being universal is preserved by isomorphisms.   □

The following is an interesting characterization of precomplete ceers.

Corollary 2.10. *Every precomplete ceer R is equal to S_v for some universal function v.*

Proof. Let R be a precomplete ceer and let S_u be the precomplete ceer determined by a universal function u. Then by Theorem 2.8, R and S_u are isomorphic. So, let π be a permutation of ω witnessing the isomorphism of R and S_u. It is straightforward to check that $v = \pi \circ u \circ \pi^{-1}$ is also universal and that $R = S_v$.   □

There are interesting extensions of Theorem 2.8, and of Corollary 2.9, due to Shavrukov [29], which we collect in the following theorem.

Theorem 2.11 [29]. *The following hold:*

(1) Any partial, and not onto monomorphism, induced by some partial computable function, from a ceer R to a precomplete ceer S can be extended to a monomorphism.

(2) Any strictly partial, and not onto monomorphism, induced by some partial computable function, between precomplete ceers R can be extended to an isomorphism.

Proof. We briefly sketch only a proof for item (1), i.e. how to show that every partial, and not onto monomorphism, from a ceer to a precomplete ceer, which is induced by some partial computable function, can be extended to a monomorphism. To prove the second item, it will be enough to combine this extension argument, with a back-and-forth argument in the style of Theorem 2.8, inserting, at odd stages, pairs that guarantee surjectivity.

Let R, S be ceers so that S is precomplete. Let φ be a partial computable function inducing a partial monomorphism from R to S: without loss of generality we may assume that the domain of φ is R-closed. Suppose we are working with suitable computable approximations $\{R_s\}$ and $\{S_s\}$ (as in Theorem 2.8) to R and S, respectively. Let F be an S-totalizer. We define an assignment $i \mapsto b_i$ such that $i \mathrel{R} j$ if and only if $b_i \mathrel{S} b_j$, and the corresponding monomorphism extends the given partial one. By the Recursion Theorem we also assume that we control the partial computable function φ_e. In the construction, at each stage $s + 1$, if i is least in its R_s-equivalence class, then we assume by induction that $\varphi_e(i)$ is still undefined by the end of stage s unless it has been already defined as $\varphi_e(i) = \varphi(i)$, for the sake of extending φ; in this regard, note that if at some stage we set $\varphi_e(i) = \varphi(i)$ then we regard $\varphi_e(i)$ as already defined, even if $\varphi(i)$ does not as yet converge, as we do so only for numbers i for which eventually $i \in \text{domain}(\varphi)$.

Pick numbers $b \mathrel{\not\!S} b'$, with $b, b' \notin [\text{range}(\varphi)]_S$. Such a pair of numbers exists because we assume that $[\text{range}(\varphi)]_S \neq \omega$ but, on the other hand, the complement of $[\text{range}(\varphi)]_S$ cannot be c.e. (see for instance Lemma 2.15 below which shows that each pair of distinct equivalence classes of a u.f.p. ceer, and a fortiori of a precomplete ceer, is effectively inseparable).

Take $b_i = F(e, i)$.

Step 0. Do nothing; $\varphi_{e,0}(i)$ is undefined for all i.

Step $s + 1$. We distinguish Cases 1. and 2., depending on whether $s + 1$ is odd or even:

(1) ($s + 1$ odd.) There are $i < j$ such that i and j R-collapse at stage $s + 1$; assume i, j are least in their R_s-equivalence classes:

    (a) if $\varphi_e(j)$ is still undefined, then set $\varphi_e(j) = b_i$: since F is an S-totalizer, this will give b_i S b_j, as $b_j = F(e, j)$ S $\varphi_e(j) = b_i$;

    (b) otherwise already $\varphi_e(j) = \varphi(j)$: set $\varphi_e(i) = \varphi(i)$, unless it has been already defined so; since φ induces a partial monomorphism, this fulfills the desired goal (notice that $\varphi(i)$ may be still undefined, but eventually it will converge).

(2) ($s + 1$ even) There are $i < j$ such that i and j are not as yet R-equivalent, but the matching b_i, b_j S-collapse:

    (a) if $\varphi_e(i)$ and $\varphi_e(j)$ are still undefined, then let $\varphi_e(i) = b$ and $\varphi_e(i) = b'$: this case cannot happen, since F is an S-totalizer, and otherwise we would get b S b_i S b_j S b';

    (b) if exactly one of $\varphi_e(i)$ and $\varphi_e(j)$ has been already defined, say $\varphi_e(i) = \varphi(i)$, then take the other one and set it equal to b: in our example, set $\varphi_e(j) = b$; again this case cannot happen, since $b \notin [\text{range}(\varphi)]_S$;

    (c) if already $\varphi_e(i) = \varphi(i)$ and $\varphi_e(j) = \varphi(j)$ have been defined, then do nothing, as φ induces a partial monomorphism.

Before leaving stage $s + 1$, we consider i such that $\varphi(i)$ converges for the first time, if any exists: if $\varphi_e(i)$ has not already been defined (otherwise it has been already stipulated that $\varphi_e(i) = \varphi(i)$), then set $\varphi_e(i) = \varphi(i)$.

Notice that the induction assumption is being preserved. This ends the construction. We skip the remaining details of the verification. □

Remark 2.12. By taking $\varphi = \varnothing$, the first item of Theorem 2.11 gives yet another proof of the universality of precomplete ceers.

2.4.1 Historical Remark

Universality of precomplete ceers was first proved by Bernardi and Sorbi in [8] and appeared before [22]. The proof in [8] used the so-called Anti Diagonal Normalization Theorem by Visser [32].

2.5 Uniformly Finitely Precomplete Ceers

The ceer $\sim_{PA}$ is not precomplete because it has a diagonal function, for instance the function induced by the connective $\neg$: we denote this function with the same symbol, namely $\neg\ulcorner\sigma\urcorner = \ulcorner\neg\sigma\urcorner$. Therefore $\sim_{PA}$ does not satisfy the Ershov Fixed Point Theorem, and thus it is not precomplete. However, although not precomplete, $\sim_{PA}$ is "locally" precomplete, i.e., every partial computable function with finite range can be totalized modulo $\sim_{PA}$ since there is some effectively found $n \geqslant 1$ such that all sentences in the range of φ are Σ_n, and thus we can totalize modulo $\sim_n$. This is exactly what led Montagna to introduce the u.f.p. ceers, see Definition 1.5(2).

Corollary 2.13. *Every precomplete ceer is u.f.p. The relation $\sim_{PA}$ is u.f.p., so there are u.f.p. ceers that are not precomplete.*

Proof. The first statement is immediate from the definitions. In order to prove that $\sim_{PA}$ is u.f.p. use the fact that, given a finite D and a sentence x, all sentences in $D \cup \{x\}$ fall into some finite level Σ_n, so that we can use a precompleteness totalizer $F_n(e, x)$ of $\sim_n$, using the fact that a totalizer for $\sim_n$ can be found uniformly in n. Some caution should be taken, since $\sim_{PA}$ and $\sim_n$ refer to different Gödel numbers. $\square$

Lemma 2.14 (Fixed Point Theorem for u.f.p. equivalence relations). *If R is u.f.p. then there exists a computable function $\mathrm{fix}(D, e)$ such that, for all D, e,*

$$\varphi_e(\mathrm{fix}(D, e)) \downarrow \in [D]_R \Rightarrow \varphi_e(\mathrm{fix}(D, e)) \, R \, \mathrm{fix}(D, e).$$

Proof. Let $f(D, e, x)$ be a totalizer of R, and let φ_u be so that for all x $\varphi_u(x) = \varphi_x(x)$. Let $s(D, e)$ be a computable function such that

$$\varphi_{s(D,e)}(z) = \varphi_e(f(D, u, z)),$$

and let $\mathrm{fix}(D, e) = f(D, u, s(D, e))$.
   Suppose that $\varphi_e(\mathrm{fix}(D, e)) \downarrow \in [D]_R$. Then

$$\varphi_e(\mathrm{fix}(D, e)) = \varphi_e(f(D, u, s(D, e)))) = \varphi_{s(D,e)}(s(D, e)) \downarrow \in [D]_R,$$

and $\varphi_u(s(D, e)) \, R \, f(D, u, s(D, e)) = \mathrm{fix}(D, e)$. $\square$

Lemma 2.15. *Every u.f.p. ceer is u.e.i.*

Proof. Let R be a u.f.p. ceer, and let $[a]_R, [b]_R$ be two distinct equivalence classes. Given c.e. sets W_u, W_v, define

$$\psi(x) = \begin{cases} b, & \text{if } (x \in W_u) \leq (x \in W_v); \\ a, & \text{if } (x \in W_u) < (x \in W_v); \\ \uparrow & \text{otherwise} \end{cases}$$

and let $n = \mathrm{fix}(\{a, b\}, e)$ be a fixed point for ψ, given by u.f.p.-ness of R where e is an index of ψ. It is clear that $n \notin W_u \cup W_v$, if $[a]_R \subseteq W_u$, $[b]_R \subseteq W_v$ and $W_u \cap W_v = \varnothing$. Since ψ is defined uniformly in the tuple (a, b, u, v), it is also clear that $n = p(a, b, u, v)$ for some computable function p. $\square$

The following theorem will be superseded by Theorem 2.34 (via Lemma 2.15). However, in order to become more acquainted with a useful proof technique, we include an outline of a direct proof here, different from the original proof given by Montagna [25].

Theorem 2.16 [25]. *Every u.f.p. ceer is universal.*

Proof. Let S be u.f.p. with totalizer f. As usual, we are assuming that S is nontrivial, and thus fix a and b with $a \not\mathrel{S} b$. In order to show that S is universal, we fix an arbitrary ceer R with $0 \not\mathrel{R} 1$ and demonstrate that $R \leqslant S$. By the Fixed Point Theorem, we assume that we control the partial computable function φ_e. Define the computable sequence y_i by $y_0 = a$, $y_1 = b$ and $y_i = f(\{y_j \mid j < i\}, e, i)$ for each $i \geqslant 2$. By our choice of whether to make $\varphi_e(i)$ converge, we can control whether y_i and y_j are S-equivalent. We show that $R \leqslant S$ via the function $i \mapsto y_i$. We will ensure in the construction that if a number k is the least number in its R-equivalence class at stage s, then $\varphi_{e,s}(k) \uparrow$.

When we witness at an odd stage $s + 1$ (we assume that R and S are approximated as in the proof of Theorem 2.8) that $i \mathrel{R} j$ for $i \neq j$ with i and j being least in their respective R_s-equivalence classes, and, say $i < j$, then we define $\varphi_{e,s}(j) = y_i$. As $f(\_, e, \_)$ is a totalizer of φ_e, it must occur that y_j becomes S-equivalent to y_i. Notice that i becomes the least number in the combined R_{s+1}-equivalence class and, as promised, that we have not yet caused $\varphi_e(i)$ to converge.

At even stages $s + 1$, we ensure that S does not collapse y_i to y_j unless already $i \mathrel{R_s} j$. We do this by threat of forcing a contradiction via the Fixed Point Theorem. Suppose i and j are the least numbers in their R-equivalence classes at an even stage $s + 1$, and the S-classes of y_i and y_j become S-equivalent at $s + 1$. Thus $\varphi_{e,s}(i) \uparrow$ and similarly $\varphi_{e,s}(j) \uparrow$. We then will cause $\varphi_{e,s+1}(i) \downarrow = a$ and $\varphi_{e,s+1}(j) \downarrow = b$, thus forcing that $a \mathrel{S} y_i \mathrel{S} y_j \mathrel{R} b$ contradicting that $a \not\mathrel{S} b$. Simply the threat of this action ensures that at no stage will it happen that $y_i \mathrel{S} y_j$ but $i \not\mathrel{R} j$. $\square$

Definition 2.17. An *extended diagonal function* for an equivalence relation R is a computable function d such that for every finite set D, we have that $x \not\mathrel{R} d(D)$ for every $x \in D$, i.e. $d(D) \notin [D]_R$.

We observe:

Corollary 2.18 [7]. *Every u.f.p. ceer R with a diagonal function has an extended diagonal function.*

Proof. Let R be a u.f.p. ceer, with a diagonal function d, and let $f(D, e, x)$ be a totalizer witnessing that R is u.f.p. By the Recursion Theorem with parameters, let $n(D)$ be a computable function such that

$$\varphi_{n(D)}(x) = d(f(D, n(D), x)):$$

then $g(D) = d(f(D, n(D), 0))$ is total, and $g(D) \notin [D]_R$: if $d(f(D, n(D), 0)) \in [D]_R$ then $\varphi_{n(D)}(0) \in [D]_R$, hence $f(D, n(D), 0) \mathrel{R} d(f(D, n(D), 0))$, contradiction. $\square$

2.6   e-Complete Ceers

The ceer $\sim_{PA}$ has an interesting additional property which is captured by the following definition, due to Montagna [25], and later independently rediscovered by Lachlan [22]. The equivalence relations described by this definition were called *uniformly finitely m-complete* by Montagna [25], and *extension complete* (or, simply, *e-complete*) by Lachlan [22]. We adopt here Lachlan's terminology.

Definition 2.19 [22,25]. An equivalence relation S is *e-complete* if for every ceer R and every pair of m-tuples $(a_1,\ldots,a_m)$, $(b_1,\ldots,b_m)$ such that the assignment $a_i \mapsto b_i$ induces a partial monomorphism from R to S, one can extend the assignment (uniformly from the two tuples and an index for R) to a computable function inducing a monomorphism. (Notice that uniformity extends also to the case in which the assignment does not provide a partial monomorphism.)

Corollary 2.20. *Every e-complete ceer is universal.*

Proof. Obvious. □

2.7   All e-Complete Ceers are Isomorphic

Finally we show that alle e-complete ceers are isomorphic.

Theorem 2.21 [22,25]. *The e-complete ceers are all isomorphic with each other.*

Proof. Let R, S be e-complete ceers. To show isomorphism, one uses a straightforward back-and-forth argument. We define an assignment $a_s \mapsto b_s$ at stages as follows.

Step 0. Do nothing.

Step $2s + 1$. Assume that we have already defined (a_i, b_i) for all $i \leqslant 2s - 1$, so that $a_i \ R \ a_j$ if and only if $b_i \ S \ b_j$. Let a_{2s} be the least such that $a_{2s} \notin \{a_i : i \leqslant 2s - 1\}$. By the uniform extension property due to the fact that S is e-complete, we can uniformly extend the finite assignment which has been defined so far, to a monomorphism, induced, say, by the computable function f. Then, let $b_{2s} = f(a_{2s})$.

Step $2s + 2$. Assume that we have already defined (a_i, b_i) for all $i \leqslant 2s$, so that $a_i \ R \ a_j$ if and only if $b_i \ S \ b_j$. Let b_{2s+1} be the least such that $b_{2s+1} \notin \{b_i : i \leqslant 2s\}$. By the uniform extension property due to the fact that R is e-complete, we can uniformly extend the finite assignment which has been defined so far, to a monomorphism, induced say, by the computable function g. Then let $a_{2s+1} = g(b_{2s+1})$. □

Theorem 2.22 [7,25]. *A ceer R is e-complete if and only if R is u.f.p. and R has a diagonal function.*

Proof. Given the fact that all e-complete ceers are isomorphic, and that there exists a ceer that is u.f.p. and with a diagonal function (namely, $\sim_{PA}$), it is enough to show that every u.f.p. ceer R with a diagonal function is e-complete as the property of being u.f.p. and having a diagonal function is invariant under computable isomorphisms. Now, by Corollary 2.18 this amounts to showing that every u.f.p. ceer R with an extended diagonal function is e-complete.

To see this, let us see that if S is any ceer, and $a_i \mapsto y_i$, for $i < m$ induces a monomorphism from S to R, then this assignment can be extended to a monomorphism. We can assume that $a_i = i$. We argue almost as in the proof of universality of u.f.p. ceers. We suppose to control, by the Recursion Theorem, a partial computable function φ_e, and define (for $i \geqslant m$),

$$y_i = f(\{y_j : j < i\} \cup \{d(\{y_j : j < i\})\}, e, i)$$

where f is an R-totalizer, and by Corollary 2.18, d is an extended diagonal function. A distinguishing difference with the proof of Theorem 2.16 is how we prevent that $y_i \ R \ y_j$ before we see that $i \ S \ j$. If we see this happen at some stage, we simply define (assume $i < j$, and j is least in its S-equivalence class at the stage, so that we assume by induction that $\varphi_e(j)$ is undefined at the given stage) $\varphi_e(j) = d(\{y_k : k < j\})$. Thus, as $\varphi_e(j) \downarrow \in \{y_k : k < i\} \cup \{d(\{y_k : k < j\})\}$,

$$d(\{y_k : k < j\}) = \varphi_e(j) \ R \ f(\{y_k : k < j\} \cup \{d(\{y_k : k < j\})\}, e, j) = y_j$$

giving a contradiction as now $d(\{y_k : k < j\}) \ R \ y_i$.   □

Lemma 2.23. *The ceer $\sim_{PA}$ is e-complete.*

Proof. By Corollary 2.13 and the presence of a diagonal function.   □

Notice that Peano Arithmetic provides examples of each one of the fundamental isomorphism types we have seen so far: in fact $\sim_{PA}$ is e-complete, whereas for instance $\sim_1$ is precomplete.

In contrast with the extension property for precomplete ceers pointed out in Theorem 2.11, and with the purpose of better distinguishing precomplete ceers from e-complete ceers, Shavrukov [29] shows

Theorem 2.24 [29]. *For every e-complete E, there is a partial non-onto monomorphism that cannot be extended to an endomorphism of E.*

Proof. Let E be e-complete, and P precomplete. We use Greek letters to denote morphisms. Let $\kappa : Id \longrightarrow P$, $\lambda : P \longrightarrow E$ be monomorphisms, and let $\eta = \lambda \circ \kappa$. Let $\theta : Id \longrightarrow E$ be given, induced by

$$t(x) = d(\{t(0), t(1), \ldots, t(x-1), x\})$$

where d is an extended diagonal function for E.

We claim that there is no endomorphism μ of E extending $\theta \circ \eta^{-1}$. Otherwise, if μ is such, let h be a computable function inducing $\mu \circ \lambda$. Then

$$\delta(x) = \text{first } y.[h(y) \, E \, t(h(x))]$$

is total and diagonal for P. For totality, notice that since μ extends $\theta \circ \eta^{-1}$, we have

$$\theta = \mu \circ \eta = \mu \circ \lambda \circ \kappa,$$

thus $\text{range}(t) \subseteq [\text{range}(h)]_E$. The remaining claim, i.e., $\delta(x) \not{P} x$ follows easily. Indeed, given x, first notice that $t(x) \not{E} x$ by definition of t; on the other hand, $h(\delta(x)) \, E \, t(h(x))$, by definition of δ; but if $x \, P \, \delta(x)$ then also $h(x) \, E \, h(\delta(x))$, as h induces a morphism: contradiction. □

Corollary 2.25. *If R is a u.f.p. ceer with a diagonal function then R has an automorphism without fixed points.*

Proof. Trivial since in this case R isomorphic to $\sim_{PA}$, for which $\neg$ induces an automorphism without fixed points. □

About fixed points of an endomorphism, Shavrukov [29] has shown that every u.f.p. ceer possesses endomorphisms with as many fixed points as we wish:

Theorem 2.26 [29]. *Let E be a u.f.p. ceer, and A a nonempty E-closed c.e. set. Then there is a computable function h, inducing and endomorphism of E such that $A = \{x : x \, E \, h(x)\}$.*

Proof. We may suppose without loss of generality that $0 \in A$. We define a computable function $h(i) = y_i$ that induces an endomorphism whose fixed points are exactly the equivalence classes of elements of A. In the rest of the proof, we say that a number is a fixed point if its equivalence class is a fixed point for the endomorphism induced by h.

The number y_i will be of the form

$$y_i = f(\{y_j : j < i\} \cup \{0, i\}, e, i),$$

where f is an E-totalizer, and e is an index such that by the Recursion Theorem we control φ_e. Since (by Lemma 2.15) the equivalence classes of E are infinite, we may suppose $f(D, z, i) \notin \{0, i\}$ for every D, z, i, and thus $y_i \neq 0, i$ for every i. At each stage, if i is least in its equivalence class and we have not previously defined $\varphi_e(i)$ to be 0 or i, then assume by induction that $\varphi_e(i)$ is undefined.

We use approximations $\{E_s\}$ to E as in Lemma 1.4, with the additional feature that if $E_{s+1} \setminus E_s \neq \varnothing$ then $s + 1 = 3t + 1$ for some t; and we use a computable approximation $\{A_s\}$ to A such that if $A_{s+1} \setminus A_s \neq \varnothing$ then $s + 1 = 3t$ for some t, and $A_{s+1} \setminus A_s \neq \varnothing$ is at most a singleton, and the approximation starts from the empty set.

The construction is by stages: at stages of the form $3t$ we make sure that all numbers in A are fixed points; at stages $3t + 1$ we make sure that h eventually induces an endomorphism; at stages $3t + 2$ we make sure that all fixed points are in A. At stage $s > 0$ we act as follows:

Stage $s = 3t$. If $i \in A_s \setminus A_{s-1}$, and $\varphi_e(i)$ is still undefined, define $\varphi_e(i) = i$.

Stage $s = 3t + 1$. If $i < j$ were least in their equivalence classes at stage $s - 1$ and they E-collapse at stage s, then we act as follows: if $\varphi_e(j)$ is still undefined, define $\varphi_e(j) = y_i$; if $\varphi_e(j)$ has been already defined (with $\varphi_e(j) \in \{0, j\}$), and if $\varphi_e(i)$ is still undefined, define $\varphi_e(i) = i$.

Stage $s = 3t + 2$. If i and y_i have become E-equivalent at the previous stage, and $\varphi_e(i)$ is still undefined, then define $\varphi_e(i) = 0$.

Notice that our action at each stage preserves the inductive assumption that $\varphi_e(i)$ is still undefined if i is least in its equivalence class unless we define $\varphi_e(i) \in \{0, i\}$. When we define $\varphi_e(i)$ we make $\varphi_e(i) \in \{y_j : j < i\} \cup \{0, i\}$ so that $\varphi_e(i) \; E \; y_i$ as f is a totalizer for E. We further observe that if $\varphi_e(i)$ is defined and $\varphi_e(i) \in \{0, i\}$, then $i \in A$ and i is a fixed point: this is trivial if $\varphi_e(i) = 0$; if $\varphi_e(i) = i$ then either $\varphi_e(i)$ has been defined at a stage $3t$, in which case the claim is trivial; or it has been defined through the second clause of a stage $3t + 1$. In this latter case, as f is a totalizer, our definition $\varphi_e(i) = i$ makes $i \; E \; y_i$; but $\varphi_e(j) \in \{0, j\}$ (where $j \; E \; i$ is the other number of the pair on which we act at the stage) and thus by induction on the stage we may assume that $j \in A$ which implies $i \in A$ as A is E-closed.

Let us now show that h induces a morphism. Assume that $i \; E \; j$, with $i < j$. Using that f is a totalizer, we get $y_i \; E \; y_j$ if we act on i, j at the stage s at which they are E-collapsed (we may again assume that they were least in their E-equivalence classes immediately before E-collapse); if we do not act on i, j, then both $\varphi_e(j)$ and $\varphi_e(i)$ have been already defined, and $\varphi_e(j) \in \{0, j\}, \varphi_e(i) \in \{0, i\}$, which, as argued above, gives $y_j \; E \; j \; E \; i \; E \; y_i$.

Finally we show that $j \in A$ if and only if j is a fixed point. If we ever define $\varphi_e(j) \in \{0, j\}$, then we have already seen that $j \in A$ and j is a fixed point. Suppose towards a contradiction that j is least so that $j \in A$ but j is not a fixed point, or vice versa. So suppose that $j \in A$ ($j \; E \; y_i$, respectively) but we never get to define $\varphi_e(j) = j$ ($\varphi_e(j) = 0$, respectively). This happens only if at the appropriate stage $3t$ ($3t + 2$, respectively), when we would like to act correspondingly, we see that $\varphi_e(j)$ has already been defined through the first clause of some step $3t + 1$, say $\varphi_e(j) = y_i$ for some $i < j$ with $i \; E \; j$. Since $i \; E \; j$, we have that $i \in A$ if and only if $j \in A$ and i is a fixed point if and only if j is a fixed point. So, $i < j$ contradicts the minimality of j. $\square$

2.8 Uniformly Effectively Inseparable Ceers

The main result of this section shows that every u.e.i. ceer is universal. To this end, we introduce a class of ceers, the *strongly uniformly m-complete (strongly u.m.c.) ceers*, and show, for any ceer R,

$$R \text{ u.e.i} \Rightarrow R \text{ strongly u.m.c.} \Rightarrow R \text{ universal.}$$

Here is the definition of a strongly u.m.c. ceer. It is a strengthening of the definition of a uniformly m-complete ceer given by Bernardi and Sorbi [8].

Namely, a nontrivial ceer R is *uniformly m-complete* (abbreviated as *u.m.c.*) if for every ceer S and every assignment $a_0 \mapsto b_0$, $a_1 \mapsto b_1$ (also denoted by $(a_0, a_1) \mapsto (b_0, b_1)$) of numbers such that $a_0 \not S a_1$ and $b_0 \not R b_1$, there exists a computable function extending the assignment and reducing S to R. It is shown in [1, Proposition 3.13] that not every u.m.c. is strongly u.m.c.

Definition 2.27. We say that a nontrivial ceer R is *strongly u.m.c.* if for every ceer S, every assignment $(a_0, a_1) \mapsto (b_0, b_1)$ can be extended uniformly (in a_0, a_1, b_0, b_1) to a total computable function f reducing S to R, provided that $a_0 \not S a_1$ and $b_0 \not R b_1$. (Note that the uniformity extends also to the cases $a_0 S a_1$ or $b_0 R b_1$; however, then no claim is made as to f reducing S to R.)

It immediately follows:

Corollary 2.28. *Every strongly u.m.c. ceer is universal.*

Proof. Straightforward.   □

Now we aim to prove that

$$R \text{ u.e.i} \Rightarrow R \text{ strongly u.m.c..}$$

For this we introduce yet another class of ceers, the weakly u.f.p. ceers, and show

$$R \text{ u.e.i} \Rightarrow R \text{ is weakly u.f.p.} \Rightarrow R \text{ strongly u.m.c..}$$

Definition 2.29. We say that a nontrivial ceer R is *weakly u.f.p.* if there exists a total computable function $f(D, e, x)$ (called a *weak totalizer for R*) such that for every finite set D, where $i \not R j$ for all distinct $i, j \in D$, and every e, x,

$$\varphi_e(x) \downarrow \in [D]_R \Rightarrow \varphi_e(x) \, R \, f(D, e, x).$$

Note that the definition differs from that of a u.f.p. ceer in that f need only satisfy the condition when $i \not R j$ for all distinct $i, j \in D$. Clearly

Corollary 2.30. *Every u.f.p. ceer is weakly u.f.p.*

Proof. Immediate.   □

A restriction of the definition is the following:

Definition 2.31. We call a nontrivial ceer *weakly n-u.f.p.* if in the definition for weakly u.f.p., we replace "finite set D" with "finite set D where $|D| \leqslant n$".

Lemma 2.32. *Each u.e.i. ceer is weakly u.f.p.*

Proof. Let R be a u.e.i. ceer. We first prove that R is weakly 2-u.f.p. To this end, assume that R is u.e.i. via the uniform productive function $p(a, b, u, v)$ as in Definition 1.5(3). We argue that R is weakly 2-u.f.p. Given any $a \neq b$, and e, we uniformly build a function $f(x) = f(\{a, b\}, e, x)$ witnessing that R is 2-u.f.p. Note that if $a = b$ then we can let f be the constant function with output a. By the

Double Recursion Theorem with parameters we build W_{a_x}, W_{b_x} for computable sequences of indices $\{a_x\}_{x \in \omega}, \{b_x\}_{x \in \omega}$, where the sequence is known to us during the construction.

Let $f(x) = p(a_x, b_x)$, where for simplicity we denote $p(a, b, \_, \_)$ by $p(\_, \_)$. Clearly f is a total computable function. Fix x, and let

$$
W_{a_x} = \begin{cases} [a]_R, & \text{if } \varphi_e(x) \not R \, b \\ [a]_R \cup \{p(a_x, b_x)\}, & \text{if } \varphi_e(x) \, R \, b, \end{cases}
$$

$$
W_{b_x} = \begin{cases} [b]_R, & \text{if } \varphi_e(x) \not R \, a \\ [b]_R \cup \{p(a_x, b_x)\}, & \text{if } \varphi_e(x) \, R \, a. \end{cases}
$$

Now assume that $a \not R \, b$, and fix e, x such that $\varphi_e(x) \downarrow \in [a]_R \cup [b]_R$. Without loss of generality suppose $\varphi_e(x) \, R \, a$. If $f(x) \not R \, a$ then $W_{a_x} \cap W_{b_x} = \varnothing$ and $p(a_x, b_x) \in W_{a_x} \cup W_{b_x}$, which contradicts p being a productive function.

Next, we show that if R is weakly 2-u.f.p. then R is weakly u.f.p. To this end, let f_i be a computable function witnessing that R is weakly i-u.f.p., for $2 \leqslant i \leqslant n$. We describe how to effectively get a function f_{n+1} witnessing that R is weakly $n+1$-u.f.p. Let e, D be given, with $|D| = i$. If $i > n+1$ or $i \leqslant 0$ then $f_{n+1}(D, e, x)$ outputs 0 for every x; if $1 \leqslant i \leqslant n$ then $f_{n+1}(D, e, x) = f_i(D, e, x)$ for every x. We assume now $D = \{d_0, \ldots, d_n\}$. By the Double Recursion Theorem, assume that we build φ_a and φ_b for some a, b. Let $E_x = \{f_n(D \smallsetminus \{d_n\}, a, x), d_n\}$, and $f_{n+1}(D, e, x) = f_2(E_x, b, x)$.

Here is how we compute $\varphi_a(x)$ and $\varphi_b(x)$. Initially both values are undefined. Step by step, we see which of the following cases happens first:

- $\varphi_e(x) \downarrow R \, d_n$: define $\varphi_b(x) = d_n$.
- $\varphi_e(x) \downarrow R \, d_i$ for some $i < n$: define $\varphi_b(x) = f_n(D \smallsetminus \{d_n\}, a, x)$ and $\varphi_a(x) = \varphi_e(x)$.
- $f_n(D \smallsetminus \{d_n\}, a, x) \, R \, d_n$: define $\varphi_a(x) = d_0$.

Clearly f_{n+1} is a total computable function, whose index can be found effectively in the indices for $f_2, \ldots, f_n$, using the fact that the fixed points in the Double Recursion Theorem can be found effectively from the parameters.

In order to see that f_{n+1} witnesses that R is weakly $n + 1$-u.f.p., fix e, D, x such that $D = \{d_0, \ldots, d_n\}$ where $d_i \not R \, d_j$ for every pair $i \neq j$, and $\varphi_e(x) \downarrow R \, d_i$ for some $i \leqslant n$. First we claim that $f_n(D \smallsetminus \{d_n\}, a, x) \not R \, d_n$: otherwise, by construction we would set $\varphi_a(x) = d_0$ unless it has previously been defined to be $\varphi_e(x) \, R \, d_i$, for some $i < n$. In either case we have $\varphi_a(x) \, R \, d_i$ for some $i < n$, which implies that $d_n \, R \, f_n(D \smallsetminus \{d_n\}, a, x) \, R \, d_i$, a contradiction.

We have thus that E_x consists of two elements that are not R-equivalent. Since $\varphi_b(x)$ is defined only when $\varphi_e(x)$ converges, it is straightforward to see that $f_{n+1}(D, e, x) \, R \, \varphi_e(x)$.   □

In the proof of Lemma 2.33 below we will use a computable infinite sequence of fixed points. This means that we wish to have an infinite sequence $\{e_i\}_{i \in \omega}$ so that we control each φ_{e_i} simultaneously. This can be done by the usual fixed point theorem, which gives us a single φ_e which we control. We simply let e_i be an index so that $\varphi_{e_i}(x) = \varphi_e(\langle i, x\rangle)$. Then by constructing the single function φ_e which we control, we simultaneously construct the infinite sequence of functions $\{\varphi_{e_i}\}_{i\in\omega}$. Of course, given the single index e we can computably list the infinite sequence $\{e_i\}_{i\in\omega}$.

Lemma 2.33. *Each weakly u.f.p. ceer is strongly u.m.c.*

Proof. We only sketch the proof, which is rather difficult. For a full and rigorous proof see [1].

Assume that R is a weakly u.f.p. ceer, with a weak totalizer f. In order to show that R is strongly u.m.c., we show in fact that for every ceer S, every assignment $(a_0', a_1') \mapsto (a_0, a_1)$ can be extended, uniformly in a_0', a_1', a_0, a_1, to a total computable function inducing a reduction from S to R, provided that $a_0' \not\mathrel{S} a_1'$ and $a_0 \not\mathrel{R} a_1$. (Uniformity extends also to the cases in which $a_0' \mathrel{S} a_1'$, or $a_0 \mathrel{R} a_1$.)

Note that by applying a computable permutation of ω, it is no loss of generality to consider an assignment $(0, 1) \mapsto (a_0, a_1)$, instead of $(a_0', a_1') \mapsto (a_0, a_1)$. Our goal (under the assumption that $0 \not\mathrel{S} 1$, and $a_0 \not\mathrel{R} a_1$) is then to extend this assignment to a total computable function yielding a reduction, by specifying a computable sequence of points $y_2, y_3, \ldots$ (we let $y_0 = a_0$, $y_1 = a_1$) where for every pair $i < k$ such that $k > 1$, we can force y_k to R-collapse to y_i, i.e., to have $y_k \mathrel{R} y_i$. The idea would be of course to mimic the proof that every u.f.p. is universal, and just define $y_i = f(\{y_j : j < i\}, e, i)$, where e is some index that we control by the Recursion Theorem. But a weak totalizer for R works only if the elements of D are pairwise R-inequivalent. Thus, if we define $y_i = f(\{y_j : j < i\}, e, i)$, and when we see $0 \mathrel{S} 2$ we force $y_2 \mathrel{R} y_0$ by making $\varphi_e(2) \downarrow = y_0$, then we would no longer be able to cause y_k to collapse to y_i ($i < k$) for any $k > 2$, because the set $\{y_j : j < k\}$ is no longer comprised of pairwise R-inequivalent elements. So the proof and the definition of y_i become more complicated.

By the Recursion Theorem we assume that we control φ_{e_i} for a computable sequence $\{e_i\}_{i\in\omega}$ of indices.

We define computable arrays $\{x_i^k, y_n\}_{i,k,n\in\omega}$, of pairwise distinct numbers in the following way:

- $x_0^k = f(\{a_0, a_1\}, e_1, k)$; notice that since a_0 and a_1 are not R-equivalent, then by suitably defining $\varphi_{e_1}(k)$ and using that f is a weak totalizer we are able to R-collapse x_0^k to either a_0 or a_1 as we wish (by making the definition $\varphi_{e_1}(k) = a_i$, we say that we *identify* x_0^k *with* a_i);
- $y_k = f(D_k \cup \{x_0^{2k}\}, e_{2k}, 0)$, where $D_k = \{x_1^k, \ldots, x_{k-1}^k\}$; notice that if $x \in D_k \cup \{x_0^{2k}\}$, $\varphi_{e_{2k}}(0)$ is still undefined, and the elements of $D_k \cup \{x_0^{2k}\}$ are eventually pairwise non-R-equivalent, then, using that f is a weak totalizer for R, we can R-collapse y_k with x by defining $\varphi_{e_{2k}}(0) = x$ (by making this definition, we say that we *identify* y_k *with* x);
- $x_i^k = f(\{y_i, x_0^{2i+1}\}, e_{2i+1}, k)$ (for $i > 0$); notice that if $x \in \{y_i, x_0^{2i+1}\}$, $\varphi_{e_{2i+1}}(k)$ is still undefined, and the elements of $\{y_i, x_0^{2i+1}\}$ are eventually non-R-equivalent, then, using that f is a weak totalizer for R, we can R-collapse

x_i^k with x by defining $\varphi_{e_{2i+1}}(k) = x$ (by making this definition, we say that we *identify* x_i^k *with* x).

Suppose now that we want to R-collapse y_k to y_i, with $i < k$, because we see $i\ S\ k$; we may also assume that i and k are least in their current S-equivalence classes, and for all $j', j < k$, we currently have $j'\ S\ j$ if and only if $y_{j'}\ R\ y_j$:

(1) if $i \leqslant 1$, then identify y_k with x_0^{2k} and x_0^{2k} with a_i;
(2) if $i > 1$, then identify y_k with x_i^k and x_i^k with y_i.

However, as already said, problems could arise if, in identifying y_k with x_i^k, elements of $D_k \cup \{x_0^{2k}\}$ were or became R-equivalent (problems of this type will be called P_a-*problems*), or, in identifying x_i^k with y_i, if y_i and x_0^{2i+1} were or became R-equivalent (P_b-*problems*). On the other hand, if these problems do not occur then our identification amount in fact to R-collapses, by properties of the weak totalizer f.

We argue that no unwanted R-collapse between two distinct elements of either $D_k \cup \{x_0^{2k}\}$ or $\{y_i, x_0^{2i+1}\}$ does take place, by *threatening*, if any such collapse occurred, to start from these two elements two parallel lines of successive identifications which propagate R and end respectively with a_0 and a_1: if the two starting elements of these two lines are (against our wishes) R-equivalent, then we would conclude that $a_0\ R\ a_1$, a contradiction. So, in fact, no unwanted R-collapse does happen.

For instance if in identifying y_k with x_i^k an unwanted R-collapse $x_r^k\ R\ x_j^k$ ($r \neq j$) occurred, with both x_r^k and x_j^k still unidentified, then our threatening action consists in identifying x_r^k with x_0^{2r+1} and x_0^{2r+1} with a_0; and also x_j^k with x_0^{2j+1} and x_0^{2j+1} with a_1. We consider a slightly more complicated situation: suppose that the identification of y_k with x_i^k, accompanied by the identification of x_i^k with y_i, faces the problem that $x_i^k\ R\ x_r^k$, with $r \neq i$. Then we can not further identify x_i^k because $\varphi_{e_{2i+1}}(k) = y_i$ has been already defined: our threatening action in this case still identifies x_r^k with x_0^{2r+1} and x_0^{2r+1} with a_0, but we now identify y_i with x_0^{2i} and x_0^{2i} with a_1.

For a given triple i, j, k, the following scheme summarizes the threatening actions one should take to face possible problems when identification is immediately possible, i.e. when the relevant numbers have not as yet been identified.

P_a: - *Problem* $x_i^k\ R\ x_0^{2k}$. Identify x_i^k with x_0^{2i+1} and x_0^{2i+1} with a_0; identify x_0^{2k} with a_1.
        - *Problem* $x_i^k\ R\ x_j^k$ ($i \neq j$). Identify x_i^k with x_0^{2i+1} and x_0^{2i+1} with a_0; identify x_j^k with x_0^{2j+1} and x_0^{2j+1} with a_1.
P_b: - *Problem* $y_i\ R\ x_0^{2i+1}$: Identify y_i with x_0^{2i} and x_0^{2i} with a_0; identify x_0^{2i+1} with a_1.

If a number x_i^k or y_i for which we threaten identification has been already identified, then we replace it with the number with which it has been identified, we threaten identification for the latter number, and so on.

Being able to prevent problems, we conclude that we are able to R-collapse y_k to y_i when we see that $i \ S \ k$. The difficult part of the verification consists in showing that we are always able to identify when we want to do so, i.e. the relevant values $\varphi_e(j)$ of the involved partial computable functions are still undefined. To show this, one can use the following facts:

i. when we choose to collapse elements in R, we always consider the y_j with j least in its current S-equivalence class. By induction on stages, we can show that we will have $\varphi_{e_1}(j)$, $\varphi_{e_{2j}}(0)$, $\varphi_{e_{2i+1}}(j)$ still undefined, allowing us the option to identify x_0^j, y_j, and any x_i^j as needed; moreover, still by induction we can show that the only identifications so far made at previous stages are those done under (1) or (2).

ii. a threatening action to solve a problem may have to face a new problem and so on, but problems alternate, i.e. the sequence of problems is such that a P_a-problem is either the last problem to occur, or it is followed by a P_b-problem; and a P_b-problem is either the last problem to occur, or it is followed by a P_a-problem; new problems refer to elements y_j with smaller j, so the sequence of threatening actions eventually terminates;

iii. when we face a P_b-problem, we deal with a y_j with a smaller j; notice that j need not be the least number in its current S-equivalence class: in this case we continue with $y_{j'}$ with j' least in its current S-equivalence class (as, $j', j < k$, we currently have that $j' \ S \ j$ implies $y_{j'} \ R \ y_j$);

iv. no threatening action does in fact take place, so no problem does in fact take place, so no new definitions of values $\varphi_e(j)$ involved in threatening actions do in fact take place (in fact the only identifications which are made are those under (1) or (2)); hence the inductive assumptions relative to values of various φ_e being undefined at the beginning of the current stage is preserved.

With the same trick, i.e., the trick of threatening to force a contradiction via suitable identifications, we argue that there is never any unwanted R-collapse between some y_k and y_i, in fact we never see y_k R-collapse to y_i before we see k S-collapse to i. □

It is now possible to close the circle, and show:

Theorem 2.34. *The following properties are equivalent for ceers:*

(i) u.e.i.
(ii) weakly u.f.p.
(iii) strongly u.m.c.

Proof. For the proof, we just need the following lemma. □

Lemma 2.35. *Every strongly u.m.c. ceer is u.e.i.*

Proof. Let R be a strongly u.m.c. ceer. Let U, V be a fixed pair of e.i. sets, and define S to be the ceer in which U and V are the only two nontrivial equivalence classes. Fix $u \in U$, $v \in V$, and given a, b, consider the assignment $(u, v) \mapsto (a, b)$. Using the fact that R is strongly u.m.c., one can uniformly extend this assignment to a computable function $f_{a,b}$. If $[a]_R \cap [b]_R = \varnothing$, then $f_{a,b}$ uniformly m-reduces the e.i. pair (U, V) to the pair $([a]_R, [b]_R)$, showing that the latter is e.i. (for this property of e.i. pairs, see, e.g., [28]). The fact that R is u.e.i. follows from the uniformity in this argument.   □

Remark 2.36. Uniformity plays a crucial role in the proof of universality for the u.e.i. ceers. Recent work has in fact shown [1] that there exist ceers yielding a partition of ω into effectively inseparable equivalence classes but they are not u.e.i. In fact the index set of the u.e.i. ceers is Σ_3^0-complete [1], but the index set of the effectively inseparable ceers is Π_4^0-complete [2].

2.9 Summarizing

Corollary 2.37 below subsumes all universality results known in the literature, including: every creative set is m-complete (Myhill [26]); every pair of effectively inseparable sets is m-complete (Smullyan [30]); all creative sequences are m-complete (Cleave [9]).

Corollary 2.37. *Every u.e.i. ceer is universal.*

Proof. Immediate by Theorem 2.34, as every strongly u.m.c. (or even u.m.c.) ceer is clearly universal: if R is a u.m.c. ceer, and S is any ceer with two distinct equivalence classes, then start off with an assignment $(a_0', a_1') \mapsto (a_0, a_1)$ with $a_0' \not\mathrel{S} a_1'$ and $a_0 \not\mathrel{R} a_1$, and extend it to a full reduction.   □

Corollary 2.38. *A ceer R is universal if and only if there exists a u.e.i. ceer S with $S \leqslant R$.*

Proof. If R is universal and S is u.e.i., then trivially $S \leqslant R$. Conversely, if S is u.e.i. and $S \leqslant R$, then R is universal, since so is S, by Corollary 2.37.   □

Corollary 2.39. *A ceer R is universal if and only if there is a c.e. set $X \subseteq \omega$ which is R-closed (i.e. so that $x \mathrel{R} y$ and $x \in X$ implies $y \in X$) and X^R is u.e.i. where $X^R = \{(i, j) : x_i \mathrel{R} x_j\}$ for a computable enumeration $X = \{x_i : i \in \omega\}$.*

Proof. If R is universal, then let S be u.e.i. with $S \leqslant R$ via a reduction f. Then let $X = \{x : (\exists y, c)[x \mathrel{R} y \,\&\, y = f(c)]\}$. Then X is chosen to have the property that $x \mathrel{R} y$ and $x \in X$ implies $y \in X$. We now show that X^R is u.e.i. Given any two numbers i, j, let c_i, c_j be so that $f(c_i) \mathrel{R} x_i$ and $f(c_j) \mathrel{R} x_j$. For any r.e. set U, let U_0 be the set $\{x : (\exists y)[f(x) \mathrel{R} y \,\&\, y \in U]\}$. If p is a uniform productive function for S, then the function $P(i, j, U, V) = i$, where i is so that $x_i = f(p(c_i, c_j, U_0, V_0))$, is a uniform productive function for X^R.

Conversely, it is clear that $X^R \leqslant R$ via the function $f(i) = x_i$. Thus if X^R is u.e.i., it is universal, and thus R is universal.   □

3   u.f.p. Ceers Which are Neither Precomplete nor e-complete

Precomplete ceers and e-complete ceers are not however the only ceers in the class of u.f.p. ceers.

Definition 3.1 [4]. An equivalence relation E is *weakly precomplete* if there exists a partial computable function fix such that, for all e,

$$\varphi_e \text{ total } \Rightarrow [\text{fix}(e)\downarrow \ \& \ \varphi_e(\text{fix}(e)) \ E \ \text{fix}(e)].$$

The following is an immediate characterization of weakly precomplete ceers.

Theorem 3.2. *Let E be a ceer. Then E is weakly precomplete if and only if E has no diagonal function.*

Proof. If E is a weakly precomplete equivalence relation then trivially E has no diagonal function. Conversely, assume that E is a ceer with no diagonal function, and let fix be the partial computable function, computed as follows: on input e, search for the first x such that $\varphi_e(x)$ is defined and $x \ E \ \varphi_e(x)$. It is now immediate to see that fix witnesses that E is weakly precomplete.   □

The following theorem and its corollary are taken from [5]. Recall from Definition 1.1 that if R, S are equivalence relations then R and S are isomorphic if and only if there is a computable function f which reduces R to S and such that for every y there exists x such that $y \ S \ f(x)$.

Theorem 3.3. *If E is a ceer such that E has an extended diagonal function, then there exist infinitely many ceers $\{E_i : i \in \omega\}$ such that, for every i, j,*

$$E \subseteq E_i \ \& \ [i \neq j \Rightarrow E_i \not\simeq E_j],$$

where $\simeq$ denotes isomorphism.

Proof. We sketch the proof: for full details see [5, Theorem 2.2]. Let E be a given ceer such that E has an extended diagonal function d.

We want to construct a countable set $\{E_i : i \in \omega\}$ of ceers such that for every i, $E \subseteq E_i$, satisfying the following requirement for each i, j, k, with $i \neq j$,

$P_{i,j,k}$: φ_k is total $\Rightarrow \varphi_k$ does not induce an isomorphism from E_i onto E_j.

Satisfaction of all requirements implies our claim, as for every isomorphism there is a total computable function inducing it.

We outline the strategy to meet $P_{i,j,k}$ in isolation, which is of course implemented at certain stages s: hence E_i and E_j have to be understood as their approximations E_i^s and E_j^s, respectively, and in particular at each such stage, $[a_0]_{E_i}$ is a finite set:

(1) choose a witness b_0 using the extended diagonal function to be E-inequivalent to every number mentioned so far;
(2) wait for a number a_0 such that $\varphi_k(a_0)\downarrow E_j b_0$;
(3) let $a_1 = d([a_0]_{E_i})$, and wait for $\varphi_k(a_1)\downarrow$;
(4) if, say, $\varphi_k(a_1) = b_1$ then E_j-collapse b_0 and b_1, and restrain $a_0 \not{E_i} a_1$.

Outcomes for the strategy to meet $P_{i,j,k}$. Here are the outcomes of the strategy:

(i) if we wait forever at (2), then we meet $P_{i,j,k}$ since φ_k, even if total, does not induce an onto morphism;
(ii) if we wait forever at (3), then we win $P_{i,j,k}$ since φ_k is not total;
(iii) if we act in (4), then we win $P_{i,j,k}$ since φ_k, even if total, does not induce a monomorphism.

The strategies can be combined by a finite priority argument. The critical part of the verification is that since b_0 is always chosen to be E-inequivalent to any number mentioned so far, and since each requirement is re-initialized if a higher-priority requirement acts, any collapse caused by the requirement $R_{i,j,k}$ cannot collapse together the elements a_0 and a_1 of a higher priority requirement.   □

Corollary 3.4. *There exist infinitely many weakly precomplete non-isomorphic u.f.p. ceers.*

Proof. Take $E =\sim_{PA}$ in the previous theorem so that E is u.f.p. and has an extended diagonal function. Then use the fact (see Lemma 1.7) that every ceer which is a nontrivial extension of a u.f.p. ceer is u.f.p. as well.   □

4   Separating u.e.i. Ceers from u.f.p. Ceers

The u.f.p. ceers are properly contained in the class of u.e.i. ceers, as shown by Andrews and Sorbi [3]:

Theorem 4.1 [3]. *There is a u.e.i. ceer which is not u.f.p.*

Proof. See [3].   □

In the same paper they show that in a sense, little is missing for a u.e.i. ceer to be u.f.p.

Theorem 4.2 [3]. *If a u.e.i. ceer has an extended diagonal function then it is u.f.p.*

Proof. See [3].   □

Figure 1 summarizes the inclusion relationships between the classes of universal ceers which have been introduced so far. The u.m.c. ceers have been defined at the beginning of Subsect. 2.8. The inside rectangular box consists of the u.f.p. ceers (which by Theorem 6.2(1) coincide with the non-trivial quotients of $\sim_{PA}$, or equivalently the nontrivial ceers that are isomorphic to ceers extending $\sim_{PA}$), and shows three disjoint regions: the precomplete ceers (all isomorphic with each other), the e-complete ceers (all isomorphic with each other), and a third unlabeled region containing by Corollary 3.4 infinitely many parwise non-isomorphic ceers.

All the inclusions shown by the picture are proper, by the above results. Not all universal ceers of course appear in one of the classes displayed in the

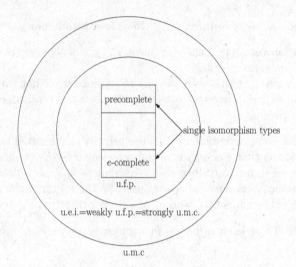

Fig. 1. Some classes of universal ceers

picture. For instance if R is a universal ceer then clearly $R \oplus \mathrm{Id}_1$ is universal but not u.m.c., where $R \oplus \mathrm{Id}_1$ is the ceer which collapses all odd numbers, and $2x\, R \oplus \mathrm{Id}_1\, 2y$ if and only if $x\, R\, y$.

The following result by Nies and Sorbi [27] shows that the class of u.e.i. contains interesting mathematical objects.

Theorem 4.3 [27]. *There is a finitely presented group D such that $=_D$ is a u.e.i. ceer.*

Proof. See [27]. □

5   A Characterization of the Universal Ceers Through a Jump operation

In this section, we look at a jump operation on ceers (due to [19]), and show that the universal ceers are exactly the ceers which are fixed points (modulo the equivalence) for this operation.

Definition 5.1 [19]. *For any ceer R, we define the jump of R to be the ceer R' so that $x\, R'\, y$ if and only if $x = y$ or $\varphi_x(x) \downarrow$, $\varphi_y(y) \downarrow$, and $\varphi_x(x)\, R\, \varphi_y(y)$.*

Notice that $(\mathrm{Id}_1)' = R_K$, that is the equivalence relation having the halting set K as its unique nontrivial equivalence class, and $(\mathrm{Id})'$ is the ceer yielding the partition $\{K_i : i \in \omega\} \cup \{\{x\} : x \notin K\}$, where $K_i = \{x : \varphi_x(x) \downarrow = i\}$.

Lemma 5.2. *The following properties hold:*

(1) $R \leqslant R'$;

(2) $R \leqslant S \Leftrightarrow R' \leqslant S'$;

(3) If R is not universal then R' is not universal.

Proof. (1) For every i, we can effectively find a number x_i so that $\varphi_{x_i}(x_i) = i$. By injectivity of s-m-n functions we may assume that the sequence (x_i) is injective. Then the map $i \mapsto x_i$ is a reduction of R to R'.

(2) Suppose $R \leqslant S$ via the function f. Given an index i, we can effectively find an index x_i so that if $\varphi_i(i) \downarrow$, then $\varphi_{x_i}(x_i) \downarrow = f(\varphi_i(i))$: as before we may assume that the sequence (x_i) is injective. Then the map $i \mapsto x_i$ gives a reduction of R' to S'.

Suppose $R' \leqslant S'$ via g. We first claim that for each x, if $\varphi_x(x) \downarrow$ then $\varphi_{g(x)}(g(x)) \downarrow$. Otherwise, we would have that the S'-class of $g(x)$ consists of a single element. But then the R'-class of x would be computable. But this is the set K_r, for $\varphi_x(x) = r$. It is a standard result that the set K_r is a complete c.e. set for any r. Thus we conclude that if $\varphi_x(x) \downarrow$ then $\varphi_{g(x)}(g(x)) \downarrow$. Now, consider the map $i \mapsto y_i$ given by taking x_i so that $\varphi_{x_i}(x_i) = i$ and letting $y_i = \varphi_{g(x_i)}(g(x_i))$. This is well-defined and gives a reduction of R to S.

(3) Suppose R' is universal. Then for any X, we have that $X' \leqslant R'$. Thus, we have that $X \leqslant R$. Thus R is universal as well. □

Note that (2) shows that the jump is an operation on degrees of ceers (where the degree of an equivalence relation is the equivalence class of the relation under the equivalence relation $\equiv$ given by $R \equiv S$ if and only if $R \leqslant S$ and $S \leqslant R$). Also, unlike most things called a jump, we can have $R' \equiv R$, for example if R is universal. Gao and Gerdes posed the question (Problem 10.2 of [19]) of whether there are non-universal ceers R so that $R' \equiv R$.

Theorem 5.3. *For every ceer E, if $E' \leqslant E$ then E is universal.*

Proof. We give an idea of the proof through a few examples. The reader interested in the full proof is invited to read [1].

Assume that h is a computable function that reduces E' to E. Let R be any ceer: we aim to show that $R \leqslant E$. We use an infinite computable set F of distinct indices (including $e_0, e_1, \ldots$), which we control. Eventually we define $g(i) = h(e_i)$, and show

$$i \, R \, j \Leftrightarrow e_i \, E' \, e_j (\Leftrightarrow h(e_i) \, E \, h(e_j))$$

i.e., g reduces R to E. For some $e \in F$, we define values $\varphi_e(e)$ during a stage of the construction, using computable approximations to R and E. The Recursion theorem will make us able to E'-collapse any pair e_i, e_j as needed.

Suppose for instance that we want to make $e_0 \, E' \, e_1$ because we see at some point that $0 \, R \, 1$. The basic module for this is the following (when in the following a new fixed point e from F is introduced, we assume that $\varphi_e(e)$ is still undefined):

(1) keep $\varphi_{e_0}(e_0)$ and $\varphi_{e_1}(e_1)$ undefined until we see $0 \ R \ 1$;
(2) define $\varphi_{e_0}(e_0) = \varphi_{e_1}(e_1) = h(e_{0,1})$ for another suitably chosen fixed point $e_{0,1} \in F$ (while keeping $\varphi_{e_{0,1}}(e_{0,1})\uparrow$).

Suppose that even later we want to E'-collapse e_1 and e_2:

(1) keep $\varphi_{e_{0,1}}(e_{0,1})$ and $\varphi_{e_2}(e_2)$ undefined, until $1 \ R \ 2$;
(2) define $\varphi_{e_2}(e_2) = h(e_2^1)$ and $\varphi_{e_{0,1}}(e_{0,1}) = \varphi_{e_2^1}(e_2^1) = h(e_{0,1,2})$ (while keeping $\varphi_{e_{0,1,2}}(e_{0,1,2})\uparrow$), where e_2^1 and $e_{0,1,2}$ are further suitably chosen fixed points.

But, using that h reduces E' to E, this implies $e_1 \ E' \ e_2$ (and thus $g(1) \ E \ g(2)$ as desired), as follows from the sequence of implications:

$$\varphi_{e_{0,1}}(e_{0,1}) = \varphi_{e_2^1}(e_2^1) \Rightarrow e_{0,1} \ E' \ e_2^1 \Rightarrow h(e_{0,1}) \ E \ h(e_2^1) \Rightarrow e_1 \ E' \ e_2.$$

Care must be taken (by carefully controlling convergence of the various computations $\varphi_e(e)$), to E-collapse only those pairs of $g(e_i)$'s which we *need* to collapse. In fact, using the power of the Recursion Theorem, we are able to prevent any E-collapse of the form $g(e_i) \ E \ g(e_j)$ if i and j have not already R-collapse; we are able to do so by simply *threatening* a contradiction. Again, we illustrate this through an example. Suppose we start from where we have just left the above computations, and assume that we see at some point $h(e_2) \ E \ h(e_3)$ but still $2 \not R 3$. Assume also for simplicity (but the general case is similar) that $\varphi_{e_2}(e_2)$ is still undefined. Then we take the following actions (using new suitable fixed points from F):

(1) define $\varphi_{e_3}(e_3) = h(e_3^1)$ and $\varphi_{e_3^1}(e_3^1) = h(e_3^2)$;
(2) stop the construction (so that $\varphi_{e_{0,1,2}}(e_{0,1,2})$ and $\varphi_{e_3^2}(e_3^2)$ remain undefined: we can do this since we have not seen as yet $2 \ R \ 3$ so we have not defined $\varphi_{e_{0,1,2}}(e_{0,1,2})$ and $\varphi_{e_3^2}(e_3^2)$).

This yields the following implications:

$$e_{0,1,2} \not E' e_3^2 \Rightarrow h(e_{0,1,2}) \not E h(e_3^2) \Rightarrow e_2^1 \not E' e_3 \Rightarrow h(e_2^1) \not E h(e_3^1) \Rightarrow e_2 \not E' e_3$$

and thus $h(e_2) \not E h(e_3)$, a contradiction.

A full proof of the theorem is just a formalization of the ideas suggested by the above examples. □

6   Characterizations of Some Classes of Universal Ceers

Bernardi and Montagna [7] use the notion of a quotient object to characterize u.f.p. ceers and precomplete ceers. Given equivalence relations R, S, we say that R is a *quotient of* S, if there is an onto morphism from S to R.

Lemma 6.1. *Let R, S be ceers with no finite classes. Then R is a quotient of S if and only if there is a ceer $S' \simeq R$ such that $S' \supseteq S$.*

Proof. Let R, S be ceers with no finite classes. If f induces an onto morphism from S to R, then define $x\ S'\ y$ if and only if $f(x)\ R\ f(y)$. The right to left implication is obvious.   □

Theorem 6.2 [7]. *The following hold:*

(1) A ceer R is u.f.p. if and only if R is a nontrivial quotient of $\sim_{PA}$.
(2) A ceer R is precomplete if and only if R is a nontrivial quotient of every universal ceer.

Proof. The two implications from right to left follow from Lemma 1.7.

We now show the implications from left to right. We begin with the first item. Let R be a u.f.p. ceer. Construct an onto morphism $\mu : \sim_{PA} \longrightarrow R$, by defining a computable h by stages. Suppose that $f(D, e, x)$ is a totalizer for R. We assume that by the Recursion Theorem we control the index e. Also, assume that we work with computable approximations $\{R_s\}$ to R, and $\sim_{PA,s}$ to $\sim_{PA}$, as in Lemma 1.4: without loss of generality we may assume that $\sim_{PA,s}$ changes only at odd stages.

At the end of stage s, suppose that we have defined a finite set of pairs $(a_0, b_0), \ldots, (a_{s-1}, b_{s-1})$ approximating a computable function h that we build and that will induce the desired onto morphism: in fact, h itself will be onto. At each stage $s + 1$ we assume by induction that if i is least such that $a_i \in [a_i]_{\sim_{PA,s}}$ then $\varphi_{e,s}(i)$ is still undefined.

Stage 0. Let $\varphi_{e,0}(x)$ be undefined for all x.

Stage $s + 1$ odd See if there are $i < j$ such that a_i, a_j become $\sim_{PA}$-equivalent. If, so, pick such a pair i, j: we may assume that j is least such that $a_j \in [a_j]_{\sim_{PA,s}}$. Define $\varphi_e(j) = b_i$.

Let now $a_s = \mu x.[x \notin \{a_i : i < s\}]$, and let $b_s = f(\{b_i : i < s\}, e, s)$.

Stage $s + 1$ even. Let $b_s = \mu x.[x \notin \{b_i : i < s\}]$, and let a_s be a number which is not $\sim_{PA}$-equivalent to any number which is already in $\{a_i : i < s\}$. We use here that $\sim_{PA}$ has an extended diagonal function. (Notice that we do not have to take any special action if b_s is already R-equivalent to some b_i with $i < s$, since we are not trying to construct a reduction but simply an (onto) R-preserving function.)

At each step the inductive assumption is preserved. It is not difficult to see that the assignment $a_s \mapsto b_s$, defines a computable function h with the desired properties.

We now turn to the second item of the statement. Let R be a precomplete ceer, and S a universal ceer: so there is a computable function f which induces a monomorphism from R to S. We want to show that there is a computable function h that induces an onto morphism from S to R. Suppose that we have already defined $h(i)$ for all $i < n$, and let e_n be a uniformly found index such that

$$\varphi_{e_n}(x) = \begin{cases} h(i) & \text{if } ((\exists i < n)[n \in [i]_S]) \leq ((\exists y)[n \ S \ f(y)]) \text{ and } i \text{ is first,} \\ y & \text{if } ((\exists y)[n \ S \ f(y)]) < ((\exists i < n)[n \in [i]_S]) \text{ and } y \text{ is first,} \\ \uparrow & \text{otherwise;} \end{cases}$$

let $f(e, z)$ be a totalizer for R, and define $h(n) = f(e_n, 0)$. For the verification, let us inductively assume that if $i, j < n$ and $i \ S \ j$ then $h(i) \ R \ h(j)$, and let $i < n$ be such that $i \ S \ n$. We want to show that $h(i) \ R \ h(n)$: by the inductive assumption, we may assume that i is least with this property. Then $\varphi_{e_n}(0)$ is defined: if it is defined through the first clause, then by the totalizer f, $h(n) \ R \ h(i)$; otherwise, let $h(n) = y$ where $f(y) \ S \ n$; but then $\varphi_{e_i}(0)$ is defined, and by minimality of i, it is defined through the second clause, so that $h(i) = z$ for some z such that $f(z) \ S \ i$. It follows that $f(y) \ S \ f(z)$, and thus $y \ R \ z$, as f induces a monomorphism. Therefore h induces a morphism: it is easy to see that this morphism is also onto. □

6.1   Extensional Formulae of Peano Arithmetic

In this section we consider ceers defined by extensional formulae of Peano Arithmetic.

Definition 6.3. *Given a formula $F(v)$ in the language of PA, let $\sim_F$ be the ceer*

$$x \sim_F y \Leftrightarrow PA \vdash F(\overline{x}) \leftrightarrow F(\overline{y}).$$

A formula $F(v)$ of PA is extensional *if for every x, y,*

$$x \sim_{PA} y \Rightarrow PA \vdash F(\overline{x}) \leftrightarrow F(\overline{y})$$

Theorem 6.4 [7]. *The u.f.p. ceers coincide with the ceers that are isomorphic to the ones induced by extensional formulas of PA.*

Proof. If R is given by an extensional formula, then $R \supseteq \sim_{PA}$, thus it is u.f.p. by Lemma 1.7.

Conversely, if R is u.f.p., then $R \simeq S$ for some ceer $S \supseteq \sim_{PA}$, by Theorem 6.2. Then, by Lemma 6.5 below, there is a formula $F(v)$ such that $S = \sim_F$. Since $S \supseteq \sim_{PA}$, F is extensional. □

Lemma 6.5 [8]. *For every ceer S there exists a Σ_1 formula $F(v)$, such that*

$$x \ S \ y \Leftrightarrow PA \vdash F(\overline{x}) \leftrightarrow F(\overline{y}).$$

Proof. Let S be a ceer. Since $\sim_L$ is precomplete, there exists a computable function f such that

$$x \ S \ y \Leftrightarrow PA \vdash \rho_{f(x)} \leftrightarrow \rho_{f(y)},$$

where $\rho_{f(x)}$ is the Σ_1 sentence with $\ulcorner_1$-code $f(x)$. Define $g(x) = \ulcorner \rho_{f(x)} \urcorner$, and let $G(u, v)$ a Σ_1 formula representing g in PA. By an argument similar to the one in the proof of Theorem 2.6, it is easy to see that can take $F(u)$ to be $(\exists v)(G(u, v) \wedge T_1(v))$. □

An important example of an extensional formula is the provability predicate $\mathrm{Pr}_{PA}(v)$, a Σ_1 formula representing the set of theorems of PA and satisfying the Hilbert-Bernays Derivability Conditions. The next lemma will be used to show that $\sim_{\mathrm{Pr}_{PA}}$ is precomplete.

Lemma 6.6 [7]. *Let $F(v)$ be a Σ_n extensional formula such that there exists $q \in \Sigma_n$ for which*

$$[\{\ulcorner F(\overline{n})\urcorner : n \in \omega\}]_{\sim_{PA}} = [\{\ulcorner p\urcorner : p \in \Sigma_n, PA \vdash q \to p\}]_{\sim_{PA}}.$$

Then $\sim_F$ is precomplete.

Proof. Let ψ be a partial computable function and let

$$\varphi(x) = \begin{cases} \text{the first } y \text{ such that } \ulcorner F(\overline{y})\urcorner \sim_{PA} x, \text{ if there is any such } y; \\ \uparrow, \qquad\qquad\qquad\qquad\qquad\qquad\qquad\qquad \text{otherwise.} \end{cases}$$

Let $\Psi(u, v)$ be a formula that represents ψ in PA, and define

$$\hat{\psi}(x) = \ulcorner \exists v(\Psi(\overline{x}, v) \land F(v)) \lor q\urcorner.$$

Clearly, $\hat{\psi}$ is total. Let now $h = \varphi \circ \hat{\psi}$. We claim that also h is total. Indeed, let x be given, and observe that

$$PA \vdash q \to (\exists v(\Psi(\overline{x}, v) \land F(v)) \lor q)$$

hence by the hypothesis there exists some z such that

$$\ulcorner (\exists v(\Psi(\overline{x}, v) \land F(v)) \lor q)\urcorner \sim_{PA} \ulcorner F(\overline{z})\urcorner.$$

This shows that $\varphi(\hat{\psi}(x))$ is defined, hence h is total.

Notice, that by the hypothesis, for every z, since $F(\overline{z})$ is provably equivalent to some sentence which is implied by q, we have $PA \vdash q \to F(\overline{z})$, and thus

$$\ulcorner F(\overline{z}) \lor q\urcorner \sim_{PA} \ulcorner F(\overline{z})\urcorner. \tag{1}$$

We now claim that h makes ψ total modulo $\sim_F$. Suppose that $\psi(x) \downarrow = y$. Then it is easy to see that

$$\hat{\psi}(x) \sim_{PA} \ulcorner F(\overline{y}) \lor q\urcorner \sim_{PA} \ulcorner F(\overline{y})\urcorner$$

(where the last equivalence is justified by (1)). Hence if $\varphi(\hat{\psi}(x)) = z$ with $\ulcorner F(\overline{z})\urcorner \sim_{PA} \hat{\psi}(x)$, then we see that $\ulcorner F(\overline{z})\urcorner \sim_{PA} \ulcorner F(\overline{y})\urcorner$, and thus

$$h(x) = \varphi(\hat{\psi}(x)) \sim_F y.$$

$\square$

Theorem 6.7 [7]. *$\sim_{\mathrm{Pr}_{PA}}$ is precomplete.*

Proof. We verify that $\mathrm{Pr}_{PA}(v)$ satisfies the hypotheses of Lemma 6.6 with the sentence $q = \neg\mathrm{Con}_{PA}$ and $n = 1$. As independently proved by Goldfarb and Friedman, see [18], for every Σ_1 sentence p such that $PA \vdash \neg\mathrm{Con}_{PA} \to p$ there is a Σ_1 sentence p' such that $PA \vdash p \leftrightarrow \mathrm{Pr}_{PA}(\ulcorner p'\urcorner)$. The other inclusion follows from the fact that for every n, $PA \vdash \neg\mathrm{Con}_{PA} \to \mathrm{Pr}_{PA}(\overline{n})$. $\square$

References

1. Andrews, U., Lempp, S., Miller, J.S., Ng, K.M., San Mauro, L., Sorbi, A.: Universal computably enumerable equivalence relations. J. Symbolic Logic **79**(1), 60–88 (2014)
2. Andrews, U., Sorbi, A.: The complexity of index sets of classes of computably enumerable equivalence relations. J. Symbolic Logic (to appear)
3. Andrews, U., Sorbi, A.: Jumps of computably enumerable equivalence relations (2016, in preparation)
4. Badaev, S.: On weakly precomplete positive equivalences. Siberian Math. J. **32**, 321–323 (1991)
5. Badaev, S., Sorbi, A.: Weakly precomplete computably enumerable equivalence relations. Mat. Log. Quart. **62**(1–2), 111–127 (2016)
6. Bernardi, C.: On the relation provable equivalence and on partitions in effectively inseparable sets. Studia Logica **40**, 29–37 (1981)
7. Bernardi, C., Montagna, F.: Equivalence relations induced by extensional formulae: classifications by means of a new fixed point property. Fund. Math. **124**, 221–232 (1984)
8. Bernardi, C., Sorbi, A.: Classifying positive equivalence relations. J. Symbolic Logic **48**(3), 529–538 (1983)
9. Cleave, J.P.: Creative functions. Z. Math. Logik Grundlag. Math. **7**, 205–212 (1961)
10. Cooper, S.B.: Computability Theory. Chapman & Hall/CRC Mathematics, Boca Raton, London, New York, Washington, DC (2003)
11. Coskey, S., Hamkins, J.D., Miller, R.: The hierarchy of equivalence relations on the natural numbers. Computability **1**, 15–38 (2012)
12. Ershov, Y.L.: Positive equivalences. Algebra Logic **10**(6), 378–394 (1973)
13. Ershov, Y.L.: Theorie der Numerierungen I. Z. Math. Logik Grundlag. Math. **19**, 289–388 (1973)
14. Ershov, Y.L.: Theorie der Numerierungen II. Z. Math. Logik Grundlag. Math. **19**, 473–584 (1975)
15. Fokina, E., Khoussainov, B., Semukhin, P., Turetsky, D.: Linear orders realized by c.e. equivalence relations. J. Symbolic Logic (to appear)
16. Fokina, E., Friedman, S., Nies, A.: Equivalence relations that are Σ_3^0 complete for computable reducibility. In: Ong, L., Queiroz, R. (eds.) WoLLIC 2012. LNCS, vol. 7456, pp. 26–33. Springer Berlin Heidelberg, Berlin, Heidelberg (2012). doi:10.1007/978-3-642-32621-9_2
17. Fokina, E.B., Friedman, S.D., Harizanov, V., Knight, J.F., McCoy, C., Montalbán, A.: Isomorphism relations on computable structures. J. Symbolic Logic **77**(1), 122–132 (2012)
18. Friedman, H.: Proof-theoretic degrees (Unpublished)
19. Gao, S., Gerdes, P.: Computably enumerable equivalence realations. Studia Logica **67**, 27–59 (2001)
20. Gravuskin, A., Jain, S., Khoussainov, A., Stephan, F.: Graphs realized by r.e. equivalence relations. Ann. Pure Appl. Logic **165**(7–8), 1263–1290 (2014)
21. Gravuskin, A., Khoussainov, A., Stephan, F.: Reducibilities among equivalence relations induced by recursively enumerable structures. Theoret. Comput. Sci. **612**(25), 137–152 (2016)
22. Lachlan, A.H.: A note on positive equivalence relations. Z. Math. Logik Grundlag. Math. **33**, 43–46 (1987)
23. Mal'tsev, A.I.: Sets with complete numberings. Algebra i Logika **2**(2), 4–29 (1963)

24. Miller III, C.F.: On Group-Theoretic Decision Problems and Their Classification. Annals of Mathematics Studies. Princeton University Press, Princeton (1971)
25. Montagna, F.: Relative precomplete numerations and arithmetic. J. Philos. Logic **11**, 419–430 (1982)
26. Myhill, J.: Creative sets. Z. Math. Logik Grundlag. Math. **1**, 97–108 (1955)
27. Nies, A., Sorbi, A.: Calibrating word problems of groups via the complexity of equivalence relations. Math. Structures Comput. Sci. (2016, to apper)
28. Rogers Jr., H.: Theory of Recursive Functions and Effective Computability. McGraw-Hill, New York (1967)
29. Shavrukov, V.Y.: Remarks on uniformly finitely positive equivalences. Math. Log. Quart. **42**, 67–82 (1996)
30. Smullyan, R.: Theory of Formal Systems. Annals of Mathematical Studies, vol. 47. Princeton University Press, Princeton (1961)
31. Soare, R.I.: Recursively Enumerable Sets and Degrees. Perspectives in Mathematical Logic, Omega Series. Springer, Heidelberg (1987)
32. Visser, A.: Numerations, λ-calculus & arithmetic. In: Seldin, J.P., Hindley, J.R. (eds.) To H. B. Curry: Essays on Combinatory Logic, Lambda Calculus and Formalism, pp. 259–284. Academic Press, London (1980)

Higher Computability

Σ_1^1 in Every Real in a Σ_1^1 Class of Reals Is Σ_1^1

Leo Harrington[1], Richard A. Shore[2(✉)], and Theodore A. Slaman[1]

[1] Department of Mathematics, University of California, Berkeley,
Berkeley, CA 94270, USA
[2] Department of Mathematics, Cornell University, Ithaca, NY 14853, USA
shore@math.cornell.edu

Abstract. We first prove a theorem about reals (subsets of $\mathbb{N}$) and classes of reals: If a real X is Σ_1^1 in every member G of a nonempty Σ_1^1 class $\mathcal{K}$ of reals then X is itself Σ_1^1. We also explore the relationship between this theorem, various basis results in hyperarithmetic theory and omitting types theorems in ω-logic. We then prove the analog of our first theorem for classes of reals: If a class $\mathcal{A}$ of reals is Σ_1^1 in every member of a nonempty Σ_1^1 class $\mathcal{B}$ of reals then $\mathcal{A}$ is itself Σ_1^1.

1 Introduction

We work in Cantor space $2^{\mathbb{N}}$ and call its members $X \subseteq \mathbb{N}$, *reals*. We think of members of Baire space $\mathbb{N}^{\mathbb{N}}$ as functions $F : \mathbb{N} \to \mathbb{N}$ (coded as real consisting of pairs of numbers). We use the standard normal form theorems for reals and classes of reals as follows: A real X is Σ_1^1 (in a real G) if it is of the form $\{n | \exists F \forall x R(F \restriction x, x, n)\}$ for a recursive (in G) predicate R. A class $\mathcal{K}$ of reals is Σ_1^1 (in G) if it is of the form $\{X | \exists F \forall x R(X \restriction x, F \restriction x, x)\}$ for a recursive (in G) predicate R. A real or class of reals is Δ_1^1 (or hyperarithmetic) (in G) if it and its complement are Σ_1^1 (in G). Our first main theorem is the following:

Theorem 2.1. *If a real X is Σ_1^1 in every member G of a nonempty Σ_1^1 class $\mathcal{K}$ of reals then X is itself Σ_1^1.*

While the statement of this theorem and certainly the proof we provide in the next section seem to have little to do with either results of hyperarithmetic theory or model theory they are all, in fact, connected along a couple of paths. Indeed, we were thinking about related matters when we proved the theorem.

A basis theorem in recursion theory typically says that every nonempty class of some sort contains a member with some property. For example, the classes may be arbitrary Σ_1^1 classes $\mathcal{K}$ of reals. One, the Gandy Basis Theorem (see Sacks (1990, III. 1.5)), says that every nonempty Σ_1^1 class of reals contains one

---

R.A. Shore—Partially supported by NSF Grant DMS-1161175. The last two authors began their work on this paper at a workshop of the Institute for Mathematical Sciences of the National University of Singapore which also partially supported them.
T.A. Slaman—Partially supported by NSF Grant DMS-1301659 and by the Institute for Mathematical Sciences of the National University of Singapore.

A. Day et al. (Eds.): Downey Festschrift, LNCS 10010, pp. 455–466, 2017.
DOI: 10.1007/978-3-319-50062-1_26

Z such that $\omega_1^Z = \omega_1^{CK}$. (For any Z, ω_1^Z is the least ordinal not recursive, or equivalently not Δ_1^1 in Z; ω_1^{CK} is ω_1^Z for Z recursive (or Δ_1^1).) Another, the Kreisel Basis Theorem (see Sacks (1990, III. 7.2)), says that if a real X is not hyperarithmetic (i.e. Δ_1^1) then $\mathcal{K}$ also contains a real Z in which X is not Δ_1^1. An equivalent version is that if X is Δ_1^1 in every member of $\mathcal{K}$ then X is Δ_1^1.

Our first theorem is the generalization of the Kreisel basis theorem where Δ_1^1 is replaced by Σ_1^1. (To see that it implies the result of Kreisel note that it says that if X and $\bar{X}$ are both Σ_1^1 (i.e. Δ_1^1) in every member of $\mathcal{K}$ then they are both Σ_1^1 (and so Δ_1^1)). Our theorem also implies the basis result of Gandy: As Kleene's O is not Σ_1^1 there is a $Z \in \mathcal{K}$ in which O is not Σ_1^1. By classical results of Spector (see Sacks (1990, II. 7.7)), this implies that $\omega_1^Z = \omega_1^{CK}$. (See Theorem 2.9.)

Sacks also provides results of hyperarithmetic theory as corollaries to Kreisel's theorem and others that, as he points out, can be viewed as omitting types theorems in ω-logic. They are also immediate consequences of our theorem as we indicate in the next section. We discus these and other related results in next section after we prove our theorem.

After his proof, Sacks (1990, p. 75) says of this connection that "The recursion theorist winding his way through a Σ_1^1 set is a brother to the model theorist threading his way through a Henkin tree." Our proof, which requires no knowledge of either hyperarithmetic theory or model theory, shows that there is another sibling traipsing (or perhaps treading carefully) through a forcing construction.

Our theorem should have been a classical one of hyperarithmetic theory. It also has analogs, both recent and classical, in other settings. When we told Stephen Simpson the result he remarked that Andrews and Miller (2015, Proposition 3.6) had recently proven the analogous result for Π_1^0 classes in place of Σ_1^1 classes. We rephrase it in our terminology as follows:

Theorem 1.1 (Andrews and Miller). *Let P be a nonempty Π_1^0 class. If X is Π_1^0 in every member of P then X is Π_1^0. (Or, equivalently, if X is Σ_1^0 in every member of P then X is Σ_1^0.)*

Their proof is a forcing proof similar to ours but using Π_1^0 classes instead of Σ_1^1 ones.

At the level of Σ_2^1 classes, a standard basis theorem gives the analogous result (as pointed out to us by John Steel). The classical result (see Moschovakis (1980, 4E.5)) is that the Δ_2^1 reals are a basis for the Σ_2^1 classes of reals. Thus if $\mathcal{K}$ is Σ_2^1 it contains a Δ_2^1 real G and, of course, any real X which is Σ_2^1 in G via Θ is itself Σ_2^1. ($X = \{n | \exists G(\Psi(G) \ \& \ \Theta(G,n))\}$ where Ψ is the Σ_2^1 formula saying G satisfies its Δ_2^1 definition.) Similar basis results hold at higher levels of the projective hierarchy assuming various set theoretic axioms. (See Moschovakis (1980, 5A.4 and 6C.6).)

About the only facts about Σ_1^1 reals and classes that we use in our proof are the standard normal form theorems mentioned at the beginning of this Introduction.

Our second main theorem is one analogous to Theorem 2.1 but at the level of classes of real.

Theorem 3.1. *If a class $\mathcal{A}$ of reals is Σ_1^1 in every member of a nonempty Σ_1^1 class $\mathcal{B}$ of reals then it is Σ_1^1.*

Our proof of this theorem requires some familiarity with effective descriptive set theory. We give some of the basic facts needed and the proof in Sect. 3.

2   The Proof for Reals

We now give the promised forcing style proof of our main theorem.

Theorem 2.1. *If a real X is Σ_1^1 in every member G of a nonempty Σ_1^1 class $\mathcal{K}$ of reals then X is itself Σ_1^1.*

Proof. We use the language of Gandy-Harrington forcing. Forcing conditions are nonempty Σ_1^1 classes $\mathcal{L}$ of reals with set containment as extension. We view the Σ_1^1 formulas $\varphi(G, n)$ as of the form $\exists F \forall x R(G \restriction x, F \restriction x, x, n)$ with R recursive. We say that $\mathcal{L} \Vdash \varphi(G, n)$ if $(\forall Z \in \mathcal{L})(\varphi(Z, n))$. If, as usual, we say $\mathcal{L} \Vdash \neg\varphi(G, n)$ if $(\forall \hat{\mathcal{L}} \subseteq \mathcal{L})(\hat{\mathcal{L}} \nVdash \varphi(G, n))$, this definition is then equivalent to $(\forall Z \in \mathcal{L})(\neg\varphi(Z, n))$. The point here is that if there is a $Z \in \mathcal{L}$ such that $\varphi(Z, n)$ then $\hat{\mathcal{L}} = \mathcal{L} \cap \{Z | \varphi(Z, n)\}$ is a nonempty extension of $\mathcal{L}$ which obviously forces $\varphi(G, n)$.

We now list all the Σ_1^1 formulas $\Theta_k(G, n)$. These are the formulas that could potentially define the reals Σ_1^1 in any G. We consider an X which is a candidate for being Σ_1^1 in every $G \in \mathcal{K}$. We build a sequence $\mathcal{L}_k$ of conditions beginning with $\mathcal{L}_0 = \mathcal{K} = \{G | \exists F_0 \forall x R_{m_0}(G \restriction x, F_0 \restriction x, x)\}$ as well as initial segments γ_k (of length at least k) of our intended G and $\delta_{i,k}$ of witnesses F_i (of length at least k) showing that $G \in \mathcal{L}_k$. More precisely, each $\mathcal{L}_k$ will be of the form $G \supset \gamma_k$ & $\forall i \leq k \exists F_i \supset \delta_{i,k} \forall x R_{m_i}(G \restriction x, F_i \restriction x, x)$ for some recursive R_{m_i} (independent of k). Thus, if we successfully continue our construction keeping $\mathcal{L}_k$ nonempty for each k then the $F_i = \lim_k \delta_{i,k}$ for $i \leq k$ will witness that $G = \lim_k \gamma_k$ is in every $\mathcal{L}_k$ as we guarantee that $R_{m_i}(\gamma_k \restriction x, \delta_{i,k} \restriction x, x)$ holds for every $i, x < k$ and every k.

We begin with $\gamma_0 = \emptyset = \delta_{0,0}$ and R_{m_0} as specified by $\mathcal{K}$. So our G will at least be in $\mathcal{K}$ as desired. Suppose we have defined γ_j and $\delta_{i,j}$ for $j, i \leq k$ and wish to define $\mathcal{L}_{k+1}$, γ_{k+1} and $\delta_{i,k+1}$ for $i \leq k+1$ so as to prevent X from being Σ_1^1 in G via Θ_k. We ask if there is an $m \in \omega$ and a nonempty $\mathcal{L} \subseteq \mathcal{L}_k$ such that

1. $m \notin X$ and $\mathcal{L} \Vdash \Theta_k(G, m)$ or
2. $m \in X$ and $\mathcal{L} \Vdash \neg\Theta_k(G, m)$.

Suppose there is such an $\mathcal{L}$ of the form $\exists F_{k+1} \forall x R_{m_{k+1}}(G \restriction x, F_{k+1} \restriction x, x)$. As $\mathcal{L} \subseteq \mathcal{L}_k$ is nonempty we can choose $\gamma_{k+1} \supset \gamma_k$ and $\delta_{i,k+1} \supset \delta_{i,k}$ for $i \leq k$ and some $\delta_{k+1,k+1}$ all of length at least $k+1$ such that $\mathcal{L}_{k+1}$ as given by $G \supset \gamma_{k+1}$ & $(\forall i \leq k+1)(\exists F_i \supset \delta_{i,k+1})(\forall x R_{m_i})(G \restriction x, F_i \restriction x, x)$ is a nonempty subclass of $\mathcal{L}$ (and so, in particular, $R_{m_i}(\gamma_{k+1} \restriction x, \delta_{i,k+1} \restriction x, x)$ for every $i, x \leq k+1$). We can now continue our induction.

Note that if we can successfully define nonempty $\mathcal{L}_k$ in this way for every k then we build a $G = \lim_k \gamma_k$ and $F_i = \lim_k \delta_{i,k}$ for each i such that $\forall x R_{m_i}(G \upharpoonright x, F_i \upharpoonright x, x)$. In particular $\forall x R_{m_0}(G \upharpoonright x, F_0 \upharpoonright x, x)$ and so $G \in \mathcal{K}$. Similarly, $G \in \mathcal{L}_k$ for every $k > 0$. If X is $\Sigma_1^1(G)$ as assumed, then $X = \{n | \Theta_k(G, n)\}$ for some k. We consider the construction at stage $k + 1$ and the $\mathcal{L}$ chosen at that stage. If we were in case (1) then as $\mathcal{L} \Vdash \Theta_k(G, m)$ and $G \in \mathcal{L}_{k+1}$, $\Theta(G, m)$ is true but $m \notin X$ for a contradiction. Similarly, if we were in case (2), as $\mathcal{L} \Vdash \neg\Theta_k(G, m)$ and $G \in \mathcal{L}_{k+1}$, $\neg\Theta(G, m)$ is true but $m \in X$ again for a contraction.

Thus we can assume that there is some first stage $k + 1$ at which there are no m and $\mathcal{L} \subseteq \mathcal{L}_k$ as required in the construction. In this case we claim that X is Σ_1^1 as desired. Indeed, we claim that X is defined as a Σ_1^1 real by $m \in X \Leftrightarrow (\exists Z \in \mathcal{L}_k)\Theta_k(Z, m)$. To see this suppose first that $(\exists Z \in \mathcal{L}_k)\Theta_k(Z, m)$. Then $\mathcal{L}$ as defined by $\mathcal{L}_k$ & $\Theta_k(G, m)$ is a nonempty Σ_1^1 class such that $\mathcal{L} \Vdash \Theta_k(G, m)$ and so we would have $m \in X$ as desired by the assumed failure of (1) at stage $k + 1$ of the construction. On the other hand, if $(\forall Z \in \mathcal{L}_k)(\neg\Theta_k(Z, m)$ then $\mathcal{L}_k \Vdash \neg\Theta_k(G, m)$ and so by the failure of (2) at stage $k + 1$ of the construction, $m \notin X$ as desired. $\qquad\square$

As usual, we may relativize the Theorem to any real C.

Our theorem easily implies several basic results of hyperarithmetic theory without any appeal to the theory of hyperarithmetic sets as used, for example, in Sacks (1990). Many of them can also be seen as consequences of type omitting theorems for certain classes of generalized logics. These type omitting arguments are also immediate consequences of our Theorem. We presented two basis theorems of this sort in the introduction and note here that the proof of Kreisel's in Sacks (1990) uses several deep facts about hyperarithmetic reals and Σ_1^1 classes.

Following his proof of the Kreisel basis theorem Sacks (1990) gives as a corollary a result of Kreisel about the intersections of all ω-models of various theories of second order arithmetic from which follow some previous specific results. We state that result now along with some similar ones earlier in Sacks's presentation. These can all be seen as type omitting arguments. After stating them, we explain a general setting which includes them all and give the relevant type omitting theorems as Corollaries of Theorem 2.1.

Theorem 2.2 (Sacks (1990, III. 4.10)). *The intersection of all ω-models of Δ_1^1 comprehension is HYP, the class of all hyperarithmetic sets or equivalently the class of all Δ_1^1 sets.*

More generally, we have the following result of Kreisel.

Theorem 2.3 (Sacks (1990, III. 7.3)). *Let K be a Π_1^1 set of axioms in the language of analysis (i.e. second order arithmetic). If a real X belongs to every countable ω-model of K then X is Δ_1^1.*

A similar result is the following.

Theorem 2.4 (Sacks (1990, III. 4.13)). *The intersection of all ω-models of Σ_1^1 choice downward closed under many-one reducibility is also HYP.*

In all of these results it is easy to see that the class of models described is Σ_1^1 and, of course, every member X of such a model is recursive in it and so any real in every such model is Σ_1^1 but these models are all trivially closed under complementation. So they all follow from our Theorem.

Moving to the type omitting point of view we, somewhat more generally, consider two sorted logics $(\mathcal{N}, \mathcal{M}, \ldots)$ in the usual sense of having two types of variables one ranging over the elements of $\mathcal{N}$ and the other over those of $\mathcal{M}$ in addition to the usual apparatus of function, relation and constant symbols of ordinary first order logic. While formally merely a version of first order logic gotten by adding on predicates for N and M, this logic can be turned into a much stronger one ($\mathcal{N}$-logic) by requiring that all models have their first sort (with some functions and relations on it as given in the structure) isomorphic to some given countable first order structure. The most common example of these logics is ω-logic where we require that $\mathcal{N}$ be isomorphic to the ordinal ω or the standard model $\mathbb{N}$ of arithmetic (depending on the language intended). Again, the most common examples are given by classes of ω-models of fragments T of second order arithmetic as mentioned above. Here, in addition to requiring that $\mathcal{N}$ be the standard model of arithmetic we intend that the elements of $\mathcal{M}$ are subsets of $\mathcal{N}$ and the membership relation $\in$ between members of $\mathcal{N}$ and those of $\mathcal{M}$ is in the language (with the usual axiom of extensionality so that the elements of $\mathcal{M}$ may be identified with true subsets of $\mathcal{N} = \mathbb{N}$). As being an $\mathcal{N}$ model, or even also satisfying some Π_1^1 theory T, is clearly Σ_1^1 in $\mathcal{N}$, we immediately get all the results from Sacks (1990) mentioned above as a corollaries of our theorem. Indeed, we have the following generalization of Kreisel's result in Sacks (1990, III. 7.3):

Theorem 2.5. *If T is a Π_1^1 set of sentences in the two sorted language of $(\mathcal{N}, \mathcal{M}, \ldots)$ and $\mathcal{N}$ is a countable structure for the appropriate sublanguage (restricted to the first sort), T has an $\mathcal{N}$-model and p is a n-type (i.e. a complete consistent set of formulas $\varphi(x)$ with n free variables in the language of $(\mathcal{N}, \mathcal{M}, \ldots)$) which is not Σ_1^1 in $\mathcal{N}$, then there is an $\mathcal{N}$-model of T not realizing p. (Note that, as types are complete sets of formulas, p being Σ_1^1 (in $\mathcal{N}$) is equivalent to its being Δ_1^1 (in $\mathcal{N}$).*

Proof. Being an $\mathcal{N}$ model of T is a Σ_1^1 in $\mathcal{N}$ property and so by our Theorem (relativized to $\mathcal{N}$) there is even an $\mathcal{N}$-model $(\mathcal{N}, \mathcal{M}, \ldots)$ of T in which p is not even Σ_1^1. (Of course, any type realized in $(\mathcal{N}, \mathcal{M}, \ldots)$ is recursive in the complete diagram of $(\mathcal{N}, \mathcal{M}, \ldots)$ and so hyperarithmetic in $(\mathcal{N}, \mathcal{M}, \ldots)$.)   □

Viewing our theorem as a type omitting argument suggests that we should be able to omit any countable sequence of types (reals) of the appropriate sort rather than just one. A simple modification of our proof gives the expected result.

Theorem 2.6. *If $\mathcal{K}$ is a nonempty Σ_1^1 class reals and X_n a countable sequence of reals none of which is Σ_1^1, then there is a $G \in \mathcal{K}$ such that no X_n is Σ_1^1 in G. Similarly if no X_n is Δ_1^1, then there is a $G \in \mathcal{K}$ such that no X_n is Δ_1^1 in G.*

Proof. Repeat the proof of the Theorem but at step $k + 1 = \langle n, j \rangle$ of the construction replace X by X_n and Θ_k by Θ_j. If we successfully pass through all steps k then the previous argument shows that no X_n is Σ_1^1 in $G \in \mathcal{K}$. On the other hand, if the construction terminates at step $k+1 = \langle n, j \rangle$ then the previous argument shows that X_n is defined as a Σ_1^1 real by $m \in X_n \Leftrightarrow (\exists Z \in \mathcal{L}_k)\Theta_j(Z, m)$ for a contradiction. For the Δ_1^1 version, simply consider the sequence Y_n where $Y_n = X_n$ if X_n is not Σ_1^1 and Y_n is the complement of X_n otherwise (i.e. X_n is not Π_1^1). As now no Y_n is $\Sigma_1^1(G)$, no X_n is $\Delta_1^1(G)$. □

This version of our Theorem also extends the analog of the result actually given by Andrews and Miller (2015, Proposition 3.6).

Of course, we can relativize this theorem as well to any real C. To give a somewhat different example of a such type omitting argument application of this last theorem we provide one for nonstandard models of ZFC for which we have uses elsewhere.

Corollary 2.7. *For every real C and reals X_n not Δ_1^1 in C, there is a countable ω-model of ZFC containing C but not containing any X_n whose well founded part consists of the ordinals less than ω_1^C, the first ordinal not recursive in C.*

Proof. Being a countable ω-model of ZFC containing (a set isomorphic to) C (under the isomorphism taking the ω of the model to true ω) is clearly a Σ_1^1 in C property. Now apply Theorem 2.6 adding on a new real $X_0 = O^C$ (i.e. Kleene's O relativized to C) to the list. It supplies a countable ω-model of ZFC containing C but not containing any of the X_n. As it contains C it contains every ordering recursive in C and so order types for every ordinal less than ω_1^C. On the other hand, if there were an ordinal in the model isomorphic to ω_1^C then, by standard results of hyperarithmetic theory, O^C would be in the model as well.

Finally, we point out that the complexity of the G of Theorem 2.6 (and hence of Corollary 2.7 as well) can be as low as possible. □

Theorem 2.8. *If $\mathcal{K}$ is a nonempty Σ_1^1 class reals and X_n a countable sequence of reals uniformly Δ_1^1 (recursive) in O none of which is Σ_1^1, then there is a $G \in \mathcal{K}$ with G Δ_1^1 (recursive) in O such that no X_n is Σ_1^1 in G. Indeed, G can be chosen to be of strictly smaller hyperdegree than O, i.e. O is not Δ_1^1 in G. As in Theorem 2.6, if we assume only that the X_n are not Δ_1^1 then we may conclude that none are Δ_1^1 in G.*

Proof. Suppose we are at step $k = \langle n, j \rangle$ of the construction. We know that either there is an $m \in X_n$ such that $(\forall Z \in \mathcal{L}_k)(\neg\Theta_k(Z, m))$ or an $m \notin X_n$ such that $(\exists Z \in \mathcal{L}_k)(\Theta_k(Z, m))$. As the X_n are uniformly Δ_1^1 (recursive) in O, and the rest of the conditions considered in the construction are either Σ_1^1 or Π_1^1, O can hyperarithmetically (recursively) decide which case to apply. As choosing the $\gamma_{k+1} \supset \gamma_k$ and $\delta_{i.k+1} \supset \delta_{i,k}$ for $i \leq k$ and so $\mathcal{L}_{k+1}$ now only require finding ones for which the corresponding Σ_1^1 class $\mathcal{L}_{k+1}$ is nonempty, this step is also recursive in O. Of course, as we can add O onto the list of X_n, we then guarantee that O is not Σ_1^1 in G and so, of course, not Δ_1^1 in G as required. □

Note that by a result of Spector's (see Sacks (1990, Theorem II.7.6(ii))) $\omega_1^{CK} < \omega_1^A$ implies that O is Δ_1^1 in A (indeed there is a pair of Σ_1^1 formulas $\varphi(X, n)$ and $\theta(X, n)$ which define O and its complement for any X with $\omega_1^X > \omega_1^{CK}$), we have the Kleene and Gandy basis theorem for Σ_1^1 classes as well.

Theorem 2.9 (Kleene and Gandy Basis Theorems). *Every nonempty Σ_1^1 class of reals $\mathcal{K}$ contains an element A recursive in and of strictly smaller hyperdegree than O. In particular, one with $\omega_1^A = \omega_1^{CK}$.*

3   The Proof for Classes of Reals

In this section we prove our result for classes of reals.

Theorem 3.1. *If a class $\mathcal{A}$ of reals is Σ_1^1 in every member of a nonempty Σ_1^1 class $\mathcal{B}$ of reals then it is Σ_1^1.*

The proof relies on several basic and important results of effective descriptive set theory. To ease reading the proof, we state the most important ones now. We state lightface versions without parameters. Relativizations to individual real parameters are routine. (Note that, when ordinals or lengths of well-ordered relations are involved, relativization to Z includes replacing ω_1^{CK} by ω_1^Z.) We don't need the full boldface versions. These facts can be found in basic books on effective descriptive set theory such as Moschovakis (1980), higher recursion theory such as Sacks (1990) or Hinman (1978) or even reverse mathematics such as Simpson (2009).

Proposition 3.2 (Codes). *We can code Δ_1^1 classes of reals $\mathcal{V}$ as either Δ_1^1 reals C (Δ_1^1 codes) or as numbers e by coding the Δ_1^1 code C as a number e (hyperarithmetic codes for Δ_1^1 reals). In either case, the property of being a code is Π_1^1 and membership of a real Z in the set coded by C or e is a Δ_1^1 relation given that C and e are codes. Similarly, membership of a number n in a Δ_1^1 real with hyperarithmetic code e is a Δ_1^1 relation. We can pass in a Δ_1^1 way between these types of codes and the syntactic ones given by the formulas required in our definition of Δ_1^1 reals and classes given that all the objects are, in fact, codes. In this situation we often abuse notation by writing $Z \in C$ to denote the assertion that Z is in the class coded by C. When C and D are both codes, we use $D \subseteq C$ to denote the assertion that $\forall Z (Z \in D \rightarrow Z \in C)$ and similarly for $D \supseteq C$. These relations are then all Π_1^1. These facts also imply that the predicate Z is $\Delta_1^1(X)$ is Π_1^1. (We also use $C \supseteq \mathcal{A}$ for an arbitrary class $\mathcal{A}$ of reals to mean that every real in the set coded by C is in $\mathcal{A}$.)*

Proposition 3.3 (Representation Theorem). *If $\mathcal{V}$ is a Π_1^1 class then there is a Δ_1^1 function $\mathcal{F}$ such that $Z \in \mathcal{V} \Leftrightarrow \mathcal{F}(Z) \in WO$ where WO is the class of reals Z which, viewed as a set of pairs of numbers, represents a well ordering. If $Z \in WO$, we write $|Z|$ for the ordinal represented by Z.*

Proposition 3.4 (Bounding). *If $\mathcal{V}$ is a Π_1^1 class of reals, $\mathcal{F}$ is as in Proposition 3.3 then $\mathcal{V}$ is Δ_1^1 if and only if there is a bound $< \omega_1^{CK}$ on the order types of $\mathcal{F}(Z)$ for $Z \in \mathcal{V}$. Moreover, if $\mathcal{V}$ contains only Δ_1^1 reals and $\mathcal{G}$ is a Δ_1^1 subset of $\mathcal{V}$ then such a bound (expressed as either a real or a number coding a recursive well-ordering) for $\{\mathcal{F}(Z)|Z \in \mathcal{G}\}$ (or even any Σ_1^1 subset of the Δ_1^1 well orderings) can be found in a Δ_1^1 way from the codes (or indices) for $\mathcal{F}, \mathcal{G}$ and $\mathcal{V}$. As a consequence we may, in this case, divide $\mathcal{V}$ into an increasing, continuous sequence $\langle \mathcal{V}_i | i < \omega_1^{CK} \rangle$ of uniformly Δ_1^1 sets given by $\mathcal{V}_i = \{Z \in \mathcal{V}| \, |\mathcal{F}(Z)| < i\}$.*

Remark 3.5. While we have not found an explicit statement in our references of the uniformity described in this bounding theorem, it can easily be deduced from the uniform version of the analogous theorem for sets of numbers (as in e.g. Sacks (1990, II. 3.4)) by translating the real codes for ordinals $< \omega_1^{CK}$ to numbers in O of at least as large a rank given by Sacks (1990, I. 4.3).

Proposition 3.6 (Gandy-Harrington Forcing). *We can define a general notion of forcing whose conditions are Σ_1^1 classes ordered by inclusion as extension. A simplified version of the proof of Theorem 2.1 that leaves out the diagonalization requirements shows that we may construct a generic G in any given Σ_1^1 class meeting any countable collection of dense sets. Thus we may use this forcing notion in any of the common ways. As usual, we will be interested in forcing over countable standard models of fragments of ZFC containing various specified reals. In addition to the typical results about forcing such as forcing equals truth, we note that, by the arguments in the proof of Theorem 2.1, a Π_1^1 sentence $\varphi(G)$ about the generic G is forced by a condition (Σ_1^1 set) $\mathcal{P}$ if and only if $\forall Z \in \mathcal{P}(\varphi(Z))$. We also note that if $\langle G_0, G_1 \rangle$ is generic then both G_0 and G_1 are generic. (See Miller (1995, Sect. 30) for more about this forcing notion and Lemma 30.3 there for this last particular fact.) Absoluteness considerations will also play a role in our applications of this forcing.*

As a notational convenience in proving our theorem, we can, by the Gandy basis theorem (Theorem 2.9) and the fact that $\omega_1^B = \omega_1^{CK}$ is a Σ_1^1 predicate (of B), assume without loss of generality that $\omega_1^B = \omega_1^{CK}$ for every $B \in \mathcal{B}$. (Note $\omega_1^B = \omega_1^{CK} \Leftrightarrow \forall e(\{e\}^B$ is a well-ordering $\rightarrow \exists i \exists f(f$ is an isomorphism of $\{i\}$ and $\{e\}^B)$.)

We begin with some crucial approximations to our class $\mathcal{A}$ and an analysis of their properties.

Notation 3.7. *We let $\mathcal{D}_B = \{C|C$ is a $\Delta_1^1(B)$ code & $C \supseteq \mathcal{A}\}$ and $\mathcal{A}_B = \{A|(\forall C \in \mathcal{D}_B)(A \in C)\}$. Similarly, we let $\mathcal{A}_0 = \{A|A$ is a member of every Δ_1^1 class containing $\mathcal{A}\}$. For $B \in \mathcal{B}$, we let $\psi_B(Z)$ be a $\Sigma_1^1(B)$ formula defining $\mathcal{A}$.*

Lemma 3.8. *For $B \in \mathcal{B}$, $\mathcal{D}_B$ is $\Pi_1^1(B)$ and $\mathcal{A}_B$ is $\Sigma_1^1(B)$.*

Proof. Fix $B \in \mathcal{B}$. For any real C, $C \in \mathcal{D}_B$ if and only if C is a $\Delta_1^1(B)$ code and $\forall Z(\psi_B(Z) \rightarrow Z \in C)$. As ψ_B is $\Sigma_1^1(B)$, both conjuncts here are $\Pi_1^1(B)$ by Proposition 3.2 and so $\mathcal{D}_B$ is $\Pi_1^1(B)$ as required. The second claim now follows directly from the definition of $\mathcal{A}_B$ and Proposition 3.2. (Rephrase the definition of $\mathcal{D}_B$ in terms of number codes to make the quantifier count work.)   □

Lemma 3.9. *For any reals A and B in $\mathcal{B}$ with $\omega_1^{A,B} = \omega_1^{CK}$, $A \in \mathcal{A} \Leftrightarrow A \in \mathcal{A}_B$.*

Proof. Clearly $A \in \mathcal{A}$ implies that $A \in \mathcal{A}_B$ for every $B \in \mathcal{B}$ by the definition of $\mathcal{A}_B$. For the other direction suppose $A \notin \mathcal{A}$. Let $\mathcal{F} \in \Delta_1^1(B)$ be as in Proposition 3.3 for the $\Pi_1^1(B)$ class $\mathcal{V} = \{Z | \neg\psi_B(Z)\}$ which is the complement of $\mathcal{A}$. Thus $\mathcal{F}(A)$ is a well ordering which is $\Delta_1^1(A, B)$ and so, by hypothesis, less than some $i < \omega_1^{CK}$. Thus the (obviously Σ_1^1) set $\{Z | \mathcal{F}(Z) < i\}$ is equal to the Π_1^1 set $\{Z | \neg\psi_B(Z)$ & $(i+1) \nleq \mathcal{F}(Z)\}$ and so is $\Delta_1^1(B)$. It is disjoint from $\mathcal{A}$ and has A as a member by our choice of i. The $\Delta_1^1(B)$ code C for its complement is then a member of $\mathcal{D}_B$ not containing A as a member. This C is then a witness that $A \notin \mathcal{A}_B$ as required. $\square$

Lemma 3.10. *$\mathcal{A}_0$ is Σ_1^1.*

Proof. Consider the real $J = \{e | e$ is a hyperarithmetic index for a Δ_1^1 code of a superset of $\mathcal{A}\} = \{e | e$ is a hyperarithmetic index for a real in $\mathcal{D}_B\}$. This real J is $\Pi_1^1(B)$ in every $B \in \mathcal{B}$ by Lemma 3.8 and Proposition 3.2. So by Theorem 2.1 (formally applied to the complements) is Π_1^1. Thus $\mathcal{A}_0$ which is the intersection of the sets coded by indices in J is Σ_1^1: $Z \in \mathcal{A}_0 \Leftrightarrow \forall e(e \in J \to Z$ is in the set coded by e). $\square$

Lemma 3.11. *If $B, C \in \mathcal{B}$ and $\omega_1^{B,C} = \omega_1^{CK}$, then $\mathcal{A}_B = \mathcal{A}_C$.*

Proof. If not, then we have, without loss of generality, an $A \in \mathcal{A}_C - \mathcal{A}_B$. So there is a code $D \in \mathcal{D}_B$ with $A \notin D$ and $A \in \mathcal{A}_C$. Now the nonempty class $\mathcal{W} = \{Z | Z \in \mathcal{A}_C$ & $Z \notin D\}$ is $\Sigma_1^1(B, C)$ by Lemma 3.8 and Proposition 3.2. Thus by the Gandy basis theorem (relative to B, C) (Theorem 2.9) there is a W in $\mathcal{W}$ and so in $\mathcal{A}_C - \mathcal{A}_B$ with $\omega_1^{W,B,C} = \omega_1^{CK}$. Lemma 3.9, however, tells us that $W \in \mathcal{A}_B \Leftrightarrow W \in \mathcal{A} \Leftrightarrow W \in \mathcal{A}_C$ for a contradiction. $\square$

Lemma 3.12. *If $B, C \in \mathcal{B}$ and $\omega_1^{B,C} = \omega_1^{CK}$, then for every $X \in \mathcal{D}_B$ there is a $Y \in \mathcal{D}_B$ and a $Z \in \mathcal{D}_C$ such that $Y \subseteq X$ and $Z \subseteq Y \subseteq Z$, i.e. Y and Z are codes for the same set.*

Proof. By Proposition 3.4, there is a uniformly Δ_1^1 continuous increasing sequences $\mathcal{D}_{B,i}$ ($i < \omega_1^{CK} = \omega_1^B$) with union $\mathcal{D}_B$. We can then set $\mathcal{A}_{B,i} = \{A | \forall C \in \mathcal{D}_{B,i}(A \in C)\}$. This sequence is clearly nested and continuous with intersection $\mathcal{A}_B$. As $\mathcal{D}_{B,i}$ and all its members are $\Delta_1^1(B)$, the $\mathcal{A}_{B,i}$ are also uniformly $\Delta_1^1(B)$ by Proposition 3.2 as we can convert to number codes. Similarly, we have $\mathcal{D}_{C,i}$ and $\mathcal{A}_{C,i}$ ($i < \omega_1^{CK} = \omega_1^C$). By Lemma 3.11 we know that for each $i < \omega_1^{CK}$ and $Z \in \overline{\mathcal{A}_{B,i}}$ there is a $j < \omega_1^{CK}$ such that $Z \in \overline{\mathcal{A}_{C,j}}$. By Proposition 3.4, there is a $k < \omega_1^{CK}$ such that for every $Z \in \overline{\mathcal{A}_{B,i}}$, $Z \in \overline{\mathcal{A}_{C,k}}$ and we can get k uniformly $\Delta_1^1(B, C)$. Of course, the analogous fact switching B and C is also true. Iterating and interleaving these $\Delta_1^1(B, C)$ functions starting with any $i < \omega_1^{CK}$ produces a $\Delta_1^1(B, C)$ increasing sequence of $k < \omega_1^{CK}$.

By Proposition 3.4, this sequence has a bound and hence a supremum $l < \omega_1^{CK}$ and $\mathcal{A}_{B,l} = \mathcal{A}_{C,l}$.

Now consider any $X \in \mathcal{D}_{B,i}$ so $X \supseteq \mathcal{A}_{B,l}$ for any $l > i$ in ω_1^{CK}. We may now choose one such that $\mathcal{A}_{B,l} = \mathcal{A}_{C,l}$. As $\mathcal{A}_{B,l} \in \Delta_1^1(B)$ and contains $\mathcal{A}$, there is a code $Y \in \mathcal{D}_B$ for it. Similarly there is a code $Z \in \mathcal{D}_C$ for $\mathcal{A}_{C,l}$. As $\mathcal{A}_{B,l} = \mathcal{A}_{C,l}$, these are then the desired Y and Z. $\square$

Lemma 3.13. *For every $B \in \mathcal{B}$, $\mathcal{A}_B = \mathcal{A}_0$.*

Proof. Fix $B \in \mathcal{B}$. Clearly, it suffices to prove that $\forall X \in \mathcal{D}_B \exists Y \in \mathcal{D}_B(X \supseteq Y$ & $\mathcal{V} = \{Z | Z \in Y\} \in \Delta_1^1)$. (This says there is, for each $X \in \mathcal{D}_B$, a Δ_1^1 code V for a Δ_1^1 class $\mathcal{V}$ contained in the class coded by X and containing $\mathcal{A}$ (as $Y \in \mathcal{D}_B$). This code shows that $\mathcal{A}_0 \subseteq \mathcal{A}_B$ by definition. On the other hand, $\mathcal{A}_B \subseteq \mathcal{A}_0$ for every B.)

Fix an $X \in \mathcal{D}_B$. Consider now the class $\mathcal{W} = \{C \in \mathcal{B} | (\forall Y \in \mathcal{D}_B)(X \supseteq Y \to \{Z | Z \in Y\} \notin \Delta_1^1(C)\}$. By Proposition 3.2, this class is $\Sigma_1^1(B)$. If it were nonempty then, by the Gandy basis theorem (relative to B) (Theorem 2.9), it would have a member C with $\omega_1^{B,C} = \omega_1^{CK}$. This would provide a counterexample to Lemma 3.12 and so $\mathcal{W}$ is empty.

We now work with a countable standard model which contains B and satisfies a fragment of ZFC sufficient to guarantee the absoluteness of Σ_1^1 formulas. Note, for example, that all reals Δ_1^1 in B (and so all in $\mathcal{D}_B$) are in this model.

Let $G \in \mathcal{B}$ be a Gandy-Harrington generic over this model as in Proposition 3.6. As $G \notin \mathcal{W}$, there is a $Y \in \mathcal{D}_B$ such that $X \supseteq Y$ and $\{Z | Z \in Y\} \in \Delta_1^1(G)$. Fix a specific Δ_1^1 definition of this class from G, i.e. $\Sigma_1^1(G)$ formulas φ and θ such that $\forall Z(\varphi(G, Z) \leftrightarrow \neg\theta(G, Z))$, $\forall Z(Z \in Y \to \varphi(G, Z))$ and $\forall Z(Z \notin Y \to \theta(G, Z))$. As G is generic we have a Σ_1^1 $\mathcal{P}$ forcing these sentences. Now consider the Σ_1^1 class $\mathcal{Q} = \{\langle C, C' \rangle | C, C' \in \mathcal{P}$ & $\exists Z(\varphi(C, Z)$ & $\theta(C', Z) \vee \varphi(C', Z)$ & $\theta(C, Z))\}$. If $\mathcal{Q}$ is nonempty then there is a Gandy-Harrington generic $\langle C, C' \rangle \in \mathcal{Q}$. Each of C and C' is in $\mathcal{P}$ and Gandy-Harrington generic by Proposition 3.6. Thus any Z witnessing that $\langle C, C' \rangle \in \mathcal{Q}$ would be a counterexample to one of the sentences above forced by $\mathcal{P}$ and hence true of C and C'. Thus $\mathcal{Q}$ is empty and so $Z \in Y \leftrightarrow (\forall C \in \mathcal{P})(\neg\theta(C, Z))$ and $Z \notin Y \leftrightarrow (\forall C \in \mathcal{P})(\neg\varphi(C, Z))$ and $\{Z | Z \in Y\}$ is Δ_1^1 as required. $\square$

We now prove our theorem on Σ_1^1 classes.

Proof of Theorem 3.1: We claim that $A \in \mathcal{A}$ if and only if $A \in \mathcal{A}_0$ (which is Σ_1^1 by Lemma 3.10) and one of the following two Σ_1^1 statements hold for a Σ_1^1 formula ψ that we will define below:

(1) $\omega_1^A = \omega_1^{CK}$ or
(2) $\omega_1^A > \omega_1^{CK} \to \psi(A)$.

Now $A \in \mathcal{A} \to A \in \mathcal{A}_0$ by the definition of $\mathcal{A}_0$. So we may assume that $A \in \mathcal{A}_0$ and show that $A \in \mathcal{A} \leftrightarrow$ (1) or (2) holds. If (1) holds then by the Gandy basis theorem (Theorem 2.9) (relative to A) we may choose a $B \in \mathcal{B}$ with

$\omega_1^{A,B} = \omega_1^{CK}$. Now by Lemma 3.9, $A \in \mathcal{A} \Leftrightarrow A \in \mathcal{A}_B$ while $A \in \mathcal{A}_B \Leftrightarrow A \in \mathcal{A}_0$ by Lemma 3.13. Thus, in this case, $A \in \mathcal{A} \Leftrightarrow A \in \mathcal{A}_0$ as required.

Assume then that (1) fails and so the hypothesis of (2) holds. We now must argue that we have a Σ_1^1 formula $\psi(A)$ that, under these assumptions, is equivalent to $A \in \mathcal{A}$. As mentioned just before Theorem 2.9, there is a pair of Σ_1^1 formulas $\varphi(X, n)$ and $\theta(X, n)$ which define O and its complement for any X with $\omega_1^X > \omega_1^{CK}$. By the Kleene basis theorem (Theorem 2.9) there is a recursive index computing a $B \in \mathcal{B}$ from O. By the hypothesis of our theorem there is a $\Sigma_1^1(B)$ formula $\psi_B(Z)$ defining $\mathcal{A}$. Thus there is a Σ_1^1 formula $\hat{\psi}(X, Z)$ which defines $\mathcal{A}$ from any X with $\omega_1^X > \omega_1^{CK}$. We now take our desired ψ to be $\hat{\psi}(A, A)$. $\square$

As a final comment, we point out that if we had only wanted to prove Theorem 3.1 in the Δ_1^1 case we would have a simple proof along the lines of the last paragraph of the proof of Lemma 3.13. This argument also gives a proof of the analog for classes of reals of the Δ_1^1 case of Theorem 2.6.

Theorem 3.14. *If $\mathcal{B}$ is a nonempty Σ_1^1 class of reals and $\mathcal{X}_n$ a countable sequence of classes of reals none of which is Δ_1^1, then there is a $G \in \mathcal{B}$ such that no $\mathcal{X}_n$ is Δ_1^1 in G.*

Proof. Note that if $\mathcal{X}_m \notin \Delta_1^1(B)$ for every $B \in \mathcal{B}$ then any $G \in \mathcal{B}$ works for $\mathcal{X}_m$. Thus we may assume that for every n there is a $B_n \in \mathcal{B}$ and φ_n and θ_n Σ_1^1 formulas with two free real variables which, with B_n for the first variable, define $\mathcal{X}_n$ and its complement. Let $G \in \mathcal{B}$ be a Gandy-Harrington generic over a countable standard model of a sufficient fragment of ZFC containing the B_n. We claim no $\mathcal{X}_n$ is $\Delta_1^1(G)$. If not, let φ and θ be $\Sigma_1^1(G)$ formulas defining some $\mathcal{X}_n$ and its complement. Let $\mathcal{P}$ be a condition which forces that $(\forall Z)(\varphi_n(B_n, Z) \rightarrow \varphi(G, Z))$, $(\forall Z)(\theta_n(B_n, Z) \rightarrow \theta(G, Z))$ and $(\forall Z(\varphi(G, Z) \leftrightarrow \neg\theta(G, Z))$. Now consider the Σ_1^1 class $\mathcal{Q} = \{\langle C, C'\rangle \,|\, C, C' \in \mathcal{P} \ \& \ \exists Z(\varphi(C, Z) \ \& \ \theta(C', Z) \lor \varphi(C', Z) \ \& \ \theta(C, Z))\}$. If $\mathcal{Q}$ is nonempty then there is a Gandy-Harrington generic $\langle C, C'\rangle \in \mathcal{Q}$. Each of C and C' is in $\mathcal{P}$ and Gandy-Harrington generic by Proposition 3.6. Thus any Z witnessing that $\langle C, C'\rangle \in \mathcal{Q}$ would be a counterexample to one of the sentences above forced by $\mathcal{P}$ and hence true of C and C'. Thus $\mathcal{Q}$ is empty and so $Z \in \mathcal{X}_n \Leftrightarrow (\forall C \in \mathcal{P})(\neg\theta(C, Z))$ and $Z \notin \mathcal{X}_n \Leftrightarrow (\forall C \in \mathcal{P})(\neg\varphi(C, Z))$ and so $\mathcal{X}_n$ is Δ_1^1 for the desired contradiction. $\square$

Corollary 3.15. *Any class $\mathcal{A}$ of reals which is Δ_1^1 in every member of a Σ_1^1 class $\mathcal{B}$ of reals is Δ_1^1.*

We do not know if the full analog of Theorem 2.6 for classes of reals, i.e. Theorem 3.14 with Δ_1^1 replaced by Σ_1^1, is also true.

References

Andrews, U., Miller, J.S.: Spectra of theories and structures. Proc. Am. Math. Soc. **143**, 1283–1298 (2015)

Hinman, P.G.: Recursion Theoretic Hierarchies. Perspectives in Mathematical Logic. Springer, Berlin (1978)

Miller, A.W.: Descriptive Set Theory and Forcing. Lecture Notes in Logic. Springer, Berlin (1995)

Moschovakis, Y.N.: Descriptive Set Theory. Studies in Logic and the Foundations of Mathematics, vol. 100. North-Holland, Amsterdam (1980)

Sacks, G.E.: Higher Recursion Theory. Perspectives in Mathematical Logic. Springer, Berlin (1990)

Simpson, S.G.: Subsystems of Second Order Arithmetic. Perspectives in Logic, 2nd edn. ASL and Cambridge University Press, New York (2009)

Turing Degree Theory, c.e. Sets

A Survey of Results on the d-c.e. and n-c.e. Degrees

Marat M. Arslanov and Iskander Sh. Kalimullin[✉]

Kazan Federal University, Kazan, Russia
{Marat.Arslanov,Iskander.Kalimullin}@kpfu.ru

Abstract. This paper is a survey on the upper semilattices of Turing and enumeration degrees of n-c.e. sets. Questions on the structural properties of these semilattices, and some model-theoretic properties are considered.

1 n-c.e. Turing Degrees

The notion of a *computably enumerable (c.e.) set*, i.e. a set of integers whose members can be effectively listed, is a fundamental one. Another way of approaching this definition is via an approximating function $\{A_s\}_{s\in\omega}$ to the set A in the following sense: we begin by guessing $x \notin A$ at stage 0 (i.e. $A_0(x) = 0$); when later x enters A at a stage $s + 1$, we change our approximation from $A_s(x) = 0$ to $A_{s+1}(x) = 1$. Note that this approximation (for fixed) x may change at most once as s increases, namely when x enters A. An obvious variation of this definition is to allow more than one change: a set A is *2-c.e.* (or *d-c.e.*) if for each x, $A_s(x)$ change at most twice as s increases. This is equivalent to requiring the set A to be the difference of two c.e. sets $B_1 - B_2$. Similarly, one can define *n-c.e.* sets by allowing n changes for each x. The last is equivalent to an existence of c.e. sets $B_1 \supseteq B_2 \supseteq \cdots \supseteq B_n$ such that

$$A = (B_1 - B_2) \cup \cdots \cup (B_{n-1} - B_n),$$

if n is even, and

$$A = (B_1 - B_2) \cup \cdots \cup (B_{n-2} - B_{n-1}) \cup B_n,$$

if n is odd.

A direct generalization of this reasoning leads to sets which are computably approximable in the following sense: for a set A there is a set of uniformly computable sequences $\{f(0, x), f(1, x), ...f(s, x), ... | x \in \omega\}$ consisting of 0 and 1 such that for any x the limit of the sequence $f(0, x), f(1, x), ...$ exists and is equal to the value of the characteristic function $A(x)$ of A. The well-known Shoenfield Limit Lemma states that the class of such sets coincides with the class of all Δ_2^0-sets. Thus, for a set A, $A \leqslant_T \emptyset'$ if and only if there is a computable function $f(s, x)$ such that $A(x) = \lim_s f(s, x)$.

A. Day et al. (Eds.): Downey Festschrift, LNCS 10010, pp. 469–478, 2017.
DOI: 10.1007/978-3-319-50062-1_27

The notion of d-c.e. and n-c.e. sets goes back to Putnam (1965) and Gold (1965) and was first investigated and generalized by Ershov (1968a, b, 1970). The arising hierarchy of sets is now known as *the Ershov difference hierarchy*. The position of a set A in this hierarchy is determined by the number of changes in the approximation of A described above, i.e. by the number of different pairs of neighboring elements of the sequence. The corresponding degree structures are denoted $\mathcal{D}_n$, the degrees of the n-c.e. sets. (So $\mathcal{D}_1 = \mathcal{R}$ the c.e. degrees.)

The degree structures $\mathcal{D}_n, n > 1$, have been intensively studied since the 1970's. It turned out that they (partially ordered by Turing reducibility) have a sufficiently rich inner structure, in many respects repeating its paramount representative, the class of c.e. degrees.

The first step toward this analysis took Barry Cooper in his Ph.D. dissertation (1971) where a 2-c.e. (d-c.e.) set whose Turing degree does not contain c.e. sets was constructed. His construction can be easily generalized to all finite levels of the difference hierarchy: for any $n < \omega$ there is a Turing degree which contains an $(n+1)-$ but not n-c.e. sets. For many years, this result remained as the only one result on Turing degrees of the n-c.e. sets until Arslanov (1985, 1988) and Downey (1989) showed that some pathological properties of c.e. degrees disappear in the difference hierarchy: Arslanov proved that for any $n > 1$ any n-c.e. degree $> \mathbf{0}$ can be cupped to $\mathbf{0}'$ by a d-c.e. degree $< \mathbf{0}'$ (in c.e. degrees this property fails by Yates (1973, unpublished), whence Downey proved that the diamond lattice is embeddable into d-c.e. degrees preserving least and greatest elements (in c.e. degrees this property also fails by Lachlan (1966). Later Cooper et al. (1991) established a nondensity result for n-c.e. degrees, $n > 1$, thus giving another difference between these two structures: for every $n > 1$, there exists a maximal n-c.e. (in fact a d-c.e.) degree in the n-c.e. degrees.

Differences between any of the other $\mathcal{D}_n$, however, seemed hard to find. The presence in the oracle of a d-c.e. set usually creates similar difficulties in the construction as in the case with n-c.e. oracles for $n > 2$, and methods which allow to cope with them for the case $n = 2$ usually allow to do so and for the case when $n > 2$ (see, for example, Cooper's (1992) proof of the splitting theorem for n-c.e. degrees). Downey (1989) even conjectured that the structures $\mathcal{D}_n$ for $n \geq 2$ might all be elementarily equivalent, i.e. all sentences (in the first order language with $\leq$) true in any $\mathcal{D}_n$ for $n \geq 2$ is true in all of them. This conjecture was refuted in Arslanov et al. (2010) where it is proved that the structures $\mathcal{D}_2$ and $\mathcal{D}_3$ differ at the $\forall\exists$-level.

Theorem 1. *Let φ be the following sentence:*

$$(\exists f, e, d)\{(f > e > d > 0) \ \&$$
$$\forall u[(u \leq f \to e \leq u \lor u \leq e) \& (u \leq e \to d \leq u \lor u \leq d)]\}.$$

Then $\mathcal{D}_3 \models \varphi$ and $\mathcal{D}_2 \models \neg\varphi$.

Remark. In Arslanov et al. (2010) we proved this statement for a slightly different φ, but later Wu and Yamaleev (2012) showed that this is the same.

In this paper we also conjectured that all the $\mathcal{D}_n$ are pairwise not elementarily equivalent, and that this level of difference ($\forall\exists$) is as small as possible in the following strongest sense: every $\exists\forall$-sentence true in any $\mathcal{D}_n$ is true in every $\mathcal{D}_m$ for $m > n$. The one quantifier theory of all the degree structures $\mathcal{D}_n, n \geq 1$, are the same since one can embed all finite (even countable) partial orderings into $\mathcal{R}$ (and so all the rest as well).

It was already mentioned that Cooper (1991) established the existence of a a d-c.e. degree $\mathbf{d}$ which is maximal in $\mathcal{D}_n, n \geq 2$. It is natural to ask, how "far" can this degree $\mathbf{d}$ be from $\mathbf{0}'$? Does it have to be a high degree? Is it possible to choose it low? Considering these questions Arslanov et al. (2000, 2004) and, independently, Downey and Yu (2004), proved that at least $\mathbf{d}$ cannot be low. Moreover, for any low d-c.e. degree $\mathbf{d}$ any c.e. degree $\mathbf{a}$ above $\mathbf{d}$ is splittable into two incomparable low d-c.e. degrees above $\mathbf{d}$. This raises the natural question: for which $n > 1$

- $\mathbf{0}'$ is splittable over any low_n d-c.e. degree?
- any c.e. degree is splittable over any low_n d-c.e. degree below?

We think these are open questions.

In general, the investigation of splitting properties of n-c.e. degrees is an important direction in the investigation of these degree structures (see, for instance, Downey and Stob (1993)). The reason is that splitting and non-splitting techniques have had a number of consequences for definability and elementary equivalence in the degree structures below $\mathbf{0}'$, major research areas in the local degree theory.

First of all, Cooper (1992) proved that any n-c.e. degree $> \mathbf{0}$ is splittable in n-c.e. degrees for any $n > 0$. The proof of this theorem is non-uniform, whereby the methods for dealing with the c.e. case (the Sacks [1963] splitting theorem) and those for the properly d-c.e. case are different. Later in Cooper and Li (2002a) they showed that there is no uniform proof of this result, and that (Cooper and Li 2002b) this non-uniformity leads to the following non-splitting theorem:

Theorem 2. *Let* $n > 1$. *There exist an n-c.e. degree* $\mathbf{a}$, *and a c.e. degree* $\mathbf{b}$, $\mathbf{0} < \mathbf{b} < \mathbf{a}$, *such that any splitting of* $\mathbf{a}$ *can not avoid the upper cone of* $\mathbf{b}$ *(as it can be done in the c.e. case by Sacks splitting theorem): for all n-c.e. degrees* $\mathbf{x}, \mathbf{y}$, *if* $\mathbf{x} \cup \mathbf{y} = \mathbf{a}$, *then either* $\mathbf{b} \leq \mathbf{x}$ *or* $\mathbf{b} \leq \mathbf{y}$.

In this work it has also been noticed that the degree $\mathbf{a}$ can be made low_3, thus not every low_3 n-c.e. degree is splittable in n-c.e. degrees avoiding upper cones of c.e. degrees below. Since in c.e. degrees such a splitting of low_3 c.e. degrees is possible by Sacks splitting theorem, it was found a nice elementary difference between the low_3 c.e. and low_3 d-c.e. degrees. Earlier Cooper (1991) have shown that in low_2 n-c.e. degrees density and splitting properties can be combined: for all low_2 n-c.e. degrees $\mathbf{a} < \mathbf{b}$, there exist n-c.e. degrees $\mathbf{x}_0, \mathbf{x}_1$ such that $\mathbf{a} < \mathbf{x}_0, \mathbf{x}_1 < \mathbf{b}$, and $\mathbf{x}_0 \cup \mathbf{x}_1 = \mathbf{b}$. This his result left open the question on the elementary equivalence of the low_2 c.e. and low_2 d-c.e. degrees. The first example which distinguishes the structures of low c.e. and low d-c.e. degrees was found by Faizrakhmanov (2010): It is known (Welch 1980) that in the c.e. degrees the following sentence holds: there are low c.e. degrees x_0, x_1 such that for any c.e. degree y there exist c.e. degrees y_0 and y_1 such that $y = y_0 \cup y_1$ and $y_0 \leq x_0$ and $y_1 \leq x_1$. Faizrakhmanov proved that for all low d-c.e. degrees x_0 and x_1 there is a low d-c.e. degree y such that $x_0 \cup x_1 \neq y$.

These results leave open the question of the elementary equivalence of the semilattices of low_n c.e. and low_n d-c.e. degrees for $n = 2$ and $n > 3$. In this connection it is interesting to note, that there is a high c.e. set whose only proper splittings are low_2 (Downey and Shore 1997).

Among important open problems in the study of structures of n-c.e. degrees for $n > 1$ the problems of definability of c.e. degrees in the broader classes of n-c.e. degrees, and the definability of m-c.e. degrees in the n-c.e. degrees for $m < n$.

It was mentioned already that any d-c.e. degree $a > 0$ is splittable into two d-c.e. degrees d_0 and d_1. If a is a *properly* (i.e. non c.e.) d-c.e. degree, then at least one of the intervals $[d_i, a], i \leq 1$, does not contain c.e. degrees. Yamaleev (2009) proved the following theorem:

Theorem 3. *Let* $a < b$ *properly d-c.e. degrees such that there are no c.e. degrees between* a *and* b. *Then* b *is splittable avoiding the upper cone of* a.

As a consequence we have the following:

Corollary 1 *(Yamaleev). Any splitting of a properly d-c.e. degree* a *into d-c.e. degrees* d_0 *and* d_1 *contains a part* $a > d_i$ *such that any d-c.e. degree* $b, a \geq b > d_i$, *is splittable avoiding the upper cone of* d_i.

This result allows to formulate questions, answers to which can set definability of c.e. degrees in the degree structures $\mathcal{D}_n, n > 1$. For instance,

Question 1. *Whether the following statement is true: for any c.e. degree* $a > 0$ *there is a splitting into d-c.e. degrees* d_0 *and* d_1 *such that each part* $a > d_i$ *contains a d-c.e. degree* $b, a \geq b > d_i$, *which is not splittable avoiding the upper cone of* d_i?

A positive answer to this question means that the c.e. degrees are definable in $\mathcal{D}_2$: a d-c.e. degree a is c.e. iff there is a splitting of a into d-c.e. degrees d_0 and d_1 such that each part $a > d_i$ contains a d-c.e. degree $b, a \geq b > d_i$, which is not splittable avoiding the upper cone of d_i.

Lachlan (unpublished) *associated* with each d-c.e. set $D = B_1 - B_2$ the following c.e. set $As(D) = \{\langle s, x \rangle : x \in D_s - D\}$. It is obvious that $As(D) \leq_T D$, D is c.e. in $As(D)$, and, therefore, if D is not c.e. then $As(D)$ is not computable. Note that for an n-c.e. set D we can similarly define an $(n-1)$ c.e. set $As(D) \leq_T D$ such that D is c.e. in $As(D)$.

It is also clear that the definition of $As(D)$ depends on enumerations of the c.e. sets B_1 and B_2 and we can uniformly determine c.e. indices for $As(D)$ from those of B_1 and B_2. But we cannot uniformly determine a c.e. index of a *non computable* $As(D)$ (in case when D is not c.e.) from indices of B_1 and B_2. Actually the following more strong claim holds:

Theorem 4 *(Downey and Stob (1993, Theorem 10.3)). There is no computable function g and partial-computable functional Φ such that if $W_e - W_i$ is non computable then $W_{g(e,i)}$ non computable and $W_{g(e,i)} = \Phi^{W_e - W_i}$.*

Nevertheless, the Turing degree of $As(D)$ is defined uniquely by the set D. It follows from the following

Proposition 1 *(Ishmukhametov (1999)). Let $D = B_1 - B_2$ be a d-c.e. set and $As(D)$ is the associated with D c.e. set. If D is c.e. in a set B, then $As(D) \leq_T B$.*

Proof. Let $D = \text{dom}(\Phi_e(B))$ for some e. For each $\langle s, x \rangle$, if $x \notin D_s$, then $\langle s, x \rangle \notin As(D)$. If $x \in D_s$, then let $t > s$ be a such stage in the enumeration of $D_0 - D_1$, that

- either $x \notin D_t$ (in this case $\langle s, x \rangle \in As(D)$)
- or $\Phi_e(B, x) \downarrow [t]$ (then $\langle s, x \rangle \notin As(D)$).

It is clear that in general the degree of $As(D)$ depends on the choice of the set D. Moreover, following statements hold:

Theorem 5. *There is a d-c.e. set D such that for any d-c.e. set $B \equiv_T D$ there is a d-e.e. set $A \equiv_T D$ such that*

(a) (Ishmukhametov 1999) $As(B) \not\leq_T As(A)$;
(b) (Wu and Yamaleev 2012) $As(A) <_T As(B)$;

Theorem 6 *(Ishmukhametov (2000)). There are d-c.e. sets A and B such that $A \equiv_T B$ and $As(B) \not\leq_T As(A), As(A) \not\leq_T As(B)$.*

If $\mathbf{d}$ is a d-c.e. degree then let

$$As(\mathbf{d}) = \{\mathbf{x} \mid As(D) \in \mathbf{x} \text{ for some d-c.e. set } D \in \mathbf{d}\}.$$

We already saw that $\mathbf{d}$ is c.e. in each degree from $As(\mathbf{d})$. The converse also holds (Fang et al. 2013): if $\mathbf{a} < \mathbf{d}$ is a c.e. degree such that $\mathbf{d}$ is c.e. in $\mathbf{a}$ then $\mathbf{d}$ contains a d-c.e. set D such that $As(D) \in \mathbf{a}$.

Proposition 2. *Let $\mathbf{a} < \mathbf{d}$ be a c.e. non computable degree. Then $As(D) \in \mathbf{a}$ for some d-c.e. set $D \in \mathbf{d}$.*

Proof. Let A be a c.e. set such that $\deg(A)$ is $\mathbf{a}$. In Arslanov et al. (1998, Theorem 4) a d-c.e. set $U \in \mathbf{a}$ such that U is c.e. in A, is constructed. It follows from Proposition 1 that $\deg(As(U)) \leq \mathbf{a}$. Now let $D = U \oplus (\omega - A)$. It is clear that D is d-c.e., $D \equiv_T U$ and $As(D) \in \mathbf{a}$.

2 n-c.e. Enumeration Degrees

A set A is *enumeration reducible to* a set B (in symbols: $A \leq_e B$), if there is an algorithm for enumerating A given any enumeration of B. Namely (see, e.g., Rogers (1967)) if there exists some computably enumerable set W, such that

$$A = \{x : (\exists u)[\langle x, u \rangle \in W \;\&\; D_u \subseteq B]\}$$

where D_u is the finite set with canonical index u (in the following we will often identify finite sets with their canonical indices). Thus, each c.e. set W can be viewed as an operator (called an *enumeration operator*), associating to each set B, the set A which is obtained from B as above. The degree structure originated by this reducibility is the structure of the *enumeration degrees*. (In the following, we will write e-reducible, e-operator, e-degree for enumeration reducible, enumeration operator, enumeration degree, respectively. We will also denote by $\deg_e(A)$ the e-degree of a set A.)

The n-c.e. e-degrees (where $n \geq 2$) form an upper semilattice $\mathcal{E}_n$ with least element $\mathbf{0}_e$ (the e-degree of the c.e. sets) and greatest element $\mathbf{0}_e'$ (the e-degree of $\overline{K}$, where K is any creative set).

For $n = 2$ the structure $\mathcal{E}_2$ is isomorphic to the upper semilattice $\mathcal{D}_1$ of c.e. Turing degrees since for d-c.e. set D we have $D \equiv_e As(D)$. In general, we can not use Lachlan's associated sets to reduce a $(n+1)$-c.e enumeration degree to some n-c.e. enumeration degree as in the n-c.e. Turing degrees. For a weaker version we can use a different method: if a $(2n+1)$-c.e. set A is not $(2n-1)$-c.e. then for the c.e. sets

$$B_1 \supseteq B_2 \supseteq \cdots \supseteq B_{2n-1} \supseteq B_{2n} \supseteq B_{2n+1}$$

such that

$$A = (B_1 - B_2) \cup \cdots \cup (B_{2n-1} - B_{2n}) \cup B_{2n+1}$$

the corresponding 3-c.e. set $C = (B_{2n-1} - B_{2n}) \cup B_{2n+1}$ is not c.e. It is easy to note that $C = A \cap B_{2n-1}$, so that $C \leq_e A$. As a consequence we have the following:

Proposition 3 *(Arslanov et al. 2001).*

1. *For every* $\mathbf{a} \in \mathcal{E}_{2n+1} \setminus \mathcal{E}_{2n}$, $n \geq 1$, *there is a nonzero enumeration degree* $\mathbf{b} \in \mathcal{E}_3$ *such that* $\mathbf{b} \leq \mathbf{a}$.
2. *(Kalimullin 2007). For every* $\mathbf{a} \in \mathcal{E}_{2n} \setminus \mathcal{E}_{2n-1}$, $n \geq 1$, *there is a nonzero enumeration degree* $\mathbf{b} \in \mathcal{E}_2$ *such that* $\mathbf{b} \leq \mathbf{a}$.

The following theorem together with the Lachlan's Non-Diamond Theorem shows that the structure $\mathcal{E}_3$ is not elementarily equivalent to $\mathcal{D}_1 \cong \mathcal{E}_2$.

Theorem 7. *There are 3-c.e. enumeration degrees* $\mathbf{a}_0 > \mathbf{0}$ *and* $\mathbf{a}_1 > \mathbf{0}$ *such that* $\mathbf{a}_0 \cup \mathbf{a}_1 = \mathbf{0}_e'$ *and* $\mathbf{a}_0 \cap \mathbf{a}_1 = \mathbf{0}_e$.

Corollary 2. *The* $\forall\exists$-*theories of* $\mathcal{D}_1 \cong \mathcal{E}_2$ *and* $\mathcal{E}_3$ *are different.*

In the proof of Theorem 7 the 3-c.e. sets $A_0 \in \mathbf{a}_0$ and $A_1 \in \mathbf{a}_1$ with corresponding 3-c.e. approximations $\{A_0^s\}_{s \in \omega}$ and $\{A_1^s\}_{s \in \omega}$ are constructed satisfying the global requirement

$$\mathcal{I} : (\forall i = 0, 1)(\forall e \in \omega)[(A_i^s - A_i^{s+1}) \cap \omega^{[e]} \neq \emptyset \Longrightarrow \omega^{[\geq e]} \restriction s \subseteq A_{1-i}],$$

where $\omega^{[e]} = \{\langle e, x \rangle : x \in \omega\}$ and $\omega^{[\geq e]} \restriction s = \{\langle i, x \rangle : i \geq e \ \& \ \langle i, x \rangle < s\}$. This requirement guarantees the condition $\deg_e(A_0) \cap \deg_e(A_1) = \mathbf{0}_e$. Moreover, the requirement $\mathcal{I}$ gives us much more.

Theorem 8 *(Kalimullin (2003)). Let $A_0 \in \mathbf{a}_0$ and $A_1 \in \mathbf{a}_1$ be Δ_2^0 sets with corresponding Δ_2^0-approximations $\{A_0^s\}_{s \in \omega}$ and $\{A_1^s\}_{s \in \omega}$ satisfying the requirement $\mathcal{I}$. Then $\mathbf{x} = (\mathbf{a}_0 \cup \mathbf{x}) \cap (\mathbf{a}_1 \cup \mathbf{x})$ for every enumeration degree $\mathbf{x}$.*

The pairs of degrees $\mathbf{a}_0, \mathbf{a}_1$ satisfying the condition $(\forall \mathbf{x})[\mathbf{x} = (\mathbf{a}_0 \cup \mathbf{x}) \cap (\mathbf{a}_1 \cup \mathbf{x})]$ were used in the definability of jump operator in the enumaration degrees (Kalimullin 2003) and the definability of the class of total enumeration degrees (Ganchev and Soskova 2015). It is easy to see that there are no m-c.e. Turing degrees $\mathbf{a}_0 > \mathbf{0}$, $\mathbf{a}_1 > \mathbf{0}$ satisfying this property.

Corollary 3. *The formula*

$$(\exists \mathbf{a}_0 > \mathbf{0})(\exists \mathbf{a}_1 > \mathbf{0})(\forall \mathbf{x})[(\mathbf{a}_0 \cup \mathbf{x}) \cap (\mathbf{a}_1 \cup \mathbf{x}) = \mathbf{x}]$$

holds in each $\mathcal{E}_n$, $n \geq 3$, and fails in each $\mathcal{D}_m$, $m \geq 1$. Thus, there is no elementarily equivalent pair $\mathcal{E}_n$ and $\mathcal{D}_m$, $n \geq 3$, $m \geq 1$.

Note that the first found formula distinguishing $\mathcal{E}_n$ and $\mathcal{D}_m$, $n \geq 3$, $m \geq 1$, is the formula

$$(\forall \mathbf{x} > \mathbf{0})(\exists \mathbf{a}_0 > \mathbf{0})(\exists \mathbf{a}_1 > \mathbf{0})[\mathbf{a}_0 \leq \mathbf{x} \ \& \ \mathbf{a}_1 \leq \mathbf{x} \ \& \ \mathbf{a}_0 \cap \mathbf{a}_1 = \mathbf{0}].$$

By Lachlan's Nonbounding Theorem the formula fails in $\mathcal{D}_1$, and hence, fails in each $\mathcal{D}_m$, $m \geq 1$. From another hand, the formula holds in $\mathcal{E}_3$ (Kalimullin 2001), and hence holds in each $\mathcal{E}_n$, $n \geq 3$, by Proposition 3.

In contrast with the situation in the Turing degrees, we know that there is no elementarily equivalent pair of upper semilattices $\mathcal{E}_n$, $\mathcal{E}_m$, $n \neq m$. To show this we need to introduce the notion of k-splittability.

Definition 1. *An enumeration degree $\mathbf{a}$ is k-splittable avoiding an enumeration degree $\mathbf{c}$ $(k \geq 1)$, if there exist enumeration degrees $\mathbf{b}_1, \mathbf{b}_2, \ldots, \mathbf{b}_k$ such that $\mathbf{a} = \mathbf{b}_1 \cup \mathbf{b}_2 \cup \cdots \cup \mathbf{b}_k$ and $\mathbf{c} \nleq \mathbf{b}_i$ for each i, $1 \leq i \leq k$.*

In particular, an enumeration degree $\mathbf{a}$ is 1-splittable avoiding $\mathbf{c}$ if and only if $\mathbf{c} \nleq \mathbf{a}$.

Theorem 9 *(Kalimullin (2002)). If $1 < m < 2k$, then every m-c.e. enumeration degree $\mathbf{a}$ is k-splittable avoiding any Δ_2^0 numeration degree $\mathbf{c} > \mathbf{0}$ via an appropriate set of m-c.e. enumeration degrees $\mathbf{b}_1, \mathbf{b}_2, \ldots, \mathbf{b}_k$.*

Due $\mathcal{E}_2 \cong \mathcal{D}_1$ the last theorem with $k = m = 2$ is the statement of Sacks Splitting Theorem. The 2-splitting still holds if $m = 3$ so that the structure $\mathcal{E}_3$ of 3-c.e. can be viewed as a natural extension of the structure of c.e. degrees preserving Sacks Splitting Theorem. By the following theorem the structure $\mathcal{E}_4$ is not such an extension.

Theorem 10 *(Kalimullin (2002)). For each $k \geq 1$ there is a $(2k)$-c.e. enumeration degree $\mathbf{a}$ which is not k-splittable avoiding some 3-c.e. enumeration degree $\mathbf{c} > \mathbf{0}$.*

Corollary 4. *For each $k \geq 1$ the formula*

$$(\exists \mathbf{a})(\exists \mathbf{c} > \mathbf{0})(\forall \mathbf{b}_1)\ldots(\forall \mathbf{b}_k)[\mathbf{c} \not\leq \mathbf{b}_1 \,\&\ldots\& \, \mathbf{c} \not\leq \mathbf{b}_k \implies \mathbf{a} \neq \mathbf{b}_1 \cup \cdots \cup \mathbf{b}_k]$$

holds in $\mathcal{E}_n$ for $n \geq 2k$ and fails in $\mathcal{E}_m$ for $m < 2k$. Thus, there is no elementarily equivalent pair $\mathcal{E}_m$ and $\mathcal{E}_n$ such that $m < 2k \leq n$.

To distinguish the upper semilattices $\mathcal{E}_{2k}$ and $\mathcal{E}_{2k+1}$ we will use the effect of the second part of Proposition 3 and the property $(\forall \mathbf{x})[\mathbf{x} = (\mathbf{a}_0 \cup \mathbf{x}) \cap (\mathbf{a}_1 \cup \mathbf{x})]$.

Theorem 11 *(Kalimullin (2007)). For each $k \geq 1$ there is a $(2k + 1)$-c.e. enumeration degree $\mathbf{a}_0$ and $\mathbf{a}_1$ such that $(\forall \mathbf{x})[\mathbf{x} = (\mathbf{a}_0 \cup \mathbf{x}) \cap (\mathbf{a}_1 \cup \mathbf{x})]$ and for every $i = 0, 1$ the enumeration degree $\mathbf{a}_i$ is not k-splittable avoiding some 3-c.e. enumeration degree $\mathbf{c}_i > \mathbf{0}$.*

Corollary 5. *For each $k \geq 1$ the formula*

$$(\exists \mathbf{a}_0)(\exists \mathbf{a}_1)(\exists \mathbf{c}_0 > \mathbf{0})(\exists \mathbf{c}_1 > \mathbf{0})(\forall \mathbf{x})(\forall \mathbf{b}_1)\ldots(\forall \mathbf{b}_k)$$
$$[[\mathbf{c}_0 \not\leq \mathbf{b}_1 \,\&\ldots\& \, \mathbf{c}_0 \not\leq \mathbf{b}_k \to \mathbf{a}_0 \neq \mathbf{b}_1 \cup \cdots \cup \mathbf{b}_k] \,\&$$
$$[\mathbf{c}_1 \not\leq \mathbf{b}_1 \,\&\ldots\& \, \mathbf{c}_1 \not\leq \mathbf{b}_k \to \mathbf{a}_1 \neq \mathbf{b}_1 \cup \cdots \cup \mathbf{b}_k] \,\&$$
$$[(\mathbf{a}_0 \cup \mathbf{x}) \cap (\mathbf{a}_1 \cup \mathbf{x}) = \mathbf{x}]]$$

holds in $\mathcal{E}_n$ for $n > 2k$ and fails in $\mathcal{E}_m$ for $m \leq 2k$.

Indeed, if this formula holds in $\mathcal{E}_{2k}$ then by Theorem 9 the corresponding $(2k)$-c.e. degrees $\mathbf{a}_0$ and $\mathbf{a}_1$ are not $(2k - 1)$-c.e. By Proposition 3 there are nonzero 2-c.e. degrees $\mathbf{d}_0 \leq \mathbf{a}_0$ and $\mathbf{d}_1 \leq \mathbf{a}_0$. Then the property $(\forall \mathbf{x})[\mathbf{x} = (\mathbf{d}_0 \cup \mathbf{x}) \cap (\mathbf{d}_1 \cup \mathbf{x})]$ holds which is impossible in $\mathcal{E}_2 \cong \mathcal{D}_1$.

Corollary 6 *(Kalimullin (2007)). There is no elementarily equivalent pair $\mathcal{E}_m$ and $\mathcal{E}_n$ with $m \neq n$.*

References

Arslanov, M.M.: Structural properties of the degrees below $\mathbf{0}'$. Sov. Math. Dokl. N.S. **283**, 270–273 (1985)

Arslanov, M.M.: On the upper semilattice of Turing degrees below $\mathbf{0}'$. Russ. Math. (Iz. VUZ) **7**, 27–33 (1988)

Arslanov, M.M., Cooper, S.B., Li, A.: There is no low maximal d.c.e. degree. Math. Logic Quart. **46**, 409–416 (2000)

Arslanov, M.M., Cooper, S.B., Li, A.: There is no low maximal d.c.e. degree - Corrigendum. Math. Logic Quart. **50**, 628–636 (2004)

Arslanov, M.M., Kalimullin, I.S., Sorbi, A.: Density results in the Δ_2^0 e-degrees. Arch. Math. Logic **40**, 597–614 (2001)

Arslanov, M.M., Kalimullin, I.S., Lempp, S.: On Downey's conjecture. J. Symb. Logic **75**, 401–441 (2010)

Arslanov, M.M., LaForte, G., Slaman, T.A.: Relative enumerability in the difference hierarchy. J. Symb. Logic **63**, 411–420 (1998)

Cooper, S.B.: Degrees of Unsolvability. Ph.D. Thesis, Leicester University, Leicester, England (1971)

Cooper, S.B.: The density of the low$_2$ n-r.e. degrees. Arch. Math. Logic **31**, 19–24 (1991)

Cooper, S.B.: A splitting theorem for the n-r.e. degrees. Proc. Am. Math. Soc. **115**, 461–471 (1992)

Cooper, S.B., Harrington, L., Lachlan, A.H., Lempp, S., Soare, R.I.: The d-r.e. degrees are not dense. Ann. Pure Appl. Logic **55**, 125–151 (1991)

Cooper, S.B., Li, A.: Non-uniformity and generalised Sacks splitting. Acta Math. Sin., Engl. Ser. **18**, 327–334 (2000a)

Cooper, S.B., Li, A.: Splitting and cone avoidance in the d.c.e. degrees. Sci. China (Ser. A) **45**, 1135–1146 (2002b)

Downey, R.G.: D.r.e. degrees and the nondiamond theorem. Bull. London Math. Soc. **21**, 43–50 (1989)

Downey, R.G., Shore, R.A.: Splitting theorems and the jump operator. Ann. Pure Appl. Logic **94**, 45–52 (1997)

Downey, R.G., Stob, M.: Splitting theorems in recursion theory. Ann. Pure Appl. Logic **65**, 1–106 (1993)

Downey, R.G., Yu, L.: There are no maximal low d.c.e. degrees. Notre Dame J. Formal Logic **45**(3), 147–159 (2004)

Ershov, Y.L.: On a hierarchy of sets I. Algebra Logika **7**(1), 47–73 (1968a)

Ershov, Y.L.: On a hierarchy of sets II. Algebra Logika **7**(4), 15–47 (1968b)

Ershov, Y.L.: On a hierarchy of sets III. Algebra Logika **9**(1), 34–51 (1970)

Faizrakhmanov, M.K.: Decomposability of low 2-computably enumerable degrees and Turing jumps in the Ershov hierarchy. Russ. Math. (Iz. VUZ) **12**, 51–58 (2010)

Fang, C., Wu, G., Yamaleev, M.M.: On a problem of Ishmukhametov. Arch. Math. Logic **52**, 733–741 (2013)

Ganchev, H.A., Soskova, M.I.: Definability via Kalimullin pairs in the structure of the enumeration degrees. Trans. Am. Math. Soc. **367**, 4873–4893 (2015)

Gold, E.M.: Limiting recursion. J. Symb. Logic **30**, 28–48 (1965)

Ishmukhametov, S.N.: On the r.e. predecessors of d.r.e. degrees. Arch. Math. Logic **38**, 373–386 (1999)

Ishmukhametov, S.T.: On relative enumerability of Turing degrees. Arch. Math. Logik Grundl. **39**, 145–154 (2000)

Kalimullin, I.S.: Elementary theories of semilattices of n-recursive enumerable degrees with respect to enumerability. Russ. Math. (Iz. VUZ) **45**(4), 22–25 (2001)

Kalimullin, I.S.: Splitting properties of n-c.e. enumeration degrees. J. Symbolic Logic **67**, 537–546 (2002)

Kalimullin, I.S.: Definability of the jump operator in the enumeration degrees. J. Math. Logic **3**(2), 257–267 (2003)

Kalimullin, I.S.: Elementary differences between the (2p)-c.e. and the (2p+1)-c.e. enumeration degrees. J. Symbolic Logic **72**, 277–284 (2007)

Lachlan, A.H.: Lower bounds for pairs of recursively enumerable degrees. Proc. London Math. Soc. **16**, 537–569 (1966)

Putnam, H.: Trial and error predicates and the solution to a problem of Mostowski. J. Symb. Logic **30**, 49–57 (1965)

Rogers Jr., H.: Theory of Recursive Functions and Effective Computability. McGraw-Hill, New York (1967)

Welch, L.: A hierarchy of families of recursively enumerable degrees and a theorem of founding minimal pairs. Ph.D. Thesis (University of Illinois, Urbana, 1980) (1980)

Wu, G., Yamaleev, M.M.: Isolation: motivations and applications. Uchenue Zapiski Kazanskogo Universiteta **154**(2), 204–217 (2012)

Yamaleev, M.M.: The splitting of 2-computably enumerable degrees avoiding the upper cones. Russ. Math. (Iz. VUZ) **6**, 76–80 (2009)

There Are No Maximal d.c.e. *wtt*-degrees

Guohua Wu[1(✉)] and Mars M. Yamaleev[2]

[1] Division of Mathematical Sciences, School of Physical and Mathematical Sciences, Nanyang Technological University, Singapore 637371, Singapore
guohua@ntu.edu.sg
[2] Institute of Mathematics and Mechanics, Kazan Federal University, 18 Kremlyovskaya Street, Kazan 420008, Russia
mars.yamaleev@ksu.ru

1 Introduction

In this article, we will study the weak-truth-table (*wtt*, for short) degrees of d.c.e. sets and show that there is no maximal d.c.e. *wtt*-degree.

Theorem 1. *For any d.c.e. wtt-degree $\mathbf{d}_{wtt} < \mathbf{0}'_{wtt}$, there is a d.c.e. wtt-degree $\mathbf{c}_{wtt}$ strictly between $\mathbf{d}_{wtt}$ and $\mathbf{0}'_{wtt}$.*

Here $\mathbf{0}'_{wtt}$ is the *wtt*-degree of K, the halting problem. Theorem 1 says that for any d.c.e. set D, if D is *wtt*-incomplete, then we can split K into c.e. sets B and C such that K cannot be *wtt*-reducible to any of $B \oplus D$ and $C \oplus D$. As K is *wtt*-equivalent to $B \sqcup C$, we have that $B \oplus C \oplus D$ is *wtt*-equivalent to K. Thus, $B \oplus D$ and $C \oplus D$ are not *wtt*-reducible to each other, so they are strictly above $\mathbf{d}_{wtt}$. Our current work shows that d.c.e. *wtt*-degrees can always split above less ones (in progress), an analogue of Ladner and Sasso's result for c.e. *wtt*-degrees in [19].

Before giving a proof of the theorem above, we first review some well-known facts of density\nondensity of Turing degrees of c.e. sets and d.c.e. sets. Recall a set $A \subseteq \mathbb{N}$ is computably enumerable (c.e. for short) if A is a domain of some partial computable function, and $D \subseteq \mathbb{N}$ is d.c.e. if D is the difference of two computably enumerable sets, i.e. $D = A - B$ for some c.e. sets A and B. The research on the structures of the c.e. degrees and the d.c.e. degrees has shown

Wu is partially supported by AcRF Tier 2 grants MOE2011-T2-1-071 (ARC 17/11, M45110030) and MOE2016-T2-1-083 from Ministry of Education of Singapore, and by AcRF Tier 1 grants, RG29/14, M4011274 and RG32/16, M4011672 from Ministry of Education of Singapore.

Yamaleev is supported by Russian Foundation for Basic Research (projects 15-41-02507, 15-01-08252), by research grant of Kazan Federal University, and by the subsidy allocated to Kazan Federal University for the project part of the state assignment in the sphere of scientific activities (project 1.2045.2014).

The original version of this chapter was revised. The spelling of the second author's name was corrected. The erratum to this chapter is available at DOI: 10.1007/978-3-319-50062-1_43

© Springer International Publishing AG 2017
A. Day et al. (Eds.): Downey Festschrift, LNCS 10010, pp. 479–486, 2017.
DOI: 10.1007/978-3-319-50062-1_28

many nice properties and also many pathological properties, which are always accompanied with new techniques of constructions.

For the c.e. degrees, Sacks proved that this structure is dense and every nonzero element splits, and Lachlan proved that the density and splitting above cannot be combined, where Lachlan developed the $0'''$ argument for the first time, an argument being called "monstrous" construction in the 1980s.

Cooper initiated the study of the structure of d.c.e. degrees in his PhD thesis [3] in 1971. Lachlan observed that the d.c.e. degrees are downwards dense and Cooper [5] proved that each nonzero d.c.e. degree splits, an analogue of Sacks splitting. Recall that a Turing degree is properly d.c.e. if it contains a d.c.e. set, but no c.e. sets. As c.e. degrees are also d.c.e., what Lachlan and Cooper needed to do in their proofs is to consider the case when the given degree is properly d.c.e. As pointed in Downey and Stob [12], and Cooper and Li [8], it is necessary to have the cases separated, as no uniform way working for both cases exists. Cooper [4] even proved that the low$_2$ d.c.e. degrees are dense.

Cooper and Yi considered the interaction between c.e. degrees and d.c.e degrees in [9] and introduced the notion of isolation. The existence of isolated degrees can be obtained from a result in Kaddah's paper [17], where she proved that low c.e. degrees branch in the d.c.e. degrees. Using this interaction phenomenon, Wu [24] provided another proof of Downey's diamond theorem, where Wu used the isolation to connect the cupping and the capping parts of the diamond embeddings. See Wu and Yamaleev's survey [25] on this topic.

Even though these two degree structures share several algebraic properties, these two structures are not elementarily equivalent. This was first proved in the 1980s by Arslanov in [2] and Downey in [11]. As to the density, Cooper, Harrington, Lachlan, Lempp and Soare proved that the d.c.e. degrees are not dense, where they constructed a maximal d.c.e. degree **d** below $0'$. Obviously, $0'$ does not split above **d**.

We will consider d.c.e. wtt-degrees in this paper, i.e. the weak-truth-table degrees of d.c.e. sets. For $A, B \subseteq \mathbb{N}$, say that A is weak-truth-table reducible to B, denoted as $A \leq_{wtt} B$, if there is a partial computable functional Φ_e and a computable function f such that (i) $A = \Phi_e^B$, and (ii) for every x, $f(x) \geq u(B; e, x)$, where $u(B; e, x)$ is the use of the computation $\Phi_e^B(x)$. We use φ_e to denote the use function of Φ_e. The wtt-reduction was proposed by Friedberg and Rogers in 1959 in [15], and is now also called bounded Turing reduction. Lachlan proved that the upper semi-lattice of c.e. wtt-degrees is distributive, providing a crucial structural difference between c.e. wtt-degrees and c.e. Turing degrees. Ladner and Sasso then gave another difference in [19] by showing that the splitting and density can be combined for the c.e. wtt-degrees. Technically, weak-truth-table degrees can be handled much easier than Turing degrees. For instance, density of the c.e. wtt-degrees can be proved by a finite injury argument, whereas the analogous result for c.e. Turing degrees requires an infinite injury priority proof.

On the other hand, some structural properties of Turing degrees can be obtained from those of wtt-degrees via the so-called contiguous degrees. Here, a c.e. Turing degree **c** is contiguous if **c** contains exactly one c.e. wtt-degree.

That is, any two c.e. sets A, B in a contiguous degree **c** are *wtt*-equivalent. Ladner and Sasso proved in [19] that any nonzero *wtt*-degree **c** has the anticupping property in the c.e. *wtt*-degrees. Thus, when **c** is contiguous, **c** also has the anticupping property in the Turing degrees, a result first proved by Yates by direct construction. This kind of "transfer" phenomenon has been further developed by Ambos-Spies in [1], Stob in [23], and Downey in [10].

In this paper, we consider the *wtt*-degrees of d.c.e. sets, and the remainder of this paper will be devoted to the proof of Theorem 1: there are no maximal d.c.e. *wtt*-degrees.

Our notation and terminology are standard and generally follow Soare [22] and Odifreddi [20]. The readers can refer Cooper's paper [6] for d.c.e. Turing degrees and Ambos-Spies' paper [1], Stob's paper [23] and Downey's paper [10] for the general idea on c.e. *wtt*-degrees.

2   Requirements and Construction

Let $K = \{e : \varphi_e(e)\}$, Turing's halting problem, and let $\{K_s : s \in \omega\}$ be a recursive enumeration of K. Note that K is *wtt*-complete among all d.c.e. sets, and we will assume that for each s, $|K_{s+1}\backslash K_s| = 1$. Let D be any d.c.e. set in $\mathbf{d}_{wtt}$, and $\{D_s : s \in \omega\}$ be a d.c.e. approximation of D. An additional addition for this approximation is: for any s, $|(D_{s+1}\backslash D_s) \cup (D_s\backslash D_{s+1})| = 1$.

For the proof of Theorem 1, we will construct c.e. sets B and C satisfying the following requirements:

$\mathcal{S}$: $K = B \sqcup C$;
$\mathcal{P}_e^B$: $K \neq \Phi_e^{B\oplus D}$, where the use function of $\Phi_e^{B\oplus D}$, i.e. φ_e, is bounded by ψ_e;
$\mathcal{P}_e^C$: $K \neq \Phi_e^{C\oplus D}$, where the use function of $\Phi_e^{C\oplus D}$, i.e. φ_e, is bounded by ψ_e.

Here $\{(\Phi_e, \psi_e) : e \in \omega\}$ is a recursive list of all pairs (Φ, ψ), Φ a partial computable functional Φ and ψ a partial computable function. As indicated at the beginning of this paper, the $\mathcal{S}$-requirement ensures that K and $B \oplus C \oplus D$ are *wtt*-equivalent. All the $\mathcal{P}_e^B$ requirements, $e \in \omega$, ensure that K is not *wtt*-reducible to $B \oplus D$, and all the $\mathcal{P}_e^C$ requirements, $e \in \omega$, ensure that K is not *wtt*-reducible to $C \oplus D$. Thus, $B \oplus D$ and $C \oplus D$ are not *wtt*-reducible to each other, which implies that both are strictly above $\mathbf{d}_{wtt}$.

The idea of satisfying the $\mathcal{S}$-requirement is standard. That is, at any stage s, we find a requirement $\mathcal{R}$ with the highest priority with k_s less than the restraint $r(\mathcal{R}, s)$, if exists. If $\mathcal{R}$ is a $\mathcal{R}_e^B$-requirement, then enumerate k_s into C. Otherwise, enumerate k_s into B. Obviously, $B \sqcup C = K$.

Before we describe how to satisfy a $\mathcal{P}$-requirement, a $\mathcal{P}_e^B$-requirement, say, we first review the idea when D is c.e., and then we show the changes we need to make for the case when D is d.c.e.

The main idea for the case *when D is c.e.* is the Sacks preservation strategy, i.e., to find a disagreement between K and $\Phi_e^{B\oplus D}$, we define expansionary stages and extend a *wtt*-reduction Δ_e at expansionary stages such that if there were infinitely many expansionary stages, then we would have $\Delta_e^D = K$, which is impossible. Here we define the length of the agreement between K and $\Phi_e^{B \oplus D}$ at stage s as

$$\ell^B(e, s) = \max\{x : \ (\forall y < x)[\Phi_e^{B \oplus D}(y)[s] \downarrow \text{ with use } \varphi_e(y) < \psi_e(y)$$

$$\text{and } \Phi_e^{B \oplus D}(y)[s] = K_s(y)]\},$$

and say that s is an expansionary stage if for any expansionary stage $t < s$, $\ell^B(e, s) > \ell^B(e, t)$. At an expansionary stage s, for any $y < \ell^B(e, s)$, if $\Delta_e^D(y)$ has no definition at stage s, then define $\Delta_e^D(y)[s] = K_s(y)$ with use $\delta_e(y)[s] = \varphi_e(y)[s]$. So, δ_e is bounded by ψ_e, and if there were infinitely many expansionary stages, then $\Phi_e^{B \oplus D}$ would be total and Δ_e^D would be defined as a total function, showing that $K \leq_{wtt} D$, which is impossible. Thus, there are only finitely many expansionary stages, and we will have some $y \leq \ell^B(e)$, the length of agreement at the last expansionary stage, such that either $\Phi_e^{B \oplus D}(y) \uparrow$ or $\Phi_e^{B \oplus D}(y) \neq K(y)$.

In this strategy, the main point is to protect computations at expansionary stages. Assume that after an expansionary stage s_1 (so a restraint is imposed on the B-part to protect computations), we see that a computation $\Phi_e^{B \oplus D}(y)$ changes, with $y < \ell^B(e, s_1)$, because of the changes of D between stages s_1 and s_2 say. We also assume that s_2 is not an expansionary stage, but we see that $\Phi_e^{B \oplus D}(y)[s_2]$ converges. The strategy says that at the next expansionary stage s_3, we will protect $\Phi_e^{B \oplus D}(y)[s_3]$. This computation $\Phi_e^{B \oplus D}(y)[s_3]$ we see at stage s_3 could be also different from $\Phi_e^{B \oplus D}(y)[s_2]$, and $\Phi_e^{B \oplus D}(y)[s_2]$ is not protected. It is okay, for D c.e., as either there are no more expansionary stages, or the changes between stages s_2 and s_3 will remain forever in D, and no more computation of $\Phi_e^{B \oplus D}(y)$ can be the same as $\Phi_e^{B \oplus D}(y)[s_2]$. Thus, among all the computations of $\Phi_e^{B \oplus D}(y)$ with use $\varphi_e(y)[s] < \psi_e(y)$, we only protect those we see at expansionary stages, and for y above, the change of D undefines $\Delta_e^D(y)$, and we will redefine $\Delta_e^D(y)$ again at the next expansionary stage. Nothing is complicated in this case.

As we are assuming that the uses of $\Phi_e^{B \oplus D}(y)$ are bounded by $\psi_e(y)$, the nature of finite injury allows to protect all computations of $\Phi_e^{B \oplus D}(y)$ we see at all stages. That is, whenever we see a new computation of $\Phi_e^{B \oplus D}(y)$, at stage s_2 above, we can protect it and redefine $\Delta_e^D(y)[s_2] = K_{s_2}(y)$. Of course, if the computation $\Phi_e^{B \oplus D}(y)$ changes between stage s_2 and s_3, then the D-changes undefine $\Delta_e^D(y)[s_2]$ again, allowing us to redefine it at stage s_3.

We adopt this idea of protecting all computations for our purpose when D is d.c.e. It can happen that a computation $\Phi_e^{B \oplus D}(y)[s]$ changes because of some enumeration of z into D, and after many stages, when z leaves D, at stage t say, the D-part of the oracle $B \oplus D$ recovers to the status at stage s, and if we protect $\Phi_e^{B \oplus D}(y)[s]$ at stage s, then we will have $\Phi_e^{B \oplus D}(y)[t] = \Phi_e^{B \oplus D}(y)[s]$. This variation of Sacks preservation strategy allows us to deal with cases when D is any Δ_2^0 set, i.e. $\mathbf{0}'_{wtt}$ splits above any other Δ_2^0 wtt-degrees.

We are now ready to provide a full construction. We first list the requirements as follows:

$$\mathcal{S} < \mathcal{P}_0^B < \mathcal{P}_0^C < \mathcal{P}_1^B < \mathcal{P}_1^C < \cdots < \mathcal{P}_e^B < \mathcal{P}_e^C < \cdots.$$

We say that a requirement $\mathcal{Q}$ has priority higher than $\mathcal{R}$ if $\mathcal{Q} < \mathcal{R}$ in the order defined above. So $\mathcal{S}$ has the highest priority, and at any stage s, we will enumerate k_s into one of B and C, but not both.

For a $\mathcal{P}$-requirement, $\mathcal{P}_e^B$ say, we call stage s a $\mathcal{P}_e^B$-*identical stage* if

1. $\ell^B(e, s) = \ell^B(e, s - 1)$,
2. for all $y \leq \ell^B(e, s)$, $\Phi_e^{B \oplus D}(y)[s]$ converges if and only if $\Phi_e^{B \oplus D}(y)[s - 1]$ converges,
3. for $\Phi_e^{B \oplus D}(y)[s]$ converges, the computation $\Phi_e^{B \oplus D}(y)[s]$ and $\Phi_e^{B \oplus D}(y)[s - 1]$ are the same.

We say that $\mathcal{P}_e^B$ *requires attention at stage* s if s is not a $\mathcal{P}_e^B$-identical stage. Construction at stage 0: For all the requirements, let the corresponding restraint as 0.

Construction at stage $s > 0$:

Step 1. Among requirements $\mathcal{P}_0^B, \mathcal{P}_0^C, \mathcal{P}_1^B, \mathcal{P}_1^C, \cdots, \mathcal{P}_s^B, \mathcal{P}_s^C$, check which one requires attention at stage s. Let it be $\mathcal{Q}[s]$, and set the corresponding restraint as s. For those y less than the length of agreement of $\mathcal{Q}[s]$, define $\Delta_e^D(y) = K_s(y)$ with use $\delta_e(y)[s] = \varphi_e(y)[s]$. Initialize all the requirements with priority lower than $\mathcal{Q}[s]$.

Step 2. Among all the requirements with priority not lower than $\mathcal{Q}[s]$, find the one with higher priority, $\mathcal{R}[s]$ say, whose restraint is larger than k_s. If $\mathcal{R}[s]$ is a $\mathcal{P}^B$-strategy, then enumerate k_s into C. Otherwise, enumerate k_s into B. Initialize all the requirements with priority lower than $\mathcal{R}[s]$.

This completes the construction of stage s.

End of construction

3 Verification

In this section, we verify that the constructed c.e. sets B and C satisfy all the requirements. The actions at step 2 of each stage s ensure that $K = B \sqcup C$, and hence

Lemma 1. *The requirement $\mathcal{S}$ is satisfied.*

Now we verify that all the $\mathcal{P}$-requirements are satisfied. The following lemma is enough to show this.

Lemma 2. *For each $e \in \omega$,*

1. $\mathcal{P}_e^B$ *can be initialized at most finitely many times;*
2. $\mathcal{P}_e^B$ *requires attention at most finitely many times;*
3. $\mathcal{P}_e^B$ *has finite restraint;*
4. *The same are true for requirement $\mathcal{P}_e^C$.*

Proof. We prove it by induction on e. So we can assume that after a stage s_0 large enough, no more $\mathcal{P}_{e'}$-requirements, with $e' < e$, requires attention, or requires further enumeration of elements of K into B. Thus, after stage s_0, $\mathcal{P}_e^B$ cannot be initialized anymore. (1) holds.

To prove (2), we assume that $\mathcal{P}_e^B$ requires attention infinitely often. Then, the bounding function ψ_e is total. As D is d.c.e., we assume that after stage s_1, D becomes fixed up to $\psi_e(0)$. According to steps 1 and 2 in every stage, after stage s_0, we protect all the computations of $\Phi_e^{B\oplus D}(0)$ whenever we have a new computation, thus, for any computation of $\Phi_e^{B\oplus D}(0)$, if it converges before stage s_1, at stage s' say, and the D-part of the use agrees with $D \upharpoonright \psi_e(0)$, then this computation will converge forever. Of course, no such a computation occurs before stage s_1, then, when $\mathcal{P}_e^B$ requires attention again, we will have a new computation of $\Phi_e^{B\oplus D}(0)$, which will be protected. In both cases, $\Phi_e^{B\oplus D}(0)$ converges. The same idea can be used to prove that $\Phi_e^{B\oplus D}$ converges at any n, by induction.

We now show that Δ_e^D is total and computes K correctly. Again, we first show that $\Delta_e^D(0)$ is defined, with $\Delta_e^D(0) = K(0)$, and the same argument can be applied to show that for any n, $\Delta_e^{D(n)}$ is defined and equals to $K(n)$.

We assume again that D has no more change below $\psi_e(0)$ after stage s_1. Then for $\Delta_e^D(0)$ defined at stage s^* with the D-part of the use agreeing with $D \upharpoonright \varphi_e(0)[s^*]$, i.e.

$$D \upharpoonright \varphi_e(0)[s^*] = D_{s_1} \upharpoonright \varphi_e(0)[s^*],$$

we have $\Delta_e^D(0) = \Delta_e^D(0)[s^*]$. To see this, assume that $s^* < t_1 < t_2 < \cdots < t_n < s_1$ be a list of stages with $D_{t_i} \upharpoonright \varphi_e(0)[s^*] = D_{s^*} \upharpoonright \varphi_e(0)[s^*]$ for each $i \in \{1, \cdots, n\}$, then $\Delta_e^D(0)[t_i] = \Delta_e^D(0)[s^*]$ with use $\varphi_e(0)[s^*]$. This actually shows that for any definition of $\Delta_e^D(0)$, which is defined at other stages, D must have changes at some number z below $\varphi_e(0)[s^*]$, before stage s^* (if $\Delta_e^D(0)$ is defined before stage s^*), or between any two stages in this list (if $\Delta_e^D(0)$ is defined between these two stages). Of course, if there is no such a stage s^*, then at stage s_1, we define $\Delta_e^D(0)$ and by the choice of s_1, D will have no change any more and hence $\Delta_e^{D(0)}$ is defined.

Now we show that $\Delta_e^D(0) = K(0)$. Note that at stage s^* above, we have $\Phi_e^{B\oplus D}(0)[s^*]$ converges, and this computation is protected since s^* onwards and hence after stage s_1, i.e. the computation $\Phi_e^{B\oplus D}(0)$ will be the same as $\Phi_e^{B\oplus D}(0)[s^*]$. Thus, $\Phi_e^{B\oplus D}(0) = \Phi_e^{B\oplus D}(0)[s^*]$, and as we assume that there are infinitely many stages $\mathcal{P}_e^B$ requires attention, we know that after stage s_1, the agreement of $\mathcal{P}_e^B$ will be always larger than 0, and hence $K(0) = \Phi_e^{B\oplus D}(0)[s^*]$ forever, and as a consequence, $\Delta_e^D(0) = K(0)$.

We can then apply the same idea and show that for any n, $\Delta_e^D(n)$ is defined and equals to $K(n)$. This shows that $K \leq_{wtt} D$ via Δ. A contradiction. Thus (2) is true for $\mathcal{P}_e^B$ requirement.

Note that (2) tells us the existence of a stage s_2, after which the $\mathcal{P}_e^B$ requirement never requires attention again, which means that after stage s_2, all stages are $\mathcal{P}_e^B$-identical, and hence no more restraint is imposed. This shows that the last restraint imposed by $\mathcal{P}_e^B$ is before stage s_2, and as a consequence, the $\mathcal{P}_e^B$ requirement has finite restraint. (3) is true.

The same argument can show that (1), (2), (3) above are also true for $\mathcal{P}_e^C$ requirement. (4) is true.

This completes the proof of Lemma 3.2, and hence the proof of Theorem 1.

4   Further Remarks

As pointed out in the introduction, we can improve Theorem 1 and show that the d.c.e. *wtt*-degrees are dense, and hence for a given Turing degree **d**, it can either contain exactly one d.c.e. *wtt*-degree, or contain infinitely many d.c.e. *wtt*-degree. We call a d.c.e. Turing degree **d** contiguous if it contains exactly one d.c.e. *wtt*-degree. A recent work of the authors shows the existence of properly d.c.e. contiguous degrees. We have seen that c.e. contiguous degrees have many unusual applications, like Downey's idea of using c.e. contiguous degrees to show the downwards density of c.e. degrees with strong anti-cupping property, and we are interested in problems of properly d.c.e. contiguous degrees, like the distribution of such degrees and how we can use such degrees to transfer properties of *wtt*-degrees to Turing degrees.

References

1. Ambos-Spies, K.: Contiguous R.E. degrees. In: Börger, E., Oberschelp, W., Richter, M.M., Schinzel, B., Thomas, W. (eds.) Computation and Proof Theory. LNM, vol. 1104, pp. 1–37. Springer, Heidelberg (1984). doi:10.1007/BFb0099477
2. Arslanov, M.: Structural properties of the degrees below 0′. Dokl. Nauk. SSSR **283**, 270–273 (1985)
3. Cooper, S.B.: Degrees of unsolvability. Ph.D. thesis, Leicester University (1971)
4. Cooper, S.B.: The density of the low₂ *n*-r.e. degrees. Arch. Math. Logic **31**, 19–24 (1991)
5. Cooper, S.B.: A splitting theorem for the *n*-r.e. degrees. Proc. Am. Math. Soc. **115**, 461–471 (1992)
6. Cooper, S.B.: Local degree theory. In: Griffor, E.R. (ed.) Handbook of Computability Theory, pp. 121–153. North-Holland, Amsterdam (1999)
7. Cooper, S.B., Harrington, L., Lachlan, A.H., Lempp, S., Soare, R.I.: The d.r.e. degrees are not dense. Ann. Pure Appl. Logic **55**, 125–151 (1991)
8. Cooper, S.B., Li, A.: Non-uniformity and generalised Sacks splitting. Acta Math. Sin. **18**, 327–334 (2002)
9. Cooper, S.B., Yi, X.: Isolated d.r.e. degrees. University of Leeds, Department of Pure Mathematics, Preprint series, vol. 17 (1995)
10. Downey, R.: Δ_2^0 degrees and transfer theorems. Ill. J. Math. **31**, 419–427 (1987)
11. Downey, R.: D.r.e. degrees and the nondiamond theorem. Bull. London Math. Soc. **21**, 43–50 (1989)
12. Downey, R., Stob, M.: Splitting theorems in recursion theory. Ann. Pure Appl. Logic **65**, 1–106 (1993)
13. Ershov, Y.L.: A hierarchy of sets, Part I (Russian). Algebra i Logika **7**, 47–73 (1968). Algebra and Logic (English translation), **7**, 24–43 (1968)

14. Ershov, Y.L.: A hierarchy of sets, Part II (Russian). Algebra i Logika **7**, 15–47 (1968). Algebra and Logic (English Translation), **7**, 212–232 (1968)
15. Friedberg, R.M., Rogers, H.: Reducibility and completeness for sets of integers. Z. Math. Logik Grundlag. Math. **5**, 117–125 (1959)
16. Ishmukhametov, S.: D.r.e. sets, their degrees and index sets. Ph.D. thesis, Novosibirsk, Russia (1986)
17. Kaddah, D.: Infima in the d.r.e. degrees. Ann. Pure Appl. Logic **62**, 207–263 (1993)
18. Lachlan, A.H.: Lower bounds for pairs of recursively enumerable degrees. Proc. London Math. Soc. **16**, 537–569 (1966)
19. Ladner, R.E., Sasso, L.P.: The weak-truth-table degrees of recursively enumerable sets. Ann. Math. Logic **8**, 429–448 (1975)
20. Odifreddi, P.: Classical Recursion Theory. Studies in Logic and the Foundations of Mathematics, vol. 143. Elsevier, New York (1999)
21. Sacks, G.E.: The recursively enumerable degrees are dense. Ann. Math. **80**, 300–312 (1964)
22. Soare, R.I.: Recursively Enumerable Sets and Degrees. Springer, Heidelberg (1987)
23. Stob, M.: Wtt-degrees and T-degrees of r.e. sets. J. Symbolic Logic **48**, 921–930 (1983)
24. Wu, G.: Isolation and lattice embeddings. J. Symbolic Logic **67**, 1055–1064 (2002)
25. Wu, G., Yamaleev, M.M.: Isolation: motivations and applications. Proc. Kazan Univ. (Phys. Math. Ser.) **52**, 204–217 (2012)

A Rigid Cone in the Truth-Table Degrees with Jump

Bjørn Kjos-Hanssen[(⊠)]

University of Hawai'i at Mānoa, Honolulu, USA
bjoern.kjos-hanssen@hawaii.edu

Abstract. The automorphism group of the truth-table degrees with order and jump is fixed on the set of degrees above the fourth jump, $\mathbf{0}^{(4)}$.

1 Introduction

A *cone* in a partial order $(D, \leq)$ is a set of the form $D(\geq a) := \{x \in D : x \geq a\}$ for some $a \in D$. A subset of S of D is *rigid* if it is fixed under the action of the automorphism group $\mathrm{Aut}(D, \leq)$, i.e., for each $x \in S$ and each $\pi \in \mathrm{Aut}(D, \leq)$, $\pi(x) = x$. We will also be interested in the case of structures $(D, \leq, \mathbf{j})$ where $\mathbf{j}$ is a unary function on D. In that case, rigidity of $S \subseteq D$ is defined with respect to $\mathrm{Aut}(D, \leq, \mathbf{j})$ rather than $\mathrm{Aut}(D, \leq)$.

It is not known whether the structure of the Turing degrees is rigid, but it is known [5] that the structure of the Turing degrees with jump contains a rigid cone. This is shown by applying a jump inversion theorem and results on initial segments. Here we show that also the structure of truth-table degrees with jump $(\mathcal{D}_{tt}, \leq, \mathbf{j})$ contains a rigid cone. For definitions relating to initial segments we refer the reader to the author's doctoral dissertation [6] and survey article [7].

Our main result is that each automorphism of the truth-table degrees with jump is equal to the identity on the cone above $\mathbf{0}^{(4)}$. This contrasts with the results of Anderson [1] that each automorphism of the truth-table degrees (not necessarily jump invariant) is equal to the identity on some cone, and each automorphism that preserves $\mathbf{0}^{(3)}$ and $\mathbf{0}^{(5)}$ is equal to the identity on the cone above $\mathbf{0}^{(5)}$. It is still open whether non-trivial automorphisms of these structures exist at all.

2 Mal'tsev Homogeneous Lattice Tables

If $(L, \leq)$ is a partial order (transitive, reflexive, antisymmetric relation) such that greatest lower bounds $\alpha \wedge \beta$ of all $\alpha, \beta \in L$ exist then $(L, \leq, \wedge)$ is called a *lower semilattice*; if least upper bounds $\alpha \vee \beta$ of all pairs $\alpha, \beta \in L$ exist, then $(L, \leq, \vee)$

B. Kjos-Hanssen—Thanks are due to Noam Greenberg for a correction to an early draft of this article. This work was partially supported by a grant from the Simons Foundation (#315188 to Bjørn Kjos-Hanssen). This material is based upon work supported by the National Science Foundation under Grant No. 1545707.

© Springer International Publishing AG 2017
A. Day et al. (Eds.): Downey Festschrift, LNCS 10010, pp. 487–500, 2017.
DOI: 10.1007/978-3-319-50062-1_29

is called an *upper semilattice* (usl). If L is both an lower semilattice and an upper semilattice then L is a *lattice*. L is called *bounded* if there exist elements $0, 1 \in L$ such that for all $\alpha \in L$, $0 \leq \alpha \leq 1$. In particular every finite lattice is bounded. If L has more than one element (so in the bounded case, $0 \neq 1$) then we say that L is *nontrivial*. A *unary algebra* is a collection of functions $f : X \to X$ on a set X, closed under composition. The *partition lattice* $\mathrm{Part}(X)$ on a set X consists of all equivalence relations (considered as sets of ordered pairs) on X, ordered by inclusion. We will be interested in the case where X is finite or countably infinite.

A lattice L_1 is a *0-1 sublattice* of a lattice L_2 if

- L_1 is a sublattice of L_2,
- both L_1 and L_2 have both a least element and a greatest element,
- the least element of L_1 equals the least element of L_2, and
- the greatest element of L_1 equals the greatest element of L_2.

A *lattice table* [9] Θ consists of

(1) a set X,
(2) a finite set of equivalence relations $\alpha_1, \ldots, \alpha_n$ on X, and
(3) an order $\leq$ given by $\alpha_i \leq \alpha_j \leftrightarrow \alpha_i \supseteq \alpha_j$ (reverse inclusion of sets of ordered pairs),

such that $\{\alpha_1, \ldots, \alpha_n\}$ ordered by *inclusion* is a 0-1 sublattice of $\mathrm{Part}(X)$. We write $\widehat{\Theta} = \{\alpha_1, \ldots, \alpha_n\}$. We think of Θ as equal to X, but endowed with additional structure. So $x \in \Theta$ means $x \in X$, etc. but for emphasis we may write $|\Theta|$ for X. Note that Θ is determined by $\widehat{\Theta}$.

Elements of $|\Theta|$ are denoted by lower-case Roman letters such as u, v, w, x, y, z, and elements of semilattices in general and $\widehat{\Theta}$ in particular by lower-case Greek letters such as α, β, γ.

If $\alpha \in \widehat{\Theta}$ and $(x, y) \in \alpha$ then we write $x \sim_\alpha y$. If Θ is a lattice table then an *endomorphism* of Θ is a map $f : \Theta \to \Theta$ preserving all equivalence relations in $\widehat{\Theta}$:

$$(\forall x, y \in \Theta)(\forall \alpha \in \widehat{\Theta})(x \sim_\alpha y \to f(x) \sim_\alpha f(y)).$$

End Θ denotes the unary algebra consisting of all endomorphisms of Θ.

$C_\Theta(x, y)$ denotes the principal equivalence relation in Θ generated by (x, y), i.e.,

$$C_\Theta(x, y) = \cap\{\alpha \in \widehat{\Theta} : (x, y) \in \alpha\}.$$

We define $\mathrm{End}_\Theta(x, y)$ to be the principal congruence relation in Θ generated by (x, y), i.e., the equivalence relation generated by all pairs $(f(x), f(y))$ for $f \in \mathrm{End}\, \Theta$.

Lemma 2.1. $\mathrm{End}_\Theta(x, y) \subseteq C_\Theta(x, y)$.

Proof. If $(u, v) \in \mathrm{End}_\Theta(x, y)$ then (u, v) is in the transitive closure of

$$\{(f(x), f(y)) \mid f \in \mathrm{End}\, \Theta\},$$

so it suffices to show each such $(f(x), f(y)) \in C_\Theta(x, y)$. For this it suffices to show $(f(x), f(y)) \in \alpha$ provided that $(x, y) \in \alpha$ for $\alpha \in \widehat{\Theta}$; this holds since $f \in \text{End } \Theta$. $\square$

Definition 2.2. *Let Θ be a lattice table. We say that Θ is Mal'tsev homogeneous if for all $x, y \in \Theta$, $C_\Theta(x, y) \subseteq \text{End}_\Theta(x, y)$ (so by Lemma 2.1, $C_\Theta(x, y) = \text{End}_\Theta(x, y)$).*

The following Proposition can readily be proved:

Proposition 2.3. Θ *is Mal'tsev homogeneous iff for all $x, y, u, v \in \Theta$ satisfying*

$$(\forall \alpha \in \widehat{\Theta})(x \sim_\alpha y \to u \sim_\alpha v),$$

there exist $n \in \omega = \{0, 1, 2, \ldots\}$, $z_1, \ldots, z_n \in \Theta$ and $f_0, \ldots, f_n \in \text{End } \Theta$ such that

$$(\forall i \leq n)(\{f_i(x), f_i(y)\} = \{z_i, z_{i+1}\})$$

where $z_0 = u$ and $z_{n+1} = v$.

The z_i are called *homogeneity interpolants*.

This notion of homogeneity is more general (weaker) than those considered in [9].

Note that if $\alpha \wedge \beta \vDash \gamma$ in $\widehat{\Theta}$ then α and β generate γ. That is, if $x \sim_\gamma y$ then there exist *meet interpolants* $z_1, \ldots, z_n$ for x, y such that $x \sim_\alpha z_1 \sim_\beta z_2 \cdots \sim_\alpha z_n \sim_\beta y$.

Definition 2.4. *If Θ is a lattice table and $Y \subseteq |\Theta|$, then for each $\alpha \in \widehat{\Theta}$, $\alpha \upharpoonright Y = \{(x, y) \in Y \times Y : (x, y) \in \alpha\}$. Let $\widehat{\Theta} \upharpoonright Y = \{\alpha \upharpoonright Y : \alpha \in \widehat{\Theta}\}$.*

If Θ_0 and Θ_1 are lattice tables, then we say that $\Theta_0 \subseteq \Theta_1$ if $|\Theta_0| \subseteq |\Theta_1|$ and $\widehat{\Theta}_1 \upharpoonright |\Theta_0| = \widehat{\Theta}_0$. Note that if $\Theta_0 \subseteq \Theta_1$ then $\widehat{\Theta}_0$ and $\widehat{\Theta}_1$ are isomorphic (nontrivial, finite) lattices.

If $\Theta_n, n \in \omega$ are lattice tables such that $\Theta_n \subseteq \Theta_{n+1}$ for each n, then $\bigcup_{n \in \omega} \Theta_n$ is the lattice table Θ such that $|\Theta| = \bigcup_{n \in \omega} |\Theta_n|$ and $\widehat{\Theta} \upharpoonright |\Theta_n| = \widehat{\Theta}_n$ for each n. In particular $\Theta_n \subseteq \Theta$ and $\widehat{\Theta}_n$ and $\widehat{\Theta}$ are isomorphic lattices for each n.

Definition 2.5. Θ *is a sequential lattice table if there exist $\Theta_n, n \in \omega$, such that $\Theta = \bigcup_{n \in \omega} \Theta_n$, and*

(1) each Θ_n is a $(0, 1, \vee)$-substructure of $\text{Part}(|\Theta_n|)$ (Θ_n is an usl table),
(2) Θ is a lattice table, and
(3) for each n, meet interpolants for elements of Θ_n exist in Θ_{n+1}.

Θ *is a sequential Mal'tsev homogeneous lattice table if in addition*

(4) Θ is Mal'tsev homogeneous, with homogeneity interpolants for elements of Θ_n appearing in Θ_{n+1} (compare [9, VII.1.1, 1.3]).

Definition 2.6 (Direct limit). *Let a sequence* $(L^i, \varphi_i)_{i \in \omega}$ *be given, where each* L^i *is a finite lattice,* $\varphi_i : L^i \to L^{i+1}$ *is a* $(0, 1, \vee)$ *homomorphism, and* $L^i \cap L^j = \emptyset$ *for* $i \neq j$.

Let $L' = \bigcup_{i \in \omega} L_i$ *as a set. Let* $\approx$ *be the equivalence relation on* L' *generated by* $a \approx \varphi_i(a)$ *for* $a \in L^i$. *Then* $L = L'/\approx$ *is an upper semilattice called the direct limit of the sequence* $(L^i, \varphi_i)_{i \in \omega}$.

Definition 2.7. *Fix finite lattices* L^0, L^1 *and a* $(0, 1, \vee)$ *homomorphism* $\varphi :$ $L^0 \to \varphi(L^0) \subseteq L^1$, *and lattice tables* Θ^0, Θ^1. *Suppose* $\Psi^i : L^i \to \widehat{\Theta}^i$, $i = 0, 1$, *are isomorphisms. For* $\alpha \in L^i$, *we write* $\sim_\alpha$ *for* $\sim_{\Psi^i \alpha}$.

We say that Θ^1 *embeds in* Θ^0 *with respect to* φ *and* Ψ_0, Ψ_1 *if there is a function* $\Theta(\varphi) : \Theta^1 \to \Theta^0$ *such that for all* $x, y \in \Theta^1$, *and all* $\alpha \in L^0$,

$$x \sim_{\varphi\alpha} y \Leftrightarrow \Theta(\varphi)(x) \sim_\alpha \Theta(\varphi)(y).$$

For our pivotal technical result, Proposition 3.2 below, Definition 2.7 plays a key role which we describe in Remark 2.8. The reader may find a still more detailed treatment in [6].

Remark 2.8. *Suppose a bounded countable upper semilattice* L *is given such that the ordering* $\leq$ *of* L *is computably enumerable. We shall need a computable sequence of lattice tables* $\Theta^0, \Theta^1, \ldots$ *such that* $\widehat{\Theta}^s$ *is isomorphic to our approximation to* L *at stage* s. *(We will start with a sequence* $(L^i, \varphi_i)_{i \in \omega}$ *having* L *as direct limit, and our approximation to* L *at stage* s *will be* L^s.) *Suppose we discover at stage* $s + 1$ *that* $\alpha \leq \beta$, *whereas at stage* s *we knew that* $\beta \leq \alpha$ *but thought that* $\alpha \not\leq \beta$. *Further suppose that we cannot ignore what was done using* Θ^s *at stage* s, *but we can let* Θ^{s+1} *be a subset of* Θ^s. *If* Θ^{s+1} *embeds into* Θ^s *with respect to* φ *(the homomorphism mapping our approximation to* L *at stage* s *to our approximation to* L *at stage* $s + 1$*) then by thinning* Θ^0 *to* $\Theta(\varphi)\Theta^1$, *we eliminate all elements* x, y *that are witnesses to the fact that* $\alpha \neq \beta$. *This allows us to identify* α *and* β, *even though so far we have been working under the assumption that* $\alpha \neq \beta$.

The full result needed for the application to the Turing degrees is contained in Theorem 2.9 and Proposition 2.10.

Theorem 2.9. *Let* L *be a bounded countable nontrivial usl and let* $(L^i, \varphi_i)_{i \in \omega}$ *be any system of nontrivial finite lattices having* L *as direct limit in the sense of Definition 2.6. Then there exists*

1. *a function* $h : \omega \to \omega$,
2. *a double sequence of finite lattice tables* $(\Theta^i_j)_{i \in \omega, j \geq h(i)}$ *with* $\Theta^i_j \subseteq \Theta^i_{j+1}$ *for each* $i \in \omega, j \geq h(i)$, *and*
3. *for each* $i \in \omega$ *an increasing function* $m_i : \omega \to \omega$ *with* $m_i(0) = h(i)$, *such that*
   1. *letting* $\Theta^i = \bigcup_{j \in \omega} \Theta^i_j$, *we have* $|\Theta^i| \supseteq |\Theta^{i+1}|$ *for each* $i \in \omega$,
   2. *for each* $i, j \geq h(i)$ *and* k *such that* $m_i(j) \leq k < m_i(j+1)$, *we have*

$$\Theta^i_k = \Theta^i_{m_i(j)},$$

3. *for each* $i \in \omega, (\Theta^i_{m_i(j)})_{j \in \omega}$ *is a sequential Mal'tsev homogeneous lattice table,*
4. *for each* $i \in \omega, \widehat{\Theta}_i$ *is isomorphic to* L^i, *and*
5. *there exist isomorphisms* $\Psi_i : L^i \to \widehat{\Theta}_i$ *such that* Θ^{i+1} *embeds in* Θ^i *with respect to* φ_i *and* Ψ_i, Ψ_{i+1}, *and the embedding is the identity map. In other words, for all* $x, y \in \Theta^{i+1}$ *and* $\alpha \in L^i$, *we have*

$$x \sim_{\Psi_i \alpha} y \leftrightarrow x \sim_{\Psi_{i+1} \varphi_i \alpha} y.$$

The essential property in Theorem 2.9, and the one that goes beyond those of [8], is (5). The following Proposition can be proved by inspecting the proof of Theorem 2.9.

Proposition 2.10 (Computability-theoretic properties). *Let* **a** *be a Turing degree and let* L *be a* $\Sigma^0_1(\mathbf{a})$*-presentable usl. Then in Theorem 2.9, we may assume that* h *is* **a***-computable; the array* $\{\Theta^i_j \mid j \geq h(i)\}$ *is* **a***-computable; each* m_i *is computable; for each* $i < \omega, (\Theta^i_{m_i(j)})_{j \in \omega}$ *is a computable sequence; each* Θ^i *is computable; and there is a computable function taking* $L^0, \ldots, L^i$ *to* Θ^i.

We now begin the development that will lead to a proof of Theorem 2.9.

If A is a unary algebra then Con A denotes the congruence lattice of A, i.e., the lattice of all equivalence relations E on X preserved by all $f \in A$, ordered by inclusion.

The following observation can be traced back to Mal'tsev [4, 10, 11].

Proposition 2.11. *For any unary algebra* A, *the dual of* Con A *is a Mal'tsev homogeneous lattice table.*

Proof. Suppose A is a unary algebra on a set X. Let Θ be the lattice table such that $\widehat{\Theta}$ is the dual of Con A. Since Con A is a 0-1 sublattice of Part(A), Θ is a lattice table.

If f is an operation in A and $\alpha \in \widehat{\Theta}$ then α is a congruence relation on A and hence $\forall x, y(x \sim_\alpha y \quad f(x) \sim_\alpha f(y))$, which means that $f \in \mathrm{End}\ \Theta$. So

$$A \subseteq \mathrm{End}\ \Theta.$$

Clearly for any unary algebras A, B on the same underlying set, we have $A \subseteq B \Rightarrow \mathrm{Con}\ A \supseteq \mathrm{Con}\ B$. Hence Con End $\Theta \subseteq$ Con $A = \widehat{\Theta}$.

If $u, v, x, y \in X, f \in \mathrm{End}\ \Theta$ and $(x, y) \in \mathrm{End}_\Theta(u, v)$ then there exist $z_1, \ldots, z_k$ such that $(z_i, z_{i+1}) = (g_i(u), g_i(v))$ for $g_i \in \mathrm{End}\ \Theta$ with $z_1 = x, z_k = y$, hence letting $w_i = f(z_i)$ and $h_i = f \circ g_i$ we have $(w_i, w_{i+1}) = (h_i(u), h_i(v)) \in \mathrm{End}_\Theta(u, v), w_1 = f(x), w_k = f(y)$, and so $(f(x), f(y)) \in \mathrm{End}_\Theta(u, v)$. Hence we have shown $\mathrm{End}_\Theta(u, v) \in \mathrm{Con}\ \mathrm{End}\ \Theta \subseteq \Theta$. Since End Θ contains the identity map, $\mathrm{End}_\Theta(u, v)$ contains (u, v). Hence $\mathrm{End}_\Theta(u, v)$ is in Θ and contains (u, v), so it contains $C_\Theta(u, v)$. So Θ is Mal'tsev homogeneous. $\square$

We recall the construction of [15].

Definition 2.12. *Let L be a nontrivial lattice. $\mathcal{A} = (A, r, h)$ is called an $L-\{1\}$-colored graph if A is a set, r is a set of size-two subsets of A, i.e., (A, r) is an undirected graph without loops, and $h : r \to L - \{1\}$ is a mapping of the set r of the edges of the graph into $L - \{1\}$.*

The map $e : L \to \mathrm{Part}(A)$ is defined by: for $\alpha \in L, e(\alpha)$ is the equivalence relation on A generated by identifying points x, y if there is a path from x to y in the graph consisting of edges all of which have color $\geq \alpha$. In this case we say that x, y are connected with color $\geq \alpha$.

Definition 2.13 (α-cells). *Let L be a nontrivial lattice and let $\alpha \in L - \{1\}$. An α-cell $\mathcal{B}_\alpha = (B_\alpha, s_\alpha, k_\alpha)$ is an $L-\{1\}$-colored graph consisting of (1) a base edge $\{x, y\}$ colored α, and (2) for each pair (α_1, α_2) of elements of L such that $\alpha_1 \wedge \alpha_2 \leq \alpha$, a chain of edges $\{x, u_1\}, \{u_1, u_2\}, \{u_2, u_3\}, \{u_3, y\}$, colored $\alpha_1, \alpha_2, \alpha_1, \alpha_2$, respectively. Here x, y, u_1, u_2, u_3 are distinct elements of B_α. The base edge and chain of edges corresponding to a particular inequality $\alpha_1 \wedge \alpha_2 \leq \alpha$ is referred to as a pentagon. So an α-cell consists of several pentagons, intersecting only in a common base edge.*

Definition 2.14 (Pudlák graphs). *Let L be a nontrivial lattice. The Pudlák graph [15] of L is an $L - \{1\}$ colored graph $\mathcal{A}^P$, defined as follows.*

1. *$\mathcal{A}_0^P$ consists of a single edge colored by $0 \in L$. (In fact, how we choose to color this one edge has no impact on later proofs.)*
2. *$\mathcal{A}_{n+1}^P$ contains $\mathcal{A}_n^P$ as a subgraph and is obtained by attaching to each edge of $\mathcal{A}_n^P$ of any color α an α-cell.*
3. *$\mathcal{A}^P = \bigcup_{n \in \omega} \mathcal{A}_n^P$.*

We will use the following modification, which contains infinitely many copies of each edge in Pudlák's graph.

1. *$\mathcal{A}_0^{(i)} = \mathcal{A}_0^P$, for each $i \in \omega$.*
2. *$\mathcal{A}_j^{(i)}$ is obtained by attaching to each edge of $\mathcal{A}_{j-1}^{(i)}$ of any color α, i many α-cells.*
3. *$\mathcal{A}_j = \mathcal{A}_j^{(j)}$.*
4. *$\mathcal{A} = \bigcup_{n \in \omega} \mathcal{A}_n = \mathcal{A}(L)$ is called the homogenized Pudlák graph of L.*

The underlying set of $\mathcal{A}_n$ is denoted by A_n.

Let $\Theta = \Theta(L)$ be the lattice table with $|\Theta| = A$, and $\widehat{\Theta} = \{e(\alpha) : \alpha \in L\}$.

Note that by definition of Θ being a lattice table, $\widehat{\Theta}$ is ordered by reverse inclusion. Similarly let Θ_n be the lattice table with $|\Theta_n| = A_n$,

$$\widehat{\Theta}_n = \{e(\alpha) \upharpoonright \Theta_n \mid \alpha \in L\}.$$

Lemma 2.15. *Let $B_0 \subseteq B_1 \subseteq A$, where A is the underlying set of $\Theta(L)$. For $i = 0, 1$, let Ξ_i be the usl table whose underlying set is B_i, and whose equivalence relations are computed using graph points belonging to B_i only. Then $\Xi_0 \subseteq \Xi_1$ in the sense of Definition 2.4.*

Proof. We have to show that if $x \sim_\alpha y$ holds in Ξ_1 then $x \sim_\alpha y$ holds in Ξ_0. The only way this could fail is if there is a path of edges between x and y leading out of Ξ_0 and then back in. We may assume that the path does not leave and re-enter Ξ_0 via the same node. So it suffices to show that any path that goes around a pentagon not contained in Ξ_0 but whose base is in Ξ_0 can be shortened to one contained in Ξ_0 with no loss of equivalence. Since the pentagons represent inequalities $\alpha \wedge \beta \leq \gamma$, any path x, u_1, u_2, u_3, y going around the pentagon in $\Xi_1 - \Xi_0$ may be replaced by the edge x, y cutting across which has equal or greater color, i.e., with no loss of equivalence. □

Lemma 2.16. $\Theta_n \subseteq \Theta_{n+1}$ *for each* n, *so* $\Theta = \bigcup_{n\in\omega} \Theta_n$.

Proof. Let $\Xi_i = \Theta_{n+i}$ for $i = 0, 1$ and apply Lemma 2.15. □

Theorem 2.17. *Let* L *be a nontrivial finite lattice.* L *is dual isomorphic to the congruence lattice of* End $\Theta(L)$. *In fact,* $e : L \to \widehat{\Theta}(L)$ *is an isomorphism, and* $\Theta(L) =$ Con End $\Theta(L)$.

Proof. Pudlák [15] assumes that L is an algebraic lattice [3], defines a certain algebra $S \subseteq$ End $\Theta^P(L)$, and shows that $e : L \to \widehat{\Theta}^P(L)$ is an isomorphism, and $\widehat{\Theta}^P(L) =$ Con S. Now trivially $\widehat{\Theta}^P(L) \subseteq$ Con End $\Theta^P(L)$ holds, and $S \subseteq$ End $\Theta^P(L)$ implies Con End $\Theta^P(L) \subseteq$ Con S. So we have $\widehat{\Theta}^P(L) =$ Con End $\Theta^P(L)$.

In fact Pudlák's proof works for our graph Θ as well, i.e., it shows that $e : L \to \widehat{\Theta}(L)$ is an isomorphism, and $\widehat{\Theta}(L) =$ Con End $\Theta(L)$. Now let L be a finite lattice. Since every finite lattice is algebraic, L is an algebraic lattice. Hence $e : L \to \widehat{\Theta}(L)$ is an isomorphism and $\widehat{\Theta}(L) =$ Con End $\Theta(L)$. □

Lemma 2.18. *The sequence* $\Theta_n(L), n \in \omega$ *has a subsequence which is a computable Mal'tsev homogeneous sequential lattice table.*

Proof. Since Θ is a congruence lattice, Θ is a Mal'tsev homogeneous lattice table. Hence as $\Theta = \bigcup_{n\in\omega} \Theta_n$, a subsequence of Θ_n, $n \in \omega$ will be a sequential Mal'tsev homogeneous lattice table. The sequence is computable since to compute an equivalence relation on elements of Θ_n, it is sufficient to consider paths in Θ_n, since $\Theta_n \subseteq \Theta_{n+1}$ by Lemma 2.16. □

From now on we will assume that in fact $\Theta_n, n \in \omega$ is itself the subsequence from Lemma 2.18. Fix nontrivial finite lattices L^0, L^1 and a $(0, 1, \vee)$-isomorphisms $\varphi : L^0 \to \varphi(L^0) \subseteq L^1$. Let the $\wedge$-isomorphism $\varphi^* : L^1 \to L^0$ be defined by $\varphi^*\beta = \bigvee\{\alpha \in L^0 \mid \varphi(\alpha) \leq \beta\}$. This φ^* is known as the Galois adjoint of φ [2].

The map φ^* has many nice properties; we list the ones we need in the following lemma.

Lemma 2.19. *1.* φ^* *is a* $(\wedge, 1)$-*homomorphism.*
2. If $\beta < 1$ *then* $\varphi^*\beta < 1$.
3. φ^* *is injective on* φL^0.
4. $\alpha \leq \varphi^*\beta \leftrightarrow \varphi^*\varphi\alpha \leq \varphi^*\beta$.

Proof. These all follow easily from the definition of φ^* and the fact that $\{\alpha \in L^0 \mid \varphi(\alpha) \leq \beta\}$ is the principal ideal generated by $\varphi^*(\beta)$, i.e., $\{\alpha \in L^0 \mid \alpha \leq \varphi^*(\beta)\}$. □

Lemma 2.20. *Let $\mathfrak{C}(\varphi)\mathcal{A}L^1$ be the graph obtained from $\mathcal{A}L^1$ by replacing each color β by $\varphi^*\beta$. Then $\mathfrak{C}(\varphi)\mathcal{A}L^1$ is isomorphic to a subgraph of $\mathcal{A}L^0$.*

Proof. Each pentagon of $\mathfrak{C}(\varphi)\mathcal{A}L^1$ represents an inequality of the form

$$\varphi^*\beta_1 \wedge \varphi^*\beta_2 \leq \varphi^*\beta,$$

for $\beta_1, \beta_2, \beta \in L^1$ satisfying $\beta_1 \wedge \beta_2 \leq \beta$. Then $\varphi^*\beta_1 \wedge \varphi^*\beta_2 = \varphi^*(\beta_1 \wedge \beta_2) \leq \varphi^*\beta$, so the represented inequality $\varphi^*\beta_1 \wedge \varphi^*\beta_2 \leq \varphi^*\beta$ holds in L^0.

Hence we can obtain an isomorphic copy of $\mathfrak{C}(\varphi)\mathcal{A}L^1$ within $\mathcal{A}L^0$ by running through the construction of $\mathcal{A}L^0$, omitting every pentagon that represents an inequality involving members of $L^0 - \varphi^*L^1$, and omitting pentagons for inequalities that are true in L^0 but not in L^1. If an edge becomes disconnected from $\mathcal{A}_0$ by such omissions then it too is omitted. Since L^1 may have many more elements than L^0, we make use of the fact that $\mathcal{A}$ contains infinitely many copies of each edge from Pudlák's original graph $\mathcal{A}^P$. Since $\varphi^*(\beta) = 1 \to \beta = 1$ by Lemma 2.19, recoloring of points is never identification of points. □

Lemma 2.21. *Let $\Theta(\varphi)$ be the isomorphism from Lemma 2.20, sending $\mathcal{A}L^1$ to a subgraph of $\mathcal{A}L^0$ isomorphic to $\mathfrak{C}(\varphi)\mathcal{A}L^1$. Then $\Theta(\varphi)\Theta L^1 \subseteq \Theta L^0$ in the sense of Definition 2.4.*

Proof. Let $\Xi_0 = \Theta(\varphi)\Theta L^1$ and $\Xi_1 = \Theta L^0$ and apply Lemma 2.15. □

Lemma 2.22. *Let Ψ_i be the map e of Definition 2.12 for $L = L^i, i = 0, 1$. Then $\Theta(L^1)$ embeds in $\Theta(L^0)$ with respect to φ and Ψ_0, Ψ_1.*

Proof. Let x, y be points in ΘL^1, i.e., in $\mathcal{A}L^1$, and let $\alpha \in L^0$. Then obviously $x \sim_{\varphi\alpha} y \to \Theta(\varphi)x \sim_{\varphi^*\varphi\alpha} \Theta(\varphi)y$. Now suppose $\Theta(\varphi)x \sim_{\varphi^*\varphi\alpha} \Theta(\varphi)y$. Then there is a path witnessing this, which by Lemma 2.20 we may assume lies within $\Theta(\varphi)\mathcal{A}L^1$. Hence the path has an inverse image path under $\Theta(\varphi)^{-1}$. This is then a path from x to y with colors β for all of which $\varphi^*\beta \geq \varphi^*\varphi\alpha$. But then $\alpha \leq \varphi^*\beta$ by Lemma 2.19(4), and so $\varphi\alpha \leq \beta$, so $x \sim_{\varphi\alpha} y$. So in fact $x \sim_{\varphi\alpha} y \leftrightarrow \Theta(\varphi)x \sim_{\varphi^*\varphi\alpha} \Theta(\varphi)y$. Colors γ of edges in $\Theta(\varphi)\mathcal{A}L^1$ are all of the form $\varphi^*(\beta)$ for some β. So $\Theta(\varphi)x \sim_{\varphi^*\varphi\alpha} \Theta(\varphi)y$ iff there is a path from $\Theta(\varphi)x$ to $\Theta(\varphi)y$, all edges of which are colored $\gamma \geq \varphi^*\varphi\alpha$, or equivalently by Lemma 2.19(4) (using $\gamma = \varphi^*\beta$), colored $\gamma \geq \alpha$. Hence equivalently $\Theta(\varphi)x \sim_\alpha \Theta(\varphi)y$. □

Proof of Theorem 2.9. Let $m_i(n)$ be the least m such that $\Theta(\varphi_i)\Theta_n^{i+1} \subseteq \Theta_m^i$. Let $h(i) = m_i(0)$, for $i \in \omega$. Let $\Theta^0 = \Theta(L^0)$ and for $i \geq 1$, denoting composition by juxtaposition,

$$\Theta^i = \Theta(\varphi_0) \cdots \Theta(\varphi_{i-1})\Theta(L^i).$$

Let $\Theta_k^i = \Theta(\varphi_0) \cdots \Theta(\varphi_{i-1})\Theta_j(L^i)$ if $k = m_0 m_1 \cdots m_{i-1}(j)$ for some j; otherwise, let $\Theta_k^i = \Theta_{k-1}^i$. The Theorem now follows easily. □

3   Initial Segments of the tt-Degrees

In the following, functions $g : \omega \to \omega$ under consideration will end up being computably bounded, hence when discussing the tt-degree $\mathbf{g}$ we may consider either g as a function or as a set (the graph of g). We recall the notion of an e-splitting tree for an element c of a finite usl [6,8], see e.g. Lerman [9, Definition VI.3.2].

Lemma 3.1. *Let $g : \omega \to \omega$. Suppose that for each e, g lies on a computable tree T_e which is e-splitting for some c for some tables with the properties of Proposition 2.10, in the sense of [8]. Then $\mathbf{g}$ is hyperimmune-free.*

Proof. For each $e \in \omega$ there exists $e^* \in \omega$ such that for all stages s and all oracles g, if $\{e^*\}_s^g(x) \downarrow$ then $\{e^*\}^g(x) = \{e\}^g(x)$ and $\{e\}_s^g(y) \downarrow$ for all $y \leq x$. If g lies on T_{e^*} then it follows that $\{e\}^g$ is total and $\{e^*\}^{T_{e^*}(\sigma)}(x) \downarrow$ for each σ of length $x + 1$. Hence $\{e\}^g = \{e^*\}^g$ is dominated by the recursive function $f(x) = \max\{\{e\}^{T_{e^*}(\sigma)}(x) : |\sigma| = x + 1\}$. $\square$

Proposition 3.2. *Let L be a $\Sigma_4^0(\mathbf{y})$-presentable upper semilattice with least and greatest element. Then there exist t, i, g such that*

1. *$t : \omega \to 2$ is $0''$-computable,*
2. *i is the characteristic function of a set I such that $I \leq_m y^{(3)}$,*
3. *$g^{(2)}(e) = t(i(0), \ldots, i(e))$ for all $e \in \omega$,*
4. *$[\mathbf{0}, \mathbf{g}]$ is isomorphic to L, and*
5. *$\mathbf{g}$ is hyperimmune-free*

Proof. The proof in [8] must be modified to employ the lattice tables of Proposition 2.10.

By Proposition 2.10, for all x, $y \in \Theta^{k+1}$ and $\alpha \in L^k$, we have [identifying the isomorphism between L^i and $\widehat{\Theta}^i$ with the identity]

$$x \sim_{\varphi_k \alpha} y \leftrightarrow x \sim_\alpha y.$$

[8, Lemma 4.1] is modified so that $\psi_{T,c}$ is $\psi_{T,\varphi_k c}$. The equivalence

$$uF_{m(i)}(c)v \leftrightarrow uG_i(c)v$$

now becomes

$$uF_{m(i)}(c)v \leftrightarrow uG_i(\varphi_k c)v$$

Just as in Lemma 2.1 it is shown that $\psi_{T,c}$ is Turing equivalent to $\psi_{T_0,c}$, it now follows that $\psi_{T,\varphi_k c}$ is Turing equivalent to $\psi_{T_0,c}$, which is what we want. $i(e) = 1$ iff the answers to the $\Pi_1^0(y^{(2)})$-question about L^e is yes.

By Lemma 3.1, $\mathbf{g}$ is hyperimmune-free. $\square$

Lemma 3.3. *If t, i, A, q satisfy*

1. *$t : \omega \to 2$ is q-computable,*
2. *i is the characteristic function of a set I such that $I \leq_m q'$,*
3. *$A(e) = t(i(0), \ldots, i(e))$ for all $e \in \omega$,*

then $A \leq_{tt} q'$.

Proof. The value of $A(e)$ is determined by the following $e + 2$ many yes-or-no questions to q': Is $i(0) = 0$? $\cdots$ Is $i(e) = 0$? and, using the answers to the first $e + 1$ many questions: Is $t(i(0), \ldots, i(e)) = 0$? □

Theorem 3.4. *Each $\Sigma_4^0(\mathbf{y})$-presentable upper semilattice with least and greatest element can be realized as an initial segment $[\mathbf{0}, \mathbf{g}]$ with $\mathbf{g}^{(2)} \leq \mathbf{y}^{(3)}$.*

Proof. Let g be as in Proposition 3.2. By Lemma 3.3 with $q = y^{(2)}$ and $A = g^{(2)}$, we have $g^{(2)} \leq_{tt} y^{(3)}$. By Proposition 3.2, L is isomorphic to $[\mathbf{0}, \mathbf{g}]_T$. Since $\mathbf{g}$ is hyperimmune-free, $[\mathbf{0}, \mathbf{g}]_T = [\mathbf{0}, \mathbf{g}]_{tt}$. □

4   Coding a Set into a Lattice

Definition 4.1. *Let L be an upper semilattice and suppose $G = \{g_i \mid i < \omega\} \subseteq L$. If there exist p, $q \in L$ such that*

$$\{g_i \mid i \in \omega\} \subseteq \{x \mid x \vee p \geq q \ \& \ (\forall y < x)(y \vee p \not\geq q)\}$$

then G is called a Slaman–Woodin set (SW-set) for p, q in L. If there exist e_0, e_1, f_0, $f_1 \in L$ such that for each $i < \omega$,

$$g_{2i+1} = (g_{2i} \vee e_1) \wedge f_1 \ \& \ g_{2i+2} = (g_{2i+1} \vee e_0) \wedge f_0,$$

then the function $i \mapsto g_i$ is called a Shore sequence for e_0, e_1, f_0, f_1 in L.

Lemma 4.2. *Let $\mathbf{a}$ be a Turing degree. Let L be a $\Sigma_1^0(\mathbf{a})$-presented upper semilattice containing elements p, q, e_0, e_1, f_0, f_1, and atoms g_i for $i \in \omega$, such that $G = \{g_i \mid i < \omega\}$ is a Slaman–Woodin set for p, q and $i \mapsto g_i$ is a Shore sequence for e_0, e_1, f_0, f_1. Then $\{\langle y, i \rangle \mid y = g_i\} \leq_T \mathbf{a}$.*

Proof

$$y = g_{2i+1} \Leftrightarrow \exists x (x = g_{2i} \ \& \ y \leq x \vee e_1 \ \& \ y \leq f_1 \ \& \ y \vee p \geq q)$$
$$y = g_{2i+2} \Leftrightarrow \exists x (x = g_{2i+1} \ \& \ y \leq x \vee e_0 \ \& \ y \leq f_0 \ \& \ y \vee p \geq q)$$

Note that the matrices of the formulas on the right hand side are positive formulas in the language with $\vee$ and $\leq$. The function $\vee$ is $\mathbf{a}$-recursive and the relation $\leq$ is $\Sigma_1^0(\mathbf{a})$. Hence the entire right hand sides are $\Sigma_1^0(\mathbf{a})$. So starting with g_0 we can find g_n, $\mathbf{a}$-recursively. □

Definition 4.3. *Let $a \in \mathcal{D}$. An usl L is said to be of degree $\mathbf{a}$ if (1) L is $\mathbf{a}$-presentable, and (2) if $\mathbf{b} \in \mathcal{D}$ and L is $\mathbf{b}$-presentable then $\mathbf{a} \leq \mathbf{b}$.*

Definition 4.4. *Given $U \subseteq \omega$ we define a lattice $L(U)$.*

It consists of $0, 1$, atoms $\{g_i : i \in \omega\}$, more atoms e_0, e_1, p, s and non-atoms f_0, $f_1 < 1$ with the properties of Lemma 4.2 (taking $q = 1$) and an additional element s with the following property for each $n \in \omega$:

$$n \in U \iff g_n \vee s = 1.$$

Remark 4.5. *Historically, the technique of enumerating the g_n $\mathbf{a}'$-recursively was first done in [17]. The idea of the improvement can be seen in [16, Lemma 1.11]. The Slaman–Woodin conditions used to combine these ideas to get the above lemma were presented in [13] with a proof appearing in [14, Lemma 2.13(i)]. The construction of $L(U)$ was presented to the author by Slaman; see also [13, Theorem 3.7].*

Remark 4.6. *In a November 2008 seminar at the University of Hawai'i we worked out some details for the proof that such a lattice $L(U)$ exists. We can make $L(U)$ a height-three lattice, i.e., every element is either 0, 1, an atom or a co-atom. The atoms are e_0, e_1, s, and the g_i. The element p may be either an atom or a co-atom, and is incomparable with all other elements except that $0 \leq p \leq 1$. The co-atoms are $e_0 \vee g_{2n+1}$, and $e_1 \vee g_{2n}$, f_0, f_1, and $g_i \vee s$ whenever $i \notin U$. These elements are incomparable except as forced by the above conditions. The point of including p and q is that $y \vee p \geq q$ is a positive statement that implies $y \nleq 0$.*

The following lemma will have many applications:

Lemma 4.7. *Let $U \subseteq \omega$.*

1. *$L(U)$ has degree $\mathbf{u}$.*
2. *If $L(U)$ is $\Sigma_1^0(\mathbf{b})$-presentable, then $U \in \Sigma_1^0(\mathbf{b})$.*
3. *If $U \in \Sigma_1^0(\mathbf{b})$ then $L(U)$ is $\Sigma_1^0(\mathbf{b})$-presentable.*

Proof. 1. The definition of $L(U)$ appeals to an oracle of degree $\mathbf{u}$ only and so $L(U)$ is $\mathbf{u}$-presentable. Suppose $L(U)$ is presented with degree $\mathbf{v}$. By Lemma 4.2, the relation $y = g_i$ is recursive in $\mathbf{v}$. Now $i \in U \leftrightarrow g_i \vee s \geq 1$, so since $\vee$ and $\geq$ are recursive in $\mathbf{v}$, $\mathbf{u} \leq \mathbf{v}$.

2. We have

$$n \in U \Leftrightarrow \exists x(x = g_n \ \& \ x \vee s \geq 1) \Leftrightarrow \forall x(x = g_n \rightarrow x \vee s \geq 1).$$

By Lemma 4.2, U is of the form $\exists x(\triangle_1^0(\mathbf{b}) \ \& \ \Sigma_1^0(\mathbf{b}))$ $U \in \Sigma_1^0(\mathbf{b})$.

3. Immediate from the fact that all clauses of the definition of $L(U)$ except "$n \in U \leftrightarrow g_n \vee s \geq 1$" are recursive. □

Proposition 4.8. *If each $\Sigma_4^0(\mathbf{x})$-presentable bounded usl is $\Sigma_4^0(\mathbf{y})$-presentable, then $\mathbf{x}^{(3)} \leq_T \mathbf{y}^{(3)}$.*

Proof. Let $\mathbf{b} = \mathbf{x}^{(3)}$. Since $L(B \oplus \overline{B})$ is $\Sigma_1^0(B)$-presentable, it is $\Sigma_1^0(\mathbf{y}^{(3)})$-presentable. Thus by Lemma 4.7(2), $B \oplus \overline{B}$ is $\Sigma_1^0(\mathbf{y}^{(3)})$ and hence $B \leq_T \mathbf{y}^{(3)}$. □

5 Proving the Main Result

Definition 5.1. *In the tt-degrees we denote the order by $\leq$. If $\mathbf{x}$, $\mathbf{y}$ are tt-degrees, we say that $\mathbf{x} \equiv_T \mathbf{y}$ if for some $X \in \mathbf{x}$ and $Y \in \mathbf{y}$, we have $X \equiv_T Y$.*

Theorem 5.2 (Mohrherr [12]). *Let $n \geq 1$ and $\mathbf{a} \geq \mathbf{0}^{(n)}$. Then for some $\mathbf{b}$, $\mathbf{a} = \mathbf{b}^{(n)}$.*

Definition 5.3 ([6]). *Suppose $\lesssim$ is a preorder (transitive and reflexive binary relation) on a countable set L and $\vee^*$ is a binary operation on L. Define an equivalence relation $\approx$ by*

$$a \approx b \iff a \lesssim b \text{ and } b \lesssim a.$$

Let $L/\approx$ be the set of $\approx$-equivalence classes.

Let $(\lesssim/\approx)$ and $(\vee^*/\approx)$ be the relation on $L/\approx$ induced by $\lesssim$ and the operation on $L/\approx$ induced by $\vee^*$, respectively.

Assume $L/\approx = \langle L/\approx, \lesssim/\approx, \vee^*/\approx \rangle$ is an upper semilattice. Assume that $\lesssim$ is $\Sigma_1^0(\mathbf{a})$ (so $\approx$ is $\Sigma_1^0(\mathbf{a})$ too) and $\vee^*$ is $\Delta_1^0(\mathbf{a})$, where $\mathbf{a}$ is a Turing degree.

If such $\lesssim$, $\vee^*$ exist then (the upper semilattice isomorphism type of) $L/\approx$ is called $\Sigma_1^0(\mathbf{a})$-presentable.

Lemma 5.4. *$[\mathbf{a}, \mathbf{b}]$ is $\Sigma_3^0(\mathbf{b})$-presentable, for any Turing degrees $\mathbf{a} \leq \mathbf{b}$.*

Proof. Let $B \in \mathbf{b}$, $A \in \mathbf{a}$, choose e such that $A = \{e\}^B$, and let

$$C = \{i \mid \{i\}^B \text{ is total and } \{e\}^B \leq_T \{i\}^B\}.$$

The set C is $\Sigma_3^0(B)$ by a standard argument, so $C = \{h(n) \mid n < \omega\}$ for some injective $h \leq_T B''$.

Let $\lesssim$ be the binary relation on ω given by $i \lesssim j \leftrightarrow \{h(i)\}^B \leq_T \{h(j)\}^B$. Recall the function $\oplus : 2^\omega \times 2^\omega \to 2^\omega$ defined by $A \oplus B = \{2x \mid x \in A\} \cup \{2x+1 \mid x \in B\}$. Let $\oplus : \omega \times \omega \to \omega$ be a total recursive function such that for each $X \subseteq \omega$ and $a, b < \omega$, if $\{a\}^X$, $\{b\}^X$ are both total and $\{a\}^X, \{b\}^X \subseteq \omega$ then $\{a \oplus b\}^X = \{a\}^X \oplus \{b\}^X$. Let $i \vee^* j = h^{-1}(h(i) \oplus h(j))$.

It is easily verified that $\lesssim$ is $\Sigma_3^0(B)$ and $\vee^* : \omega \times \omega \to \omega$ is B''-recursive, and $\lesssim$ and $\vee^*$ have the required properties. $\square$

Proposition 5.5. *For each $\mathbf{g}$, $[\mathbf{0}, \mathbf{g}]$ is $\Sigma_3^0(\mathbf{g})$-presentable.*

Proof. An analysis of the definition of tt-reducibility similar to Lemma 5.4. $\square$

Corollary 5.6. *Each upper semilattice with least and greatest element that can be realized as an initial segment $[\mathbf{0}, \mathbf{g}]$ with $\mathbf{g}^{(2)} \leq \mathbf{y}^{(3)}$ is $\Sigma_4^0(\mathbf{y})$-presentable.*

Theorem 5.7. *For any y, the upper semilattices with least and greatest element that can be realized as initial segments $[\mathbf{0}, \mathbf{g}]$ with $\mathbf{g}^{(2)} \leq \mathbf{y}^{(3)}$ are exactly the $\Sigma_4^0(\mathbf{y})$-presentable ones.*

Proof. This is immediate from Corollary 5.6 and Theorem 3.4. $\square$

Theorem 5.8. *Let π be an automorphism of the truth-table degrees with jump and let $\mathbf{x} \geq \mathbf{0}^{(3)}$. Then $\pi(\mathbf{x}) \equiv_T \mathbf{x}$.*

Proof. By Theorem 5.2 there is a $\mathbf{y}$ such that $\mathbf{x} = \mathbf{y}^{(3)}$. The initial segments $[\mathbf{0}, \mathbf{y}']$ and $[\mathbf{0}, \pi(\mathbf{y}')]$ are jump-isomorphic via π, so by Theorem 5.7, the $\Sigma_4^0(\mathbf{y})$- and $\Sigma_4^0(\pi(\mathbf{y}))$-presentable bounded usls coincide. Hence by Proposition 4.8,

$$\pi(\mathbf{y})^{(3)} \equiv_T \mathbf{y}^{(3)}$$

and so

$$\pi(\mathbf{x}) = \pi(\mathbf{y}^{(3)}) = \pi(\mathbf{y})^{(3)} \equiv_T \mathbf{y}^{(3)} = \mathbf{x}. \qquad \square$$

Lemma 5.9. $\mathbf{a} \equiv_T \mathbf{b} \Rightarrow \mathbf{a}' = \mathbf{b}'$.

Theorem 5.10. *Let π be an automorphism of the truth-table degrees with jump and let $\mathbf{x} \geq \mathbf{0}^{(4)}$. Then $\pi(\mathbf{x}) = \mathbf{x}$.*

Proof. By Theorem 5.2, there is a $\mathbf{y}$ such that $\mathbf{x} = \mathbf{y}^{(4)}$. Let $\mathbf{z} = \mathbf{y}^{(3)}$, so $\mathbf{x} = \mathbf{z}'$ and $\mathbf{z} \geq \mathbf{0}^{(3)}$. By Theorem 5.8, $\pi(\mathbf{z}) \equiv_T \mathbf{z}$ and by Lemma 5.9, $\mathbf{a} \equiv_T \mathbf{b} \Rightarrow \mathbf{a}' = \mathbf{b}'$. Hence

$$\pi(\mathbf{x}) = \pi(\mathbf{z}') = \pi(\mathbf{z})' = \mathbf{z}' = \mathbf{x}. \qquad \square$$

References

1. Anderson, B.A.: Automorphisms of the truth-table degrees are fixed on a cone. J. Symb. Log. **74**(2), 679–688 (2009)
2. Gierz, G., Hofmann, K.H., Keimel, K., Lawson, J.D., Mislove, M., Scott, D.S.: Continuous Lattices and Domains. Encyclopedia of Mathematics and its Applications, vol. 93. Cambridge University Press, Cambridge (2003)
3. Grätzer, G.: General Lattice Theory. Birkhäuser, Basel (2003). With appendices by Davey, B.A., Freese, R., Ganter, B., Greferath, M., Jipsen, P., Priestley, H.A., Rose, H., Schmidt, E.T., Schmidt, S.E., Wehrung, F., Wille, R., Reprint of the 1998 2nd edn. [MR1670580]
4. Grätzer, G.: Universal Algebra, 2nd edn. Springer, New York (2008). With appendices by Grätzer, Jónsson, B., Taylor, W., Quackenbush, R.W., Wenzel, G.H., Grätzer, Lampe, W.A
5. Jockusch Jr., C.G., Solovay, R.M.: Fixed points of jump preserving automorphisms of degrees. Israel J. Math. **26**(1), 91–94 (1977)
6. Kjos-Hanssen, B.: Lattice initial segments of the Turing degrees. ProQuest LLC, Ann Arbor, MI, 2002. Thesis (Ph.D.)-University of California, Berkeley (2002)
7. Kjos-Hanssen, B.: Local initial segments of the Turing degrees. Bull. Symb. Log. **9**(1), 26–36 (2003)
8. Lachlan, A.H., Lebeuf, R.: Countable initial segments of the degrees of unsolvability. J. Symb. Log. **41**(2), 289–300 (1976)
9. Lerman, M.: Degrees of Unsolvability: Local and Global Theory. Perspectives in Mathematical Logic. Springer, Berlin (1983)
10. Mal'cev, A.I.: On the general theory of algebraic systems. Mat. Sb. N.S. **35**(77), 3–20 (1954)
11. Mal'cev, A.I.: On the general theory of algebraic systems. Am. Math. Soc. Transl. **2**(27), 125–142 (1963)

12. Mohrherr, J.: Density of a final segment of the truth-table degrees. Pac. J. Math. **115**(2), 409–419 (1984)
13. Nies, A., Shore, R.A., Slaman, T.A.: Definability in the recursively enumerable degrees. Bull. Symb. Log. **2**(4), 392–404 (1996)
14. Nies, A., Shore, R.A., Slaman, T.A.: Interpretability and definability in the recursively enumerable degrees. Proc. Lond. Math. Soc. **77**(2), 241–291 (1998)
15. Pudlák, P.: A new proof of the congruence lattice representation theorem. Algebra Univers. **6**(3), 269–275 (1976)
16. Shore, R.A.: The theory of the degrees below $0'$. J. Lond. Math. Soc. **24**(1), 1–14 (1981)
17. Shore, R.A.: On homogeneity and definability in the first-order theory of the Turing degrees. J. Symb. Log. **47**(1), 8–16 (1982)

Asymptotic Density and the Theory of Computability: A Partial Survey

Carl G. Jockusch Jr.$^{(\boxtimes)}$ and Paul E. Schupp

Department of Mathematics, University of Illinois at Urbana-Champaign,
1409 West Green Street, Urbana, IL 61801, USA
{jockusch,schupp}@math.uiuc.edu
http://www.math.uiuc.edu/~jockusch/

This paper is dedicated to Rod Downey in honor of his important contributions to computability theory.

Keywords: Asymptotic density · Generic computability · Coarse computability · Generic-case complexity

2010 Mathematics Subject Classification: 03D30 · 03D25 · 03D28 · 03D32 · 03D40

1 Introduction

The purpose of this paper is to survey recent work on how classical asymptotic density interacts with the theory of computability. We have tried to make the survey accessible to those who are not specialists in computability theory and we mainly state results without proof, but we include a few easy proofs to illustrate the flavor of the subject.

In complexity theory, classes such as $\mathcal{P}$ and $\mathcal{NP}$ are defined by using worst-case measures. That is, a problem belongs to the class if there is an algorithm solving it which has a suitable bound on its running time over *all* instances of the problem. Similarly, in computability theory, a problem is classified as computable if there is a single algorithm which solves all instances of the given problem.

There is now a general awareness that worst-case measures may not give a good picture of a particular algorithm or problem since hard instances may be very sparse. The paradigm case is Dantzig's Simplex Algorithm (see [6]) for linear programming problems. This algorithm runs many hundreds of times every day for scheduling and transportation problems, almost always very quickly. There are clever examples of Klee and Minty [21] showing that there exist instances for which the Simplex Algorithm must take exponential time, but such examples are not encountered in practice.

Observations of this type led to the development of *average-case complexity* by Gurevich [12] and by Levin [23] independently. There are different approaches

The authors would like to thank the Simons Foundation for its support.

© Springer International Publishing AG 2017
A. Day et al. (Eds.): Downey Festschrift, LNCS 10010, pp. 501–520, 2017.
DOI: 10.1007/978-3-319-50062-1_30

to the average-case complexity, but they all involve computing the expected value of the running time of an algorithm with respect to some measure on the set of inputs. Thus the problem must be decidable and one still needs to know the worst-case complexity.

Another example of hard instances being sparse is the behavior of algorithms for decision problems in group theory used in computer algebra packages. There is often some kind of an easy "fast check" algorithm which quickly produces a solution for "most" inputs of the problem. This is true even if the worst-case complexity of the particular problem is very high or the problem is even unsolvable. Thus many group-theoretic decision problems have a very large set of inputs where the (usually negative) answer can be obtained easily and quickly.

Such examples led Kapovich et al. [20] to introduce generic-case complexity as a complexity measure which is often more useful and easier to work with than either worst-case or average-case complexity. In generic-case complexity, one considers algorithms which answer correctly within a given time bound on a set of inputs of asymptotic density 1. They showed that many classical decision problems in group theory resemble the situation of the Simplex Algorithm in that hard instances are very rare. For example, consider the word problem for one-relator groups. In the 1930's Magnus (see [24]) showed that this problem is decidable but we still have no idea of the possible worst-case complexities over the whole class of one-relator groups. However, for *every* one-relator group with at least three generators, the word problem is generically linear time by Example 4.7 of [20]. Also, in the famous groups of Novikov [31] and Boone (see [33]) with undecidable word problem, the word problem has linear time generic-case complexity by Example 4.6 of [20].

Although it focused on complexity, the paper [20] introduced a general definition of generic computability in Sect. 9.

Let Σ be a nonempty finite alphabet and let Σ^* denote the set of all finite words on Σ. The *length*, $|w|$, of a word w is the number of letters in w. Let S be a subset of Σ^*. For every $n \geq 0$ let $S{\upharpoonright}n$ denote the set of all words in S of length less than or equal to n. In this situation we can copy the classical definition of asymptotic density from number theory.

Definition 1.1. For every $n \geq 0$, the *density of S up to n* is

$$\rho_n(S) = \frac{|S{\upharpoonright}n|}{|\Sigma^*{\upharpoonright}n|}$$

The *density* of S is

$$\rho(S) = \lim_{n \to \infty} \rho_n(S)$$

if this limit exists.

Definition 1.2. Let $S \subseteq \Sigma^*$. We say that S is *generic* if $\rho(S) = 1$ and S is *negligible* if $\rho(S) = 0$.

It is clear that S is generic if and only if its complement $\overline{S} = \Sigma^* \backslash S$ is negligible. Also, the intersection (union) of finitely many generic (negligible) sets is generic (negligible). This notion of genericity should not be confused with notions of genericity from forcing in computability theory and set theory. The latter are related to Baire category rather than density.

Definition 1.3 ([20]). Let S be a subset of Σ^* with characteristic function χ_S. A set S is *generically computable* if there exists a *partial computable function* φ such that $\varphi(x) = \chi_S(x)$ whenever $\varphi(x)$ is defined (written $\varphi(x) \downarrow$) and the domain of φ is generic in Σ^*.

We stress that *all* answers given by φ must be correct even though φ need not be everywhere defined, and, indeed, we do not require the domain of φ to be computable. In studying complexity we can clock the partial algorithm and consider it as not answering if it does not answer within the allotted amount of time.

To illustrate that even undecidable problems may be generically easy, we consider the *Post Correspondence Problem* (PCP). Fix a finite alphabet Σ of size $k \geq 2$. A typical instance of the problem consists of a finite sequence of pairs of words $(u_1, v_1), (u_2, v_2), \ldots, (u_n, v_n)$, where $u_i, v_i \in \Sigma^*$ for $1 \leq i \leq n$. The problem is to determine whether or not there is a finite nonempty sequence of indices $i_1, i_2, \ldots, i_k$ such that

$$u_{i_1} u_{i_2} \ldots u_{i_k} = v_{i_1} v_{i_2} \ldots v_{i_k}$$

holds.

In other words, can finitely many u's be concatenated to give the same word as the corresponding concatenation of v's? Emil Post proved in 1946 [32] that this problem is unsolvable for each alphabet Σ of size at least 2 and this result has been used to show that many other problems are unsolvable. Our exposition of a fast generic algorithm for the PCP follows the book [29] by Myasnikov, Shpilrain, and Ushakov.

The generic algorithm works as follows. Say that two words u and v are *comparable* if either is a prefix of the other. Given an instance $(u_1, v_1), (u_2, v_2), \ldots, (u_n, v_n)$ of the PCP determine whether or not u_i and v_i are comparable for some i between 1 and n. If not, output "no". Otherwise, give no output.

If the given instance has a solution $u_{i_1} \ldots u_{i_n} = v_{i_1} \ldots v_{i_n}$, then u_{i_1} and v_{i_1} must be comparable. Hence the above algorithm never gives a wrong answer.

We now show that the algorithm gives an answer with density 1 on the natural stratification of instances of the problem. Let I_s be the set of instances $(u_1, v_1), (u_2, v_2), \ldots, (u_n, v_n)$ where $n \leq s$ and each word u_i, v_i has length at most s. Each I_s is finite, each $I_j \subseteq I_{j+1}$ and every instance of the PCP belongs to some I_s. Let D_s be the set of instances in I_s for which the algorithm gives an output.

Claim 1.4. $\lim_s \frac{|D_s|}{|I_s|} = 1$

Proof. Put the uniform measure on I_s and let an element $(u_1, v_1), (u_2, v_2), \ldots,$ (u_n, v_n) of I_s be chosen uniformly at random. To prove the claim, we show that the probability that the algorithm diverges on a random element of I_s approaches 0 as s approaches infinity.

For any fixed values of $v_1, u_2, \ldots, v_n$ the conditional probability that u_1 is a prefix of v_1 is at most $\frac{s+1}{2^s}$ since there at least 2^s words on Σ of length s and at most $s + 1$ of these are prefixes of v_1.

Hence, the probability that u_1 is a prefix of v_1 is at most $\frac{s+1}{2^s}$, and the probability that some u_i is comparable with v_i is at most $\frac{2s(s+1)}{2^s}$. So the probability that the algorithm gives no answer on the given instance is at most $\frac{2s(s+1)}{2^s}$, which tends to 0 as s approaches infinity. $\qquad\qquad\square$

The generic algorithm we described works in quadratic time, so the generic-case complexity of the Post Correspondence Problem is at most quadratic time.

From now on we mainly consider subsets of the set $\mathbb{N} = \{0, 1, \ldots\}$ of natural numbers, which we identify with the set ω of finite ordinals, In terms of the preceding definitions, we are using the 1-element alphabet $\Sigma = \{1\}$ and identifying $n \in \omega$ with its unary representation $1^n \in \{1\}^*$. In this context, we are using classical asymptotic density. If $A \subseteq \mathbb{N}$, then, for $n \geq 1$, the *density of A below* n is

$$\rho_n(A) = \frac{|\{m \in A : m < n\}|}{n}$$

The *(asymptotic) density* $\rho(A)$ of A is $\lim_{n\to\infty} \rho_n(A)$ if this limit exists.

While the limit for density does not exist in general, the *upper density*

$$\overline{\rho}(A) = \limsup_n \{\rho_n(A)\}$$

and the *lower density*

$$\underline{\rho}(A) = \liminf_n \{\rho_n(A)\}$$

always exist.

We use φ_e for the unary partial function computed by the e-th Turing machine. Let W_e be the domain of φ_e. We identify a set $A \subseteq \omega$ with its characteristic function χ_A.

First observe that *every* Turing degree contains a generically computable set. Let $A \subseteq \mathbb{N}$. Let $C(A) = \{2^n : n \in A\}$. Then $C(A)$ is generically computable since the set of powers of 2 is computable and has density 0. All the information about A is in a set of density 0. When given m, the generic algorithm checks if m is a power of 2. If not, the algorithm answers $m \notin C(A)$ and otherwise does not answer. This example shows that one partial algorithm can generically compute uncountably many different sets.

The following sets R_k are extremely useful.

Definition 1.5 ([19], Definition 2.5).

$$R_k = \{m : 2^k | m, 2^{(k+1)} \nmid m\}.$$

For example, R_0 is the set of odd nonnegative integers. Note that $\rho(R_k) = 2^{-(k+1)}$. The collection of sets $\{R_k\}$ forms a partition of $\omega - \{0\}$ since these sets are pairwise disjoint and $\bigcup_{k=0}^{\infty} R_k = \omega - \{0\}$.

From the definition of asymptotic density it is clear that we have *finite additivity* for densities. Of course we do not have countable additivity for densities in general, since ω is a countable union of singletons. However, we do have countable additivity in the situation where the "tails" of a sequence contribute vanishingly small density to the union of a sequence of sets.

Lemma 1.6 *([19], Lemma 2.6, Restricted countable additivity).* *If $\{S_i\}, i = 0, 1, \ldots$ is a countable collection of pairwise disjoint subsets of ω such that each $\rho(S_i)$ exists and $\overline{\rho}(\bigcup_{i=N}^{\infty} S_i) \to 0$ as $N \to \infty$, then*

$$\rho(\bigcup_{i=0}^{\infty} S_i) = \sum_{i=0}^{\infty} \rho(S_i).$$

Definition 1.7 ([19], Definition 2.7). If $A \subseteq \omega$ then $\mathcal{R}(A) = \bigcup_{n \in A} R_n$.

Our sequence $\{R_n\}$ satisfies the hypotheses of Lemma 1.6, so we have the following corollary.

Corollary 1.8 *([19], Corollary 2.8).* $\rho(\mathcal{R}(A)) = \sum_{n \in A} 2^{-(n+1)}$.

This gives an explicit construction of sets with pre-assigned densities. and shows that every real number $r \in [0, 1]$ is a density.

Proposition 1.9 *([19], Observation 2.11). Every nonzero Turing degree contains a set which is not generically computable since the set $\mathcal{R}(A)$ is generically computable if and only if A is computable.*

Proof. It is clear that $\mathcal{R}(A)$ is Turing equivalent to A. If $\mathcal{R}(A)$ is generically computable by a partial algorithm φ, to compute $A(n)$ search for $k \in R_n$ with $\varphi(k) \downarrow$ and output $\varphi(k)$. Since R_n has positive density, this procedure must eventually answer, and the answer is correct because φ never gives a wrong answer. $\square$

Recall that a set A is *immune* if A is infinite and A does not have any infinite c.e. subset and A is *bi-immune* if both A and its complement $\overline{A}$ are immune. It is clear that no bi-immune set can be generically computable.

Now the class of bi-immune sets is both comeager and of measure 1. This is clear by countable additivity since the family of sets containing a given infinite set is of measure 0 and nowhere dense. Thus the family of generically computable sets is both meager and of measure 0.

There are numerous interactions between the area of this paper and effective randomness. For information on the latter see, for example, [7].

2   Densities and C.E. Sets

Observe that a set A is generically computable if and only if there exist c.e. sets $B \subseteq A$ and $C \subseteq \overline{A}$ such that $B \cup C$ has density 1. In particular, every c.e. set of density 1 is generically computable. This suggests the question of how well c.e. sets can be approximated by computable subsets in general. The following definition gives two ways to measure how good an approximation is.

Definition 2.1 ([8], Definition 3.1). Let $A, B \subseteq \omega$.

(i) Define $d(A, B) = \underline{\rho}(A \triangle B)$, the lower density of the symmetric difference of A and B.
(ii) Define $D(A, B) = \overline{\rho}(A \triangle B)$, the upper density of the symmetric difference of A and B.

To our knowledge the first result on approximating c.e. sets by computable subsets is a result of Barzdin' [3] from 1970 showing that for every c.e. set A and every real number $\epsilon > 0$, there is a computable set $B \subseteq A$ such that $d(A, B) < \epsilon$. We thank Evgeny Gordon for bringing this result to our attention. The following result of Downey, Jockusch, and Schupp improves Barzdin's result from d to D.

Theorem 2.2 ([8], Corollary 3.10). *For every c.e. set A and real number $\epsilon > 0$, there is a computable set $B \subseteq A$ such that $D(A, B) < \epsilon$.*

Jockusch and Schupp ([19], Theorem 2.22) showed that there is a c.e. set of density 1 which does not have any computable subset of density 1. It turns out that this property characterizes an important class of c.e. degrees, where a c.e. degree is one which contains a c.e. set. Recall that if $\mathbf{a}$ is a Turing degree with $A \in \mathbf{a}$, then the *jump* of $\mathbf{a}$, denoted $\mathbf{a}'$, is the Turing degree of the halting problem for machines with an oracle for A. If $\mathbf{a}$ is a c.e. degree then $\mathbf{0}' \leq \mathbf{a}' \leq \mathbf{0}''$. A degree $\mathbf{a}$ is *low* if $\mathbf{a}' = \mathbf{0}'$, that is, $\mathbf{a}'$ is as low as possible. A degree $\mathbf{a}$ is *high* if $\mathbf{a}' \geq \mathbf{0}''$.

Downey et al. [8] proved the following characterization of non-low c.e. degrees.

Theorem 2.3 ([8], Corollary 4.4). *Let $\mathbf{a}$ be a c.e. degree. Then $\mathbf{a}$ is not low if and only if $\mathbf{a}$ contains a c.e. set A of density 1 with no computable subset of density 1.*

With Eric Astor they also proved the following result.

Theorem 2.4 ([8], Corollary 4.2). *There is a c.e. set A of density 1 such that the degrees of subsets of A of density 1 are exactly the high degrees.*

One of the striking things to emerge from considering density and computability is that there is a very tight connection between the positions of sets in the arithmetical hierarchy and the complexity of their densities as real numbers.

Fix a computable bijection between the rationals and $\mathbb{N}$, so we can classify sets of rationals in the arithmetical hierarchy.

Definition 2.5. Define a real number r to be *left-Σ_n^0* if its corresponding lower cut in the rationals, $\{q \in \mathbb{Q} : q < r\}$, is Σ_n^0. Define "left-Π_n^0" analogously.

Jockusch and Schupp [19] proved that a real number $r \in [0, 1]$ is the density of a computable set if and only if r is a Δ_2^0 real. Downey et al. [8] carried this much further and proved the following results.

Theorem 2.6 *([19], Theorem 2.21, [8] Corollary 5.4, Theorems 5.6, 5.7, and 5.13). Let r be a real number in the interval $[0,1]$ and suppose that $n \geq 1$. Then the following hold:*

(i) r is the density of some set in Δ_n^0 if and only if r is left-Δ_{n+1}^0.
(ii) r is the lower density of some set in Δ_n^0 if and only if r is left-Σ_{n+1}^0.
(iii) r is the upper density of some set in Δ_n^0 if and only if r is left-Π_{n+1}^0.
(iv) r is the lower density of some set in Σ_n^0 if and only if r is left-Σ_{n+2}^0.
(v) r is the upper density of some set in Σ_n^0 if and only if r is left-Π_{n+1}^0.
(vi) r is the density of some set in Σ_n^0 if and only if r is left-Π_{n+1}^0.

This result follows by relativization from characterizing the densities, upper densities, and lower densities of the computable and c.e. sets.

2.1 Asymptotic Density and the Ershov Hierarchy

The correlation of densities and position in the arithmetical hierarchy is further clarified by considering densities of sets in the Ershov Hierarchy. The Shoenfield Limit Lemma shows that a set A is Δ_2^0 exactly if there is a computable function g such that for all x, $A(x) = lim_s g(x, s)$. Roughly speaking, the Ershov Hierarchy classifies Δ_2^0 sets by the number of s with $g(x, s) \neq g(x, s + 1)$. A set A is n-c.e. if there exists a computable function g as above such that, for all x, $g(x, 0) = 0$ and there are at most n values of s such that $g(x, s) \neq g(x, s + 1)$.

The 1-c.e. sets are just the c.e. sets. The 2-c.e. sets, also called the d.c.e. sets, are sets which are the differences of two c.e. sets. Since the densities of c.e. sets are precisely the left-Π_2^0 reals in the unit interval, one is led to suspect that the densities of the 2-c.e. sets should be exactly the differences of two left-Π_2^0 reals which are in the unit interval. This is true but there is something to prove since the difference of A and B may have a density even though A and B do not have densities. Let $\mathcal{D}_2$ denote the set of reals which are the difference of two left Π_2^0 reals. Downey et al. [9] proved the following results.

Theorem 2.7 *([9], Corollary 4.3). For every $n \geq 2$, the densities of the n-c.e. sets coincide with the reals in $\mathcal{D}_2 \cap [0, 1]$.*

It follows that there is a real r which is the density of a 2-c.e. set but not of any c.e. or co-c.e. set.

Say that a Δ_2^0 set A is *f-c.e.* if there is a computable function g such that, for all x, $g(x, 0) = 0$, $A(x) = \lim_s g(x, s)$, and $|\{s : g(x, s) \neq g(x, s+1)\}| \leq f(x)$.

Theorem 2.8 *([9], Corollary 5.2). Let f be any computable, nondecreasing, unbounded function. If A is a Δ_2^0 set that has a density, then the density of A is the same as the density of a set B such that B is f-c.e.*

2.2 Bi-immunity and Absolute Undecidability

If A is bi-immune then any c.e. set contained in either A or $\overline{A}$ is finite so being bi-immune is an extreme non-computability condition. Jockusch [18] proved that there are nonzero Turing degrees which do not contain any bi-immune sets. This raises the natural question of how strong a non-computability condition can be pushed into every non-zero degree. Miasnikov and Rybalov [28] defined a set A to be *absolutely undecidable* if every partial computable function which agrees with A on its domain has a domain of density 0. We might suggest the term *densely undecidable* as a synonym for "absolutely undecidable", since being absolutely undecidable is a weaker condition than being bi-immune. The following beautiful and surprising result is due to Bienvenu et al. [4].

Theorem 2.9 *([4]). Every nonzero Turing degree contains an absolutely undecidable set.*

The theorem was proved using the Hadamard error-correcting code, which the authors of [4] rediscovered to prove the result.

3 Coarse Computability

The following definitions suggest another quite reasonable concept of "imperfect computability".

Definition 3.1 ([19], Definition 2.12). Two sets A and B are *coarsely similar*, which we denote by $A \sim_c B$, if their symmetric difference $A \triangle B = (A \setminus B) \cup (B \setminus A)$ has density 0. If B is any set coarsely similar to A then B is called a *coarse description* of A.

It is easy to check that $\sim_c$ is an equivalence relation. Any set of density 1 is coarsely similar to ω, and any set of density 0 is coarsely similar to $\emptyset$.

Definition 3.2 ([19], Definition 2.13). A set A is *coarsely computable* if A is coarsely similar to a computable set. That is, A has a computable coarse description.

We can think of coarse computability in the following way: The set A is coarsely computable if there exists a *total* algorithm φ which may make mistakes on membership in A but the mistakes occur only on a negligible set. A generic algorithm is always correct when it answers and almost always answers, while a coarse algorithm always answers and is almost always correct. Note that all sets of density 1 or of density 0 are coarsely computable.

Using the Golod-Shafarevich inequality, Miasnikov and Osin [27] constructed finitely generated, computably presented groups whose word problems are not generically computable. Whether or not there exist finitely presented groups whose word problem is not generically computable is a difficult open question. The situation for coarse computability is very different.

Observation 3.3 ([19], Observation 2.14). The word problem of any finitely generated group $G = \langle X : R \rangle$ is coarsely computable.

Proof. If G is finite then the word problem is computable. If G is an infinite group, the set of words on $X \cup X^{-1}$ which are not equal to the identity in G has density 1 and hence is coarsely computable. (See, for example, [35].)   □

It is easy to check that the family of coarsely computable sets is meager and of measure 0. In fact, if A is coarsely computable, then A is neither 1-generic nor 1-random. This is a consequence of the fact that if A is 1-random and C is computable, then the symmetric difference $A \triangle C$ is also 1-random, and the analogous fact also holds for 1-genericity. The result now follows because 1-random sets have density 1/2 [30], and 1-generic sets have upper density 1.

Proposition 3.4 *([19], Proposition 2.15). There is a c.e. set which is coarsely computable but not generically computable.*

Proof. Recall that a c.e. set A is *simple* if $\overline{A}$ is immune. It suffices to construct a simple set A of density 0, since any such set is coarsely computable but not generically computable. This is done by a slight modification of Post's simple set construction. Namely, for each e, enumerate W_e until, if ever, a number $>e^2$ appears, and put the first such number into A. Then A is simple, and A has density 0 because for each e, it has at most e elements less than e^2.   □

The following construction shows that c.e. sets may be neither generically nor coarsely computable.

Theorem 3.5 *([19], Theorem 2.16). There exists a c.e. set which is not coarsely similar to any co-c.e. set and hence is neither coarsely computable nor generically computable.*

Proof. Let $\{W_e\}$ be a standard enumeration of all c.e. sets. Let

$$A = \bigcup_{e \in \omega} (W_e \cap R_e)$$

Clearly, A is c.e. We first claim that A is not coarsely similar to any co-c.e. set and hence is not coarsely computable. Note that

$$R_e \subseteq A \triangle \overline{W_e}$$

since if $n \in R_e$ and $n \in A$, then $n \in (A \setminus \overline{W_e})$, while if $n \in R_e$ and $n \notin A$, then $n \in (\overline{W_e} \setminus A)$. So, for all e, $(A \triangle \overline{W_e})$ has positive lower density, and hence A is not coarsely similar to $\overline{W_e}$. It follows that A is not coarsely computable. Of course, this construction is simply a diagonal argument, but instead of using a single witness for each requirement, we use a set of witnesses of positive density.

Suppose now for a contradiction that A were generically computable. Let W_a, W_b be c.e. sets such that $W_a \subseteq A$, $W_b \subseteq \overline{A}$, and $W_a \cup W_b$ has density 1. Then A would be generically similar to $\overline{W_b}$ since

$$A \triangle \overline{W_b} \subseteq \overline{W_a \cup W_b}$$

and $\overline{W_a \cup W_b}$ has density 0. This shows that A is not generically computable. □

We introduce the following construction which will be used repeatedly.

Definition 3.6 ([19]). Let $\mathcal{I}_0 = \{0\}$ and for $n > 0$ let I_n be the interval $[n!, (n+1)!)$. For $A \subseteq \omega$, let $\mathcal{I}(A) = \cup_{n \in A} I_n$.

Theorem 3.7 ([19], *proof of Theorem 2.20*). *For all A, the set $\mathcal{I}(A)$ is coarsely computable if and only if A is computable.*

Proof It is clear that $\mathcal{I}(A) \equiv_T A$, so it suffices to show that if A is not computable then $\mathcal{I}(A)$ is not coarsely computable. If $\mathcal{I}(A)$ is coarsely computable, we can choose a computable set C such that $\rho(C \triangle \mathcal{I}(A)) = 0$. The idea is now that we can show that A is computable by using "majority vote" to read off from C a set D which differs only finitely from A. Specifically, define

$$D = \{n : |I_n \cap C| > (1/2)|I_n|\}.$$

Then D is a computable set and we claim that $A \triangle D$ is finite. To prove the claim, assume for a contradiction that $A \triangle D$ is infinite. If $n \in A \triangle D$, then more than half of the elements of I_n are in $C \triangle \mathcal{I}(A)$. It follows that, for $n \in A \triangle D$,

$$\rho_{(n+1)!}(C \triangle \mathcal{I}(A)) \geq \frac{1}{2} \frac{|I_n|}{(n+1)!} = \frac{1}{2} \frac{(n+1)! - n!}{(n+1)!} = \frac{1}{2}(1 - \frac{1}{n+1}).$$

As the above inequality holds for infinitely many n, it follows that $\overline{\rho}(C \triangle \mathcal{I}(A)) \geq 1/2$, in contradiction to our assumption that $\rho(C \triangle \mathcal{I}(A)) = 0$. It follows that $A \triangle D$ is finite and hence A is computable. □

A similar argument shows that if A is not computable then $\mathcal{I}(A)$ is also not generically computable. We thus have the following result.

Theorem 3.8 ([19], *Theorem 2.20*). *Every nonzero Turing degree contains a set which is neither coarsely computable nor generically computable.*

Since $\mathcal{R}(A)$ is generically computable if and only if A is computable, it seems natural to ask about the coarse computability of $\mathcal{R}(A)$. Post's Theorem shows that the sets Turing reducible to $0'$ are precisely the sets which are Δ_2^0 in the arithmetical hierarchy. Using the limit lemma one can prove the following result.

Theorem 3.9 ([19], *Theorem 2.19*). *For all A, the set $\mathcal{R}(A) = \cup_{n \in A} R_n$ is coarsely computable if and only if $A \leq_T 0'$.*

In particular, if A is any noncomputable set Turing reducible to $0'$ then $\mathcal{R}(A)$ is coarsely computable but not generically computable.

4   Computability at Densities Less Than 1

Generic and coarse computability are computabilites at density 1. Downey et al. [8] took the natural step of considering computability at densities less than 1.

Definition 4.1 ([8], Definition 5.9). If $r \in [0,1]$, a set A is *partially computable at density* r if there exists a partial computable function φ agreeing with $A(n)$ whenever $\varphi(n) \downarrow$ and with the lower density of domain(φ) greater than or equal to r.

A natural first question is: Are there sets which are computable at all densities $r < 1$ but are not generically computable? Actually, we have already seen that every nonzero Turing degree contains such sets. Any set of the form $\mathcal{R}(A)$ is partially computable at all densities less than 1, as Asher Kach observed. Note that for any $t \geq 0$, the set $\bigcup R_k$ where $k \leq t$ and $k \in A$ is a computable set whose symmetric difference with $\mathcal{R}(A)$ is contained in $\bigcup \{R_k : k > t\}$, and the latter set has density 2^{-t-1}. Furthermore, $\mathcal{R}(A)$ is generically computable if and only if A is computable.

This "approachability" phenomenon holds very generally.

Definition 4.2 ([8], Definition 6.9). If $A \subseteq \omega$, the *partial computability bound* of A is
$\alpha(A) := sup\{r : A \text{ is computable at density } r\}$.

Theorem 4.3 ([8], *Theorem 6.10*). *If* $r \in [0,1]$, *then there is a set* A *of density* r *with* $\alpha(A) = r$.

Proof. Let $.b_0 b_1...$ be the binary expansion of r. By Corollary 1.8 the set $D = \bigcup_{b_i=1} R_i$ has density r. We let $A = D \cup S$ where S is a simple set of density 0 (Proposition 3.4). If $s < r$ we can take enough digits of the expansion of r so that if $t = .b_1 \ldots b_n$ then $s < t < r$. The set C which is the union of the R_j where $j \leq n, b_j = 1$ is a computable subset of A of density t so A is computable at density t. Since we can take t arbitrarily close to r, it follows that $\alpha(A) \geq r$. To show that $\alpha(A) \leq r$, assume that φ is a computable partial function which agrees with A on its domain W. We must show that $\underline{\rho}(W) \leq r$. For $i \in \{0,1\}$, let $T_i = \{n : \varphi(n) = i\}$, so $W = T_0 \cup T_1$. Then T_0 is c.e. and $T_0 \subseteq \overline{A} \subseteq \overline{S}$, so T_0 is finite because S is simple. Also $T_1 \subseteq A$, so $\underline{\rho}(T_1) \leq \underline{\rho}(A) = r$, so $\underline{\rho}(W) \leq r$, as needed to complete the proof.   □

In analogy with partial computability at densities less than 1, Hirschfeldt et al. [15] introduced the analogous concepts for coarse computability. We define

$$A \triangledown C = \{n : A(n) = C(n)\}$$

and call $A \triangledown C$ the *symmetric agreement* of A and C. Of course, the symmetric agreement of A and C is the complement of the symmetric difference of A and C.

Definition 4.4 ([15], Definition 1.5). A set A is *coarsely computable at density* r if there is a computable set C such that the lower density of the symmetric agreement of A and C is at least r, that is

$$\underline{\rho}(A \triangledown C) \geq r$$

Definition 4.5 ([15], Definition 1.6). If $A \subseteq \mathbb{N}$, the *coarse computability bound* of A is

$$\gamma(A) := \sup\{r : A \text{ is coarsely computable at density } r\}$$

Proposition 4.6 *([15], Lemma 1.7). For every set A, $\alpha(A) \leq \gamma(A)$.*

This result follows easily from Theorem 2.2.

The next result is due to Greg Igusa and shows that this is the *only* restriction on the values taken simultaneously by α and γ.

Theorem 4.7 *(Igusa, personal communication). If r and s are real numbers with $0 \leq r \leq s \leq 1$, there is a set A such that $\alpha(A) = r$ and $\gamma(A) = s$.*

The coarse computability bound of every 1-random set A is $1/2$. This is because for every computable set C, the set $A \triangledown C$ is also 1-random and so has density $1/2$.

Recall that we defined the distance function $D(A, B) = \overline{\rho}(A \triangle B)$. It is easily seen that D satisfies the triangle inequality and hence is a pseudometric on Cantor space 2^ω. Since $D(A, B) = 0$ exactly when A and B are coarsely similar, D is actually a metric on the space $\mathcal{S}$ of coarse similarity classes.

Note that A is coarsely computable at density 1 if and only if A is coarsely computable. To exhibit many sets with $\gamma = 1$ which are not coarsely computable, again consider sets of the form $\mathcal{R}(A) = \bigcup_{n \in A} R_n$. Essentially the same argument as before shows that $\gamma(\mathcal{R}(A)) = 1$ for every A. For each k, use the finite list of which $i \leq k$ are in A, to answer correctly on $\bigcup_{i=0}^{k} R_i$ and answer "yes" on all R_l with $l > k$. This algorithm is correct with density at least $1 - \frac{1}{2^{k+1}}$.

Lemma 4.8 *([15]). For $A \subseteq \omega$, $\underline{\rho}(A) = 1 - \overline{\rho}(\overline{A})$*

For each n, $\rho_n(A) = 1 - \rho_n(\overline{A})$, so the lemma follows by taking the least upper bound of both sides. As a corollary we have

$$\underline{\rho}(A \triangledown C) = 1 - D(A, C).$$

So, $\gamma(A) = 1$ if and only if A is a limit of computable sets in the pseudo-metric D. In general, $\gamma(A) = r$ means that the distance from A to the family $\mathcal{C}$ of computable sets is $1 - r$.

Theorem 4.9 *([15], Theorems 3.1 and 3.4). For every $r \in (0, 1]$ there is a set A with $\gamma(A) = r$ such that A is not coarsely computable at density r, and a set B such that $\gamma(B) = r$ and B is coarsely computable at density r.*

We have seen that if A is not Δ_2^0 then $\mathcal{R}(A)$ is Turing equivalent to A, and $\gamma(\mathcal{R}(A)) = 1$, but $\mathcal{R}(A)$ is not coarsely computable. Also, every non-zero c.e. degree contains a c.e. set A which is generically computable but not coarsely computable ([8], Theorem 4.5). So the question is whether or not *every* nonzero Turing degree contains a set A such that $\gamma(A) = 1$ but A is not coarsely computable. The following result gives a negative answer. The proof includes a crucial lemma due to Joe Miller.

Theorem 4.10 *([15], Theorem 5.12). If A is computable from a Δ_2^0 1-generic set and $\gamma(A) = 1$, then A is coarsely computable.*

Theorem 4.11 *([15], Theorem 2.1). Every nonzero (c.e.) degree contains a (c.e.) set B such that $\alpha(B) = 0$ and $\gamma(B) = \frac{1}{2}$.*

Proof. Given A, let $B = \mathcal{I}(A)$. The majority vote argument about $\mathcal{I}(A)$ in the proof of Theorem 3.7 actually shows that if A is not computable then $\gamma(\mathcal{I}(A)) \leq \frac{1}{2}$. If E is the set of even numbers, then $E \triangledown \mathcal{I}(A)$ has density $1/2$, so $\gamma(\mathcal{I}(A)) \geq \frac{1}{2}$. Also, it is easily seen $\alpha(\mathcal{I}(A)) = 0$ if A is noncomputable. □

We observe that large classes of degrees contain sets A with $\gamma(A) = 0$.

A set $S \subseteq 2^{<\omega}$ of finite binary strings is *dense* if every string has some extension in S. Kurtz [22] defined a set A to be *weakly 1-generic* if A meets every dense c.e. set S of finite binary strings.

Theorem 4.12 *([15], proof of Theorem 2.1). If A is a weakly 1-generic set, then $\gamma(A) = 0$.*

Proof. If f is a computable function then, for each $n, j > 0$, define

$$S_{n,j} = \left\{ \sigma \in 2^{<\omega} : |\sigma| \geq j \ \& \ \rho_{|\sigma|}(\{k < |\sigma| : \sigma(k) = f(k)\}) < \frac{1}{n} \right\}.$$

Each set $S_{n,j}$ is computable and dense. A meets each $S_{n,j}$ since A is weakly 1-generic. Thus $\{k : f(k) = A(k)\}$ has lower density 0. □

Let D_n be the finite set with canonical index n, so $n = \sum\{2^i : i \in D\}$.

Recall that a set A is *hyperimmune* if A is infinite and there is no computable function f such that the sets $D_{f(0)}, D_{f(1)}, \ldots$ are pairwise disjoint and all intersect A, where D_n is the finite set with canonical index n. A degree $\mathbf{a}$ is called *hyperimmune* if it contains a hyperimmune set and otherwise *hyperimmune-free*. Kurtz [22] proved that the weakly 1-generic degrees coincide with the hyperimmune degrees. We thus have the following corollary.

Corollary 4.13 *([15], Theorem 2.2). Every hyperimmune degree contains a set A with $\gamma(A) = 0$.*

A degree $\mathbf{a}$ is called *PA* if every infinite computable tree of binary strings has an infinite $\mathbf{a}$-computable path.

Proposition 4.14 *([1], Proposition 1.8). If **a** is PA, then **a** contains a set A with $\gamma(A) = 0$.*

Proof. It is straightforward to construct an infinite computable tree T of binary strings such that the paths through T are exactly the sets X which, on every interval I_n, disagree with the partial computable function φ_n on all arguments where the latter is defined. Then an easy argument shows that $\gamma(X) = 0$ for every path X through T, and T has an **a**-computable path since **a** is PA. □

It is easily seen that $\alpha(\mathcal{I}(A)) = 0$ whenever A is noncomputable, and hence every nonzero degree contains a set B such that $\alpha(B) = 0$. In view of the preceding results on hyperimmune and PA degrees it is natural to ask whether *every* nonzero degree contains a set B such that $\gamma(B) = 0$.

This question is investigated and answered in the negative in Andrews et al. [1], where the following definition was introduced.

Definition 4.15 ([1]). If **d** is a Turing degree,

$$\Gamma(\mathbf{d}) = \inf\{\gamma(A) : A \leq_T \mathbf{d}\}$$

Recall that the majority vote argument shows that if A is any noncomputable set then $\gamma(\mathcal{I}(A)) \leq 1/2$. Therefore if a Turing degree has a Γ-value greater than $1/2$ then it is computable and so has Γ-value 1.

We call a function g a *trace* of a function f if $f(n) \in D_{g(n)}$ for every n.

Definition 4.16 (Terwijn and Zambella [34]). A set A is *computably traceable* if there is a computable function p with the property that every A-computable function f has a computable trace g such that $(\forall n)[|D_{g(n)}| \leq p(n)]$. (Note that p is independent of f.)

Theorem 4.17 *([1], Theorem 1.10). If A is computably traceable, then A is coarsely computable at density $\frac{1}{2}$.*

The proof is a probabilistic argument. Since the computably traceable sets are closed downwards under Turing reducibility, it follows easily that $\Gamma(\mathbf{a}) = \frac{1}{2}$ for every degree $\mathbf{a} > \mathbf{0}$ which contains a computably traceable set.

Theorem 4.18 *([1], Theorem 1.12). If A is a 1-random set of hyperimmune-free Turing degree and $B \leq_T A$, then B is coarsely computable at density $\frac{1}{2}$.*

In summary, we know the following.

- $\Gamma(\mathbf{0}) = 1$.
- If $\mathbf{a} > \mathbf{0}$, then $\Gamma(\mathbf{a}) \leq \frac{1}{2}$.
- If $\mathbf{a}$ is hyperimmune or PA, then $\Gamma(\mathbf{a}) = 0$.
- If $\mathbf{a}$ is computably traceable and nonzero, then $\Gamma(\mathbf{a}) = \frac{1}{2}$.
- If $\mathbf{a}$ is both 1-random and hyperimmune-free, then $\Gamma(\mathbf{a}) = \frac{1}{2}$.

The following question was raised in [1].

Question 4.19. What is the range of Γ? Does it equal $\{0, \frac{1}{2}, 1\}$?

Monin [25] has recently announced the remarkable result that $\Gamma(\mathbf{d})$ is equal to 0, 1/2 or 1 for every degree $\mathbf{d}$. Together with the results just above, this gives a positive answer to the second half of the above question, and thus a natural trichotomy of the Turing degrees according to their Γ-values. In contrast, Matthew Harrison-Trainor [13] has just announced that the range of the analogue for Γ for many-one degrees is $[0, 1/2] \cup \{1\}$.

Monin and Nies [26] have also recently extended and unified some of the above results on Γ using Schnorr randomness. In particular they showed the existence of degrees $\mathbf{a}$ with $\Gamma(\mathbf{a}) = \frac{1}{2}$ which are neither computably traceable nor 1-random. They also gave a new proof of Liang Yu's unpublished result that there are degree $\mathbf{a}$ with $\Gamma(\mathbf{a}) = 0$ such that $\mathbf{a}$ is neither hyperimmune nor PA.

5 Generic and Coarse Reducibility and Their Corresponding Degrees

One might first consider relative generic computability: That is, what sets are generically computable by Turing machines with a full oracle for a set A? Say that a set B is *generically A-computable* if there is a generic computation of B using a *full* oracle for A. It is easy to see that this notion is not transitive because we start with full information but compute only partial information. For example, let $A = \emptyset$ and let $B = \{2^n : n \in C\}$ where C is any set which is not generically computable. Then B is generically A-computable and C is generically B-computable, but C is not generically A-computable. The following is a remarkable and surprising result of Igusa [16] showing there are no minimal pairs for this non-transitive notion of relative generic computability.

Theorem 5.1 *([16], Theorem 2.1). For any noncomputable sets A and B there is a set C which is not generically computable but which is both generically A-computable and generically B-computable.*

Generic reducibility (denoted $\leq_g$) was introduced by Jockusch and Schupp [19] (Sect. 4), and we review the definition here. A *generic description* of a set A is a partial function θ which agrees with A on its domain and has a domain of density 1. Note that A is generically computable if and only if A has a partial computable generic description. The basic idea is then that $B \leq_g A$ if and only if there is an effective procedure which, from any generic description of A, computes a generic description of B. Since computing a partial function is tantamount to enumerating its graph, this is made precise using enumeration operators. These are similar to Turing reductions but use only *positive* oracle information and also output only positive information. An *enumeration operator* is a c.e. set W of pairs $\langle n, D \rangle$ where $n \in \omega$ and D is a finite subset of ω. (Here we identify finite sets with their canonical indices and pairs with their codes in saying that W is c.e. The membership of $\langle n, D \rangle$ in W means intuitively that from the positive information that D is a subset of the oracle, W computes that n belongs to the output.) Hence if W is an enumeration operator and $X \subseteq \omega$, define

$$W^X := \{n : (\exists D)[\langle n, D \rangle \in W \ \& \ D \subseteq X]\}$$

Note that from any enumeration of X one may effectively obtain an enumeration of W^X. If θ is a partial function, let $\gamma(\theta) = \{\langle a, b \rangle : \theta(a) = b\}$, so $\gamma(\theta)$ is a set of natural numbers coding the graph of θ. We can now state our formal definition of generic reducibility.

Definition 5.2. The set B is *generically reducible* to the set A (written $B \leq_g A$) if there is a fixed enumeration operator W such that, for every generic description θ of A, $W^{\gamma(\theta)} = \gamma(\delta)$ for some generic description δ of B.

This reducibility is also called "uniform generic reducibility" and denoted $\leq_{ug}$. (There is also a nonuniform version, $\leq_{ng}$, of generic reducibility which we do not consider in this survey.)

It is easily seen that $\leq_g$ is transitive since the maps induced by enumeration operators are closed under composition.

Definition 5.3. The *generic degree* of A is $\{B : B \leq_g A \ \& \ A \leq_g B\}$.

We have seen that the map $\widehat{\mathcal{R}}$ which sends the Turing degree of A to the generic degree of $\mathcal{R}(A)$ embeds the Turing degrees into the generic degrees, since any generic algorithm for $\mathcal{R}(A)$ will compute A, and the proof of this is uniform. The generic degrees have a least degree under the ordering induced by $\leq_g$, and this least degree consists of the generically computable sets.

Define B to be *enumeration reducible* to A (written $B \leq_e A$) if there is an enumeration operator W such that $W^A = B$.

Enumeration reducibility leads analogously to the *enumeration degrees*, i.e. equivalence classes under the equivalence relation $A \leq_e B$ and $B \leq_e A$. The Turing degrees can be embedded in the enumeration degrees by the map which takes the Turing degree of A to the enumeration degree of $A \oplus \overline{A}$. An enumeration degree $\mathbf{a}$ is called *quasi-minimal* if it is nonzero and no nonzero enumeration degree $\mathbf{b} \leq \mathbf{a}$ is in the range of this embedding. The following definition is analogous:

Definition 5.4 ([17]). A generic degree $\mathbf{a}$ is *quasi-minimal* if it is nonzero and no nonzero generic degree $\mathbf{b} \leq \mathbf{a}$ is in the range of the embedding $\widehat{\mathcal{R}}$ of the Turing degrees into the generic degrees defined above.

The following result gives a connection between quasi-minimality for enumeration degrees and generic degrees.

Lemma 5.5 ([19], *Lemma 4.9*). *If A is a set of density 1 which is not generically computable and the enumeration degree of A is quasi-minimal, then the generic degree of A is also quasi-minimal.*

It is shown in the proof of Theorem 4.8 of [19] that there is a set A which meets the hypotheses of the lemma. It follows that there exist quasi-minimal generic degrees which contain sets of density 1.

It is therefore natural to consider generic degrees which are *density*-1, that is, generic degrees which contain a set of density 1 [17].

A *hyperarithmetical* set is a set computable from any set that can be obtained by iterating the jump operator through the computable ordinals. The class of such sets coincides with the class of Δ_1^1 sets, which are those sets which can be defined by a prenex formula of second-order arithmetic with all set quantifiers universal and also by a prenex formula with all set quantifiers existential. Igusa [17] proves the following striking characterization.

Theorem 5.6 *([17], Theorem 2.15). A set A is hyperarithmetical if and only if there is a density-1 set B such that $\mathcal{R}(A) \leq_g B$.*

Cholak and Igusa [5] consider the question of whether or not every non-zero generic degree bounds a non-zero density-1 generic degree. By the results of [17] a positive answer would show that there are no minimal generic degrees and a negative answer would show that there are minimal pairs in the generic degrees. However, it is not yet known whether or not there are minimal degrees or minimal pairs in the generic degrees.

Recall that a *coarse description* of a set A is a set C which agrees with A on a set of density 1. Hirschfeldt et al. [14] introduced both uniform and nonuniform versions of coarse reducibility and their corresponding degrees.

Definition 5.7 ([14], Definition 2.1). A set A is *uniformly coarsely reducible* to a set B, written $A \leq_{uc} B$, if there is a fixed oracle Turing machine M which, given any coarse description of B as an oracle, computes a coarse description of A. A set A is *nonuniformly coarsely reducible* to a set B, written $A \leq_{nc} B$ if every coarse description of B computes a coarse description of A.

These coarse reducibilities induce respective equivalence relations $\equiv_{uc}$ and $\equiv_{nc}$.

Definition 5.8 ([14]). The *uniform coarse degree* of A is $\{B : B \equiv_{uc} A\}$ and the *nonuniform coarse degree* of A is $\{B : B \equiv_{nc} A\}$.

We can embed the Turing degrees into both the nonuniform and the uniform coarse degrees. We have already seen that the function $\mathcal{I}$ induces an embedding of the Turing degrees into the nonuniform coarse degrees since $\mathcal{I}(A) \leq_T A$ and each coarse description of $\mathcal{I}(A)$ computes A, but the adjustments needed to compute A depend on the coarse description used.

To construct an embedding of the Turing degrees into the uniform coarse degrees we need more redundancy. The following map is slightly different from but equivalent to the map used in [14], Proposition 2.3.

Proposition 5.9 *([14]). Define $\mathcal{E}(A) = \mathcal{I}(\mathcal{R}(A))$. The function $\mathcal{E}$ induces an embedding of the Turing degrees into the uniform coarse degrees.*

Recall that a set X is *autoreducible* if there exists a Turing functional Φ such that for every $n \in \omega$ we have $\Phi^{X \setminus \{n\}}(n) = X(n)$. Equivalently, we could require that Φ not ask whether its input belongs to its oracle. Figueira et al. [11] showed that no 1-random set is autoreducible and it is not difficult to show that no 1-generic set is autoreducible.

Dzhafarov and Igusa [10] study various notions of "robust information coding" and introduced uniform "mod-finite", "co-finite" and "use-bounded from below" reducibilities. Using the relationships between these reducibilities and generic and coarse reducibility, Igusa proved the following result.

Theorem 5.10 (see [14], *Theorem 2.7*). *If $\mathcal{E}(X) \leq_{uc} \mathcal{I}(X)$ then X is autoreducible. Therefore if A is 1-random or 1-generic then $\mathcal{E}(X) \leq_{nc} \mathcal{I}(X)$ but $\mathcal{E}(X) \nleq_{uc} \mathcal{I}(X)$.*

There are striking connections between coarse degrees and algorithmic randomness. The paper [14] shows the following.

Theorem 5.11 ([14], *Corollary 3.3*). *If X is weakly 2-random then $\mathcal{E}(A) \nleq_{nc} X$ for every noncomputable set A, so the degree of X is quasi-minimal (in the obvious sense) in both the uniform and nonuniform coarse degrees.*

For the uniform coarse degrees, this result was strengthened by independently motivated work by Cholak and Igusa [5].

Theorem 5.12 ([5]). *If A is either 1-random or 1-generic, then the degree of A is quasiminimal in the uniform coarse degrees.*

Theorem 5.13 ([14], *Corollary 5.3*). *If Y is not coarsely computable and X is weakly 3-random relative to Y, then their nonuniform coarse degrees form a minimal pair for both uniform and nonuniform coarse reducibility.*

Astor et al. [2] introduced "dense computability" as a weakening of both generic and coarse computability.

Definition 5.14 ([2]). A set A is *densely computable* (or *weakly partially computable*) if there is a partial computable function φ such that $\underline{\rho}(\{n : \varphi(n) = A(n)\}) = 1$.

In other words, the partial computable function may diverge on some arguments and give wrong answers on others but agrees with the characteristic function of A on a set of density 1. It is obvious that every generically computable set and every coarsely computable set is densely computable. Note that if A is generically computable but not coarsely computable and B is coarsely computable but not generically computable then $A \oplus B$ is neither generically computable nor coarsely computable, where, as usual, $A \oplus B = \{2n : n \in A\} \cup \{2n + 1 : n \in B\}$. But $A \oplus B$ is densely computable by using the generic algorithm on even numbers and the coarse algorithm on odd numbers. Thus dense computability is strictly weaker than the disjunction of coarse computability and generic computability.

We can consider weak partial computability at densities less than 1.

Definition 5.15 ([2]). Let $r \in [0,1]$. A set A is *weakly partially computable* at density r if there exists a partial computable function such that $\underline{\rho}(\{n : \varphi(n) = A(n)\}) \geq r$. Let

$$\delta(A) = sup\{r : A \text{ is weakly partially computable at density } r\}.$$

It is easy to show the following.

Lemma 5.16 *([2]). For all $A, \delta(A) = \gamma(A)$.*

Proof. If A is weakly partially computable at density r by a partial computable function φ, then by Theorem 2.2 $\text{dom}(\varphi)$ has a computable subset C such that $\underline{\rho}(C) > \underline{\rho}(\text{dom}(\varphi)) - \epsilon$. Let h be the total computable function defined by $h(n) = \varphi(n)$ if $n \in C$ and $h(n) = 0$ otherwise. Since $A \cap C \subseteq \{n : A(n) = \varphi(n)\}$ it follows that A is coarsely computable at density $r - \epsilon$. So $\gamma(A) \geq \delta(A)$. Since $\delta(A) \geq \gamma(A)$ by definition, the two are equal. $\square$

Definition 5.17. A partial function Θ is a *dense description of A* if $\{n : \Theta(n) = A(n)\}$ has density 1.

Using dense descriptions one can define dense reducibility and dense degrees as in [2].

References

1. Andrews, U., Cai, M., Diamondstone, D., Jockusch, C., Lempp, S.: Asymptotic density, computable traceability, and 1-randomness. Fundam. Math. **234**, 41–53 (2016)
2. Astor, E., Hirschfeldt, D., Jockusch, C.: Dense computability, upper cones and minimal pairs. In preparation
3. Barzdin, J.: On a frequency solution to the problem of occurrence in a recursively enumerable set. Proc. Steklov Inst. Math. **133**, 49–56 (1973)
4. Bienvenu, L., Day, A., Hölzl, R.: From bi-immunity to absolute undecidability. J. Symb. Log. **78**(4), 1218–1228 (2013)
5. Cholak, P., Igusa, G.: Bounding a density-1 and quasiminimality in the generic degrees. Preprint
6. Cormen, T.H., Leiserson, C.E., Rivest, R.L., Stein, C.: Introduction to Algorithms, 2nd edn. MIT Press and McGraw-Hill (2001). Section 29.3: The simplex algorithm, pp. 790–804
7. Downey, R.G., Hirschfeldt, D.R.: Algorithmic Randomness and Complexity. Theory and Applications of Computability. Springer, Springer, New York (2010)
8. Downey, R.G., Jockusch Jr., C.G., Schupp, P.E.: Asymptotic density and computably enumerable sets. J. Math. Log. **13** (2013). 1350005, 43 pp.
9. Downey, R.G., Jockusch Jr., C.G., McNicholl, T.H., Schupp, P.E.: Asymptotic density and the Ershov Hierarchy. Math. Log. Q. **61**, 189–195 (2015)
10. Dzhafarov, D.D., Igusa, G.: Notions of robust information coding, computability. To appear
11. Figueira, S., Miller, J.S., Nies, A.: Indifferent sets. J. Log. Comput. **19**, 425–443 (2009)

12. Gurevich, Y.: Average case completeness. J. Comput. Syst. Sci. **42**, 346–398 (1991)
13. Harrison-Trainor, M.: The Gamma questions for many-one degrees. Preprint. arXiv: 1606.05701
14. Hirschfeldt, D., Jockusch, C., Kuyper, R., Schupp, P.: Coarse reducibility and algorithmic randomness. J. Symb. Log. **81**, 1028–1046 (2016)
15. Hirschfeldt, D.R., Jockusch, C.G., McNicholl, T., Schupp, P.E.: Asymptotic density and the coarse computability bound. Computability **5**, 13–27 (2016)
16. Igusa, G.: Nonexistence of minimal pairs for generic computability. J. Symb. Log. **78**, 51–522 (2012)
17. Igusa, G.: The generic degrees of density-1 sets and a characterization of the hyperarithmetic reals. J. Symb. Log. **80**, 1290–1314 (2015)
18. Jockusch, C.: The degrees of bi-immune sets. Zeitschr. f. math. Logik und Grundlagen d. Math. **15**, 135–140 (1969)
19. Jockusch, C.G., Schupp, P.E.: Generic computability, Turing degrees, and asymptotic density. J. Lond. Math. Soc. **85**, 472–490 (2012)
20. Kapovich, I., Myasnikov, A., Schupp, P., Shpilrain, V.: Generic-case complexity, decision problems in group theory and random walks. J. Algebra **264**, 665–694 (2003)
21. Klee, V., Minty, G.: How good is the simplex algorithm? Inequalities, III (Proceedings of Third Symposium, University of California, Los Angeles, 1969; dedicated to the memory of Theodore S. Motzkin), pp. 159–175. Academic Press, New York (1972)
22. Kurtz, S.A.: Notions of weak genericity. J. Symb. Log. **48**, 764–770 (1983)
23. Levin, L.: Average case complete problems. SIAM J. Comput. **15**, 285–286 (1986)
24. Lyndon, R.C., Schupp, P.E.: Combinatorial Group Theory. Classics in Mathematics. Springer, Heidelberg (2000)
25. Monin, B.: Asymptotic density, error-correcting codes. https://www.lacl.fr/benoit.monin/ressources/papers/resolution_gamma.pdf
26. Monin, B., Nies, A.: A unifying approach to the Gamma question. In: 30th Annual ACM/IEEE Symposium on Logic in Computer Science, LICS 2015, Kyoto, pp. 585–596, 6–10 July 2015
27. Myasnikov, A., Osin, D.: Algorithmically finite groups. J. Pure Appl. Algebra **215**, 2789–2796 (2011)
28. Myasnikov, A., Rybalov, A.: Generic complexity of undecidable problems. J. Symb. Log. **73**, 656–673 (2008)
29. Myasnikov, A., Shpilrain, V., Ushakov, A.: Non-commutative Cryptography and Complexity of Group-Theoretic Problems. Mathematical Surveys and Monographs, vol. 177. American Mathematical Society (2011)
30. Nies, A.: Computability and Randomness. Oxford University Press, Oxford (2009)
31. Novikov, P.S.: On the algorithmic unsolvability of the word problem in group theory. Trudy Math. Inst. Steklov **44**, 3–143 (1955)
32. Post, E.: A variant of a recursively unsolvable problem. Bull. Am. Math. Soc. **52**, 264–268 (1946)
33. Rotman, J.: An Introduction to the Theory of Groups. Graduate Texts in Mathematics, 4th edn. Springer, New York (1995)
34. Terwijn, S.A., Zambella, D.: Algorithmic randomness and lowness. J. Symb. Log. **66**, 1199–1205 (2001)
35. Woess, W.: Cogrowth of groups and simple random walks. Arch. Math. **41**, 363–370 (1983)

On Splits of Computably Enumerable Sets

Peter A. Cholak$^{(\boxtimes)}$

Department of Mathematics, University of Notre Dame,
Notre Dame, IN 46556-5683, USA
Peter.Cholak.1@nd.edu
http://www.nd.edu/~cholak

Abstract. Our focus will be on the computably enumerable (c.e.) sets
and trivial, non-trivial, Friedberg, and non-Friedberg splits of the c.e.
sets. Every non-computable set has a non-trivial Friedberg split. Moreover, this theorem is uniform. V. Yu. Shavrukov recently answered the
question which c.e. sets have a non-trivial non-Friedberg splitting and
we provide a different proof of his result. We end by showing there is
no uniform splitting of all c.e. sets such that all non-computable sets
are non-trivially split and, in addition, all sets with a non-trivial non-
Friedberg split are split accordingly.

2010 Mathematics Subject Classification: Primary 03D25

1 Trivial Splits

Given a c.e. set A, a *split* of A is a pair of c.e. sets A_0, A_1 such that $A_0 \sqcup A_1 = A$,
$\sqcup$ is disjoint union. If one of A_0 or A_1 is computable the splitting is *trivial*. If A_0
is computable then $A = A_0 \sqcup (\overline{A_0} \cap A)$.

It is straightforward to see that any splitting of a computable set is trivial.
Given a c.e. set A, letting $A_0 = \emptyset$ and $A_1 = A$, provides a trivial splitting of
A. We would like to avoid splits where one of the sets is finite. It is known that
every infinite c.e. set A has an infinite computable subset R. This provides a
trivial splitting of A, $A = R \sqcup (A \cap \overline{R})$, into two infinite c.e. sets assuming A is
not computable.

Given this, Myhill asked

Question 1.1 (Myhill [9]). *Does every non-computable c.e. set have a non-trivial splitting?* .

Myhill's Question was answered positively by Friedberg [5].

Draft as of August 7, 2016. We want to thank V. Yu. Shavrukov for allowing us
to include his result, Theorem 3.8. Without it, this paper would look very different.
This research was started while Cholak participated in the Buenos Aires Semester in
Computability, Complexity and Randomness, 2013. Thanks to Rachel Epstein, Greg
Igusa, Nathan Pierson, Mike Stob, and the referees for comments and suggestions.
My interest in Friedberg splits was sparked in 1989 by Rod Downey. I cannot forgive
him.

© Springer International Publishing AG 2017
A. Day et al. (Eds.): Downey Festschrift, LNCS 10010, pp. 521–535, 2017.
DOI: 10.1007/978-3-319-50062-1_31

2   Friedberg Splits

Most of this section is known but we wanted to provide an explicit proof of Corollary 2.7. This corollary will be useful later. One of the focuses of this paper is splitting procedures that always produce a non-trivial split when possible.

At this point we will fix the standard uniform enumeration $W_{e,s}$ of all c.e. sets with the convention that at stage s, there is at most one pair e, x where x enters W_e at stage s. Some details on how we can effectively achieve this enumeration can be found in Soare [12, Exercise I.3.11].

Every c.e. set has an index according to this fixed enumeration. For the sets that we construct we have to appeal to Kleene's Recursion Theorem to find this index. Moreover, by the standard trick of slowing down or pausing our construction, we can assume the enumerations of our fixed point W_e and our constructed set A are the same. Our construction, at times, will construct sets other than A. While we will focus on the constructed sets, the actual outcome of our constructions will be a uniform enumeration of all constructed sets. We will be using Kleene's Recursion Theorem with parameters to get a function from each constructed set to an index with the same enumeration for that set in the above enumeration.

By the Padding Lemma, we know that each c.e. set A has infinitely many indices. By Rice's Theorem, we know that for a given c.e. set A the set of indices (in this fixed enumeration) for A is not computable.

Also at this point we will fix the convention that A, B, W, X and Y always refer to c.e. sets with some fixed index in our given enumeration. Now we need the following.

Definition 2.1. A split $A_0 \sqcup A_1 = A$ is a *Friedberg split* of A iff, for all W, if $W - A$ is not c.e. then both $W - A_0$ and $W - A_1$ are not c.e. sets.

Lemma 2.2. *If A is not computable and $A_0 \sqcup A_1$ is a Friedberg split then the split is not trivial.*

Proof. $\mathbb{N} - A$ is not c.e. so $\mathbb{N} - A_0$ and $\mathbb{N} - A_1$ are not c.e. and hence A_0 and A_1 are not computable. $\square$

Definition 2.3. For $A = W_e$ and $B = W_i$,

$$A \backslash B = \{x | \exists s [x \in (W_{e,s} - W_{i,s})]\}$$

and $A \searrow B = A \backslash B \cap B$. (This is with respect to our given enumeration and hence this definition depends on our chosen enumeration.)

By the above definition, $A \backslash B$ is a c.e. set. $A \backslash B$ is the set of balls that enter A before they enter B. If $x \in A \backslash B$ then x may or may not enter B and if x does enter B, it only does so after x enters A (in terms of our enumeration). Since the intersection of two c.e. sets is c.e., $A \searrow B$ is a c.e. set. The c.e. set $A \searrow B$ is the c.e. set of balls that first enter A and then enter B (under the above enumeration). So $A \backslash B$ reads "A before B" and $A \searrow B$ reads "A before B and then B".

Note that for all W, $W \backslash A = (W - A) \sqcup (W \searrow A)$. Since $W \backslash A$ is a c.e. set, if $W - A$ is not a c.e. set then $W \searrow A$ must be infinite. (This happens for all enumerations not just our given enumeration.)

Lemma 2.4 (Friedberg). *Assume $A = A_0 \sqcup A_1$, and, for all e, if $W_e \searrow A$ is infinite then both $W_e \searrow A_0$ and $W_e \searrow A_1$ are infinite. Then $A_0 \sqcup A_1$ is a Friedberg split of A.*

Proof. Assume that $W - A$ is not a c.e. set but $X = W - A_0$ is a c.e. set. $X - A = (W - A_0) - A = W - A$ is not a c.e. set. So $X \searrow A$ is infinite which implies that $X \searrow A_0$ is infinite but $X \searrow A_0 = (W - A_0) \searrow A_0 = \emptyset$. Contradiction. $\qquad\qquad\square$

Friedberg invented the priority method to split every c.e. set into two disjoint c.e. sets while meeting the hypothesis of the above lemma.

Theorem 2.5 (Friedberg). *Every non-computable set A has a Friedberg split.*

Proof. When a ball x enters A at stage s we add it to one of A_0 or A_1 but which one x enters is determined by priority. Our requirements are:

$$\mathcal{P}_{e,i,k}: \qquad \text{if } W_e \searrow A \text{ is infinite then } |W_e \searrow A_i| \geq k.$$

We say x meets $\mathcal{P}_{e,i,k}$ at stage s if $|W_e \searrow A_i| = k - 1$ by stage $s - 1$ and if we add x to A_i at stage s then $|W_e \searrow A_i| = k$ at stage s. Find the smallest $\langle e, i, k \rangle$ that x can meet and add x to A_i at stage s. If no such triple can be found, add x to A_0 at stage s. It is not hard to show that all the $\mathcal{P}_{e,i,k}$ are met. $\qquad\square$

Observe that the procedure in Theorem 2.5 is uniform. Given this we made the following definition and corollary.

Definition 2.6. A computable function h is a *splitting procedure* iff, for all e, if $h(e) = \langle e_0, e_1 \rangle$ then $W_{e_0} \sqcup W_{e_1}$ is a split of A and if W_e is not computable then this split is not trivial. If h is a splitting procedure, we say that $h(e)$ gives a split of W_e or splits W_e.

Corollary 2.7 (of Friedberg's Proof). *There is a splitting procedure h such that if W_e is not computable then $h(e)$ gives a Friedberg split of W_e.*

3   Non-trivial Non-Friedberg Splits

The above section brings us to the following question:

Question 3.1. *When does a c.e. set have a non-trivial non-Friedberg split?*

This question was first asked, in a different form, as Question 1.4 in Cholak [1]. In [1], it was asked if there is a definable collection of c.e. sets such that for each set A in this collection the Friedberg splits of A are a proper subclass of the non-trivial splits of A. This question later appeared, in yet a different form, as Question 4.6 in the first unpublished version of Cholak et al. [3]. There it was suggested to compare the class of all c.e. sets all of whose non-trivial splits are Friedberg with the $\mathcal{D}$-maximal sets (defined below). As we will see in Theorem 3.8 every form of this question was answered by [11]. Shavrukov showed that a c.e. set A has a non-trivial non-Friedberg split iff A is not $\mathcal{D}$-maximal.

3.1   There Are c.e Sets with Non-trivial Non-Friedberg Splits

Let R be an infinite, coinfinite, computable set. There is a non-computable c.e. subset of R, call this set K_R. There is a non-computable c.e. subset of $\overline{R}$, call this set $K_{\overline{R}}$. Let $A = K_R \sqcup K_{\overline{R}}$. Then $K_R \sqcup K_{\overline{R}}$ is a non-trivial split of A. $R - A = R - K_R$ is not c.e. but $R - K_{\overline{R}} = R$ is a c.e. set. So this split is not Friedberg. Please note that the set A and its non-trivial non-Friedberg split are built simultaneously.

See Theorem 3.8 for more examples of sets with non-trivial non-Friedberg Splits. There are published examples of sets with non-trivial non-Friedberg splits. In Sect. 3.2 of Cholak et al. [3], a number of such sets are constructed. But, like in the construction in the above paragraph and Theorem 3.8, for the examples in [3] the set A and its non-trivial non-Friedberg split are built simultaneously.

3.2   There Are c.e Sets Without Non-trivial Non-Friedberg Splits

For this we need the following definitions:

Definition 3.2. 1. $\mathcal{D}(A) = \{B | B - A \text{ is a c.e. set}\}$.
2. W is *complemented* modulo $\mathcal{D}(A)$ iff there is a c.e. Y such that $W \cup Y \cup A = \mathbb{N}$ and $(W \cap Y) - A$ is a c.e. set.
3. A is $\mathcal{D}$-*hhsimple* iff, for every c.e. W, W is complemented modulo $\mathcal{D}(A)$.
4. A c.e. set W is 0 modulo $\mathcal{D}(A)$ iff $W \in \mathcal{D}(A)$.
5. A c.e. set W is 1 modulo $\mathcal{D}(A)$ iff there is a Y such that $Y \cap A = \emptyset$ and $W \cup Y \cup A = \mathbb{N}$.
6. A non-computable set A is $\mathcal{D}$-*maximal* iff for every W, W is complemented modulo $\mathcal{D}(A)$ and either 0 or 1 modulo $\mathcal{D}(A)$.

Assume W is 0 modulo $\mathcal{D}(A)$. WLOG we can assume $W \cap A = \emptyset$. Then $W \cup \mathbb{N} \cup A = \mathbb{N}$ and $\mathbb{N} \cap W = W$ is disjoint from A. So W is complemented modulo $\mathcal{D}(A)$. If $W - A$ is not c.e. then W is not 0 modulo $\mathcal{D}(A)$. A c.e. set W is 0 modulo $\mathcal{D}(A)$ iff $W - A$ is a c.e. set. The set W is 1 modulo $\mathcal{D}(A)$ as witnessed by Y iff W is complemented by Y modulo $\mathcal{D}(A)$ and Y is 0 modulo $\mathcal{D}(A)$. We will not go through the details but the property of a set A being $\mathcal{D}$-maximal is definable in the c.e. sets, $\mathcal{E}$.

Lemma 3.3 (Cholak, Downey, Herrmann). *All non-trivial splits of a $\mathcal{D}$-maximal set A are Friedberg.*

Proof. Let $A_0 \sqcup A_1 = A$ be a non-trivial split of A. Assume that $W - A$ is not a c.e. set. So $W \cup A$ is 1 modulo $\mathcal{D}(A)$. Then, for some Y, $W \cup A \cup Y = \mathbb{N}$ and $Y \cap A = \emptyset$. If $W - A_0$ is c.e. then $A_0 \sqcup ((W - A_0) \cup A_1 \cup Y) = \mathbb{N}$ and hence A_0 is computable. Contradiction.   □

This result and the above proof explicitly appears in an earlier unpublished version of Cholak et al. [3] but not in the published version. It was first implicitly mentioned in Cholak et al. [2]. It follows a similar result about maximal sets in Downey and Stob [4].

3.3   The Herrmann and Kummer Splitting Theorem

Shortly we will need the following theorem.[1]

Theorem 3.4 (Herrmann and Kummer Splitting Theorem). *Let A and B be c.e. sets such that $A \subseteq B$ and B is non-complemented modulo $\mathcal{D}(A)$. Then there are B_0 and B_1 such that B_i is non-complemented modulo $\mathcal{D}(A)$ and $B_0 \sqcup B_1 = B$.*

Proof. As balls x enter B they will be enumerated into either B_0 or B_1. So $B = B_0 \sqcup B_1$. Let Y_e, Z_j be two listings of all c.e. sets. We need to meet the requirements:

$$\mathcal{R}_{e,j,i}: \qquad \text{either } B_i \cup A \cup Y_e \neq \mathbb{N} \text{ or } (B_i \cap Y_e) - A \neq Z_j.$$

If we fail to meet this requirement then Y_e and Z_j witness that B_i is complemented modulo $\mathcal{D}(A)$.

We need a *disagreement* function. Let $l(e, j, i, s)$ be the least $x \leq s$ such that either $x \notin B_{i,s} \cup A_s \cup Y_{e,s}$, or $x \in ((B_{i,s} \cap Y_{e,s}) - A_s)$ iff $x \notin Z_{j,s}$.

If x does not exist, let $l(e, j, i, s) = s$. The $\lim_s l(e, j, i, s)$ exists iff we will have meet $\mathcal{R}_{e,j,i}$.

We will use l to define a *restraint* function, $r(e, j, i, -1) = \langle e, j, i \rangle$ and $r(e, j, i, s)$ is the max of $r(e, j, i, s - 1)$ and $l(e, j, i, s)$. Again, the $\lim_s r(e, j, i, s)$ exists iff we will have meet $\mathcal{R}_{e,j,i}$. Moreover $r(e, j, i, s)$ is a non-decreasing function in s.

When a ball x enters B at stage s find the least $\langle e, j, i \rangle$ such that $x \leq r(e, j, i, s)$ and add x to B_i.

Let $\langle e, j, i \rangle$ be the least triple such that $\lim_s r(e, j, i, s)$ does not exist. Let x be such that for all $\langle e', j', i' \rangle < \langle e, j, i \rangle$, $\lim_s l(e', j', i', s) < x$. Assume $i = 0$. Then B_1 is computable (for all $y > x$, after $r(e, j, 0, s) > y$, y cannot enter B_1), Y_e and Z_j witness that B_0 is complemented modulo $\mathcal{D}(A)$. Now $Y = Y_e \cap \overline{B_1}$ and Z_j witness that B is complemented modulo $\mathcal{D}(A)$. Contradiction. Similarly if $i = 1$.   □

---

[1] The Herrmann and Kummer Splitting Theorem appears, in a very different form, in Herrmann and Kummer [7]. This theorem appears in the only if direction of the proof of Theorem 2.4 of Herrmann and Kummer [7] starting on page 63 from the first full paragraph on that page. It is interesting enough to be isolated in its own right as a theorem.

This construction is uniform. Given an index for B we can uniformly get a split of B via the above theorem. Assume B is 0 modulo $\mathcal{D}(A)$ witnessed by the c.e. set $Z = B - A$. Then $\mathbb{N}$ and Z witness that B and any splits of B are complemented modulo $\mathcal{D}(A)$. Let e' and j' be the least such that $Y_{e'} = \mathbb{N}$ and $Z_{j'} = Z$, and $l(e', j', i, s) = s$ (this last item just takes playing a little with the enumeration of these sets). For some $e \le e', j \le j'$ and i, $\lim_s r(e, j, i, s)$ does not exist and the argument above shows that the split is trivial. So if $B \subseteq A$ this split will be trivial. So this theorem does not give rise to a splitting procedure.

If B is not complemented modulo $\mathcal{D}(A)$ then it is open if the above split (as given above) is always Friedberg. We conjecture yes with the following evidence: We can combine the requirements $\mathcal{P}$ from the proof of Theorem 2.5 with the one here to force the split to be a Friedberg split.

We also want to point out that the Herrmann and Kummer Splitting Theorem is very similar to the Owings Splitting Theorem. B is *non complemented modulo* A iff $B - A$ is not co-c.e. iff $\overline{B} \cup A$ is not c.e. The following theorem is an easy corollary of the Owings Splitting Theorem, [10]. Also see Soare [12, X.2.5].

Theorem 3.5. (Owings). *Let A and B be c.e. sets such that $A \subseteq B$ and B is non-complemented modulo A. Then there are B_0 and B_1 such that B_i is non-complemented modulo A and $B_0 \sqcup B_1 = B$.*

We are not going to provide a proof. The standard proof is Soare [12, X.2.5]. What is not clear is whether this standard proof always provides a Friedberg split and, if $B \subseteq A$, whether the resulting split is non-trivial. We can arrange the enumeration (let $W_0 = \mathbb{N}$) such that if $B \subseteq A$ then the resulting split is non-trivial. But it is open what occurs when we use the standard enumeration. So it is unknown if the Owings Splitting Theorem gives a splitting procedure.

The Owings and the Herrmann and Kummer Splitting theorems are like Friedberg's in that all three are uniform, but unlike Friedberg's in that they do not necessarily provide non-trivial splits when possible. Herrmann and Kummer Splitting Theorem does not give rise to a splitting procedure. It is open if the Ownings Splitting Theorem gives rise to a splitting procedure. Friedberg Splitting Theorem does give rise to a splitting procedure.

There is one more (little) known splitting theorem, Hammond [6], which extends all three of the splitting theorems above discussed in this subsection. Let $\mathcal{E}$ be the collection of c.e. sets with inclusion, intersection, union, $\emptyset$ and $\mathbb{N}$; this is called the lattice of c.e. sets. An *ideal* of $\mathcal{E}$ is a collection of sets $\mathcal{I}$ such that $\emptyset \in \mathcal{I}$ and $\mathcal{I}$ is closed under subset and inclusion. An ideal $\mathcal{I}$ is Σ_3^0 if the relation $W_e \in \mathcal{I}$ is Σ_3^0. $\mathcal{F}$, collection of all finite sets, is an Σ_3^0 ideal. For any A, so are $\mathcal{S}(A) = \{B | B \subseteq A\}$ and $\mathcal{D}(A)$. W is *complemented modulo* $\mathcal{I}$ iff there is a Y such that $W \cup Y = \mathbb{N}$ and $W \cap Y$ is in $\mathcal{I}$. For any A, the Friedberg, Ownings, and Herrmann and Kummer Splitting Theorems, respectively, imply any B which is non-complemented modulo $\mathcal{F}$, $\mathcal{S}(A)$, or $\mathcal{D}(A)$ can be split into B_0 and B_1 such that each B_i is non-complemented modulo $\mathcal{F}$, $\mathcal{S}(A)$, or $\mathcal{D}(A)$.

Theorem 3.6 (Hammond [6]). *Let $\mathcal{I}$ be any Σ_3^0 ideal. If B is non-complemented modulo $\mathcal{I}$ then B can be split into B_0 and B_1 such that each B_i is non-complemented modulo $\mathcal{I}$.*

We will not include a proof here. Unlike the other three splitting theorems discussed here the proof is not finite injury. It is uniform in $\mathcal{I}$. Since $\mathcal{I}$ can equal $\mathcal{D}(A)$, it does not always give raise to a splitting procedure. What happens when $\mathcal{I}$ is $\mathcal{S}(A)$ is open.

3.4 Shavrukov's Result

First we need to use the Herrmann and Kummer Splitting Theorem for the following corollary. The proof is not uniform.

Corollary 3.7. *For all non-computable non-$\mathcal{D}$-maximal A, there are disjoint X_0 and X_1 such that $X_i - A$ is not c.e. and $A \subseteq X_0 \sqcup X_1$.*

Proof. When A is not $\mathcal{D}$-hhsimple there is a c.e. X such that $A \subseteq X$ and X is not complemented modulo $\mathcal{D}(A)$. Apply the above Herrmann and Kummer Splitting Theorem to get $X_0 \sqcup X_1 = X$ where the X_is are also not complemented modulo $\mathcal{D}(A)$. If $X_i - A$ is c.e. then X_i is 0 and hence complemented modulo $\mathcal{D}(A)$. Therefore $X_i - A$ is not a c.e. set.

Otherwise A is $\mathcal{D}$-hhsimple but not $\mathcal{D}$-maximal. So there must be a c.e. superset W of A which is not 0 or 1. So $W - A$ is not a c.e. set. There is a Y such that $W \cup Y = \mathbb{N}$, $(W \cap Y) - A$ is c.e. but $Y - A$ is not a c.e. set.

Let $X_0 = W \backslash Y$ and $X_1 = Y \backslash W$. Now $W = X_0 \cup (W \cap Y)$. So $W - A = (X_0 - A) \cup ((W \cap Y) - A)$. The set $(W \cap Y) - A$ is known to be c.e., so if $X_0 - A$ is c.e. then so is $W - A$. Therefore $X_0 - A$ is not a c.e. set. $Y = X_1 \cup (W \cap Y)$. So $Y - A = (X_1 - A) \cup ((W \cap Y) - A)$. $(W \cap Y) - A$ is known to be c.e., so if $X_1 - A$ is c.e. then so is $Y - A$. Therefore $X_1 - A$ is not a c.e. set.   □

Theorem 3.8 (Shavrukov). *All c.e. non-computable non-$\mathcal{D}$-maximal sets A have non-trivial non-Friedberg splits.*

Proof By the above corollary, there are disjoint X_0 and X_1 such that $\overline{X_i - A}$ is not c.e. and $A \subseteq X_0 \sqcup X_1$. If $X_i \cap A$ were computable then $X_i - A = X_i \cap \overline{(X_i \cap A)}$ is c.e. Therefore $X_0 \cap A, X_1 \cap A$ is a non-trivial split of A. $X_0 - A$ is not c.e. but $X_0 - (X_1 \cap A) = X_0$ is a c.e. set. Hence $X_0 \cap A, X_1 \cap A$ is a non-trivial non-Friedberg split.   □

Corollary 3.9 (Shavrukov). *All of A's non-trivial splits are Friedberg iff A is $\mathcal{D}$-maximal.*

Again we want to thank V. Yu. Shavrukov for allowing us to include his results. The proof we presented here is very different than Shavrukov's, see [11]. Shavrukov's proof used the fact that every $\mathcal{D}$-hhsimple is not a diagonal. For the definition of a diagonal set see Kummer [8] and Herrmann and Kummer [7].

4   Uniform Non-trivial Non-Friedberg Splits

The question we will answer in this section follows:

Question 4.1. *Is there a splitting procedure h such that all non-$\mathcal{D}$-maximal sets W_e are split by $h(e)$ into a non-trivial non-Friedberg split?*

The answer is no by the following theorem:

Theorem 4.2. *For every total computable h there is an e such that W_e is not computable and $h(e) = \langle e_0, e_1 \rangle$ then either*

(1) W_{e_0}, W_{e_1} is not a split of W_e,
(2) $W_{e_0} \sqcup W_{e_1}$ is a trivial split of W_e, or
(3) $W_{e_0} \sqcup W_{e_1}$ is a Friedberg split of W_e and W_e is not $\mathcal{D}$-maximal.

Moreover given an index for h we can effectively find e.

Hence if h is a splitting procedure then Case (3) applies. Actually, Case (3) applies infinitely often.

Corollary 4.3. *Let h be a splitting procedure. Then there is an infinite set J of indices that, for all $e \in J$, W_e has a non-Friedberg split but the split given by $h(e)$ is a Friedberg split.*

Proof of the Corollary. Let $h_0 = h$ and apply Theorem 4.2 to get e_0. Only Case (3) can apply. So W_{e_0} has a non-trivial non-Friedberg split but $h(e_0)$ gives a Friedberg split. Inductively, assume for all $j \le i$, that h_j and distinct e_j exist and that Case (3) applies to W'_{e_j}. Let $W_{a_i} \sqcup W_{b_i}$ be a non-trivial non-Friedberg split of W_{e_i}. Let $h_{i+1}(e_i) = \langle a_i, b_i \rangle$ and if $e \ne e_i$ let $h_{i+1}(e) = h_i(e)$. Apply Theorem 4.2 to h_{i+1} to effectively get an e_{i+1}. Case (3) applies to e_{i+1} and $e_{i+1} \ne e_j$, for all $j \le i$. Let J be the infinite set $\{e_i | i \in \omega\}$. $\square$

We can create a splitting procedure that is correct on infinite many indices of a non-$\mathcal{D}$-maximal set. Take $A_0 \sqcup A_1 = A = W_a \sqcup W_b = W_c$ to be a non-trivial non-Friedberg splitting of A. Using the padding lemma, let I be an infinite computable set of indices for A. Define $h(e)$ to be $\langle a, b \rangle$ if $e \in I$ and $h_F(e)$ otherwise, where h_F is from Corollary 2.7. By Rice's Theorem, I is not all indices for A. But the following is open.

Question 4.4. *Is there a splitting procedure h and a c.e. set A with a non-trivial non-Friedberg split such that, if $W_e = A$ then $h(e)$ gives a a non-trivial non-Friedberg split of $W_e = A$?*

5   Proof of Theorem 4.2

The goal of the rest of the paper is to provide a proof of the above Theorem 4.2. Assume that we are given h and we will construct A. Via the Recursion Theorem we can assume that $W_e = A$. Also assume that $h(e) = \langle e_0, e_1 \rangle$.

For our proof we will work using an oracle for certain Π_2^0 questions. Certainly $\mathbf{0}''$ works but is overkill. The index set of all infinite c.e. sets works nicely. We will use a tree argument to provide answers to our Π_2^0 questions. The tree will also provide a framework for our construction.

We will build A in pieces. First we will construct a Δ_3^0 list of pairwise disjoint computable sets R such that every c.e. set or it's complement will be in the union of finitely many of these computable sets and the union of all them is $\mathbb{N}$. Inside each of these computable sets we will build a piece of A. The default is that A will be maximal inside each R but finite or cofinite inside R are also possible. The construction will ensure that the union of these pieces is a c.e. set A. If A is maximal in only finitely many of these computable sets then A will turn out to be $\mathcal{D}$-maximal.

We will *try* to construct infinite, coinfinite, computable sets R_i such that, for all j, either

$$(5.0.1) \qquad W_j \subseteq^* \bigsqcup_{i \leq j} R_i \cup A,$$

or

$$(5.0.2) \qquad W_j \cup \bigsqcup_{i \leq j} R_i \cup A =^* \mathbb{N}.$$

(We will remind the reader that $X =^* Y$ iff $(X - Y) \sqcup (Y - X)$ is finite.) Since these sets are meant to be computable we also have to build $\overline{R}_i$ while we are building R_i. Assume that we have built the sets R_i up to j. The balls in $\bigcap_{i<j} \overline{R}_i$ have not yet been added to R_j or A. So our construction will ensure $(\bigcap_{i<j} \overline{R}_i) = (\bigcap_{i<j} \overline{R}_i) \backslash A$ is infinite. To build R_j ask if

$$(5.0.3) \qquad P_j = (W_j \cap \bigcap_{i<j} \overline{R}_i) \backslash A$$

is infinite. This is a Π_2^0 question. If P_j is infinite, we will build $\overline{R}_j$ as a subset of W_j, so that Eq. 5.0.2 is satisfied. When we add balls from the set $\bigcap_{i<j} \overline{R}_i$ to R_j, we will make sure that there is at least one ball in $W_j \cap \bigcap_{i<j} \overline{R}_i$ currently uncommitted. We will add that ball to $\overline{R}_j$ and the rest of the balls under consideration to R_j. We will do this infinitely often. In this case, we satisfy Eq. 5.0.2. If P_j is finite, then, since $(\bigcap_{i<j} \overline{R}_i) \backslash A$ is infinite, we just build R_j and $\overline{R}_j$ to be infinite within $\bigcap_{i<j} \overline{R}_i$ and Eq. 5.0.1 is satisfied.

Now inside each R_i we will build A to be finite, cofinite, or maximal depending on various outcomes. The default will be for A to be maximal in R_i. To do this we use the construction presented in Soare [12, X.3.3] as a guide to work inside R_i. We will go over the details later. Since maximal sets are not computable, A will not be computable. Assume that A is maximal inside R_i and R_l, where $l \neq i$, then, since $A \cap R_l$ is a non-computable subset of $A \cap \overline{R}_i$, $A = (A \cap R_i) \sqcup (A \cap \overline{R}_i)$ is a non-trivial non-Friedberg split of A. The details follow the construction in Subsect. 3.1. Now by Theorem 3.8, A is not $\mathcal{D}$-maximal. If $W_{e_0} \sqcup W_{e_1}$ is not a split of A then we are done. So we may as well assume that $W_{e_0} \sqcup W_{e_1}$ is a split of A.

We will now consider how this split behaves inside each R_i. Since A is maximal inside R_i there are two choices either the split is trivial or Friedberg. We are going to ask an infinite series of questions designed to tell if the split inside R_i is trivial. The questions are is "$W_k \sqcup (W_{e_0} \cap R_i) = R_i$" and is "$W_k \sqcup (W_{e_1} \cap R_i) = R_i$", for all k. Again these questions are Π^0_2. A positive answer will tell us the split is trivial inside R_i and which set $W_{e_0} \cap R_i$ or $W_{e_1} \cap R_i$ is computable.

Assume that we get a positive answer and the information that the set $W_{e_0} \cap R_i$ is computable. In this case we will take the following action: Dump almost all of R_l, for $l < i$, into A and, for $l > i$, stop adding balls from R_l into A. In fact, stop building R_l. In this case, A is computable outside R_i and hence W_{e_0} must also be computable. So $W_{e_0} \sqcup W_{e_1}$ is a trivial split of A. We act similarly if $W_{e_1} \cap R_i$ is computable.

If none of the answers to these questions for each R_i is positive then $W_{e_0} \sqcup W_{e_1}$ is a non-trivial split of A. We know inside each R_i the split is Friedberg. We must show that globally the split is Friedberg. Let's consider W_j. If Eq. 5.0.1 holds, then $W_j - A \subseteq^* \bigsqcup_{i \leq j} R_i$. So, if $W_j - A$ is not a c.e. set neither are $W_j - W_{e_0}$ and $W_j - W_{e_1}$. So assume Eq. 5.0.2 holds, $W_j - A$ is not a c.e. set, but $W_j - W_{e_0}$ is a c.e. set. For any $n > j$, $(W_j - A) \cap R_n$ cannot be a c.e. set. But $(W_j - W_{e_0}) \cap R_n$ is a c.e. set. This contradicts that our split is Friedberg inside R_n. A similar argument works if Eq. 5.0.2 holds, $W_j - A$ is not a c.e. set, but $W_j - W_{e_1}$ is a c.e. set. Our split is a Friedberg split.

With one positive answer, we must take action to ensure that our given split is trivial. One positive answer is a Σ^0_3 event. If all questions have negative answers then we have a Π^0_3 event and, in this case, our split is a Friedberg split.

5.1 Coding Our Π^0_2 Questions via a Tree

We will work with the tree, $2^{<\omega}$. We consider the tree to grow downward. At the empty node, λ, we will construct A and $\overline{R}_\lambda = \tilde{R}_\lambda = \mathbb{N}$. At nodes α of length $i^2 > 0$ we will construct R_α and $\tilde{R}_\alpha$ ($\overline{R}_\alpha = \sqcup_{\beta \subset \alpha} R_\beta \sqcup \tilde{R}_\alpha$.) We will call such nodes R-nodes. The idea is that if f is the true path, $|\alpha| = i^2$, and $\alpha \prec f$, then $R_\alpha = R_i$ and $\bigsqcup_{\beta \subset \alpha} R_\beta \sqcup \tilde{R}_\alpha =^* \mathbb{N}$. (We will start indexing the R_i at 1.)

Since we need to ask questions about the potential R_i's we need the indices for the R_α's. So the real outcome of our construction is a pair of functions g and $\tilde{g}$ such that $W_{g(\lambda)} = A$, $W_{g(\alpha)} = R_\alpha$, and $W_{\tilde{g}(\alpha)} = \tilde{R}_\alpha$, for all α. Via the Recursion Theorem, we can assume we know g and $\tilde{g}$ prior to the construction. We will use this knowledge to code our questions into the tree.

Let $|\gamma| = j^2 - 1$. Let $\delta \subset \gamma$ such that $|\delta| = (j-1)^2$. At γ we will code the question "Is $(W_j \cap \tilde{R}_\delta) \backslash A$ infinite?". *Strictly* between two R-nodes of length j^2

and $(j+1)^2$ there are $((j+1)^2-1)-j^2 = 2j$ nodes. If $|\gamma| = j^2+2k-2, 1 \le k \le j$, $\beta \preceq \gamma$, and $|\beta| = k^2$, then at γ code the question "Does $W_j \sqcup (W_{e_0} \cap R_\beta) = R_\beta$?". If $|\gamma| = j^2 + 2k - 1, 1 \le k \le j$, $\beta \preceq \gamma$, and $|\beta| = k^2$, then at γ code the question "Does $W_j \sqcup (W_{e_1} \cap R_\beta) = R_\beta$?". (The only difference in these two sentences is the length of γ differences by 1 and the second uses W_{e_1} rather than W_{e_0}.)

Via the use of the Recursion Theorem, as we discussed two paragraphs above, these are uniformly Π_2^0 questions. There is a uniform reduction from these questions to the index set of infinite c.e. sets or INF. So uniformly, for all γ, we can associate a c.e. *chip* set C_γ such that C_γ is infinite iff the question coded at γ has a positive answer.

Earlier we have called some nodes R-nodes. These were the nodes whose length is a prefect square. Other than the empty node, we will call the remaining nodes A-nodes; they provide answers to questions coded at α's predecessor, $\alpha^- = \gamma$. We call an A-node α *positive* iff $\alpha^\smallfrown 1 = \gamma$. Otherwise an A-node is negative.

We will inductively define the *true path*, f. λ is on f. Assume that $\alpha \preceq f$. If α is a positive A-node then $f = \alpha$. Otherwise, $\alpha^\smallfrown 1 \prec f$ iff C_α is infinite and $\alpha^\smallfrown 0 \prec f$ iff C_α is finite. Since nodes of length 0 and 1 are not A-nodes, there is always an R-node on the true path. Either all the A-nodes on f are negative or f is finite and ends with a positive A-node.

A key to the construction is the approximation to the true path at stage s, f_s. Define $f_0 = \lambda$, the empty node. Assume that $\alpha \subseteq f_{s+1}$ and $|\alpha| < s^2$. If α is a positive A-node, let $f_{s+1} = \alpha$. Assume that α is not a positive A-node. Let t be the greatest stage less than $s + 1$ such that $\alpha \subseteq f_t$. If no such stage exists, let $t = 0$. If $C_{\alpha,t} \neq C_{\alpha,s+1}$ then let $\alpha^\smallfrown 1 \subseteq f_{s+1}$. Otherwise, $\alpha^\smallfrown 0 \subseteq f_{s+1}$.

Since nodes of length s^2 are R-nodes, for $s > 0$, f_s always ends in an R-node or a positive A-node. We say $\alpha <_L \beta$ (or α is to the left of β) iff $\alpha \subsetneq \beta$ or there is a γ such that $\gamma^\smallfrown 1 \subseteq \alpha$ and $\gamma^\smallfrown 0 \subseteq \beta$. By induction on l, we can show that $\liminf_s f_s \upharpoonright l = f \upharpoonright l$ (the lim inf is measured w.r.t. $<_L$). So, $\liminf_s f_s = f$. If $f_s <_L \alpha$ then there is always a least (in terms of length) R-node or positive A-node, β, such that $\beta \subseteq f_s$ and $\beta <_L \alpha$.

5.2 Action on the Tree

We will use the tree and f_s to construct A, R_α, and $\tilde{R}_\alpha$, for all α. We think of this construction as a pinball machine. Integers or balls enter at top node, λ, and move downwards and leftwards. The position of a ball, x, at the end of stage s is given by the function $\alpha(x, s)$. The movement on the tree is done such that the $\lim_s \alpha(x, s)$ exists. Let $\alpha(x) = \lim_s \alpha(x, s)$. Initially, $\alpha(x, s)$ is not defined (so x is not on the machine) and, unless explicitly changed, $\alpha(x, s)$ remains the same from stage to stage. For the balls on the machine, at every stage s, $\alpha(x, s)$ is always an R-node or a positive A-node, and $|\alpha(x, s)| \le x^2$. (The bound x^2 was chosen since balls can only rest at R-nodes or positive A nodes and the length of R-nodes are perfect squares.) If a ball x enters A at stage s, x is removed from the tree at stage s and $\alpha(x, s)$ is undefined again.

Entering the machine and leftward movement is determined by f_{s+1}. Downward movement will be discussed later. Let β be the R-node of length 1 such that $\beta \subseteq f_{s+1}$. Let $\alpha(s, s+1) = \beta$. So all the balls on the machine at stage s are less than s. Assume that $\alpha(x, s) = \alpha$ and $f_{s+1} <_L \alpha$. Then there is always a least (in terms of length) R-node or positive A-node, β, such that $\beta \subseteq f_{s+1}$, $\beta \not\subseteq \alpha$, and $\beta <_L \alpha$. Let $\alpha(x, s+1) = \beta$. Since $|\alpha| \leq x^2 + 1$, the same is true for β. A ball x can only move leftward finitely many times. Since $\liminf_s f_s = f$, $\alpha(x) <_L f$ or $\alpha(x) \subseteq f$.

Assume that α is an R-node. So the length of α is j^2 for some j. Either $\alpha = \alpha^{-\smallfrown}1$ or $\alpha = \alpha^{-\smallfrown}0$. At $\gamma = \alpha^-$ we asked the question "Is $(W_j \cap \tilde{R}_\delta)\backslash A$ infinite?", where δ is the greatest proper R-subnode of γ. If α ends with a 1, then α believes this set is infinite. If α ends with a 0 then α believes this set is finite. If α ends with a 1 let $P_\alpha = (W_j \cap \tilde{R}_\delta)\backslash A$. Otherwise, let $P_\alpha = \tilde{R}_\delta\backslash A$. We also defined P_α for positive A-nodes to be $P_\alpha = \tilde{R}_\delta\backslash A$, where $\delta \subset \alpha$ is the greatest R-node contained in α. α wants all balls in P_α to go though α. Moreover the construction of A inside R_α requires that α see fresh balls in P_α. So these α are allowed to pull balls in P_α.

We will now work on the remaining movement, pulling, on our pinball machine. An R-node or positive A-node α is allowed to pull balls from subnodes of α or nodes to the right of α. Pulling will be downward or leftward movement. The only downward movement allowed is done via pulling. When α can pull balls is controlled by f_s. When $\alpha \subseteq f_s$, α puts a request coded by s on a list denoted by $\mathcal{P}_\alpha$ at stage s. α can only pull balls when there is an unfulfilled request on the list. If α takes action (as described below) at stage s then the least request on $\mathcal{P}_\alpha$ has been fulfilled. If $f_s <_L \alpha$ then all the current requests at stage s on $\mathcal{P}_\alpha$ are declared fulfilled.

Let α be an R-node or A-node of length l and assume that there is an unfulfilled request on $\mathcal{P}_\alpha$ at stage s. Assume that there are two different balls, x_0 and x_1, such that $x_i > l$, $x_i \in P_{\alpha,s}$, and either $\alpha <_L \alpha(x_i, s)$ (x_i is to the right of α) or $\alpha(x_i, s) \subset \alpha$ (x_i is above α). For leftmost α and the least such pair, at stage $s + 1$, take the following action: Let $\alpha(x_i, s+1) = \alpha$ and, if α is a R-node, then put x_0 into $R_{\alpha,s+1}$ and put x_1 into $\tilde{R}_{\alpha,s+1}$. For all balls y, such that $|\alpha|^2 < y < \max x_i$, $y \in \tilde{R}_{\delta,s}$ (using the above notation for δ), and either $\alpha <_L \alpha(x_i, s)$ or $\alpha(x_i, s) \subset \alpha$, let $\alpha(y, s+1) = \alpha$ and, if α is R-node, then add y to $R_{\alpha,s+1}$. This request is declared fulfilled.

There is just a little more to the construction of R_α. In the next section we will discuss the construction of A inside $R_\alpha\backslash A$. Recall earlier that we said that if a ball enters A it is removed from the machine. That means that none of the above balls added to R_α and $\tilde{R}_\alpha$ are in A. To make sure that R_α is computable when $\alpha \subset f$ we must be sure that almost all balls from $\tilde{R}_\delta$ enter R_α or $\tilde{R}_\alpha$. Because of the construction to the right of the true path, infinitely many balls in $\tilde{R}_\delta$ might enter A before they enter R_α or $\tilde{R}_\alpha$. The balls we are talking about are in the c.e. set $(\tilde{R}_\delta \smallsetminus A)\backslash(R_\alpha \sqcup \tilde{R}_\alpha)$. The above action cannot add these balls to R_α or $\tilde{R}_\alpha$. So we will simply add these balls to R_α. So the above set is equal to $A \smallsetminus R_\alpha$.

Let's see inductively that for $\alpha \subset f$ and α is an R-node, that $R_\alpha \backslash A$ is infinite, $\tilde{R}_\alpha \backslash A$ is infinite, $\bigsqcup_{\beta \subseteq \alpha} R_\beta \sqcup \tilde{R}_\alpha =^* \mathbb{N}$, and $A \searrow R_\alpha$ is computable. Let δ be the greatest proper $\tilde{R}$-subnode of α. If no such node exists let $\delta = \lambda$. So by our inductive hypothesis $\tilde{R}_\delta \backslash A$ is infinite. Moreover, by the movement on the tree, only finite many of these balls are ever to the left of α. Ignore those balls. Since α is on the true path, infinitely many requests are placed on $\mathcal{P}_\alpha$ and only finitely many of them are fulfilled because $f_s <_L \alpha$. We claim all of the remaining requests are fulfilled. If not then all but finitely balls of P_α can be pulled by α and α will eventually pull two balls fulfilling the desired request. So the action discussed two paragraphs above occurs infinitely often. We have ensured that $R_\alpha \backslash A$ and $\tilde{R}_\alpha \backslash A$ are infinite. The sets R_β, for $\beta \subseteq \alpha$, and $\tilde{R}_\alpha$ are all pairwise disjoint. By the action in the above paragraph the union of all these sets is almost everything. Since the disjoint union of these sets is almost everything, we also have that $A \searrow R_\alpha$ is computable. Moreover if α ends with a 1 then $\tilde{R}_\alpha \subseteq W_j$, where $|\alpha| = j^2$, and hence $W_j \cup A \cup \bigsqcup_{\beta \subseteq \alpha} R_\beta =^* \mathbb{N}$ and Eq. 5.0.2 holds. If α ends with a 0 then $W_j \subseteq^* A \cup \bigsqcup_{\beta \subseteq \alpha} R_\beta$ and Eq. 5.0.1 holds.

Assume f is finite. So $\alpha = f$ is a positive A node. Let γ be the greatest R-subnode of α. Let Z be the set of x such that there is a stage s where $\alpha(x,s) = \alpha$. Z is a c.e. set. Because $\alpha = f$ for almost all balls x in Z, $\alpha(x) = \alpha$. Almost all of the balls in Z never enter A. Z is the end of the line. Recall that $P_\alpha = \tilde{R}_\delta \backslash A$. By the pulling action almost all of the balls in P_α will enter Z. By the above paragraph, $\bigsqcup_{\beta \subseteq \delta} R_\beta \sqcup \tilde{R}_\delta =^* \mathbb{N}$. So Z and $A \searrow \tilde{R}_\delta$ are computable sets.

5.3 The Construction of A

We will build A to be maximal inside $R_\alpha \backslash A$. Since $\alpha \subset f$, $R_\alpha \backslash A$ is an infinite computable set. Let $R = R_\alpha \backslash A$. We build $A \cap R$ stagewise based on the construction of a maximal set from Soare [12, Theorem X.3.3].

The main requirement is to ensure that, for all e,

$$\mathcal{M}_e: \qquad W_e \cap R \subseteq^* A \cap R \text{ or } (W_e \cap R) \cup (A \cap R) = R.$$

$\sigma(e,x,s) = \{i : i \le e \wedge x \in W_{i,s}\}$ is the e-state of x at stage s. We will have a series of markers Γ_n^α with $a_n^{\alpha,s}$ denoting the position of Γ_n^α at stage s and such that $\overline{A}_s \cap R = \{a_0^{\alpha,s} < a_1^{\alpha,s} \ldots\}$. Each marker Γ_e wants to move to maximize the e-state of $a_e^\alpha = \lim a_e^{\alpha,s}$.

Initially, we let $A_0 \cap R = \emptyset$ and define the $a_n^{\alpha,0}$ accordingly. At stage $s+1$, if there is a least e such that for some least i, $e < i < s$ and $\sigma(e, a_i^{\alpha,s}, s) > \sigma(e, a_e^{\alpha,s}, s)$, then we dump $a_e^{\alpha,s}, a_{e+1}^{\alpha,s}, \ldots a_{i-1}^{\alpha,s}$ into A at stage $s+1$. So $a_e^{\alpha,s+1} = a_i^{\alpha,s}$. Let's call this dumping the *original dumping*. If e does not exist do nothing.

Certain positive A-nodes γ below α can also dump balls from $R = R_\alpha \backslash A$ into A. Let γ be a positive A-node such that $\alpha \subset \gamma$ and at γ^- is coded the question "Is $W_j \sqcup (W_{e_0} \cap R_\beta) = R_\beta$ infinite'?" or "Is $W_j \sqcup (W_{e_1} \cap R_\beta) = R_\beta$ infinite?", for some j and some $\beta \neq \alpha$. γ believes that our split is trivial inside some R_β and wants to dump almost all of R_α into A. Let $t_{\gamma,s}$ be the maximum of $|\gamma|$ and the greatest stage t such that $t \le s$ and $f_t <_L \gamma$. Assume $\gamma \subset f_{s+1}$, dump $a_s^{\alpha, t_{\gamma,s}}$

into A at stage $s+1$ (if the above movement of balls at stage $s+1$ has already forced $a_s^{\alpha,t_{\gamma,s}} \neq a_{s+1}^{\alpha,t_{\gamma,s}}$ that is enough). Let's call this dumping, *extra dumping*.

The positive A-nodes to the left or to the right of the true path only dump $a_s^{\alpha,e}$ finitely often. The ones to the left of the true path are only on f_s finitely often and hence only dump finitely many balls from $R_\alpha \backslash A$ into A. If $f <_L \gamma$ then $\lim_s t_{\gamma,s}$ goes to infinity and γ can only dump each $a_s^{\alpha,e}$ into A finitely often.

Assume $\gamma = f_s$ is a positive A-node and $\alpha \subset \gamma$ and at γ^- is coded the question "Is $W_j \sqcup (W_{e_0} \cap R_\beta) = R_\beta$ infinite?" or "Is $W_j \sqcup (W_{e_1} \cap R_\beta) = R_\beta$ infinite?", for some j and some $\beta \neq \alpha$. Then $\lim_s t_{\gamma,s}$ exists and almost all balls in R_α are dumped into A, i.e. $(R_\alpha \backslash A) \subseteq^* A$. For the rest of this section we will assume the above assumption is false.

So the extra dumping at most dumps each $a_s^{\alpha,e}$ into A finitely often. Assume that $a_s^{\alpha,e}$ will not be dumped after stage s via our extra dumping. Since the original dumping only dumps to increase the e-state and there are 2^e many e-states, the original dumping only dumps $a_s^{\alpha,e}$ finitely often. Hence $\lim_s a_s^{\alpha,e}$ exists and equals $a^{\alpha,e}$.

Now we are in a position to show that the requirements $\mathcal{M}_e$ are met. Assume that $\mathcal{M}_i$ holds for $i < e$ and there is an $(e-1)$-state τ such that almost all of $R - A$ are in state τ. Assume all balls greater than k in $R - A$ are in state τ. Let

$$M = \{x : \exists s, n[\sigma(e-1,x,s) = \tau \wedge n \geq k \wedge x = a_s^{\alpha,n}]\}.$$

So $R - A \subseteq^* M$. Assume $(M \cap W_e) \backslash A$ is finite. Then $W_e \cap R \subseteq^* A \cap R$ and almost all balls in $R - A$ are in e-state τ. Now assume $(M \cap W_e) \backslash A$ is infinite. Let $n \geq k$ and $\sigma(e, a_s^{\alpha,n}, s) = \tau$. Since eventually there will be an m and stage t where $\sigma(e, a_t^{\alpha,m}, t) = \tau \cup \{e\}$, $a^{\alpha,n} \neq a_s^{\alpha,n}$. So $R - A \subseteq^* W_e$. So, A is maximal inside R_α.

5.4   Putting It All Together

Recall that if $\alpha = f$ is a positive A-node then, for some j, some i, and some $\beta \subset f$, W_j witnesses that $W_{e_i} \cap A$ is a computable subset of R_β. In this case, a Σ_3^0 event occurs. By the work in the above paragraph, we know that $A \cap R_\beta$ is maximal in R_β and hence $A \cap R_\beta$ is not computable. So A is not computable. So if $W_{e_0} \sqcup W_{e_1}$ is not a split of A we are done. Assume otherwise. By our assumption, inside R_β, $W_{e_0} \sqcup W_{e_1}$ is the trivial split. Let δ be the greatest R-subnode of α. By work in the last paragraph of Sect. 5.2, there is a set Z, such that $\bigsqcup_{\gamma \subseteq \delta} R_\gamma \sqcup (A \diagdown \tilde{R}_\delta) \sqcup Z =^* \mathbb{N}$ and $Z \diagdown A = \emptyset$. Now, by the above section, for all γ, such that $\gamma \subset \alpha$ and $\gamma \neq \beta$, $A \cap R_\gamma =^* R_\gamma$. Therefore outside of R_β, A is computable; i.e. $A \cap \overline{R_\beta}$ is computable. Any split of a computable set is trivial. Therefore, $W_{e_0} \sqcup W_{e_1}$ is a trivial split of A. So, by Theorem 4.2, A is $\mathcal{D}$-maximal.

For the remaining part of this paper, assume that f is an infinite path through $2^{<\omega}$. So a Π_3^0 event occurs. In this case, by the above section, for all $\alpha \subset f$, where α is an R-node, $A \cap R_\alpha$ is not computable. So A is not computable. If $W_{e_0} \sqcup W_{e_1}$ is not a split of A we are done. Assume otherwise. Let α be any R-node where $\alpha \subset f$. Let $A = (A \cap R_\alpha) \sqcup (A \cap \overline{R_\alpha})$. There is an R-node $\beta \neq \alpha$ on the true path. $A \cap R_\beta$ is also not computable. Hence, $(A \cap \overline{R_\alpha})$ is not computable and $A = (A \cap R_\alpha) \sqcup (A \cap \overline{R_\alpha})$ is a non-trivial split of A. The split is not Friedberg, since $R_\alpha - A$ is not a c.e. set but $R_\alpha - (A \cap \overline{R_\alpha}) = R_\alpha$ is computable. Therefore, by Theorem 4.2, A is not $\mathcal{D}$-maximal.

It just remains to show that $W_{e_0} \sqcup W_{e_1}$ is a Friedberg split of A. We know that for all $\gamma \subset f$, $W_{e_0} \sqcup W_{e_1}$ is a Friedberg split of A inside R_γ. Since splits of maximal sets are either trivial or Friedberg, otherwise the above Σ_3^0 event occurs. We must show that globally the split is Friedberg. Let's consider W_j. Let $\alpha \subset f$ such that $|\alpha| = j^2$. By the work in the second to last paragraph of Sect. 5.2, either $W_j \subseteq^* A \cup \bigsqcup_{\beta \subseteq \alpha} R_\beta$ or $W_j \cup A \cup \bigsqcup_{\beta \subseteq \alpha} R_\beta =^* \mathbb{N}$. In the first case, if $W_j - A$ is not a c.e. set neither are $W_j - W_{e_0}$ and $W_j - W_{e_1}$. Assume that $W_j \cup A \cup \bigsqcup_{\beta \subseteq \alpha} R_\beta =^* \mathbb{N}$. Furthermore, assume $W_j - A$ is not a c.e. set, but $W_j - W_{e_0}$ is a c.e. set. For any γ, where $\alpha \subset \gamma \subset f$, $(W_j - A) \cap R_\gamma$ cannot be a c.e. set since this set contains $R_\gamma - A$ and A is maximal inside R_γ. But, since $W_j - W_{e_0}$ is a c.e. set, $(W_j - W_{e_0}) \cap R_\gamma$ is a c.e. set. This contradicts the fact that our split is Friedberg inside R_γ. A similar argument works if $W_j - A$ is not a c.e. set, but $W_j - W_{e_1}$ is a c.e. set. So our split is a Friedberg split. $\square$

References

1. Cholak, P.: Some recent research directions in the computably enumerable sets. To appear in the book "The Incomputable" in the Springer/CiE book series (2013)
2. Cholak, P., Downey, R., Herrmann, E.: Some orbits for $\mathcal{E}$. Ann. Pure Appl. Log. **107**(1–3), 193–226 (2001). doi:10.1016/S0168-0072(00)00060-9. ISSN: 0168-0072
3. Cholak, P.A., Gerdes, P., Lange, K.: $\mathcal{D}$-maximal sets. J. Symb. Log. **80**(4), 1182–1210 (2015). The Url is to the first unpublished version of this paper. The first and last citation of this paper are actually to this version. arxiv:1401.1266v1. ISSN: 0022-4812
4. Downey, R.G., Stob, M.: Automorphisms of the lattice of recursively enumerable sets: orbits. Adv. Math. **92**, 237–265 (1992)
5. Friedberg, R.M.: Three theorems on recursive enumerationI. I. Decomposition. II. Maximal set. III. Enumeration without duplication. J. Symb. Logic **23**, 309–316 (1958). ISSN: 0022-4812
6. Hammond, T.: Friedberg splittings in Σ_3^0 quotient lattices of $\mathcal{E}$. J. Symb. Log. **64**(4), 1403–1406 (1999). doi:10.2307/2586786. ISSN: 0022-4812
7. Herrmann, E., Kummer, M.: Diagonals and $\mathcal{D}$-maximal sets. J. Symb. Log. **59**(1), 60–72 (1994)
8. Kummer, M.: Diagonals and semihyperhypersimple sets. J. Symb. Log. **56**(3), 1068–1074 (1991)
9. Myhill, J.: Problems. J. Symb. Log. **21**, 215 (1956). This question was the eighth problem appearing in this section
10. Owings, J.C.: Recursion, metarecursion, and inclusion. J. Symb. Log. **32**, 173–178 (1967)
11. Shavrukov, V.Y.: Friedberg splittings and $\mathcal{D}$-maximal sets. A letter from Shavrukov, V.Y., to Cholak, P. and Schmerl, J., 16 May 2015
12. Soare, R.I.: Recursively Enumerable Sets and Degrees. Perspectives in Mathematical Logic, Omega Series. Springer, Heidelberg (1987)

1-Generic Degrees Bounding Minimal Degrees Revisited

C.T. Chong[(⊠)]

Department of Mathematics, National University of Singapore,
Singapore 119076, Singapore
chongct@math.nus.edu.sg

Abstract. We show that over the base system $P^- + \Sigma_2$-bounding, the existence of a 1-generic degree $< 0''$ bounding a minimal degree is equivalent to Σ_2-induction.

1 Introduction

In considering a suitable topic on which to write for the *Festschrift*, I looked for one that would represent a common research interest between Rod Downey and me. Surprisingly, although Downey spent more than three years (1983–1986) of his early career in NUS where we were colleagues and had mathematical discussions quite regularly, there was no joint paper written until several years after he moved to New Zealand. During one of his fairly frequent visits to Singapore, we thought it would be a good idea to work on a problem together. That collaboration resulted in two papers published on the ordering relation between 1-generic degrees and minimal degrees (Chong and Downey [1,2]), and in essence is the subject of this paper, now viewed from a different perspective.

Let $n \geq 1$. A set G of natural numbers is *n-generic* if for each Σ_n-definable set Y of binary strings, there is an initial segment σ of G such that either $\sigma \in Y$ or no extension of σ belongs to Y. The Turing degree of an n-generic set is called an n-generic degree. A set A has minimal degree if every set of strictly lower Turing degree is recursive. Methodologically these sets are constructed differently: the former uses Cohen forcing while the latter uses perfect set forcing introduced by Spector. The class of n-generic sets and the class of sets of minimal degree are mutually exclusive. In fact, no n-generic set is recursive in a set of minimal degree. Jockusch [8] proved the converse for $n \geq 2$, that no n-generic set bounds a set of minimal degree, so that G and A are Turing incomparable. This incomparability extends to $n = 1$ when $G \leq_T \varnothing'$ (Chong and Jockusch [3]). Haught [7] strengthened this result by showing that every nonzero Turing degree below a 1-generic degree $< 0'$ is 1-generic. The question then turned to whether a 1-generic degree in general could bound a minimal degree. In Chong

This research was partially supported by grant C146-000-042-001 at the National University of Singapore. The author wishes to thank the referee for a comment on the proof of Theorem 4 in [2] (see Footnote 1 in Sect. 3).

A. Day et al. (Eds.): Downey Festschrift, LNCS 10010, pp. 536–546, 2017.
DOI: 10.1007/978-3-319-50062-1_32

and Downey [2] we constructed a 1-generic set $G \leq_T \varnothing''$ bounding a set of minimal degree (in fact the minimal degree is recursive in $\varnothing'$). This was also independently obtained by Kumabe [9]. In [1] we showed the existence of a minimal degree below $\mathbf{0}'$ not bounded by a 1-generic degree. Thus the ordering relation between 1-generic degrees and minimal degrees is fairly complex. This complexity—at least at the $\mathbf{0}''$ level—turns out to depend on the strength of the underlying mathematical theory, i.e. the strength of mathematical induction allowed. Specifically, we show in this paper that Σ_2-induction implies the existence of a 1-generic degree $< \mathbf{0}''$ bounding a minimal degree, and that the theorem fails in the weaker theory where only Δ_2-induction is assumed. Indeed, over the base system $P^- + \Sigma_2$-bounding, the statement asserting the existence of a 1-generic degree $< \mathbf{0}''$ bounding a minimal degree is equivalent to Σ_2-induction.

2 Preliminaries

Let P^- be the set of axioms of Peano arithmetic minus the scheme of mathematical induction. $I\Sigma_n$ is induction for Σ_n-formulas and $B\Sigma_n$ denotes the principle of Σ_n-bounding, which states that every Σ_n-definable function maps a finite set of numbers into a bounded set. Slaman [11] showed that over P^-, $B\Sigma_n$ is equivalent to induction scheme for provably Δ_n-formulas. It is known that $I\Sigma_n$ is strictly stronger than $B\Sigma_n$. The subject of recursion theory in the context of fragments of Peano arithmetic has been extensively investigated (see Chong, Li and Yang [4] for an exposition). In this paper the focus is on $n = 2$. We first state a theorem whose proof is a straightforward adaptation of that in Sacks [10]:

Theorem 2.1. $P^- + I\Sigma_2 \rightarrow$ *There is a minimal degree* $< \mathbf{0}'$.

On the other hand, whether there is a minimal degree $< \mathbf{0}'$ under the weaker system $P^- + B\Sigma_2$ remains a major open problem. In many respects, this is the arithmetic version of the long standing question of whether there is a minimal α-degree for admissible ordinals α such as $\alpha = \aleph_\omega^L$. In spite of this knowledge gap, it still makes sense to ask about the ordering relation between 1-generic degrees and minimal degrees for models of this system and seek a satisfactory resolution.

Unless otherwise stated, for the rest of this paper we take $\mathcal{M} = (M, +, \cdot)$ to be a model of $P^- + B\Sigma_2 + \neg I\Sigma_2$. Such models are called $B\Sigma_2$-models for short. Let $2^{<M}$ be the collection of $\mathcal{M}$-finite binary strings. Let $G \subseteq M$. $Y \subseteq 2^{<M}$ is G-dense if every initial segment of G is extended by some member of Y.

Definition 2.2. *A set* $G \subset M$ *is 1-generic if whenever* $Y \subseteq 2^{<M}$ *is G-dense then it contains an initial segment of G.*

We say that G is *regular* if $G \restriction s$ is $\mathcal{M}$-finite for every $s \in M$. It is not difficult to verify that every 1-generic set is regular. An example, albeit misleading, of a nonregular set is a *cut*: a bounded set that is closed downwards as well as under the successor function. The set of natural numbers is a cut in $\mathcal{M}$ though not necessarily definable in the model. By the regularity of G, one can show that G is GL_1, in the sense that G' is pointwise Turing equivalent to $G \oplus \varnothing'$. Furthermore, if $G \leq_T \varnothing'$ then it is low $(G' \equiv_T \varnothing')$ and so $I\Sigma_1$ holds relative to G.

Proposition 2.3. *In $\mathcal{M}$ there is a Σ_2-definable cut I and a Σ_2-definable cofinal increasing function $g : I \to M$.*

We fix I and g to be as above and call I a Σ_2-cut.

The next proposition was proved in Chong and Yang [6] and is a basic fact about regular sets below $\varnothing''$ to be used in this paper.

Proposition 2.4. *Every regular set recursive in $\varnothing''$ is recursive in $I \oplus \varnothing'$. The conclusion fails for nonregular sets.*

Computation involving I as part of an oracle has a particularly simple representation. Fix an upper bound a of I. Without loss of generality, one may identify a positive condition about I to be a point $k \in I$ and a negative condition about I to be a point $l \leq a$ in $\overline{I}$. This is possible since any $\mathcal{M}$-finite subset of I is bounded above by a $k \in I$, and any $\mathcal{M}$-finite subset of $\overline{I}$ is bounded below by an $l \in \overline{I}$, and any reduction procedure that uses a negative condition $u > a$ may be replaced by one that uses the negative condition $u = a$. Then if $G = \Phi^{I \oplus \varnothing'}$, one has for each $i \in I$, a $j \in I$ and a pair $(k, l) \in I \times \overline{I}$ such that $G \restriction g(i) = \Phi^{(k,l) \oplus \varnothing'} \restriction g(j)$.

A property about 1-generic sets below $\varnothing'$ which was used in Haught [7] states:

(*) If G is 1-generic $< \varnothing'$ and $\{\gamma_i\}_{i \in \omega} \subseteq 2^{<\omega}$ is a recursive approximation of G (i.e. $\lim_i \gamma_i(x) = G(x)$ for all x), then every infinite Σ_1-subset of ω contains an i such that γ_i is an initial segment of G (written $\gamma_i \prec G$).

This property continues to hold for 1-generic sets $<_T \varnothing'$ in models of $P^- + B\Sigma_2$. An analog of this for $G \leq_T I \oplus \varnothing'$ is the following:

Lemma 2.5. *Let a be an upper bound of a Σ_2-cut I. If $G \leq_T I \oplus \varnothing'$ is 1-generic, then there is an $\mathcal{M}$-finite set $X \subset a \times a$ and a uniform r. e. sequence of sets $\{X_{k,l}\}_{(k,l) \in X}$, where $X_{k,l} \subset 2^{<M}$, with the following properties:*

(i) *For each $(k, l) \in X$, there is a $\gamma \in X_{k,l}$ such that $\gamma \prec G$;*
(ii) $\max_{(k,l) \in X} \{\gamma : \gamma \in X_{k,l} \wedge \gamma \prec G\}$ *is unbounded in M.*
(iii) *If S is an $\mathcal{M}$-infinite Σ_1-subset of $\{(k, l, s) : (k, l) \in X \wedge s \in M\}$, then for each (k, l), either there is an s where $(k, l, s) \in S$ and an initial segment of G which is enumerated in $X_{k,l}$ at stage s, or there is an initial segment of G for which none of its extensions is in $X_{k,l}$ at any stage s where $(k, l, s) \in S$.*

Proof. Let $\varnothing'$ be approximated by the recursive sequence $\{\sigma_s : s \in M\}$ and assume $\Phi_G^{I \oplus \varnothing'} = G$. Fix $k_0 \in I$ and $l_0 \in \overline{I}$ such that $\Phi_G^{(k_0, l_0) \oplus \varnothing'} \succ \gamma_0$ for some $\varnothing \prec \gamma_0 \prec G$. Let $X = \{(k, l) \in a \times a : k_0 \leq k < l \leq l_0\}$, let

$$X_{k,l} = \{\gamma : \gamma \succeq \gamma_0 \wedge \exists s (\Phi_G^{(k,l) \oplus \sigma_s}[\|\sigma_s\|] = \gamma\}.$$

Clearly (i) holds since $\gamma_0 \in X_{k,l}$ for each $(k,l) \in X$. Since $\Phi_G^{I \oplus \varnothing'} = G$, there are cofinally many $k \in I$ and corresponding $l(k) \in a \setminus I$ such that, by the 1-genericity of G, $X_{k,l(k)}$ contains a $\gamma \prec G$ properly extending γ_0. Furthermore, cofinally many initial segments of G are enumerated into $\bigcup_{(k,l) \in X} X_{k,l}$, giving (ii). It is straightforward to verify that (iii) also follows from the 1-genericity of G. $\square$

3   An Existence Theorem Under Σ_2-Induction

The proof of the following theorem follows from the observation that the construction in [2] (proof of Theorem 4) may be implemented in a model of $P^- + I\Sigma_2$. We give a sketch of the main idea and leave the details to the reader.

Theorem 3.1. *In a model of $P^- + I\Sigma_2$, there is a 1-generic degree $< \mathbf{0}''$ bounding a minimal degree.*

Proof. Let $\mathcal{M} = (M, +, \cdot)$ be a model of $P^- + I\Sigma_2$ and $A \subset M$. We say that $Y \subseteq 2^{<M}$ is Σ_1-*dense* in A if it is Σ_1, contains no initial segment of A, and the following holds: For any Σ_1 $W \subseteq 2^{<M}$ such that every initial segment of A is extended by some member of W, and no member of W is an initial segment of A, there is a $\sigma \in Y$ extended by a $\tau \in W$. We say in this case that A has a Σ_1-dense set. The proof of the next lemma is an adaptation of the classical construction of a set of minimal degree below $\varnothing'$, originally due to Sacks [10].

Lemma 3.2. *There is a set $A < \varnothing'$ of minimal degree that has no Σ_1-dense set.*

Lemma 3.3. *Any $A < \varnothing'$ with no Σ_1-dense set is recursive in a 1-generic set below $\varnothing''$.*

Let A be the set of minimal degree constructed in Lemma 3.2. Define a partial recursive functional $\Phi : 2^{<M} \to 2^{<M}$ and a 1-generic G such that $\sigma \prec \tau \to \Phi(\sigma) \preceq \Phi(\tau)$ (if both are defined) and $\Phi(G) = A$. Let $S \subseteq 2^{<M}$ be Σ_1. Φ and G will satisfy three conditions:

(i) For any Σ_1 $S \subseteq 2^{<M}$, if no initial segment of A belongs to $Y = \{\Phi(\sigma) : \sigma \in S\}$, then there is a $\gamma \prec G$ with no extension in S;

(ii) If cofinally many initial segments of G are extended by strings in S (S is Σ_1), then there is a $\sigma \in S$ such that $\sigma \prec G$ and $\Phi(\sigma) \prec A$;

(iii) $\gamma \prec G \to \Phi(\gamma) \prec A$, and cofinally many initial segments of A are of the form $\Phi(\gamma)$ for some $\gamma \prec G$.

Conditions (i)–(iii) ensure that G is 1-generic and $\Phi(G) = A$. Satisfying (i) is worth particular note since one appeals to Lemma 3.3 to succeed. Suppose S is Σ_1 and every initial segment of A is extended by some member of $Y = \{\Phi(\sigma) : \sigma \in S\}$, but no member of Y is an initial segment of A. Since A has no Σ_1-dense set, there is a Y^* witnessing the failure of Y being Σ_1-dense in A. Thus every initial segment of A is extended by some member of Y^* while no member of Y

extends a member of Y^*. The idea is then to define Φ and G so that for some $\gamma \prec G$, if $\sigma \succ \gamma$ and $\Phi(\sigma)$ is defined, then $\Phi(\sigma) \in Y^*$. It follows that (i) is satisfied for S.

There are two main issues to resolve to make this work. Firstly the set G is chosen only after Φ is defined. Thus it is not possible to arrange at the beginning of the construction which S will meet the hypothesis of (i) and hence which Y to search for a Y^*. Secondly given an index for Y it is not possible to recursively find an index for Y^*. The strategy is to create "sufficient space" for different guesses of where G will be and which of (i)–(ii) will S fall under. This problem is addressed by introducing, for every string σ potentially in the domain of Φ, a recursive sequence of M-many pairwise incompatible extensions $\sigma^{u,j}$ (for $u \in M, j < 2$) of σ such that if $\sigma' \succ \sigma^{u,j}$ and $\Phi(\sigma')$ is defined, then it is a substring of a string in Y_u (the uth Σ_1-set of strings). More precisely, tag each string that may be in the domain of Φ with an $\mathcal{M}$-finite function f. The function that is tagged to a string may change at different stages, but will eventually stabilize. The requirement is that if γ is (eventually) tagged with f and $\Phi(\gamma)$ is defined, then for any $\sigma' \succ \gamma$, if $\Phi(\sigma')$ is defined then it is a substring of a member of $Y_{f(x)}$ for each $x \in \mathrm{Dom}(f)$. It follows that if γ has a final tag f, $Y^* = Y_{f(x')}$ for some $x' \in \mathrm{Dom}(f)$ and γ is chosen to be an initial segment of G, then (i) is satisfied.

Of course there is tension between the attempt to satisfy (i) and the attempt to satisfy (ii), (iii). This is due to the fact that, on the one hand, if σ is tagged with f, then $\Phi(\sigma)$ is not defined until a τ is enumerated with the property that for each x in the domain of f, τ is a substring of a string in $Y_{f(x)}$, making τ a candidate to be chosen as $\Phi(\sigma)$, while on the other hand, to satisfy (ii) or (iii), if $\sigma \in S$ is seen to extend an initial segment of G, then the demand of 1-genericity would require σ to be chosen as an initial segment of G (and hence not to wait for the enumeration of a τ that is a substring of a string in $Y_{f(x)}$ for each x in the domain of f). These are competing requirements and are resolved by an ordering of priorities, as is standard in a recursion-theoretic construction.

By invoking Σ_2-induction, one shows that every string σ has a final tag in the limit.[1] A $\varnothing''$-recursive increasing sequence of strings $\{G_s\}_{s \in M}$ is defined by induction on s. An analysis of the construction in [2] shows that G_s is defined for each $s \in M$, if $\mathcal{M}$ is a model of $P^- + I\Sigma_2$. Let $G = \bigcup_s G_s$. Then G is recursive in $\varnothing''$. The 1-genericity of G follows from the verification that (i)–(iii) are satisfied. Corollary 4.3 in the next section implies that the construction fails in any model of a weaker system.   □

---

[1] The referee has pointed out that in the proof of Theorem 4 in [2], Subcase (1) of Case B (in the construction of the 1-generic set G) has not considered the possibility that $\tau_s^{\varnothing,j}(\sigma)$ is not $(n+1)$-attended for all $t > s$ but notes, however, that as the tag assigned to $\tau_s^{\varnothing,j}(\sigma)$ is the same as that assigned to σ (except for elements that are mapped to $\varnothing$), this tag will eventually be confirmed and the construction as presented succeeds. This observation applies to any model of Σ_2 induction.

4 Nonexistence of a 1-Generic Degrees Bounding a Minimal Degree

For the rest of the paper, we work within a $B\Sigma_2$ model $\mathcal{M}$. The following extends the result in Chong and Jockusch [3] to $B\Sigma_2$-models beyond the degree of the halting set.

Theorem 4.1. *Let $G \leq_T I \oplus \varnothing'$ be 1-generic. If $\varnothing <_T B \leq_T G$, then B computes a 1-generic set.*

Corollary 4.2. *No 1-generic set recursive in $I \oplus \varnothing'$ bounds a set of minimal degree.*

Lemma 2.4, Corollary 4.2 and Theorem 3.1 together yield the following characterization.

Corollary 4.3. *Over the base system $P^- + B\Sigma_2$, the following are equivalent:*

(i) *Σ_2-induction;*
(ii) *There is a 1-generic degree $< \mathbf{0}''$ bounding a minimal degree.*

Proof of Theorem 4.1.
Let $\Phi_G^{I \oplus \varnothing'} = G$ and $\Phi_B^G = B$. Let X and $\{X_{k,l}\}_{(k,l) \in X}$ be as in Lemma 2.5. We construct a 1-generic set $D \leq_T B$ following the approach in [3]. The absence of Σ_2-induction adds complexity to the construction and verification. Define a partial recursive functional θ on $2^{<M}$ so that $\theta(B) = D$ (i.e. cofinally many initial segments of B are in the domain of θ and are mapped onto cofinally many initial segments of D).

A tree T of height s is an $\mathcal{M}$-finite function with domain $2^{<s}$ such that $\sigma \prec \tau$ if and only if $T(\sigma) \prec T(\tau)$ for $\sigma, \tau \in 2^{<s}$. T is Φ_B-splitting if $\Phi_B^{T(\gamma)}$ and $\Phi_B^{T(\gamma')}$ are incompatible if γ, γ' are incompatible strings in the domain of T. The proof of the following claim, which applies to an arbitrary 1-generic set, is similar to that in [3].

Claim 1. For every s and $\gamma \prec G$, there is a Φ_B-splitting tree T of height s with $T(\varnothing) \succeq \gamma$.

The construction of the set D is split into two cases:

Case 1. There is a $(k,l) \in X$ such that $X_{k,l} \cap G$ is unbounded.

Since $\{\sigma_s : s \in M\}$ is a recursive approximation of $\varnothing'$, we have $\Phi_G^{(k,l) \oplus \varnothing'} = G$ and hence $G \leq_T \varnothing'$. In the construction below, we will avoid using this fact so that the argument is applicable to Case 2. Define the partial recursive functional θ as follows:

Fix γ_0 as in Lemma 2.5. Let $\theta_1 = T_1 = \gamma_1 = \varnothing$. Suppose θ_s, T_s and γ_s are defined, where T_s is a Φ_B-splitting tree of height s and $T_s(\varnothing) = \gamma_s$. Recursively search for the least $\gamma \in X_{k,l}$, denoted γ_{s+1}, such that $|\gamma| > |\gamma_s|$ and a least Φ_B-splitting tree, denoted T_{s+1}, with the following properties:

(1) T_{s+1} has height $s + 1$;
(2) $T_{s+1}(\varnothing) = \gamma_{s+1}$;
(3) $|T_{s+1}(\varnothing)| > |T_s(\tau)|$ for every τ in the domain of T_s.

Let ν_s be the string in the domain of θ_s with the greatest length extended by $\Phi_B^{T_{s+1}(\varnothing)}$. Let $\theta_{s+1}(\nu) = \theta_s(\nu)$ for $\nu \in \mathrm{Dom}(\theta_s)$. For $\tau \in 2^{<s}$, let

$$\theta_{s+1}(\Phi_B^{T_{+1s}(\tau)}) = \theta_s(\nu_s)^\frown \tau.$$

Let $\theta = \bigcup_s \theta_s$. Note that θ is consistent, meaning if $\nu \prec \nu'$ are in the domain of θ, then $\theta(\nu) \preceq \theta(\nu')$. Furthermore, cofinally many initial segments of G are extended by strings of the form $T_s(\varnothing)$ since $\Phi_G^{(k,l)\oplus\varnothing'} = G$. By the 1-genericity of G, cofinally many such $T_s(\varnothing)$'s are initial segments of G. Hence $\Phi_B^{T_s(\varnothing)} \prec B$. Since $\Phi_B^{\Phi_G^{(k,l)\oplus\varnothing'}} = B$, we see that θ is total on B. Thus $\theta(\Phi_B^G) = \theta(B)$ which we denote as D.

The construction gives the following useful fact which is proved by induction on s (cf. Lemma 2 (iii) of [3]):

Claim 2. If $\delta \succ \theta_s(\Phi_B^{T_s(\tau)})$, then either $\delta = \theta_s(\Phi_B^{T_s(\tau')})$ for some $\tau' \in \mathrm{Dom}(T_s)$, or δ extends a maximal string in $\theta_s(\Phi_B^{T_s})$.

We show that $D = \theta(B)$ is 1-generic. Suppose $\Delta = \{\delta_s : s \in M\}$ is a Σ_1-set of strings such that every initial segment of D is extended by some member of Δ. Assume for the sake of contradiction that there is no s with $\delta_s \prec D$.

Claim 3. There is an s_0 and a γ_Δ such that $\gamma_\Delta = T_{s_0}(\varnothing) \prec G$ and $\theta(\Phi_B^{T_s(\tau)}) \notin \Delta$ for any $s \geq s_0$ and $\tau \in \mathrm{Dom}(T_s)$ such that $T_s(\tau) \succ T_{s_0}(\varnothing)$.

Since $\theta(\Phi_B^G) = D$ and there is no initial segment of D that belongs to Δ, the set $\{T_s(\tau) : s \in M \wedge \theta(\Phi_B^{T_s(\tau)}) \in \Delta\}$ is Σ_1 and has no intersection with G. Hence by the 1-genericity of G there is an s_0 such that $T_{s_0}(\varnothing) \prec G$ and for any $s \geq s_0$ and τ, if $T_s(\tau) \succ T_{s_0}(\varnothing)$ then $\theta(\Phi_B^{T_s(\tau)}) \notin \Delta$. Let $\gamma_\Delta = \Phi_B^{T_{s_0}(\varnothing)}$.

Define

$$U_1 = \{T_s(\tau) : s \geq s_0 \wedge \exists t(|\delta_t| < s \wedge \theta_s(\Phi_B^{T_s(\tau)}) \text{ is maximal in } \theta_s(\Phi_B^{T_s}) \wedge \theta_s(\Phi_B^{T_s(\tau)}) \prec \delta_t)\}.$$

Claim 4. There is an $s_1 \geq s_0$ and a $\gamma_{U_1} \prec G$ such that $\gamma_\Delta \prec \gamma_{U_1}$ and for all $s \geq s_1$ and τ, if $T_s(\tau) \succ \gamma_{U_1}$ then it is not in U_1.

Let $s \geq s_0$. Suppose $T_s(\tau) \in U_1$ is an initial segment of G that extends γ_Δ, and δ_t is as given in U_1. Then there is a (least) $s' > s$ such that $T_{s'}(\varnothing) \succ T_s(\tau)$. Now the construction at stage s' ensures that every string of length less than s' is the image of some $\theta_{s'}(\Phi_B^{T_{s'}(\tau')})$. This applies in particular to δ_t. But then δ_t is in the range of θ, contradicting the choice of s_0 for Claim 3. Thus no initial segment of G belongs to U_1. By the 1-genericity of G, there is an initial segment $\gamma_{U_1} \succ \gamma_\Delta$ of G such that $\gamma_{U_1} = T_{s_1}(\varnothing)$ for some $s_1 > s_0$, and no extension of γ_{U_1} belongs to U_1. Hence Claim 4 holds.

Let $\gamma \prec G$ be chosen so that $\gamma = T_s(\varnothing)$ for some $s > s_1$, $\gamma_{U_1} \prec \gamma$, $|\gamma| > s_1$, $\theta(\Phi_B^\gamma)$ is defined and has length greater than $\max\{|\theta_{s_1}(\nu)| : \nu \in \mathrm{Dom}(\theta_{s_1})\}$. Then $\theta(\Phi_B^\gamma)$ is an initial segment of D. By assumption on Δ there is a $\delta_t \in \Delta$ extending $\theta(\Phi_B^\gamma)$. Choose $s_2 > s_1$ such that $s_2 > \max\{|\delta_t|, t\}$ and $\theta_{s_2}(\Phi_B^\gamma)$ is defined. By Claim 2, at stage s_2 either $\delta_t = \theta_{s_2}(\nu)$ for some $\nu \in \mathrm{Dom}(\theta_{s_2})$, or δ_t extends a maximal string $\theta_{s_2}(\nu)$ for some ν in $\mathrm{Dom}(\theta_{s_2})$.

Assume $\delta_t = \theta_{s_2}(\nu) = \theta_{s_2}(\Phi_B^{T_{s_2}(\tau)})$. As $|\theta_{s_2}(\Phi_B^\gamma)| > \max\{|\theta_{s_1}(\nu)| : \nu \in \mathrm{Dom}(\theta_{s_1})\}$, the same is true of δ_t and hence $\delta_t \notin \mathrm{Range}(\theta_{s_1}) \supset \mathrm{Range}(\theta_{s_0})$. However, the choice of γ implies that $T_{s_2}(\tau) \succ T_{s_0}(\varnothing)$, since $\delta_t \succ \gamma_U \succ \gamma_\Delta$. By Claim 3 we see that $\delta_t \notin \mathrm{Range}(\theta)$ and this is a contradiction.

Now suppose that $\delta_t \succ \theta_{s_2}(\nu)$ where $\theta_{s_2}(\nu)$ is a maximal string in the range of θ_{s_2}. Then $\theta_{s_2}(\nu) \succ \theta_{s_2}(\Phi_B^\gamma)$ and this implies that $\theta_{s_2}(\nu)$ is not in the range of θ_{s_1} since $|\theta_{s_2}(\Phi_B^\gamma)|$, hence $|\theta_{s_2}(\nu)|$, is longer than every string enumerated in the range of θ_{s_1}. Thus $\nu \notin \mathrm{Dom}(\theta_{s_1})$. Let $\nu = \Phi_B^{T_{s_2}(\tau')}$. Then $T_{s_2}(\tau') \succ \gamma_U$. This implies that $T_{s_2}(\tau') \in U_1$ which contradicts the choice of γ_{U_1} for Claim 4.

Case 2. For all $(k, l) \in X$, $X_{k,l} \cap G$ is bounded.

We leave the proof of the next claim to the reader.

Claim 5. For each $(k, l) \in X$, $\Phi_G^{(k,l) \oplus \varnothing'}$ is (possibly $\mathcal{M}$-finite and in any case) recursive in $\varnothing'$ but not equal to G. Furthermore by Lemma 2.5 (iii), $\bigcup_{(k,l) \in X} \{\gamma \in X_{k,l} : \gamma \prec G \wedge \gamma \prec \Phi_G^{(k,l) \oplus \varnothing'}\}$ is unbounded.

We now define the partial recursive functional θ. Due to Claim 5, the definition of θ will not focus on just one (k, l), but all $(k, l) \in X$. Denote by θ_s the subset of θ that is defined after s steps of computation.

Let γ_0 be as defined in Lemma 2.5. Let $\theta_0(\Phi_B^{\gamma_0}) = \varnothing$. For each $(k, l) \in X$, let $\gamma_{1,k,l} = \gamma_0$. Let $T_{1,k,l}$ be the first Φ_B-splitting tree of height 1 enumerated at the least stage $t_{1,k,l} \geq s$ (if it exists) such that $T_{1,k,l}(\varnothing) = \gamma_0$. Suppose $s \geq 1$ and $\gamma_{s,k,l} \in X_{k,l}$ and $T_{s,k,l}$ are defined at stage $t_{s,k,l}$, where $T_{s,k,l}$ is a Φ_B-splitting tree of height s with $T_{s,k,l}(\varnothing) = \gamma_{s,k,l}$. For each (k, l), recursively search for the least stage $t_{s+1,k,l} \geq \max\{t_{s,k,l}, s+1\}$ where a $\gamma \in X_{k,l}$ is enumerated satisfying the following conditions:

(4) $|\gamma| > |\gamma_{s,k,l}|$;

(5) There is a Φ_B-splitting tree T of height $s + 1$ such that $T(\varnothing) = \gamma$;

(6) $|T(\varnothing)| > |T_{s,k,l}(\tau)|$ for every τ in the domain of $T_{s,k,l}$.

(7) Φ_B^γ is not a substring of any $\theta_t(\Phi_B^{\gamma'})$ that is defined for some γ' and $t \leq t_{s+1,k,l}$.

Let $\gamma_{s+1,k,l}$ be the least such γ and let $T_{s+1,k,l}$ be its corresponding least tree T. Let $\nu_{s,k,l}$ be the string in $\mathrm{Dom}(\theta_{t_{s,k,l}})$ with the greatest length extended by $\Phi_B^{T_{s+1,k,l}(\varnothing)}$. For $\tau \in 2^{<s}$, let

$$\theta_{t_{s+1,k,l}}(\Phi_B^{T_{s+1,k,l}(\tau)}) = \theta_{t_{s,k,l}}(\nu_{s,k,l})^\frown \tau.$$

Define $\theta_{t_{s+1,k,l}}(\nu) = \theta_t(\nu)$ if $t < t_{s+1,k,l}$ and $\theta_t(\nu)$ is defined. Condition (7) guarantees that θ is consistent. Clearly θ is a partial recursive functional. Note by Claims 1 and 5 that the domain of θ is unbounded. In particular, for each $s \geq 1$ there is a (k,l) such that $t_{s+1,k,l}$ is defined. Furthermore, Lemma 2.5 (iii) ensures that for cofinally many s, there is a $(k,l) \in X$ such that $T_{s,k,l}(\varnothing) \prec G$. This gives

Claim 6. θ is total on $\Phi_B^G = B$, i.e. $\theta(\Phi_G^\gamma)$ is defined for cofinally many $\gamma \prec G$ of the form $T_{s,k,l}(\varnothing)$.

As before we show that $\theta(B) = D$ is 1-generic. Let Δ be a Σ_1-set of strings such that every initial segment of D is extended by some member of Δ. We claim that $\delta \prec D$ for some $\delta \in \Delta$. Assume for the sake of contradiction that no $\delta \in \Delta$ is an initial segment of D. The following is an analog of Claim 3.

Claim 7. There exist $\gamma_\Delta \prec G$, (s_3, k_3, l_3) such that $\gamma_\Delta = T_{s_3,k_3,l_3}(\varnothing)$ with the property that $\theta_{t_{s_3,k_3,l_3}}(\Phi_B^{T_{s_3,k_3,l_3}(\tau)})$ is defined for $\tau \in \mathrm{Dom}(T_{s_3,k_3,l_3})$, and for all $\gamma \succ \gamma_\Delta$ and $s \geq s_3$, if $\theta_{t_{s,k,l}}(\Phi_B^\gamma)$ is defined for some $(k,l) \in X$ then it is not in Δ.

Since no initial segment of D belongs to Δ, the Σ_1-definable set

$$\{\gamma : \exists(k,l)\exists s((k,l) \in X \wedge \theta_{t_{s,k,l}}(\Phi_B^\gamma) \in \Delta)\}$$

has empty intersection with G. Hence by the 1-genericity of G there is a $\gamma_\Delta \prec G$ for which no extension belongs to the set. We may choose γ_Δ, (s_3, k_3, l_3) such that $\gamma_\Delta = T_{s_3,k_3,l_3}(\varnothing)$ (hence $\theta_{t_{s_3,k_3,l_3}}(\Phi_B^{\gamma_\Delta})$ is defined). Then γ_Δ and (s_3, k_3, l_3) satisfy the requirements of Claim 7.

Let

$$U_2 = \{T_{s,k,l}(\tau) : s \geq s_3 \wedge \exists t(|\delta_t| < s \wedge \theta_{t_{s,k,l}}(\Phi_B^{T_{s,k,l}(\tau)}) \text{ is maximal in}$$
$$\theta_{t_{s,k,l}}(\Phi_B^{T_{s,k,l}}) \wedge \theta_{t_{s,k,l}}(\Phi_B^{T_{s,k,l}(\tau)}) \prec \delta_t)\}.$$

Claim 8. There is an (s_4, k_4, l_4), where $s_4 \geq s_3$, and a $\gamma_{U_2} \prec G$ such that $\gamma_\Delta \prec \gamma_{U_2}$, $T_{s_4,k_4,l_4}(\varnothing) = \gamma_{U_2}$, and for all $s \geq s_4$, $(k,l) \in X$, if $T_{s,k,l}(\tau) \succ \gamma_{U_2}$ then $T_{s,k,l}(\tau)$ is not in U_2.

Suppose $s \geq s_3$ and $\gamma_\Delta \prec T_{s,k,l}(\tau) \prec G$ where $T_{s,k,l}(\tau)$ belongs to U_2. By Claim 6 there is an $s' > s$ and $(k',l') \in X$ such that $T_{s,k,l}(\tau) \prec T_{s',k',l'}(\varnothing) \prec G$. Let δ_t be as given in U_2 for $T_{s,k,l}(\tau)$ so that $|\delta_t| < s < s'$. Then the definition of θ implies that δ_t is in the range of $\theta_{t_{s',k',l'}}$, contradicting the choice of γ_Δ. Hence there is an initial segment of G extending γ_Δ for which no extension belongs to U_2. Let (s_4, k_4, l_4) and $\gamma_\Delta \prec \gamma_{U_2} \prec G$ be chosen so that $T_{s_4,k_4,l_4}(\varnothing) = \gamma_{U_2}$ to satisfy the prescribed requirement.

Now let $\gamma \prec G$ be a string extending γ_{U_2} satisfying the following for some $(k, l) \in X$ and (least) $s > s_4$:

(8) $|\gamma| > s_4$;
(9) $\gamma = T_{s,k,l}(\varnothing)$;
(10) $\theta_{t_{s,k,l}}(\Phi_B^\gamma)$ is defined and has length greater than $\max\{|\theta_{t_{s_4,k_4,l_4}}(\nu)| : \nu \in \mathrm{Dom}(\theta_{t_{s_4,k_4,l_4}})\}$.

Fix (s, k, l). Then $\theta(\Phi_B^\gamma)$ is an initial segment of D and there is a $\delta_t \in \Delta$ extending it. Choose $s_5 > s_4$ such that $s_5 > \max\{|\delta_t|, t\}$ and $\theta_{t_{s_5,k,l}}(\Phi_B^\gamma)$ is defined. Then the construction ensures that an analog of Claim 2 holds here: either $\delta_t = \theta_{t_{s_5,k,l}}(\nu)$ for some ν in the domain of $\theta_{t_{s_5,k,l}}$, or δ_t extends a maximal string in the range of $\theta_{t_{s_5,k,l}}$.

Assume first that $\delta_t = \theta_{t_{s_5,k,l}}(\nu) = \theta_{t_{s_5,k,l}}(\Phi_B^{T_{s_5,k,l}(\tau)})$. Then $T_{s_5,k,l}(\varnothing) \succeq \gamma$. As $|\theta_{t_{s,k,l}}(\Phi_B^\gamma)| > \max\{|\theta_{t_{s_4,k_4,l_4}}(\nu)| : \nu \in \mathrm{Dom}(\theta_{t_{s_4,k_4,l_4}})\}$, the same is true of δ_t. It follows that $\delta_t \notin \mathrm{Range}(\theta_{t_{s_4,k_4,l_4}}) \supset \mathrm{Range}(\theta_{t_{s_3,k_3,l_3}})$. However, the choice of γ implies that $T_{s_5,k,l}(\tau) \succ \gamma_\Delta$, since $\delta_t \succ \gamma \succ \gamma_{U_2} \succ \gamma_\Delta$. By Claim 7 we see that $\delta_t \notin \mathrm{Range}(\theta)$ and this is a contradiction.

Now suppose that $\delta_t \succ \theta_{t_{s_5,k,l}}(\nu)$, a maximal string in the range of $\theta_{t_{s_5,k,l}}$. Then $\theta_{t_{s_5,k,l}}(\nu) \succ \theta_{t_{s_5,k,l}}(\Phi_B^\gamma)$ and this implies that $\theta_{t_{s_5,k,l}}(\nu)$ is not in the range of $\theta_{t_{s_4,k_4,l_4}}$ since $|\theta_{t_{s_5,k,l}}(\Phi_B^\gamma)|$, hence $|\theta_{t_{s_5,k,l}}(\nu)|$, is larger than the length of every string enumerated in the range of $\theta_{t_{s_4,k_4,l_4}}$. Thus $\nu \notin \mathrm{Dom}(\theta_{t_{s_4,k_4,l_4}})$. Let $\nu = \Phi_B^{T_{s_5,k,l}(\tau')}$. Then $T_{s_5,k,l}(\tau') \succ \gamma_{U_2}$. This implies that $T_{s_5,k,l}(\tau') \in U_2$ which contradicts the choice of γ_{U_2}.

This completes the proof of Theorem 4.1.   □

Remark 1. It is not known if in a $B\Sigma_2$ model there is a 1-generic degree bounding a minimal degree. While one can construct a set of minimal degree with no Σ_1-dense set in a countable $B\Sigma_2$ model, such a set need not be definable. The proof of Theorem 3.1 shows that if A is a set of minimal degree that has no Σ_1-dense set, and $\mathcal{M} \models P^- + I\Sigma_2(A)$, then A is bounded by a 1-generic $G \leq_T A \oplus \varnothing''$. The argument breaks down in the absence of $I\Sigma_2(A)$.

Remark 2. A simpler reverse mathematics type question, but not necessarily easier, is whether over a countable $B\Sigma_2$ model $\mathcal{M}$, there is a set A of minimal degree such that $\mathcal{M}[A] \models B\Sigma_2$.

References

1. Chong, C.T., Downey, R.G.: Degrees bounding minimal degrees. Math. Proc. Camb. Philos. Soc. **105**, 211–222 (1989)
2. Chong, C.T., Downey, R.G.: Minimal degrees recursive in 1-generic degrees. Ann. Pure Appl. Logic **48**, 215–225 (1990)
3. Chong, C.T., Jockusch, C.G.: Minimal degrees and 1-generic sets below 0′. In: Börger, E., Oberschelp, W., Richter, M.M., Schinzel, B., Thomas, W. (eds.) Computation and Proof Theory. LNM, vol. 1104, pp. 63–77. Springer, Heidelberg (1984). doi:10.1007/BFb0099479

4. Chong, C.T., Li, W., Yang, Y.: Nonstandard models in recursion theory and reverse mathematics. Bull. Symbolic Logic **20**, 170–200 (2014)
5. Chong, C.T., Mourad, K.J.: The degree of a Σ_n-cut. Ann. Pure Appl. Logic **48**, 227–235 (1990)
6. Chong, C.T., Yang, Y.: The jump of a Σ_n cut. J. Lond. Math. Soc. **75**, 690–704 (2007)
7. Haught, C.A.: The degrees below a 1-generic degree $<\mathbf{0}'$. J. Symbolic Logic **51**, 770–777 (1986)
8. Jockusch, C.G.: Degrees of Generic Sets. In: Drake, F.R., Wainer, S.S. (eds.) Recursion Theory: Its Generalisations and Applications, pp. 110–139. Cambridge University Press, Cambridge (1980)
9. Kumabe, M.: A 1-generic degree which bounds a minimal degree. J. Symbolic Logic **55**, 733–743 (1990)
10. Sacks, G.E.: On the degrees less than $\mathbf{0}'$. Ann. Math. **77**, 211–231 (1963)
11. Slaman, T.A.: Σ_n-bounding and Δ_n-induction. Proc. Am. Math. Soc. **132**, 2449–2456 (2004)

Nondensity of Double Bubbles
in the D.C.E. Degrees

Uri Andrews[1], Rutger Kuyper[1], Steffen Lempp[1],
Mariya I. Soskova[2], and Mars M. Yamaleev[3(✉)]

[1] Department of Mathematics, University of Wisconsin–Madison,
Madison, WI 53706, USA
{andrews,lempp}@math.wisc.edu, mail@rutgerkuyper.com
[2] Faculty of Mathematics and Computer Science, Sofia University,
5 James Bourchier Blvd., 1164 Sofia, Bulgaria
msoskova@fmi.uni-sofia.bg
[3] Lobachevsky Institute of Mathematics and Mechanics, Kazan Federal University,
18 Kremlyovskaya Street, Kazan 420008, Russia
mars.yamaleev@kpfu.ru
http://www.math.wisc.edu/~andrews/, http://rutgerkuyper.com,
http://www.math.wisc.edu/~lempp/,
https://www.fmi.uni-sofia.bg/fmi/logic/msoskova/,
http://kpfu.ru/Mars.Yamaleev&p_lang=2

Abstract. In this paper, we show that the so-called "double bubbles" are not downward dense in the d.c.e. degrees. Here, a pair of d.c.e. degrees $d_1 > d_2 > 0$ forms a *double bubble* if all d.c.e. degrees below d_1 are comparable with d_2.

Keywords: Ershov hierarchy · d.c.e. sets · Lachlan sets · exact degrees

2010 Mathematics Subject Classification: Primary 03D28.

1 Introduction

In this paper, we study a fundamental structural property of the d.c.e. degrees. The d.c.e., and more generally the n-c.e., sets and degrees were introduced by

This research was carried out while Yamaleev was visiting the University of Wisconsin under binational NSF grant DMS-1101123 entitled "Collaboration in Computability". Kuyper's research was supported by John Templeton Foundation grant 15619: "Mind, Mechanism and Mathematics: Turing Centenary Research Project". Lempp's research was partially supported by AMS-Simons Foundation Collaboration Grant 209087. Soskova's research was supported by Sofia University Science Fund Grant 54/12.04.2016 and by the L'Oréal-UNESCO program "For women in science". Yamaleev's research was supported by the Russian Foundation for Basic Research (projects 15-41-02507, 15-01-08252), by the Russian Government Program of Competitive Growth of Kazan Federal University, and by the subsidy allocated to Kazan Federal University for the project part of the state assignment in the sphere of scientific activities (project 1.2045.2014).

© Springer International Publishing AG 2017
A. Day et al. (Eds.): Downey Festschrift, LNCS 10010, pp. 547–562, 2017.
DOI: 10.1007/978-3-319-50062-1_33

Putnam [13] and Gold [8] as a generalization of the c.e. sets and degrees. A set A is n-c.e. if it has an approximation that can change the value of $A(x)$ at most n times for every natural number x, starting with $x \notin A$. When $n = 1$, we obtain the c.e. sets, and when $n = 2$, we obtain the difference of c.e. sets—the d.c.e. sets. Later on, this hierarchy was extended by Ershov [5–7] to arbitrary computable ordinals. The difference hierarchy gives rise to a corresponding nested hierarchy of degree structures, all contained in the Δ_2^0-Turing degrees. Naturally, one wonders if these structures are different. Lachlan showed that every nonzero n-c.e. degree bounds a nonzero c.e. degree, thus the structures of the n-c.e. degrees are different from that of the Δ_2^0-degrees, which contains minimal elements. Lachlan's proof relies on a particular set that is fairly easy to define: If A is d.c.e. and $\{A_s\}_{s<\omega}$ is a d.c.e. approximation to A, then the *Lachlan set* $L(A)$ is the c.e. set of stages s at which some element x enters A which then later leaves A.

Next, Arslanov [1] found an elementary difference between the structures of the c.e. degrees and the d.c.e degrees: Cooper and Yates (see [3,12]) had constructed a noncuppable nonzero c.e. degree, whereas Arslanov [1] showed that every nonzero d.c.e. degree is cuppable. Downey [4] found a further difference between the two structures: He showed that the diamond can be embedded into the d.c.e. degrees, in contrast to the Lachlan Non-Diamond Theorem for the c.e. degrees [10]. Downey's work lead him to conjecture that for any distinct $n, m > 1$, the structures of the n-c.e. degrees and the m-c.e. degrees are elementarily equivalent.

Arslanov, Kalimullin and Lempp [2] disproved this conjecture. They showed that the structure of d.c.e. degrees contains special pairs of degrees which they informally called a *double bubble*. They used this notion and a generalization of this notion to the 3-c.e. degrees to refute Downey's Conjecture by showing that the partial orders of the d.c.e. and the 3-c.e. Turing degrees are not elementarily equivalent. A pair of d.c.e. degrees $\mathbf{d}_1 > \mathbf{d}_2 > \mathbf{0}$ forms a *double bubble* if all d.c.e. degrees below $\mathbf{d}_1$ are comparable with $\mathbf{d}_2$. We call $\mathbf{d}_2$ the *middle* of the bubble and $\mathbf{d}_1$ the *top* of the bubble. Double bubbles play an important role in the study of the properly d.c.e. degrees. They have many nontrivial properties and have sparked a lot of interest. It is easy to see, by relativizing the Sacks Splitting Theorem [14], that the top of a bubble must always be properly d.c.e. On the other hand, it was shown in [2] that the middle is always a c.e. degree. A more elaborate property is related to the notion of an *exact d.c.e. degree*. Exact degrees were introduced and first studied by Ishmukhametov [9]; a d.c.e. degree $\mathbf{d}$ is called *exact* if all Lachlan sets of d.c.e. members of $\mathbf{d}$ have the same degree. It follows from [2] that the top of every bubble is an exact degree and the middle of the bubble is the degree of the Lachlan set of any member of the top of the bubble (a proof can be found, e.g., in [16]).

Liu, Wu and Yamaleev [11] investigated the possibility of combining the construction of a double bubble in the d.c.e. degrees with other properties, such as upward and downward density in the c.e. degrees. They noted that a positive answer to the full density question would allow us to define the c.e. degrees within the d.c.e. degrees.[1] If $\mathbf{d}$ is a properly d.c.e. degree and $\mathbf{d_1}, \mathbf{d_2}$ form a nontrivial splitting of $\mathbf{d}$ in the d.c.e. degrees, then at least one of the intervals $(\mathbf{d_1}, \mathbf{d})$ or $(\mathbf{d_2}, \mathbf{d})$ must be free of c.e. degrees and hence bubbles, otherwise $\mathbf{d}$ would be c.e. On the other hand, if every nonempty interval of c.e. degrees contained a bubble, then, by the Sacks Splitting Theorem, every nonzero c.e. degree $\mathbf{c}$ would have a nontrivial splitting $\mathbf{c_1}, \mathbf{c_2}$, such that both intervals $(\mathbf{c_1}, \mathbf{c})$ and $(\mathbf{c_2}, \mathbf{c})$ contain a bubble. Liu, Wu and Yamaleev [11] showed that exact degrees are downward dense in the c.e. degrees and left the downward density of double bubbles as an open question. In this paper, we show that double bubbles are not downward dense in the d.c.e. degrees. Of course, it suffices to show that double bubbles do not necessarily exist below any nonzero c.e. degree:

Theorem 1.1. *There exists a c.e. degree* $\mathbf{a}$ *such that there are no d.c.e. degrees* $\mathbf{d_2} < \mathbf{d_1} \leq \mathbf{a}$ *which form a double bubble in the d.c.e. degrees.*

The rest of this paper is devoted to the proof of this theorem. Our notation and terminology is standard and generally follows Soare [15]. We also use standard notation and terminology for priority constructions.

2 Strategies

2.1 Requirements

Recall that the top of a double bubble is always an exact degree. We will give a more formal definition of what this means. Fix a d.c.e. set D and a d.c.e. enumeration $\{D_s\}_{s \in \omega}$ of D. For technical reasons, we will assume from now on that at any stage, any set D changes at at most one number, thus $|D_s \triangle D_{s-1}| \leq 1$ for any $s \in \omega$. We define the partial computable function $s^D(x)$ as the stage of entry of x into D, i.e., $s^D(x) \downarrow = s$ is defined if x is enumerated into D at stage s. If x is never enumerated into D then $s^D(x) \uparrow$. The *Lachlan set of D with respect to the enumeration* $\{D_s\}_{s < \omega}$ is defined as

$$L(D) = \{s \mid (\exists x)(s^D(x) \downarrow = s \ \& \ x \notin D)\}.$$

Although it may not be immediately obvious from the definition, every Lachlan set is c.e. This follows from the fact that it is defined with respect to a d.c.e. approximation and so '$x \notin D$' can be substituted by '$(\exists t > s)(x \notin D_t)$'. Furthermore, it is not difficult to see that the degree of $L(D)$ does not depend on the particular choice of a d.c.e. approximation for D.

If $\mathbf{d} = \deg(D)$ is a d.c.e. degree, then the *set of Lachlan degrees of* $\mathbf{d}$ is the set

$$L[\mathbf{d}] = \{\deg(L(B)) \mid B \in \mathbf{d} \text{ and } B \text{ is d.c.e.}\}.$$

A d.c.e. degree $\mathbf{d}$ is *exact* if $|L[\mathbf{d}]| = 1$.

---

[1] In fact, this idea goes back to Arslanov, who noted it in private communication with Shore. Later, he publicized this idea in conference talks.

Fix a double bubble $\mathbf{d}_1 > \mathbf{d}_2 > \mathbf{0}$. If $D \in \mathbf{d}_1$ is d.c.e., then $L(D) \in \mathbf{d}_2$ (by Ishmukhametov [9, Proposition 1.2] and Arslanov/Kalimullin/Lempp [2, Theorem 5]). So in order to prove the theorem, we must construct a non-computable c.e. set A such that for any noncomputable d.c.e. set $D \leq_T A$, if $\mathbf{0} < \deg(L(D)) < \deg(D)$, then there is a d.c.e. set $E \leq_T D$ that is Turing incomparable with $L(D)$. Fix a computable listing of all tuples $\langle \Phi, \Psi, \Theta, \Omega, D \rangle$ of partial computable functionals $\Phi, \Psi, \Theta, \Omega$ and d.c.e. sets D. It suffices to build a c.e. set A satisfying the following list of requirements:

$$\mathcal{P}_\Theta : A \neq \Theta;$$
$$\mathcal{R}_{\Phi,D} : D = \Phi^A \Rightarrow$$
$$\exists E \, \exists \Lambda_{\Phi,D} \, (E = \Lambda_{\Phi,D}^D \wedge E \mid_T L(D)) \vee D \leq_T L(D) \vee L(D) \leq_T \emptyset,$$

where each $\mathcal{R}$-requirement has its own infinite list of subrequirements:

$$\mathcal{T}_\Psi : E = \Psi^{L(D)} \Rightarrow \exists \Gamma_\Psi \, (D = \Gamma_\Psi^{L(D)});$$
$$\mathcal{S}_\Omega : L(D) = \Omega^E \Rightarrow \exists \Delta_\Omega \, (L(D) = \Delta_\Omega) \vee \exists \Gamma_\Omega \, (D = \Gamma_\Omega^{L(D)}).$$

(We will usually suppress the subscripts on the functionals above when they are clear from the context.) We will construct A using a tree of strategies and the gap/co-gap method. The proof will be a $\mathbf{0}'''$-priority argument. We will first describe the intuition behind the construction, starting with each strategy in isolation.

2.2   Strategies in Isolation

Recall our convention that at every stage, any of the given sets can change at at most one element.

The basic $\mathcal{P}$-strategy. The basic $\mathcal{P}$-strategy is a variant of the standard Friedberg–Muchnik strategy. We choose a fresh witness a, wait for a stage s such that $\Theta(a)[s] \downarrow = 0$, and enumerate a into A.

The basic $\mathcal{R}$-strategy. An $\mathcal{R}$-strategy ρ serves as the mother strategy for all of its substrategies. It monitors the length of agreement between D and Φ^A. At non-expansionary stages, it takes the finitary outcome fin. At expansionary stages, it makes progress towards building the functional Λ so that $\Lambda^D = E$ and takes its infinite outcome ∞, allowing its $\mathcal{S}$- and $\mathcal{T}$-substrategies to act.

The basic $\mathcal{T}$-strategy. A $\mathcal{T}$-strategy τ is a child strategy of some $\mathcal{R}$-strategy. In isolation, it checks the length of agreement between E and $\Psi^{L(D)}$. At expansionary stages, τ builds Γ_τ so that $\Gamma_\tau^{L(D)} = D$. The strategy has two possible outcomes, Γ and fin.

The basic $\mathcal{S}$-strategy. An $\mathcal{S}$-strategy σ, say, is a child strategy of some $\mathcal{R}$-strategy. In isolation, it checks the length of agreement between $L(D)$ and Ω^E, and if the stage is expansionary, then σ first tries to build Δ_σ so that $\Delta_\sigma = L(D)$.

This strategy exhibits an interesting behavior in response to other strategies. We will describe this in the next subsection and see how this response may cause σ to build a backup functional Γ_σ so that $\Gamma_\sigma^{L(D)} = D$ at expansionary stages. The strategy has three possible outcomes, Γ, Δ, and fin.

2.3   Interactions Between Strategies

In this section, we consider nontrivial interactions between strategies and describe how to overcome the corresponding problems. Since all problems begin when a $\mathcal{P}$-strategy enumerates an element into A, we will always assume that there is a $\mathcal{P}$-strategy below the other strategies we consider.

A $\mathcal{T}$-strategy τ below its mother $\mathcal{R}$-strategy ρ. The nontrivial case is when a $\mathcal{P}$-strategy π below the Γ-outcome of τ acts. Let us consider τ in more detail. For every x, we need to correctly define $\Gamma^{L(D)}(x) = D(x)$. We pick a big $y = y_x$ first and wait until the length of agreement between $\Psi^{L(D)}$ and E is larger than y. At the first expansionary stage s at which this happens, we define $\Gamma^{L(D)}(x)[s] = D(x)[s]$ with use $\gamma(x)[s] = s > \psi(y)[s]$. From now on (assuming τ is along the true path), the equality between $\Gamma^{L(D)}(x)$ and $D(x)[s]$ can be broken only if a witness a of the $\mathcal{P}$-strategy $\pi \supseteq \tau^\frown\Gamma$ is enumerated into A. It is worth noting that a must have been chosen before stage s, and so this can happen at most finitely many times (since all new witnesses of $\mathcal{P}$-strategies after initialization will be chosen big enough and there are only finitely many old witnesses).

The change in A allows a change in D on any x with Φ-use $\varphi(x)[s] \geq a$. We have the following possible cases:

(1) x enters D but there is no change in $L(D) \restriction (\gamma(x) + 1)$: Then we enumerate $y = y_x$ into E and we initialize all strategies below τ. So we have $1 = E(y) \neq \Psi^{L(D)}(y) = \Psi^{L(D)}(y)[s] = 0$, and τ wins. Initialized strategies must pick fresh witnesses, so from this moment on only strategies of higher priority than τ can enumerate numbers into A that allow changes of $\Psi^{L(D)}(y)[s]$. Indeed, if $\Psi^{L(D)}(y)[s]$ changes at a stage $s_1 > s$, then a number x_1 is extracted from D where $s^D(x_1) \leq \psi(y) < s$. It follows that some $a_1 \leq \varphi(x_1)[s] < s$ entered A after stage s, so a_1 must have been chosen before stage s.

(2) x enters or leaves D and there is a change in $L(D) \restriction (\gamma(x) + 1)$: In this case, we can update $D(x) = \Gamma^{L(D)}(x)$ with new big use $\gamma(x)$. Note that a new update of $\Gamma^{L(D)}(x)$ can only be caused by a number $a_1 < a$ entering A. This is because when a is enumerated into A by a $\mathcal{P}$-strategy, we initialize all lower-priority strategies, and hence all strategies with witnesses greater than a. New witnesses will be greater than the current use $\varphi(x)$ and will not be able to change computations related to x. So an increase in $\varphi(x)$ can only be caused by the enumeration of some $a_1 < a$, and as we noted above, this can happen at most finitely often.

Note that if x leaves D, then there must be a change in $L(D) \restriction (\gamma(x) + 1)$. This is because we defined $\Gamma^{L(D)}(x)$ correctly at stage s, when we have that x

is already in D, and so $s^D(x) < s = \gamma(x)$. It follows that the two cases above exhaust all possibilities.

In what follows, when we build a functional Γ, we can think of it as *opening a gap* and allowing for some number a to enter A. Hence, either a gap will be *closed successfully*, namely, at some point we have case (1) and a diagonalization at τ, or all gaps will be *closed unsuccessfully*, namely, we always have case (2), in which case we will correctly reduce D to $L(D)$. In the construction, we will create a link from τ to ρ. This link allows us to jump directly from ρ to τ and decide whether we want to enumerate y into E while keeping $E = \Lambda^D_\rho$ correct. So we will enumerate a number into or extract a number from E only when we come to a substrategy of ρ using a link (if there is no link at ρ then we change E at ρ); otherwise, we will not need to change E at ρ, since at ρ we will not be in a position in which we must change E back due to D returning to an old initial segment (except for the situation when some $\mathcal{P}$-strategy between ρ and τ enumerates a small number into A, which allows a D-change which can force us to change E back at ρ but also causes τ to be initialized).

A $\mathcal{T}$-strategy τ below an $\mathcal{S}$-strategy σ below their mother $\mathcal{R}$-strategy ρ. The real conflict, which also causes this priority argument to be a $0'''$-argument rather than just an infinite-injury argument, first arises in the following scenario: Suppose we have an $\mathcal{R}$-strategy ρ with an $\mathcal{S}$-substrategy σ and a $\mathcal{T}$-substrategy τ below such that τ is below the finite outcome of σ. Furthermore, assume we have three $\mathcal{P}$-strategies π_2, π_1 and π_0 below the Γ-outcome of τ, the Δ-outcome of σ and the Γ-outcome of σ, respectively. Suppose now the following sequence of events:

First, the $\mathcal{P}$-strategy π_2 enumerates a witness a_2 into A, allowing a number x to enter D and causing τ to enumerate a number $y = y_x$ into E in order to diagonalize τ. Next, the $\mathcal{P}$-strategy π_1 enumerates a witness $a_1 < a_2$ into A, allowing x to leave D, which would normally force y to be extracted from E in order to keep Λ correct. However, for the stage $s^D(x)$ at which x entered D, $s^D(x)$ will enter $L(D)$ when x leaves D, while σ has possibly already defined $\Delta(s^D(x)) = 0$, which cannot be corrected. We resolve this conflict by threatening to let σ build a Turing functional $\Gamma^{L(D)} = D$ to permanently satisfy ρ.

However, letting σ build Γ (and taking an infinite Γ-outcome to the left of the infinite Δ-outcome) creates a new problem: Suppose our $\mathcal{P}$-strategy π_0 below the Γ-outcome of σ next enumerates a number a_0 into A, allowing D to change at a number on which $\Gamma^{L(D)}$ is already defined and now possibly wrong. The strategy for σ can use the following procedure to force an $L(D)$-change and correct $\Gamma^{L(D)}$: Before letting π_0 choose its witness a_0, we have a number x from some $\mathcal{P}$-strategy π_1 ready that just left D and caused the function Δ of σ to be incorrect. We will have a link from ρ to σ so that we can visit σ directly before ρ has a chance to extract y from E, allowing Λ^D to be temporarily incorrect. If the functional Δ is now wrong on $s^D(x)$, then we create a second link from ρ to σ and move to outcome Γ, only then allowing a_0 to be enumerated in A. Suppose that this causes a change in $D(x')$.

(1) If x' enters D, then there need not be any $L(D)$-change and thus $\Gamma^{L(D)}(x')$ may now be incorrect. If $\Gamma^{L(D)}(x')$ is defined, then this means that x' is small enough to allow us to preserve y in E while still keeping Λ^D correct. This causes a permanent disagreement between Ω^E and $L(D)$ at $s^D(x)$, since the old definition of $\Omega^E(s^D(x)) = 0$ is still valid while $s^D(x) \in L(D)$; this disagreement can only be undone by an action of a strategy of higher priority than σ, since σ can now switch to a permanent finitary diagonalization outcome unless initialized later.

(2) If x' leaves D (and had previously entered D at a stage $s^D(x')$), then it will follow from the way we construct Γ that $\gamma(x') \geq s^D(x')$. So x' leaving D will cause $s^D(x')$ to enter $L(D)$ and allow $\Gamma^{L(D)}(x')$ to be corrected.

Similarly to the previous case, we *open a second gap* when we allow the number a_0 to enter A. Either one of these gaps will be *closed successfully* (i.e., at some point, we have case (1)) and we have a permanent win at σ, or all gaps will be *closed unsuccessfully* (i.e., we always have case (2)), then we correctly reduce D to $L(D)$. Again, in the construction, we will create a link from σ to ρ since we jump from ρ to σ when we need to decide whether to enumerate y into E or not, and the link allows us to keep $E = \Lambda_\rho^D$ correct.

2.4   Several $\mathcal{R}$-Strategies

Now we consider several $\mathcal{R}$-strategies with their substrategies. In our intuitive analysis, we restrict ourselves to two $\mathcal{R}$-strategies ρ_0 and ρ_1. Assume that we have $\rho_0 \subset \rho_1$, and that they have substrategies σ_0 and σ_1, respectively (also assume that σ_0 and σ_1 have Γ-outcome). The conceivable relative priorities for these strategies are as follows:

(1) $\rho_0 \subset \sigma_0 \subset \rho_1 \subset \sigma_1$,
(2) $\rho_0 \subset \rho_1 \subset \sigma_1 \subset \sigma_0$, and
(3) $\rho_0 \subset \rho_1 \subset \sigma_0 \subset \sigma_1$.

The third case could produce non-nested links; so we disallow it as follows: When σ_0 changes the global outcome of ρ_0 along the true path, we introduce another version of ρ_1, say, an $\mathcal{R}$-strategy ρ_1', first, and only allow substrategies of ρ_1' but not of ρ_1 below ρ_1'. This reduces the third case above to the first, in the usual manner of $\mathbf{0}'''$-arguments.

In the first case, there is no real conflict, since ρ_1 already knows that σ_0 will build its Γ, which permanently satisfies ρ_0. In the second case, there may be links from ρ_0 directly to σ_0, over ρ_1 and σ_1; but if σ_0 truly has Γ-outcome, then we again introduce another version of ρ_1, say, an $\mathcal{R}$-strategy ρ_1', below σ_0 and only allow substrategies of ρ_1' but not of ρ_1 below σ_0.

3 Construction

3.1 Outcomes and the Tree of Strategies

Throughout the construction, we insert comments in brackets which we hope will help the reader connect the formal construction back to the intuition given above.

Let $\text{ListFunc} = \{\langle \Phi, \Psi, \Theta, \Omega, D \rangle\}$ be the above-mentioned computable listing of all tuples of p.c. functionals $\Phi, \Psi, \Theta, \Omega$ and d.c.e. sets D.

Let ListReq be a computable listing of all requirements defined as follows: First, we fix the least element $\langle \Phi_0, \Psi_0, \Theta_0, \Omega_0, D_0 \rangle$ in ListFunc. Then we set

$$\text{ListReq} = \{\mathcal{P}_\Theta \mid \langle \Phi_0, \Psi_0, \Theta, \Omega_0, D_0 \rangle \in \text{ListFunc}\} \cup$$
$$\{\mathcal{R}_{\Phi,D} \mid \langle \Phi, \Psi_0, \Theta_0, \Omega_0, D \rangle \in \text{ListFunc}\} \cup$$
$$\{\mathcal{T}_{\Phi,D,\Psi} \mid \langle \Phi, \Psi, \Theta_0, \Omega_0, D \rangle \in \text{ListFunc}\} \cup$$
$$\{\mathcal{S}_{\Phi,D,\Omega} \mid \langle \Phi, \Psi_0, \Theta_0, \Omega, D \rangle \in \text{ListFunc}\}$$

For $\mathcal{X} \in \text{ListReq}$, let $\text{ind}(\mathcal{X}) \in \text{ListFunc}$ be the corresponding tuple for $\mathcal{X}$. Now we say that $\mathcal{X} < \mathcal{Y}$ for $\mathcal{X}, \mathcal{Y} \in \text{ListReq}$ if and only if either $\text{ind}(\mathcal{X}) \neq \text{ind}(\mathcal{Y})$ and $\text{ind}(\mathcal{X})$ is listed before $\text{ind}(\mathcal{Y})$ in ListFunc, or if $\text{ind}(\mathcal{X}) = \text{ind}(\mathcal{Y})$ and $\mathcal{X} \neq \mathcal{Y}$, then we use the following ordering of the requirements when we compare $\mathcal{X}$ and $\mathcal{Y}$: $\mathcal{P} < \mathcal{R} < \mathcal{T} < \mathcal{S}$.

The strategies for these requirements have the following outcomes, where $L = \{d, \infty, \Gamma, \Delta, \text{fin}\}$ is the set of outcomes (we have added one more outcome d that was not mentioned in the intuition, meant to isolate the situation when a strategy has a permanent win by diagonalization, i.e., by successfully closing a gap):

- A $\mathcal{P}$-strategy has two possible outcomes: $d <_L \text{fin}$.
- An $\mathcal{R}$-strategy has two possible outcomes: $\infty <_L \text{fin}$.
- A $\mathcal{T}$-strategy has three possible outcomes: $d <_L \Gamma <_L \text{fin}$.
- An $\mathcal{S}$-strategy has four possible outcomes: $d <_L \Gamma <_L \Delta <_L \text{fin}$.

The *tree of strategies* $T \subset L^{<\omega}$ is defined by induction as follows. When we assign a requirement to some node, then the strategy of this requirement will work at this node. For the empty node λ, we set $\text{ListReq}_\lambda = \text{ListReq}$. Given a node $\xi \in T$, we assign to it the highest-priority (sub)requirement from ListReq_ξ. If it is a subrequirement of an $\mathcal{R}$-requirement, then there will be a longest strategy $\rho \subset \xi$ assigned to the corresponding $\mathcal{R}$-requirement, and we call ξ a *child node* of ρ and ρ the *mother node* of ξ. Depending on the requirement assigned to ξ, we next define the list of requirements yet to be satisfied as follows.

- If it is a $\mathcal{P}_\Theta$-requirement, then define

$$\text{ListReq}_{\xi^\frown d} = \text{ListReq}_{\xi^\frown \text{fin}} = \text{ListReq}_\xi - \{\mathcal{P}_\Theta\}.$$

- If it is an $\mathcal{R}_{\Phi,D}$-requirement, then define

$$\text{ListReq}_{\xi^\frown\infty} = \text{ListReq}_\xi - \{\mathcal{R}_{\Phi,D}\}, \text{ and}$$

$$\text{ListReq}_{\xi^\frown\text{fin}} = \text{ListReq}_\xi - \{\mathcal{R}_{\Phi,D}\}$$
$$- \{\mathcal{T}_{\Phi,D,\Psi} \mid \text{ind}(\mathcal{T}_{\Phi,D,\Psi}) \in \text{ListFunc}\}$$
$$- \{\mathcal{S}_{\Phi,D,\Omega} \mid \text{ind}(\mathcal{S}_{\Phi,D,\Omega}) \in \text{ListFunc}\}.$$

[So, for the infinite outcome, we remove only $\mathcal{R}_{\Phi,D}$, but for the finite outcome, we remove $\mathcal{R}_{\Phi,D}$ and all its subrequirements.]

- If it is a $\mathcal{T}_{\Phi,D,\Psi}$-subrequirement and the mother node of ξ is ρ, then define

$$\text{ListReq}_{\xi^\frown\Gamma} = \text{ListReq}_{\rho^\frown\text{fin}}, \text{ and}$$

$$\text{ListReq}_{\xi^\frown d} = \text{ListReq}_{\xi^\frown\text{fin}} = \text{ListReq}_\xi - \{\mathcal{T}_{\Phi,D,\Psi}\}.$$

- If it is an $\mathcal{S}_{\Phi,D,\Omega}$-subrequirement and the mother node of ξ is ρ, then define

$$\text{ListReq}_{\xi^\frown\Gamma} = \text{ListReq}_{\xi^\frown\Delta} = \text{ListReq}_{\rho^\frown\text{fin}}, \text{ and}$$

$$\text{ListReq}_{\xi^\frown d} = \text{ListReq}_{\xi^\frown\text{fin}} = \text{ListReq}_\xi - \{\mathcal{S}_{\Phi,D,\Omega}\}.$$

[Namely, in both these cases, under outcomes d and fin, we remove only the subrequirement itself, whereas under outcomes Γ and Δ, we consider the same list of requirements as under the finite outcome of the mother node.]

Now we define the expansionary stages. A stage s is called a ξ-*stage* if the node ξ is visited at this stage. The *length of agreement functions* for an $\mathcal{R}$-strategy ρ and for an $\mathcal{S}$-strategy σ are defined as follows:

$$l(\rho)[s] = \max\left\{t < s \mid \forall x < t\,(D(x)[s] = \Phi_\rho^A(x) \downarrow [s])\right\},$$

$$l(\sigma)[s] = \max\left\{t < s \mid \forall z < t\,(L(D)(z)[s] = \Omega_\sigma^E(z) \downarrow [s])\right\}.$$

A stage s is ξ-*expansionary* (for $\xi \in \{\rho, \sigma\}$) if $l(\xi)[s] > s^-[s]$, where $s^-[s] = s_\xi^-[s]$ is the previous ξ-expansionary stage at stage s, and where stage 0 is always ξ-expansionary. (Note that we will not need the notion of a ξ-expansionary stage for $\mathcal{T}$-strategies ξ, where we will use a version of the Sacks coding strategy instead.) During the construction, *initializing* a node will mean canceling its satisfaction, its witnesses, and its associated functionals.

3.2 Full Construction

We build a computable approximation TP_s to the true path TP at each stage. Meanwhile we define approximations of all sets and functionals at these stages (keeping sets and functionals the same unless we redefine them explicitly). The construction proceeds as follows.

Stage $s = 0$. We set $A_0 = \emptyset$ and initialize all strategies.

Stage $s+1$. We work in substages $t \leq s+1$, possibly skipping over some substages. Let $TP_{s,0} = \lambda$. Given $TP_{s+1,t}$ at a substage $t+1$, we define $TP_{s+1,t'}$ for some $t' > t$ (usually $t' = t+1$). After we define $TP_{s+1,t+1}$, if $t < s$, then we proceed to substage $t+2$ unless explicitly stated otherwise. If $t = s$, then we define $TP_{s+1} = TP_{s+1,t+1}$, proceed to the next stage, and initialize all nodes $\xi \not\leq TP_{s+1}$. At substage $t+1$, the construction depends on the requirement assigned to $TP_{s+1,t}$:

Case 1: $TP_{s+1,t} = \pi$ *is a* $\mathcal{P}$-*strategy:* Go to the first subcase which applies.
$\quad$ $\pi1$. If no witness is defined for π, then define $a = a_\pi$ to be a big number and
$\qquad$ let $TP_{s+1,t+1} = \pi\,\widehat{}\,\text{fin}$.
$\quad$ $\pi2$. Otherwise, if $\Theta(a)[s] \uparrow$ or $\Theta(a)[s] \neq 0$, then define $TP_{s+1,t+1} = \pi\,\widehat{}\,\text{fin}$.
$\quad$ $\pi3$. Otherwise, if $\Theta(a)[s] = 0$ and $a \notin A_s$, then enumerate a into A and define
$\qquad$ $TP_{s+1,t+1} = \pi\,\widehat{}\,d$.
$\quad$ $\pi4$. Otherwise, $\Theta(a)[s] = 0$ and $a \in A_s$, so define $TP_{s+1,t+1} = \pi\,\widehat{}\,d$.

Case 2: $TP_{s+1,t} = \rho$ *is an* $\mathcal{R}$-*strategy:* Go to the first subcase which applies. [The goal of ρ is to use links and to define the reduction $E \leq_T D$.]
$\quad$ $\rho1$. If stage s is not ρ-expansionary, then define $TP_{s+1,t+1} = \rho\,\widehat{}\,\text{fin}$.
$\quad$ $\rho2$. Otherwise, fix the previous ρ-expansionary stage $s^-[s+1] = \tilde{s}$ and consider the following subcases.
$\quad$ $\rho2.1$. If there is no link to ρ, then extract the necessary numbers from $E = E_\rho$
$\qquad$ in order to keep $E(y)[s+1] = \Lambda_\rho^D(y)[s]$ correct for all $y \in \text{dom}(\Lambda_\rho^D[s])$,
$\qquad$ define $\Lambda_\rho^D(y_0)[s+1] = E(y_0)[s+1]$ for the least $y_0 \notin \text{dom}(\Lambda_\rho^D[s])$, with
$\qquad$ use $\lambda_\rho(y_0)[s+1] = y_0$. Let $TP_{s+1,t+1} = \rho\,\widehat{}\,\infty$. [It is easy to see that each
$\qquad$ $\lambda_\rho(y_0)$ will not increase and is bigger than x_0, the number to which it is
$\qquad$ potentially related by some $\mathcal{T}$-strategy.]
$\quad$ $\rho2.2$. Otherwise, travel the link to the child node η, say, which created the
$\qquad$ link, define $TP_{s+1,|\eta|} = \eta$, and proceed to substage $|\eta| + 1$.

Case 3: $TP_{s+1,t} = \tau$ *is a* $\mathcal{T}$-*strategy:* The strategy works in cycles. Let ρ be the mother node of τ. Proceed as in the first subcase which applies. [The goal of τ is to diagonalize against $E = \Psi^{L(D)}$ or to define a reduction $D \leq_T L(D)$.]
$\quad$ $\tau1$. If τ is visited through a link, then the link must be from the mother
$\qquad$ node ρ. Cancel this link and consider the following subcases. [This means
$\qquad$ that at the previous τ-stage we had outcome Γ, and now we either diagonalize or extend the Γ-functional.]
$\quad$ $\tau1.1$. If there is x such that $\Gamma^{L(D)}(x)[s]$ is defined and such that $D(x)[s] \neq$
$\qquad$ $\Gamma^{L(D)}(x)[s]$, then put y_x into E and define $TP_{s+1,t+1} = \tau\,\widehat{}\,d$. Declare τ
$\qquad$ *satisfied*. [This is Case (1) in the intuition, which allows us to change
$\qquad$ $E(y_x)$ since $\Lambda^D(y_x)$ becomes undefined because of x, hence it allows a
$\qquad$ diagonalization at τ.]
$\quad$ $\tau1.2$. Otherwise, let x be the greatest opened cycle (if there is none, set $x = 0$).
$\qquad$ Open cycle $x+1$ and, for all $u \leq x$, define $\Gamma_\tau^{L(D)}(u)[s+1] = D(u)[s+1]$
$\qquad$ (if $\Gamma_\tau^{L(D)}(u)[s] \uparrow$) with use $\gamma(u) = s+1$, and define $TP_{s+1,t+1} = \tau\,\widehat{}\,\text{fin}$.
$\qquad$ [This is Case (2) in the intuition.]
$\quad$ $\tau2$. Otherwise, if τ is satisfied, then define $TP_{s+1,t+1} = \tau\,\widehat{}\,d$.

$\tau 3$. Otherwise, let x be the greatest opened cycle (if there is none, set $x = 0$).
Consider the following subcases.

$\tau 3.1$. If there is no attacker $y = y_x$, then choose y big, and define $TP_{s+1,t+1} = \tau\hat{\ }fin$.

$\tau 3.2$. Otherwise, if $E(y)[s] \neq \Psi_\tau^{L(D)}(y)[s]$, then define $TP_{s+1,t+1} = \tau\hat{\ }fin$.

$\tau 3.3$. Otherwise, we have $E(y)[s] = \Psi_\tau^{L(D)}(y)[s] \downarrow$, so we create a link with ρ and define $TP_{s+1,t+1} = \tau\hat{\ }\Gamma$.

Case 4: $TP_{s+1,t} = \sigma$ is an S-strategy: The strategy works in cycles. Let ρ be the mother node of σ. Go to the first subcase which applies. [The goal of σ is to diagonalize against $L(D) = \Omega^E$ or to either define a reduction $L(D) \leq_T \emptyset$ or to define a reduction $D \leq_T L(D)$. Also note that cycles here are analogues of cycles in $\mathcal{T}$-strategies; however, inside these cycles, we use σ-expansionary stages, which could be considered as inner cycles.]

$\sigma 1$. If σ is visited through a link, then the link must be from the mother node ρ, and before creating the link at the previous σ-expansionary stage $s^- = s_\sigma^-[s]$, the node σ had either outcome Δ or Γ. Cancel this link and consider the following subcases.

$\sigma 1.1$. If the previous outcome was Δ, then if there is some $z \leq s^-[s^-[s]]$ with $\Delta_\sigma(z)[s] \neq L(D)(z)[s]$, then keep $\Lambda_\rho^D(y)$ undefined (where $y = y_x$ is the number which entered E earlier due to x entering D but now x has left D again, and $z = s^D(x)$), create a link between σ and ρ again, and define $TP_{s+1,t+1} = \sigma\hat{\ }\Gamma$. [The first gap is closed successfully.]

$\sigma 1.2$. Otherwise, if the previous outcome was Δ, but the functional Δ_σ agrees with $L(D)$ on its domain, then, for all $z \leq s^-[s]$, define $\Delta_\sigma(z)[s+1] = L(D)(z)[s+1]$, and define $TP_{s+1,t+1} = \sigma\hat{\ }fin$. [The first gap is closed unsuccessfully.]

$\sigma 1.3$. Otherwise, the previous outcome was Γ. If there is an opened cycle x such that $\Gamma_\sigma^{L(D)}(x)[s]$ is defined and $D(x)[s] \neq \Gamma_\sigma^{L(D)}(x)[s]$, then declare σ *satisfied*, keep E unchanged, and redefine $\Lambda_\rho^D(y) = E(y)$ with old use $\lambda(y)$ for all y (which is possible due to the fresh $x \in D$), and define $TP_{s+1,t+1} = \sigma\hat{\ }d$. [The second gap is closed successfully. This is Case (1) in the intuition which allows to keep $E(y)$ unchanged since $\Lambda^D(y)$ became undefined because of x, hence it allows diagonalization at σ.]

$\sigma 1.4$. Otherwise, the previous outcome was Γ, and $\Gamma^{L(D)}$ is correct on its domain. Let x be the greatest opened cycle (if there is none, set $x = 0$). Then open cycle $x + 1$, and for all $u \leq x$, define $\Gamma_\sigma^{L(D)}(u)[s+1] = D(u)[s+1]$ (if $\Gamma_\sigma^{L(D)}(u)[s] \uparrow$) with use $\gamma(u) = s+1$, cancel Δ, and define $TP_{s+1,t+1} = \sigma\hat{\ }fin$. [The second gap is closed unsuccessfully. This is Case (2) in the intuition.]

$\sigma 2$. Otherwise, if there is an opened cycle x such that σ is satisfied at cycle x, then define $TP_{s+1,t+1} = \sigma\hat{\ }d$.

$\sigma 3$. Otherwise, if stage s is not σ-expansionary, or if this is the first visit of σ after initialization, then $TP_{s+1,t+1} = \sigma\hat{\ }fin$.

$\sigma 4$. Otherwise, s is a σ-expansionary stage, then fix the previous expansionary stage $s_\sigma^-[s+1] = s$, create a link with ρ, and define $TP_{s+1,t+1} = \sigma\hat{\ }\Delta$. [Note that this feature introduces a small delay into the definition of Δ.]

4   Verification

Define the *true path* $TP = \liminf_s TP_s$. We will show that TP exists and that each requirement is satisfied by some node along TP.

Lemma 4.1. *The true path TP exists.*

Proof. This is clear by definition since the tree is finitely branching and $|TP_s| = s$ for all $s \in \omega$.   □

Lemma 4.2. *Each node along the true path TP is initialized only finitely often.*

Proof. At stage s, we initialize only the nodes $\xi \not\preceq TP_s$. So eventually, every node along TP will not be initialized.   □

Lemma 4.3. *There are no nodes along TP which are part of a permanent link.*

Proof. A link connects a mother node and one of its children. So, if a mother node is along TP and is part of a link, then when we visit the mother node, we travel the link to its child node and cancel the link if the child node is a $\mathcal{T}$-substrategy, or cancel the link after traveling it at most twice if the child node is an $\mathcal{S}$-substrategy.   □

Lemma 4.4. *For every $\mathcal{R}$-strategy ρ along the true path and with true outcome ∞, there are infinitely many stages at which it does not travel links and $\rho^\frown\infty$ is visited. More generally, each node on TP is visited infinitely often.*

Proof. Suppose ρ travels a link to a child strategy ξ. It follows that ξ has not yet been satisfied, and so all strategies below $\xi^\frown d$ are in initial state. By construction, when we previously visited ξ it took an infinite outcome Γ or Δ, and so all strategies extending $\xi^\frown fin$ are also in initial state. If ξ is a $\mathcal{T}$-strategy τ, the link is canceled and τ has either outcome d for the first time since initialization or outcome fin, visiting in each case strategies in their initial state. If ξ is an $\mathcal{S}$-strategy σ, the link is canceled and σ has either outcome fin, visiting strategies in their initial state, or creates a second link and has outcome Γ. At the next expansionary ρ-stage, the second link is traveled and canceled, and σ has outcome d or fin, visiting in each case strategies in their initial state. No strategy in its initial state can create a link on its first visit. So, when ρ is next visited, it will not travel a link, and at its next expansionary stage, ρ will have outcome ∞.

The second part of this lemma is now an easy induction. Consider a node $\eta \in TP$ and assume that lemma holds for all $\xi \subset \eta$. The case of the empty node η is trivial, so let $\eta = \xi^\frown o$. The case when ξ is an $\mathcal{R}$-strategy and $o = \infty$ was just proved. If $o = fin$ or $o = d$, then η is visited at all but finitely many ξ-stages. If $o = \Gamma$ or $o = \Delta$, then unless ξ has outcome o at infinitely many stages, η is not along the true path, contradicting our choice of η.   □

Lemma 4.5. *Each $\mathcal{P}$-requirement is satisfied by a node along the true path TP.*

Proof. Consider a requirement $\mathcal{P}_\Theta$. By the definition of the tree of strategies, we can choose a node $\pi \subset TP$ assigned to $\mathcal{P}_\Theta$ of maximal length. By Lemma 4.4, π is visited at infinitely many stages. By Lemma 4.2, we fix a stage s_0 such that π is not initialized after stage s_0 (even though links starting at $\rho \subset \pi$ to some child node $\tau \supset \pi$ of ρ may be traveled after stage s_0, this would not initialize π). It follows that π has a final witness a. Now, clearly, the requirement will be satisfied: Either $\Theta(a) \neq 0$ and $a \notin A$; or at some stage $s_1 > s_0$, $\Theta(a)[s_1] = 0$ and when we next visit π, we enumerate a into A. $\qquad\square$

Lemma 4.6. *If a $\mathcal{T}_{\Phi,D,\Psi}$-strategy of maximal length along TP has outcome d or fin, then its requirement is satisfied.*

Proof. By the definition of the tree of strategies, we can choose a node $\tau \subset TP$ assigned to $\mathcal{T}_{\Phi,D,\Psi}$ of maximal length. By Lemma 4.2, we fix a stage s_0 such that τ is not initialized after stage s_0. Now consider the following two cases:

$\tau^\frown fin \subset TP$: Then fix a stage $s_1 > s_0$ such that τ does not take an outcome to the left of fin after stage s_1. Now let x be the greatest opened cycle, so for all τ-stages $t > s_1$ we have $0 = E(y_x) = E(y_x)[t] \neq \Psi^{L(D)}(y_x)[t]$ (otherwise, we would have outcome Γ at least once). Hence, $\mathcal{T}_{\Phi,D,\Psi}$ is satisfied.

$\tau^\frown d \subset TP$: Then fix a stage $s_1 > s_0$ such that τ takes outcome d at stage s_1 and diagonalizes via cycle x. Then we have that $1 = E(y_x) = E(y_x)[s_1 + 1] \neq \Psi^{L(D)}(y_x)[s_1] = 0$. Furthermore, for all τ-stages $t > s_1$, the only way this can change is if $L(D) \upharpoonright (\psi(y_x)[s_1] + 1)$ changes. However, this can happen only if a number x_0 leaves D after stage s_1 and $s^D(x_0) \leq \psi(y_x)[s_1] < s_1$, but this means that $A \upharpoonright (\varphi(x_0)[s_1] + 1)$ has changed after stage s_1, and this can happen only due to a node $\xi < \tau$ (and so τ would be initialized in that case). Indeed, after stage s_1, τ will always take outcome d, below which every $\mathcal{P}$-strategy will choose a fresh witness greater than $s_1 > \varphi(x_0)[s_1]$. (Moreover, since we visited τ at stage s_1, there was no link which crossed over $\tau^\frown fin$ at stage s_1. Also, if new links cross over $\tau^\frown d$ later, they don't affect our restraint on A.) So $\mathcal{T}_{\Phi,D,\Psi}$ is satisfied. $\qquad\square$

Lemma 4.7. *If an $\mathcal{S}_{\Phi,D,\Omega}$-strategy of maximal length along TP has outcome d or fin, then its requirement is satisfied.*

Proof. By the definition of the tree of strategies, we can choose a node $\sigma \subset TP$ assigned to $\mathcal{S}_{\Phi,D,\Omega}$ of maximal length. By Lemma 4.2, we fix a stage s_0 such that σ is not initialized after stage s_0. Now consider the following two cases:

$\sigma^\frown fin \subset TP$: Then fix a stage $s_1 > s_0$ such that σ does not take an outcome to the left of fin after stage s_1. Then we never see another σ-expansionary stage, and so $\mathcal{S}_{\Phi,D,\Omega}$ is satisfied vacuously.

$\sigma^\frown d \subset TP$: Then fix a stage $s_1 > s_0$ such that σ first takes outcome d at a σ-stage $\geq s_1$. Assume this was due to cycle x, so by case $\sigma 1.3$ of the construction, we have that $D(x)[s_1] \neq \Gamma^{L(D)}(x)[s_1]$. We first argue that $x \in D[s_1]$, since otherwise, $\Gamma^{L(D)}(x)[s_1] = 1$ with use $\gamma(x) \geq s^D(x)$, but after $\Gamma^{L(D)}(x)$ was defined, x has left D and so $s^D(x)$ has entered $L(D)$, destroying the computation $\Gamma^{L(D)}(x)$. This allows us to redefine Λ as described in $\sigma 1.3$.

Now consider the number z which caused σ to proceed to case $\sigma 1.1$; it must be of the form $z = s^D(x')$ for some x' which had left D already at a stage $s' < s_1$, say, causing z to enter $L(D)$ while $\Delta(z) = 0$. But just before stage s', we had $L(D)(z) = 0 = \Omega^E(z)$ since the σ-stage before stage s' was σ-expansionary; and since E was not allowed to change until stage s_1, we will have $\Omega^E(z)[s_1] = 0$ while $z \in L(D)$. By initialization, we then have that $\Omega^E(z) = 0$ is preserved, so $\mathcal{S}_{\Phi,D,\Omega}$ is satisfied.   □

Lemma 4.8. *If a $\mathcal{T}_{\Phi,D,\Psi}$-strategy of maximal length along TP has outcome Γ, then it correctly builds a Γ-functional.*

Proof. By the definition of the tree of strategies, we can choose a node $\tau \subset TP$ assigned to $\mathcal{T}_{\Phi,D,\Psi}$ of maximal length. By Lemma 4.2, we fix a stage s_0 such that τ is not initialized after stage s_0. Now we prove that $\Gamma^{L(D)} = D$. It is clear by the construction (case $\tau 1.2$) that for any x, there are infinitely many stages $s \geq s_0$ at which we have $\Gamma^{L(D)}(x)[s] \downarrow = D(x)[s]$. It remains to show that $\gamma(x)$ is bounded. So fix x and assume that $\gamma(x)[s] = s$ is defined at stage s (via case $\tau 1.2$). Since we traveled the link, the numbers below outcome *fin* will be chosen big, in particular, any witness a chosen below it is bigger than the use $\varphi(x)$ (namely, from now on, it is bigger than $s^-[s+1] > \varphi(x)[s]$, where $s^-[s+1] = s$ is that ρ-expansionary stage). Hence, only numbers of $\mathcal{P}$-strategies below the Γ-outcome can change $D \upharpoonright (x+1)$ (and so change $L(D) \upharpoonright (\gamma(x)+1)$). Note, however, that the enumeration of each such number a initializes all lower-priority strategies, so we will have fewer and fewer such numbers a, and so $\gamma(x)$ will be increased only finitely often. Hence, $\Gamma^{L(D)} = D$.   □

Lemma 4.9. *If an $\mathcal{S}_{\Phi,D,\Omega}$-strategy of maximal length along TP has outcome Γ or Δ, then it correctly builds a Γ- or Δ-functional.*

Proof. By the definition of the tree of strategies, we can choose a node $\tau \subset TP$ assigned to $\mathcal{S}_{\Phi,D,\Omega}$ of maximal length. By Lemma 4.2, we fix a stage s_0 such that σ is not initialized after stage s_0. If $\sigma^\frown\Gamma \subset TP$, then the proof is similar to the proof in Lemma 4.8. So let $\sigma^\frown\Delta \subset TP$. Then, after some stage $s_1 > s_0$, we never go to the left of outcome Δ. By the construction, this means that we can only have cases $\sigma 1.2$, $\sigma 3$, and $\sigma 4$; and cases $\sigma 4$ and then $\sigma 1.2$ must occur infinitely often. Hence, $\Delta = L(D)$, and the lemma is proved.   □

Lemma 4.10. *Each $\mathcal{R}$-requirement is satisfied by a node along the true path TP.*

Proof. Consider a requirement $\mathcal{R}_{\Phi,D}$. By the definition of the tree of strategies, we can choose a node $\rho \subset TP$ assigned to $\mathcal{R}_{\Phi,D}$ of maximal length. Clearly, ρ cannot be strictly between two fixed nodes forming a link created and canceled infinitely often (otherwise, ρ would not have maximal length). By Lemma 4.2, fix a stage s_0 such that ρ is not initialized after stage s_0. If ρ has finitely many ρ-expansionary stages, then $\mathcal{R}$ is satisfied vacuously; otherwise, assume that ρ has final versions of its set $E = E_\rho$ and functional $\Lambda = \Lambda_\rho$.

We prove that Λ^D is total and correctly computes E. Fix a natural number y and suppose inductively that $\Lambda^D(z) \downarrow = E(z)$ for all $z < y$ at all stages s starting at some stage $s_1 \geq s_0$. By Lemma 4.4, there are infinitely many stages at which case $\rho 2.1$ is executed. At such stages $t > s_1$, the strategy ρ ensures that $\Lambda^D(y) = E(y)$ with use y. As D is d.c.e, it follows that $D \upharpoonright (y+1)$ will eventually stop changing, after stage s_2, say, and hence $\Lambda^D(y)$ is defined. On the other hand, $E(y)$ can change at most twice: $E(y) = 0$ holds at all stages unless $y = y_x$ is a number used by a specific $\mathcal{T}$-strategy τ in relation to some number $x < y$. In this case, τ is the only strategy that can enumerate y into E, and this happens under case $\tau 1.1$, when $D(x)$ also changes and τ is declared satisfied. The change in D at x allows ρ to correct $\Lambda^D(y)$. The strategy τ then has outcome d until (if ever) it is initialized and so it will never deal with the number y again. After that, $E(y)$ can change only once, if it is extracted from E by ρ under case $\rho 2.1$. This happens if $D \upharpoonright (y+1)$ has reverted to an old state, and so x was previously enumerated into D when y was enumerated in E, but now $D(x) = 0$ again. Note that $D(x) = 0$ at all future stages, and so $\Lambda^D(y) = E(y)$ will remain true at all future stages.

If all substrategies of ρ along the true path have finite outcomes, then it follows from Lemmas 4.6 and 4.7 that E is Turing incomparable to $L(D)$. Otherwise, it follows from Lemmas 4.8 and 4.9 that either $L(D)$ is computable or $L(D) \equiv_T D$. In both cases, $\mathcal{R}$ is satisfied. $\qquad\square$

References

1. Arslanov, M.M.: The lattice of the degrees below $0'$, Izv. Vyssh. Uchebn. Zaved. Mat. **7**, 27–33 (1988). MR 968729
2. Arslanov, M.M., Kalimullin, I.S., Lempp, S.: On Downey's conjecture. J. Symbolic Logic **75**(2), 401–441 (2010). MR 2648149
3. Cooper, S.B.: On a theorem of C. E. M. Yates, handwritten notes
4. Downey, R.G.: D.r.e. degrees and the nondiamond theorem. Bull. London Math. Soc. **21**(1), 43–50 (1989). MR 967789
5. Ershov, Y.L.: A certain hierarchy of sets. I. Algebra i Logika **7**(1), 47–74 (1968). MR 0270911
6. Ershov, Y.L.: A certain hierarchy of sets. II. Algebra i Logika **7**(4), 15–47 (1968). MR 0270912
7. Ershov, Y.L.: A certain hierarchy of sets. III. Algebra i Logika **9**, 34–51 (1970). MR 0299478
8. Gold, E.M.: Limiting recursion. J. Symbolic Logic 30, 28–48 (1965). MR 0239972
9. Ishmukhametov, S.T.: On the r.e. predecessors of d.r.e. degrees. Arch. Math. Logic **38**(6), 373–386 (1999). MR 1711400
10. Lachlan, A.H.: Lower bounds for pairs of recursively enumerable degrees. Proc. London Math. Soc. **16**(3), 537–569 (1966). MR 0204282
11. Liu, J., Wu, G., Yamaleev, M.M.: Downward density of exact degrees. Lobachevskii. J. Math. **36**(4), 389–398 (2015). MR 3431199
12. Miller, D.P.: High recursively enumerable degrees, the anti-cupping property, Logic Year 1979–80: University of Connecticut, pp. 230–245 (1981)
13. Putnam, H.W.: Trial and error predicates and the solution to a problem of Mostowski. J. Symbolic Logic **30**, 49–57 (1965). MR 0195725

14. Sacks, G.E.: On the degrees less than $0'$. Ann. Math. **77**(2), 211–231 (1963). MR 0146078
15. Soare, R.I.: Recursively enumerable sets and degrees: A study of computable functions and computably generated sets. Perspectives in Mathematical Logic, Springer, Berlin (1987). MR 882921
16. Guohua, W., Yamaleev, M.M.: Isolation: motivations and applications. Uchenye Zapiski Kazanskogo Universiteta **154**, 204–217 (2012)

On the Strongly Bounded Turing Degrees
of the Computably Enumerable Sets

Klaus Ambos-Spies[✉]

Institut für Informatik, Heidelberg University, INF 205, 69120 Heidelberg, Germany
ambos@math.uni-heidelberg.de

Abstract. We introduce and discuss some techniques designed for the study of the strongly bounded Turing degrees of the computably enumerable sets, i.e., of the computable Lipschitz degrees and of the identity bounded Turing degrees of c.e. sets. In particular we introduce some tools which allow the transfer of certain facts on the weak truth-table degrees to these degree structures. Using this approach we show that the first order theories of the partial orderings $(R_{cl}, \leq)$ and $(R_{ibT}, \leq)$ of the c.e. cl- and ibT-degrees are not $\aleph_0$-categorical and undecidable. Moreover, various other results on the structure of the partial orderings $(R_{cl}, \leq)$ and $(R_{ibT}, \leq)$ are obtained along these lines.

1 Introduction

A *computable Lipschitz* reduction, cl-reduction for short, is a Turing reduction in which, on an input x, the queries are not greater than $x + c$ for some constant c. By this strong bound on the use function, this reducibility, introduced by Downey, Hirschfeldt and Laforte [15,16], is not only a measure of relative computability but also a measure of the Kolmogorov complexity. Namely, if A is cl-reducible to B, $A \leq_{cl} B$, then A is Turing reducible (in fact weak truth-table) reducible to B and, for any $n \geq 0$, the Kolmogorov complexity of the initial segment $A \restriction n$ of length n is bounded by the Kolmogorov complexity of the corresponding initial segment $B \restriction n$ up to an additive constant. So this reducibility proved to be a useful tool in the theory of algorithmic randomness (see the recent monographs by Downey and Hirschfeldt [14] and by Nies [25] for details). The special case of a cl-reduction where the additive constant is zero, i.e., where the input itself is a bound on the oracle queries, has been independently introduced by Soare [28] in a different context, and is called an *identity bounded Turing* reduction (ibT-reduction, for short) there. In the following we refer to cl- and ibT-reducibility as the *strongly bounded Turing* reducibilities.

In this paper we study the partial orderings of the degrees induced by the strongly bounded Turing reducibilities on the computably enumerable (c.e.) sets, i.e., the partial orderings of the c.e. cl- and ibT-degrees, $(R_{cl}, \leq)$ and $(R_{ibT}, \leq)$. These partial orderings have some quite surprising properties which distinguish them from the partial orderings of the c.e. weak truth-table (wtt) and Turing (T) degrees. So Barmpalias [7] has shown that the partial ordering of the c.e.

© Springer International Publishing AG 2017
A. Day et al. (Eds.): Downey Festschrift, LNCS 10010, pp. 563–598, 2017.
DOI: 10.1007/978-3-319-50062-1_34

cl-degrees does not have any maximal elements hence, in particular, no greatest element, and this observation applies to the c.e. ibT-degrees too; and Barmpalias [7] and Fan and Lu [19] have shown that there are *maximal pairs* in these partial orderings, i.e., c.e. degrees **a** and **b** with no c.e. degree **c** above them. So, in particular, for the strongly bounded Turing reducibilities $r = \text{ibT}, \text{cl}$, $(R_r, \leq)$ is not an upper semilattice (and, as for most of the c.e. degree structures, it is neither a lower semilattice as shown by Downey and Hirschfeldt [14], Theorem 9.5.2). Moreover, in contrast to the c.e. wtt- and T-degrees, the partial orderings of the c.e. ibT- and cl-degrees are not dense (Barmpalias and Lewis [9] and Day [12], respectively). More recently, Ambos-Spies, Bodewig, Kräling and Yu [4] have shown that these partial orderings are nondistributive.

The goal of this paper is twofold. First we want to give some more facts on the strongly bounded Turing reducibilities and their c.e. degrees. Second we want to introduce and discuss some tools for studying these reducibilities and degrees. Though the study of the partial orderings $(R_r, \leq)$ for $r = \text{ibT}, \text{cl}$ uses some of the standard methods developed for studying computably enumerable sets and their degrees, like priority arguments, there are also some specific methods taylored for the strongly bounded Turing reducibilities. For example, the fact that there are no minimal nonzero degrees is trivial in this setting, since - as observed by Downey et al. [15] - $2A <_{\text{cl}} A$ for any noncomputable set A where $2A = \{2x : x \in A\}$; and - as independently observed in Ambos-Spies et al. [5] and Bélanger [10] - Barmpalias's result that there are no maximal c.e. cl-degrees can be proven by a simple shift argument too (see Theorem 3.4 below).

Expanding some work of Ambos-Spies, Ding, Fan and Merkle [5] we first explore some basic properties of computable shifts (i.e., strictly increasing computable functions) and give some applications of this notion (Sects. 3 and 4). For instance we show that computable invariance of the strongly bounded Turing reducibilities fails as badly as possible: for any noncomputable c.e. set A there are computable permutations p and $\hat{p}$ such that $p(A) <_r A <_r \hat{p}(A)$ (for $r = \text{ibT}, \text{cl}$). Moreover, we show that the bounded shifts f (i.e., shifts where the difference $f(x) - x$ is bounded) induce automorphisms of the partial ordering $(R_{\text{ibT}}, \leq)$ of the c.e. ibT-degrees (and of the partial ordering of the ibT-degrees in general) whence this partial ordering is not rigid. On the other hand we show that no unbounded shift induces an automorphism of $(R_{\text{ibT}}, \leq)$ or $(R_{\text{cl}}, \leq)$ whence we leave open the question whether $(R_{\text{cl}}, \leq)$ is rigid or not.

Next, in Sect. 5, we exploit the close relation between the strongly bounded Turing reducibilities and the permitting technique in order to give a very useful Representation Lemma for reductions of this type. As first applications of this lemma, (1) we observe that, for ibT-equivalent c.e. sets A and B, the sets A, $A \cup B$ and $A \cap B$ are ibT-equivalent whence, for disjoint c.e. sets A and B such that $A \leq_{\text{ibT}} B$, it holds that $A <_{\text{ibT}} B$; (2) we show that the joins in the strongly bounded Turing degrees of c.e. sets which are represented by c.e. set splittings are locally distributive. Either of these observations is exploited in the following.

In Sect. 6 we combine the preceding observations on shifts, representability and splittings in order to give a quite simple proof for a first global result on the theory of the c.e. ibT-degrees: the first order theory $\text{Th}(\mathbf{R}_{\text{ibT}}, \leq)$ of the partial ordering of c.e. ibT-degrees realizes infinitely many 2-types. So $\text{Th}(\mathbf{R}_{\text{ibT}}, \leq)$ is not $\aleph_0$-categorical. The proof is based on the analysis of the degrees which cup to a given a degree in this degree structure. (These results are also used in Ambos-Spies, Bodewig, Fan, and Kräling [3], where it is shown, that some of the cupping properties of the c.e. ibT-degrees which are proven here do not carry over to the c.e. cl-degrees, thereby giving the so far only known elementary difference between these degree structures.)

In the remainder of the paper we explore interactions among ibT-reducibility, cl-reducibility and wtt-reducibility on the c.e. sets, and we show how these relations can be used in order to transfer results on the weak truth-table degrees to the strongly bounded Turing degrees. Some basic observations are provided in Sect. 7. In particular, there we show that joins and meets in the c.e. ibT-degrees are preserved in the c.e. cl-degrees, and that joins and meets in the c.e. cl-degrees are preserved in the c.e. wtt-degrees. A typical consequence of these preservation lemmas is the fact that minimal pairs in the c.e. ibT-, cl-, and wtt-degrees coincide (as observed in [5] already); and so do the (non)bounding and (non)top degrees, i.e., the degrees which (do not) bound a minimal pair and which are (not) the join of a minimal pair, see Sect. 8. There we also give an Embedding Lemma which shows that the nondistributive partial orderings $(\mathbf{R}_r, \leq)$ ($r = \text{ibT}, \text{cl}$) provide sufficient local distributivity such that, any sequence of n degrees which are pairwise minimal pairs and which can be represented by pairwise disjoint c.e. sets generates the n-atom Boolean algebra. So, in particular, any finite distributive lattice can be embedded into $(\mathbf{R}_r, \leq)$.

Sections 9 and 10 contain our main results: the first order theories $\text{Th}(\mathbf{R}_{\text{ibT}}, \leq)$ and $\text{Th}(\mathbf{R}_{\text{cl}}, \leq)$ realize infinitely many 1-types - hence are not $\aleph_0$-categorical - and are undecidable. Either result is based on the proof of a corresponding result for the c.e. wtt-degrees obtained by Ambos-Spies and Soare [6] and by Lempp and Nies [23], respectively. The core of either of these previous proofs is some technical theorem, on nonbounding and nontop degrees, respectively, which is proven by a very sophisticated $0'''$-priority argument. Then, in either case, distributivity of the upper semilattice of the c.e. wtt-degrees is used in order to get the desired result on the theory $\text{Th}(\mathbf{R}_{\text{wtt}}, \leq)$. By using the transfer tools developed in the preceding sections together with the local distributivity results obtained there, we can argue that the original technical theorems in [6] and [23] suffice to carry over the 1-types and undecidability results for the wtt-degrees there to the strongly bounded Turing degrees.

The paper is rounded off by some observations on the relations among the strongly bounded Turing reducibilities and the classical strong reducibilities of truth-table type (given in the initial Sect. 2) and some open problems (Sect. 11).

We conclude this section with introducing the basic notions studied in this paper. In general our notation is standard; unexplained notation can be found in Downey and Hirschfeldt [14], Nies [25] or Soare [27].

A set A is *computable Lipschitz* (cl for short) *reducible* to a set B, $A \leq_{\text{cl}} B$, if there is a Turing functional Ψ such that $A = \Psi^B$ and such that, for some constant $k \geq 0$, the greatest oracle query in the computation of $\Psi^B(x)$ (if any) is $\leq x+k$. In this case we also say that A is $(i+k)$-*bounded Turing* $((i+k)\text{bT}$ for short) reducible to B $(k \geq 0)$ and we write $A \leq_{(i+k)\text{bT}} B$. A is *identity bounded Turing* (ibT for short) reducible to B, $A \leq_{\text{ibT}} B$, if A is $(i+0)\text{bT}$-reducible to B. Correspondingly we call a Turing functional Ψ an $(i+k)\text{bT}$-*functional* if, for any oracle B and any input x the greatest oracle query in the computation of $\Psi^B(x)$ is bounded by $x+k$ $(k \geq 0)$; we call Ψ a cl-*functional* if Ψ is an $(i+k)\text{bT}$-functional for some $k \geq 0$; and we call Ψ an ibT-*functional* if Ψ is an $(i+0)\text{bT}$-functional. Note that, for $r = \text{cl}, \text{ibT}, (i+k)\text{bT}$, $A \leq_r B$ iff there is an r-functional Ψ such that $A = \Psi^B$. (For instance, for $r = \text{ibT}$, if $A \leq_{\text{ibT}} B$ via the Turing functional Ψ then $A = \hat{\Psi}^B$ for the ibT-functional $\hat{\Psi}$ obtained from Ψ by suppressing all queries which are greater than the input and by making the computation divergent if there is such a query in the computation of Ψ.) Obviously, for $r = \text{cl}, \text{ibT}$, r-reducibility is transitive whence the r-degree

$$deg_r(A) = \{B : A =_r B\}$$

is well defined and $\leq_r$ induces a partial ordering $\leq$ on the r-degrees. We call an r-degree *computably enumerable* if it contains a c.e. set, and we let $(R_r, \leq)$ denote the partial ordering of the c.e. r-degrees. Note that (just as for $r = \text{wtt}, \text{T}$) $\mathbf{0} = \{A : A \text{ is computable}\}$ is the least c.e. r-degree. Moreover, in the following we will tacitly use that, for $r = \text{cl}, \text{ibT}, (i+k)\text{bT}$, r-reducibility is invariant under finite variants, i.e., that $\hat{A} \leq_r \hat{B}$ for any sets $\hat{A}$ and $\hat{B}$ such that $\hat{A} =^* A$, $\hat{B} =^* B$ and $A \leq_r B$.

2   Comparing Reducibilities

The strongly bounded Turing reducibilities are related to the other common strong reducibilities refining weak truth-table reducibility as follows.

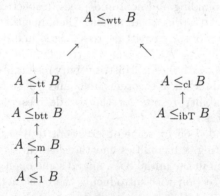

Diagram 1

Note that the strong reducibilities of truth-table type on the left hand side are on one hand more flexible than the strongly bounded Turing reducibilities on the right hand side since their use functions are only computably bounded and not necessarily bounded by the identity function (or the identity function plus a constant). On the other hand, the strongly bounded Turing reducibilities are more flexible in the use of the oracle since - in contrast to a truth-table reduction - the result of the reduction is not completely determinded by the answers of the oracle queries. So it is not surprising that in the above diagram only the indicated relations hold, even if we consider only c.e. sets. For the strictness of the implications on the left hand side see e.g. Odifreddi [26]. The strictness of the implications on the right hand side have been established by Barmpalias and Lewis [9] and Downey, Hirschfeldt and LaForte [15], respectively.

Theorem 2.1 (Barmpalias and Lewis [9], Downey et al. [15]). *Let A be a noncomputable set.*

(i) For $A + 1 = \{x + 1 : x \in A\}$, $A + 1 <_{\text{ibT}} A$ and $A + 1 =_{\text{cl}} A$.
(ii) For $2A = \{2x : x \in A\}$, $2A <_{\text{cl}} A$ and $2A =_{\text{wtt}} A$.

(Note that, for a c.e. set A, the sets $A + 1$ and $2A$ are c.e. too.)

PROOF (SKETCH). We sketch the proof of (ii). The proof of (i) is similar. Obviously, for any set A, $2A \leq_{\text{ibT}} A$, hence $2A \leq_{\text{cl}} A$, and $A \leq_{\text{wtt}} 2A$. So, given A such that $A \leq_{\text{cl}} 2A$, it suffices to show that A is computable. For this we show that A is selfreducible, i.e., that $A(x)$ can be computed from $A \upharpoonright x$ uniformly in x. Obviously, this implies that A is computable.

So fix $k \geq 0$ such that $A \leq_{(i+k)\text{bT}} 2A$, fix an $(i+k)$bT-functional Ψ such that $A = \Psi^{2A}$, and fix y_0 such that $y + k < 2y$ for $y \geq y_0$. Then, for $x \geq y_0$, $A(x)$ can be computed from $A \upharpoonright x$ by answering the oracle queries in the computation of $\Psi^{2A}(x)$ using $A \upharpoonright x$ as an oracle. Namely, for any oracle query y, $y \in 2A$ if and only if y is even and $y/2 \in A$. But, since the functional Ψ is $id + k$-bounded, $y \leq x + k < 2x$ whence $y/2 < x$. $\qquad\square$

In order to complete the analysis of the relations among the strong reducibilities it remains to show that, in Diagram 1, the reducibilities on the left hand side are incomparable with the reducibilities on the right hand side. We do this by constructing a pair of c.e. sets such that the relation between these sets with respect to the truth-table reducibilities is just the opposite of the relation between these sets with respect to the strongly bounded Turing reducibilities.

Theorem 2.2. *There are noncomputable c.e. sets A and B such that*

$$A <_r B \text{ for } r \in \{1, \text{m}, \text{btt}, \text{tt}\} \tag{1}$$

whereas

$$B <_{r'} A \text{ for } r' \in \{\text{ibT}, \text{cl}\}. \tag{2}$$

PROOF. (SKETCH). By a finite injury argument enumerate c.e. sets A and B such that

$$x \in A_{at\ s} \Leftrightarrow 2x \in B_{at\ s} \tag{3}$$

and

$$2x + 1 \in B_{at\ s} \Rightarrow x \in A_{at\ s} \tag{4}$$

hold and such that the requirements

$$\Re_e : \Phi_e \text{ is a tt-reduction (i.e., total)} \Rightarrow \exists\, x\ (B(2x+1) \neq \Phi_e^A(2x+1))$$

are met where $\{\Phi_e\}_{e\geq 0}$ is a computable enumeration of the Turing functionals.

Note that this guarantees that A and B have the required properties. Namely, for a proof of (1) it suffices to show that $A \leq_1 B$ and $B \not\leq_{tt} A$. But the former is immediate by (3) while the latter is guaranteed by the requirements $\Re_e$ ($e \geq 0$). For a proof of (2) it suffices to show that $B \leq_{ibT} A$ and $A \not\leq_{cl} B$. The former is immediate by (3) and (4). (For odd numbers this follows from (4) by permitting; see Sect. 5 for more on the permitting technique.) For a proof of the latter note that, by (1), B is noncomputable hence so is A by $B \leq_{ibT} A$. So, by Theorem 2.1, $A \not\leq_{cl} 2A$. Since, by (3) and (4), $B \leq_{ibT} 2A$ (hence $B \leq_{cl} 2A$), it follows that $A \not\leq_{cl} B$.

In the remainder of the proof we explain the strategy for meeting requirement $\Re_e$. Reserve an infinite computable set R_e of diagonalization candidates. Pick $x \in R_e$ minimal such that x is not restrained by any higher priority requirement and such that x has not been enumerated into A by some previous attack on $\Re_e$ (which later became injured by a higher priority requirement). Wait for a stage s such that $\Phi_{e,s}^{A_s \cup \{x\}}(2x+1)$ is defined, say $\Phi_{e,s}^{A_s \cup \{x\}}(2x+1) = i$. (Note that if no such stage s exists then Φ_e is not total hence requirement $\Re_e$ is trivially met.) Then, at stage $s+1$, put x into A and $2x$ into B, and, if $i = 0$, put $2x+1$ into B too. Moreover, by imposing a restraint on A, preserve the computation $\Phi_{e,s}^{A_{s+1}}(2x+1) = \Phi_{e,s}^{A_s \cup \{x\}}(2x+1)$. Note that this action is compatible with (3) and (4) and it ensures that $B(2x+1) \neq \Phi^A(2x+1)$ unless the restraint imposed at stage $s+1$ is injured by some higher priority requirement later. $\square$

3   Bounded and Computable Shifts

In [5] Ambos-Spies, Ding, Fan and Merkle have extended the observations made in Theorem 2.1 by introducing finite and computable shifts. We shortly review the basic definitions and results here. Moreover, we extend some of the observations in [5] by proving some noncupping properties of shifts (see parts (iii) of Lemmas 3.2 and 3.3 below) which will be used in the following.

Definition 3.1. *(a) A (computable) shift f is a strictly increasing (computable) function $f : \omega \to \omega$. A shift f is nontrivial if $f(x) > x$ for some (hence for almost all) x; and a shift f is unbounded if, for any number k, there is a number x such that $f(x) - x > k$ (and f is bounded otherwise).*

(b) For any set A and any shift f, the f-shift A_f of A is defined by

$$A_f = f(A) = \{f(x) : x \in A\}.$$

Note that, for any shift f, $x \leq f(x)$ and $f(x) - x$ is nondecreasing in x. So a shift f is unbounded if and only

$$\lim_{n \to \infty} (f(n) - n) = \sup_{n \to \infty} (f(n) - n) = \infty.$$

For $k \geq 0$ let

$$A + k = \{x + k : x \in A\} \quad \text{and} \quad A - k = \{x - k : x \in A \ \& \ x \geq k\}.$$

Then $A + k = A_f$ for the bounded shift $f(x) = x + k$. In fact, for any bounded shift f, $A_f =^* A + k$ for some $k \geq 0$. So, in particular, any bounded shift f is computable. Moreover, obviously, $A + (k + 1) = (A + k) + 1$ whence the observation in Theorem 2.1 that, for noncomputable A, $A + 1 <_{\mathrm{ibT}} A$ extends to all nontrivial bounded shifts.

Lemma 3.2 (Bounded-Shift Lemma). *Let f be a nontrival bounded shift and let A be a noncomputable (not necessarily c.e.) set.*

(i) $A_f =_m A$ (in fact $A \leq_1 A_f$ and $A_f \leq_m A$) and $A_f =_{\mathrm{cl}} A$.
(ii) $A_f <_{\mathrm{ibT}} A$.
(iii) For any (not necessarily c.e.) set B such that $A \leq_{\mathrm{ibT}} A_f \cup B$, $A \leq_{\mathrm{ibT}} B$.

PROOF. Part (i) is straightforward, and, obviously, $A_f \leq_{\mathrm{ibT}} A$. By the latter, part (ii) follows from (iii) by letting B be the empty set. This leaves (iii). The proof of this part is similar to the proof of (the nontrivial parts of) Theorem 2.1.

Fix B and assume that $A = \Psi^{A_f \cup B}$ for the ibT-functional Ψ. Moreover, by nontriviality of f, fix y_0 such that $y < f(y)$ for all $y \geq y_0$. In order to show $A \leq_{\mathrm{ibT}} B$, we give an inductive procedure for computing $A(x)$ from $B \upharpoonright x + 1$ (uniformly in x). Given x and the initial segment $B \upharpoonright x + 1$ of B as an oracle, it suffices to compute $(A_f \cup B)(y)$ for the queries $y \geq y_0$ occuring in the computation of $\Psi^{A_f \cup B}(x)$ in order of appearance. Fix such a query y. Since Ψ is an ibT-functional, $y \leq x$. So, using $B \upharpoonright x + 1$ as an oracle, we can decide whether $y \in B$. It remains to decide whether $y \in A_f$. For this sake, first decide whether y is in the range of f and, if so, compute the unique x' such that $f(x') = y$. (Note that this can be done since f is a computable shift.) Now, if x' does not exist then $A_f(y) = 0$. Otherwise, $A(x') = A_f(y)$, and, by $y \geq y_0$, $x' < y \leq x$. So $A(x')$ can be computed from $B \upharpoonright x' + 1$ (hence from $B \upharpoonright x + 1$) by inductive hypothesis. □

For unbounded computable shifts f, we obtain an analog of Lemma 3.2 with cl and wtt in place of ibT and cl, respectively.

Lemma 3.3 (Computable-Shift Lemma). *Let f be a computable shift and let A be a noncomputable (not necessarily c.e.) set.*

(i) $A_f =_m A$ (in fact $A \leq_1 A_f$ and $A_f \leq_m A$) and $A_f =_{\mathrm{wtt}} A$.
(ii) $A_f \leq_{\mathrm{ibT}} A$. Moreover, if f is unbounded then $A \not\leq_{\mathrm{cl}} A_f$ (whence $A_f <_{\mathrm{ibT}} A$ and $A_f <_{\mathrm{cl}} A$).
(iii) If f is unbounded then, for any (not necessarily c.e.) set B such that $A \leq_{\mathrm{cl}} A_f \cup B$, $A \leq_{\mathrm{cl}} B$.

The proof of the Computable-Shift Lemma is similar to the proof of the Bounded-Shift Lemma and is left to the reader.

As Theorem 2.1 above shows, computable shifts witness the fact that the partial orderings of the c.e. ibT-degrees and the c.e. cl-degrees do not have minimal nonzero elements. As Ambos-Spies et al. [5] and, independently, Bélanger [10] have shown, shifts can be also applied in order to show that these orderings do not have maximal elements. In case of the c.e. ibT-degrees this is straightforward. Given a noncomputable c.e. set A, $A <_{\text{ibT}} A - 1$. Namely, by $A =^* (A - 1) + 1$, A is ibT equivalent to the nontrival bounded shift $(A - 1) + 1$ of $A - 1$ whence $A <_{\text{ibT}} A - 1$ by (part (ii) of) the Bounded-Shift Lemma. For the cl-degrees the nonexistence of maximal c.e. degrees was shown by Barmpalias [7]. Barmpalias's original proof was based on a quite sophisticated nonuniform argument. By using the fact, that any infinite c.e. set contains an infinite computable subset, however, there is a quite simple proof of this fact using shifts.

Theorem 3.4 (Barmpalias [7]). *The partial ordering* $(\mathbf{R}_{\text{cl}}, \leq)$ *of the c.e. cl-degrees does not have maximal elements. I.e., for any c.e. set A there is a c.e. set $\hat{A}$ such that $A <_{\text{cl}} \hat{A}$. Moreover, $\hat{A}$ can be chosen so that $A <_{\text{ibT}} \hat{A}$ too.*

PROOF OF THEOREM 3.4 (AMBOS-SPIES ET AL. [5] AND BÉLANGER [10]). Let A be any c.e. set. If A is computable then, for any noncomputable c.e. set $\hat{A}$, $A <_r \hat{A}$ for $r = \text{ibT}, \text{cl}$. So w.l.o.g. we may assume that A is noncomputable, hence infinite, and we may fix an infinite computable subset C of A. Then, obviously, $A =_{\text{ibT}} A \setminus C = A \cap \overline{C}$, where $\overline{C}$ is the complement of C. Moreover, by noncomputability of A, $\overline{C}$ is infinite.

Now, intuitively, we can effectively *compress* the set $A \setminus C$ to a set $\hat{A}$ by using the space in C. Then $A \setminus C$ will be a computable unbounded shift of $\hat{A}$ whence, by the Computable-Shift Lemma, $(A =_r) A \setminus C <_r \hat{A}$ for $r = \text{ibT}, \text{cl}$.

More formally, let $f : \omega \to \overline{C}$ enumerate $\overline{C}$ in order of magnitude, and let $\hat{A} = f^{-1}(A \cap \overline{C})$ be the preimage of $A \cap \overline{C}$ under f. Then f is a computable shift and $\hat{A}_f = A \cap \overline{C}$. Moreover, by infinity of C, f is unbounded. So, for $r = \text{ibT}, \text{cl}$, $A \cap \overline{C} <_r \hat{A}$ by the Computable-Shift Lemma. By $A =_{\text{ibT}} A \cap \overline{C}$ this implies the claim. □

Note that the first step in the above proof shows that any noncomputable c.e. set A is ibT-equivalent to a non-simple c.e. set (namely, to $A \setminus C$ where C is any infinite computable subset of A):

Proposition 3.5. *For any (noncomputable) c.e. set A there is a c.e. set $\hat{A}$ such that $A =_{\text{ibT}} \hat{A}$ and $\hat{A}$ is not simple.*

The dual of this proposition fails: as shown in Ambos-Spies [2] there are noncomputable c.e. sets which are not cl-equivalent to any simple set.

The argument used in the proof of Theorem 3.4 can be easily modified in order to show that the strongly bounded Turing reducibilities are not computably invariant. In fact, computable invariance fails for the ibT- and cl-degrees of the computably enumerable sets almost as badly as possible.

Theorem 3.6. *Let A be any noncomputable c.e. set.*

(i) There is a computable permutation p such that, for $r \in \{\text{ibT}, \text{cl}\}$, $p(A) <_r A$. In fact, for any computable shift f, there is a computable permutation p such that $p(A) =_{\text{ibT}} A_f$.

(ii) There is a computable permutation p such that, for $r \in \{\text{ibT}, \text{cl}\}$, $A <_r p(A)$.

PROOF. (i). Note that the first part follows from the second part by the Computable-Shift Lemma. So, given a computable shift f, it suffices to define a computable permutation p such that $p(A) =_{\text{ibT}} f(A)$. (Note that $f(A) = A_f$.)

Let B be an infinite computable subset of A and let $C = \overline{f(\overline{B})}$. Then C is infinite and computable too. Hence the one-to-one functions b and c enumerating B and C, respectively, in order of magnitude are computable. It follows that p defined by $p(b(n)) = c(n)$ and $p(x) = f(x)$ for $x \in \overline{B}$ is a computable permutation. Moreover, $f(A)$ is the disjoint union of $f(A\backslash B)$ and the computable set $f(B)$, while $p(A)$ is the disjoint union of $p(A\backslash B) = f(A\backslash B)$ and the computable set $p(B)$. So $f(A) =_{\text{ibT}} f(A\backslash B) =_{\text{ibT}} p(A)$.

(ii). Fix a pair of disjoint infinite computable subsets B_0 and B_1 of A, let $B = B_0 \cup B_1$, let $f : \omega \to \overline{B_0}$ enumerate the complement $\overline{B_0}$ of B_0 in order of magnitude, and let

$$\hat{A} = f^{-1}(A\backslash B) = \{x : f(x) \in A\backslash B\}.$$

Then f is an unbounded computable shift and $A\backslash B$ is the f-shift of the c.e. set $\hat{A}$, hence $A\backslash B <_r \hat{A}$ by the Computable-Shift Lemma ($r \in \{\text{ibT}, \text{cl}\}$). Since A is the disjoint union of $A\backslash B$ and the computable set B, it follows that $A <_r \hat{A}$ holds.

So it suffices to define a computable permutation p such that $p(A) =_{\text{ibT}} \hat{A}$. For the definition of p first note that $\overline{f^{-1}(\overline{B})} = f^{-1}(B) = f^{-1}(B_1)$ is infinite and computable. Hence the one-to-one functions b and c enumerating B and $C = \overline{f^{-1}(\overline{B})}$, respectively, in order of magnitude are computable. It follows that p defined by $p(b(n)) = c(n)$ and $p(x) = f^{-1}(x)$ for $x \in \overline{B}$ is a computable permutation. Moreover, $p(A)$ is the disjoint union of $p(A\backslash B) = f^{-1}(A\backslash B) = \hat{A}$ and the computable set $p(B) = C$. So $p(A) =_{\text{ibT}} \hat{A}$. □

It is natural to ask whether the previous theorem can be extended by showing that, for any noncomputable c.e. set A, there is a computable permutation p such that, for $r \in \{\text{ibT}, \text{cl}\}$, the c.e. sets A and $p(A)$ are r-incomparable. Though, for some c.e. sets A - for instance for any m-complete set A (see the remark following Theorem 5.5 below) - such a permutation p can be found, in general this is impossible: by a finite injury argument we can construct a noncomputable c.e. set A such that, for any permutation p, $p(A) \leq_{\text{ibT}} A$ or $A \leq_{\text{ibT}} p(A)$ holds. (We omit the proof here.)

4 Shifts and Automorphisms

Here we give another interesting application of computable shifts. We show that the bounded shifts induce automorphisms of the partial ordering $(\mathbf{R}_{\text{ibT}}, \leq)$ of the c.e. ibT-degrees. So this degree structure is not rigid. This approach for defining nontrivial automorphisms, however, cannot be extended to the structure of the c.e. cl-degrees. Namely, as we will show too, no unbounded computable shift f induces an automorphism of $(\mathbf{R}_{cl}, \leq)$ or $(\mathbf{R}_{ibT}, \leq)$. So the question, whether the partial ordering of the c.e. cl-degrees is rigid too, is left open.

Recall that an *(order) automorphism* of a partial ordering $(P, \leq)$ is a surjective map $f : P \to P$ such that, for any $x, y \in P$, $x \leq y$ if and only if $f(x) \leq f(y)$. In order to show that the bounded shifts induce automorphisms of $(\mathbf{R}_{ibT}, \leq)$, we first observe that any computable shift is ibT-degree invariant and preserves the ordering $\leq_{ibT}$ and then show that the function on the c.e. ibT-degrees induced by a bounded shift is surjective.

Lemma 4.1 (Invariance Lemma for Computable Shifts). *Let f be a computable shift. Then, for any c.e. sets A and B,*

$$A \leq_{ibT} B \; \Leftrightarrow \; A_f \leq_{ibT} B_f. \tag{5}$$

We omit the straightforward proof.

By the Invariance Lemma, any computable shift f defines a function $\mathbf{f} :$ $\mathbf{R}_{ibT} \to \mathbf{R}_{ibT}$ on the c.e. ibT-degrees where $\mathbf{f}(deg_{ibT}(A)) = deg_{ibT}(A_f)$. For bounded shifts this function is on-to.

Lemma 4.2 (Surjectivity Lemma for Bounded Shifts). *For $k \geq 0$ let $\mathbf{f}_k : \mathbf{R}_{ibT} \to \mathbf{R}_{ibT}$ be defined by $\mathbf{f}_k(\mathbf{a}) = \mathbf{a} + k$ where $\mathbf{a} + k = deg_{ibT}(A + k)$ for any c.e. set $A \in \mathbf{a}$. Then $\mathbf{f}_k$ is well defined and surjective.*

PROOF. Fix $k \geq 0$. Since $\mathbf{f}_k$ is well defined by Lemma 4.1, it suffices to show that, for any c.e. set A, there is a c.e. set B such that $\mathbf{f}_k(deg_{ibT}(B)) = deg_{ibT}(A)$, i.e., $A =_{ibT} B + k$. But this is true for $B = A - k$ since $A =^* (A - k) + k$. □

Theorem 4.3 (Automorphism Theorem). *Let $k \geq 0$. The function $\mathbf{f}_k :$ $\mathbf{R}_{ibT} \to \mathbf{R}_{ibT}$ induced by the bounded shift $f(x) = x + k$ is an automorphism of $(\mathbf{R}_{ibT}, \leq)$.*

PROOF. By Lemma 4.1, $\mathbf{a} + k$ hence $\mathbf{f}_k$ is well defined and satisfies

$$\forall \, \mathbf{a}, \mathbf{b} \in \mathbf{R}_{ibT} \; [\mathbf{a} \leq \mathbf{b} \Leftrightarrow \mathbf{f}_k(\mathbf{a}) \leq \mathbf{f}_k(\mathbf{b})], \tag{6}$$

while, by Lemma 4.2, $\mathbf{f}_k$ is surjective. □

Corollary 4.4. *The partial ordering $(\mathbf{R}_{ibT}, \leq)$ of the c.e. ibT-degrees is not rigid. In fact, the automorphism group of $(\mathbf{R}_{ibT}, \leq)$ is infinite.*

PROOF. By Theorem 4.3, for any $k \geq 0$, $\mathbf{f}_k$ is an automorphism of $(\mathbf{R}_{ibT}, \leq)$. On the other hand, by definition of $\mathbf{f}_1$ and by the Bounded-Shift Lemma,

$$\forall \, \mathbf{a} > \mathbf{0} \; (\mathbf{f}_1(\mathbf{a}) = \mathbf{a} + 1 < \mathbf{a}). \tag{7}$$

So $\mathbf{f}_1$ is a nontrivial automorphism whence $(\mathbf{R}_{\mathrm{ibT}}, \leq)$ is not rigid. Since $A + (k+1) = (A+k)+1$ and since, by definition of $\mathbf{f}_k$, $\mathbf{f}_{k+1}(\mathbf{a}) = \mathbf{f}_1(\mathbf{f}_k(\mathbf{a}))$, it follows from (7) that, for any nonzero c.e. ibT-degree $\mathbf{a}$,

$$\cdots < \mathbf{f}_3(\mathbf{a}) < \mathbf{f}_2(\mathbf{a}) < \mathbf{f}_1(\mathbf{a}) < \mathbf{f}_0(\mathbf{a}) = \mathbf{a}.$$

So the automorphisms $\mathbf{f}_k$, $k \geq 0$, are pairwise different.   □

Note that the above observations apply to non-c.e. sets too. I.e., the (nontrivial) bounded shifts induce (nontrivial) automorphisms of the global degree structure $(\mathbf{D}_{\mathrm{ibT}}, \leq)$ too. So $(\mathbf{D}_{\mathrm{ibT}}, \leq)$ is not rigid too. Also note that the nontrivial automorphisms $\mathbf{f}_k$ of $(\mathbf{R}_{\mathrm{ibT}}, \leq)$ ($k \geq 1$) are *push-down* automorphisms, i.e., map any nonzero degree to a strictly lesser degree. Correspondingly, the inverse automorphisms $\mathbf{f}_k^{-1}$ (which, of course, are induced by the functions mapping a c.e. set A to $A - k$) are *push-up* automorphisms, i.e., map any nonzero degree to a strictly greater degree. It is an interesting open question whether there are also automorphisms of $(\mathbf{R}_{\mathrm{ibT}}, \leq)$ which map some degree to an incomparable one. (Here it might be of interest to note that for the c.e. Turing degrees - where it is still open whether nontrivial automorphisms exist - any nontrivial automorphism $\mathbf{f}$ (if there is any) has to move some degree $\mathbf{a}$ to a degree $\mathbf{f}(\mathbf{a})$ which is incomparable with $\mathbf{a}$, whence, in particular, there are no push-up or push-down automorphisms of this structure. This follows from the fact that the anti-chain of maximal contiguous is an automorphism base (Cholak, Downey, and Walk [11]) and definable (Downey and Lempp [17]).)

Note that the bounded shifts induce the trivial automorphism on $(\mathbf{R}_{\mathrm{cl}}, \leq)$ since $A + k =_{\mathrm{cl}} A$ for any $k \geq 0$ and any set A. One might guess, however, that the unbounded computable shifts may induce nontrivial automorphisms of the partial ordering of the c.e. cl-degrees. In the remainder of this section we show that this is not the case: no unbounded computable shift induces an automorphism of $(\mathbf{R}_{\mathrm{cl}}, \leq)$ or $(\mathbf{R}_{\mathrm{ibT}}, \leq)$. So we have to leave open the question of whether there are nontrivial automorphisms of $(\mathbf{R}_{\mathrm{cl}}, \leq)$.

Though, as one can easily show, linear shifts $f(x) = k \cdot x + k'$ ($k \geq 1, k' \geq 0$) are invariant under cl-equivalence and preserve cl-reducibility, in general the analog of the invariance lemma for cl-reducibility fails for hyper-linear computable shifts. We demonstrate this by considering the quadratic shift $f(x) = x^2$. Then, for any noncomputable c.e. set A, the f-shifts of the cl-equivalent sets A and $B = A + 1$ are not cl-equivalent, since $B_f = (A + 1)_f = (A_f)_g$ for some unbounded computable shift g (namely for g defined by $g(0) = 1$ and $g((n+1)^2 - m) = (n+2)^2 - m$ for $n \geq 0$ and $0 \leq m < (n+1)^2 - n^2$) hence $B_f <_{\mathrm{cl}} A_f$ by the Computable-Shift Lemma.

In order to show that *no* unbounded computable shift f induces an automorphism of $(\mathbf{R}_{\mathrm{cl}}, \leq)$ or $(\mathbf{R}_{\mathrm{ibT}}, \leq)$, we show that the function $\mathbf{f}$ induced by f on the c.e. cl-degrees (if well defined) respectively on the c.e. ibT-degrees is not on-to. This is an immediate consequence of the following stronger result where $\#_A$ denotes the *census function* of set A, i.e., $\#_A(n) = |A \upharpoonright n|$.

Lemma 4.5. *Let g be a nondecreasing computable function such that, for $n \geq 0$, $g(n) \leq n$ and such that*

$$\lim_{n \to \infty} (n - g(n)) = \infty. \tag{8}$$

There is a c.e. set A such that, for all c.e. sets B with $\#_B \leq g$, $A \not\leq_{\text{cl}} B$.

PROOF. Since the argument is similar to the construction of a maximal pair in the c.e. ibT-degrees given in Ambos-Spies et al. [5], we only sketch the proof. We effectively enumerate the desired c.e. set A in stages, and let A_s denote the finite part of A enumerated by the end of stage s ($s \geq 0$). Fix a computable enumeration $\{(V_e, \Psi_e)\}_{e \geq 0}$ of all pairs (V, Ψ) such that V is a c.e. set and Ψ is a cl-functional where w.l.o.g. Ψ_e is an $(i + e)$bT-functional. Then it suffices to meet the requirements

$$\mathfrak{R}_e : \#_{V_e} \leq g \Rightarrow A \neq \Psi_e^{V_e}$$

for all $e \geq 0$.

Now, by choice of g, there is a computable ascending sequence $x_0 = 0 < x_1 < x_2 < x_3 < \dots$ such that, for $e \geq 0$,

$$x_{e+1} - x_e > g(x_{e+1} + e). \tag{9}$$

The numbers in the interval $I_e = [x_e, x_{e+1})$ are used as potential diagonalization candidates for meeting requirement $\mathfrak{R}_e$ as follows. If, at the end of stage s, $A_s(y) = \Psi_{e,s}^{V_{e,s}}(y)$ for all numbers $y \in I_e$ then, for the least $y \in I_e$ such that $y \notin A_s$ (if any), put y into A_{s+1}. This ensures that $A(y) \neq \Psi_e^{V_e}(y)$ unless a number $z \leq y + e$ is enumerated into V_e after stage s. So we can argue that $A(y) \neq \Psi_e^{V_e}(y)$ for some $y \in I_e$ (whence requirement $\mathfrak{R}_e$ is met) unless we eventually put all elements of I_e into A thereby forcing $|I_e| = x_{e+1} - x_e$ numbers $z < x_{e+1} + e$ into V_e. But if the latter happens then, by (9), $\#_{V_e}(x_{e+1} + e) > g(x_{e+1} + e)$, whence the hypothesis of requirement $\mathfrak{R}_e$ fails (and $\mathfrak{R}_e$ is trivially met). □

Theorem 4.6. *Let f be an unbounded computable shift. There is a c.e. set A such that, for all c.e. sets B, $A \not\leq_{\text{cl}} B_f$ (hence $\deg_r(A) \notin \{\deg_r((W_e)_f) : e \geq 0\}$ for $r = \text{ibT, cl}$).*

PROOF. By choice of f, for any c.e. set B, $\#_{B_f} \leq g$ for $g = \#_{\omega_f}$, and g satisfies the hypotheses of Lemma 4.5. □

Remark 4.7. *For unbounded computable shifts f of sufficiently fast growth rate like $f(x) = x^2$, namely those computable shifts f satisfying*

$$\lim_{n \to \infty} f(n+1) - f(n) = \infty, \tag{10}$$

Theorem 4.6 can be improved by showing that there is a c.e. set A such that $B_f \leq_{\text{ibT}} A$ for all c.e. sets B. Namely, as one can easily check, it suffices to let

$$A = \{f(n) + e : e \geq 0 \ \& \ f(n) + e < f(n+1) \ \& \ f(n+1) \in W_e\}.$$

For unbounded linear shifts, however, Theorem 4.6 cannot be improved in this way. We demonstrate this for the shift $f(n) = 2n$ (the general case is similar). Given any c.e. set A and $r \in \{\mathrm{ibT}, \mathrm{cl}\}$, we have to show that there is a c.e. set B such that $B_f \not\leq_r A$.

By Theorem 3.4, fix a c.e. set $\hat{A}$ such that $A <_r \hat{A}$. Then, for the even part $\hat{A}_0 = \hat{A} \cap \{2n : n \geq 0\}$ and odd part $\hat{A}_1 = \hat{A} \cap \{2n+1 : n \geq 0\}$ of $\hat{A}$, $deg_r(\hat{A}) = deg_r(\hat{A}_0) \vee deg_r(\hat{A}_1)$ (see the Splitting Lemma below). So, for some $i \leq 1$, $\hat{A}_i \not\leq_r A$. Fix such an i and define the c.e. set B by $B = \{n : 2n+i \in \hat{A}\}$. Then $B_f = \{2n : 2n+i \in \hat{A}\}$. So, for $i = 0$, $\hat{A}_i = B_f$, while, for $i = 1$, $\hat{A}_i = B_f + 1 \leq_r B_f$. In either case, it follows from $\hat{A}_i \not\leq_r A$ that $B_f \not\leq_r A$ holds.

5   Permitting and Splitting

For the further analysis of the strongly bounded Turing reducibilities on the c.e. sets we will need some observations relating the permitting technique and splittings of c.e. sets to these reducibilities.

A fundamental quite common technique for constructing a c.e. set A below a given c.e. set B which actually gives an ibT-reduction (but, in general, not a tt-reduction, hence not an m-reduction) is the *permitting method*. By an obvious generalization we obtain a basic technique for obtaining $(i+k)$bT-reductions.

Definition 5.1. *Let A and B be c.e. sets and $k \geq 0$. Then $A \leq_T B$ by k-permitting if there are computable enumerations $\{A_s\}_{s \geq 0}$ and $\{B_s\}_{s \geq 0}$ of A and B, respectively, such that*

$$\forall x \forall s (x \in A_{at\ s} \Rightarrow \exists y \leq x + k (y \in B_{at\ s})) \tag{11}$$

holds. In particular, $A \leq_T B$ by permitting if $A \leq_T B$ by 0-permitting.

Proposition 5.2. *Let A and B be c.e. sets such that $A \leq_T B$ by k-permitting. Then $A \leq_{(i+k)\mathrm{bT}} B$. In particular, if A and B are c.e. sets such that $A \leq_T B$ by permitting then $A <_{\mathrm{ibT}} B$.*

Conversely, any $(i+k)$bT-reduction from a c.e. set A to a c.e. set B can be represented by a k-permitting if we replace A and B by some ibT equivalent subsets. In the following we state this observation in a somewhat more general form.

Lemma 5.3 (Representation Lemma). *Let A and $B_0, \ldots, B_m$ $(m \geq 0)$ be noncomputable c.e. sets and let $k \geq 0$ such that $B_j \leq_{(i+k)\mathrm{bT}} A$ for $j \leq m$. There are c.e. subsets $\hat{A}$ of A and $\hat{B}_j$ of B_j, and computable one-to-one functions $a(n)$ and $b_j(n)$ enumerating $\hat{A}$ and $\hat{B}_j$, respectively, such that, for $j \leq m$,*

(i) $\hat{A} =_{\mathrm{ibT}} A$ and $\hat{B}_j =_{\mathrm{ibT}} B_j$ and
(ii) $\forall n (a(n) \leq \min\{b_0(n), \ldots, b_m(n)\} + k)$.

Note that (in the context of the above lemma) $\hat{B}_j \leq_T \hat{A}$ by k-permitting via the enumerations $\{\hat{B}_{j,s}\}_{s \geq 0}$ and $\{\hat{A}\}_{s \geq 0}$ given by $\hat{B}_{j,s} = \{b_j(0), \ldots, b_j(s-1)\}$ and $\hat{A}_s = \{a(0), \ldots, a(s-1)\}$.

PROOF OF LEMMA 5.3. Fix computable enumerations $\{A_s\}_{s \geq 0}$ and $\{B_{j,s}\}_{s \geq 0}$ of A and B_j, respectively, such that $A_{s+1} \neq A_s$ and $B_{j,s+1} \neq B_{j,s}$ ($s \geq 0$), and let Φ_j be an $(i+k)$bT-functional such that $B_j = \Phi_j^A$ ($j \leq m$). Define the *length (of agreement) function* l by

$$l(s) = max\{x \leq s : \forall\, y < x \,\forall\, j \leq m \,(B_{j,s}(y) = \Phi_{j,s}^{A_s}(y))\}.$$

Then

$$\lim_{s \to \omega} l(s) = \omega$$

whence there are infinitely many *expansionary stages* s, i.e., stages s such that

$$\forall t < s \,(l(t) < l(s)).$$

Call an expansionary stage s *critical* if, for some $j \leq m$, there is a number $x < l(s)$ such that $x \in B_j \setminus B_{j,s}$, and say that criticalness of s is witnessed by t if $t > s$ and $B_{j,s} \upharpoonright l(s) \neq B_{j,t} \upharpoonright l(s)$. Note that if criticalness of s is witnessed by t then criticalness of s is witnessed by all $t' \geq t$. Moreover, by noncomputability of the sets B_j, there are infinitely many critical expansionary stages. Finally, the set of all pairs (s,t) such that s is critical and criticalness of s is witnessed by t is computable. So we can define a computable ascending sequence of expansionary stages $s_0 < s_1 < s_2 < \ldots$ such that s_n is critical and s_{n+1} witnesses criticalness of s_n.

Now let

$$a(n) = \mu\, x\, (x \in A_{s_{n+1}} \setminus A_{s_n}) \ \& \ b_j(n) = \mu\, x\, (x \in B_{j,s_{n+1}} \setminus B_{j,s_n}) \qquad (12)$$

and let $\hat{A} = \{a(n) : n \geq 0\}$ and $\hat{B}_j = \{b_j(n) : n \geq 0\}$. Then, obviously, a and b_j are computable one-to-one functions enumerating $\hat{A}$ and $\hat{B}_j$, respectively, and $\hat{A} \subseteq A$ and $\hat{B}_j \subseteq B_j$. So it only remains to show that (i) and (ii) hold.

For a proof of (ii), given n and $j \leq m$ such that $b_j(n) = min\{b_0(n), \ldots, b_m(n)\}$, it suffices to show that

$$a(n) \leq b_j(n) + k. \qquad (13)$$

Now, by definition of b_j and by choice of j,

$$b_j(n) \in B_{j,s_{n+1}} \setminus B_{j,s_n} \ \& \ b_j(n) < l(s_n).$$

So

$$0 = B_{j,s_n}(b_j(n)) = \Phi_{j,s_n}^{A_{s_n}}(b_j(n))$$

and, since s_{n+1} is expansionary,

$$1 = B_{j,s_{n+1}}(b_j(n)) = \Phi_{j,s_{n+1}}^{A_{s_{n+1}}}(b_j(n)).$$

Since Φ_j is an $(i+k)$bT-functional it follows that

$$A_{s_n} \upharpoonright b_j(n) + k + 1 \neq A_{s_{n+1}} \upharpoonright b_j(n) + k + 1.$$

So, by definition of a, (13) holds.

Finally, for a proof of (i), let

$$\hat{A}_n = \{a(0), \ldots, a(n-1)\} \ \& \ \hat{B}_{j,n} = \{b_j(0), \ldots, b_j(n-1)\}.$$

Then

$$\mu\, x\ (x \in \hat{A}_{n+1} \setminus \hat{A}_n) = \mu\, x\ (x \in A_{s_{n+1}} \setminus A_{s_n})$$

and

$$\mu\, x\ (x \in \hat{B}_{j,n+1} \setminus \hat{B}_{j,n}) = \mu\, x\ (x \in B_{j,s_{n+1}} \setminus B_{j,s_n}).$$

So $\hat{A} =_{\mathrm{ibT}} A$ and $\hat{B}_j =_{\mathrm{ibT}} B_j$ by permitting. $\square$

If we split a c.e. set A into two disjoint c.e. sets A_0 and A_1 then the Turing degree of A is the least upper bound (join) of the Turing degrees of the parts A_0 and A_1. This simple but fundamental observation carries over not only to wtt reducibility but to the strongly bounded Turing reducibilities too.

Lemma 5.4 (Splitting Lemma). *Let $A_0, \ldots, A_m$ $(m \geq 1)$ be pairwise disjoint c.e. sets and let $A = A_0 \cup \cdots \cup A_m$. Then, for $r \in \{\mathrm{ibT}, \mathrm{cl}\}$,*

$$deg_r(A) = deg_r(A_0) \vee \cdots \vee deg_r(A_m).$$

PROOF. For $j \leq m$, $A_j \leq_{\mathrm{ibT}} A_0 \cup \cdots \cup A_m = A$ by permitting (where the enumerations $\{A_s\}_{s \geq 0}$ and $\{A_{i,s}\}_{s \geq 0}$ of the sets A and A_i, respectively, are chosen so that $A_s = A_{0,s} \cup \cdots \cup A_{m,s}$). So, given B such that $A_j \leq_{(i+k_j)\mathrm{bT}} B$, it suffices to show that $A_0 \cup \cdots \cup A_m \leq_{(i+k)\mathrm{bT}} B$ where $k = max\ k_j$. But this is obviously true. $\square$

As a first application of the Splitting Lemma we show that the observation (made in Sect. 3) that, for $r = \mathrm{ibT}, \mathrm{cl}$ and for any noncomputable c.e. set A, there are c.e. sets C and D such that $D <_r A <_r C$ can be extended, by showing that there is also a c.e. set B such that $A|_r B$.

Theorem 5.5. *Let A be a noncomputable c.e. set. There is a c.e. set B such that, for $r = \mathrm{ibT}, \mathrm{cl}$, A and B are r-incomparable. Moreover, if A is m-complete then the set B can be chosen to be m-complete too.*

PROOF. By Theorem 3.4 fix a c.e. set C such that, for $r = \mathrm{ibT}, \mathrm{cl}$, $A <_r C$. By Sacks's Splitting Theorem split C into c.e. sets C_0 and C_1 such that, for $i \leq 1$, $A \not\leq_T C_i$ hence $A \not\leq_r C_i$. Then, by the Splitting Lemma, $C_i|_r A$ for some i. By symmetry assume that $i = 0$. So the first part of the claim holds for $B = C_0$.

For a proof of the second part, assume that A is m-complete and fix C, C_0 and C_1 as above. In order to get an m-complete set B with $A|_r B$, fix an infinite computable subset D of C_1, let d be the computable shift enumerating D in order of magnitude, and let $B = C_0 \cup A_d$. Note that C_0 and D are disjoint and $A_d \subseteq D$. So $A \leq_m B$ via d whence B is m-complete. Moreover, since C_0 is noncomputable, the computable shift d is unbounded. So, by the Computable-Shift Lemma and by choice of C_0, $A \not\leq_r B$. Finally, since $C_0 \not\leq_r A$ and, by the Splitting Lemma (applied to the splitting $B = C_0 \cup A_d$), $C_0 \leq_r B$, it follows that $B \not\leq_r A$.   □

Note that, by Myhill's Theorem, for m-complete A, the r-incomparable set B provided by Theorem 5.5 is computably isomorphic to A. So, for any m-complete set A, there is a computable permutation p such that $A|_r p(A)$ for $r = \text{ibT}, \text{cl}$ (compare with the remark following Theorem 3.6).

For the Turing degrees (or wtt-degrees) any join $deg_T(A) = deg_T(B) \vee deg_T(C)$ can be represented by a splitting, i.e., there are c.e. sets $\hat{A} \in deg_T(A), \hat{B} \in deg_T(B), \hat{C} \in deg_T(C)$ such that $\hat{A} = \hat{B} \cup \hat{C}$ and $\hat{B} \cap \hat{C} = \emptyset$. (For instance, let $\hat{A} = B \oplus C$, $\hat{B} = B \oplus \emptyset$ and $\hat{C} = \emptyset \oplus C$.) The corresponding observation for the strongly bounded Turing degrees fails. This can be shown by considering distributivity properties.

As Lachlan has shown (see Stob [29]), the upper semilattice of the c.e. wtt-degrees is distributive, i.e., satisfies the following distributivity law for upper semilattices:

$$\forall\, a_0, a_1, b\, [b \leq a_0 \vee a_1 \Rightarrow \exists\, b_0 \leq a_0, b_1 \leq a_1\, (b = b_0 \vee b_1)]. \tag{14}$$

Note that a lattice is distributive in the common sense if and only if it satisfies (14). Moreover, no nondistributive lattice can be embedded into any distributive upper semilattice.

Now, Ambos-Spies, Bodewig, Kräling and Yu [4] have shown that the non-modular five-element lattice N_5 can be embedded into the partial orderings of the c.e. ibT- and cl-degrees. So there are joins in these degree structures for which (14) fails. Hence, in order to argue that there are joins in the strongly bounded Turing degrees of c.e. sets which cannot be represented by c.e. splittings, it suffices to show that the joins represented by such splittings have the distributive splitting property. (This observation will be very useful in the following.)

Lemma 5.6 (Distributivity Lemma). *Let $A_0, \ldots, A_m$ $(m \geq 1)$ be pairwise disjoint c.e. sets, let $A = A_0 \cup \cdots \cup A_m$, and let B be a c.e. set such that $B \leq_{(i+k)\text{bT}} A$ $(k \geq 0)$. There is a splitting $B = B_0 \cup \cdots \cup B_m$ of B into pairwise disjoint c.e. sets B_j such that $B_j \leq_{(i+k)\text{bT}} A_j$ $(j \leq m)$.*

PROOF. For computable B the claim is trivial. So w.l.o.g. assume that B is noncomputable. Apply the Representation Lemma to A and the single set B. This yields c.e. sets $\hat{A}$ and $\hat{B}$ and computable one-to-one functions $a(n)$ and $b(n)$ enumerating $\hat{A}$ and $\hat{B}$, respectively, such that $\hat{A} \subseteq A$, $\hat{B} \subseteq B$, $\hat{A} =_{\text{ibT}} A$, $\hat{B} =_{\text{ibT}} B$, and $a(n) \leq b(n) + k$ (for $n \geq 0$) hold. Moreover, by (12) in the proof of the Representation Lemma, there is a computable enumeration $\{\tilde{B}_s\}_{s \geq 0}$ of B such that $b(s)$ is the least element of $\tilde{B}_{s+1} \setminus \tilde{B}_s$ $(s \geq 0)$.

Then the sets B_j defined by

$$B_0 = \tilde{B}_0 \cup \bigcup_{\{s:a(s)\in A_0\}} (\tilde{B}_{s+1} \setminus \tilde{B}_s)$$

and

$$B_j = \bigcup_{\{s:a(s)\in A_j\}} (\tilde{B}_{s+1} \setminus \tilde{B}_s)$$

for $0 < j \le m$ have the required properties. Namely, obviously, the sets B_j are c.e. and pairwise disjoint, and $B = B_0 \cup \cdots \cup B_m$. Finally, $B_j \le_{(i+k)\text{bT}} A_j$ by k-permitting (in case of $j = 0$ up to a finite variant).   □

We conclude this section with some observations on ibT-reductions among disjoint c.e. sets and on the relations between ibT-equivalent c.e. sets.

Lemma 5.7 (Disjoint Sets Lemma). *Let D and E be disjoint noncomputable c.e. sets such that $D \le_{\text{ibT}} E$. Then $D \le_{\text{ibT}} E + 1$.*

PROOF. By the Representation Lemma (and the Invariance Lemma) w.l.o.g. we may assume that there are one-to-one computable functions d and e which enumerate D and E, respectively, such that $d(n) \ge e(n)$ for all n. Since D and E are disjoint, the latter implies $d(n) > e(n)$, i.e., $d(n) \ge e(n)+1$. Since $e(n)+1$ is a computable one-to-one enumeration of $E + 1$, it follows that $D \le_{\text{ibT}} E + 1$ by permitting.   □

Note that, by the Disjoint Sets Lemma and by the Bounded-Shift Lemma, for any noncomputable c.e. sets D and E such that $D =_{\text{ibT}} E$, $D \cap E \ne \emptyset$. This can be improved as follows.

Lemma 5.8 (Equivalent Sets Lemma). *Let A and B be c.e. sets such that $A =_{\text{ibT}} B$. Then $A =_{\text{ibT}} A \cup B =_{\text{ibT}} A \cap B$.*

PROOF. A proof can be obtained by giving a symmetric version of the representation lemma for ibT-equivalent sets. Here we give a direct proof.

Let $\{A_s\}_{s\ge 0}$ and $\{B_s\}_{s\ge 0}$ be computable enumerations of A and B, respectively, and fix ibT-functionals $\hat{\Phi}_e$ and $\hat{\Phi}_{e'}$ such that

$$A = \hat{\Phi}_e^B \ \& \ B = \hat{\Phi}_{e'}^A. \tag{15}$$

Obviously, by $B \le_{\text{ibT}} A$, $A \cup B \le_{\text{ibT}} A$ and $A \cap B \le_{\text{ibT}} A$. So it suffices to show that $A \le_{\text{ibT}} A \cup B$ and $A \le_{\text{ibT}} A \cap B$.

For a proof of $A \le_{\text{ibT}} A \cup B$ it suffices to compute $A(x)$ from $(A \cup B) \restriction x+1$. This is done inductively. So, given x, by inductive hypothesis and by $B \le_{\text{ibT}} A$, it suffices to compute $A(x)$ from $A \restriction x$, $B \restriction x$ and $(A \cup B) \restriction x+1$. This is done as follows. First check whether $x \in A \cup B$. If not then, obviously, $x \notin A$. So, for the remainder of the argument assume that $x \in A \cup B$. Compute s minimal such that $x \in A_s$ or $x \in B_s$. In the former case, $x \in A$. In the latter case,

$B \upharpoonright x + 1 = (B \upharpoonright x) \cup \{x\}$. Hence, by (15), $A(x) = \hat{\Phi}_e^{(B \upharpoonright x) \cup \{x\}}(x)$ where the right hand side can be computed from $B \upharpoonright x$.

For a proof of $A \leq_{\mathrm{ibT}} A \cap B$ it suffices to compute $A(x)$ from $(A \cap B) \upharpoonright x + 1$. Again, this is done inductively, and, for given x, we may argue that it suffices to compute $A(x)$ from $A \upharpoonright x$, $B \upharpoonright x$ and $(A \cap B) \upharpoonright x + 1$. The computation of $A(x)$ from these parameters is as follows. First check whether $x \in A \cap B$. If so, x is in A. So, for the following, assume that $x \notin A \cap B$. Then (from $A \upharpoonright x$ and $B \upharpoonright x$) compute s minimal such that (i) $x \in A_s$ or (ii) $x \in B_s$ or (iii) $0 = \hat{\Phi}_{e,s}^{B \upharpoonright x}(x) = \hat{\Phi}_{e',s}^{A \upharpoonright x}(x)$. (Note that such an s must exist since if (i) and (ii) fail for all s then $A(x) = B(x) = 0$. Hence $A \upharpoonright x + 1 = A \upharpoonright x$ and $B \upharpoonright x + 1 = B \upharpoonright x$ whence, by (15), $0 = A(x) = \hat{\Phi}_e^{B \upharpoonright x + 1}(x) = \hat{\Phi}_e^{B \upharpoonright x}(x)$ and $0 = B(x) = \hat{\Phi}_{e'}^{A \upharpoonright x + 1}(x) = \hat{\Phi}_{e'}^{A \upharpoonright x}(x)$. So (iii) holds for all sufficiently large s.) Now if (i) holds then trivially $x \in A$, if (ii) holds then $x \notin A$ (since $x \notin A \cap B$), and if (iii) holds then $x \notin A$ too. The latter is shown as follows. For a contradiction assume that $x \in A$. Then, by $x \notin A \cap B$, x is not in B. So $B \upharpoonright x + 1 = B \upharpoonright x$ whence, by (15), $1 = A(x) = \hat{\Phi}_e^{B \upharpoonright x}(x)$. But this implies that $0 \neq \hat{\Phi}_{e,s}^{B \upharpoonright x}(x)$ for all $s \geq 0$ contrary to the assumption that (iii) holds.

This completes the proof of Lemma 5.8.   □

6   C.E. ibT-Degrees: Types and Definability

Here we give another application of bounded shifts to the partial ordering of the c.e. ibT-degrees by using some of the observations made in the preceding section. We show that, for certain noncomputable c.e. sets A, the ibT-degrees of the bounded shifts $A + k$ ($k \geq 1$) can be defined from $deg_{\mathrm{ibT}}(A)$ in the partial ordering $(\mathbf{R}_{\mathrm{ibT}}, \leq)$ by first order formulas $\varphi_k(x, y)$. By the Bounded-Shift Lemma, this implies that the first order theory of $(\mathbf{R}_{\mathrm{ibT}}, \leq)$, $\mathrm{Th}(\mathbf{R}_{\mathrm{ibT}}, \leq)$, realizes infinitely many 2-types. So, by Ryll-Nardzewski's Theorem, $\mathrm{Th}(\mathbf{R}_{\mathrm{ibT}}, \leq)$ is not $\aleph_0$-categorical. By other, more involved methods, in Sect. 9 we will obtain the corresponding result for the cl-degrees too. In fact, there we will show that both, $\mathrm{Th}(\mathbf{R}_{\mathrm{ibT}}, \leq)$ and $\mathrm{Th}(\mathbf{R}_{\mathrm{cl}}, \leq)$, realize infinitely many 1-types.

Definition 6.1. *A set A is* scattered *if $A \subseteq B$ for some computable set B such that*

$$\forall x \in B \ (x + 1 \notin B).$$

A c.e. r-degree $\mathbf{a}$ is scattered *if there is a scattered c.e. set $A \in \mathbf{a}$.*

Note that, for any c.e. set A, the c.e. sets $2A$ and $(2A) + 1$ are scattered and wtt-equivalent to A. So, in particular, any c.e. wtt-degree is scattered. On the other hand, there is a c.e. cl-degree $\mathbf{a}$ such that no c.e. cl-degree $\mathbf{b} \geq \mathbf{a}$ is scattered. This follows from Lemma 4.5 since, for any scattered set A, $\#_A(2n) \leq n$.

For defining $\mathbf{a} + 1$ from $\mathbf{a}$ for scattered $\mathbf{a}$ in the partial ordering of the c.e. ibT-degrees we use the following notion.

Definition 6.2. *Let $(P, \leq)$ be a partial ordering, and let $a, b \in P$ such that $b \leq a$. b* cups *to a if there is some $c < a$ such that the least upper bound $b \vee c$ of b and c exists and $a = b \vee c$.*

Lemma 6.3. *Let* $\mathbf{a}$ *and* $\mathbf{b}$ *be c.e. ibT-degrees such that* $\mathbf{b} \leq \mathbf{a}$. *Then*

$$\mathbf{b} \not\leq \mathbf{a} + 1 \ \Rightarrow \ \mathbf{b} \ cups \ to \ \mathbf{a} \tag{16}$$

holds. Moreover, if $\mathbf{a}$ *is scattered then the converse of* (16) *holds too.*

Note that (the first part of) Lemma 6.3 and the Bounded-Shift Lemma imply that, for any c.e. ibT-degree $\mathbf{a} > \mathbf{0}$, there is a c.e. ibT-degree $\mathbf{c} < \mathbf{a}$ (namely $\mathbf{c} = \mathbf{a}+1$) which bounds the degrees which do not cup to $\mathbf{a}$. By observing that the corresponding fact fails for the c.e. cl-degrees, Ambos-Spies, Bodewig, Fan, and Kräling [3] have shown that the partial orderings of the c.e. ibT-degrees and of the c.e. cl-degrees, i.e., $(\mathbf{R}_{ibT}, \leq)$ and $(\mathbf{R}_{cl}, \leq)$, are not elementarily equivalent.

PROOF OF LEMMA 6.3. For a proof of (16) fix c.e. sets A and B such that $A \in \mathbf{a}$, A is scattered, $B \leq_{ibT} A$ and $B \not\leq_{ibT} A + 1$. It suffices to give a c.e. set C such that $C <_{ibT} A$ and

$$deg_{ibT}(A) = deg_{ibT}(B) \vee deg_{ibT}(C) \tag{17}$$

holds.

By the Representation Lemma, w.l.o.g. we may assume that there are computable one-to-one enumeration functions $a(n)$ and $b(n)$ of A and B, respectively, such that $a(n) \leq b(n)$. Split B into the disjoint c.e. sets

$$B_0 = \{b(n) : a(n) = b(n)\} \ \text{and} \ B_1 = \{b(n) : a(n) < b(n)\}, \tag{18}$$

and let

$$C = \{a(n) : a(n) < b(n)\}.$$

Then A is the disjoint union of B_0 and C. So, by the Splitting Lemma, $C \leq_{ibT} A$ and

$$deg_{ibT}(A) = deg_{ibT}(B_0) \vee deg_{ibT}(C).$$

Since, by (18), $B_0 \leq_{ibT} B$, this implies (17).

It remains to show that $C \neq_{ibT} A$. For a contradiction assume that $C -_{ibT} A$. Then $B_0 \leq_{ibT} C$. Since B_0 and C are disjoint it follows with the Disjoint Sets Lemma that $B_0 \leq_{ibT} C + 1$. So, by the Invariance Lemma, $B_0 \leq_{ibT} A + 1$. Since, on the other hand, by definition of B_1, $B_1 \leq_{ibT} A + 1$ too, it follows with the Splitting Lemma that $B \leq_{ibT} A + 1$. But this contradicts the choice of B.

The proof that the implication in (16) can be reversed, which is based on the assumption that $\mathbf{a}$ is scattered, is by contraposition. Since the class of the ibT-degrees below $\mathbf{a}$ which does not cup to $\mathbf{a}$ is closed downwards, it suffices to show that $\mathbf{a} + 1$ does not cup to $\mathbf{a}$. So fix c.e. sets A and C such that $A \in \mathbf{a}$, A is scattered, and $C \leq_{ibT} A$, and assume

$$deg_{ibT}(A) = deg_{ibT}(A + 1) \vee deg_{ibT}(C). \tag{19}$$

We have to show that $A \leq_{ibT} C$.

By the Representation Lemma, w.l.o.g. we may assume that there are computable one-to-one enumeration functions $a(n)$ and $c(n)$ of A and C, respectively, such that $a(n) \leq c(n)$. Split C into the disjoint c.e. sets

$$C_0 = \{c(n) : a(n) = c(n)\} \text{ and } C_1 = \{c(n) : a(n) < c(n)\}.$$

Then, by the Splitting Lemma, $deg_{\mathrm{ibT}}(C) = deg_{\mathrm{ibT}}(C_0) \vee deg_{\mathrm{ibT}}(C_1)$. Since $C_1 \leq_{\mathrm{ibT}} A + 1$ by permitting, it follows with (19) that

$$deg_{\mathrm{ibT}}(A) = deg_{\mathrm{ibT}}(A + 1) \vee deg_{\mathrm{ibT}}(C_0).$$

Moreover, since C_0 is contained in A and since A is scattered, it follows that C_0 and $A + 1$ are disjoint. So, by the Splitting Lemma,

$$A =_{\mathrm{ibT}} A + 1 \cup C_0$$

whence $A \leq_{\mathrm{ibT}} C_0$ by the Bounded-Shift Lemma. Since, by the Splitting Lemma, $C_0 \leq_{\mathrm{ibT}} C$, it follows that $A \leq_{\mathrm{ibT}} C$. □

Theorem 6.4. *The first order theory* $\mathrm{Th}(\mathbf{R}_{\mathrm{ibT}}, \leq)$ *of the partial ordering of the c.e. ibT-degrees realizes infinitely many 2-types. So* $\mathrm{Th}(\mathbf{R}_{\mathrm{ibT}}, \leq)$ *is not* $\aleph_0$-*categorical.*

PROOF. It suffices to give first order formulas $\varphi_k(x, y)$ with two free variables x, y in the language of partial orderings ($k \geq 1$) such that, for the sets

$$\mathbf{D}_k = \{(\mathbf{a}, \mathbf{b}) \in \mathbf{R}_{\mathrm{ibT}}^2 : (\mathbf{R}_{\mathrm{ibT}}, \leq) \vDash \varphi_k(\mathbf{a}, \mathbf{b})\},$$

we have $\mathbf{D}_k \neq \mathbf{D}_{k'}$ for any $k, k' \geq 1$ such that $k \neq k'$.

Let the formula $\varphi_1(x, y)$ express that y is the greatest element $\leq x$ such that y does not cup to x, and, for $k \geq 2$, define φ_k by

$$\varphi_k(x, y) \equiv \exists y_1, \ldots, y_{k-1} \, (\varphi_1(x, y_1) \,\&\, \varphi_1(y_1, y_2) \,\&\, \ldots \,\&\, \varphi_1(y_{k-1}, y)).$$

Then, by Lemma 6.3, for any scattered c.e. ibT-degree $\mathbf{a}$,

$$(\mathbf{R}_{\mathrm{ibT}}, \leq) \vDash \varphi_1(\mathbf{a}, \mathbf{b}) \,\Leftrightarrow\, \mathbf{b} = \mathbf{a} + 1.$$

Since, for any scattered degree $\mathbf{a}$, the degree $\mathbf{a} + 1$ is scattered too, it follows by a straightforward induction on $k \geq 1$ that

$$(\mathbf{R}_{\mathrm{ibT}}, \leq) \vDash \varphi_k(\mathbf{a}, \mathbf{b}) \,\Leftrightarrow\, \mathbf{b} = \mathbf{a} + k.$$

Since, by the Bounded-Shift Lemma, for $\mathbf{a} > \mathbf{0}$, the degrees $\mathbf{a} + k$ and $\mathbf{a} + k'$ differ for $k \neq k'$, this implies the claim. □

7 Relations Among the ibT-, cl- and wtt-Degrees

In the remainder of this paper we will show how to transfer certain results about the c.e. wtt (and, to a lesser extent, T) degrees to the strongly bounded Turing degrees. In order to do so we first explore some basic relations among the strongly bounded Turing reducibilities and weak truth-table reducibility. Some of the material in this section is taken from Ambos-Spies et al. [5].

If a reducibility r is stronger than a reducibility r' on the c.e. sets, i.e., if for any c.e. sets A and B, $A \leq_r B$ implies $A \leq_{r'} B$ then, obviously, $deg_r(A) \leq deg_r(B)$ implies $deg_{r'}(A) \leq deg_{r'}(B)$. In general, however, this does not imply that any join $deg_r(A) \vee deg_r(B) = deg_r(C)$ in the c.e. r-degrees yields the corresponding join $deg_{r'}(A) \vee deg_{r'}(B) = deg_{r'}(C)$ in the c.e. r'-degrees. I.e., joins in the c.e. r-degrees may not be preserved in the c.e. r'-degrees, and similarly for meets.

If, for example, we let $r = $ wtt and $r' = $ T then joins are preserved since for any sets A and B, the effective disjoint union $A \oplus B$ of A and B represents the join of the degrees of the degrees of A and B in both, in the wtt-degrees and in the T-degrees. Meets in the c.e. wtt-degrees, however, are not preserved in the c.e. T-degrees. For instance, as observed by Downey and Stob [18], there are noncomputable c.e. sets A and B such that $deg_{\text{wtt}}(A) \wedge deg_{\text{wtt}}(B) = \mathbf{0}$ but $A =_T B$ (whence $deg_T(A) \wedge deg_T(B) = deg_T(A) > \mathbf{0}$).

As we will show in the following, however, joins and meets in the c.e. ibT-degrees are preserved in the c.e. cl-degrees and joins and meets in the c.e. cl-degrees are preserved in the c.e. wtt-degrees.

For a proof of the former, we start with some observations on the convertibility of cl-reductions into ibT-reductions.

Proposition 7.1. *Let $k \geq 0$ and let A and B be c.e. sets such that $A \leq_{(i+k)\text{bT}} B$. Then, for any $k', k'' \geq 0$ such that $k \leq k' + k''$, $A + k' \leq_{\text{ibT}} B - k''$. So, in particular, for $k' \geq k$, $A + k' \leq_{\text{ibT}} B$ and $A \leq_{\text{ibT}} B - k'$.*

PROOF. Straightforward.   □

Lemma 7.2 (cl-ibT-**Conversion Lemma**). *(a) Let $A, B_0, \ldots, B_n$ be c.e. sets such that $A \leq_{\text{cl}} B_0, \ldots, B_n$. There is a c.e. set $\hat{A} =_{\text{cl}} A$ such that $\hat{A} \leq_{\text{ibT}} B_0, \ldots, B_n$.*

(b) Let $A_0, \ldots, A_n, B$ be c.e. sets such that $A_0, \ldots, A_n \leq_{\text{cl}} B$. There is a c.e. set $\hat{B} =_{\text{cl}} B$ such that $A_0, \ldots, A_n \leq_{\text{ibT}} \hat{B}$.

PROOF. For a proof of (a), fix k minimal such that $A \leq_{(i+k)\text{bT}} B_0, \ldots, B_n$ and let $\hat{A} = A + k$. Then, by the Bounded-Shift Lemma, $\hat{A} =_{\text{cl}} A$, and, by Proposition 7.1, $\hat{A} \leq_{\text{ibT}} B_0, \ldots, B_n$.

The proof of (b) is similar: Given k such that $A_0, \ldots, A_n \leq_{(i+k)\text{bT}} B$, the set $\hat{B} = B - k$ will have the required properties.   □

In [5] it is concluded from Lemma 7.2(b) that, for any c.e. sets A and B, $(deg_{\text{ibT}}(A), deg_{\text{ibT}}(B))$ is a maximal pair in the c.e. ibT-degrees if and only if

$(deg_{cl}(A), deg_{cl}(B))$ is a maximal pair in the c.e. cl-degrees. By a similar argument we obtain the preservation of ibT-meets and ibT-joins in the cl-degrees.

Lemma 7.3 (ibT-**Meet Lemma**). *Let $A, B_0, \ldots, B_n$ $(n \geq 0)$ be c.e. sets such that*

$$deg_{ibT}(A) = deg_{ibT}(B_0) \wedge \cdots \wedge deg_{ibT}(B_n). \tag{20}$$

Then

$$deg_{cl}(A) = deg_{cl}(B_0) \wedge \cdots \wedge deg_{cl}(B_n). \tag{21}$$

PROOF. Since A is a lower bound of $B_0, \ldots, B_n$ with respect to ibT-reducibility and since ibT-reducibility is stronger than cl-reducibility, A is a lower bound of $B_0, \ldots, B_n$ with respect to cl-reducibility too. So, given a c.e. set C such that $C \leq_{cl} B_0, \ldots, B_n$, it suffices to show that $C \leq_{cl} A$. By the first part of the cl-ibT-Conversion Lemma, there is a c.e. set $\hat{C}$ such that $\hat{C} =_{cl} C$ and $\hat{C} \leq_{ibT} B_0, \ldots, B_n$. It follows by (20) that $\hat{C} \leq_{ibT} A$. So, by $\hat{C} =_{cl} C$, $C \leq_{cl} A$. □

Lemma 7.4 (ibT-**Join Lemma**). *Let $A, B_0, \ldots, B_n$ $(n \geq 0)$ be c.e. sets such that*

$$deg_{ibT}(A) = deg_{ibT}(B_0) \vee \cdots \vee deg_{ibT}(B_n).$$

Then

$$deg_{cl}(A) = deg_{cl}(B_0) \vee \cdots \vee deg_{cl}(B_n).$$

PROOF. This easily follows from the second part of the cl-ibT-Conversion Lemma just as the ibT-Meet Lemma followed from the first part. □

Having shown that joins and meets in the c.e. ibT-degrees are preserved in the c.e. cl-degrees, we now turn to the corresponding question for the cl-degrees and the wtt-degrees. Again we start with a conversion lemma.

Lemma 7.5 (wtt-ibT-**Conversion Lemma**). *(a) Let $A, B_0, \ldots, B_n$ be c.e. sets such that $A \leq_{wtt} B_0, \ldots, B_n$. There is a c.e. set $\hat{A} =_{wtt} A$ such that $\hat{A} \leq_{ibT} B_0, \ldots, B_n$.*

(b) Let A, B be c.e. sets such that $A \leq_{wtt} B$. There is a c.e. set $\hat{B} =_{wtt} B$ such that $A \leq_{ibT} \hat{B}$.

PROOF. (a) Fix wtt-reductions $A = \Gamma_j^{B_j}$, let f_j be computable bounds on the use functions of theses reductions, let f be any strictly increasing computable function which dominates the functions f_j $(j \leq n)$, and let $\hat{A} = A_f$ be the f-shift of A. Then, as one can easily check, $\hat{A}$ has the required properties.

(b) Since the claim is trivial for computable A, w.l.o.g. we may assume that A is not computable hence infinite. Fix an infinite computable subset C of A, let f enumerate C in order of magnitude (note that f is an unbounded computable shift), and let $\hat{B} = (A \setminus C) \cup B_f$. Again, one can easily check, that $\hat{B}$ has the required properties. □

Since the partial ordering of the c.e. wtt-degrees is an upper semilattice, it follows from the existence of maximal pairs in the c.e. ibT- and cl-degrees, that in the second part of the wtt-ibT-Conversion Lemma we cannot replace the set A by a pair of sets A_0, A_1 (even if we replace ibT by cl). So, here the conversion lemma implies the corresponding meet lemma but not the corresponding join lemma.

Lemma 7.6 (cl-Meet Lemma). *Let* $A, B_0, \ldots, B_n$ *(n $\geq$ 0) be c.e. sets such that*

$$deg_{cl}(A) = deg_{cl}(B_0) \wedge \cdots \wedge deg_{cl}(B_n). \tag{22}$$

Then

$$deg_{wtt}(A) = deg_{wtt}(B_0) \wedge \cdots \wedge deg_{wtt}(B_n). \tag{23}$$

PROOF. This follows from the first part of the wtt-ibT-Conversion Lemma just as the ibT-Meet Lemma followed from the first part of the cl-ibT-Conversion Lemma. □

Finally we give the dual of Lemma 7.6 for joins which is due to Ambos-Spies, Bodewig, Kräling and Yu, and which is published here first. As pointed out above, here we cannot use a general conversion lemma for the wtt- and cl-reducibilities but we have to use a somewhat more sophisticated argument.

Lemma 7.7 (cl-Join Lemma; Ambos-Spies, Bodewig, Kräling and Yu). *Let* $A, B_0, \ldots, B_n$ *(n $\geq$ 0) be c.e. sets such that*

$$deg_{cl}(A) = deg_{cl}(B_0) \vee \cdots \vee deg_{cl}(B_n). \tag{24}$$

Then

$$deg_{wtt}(A) = deg_{wtt}(B_0) \vee \cdots \vee deg_{wtt}(B_n). \tag{25}$$

PROOF. Since cl-reducibility is stronger than wtt-reducibility (and since the partial ordering of the c.e. wtt-degrees is an upper semilattice), (24) implies

$$deg_{wtt}(A) \geq deg_{wtt}(B_0) \vee \cdots \vee deg_{wtt}(B_n).$$

So, by

$$deg_{wtt}(B_0) \vee \cdots \vee deg_{wtt}(B_n) = deg_{wtt}(B_0 \oplus \cdots \oplus B_n),$$

it suffices to prove

$$A \leq_{wtt} B_0 \oplus \cdots \oplus B_n. \tag{26}$$

W.l.o.g. we may assume that the sets $A, B_0, \ldots, B_n$ are not computable. By Proposition 3.5 we may assume that A is not simple. So we may fix an infinite computable subset R of $\overline{A}$, and let r be the one-to-one computable function enumerating R in order of magnitude. Note that, by noncomputability of A, r is an unbounded computable shift. Next, by (the $\geq$ part of) (24), fix $k \geq 0$ such that $B_i \leq_{(i+k)bT} A$ $(i \leq n)$. Then, by the Representation Lemma, w.l.o.g. we may assume that there are one-to-one computable enumeration functions a and b_i of A and B_i, respectively, such that

$$\forall\, s \geq 0\ [a(s) \leq \min(b_0(s), \ldots, b_n(s)) + k]. \tag{27}$$

(Note that in the Representation Lemma, A is replaced by an ibT-equivalent subset of A. So this is compatible with the assumption that $A \cap R = \emptyset$.)

Now, for showing (26), we split A into the c.e. sets

$$C_0 = \{a(s) : \min(b_0(s), \ldots, b_n(s)) < r(a(s))\}$$

and

$$C_1 = \{a(s) : \min(b_0(s), \ldots, b_n(s)) \geq r(a(s))\}.$$

Note that $C_0 \leq_{\mathrm{wtt}} B_0 \oplus \cdots \oplus B_n$. Namely, given x, using $B_0 \restriction r(x), \ldots, B_n \restriction r(x)$ as oracles we can compute the finitely many stages s such that $b_i(s) < r(x)$ for some $i \leq n$; and $x \in C_0$ iff $x = a(s)$ for one of theses stages s. So it suffices to argue that, by (the $\leq$ part of) (24), the part C_1 of A can be neglected.

For this sake it suffices to show that

$$\forall\, i \leq n\ (B_i \leq_{\mathrm{cl}} C_0 \cup A_r) \tag{28}$$

holds. Then, by (24), $A \leq_{\mathrm{cl}} C_0 \cup A_r$ too whence, by the Computable-Shift Lemma, $A \leq_{\mathrm{cl}} C_0$ hence $A \leq_{\mathrm{wtt}} C_0$. By $C_0 \leq_{\mathrm{wtt}} B_0 \oplus \cdots \oplus B_n$, this implies (26).

So it only remains to prove (28). Fix $i \leq n$ and $x \geq 0$. Since C_0 and A_r are computably separated by $\overline{R}$, it suffices to (uniformly) compute $B_i(x)$ using $C_0 \restriction (x+k+1)$ and $A_r \restriction (x+1)$ as oracles. But this can be done by the following observations. By (27), $x \in B_i$ if and only if there is a number $y \leq x + k$ such that $y \in A$ and, for the unique s such that $y = a(s)$, $x = b_i(s)$. So, by definition of C_0, $x \in B_i$ if and only if

(i) there is a number $y \leq x + k$ in C_0 such that, for the unique s such that $y = a(s)$, $x = b_i(s)$ or

(ii) there is a number $y \leq x + k$ such that $r(y) \leq x$, $r(y) \in A_r$ and, for the unique s such that $y = a(s)$, $x = b_i(s)$. $\square$

In the remainder of the paper we use the above introduced tools (i.e., the join and meet preservation lemmas and the conversion lemmas together with the splitting and distributivity lemmas) in order to carry over results on the c.e. wtt-degrees to the c.e. ibT- and cl-degrees. We conclude this section with a first example of such a transfer.

Theorem 7.8 (Downey and Hirschfeldt [14]). *For $r \in \{\mathrm{ibT}, \mathrm{cl}\}$, there are c.e. r-degrees* $\mathbf{a}$ *and* $\mathbf{b}$ *such that* $\mathbf{a} \wedge \mathbf{b}$ *does not exist.*

This theorem was originally proven by Downey and Hirschfeldt by a direct construction. Alternatively we can use the fact that Jockusch [21] has shown that there is a pair of c.e. sets A and B such that $deg_{\mathrm{wtt}}(A) \wedge deg_{\mathrm{wtt}}(B)$ does not exist. Namely, by Lemmas 7.6 and 7.3, Jockusch's result implies that $deg_{\mathrm{cl}}(A) \wedge deg_{\mathrm{cl}}(B)$ and $deg_{\mathrm{ibT}}(A) \wedge deg_{\mathrm{ibT}}(B)$ do not exist too.

8 Minimal Pairs, Embedding Distributive Lattices, and Nonbounding Degrees

In this section we transfer some results on minimal pairs and lattice embeddings from the c.e. weak truth-table degrees to the c.e. strongly bounded Turing degrees. These results will be used in the proofs of some global results on the theories of the c.e. ibT- and cl-degrees given in the subsequent sections.

We first observe that minimal pairs in the c.e. wtt-, cl- and ibT-degrees coincide. Recall that $(\mathbf{a}, \mathbf{b})$ is a *minimal pair* if $\mathbf{a}, \mathbf{b} > \mathbf{0}$ and $\mathbf{a} \wedge \mathbf{b} = \mathbf{0}$.

Lemma 8.1 (Minimal-Pair Lemma [5]). *For c.e. sets A and B the following are equivalent.*

(i) The pair $(deg_{\mathrm{wtt}}(A), deg_{\mathrm{wtt}}(B))$ is a minimal pair of c.e. wtt-degrees.
(ii) The pair $(deg_{\mathrm{cl}}(A), deg_{\mathrm{cl}}(B))$ is a minimal pair of c.e. cl-degrees.
(iii) The pair $(deg_{\mathrm{ibT}}(A), deg_{\mathrm{ibT}}(B))$ is a minimal pair of c.e. ibT-degrees.

PROOF. The implications $(i) \Rightarrow (ii)$ and $(ii) \Rightarrow (iii)$ are immediate by the fact that $\leq_{\mathrm{cl}}$ is stronger than $\leq_{\mathrm{wtt}}$ and $\leq_{\mathrm{ibT}}$ is stronger than $\leq_{\mathrm{cl}}$. The implication $(iii) \Rightarrow (ii)$ and $(ii) \Rightarrow (i)$ hold by the ibT-Meet Lemma and the cl-Meet Lemma, respectively.   □

It might be of interest to note that we cannot expand Lemma 8.1 by adding Turing reducibility: As mentioned before, there is a wtt-minimal pair (A, B) such that $A =_{\mathrm{T}} B$ whence the pair $(deg_{\mathrm{T}}(A), deg_{\mathrm{T}}(B))$ of c.e. Turing degrees is not minimal (see Downey and Stob [18]). For *halves* of minimal pairs, however, Ambos-Spies [1] has shown that a c.e. set A is half of a wtt-minimal pair if and only if A is half of a T-minimal pair. So a c.e. set A is half of a T-minimal pair iff A is half of a wtt-minimal pair iff A is half of a cl-minimal pair iff A is half of an ibT-minimal pair.

Next we show that, for the strongly bounded Turing reducibilities $r = \mathrm{ibT}, \mathrm{cl}$, any finite distributive lattice can be embedded into the partial ordering $(\mathbf{R}_r, \leq)$ of the c.e. r-degrees by a map which preserves the least element. Since any finite distributive lattice can be (0-preserving) embedded into some finite Boolean algebra, it suffices to show that, for $n \geq 2$, the n-atom Boolean algebra $\mathcal{B}_n$ can be embedded into $(\mathbf{R}_r, \leq)$ by a map which preserves the least element. In a distributive upper semilattice (of c.e. degrees), like the upper semilattice $(\mathbf{R}_{\mathrm{wtt}}, \leq)$ of the c.e. wtt-degrees, any finite anti-chain of n degrees $\mathbf{a}_0, \dots, \mathbf{a}_{n-1}$ which are pairwise minimal pairs generates the n-atom Boolean algebra $\mathcal{B}_n$. Though, for $r = \mathrm{ibT}, \mathrm{cl}$, $(\mathbf{R}_r, \leq)$ is neither an upper semilattice nor distributive, by the Splitting and Distributivity Lemmas we still obtain the corresponding result here, provided that the degrees $\mathbf{a}_0, \dots, \mathbf{a}_{n-1}$ can be represented by mutually disjoint c.e. sets.

Lemma 8.2 (Embedding Lemma). *Let $r \in \{\mathrm{ibT}, \mathrm{cl}\}$ and let $A_0, \dots, A_{n-1}$ $(n \geq 2)$ be noncomputable pairwise disjoint c.e. sets such that*

$$\forall\, i, j < n\ (i \neq j \Rightarrow deg_r(A_i) \wedge deg_r(A_j) = \mathbf{0}). \tag{29}$$

Then $f_r : \text{POWER}(\{0, \ldots, n-1\}) \to \mathbf{R}_r$ defined by

$$f_r(\alpha) = deg_r(A_\alpha) \quad \text{where} \quad A_\alpha = \bigcup_{i \in \alpha} A_i$$

defines a lattice embedding of the n-atom Boolean algebra

$$\mathcal{B}_n = \text{POWER}(\{0, \ldots, n-1\}), \subseteq)$$

into the partial ordering $(\mathbf{R}_r, \leq)$ of the c.e. r-degrees which preserves the least element.

PROOF. Since $\leq_{ibT}$ is stronger than $\leq_{cl}$ and since, by the ibT-Meet Lemma and the ibT-Join Lemma, joins and meets in the c.e. ibT-degrees are preserved in the c.e. cl-degrees, given $\alpha, \beta \subseteq \{0, \ldots, n-1\}$ it suffices to show

$$\alpha \subseteq \beta \Rightarrow A_\alpha \leq_{ibT} A_\beta \quad \text{(ordering)} \tag{30}$$

$$\alpha \not\subseteq \beta \Rightarrow A_\alpha \not\leq_{cl} A_\beta \quad \text{(non-ordering)} \tag{31}$$

$$deg_{ibT}(A_\alpha) \vee deg_{ibT}(A_\beta) = deg_{ibT}(A_{\alpha \cup \beta}) \quad \text{(joins)} \tag{32}$$

$$deg_{ibT}(A_\alpha) \wedge deg_{ibT}(A_\beta) = deg_{ibT}(A_{\alpha \cap \beta}) \quad \text{(meets)} \tag{33}$$

Moreover, by the Minimal-Pair Lemma, we may assume that (29) holds for both, $r = ibT$ and $r = cl$.

Now, (30) and (32) are immediate by the Splitting Lemma.

For a proof of (31), given α and β such that $\alpha \not\subseteq \beta$, fix $i \in \alpha \setminus \beta$. By (30), it suffices to show that $A_i \not\leq_{cl} A_\beta$. For a contradiction assume that $A_i \leq_{cl} A_\beta$. Then by the Distributivity Lemma, there is a splitting of A_i into pairwise disjoint c.e. sets $A_{i,j}$, $j \in \beta$, such that $A_{i,j} \leq_{cl} A_j$; and, by the Splitting Lemma,

$$deg_{cl}(A_i) = \vee_{j \in \beta} deg_{cl}(A_{i,j}). \tag{34}$$

Hence, for $j \in \beta$, $A_{i,j} \leq_{cl} A_i, A_j$, and, by $i \neq j$, (A_i, A_j) is a cl-minimal pair. So $A_{i,j}$ is computable. It follows with (34) that A_i is computable too. But this contradicts the choice of A_i.

Finally, for a proof of (33), by (30) it suffices to show, that, for any c.e. set B,

$$[B \leq_{ibT} A_\alpha \ \& \ B \leq_{ibT} A_\beta] \Rightarrow B \leq_{ibT} A_{\alpha \cap \beta}$$

holds. So fix B such that $B \leq_{ibT} A_\alpha$ and $B \leq_{ibT} A_\beta$. By the former and by the Distributivity Lemma, there are pairwise disjoint c.e. sets $B_i \leq_{ibT} A_i, i \in \alpha$, such that $B = \bigcup_{i \in \alpha} B_i$. It follows by the Splitting Lemma that $B_i \leq_{ibT} B$ whence, by $B \leq_{ibT} A_\beta$, $B_i \leq_{ibT} A_\beta$. So, again by the Distributivity Lemma, there are pairwise disjoint c.e. sets $B_{i,j} \leq_{ibT} A_j$ ($i \in \alpha$, $j \in \beta$) such that $B_i = \bigcup_{j \in \alpha} B_{i,j}$. Hence, by the Splitting Lemma, $B_{i,j} \leq_{ibT} A_i, A_j$ and B is the disjoint union

$$B = \bigcup_{(i,j) \in \alpha \times \beta} B_{i,j}.$$

By the former, for $i \neq j$, $B_{i,j}$ is computable since (A_i, A_j) is an ibT-minimal pair. So

$$B =_{\text{ibT}} \bigcup_{i \in \alpha \cap \beta} B_{i,i}$$

where $B_{i,i} \leq_{\text{ibT}} A_i$. Hence, by the Splitting Lemma, $B \leq_{\text{ibT}} A_{\alpha \cap \beta}$.   □

Theorem 8.3 *Let $r \in \{\text{ibT}, \text{cl}\}$. Any finite distributive lattice $\mathcal{L}$ is embeddable (as a lattice) into the partial ordering of the c.e. r-degrees by a map which preserves the least element.*

PROOF. For any finite distributive lattice $\mathcal{L}$ there is some $n \geq 2$ such that $\mathcal{L}$ can be embedded into the n-atom Boolean algebra $\mathcal{B}_n$ by a map which preserves the least element. So, given $n \geq 2$ and $r \in \{\text{ibT}, \text{cl}\}$, it suffices to embed $\mathcal{B}_n$ into the partial ordering $(\mathbf{R}_r, \leq)$ by a map which preserves the least element. By the Embedding Lemma, the latter can be done by giving pairwise disjoint c.e. sets $A_0, \ldots, A_{n-1}$ which are pairwise r-minimal pairs.

The existence of such sets can be shown by various results in the literature. For instance, Thomason [30] has shown that, for any $n \geq 0$, there are c.e. T-degrees $\mathbf{a_i}$ $(i < n)$ which are pairwise T-minimal pairs. Since any T-minimal pair is a fortiori an r-minimal pair for $r \in \{\text{ibT}, \text{cl}\}$, it suffices to choose any pairwise disjoint c.e. sets $A_i \in \mathbf{a}_i$ $(i < n)$.   □

We close this section with some results on bounds of minimal pairs. A c.e. r-degree $\mathbf{a}$ is called *bounding* if there is a minimal pair $(\mathbf{a}_0, \mathbf{a}_1)$ of c.e. r-degrees such that $\mathbf{a}_0, \mathbf{a}_1 \leq \mathbf{a}$, and $\mathbf{a}$ is called *nonbounding* otherwise. A c.e. r-degree $\mathbf{a}$ is called a *top* (of a minimal pair) if there is a minimal pair $(\mathbf{a}_0, \mathbf{a}_1)$ of c.e. r-degrees such that $\mathbf{a}_0 \vee \mathbf{a}_1 = \mathbf{a}$, and $\mathbf{a}$ is called a *nontop* otherwise.

Lemma 8.4 (Nonbounding Lemma). *For any noncomputable c.e. set A the following are equivalent.*

(i) $deg_{\text{wtt}}(A)$ is (non)bounding.
(ii) $deg_{\text{cl}}(A)$ is (non)bounding.
(iii) $deg_{\text{ibT}}(A)$ is (non)bounding.

PROOF. It suffices to consider the case of bounding degrees.

For the proof of $(i) \Rightarrow (ii)$ assume that $deg_{\text{wtt}}(A)$ is bounding. Then there are noncomputable c.e. sets $B, C \leq_{\text{wtt}} A$ such that $(deg_{\text{wtt}}(B), deg_{\text{wtt}}(C))$ is a minimal pair. By the wtt-ibT-Conversion Lemma there are c.e. sets $\hat{B}, \hat{C} \leq_{\text{cl}} A$ such that $\hat{B} =_{\text{wtt}} B$ and $\hat{C} =_{\text{wtt}} C$. By the latter, $(deg_{\text{wtt}}(\hat{B}), deg_{\text{wtt}}(\hat{C}))$ is a minimal pair too. It follows by (the trivial direction of) the Minimal-Pair Lemma that $deg_{\text{cl}}(A)$ bounds the minimal pair $(deg_{\text{cl}}(\hat{B}), deg_{\text{cl}}(\hat{C}))$.

The proof of $(ii) \Rightarrow (iii)$ is similar to the proof of $(i) \Rightarrow (ii)$ (now applying the cl-ibT-Conversion Lemma).

Finally, for a proof of $(iii) \Rightarrow (i)$, assume that $deg_{\text{ibT}}(A)$ is bounding. Fix noncomputable c.e. sets $B, C \leq_{\text{ibT}} A$ such that $(deg_{\text{ibT}}(B), deg_{\text{ibT}}(C))$ is a minimal pair. Then a fortiori $B, C \leq_{\text{wtt}} A$ and by (the nontrivial direction of) the Minimal-Pair Lemma, $(deg_{\text{wtt}}(B), deg_{\text{wtt}}(C))$ is a minimal pair too. So $deg_{\text{wtt}}(A)$ is bounding.   □

Lemma 8.5 (Nontop Lemma). *For any noncomputable c.e. set A the following are equivalent.*

(i) $deg_{\mathrm{wtt}}(A)$ *is a (non)top.*
(ii) $deg_{\mathrm{cl}}(A)$ *is a (non)top.*
(iii) $deg_{\mathrm{ibT}}(A)$ *is a (non)top.*

PROOF. It suffices to consider the case of tops.

$(i) \Rightarrow (iii)$. Assume that $deg_{\mathrm{wtt}}(A)$ is a top, say $deg_{\mathrm{wtt}}(A) = deg_{\mathrm{wtt}}(B) \vee deg_{\mathrm{wtt}}(C)$ for the wtt-minimal pair $(deg_{\mathrm{wtt}}(B), deg_{\mathrm{wtt}}(C))$. Then $A \leq_{\mathrm{wtt}} B \oplus C$. So, by distributivity, we may split A into c.e. sets A_0 and A_1 such that $A_0 \leq_{\mathrm{wtt}} B$ and $A_1 \leq_{\mathrm{wtt}} C$ (see Stob [29]). Then $(deg_{\mathrm{wtt}}(A_0), deg_{\mathrm{wtt}}(A_1))$ is a minimal pair again. So, by the Minimal-Pair Lemma, $(deg_{\mathrm{ibT}}(A_0), deg_{\mathrm{ibT}}(A_1))$ is a minimal pair, and, by the Splitting Lemma, $deg_{\mathrm{ibT}}(A) = deg_{\mathrm{ibT}}(A_0) \vee deg_{\mathrm{ibT}}(A_1)$. So $deg_{\mathrm{ibT}}(A)$ is a top too.

$(iii) \Rightarrow (ii)$. Assume that $deg_{\mathrm{ibT}}(A)$ is a top, say $deg_{\mathrm{ibT}}(A) = deg_{\mathrm{ibT}}(B) \vee deg_{\mathrm{ibT}}(C)$ for the ibT-minimal pair $(deg_{\mathrm{ibT}}(B), deg_{\mathrm{ibT}}(C))$. Then, by the Minimal Pair Lemma, $(deg_{\mathrm{cl}}(B), deg_{\mathrm{cl}}(C))$ is a cl-minimal pair and, by the ibT-Join Lemma, $deg_{\mathrm{cl}}(A) = deg_{\mathrm{cl}}(B) \vee deg_{\mathrm{cl}}(C)$. So $deg_{\mathrm{cl}}(A)$ is a top too.

$(ii) \Rightarrow (i)$. The proof that, for any top $deg_{\mathrm{cl}}(A)$, $deg_{\mathrm{wtt}}(A)$ is a top is similar to the proof of the implication $(iii) \Rightarrow (ii)$ (using the cl-Join Lemma in place of the ibT-Join Lemma). □

9   1-Types

In [6] Ambos-Spies and Soare have shown that the theory of the c.e. wtt-degrees realizes infinitely many 1-types. They showed that, for any number $n \geq 2$, there is an n-bounding c.e. wtt-degree which is not $(n + 1)$-bounding (hence not m-bounding for $m > n$), where a c.e. r-degree **b** is n-*bounding* if there are n c.e. r-degrees $\mathbf{a}_0, \ldots, \mathbf{a}_{n-1} < \mathbf{b}$ which are pairwise minimal pairs. Ambos-Spies and Soare deduce this result from the following technical lemma (which is proven by a quite sophisticated $0'''$-priority argument) by exploiting distributivity of the upper semilattice of the c.e. wtt-degrees.

Lemma 9.1 (Ambos-Spies and Soare [6]). *For any* $n \geq 2$ *there are c.e. wtt-degrees* $\mathbf{a}_0, \ldots, \mathbf{a}_{n-1} > \mathbf{0}$ *such that*

$$\forall\, i, j < n\ (i \neq j \Rightarrow \mathbf{a}_i \wedge \mathbf{a}_j = \mathbf{0}) \tag{35}$$

$$\forall\, i < n\ (\mathbf{a}_i\ \text{is nonbounding}) \tag{36}$$

In fact, Ambos-Spies and Soare [6] proved a stronger version of this lemma where the degrees $\mathbf{a}_i$ (and their joins) are chosen to be contiguous, i.e., where the Turing degree $\hat{\mathbf{a}}_i$ which contains $\mathbf{a}_i$ contains only one c.e. wtt-degree (namely $\mathbf{a}_i$). This allowed them to argue that, locally, there is enough agreement between the distributive structure of the c.e. wtt-degrees and the nondistributive structure of the c.e. T-degrees in order to get the corresponding result on one-types for the weaker Turing reducibility too. (The transfer of results on the (c.e.) wtt-degrees to the (c.e.) Turing degrees via contiguous degrees has its origin in Ladner and Sasso [22]. For more applications of this transfer technique see Downey [13].)

Here we proceed in a similar fashion: we will show that our preceding observations on the relations between the strongly bounded Turing reducibilities and weak truth-table reducibility together with the local distributivity phenomena in the c.e. ibT- and cl-degrees established in Sect. 5 allow us to deduce the existence of infinitely many 1-types in the strongly bounded Turing degrees of the c.e. sets from Lemma 9.1 too.

Theorem 9.2. *For* $r \in \{\text{ibT}, \text{cl}\}$, *the first order theory* $\text{Th}(\mathbf{R}_r, \leq)$ *of the partial ordering of c.e.* r-degrees realizes infinitely many 1-types. So $\text{Th}(\mathbf{R}_r, \leq)$ is not $\aleph_0$-categorical.

PROOF. Given $n \geq 2$ it suffices to show that there is a c.e. r-degree $\mathbf{b}$ which is n-bounding but not $(n+1)$-bounding (hence not m-bounding for all $m > n$). By Lemma 9.1 fix c.e. wtt-degrees $\mathbf{a}_0, \ldots, \mathbf{a}_{n-1} > \mathbf{0}$ such that (35) and (36) hold, choose pairwise disjoint c.e. sets $A_i \in \mathbf{a}_i$ $(i < n)$, let $B = A_0 \cup \cdots \cup A_{n-1}$, and let $\mathbf{b}$ be the r-degree of B.

Then $\mathbf{b}$ is n-bounding since, by the Splitting Lemma, $\mathbf{b}$ bounds the c.e. r-degrees $\hat{\mathbf{a}}_i = deg_r(A_i)$ $(i < n)$, and, by the Minimal-Pair Lemma and by (35), the degrees $\hat{\mathbf{a}}_i$ are pairwise minimal pairs.

It remains to show that $\mathbf{b}$ is not $(n+1)$-bounding. For a contradiction assume that there are noncomputable c.e. sets $C_0, \ldots, C_n \leq_r B$ such that

$$\forall \, j, j' \leq n \,(j \neq j' \Rightarrow deg_r(C_j) \wedge deg_r(C_{j'}) = \mathbf{0}). \qquad (37)$$

By the Distributivity Lemma, split C_j into pairwise disjoint c.e. sets $C_{j,i}$ such that $C_{j,i} \leq_r A_i$ $(j \leq n, i < n)$. Then, by the Splitting Lemma and by noncomputability of C_j, $C_{j,i} \leq_r C_j$ for all $i < n$ and there is some $i < n$ such that $C_{j,i}$ is noncomputable. Let i_j be the least such i. By the pigeon hole principle, fix $j \neq j'$ such that $i_j = i_{j'}$. It follows by (37) that $(deg_r(C_{j,i_j}), deg_r(C_{j',i_j}))$ is a minimal pair bounded by the r degree $\hat{\mathbf{a}}_{i_j}$. So $\hat{\mathbf{a}}_{i_j}$ is bounding whence, by the Nonbounding Lemma, the wtt-degree $\mathbf{a}_{i_j}$ is bounding too. But this contradicts (36). □

10   Undecidability

Our final result is the undecidability of the first order theories of the partial orderings of the c.e. ibT- and cl-degrees. Just as in case of the preceding theorem on 1-types, our proof will be based on a proof of the undecidability of the theory of the partial ordering of the c.e. wtt-degrees, namely on the proof of the undecidability of the Π_4-theory of this structure given in Lempp and Nies [23]. We will use the main technical lemma on the c.e. wtt-degrees of [23] together with a sufficient condition for a partial ordering to have an undecidable theory given there too. We first state these results from [23].

Lemma 10.1 (Main Lemma of Lempp and Nies [23]). *Let $n \geq 1$. There are noncomputable c.e. sets A_i, B_j and $D_{i,j}$ $(i, j < n)$ such that*

$$\forall\, i, j < n\ (D_{i,j} \leq_{\mathrm{wtt}} A_i, B_j), \tag{38}$$

$$\forall\, i, j < n\ (deg_{\mathrm{wtt}}(A_i)\ and\ deg_{\mathrm{wtt}}(B_j)\ are\ nontops), \tag{39}$$

$$\forall\, i, i' < n\ (i \neq i' \Rightarrow (deg_{\mathrm{wtt}}(A_i), deg_{\mathrm{wtt}}(A_{i'}))\ is\ a\ \text{wtt-minimal pair}), \tag{40}$$

and

$$\forall\, j, j' < n\ (j \neq j' \Rightarrow (deg_{\mathrm{wtt}}(B_j), deg_{\mathrm{wtt}}(B_{j'}))\ is\ a\ \text{wtt-minimal pair}). \tag{41}$$

Lemma 10.2 (Undecidability Lemma; [23]). *Let $\mathcal{P} = (P, \leq)$ be a partial ordering with least element 0. Suppose that there is a first order formula $\varphi(x, y)$ in the language of partial orderings such that, for any $n \geq 1$, there are elements $a_i, b_j, d_{i,j}$ of P $(i, j < n)$ such that*

(i) $0 < d_{i,j} \leq a_i, b_j$,
(ii) for any $I \subseteq \{0, \ldots, n-1\} \times \{0, \ldots, n-1\}$, the supremum of the elements $d_{i,j}$ of P with $(i, j) \in I$, $\bigvee_{(i,j) \in I} d_{i,j}$, exists in $(P, \leq)$,
(iii) for $\hat{d}_{i,j} = \bigvee_{(i',j') \neq (i,j)} d_{i',j'}$, the infimum $a_i \wedge b_j \wedge \hat{d}_{i,j}$ of $a_i, b_j, \hat{d}_{i,j}$ exists and $a_i \wedge b_j \wedge \hat{d}_{i,j} = 0$, and
(iv) there are elements $\hat{a}, \hat{b} \in P$ such that, for $A = \{a_0, \ldots, a_{n-1}\}$ and $B = \{b_0, \ldots, b_{n-1}\}$,

$$\mathcal{P} \vDash \varphi(a, \hat{a}) \Leftrightarrow a \in A\ and\ \mathcal{P} \vDash \varphi(b, \hat{b}) \Leftrightarrow b \in B.$$

Then the first order theory $\mathrm{Th}(P, \leq)$ is undecidable.

Actually, Lemma 10.2 is stated in [23] (see Theorem 2.1 there) only for upper semilattices (and there it is stated as a criterion for proving Π_4-$\mathrm{Th}(P, \leq)$ to be undecidable by imposing some bound on the complexity of the formula φ). But the existence of the joins required in the proof of Theorem 2.1 of [23] is guaranteed by clause (ii) above (which is not present in [23]). The idea of the proof of Lemma 10.2 is as follows. Given a finite bipartite graph G with left side $\{0, \ldots, n-1\}$, right side $\{0', \ldots, (n-1)'\}$, and edge relation $E \subseteq \{0, \ldots, n-1\} \times \{0', \ldots, (n-1)'\}$, G can be defined in $(P, \leq)$ with parameters $\hat{a}$, $\hat{b}$, and $\hat{c} = \bigvee_{(i,j) \in E} d_{i,j}$ (which exists by (ii)). Namely, by representing the left and right parts of the vertex set of G by $A = \{a_0, \ldots, a_{n-1}\}$ and $B = \{b_0, \ldots, b_{n-1}\}$, respectively, these parts are definable from $\hat{a}$ and $\hat{b}$ by the formula φ. Finally, the edge relation becomes definable from $\hat{c}$ by

$$(i, j) \in E \Leftrightarrow \exists\, u \leq \hat{c}\ (u \neq 0\ \&\ u \leq a_i, b_j).$$

Since the theory of finite bipartite graphs with left and right sides of the same size is hereditarily undecidable, the above interpretation implies undecidability of $\mathrm{Th}(P, \leq)$. For details see [23].

In order to derive the necessary facts on the c.e. ibT- and cl-degrees from Lemma 10.1 which will allow us to argue that the partial orderings $(\mathbf{R}_{ibT}, \leq)$ and $(\mathbf{R}_{cl}, \leq)$ satisfy the hypotheses of Lemma 10.2, we have to provide sufficient distributivity in these structures. So we will show first that (degree) splittings of the ibT- and cl-degrees of sufficiently scattered c.e. sets are distributive.

Definition 10.3. *A set A is* strongly scattered *if there is a computable set R and a nondecreasing and unbounded computable function $l : \omega \rightarrow \omega$ such that $A \subseteq R$ and*

$$\forall n \in R \left([n - l(n), n + l(n)] \cap R = \{n\} \right) \tag{42}$$

holds. A c.e. r-degree $\mathbf{a}$ is strongly scattered *if there is a strongly scattered c.e. set $A \in \mathbf{a}$.*

Note that, for any set A and any unbounded computable shift f, the f-shift A_f of A is strongly scattered. So, by the Computable-Shift Lemma, any c.e. wtt-degree is strongly scattered.

Lemma 10.4. *Let $r \in \{ibT, cl\}$. Suppose that A, B_0, B_1 are c.e. sets such that A is strongly scattered and $deg_r(A) = deg_r(B_0) \vee deg_r(B_1)$. There are disjoint c.e. sets $\hat{B}_i$ such that*

$$\hat{B}_i \leq_r B_i \ (i \leq 1) \tag{43}$$

and

$$A =_r \hat{B}_0 \cup \hat{B}_1. \tag{44}$$

PROOF. We give the proof for $r = cl$. The proof for $r = ibT$ is similar.

Fix a computable set R and a nondecreasing and unbounded computable function $l : \omega \rightarrow \omega$ such that $A \subseteq R$ and (42) holds, and fix $k \geq 0$ minimal such that $B_0, B_1 \leq_{(i+k)bT} A$. W.l.o.g. we may assume that $l(n) > k$ (by letting $l(n) = k + 1$ for the finitely many numbers n such that $l(n) \leq k$ and by omitting these numbers n from R and A). Moreover, since any c.e. subset of A will be strongly scattered via R and l too, by the Representation Lemma, we may assume that there are computable one-to-one functions a, b_0 and b_1 enumerating A, B_0 and B_1, respectively, such that, for $b(n) = min(b_0(n), b_1(n))$,

$$\forall n \ (a(n) \leq b(n) + k). \tag{45}$$

Note that by a being a one-to-one enumeration of A and by choice of R and l, for any numbers n and n',

$$n \neq n' \Rightarrow [a(n) - k, a(n) + l(a(n))] \cap [a(n') - k, a(n') + l(a(n'))] = \emptyset \tag{46}$$

holds.

Now, let

$$\hat{B}_0 = \{b_0(n) : n \geq 0 \ \& \ b_0(n) \leq b_1(n) \ \& \ b_0(n) < a(n) + l(a(n))\},$$

$$\hat{B}_1 = \{b_1(n) : n \geq 0 \ \& \ b_1(n) < b_0(n) \ \& \ b_1(n) < a(n) + l(a(n))\},$$

and
$$\hat{A} = \{a(n) + l(a(n)) : n \geq 0\}.$$

Note that, by definition of the sets $\hat{B}_i$ and by (45), for any n such that $b_i(n) \in \hat{B}_i$, $b_i(n) \in [a(n)-k, a(n)+l(a(n))-1]$ and $b_{1-i}(n) \notin \hat{B}_{1-i}$. So, by (46), the sets $\hat{B}_0$, $\hat{B}_1$, $\hat{A}$ are pairwise disjoint. Moreover, $\hat{A} = A_f$ for the computable unbounded shift f defined by $f(n) = n + l(n)$, and $\hat{B}_i \leq_{\text{ibT}} B_i$ by permitting whence (43) holds.

For a proof of (44), note that, by the Splitting Lemma and by (43),

$$deg_{\text{cl}}(\hat{B}_0 \cup \hat{B}_1) = deg_{\text{cl}}(\hat{B}_0) \vee deg_{\text{cl}}(\hat{B}_1)$$
$$\leq deg_{\text{cl}}(B_0) \vee deg_{\text{cl}}(B_1) = deg_{\text{cl}}(A).$$

It remains to show that $A \leq_{\text{cl}} \hat{B}_0 \cup \hat{B}_1$. Since $\hat{A}$ is a computable unbounded shift of A, by the Computable-Shift Lemma it suffices to show that $A \leq_{\text{cl}} \hat{B}_0 \cup \hat{B}_1 \cup \hat{A}$. In fact, since $deg_{\text{cl}}(A) = deg_{\text{cl}}(B_0) \vee deg_{\text{cl}}(B_1)$, it suffices to show that $B_i \leq_{\text{cl}} \hat{B}_0 \cup \hat{B}_1 \cup \hat{A}$ $(i = 0, 1)$. But the latter follows from the fact that if $x = b_i(n)$ enters B_i at stage n then $b_0(n) \leq b_i(n)$ and $b_0(n)$ enters $\hat{B}_0$ at stage n or $b_1(n) \leq b_i(n)$ and $b_1(n)$ enters enters $\hat{B}_1$ at stage n or $a(n) + l(a(n)) \leq b_i(n)$ and $a(n) + l(a(n))$ enters $\hat{A}$ at stage n. So $B_i \leq_{\text{ibT}} \hat{B}_0 \cup \hat{B}_1 \cup \hat{A}$ by permitting. $\square$

Theorem 10.5. *Let $r \in \{\text{ibT}, \text{cl}\}$. The first order theory $\text{Th}(\mathbf{R}_r, \leq)$ of the partial ordering of the c.e. r-degrees is undecidable.*

PROOF. Let $\varphi(x, y)$ express that x is minimal with the property that $0 < x < y$ and there is a z such that $0 < z < y$ and $0 = x \wedge z$ and $y = x \vee z$. (i.e., x is a minimal element of the set of elements of the open interval $(0, y)$ which possess a complement in the closed interval $[0, y]$). Then, given $n \geq 1$, it suffices to give c.e. r-degrees $\mathbf{a}_i$, $\mathbf{b}_j$ and $\mathbf{d}_{i,j}$ $(i, j < n)$ and $\hat{\mathbf{a}}$ and $\hat{\mathbf{b}}$ satisfying the conditions $(i) - (iv)$ of Lemma 10.2 in the partial ordering $(\mathbf{R}_r, \leq)$.

By Lemma 10.1 fix noncomputable c.e. sets A_i, B_j and $D_{i,j}$ $(i, j < n)$ such that (38) to (41) hold. Since, for any c.e. set C and any infinite computable set R, there is a c.e. set $\hat{C} \subseteq R$ which is wtt-equivalent to C, w.l.o.g. we may assume that there are pairwise disjoint, infinite computable sets R_i^A, R_j^B and $R_{i,j}^D$ such that $A_i \subseteq R_i^A$, $B_j \subseteq R_j^B$ and $D_{i,j} \subseteq R_{i,j}^D$ $(i, j < n)$ and such that there is a nondecreasing and unbounded computable function l such that (42) holds for

$$R = \bigcup_{i < n} R_i^A \cup \bigcup_{j < n} R_j^B \cup \bigcup_{i,j < n} R_{i,j}^D.$$

Moreover, since (by (38)) $D_{i,j} \leq_{\text{wtt}} A_i, B_j$, as in the proof of the wtt-ibT-Conversion Lemma, for any sufficiently fast growing computable shift f, $D_{i,j} =_{\text{wtt}} (D_{i,j})_f \leq_{\text{ibT}} A_i, B_j$. So, by choosing f so that $f(R_{i,j}^D) \subseteq R_{i,j}^D$, w.l.o.g. we may assume that

$$\forall i, j < n \ (D_{i,j} \leq_{\text{ibT}} A_i, B_j). \tag{47}$$

Finally, let $A = A_0 \cup \cdots \cup A_{n-1}$ and $B = B_0 \cup \cdots \cup B_{n-1}$.

Then define the desired r-degrees $\mathbf{a}_i$, $\mathbf{b}_j$, $\mathbf{d}_{i,j}$, $\hat{\mathbf{a}}$ and $\hat{\mathbf{b}}$ by $\mathbf{a}_i = deg_r(A_i)$, $\mathbf{b}_j = deg_r(B_j)$, $\mathbf{d}_{i,j} = deg_r(D_{i,j})$, $\hat{\mathbf{a}} = deg_r(A)$ and $\hat{\mathbf{b}} = deg_r(B)$ $(i,j < n)$, and call the given sets the canonical representatives of the thus defined r-degrees.

For verifying conditions $(i) - (iv)$ of Lemma 10.2 we start with some observations.

Since the sets A_i, B_j and $D_{i,j}$ are pairwise disjoint, it follows by the Splitting Lemma that, for any finite collection $S_1, \ldots, S_m$ $(m \geq 1)$ of these sets,

$$deg_r(S_1 \cup \cdots \cup S_m) = deg_r(S_1) \vee \cdots \vee deg_r(S_m).$$

Moreover, since $S_1 \cup \cdots \cup S_m$ is contained in R, it follows that $S_1 \cup \cdots \cup S_m$ (hence $deg_r(S_1 \cup \cdots \cup S_m)$) is strongly scattered. So, for any nonempty collection $\mathbf{S}$ of the degrees $\mathbf{a}_i, \mathbf{b}_j, \mathbf{d}_{i,j}$, the join of $\mathbf{S}$ exists and the join is strongly scattered and represented by the union of the canonical representatives of the members of $\mathbf{S}$. We will tacitly use these facts in the following.

Next we observe that, by Lemmas 8.1 and 8.5, in $(39) - (41)$ we may replace wtt-reducibility by r-reducibility:

$$\forall \, i,j < n \; (deg_r(A_i) \text{ and } deg_r(B_j) \text{ are nontops}), \tag{48}$$

$$\forall \, i,i' < n \; (i \neq i' \Rightarrow (deg_r(A_i), deg_r(A_{i'})) \text{ is an } r\text{-minimal pair}), \tag{49}$$

and

$$\forall \, j,j' < n \; (j \neq j' \Rightarrow (deg_r(B_j), deg_r(B_{j'})) \text{ is an } r\text{-minimal pair}). \tag{50}$$

(So, by (47) above, Lemma 10.1 holds for r-reducibility in place of wtt-reducibility for the sets A_i, B_j and $D_{i,j}$ chosen above.) Moreover, by (49) and (50) and by the Embedding Lemma, for any nonempty α which is strictly contained in $\{0, \ldots, n-1\}$ and for $\overline{\alpha} = \{0, \ldots, n-1\} \setminus \alpha$,

$$\mathbf{0} < deg_r(A_\alpha), deg_r(A_{\overline{\alpha}}) \; \& \; \hat{\mathbf{a}} = deg_r(A_\alpha) \vee deg_r(A_{\overline{\alpha}}) \; \& \; deg_r(A_\alpha) \wedge deg_r(A_{\overline{\alpha}}) = \mathbf{0} \tag{51}$$

and

$$\mathbf{0} < deg_r(B_\alpha), deg_r(B_{\overline{\alpha}}) \; \& \; \hat{\mathbf{b}} = deg_r(B_\alpha) \vee deg_r(B_{\overline{\alpha}}) \; \& \; deg_r(B_\alpha) \wedge deg_r(B_{\overline{\alpha}}) = \mathbf{0} \tag{52}$$

where $A_\alpha = \bigcup_{i \in \alpha} A_i$ and $B_\alpha = \bigcup_{j \in \alpha} B_j$.

We are now ready to establish conditions (i) - (iv) of Lemma 10.2. Condition (i) is immediate by noncomputability of the sets $D_{i,j}$ and by (47); and condition (ii) is immediate by the preceding observations on joins.

For a proof of (iii) fix $i,j < n$ and let $\hat{\mathbf{d}}_{i,j} = \vee_{(i',j') \neq (i,j)} \mathbf{d}_{i',j'}$. Then $\hat{\mathbf{d}}_{i,j}$ is represented by the set $\bigcup_{(i',j') \neq (i,j)} D_{i',j'}$. So, in order to show that $\mathbf{a}_i \wedge \mathbf{b}_j \wedge \hat{\mathbf{d}}_{i,j} = \mathbf{0}$, it suffices to show that any given c.e. set E with

$$E \leq_r A_i, B_j, \bigcup_{(i',j') \neq (i,j)} D_{i',j'}$$

is computable. Now since

$$\bigcup_{(i',j')\neq(i,j)} D_{i',j'} = \bigcup_{i'\neq i,j'<n} D_{i',j'} \cup \bigcup_{j'\neq j} D_{i,j'},$$

by the Splitting Lemma, E can be split into disjoint c.e. sets

$$E_0 \leq_r E, \quad \bigcup_{i'\neq i,j'<n} D_{i',j'} \tag{53}$$

and

$$E_1 \leq_r E, \quad \bigcup_{j'\neq j} D_{i,j'}, \tag{54}$$

and it suffices to show that E_0 and E_1 are computable. This is done as follows. Note that, by (47), $\bigcup_{i'\neq i,j'<n} D_{i',j'} \leq_r A_{\{0,\dots,n-1\}\setminus\{i\}}$ and, by choice of E, $E \leq_r A_i$. So E_0 is computable by (51) and (53). Computability of E_1 follows in a similar way from (52) and (54) by observing that $\bigcup_{j'\neq j} D_{i,j'} \leq_r B_{\{0,\dots,n-1\}\setminus\{j\}}$ and $E \leq_r B_j$.

Finally, for a proof of (iv) it suffices to show that

$$(\mathbf{R}_r, \leq) \vDash \varphi(\mathbf{a}, \hat{\mathbf{a}}) \Leftrightarrow \mathbf{a} \in \{\mathbf{a}_0, \dots, \mathbf{a}_{n-1}\}.$$

(The proof of the corresponding claim for $\mathbf{b}$, $\hat{\mathbf{b}}$, $\mathbf{b}_0, \dots, \mathbf{b}_{n-1}$ in place of $\mathbf{a}$, $\hat{\mathbf{a}}$, $\mathbf{a}_0, \dots, \mathbf{a}_{n-1}$ is symmetric.) Note that, by (51), $\mathbf{0} < \mathbf{a}_i < \hat{\mathbf{a}}$ and $\mathbf{a}_i$ has a complement in $[\mathbf{0}, \hat{\mathbf{a}}]$. So, by definition of φ, it suffices to show that, for any c.e. r-degrees $\mathbf{c}_0$ and $\mathbf{c}_1$ in $(\mathbf{0}, \hat{\mathbf{a}})$ such that $\mathbf{c}_0 \vee \mathbf{c}_1 = \hat{\mathbf{a}}$ and $\mathbf{c}_0 \wedge \mathbf{c}_1 = \mathbf{0}$, there is some $i < n$ such that $\mathbf{a}_i \leq \mathbf{c}_0$.

Since $\hat{\mathbf{a}} = \mathbf{c}_0 \vee \mathbf{c}_1$ and $\hat{\mathbf{a}}$ is strongly scattered, it follows by Lemma 10.4 and the Distributivity Lemma, that, for any c.e. r-degree $\mathbf{x} \leq \hat{\mathbf{a}}$ there are c.e. r-degrees $\mathbf{x}_0 \leq \mathbf{c}_0$ and $\mathbf{x}_1 \leq \mathbf{c}_1$ such that $\mathbf{x} = \mathbf{x}_0 \vee \mathbf{x}_1$. So, in particular, for $i < n$, there are c.e. r-degrees $\mathbf{a}_{i,0} \leq \mathbf{c}_0$ and $\mathbf{a}_{i,1} \leq \mathbf{c}_1$ such that $\mathbf{a}_i = \mathbf{a}_{i,0} \vee \mathbf{a}_{i,1}$. Note that, by the former and by $\mathbf{c}_0 \wedge \mathbf{c}_1 = \mathbf{0}$, $\mathbf{a}_{i,0} \wedge \mathbf{a}_{i,1} = \mathbf{0}$. Since $\mathbf{a}_i$ is a nontop it follows that $\mathbf{a}_{i,0} = \mathbf{a}_i$ or $\mathbf{a}_{i,1} = \mathbf{a}_i$. Now, if the former happens for some $i < n$ then we are done since $\mathbf{a}_{i,0} \leq \mathbf{c}_0$. Otherwise, however, $\mathbf{a}_i \leq \mathbf{c}_1$ for all $i < n$ whence $\hat{\mathbf{a}} = \mathbf{a}_0 \vee \cdots \vee \mathbf{a}_{n-1} \leq \mathbf{c}_1$ contradicting the choice of $\mathbf{c}_1$. So this case cannot occur.

This completes the proof of the theorem.   □

11   Open Problems

It is natural to ask whether the undecidability theorem for the first order theory of the partial orderings of the c.e. ibT- and cl-degrees can be extended to show that these theories are equivalent to true arithmetic. Moreover, our proofs for undecidability and not-$\aleph_0$-categoricity of $\mathrm{Th}(\mathbf{R}_{ibT}, \leq)$ and $\mathrm{Th}(\mathbf{R}_{cl}, \leq)$ (Theorems 9.2 and 10.5) are based on some quite technical results on the c.e. wtt-degrees. Only in case of ibT we could give a fairly simple proof that

Th(R_{ibT}, $\leq$) is not $\aleph_0$-categorical (see Theorem 6.4). This leads to the question whether there are less involved proofs for the other results too. Finally, in contrast to the ibT-degrees where the existence of nontrivial automorphisms can be very easily shown (see Sect. 4), the question whether the partial ordering (R_{cl}, $\leq$) of the c.e. cl-degrees is rigid or not seems to be much more challenging.

Note. The main results of this paper have been presented at the Maltsev Meeting 2010 in Novosibirsk.

References

1. Ambos-Spies, K.: Cupping and noncapping in the r.e. weak truth-table and Turing degrees. Arch. Math. Logik Grundlag. **25**(3–4), 109–126 (1985)
2. Ambos-Spies, K.: On the strongly bounded Turing degrees of simple sets. Logic Comput. Hierarchies Ontos Math. Log. **4**, 23–78 (2014). De Gruyter, Berlin
3. Ambos-Spies, K., Bodewig, P., Fan, Y., Kräling, T.: The partial orderings of the computably enumerable ibT-degrees and cl-degrees are not elementarily equivalent. Ann. Pure Appl. Logic **164**(5), 577–588 (2013)
4. Ambos-Spies, K., Bodewig, Thorsten, P., Liang, Y.: Nondistributivity in the strongly bounded Turing degrees of c.e. sets. (to appear)
5. Ambos-Spies, K., Ding, D., Fan, Y., Merkle, W.: Maximal pairs of computably enumerable sets in the computably Lipschitz degrees. Theory Comput. Syst. **52**(1), 2–27 (2013)
6. Ambos-Spies, K., Soare, R.I.: The recursively enumerable degrees have infinitely many one-types. Ann. Pure Appl. Logic **44**(1–2), 1–23 (1989)
7. Barmpalias, G.: Computably enumerable sets in the solovay and the strong weak truth table degrees. In: Cooper, S.B., Löwe, B., Torenvliet, L. (eds.) CiE 2005. LNCS, vol. 3526, pp. 8–17. Springer, Heidelberg (2005). doi:10.1007/11494645_2
8. Barmpalias, G., Downey, R., Greenberg, N.: Working with strong reducibilities above totally ω-c.e. and array computable degrees. Trans. Amer. Math. Soc. **362**(2), 777–813 (2010)
9. Barmpalias, G., Lewis, A.E.M.: The ibT degrees of computably enumerable sets are not dense. Ann. Pure Appl. Logic **141**(1–2), 51–60 (2006)
10. Bélanger, D.R.: Structures of some strong reducibilities. In: Ambos-Spies, K., Löwe, B., Merkle, W. (eds.) CiE 2009. LNCS, vol. 5635, pp. 21–30. Springer, Heidelberg (2009). doi:10.1007/978-3-642-03073-4_3
11. Cholak, P., Downey, R., Walk, S.: Maximal contiguous degrees. J. Symbolic Logic **67**(1), 409–437 (2002)
12. Day, A.R.: The computable Lipschitz degrees of computably enumerable sets are not dense. Ann. Pure Appl. Logic **161**(12), 1588–1602 (2010)
13. Downey, R.G.: Δ_2^0 degrees and transfer theorems. Illinois J. Math. **31**(3), 419–427 (1987)
14. Downey, R.G., Hirschfeldt, D.R.: Algorithmic Randomness and Complexity. Theory and Applications of Computability. Springer, New York (2010)
15. Downey, R.G., Hirschfeldt, D.R., Forte, G.: Randomness and reductibility. In: Sgall, J., Pultr, A., Kolman, P. (eds.) MFCS 2001. LNCS, vol. 2136, pp. 316–327. Springer, Heidelberg (2001). doi:10.1007/3-540-44683-4_28
16. Downey, R.G., Hirschfeldt, D.R., LaForte, G.: Randomness and reducibility. J. Comput. System Sci. **68**(1), 96–114 (2004)

17. Downey, R.G., Lempp, S.: Contiguity and distributivity in the enumerable Turing degrees. J. Symbolic Logic 62(4), 1215–1240 (1997)
18. Downey, R.G., Stob, M.: Structural interactions of the recursively enumerable T- and w-degrees. Special Issue: Second Southeast Asian Logic Conference, Bangkok (1984). Ann. Pure Appl. Logic 31(2–3), 205–236 (1986)
19. Fan, Y., Lu, H.: Some properties of sw-reducibility. J. Nanjing Univ. (Mathematical Biquarterly) 2, 244–252 (2005)
20. Fan, Y., Yu, L.: Maximal pairs of c.e. reals in the computably Lipschitz degrees. Ann. Pure Appl. Logic 162(5), 357–366 (2011)
21. Jockusch, C.G.: Three easy constructions of recursively enumerable sets. In: Lerman, M., Schmerl, J.H., Soare, R.I. (eds.) Logic Year 1979–80. LNM, vol. 859, pp. 83–91. Springer, Heidelberg (1981). doi:10.1007/BFb0090941
22. Ladner, R.E., Sasso Jr., L.P.: The weak truth table degrees of recursively enumerable sets. Ann. Math. Logic 8(4), 429–448 (1975)
23. Lempp, S., Nies, A.: The undecidability of the Π_4-theory for the r.e. wtt and Turing degrees. J. Symbolic Logic 60(4), 1118–1136 (1995)
24. Lewis, A.E.M., Barmpalias, G.: Random reals and Lipschitz continuity. Math. Structures Comput. Sci. 16(5), 737–749 (2006)
25. André, N.: Computability and Randomness. Oxford University Press, Oxford (2009)
26. Odifreddi, P.G.: Classical Recursion Theory. Studies in Logic and the Foundations of Mathematics, vol. 2. North-Holland Publishing Co., Amsterdam (1999)
27. Soare, R.I.: Recursively Enumerable Sets and Degrees. A study of computable functions and computably generated sets. Perspectives in Mathematical Logic. Springer, Berlin (1987)
28. Soare, R.I.: Computability theory and differential geometry. Bull. Symbolic Logic 10(4), 457–486 (2004)
29. Stob, M.: wtt-degrees and T-degrees of r.e. sets. J. Symbolic Logic 48, 921–930 (1983)
30. Thomason, S.K.: Sublattices of the recursively enumerable degrees. Z. Math. Logik Grundlagen Math. 17, 273–280 (1971)

Permutations of the Integers Induce only the Trivial Automorphism of the Turing Degrees

Bjørn Kjos-Hanssen[✉]

Department of Mathematics, University of Hawai'i at Mānoa, Honolulu, USA
bjoern.kjos-hanssen@hawaii.edu

Abstract. In the 1960s, Clement F. Kent showed that there are continuum many permutations of ω that map computable sets to computable sets. Thus these permutations preserve the bottom Turing degree $\mathbf{0}$. We show that a permutation of ω cannot induce any nontrivial automorphism of the Turing degrees of members of 2^ω, and in fact any permutation that induces the trivial automorphism must be computable.

1 Introduction

Let $\mathscr{D}_T$ denote the set of Turing degrees and let $\leq$ denote its ordering. This article gives a partial answer to the following famous question.

Question 1. Does there exist a nontrivial automorphism of $\mathscr{D}_T$?

Definition 1. A bijection $\pi : \mathscr{D}_T \to \mathscr{D}_T$ is an *automorphism* of $\mathscr{D}_T$ if for all $\mathbf{x}, \mathbf{y} \in \mathscr{D}_T$, $\mathbf{x} \leq \mathbf{y}$ iff $\pi(\mathbf{x}) \leq \pi(\mathbf{y})$. If moreover there exists an $\mathbf{x}$ with $\pi(\mathbf{x}) \neq \mathbf{x}$ then π is *nontrivial*.

Question 1 has a long history. Already in 1977, Jockusch and Solovay [3] showed that each jump-preserving automorphism of the Turing degrees is the identity above $\mathbf{0}^{(4)}$. Nerode and Shore 1980 [8] showed that each automorphism (not necessarily jump-preserving) is equal to the identity on some cone. Slaman and Woodin [11] showed that each automorphism is equal to the identity on the cone above $\mathbf{0}''$. Cooper (around 1999) worked on a construction of a nontrivial automorphism, induced by a continuous function on 2^ω, but that project was not completed and so the problem of existence of a nontrivial automorphism is still open.

Was it ever plausible that a permutation would induce an automorphism? Haught and Slaman [2] used permutations of the integers to obtain automorphisms of the polynomial-time Turing degrees in an ideal (below a fixed set).

This work was partially supported by a grant from the Simons Foundation (#315188 to Bjørn Kjos-Hanssen). This material is based upon work supported by the National Science Foundation under Grant No. 1545707. The author acknowledges the support of the Institut für Informatik at the University of Heidelberg, Germany during the workshop on *Computability and Randomness*, June 15 – July 9, 2015.

© Springer International Publishing AG 2017
A. Day et al. (Eds.): Downey Festschrift, LNCS 10010, pp. 599–607, 2017.
DOI: 10.1007/978-3-319-50062-1_35

Theorem 2 (Haught and Slaman [2]). *There is a permutation of $2^{<\omega}$, or equivalently of ω, that induces a nontrivial automorphism of*

$$(\text{PTIME}^A, \leq_{\text{pT}}).$$

for some A.

Caveat: the automorphism is probably not in the ideal itself.

Our proof below shows, informally speaking, that any ideal, for any reducibility, where the degrees are sufficiently closed under iterated exponential time reductions, will have no nontrivial automorphism induced by a permutation belonging to the ideal. This includes T, wtt, tt, EXPTIME, and ELEMENTARY reducibilities. But the argument only works when the permutation belongs to the ideal.

Our result can be seen as a contrast to the following work of Kent.

Definition 3. $A \subset \omega$ is *cohesive* if for each recursively enumerable set W_e, either $A \cap W_e$ is finite or $A \cap (\omega \setminus W_e)$ is finite.

Theorem 4 (Kent [9, Theorem 12.3.IX], [4,5]). *There exists a permutation f such that*

(i) *for all recursively enumerable B, $f(B)$ and $f^{-1}(B)$ are recursively enumerable (and hence for all recursive A, $f(A)$ and $f^{-1}(A)$ are recursive);*
(ii) *f is not recursive.*

Proof. Kent's permutation is just any permutation of a cohesive set (and the identity off the cohesive set). □

2   Universal Algebra Setup

Definition 5. The *pullback* of $f : \omega \to \omega$ is $f^* : \omega^\omega \to \omega^\omega$ given by

$$f^*(A)(n) = A(f(n)).$$

We often write $F = f^*$. Given a set $S \subseteq \omega$, let $\mathscr{D}_S = S^\omega / \equiv_T$. Thus the elements of $\mathscr{D}_S$ are of the form

$$[g]_S = \{ h \in S^\omega \mid h \equiv_T g \}, \qquad g \in S^\omega.$$

Given $F : S^\omega \to S^\omega$ for which

$$A \equiv_T B \implies F(A) \equiv_T F(B),$$

we may define $F_S : \mathscr{D}_S \to \mathscr{D}_S$ by

$$F_S([A]_S) = [F(A)]_S.$$

If $F = f_S^*$ then we say that F_S and F are both *induced* by f.

In Definition 5 we are mostly interested in the case where f is a bijection, but the definition does not require it.

Lemma 6. *For each $f : \omega \to \omega$ and each $S \subseteq \omega$, the pullback f^* maps S^ω into S^ω.*

Proof

$$A \in S^\omega, n \in \omega \implies f^*(A)(n) = A(f(n)) \in S.$$ □

In light of Lemma 6, we can define:

Definition 7. $f_S^* : \mathscr{D}_S \to \mathscr{D}_S$ is the map given by

$$f_S^*([g]_S) = [f^*(g)]_S.$$

Our main result concerns $\mathscr{D}_S$ with $S = 2 = \{0, 1\}$. The easier corresponding result for $S = \omega$ is Theorem 8.

Theorem 8. *Let $f : \omega \to \omega$ be a bijection and let f^* be its pullback. If f_S^* is an automorphism of $\mathscr{D}_S$ for some infinite computable set S, then f is computable.*

Proof. Let $\eta : \omega \to S$ be a computable bijection between ω and S. Then for all $x \in \omega$,

$$f^*(\eta \circ f^{-1})(x) = (\eta \circ f^{-1})(f(x)) = \eta(f^{-1}(f(x))) = \eta(x).$$

Since $\eta \in S^\omega$ is computable and f_S^* is an automorphism, $\eta \circ f^{-1} \in S^\omega$ must be computable. Hence f is computable.   □

3   Permutations Preserve Randomness

Theorem 9. *If B is $f\text{-}\mu_p$-random, $F = f^*$ and $A = F(B)$ or $A = F^{-1}(B)$, then A is $f\ \mu_p$ random.*

Proof. First note that $f^{-1}\text{-}\mu_p$-randomness is the same as $f\text{-}\mu_p$-randomness since $f \equiv_T f^{-1}$. Thus the result for $A = F^{-1}(B)$ follows from the result for $A = F(B)$. So suppose $A = F(B)$ and A is not $f\text{-}\mu_p$-random. So $A \in \cap_n U_n$ where $\{U_n\}_n$ is an $f\text{-}\mu_p$-ML test. Then

$$B \in \{X \mid F(X) \in \cap_n U_n\} = \cap_n V_n$$

where

$$V_n = \{X \mid F(X) \in U_n\} = F^{-1}(U_n)$$

We claim that V_n is $\Sigma_1^0(f)$ (uniformly in n) and $\mu_p(V_n) = \mu_p(U_n)$. Write $U_n = \cup_k[\sigma_k]$ where the strings σ_k are all incomparable. Then

$$V_n = \cup_k F^{-1}([\sigma_k])$$

and

$$\mu_p[\sigma_k] = \mu_p F^{-1}([\sigma_k])$$

and the $F^{-1}([\sigma_k])$, $k \in \omega$ are still disjoint and clopen. (If we think of $\sigma \in 2^{<\omega}$ as a partial function from ω to 2 then

$$F^{-1}([\sigma]) = \{X \mid F(X) \in [\sigma]\}$$
$$= \{X \mid X(f(n)) = \sigma(n), n < |\sigma|\} = [\{\langle f(n), \sigma(n)\rangle \mid n < |\sigma|\}].$$

Thus $\{V_n\}_n$ is another f-μ_p-ML test, and so B is not f-μ_p-random, which completes the proof.   □

Theorem 10. $\mu_p(\{A : A \geq_T p\}) = 1$, *in fact if A is μ_p-ML-random then A computes p.*

Proof. Kjos-Hanssen [6] showed that each Hippocratic μ_p-random set computes p. In particular, each μ_p-random set computes p.   □

4   Cones Have Small Measure

Definition 11 (Bernoulli measures). For each $n \in \omega$,

$$\mu_p(\{X \in 2^\omega : X(n) = 1\}) = p$$

and $X(0), X(1), X(2), \ldots$ are mutually independent random variables.

Ben Miller proved the following extension of the Lebesgue Density Theorem to Bernoulli measures and beyond [7, Proposition 2.10].

Definition 12. An *ultrametric* space is a metric space with metric d satisfying the strong triangle inequality

$$d(x,y) \leq \max\{d(x,z), d(z,y)\}.$$

Definition 13. A *Polish space* is a separable completely metrizable topological space.

Definition 14. In a metric space, $B(x, \varepsilon) = \{y : d(x,y) < \varepsilon\}$.

Theorem 15 ([7, Proposition 2.10]). *Suppose that X is a Polish ultrametric space, μ is a probability measure on X, and $\mathcal{A} \subseteq X$ is Borel. Then*

$$\lim_{\varepsilon \to 0} \frac{\mu(\mathcal{A} \cap B(x,\varepsilon))}{\mu(B(x,\varepsilon))} = 1$$

for μ-almost every $x \in \mathcal{A}$.

Definition 16. For any measure μ define the conditional measure by

$$\mu(\mathcal{A} \mid \mathcal{B}) = \frac{\mu(\mathcal{A} \cap \mathcal{B})}{\mu(\mathcal{B})}.$$

A measurable set $\mathcal{A}$ has density d at X if

$$\lim_n \mu_p(\mathcal{A} \mid [X \upharpoonright n]) = d.$$

Let $\Xi(\mathcal{A}) = \{X : \mathcal{A} \text{ has density 1 at } X\}$.

Theorem 17 (Lebesgue Density Theorem for μ_p). *For Cantor space with Bernoulli(p) product measure μ_p, the Lebesgue Density Theorem holds:*

$$\lim_{n \to \infty} \frac{\mu_p(\mathcal{A} \cap [x \upharpoonright n])}{\mu_p([x \upharpoonright n])} = 1$$

for μ-almost every $x \in \mathcal{A}$.

If $\mathcal{A}$ is measurable then so is $\Xi(\mathcal{A})$. Furthermore, the measure of the symmetric difference of $\mathcal{A}$ and $\Xi(\mathcal{A})$ is zero, so $\mu(\Xi(\mathcal{A})) = \mu(\mathcal{A})$.

Proof. Consider the ultrametric $d(x, y) = 2^{-\min\{n : x(n) \neq y(n)\}}$. It induces the standard topology on 2^ω. Apply Theorem 15. ☐

Sacks [10] and de Leeuw, Moore, Shannon, and Shapiro [1] showed that each cone in the Turing degrees has measure zero. Here we use Theorem 17 to extend this to μ_p.

Theorem 18. *If $\mu_p(\{X : W_e^X = A\}) > 0$ then A is c.e. in p.*

Proof. Suppose $\mu_p(\{X : W_e^X = A\}) > 0$. Then $S := \{X \mid W_e^X = A\}$ has positive measure, so $\Xi(S)$ has positive measure, and hence by Theorem 15 there is an X such that S has density 1 at X. Thus, there is an n such that $\mu_p(S \mid [X \upharpoonright n]) > \frac{1}{2}$. Let $\sigma = X \upharpoonright n$. We can now enumerate A using p by taking a "vote" among the sets extending σ. More precisely, $n \in A$ iff

$$\mu_p(\{Y : \sigma \prec Y \wedge n \in W_e^Y\}) > \frac{1}{2},$$

and the set of n for which this holds is clearly c.e. in p. ☐

Theorem 19. *Each cone strictly above p has μ_p-measure zero:*

$$\mu_p(\{A : A \geq_T q\}) = 1 \quad \implies \quad q \leq_T p.$$

Proof. If A can compute q then A can enumerate both q and the complement of q. Hence by Theorem 18, q is both c.e. in p and co-c.e. in p; hence $q \leq_T p$. ☐

5   Main Result

We are now ready to prove our main result Theorem 20 that no nontrivial automorphism of the Turing degrees is induced by a permutation of ω.

Theorem 20. *If π is an automorphism of $\mathscr{D}_2$ which is induced by a permutation of ω then $\pi(\mathbf{p}) = \mathbf{p}$ for each $\mathbf{p} \in \mathscr{D}_T$.*

Proof. Fix a permutation $f : \omega \to \omega$ and let $F = f^* \upharpoonright 2^\omega$. Let B be f-μ_p-random. We claim that B computes $F(p)$.

By Theorem 10, for any f-μ_p random A, we have $p \leq_T A$, hence $F(p) \leq_T F(A)$. So it suffices to represent B as $F(A)$.

Now $B = F(F^{-1}(B))$. Let $A = F^{-1}(B)$. By Theorem 9, A is f-μ_p-random. Thus every f-μ_p-random computes $F(p)$.

Thus we have completed the proof of our claim that μ_p-almost every real computes $F(p)$.

By the Sacks / de Leeuw result, Theorem 19, it follows that $F(p) \leq_T p$.

By considering the inverse f^{-1} we also obtain $F^{-1}(p) \leq_T p$ and hence $p \leq_T F(p)$. So $F(p) \equiv_T p$ and F induces the identity automorphism. $\square$

$$F(A + n) \xrightarrow{\;\mathbb{P} \geq 1-\varepsilon\;} \Phi^{A+n}$$

$$\mathbb{P}=1 \Big| \qquad\qquad \Big| \therefore \mathbb{P} \geq 1-2\varepsilon$$

$$F(A - n) \xrightarrow{\;\mathbb{P} \geq 1-\varepsilon\;} \Phi^{A-n}$$

Fig. 1. = means equal and − means a Hamming distance of 1.

6   Computing the Permutation

Theorem 21. *Let $f : \omega \to \omega$ be a permutation. Let $F = f^*$ be its pullback (Definition 5) to 2^ω. If for positive Lebesgue measure many G, $F(G) \leq_T G$, then f is recursive.*

Proof. By the Lebesgue Density Theorem we can get a Φ and a σ such that, if μ_σ denotes conditional probability on σ and $E = \{A : F(A) = \Phi^A\}$, then

$$\mu_\sigma(E) \geq 95\%.$$

For simplicity let us write $p_n(A) = A+n = A \cup \{n\}$ and $m_n(A) = A-n = A \setminus \{n\}$. Then $p_n^{-1}E = \{A : p_n(A) \in E\}$. Note that

$$E \subseteq p_n^{-1}(E) \cup m_n^{-1}(E)$$

and

$$E^c \subseteq p_n^{-1}(E^c) \cup m_n^{-1}(E^c)$$

Then

$$\mu_\sigma(E) \leq \mu_\sigma(p_n^{-1}(E) \cup m_n^{-1}(E)) \leq \mu_\sigma(p_n^{-1}(E)) + \mu_\sigma(m_n^{-1}(E))$$

We now have

$$\mu_\sigma\{A : F(A+n) = \Phi^{A+n}\} \geq 90\%$$

and

$$\mu_\sigma\{A : F(A-n) = \Phi^{A-n}\} \geq 90\%;$$

Indeed, the events $A \in m_n^{-1}(E)$, $A \in p_n^{-1}(E)$ are each independent of the event $n \in A$, so for $n > |\sigma|$,

$$95\% \leq \mu_\sigma(E) = \mu_\sigma(\{A : A \in E \text{ and } (n \in A \text{ or } n \notin A)\})$$
$$= \mu_\sigma(\{A : A \in E \text{ and } n \in A\}) + \mu_\sigma(\{A : A \in E \text{ and } n \notin A\})$$
$$= \mu_\sigma(E \mid n \in A)\,\mu_\sigma(n \in A) + \mu_\sigma(E \mid n \notin A)\,\mu_\sigma(n \notin A)$$
$$= \mu_\sigma(p_n^{-1}(E) \mid n \in A)\,\mu_\sigma(n \in A) + \mu_\sigma(m_n^{-1}(E) \mid n \notin A)\,\mu_\sigma(n \notin A)$$
$$= \frac{1}{2}\left(\mu_\sigma(p_n^{-1}(E) \mid n \in A) + \mu_\sigma(m_n^{-1}(E) \mid n \notin A)\right) = \frac{1}{2}\left(\mu_\sigma(p_n^{-1}(E)) + \mu_\sigma(m_n^{-1}(E))\right),$$

which gives

$$1.9 \leq \mu_\sigma(p_n^{-1}(E)) + \mu_\sigma(m_n^{-1}(E)) \leq 1 + \min\{\mu_\sigma(p_n^{-1}(E)), \mu_\sigma(m_n^{-1}(E))\}.$$

Also $F(A-n)$ and $F(A+n)$ differ in exactly one bit, namely $f^{-1}(n)$, for all A:

$$F(A-n)(b) \neq F(A+n)(b) \iff (A-n)(f(b)) \neq (A+n)(f(b))$$
$$\iff n = f(b) \iff b - f^{-1}(n),$$

that is

$$\{A : (\forall b)(F(A+n)(b) \neq F(A-n)(b) \leftrightarrow b = f^{-1}(n))\} = 2^\omega.$$

See Fig. 1. Let $D_{n,b} = \{A : \Phi^{A+n}(b) \downarrow \neq \Phi^{A-n}(b) \downarrow\}$. For $n > |\sigma|$,

$$\mu_\sigma\left(D_{n,f^{-1}(n)} \setminus \bigcup_{b \neq f^{-1}(n)} D_{n,b}\right) = \mu_\sigma\{A : (\forall b)(A \in D_{n,b} \leftrightarrow b = f^{-1}(n))\} \geq 80\%$$

since

$$\mu_\sigma\{A : \neg(\forall b)(A \in D_{n,b} \quad \leftrightarrow \quad b = f^{-1}(n))\}$$
$$\leq \mu_\sigma(\neg p_n^{-1}(E)) + \mu_\sigma(\neg m_n^{-1}(E)) \leq 10\% + 10\% = 20\%.$$

Therefore, given any n, we can compute $f^{-1}(n)$: enumerate computations until we have found some bit b such that

$$\mu_\sigma D_{n,b} \geq 80\%.$$

Then $b = f^{-1}(n)$.

Thus f^{-1} is computable and hence so is f. □

Theorem 22. *If π is an automorphism of $\mathscr{D}_T$ which is induced by a permutation f of ω then f is recursive.*

Proof. By Theorem 20, $f^*(G) \equiv_T G$ for each $G \in 2^\omega$. By Theorem 21, f is recursive. □

7 Measure-Preserving Homeomorphisms of the Cantor Set

Proposition 23. *A permutation of ω induces a homeomorphism of 2^ω that is μ_p-preserving for each p.*

Proposition 24. *There exist homeomorphisms of 2^ω that are μ_p-preserving for each p, but are not induced by a permutation.*

Proof. Map

$$[1] \mapsto [111] \cup [001] \cup [101] \cup [110]$$

(more generally, any collection of cylinders of strings of length 3 including 2 strings of Hamming weight 2 and 1 of Hamming weight 1).

Another way to express this is that the homeomorphism preserves the fraction of 1 s in a certain sense.

More precisely,

$$100 \mapsto 001,$$
$$101 \mapsto 101,$$
$$110 \mapsto 110,$$
$$111 \mapsto 111.$$

□

Theorem 25. *Suppose φ is a homeomorphism of 2^ω which is μ_p-preserving for all p (it suffices to require this for infinitely many p, or for a single transcendental p). Suppose φ induces an automorphism π of the Turing degrees. Then $\pi = \mathrm{id}$.*

We omit the proof, which is along similar lines to that of Theorem 20. There are other kinds of functions that one may wonder whether induce nontrivial automorphisms of the Turing degrees. We close with an easy example.

Theorem 26. *A polynomial cannot restrict to a homeomorphism of $[0,1]$ inducing a nontrivial automorphism of $\mathscr{D}_T$.*

Proof. A polynomial that maps all computable points to computable points must be computable. This follows from the effectivity in the unisolvence theorem, in which the relevant matrix is the Vandermonde matrix. □

References

1. de Leeuw, K., Moore, E.F., Shannon, C.E., Shapiro, N.: Computability by probabilistic machines. In: Automata Studies, Annals of Mathematics Studies, vol. 34, pp. 183–212. Princeton University Press, Princeton, N.J. (1956)
2. Haught, C.A., Slaman, T.A.: Automorphisms in the PTIME-Turing degrees of recursive sets. Ann. Pure Appl. Logic **84**(1), 139–152 (1997). Fifth Asian Logic Conference, Singapore (1993)
3. Jockusch Jr., C.J., Solovay, R.M.: Fixed points of jump preserving automorphisms of degrees. Israel J. Math. **26**(1), 91–94 (1977)
4. Kent, C.F.: Constructive analogues of the group of permutations of the natural numbers. Trans. Amer. Math. Soc. **104**, 347–362 (1962)
5. Kent, C.F.: Algebraic structure of some groups of recursive permutations. Thesis (Ph.D.). Massachusetts Institute of Technology (1960)
6. Kjos-Hanssen, B.: The probability distribution as a computational resource for randomness testing. J. Log. Anal. **2**, Paper 10, 13 (2010)
7. Miller, B.: The existence of measures of a given cocycle. I. Atomless, ergodic σ-finite measures. Ergodic Theo. Dynam. Syst. **28**(5), 1599–1613 (2008)
8. Nerode, A., Shore, R.A.: Reducibility orderings: theories, definability and automorphisms. Ann. Math. Logic **18**(1), 61–89 (1980)
9. Rogers Jr., H.: Theory of Recursive Functions and Effective Computability. McGraw-Hill Book Co., New York-Toronto, Ont.-London (1967)
10. Sacks, G.E.: Degrees of Unsolvability. Princeton University Press, Princeton, N.J. (1963)
11. Slaman, T.A.: Global properties of the turing degrees and the turing jump. In: Computational Prospects of Infinity. Part I. Tutorials. Lect. Notes Ser. Inst. Math. Sci. Natl. Univ. Singap., vol. 14, pp. 83–101. World Sci. Publ., Hackensack, NJ (2008)

Algorithmic Randomness

On the Reals Which Cannot Be Random

Liang Yu[1(✉)] and Yizheng Zhu[2]

[1] Institute of Mathematics, Nanjing University, 22 Hankou Road,
Nanjing 210093, People's Republic of China
yuliang.nju@gmail.com
[2] Institut für Mathematische Logik, Universität Münster, Münster, Germany
zhuyizheng@gmail.com

Abstract. We investigate which reals can never be L-random. That is
to give a description of the reals which are always belong to some $L[\lambda]$-
null set for any continuous measure λ. Among other things, we prove
that NCR_L is an L-cofinal subset of Q_3 under $ZFC + PD$.

1 Introduction

This paper is inspired by the work of Reimann and Slaman [11,12].

A real x is called *never continuous random* (NCR_1) if there is no continuous
measure λ so that x is Martin-Löf random with respect to λ. In both papers,
a fairly clear description of NCR_1 was given. For example, they proved that
NCR_1 is a subset of the collection of hyperarithmetic reals and contains all the
reals which belong to some countable Π_1^0-set.

Martin-Löf randomness may be a "real" randomness notion from a *computability theorist* point view. In this paper, we investigate L-randomness, the
randomness relative to constructibility, which may be viewed as an "actual"
randomness notion.

The L-randomness notion was introduced by Solovay in his celebrated paper
[16]. A real r is L-random if it does not belong to any Borel null set which has
a Borel code in L. We may generalize this notion to any continuous measure λ
and introduce $L[\lambda]$-randomness. Then a notion of *never L-continuous random*
(NCR_L) can be naturally defined. The target of this paper is to give a description
of NCR_L.

It turns out that NCR_L becomes interesting only under certain large cardinal
assumptions. If people think of that Π_2^1-ness and Σ_3^1 "correspond" to Π_1^0-ness and
Σ_1^1-ness respectively under PD, then many results in [11,12] can be lifted.

We organize the paper as follows: In Sect. 2, we give non self-contained preliminaries for the further reading. In Sect. 3, we investigate NCR_L under certain
fairly weak set theory assumptions (not stronger than the existence of an inaccessible cardinal). In Sect. 4, we give a description of NCR_L under PD.

Yu was partially supported by National Natural Science Fund of China
grant 11322112 and Humboldt foundation. Both authors thank Professor Ambos-
Spies from Heidelberg University and Professor Schindler from University of Münster
for their hospitality.

© Springer International Publishing AG 2017
A. Day et al. (Eds.): Downey Festschrift, LNCS 10010, pp. 611–622, 2017.
DOI: 10.1007/978-3-319-50062-1_36

2   Preliminaries

Since a lot of facts from set theory, recursion theory and algorithmic randomness theory are needed, we feel that it is unlikely to give a self-contained preliminary. We mostly follow standard terminology and notations from the standard references like [3, 5, 15] to make the paper accessible to readers.

We identify an open set in 2^ω with its representation, a subset of $2^{<\omega}$. For a finite string $\sigma \in 2^{<\omega}$, we use $[\sigma]$ to denote the basic open set $\{x \in 2^\omega \mid x \succ \sigma\}$.

First note that if λ is a finite Borel measure, then it is uniquely determined by a measure $\tilde{\lambda}$ over open sets. Throughout this paper, we only consider Borel measures. So they all have standard representations.

Definition 2.1. *For any measure λ over 2^ω, we use $\hat{\lambda} \in \mathbb{Q}^2 \times 2^{<\omega}$ to denote its standard representation $\{(p, q, \sigma) \mid \lambda(\sigma) \in [p, q]\}$.*

From now on, *we identify a Borel measure with its representation.*

Definition 2.2. *A probability measure λ over 2^ω is a Borel measure so that*

(1) $\lambda(2^\omega) = 1$; *and*
(2) *For any $\sigma \in 2^{<\omega}$, $\lambda([\sigma]) = \lambda([\sigma^\frown 0]) + \lambda([\sigma^\frown 1])$.*

Definition 2.3. *A continuous measure λ over 2^ω is a probability Borel measure so that for any real x, $\lambda(\{x\}) = 0$.*

Note that a probability measure λ is continuous if and only if for any n, there is some m so that for any $\sigma \in 2^m$, $\lambda([\sigma]) \le 2^{-n}$. So we have the following result.

Lemma 2.4. *The set $\{\hat{\lambda} \mid \hat{\lambda}$ represents a continuous measure$\}$ is Δ_1^1.*

Definition 2.5. *For any real x and measure λ, a real r is $L[\lambda \oplus x]$-λ-random if for any λ-null Borel set A which has a Borel code in $L[\hat{\lambda} \oplus x]$, $x \notin A$.*

If $x \in L[\hat{\lambda}]$ and r is $L[\lambda \oplus x]$-λ-random, we simply say that r is $L[\lambda]$-random. Further more, if λ is Lesbegue measure and r is $L[\lambda]$-random, then we simply say that r is L-random.

Definition 2.6

$$NCR_L = \{x \mid \text{For any continuous measure } \lambda, x \text{ is not } L[\lambda]\text{-random.}\}.$$

Lemma 2.7. NCR_L *is Π_3^1.*

Proof. $x \in NCR_L$ if and only if for any $\hat{\lambda}$, if $\hat{\lambda}$ represents a continuous measure, then there is a Borel set A having a Borel code in $L[\hat{\lambda}]$ so that $\lambda(A) = 0$ and $x \in A$. By Lemma 2.4 and some well known descriptive set theory result (see [10] or [2]), NCR_L is Π_3^1. $\square$

The following proposition is routine.

Proposition 2.8. *If λ is a continuous measure, then r is $L[\lambda \oplus x]$-λ-random if and only if for any $\mathbf{\Pi}_2^0$-λ-null set A having a Borel code in $L[\hat{\lambda} \oplus x]$, $x \notin A$.*

Fix a real x and continuous measure λ, let $\mathbb{P}_{\lambda,x} = (\mathbf{P}_{\lambda,x}, \leq)$ be a λ-x-Solovay forcing so that

(1) $P \in \mathbf{P}_{\lambda,x}$ if and only if P is a closed non-λ-null set in $L[\hat{\lambda} \oplus x]$; and
(2) For two conditions P_0 and P_1, $P_0 \subseteq P_1$ if and only if $P_0 \leq P_1$.

If $x \in L[\hat{\lambda}]$, then simply use $\mathbb{P}_\lambda$ to denote $\mathbb{P}_{\lambda,\hat{\lambda}}$.

$\mathbb{P}_{\lambda,x}$ has almost all the properties of classical Solovay forcing. For example, it is c.c.c and has the homogeneity property.

The following proposition is obvious.

Proposition 2.9. *Fix a real x and continuous measure λ, a real r is $L[\lambda \oplus x]$-λ-random if and only if r is a $\mathbb{P}_{\lambda,x}$-generic real over $L[\hat{\lambda} \oplus x]$.*

3   Basic Results

In this section, we investigate NCR_L under weak set theoretic hypotheses (not stronger than the existence of an inaccessible cardinal).

The following result can be viewed as a set theoretical version of Demuth's theorem (see [9]).

Theorem 3.1. *For any real x, continuous measure λ, $L[\lambda \oplus x]$-λ-random real r, if $z \in L[\hat{\lambda} \oplus x \oplus r] \setminus L[\hat{\lambda} \oplus x]$, then z is $L[\lambda \oplus x \oplus \rho]$-$\rho$-random with respect to some continuous measure $\rho \in L[\hat{\lambda} \oplus x]$. In particular, if r is L-random and $z \in L[r] \setminus L$, then z is $L[\rho]$-random with respect to some continuous measure $\rho \in L[r]$.*

Proof. Suppose that r is $L[\lambda \oplus x]$-λ-random and $z \in L[\hat{\lambda} \oplus x \oplus r] \setminus L[x \oplus \hat{\lambda}]$. Then there is a condition $P \in \mathbb{P}_{\lambda,x}$ such that $P \Vdash \dot{z} \in 2^\omega$ and $r \in P$. Since $\mathbb{P}_{\lambda,x}$ is c.c.c, there is a sequence of conditions $\{P_n^i \mid i, n \in \omega\} \in L$ below P so that

- $\forall i \forall n \exists j_i (P_n^i \Vdash \dot{x}(\check{n}) = \check{j}_i)$; and
- For all n, $\{P_n^i\}_{i \in \omega}$ is a maximal antichain below P.

Note that for each n, there is only one k_n such that $r \in [P]_n^{k_n}$. Then there is a function $f \in L[\hat{\lambda} \oplus x]$ such that for any $i \neq k_n$, $r \restriction f(\langle i, n \rangle) \notin P_n^i$. Since random forcing is dominated (i.e. $2^{<\omega} \Vdash \forall f \exists g \in L[\hat{\lambda} \oplus \check{x}] \forall n (f(n) \leq g(n))$), there is a function $g \in L[\hat{\lambda} \oplus x]$ such that g dominates f. Hence we may code the sequence $\{P_n^i \mid i, n \in \omega\}$ and the relation $P_n^i \Vdash \dot{z}(\check{n}) = \check{j}_i$ into a single real $t = \{\langle i, n, \sigma, j_i \rangle \mid \sigma \in P_n^i \wedge P_n^i \Vdash \dot{z}(\check{n}) = \check{j}_i\} \in L[\hat{\lambda} \oplus x]$. Now for each n, we $r \oplus t \oplus g$-recursively find an i such that $r \restriction g(\langle i, n \rangle) \in P_n^i$. Then $i = k_n$ as above. Then there is a unique j_{k_n} such that $\langle k_n, n, r \upharpoonright g(\langle k_n, n \rangle)), j_{k_n} \rangle \in t$. So $z(n) = j_{k_n}$. In other words,

$$z = \Psi^{r \oplus t \oplus g}$$

for some Turing functional Ψ. Again since random forcing only adds dominated functions, there is a function $h_0 \in L[\hat\lambda \oplus x]$ that dominates the use function of $\Psi^{r \oplus t \oplus g}$.

By the dominated property again and the fact that $z \notin L[\hat\lambda \oplus x]$, we may assume that there is a non-decreasing function $h_1 \in \omega^\omega \cap L[\lambda \oplus x]$ so that

- $\lim_{n \to \omega} h_1(n) = \infty$; and
- $h_1(0) = 0$; and
- $\forall n (\lambda(\{y \mid z \upharpoonright n = \Psi^{y \oplus t \oplus g} \upharpoonright n\}) \le 2^{-h_1(n)})$.

For any $\tau \in 2^{<\omega}$, let

$$C(\tau) = \{\sigma \mid \sigma \in 2^{h_0(|\tau|)} \wedge \Psi^{\sigma \oplus t \upharpoonright h_0(|\tau|) \oplus g \upharpoonright h_0(|\tau|)}[h_0(|\tau|)] \succeq \tau\}.$$

Inductively define $\rho \in L[\hat\lambda \oplus x]$ as follows:

$$\rho(\emptyset) = 1, \text{ and}$$

$$\rho(\tau^\frown i) = \begin{cases} \lambda(\bigcup_{\sigma \in C(\tau^\frown i)}[\sigma]), & \forall \tau' \preceq \tau (\rho(\tau') \le 2^{-h_1(|\tau'|)}); \\ \frac{\rho(\tau)}{2}, & \text{Otherwise.} \end{cases}$$

Note that for any τ,

$$C(\tau) = C(\tau^\frown 0) \cup C(\tau^\frown 1), \text{ and } C(\tau^\frown 0) \cap C(\tau^\frown 1) = \emptyset.$$

Since λ is a probability measure, for any τ with the property that $\forall \tau' \preceq \tau (\rho(\tau') \le 2^{-h_1(|\tau'|)})$, it must be that $\rho(\tau) = \rho(\tau^\frown 0) + \rho(\tau^\frown 1)$. Then one may easily check that ρ is induces a probability measure. Since the limit of h_1 is infinite and λ is continuous, ρ must be continuous.

Now suppose that $\{U_n\}_{n \in \omega}$ in $L[\hat\lambda \oplus x]$ is a descending sequence of open sets so that $z \in \bigcap_{n \in \omega} U_n$ and $\rho(\bigcap_{n \in \omega} U_n) = 0$. Define a sequence of open sets $\{\hat U_n\}_{n \in \omega}$ so that for any n,

$$\tau \in \hat U_n \text{ iff } \forall \tau' \preceq \tau (\rho(\tau') \le 2^{-h_1(|\tau'|)}).$$

Then $\forall n (\hat U_n \subseteq U_n)$ and $z \in \bigcap_{n \in \omega} \hat U_n$.

Now for every n, let $V_n = \{\sigma \mid \exists \tau \in \hat U_n (\sigma \in C(\tau))\}$. Note that for every n,

$$\lambda(V_n) = \sum_{\exists \tau \in \hat U_n (\sigma \in C(\tau))} \lambda([\sigma]) = \sum_{\tau \in \hat U_n} \rho(\tau) = \rho(\hat U_n) \le \rho(U_n).$$

Moreover since $z \in \bigcap_{n \in \omega} \hat U_n$, by the definition of V_n, we have that $r \in V_n$ for every n.

So $\{V_n\}_{n \in \omega}$ in $L[\hat\lambda \oplus x]$ is a descending sequence of open sets so that $r \in \bigcap_{n \in \omega} V_n$ and $\rho(\bigcap_{n \in \omega} V_n) = 0$. Then r is not $L[\hat\lambda \oplus x]$-λ-random, a contradiction. Hence z must be $L[\lambda \oplus x]$-ρ-random and so $L[\lambda \oplus x \oplus \rho]$-$\rho$-random. $\square$

We use $x \equiv_L y$ to denote $x \in L[y]$ and $y \in L[x]$. Then immediately, we have the following result.

Corollary 3.2. *If $x \in NCR_L$ and $x \equiv_L y$, then $y \in NCR_L$.*

Obviously if $2^\omega \subseteq L$, then $NCR_L = 2^\omega$.

Proposition 3.3

(1) If $NCR_L \neq 2^\omega$, then NCR_L is not Π_2^1.
(2) If $V = L[r]$ for some L-random real r, then NCR_L is Σ_2^1.
(3) If $(\aleph_1)^{L[x]}$ is countable for any real x, then NCR_L is thin; and NCR_L is Σ_2^1 if and only if $NCR_L \subseteq L$.

Proof. (1) If $NCR_L \neq 2^\omega$ and NCR_L is Π_2^1, then $2^\omega \setminus NCR_L$ is a nonempty Σ_2^1-set and so must a contain a real in L, which is a contradiction to the Shoenfield's absolutness.

(2) If $V = L[r]$ for some L-random real r, then by Theorem 3.1, $NCR_L = 2^\omega \cap L$ and so must be Σ_2^1.

(3) Suppose that for any real x, $(\aleph_1)^{L[x]}$ is countable and NCR_L is not thin. Then there is a perfect tree $T \subseteq 2^{<\omega}$ so that $[T] \subseteq NCR_L$. Define a continuous measure ρ "focusing on T" in $L[T]$ as follows:

$$\rho(\emptyset) = 1, \text{ and}$$

$$\rho(\sigma^\frown i) = \begin{cases} \frac{\rho(\sigma)}{2}, & \sigma^\frown i \in T \wedge \sigma^\frown(1-i) \in T; \\ \rho(\sigma), & \sigma^\frown i \in T \wedge \sigma^\frown(1-i) \notin T; \\ 0, & \text{Otherwise}; \end{cases}$$

Then $\rho([T]) = 1$. Since $2^\omega \cap L[T]$ is countable, there must be some $L[T]$-ρ-random real $r \in [T]$, a contradiction.

Now if NCR_L is Σ_2^1, then $NCR_L \subseteq L$ since $2^\omega \cap L$ is the largest Σ_2^1-thin set. If $NCR_L \subseteq L$, then $NCR_L = 2^\omega \cap L$ and so must be Σ_2^1.   □

The following fact gives a plenty of examples of NCR_L.

Proposition 3.4. *Suppose that for any real x, $(\aleph_1)^{L[x]}$ is countable. If $A \subseteq 2^\omega$ is a Π_2^1-thin set, then $A \subsetneq NCR_L$.*

Proof. Suppose that φ is a Π_2^1-formula and x is a real so that $\varphi(x)$ and the set $\{y \mid \varphi(y)\}$ is countable. Suppose that there is a continuous measure ρ so that x is $L[\rho]$-random. Note that $L[\rho \oplus x] \models \varphi(x)$ by the Shoenfield absoluteness. Then $p \Vdash \varphi(\dot{x})$ for some condition $p \in \mathbf{P}_\rho$. Then by the homogeneity of random forcings, for any $L[\rho]$-random real $y \in p$, $L[\rho \oplus y] \models \varphi(y)$. By Shoenfield absoluteness again, $\varphi(y)$ is true. Since there are countably many reals in $L[\rho]$, there must be ρ-conull many $L[\rho]$-random reals. So $\{y \mid \varphi(y)\}$ has a perfect subset, a contradiction.   □

We don't know whether the assumption of Proposition 3.4 can be weakened to a non-large cardinal one. But note that if r is an L-random, then $\{x \mid x \in L[r]$ is L-random$\}$ is a Π_2^1 thin set (see Theorem 3.2.17 in [1]).

We also remark that the union of all Π_2^1-thin sets is a proper subset of NCR_L. Actually Friedman proved [4] that there is a Δ_3^1 real $x \in L$ which does not belong to any Π_2^1-countable set.

4   Under PD

Throughout this section, we assume that ZFC+projective determinacy, PD.

By (3) of Proposition 3.3, NCR_L is a Π_3^1 countable set.

We need the following theorem which can be found in [5] (Exercise 18.6).

Theorem 4.1 (Kunen). *If κ is weakly compact and $|(\kappa^+)^L| = \kappa$, then $0^\sharp$ exists.*

Definition 4.2. *Let $j : 2^\omega \to Ord$ be a function so that $\forall x (j(x) = (\kappa^+)^{L[x]})$ where $\kappa = \aleph_1$[1].*

Lemma 4.3 (Simpson [14]). *The function j has the following property:*

(1) $x \in L[y] \to j(x) \leq j(y)$; *and*
(2) $x \in L[y] \to (j(x) < j(y) \leftrightarrow x^\sharp \in L[y])$.

(1) is obvious. To see (2), for any real x, note that κ is weakly compact in $L[x]$. So if $x \leq_L y$ and $j(x) < j(y)$, then $L[y] \models |(\kappa^+)^{L[x]}| = \kappa$. Then by Theorem 4.1 relative to x, $L[y] \models x^\sharp$ exists. So $x^\sharp \in L[y]$. Another direction of (2) is obvious.

Proposition 4.4. $0^\sharp \in NCR_L$.

Proof. This follows from Proposition 3.4 since $0^\sharp$ is a Π_2^1-singleton.

We give alternative proof that is forcing-free. By Theorem 4.1 again, for any continuous measure $\lambda \not\geq_L 0^\sharp$, $j(\lambda) < j(0^\sharp) \leq j(\lambda \oplus 0^\sharp)$. So by (2) above, $\lambda \oplus 0^\sharp \geq_L \lambda^\sharp$. Let $\kappa = \aleph_1$, then $(\kappa^+)^{L[\lambda]} < (\kappa^+)^{L[\lambda \oplus 0^\sharp]}$. Since random forcing does not collapse cardinals, $0^\sharp$ cannot be $L[\lambda]$-random.   □

The general form of Proposition 4.4 will be proved in Lemma 4.21 and Theorem 4.22, using the covering property for the core model below one Woodin cardinal.

So by Proposition 4.4, and (1) and (3) of Proposition 3.3, we have the following corollary.

Corollary 4.5. NCR_L *is neither Π_2^1 nor Σ_2^1.*

However, NCR_L is not closed under Δ_3^1-equivalence relations. For example, there is an L-random real r Turing below $0^\sharp$. Then the real r must be Δ_3^1.

Let C_3 be the largest countable Π_3^1-set.

The existence of C_3 was proved by Kechris [7]. By the discussion above, $NCR_L \subset C_3$. We will show that NCR_L lives inside the "bottom" of C_3.

Definition 4.6. $Q_3 = \{x \mid \exists \alpha < \omega_1 \forall z (\omega_1^z > \alpha \to x \leq_{\Delta_3^1} z)\}$, *where ω_1^z is the least non-z-recursive ordinal.*

---

[1] The function j was introduced in [14]. We use κ to denote $\aleph_1$ in V to avoid any confusion.

In [8], it was proved that Q_3 is a Π^1_3 countable set which is downward closed under $\leq_{\Delta^1_3}$. One may also relativize the definition of Q_3 to any real x and obtain $Q_3(x)$. Then it induces a reduction $y \leq_{Q_3} x$ iff $y \in Q_3(x)$. Just like in the higher recursion theory, any two reals in C_3 are Q_3-comparable, and C_3 is closed under Q_3-equivalence relation, and every real in the least Q_3-degree above $\mathbf{0}$ is a Π^1_3-singleton.

We choose a representative $y_{0,3}$ as a Q_3-complete real so that every nonempty Σ^1_3 set contains a real recursive in $y_{0,3}$. Note that $y_{0,3}$ is far more complex than the Π^1_3-complete real which actually belongs to Q_3 (see [8]).

Lemma 4.7. *For any real x, there is a real $y \geq_T x$ such that there is a continuous measure $\rho \leq_T y$ so that y is $L[\rho]$-random.*

Proof. For any x, let r be $L[x]$-random.

Let $\rho \leq_T x$ be a continuous measure so that

$$\rho(\emptyset) = 1, \text{ and}$$

$$\rho(\sigma^\frown i) = \begin{cases} \frac{\rho(\sigma)}{2}, & |\sigma| \text{ is odd,} \\ \rho(\sigma), & |\sigma| \text{ is even} \wedge i = x(\frac{|\sigma|}{2}), \\ 0, & \text{Otherwise.} \end{cases}$$

Since r is $L[x]$-random, it is not difficult to see that $y = x \oplus r$ is $L[\rho]$-random. $\square$

Definition 4.8. *Let NCR^T_L be the set of reals r so that there is no continuous measure $\rho \leq_T r$ so that r is $L[\rho]$-random.*

Then NCR^T_L is a Σ^1_2-set and so $NCR^T_L \neq NCR_L$. By Lemma 4.7, $2^\omega \setminus NCR^T_L$ has cofinally many L-degrees and so contains an upper cone of L-degrees.

Proposition 4.9. *Every Sacks generic real g over L belongs to NCR^T_L.*

Proof. For a contradiction, suppose that g is a Sacks generic real over L and g is $L[\rho]$-random with respect to some continuous measure $\rho \in L[r]$. Then $\rho \in L$.

It is well known (see [1]) that for any function $f \in L[r] \cap \omega^\omega$, there is a function $t \in L \cap (\mathcal{P}_{<\omega}(\omega))^\omega$, where $\mathcal{P}_{<\omega}$ is the collection of finite subsets of ω, so that for any n,

- $f(n) \in t(n)$; and
- $|t(n)| < n$.

Since ρ is a continuous measure, there is a function $\hat{h} \in L[g] \cap \omega^\omega$ so that for any n, $\rho([x \upharpoonright \hat{h}(n)]) < 2^{-n}$. By the dominated property of Sacks forcing, there is a function $h \in L \cap \omega^\omega$ that dominates $\hat{h}$.

Then let $f \in L[g]$ be a function so that $f(n) = g \upharpoonright h(n)$. Then let t be as above.

Now for each n, let $V_n = \{\sigma \in 2^{h(n)} \cap t(n) \mid \rho([\sigma]) < 2^{-n}\}$. Then $\{V_n\}_{n \in \omega}$ is a sequence in L so that

- $\forall n(\rho(V_n) \leq n \cdot 2^{-n})$; and
- $g \in \bigcap_{n \in \omega} V_n$.

So g is not $L[\rho]$-random, a contradiction. □

Corollary 4.10. NCR_L^T *contains a perfect subset.*

Actually by the proof above, for any real $x \in P_2$, $NCR_L^T \cap \{y \mid x \in L[y]\}$ contains a perfect subset.

Let

$$D = \{y_0 \mid \forall y(y \geq_T y_0 \rightarrow y \notin NCR_L^T)\}.$$

By PD and Lemma 4.7, D is a nonempty Π_2^1-set and so the Q_3-complete real $y_{0,3} \in D$.

By relativizing the discussion above, we have the following lemma.

Lemma 4.11. *For any real z and real $x \geq_T y_{0,3}^z$, where $y_{0,3}^z$ is the Q_3-jump relative to z, there is a continuous measure $\rho \leq_T x \oplus z$ so that x is $L[z \oplus \rho]$-ρ-random.*

We need the following Posner-Robinson Theorem.

Theorem 4.12 *(Woodin). If $x \notin Q_3$, then there is a real z so that $x \oplus z \geq_T y_{0,3}^z$.*

Woodin's proof remains unpublished, though it is confirmed by him. Hopefully we may figure it out in the near future.

Lemma 4.13. *For any $x \notin Q_3$, there is a real z so that $x \oplus z$ is $L[z \oplus \rho]$-random with respect to some continuous measure $\rho \leq_T x \oplus z$. Further more, x must be $L[\rho_0]$-random with respect to some continuous measure ρ_0.*

Proof. Suppose that $x \notin Q_3$, then by Posner-Robinson Theorem 4.12, there is a real z so that $x \oplus z \geq_T y_{0,3}^z$. By Lemma 4.11, $x \oplus z$ is $L[z \oplus \rho]$-ρ-random with respect to some measure $\rho \leq_T x \oplus z$.

Obviously $x \not\leq_L \rho \oplus z$. Then by Theorem 3.1, x is $L[z \oplus \rho \oplus \rho_0]$-$\rho_0$-random with respect to some continuous measure $\hat{\rho}_0 \in L[z \oplus \hat{\rho} \oplus \hat{\rho}_0]$. In particular, x is $L[\rho_0]$-random. □

In summary, we have the following theorem.

Theorem 4.14. $NCR_L \subseteq Q_3$.

Now we want to know how "large" is NCR_L.

Proposition 4.15. *For every Δ_3^1-real x, there is a Π_2^1-singleton z so that $x \leq_T z$.*

Proof. If x is Δ_3^1, then $\{x\}$ is also Δ_3^1. So by 6E.14 in [10], there is a Π_2^1-set A and a recursive function $f : \omega^\omega \rightarrow 2^\omega$ so that $f(A) = \{x\}$ and f is $1-1$ over A. Obviously there is a $z \in \omega^\omega$ so that $A = \{z\}$. So $x \leq_T z$ and z is a Π_2^1-singleton. □

So by Proposition 3.4, $NCR_L \cap \Delta_3^1$ is L-cofinal in the collection of Δ_3^1-reals.

Definition 4.16. $P_2 = \{x \mid \forall y(j(x) \leq j(y) \to x \in L[y])\}$.

Lemma 4.17

(1) $0 \in P_2$.
(2) $x \in P_2 \wedge z \equiv_L x \implies z \in P_2$.
(3) $x \in P_2 \implies x^\sharp \in P_2$.
(4) $P_2 \subseteq NCR_L$.

Proof. (1) is obvious.

If $z \equiv_L x$, then $j(x) = j(z)$. So (2) is true.

If $x \in P_2$ and $j(x^\sharp) \leq j(y)$, then $j(x) < j(y)$. So $x \in L[y]$. By Lemma 4.3, $x^\sharp \leq_L y$. So (3) is proved.

To see (4), suppose that $x \in P_2 \setminus NCR_L$, then x is $L[\lambda]$-random for some continuous measure λ. Since random forcing preserves all cardinals, we have that $j(x) \leq j(\lambda)$. Since $x \in P_2$, we also have that $x \in L[\hat\lambda]$, which is a contradiction. $\square$

From this point on, we require basic knowledge of inner model theory in the region of one Woodin cardinal. We shall follow the notations in [20]. Briefly speaking, a premouse is a model of the form $\mathcal{M} = J_\alpha^{\vec{E}}$ with certain fine structural properties and coherency properties, where $\vec{E}$ codes a sequence of extenders. $\rho(\mathcal{M})$ denotes the ultimate projectum of $\mathcal{M}$. $o(\mathcal{M})$ denotes the height of $\mathcal{M}$. $\mathcal{M}|\xi$ denotes the initial segment of $\mathcal{M}$ of height ξ, that is, $J_\alpha^{\vec{E}}|\xi = J_\xi^{\vec{E}}$. $\mathcal{M} \trianglelefteq \mathcal{N}$ means $\mathcal{M}$ is an initial segment of $\mathcal{N}$. A normal iteration tree $\mathcal{T}$ on a premouse $\mathcal{M}$ consists of premice $(\mathcal{M}_\alpha : \alpha < \lambda)$, extenders $(E_\alpha : \alpha < \lambda)$, a tree order T on λ, and a set $D \subseteq \lambda$. Here, E_α is an extender on the $\mathcal{M}_\alpha$-sequence, $\mathcal{M}_{\alpha+1}$ is the fine-structural ultrapower of an initial segment of $\mathcal{M}_{T\text{-pred}(\alpha+1)}$ according to E_α, $\alpha < \beta \to \mathrm{lh}(E_\alpha) < \mathrm{lh}(E_\beta)$, D is the set of dropping points. If $\alpha <^T \beta$ and $D \cap (\alpha, \beta]_T = \emptyset$ then we have an iteration map $i_{\alpha\beta} : \mathcal{M}_\alpha \to \mathcal{M}_\beta$.

If $\mathcal{T}$ is an iteration tree of limit length, in order to continue the iteration, we need to find a cofinal branch through $\mathcal{T}$. The branch choice is at the level of one Woodin cardinal is handled in [20, Theorem 6.10]. We denote by $\mathcal{M}(\mathcal{T})$ the common part of $\mathcal{T}$, $\delta(\mathcal{T})$ the sup of lengths of extenders used in $\mathcal{T}$, as in [20, Definition 6.9]. If b is a cofinal branch through $\mathcal{T}$, then $\mathcal{M}_b^{\mathcal{T}}$ is the (not necessarily wellfounded) direct limit of models along b, and $\mathcal{Q}(b, \mathcal{T})$ is the least initial segment of $\mathcal{M}_b^{\mathcal{T}}$ which either projects across $\delta(\mathcal{T})$ or defines a failure of Woodinness of $\delta(\mathcal{T})$ at the next level[2]. By the proof of [20, Corollary 6.14], if b, c are both cofinal branches through $\mathcal{T}$ and $\mathcal{Q}(b, \mathcal{T}) = \mathcal{Q}(c, \mathcal{T})$, then $b = c$.

If there is no inner model with a Woodin cardinal, the core model K exists. K is defined initially by Steel in [19] as an inner model of V_Ω for a measurable cardinal Ω, and later by Jensen-Steel in [6] from ZFC alone. The iteration strategy for K or any mouse $\mathcal{M}$ is simply as follows: if $\mathcal{T}$ is an iteration tree of limit length, choose the unique branch b through $\mathcal{T}$ such that $Q(b, \mathcal{T})$ exists and $Q(b, \mathcal{T}) = J_\xi[\mathcal{M}(\mathcal{T})]$ for some ordinal ξ. Such an iteration strategy is called the

---

[2] This is slightly different from [20, Definition 6.11]. In our situation, b needs not be wellfounded.

L-guided strategy, in the sense that $Q(b, \mathcal{T})$ is an initial segment of $L[\mathcal{M}(\mathcal{T})]$. If N is a model of ZFC and "no inner model with a Woodin cardinal", K^N denotes the core model defined in N.

If $\mathcal{M}$ is a sound premouse projecting to ω, its master code is (modulo arithmetic equivalence) the first order theory of $\mathcal{M}$, coded into a real.

We use M_1 to denote the structure introduced by Steel in [17] which is the least inner model containing a Woodin cardinal. A master code in M_1 is a master code of $\mathrm{M}_1|\alpha$ for some α so that $\rho(\mathrm{M}_1|\alpha) = \omega$. Note that every real in M_1 is recursive in a master code in M_1.

The connection between the theory of M_1 and Q_3-theory was built in [18].

Theorem 4.18 (Steel [18]). $2^\omega \cap \mathrm{M}_1 = Q_3$. *Moreover, if N is a proper class inner model with a Woodin cardinal, then $Q_3 \subseteq N$.*

Definition 4.19. *Let*

$$\tilde{P}_2 = \{x \mid \exists y \in \mathrm{M}_1(x \equiv_L y \wedge y \text{ is a master code in } \mathrm{M}_1)\}.$$

By Theorem 4.18, we have the following result.

Corollary 4.20. *For any real $x \in Q_3$, there is a real $y \in \tilde{P}_2$ so that $x \leq_L y$. Namely $\tilde{P}_2$ is L-cofinal in Q_3.*

Actually $\tilde{P}_2$ is contained in P_2.

Lemma 4.21. $\tilde{P}_2 \subseteq P_2$.

Proof. Suppose $x \in \tilde{P}_2$. Without loss of generality, x is the master code of $\mathrm{M}_1|\alpha$, where $\rho(\mathrm{M}_1|\alpha) = \omega$. Let $\mathcal{M} = \mathrm{M}_1|\alpha + 1$, so that $\rho(\mathcal{M}) = \omega$ and $x \in \mathcal{M}$. Suppose towards a contradiction that for some y, $j(x) \leq j(y)$ but $x \notin L[y]$. By Theorem 4.18, $L[y]$ does not have an inner model with a Woodin cardinal. So $K^{L[y]}$ exists. Let $\kappa = \aleph_1$. Using the fact κ is weakly compact in $L[y]$, by Schimmerling-Steel [13], $(\kappa^+)^{K^{L[y]}} = (\kappa^+)^{L[y]}$. So $x \notin L[y]$ but $(\kappa^+)^{K^{L[y]}} \geq (\kappa^+)^{L[x]}$. We shall derive a contradiction by comparing $K^{L[y]}$ versus $\mathcal{M}$.

The comparison takes place in $L[x, y]$, using L-guided iteration strategies. Both $K^{L[y]}$ and $\mathcal{M}$ are Ord $+1$-iterable in $L[x, y]$. The fact that $x \in \mathcal{M} \setminus K^{L[y]}$ implies that the $K^{L[y]}$-side comes out strictly shorter. Let $(\mathcal{T}, \mathcal{U})$ be the padded normal trees[3] on the $K^{L[y]}$-side and $\mathcal{M}$-side respectively, both of length Ord $+1$. For $\alpha \leq \beta \leq \infty$, let $\mathcal{M}_\alpha^{\mathcal{T}}$ be the α-th model of $\mathcal{T}$ and $i_{\alpha\beta}^{\mathcal{T}}$ (if exists) be the iteration map from $\mathcal{M}_\alpha^{\mathcal{T}}$ to $\mathcal{M}_\beta^{\mathcal{T}}$; similar notations apply to the $\mathcal{U}$-side. Let $\mathcal{P}$ be the last model of $\mathcal{T}$. So the iteration map $i_{0\infty}^{\mathcal{T}} : K^{L[y]} \to \mathcal{P}$ exists, while the main branch of $\mathcal{U}$ drops. Usual arguments (e.g. [19, Sect. 3]) show that there is an $L[x, y]$-definable closed unbounded proper class Γ such that for every $\xi \in \Gamma$,

(1) ξ belongs to the main branches of $\mathcal{T}$ and $\mathcal{U}$,
(2) $i_{0\xi}^{\mathcal{T}}(\xi) = \xi$, $i_{\xi\infty}^{\mathcal{T}} \restriction (\xi + 1) = \mathrm{id}$,

---

[3] i.e. $E_\alpha^{\mathcal{T}}$ might be empty, in which case we do nothing and put $\mathcal{M}_{\alpha+1}^{\mathcal{T}} = \mathcal{M}_\alpha^{\mathcal{T}}$, and similarly for $\mathcal{U}$.

(3) there is no drop on the main branch of $\mathcal{U}$ in the interval $[\xi, \infty)$, i.e., $i^{\mathcal{U}}_{\xi\infty}$ exists,

(4) $i^{\mathcal{U}}_{\xi\infty} \upharpoonright \xi = id$,

(5) if $\bar{\xi} < \xi$, then $o(\mathcal{M}^{\mathcal{U}}_{\bar{\xi}}) < \xi$,

(6) $\mathcal{M}^{\mathcal{T}}_{\xi} | (\xi^+)^{\mathcal{M}^{\mathcal{T}}_{\xi}} \trianglelefteq \mathcal{M}^{\mathcal{U}}_{\xi}$.

Obviously, Γ contains every (x, y)-indiscernible, and in particular, $\kappa \in \Gamma$.

Consider the set A, where $z \in A$ iff z codes $(\mathcal{S}_z, \mathcal{N}_z, \alpha_z)$, $\mathcal{S}_z$ is a countable L-guided normal iteration tree on $\mathcal{M}$, $\mathcal{N}_z$ is the last model of $\mathcal{S}_z$, α_z is an ordinal in $\mathcal{N}_z$. A is $\Sigma^1_2(x)$, equipped with the $\Sigma^1_2(x)$ wellfounded relation $<^*$ defined as follows: $z <^* z'$ iff $\mathcal{S}_z$ is an initial segment of $\mathcal{S}_{z'}$, $\mathcal{N}_z$ is on the branch of $\mathcal{S}_{z'}$ leading from $\mathcal{M}$ to $\mathcal{N}_{z'}$, there is no drop on the branch of $\mathcal{S}_{z'}$ from $\mathcal{N}_z$ to $\mathcal{N}_{z'}$, and letting $k : \mathcal{N}_z \to \mathcal{N}_{z'}$ be the iteration map encoded in $\mathcal{S}_{z'}$, then $k(\alpha_z) > \alpha_{z'}$. Every $\Sigma^1_2(x)$ subset of ω^ω is ω_1-Suslin as witnessed by a tree on $\omega \times \omega_1$ in $L[x]$. By Kunen-Martin, the rank of $<^*$ is smaller than $j(x)$. By definition, $o(\mathcal{M}^{\mathcal{U}}_\kappa)$ is smaller than the rank of $<^*$, hence smaller than $j(x)$.

However, the main branch of $\mathcal{T}$ does not drop, $i^{\mathcal{T}}_{0\kappa}(\kappa) = \kappa$, and $\mathcal{M}^{\mathcal{T}}_\kappa | (\kappa^+)^{\mathcal{M}^{\mathcal{T}}_\kappa} \trianglelefteq \mathcal{M}^{\mathcal{U}}_\kappa$, implying that $(\kappa^+)^{K^{L[y]}} \leq (\kappa^+)^{\mathcal{M}^{\mathcal{T}}_\kappa} \leq o(\mathcal{M}^{\mathcal{U}}_\kappa) < j(x)$, a contradiction. $\qquad\square$

By Lemma 4.21, we have the following theorem.

Theorem 4.22

(1) $\tilde{P}_2 \subseteq NCR_L$.

(2) NCR_L is L-cofinal in Q_3.

(3) NCR_L is not Σ^1_3.

Proof. (1) follows from Lemmas 4.17 and 4.21.

(2) follows from (1) and Corollary 4.20.

For (3), suppose that NCR_L is Σ^1_3. Then by (2), $Q_3 = \{x \mid \exists y \in NCR_L(x \in L[y])\}$. So Q_3 is a Σ^1_3-set, a contradiction. $\qquad\square$

We don't know whether the converse of Lemma 4.21 is true.

Conjecture 4.23. $P_2 = \tilde{P}_2$.

Though a number of results concerning the structure of NCR_L are proved in this paper, the picture of NCR_L still remains vague for us.

Question 4.24. *Assuming $ZFC + PD$, give a clearer description of NCR_L?*

References

1. Bartoszyński, T., Judah, H.: Set Theory: On the Structure of the Real Line. A K Peters Ltd., Wellesley, MA (1995)
2. Chong, C.T., Liang, Y.: Recursion Theory: Computational Aspects of Definability. De Gruyter Series in Logic and Its Applications, vol. 8. De Gruyter, Berlin (2015). With an interview with Gerald E, Sacks

3. Downey, R.G., Hirschfeldt, D.R.: Algorithmic Randomness and Complexity. Theory and Applications of Computability. Springer, New York (2010)
4. Friedman, S.D.: Fine Structure and Class Forcing. De Gruyter Series in Logic and Its Applications. De Gruyter (2000)
5. Jech, T.: Set Theory. Springer Monographs in Mathematics. Springer, Berlin (2003)
6. Jensen, R., Steel, J.: K without the measurable. J. Symbolic Logic **78**(3), 708–734 (2013)
7. Kechris, A.S.: The theory of countable analytical sets. Trans. Amer. Math. Soc. **202**, 259–297 (1975)
8. Kechris, A.S., Martin, D.A., Solovay, R.M.: Introduction to Q-theory. In: Kechris, A.S., Martin, D.A., Moschovakis, Y.N. (eds.) Cabal Seminar 79–81. Lecture Notes in Mathematics, vol. 1019, pp. 199–282. Springer, Berlin (1983)
9. Kučera, A., Nies, A., Porter, C.P.: Demuth's path to randomness. Bull. Symb. Log. **21**(3), 270–305 (2015)
10. Moschovakis, Y.N.: Descriptive Set Theory. Mathematical Surveys and Monographs, vol. 155, 2nd edn. American Mathematical Society, Providence, RI (2009)
11. Reimann, J., Slaman, T.A.: Measures and their random reals. Trans. Amer. Math. Soc. **367**(7), 5081–5097 (2015)
12. Reimann, J., Slaman, T.A.: Randomness for continuous measures (2016)
13. Schimmerling, E., Steel, J.R.: The maximality of the core model. Trans. Amer. Math. Soc. **351**(8), 3119–3141 (1999)
14. Simpson, S.G.: Minimal covers and hyperdegrees. Trans. Amer. Math. Soc. **209**, 45–64 (1975)
15. Soare, R.I.: Recursively Enumerable Sets and Degrees. Springer, Berlin (1987)
16. Solovay, R.M.: A model of set-theory in which every set of reals is Lebesgue measurable. Ann. of Math. (2) **92**, 1–56 (1970)
17. Steel, J.R.: Inner models with many Woodin cardinals. Ann. Pure Appl. Logic **65**(2), 185–209 (1993)
18. Steel, J.R.: Projectively well-ordered inner models. Ann. Pure Appl. Logic **74**(1), 77–104 (1995)
19. Steel, J.R.: The Core Model Iterability Problem. Lecture Notes in Logic, vol. 8. Springer, Berlin (1996)
20. Steel, J.R.: An outline of inner model theory. In: Handbook of Set Theory. vols. 1, 2, 3, pp. 1595–1684. Springer, Dordrecht (2010)

A Note on the Differences of Computably Enumerable Reals

George Barmpalias[1,2(✉)] and Andrew Lewis-Pye[3]

[1] State Key Lab of Computer Science, Institute of Software,
Chinese Academy of Sciences, Beijing, China
barmpalias@gmail.com
[2] School of Mathematics, Statistics and Operations Research,
Victoria University of Wellington, Wellington, New Zealand
[3] Department of Mathematics, Columbia House, London School of Economics,
Houghton Street, London WC2A 2AE, UK
A.Lewis7@lse.ac.uk
http://barmpalias.net, http://aemlewis.co.uk

Abstract. We show that given any non-computable left-c.e. real α there exists a left-c.e. real β such that $\alpha \neq \beta + \gamma$ for all left-c.e. reals and all right-c.e. reals γ. The proof is non-uniform, the dichotomy being whether the given real α is Martin-Löf random or not. It follows that given any universal machine U, there is another universal machine V such that the halting probability Ω_U of U is not a translation of the halting probability Ω_V of V by a left-c.e. real. We do not know if there is a uniform proof of this fact.

1 Introduction

The reals which have a computably enumerable left or right Dedekind cut, also known as c.e. reals, play a ubiquitous role in computable analysis and algorithmic randomness. The differences of c.e. reals, also known as d.c.e. reals, form a field under the usual addition and multiplication, as was demonstrated by Ambos-Spies, Weihrauch, and Zheng [ASWZ00]. Raichev [Rai05] and Ng [Ng06] showed that this field is real-closed. Downey, Wu and Zheng [DWZ04] studied the Turing degrees of d.c.e. reals. Clearly d.c.e. reals are Δ_2^0 since they can be computably approximated. Downey, Wu and Zheng [DWZ04] showed that every real which is truth-table reducible to the halting problem is Turing equivalent to a d.c.e. real. However they also showed that there are Δ_2^0 degrees which do not contain any d.c.e. reals. In this strong sense, d.c.e. reals form a strict subclass of the Δ_2^0 reals.

Barmpalias was supported by the 1000 Talents Program for Young Scholars from the Chinese Government, and the Chinese Academy of Sciences (CAS) President's International Fellowship Initiative No. 2010Y2GB03. Additional support was received by the CAS and the Institute of Software of the CAS. Partial support was also received from a Marsden grant of New Zealand and the China Basic Research Program (973) grant No. 2014CB340302. Lewis-Pye was supported by a Royal Society University Research Fellowship.

© Springer International Publishing AG 2017
A. Day et al. (Eds.): Downey Festschrift, LNCS 10010, pp. 623–632, 2017.
DOI: 10.1007/978-3-319-50062-1_37

Despite this considerable body of work on d.c.e. reals, the following rather basic question does not have an answer in the current literature. Given a non-computable c.e. real α, is there a c.e. real β such that $\alpha - \beta$ is not a c.e. real? The answer is, perhaps unsurprisingly, positive. We say that a real is left-c.e. or right-c.e. if its left or right Dedekind cut respectively is computably enumerable.

Theorem 1.1. *If α is a non-computable left-c.e. real there exists a left-c.e. real β such that $\alpha \neq \beta + \gamma$ for all left-c.e. and all right-c.e. reals γ.*

An interesting aspect of Theorem 1.1 is that its proof depends crucially on the well-developed theory of Martin-Löf random left-c.e. reals, and in particular the methodology developed by Downey, Hirschfeldt and Nies in [DHN02]. The proof is nonuniform and one has to consider separately the case where α is Martin-Löf random and the case where it is not. We do not know if there is a uniform proof of Theorem 1.1, in the sense that from a left-c.e. approximation to a non-computable real α we can compute a left-c.e. approximation to a real β such that $\alpha \neq \beta + \gamma$ for all left-c.e. and all right-c.e. reals γ.

Let us focus on the connection with the theory of Martin-Löf random left-c.e. reals, as it is crucial in both of the two cases. It follows from the work of Downey, Hirschfeldt and Nies [DHN02] that:

> if α, β are left-c.e. reals and α is Martin-Löf random while β is not, then $\alpha - \beta$ is a Martin-Löf random left-c.e. real. $\qquad$ (1.0.1)

This, in particular, means that in Theorem 1.1, α is Martin-Löf random if and only if β is Martin-Löf random. Moreover we can use this fact in order to reduce Theorem 1.1 to the following special case, which we prove in Sect. 3.

Lemma 1.2. *If α is a left-c.e. real which is neither computable nor Martin-Löf random, then there exists a left-c.e. real β (also not Martin-Löf random) such that $\alpha - \beta$ is neither a left-c.e. real nor a right-c.e. real.*

Let us now see how Theorem 1.1 can be derived from this special case. First, assume that the given α is Martin-Löf random. Lemma 1.2 implies the existence of two left-c.e. reals δ_0, δ_1 which are not Martin-Löf random and such that $\delta := \delta_0 - \delta_1$ is neither a left-c.e. nor a right-c.e. real. Indeed, we can start with any non-computable left-c.e. real δ_0 which is not Martin-Löf random (such as the halting problem) and apply Lemma 1.2 in order to get δ_1 with the required properties. Note that δ_1 is necessarily not Martin-Löf random, because otherwise, given that δ_0 is not Martin-Löf random, it would follow from (1.0.1) that $\delta_0 - \delta_1$ would be a right-c.e. real. To establish Theorem 1.1 for this case, we choose $\beta = \alpha + \delta$. First note that $\alpha - \beta$ is not a left-c.e. real or a right-c.e. real, by the choice of δ. Second, $\beta = (\alpha - \delta_1) + \delta_0$ and $\alpha - \delta_1$ is Martin-Löf random by (1.0.1), since α is Martin-Löf random. Then β is a Martin-Löf random left-c.e. real as the sum of a Martin-Löf random left-c.e. real and another left-c.e. real (a result that was originally proved by Demuth [Dem75]). The case of Theorem 1.1 when α is not Martin-Löf random is exactly Lemma 1.2. We note that, as will become apparent in Sect. 3, the proof of this case also makes essential use of (1.0.1).

A subclass of the left-c.e. reals are the characteristic functions of c.e. sets (viewed as binary expansions). These reals were called *strongly left-c.e. reals* by Downey, Hirschfeldt and Nies [DHN02] and are highly non-random reals. It will be clear from the discussion of Sect. 2 that in Theorem 1.1 we cannot (in general) choose the real β to be strongly left-c.e. as in that case, if the given α is Martin-Löf random, then $\alpha - \beta$ is a left-c.e. real. However the following can be proved using standard finite injury methods.

Proposition 1.3 (Properly d.c.e. reals). *There exist strongly left-c.e. reals α, β such that $\alpha - \beta$ is not a left-c.e. real and is not a right-c.e. real.*

We conclude this discussion with a corollary of Theorem 1.1 in terms of halting probabilities. The cumulative work of Solovay [Sol75], Calude, Hertling, Khoussainov and Wang [CHKW01] and Kučera and Slaman [KS01] has shown that the Martin-Löf random left-c.e. reals are exactly the halting probabilities of universal machines. This class remains the same whether we consider prefix-free machines or plain Turing machines. Here we consider Turing machines operating on strings, and given an effective list of all Turing machines (M_e), a Turing machine U is called *universal* if there exists a computable function $e \mapsto \sigma_e$ from numbers to strings such that $U(\sigma_e * \tau) = M_e(\tau)$ for all e and all strings τ. A similar definition applies to universal prefix-free machines, restricted to Turing machines with prefix-free domain.

Halting probabilities, or equivalently Martin-Löf random left-c.e. reals, are all similar in the sense that they all have the same degree with respect to a wide variety of degree structures (see Downey and Hirschfeldt [DH10, Chap. 9]). A number of results have been established, however, which show that halting probabilities may differ in certain ways, depending on the universal machine used. For example, Figueira, Stephan, and Wu [FSW06] showed that for each universal machine U there exists universal machine V such that Ω_U and Ω_V have incomparable truth-table degrees. Their proof consists of considering $\Omega_V = \Omega_U + X$ for a creative set X like the halting problem, and then using the fact from [Ben88, CN97] that no Martin-Löf random real truth-table computes a creative set. Recall that the use of an oracle computation of a set A from a set B is an upper bound (as a function of n) on the largest position in the oracle B queried in the computation of the first n bits of A. Frank Stephan (see [BDG10, Sect. 6]) showed that for each universal machine U there exists universal machine V such that Ω_U cannot compute Ω_V with use $n + c$ for any constant c. Recently Barmpalias and Lewis-Pye have improved the use-bound in this statement to $n + \log n$, while they also showed that Ω_U, Ω_V can be computed from each other with use $n + 2 \log n$, for any universal machines U, V. Along these lines, we can formulate Theorem 1.1 as follows.

Corollary 1.4. *For each universal by adjunction machine U_0 there exists another universal by adjunction machine U_1 such that for all left-c.e. and all right-c.e. reals β we have $\Omega_{U_0} \neq \Omega_{U_1} + \beta$.*

This shows that halting probabilities are not always translations of the halting probability of a fixed universal machine by a left-c.e. or a right-c.e. real.

2 Overview of Martin-Löf random left-c.e. Reals

Some familiarity with the basic concepts of algorithmic information theory and the basic methods of computability theory would be helpful for the reader. For such background we refer to one of the monographs [LV97, DH10, Nie09], where the latter two are more focused on computability theory aspects of algorithmic randomness. The theory of left-c.e. reals has grown into a significant part of modern algorithmic randomness, and is best presented in [DH10, Chaps. 5 and 9]. The present section is an original presentation of some facts regarding Martin-Löf random reals that stem from [Sol75, CHKW01, KS01] and are further elaborated on in [DHN02], which are essential for the proof of Theorem 1.1. Moreover, some of these facts are not given explicitly in the sources above, but can be recovered from the proofs.

The systematic study of Martin-Löf random c.e. reals started with Solovay in [Sol75], who showed that Chaitin's halting probability of a prefix-free machine (a well known Martin-Löf random left-c.e. real) has maximum degree in a degree structure that measures the hardness of approximating left-c.e. reals by increasing sequences of rationals. This result was complemented by the work of Calude, Hertling, Khoussainov and Wang [CHKW01] and Kučera and Slaman [KS01], who showed that these maximally hard to approximate left-c.e. reals are exactly the halting probabilities of universal machines, which also coincide with the Martin-Löf random left-c.e. reals. The degree structure introduced in [Sol75] is now known as the *Solovay degrees of left-c.e. reals* and was extensively studied in [DHN02]. An increasing computable sequence of rationals (α_i) that converges to a real α is called a *left-c.e. approximation to* α, denoted $(\alpha_s) \to \alpha$. The Solovay reducibility $\beta \leq_S \alpha$ between left-c.e. reals α, β can be defined equivalently by any of the following clauses:

(a) there exists a rational q such that $q\alpha - \beta$ is left-c.e.
(b) there exist a rational q and $(\alpha_s) \to \alpha$, $(\beta_s) \to \beta$ such that $\beta - \beta_s < q \cdot (\alpha - \alpha_s)$ for all s;
(c) there exist a rational q and $(\alpha_s) \to \alpha$, $(\beta_s) \to \beta$ such that $\beta_{s+1} - \beta_s < q \cdot (\alpha_{s+1} - \alpha_s)$ for all s.

Note that the set of rationals q for which one of the above clauses holds is upward closed - if the clause holds for the rational q then it also holds for all rationals $q' > q$. Although it is not explicitly stated in [DHN02], it follows from the proofs that when $\beta \leq_S \alpha$, the infimums of the rationals q for which the clauses (a), (b) and (c) hold are equal.

Kučera and Slaman [KS01] proved that:

> if (α_s), (β_s) are left-c.e. approximations to α, β respectively and if α is Martin-Löf random, then $\liminf_s \left[(\alpha - \alpha_s)/(\beta - \beta_s) \right] > 0$. $\qquad$ (2.0.1)

In this sense, Martin-Löf random left-c.e. reals can only have slow left-c.e. approximations, compared to any other left-c.e. real and any left-c.e. approximation to it. Downey, Hirschfeldt and Nies [DHN02] showed that any left-c.e.

approximation to a non-random left-c.e. real is considerably faster than every left-c.e. approximation to any Martin-Löf random real, in the sense that:

if (α_s), (β_s) are left-c.e. approximations to α, β respectively, β is Martin-Löf random and α is not Martin-Löf random, then $\liminf_s \left[(\alpha - \alpha_s)/(\beta - \beta_s) \right] = 0.$ (2.0.2)

Demuth [Dem75] showed that if α, β are left-c.e. reals and at least one of them is Martin-Löf random, then $\alpha + \beta$ is also Martin-Löf random. Downey, Hirschfeldt and Nies [DHN02] proved that the converse also holds, i.e.:

if α, β are left-c.e. reals and $\alpha + \beta$ is Martin-Löf random then at least one of α, β is Martin-Löf random. (2.0.3)

We conclude our overview with a proof of (1.0.1) which is essential for the proof of Theorem 1.1, but which is not stated or proved in [DHN02] (although it follows easily from the arguments in that paper). We need the following fact which was proved in [DHN02] (but stated in a weaker form) and which is also related to the above discussion regarding clauses (a)–(c).

Lemma 2.1 (Downey, Hirschfeldt and Nies [DHN02]). *Suppose that α, β have left-c.e. approximations $(\alpha_s), (\beta_s)$ such that $\forall s \left(\alpha - \alpha_s < q \cdot (\beta - \beta_s) \right)$ for some rational $q > 0$. If $p > q$ is another rational, then there exists a left-c.e. approximation (γ_s) to α such that $\forall s \left(\gamma_{s+1} - \gamma_s < p \cdot (\beta_{s+1} - \beta_s) \right)$.*

Now for (1.0.1), assume that α is Martin-Löf random and β is not Martin-Löf random. By (2.0.2) for each left-c.e. approximation (α_s) to α there exists a left-c.e. approximation (β_s) to β such that $\beta - \beta_s < 2^{-1} \cdot (\alpha - \alpha_s)$ for all s. Then by Lemma 2.1 there exists a left-c.e. approximation $(\gamma_s) \to \beta$ such that $\gamma_{s+1} - \gamma_s < \alpha_{s+1} - \alpha_s$ for all s. This means that the approximation $(\alpha_s - \gamma_s)$ to $\alpha - \beta$ is an increasing left-c.e. approximation. So $\alpha - \beta$ is a left-c.e. real. It remains to show that $\alpha - \beta$ is Martin-Löf random. Since β is not Martin-Löf random, by (2.0.3) it suffices to show that $(\alpha - \beta) + \beta$ is Martin-Löf random. The latter follows from the hypothesis that α is Martin-Löf random.

3   Proof of Lemma 1.2

We can use a priority injury construction. Let $(\gamma_s^i), (\delta_s^i)$ be an effective list of all increasing and decreasing computable sequences of rationals in $(0, 1)$ respectively. Let γ^i be the limit of (γ_s^i) and let δ^i be the limit of (δ_s^i). Given α as in the statement of the lemma, it suffices to construct a left-c.e. real β such that the following conditions are met:

$$\mathcal{L}_i: \ \alpha - \beta \neq \gamma^i \quad \text{and} \quad \mathcal{R}_i: \ \alpha - \beta \neq \delta^i.$$

Given an increasing computable sequence of rationals (α_s) that coverges to α, our construction will define an increasing sequence of rationals (β_s) converging to β such that the above requirements are met. We list the requirements in order of priority as $\mathcal{L}_0, \mathcal{R}_0, \mathcal{L}_1, \ldots$.

Parameters of the construction. Let $\beta_0 = 0$. The strategy for $\mathcal{L}_i$ will use a dynamically defined parameter c_i and the strategy for $\mathcal{R}_i$ will use a similar parameter d_i. Let $c_i[0] = d_i[0] = 0$. We say that stage $s+1$ is $\mathcal{L}_i$-expansionary if $|\alpha_{s+1} - \beta_{s+1} - \gamma^i_{s+1}| < 2^{-c_i[s]}$. Similarly, stage $s+1$ is $\mathcal{R}_i$-expansionary if $|\alpha_{s+1} - \beta_{s+1} - \delta^i_{s+1}| < 2^{-d_i[s]}$. The strategy for each requirement $\mathcal{L}_i$ will define a left-c.e. real β^i, which will be its contribution toward the global left-c.e. real β. Formally, given the approximations (β^i_s) defined by the requirements $\mathcal{L}_i$ respectively, for each s we define:

$$\beta_s = \sum_{i \leq s} \beta^i_s.$$

If $s + 1$ is $\mathcal{L}_i$-expansionary we let $c_i[s + 1] = c_i[s] + 1$, and otherwise we let $c_i[s+1] = c_i[s]$. Similarly, if $s+1$ is $\mathcal{R}_i$-expansionary we let $d_i[s+1] = d_i[s]+1$, and if not we let $d_i[s+1] = d_i[s]$. This completes the definition of the parameters c_i, d_i throughout the stages of the construction. At each stage $s + 1$ the strategy for $\mathcal{R}_i$ imposes an automatic restraint on the strategies for $\mathcal{L}_j$ of lower priority, which prohibits any increase of β by more than $2^{-d_i[s+1]}$. All of the strategies for the $\mathcal{L}_i$ requirements will use a fixed Martin-Löf random left-c.e. real $\eta \in (0,1)$ and an increasing computable rational approximation (η_s) to η. The strategy for each $\mathcal{L}_i$ has an extra parameter q_i, which is updated during the stages s and which dictates the scale at which η is going to affect the growth of (β_s). At stage $s + 1$ we define $q_0[s + 1] = \frac{1}{2}$, and for $i > 0$ we define $q_i[s + 1]$ to be the least of all $2^{-i-d_j[s+1]-1}$ for $j < i$.

Construction of (β_s). At each stage $s + 1$ and each $i \leq s$, if $s + 1$ is $\mathcal{L}_i$-expansionary we define $\beta^i_{s+1} = \beta^i_s + q_i[s + 1] \cdot (\eta_{s+1} - \eta_t)$, where t is the largest $\mathcal{L}_i$-expansionary stage before $s+1$ if there is such, and where $t = 0$ otherwise. If $s + 1$ is not $\mathcal{L}_i$-expansionary, we define $\beta^i_{s+1} = \beta^i_s$. This completes the definition of (β_s).

Verification. First we verify that (β_s) reaches a finite limit β. Let β^i be the limit of β^i_s as $s \to \infty$ and note that for each i:

$$\beta^i \leq 2^{-i-1} \cdot \eta < 2^{-i-1} \quad \text{so} \quad \beta = \sum_i \beta^i < 1.$$

Recall the dynamic definition of $c_i[s]$ and $d_i[s]$. It follows that if $c_i[s]$ reaches a limit, requirement $\mathcal{L}_i$ is met. Similarly, if $d_i[s]$ reaches a limit, requirement $\mathcal{R}_i$ is met. We prove both of these statements by induction. Suppose that the claim holds for all $i < n$. Also let s_0 be a stage such that $c_i[s] = c_i[s_0]$ and $d_i[s] = d_i[s_0]$ for all $i < n$ and all $s > s_0$. Then by definition $q_n[s] = q_n[s_0]$ for all $s > s_0$. Let q_n denote the limit $q_n[s_0]$ of $q_n[s]$ from now on. If $c_n[s]$ does not reach a limit, then there are infinitely many $\mathcal{L}_n$-expansionary stages, which implies that $\alpha - \beta = \gamma^n$. Moreover if (t_j) is a monotone enumeration of the $\mathcal{L}_n$-expansionary stages, then $\beta_{t_{s+1}} - \beta_{t_s} > q_n \cdot (\eta_{t_{s+1}} - \eta_{t_s})$ for all s. Since η is Martin-Löf random, this means that β is also Martin-Löf random. But by hypothesis α is not Martin-Löf random, so $\alpha - \beta$ is a Martin-Löf random right-c.e. real. This contradicts the fact that $\alpha - \beta = \gamma$ since right-c.e. reals which have a left-c.e. approximation are computable. It follows that there are only finitely many $\mathcal{L}_n$-expansionary stages, which implies that $c_n[s]$ reaches a limit. Let $s_1 > s_0$ be a stage such that $c_n[s] = c_n[s_1]$ for all $s > s_1$.

It remains to show that $d_n[s]$ reaches a limit. Towards a contradiction, suppose that this is not the case, so that there are infinitely many $\mathcal{R}_n$-expansionary stages. Then it follows that $\alpha - \beta = \delta^n$. Let (t_k) be a computable enumeration of all $\mathcal{R}_n$-expansionary stages. Then $d_n[t_k] = k$ for all k. For each $i > n$ and each k we have $\beta^i - \beta^i_{t_k} \leq 2^{-i-k-1}$ which means that for k large enough that $t_k > s_1$:

$$\beta - \beta_{t_k} < \sum_{i>n}(\beta^i - \beta^i_{t_k}) \leq \sum_{i>n} 2^{-i-k-1} \leq 2^{-k-1}.$$

This means that β is a computable real. Since $\alpha = \delta^n + \beta$ and δ^n is a right-c.e. real, it follows that α is a right-c.e. real. Since α also a left-c.e. real, it must therefore be computable, contrary to hypothesis. So we may conclude that there are finitely many $\mathcal{R}_n$-expansionary stages, which establishes that $\mathcal{R}_n$ is met and d_n reaches a limit. This concludes the induction step and the proof that the constructed real β meets the requirements $\mathcal{L}_n$ and $\mathcal{R}_n$ for all n.

Remark. The reader may wonder why a uniform argument for Theorem 1.1 might not work, i.e. why we needed to divide into two cases according to whether the given real is Martin-Löf random or not. While it is not easy to explain why some things do not work, the immediate answer is that in a construction such as the argument above, if we did not assume that the given real is not Martin-Löf random or we did not code randomness into the real we construct, we do not see a way to argue that requirements $\mathcal{L}_i$ act only finitely often. More generally, if a direct standard uniform construction worked, in our view we could use it to show that *given a left ce real α we can find a left ce real β such that $2\alpha - \beta$ is not left-c.e. and $\alpha - \beta$ not a right-c.e. real*. However we know that this is not possible by one of the results in [BLP16]. This non-uniformity seems to relate to the non-uniformities in the characterization of the halting probabilities in [Sol75, CHKW01, KS01] that we discussed in Sect. 1. Showing that such non-uniformities are necessary may be an interesting exercise.

4   Proof of Proposition 1.3

We can use a standard priority injury construction. Let $(\gamma^i_s), (\delta^i_s)$ be an effective list of all increasing and decreasing computable sequences of rationals in $(0, 1)$ respectively. Moreover let γ^i be the limit of (γ^i_s) and let δ^i be the limit of (δ^i_s). It suffices to satisfy the following conditions.

$$\mathcal{L}_i: \ \alpha - \beta \neq \gamma^i \quad \text{and} \quad \mathcal{R}_i: \ \alpha - \beta \neq \delta^i$$

Our construction will define increasing sequences $(\alpha_s), (\beta_s)$ of rationals which converge to α, β respectively. Let $\alpha_0 = \beta_0 = 0$. Strategies $\mathcal{L}_i$ will use a parameter

c_i which takes values from $\mathbb{N}^{[2i]}$ (i.e. the even numbers) and strategies $\mathcal{R}_i$ will use a parameter d_i which takes values from $\mathbb{N}^{[2i+1]}$. We say that $\mathcal{L}_i$ requires attention at stage $s + 1$ if either c_i is undefined, or $c_i[s]$ is defined and $|\alpha_s - \beta_s - \gamma_{s+1}^i| < 2^{-c_i[s]-3}$. Similarly we say that $\mathcal{R}_i$ requires attention at stage $s + 1$ if either d_i is undefined, or $d_i[s]$ is defined and $|\alpha_s - \beta_s - \delta_{s+1}^i| < 2^{-d_i[s]-3}$. Strategy $\mathcal{L}_i$ will impose a restraint ℓ_i on α while strategy $\mathcal{R}_i$ will impose a restraint r_i on β. The parameters ℓ_i, r_i will be defined (and possibly redefined) dynamically during the construction, before reaching a limit. We list the requirements in order of priority as $\mathcal{L}_0, \mathcal{R}_0, \mathcal{L}_1, \ldots$ and construct α, β as c.e. sets A, B with characteristic sequences the binary expansions of α, β. In this way, the restraints ℓ_i, r_i will apply to the enumerations into A and B respectively. Note that enumerating a number n into A increases $\alpha - \beta$ by 2^{-n} while enumerating n into B decreases $\alpha - \beta$ by 2^{-n}. Initializing requirement $\mathcal{L}_i$ at stage $s + 1$ means to let $c_i[s+1], \ell_i[s+1]$ be undefined. Similarly, initializing $\mathcal{R}_i$ at stage $s + 1$ means to let $d_i[s+1], r_i[s+1]$ be undefined. If $c_i[s]$ is defined and $\mathcal{L}_i$ is not initialized at stage $s + 1$ then we automatically assume that $c_i[s] = c_i[s+1]$. Similarly, if $d_i[s]$ is defined and $\mathcal{R}_i$ is not initialized at stage $s + 1$ then we automatically assume that $d_i[s] = d_i[s+1]$.

At stage $s+1$ let i be the least number $\leq s$ such that $\mathcal{L}_i$ or $\mathcal{R}_i$ requires attention. If there is no such number, go to the next stage. Otherwise, first assume that $\mathcal{L}_i$ requires attention at stage $s + 1$. If $c_i[s]$ is not defined, let $c_i[s + 1]$ be the least number in $\mathbb{N}^{[2i]}$ which is larger than any value of any parameter defined so far in the construction (in particular larger than all previous values of c_i and larger than any restraint r_j on β which is currently defined). If, on the other hand $c_i[s]$ is defined, then enumerate it into B, define $\ell_i[s + 1] = c_i[s] + 3$ and initialize all $\mathcal{L}_{j+1}, \mathcal{R}_j$ for all $j \geq i$. In this latter case we say that $\mathcal{L}_i$ acts at stage $s + 1$.

Second, assume that $\mathcal{R}_i$ requires attention at stage $s + 1$. If $d_i[s]$ is not defined, let $d_i[s + 1]$ be the least number in $\mathbb{N}^{[2i+1]}$ which is larger than any value of any parameter defined so far in the construction (in particular larger than all previous values of d_i and larger than any restraint ℓ_j on α which is currently defined). If, on the other hand $d_i[s]$ is defined, then enumerate it into A, define $r_i[s + 1] = d_i[s] + 3$ and initialize all $\mathcal{L}_j, \mathcal{R}_j$ for all $j \geq i$. In this latter case we say that $\mathcal{R}_i$ acts at stage $s + 1$.

The construction defined computable enumerations of the sets A, B which in turn define computable non-decreasing rational approximations $(\alpha_s), (\beta_s)$ to the reals α, β. Since A, B are c.e. and no c.e. set is Martin-Löf random, we immediately get that α, β are not random. It remains to show that α, β meet the requirements $\mathcal{L}_i$ and $\mathcal{R}_i$. Note that if $\mathcal{L}_i$ acts at stage $s + 1$ and is not initialized at any later stage, then it will not require attention at any later stage. Indeed, in this case no higher priority requirement will act at later stages, and both $c_i[t]$ and $\ell_i[t + 1]$ remain constant for all $t \geq s$. Let c_i, ℓ_i denote their final values respectively. Since $\mathcal{L}_i$ required attention at stage $s+1$ we have $|\alpha_s - \beta_s - \gamma_{s+1}^i| < 2^{-c_i-3}$. Moreover $\beta_{s+1} - \beta_s = 2^{-c_i}$ and $\alpha_s = \alpha_{s+1}$. So $\alpha_{s+1} - \beta_{s+1} < \gamma_{s+1}^i - 2^{-c_i-1}$ and since $\ell_i = c_i + 3$ we have $\alpha_t - \alpha_{s+1} < 2^{-c_i-2}$ for all $t > s$. Therefore $\alpha_t - \beta_t < \gamma_t^i - 2^{-c_i-2}$ for all $t > s$ and $\mathcal{L}_i$ will not require attention at any stage after s. Moreover we also get that $\alpha - \beta \leq \gamma^i - 2^{-c_i-2}$ which means that in this case condition $\mathcal{L}_i$ is met. We have shown that:

> If $\mathcal{L}_i$ *acts* at stage $s + 1$ and is not initialized at any later stage, then it will not require attention at any later stage and is satisfied. $\qquad$ (4.0.1)

An entirely similar argument shows that:

> If $\mathcal{R}_i$ *acts* at stage $s + 1$ and is not initialized at any later stage, then it will not require attention at any later stage and is satisfied. $\qquad$ (4.0.2)

It remains to use (4.0.1) and (4.0.2) inductively in order to show that $\alpha - \beta$ meets $\mathcal{L}_i, \mathcal{R}_i$ for all i. Note that $\mathcal{L}_0$ cannot be initialized. So c_0 will be defined and remain constant for the rest of the stages. If $\mathcal{L}_0$ never acts, then it does not require attention after the first time that it required (and received) attention. This means that $|\alpha_s - \beta_s - \gamma^0| \geq 2^{-c_i-3}$ for all but finitely many stages s, so $\alpha - \beta \neq \gamma^i$. If it does act at some stage, then by (4.0.1) it is satisfied and never requires attention at any later stage. Now inductively assume that the same is true for all $\mathcal{L}_i, \mathcal{R}_i$, $i < e$. Then consider a stage s_0 after which none of $\mathcal{L}_i, \mathcal{R}_i$, $i < e$ acts or requires attention. Then the same argument shows that $\mathcal{L}_e$ does not act or require attention after a certain stage, and is met. The same argument applies to $\mathcal{R}_e$ through property (4.0.2), and this concludes the induction step. We can conclude that $\alpha - \beta$ meets $\mathcal{L}_i, \mathcal{R}_i$ for all i.

Remark. The referee has pointed out that a proof of Proposition 1.3 may be given without a direct construction. Consider two c.e. sets A, B such that $A - B$ has properly d.c.e. degree, i.e. there is no c.e. set which is Turing equivalent to $A - B$. Such c.e. sets were originally constructed in Cooper [Coo71], and the standard construction gives $B \subseteq A$. Let α, β be the reals in $(0,1)$ whose binary expansions are the characteristic sequences of A, B respectively. Then the binary expansion of $\alpha - \beta$ is the characteristic sequence of $A - B$. If $\alpha - \beta$ had a left-c.e. or a right-c.e. approximation, then $A - B$ would be Turing equivalent to the left or the right Dedekind cut of $\alpha - \beta$ which would be a c.e. set. This would contradict the choice of $A - B$. Hence α, β have the required properties.

References

[ASWZ00] Ambos-Spies, K., Weihrauch, K., Zheng, X.: Weakly computable real numbers. J. Complex. **16**(4), 676–690 (2000)

[BDG10] Barmpalias, G., Downey, R., Greenberg, N.: Working with strong reducibilities above totally ω-c.e. and array computable degrees. Trans. Am. Math. Soc. **362**(2), 777–813 (2010)

[Ben88] Bennett, C.H.: Logical depth and physical complexity. In: The Universal Turing Machine: A Half-Century Survey, pp. 227–257. Oxford University Press (1988)

[BLP16] Barmpalias, G., Lewis-Pye, A.: Differences of halting probabilities. arXiv:1604.00216 [cs.CC], April 2016

[CHKW01] Calude, C., Hertling, P., Khoussainov, B., Wang, Y.: Recursively enumerable reals and Chaitin Ω numbers. Theor. Comput. Sci. **255**(1–2), 125–149 (2001)

[CN97] Calude, C., Nies, A.: Chaitin Ω numbers, strong reducibilities. J. UCS **3**(11), 1162–1166 (1997)

[Coo71] Cooper, S.B.: Degrees of Unsolvability. PhD thesis, Leicester University (1971)

[Dem75] Demuth, O.: On constructive pseudonumbers. Comment. Math. Univ. Carolinae **16**, 315–331 (1975). In Russian

[DH10] Downey, R.G., Hirshfeldt, D.R.: Algorithmic Randomness and Complexity. Springer, New York (2010)

[DHN02] Downey, R.G., Hirschfeldt, D.R., Nies, A.: Randomness, computability, and density. SIAM J. Comput. **31**, 1169–1183 (2002)

[DWZ04] Downey, R.G., Guohua, W., Zheng, X.: Degrees of d.c.e. reals. Math. Log. Q. **50**(4–5), 345–350 (2004)

[FSW06] Figueira, S., Stephan, F., Guohua, W.: Randomness and universal machines. J. Complex. **22**(6), 738–751 (2006)

[KS01] Kučera, A., Slaman, T.: Randomness and recursive enumerability. SIAM J. Comput. **31**(1), 199–211 (2001)

[LV97] Li, M., Vitányi, P.: An Introduction to Kolmogorov Complexity and Its Applications. Graduate Texts in Computer Science, 2nd edn. Springer, New York (1997)

[Ng06] Ng, K.-M.: Some properties of d.c.e. reals and their Degrees. M.Sc. thesis, National University of Singapore (2006)

[Nie09] Nies, A.: Computability and Randomness. Oxford University Press, 444 pp. (2009)

[Rai05] Raichev, A.: Relative randomness and real closed fields. J. Symb. Log. **70**(1), 319–330 (2005)

[Sol75] Solovay, R.: Handwritten manuscript related to Chaitin's work. IBM Thomas J. Watson Research Center, Yorktown Heights, 215 pages (1975)

Effective Bi-immunity and Randomness

Achilles A. Beros, Mushfeq Khan, and Bjørn Kjos-Hanssen[✉]

Department of Mathematics, University of Hawai'i at Mānoa,
Honolulu, HI 96822, USA
{beros,khan}@math.hawaii.edu, bjoern.kjos-hanssen@hawaii.edu

Abstract. We study the relationship between randomness and effective bi-immunity. Greenberg and Miller have shown that for any oracle X, there are arbitrarily slow-growing DNR functions relative to X that compute no Martin-Löf random set. We show that the same holds when Martin-Löf randomness is replaced with effective bi-immunity. It follows that there are sequences of effective Hausdorff dimension 1 that compute no effectively bi-immune set.

We also establish an important difference between the two properties. The class Low(MLR, EBI) of oracles relative to which every Martin-Löf random is effectively bi-immune contains the jump-traceable sets, and is therefore of cardinality continuum.

1 Introduction

Let W_0, W_1, W_2, ... be an effective enumeration of the recursively enumerable (or r.e.) sets of natural numbers. An infinite set A of natural numbers is said to be *immune* if it contains no infinite r.e. subset. It is said to be *effectively immune* when there is a recursive function f such that for all e, if W_e is a subset of A, then $|W_e| \leq f(e)$. The interest in sets whose immunity is effectively witnessed in this manner originally arose in the search for a solution to Post's problem.

The complement of an effectively immune set, if it is r.e., is called *effectively simple*. Smullyan [15] appears to be the first to explicitly isolate the notion, observing that Post's construction [13] of a simple set[1] actually produces an effectively simple set. Sacks [14] established the existence of a simple set that is not effectively simple. Subsequently, Martin [11] showed that every effectively simple set is Turing complete, that is, it computes the halting problem, and thus cannot constitute a solution to Post's problem.

A key result that establishes the significance of the notion of effective immunity outside the context of the co-r.e. sets is a theorem by Jockusch [6] that says that the Turing degrees of the effectively immune sets coincide with those of the diagonally nonrecursive (or DNR) functions. Recently, Jockusch and Lewis [7] have shown that every DNR function computes a *bi-immune* set, i.e., one

---

This work was partially supported by a grant from the Simons Foundation (#315188 to Bjørn Kjos-Hanssen). This material is based upon work supported by the National Science Foundation under Grant No. 1545707.

[1] A simple set is an r.e. set whose complement is immune.

© Springer International Publishing AG 2017
A. Day et al. (Eds.): Downey Festschrift, LNCS 10010, pp. 633–643, 2017.
DOI: 10.1007/978-3-319-50062-1_38

such that both it and its complement are immune. They left open the question of whether the result could be extended to show that every DNR function computes an *effectively bi-immune set*.

Definition 1.1. A set X is *effectively bi-immune* (or EBI) if X and its complement, $\bar{X}$, are both effectively immune. If f is a recursive function that witnesses the effective immunity of both X and $\bar{X}$, we say it is *effectively bi-immune via* f, or f-EBI.

The first author has provided a negative answer [3] to Jockusch and Lewis's question. To summarize: every DNR function computes an effectively immune set (in fact, of the same Turing degree), and a bi-immune set, but not every DNR function computes an EBI set. In Sect. 2, we provide a short proof of the main result from [3] that builds on previous work by Ambos-Spies, Kjos-Hanssen, Lempp, and Slaman [1].

Every Martin-Löf random set is EBI. How close are these properties? Greenberg and Miller [5] have shown that there are sets of effective Hausdorff dimension 1 that compute no Martin-Löf random set. The main result of Sect. 3 shows that there are sets of the former type that compute no EBI set, and is a possible strengthening of the Greenberg-Miller result.

It is not known whether every EBI set computes a Martin-Löf random set. However, existing results imply that the Turing degrees of these two classes do not coincide. Barmpalias, Lewis, and Ng [2] have shown that every PA degree is the join of two Martin-Löf random degrees. The join of two EBI sets is easily seen to be EBI, and so EBI sets are present in every PA degree, in particular, the incomplete ones. Such a degree cannot contain a Martin-Löf random set, by a theorem of Stephan [16].

2   Computing Recursively Bounded DNR Functions

The following has been obtained independently by Sanjay Jain and Ludovic Patey:

Theorem 2.1. *Every EBI set uniformly computes a recursively bounded DNR function. Moreover, a recursive bound for the DNR function can be obtained uniformly from a witnessing function for the EBI set.*

Proof. Let γ be any recursive bijection from ω to the collection of finite subsets of ω, and for an infinite set $Y \subseteq \omega$, let Y_n denote the set consisting of the first n elements of Y.

Suppose X is effectively bi-immune via f. Let h be a recursive function such that for all n,

$$W_{h(n)} = \begin{cases} \gamma(\varphi_n(n)), & \text{if } \varphi_n(n) \downarrow \\ \emptyset, & \text{otherwise.} \end{cases}$$

Now let $g(n) = \gamma^{-1}(X_{f(h(n))+1})$ and let $\bar{g}(n) = \gamma^{-1}(\bar{X}_{f(h(n))+1})$. We claim that both g and $\bar{g}$ are DNR. Suppose that for some e, $\varphi_e(e) = g(e)$.

Then $X_{f(h(e))+1} = W_{h(e)}$. But then $W_{h(e)} \subset X$ and $|W_{h(e)}| > f(h(e))$, a contradiction. The argument for $\bar{g}$ is identical.

Finally, let $\tilde{g} = \min(g, \bar{g})$. Clearly, $\tilde{g}$ is DNR and recursive in X. Given any $n \in \omega$, the largest elements in $X_{f(h(n))+1}$ and $\bar{X}_{f(h(n))+1}$ cannot both be larger than $2f(h(n)) + 1$. Thus, letting

$$\pi(n) = \max\{\gamma^{-1}(D) : \max(D) \leq 2f(h(n)) + 1\},$$

we have $\tilde{g}(n) \leq \pi(n)$.   □

Ambos-Spies et al. [1] have shown that there is a DNR function that computes no recursively bounded DNR function, and so we reprove the main result of [3]:

Corollary 2.2. *There is a* DNR *function that computes no* EBI *set.*

It is worth noting that the construction in [3] achieves significantly more than was claimed in that paper. It partially relativizes, and Turing reduction can be replaced with recursive enumeration:

Theorem 2.3 (Beros). *For any set A, there is a function f that is* DNR *relative to A, such that no* EBI *set is r.e. in f.*

3   Slow-Growing DNR Functions

In the language of mass problems, Theorem 2.1 says that the problem of computing a recursively bounded DNR is strongly (or Medvedev) reducible to that of computing an EBI set. One might wonder if the reverse is true, that is, if the two mass problems can be shown to be equivalent. Failing that, one might hope to show that sufficiently slow-growing DNR functions suffice. More precisely, perhaps there is a slow enough recursive bound g such that all g-bounded DNR functions compute EBI sets. Khan and Miller have shown [9] that by varying g, one can obtain a proper hierarchy of mass problems of recursively bounded DNR functions. Our main result in this section settles these questions.

Definition 3.1. An *order function* is a recursive, unbounded and nondecreasing function from ω to $\omega \backslash \{0, 1\}$.

Theorem 3.2. *For each order function g, for each oracle X, there is a g-bounded function f that is* DNR *relative to X and that computes no* EBI *set.*

In other words, there are arbitrarily slow-growing DNR functions relative to any oracle that compute no effectively bi-immune set. On the other hand, sufficiently slow-growing DNR functions are known to compute sets of effective Hausdorff dimension 1:

Theorem 3.3 (Greenberg and Miller [5]). *There is an order function h such that every h-bounded* DNR *function computes a set of effective Hausdorff dimension 1.*

Together, these theorems imply the following:

Corollary 3.4. *There is a real of effective Hausdorff dimension 1 that computes no EBI set.*

In order to prove Theorem 3.2, we force with bushy trees.

3.1 Definitions and Combinatorial Lemmas

The following definitions can also be found in [5,9].

Definition 3.5. Given $\sigma \in \omega^{<\omega}$, we say that a tree $T \subseteq \omega^{<\omega}$ is *n-bushy above* σ if every element of T is comparable with σ, and for every $\tau \in T$ that extends σ and is not a leaf of T, τ has at least n immediate extensions in T. We refer to σ as the *stem* of T.

Definition 3.6. Given $\sigma \in \omega^{<\omega}$, we say that a set $B \subseteq \omega^{<\omega}$ is *n-big above* σ if there is a finite n-bushy tree T above σ such that all its leaves are in B. If B is not n-big above σ then we say that B is *n-small above* σ.

Proofs of the following lemmas can be found in [5,9].

Lemma 3.7 (Smallness preservation property). *Suppose that B and C are subsets of $\omega^{<\omega}$ and that $\sigma \in \omega^{<\omega}$. If B and C are respectively n- and m-small above σ, then $B \cup C$ is $(n+m-1)$-small above σ.*

Lemma 3.8 (Small set closure property). *Suppose that $B \subset \omega^{<\omega}$ is n-small above σ. Let $C = \{\tau \in \omega^{<\omega} : B \text{ is } n-big \text{ above } \tau\}$. Then C is n-small above σ. Moreover C is n-closed, meaning that if C is n-big above a string ρ, then $\rho \in C$.*

3.2 Proof of Theorem 3.2

For an order function g, let $g^{<\omega}$ denote the set of strings in $\omega^{<\omega}$ whose entries are pointwise bounded by g. We define g^{ω} analogously.

We work entirely in $g^{<\omega}$, forcing with conditions of the form (σ, B), where $\sigma \in g^{<\omega}$ and $B \subset g^{<\omega}$ and B is $g(|\sigma|)$-small above σ. A condition (σ, B) *extends* another condition (τ, C) if $\sigma \succeq \tau$ and $C \subseteq B$. Let $\mathbb{P}$ denote this partial order of conditions. Let $[\sigma]$ denote the elements of g^{ω} that extend σ, and let $[B]^{\prec}$ denote the set of elements of g^{ω} that extend an element of B.

For a functional Γ and a recursive function q, let $\mathcal{H}_{\Gamma,q}$ be the set of all conditions (σ, B) such that if $f \in [\sigma] \setminus [B]^{\prec}$, then Γ^f is not effectively bi-immune as witnessed by the function q.

We assume that for all $f \in g^{\omega}$, for any functional Γ, the domain of Γ^f is an initial segment of ω.

Lemma 3.9. $\mathcal{H}_{\Gamma,q}$ *is dense in* $\mathbb{P}$.

Proof. Let (σ, B) be any condition and suppose B is k-small (and k-closed) above σ. By suitably extending σ, we may assume that $g(|\sigma|) \geq 8k$.

Suppose first that there is a $\tau \notin B$ extending σ such that for some m,

$$C_m = \{\rho : \Gamma^\rho(m) \downarrow\}$$

is $7k$-small above τ. Then $(\tau, B \cup C_m)$ is a condition extending (σ, B), and for every $f \in [\sigma] \backslash [B]^{\prec}$, Γ^f is not total. So we assume from now on that for every $\tau \notin B$ extending σ and every $m \in \omega$, C_m is $7k$-big above τ.

It now follows that for every $\tau \notin B$ extending σ, there is an infinite exactly $6k$-bushy tree T_τ without leaves above τ such that for every $f \in [T_\tau]$, Γ^f is total: Let S_0 consist of τ and its initial segments. Next, suppose we have already constructed a finite tree S_n that is exactly $6k$-bushy above τ and such that for each leaf ρ of this tree, $\rho \notin B$ and Γ^ρ is defined up to $n - 1$. By our assumption above, C_n is $7k$-big above each leaf, so $C_n \backslash B$ is $6k$-big above each leaf by Lemma 3.7. For a leaf ρ of S_n, let A_ρ be a finite exactly $6k$-bushy tree above τ with leaves in $C_n \backslash B$. We construct S_{n+1} by appending A_ρ to each leaf ρ of S_n. Finally, let $T_\tau = \bigcup_{n \in \omega} S_n$.

Definition 3.10. Let τ be any extension of σ that is not in B. We say τ *admits fusion* if for infinitely many $m \in \omega$, for some $i \in \{0, 1\}$,

$$\Delta_{\tau, m, i} = \{\rho \in T_\tau : \Gamma^\rho(m) = i\}$$

is $4k$-big above τ.

Claim 3.11. *If τ admits fusion, then there is a subtree T' of T_τ which is $2k$-bushy above τ and for infinitely many $m \in \omega$ there is an $i \in \{0, 1\}$ with $\Gamma^f(m) = i$ for all $f \in [T']$.*

Proof. Let $I_0 \subseteq \omega$ be such that for all $l \in I_0$, either $\Delta_{\tau, l, 0}$ or $\Delta_{\tau, l, 1}$ is $4k$-big above τ, and let Δ_l denote whichever one is. Let S_0 consist of τ and its initial segments, and note that S_0 is $2k$-bushy above τ.

Next, suppose that we have constructed a finite tree $S_k \subseteq T_\tau$, $2k$-bushy above τ, and a subset I_k of ω such that:

(1) There are $n_0 < n_1 < \cdots < n_{k-1}$ such that for each $i < k$, $\Gamma^\rho(n_i)$ is constant as ρ ranges over the leaves of S_k.
(2) For all $l \in I_k$, there is a tree which is $4k$-bushy above τ and contains S_k, whose leaves are in Δ_l.

Let n be the least element in I_k greater than n_{k-1}, and let C be a finite $4k$-bushy tree above τ containing S_k whose leaves are in Δ_n. Now for any $l > n$ in I_k, if F_l is any $4k$-bushy tree above τ containing S_k with leaves in Δ_l, then $F_l \cap C$ is a $2k$-bushy tree above τ that contains S_k. To see this, let $\rho \in F_l \cap C$. If ρ has an immediate extension in $F_l \cap C$, then ρ has $4k$ many extensions in each of F_l and C. But T_τ is exactly $6k$-bushy above τ, so at least $2k$ of these must be in $F_l \cap C$.

It follows from the pigeonhole principle (note that C is finite) that there is an infinite subset I_{k+1} of I_k, such that for all $l \in I_{k+1}$, there are $4k$-bushy subtrees above τ with leaves in Δ_l that intersect C in the *same* $2k$-bushy subtree S_{k+1} above τ that contains S_k.

This completes the definition of the sequence $\langle S_k \rangle_{k \in \omega}$. Let $T' = \bigcup_{k \in \omega} S_k$. Then T' is as desired. □

Case 1: Some $\tau \succeq \sigma$ admits fusion. We begin by extending σ to τ obtaining the condition (τ, B) (note that τ is by definition not in B). Claim 3.11 implies that, uniformly in k, we can find a finite $2k$-bushy tree R_k above τ such that for at least k distinct inputs m, $\Gamma^\rho(m)$ is constant as ρ ranges over the leaves of R_k.

For $i \in \{0,1\}$, let W_{e_i} be the r.e. set defined as follows: If $m = \max(q(e_0), q(e_1))^2$, let

$$W_{e_i} = \{n : \Gamma^\rho(n) = i \text{ for each leaf } \rho \text{ of } R_{2m+1}\}.$$

It must now be the case that for some $i \in \{0,1\}$, $|W_{e_i}| > m$. Suppose $i = 0$ (the argument for the other case is symmetric). Let $\rho \succeq \tau$ be a string in $R_{2m+1} \backslash B$ (note that B is k-small above τ). Then (ρ, B) is a condition, and for all $f \in [\rho] \backslash [B]^{\prec}$, if Γ^f is total, then W_{e_0} is contained in its complement.

Case 2: No extension of σ admits fusion. This means that for every extension τ of σ such that $\tau \notin B$, there is an $m_\tau \in \omega$ such that for all $l \geq m_\tau$, both $\Delta_{\tau, l, 0}$ and $\Delta_{\tau, l, 1}$ are $4k$-small above τ. Recall that T_τ is exactly $6k$-bushy above τ. Therefore, $\Delta_{\tau, l, 0} \cup \Delta_{\tau, l, 1}$ is $6k$-big above τ. By Lemma 3.7, if one of these sets is $2k$-small above τ, the other is $4k$-big, so both must be $2k$-big above τ.

Let S_0 consist of σ and its initial segments. Note that no leaf of S_0 is in B.

Proceeding by induction, suppose we have constructed a finite k-bushy tree $S_k \subseteq T_\sigma$ above σ with the following properties:

(1) There are $n_0 < n_1 < \cdots < n_{k-1}$ such that for every leaf ρ of S_k, $\Gamma^\rho(n_i) = 0$ for each $i < k$.
(2) None of the leaves of S_k is in B.

Let $n_k = \max\{m_\tau : \tau \text{ a leaf of } S_k\} + 1$. By the observation above, for each leaf τ of S_k, $\Delta_{\tau, n_k, 0}$ is $2k$-big above τ, so $\Delta_{\tau, n_k, 0} \backslash B$ is k-big above τ. Let F_τ be a finite k-bushy tree above τ with leaves in $\Delta_{\tau, n_k, 0} \backslash B$, and let S_{k+1} be obtained from S_k by extending each leaf τ of S_k by F_τ.

Finally, let $T' = \bigcup_{k \in \omega} S_k$. Then for all k, for every $f \in [T']$, $\Gamma^f(n_k) = 0$. A strategy similar to the one employed in case 1 now diagonalizes against the pair (Γ, q). This concludes the proof of the lemma. □

To conclude the proof of Theorem 3.2, let B_{DNR^X} be the set of finite strings that cannot be extended to a DNR relative to X and let $\mathcal{G}$ be any filter containing $(\langle\rangle, B_{\mathrm{DNR}^X})$ that meets $\mathcal{H}_{\Gamma, q}$ for each functional Γ and recursive function q. Then $f_{\mathcal{G}}$ is a g-bounded DNR relative to X and does not compute an effectively bi-immune set.

---

[2] We use the recursion theorem here.

4   Traceability and Lowness

There is more than one way to define effective immunity relative to an oracle. We focus on a partial relativization, motivated by the fact that under this definition, a Martin-Löf random set relative to any oracle X will be effectively immune relative to X via the function $h(e) = e + c$ for some $c \in \omega$.

Definition 4.1. An infinite set R is *effectively immune relative to G* if there is a recursive function h such that for all e, if $W_e^G \subseteq R$ then $|W_e^G| \leq h(e)$.

Definition 4.2. A set $G \in$ Low(MLR,EBI) if each MLR set R is EBI relative to G.

Definition 4.3. A recursive enumerable (r.e.) *trace* T is a sequence of sets $T^{[e]} = W_{g(e)}, e \in \omega$ such that $|W_{g(e)}| \leq h(e)$ for all e, where g and h are recursive functions. For a function f, we say that T *traces* f on input n if $f(n) \in T^{[n]}$. A set G is *jump traceable* if there is a r.e. trace T such that for all e, if $\varphi_e^G(e) \downarrow$ then $\varphi_e^G(e) \in T^{[e]}$.

Theorem 4.4 gives a contrast between MLR and EBI.

Theorem 4.4. *Each jump traceable Turing degree is* Low(MLR, EBI).

Proof. Let G be jump traceable via h, and let J^G denote the diagonal partial recursive function relative to G.

We define a recursive function f knowing its index in advance by the recursion theorem. Let φ be the function partial recursive in G that on input e, waits for W_e^G to enumerate at least $f(e) + 1$ elements, and then outputs the natural number that encodes the finite set B_e consisting of the first $f(e) + 1$ of these. Next, let p be a recursive function such that $J^G \circ p = \varphi$. Note that p can be obtained uniformly from an index for f. Now define f so that

$$h(p(e)) \, 2^{-(f(e)+1)} \leq 2^{-e}.$$

We have an r.e. trace $S^{[p(e)]}$ for (the code for) B_e, and there are at most $h(p(e))$ many elements in it. Let $B_e^{(i)}$ denote the ith candidate for B_e if it exists, for $i < h(p(e))$. Then let

$$U_c = \{A : (\exists e > c)(\exists i < h(p(e)))(B_e^{(i)} \downarrow \subseteq A)\}$$

Then

$$\mu(U_c) \leq \sum_{e>c} h(p(e)) \, 2^{-(f(e)+1)} \leq \sum_{e>c} 2^{-e} = 2^{-c}.$$

If A is MLR then there exists c such that for all $e > c$ and i, it is not the case that $B_e^{(i)} \downarrow \subseteq A$. Thus for all $e \geq c$, if $W_e^G \subseteq A$ then W_e^G has size at most $f(e)$. Thus A is EBI relative to G. $\square$

5   Canonical Immunity

It is natural to next consider lowness notions associated with Schnorr randomness. This idea leads us to a new notion of immunity.

A *canonical numbering of the finite sets* is a surjective function $D : \omega \to \{A : A \subseteq \omega$ and A is finite$\}$ such that $\{(e, x) : x \in D(e)\}$ is recursive and the cardinality function $e \mapsto |D(e)|$, or equivalently, $e \mapsto \max D(e)$, is also recursive. We write $D_e = D(e)$.

Definition 5.1. R is *canonically immune* if R is infinite and there is a recursive function h such that for each canonical numbering of the finite sets D_e, $e \in \omega$, we have that for all but finitely many e, if $D_e \subseteq R$ then $|D_e| \le h(e)$.

Theorem 5.2. *Schnorr randoms are canonically immune.*

Proof. Fix a canonical numbering of the finite sets, $\{D_e\}_{e \in \omega}$. Define $U_c = \{X : (\exists e > c)(|D_e| \ge 2e \wedge D_e \subset X)\}$. Since $e \mapsto |D_e|$ is recursive, $\mu(U_c)$ is recursive and bounded by 2^{-c}. Thus, the sequence $\{U_c\}_{c \in \omega}$ is a Schnorr test. If A is a Schnorr random, then $A \in U_c$ for only finitely many $c \in \omega$. We conclude that A is canonically immune. $\qquad\square$

Theorem 5.3. *Each canonically immune set is immune.*

Proof. Suppose A has an infinite recursive subset R. Let h be any recursive function. Let R_n denote the set of the first n elements of R, and let $\{D_e : e \in \omega\}$ be a canonical numbering of the finite sets such that $D_{2n} = R_{h(2n)+1}$ for all $n \in \omega$. For all n, $D_{2n} \subseteq R \subseteq A$ and $|D_{2n}| = h(2n) + 1 > h(2n)$, and so h does not witness the canonical immunity of A. $\qquad\square$

We now show that canonically immune is the "correct" analogue of effectively immune.

Definition 5.4 (Kjos-Hanssen, Merkle, and Stephan [10]). A function is *strongly nonrecursive* (SNR) if it differs from each recursive function on all but finitely many inputs.

Theorem 5.5. *Each canonically immune set computes a strongly nonrecursive function.*

Proof. Let R be canonically immune as witnessed by the recursive function h. Define $f(e)$ to be (a code for) the first $h(2e) + 1$ many elements of R, and note that f is recursive in R. We claim that f is strongly nonrecursive.

Suppose that the recursive function g is infinitely often equal to f. Let $\{D_e : e \in \omega\}$ be any canonical numbering of finite sets such that for all e, D_{2e} is the finite set coded by $g(e)$. We now have that for infinitely many e, D_{2e} is the set consisting of the first $h(2e) + 1$ many elements of R, a contradiction. $\qquad\square$

Interestingly, Theorem 5.5 shows that we can strengthen "$D_e \subseteq R$" to "D_e is an initial segment of R".

Corollary 5.6. *The following are equivalent for an oracle A:*

(1) *A computes a canonically immune set,*
(2) *A computes an SNR function,*
(3) *A computes an infinite subset of a Schnorr random.*

Proof. (1) implies (2) is proved in Theorem 5.5. (2) implies (3) follows from older results: each SNR either is high or computes a DNR [10], hence either computes a Schnorr random [12] or computes an infinite subset of an MLR [4], hence either way computes an infinite subset of a Schnorr random. (3) implies (1) is proved in Theorem 5.2. □

6   A Class Between EI and EBI

Theorem 6.1. *There is a bi-immune set such that it is effectively immune while its complement is not.*

Proof. We build a set A in stages by describing its characteristic function, g.

Stage 0: Define g_0 to be the function with empty domain.

Stage 2e + 1: Define $g_{2e+1} \upharpoonright \text{dom}(g_{2e}) = g_{2e}$. Let $m = min(\mathbb{N}\backslash\text{dom}(g_e))$ and set $g_{2e+1}(m) = 1$. If $|W_e| > 2e + 1$ and there is no $a \in W_e \cap \text{dom}(g_{2e})$ such that $g_{2e}(a) = 0$, pick $x \in W_e\backslash\text{dom}(g_{2e})$ and set $g_{2e+1}(x) = 0$. If W_e is infinite, select a $y \neq x$ such that $y \in W_e\backslash\text{dom}(g_e)$ and set $g_{2e+1}(y) = 1$.

Stage 2e + 2: Define $g_{2e+1} \upharpoonright \text{dom}(g_{2e}) = g_{2e}$. If ϕ_e is total, pick an r.e. set W_a such that $\phi_e(a) < |W_a| < \infty$ and $|W_a| \cap \text{dom}(g_{2e+1}) = \emptyset$. Set $g_{2e+2}(x) = 0$ for all $x \in W_a$.

Notice that there are no more than $2s$ elements x such that $g_s(x) = 1$. So either it is possible to pick an x as in the odd stages, $2e + 1$, whenever $|W_e| > 2e + 1$, or there is already an element of the domain of g_{2e} which is in W_e on which g_{2e} takes the value 0. Let $g = \bigcup_{s\in\mathbb{N}} g_s$. Observe that g is total and $\{0,1\}$-valued. Let A be the set whose characteristic function is g. The effective immunity of A is witnessed by $f(x) = 2x + 1$ and $\overline{A}$ is clearly immune, however, for any total function h there is an r.e. set W_a such that $h(a) < |W_a|$ and $W_a \subseteq \overline{A}$. Thus, $\overline{A}$ is not effectively immune. □

7   Boldface Complexity

Theorem 7.1. *Let f be a recursive function. The class of reals that are effectively immune via f is closed.*

Proof. Suppose that A is not effectively immune via f. Then there is a e such that W_e is a subset of A and $|W_e| > f(e)$. If W_e is a finite set, then there is an initial segment σ of A such that W_e is contained in any set whose characteristic function extends σ, and so no extension of σ is effectively immune via f. So suppose that W_e is infinite. By the recursion theorem, there exists an e' such that $W_{e'}$ consists of the first $f(e') + 1$ elements of W_e. Thus, in this case there is also an initial segment σ of A such that any set whose characteristic function extends σ contains $W_{e'}$, and is therefore not effectively immune via f. □

Recall that a set of reals is F_σ if it is a countable union of closed sets.

Corollary 7.2. *The class of* EBI *reals is* F_σ.

Additionally, the class is no simpler:

Theorem 7.3. *The class of* EBI *reals is Wadge complete for* F_σ.

Proof. Let $A \subset 2^\omega$ be the set of reals that are eventually zero. It is well-known that A is Wadge complete for the F_σ sets (see, for example, [8], Exercise 21.17). We construct a continuous $h : 2^\omega \to 2^\omega$ such that $X \in A$ iff $h(X)$ is EBI, showing that A is Wadge reducible to the class of EBI reals.

We first define a function $f : 2^{<\omega} \to 2^{<\omega}$ recursively. Let $e_0, e_1, \ldots$ be an increasing list of all codes for total functions. Also, for each $\tau \in 2^{<\omega}$, let g_τ be an EBI real which has τ as an initial segment and let α_τ^i be an extension of τ such that no real extending α_τ^i is ϕ_{e_i}-EBI. Note that α_τ^i exists by the argument in the proof of Theorem 7.1.

Let $f(\langle\rangle) = \langle\rangle$. Suppose $\sigma \in 2^{<\omega}$, let $n = |\sigma|$ and k be the number of bits of σ which are 1. Given $f(\sigma) = \tau$, we define $f(\sigma\hat{\,}0) = g_\tau \upharpoonright (|\tau| + 1)$ and $f(\sigma\hat{\,}1) = \alpha_\tau^k$. Finally, define h so that the nth bit of $h(x)$ is the nth bit of $f(x \upharpoonright (n+1))$. $\square$

Acknowledgements. The authors would like to thank Uri Andrews, Daniel Turetsky, Linda Westrick, Rohit Nagpal, and Ashutosh Kumar for helpful discussions.

References

1. Ambos-Spies, K., Kjos-Hanssen, B., Lempp, S., Slaman, T.A.: Comparing DNR and WWKL. J. Symbolic Logic **69**(4), 1089–1104 (2004)
2. Barmpalias, G., Lewis, A.E.M., Ng, K.M.: The importance of Π_1^0 classes in effective randomness. J. Symbolic Logic **75**(1), 387–400 (2010)
3. Beros, A.A.: A DNC function that computes no effectively bi-immune set. Arch. Math. Logic **54**(5–6), 521–530 (2015)
4. Greenberg, N., Miller, J.S.: Lowness for Kurtz randomness. J. Symbolic Logic **74**(2), 665–678 (2009)
5. Greenberg, N., Miller, J.S.: Diagonally non-recursive functions and effective Hausdorff dimension. Bull. Lond. Math. Soc. **43**(4), 636–654 (2011)
6. Jockusch Jr., C.G.: Degrees of functions with no fixed points. In: Logic, methodology and philosophy of science, VIII (Moscow, 1987), vol. 126. Studies in Logic and the Foundations of Mathematics, pp. 191–201, North-Holland, Amsterdam (1989)
7. Jockusch Jr., C.G., Lewis, A.E.M.: Diagonally non-computable functions and bi-immunity. J. Symbolic Logic **78**(3), 977–988 (2013)
8. Kechris, A.S.: Classical Descriptive Set Theory. Graduate Texts in Mathematics, vol. 156. Springer, New York (1995)
9. Khan, M., Miller, J.S.: Forcing with bushy trees. http://www.math.hawaii.edu/khan/bushy_trees.pdf
10. Kjos-Hanssen, B., Merkle, W., Stephan, F.: Kolmogorov complexity and the recursion theorem. Trans. Amer. Math. Soc. **363**(10), 5465–5480 (2011)
11. Martin, D.A.: Completeness, the recursion theorem, and effectively simple sets. Proc. Am. Math. Soc. **17**(4), 838–842 (1966)
12. Nies, A., Stephan, F., Terwijn, S.A.: Randomness, relativization and Turing degrees. J. Symbolic Logic **70**(2), 515–535 (2005)

13. Post, E.L.: Recursively enumerable sets of positive integers and their decision problems. Bull. Amer. Math. Soc. **50**, 284–316 (1944)
14. Sacks, G.E.: A simple set which is not effectively simple. Proc. Am. Math. Soc. **15**(1), 51–55 (1964)
15. Smullyan, R.M.: Effectively simple sets. Proc. Am. Math. Soc. **15**(6), 893–895 (1964)
16. Stephan, F.: Martin-Löf random and PA-complete sets. In: Logic Colloquium 2002, vol. 27. Lecture Notes in Logic, pp. 342–348. Association for Symbolic Logic, La Jolla, CA (2006)

On Work of Barmpalias and Lewis-Pye: A Derivation on the D.C.E. Reals

Joseph S. Miller[✉]

Department of Mathematics, University of Wisconsin–Madison,
480 Lincoln Dr., Madison, WI 53706, USA
jmiller@math.wisc.edu

Let α and β be (Martin-Löf) random left-c.e. reals with left-c.e. approximations $\{\alpha_s\}_{s\in\omega}$ and $\{\beta_s\}_{s\in\omega}$. To compare the rates of convergence, consider[1]

$$\frac{\partial\alpha}{\partial\beta} = \lim_{s\to\infty} \frac{\alpha - \alpha_s}{\beta - \beta_s}. \tag{1}$$

Barmpalias and Lewis-Pye [2] recently proved that this limit exists and is independent of the choice of approximations to α and β. Furthermore, they showed that $\alpha - \beta$ is random if and only if $\partial\alpha/\partial\beta \neq 1$, and that

$$\frac{\partial\alpha}{\partial\beta} = \sup\{c \in \mathbb{Q}\colon \alpha - c\beta \text{ is a left-c.e. real}\}$$

$$= \inf\{c \in \mathbb{Q}\colon \alpha - c\beta \text{ is a right-c.e. real}\} \tag{2}$$

These are beautiful results that clarify the behavior of random left-c.e. reals. It has long been understood that all random left-c.e. reals are "essentially interchangeable". One of the key arguments for this heuristic was given by Kučera and Slaman [8], who showed that, up to multiplicative constants, we cannot approximate one random left-c.e. real faster than another (see Lemma 1.1). The convergence of (1) shows more: all approximations to random left-c.e. reals converge in essentially the same way. This not only solidifies our belief that that random left-c.e. reals are interchangeable, but ironically, it gives us a useful way to contrast them. For example, it follows that $\partial\alpha/\partial\beta > 1$ if and only if $\alpha - \beta$ is a random left-c.e. real and $\partial\alpha/\partial\beta < 1$ if and only if $\alpha - \beta$ is a random right-c.e. real.

This note has three main purposes. The first two go hand in hand: to give relatively short, self-contained proofs of the results of Barmpalias and Lewis-Pye, and to extend them to the d.c.e. reals. This extension is easy; the main technical breakthrough is the convergence of (1). However, extending to the d.c.e. reals gives us a clearer picture.

2010 Mathematics Subject Classification. Primary 03D32; Secondary 68Q30, 13N15.

The author was partially supported by grant #358043 from the Simons Foundation.

[1] For reasons that will become clear, we use different notation than Barmpalias and Lewis-Pye [2]. They write $\mathcal{D}(\alpha, \beta)$ instead of $\partial\alpha/\partial\beta$.

© Springer International Publishing AG 2017
A. Day et al. (Eds.): Downey Festschrift, LNCS 10010, pp. 644–659, 2017.
DOI: 10.1007/978-3-319-50062-1_39

Fix a random left-c.e. real Ω with left-c.e. approximation $\{\Omega_s\}_{s\in\omega}$. We will use this as the benchmark against which we measure the convergence of other d.c.e. reals. If α is a d.c.e. real with d.c.e. approximation $\{\alpha_s\}_{s\in\omega}$, let

$$\partial\alpha = \frac{\partial\alpha}{\partial\Omega} = \lim_{s\to\infty}\frac{\alpha-\alpha_s}{\Omega-\Omega_s}.$$

We show that $\partial\alpha = 0$ if and only if α is nonrandom, $\partial\alpha > 0$ if and only if α is a random left-c.e. real, and $\partial\alpha < 0$ if and only if α is a random right-c.e. real. Note that implicit in this case breakdown is the fact, due to Rettinger and Zheng [12], that random d.c.e. reals must either be left-c.e. or right-c.e. (see Remark 1.4).

As we have telegraphed by our choice of notation (and the title of the paper), ∂ acts somewhat like differentiation. This should not be surprising; $\partial\alpha$ is, after all, defined as the limit of a difference quotient and is meant to capture the rate of convergence of $\{\alpha_s\}_{s\in\omega}$ to α. In fact, ∂ is a derivation on the field of d.c.e. reals.[2] In other words, ∂ preserves addition and satisfies the Leibniz law:

$$\partial(\alpha\beta) = \alpha\,\partial\beta + \beta\,\partial\alpha.$$

Furthermore, if $f\colon \mathbb{R} \to \mathbb{R}$ is a computable function that is differentiable at α, then $\partial f(\alpha) = f'(\alpha)\,\partial\alpha$. This allows us to apply basic identities from calculus, so for example, $\partial\alpha^n = n\alpha^{n-1}\,\partial\alpha$ and $\partial e^\alpha = e^\alpha\,\partial\alpha$. Since $\partial\Omega = 1$, we have $\partial e^\Omega = e^\Omega$.

The third purpose of this note is to investigate the nonrandom d.c.e. reals. Given a derivation on a field, the elements that it maps to zero also form a field: the *field of constants*. In our case, these are the nonrandom d.c.e. reals. We show that, in fact, the nonrandom d.c.e. reals form a *real closed field*. It was not even previously known that the nonrandom d.c.e. reals are closed under addition, and indeed, in Remark 3.2, we note that it is easy to prove the convergence of (1) from this fact. In contrast, it has long been known that the nonrandom left-c.e. reals are closed under addition (Demuth [5] and Downey, Hirschfeldt, and Nies [7]). While also nontrivial, this fact seems to be easier to prove. Towards understanding this difference, we show that the real closure of the nonrandom left-c.e. reals is strictly smaller than the field of nonrandom d.c.e. reals. In particular, there are nonrandom d.c.e. reals that cannot be written as the difference of nonrandom left-c.e. reals; despite being nonrandom, they carry some kind of intrinsic randomness.

We should compare the results above to the work on the Solovay degrees of left-c.e. reals. Solovay [13] introduced Solovay reducibility in his study of the halting probability of a universal prefix-free machine, the standard example of a random left-c.e. real [4]. As can be seen in Fig. 1, the Solovay degrees are complementary to ∂; on the one hand, all random left-c.e. reals are Solovay equivalent [8],[3] while on the other hand, ∂ maps all nonrandom d.c.e. reals to 0

---

[2] However, we will show that ∂ maps outside of the d.c.e. reals, so it does not make them a differential field.

[3] In fact, Rettinger and Zheng [12,14] extended Solovay reducibility to the d.c.e. reals and showed that their notion retains this basic property, putting all randoms in the top degree.

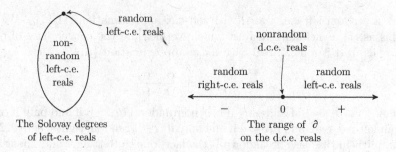

Fig. 1. Two ways to measure the randomness of effective reals

and distinguishes the random left-c.e. (and right-c.e.) reals. There is significant overlap, however, in what the two approaches tell us about the random left-c.e. reals. For example, in their work on Solovay degrees, Downey, Hirschfeldt, and Nies [7] showed that a left-c.e. real β is random if and only if

for every left-c.e. real α, there is a $c \in \omega$ and a left-c.e. real γ such that $c\beta = \alpha + \gamma$.

This follows easily from the work above: If β is random, then $\partial\beta > 0$. So given any left-c.e. real α, take c large enough that $c\partial\beta > \partial\alpha$. Then let $\gamma = c\beta - \alpha$ and note that $\partial\gamma > 0$, so it is left-c.e. For the other direction, if β is not random and α is, then for any c and any left-c.e. real γ, we have $\partial(c\beta) = 0 < \partial\alpha + \partial\gamma$.

1   Preliminaries

We assume that the reader is familiar with the basics of computability theory and effective randomness. See Downey and Hirschfeldt [6] and Nies [10] for background, including past work on random left-c.e. reals.

1.1   Left-c.e Reals

Let $\{\alpha_s\}_{s \in \omega}$ be a computable nondecreasing sequence of rationals converging to α. We say that $\{\alpha_s\}_{s \in \omega}$ is a *left-c.e. approximation* of the *left-c.e. real* α.[4] We define *right-c.e.* approximations and reals similarly. It is easy to see that a real is computable if and only if it is both a left-c.e. and a right-c.e. real.

As we have already hinted, the random left-c.e. reals are an interesting class. The key steps in understanding this class were made by Chaitin [4], Solovay [13], Calude, Hertling, Khoussainov, and Wang [3], and Kucera and Slaman [8]. Together, they showed that the following are equivalent:

---

[4] There is not broad agreement in the literature on what to call left-c.e. reals. They are often called "c.e. reals", as in Downey, Hirschfeldt, and Nies [7], or "left computable", as in Ambos-Spies, Weihrauch, and Zheng [1]. Several other names have been used, including "lower semicomputable". Both Downey and Hirschfeldt [6] and Nies [10] use "left-c.e.", so perhaps a consensus is forming.

○ α is a random left-c.e. real,
○ α is the halting probability of a universal prefix-free machine,
○ Any left-c.e. approximation to α converges at least as slowly as any left-c.e. approximation to any other left-c.e. real.

The last of these conditions is made precise in the next lemma. It is somewhat stronger than saying that α is "Solovay complete", but since we do not need Solovay reducibility below, we will leave this hair unsplit.

Lemma 1.1. (Kučera and Slaman [8]). *Let α and β be a left-c.e. reals with left-c.e. approximations $\{\alpha_s\}_{s\in\omega}$ and $\{\beta_s\}_{s\in\omega}$. If β is random, then there is a $c \in \omega$ such that*

$$(\forall k)\; \alpha - \alpha_k \leqslant c\,(\beta - \beta_k).$$

Proof. We define a Martin-Löf test $\{U_n\}_{n\in\omega}$. Fix n. We will build U_n in stages. At stage t, we will define $s(t)$ and put $[\beta_{s(t)}, \beta_{s(t)} + 2^{-n}(\alpha_{t+1} - \alpha_t)]$ into U_n. First, let $s(0) = 0$ and put $[\beta_0, \beta_0 + 2^{-n}(\alpha_1 - \alpha_0)]$ into U_n. At stage $t + 1$, define $s(t + 1) > s(t)$ such that $\beta_{s(t+1)}$ is no longer in the previous interval added to U_n. In other words, we have $\beta_{s(t+1)} > \beta_{s(t)} + 2^{-n}(\alpha_{t+1} - \alpha_t)$. Add the corresponding interval to U_n and complete the stage. Note that $\mu(U_n) \leqslant \sum_{t\in\omega} 2^{-n}(\alpha_{t+1} - \alpha_t) = 2^{-n}(\alpha - \alpha_0)$, so $\{U_n\}_{n\in\omega}$ is a Martin-Löf test (perhaps offset by a constant).

By assumption, β is random, so take n such that $\beta \notin U_n$. For this n, we add infinitely many intervals to U_n. Note that these intervals are all disjoint. In particular, for any k, we add disjoint intervals of lengths $2^{-n}(\alpha_{k+1}-\alpha_k), 2^{-n}(\alpha_{k+2}-\alpha_{k+1}),\ldots$ between $\beta_{s(k)}$ and β. Therefore, $\beta - \beta_k \geqslant \beta - \beta_{s(k)} \geqslant 2^{-n}(\alpha - \alpha_k)$. $\square$

The next lemma is the main technical tool used in the rest of the paper.

Lemma 1.2 (Barmpalias and Lewis-Pye [2]). *Let α and β be left-c.e. reals with left-c.e. approximations $\{\alpha_s\}_{s\in\omega}$ and $\{\beta_s\}_{s\in\omega}$. If β is random, then*

$$\lim_{s\to\infty} \frac{\alpha - \alpha_s}{\beta - \beta_s}\; exists.$$

Proof. Assume, for a contradiction, that the limit fails to exists. By Lemma 1.1, $\limsup_{s\to\infty}(\alpha - \alpha_s)/(\beta - \beta_s) < \infty$. On the other hand, all of the terms in the sequence are non-negative, so $\liminf_{s\to\infty}(\alpha - \alpha_s)/(\beta - \beta_s) \geqslant 0$. Therefore, there must be $c, d \in \mathbb{Q}$ such that

$$\liminf_{s\to\infty} \frac{\alpha - \alpha_s}{\beta - \beta_s} < c < d < \limsup_{s\to\infty} \frac{\alpha - \alpha_s}{\beta - \beta_s}.$$

In particular, there are infinitely many s such that $\alpha_s - d\beta_s < \alpha - d\beta$ and infinitely many t such that $\alpha_t - c\beta_t > \alpha - c\beta$. Fix such stages $s < t$. So

$$\alpha_t - c\beta_t > \alpha - c\beta = \alpha - d\beta + (d - c)\beta > \alpha_s - d\beta_s + (d - c)\beta.$$

Rearranging, we have

$$\beta < \frac{\alpha_t - \alpha_s + d\beta_s - c\beta_t}{d - c}.$$

The idea of the proof is to use such upper bounds to cover β with a Solovay test. The difficulty is that we cannot effectively determine which stages s and t satisfy our requirements, so we guess and update our guesses dynamically.

At stage t of the construction, first search for the largest $u < t$ such that $\alpha_u - c\beta_u \geqslant \alpha_t - c\beta_t$. If no such u exists, let $u = -1$. Now take the largest $s \in (u, t]$ minimizing $\alpha_s - d\beta_s$. We say that t is *absorbed* by s and we tentatively guess that s and t will give us an upper bound of β as described above (even though we may know better, for example, when $s = t$). We would like to add the interval

$$\left(\beta_s, \frac{\alpha_t - \alpha_s + d\beta_s - c\beta_t}{d - c} \right) \tag{3}$$

to the Solovay test, but this might cost too much, so we act more conservatively. First note that if $s = t$, then (3) is the empty interval (β_s, β_s), so we can "add" it to the Solovay test for free. Now consider $s < t$. Let $v \in [s, t)$ be the largest stage that has previously been absorbed by s. (It is not hard to see from our choice of s that s must have absorbed itself, so v is well-defined.) We claim that $\alpha_v - c\beta_v \geqslant \alpha_{t-1} - c\beta_{t-1}$. If not, then it must be the case that $s \leqslant v < t - 1$ and

$$\alpha_v - c\beta_v < \alpha_{t-1} - c\beta_{t-1} < \alpha_t - c\beta_t.$$

(If the second inequality were false, then we would have picked $u = t - 1$ and $s = t$.) The fact that both v and t are absorbed by s implies that $t - 1$ should have also been absorbed by s, contradicting the choice of v.

Now assume inductively that our Solovay test contains the interval

$$\left(\beta_s, \frac{\alpha_v - \alpha_s + d\beta_s - c\beta_v}{d - c} \right).$$

We extend this to the desired interval from (3), which adds measure

$$\frac{(\alpha_t - c\beta_t) - (\alpha_v - c\beta_v)}{d - c} \leqslant \frac{(\alpha_t - c\beta_t) - (\alpha_{t-1} - c\beta_{t-1})}{d - c}$$

$$\leqslant \frac{(\alpha_t - \alpha_{t-1}) - c(\beta_t - \beta_{t-1})}{d - c} \leqslant \frac{\alpha_t - \alpha_{t-1}}{d - c}.$$

Hence the total weight of the Solovay test is bounded by $\alpha/(d - c)$.

What remains is to prove that β is captured by the Solovay test. Pick s_0 to be the largest stage minimizing $\alpha_{s_0} - d\beta_{s_0}$, and $t_0 > s_0$ to be the least stage maximizing $\alpha_{t_0} - c\beta_{t_0}$ among stages greater than s_0. Note that t_0 is absorbed by s_0, so the corresponding interval is in the Solovay test. Also, it must be the case that $\alpha_{s_0} - d\beta_{s_0} < \alpha - d\beta$ and $\alpha_{t_0} - c\beta_{t_0} > \alpha - c\beta$, so β is contained in this interval. Now, pick $s_1 \geqslant t_0$ to be the greatest stage minimizing $\alpha_{s_1} - d\beta_{s_1}$ and $t_1 > s_1$ to be the least stage maximizing $\alpha_{t_1} - c\beta_{t_1}$. Again, β is contained in the corresponding interval, which in turn, is in the Solovay test. Continuing in this way, β fails the Solovay test, which is a contradiction. $\square$

1.2   D.C.E Reals

If β and γ are left-c.e. reals, we call $\alpha = \beta - \gamma$ a *d.c.e. real*.[5] Let $\{\beta_s\}_{s\in\omega}$ and $\{\gamma_s\}_{s\in\omega}$ be left-c.e. approximations of β and γ, respectively. If we set $\alpha_s = \beta_s - \gamma_s$, then not only do we have $\lim_{s\to\infty} \alpha_s = \alpha$, but the *variation* of the approximation is finite, i.e.,

$$\sum_{s\in\omega} |\alpha_{s+1} - \alpha_s| = \sum_{s\in\omega} |(\beta_{s+1} - \beta_s) - (\gamma_{s+1} - \gamma_s)|$$

$$\leqslant \sum_{s\in\omega} |\beta_{s+1} - \beta_s| + \sum_{s\in\omega} |\gamma_{s+1} - \gamma_s| = \beta + \gamma < \infty.$$

This characterizes the d.c.e. reals.

Proposition 1.3 (Ambos-Spies, Weihrauch, and Zheng [1]). *A real α is d.c.e. if and only if it is the limit of a computable sequence $\{\alpha_s\}_{s\in\omega}$ of rationals such that*

$$\sum_{s\in\omega} |\alpha_{s+1} - \alpha_s| < \infty.$$

In this case, we call $\{\alpha_s\}_{s\in\omega}$ a d.c.e. approximation *of α.*

Proof. We proved one direction above. Now assume that α is the limit of a sequence $\{\alpha_s\}_{s\in\omega}$ with finite variation. Let $\beta = \alpha_0 + \sum\{\alpha_{s+1} - \alpha_s : \alpha_{s+1} - \alpha_s \geqslant 0\}$ and $\gamma = \sum\{\alpha_s - \alpha_{s+1} : \alpha_{s+1} - \alpha_s < 0\}$. Since $\{\alpha_s\}_{s\in\omega}$ has finite variation, both β and γ are finite. It should be clear that they are left-c.e. reals and that $\alpha = \beta - \gamma$.   □

It is evident that the d.c.e. reals are closed under addition and subtraction and not too hard to see that they form a field [1]. Ng [9] and Raichev [11] independently proved that they actually form a *real closed field*; this just means that the real roots of a polynomial whose coefficients are d.c.e. reals must also be d.c.e. reals.

Rettinger and Zheng [12] observed that d.c.e. approximations of random reals are severely limited.

Remark 1.4 (Rettinger and Zheng [12]). Let $\{\alpha_s\}_{s\in\omega}$ be a d.c.e. approximation of α. Consider the Solovay test $\{[\alpha_s, \alpha_{s+1}] : \alpha_s < \alpha_{s+1}\}$; note that it has finite weight because $\{\alpha_s\}_{s\in\omega}$ has finite variation. If there are infinitely many s such

---

[5] D.c.e. is short for "difference of computably enumerable", which is admittedly an imperfect name because it is too easy to confuse d.c.e. *reals* with d.c.e. *sets*. As with "left-c.e.", various other terms have been used in the literature. Many sources, including Ambos-Spies, Weihrauch, and Zheng [1], call them "weakly computable" real numbers, which is not particularly descriptive. On the other hand, Downey and Hirschfeldt [6] call them "left-d.c.e.", while admitting that "d.l.c.e." would make somewhat more sense. Indeed, Nies [10] calls them "difference left-c.e.".

that $\alpha_s < \alpha$ and infinitely many t such that $\alpha_t > \alpha$, then α would be covered by the test, hence it would be nonrandom.

Now assume that α is random. We know that all but finitely many of the elements of the approximation fall on the same side of α. Assume, for the sake of argument, that there is an $s* \in \omega$ such that $(\forall s \geqslant s*)\ \alpha_s < \alpha$. Then $\alpha_s^* = \max_{s* \leqslant t \leq s} \alpha_t$ is a left-c.e. approximation of α, so α is a left-c.e. real. Similarly, if we assume that almost all elements of the approximation are greater than α, then α is a right-c.e. real. Note that α cannot be both a left-c.e. real and a right-c.e. real or it would be computable, and hence not random. So if we know that α is a random left-c.e. real, then we know that $\alpha_s < \alpha$ for almost all s. $\square$

Proposition 1.5 (Rettinger and Zheng [12]). *Random d.c.e. reals are either left-c.e. reals or right-c.e. reals.*

We finish with what is essentially the converse of Remark 1.4: nonrandom d.c.e. reals have "properly" d.c.e. approximations.

Lemma 1.6. *Let α be a nonrandom d.c.e. real. There is a d.c.e. approximation $\{\alpha_s\}_{s\in\omega}$ of α such that there are infinitely many s for which $\alpha_s < \alpha$ and infinitely many t for which $\alpha_t > \alpha$.*

Proof. Let $\{\alpha_s^*\}_{s\in\omega}$ be a d.c.e. approximation of α. Let $\{[c_n, d_n]\}_{n\in\omega}$ be a Solovay test that covers α, viewed as a sequence of rational intervals. We define our new approximation of α as follows. At stage s, check if α_s^* is contained in some *unused* interval $[c_n, d_n]$ for $n \leqslant s$. If so, mark that interval *used* and let $\alpha_{4s} = \alpha_{4s+3} = \alpha_s^*$, $\alpha_{4s+1} = c_n$, and $\alpha_{4s+2} = d_n$. Otherwise, let $\alpha_{4s} = \cdots = \alpha_{4s+3} = \alpha_s^*$.

Note that the variation of $\{\alpha_s\}_{s\in\omega}$ is bounded by the variation of $\{\alpha_s^*\}_{s\in\omega}$ plus the extra variation added when intervals are used. When an interval $[c_n, d_n]$ is used, it adds $2|d_n - c_n|$ to the variation. Each interval in the Solovay test is used at most once, so the contribution of all such intervals is bounded by twice the weight of the test. So $\{\alpha_s\}_{s\in\omega}$ has finite variation, which implies that it converges. Since there is a subsequence converging to α, this must be the limit. Therefore, $\{\alpha_s\}_{s\in\omega}$ is a d.c.e. approximation of α.

Now note that if an interval in the Solovay test contains α, then it will eventually be used. If such an interval is used at stage s, then $\alpha_{4s+1} < \alpha$ and $\alpha_{4s+2} > \alpha$. Since there are infinitely many such intervals, the lemma is proved. $\square$

2   A Derivation on the D.C.E Reals

As before, fix a random left-c.e. real Ω with left-c.e. approximation $\{\Omega_s\}_{s\in\omega}$.

Definition 2.1. If α is a d.c.e. real with d.c.e. approximation $\{\alpha_s\}_{s\in\omega}$, let

$$\partial\alpha = \lim_{s\to\infty} \frac{\alpha - \alpha_s}{\Omega - \Omega_s}.$$

To justify this definition, we must prove that the limit is independent of the choice of approximation. Before we have done so, it will be convenient to write $\partial\{\alpha_s\}$ instead of $\partial\alpha$. In light of the results from the previous section, the basic properties of $\partial\{\alpha_s\}$ are now fairly easy to prove.

First, note that we get linearity, a fact also observed by Barmpalias and Lewis-Pye [2]: if α and β are d.c.e. reals with d.c.e. approximations $\{\alpha_s\}_{s\in\omega}$ and $\{\beta_s\}_{s\in\omega}$, respectively, then

$$\partial\{\alpha_s + \beta_s\} = \lim_{s\to\infty} \frac{(\alpha + \beta) - (\alpha_s + \beta_s)}{\Omega - \Omega_s}$$

$$= \lim_{s\to\infty} \frac{\alpha - \beta_s}{\Omega - \Omega_s} + \lim_{s\to\infty} \frac{\beta - \beta_s}{\Omega - \Omega_s} = \partial\{\alpha_s\} + \partial\{\beta_s\}.$$

Similarly, if c is rational, then $\partial\{c\alpha_s\} = c\,\partial\{\alpha_s\}$.

Lemma 2.2. *Let α be a d.c.e. real with d.c.e. approximation $\{\alpha_s\}_{s\in\omega}$.*

(a) $\partial\{\alpha_s\}$ *converges.*
(b) *If $\partial\{\alpha_s\} > 0$, then α is a left-c.e. real.*
(c) *If $\partial\{\alpha_s\} < 0$, then α is a right-c.e. real.*
(d) *If $\alpha = 0$, then $\partial\{\alpha_s\} = 0$.*
(e) *If $\{\alpha_s^*\}_{s\in\omega}$ is another d.c.e. approximation of α, then $\partial\{\alpha_s\} = \partial\{\alpha_s^*\}$.*

Proof. As in the proof of Proposition 1.3, let β and γ be left-c.e. reals with left-c.e. approximations $\{\beta_s\}_{s\in\omega}$ and $\{\gamma_s\}_{s\in\omega}$ such that $\alpha_s = \beta_s - \gamma_s$ for all s. Then $\alpha = \beta - \gamma$ and $\partial\{\alpha_s\} = \partial\{\beta_s\} - \partial\{\gamma_s\}$. Both $\partial\{\beta_s\}$ and $\partial\{\gamma_s\}$ converge by Lemma 1.2, so $\partial\{\alpha_s\}$ also converges. This proves (a).

For (b), note that if $\partial\{\alpha_s\} > 0$, then there is an $s* \in \omega$ such that $(\forall s \geqslant s*)\ \alpha_s < \alpha$. Hence by the argument in Remark 1.4, α is a left-c.e. real. Part (c) is proved similarly.

To prove (d), assume that $\alpha = 0$ but $\partial\{\alpha_s\} \neq 0$. Pick an integer c such that $\partial\{\Omega_s + c\alpha_s\} = \partial\{\Omega_s\} + c\,\partial\{\alpha_s\} = 1 + c\,\partial\{\alpha_s\} < 0$. But $\{\Omega_s + c\alpha_s\}_{s\in\omega}$ is a d.c.e. approximation of $\Omega + c\cdot 0 = \Omega$, so by part (c), Ω is a right-c.e. real. This implies that Ω is computable, which is a contradiction.

Finally, to prove (e), note that $\partial\{\alpha_s\} - \partial\{\alpha_s^*\} = \partial\{\alpha_s - \alpha_s^*\} = 0$, because $\{\alpha_s - \alpha_s^*\}_{s\in\omega}$ is a d.c.e. approximation of 0. □

Theorem 2.3. *Let α be a d.c.e. real.*

(a) $\partial\alpha$ *converges and does not depend on the d.c.e. approximation of α.*
(b) $\partial\alpha = 0$ *if and only if α is not random.*
(c) $\partial\alpha > 0$ *if and only if α is a random left-c.e. real.*
(d) $\partial\alpha < 0$ *if and only if α is a random right-c.e. real.*
(e) $\partial\alpha = \sup\{c \in \mathbb{Q}\colon \alpha - c\Omega \text{ is left-c.e.}\}$
$= \inf\{c \in \mathbb{Q}\colon \alpha - c\Omega \text{ is right-c.e.}\}.$

Proof. Part (a) is immediate from the previous lemma. Now assume that α is not random. Let $\{\alpha_s\}_{s\in\omega}$ be the approximation guaranteed by Lemma 1.6. So there are infinitely many s for which $\alpha - \alpha_s > 0$ and infinitely many t for which $\alpha - \alpha_t < 0$. This implies that $\partial\alpha = 0$.[6] On the other hand, if α is random, then by Proposition 1.5, it must be either a left-c.e. real or a right-c.e. real. Assume that α is a random left-c.e. real. By Lemma 1.1, there is a $c \in \omega$ such that

$$(\forall s)\ \Omega - \Omega_s \leqslant c\,(\alpha - \alpha_s)\,.$$

This implies that $\partial\alpha > 1/c > 0$. Similarly, if α is a random right-c.e. real, then $\partial\alpha < 0$. This proves part (b) and the "if" directions of parts (c) and (d). The "only if" directions also follow. For example, if $\partial\alpha > 0$, then α is random by (b) and left-c.e. by the previous lemma.

Finally, (e) follows from parts (c) and (d) and the fact that $\partial(\alpha - c\Omega) = \partial\alpha - c$.
□

We have now recovered the work of Barmpalias and Lewis-Pye [2] that was discussed in the introduction. Note that we have lost nothing by working with Ω as a fixed benchmark; it is easy to see that if β is a random d.c.e. real, then

$$\frac{\partial\alpha}{\partial\beta} = \frac{\partial\alpha/\partial\Omega}{\partial\beta/\partial\Omega}.$$

Therefore, $\partial\alpha/\partial\beta$ is not ambiguous: it can either be defined as in equation (1), or as a ratio of derivations as in Definition 2.1.

Next, we show that ∂ is a derivation on the field of d.c.e. reals; in other words, that it respects addition and satisfies the Leibniz law.

Theorem 2.4. *Let* α, β *be d.c.e. reals.*

(a) $\partial(\alpha + \beta) = \partial\alpha + \partial\beta$.
(b) $\partial(\alpha\beta) = \alpha\,\partial\beta + \beta\,\partial\alpha$.

Proof. We proved (a) above. The proof for (b) is standard and simple:

$$\partial(\alpha\beta) = \lim_{s\to\infty} \frac{\alpha\beta - \alpha_s\beta_s}{\Omega - \Omega_s}$$
$$= \lim_{s\to\infty} \alpha\left(\frac{\beta - \beta_s}{\Omega - \Omega_s}\right) + \lim_{s\to\infty} \beta_s\left(\frac{\alpha - \alpha_s}{\Omega - \Omega_s}\right) = \alpha\,\partial\beta + \beta\,\partial\alpha. \quad □$$

We also get the following version of the chain rule.

---

[6] An alternate proof might appeal to those familiar with Solovay reducibility: we can show that if $\partial\alpha \neq 0$, then we can extract good approximations of Ω from good approximations of α; hence, if α were not random, then we could derandomize Ω.

Theorem 2.5. *Let* $f \colon \mathbb{R} \to \mathbb{R}$ *be a computable function. If* f *is differentiable at the d.c.e. real* α, *then*

(a) $f(\alpha)$ *is d.c.e., and*
(b) $\partial f(\alpha) = f'(\alpha)\,\partial\alpha$.

Proof. Let $\{\alpha_s\}_{s\in\omega}$ be a d.c.e. approximation of α. If f were sufficiently nice, then $\{f(\alpha_s)\}_{s\in\omega}$ would be a d.c.e. approximation of $f(\alpha)$. In particular, it would be enough to assume that f is Lipschitz in some neighborhood of α, which is true for any continuously differentiable function. For the stated generality, assume only that f is differentiable at α. Hence it is continuous at α and there is an $\epsilon > 0$ and a $c \in \omega$ such that

$$(\forall x \in \mathbb{R})\ |\alpha - x| < \epsilon \implies |f(\alpha) - f(x)| < c|\alpha - x|.$$

Fix N large enough that $(\forall n \geqslant N)\ |\alpha - \alpha_n| < \epsilon$. Let $n(0) = N$. If $n(s)$ has been defined, let $n(s+1) > n(s)$ be chosen so that $|f(\alpha_{n(s+1)}) - f(\alpha_{n(s)})| < c|\alpha_{n(s+1)} - \alpha_{n(s)}|$. Note that $n(s+1)$ exists because $\{\alpha_s\}_{s\in\omega}$ converges to α and f is continuous at α. In this way, we get an approximation $\{f(\alpha_{n(s)})\}_{s\in\omega}$ of $f(\alpha)$. It is a d.c.e. approximation because its variation is at most c times the variation of $\{\alpha_{n(s)}\}_{s\in\omega}$. This proves (a).

For (b), let $\{\alpha_s^*\}_{s\in\omega} = \{\alpha_{n(s)}\}_{s\in\omega}$ be the d.c.e. approximation of α from the previous paragraph. Then

$$\partial f(\alpha) = \lim_{s\to\infty} \frac{f(\alpha) - f(\alpha_s^*)}{\Omega - \Omega_s}$$

$$= \left(\lim_{s\to\infty} \frac{f(\alpha) - f(\alpha_s^*)}{\alpha - \alpha_s^*} \right) \left(\lim_{s\to\infty} \frac{\alpha - \alpha_s^*}{\Omega - \Omega_s} \right)^* = f'(\alpha)\,\partial\alpha. \qquad \square$$

The previous theorem allows us to apply basic identities from calculus, so for example, $\partial e^\Omega = e^\Omega$.

As already noted, ∂ does not make the d.c.e. reals into a differential field; it is straightforward to show that ∂ maps outside of the d.c.e. reals, though we do not know its range.

Proposition 2.6. *If* β *is a positive* Δ_2^0 *real, then there is a left-c.e. real* α *such that* $\partial\alpha = \beta$.

Proof. Let $\{\beta_s\}_{s\in\omega}$ be an approximation of β; we may assume that is consists only of positive rationals. Define a left-c.e. approximation $\{\alpha_s\}_{s\in\omega}$ as follows: let $\alpha_0 = 0$ and $\alpha_{s+1} = \alpha_s + \beta_s(\Omega_{s+1} - \Omega_s)$. The fact that $\{\beta_s\}_{s\in\omega}$ is bounded above implies that $\alpha = \lim_{s\to\infty} \alpha_s$ is finite. We must show that $\partial\alpha = \beta$. Fix $\epsilon > 0$ and take N such that $(\forall s \geqslant N)\ |\beta - \beta_s| < \epsilon$. Then, for any $n \geqslant N$,

$$|\beta(\Omega - \Omega_n) - (\alpha - \alpha_n)| \leqslant \sum_{s \geqslant n} |\beta(\Omega_{s+1} - \Omega_s) - (\alpha_{s+1} - \alpha_s)|$$

$$= \sum_{s \geqslant n} |\beta - \beta_s|(\Omega_{s+1} - \Omega_s) \leqslant \epsilon(\Omega - \Omega_n).$$

For such n,

$$\left| \beta - \frac{\alpha - \alpha_n}{\Omega - \Omega_n} \right| \leqslant \epsilon.$$

But $\epsilon > 0$ was arbitrary, so $\partial \alpha = \beta$. $\qquad\square$

In the same way, every negative Δ_2^0 real is $\partial \alpha$ for some right-c.e. real α. So the range of ∂ contains the Δ_2^0 reals, which is a proper superset of the d.c.e. reals.

Question 2.7. What is the range of ∂ on the d.c.e. reals?

3   The Field of Nonrandom D.C.E. Reals

We finish with an exploration of the nonrandom d.c.e. reals, in part as an application of the work above. First, it is easy to see that if ∂ is a derivation on a field, then its kernel—in this case the nonrandom d.c.e. reals—is also a field. It is called the *field of constants*. With a little more work, we can show:

Corollary 3.1. *The nonrandom d.c.e. reals form a real closed field.*

Proof. Let α and β be nonrandom d.c.e. reals. Then $\partial(\alpha + \beta) = \partial \alpha + \partial \beta = 0$, so $\alpha + \beta$ is not random. It is similarly easy to see that $\alpha - \beta$, $\alpha\beta$ and α/β are not random. So the nonrandom d.c.e. reals form a field.

Now let $p(x)$ be a polynomial whose coefficients are nonrandom d.c.e. reals. Assume that α is a real root of $p(x)$. As mentioned above, the d.c.e. reals form a real closed field [9,11], so α must be a d.c.e. real. We need to show that α is nonrandom. We may assume that α has multiplicity one as a root of $p(x)$; otherwise, we could replace $p(x)$ with the greatest common divisor of $p(x)$ and $p'(x)$, which also has coefficients in the field of nonrandom d.c.e. reals. This ensures that $p'(\alpha) \neq 0$. Now note that $\partial p(\alpha) = p'(\alpha)\,\partial \alpha$. (This does not follow from Theorem 2.5 because $p(x)$ may not be a computable function, but it can be shown by an easy induction using parts (a) and (b) of Theorem 2.4.) Therefore, we have

$$\partial \alpha = \frac{\partial p(\alpha)}{p'(\alpha)} = \frac{\partial 0}{p'(\alpha)} = 0,$$

so α is nonrandom. $\qquad\square$

The nonrandom d.c.e. reals were not previously known to be a field. In particular, it was not previously known that the sum of nonrandom d.c.e. reals is nonrandom. This was, however, known for left-c.e. reals. It was first claimed by Demuth [5] and later independently proved by Downey, Hirschfeldt, and Nies [7].

Remark 3.2. The fact that the sum of nonrandom d.c.e. reals is itself nonrandom is not, apparently, a trivial generalization of the corresponding fact for left-c.e. reals. To back up this claim, we use the fact to give a short (albeit circular) proof of Lemma 1.2. As in the actual proof, if $\lim_{s\to\infty}(\alpha - \alpha_s)/(\beta - \beta_s)$ does not exist, then there are rationals $c < d$ such that there are infinitely many s for which $\alpha_s - d\beta_s < \alpha - d\beta$ and infinitely many t for which $\alpha_t - c\beta_t > \alpha - c\beta$. Note that if $\alpha_s - d\beta_s < \alpha - d\beta$, then

$$\alpha_s - c\beta_s = \alpha_s - d\beta_s + (d - c)\beta_s < \alpha - d\beta + (d - c)\beta = \alpha - c\beta.$$

Similarly, if $\alpha_t - c\beta_t > \alpha - c\beta$, then $\alpha_t - d\beta_t > \alpha - d\beta$. Therefore, by Remark 1.4, both $\alpha - c\beta$ and $\alpha - d\beta$ are nonrandom, so their difference $(d - c)\beta$ is nonrandom. But this implies that β is nonrandom, which is a contradiction. $\qquad\square$

This leads to a natural question: why is it (apparently) harder to prove things about nonrandom d.c.e. reals than nonrandom left-c.e. reals? One immediate answer is that there are nonrandom d.c.e. reals that can only be expressed as a difference of *random* left-c.e. reals. Although they are nonrandom, such d.c.e. reals have an intrinsic randomness. This property can also be captured by looking at the variation of d.c.e. approximations.

Definition 3.3. Call a d.c.e. real α *variation nonrandom* if it has a d.c.e. approximation $\{\alpha_s\}_{s\in\omega}$ such that the variation $\sum_{s\in\omega}|\alpha_{s+1} - \alpha_s|$ is not random. Otherwise, call α *variation random*.

Proposition 3.4. *The following are equivalent for a d.c.e. real α:*

- α *is variation nonrandom,*
- *There are nonrandom left-c.e. reals β and γ such that $\alpha = \beta - \gamma$.*

Proof. First, assume that α is variation nonrandom, as witnessed by the d.c.e. approximation $\{\alpha_s\}_{s\in\omega}$. Let $\alpha*$ be the variation of this approximation, with the natural left-c.e. approximation $\{\alpha_s^*\}_{s\in\omega}$. Following Proposition 1.3, let $\beta_{s+1} = \alpha_{s+1} - \alpha_s$ if this is positive; otherwise let $\gamma_{s+1} = \alpha_s - \alpha_{s+1}$. Let $\beta_0 = \alpha_0$, and set all remaining values of β_s and γ_s to 0. Thus β and γ are left-c.e. reals and $\alpha = \beta - \gamma$. Note that

$$\beta_{s+1} - \beta_s \leqslant |\alpha_{s+1} - \alpha_s| = \alpha_{s+1}^* - \alpha_s^*,$$

for all s. So $\beta - \beta_s \leqslant \alpha^* - \alpha_s^*$, which means that $\partial\beta \leqslant \partial\alpha^* = 0$. But $\partial\beta \geqslant 0$ since β is left-c.e., so β is not random. A similar argument works for γ.

Now assume that $\alpha = \beta - \gamma$, where β and γ are nonrandom left-c.e. reals with left-c.e. approximations $\{\beta_s\}_{s\in\omega}$ and $\{\gamma_s\}_{s\in\omega}$. Let $\alpha_s = \beta_s - \gamma_s$, so $\{\alpha_s\}_{s\in\omega}$ is a d.c.e. approximation of α. As before, let $\alpha*$ be the variation of this approximation and $\{\alpha_s^*\}_{s\in\omega}$ the natural left-c.e. approximation to α^*. Then

$$\alpha_{s+1}^* - \alpha_s^* = |\alpha_{s+1} - \alpha_s| \leqslant (\beta_{s+1} - \beta_s) + (\gamma_{s+1} - \gamma_s),$$

for all s. So $\alpha^* - \alpha_s^* \leqslant (\beta - \beta_s) + (\gamma - \gamma_s)$. This means that $\partial\alpha^* \leqslant \partial\beta + \partial\gamma = 0$, so α^* is nonrandom. $\qquad\square$

Next, we will show that variation randomness is a nontrivial notion. So as promised, there is a nonrandom d.c.e. real that cannot be written as the difference of nonrandom left-c.e. reals.

Theorem 3.5. *There is a nonrandom, variation random d.c.e. real.*

Proof. Let $\{\beta_{0,s}\}_{s\in\omega}, \{\beta_{1,s}\}_{s\in\omega}, \ldots$ be an effective list of rational sequences that contains d.c.e. approximations of every d.c.e. real, with every possible variation. This is possible because we can pad a partial computable sequence of rationals by repeating the last value until a new convergence is seen, and this process does not change the variation.

We build a nonrandom d.c.e. real α such that, for each $e \in \omega$:

R_e: If $\{\beta_{e,s}\}_{s\in\omega}$ is a d.c.e. approximation of α,
then its variation is random.

The construction uses the infinite injury priority method. Strategies are organized on a tree, with the eth level containing strategies for R_e. Each node on the tree has outcomes $\infty < \cdots < w_2 < w_1 < w_0$. Strategies will update the global value of α as they are executed; we start with $\alpha = 1/2$. To each node σ on the priority tree, we assign a rational parameter $\epsilon_\sigma > 0$, in an effective way, such that the total sum of these parameters is bounded by 1. They will be used to meet the global requirements that α is nonrandom and d.c.e.

We are ready to describe the behavior of a node σ on level e of the tree. Let $\epsilon = \epsilon_\sigma$. The goal of σ is to make sure that, at any stage, the error in the current approximation to the variation of $\{\beta_{e,s}\}_{s\in\omega}$ is at least ϵ times the error in the current approximation of Ω. That will ensure that the variation is random. To force the variation to increase, σ will move α back and forth, subject to restraints imposed by other nodes, each time waiting for $\beta_{e,s}$ to get close to α before moving again.

If σ is visited at stage s, it runs the following algorithm, picking up where it left off after the last visit:

(1) Impose the restraint $(\alpha - \epsilon/2, \alpha + \epsilon/2)$. Let $t = 0$.
(2) End the substage with outcome ∞.
(3) Let (a, b) be the intersection of all current restraints. Let c be the current value of α (which will be in the interval (a, b)). Pick $n \in \omega$ and a rational $\delta < b - c$ such that $n\delta = \epsilon(\Omega_{t+1} - \Omega_t)$. Run the following loop n times:
    (a) Let σw_m be the rightmost unvisited child. Move α to c, if it is not already there. Establish the restraint $(\alpha - \delta/8, \alpha + \delta/8)$.
    (b) If $\beta_{e,s}$ is within $\delta/8$ of α, then cancel the restraint from (3a) and all restraints imposed by nodes extending σw_m, including itself, (these nodes will never again be visited); continue the algorithm. Otherwise, end the substage with outcome w_m; the next time σ is visited, repeat this step.
    (c) Let σw_m be the rightmost unvisited child. Move α to $c + \delta$. Establish the restraint $(\alpha - \delta/8, \alpha + \delta/8)$.
    (d) If $\beta_{e,s}$ is within $\delta/8$ of α, then cancel the restraint from (3c) and all restraints imposed by nodes extending σw_m; continue the algorithm. Otherwise, end the substage with outcome w_m; the next time σ is visited, repeat this step.

(4) Execute steps (3a) and (3b) one more time.
(5) Increment t and go to (2).

At stage s of the construction, we execute the algorithm above, starting at the root node and following the outcomes until we get to a node at level s of the tree. Let α_s be the value of α at the end of the stage. Note that α always respects all current restraints. In particular, any new restraints that are imposed while we wait in steps (3b) or (3d) for σ are canceled before we move α again for the sake of σ.

We must show that $\{\alpha_s\}_{s\in\omega}$ is a d.c.e. approximation. Let us look at how much σ can move α. Fix a value of t and the corresponding n and δ from step (3). When we transition from (3b) to (3c), we move α by at most $9\delta/8$. The same holds for the transition from (3d) to (3a). There are a total of $2n$ such transitions for t, so α is moved by at most

$$2n \cdot 9\delta/8 = 9/4 \cdot n\delta = 9/4 \cdot \epsilon(\Omega_{t+1} - \Omega_t),$$

where $\epsilon = \epsilon_\sigma$. Over all t, the algorithm for σ moves α by at most $9/4 \cdot \epsilon\Omega \leqslant 9/4 \cdot \epsilon_\sigma$. So in total, α is moved by at most $9/4$. Therefore, $\{\alpha_s\}_{s\in\omega}$ is a d.c.e. approximation converging to a d.c.e. real, which of course we call α.

Next, it is not hard to see that α is nonrandom. Each σ that is visited imposes a restraint in step (1). Put the *closure* of this restraint into a Solovay test; it has length ϵ_σ, so the total weight of the test is bounded by 1. If σ is on the true path, this restraint is never canceled, hence all future approximations of α must respect it. This means that (in the limit) α must be in the closure of the restraint. There are infinitely many nodes on the true path, so α is covered by the Solovay test.

Finally, we must prove that each R_e is satisfied. Assume that $\{\beta_{c,s}\}_{s\in\omega}$ is a d.c.e. approximation of α. Let β_e^* be the variation of $\{\beta_{e,s}\}_{s\in\omega}$, and let $\{\beta_{e,s}^*\}_{s\in\omega}$ be its natural left-c.e. approximation. Let σ be the node at level e of the true path and let $\epsilon = \epsilon_\sigma$. Fix t and the corresponding n and δ. Every time we leave (3b), $\beta_{e,s}$ is within $\delta/8$ of α, which is within $\delta/8$ of c. Every time we leave (3d), $\beta_{e,s}$ is within $\delta/8$ of α, which is within $\delta/8$ of $c + \delta$. So every transition adds at least $\delta/2$ to the variation of $\{\beta_{e,s}\}_{s\in\omega}$. By assumption, the algorithm for σ does not get stuck in steps (3b) or (3d), so there are $2n$ such transitions. Therefore, at least $2n \cdot \delta/2 = n\delta = \epsilon(\Omega_{t+1} - \Omega_t)$ is added to β_e^* for this t. But t is always less than s, the current stage, so this increase in the variation happens after stage t. This means that

$$\beta_e^* - \beta_{e,t}^* \geqslant \epsilon(\Omega - \Omega_t),$$

for all t. Therefore, $\partial\beta_e^* \geqslant \epsilon > 0$, so β_e^* is random and R_e is satisfied. $\square$

We finish by arguing that the nonrandom, variation random d.c.e. reals cannot be generated in any reasonably way from nonrandom left-c.e. reals. This is because the variation nonrandom reals form a robust class with a lot of closure. We will see that it is a real closed field, making it the real closure of the nonrandom left-c.e. reals. Furthermore, the field of variation nonrandom d.c.e. reals is closed under the application of sufficiently well-behaved computable functions.

Lemma 3.6. *Assume that $\alpha_1, \ldots, \alpha_n$ are variation nonrandom d.c.e. reals and $f \colon \mathbb{R}^n \to \mathbb{R}$ is a computable function. Let $\beta = f(\alpha_1, \ldots \alpha_n)$. If either*

(a) *f is Lipschitz in a neighborhood of $(\alpha_1, \ldots \alpha_n)$, or*
(b) *f is differentiable at $(\alpha_1, \ldots \alpha_n)$,*

then β is variation nonrandom

Proof. (a) Let $\{\alpha_{1,s}\}_{s \in \omega}, \ldots, \{\alpha_{n,s}\}_{s \in \omega}$ be d.c.e. approximations of $\alpha_1, \ldots, \alpha_n$ that have nonrandom variations $\alpha_1^*, \ldots, \alpha_n^*$. Let $\{\beta_s\}_{s \in \omega}$ be an approximation of β such that β_s is within 2^{-s-1} of $f(\alpha_{1,s}, \ldots, \alpha_{n,s})$. By the Lipschitz assumption, there is a $c \in \omega$ such that

$$
\begin{aligned}
|\beta_{s+1} - \beta_s| &\leqslant 2^{-s} + |f(\alpha_{1,s+1}, \ldots, \alpha_{n,s+1}) - f(\alpha_{1,s}, \ldots, \alpha_{n,s})| \\
&\leqslant 2^{-s} + c \, \|(\alpha_{1,s+1}, \ldots, \alpha_{n,s+1}) - (\alpha_{1,s}, \ldots, \alpha_{n,s})\|_2 \\
&\leqslant 2^{-s} + c \, |\alpha_{1,s+1} - \alpha_{1,s}| + \cdots + c \, |\alpha_{n,s+1} - \alpha_{n,s}|.
\end{aligned}
$$

This proves that $\{\beta_s\}_{s \in \omega}$ has finite variation; call it β^*. Furthermore, assuming the natural approximations for β^* and $\alpha_1^*, \ldots, \alpha_n^*$, we have

$$
\beta^* - \beta_s^* \leqslant 2^{-s+1} + c \, (\alpha_1^* - \alpha_{1,s}^*) + \cdots + c \, (\alpha_n^* - \alpha_{n,s}^*).
$$

Using the fact that $\{-2^{-s+1}\}_{s \in \omega}$ is a d.c.e. approximation of 0, we have $\partial \beta^* \leqslant \partial(0 + c\alpha_1^* + \cdots + c\alpha_n^*) = 0$, so β is a variation nonrandom d.c.e. real.

The argument for (b) is similar, but now the d.c.e. approximation to β must be defined using the method in the proof of Theorem 2.5(a). $\qquad\square$

Proposition 3.7. *The variation nonrandom d.c.e. reals form a real closed field.*

Proof. Closure under addition and subtraction are obvious. Multiplication and division are computable and locally Lipschitz, so by the previous lemma, the variation nonrandom d.c.e. reals form a field.

Now let $p(x)$ be a polynomial whose coefficients are variation nonrandom d.c.e. reals. Assume that α is a real root of $p(x)$. We need to show that α is variation nonrandom. As in Corollary 3.1, we may assume that α has multiplicity one as a root of $p(x)$, so $p'(\alpha) \neq 0$. We now essentially follow the proof of Theorem 2.9 in Raichev [11]. Say that $p(x) = \gamma_0 + \gamma_1 x + \cdots + \gamma_n x^n$ and let $f(x, y_0, \ldots, y_n) = y_0 + y_1 x + \cdots + y_n x^n$. So $f(\alpha, \gamma_1, \ldots, \gamma_n) = 0$ and $(\partial_x f)(\alpha, \gamma_1, \ldots, \gamma_n) = p'(x) \neq 0$. By the implicit function theorem, there is an open rational ball V containing $(\gamma_1, \ldots, \gamma_n)$ and an open rational interval U containing α such that $f(x, y_1, \ldots, y_n)$ has a unique root $g(y_1, \ldots, y_n) \in U$ for every $(y_1, \ldots, y_n) \in V$. Furthermore, g is continuously differentiable, hence Lipschitz in a neighborhood of $(\gamma_1, \ldots, \gamma_n)$. By taking V to be small enough to ensure that $g(y_1, \ldots, y_n)$ is a multiplicity one root of $f(x, y_1, \ldots, y_n)$ for every $(y_1, \ldots, y_n) \in V$, it is not hard to see that $g \colon V \to \mathbb{R}$ is computable. Therefore, α is a variation nonrandom d.c.e. real. $\qquad\square$

References

1. Ambos-Spies, K., Weihrauch, K., Zheng, X.: Weakly computable real numbers. J. Complexity **16**(4), 676–690 (2000)
2. Barmpalias, G., Lewis-Pye, A.: Differences of halting probabilities (submitted)
3. Calude, C.S., Hertling, P.H., Khoussainov, B., Wang, Y.: Recursively enumerable reals and Chaitin Ω numbers. Theoret. Comput. Sci. **255**(1–2), 125–149 (2001)
4. Chaitin, G.J.: A theory of program size formally identical to information theory. J. Assoc. Comput. Mach. **22**, 329–340 (1975)
5. Demut, O.: Constructive pseudonumbers. Comment. Math. Univ. Carolinae **16**, 315–331 (1975)
6. Rodney, G.: Algorithmic Randomness and Complexity. Theory and Applications of Computability. Springer, New York (2010)
7. Downey, R.G., Hirschfeldt, D.R., Nies, A.: Randomness, computability, and density. SIAM J. Comput. **31**(4), 1169–1183 (2002)
8. Kučera, A., Slaman, T.A.: Randomness and recursive enumerability. SIAM J. Comput. **31**(1), 199–211 (2001)
9. Ng, K.M.: Some properties of d.c.e. reals and their degrees. Master's thesis, National University of Singapore (2006)
10. Nies, A.: Computability and Randomness. Oxford Logic Guides, vol. 51. Oxford University Press, Oxford (2009)
11. Raichev, A.: Relative randomness and real closed fields. J. Symbolic Logic **70**(1), 319–330 (2005)
12. Rettinger, R., Zheng, X.: Solovay reducibility on d-c.e real numbers. In: Wang, L. (ed.) COCOON 2005. LNCS, vol. 3595, pp. 359–368. Springer, Heidelberg (2005). doi:10.1007/11533719_37
13. Solovay, R.M.: Draft of paper (or series of papers) related to Chaitin's work. IBM Thomas J. Watson Research Center, Yorktown Heights, NY, 215 pages (1975)
14. Zheng, X., Rettinger, R.: On the extensions of Solovay-Reducibility. In: Chwa, K.-Y., Munro, J.I.J. (eds.) COCOON 2004. LNCS, vol. 3106, pp. 360–369. Springer, Heidelberg (2004). doi:10.1007/978-3-540-27798-9_39

Turing Degrees and Muchnik Degrees
of Recursively Bounded DNR Functions

Stephen G. Simpson[✉]

Department of Mathematics, Vanderbilt University, Nashville, TN 37240, USA
sgslogic@gmail.com
http://www.math.psu.edu/simpson

In honor of Rod Downey's 60th birthday.

1 Introduction

Let φ_i, $i \in \mathbb{N}$ be a standard enumeration of the 1-place partial recursive functions $\varphi : \subseteq \mathbb{N} \to \mathbb{N}$. A function $Y : \mathbb{N} \to \mathbb{N}$ is said to be *diagonally nonrecursive* (with respect to the given enumeration), abbreviated *DNR*, if $\forall i \, (Y(i) \neq \varphi_i(i))$. Such a Y is said to be *recursively bounded* if there exists a recursive function $p : \mathbb{N} \to \mathbb{N}$ such that $\forall i \, (Y(i) < p(i))$. In this situation it is known that the growth rate of p has a strong influence on the Turing degree of Y. For example, it follows from [1] (see also [15, §10]) that the Turing degrees of elementary-recursively bounded DNR functions form a proper subclass of the Turing degrees of primitive-recursively bounded DNR functions. Additional results in this vein may be found in [11, Chap. 3], and still other results may be obtained by translating theorems about partial randomness [8,9] into the context of recursively bounded DNR functions [12,13].

In this note we exposit two striking results along these lines due to Joseph S. Miller. Roughly speaking, the results are as follows. Let $p : \mathbb{N} \to \mathbb{N}$ be a nondecreasing recursive function such that $p(0) \geq 2$.

1. If $\sum_i p(i)^{-1} < \infty$, then every Martin-Löf random real computes a p-bounded DNR function.
2. If $\sum_i p(i)^{-1} = \infty$, then no Martin-Löf random real computes a p-bounded DNR function unless it is Turing complete.

Note that 2 may be viewed as a vast generalization of a theorem of Stephan [20]. Combining results 1 and 2, we see that $\sum_i p(i)^{-1} < \infty$ if and only if the Turing upward closure of the set of p-bounded DNR functions is of full measure.

In order to formulate results 1 and 2 precisely, we find it convenient to replace the class DNR by the closely related class LDNR of *linearly DNR* functions. As a by-product of this move, we use LDNR to identify some specific, natural Muchnik degrees in $\mathcal{E}_w$ which are associated with 1 and 2.

In our exposition of Miller's results, we draw heavily on the ideas of Bienvenu and Porter [3]. Of course [3] contains many other interesting results concerning

This research was partially supported by Simons Foundation Collaboration Grant 276282.

other topics such as shift-complexity. Our intention here is to break down the proofs of Miller's results into easily manageable components.

2   When $\sum_i p(i)^{-1} < \infty$

Let $\mathbb{N} = \{0, 1, 2,, \ldots\}$ = the natural numbers. Let MLR $= \{X \in \{0,1\}^{\mathbb{N}} \mid X$ is Martin-Löf random$\}$. The following theorem is a slight generalization of [3, Theorem 7.6(i)]. See also Kurtz's earlier result in [10, Proposition 3].

Definition 2.1. Given a function $p : \mathbb{N} \to \mathbb{N}$, we write

$$\prod p = \prod_{i=0}^{\infty} \{j \mid j < p(i)\} = \{Y \in \mathbb{N}^{\mathbb{N}} \mid \forall i \, (Y(i) < p(i))\}$$

denoting the set of p-bounded functions.

Theorem 2.2 (Miller). Let $p : \mathbb{N} \to \mathbb{N}$ be a recursive function such that $\forall i \, (p(i) \geq 2)$ and $\sum_{i=0}^{\infty} p(i)^{-1} < \infty$. Let $\psi : \subseteq \mathbb{N} \to \mathbb{N}$ be a partial recursive function. Then $(\forall X \in \text{MLR}) \, (\exists Y \leq_{\text{T}} X) \, (Y \in \prod p$ and $Y \cap \psi = \emptyset)$.

Proof. For each i let $q(i)$ be such that $2^{q(i)} \leq p(i) < 2^{q(i)+1}$. Note that $q : \mathbb{N} \to \mathbb{N}$ is recursive and $\forall i \, (q(i) \geq 1)$. For all $X \in \{0,1\}^{\mathbb{N}}$ define $\Psi^X : \mathbb{N} \to \mathbb{N}$ by $\Psi^X(i) = \sum_{j < q(i)} X(j) 2^j < 2^{q(i)}$. Let $U_i = \{X \mid \Psi^X(i) = \psi(i)\}$, and let λ be the fair coin probability measure on $\{0,1\}^{\mathbb{N}}$. Clearly U_i is uniformly Σ_1^0 and $\lambda(U_i) \leq 2^{-q(i)}$, hence $\sum_i \lambda(U_i) \leq \sum_i 2^{-q(i)} = 2 \sum_i 2^{-q(i)-1} < 2 \sum_i p(i)^{-1} < \infty$. Hence by Solovay's Lemma [16, Lemma 3.5] we have $(\forall X \in \text{MLR}) \, \exists n \, (\forall i \geq n) \, (X \notin U_i)$, i.e., $(\forall X \in \text{MLR}) \, \exists n \, (\forall i \geq n) \, (\Psi^X(i) \neq \psi(i))$. Given $X \in \text{MLR}$, fix such an n and define $Y : \mathbb{N} \to \mathbb{N}$ by

$$Y(i) = \begin{cases} 1 & \text{if } i < n \text{ and } \psi(i) = 0, \\ 0 & \text{if } i < n \text{ and } \psi(i) \neq 0, \\ \Psi^X(i) & \text{if } i \geq n. \end{cases}$$

Then Y differs at most finitely from Ψ^X, hence $Y \leq_{\text{T}} X$, and it is also clear that $\forall i \, (Y(i) < 2^{q(i)} \leq p(i)$ and $Y(i) \neq \psi(i))$. $\square$

Definition 2.3. Let DNR $= \{Y \in \mathbb{N}^{\mathbb{N}} \mid \forall i \, (Y(i) \neq \varphi_i(i))\}$ where $\varphi_e, e \in \mathbb{N}$ is some fixed standard enumeration of the 1-place partial recursive functions. Given $p : \mathbb{N} \to \mathbb{N}$, let $\text{DNR}_p = \text{DNR} \cap \prod p$, and let $\text{DNR}_{\text{REC}} = \{Y \mid Y \in \text{DNR}_p$ for some recursive function $p\}$.

Corollary 2.4. Let $p : \mathbb{N} \to \mathbb{N}$ be a recursive function such that $\forall i \, (p(i) \geq 2)$ and $\sum_{i=0}^{\infty} p(i)^{-1} < \infty$. Then $(\forall X \in \text{MLR}) \, (\exists Y \in \text{DNR}_p) \, (Y \leq_{\text{T}} X)$.

Proof. This is the special case of Theorem 2.2 with $\psi(i) \simeq \varphi_i(i)$. $\square$

3   When $\sum_i p(i)^{-1} = \infty$

The following definition and theorem are slight generalizations of [3, Definition 4.1(i), Theorem 5.3].

Definition 3.1

1. We write $\mathbb{N}^* = \bigcup_{n=0}^{\infty} \mathbb{N}^n$ denoting the set of finite sequences of natural numbers. We use σ as a variable ranging over $\mathbb{N}^*$. Let $[0, 1]$ denote the unit interval in the real line, and let $\mathbb{Q}$ denote the set of rational numbers.
2. A *continuous semimeasure on* $\mathbb{N}^*$ is a function $M : \mathbb{N}^* \to [0, 1]$ such that $\forall \sigma \, (M(\sigma) \geq \sum_{i \in \mathbb{N}} M(\sigma^\frown \langle i \rangle))$.
3. A continuous semimeasure M on $\mathbb{N}^*$ is said to be *left recursively enumerable*, abbreviated *left r.e.*, if there exists a recursive function $(s, \sigma) \mapsto M_s(\sigma) : \mathbb{N} \times \mathbb{N}^* \to \mathbb{Q}$ such that $\forall \sigma \, (M(\sigma) = \lim_s M_s(\sigma)$ and $\forall s \, (0 \leq M_s(\sigma) \leq M_{s+1}(\sigma)))$. We may safely assume that $\forall s \, (M_s$ is a continuous semimeasure on $\mathbb{N}^*$ and $\{\sigma \mid M_s(\sigma) > 0\}$ is finite).
4. A left r.e. continuous semimeasure M on $\mathbb{N}^*$ is said to be *universal* if for all left r.e. continuous semimeasures $\overline{M}$ on $\mathbb{N}^*$ we have $\exists c \, \forall \sigma \, (\overline{M}(\sigma) < c \cdot M(\sigma))$. It is straightforward to prove the existence of such an M.
5. Throughout this note we let M denote a fixed universal left r.e. continuous semimeasure on $\mathbb{N}^*$, and we fix $M_s(\sigma)$ as above. Our definitions and results will not depend on the choice of M and $M_s(\sigma)$.
6. Given $Q \subseteq \mathbb{N}^{\mathbb{N}}$ and $n \in \mathbb{N}$, let $Q{\upharpoonright}n = \{Y{\upharpoonright}n \mid Y \in Q\}$. Note that $Q{\upharpoonright}n$ is a subset of $\mathbb{N}^n$, which is a prefix-free subset of $\mathbb{N}^*$. For any prefix-free set $S \subseteq \mathbb{N}^*$ let $M(S) = \sum_{\sigma \in S} M(\sigma)$.
7. A set $Q \subseteq \mathbb{N}^{\mathbb{N}}$ is said to be *deep* if there exists a recursive function $r : \mathbb{N} \to \mathbb{N}$ such that $\forall n \, (M(Q{\upharpoonright}r(n)) \leq 2^{-n})$.

Theorem 3.2 ([3, Theorem 5.3]). *Let* $p : \mathbb{N} \to \mathbb{N}$ *be a recursive function, and let* $Q \subseteq \prod p$ *be deep and* Π_1^0. *Then* $(\forall X \in \mathrm{MLR}) \, (\forall Y \in Q) \, (Y \leq_{\mathrm{T}} X \Rightarrow 0' \leq_{\mathrm{T}} X)$.

Proof. A *difference test* is a pair of sequences U_n, V_n, $n \in \mathbb{N}$ of uniformly Σ_1^0 subsets of $\{0, 1\}^{\mathbb{N}}$ such that $\forall n \, (\lambda(U_n \setminus V_n) \leq 2^{-n})$. A real $X \in \{0, 1\}^{\mathbb{N}}$ is said to be *difference random* [6] if for all such difference tests we have $\exists n \, (X \notin U_n \setminus V_n)$. We shall use the following result of Franklin and Ng [6]: X is difference random if and only if X is Martin-Löf random and $\not\geq_{\mathrm{T}} 0'$.

Let p and Q be as in the hypothesis of Theorem 3.2. Let r be a recursive function such that $\forall n \, (M(Q{\upharpoonright}r(n)) \leq 2^{-n})$. Since p and r are recursive and Q is a Π_1^0 subset of $\prod p$, it follows by König's Lemma that $Q{\upharpoonright}r(n)$ is Π_1^0 uniformly in n. Given a partial recursive functional Φ, consider the left r.e. continuous semimeasure M_Φ on $\mathbb{N}^*$ given by $M_\Phi(\sigma) = \lambda(\{X \in \{0, 1\}^{\mathbb{N}} \mid \Phi^X{\upharpoonright}|\sigma| = \sigma\})$. Since M is a universal left r.e. continuous semimeasure on $\mathbb{N}^*$, let c_Φ be a constant such that $\forall \sigma \, (M_\Phi(\sigma) \leq c_\Phi \cdot M(\sigma))$. Let

$$U_n = \{X \in \{0, 1\}^{\mathbb{N}} \mid (\forall i < r(n)) \, (\Phi^X(i) \downarrow)\}$$

and let $V_n = \{X \in U_n \mid \Phi^X \upharpoonright r(n) \notin Q \upharpoonright r(n)\}$. Then U_n and V_n are uniformly Σ_1^0 and $\lambda(U_n \setminus V_n) = M_\Phi(Q \upharpoonright r(n)) \leq c_\Phi \cdot M(Q \upharpoonright r(n)) \leq c_\Phi \cdot 2^{-n}$. We now see that if $\Phi^X \in Q$ then X is not difference random, so by [6] $X \in \mathrm{MLR}$ implies $0' \leq_T X$. Since Φ is an arbitrary partial recursive functional, Theorem 3.2 is proved. $\square$

Theorem 3.3 ([3, Theorem 7.6(ii)]). Let p be a recursive function such that $\sum_{i=0}^{\infty} p(i)^{-1} = \infty$. Then, we can effectively find a partial recursive function $\psi : \subseteq \mathbb{N} \to \mathbb{N}$ such that the Π_1^0 set $Q = \{Y \in \prod p \mid Y \cap \psi = \emptyset\}$ is deep.

Proof. We may safely assume that $p(i) > 0$ for all i, because otherwise $Q = \emptyset$. Since p is recursive and $\sum_i p(i)^{-1} = \infty$, let $r : \mathbb{N} \to \mathbb{N}$ be recursive such that $\sum_{r(n) \leq i < r(n+1)} p(i)^{-1} > 2^n$ holds for all n. We shall have $\psi = \bigcup_s \psi_s$ where ψ_s is defined recursively by stages, as follows.

Stage 0. Let $\psi_0 = \emptyset$.

Stage $s + 1$. Let $Q_s = \{Y \in \prod p \mid Y \cap \psi_s = \emptyset\}$ and let $n = (s + 1)_0 = $ the largest n such that 2^n is a divisor of $s + 1$. There are three cases.

Case 1. If $M_s(Q_s \upharpoonright r(n + 1)) \leq 2^{-n}$ then do nothing, i.e., $\psi_{s+1} = \psi_s$.

Case 2. Otherwise, if $\{i \mid r(n) \leq i < r(n+1)\} \subseteq \mathrm{dom}(\psi_s)$ then again do nothing, i.e., $\psi_{s+1} = \psi_s$.

Case 3. Otherwise, pick an i such that $r(n) \leq i < r(n + 1)$ and $i \notin \mathrm{dom}(\psi_s)$. For each $j < p(i)$ let $Q_s^j = \{X \in Q_s \mid X(i) = j\}$. Thus $Q_s = \bigcup_{j<p(i)} Q_s^j$ and $Q_s \upharpoonright r(n + 1) = \bigcup_{j<p(i)} Q_s^j \upharpoonright r(n + 1)$ and these unions are disjoint unions. Since $M_s(Q_s \upharpoonright r(n+1)) > 2^{-n}$, there is at least one $j < p(i)$ such that $M_s(Q_s^j \upharpoonright r(n+1)) > 2^{-n} p(i)^{-1}$. Pick such a j and let $\psi_{s+1} = \psi_s \cup \{\langle i, j \rangle\}$.

In Case 3 we have $Q_{s+1} = Q_s \setminus Q_s^j$, hence $Q_{s+1} \upharpoonright r(n + 1) = Q_s \upharpoonright r(n + 1) \setminus Q_s^j \upharpoonright r(n + 1)$, hence

$$M(Q_{s+1} \upharpoonright r(n + 1)) = M(Q_s \upharpoonright r(n + 1)) - M(Q_s^j \upharpoonright r(n + 1))$$
$$\leq M(Q_s \upharpoonright r(n + 1)) - M_s(Q_s^j \upharpoonright r(n + 1)) \quad (1)$$
$$< M(Q_s \upharpoonright r(n + 1)) - 2^{-n} p(i)^{-1}.$$

But $M(Q_0 \upharpoonright r(n + 1)) \leq 1 < \sum_{r(n) \leq i < r(n+1)} 2^{-n} p(i)^{-1}$, so from (1) we see that for each n Case 3 holds at fewer than $r(n + 1) - r(n)$ many stages $s + 1$ with $(s + 1)_0 = n$, and Case 2 never holds. Hence Case 1 holds at stage $s + 1$ for all sufficiently large s such that $(s + 1)_0 = n$, hence $M_s(Q_s \upharpoonright r(n + 1)) \leq 2^{-n}$ for all such s, so letting $Q = \bigcap_s Q_s$ we have $M(Q \upharpoonright r(n + 1)) \leq 2^{-n}$, Q.E.D. $\square$

Theorem 3.4 (Miller). Let p be a recursive function such that $\sum_{i=0}^{\infty} p(i)^{-1} = \infty$. Then, we can find a partial recursive function $\psi : \subseteq \mathbb{N} \to \mathbb{N}$ such that $(\forall X \in \mathrm{MLR}) (\forall Y \in \prod p)$ (if $Y \cap \psi = \emptyset$ and $Y \leq_T X$ then $0' \leq_T X$).

Proof. This is immediate from Theorems 3.2 and 3.3. $\square$

Corollary 3.5 (Stephan [20]). *If $X \in \mathrm{MLR}$ is of PA-degree, then $0' \leq_T X$.*

Proof. Applying Theorem 3.4 with $p(i) = 2$ for all i, we obtain a disjoint pair of recursively enumerable sets $A_0 = \{i \mid \psi(i) = 0\}$ and $A_1 = \{i \mid \psi(i) = 1\}$ with the following property: $(\forall Y \in \{0,1\}^{\mathbb{N}})$ (if Y separates A_0 from A_1 then $(\forall X \in \mathrm{MLR})(Y \leq_T X \Rightarrow 0' \leq_T X)$). The corollary follows, because any X which is of PA-degree computes a separating function for any disjoint pair of recursively enumerable sets. □

4   Linear Universality

Despite Theorems 3.2 and 3.4, it is not clear whether the following holds:

If $p : \mathbb{N} \to \mathbb{N}$ is nondecreasing and recursive and $\sum_{i=0}^{\infty} p(i)^{-1} = \infty$, then $(\forall X \in \mathrm{MLR})(\forall Y \in \mathrm{DNR}_p)(Y \leq_T X \Rightarrow 0' \leq_T X)$.

The difficulty here is that, depending on our choice of a standard enumeration of the partial recursive functions, there may or may not exist a one-to-one recursive function $r : \mathbb{N} \to \mathbb{N}$ such that $\forall i \, (\psi(i) \simeq \varphi_{r(i)}(r(i)))$ and $\sum_{i=0}^{\infty} p(r(i))^{-1} - \infty$. See also the remarks of Bienvenu and Porter concerning their [3, Definition 7.5].

However, as we shall explain in this section and the next, the statement displayed above holds if we replace DNR functions by *linearly DNR* functions.

Definition 4.1. Let $\psi : \subseteq \mathbb{N} \to \mathbb{N}$ be a partial recursive function. We say that ψ is *universal* if for all partial recursive functions $\varphi : \subseteq \mathbb{N} \to \mathbb{N}$ there exists a recursive function $r : \mathbb{N} \to \mathbb{N}$ such that $\forall i \, (\varphi(i) \simeq \psi(r(i)))$. We say that ψ is *linearly universal* if it is "universal via linear functions," i.e., for all partial recursive functions $\varphi : \subseteq \mathbb{N} \to \mathbb{N}$ there exist constants $a, b \in \mathbb{N}$ such that $\forall i \, (\varphi(i) \simeq \psi(ai + b))$.

Example 4.2. Let φ_e, $e \in \mathbb{N}$ be a standard enumeration of the 1-place partial recursive functions. The partial recursive function ψ defined by $\psi(i) \simeq \varphi_i(i)$ is universal. The partial recursive function ψ defined by $\psi(2^e(2i+1)) \simeq \varphi_e(i)$ is linearly universal.

Lemma 4.3. *If a partial recursive function $\psi : \subseteq \mathbb{N} \to \mathbb{N}$ is linearly universal, then it is "uniformly linearly universal." More precisely, there exist primitive recursive functions $a, b : \mathbb{N} \to \mathbb{N} \setminus \{0\}$ such that $\forall e \forall i \, (\varphi_e(i) \simeq \psi(a(e)i + b(e)))$.*

Proof. Fix an index $\widehat{e}$ such that $\forall e \forall i \, (\varphi_{\widehat{e}}(2^e(2i+1)) \simeq \varphi_e(i))$. Since ψ is linearly universal, fix constants $\widehat{a}, \widehat{b} \in \mathbb{N}$ such that $\forall i \, (\varphi_{\widehat{e}}(i) \simeq \psi(\widehat{a}i + \widehat{b}))$. For all e and all i we have $\varphi_e(i) \simeq \varphi_{\widehat{e}}(2^e(2i+1)) \simeq \psi(\widehat{a}2^e(2i+1) + \widehat{b})$, so we may take $a(e) = 2^{e+1}\widehat{a}$ and $b(e) = 2^e\widehat{a} + \widehat{b}$. Since $\varphi_{\widehat{e}}$ is not a constant function, we have $\widehat{a} > 0$, hence $a(e) > 0$ and $b(e) > 0$ for all e. □

The next two theorems improve the conclusions of Theorems 3.3 and 3.4 by saying that they hold for any ψ which is linearly universal.

Theorem 4.4 ([3, Theorem 7.6(ii)]). Let p be a nondecreasing recursive function such that $\sum_{i=0}^{\infty} p(i)^{-1} = \infty$. Let ψ be a partial recursive function which is linearly universal. Then the Π_1^0 set $Q = \{Y \in \prod p \mid Y \cap \psi = \emptyset\}$ is deep.

Proof. Let $\overline{e} \in \mathbb{N}$ be given. Since ψ is linearly universal, let $\overline{a} = a(\overline{e})$ and $\overline{b} = b(\overline{e})$ where $a, b : \mathbb{N} \to \mathbb{N}$ are fixed primitive recursive functions as given by Lemma 4.3. Thus we have $\varphi_{\overline{e}}(i) \simeq \psi(\overline{a}i + \overline{b})$ for all i. Define $\overline{p} : \mathbb{N} \to \mathbb{N}$ by $\overline{p}(i) = p(\overline{a}i + \overline{b})$. Since p is recursive and nondecreasing with $\sum_i p(i)^{-1} = \infty$, we claim that $\overline{p}$ is likewise recursive and nondecreasing with $\sum_i \overline{p}(i)^{-1} = \infty$. To see this, note that for all i and j we have $\overline{a}i + \overline{b} \leq \overline{a}i + \overline{b} + j$, hence $\overline{p}(i) = p(\overline{a}i + \overline{b}) \leq p(\overline{a}i + \overline{b} + j)$, hence $\overline{p}(i)^{-1} \geq p(\overline{a}i + \overline{b} + j)^{-1}$, hence $\overline{a}\,\overline{p}(i)^{-1} \geq \sum_{j < \overline{a}} p(\overline{a}i + \overline{b} + j)^{-1}$, hence $\overline{a} \sum_i \overline{p}(i)^{-1} \geq \sum_i \sum_{j < \overline{a}} p(\overline{a}i + \overline{b} + j)^{-1} = \sum_{j \geq \overline{b}} p(j)^{-1} = \infty$, hence $\sum_i \overline{p}(i)^{-1} = \infty$ as claimed. But then, applying Theorem 3.3 to $\overline{p}$, we can effectively find a partial recursive function $\overline{\psi} : \mathbb{N} \to \mathbb{N}$ such that the Π_1^0 set $\overline{Q} = \{\overline{Y} \in \prod \overline{p} \mid \overline{Y} \cap \overline{\psi} = \emptyset\}$ is deep.

Our construction of $\overline{\psi}$ given $\overline{e}$ is uniform in the following sense: there is a primitive recursive function which maps an arbitrary $\overline{e}$ to an index of the corresponding partial recursive function $\overline{\psi}$. Therefore, by the Recursion Theorem (a.k.a., the Recursion-Theoretic Fixed Point Theorem, see [14, §11.2]) we can find an $\overline{e}$ which is an index of the corresponding $\overline{\psi}$. For this $\overline{e}$ and for all i we have $\overline{\psi}(i) \simeq \varphi_{\overline{e}}(i) \simeq \psi(\overline{a}i + \overline{b})$. Thus the recursive functional $Y \mapsto \overline{Y}$ given by $\overline{Y}(i) = Y(\overline{a}i + \overline{b})$ maps Q into $\overline{Q}$. Since $\overline{Q}$ is deep, it follows by [3, Theorem 6.4] that Q is deep, Q.E.D.   □

Theorem 4.5 (essentially due to Miller). Let $p : \mathbb{N} \to \mathbb{N}$ be a nondecreasing recursive function such that $\sum_{i=0}^{\infty} p(i)^{-1} = \infty$. Let $\psi : \subseteq \mathbb{N} \to \mathbb{N}$ be a partial recursive function which is linearly universal. Then

$$(\forall X \in \mathrm{MLR})\,(\forall Y \in \prod p)\,(\text{if } Y \cap \psi = \emptyset \text{ and } Y \leq_{\mathrm{T}} X \text{ then } 0' \leq_{\mathrm{T}} X).$$

Proof. This is immediate from Theorems 3.2 and 4.4.   □

5   Some Muchnik Degrees in $\mathcal{E}_{\mathrm{w}}$

Recall from [17–19] that $\mathcal{E}_{\mathrm{w}}$ is the lattice of Muchnik degrees of nonempty Π_1^0 subsets of $\{0,1\}^{\mathbb{N}}$. Recall also from [15, 17–19] that $\mathbf{r}_1 = \deg_{\mathrm{w}}(\mathrm{MLR}) \in \mathcal{E}_{\mathrm{w}}$ and $\mathbf{0} < \mathbf{r}_1 < \mathbf{1}$, where $\mathbf{0}$ and $\mathbf{1}$ are the bottom and top Muchnik degrees in $\mathcal{E}_{\mathrm{w}}$. The purpose of this section is to define and discuss some specific, natural Muchnik degrees in $\mathcal{E}_{\mathrm{w}}$ which are associated with Theorems 2.2 and 4.5.

Definition 5.1. A function $Y : \mathbb{N} \to \mathbb{N}$ is said to be *linearly DNR* if $Y \cap \psi = \emptyset$ for some linearly universal partial recursive function $\psi : \subseteq \mathbb{N} \to \mathbb{N}$. We write $\mathrm{LDNR} = \{Y \in \mathbb{N}^{\mathbb{N}} \mid Y \text{ is linearly DNR}\}$ and $\mathrm{LDNR}_{\mathrm{REC}} = \{Y \in \mathrm{LDNR} \mid Y \in \prod p \text{ for some recursive function } p\}$. Given $p : \mathbb{N} \to \mathbb{N}$ let $\mathbf{d}_p = \deg_{\mathrm{w}}(\mathrm{LDNR}_p)$ be the Muchnik degree of $\mathrm{LDNR}_p = \mathrm{LDNR} \cap \prod p$.

Remark 5.2

1. It is easy to see that $\deg_w(\mathrm{LDNR}) = \deg_w(\mathrm{DNR}) = \mathbf{d}$ and $\deg_w(\mathrm{LDNR_{REC}}) = \deg_w(\mathrm{DNR_{REC}}) = \mathbf{d_{REC}}$, and by [1,15,17] these Muchnik degrees belong to $\mathcal{E}_w$ and we have $\mathbf{0} < \mathbf{d} < \mathbf{d_{REC}} < \mathbf{r_1}$. Moreover $\mathbf{d} = \inf_p \mathbf{d}_p$ where p ranges over all functions, and $\mathbf{d_{REC}} = \inf_p \mathbf{d}_p$ where p ranges over all recursive functions.

2. Note that LDNR and $\mathrm{LDNR_{REC}}$ are independent of the choice of a standard enumeration of the partial recursive functions. Moreover, LDNR_p and $\mathbf{d}_p$ are also independent of this choice, provided p is nondecreasing. In particular, the Muchnik degree $\mathbf{d}_p$ is specific and natural[1] provided p is specific, natural, and nondecreasing. This would not be the case if we had based our definition of $\mathbf{d}_p$ on DNR instead of LDNR. By using LDNR instead of DNR, we can now sharpen the observations in [15, §10].

3. Let p be nondecreasing and unbounded such that $p(0) \geq 2$. Let ψ be a linearly universal partial recursive function. Is the Muchnik degree of $Q = \{Y \in \prod p \mid Y \cap \psi = \emptyset\}$ independent of the choice of ψ? If so, then we could define $\mathbf{d}_p$ more simply as $\mathbf{d}_p = \deg_w(Q)$. Our actual definition of $\mathbf{d}_p$ circumvents this question, at the cost of extra complication.

4. Clearly $\forall i\,(p(i) \leq q(i))$ implies $\mathbf{d}_q \leq \mathbf{d}_p$. There are many open questions here concerning specific, natural Muchnik degrees in $\mathcal{E}_w$. For instance, letting $p(i) = \max(i^2, 1)$ and $q(i) = \max(i^3, 1)$, do we have $\mathbf{d}_q < \mathbf{d}_p$?

Lemma 5.3. The predicates "φ_e is linearly universal" and "Y is linearly DNR" are Σ_3^0.

Proof. Fix an index $\widehat{e}$ such that $\varphi_{\widehat{e}}$ is linearly universal. Then for all e, φ_e is linearly universal if and only if $\exists a\,\exists b\,\forall i\,(\varphi_{\widehat{e}}(i) \simeq \varphi_e(ai+b))$. A Tarski/Kuratowski computation [14, §14.3] shows that this predicate is Σ_3^0. Moreover, $Y \in \mathrm{LDNR}$ if and only if $\exists e\,(\varphi_e$ is linearly universal and $Y \cap \varphi_e = \emptyset)$, which is again Σ_3^0. $\square$

Theorem 5.4. Let $p : \mathbb{N} \to \mathbb{N}$ be a nondecreasing recursive function such that $p(0) \geq 2$. Then $\mathbf{d}_p \in \mathcal{E}_w$. Moreover, $\mathbf{d}_p \leq \mathbf{r_1}$ if and only if $\sum_{i=0}^{\infty} p(i)^{-1} < \infty$, and $\mathbf{d}_p \geq \mathbf{r_1}$ if and only if p is bounded, in which case $\mathbf{d}_p = \mathbf{1}$.

Proof. Lemma 5.3 implies that LDNR_p is Σ_3^0, and our assumption $\forall i\,(p(i) \geq 2)$ implies that LDNR_p includes a nonempty Π_1^0 subset of $\{0,1\}^{\mathbb{N}}$. It follows by the Σ_3^0 Embedding Lemma (see [17, Lemma 3.3] or [18, §3.3]) that $\mathrm{LDNR}_p \equiv_w D_p$ for some nonempty Π_1^0 set $D_p \subseteq \{0,1\}^{\mathbb{N}}$. Thus $\mathbf{d}_p = \deg_w(D_p) \in \mathcal{E}_w$. Theorem 2.2 tells us that $\sum_i p(i)^{-1} < \infty$ implies $\mathbf{d}_p \leq \mathbf{r_1}$. Theorem 4.5 tells us that $\sum_i p(i)^{-1} = \infty$ implies $\mathbf{d}_p \not\leq \mathbf{r_1}$. A theorem of Jockusch [10, Theorem 5] says that if p is bounded then $\mathbf{d}_p = \mathbf{1}$. A theorem of Greenberg and Miller [7] says that if p is unbounded then $\mathbf{d}_p \not\geq \mathbf{r_1}$. $\square$

Definition 5.5. An *order function* is an unbounded nondecreasing recursive function $p : \mathbb{N} \to \mathbb{N}$ such that $p(0) \geq 2$. Let us say that p is *slow-growing* if

---

[1] For an explanation of what we mean by *specific and natural*, see [19, footnote 2].

$\sum_i p(i)^{-1} = \infty$, otherwise *fast-growing*. Define $\text{LDNR}_{\text{slow}} = \bigcup_p \text{LDNR}_p$ and $\mathbf{d}_{\text{slow}} = \deg_{\text{w}}(\text{LDNR}_{\text{slow}}) = \inf_p \mathbf{d}_p$ where p ranges over all slow-growing order functions. We could define $\text{LDNR}_{\text{fast}}$ and $\mathbf{d}_{\text{fast}}$ similarly, but this would give us nothing new, because we would have $\text{LDNR}_{\text{fast}} = \text{LDNR}_{\text{REC}}$ and $\mathbf{d}_{\text{fast}} = \mathbf{d}_{\text{REC}}$.

Theorem 5.6. For each slow-growing order function p, we have $\mathbf{d}_p \in \mathcal{E}_{\text{w}}$ and $\mathbf{d}_{\text{REC}} < \mathbf{d}_p < 1$ and $\mathbf{d}_p$ is incomparable with $\mathbf{r}_1$. And similarly, we have $\mathbf{d}_{\text{slow}} \in \mathcal{E}_{\text{w}}$ and $\mathbf{d}_{\text{REC}} < \mathbf{d}_{\text{slow}} < 1$ and $\mathbf{d}_{\text{slow}}$ is incomparable with $\mathbf{r}_1$.

Proof. The statements concerning $\mathbf{d}_p$ follow directly from Theorem 5.4. To obtain the same conclusions for $\mathbf{d}_{\text{slow}}$, first imitate the proof of Lemma 5.3 to show that $\text{LDNR}_{\text{slow}}$ is Σ_3^0, then imitate the proof of Theorem 5.4. □

Remark 5.7. Given an order function p, Khan [11, Theorems 3.13 and 3.15] has shown how to construct order functions p^+ and p^- such that $\mathbf{d}_{p^+} < \mathbf{d}_p < \mathbf{d}_{p^-}$. If p is a slow-growing order function, it should be possible to construct a slow-growing order function p^+ such that $\mathbf{d}_{p^+} < \mathbf{d}_p$. This would imply that $\mathbf{d}_{\text{slow}} < \mathbf{d}_p$ for all slow-growing order functions p.

References

1. Ambos-Spies, K., Kjos-Hanssen, B., Lempp, S., Slaman, T.A.: Comparing DNR and WWKL. J. Symbolic Logic **69**(4), 1089–1104 (2004)
2. Beckmann, A., Mitrana, V., Soskova, M. (eds.): CiE 2015. LNCS, vol. 9136. Springer, Heidelberg (2015)
3. Bienvenu, L., Porter, C.P.: Deep Π_1^0 classes. Bull. Symbolic Logic **22**(2), 249–286 (2016)
4. Chatzidakis, Z., Koepke, P., Pohlers, W. (eds.): Logic Colloquium 2002. Lecture Notes in Logic, no. 27. Association for Symbolic Logic, VIII + 359 pages (2006)
5. Fenstad, J.-E., Frolov, I.T., Hilpinen, R. (eds.): Logic, Methodology and Philosophy of Science VIII. Studies in Logic and the Foundations of Mathematics, no. 126. North-Holland, XVII + 702 pages (1989)
6. Franklin, J.N.Y., Ng, K.M.: Difference randomness. Proc. Am. Math. Soc. **139**(1), 345–360 (2011)
7. Greenberg, N., Miller, J.S.: Diagonally non-recursive functions and effective Hausdorff dimension. Bull. Lond. Math. Soc. **43**(4), 636–654 (2011)
8. Higuchi, K., Hudelson, P.W.M., Simpson, S.G., Yokoyama, K.: Propagation of partial randomness. Ann. Pure Appl. Logic **165**(2), 742–758 (2014). http://dx.doi.org/10.1016/j.apal.2013.10.006
9. Hudelson, P.W.M.: Mass problems and initial segment complexity. J. Symbolic Logic **79**(1), 20–44 (2014)
10. Jockusch Jr., C.G.: Degrees of functions with no fixed points. In: [5], pp. 191–201 (1989)
11. Khan, M.: Some results on algorithmic randomness and computability-theoretic strength. Ph.D. thesis, University of Wisconsin, VII + 93 pages (2014)
12. Kjos-Hanssen, B., Merkle, W., Stephan, F.: Kolmogorov complexity and the recursion theorem. Trans. Am. Math. Soc. **363**(10), 5465–5480 (2011)
13. Kjos-Hanssen, B., Simpson, S.G.: Mass problems and Kolmogorov complexity. 4 October 2006. Preprint, 1 page (2006)

14. Rogers Jr., H.: Theory of Recursive Functions and Effective Computability. McGraw-Hill, XIX + 482 pages (1967)
15. Simpson, S.G.: Mass problems and randomness. Bull. Symbolic Logic 11(1), 1–27 (2005)
16. Simpson, S.G.: Almost everywhere domination and superhighness. Math. Logic Q. 53(4–5), 462–482 (2007)
17. Simpson, S.G.: An extension of the recursively enumerable Turing degrees. J. Lond. Math. Soc. 75(2), 287–297 (2007)
18. Simpson, S.G.: Mass problems associated with effectively closed sets. Tohoku Math. J. 63(4), 489–517 (2011)
19. Simpson, S.G.: Degrees of unsolvability: a tutorial. In: [2], pp. 83–94 (2015)
20. Stephan, F.: Martin-Löf random and PA-complete sets. In: [4], pp. 341–347 (2006)

Algorithmic Statistics: Forty Years Later

Nikolay Vereshchagin[1,2,3] and Alexander Shen[4(✉)]

[1] National Research University Higher School of Economics, Moscow, Russia
[2] Yandex, Moscow, Russia
[3] Moscow State University, Moscow, Russia
[4] LIRMM, CNRS & Univ. Montpellier,
161 Rue Ada, 34095 Montpellier, France
alexander.shen@lirmm.fr

Abstract. Algorithmic statistics has two different (and almost orthogonal) motivations. From the philosophical point of view, it tries to formalize how the statistics works and why some statistical models are better than others. After this notion of a "good model" is introduced, a natural question arises: it is possible that for some piece of data there is no good model? If yes, how often these bad (*non-stochastic*) data appear "in real life"?

Another, more technical motivation comes from algorithmic information theory. In this theory a notion of complexity of a finite object (=amount of information in this object) is introduced; it assigns to every object some number, called its *algorithmic complexity* (or *Kolmogorov complexity*). Algorithmic statistic provides a more fine-grained classification: for each finite object some curve is defined that characterizes its behavior. It turns out that several different definitions give (approximately) the same curve.

Road-map: Sect. 2 considers the notion of (α, β)-stochasticity; Sect. 3 considers two-part descriptions and the so-called "minimal description length principle"; Sect. 4 gives one more approach: we consider the list of objects of bounded complexity and measure how far some object is from the end of the list, getting some natural class of "standard descriptions" as a by-product; finally, Sect. 5 establishes a connection between these notions and resource-bounded complexity. The rest of the paper deals with an attempts to make theory close to practice by considering restricted classes of descriptions (Sect. 6) and strong models (Sect. 7).

In this survey we try to provide an exposition of the main results in the field (including full proofs for the most important ones), as well as some historical comments. We assume that the reader is familiar with the main notions of algorithmic information (Kolmogorov complexity) theory. An exposition can be found in [42, Chaps. 1, 3, 4] or [22, Chaps. 2, 3], see also the survey [36].

A short survey of main results of algorithmic statistics was given in [41] (without proofs); see also the last chapter of the book [42].

The work was in part funded by RFBR according to the research project grant 16-01-00362-a (N.V.) and by RaCAF ANR-15-CE40-0016-01 grant (A.S.).

© Springer International Publishing AG 2017
A. Day et al. (Eds.): Downey Festschrift, LNCS 10010, pp. 669–737, 2017.
DOI: 10.1007/978-3-319-50062-1_41

1   Statistical Models

Let us start with a (very rough) scheme. Imagine an experiment that produces some bit string x. We know nothing about the device that produced this data, and cannot repeat the experiment. Still we want to suggest some statistical model that fits the data ("explains" x in a plausible way). This model is a probability distribution on some finite set of binary strings containing x. What do we expect from a reasonable model?

There are, informally speaking, two main properties of a good model. First, the model should be "simple". If a model contains so many parameters that it is more complicated than the data itself, we would not consider it seriously. To make this requirement more formal, one can use the notion of Kolmogorov complexity.[1] Let us assume that measure P (used as a model) has finite support and rational values. Then P can be considered as a finite (constructive) object, so we can speak about Kolmogorov complexity of P. The requirement then says that complexity of P should be much smaller than the complexity of the data string x itself.

For example, if a data string x contains n bits, we may consider a model that corresponds to n independent fair coin tosses, i.e., the uniform distribution P on the set of all n-bit strings. Such a distribution is a constructive object that is completely determined by the value of n, so its complexity is $O(\log n)$, while the complexity of most n-bit strings is close to n (and therefore is much larger than the complexity of P, if n is large enough).

Still this simple model looks unacceptable if, for example, the sequence x consists of n zeros, or, more generally, if the frequency of ones in x deviates significantly from $1/2$, or if zeros and ones alternate. This feeling was one of the motivations for the development of algorithmic randomness notions: why some bit sequences of length n look plausible as outcomes of n fair coin tosses while other do not, while all the n-bit sequences have the same probability 2^{-n} according to the model? This question does not have a clear answer in the classical probability theory, but the algorithmic approach to randomness says that plausible strings should be incompressible: the complexity of such a string (the minimal length of a program producing it) should be close to its length.

This answer works for a uniform distribution on n-bit strings; for arbitrary P it should be modified. It turns out that for arbitrary P we should compare the complexity of x not with its length but with the value $(-\log P(x))$ (all logarithms are binary); if P is the uniform distribution on n-bit strings, the value of $(-\log P(x))$ is n for all n-bit strings x. Namely, we consider the difference between $(-\log P(x))$ and complexity of x as *randomness deficiency* of x with respect to P. We discuss the exact definition in the next section, but let us note here that this approach looks natural: different data strings require different models.

---

[1] We assume that the reader is familiar with basic notions of algorithmic information theory (complexity, a priory probability). See [36] for a concise introduction, and [22, 42] for more details.

Disclaimer. The scheme above is oversimplified in many aspects. First, it rarely happens that we have no a priori information about the experiment that produced the data. Second, in many cases the experiment can be repeated (the same experimental device can be used again, or a similar device can be constructed). Also we often deal with a data stream: we are more interested, say, in a good prediction of oil prices for the next month than in a construction of model that fits well the prices in the past. All these aspects are ignored in our simplistic model; still it may serve as an example for more complicated cases. One should stress also that algorithmic statistics is more theoretical than practical: one of the reasons is that complexity is a non-computable function and is defined only asymptotically, up to a bounded additive term. Still the notions and results from this theory can be useful not only as philosophical foundations of statistics but as a guidelines when comparing statistical models in practice (see, for example, [32]).

More practical approach to the same question is provided by machine learning that deals with the same problem (finding a good model for some data set) in the "real world". Unfortunately, currently there is a big gap between the algorithmic statistics and machine learning: the first one provides nice results about mathematical models that are quite far from practice (see the discussion about "standard models" below), while machine learning is a tool that sometimes works well without any theoretical reasons. There are some attempts to close this gap (by considering models from some class or resource-bounded versions of the notions), but much more remains to be done.

A Historical Remark. The principles of algorithmic statistics are often traced back to Occam's razor principle often stated as "Don't multiply postulations beyond necessity" or in a similar way. Poincare writes in his *Science and Method'* (Chap. 1, *The choice of facts*) that "this economy of thought, this economy of effort, which is, according to Mach, the constant tendency of science, is at the same time a source of beauty and a practical advantage". Still the mathematical analysis of these ideas became possible only after a definition of algorithmic complexity was given in 1960s (by Solomonoff, Kolmogorov and then Chaitin): after that the connection between randomness and incompressibility (high complexity) became clear. The formal definition of (α, β)-stochasticity (see the next section) was given by Kolmogorov (the authors learned it from his talk given in 1981 [17], but most probably it was formulated earlier in 1970s; the definition appeared in print in [34]). For the other related approaches (the notions of logical depth and sophistication, minimal description length principle) see the discussion in the corresponding sections (see also [22, Chap. 5].)

2 (α, β)-stochasticity

2.1   Prefix Complexity, a Priori Probability and Randomness Deficiency

Preparing for the precise definition of (α, β)-stochasticity, we need to fix the version of complexity used in this definition. There are several versions (plain and prefix complexities, different types of conditions), see [42, Chap. 6]. For most of the results the choice between these versions is not important, since the difference between the different versions is small (at most $O(\log n)$ for strings of length n), and we usually allow errors of logarithmic size in the statements.

We will use the notion of *conditional prefix complexity*, usually denoted by $K(x|c)$. Here x and c are finite objects; we measure the complexity of x when c is given. This complexity is defined as the length of the minimal prefix-free program that, given c, computes x.[2] The advantage of this definition is that it has an equivalent formulation in terms of a priori probability [42, Chap. 4]: if $\mathbf{m}(x|c)$ is the conditional a priori probability, i.e., the maximal lower semicomputable function of two arguments x and c such that $\sum_x \mathbf{m}(x|c) \leqslant 1$ for every c, then

$$K(x|c) = -\log \mathbf{m}(x|c) + O(1).$$

In particular, if a probability distribution P with finite support and rational values (we consider only distributions of this type) is considered as a condition, we may compare $\mathbf{m}$ with function $(x, P) \mapsto P(x)$ and conclude that $\mathbf{m}(x|P) \geqslant P(x)$ up to an $O(1)$-factor, so $K(x|P) \leqslant -\log P(x)$. So if we define the randomness deficiency as

$$d(x|P) = -\log P(x) - K(x|P),$$

we get a non-negative (up to $O(1)$ additive term) function. One may also explain in a different way why $K(x|P) \leqslant -\log P(x)$: this inequality is a reformulation of a standard result from information theory (Shannon–Fano code, Kraft inequality).

Why do we define the deficiency in this way? The following proposition provides some additional motivation.

Proposition 1. The function $d(x|P)$ is (up to $O(1)$-additive term) the maximal lower semicomputable function of two arguments x and P such that

$$\sum_x 2^{d(x|P)} \cdot P(x) \leqslant 1 \qquad (*)$$

for every P.

Here x is a binary string, and P is a probability distribution on binary strings with finite support and rational values. By lower semicomputable functions we mean functions that can be approximated from below by some algorithm (given x and P, the algorithm produces an increasing sequence of rational numbers that converges to $d(x|P)$; no bounds for the convergence speed are required). Then, for a given P, the function $x \mapsto 2^{d(x|P)}$ can be considered as a random variable on the probability space with distribution P. The requirement $(*)$ says that its expectation is at most 1. In this way we guarantee (by Markov inequality) that

---

[2] We do not go into details here, but let us mention one common misunderstanding: the set of programs should be prefix-free for each c, but these sets may differ for different c and the union is not required to be prefix-free.

only a P-small fraction of strings have large deficiency: the P-probability of the event $d(x|P) > c$ is at most 2^{-c}. It turns out that there exists a maximal function d satisfying $(*)$ up to $O(1)$ additive term, and our formula gives the expression for this function in terms of prefix complexity.

Proof. The proof uses standard arguments from Kolmogorov complexity theory. The function $K(x|P)$ is upper semicomputable, so $d(x|P)$ is lower semicomputable. We can also note that

$$\sum_x 2^{d(x|P)} \cdot P(x) = \sum_x \frac{\mathbf{m}(x|P)}{P(x)} \cdot P(x) = \sum_x \mathbf{m}(x|P) \leqslant 1,$$

so the deficiency function satisfies $(*)$.

To prove the maximality, consider an arbitrary function $d'(x|P)$ that is lower semicomputable and satisfies $(*)$. Then consider a function $m(x|P) = 2^{d'(x|P)} \cdot P(x)$ (the function equals 0 if x is not in the support of P). Then m is lower semicomputable, $\sum_x m(x|P) \leqslant 1$ for every P, so $m(x|P) \leqslant \mathbf{m}(x|P)$ up to $O(1)$-factor; this implies that $d'(x|P) \leqslant d(x|P) + O(1)$. □

For the case where P is the uniform distribution on n-bit strings, using P as a condition is equivalent to using n as the condition, so

$$d(x|P) = n - K(x|n)$$

in this case, and small deficiency means that complexity $K(x|n)$ is close to the length n, so x is incompressible.[3]

2.2 Definition of Stochasticity

Definition 1. A string x is called (α, β)-stochastic if there exists some probability distribution P (with rational values and finite support) such that $K(P) \leqslant \alpha$ and $d(x|P) \leqslant \beta$.

By definition every (α, β)-stochastic string is (α', β')-stochastic for $\alpha' \geqslant \alpha$, $\beta' \geqslant \beta$. Sometimes we say informally that a string is "stochastic" meaning that it is (α, β)-stochastic for some reasonably small thresholds α and β (for example, one can consider $\alpha, \beta = O(\log n)$ for n-bit strings).

Let us start with some simple remarks.

- Every simple string is stochastic. Indeed, if P is concentrated on x (singleton support), then $K(P) \leqslant K(x)$ and $d(x|P) = 0$ (in both cases with $O(1)$-precision), so x is always $(K(x) + O(1), O(1))$-stochastic.

---

[3] Initially Kolmogorov suggested to consider $n - C(x)$ as "randomness deficiency" in this case, where C stands for the plain (not prefix) complexity. One may also consider $n - C(x|n)$. But all three deficiency functions mentioned are close to each other for strings x of length n; one can show that the difference between them is bounded by $O(\log d)$ where d is any of these three functions. The proof works by comparing the expectation and probability-bounded characterizations as explained in [9].

- On the other end of the spectrum: if P is a uniform distribution on n-bit strings, then $K(P) = O(\log n)$, and most strings of length n have $d(x|P) = O(1)$, so most strings of length n are $(O(\log n), O(1))$-stochastic. The same distribution also witnesses that every n-bit string is $(O(\log n), n + O(1))$-stochastic.

- It is easy to construct stochastic strings that are between these two extreme cases. Let x be an incompressible string of length n. Consider the string $x0^n$ (the first half is x, the second half is zero string). It is $(O(\log n), O(1))$-stochastic: let P be the uniform distribution on all the strings of length $2n$ whose second half contains only zeros.

- For every distribution P (with finite support and rational values, as usual) a random sampling according to P gives us a $(K(P), c)$-stochastic string with probability at least $1 - 2^{-c}$. Indeed, the probability to get a string with deficiency greater than c is at most 2^{-c} (Markov inequality, see above).

After these observations one may ask whether non-stochastic strings exists at all — and how they can be constructed? A non-stochastic string should have non-negligible complexity (our first observation), but a standard way to get strings of high complexity, by coin tossing or other random experiment, can give only stochastic strings (our last observation).

We will see that non-stochastic strings do exist in the mathematical sense; however, the question whether they appear in the "real world", is philosophical. We will discuss both questions soon, but let us start with some mathematical results.

First of all let us note that with logarithmic precision we may restrict ourselves to uniform distributions on finite sets.

Proposition 2. Let x be an (α, β)-stochastic string of length n. Then there exist a finite set A containing x such that $K(A) \leqslant \alpha + O(\log n)$ and $d(x|U_A) \leqslant \beta + O(\log n)$, where U_A is the uniform distribution on A.

Since $K(A) = K(U_A)$ (with $O(1)$-precision, as usual), this proposition means that we may consider only uniform distributions in the definition of stochasticity, and get an equivalent (up to logarithmic change in the parameters) definition. According to this modified definition, a string x in (α, β)-stochastic if there exists a finite set A such that $K(A) \leqslant \alpha$ and $d(x|A) \leqslant \beta$, where $d(x|A)$ is now defined as $\log \#A - K(x|A)$. Kolmogorov originally proposed the definition in this form (but used plain complexity).

Proof. Let P be the (finite) distribution that exists due to the definition of (α, β)-stochasticity of x. We may assume without loss of generality that $\beta \leqslant n$ (as we have seen, all strings of length n are $(O(\log n), n + O(1))$-stochastic, so for $\beta > n$ the statement is trivial). Consider the set A formed by all strings that have sufficiently large P-probability. Namely, let us choose minimal k such that $2^{-k} \leqslant P(x)$ and consider the set A of all strings such that $P(x) \geqslant 2^{-k}$. By construction A contains x. The size of A is at most 2^k, and $-\log P(x) = k$

with $O(1)$-precision. According to our assumption, $d(x\,|\,P) = k - \mathrm{K}(x\,|\,P) \leqslant n$, so $k = d(x\,|\,P) + \mathrm{K}(x\,|\,P) \leqslant O(n)$. Then

$$\mathrm{K}(x\,|\,A) \geqslant \mathrm{K}(x\,|\,P,k) \geqslant \mathrm{K}(x\,|\,P) - O(\log n),$$

since A is determined by P, k, and the additional information in k is $O(\log k) = O(\log n)$ since $k = O(n)$ by our assumption. So the deficiency may increase only by $O(\log n)$ when we replace P by U_A, and

$$\mathrm{K}(A) \leqslant \mathrm{K}(P,k) \leqslant \mathrm{K}(P) + O(\log n)$$

for the same reasons.   □

Remark 1. Similar argument can be applied if P is a computable distribution (may be, with infinite support) computed by some program p, and we require $\mathrm{K}(p) \leqslant \alpha$ and $-\log P(x) - \mathrm{K}(x\,|\,p) \leqslant \beta$. So in this way we also get the same notion (with logarithmic precision). It is important, however, that program p *computes* the distribution P (given some point x and some precision $\varepsilon > 0$, it computes the probability of x with error at most ε). It is *not* enough for P to be an output distribution for a randomized algorithm p (in this case P is called the semimeasure lower *semicomputed* by p; note that the sum of probabilities may be strictly less than 1 since the computation may diverge with positive probability). Similarly, it is very important in the version with finite sets A (and uniform distributions on them) that the set A is considered as a finite object: A is simple if there is a short program that prints the list of all elements of A. If we allowed the set A to be presented by an algorithm that enumerates A (but never says explicitly that no more elements will appear), then situation would change drastically: for every string of complexity k the finite set S_k of strings that have complexity at most k, would be a good explanation for x, so all objects would become stochastic.

2.3   Stochasticity Conservation

We have defined stochasticity for binary strings. However, the same definition can be used for arbitrary finite (constructive) objects: pairs of strings, tuples of strings, finite sets of strings, graphs, etc. Indeed, complexity can be defined for all these objects as the complexity of their encodings; note that the difference in complexities for different encodings is at most $O(1)$. The same can be done for finite sets of these objects (or probability distributions), so the definition of (α, β)-stochasticity makes sense.

One can also note that computable bijection preserves stochasticity (up to a constant that depends on the bijection, but not on the object). In fact, a stronger statement is true: every total computable mapping preserves stochasticity. For example, consider a stochastic pair of strings (x, y). Does it imply that x (or y) is stochastic? It is indeed the case: if P is a distribution on pairs that is a reasonable model for (x, y), then its projection (marginal distribution on the first components) should be a reasonable model for x. In fact, projection can be replaced by any *total* computable mapping.

Proposition 3. Let F be a total computable mapping whose arguments and values are strings. If x is (α, β)-stochastic, then $F(x)$ is $(\alpha + O(1), \beta + O(1))$-stochastic. Here the constant in $O(1)$ depends on F but not on x, α, β.

Proof. Let P be the distribution such that $K(P) \leqslant \alpha$ and $d(x|P) \leqslant \beta$; it exists according to the definition of stochasticity. Let $Q = F(P)$ be the image distribution. In other words, if ξ is a random variable with distribution P, then $F(\xi)$ has distribution Q. It is easy to see that $K(Q) \leqslant K(P) + O(1)$, where the constant depends only on F. Indeed, Q is determined by P and F in a computable way. It remains to show that $d(F(x)|Q) \leqslant d(x|P) + O(1)$.

The easiest way to show this is to recall the characterization of deficiency as the maximal lower semicomputable function such that

$$\sum_u 2^{d(u|S)} S(u) \leqslant 1$$

for every distribution S. We may consider another function d' defined as

$$d'(u|S) = d(F(u)|F(S))$$

It is easy to see that

$$\sum_u 2^{d'(u|S)} S(u) = \sum_u 2^{d(F(u)|F(S))} S(u) = \sum_v 2^{d(v|F(S))} \cdot [F(S)](v) \leqslant 1$$

(in the second equality we group all the values of u with the same $v = F(u)$). Therefore the maximality of d guarantees that $d'(u|S) \leqslant d(u|S) + O(1)$, so we get the required inequality.

This proof can be also rephrased using the definition of stochasticity with a priori probability. We need to show that for $y = P(x)$ and $Q = F(P)$ we have

$$\frac{\mathbf{m}(y|Q)}{Q(y)} \leqslant O(1) \cdot \frac{\mathbf{m}(x|P)}{P(x)}$$

or

$$\frac{\mathbf{m}(F(x)|F(P)) \cdot P(x)}{Q(F(x))} \leqslant O(\mathbf{m}(x|P)).$$

It remains to note that the left hand side is a lower semicomputable function of x and P whose sum over all x (for every P) is at most 1. Indeed, if we group all terms with the same $F(x)$, we get the sum $\sum_y \mathbf{m}(y|F(P)) \leqslant 1$, since the sum of $P(x)$ over all x with $F(x) = y$ equals $Q(y)$. $\square$

Remark 2. In this proof it is important that we use the definition with distributions. If we replace is with the definition with finite sets, the results remains true with logarithmic precision, but the argument becomes more complicated, since the image of the uniform distribution may not be a uniform distribution. So if a set A is a good model for x, we should not use $F(A)$ as a model for $F(x)$. Instead, we should look at the maximal k such that $2^k \leqslant \#F^{-1}(y)$, and consider the set of all y' that have at least 2^k preimages in A.

Remark 3. It is important in Proposition 3 that F is a total function. If x is some non-stochastic object and x^* is the shortest program for x, then x^* is incompressible and therefore stochastic. Still the interpreter (decompressor) maps x^* to x. We discuss the case of non-total F below, see Sect. 5.4.

Remark 4. A similar argument shows that $d(F(x)|F(P)) \leqslant d(x|P) + \mathrm{K}(F) + O(1)$ (for total F), so both $O(1)$-bounds in Proposition 3 may be replaced by $\mathrm{K}(F) + O(1)$ where $O(1)$-constant does not depend on F anymore.

2.4 Non-stochastic Objects

Note that up to now we have not shown that non-stochastic objects exist at all. It is easy to show that they exist for rather large values of α and β (linearly growing with n).

Proposition 4 ([34]). For some c and all n:

(1) if $\alpha + 2\beta < n - c\log n$, then there exist n-bit strings that are not (α, β)-stochastic;

(2) however, if $\alpha + \beta > n + c\log n$, then every n-bit string is (α, β)-stochastic.

Note that the term $c\log n$ allows us to use the definition with finite sets (i.e., uniform distributions on finite sets) instead of arbitrary finite distributions, since both versions are equivalent with $O(\log n)$-precision.

Proof. The second part is obvious (and is added just for comparison): if $\alpha + \beta = n$, then all n-bit strings can be split into 2^α groups of size 2^β each. Then the complexity of each group is $\alpha + O(\log n)$, and the randomness deficiency of every string in the corresponding group is at most $\beta + O(1)$. It is slightly bigger than the bounds we need, but we have reserve $c\log n$, and α and β can be decreased, say, by $(c/2)\log n$ before using this argument.

The first part: Consider all finite sets A of strings that have complexity at most α and size at most $2^{\alpha+\beta}$. Since $\alpha + (\alpha + \beta) < n$, they cannot cover all n-bit strings. Consider then the first (say, in the lexicographical order) n-bit string u not covered by any of these sets. What is the complexity of u? To specify u, it is enough to give n, α, β and the program of size at most α (from the definition of Kolmogorov complexity) that has maximal running time among programs of that size. Then we can wait until this program terminates and look at the outputs of all programs of size at most α after the same number of steps, select sets of strings of size at most $\alpha + \beta$, and take the first u not covered by these sets. So the complexity of u is at most $\alpha + O(\log n)$ (the last term is needed to specify n, α, β). The same is true for conditional complexity with arbitrary condition, since it is bounded by the unconditional complexity. So the randomness deficiency of u in every set A of size $2^{\alpha+\beta}$ is at least $\beta - O(\log n)$. We see that u is not $(\alpha, \beta - O(\log n))$-stochastic. Again the $O(\log n)$-term can be compensated by $O(\log n)$-change in β (we have $c\log n$ reserve for that). $\square$

Remark 5. There is a gap between lower and upper bounds provided by Proposition 4. As we will see later, the upper bound (2) is tight with $O(\log n)$-precision, but we need more advanced technique (properties of two-part descriptions, Sect. 3) to prove this.

Proposition 4 shows that non-stochastic objects exist for rather large values of α and β (proportional to n). This, of course, is a mathematical existence result; it does not say anything about the possibility to observe non-stochastic objects in the "real world". As we have discussed, random sampling (from a simple distribution) may produce a non-stochastic object only with a negligible probability; *total* algorithmic transformations (defined by programs of small complexity) also cannot not create non-stochastic object from stochastic ones. What about non-total algorithmic transformations? As we have discussed in Remark 3, a non-total computable transformation may transform a stochastic object into a non-stochastic one, but does it happen with non-negligible probability?

Consider a randomized algorithm that outputs some string. It can be considered as a deterministic algorithm applied to random bit sequence (generated by the internal coin of the algorithm). This deterministic algorithm may be non-total, so we cannot apply the previous result. Still, as the following result shows, randomized algorithms also generate non-stochastic objects only with small probability.

To make this statement formal, we consider the sum of $\mathbf{m}(x)$ over all non-stochastic x of length n. Since the a priori probability $\mathbf{m}(x)$ is the upper bound for the output distribution of any randomized algorithm, this implies the same bound (up to $O(1)$-factor) for every randomized algorithm. The following theorem gives an upper bound for this sum:

Proposition 5. (see [30], Sect. 10)

$$\sum \{\, \mathbf{m}(x) \mid x \text{ is a } n\text{-bit string that is not } (\alpha, \alpha)\text{-stochastic} \,\} \leqslant 2^{-\alpha + O(\log n)}$$

for every n and α.

Proof. Consider the sum of $\mathbf{m}(x)$ over *all* strings of length n. This sum is some real number $\omega \leqslant 1$. Let $\tilde{\omega}$ be the number represented by first α bits in the binary representation of ω, minus $2^{-\alpha}$. We may assume that $\alpha \leqslant O(n)$, otherwise all strings of length n are (α, α)-stochastic.

Now construct a probability distribution as follows. All terms in a sum for ω are lower semicomputable, so we can enumerate increasing lower bounds for them. When the sum of these lower bounds exceeds $\tilde{\omega}$, we stop and get some measure P with finite support and rational values. Note that we have a measure, not a distribution, since the sum of $P(x)$ for all x is less than 1 (it does not exceed ω). So we normalize P (by some factor) to get a distribution $\tilde{P}$ proportional to P. The complexity of $\tilde{P}$ is bounded by $\alpha + O(\log n)$ (since $\tilde{P}$ is determined by $\tilde{\omega}$ and n). Note that the difference between P (without normalization factor) and a priori probability $\mathbf{m}$ (the sum of differences over all strings of length n) is bounded by $O(2^{-\alpha})$. It remains to show that for $\mathbf{m}$-most strings the distribution $\tilde{P}$ is a good model.

Let us prove that the sum of a priori probabilities of all n-bit strings x that have $d(x \mid \tilde{P}) > \alpha + c \log n$ is bounded by $O(2^{-\alpha})$, if c is large enough. Indeed, for those strings we have

$$- \log \tilde{P}(x) - \mathrm{K}(x \mid \tilde{P}) > \alpha + c \log n.$$

The complexity of $\tilde{P}$ is bounded by $\alpha + O(\log n)$ and therefore $\mathrm{K}(x)$ exceeds $\mathrm{K}(x \mid \tilde{P})$ at most by $\alpha + O(\log n)$, so $- \log \tilde{P}(x) - \mathrm{K}(x) > 1$ (or $\tilde{P}(x) < \mathbf{m}(x)/2$) for those strings, if c is large enough (it should exceed the constants hidden in $O(\log n)$ notation). The difference 1 is enough for the estimate below, but we could have arbitrary constant or even logarithmic difference by choosing larger value of c.

Prefix complexity can be defined in terms of a priori probability, so we get

$$\log(\mathbf{m}(x)/\tilde{P}(x)) > 1$$

for all x that have deficiency exceeding $\alpha + c \log n$ with respect to $\tilde{P}$. The same inequality is true for P instead of $\tilde{P}$, since P is smaller. So for all those x we have $P(x) < \mathbf{m}(x)/2$, or $(\mathbf{m}(x) - P(x)) > \mathbf{m}(x)/2$. Recalling that the sum of $\mathbf{m}(x) - P(x)$ over all x of length n does not exceed $O(2^{-\alpha})$ by construction of $\tilde{\omega}$, we conclude that the sum of $\mathbf{m}(x)$ over all strings of randomness deficiency (with respect to $\tilde{P}$) exceeding $\alpha + c \log n$ is at most $O(2^{-\alpha})$.

So we have shown that the sum of $\mathbf{m}(x)$ for all x of length n that are not $(\alpha + O(\log n), \alpha + O(\log n))$-stochastic, does not exceed $O(2^{-\alpha})$. This differs from our claim only by $O(\log n)$-change in α.   □

Bruno Bauwens noted that this argument can be modified to obtain a stronger result where (α, α)-stochasticity is replaced by $(\alpha + O(\log n), O(\log n))$-stochasticity. Instead of one measure P, one should consider a family of measures. Let us approximate ω and look when the approximations cross the thresholds corresponding to k first bits of the binary expansion of ω. In this way we get $P = P_1 + P_2 + \ldots + P_\alpha$, where P_i has total weight at most 2^{-i}, and complexity at most $i + O(\log n)$. Let us show that all strings x where $P(x)$ is close to $\mathbf{m}(x)$ (say, $P(x) \geqslant \mathbf{m}(x)/2$) are $(\alpha + O(\log n), O(\log n))$-stochastic, namely, one of the measures P_i multiplied by 2^i is a good explanations for them. Indeed, for such x and some i the value of $P_i(x)$ coincides with $\mathbf{m}(x)$ up to polynomial (in n) factor, since the sum of all P_i is at least $\mathbf{m}(x)/2$. On the other hand, $\mathbf{m}(x \mid 2^i P_i) \leqslant 2^i \mathbf{m}(x) \approx 2^i P_i(x)$, since the complexity of $2^i P_i$ is at most $i + O(\log n)$. Therefore the ratio $\mathbf{m}(x \mid P_i)/(2^i P_i(x))$ is polynomially bounded, and the model $2^i P_i$ has deficiency $O(\log n)$. This better bound also follows from the Levin's explanation, see below.

This result shows that non-stochastic objects rarely appear as outputs of randomized algorithms. There is an explanation of this phenomenon (that goes back to Levin): non-stochastic objects provide a lot of information about halting problem, and the probability of appearance of an object that has a lot of information about some sequence α, is small (for any fixed α). We discuss this argument below, see Sect. 4.6.

It is natural to ask the following general question. For a given string x, we may consider the set of all pairs (α, β) such that x is (α, β)-stochastic. By definition, this set is upwards-closed: a point in this set remains in it if we increase α or β, so there is some boundary curve that describes the trade-off between α and β. What curves could appear in this way? To get an answer (to characterizes all these curves with $O(\log n)$-precision), we need some other technique, explained in the next section.

3 Two-Part Descriptions

Now we switch to another measure of the quality of a statistical model. It is important both for philosophical and technical reasons. The philosophical reason is that it corresponds to the so-called "minimal description length principle". The technical reason is that it is easier to deal with; in particular, we will use it to answer the question asked at the end of the previous section.

3.1 Optimality Deficiency

Consider again some statistical model. Let P be a probability distribution (with finite support and rational values) on strings. Then we have

$$K(x) \leqslant K(P) + K(x|P) \leqslant K(P) + (-\log P(x))$$

for arbitrary string x (with $O(1)$-precision). Here we use that (with $O(1)$-precision):

- $K(x|P) \leqslant -\log P(x)$, as we have mentioned;
- the complexity of the pair is bounded by the sum of complexities: $K(u, v) \leqslant K(u) + K(v)$;
- $K(v) \leqslant K(u, v)$ (in our case, $K(x) \leqslant KP(x, P)$).

If P is a uniform distribution on some finite set A, this inequality can be explained as follows. We can specify x in two steps:

- first, we specify A;
- then we specify the ordinal number of x in A (in some natural ordering, say, the lexicographic one).

In this way we get $K(x) \leqslant K(A) + \log \#A$ for every element x of arbitrary finite set A. This inequality holds with $O(1)$-precision. If we replace the prefix complexity by the plain version, we can say that $C(x) \leqslant C(A) + \log \#A$ with precision $O(\log n)$ for every string x of length at most n: we may assume without loss of generality that both terms in the right hand side are at most n, otherwise the inequality is trivial.

The "quality" of a statistical model P for a string x can be measured by the difference between sides of this inequality: for a good model the "two-part description" should be almost minimal. We come to the following definition:

Definition 2. The *optimality deficiency* of a distribution P considered as the model for a string x is the difference

$$\delta(x, P) = (K(P) + (-\log P(x))) - K(x).$$

As we have seen, $\delta(x, P) \geqslant 0$ with $O(1)$-precision.

If P is a uniform distribution on a set A, the optimality deficiency $\delta(x, P)$ will also be denoted by $\delta(x, A)$, and

$$\delta(x, A) = (K(A) + \log \#A) - K(x).$$

The following proposition shows that we may restrict our attention to finite sets as models (with $O(\log n)$-precision):

Proposition 6. Let P be a distribution considered as a model for some string x of length n. Then there exists a finite set A such that

$$K(A) \leqslant K(P) + O(\log n); \quad \log \#A \leqslant -\log P(x) + O(1) \qquad (*)$$

This proposition will be used in many arguments, since it is often easier to deal with sets as statistical models (instead of distributions). Note that the inequalities $(*)$ evidently imply that

$$\delta(x, A) \leqslant \delta(x, P) + O(\log n),$$

so arbitrary distribution P may be replaced by a uniform one (U_A) with a logarithmic-only change in the optimality deficiency.

Proof. We use the same construction as in Proposition 2. Let 2^{-k} be the maximal power of 2 such that $2^{-k} \leqslant P(x)$, and let $A = \{x \mid P(x) \geqslant 2^{-k}\}$. Then $k = -\log P(x) + O(1)$. We may assume that $k = O(n)$: if k is much bigger than n, then $\delta(x, P)$ is also bigger than n (since the complexity of x is bounded by $n + O(\log n)$), and in this case the statement is trivial (let A be the set of all n-bit strings).

Now we see that that A is determined by P and k, so $K(A) \leqslant K(P) + K(k) \leqslant K(P) + O(\log n)$. Note also that $\#A \leqslant 2^k$, so $\log \#A \leqslant -\log P(x) + O(1)$. $\qquad\square$

Let us note that in a more general setting [25] where we consider several strings as outcomes of the repeated experiment (with independent trials) and look for a model that explains all of them, a similar result is not true: not every probability distribution can be transformed into a uniform one.

3.2   Optimality and Randomness Deficiencies

Now we have two "quality measures" for a statistical model P: the randomness deficiency $d(x \mid P)$ and the optimality deficiency $\delta(x, P)$. They are related:

Proposition 7

$$d(x \mid P) \leqslant \delta(x, P)$$

with $O(1)$-precision.

Proof. By definition

$$d(x\,|\,P) = -\log P(x) - \mathrm{K}(x\,|\,P);$$
$$\bullet \qquad \delta(x,P) = -\log P(x) + \mathrm{K}(P) - \mathrm{K}(x).$$

It remains to note that $\mathrm{K}(x) \leqslant \mathrm{K}(x,P) \leqslant \mathrm{K}(P) + \mathrm{K}(x\,|\,P)$ with $O(1)$-precision. □

Could $\delta(x,P)$ be significantly larger than $d(x\,|\,P)$? Look at the proof above: the second inequality $\mathrm{K}(x,P) = \mathrm{K}(P) + \mathrm{K}(x\,|\,P)$ is an equality with logarithmic precision. Indeed, the exact formula (Levin–Gács formula for the complexity of a pair with $O(1)$-precision) is

$$\mathrm{K}(x,P) = \mathrm{K}(P) + \mathrm{K}(x\,|\,P,\mathrm{K}(P)).$$

Here the term $\mathrm{K}(P)$ in the condition changes the complexity by $O(\log \mathrm{K}(P))$, and we may ignore models P whose complexity is much greater than the complexity of x.

On the other hand, in the first inequality the difference between $\mathrm{K}(x,P)$ and $\mathrm{K}(x)$ may be significant. This difference equals $\mathrm{K}(P\,|\,x)$ with logarithmic accuracy and, if it is large, then $\delta(x,P)$ is much bigger than $d(x\,|\,P)$. The following example shows that this is possible. In this example we deal with sets as models.

Example 1. Consider an incompressible string x of length n, so $\mathrm{K}(x) = n$ (all equalities with logarithmic precision). A good model for this string is the set A of all n-bit strings. For this model we have $\#A = 2^n$, $\mathrm{K}(A) = 0$ and $\delta(x,A) = n+0-n = 0$ (all equalities have logarithmic precision). So $d(x\,|\,P) = 0$, too. Now we can change the model by excluding some other n-bit string. Consider a n-bit string y that is incompressible and independent of x: this means that $\mathrm{K}(x,y) = 2n$. Let A' be $A \setminus \{y\}$.

The set A' contains x (since x and y are independent, y differs from x). Its complexity is n (since it determines y). The optimality deficiency is then $n+n-n = n$, but the randomness deficiency is still small: $d(x\,|\,A') = \log \#A' - \mathrm{K}(x\,|\,A') = n - n = 0$ (with logarithmic precision). To see why $\mathrm{K}(A'\,|\,x) = n$, note that x and y are independent, and the set A' has the same information as (n,y).

One of the main results of this section (Theorem 3) clarifies the situation: it implies that if optimality deficiency of a model is significantly larger than its randomness deficiency, then this model can be improved and another model with better parameters can be found. More specifically, the complexity of the new model is smaller than the complexity of the original one while both the randomness deficiency and optimality deficiency of the new model are not worse than the randomness deficiency of the original one. This is one of the main results of algorithmic statistics, but first let us explore systematically the properties of two-part descriptions.

3.3   Trade-off Between Complexity and Size of a Model

It is convenient to consider only models that are sets (=uniform distribution on sets). We will call them *descriptions*. Note that by Propositions 2 and 6 this restriction does not matter much since we ignore logarithmic terms. For a given string x there are many different descriptions: we can have a simple large set containing x, and at the same time some more complicated, but smaller one. In this section we study the trade-off between these two parameters (complexity and size).

Definition 3. A finite set A is an $(i * j)$-description[4] of x if $x \in A$, complexity $K(A)$ is at most i, and $\log \#A \leqslant j$. For a given x we consider the set P_x of all pairs (i, j) such that x has some $(i * j)$-description; this set can be called the *profile* of x.

Informally speaking, an $(i * j)$-description for x consists of two parts: first we spend i bits to specify some finite set A and then j bits to specify x as an element of A. ·

What can be said about P_x for a string x of length n and complexity $k = K(x)$? By definition, P_x is closed upwards and contains the points $(0, n)$ and $(k, 0)$. Here we omit terms $O(\log n)$: more precisely, we have a $(O(\log n) * n)$-description that consists of all strings of length n, and a $((k + O(1)) * 0)$-description $\{x\}$. Moreover, the following proposition shows that we can move the information from the second part of the description into its first part (leaving the total length almost unchanged). In this way we make the set smaller (the price we pay is that its complexity increases).

Proposition 8 ([13, 15, 35]). Let x be a string and A be a finite set that contains x. Let s be a non-negative integer such that $s \leqslant \log \#A$. Then there exists a finite set A' containing x such that $\#A' \leqslant \#A/2^s$ and $K(A') \leqslant K(A) + s + O(\log s)$.

Proof. List all the elements of A in some (say, lexicographic) order. Then we split the list into 2^s parts (first $\#A/2^s$ elements, next $\#A/2^s$ elements etc.; we omit evident precautions for the case when $\#A$ is not a multiple of 2^s). Then let A' be the part that contains x. It has the required size. To specify A', it is enough to specify A and the part number; the latter takes at most s bits. (The logarithmic term is needed to make the encoding of the part number self-delimiting.)   □

This statement can be illustrated graphically. As we have said, the set P_x is "closed upwards" and contains with each point (i, j) all points on the right (with bigger i) and on the top (with bigger j). It contains points $(0, n)$ and $(K(x), 0)$; Proposition 8 says that we can also move down-right adding $(s, -s)$ (with logarithmic precision). We will see that movement in the opposite direction is not always possible. So, having two-part descriptions with the same total

---

[4] This notation may look strange; however, we speak so often about finite sets of complexity at most i and cardinality at most 2^j that we decided to introduce some short name and notation for them.

length, we should prefer the one with bigger set (since it always can be converted into others, but not vice versa).

The boundary of P_x is some curve connecting the points $(0, n)$ and $(k, 0)$. This curve (introduced by Kolmogorov in 1970s, see [16]) never gets into the triangle $i + j < K(x)$ and always goes down (when moving from left to right) with slope at least -1 or more.

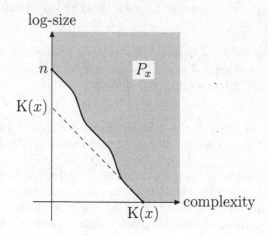

Fig. 1. The set P_x and its boundary curve

This picture raises a natural question: which boundary curves are possible and which are not? Is it possible, for example, that the boundary goes along the dotted line on Fig. 1? The answer is positive: take a random string of desired complexity and add trailing zeros to achieve desired length. Then the point $(0, K(x))$ (the left end of the dotted line) corresponds to the set A of all strings of the same length having the same trailing zeros. We know that the boundary curve cannot go down slower than with slope -1 and that it lies above the line $i + j = K(x)$, therefore it follows the dotted line (with logarithmic precision).

A more difficult question: is it possible that the boundary curve starts from $(0, n)$, goes with the slope -1 to the very end and then goes down rapidly to $(K(x), 0)$ (Fig. 2, the solid line)? Such a string x, informally speaking, would have essentially only two types of statistical explanations: a set of all strings of length n (and its parts obtained by Proposition 8) and the exact description, the singleton $\{x\}$.

It turns out that not only these two opposite cases are possible, but also all intermediate curves (provided they decrease with slope -1 or faster, and are simple enough), at least with logarithmic precision. More precisely, the following statement holds:

Theorem 1 ([43]). *Let $k \leqslant n$ be two integers and let $t_0 > t_1 > \ldots > t_k$ be a strictly decreasing sequence of integers such that $t_0 \leqslant n$ and $t_k = 0$; let m be the complexity of this sequence. Then there exists a string x of complexity*

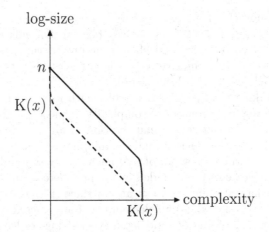

Fig. 2. Two opposite possibilities for a boundary curve

$k + O(\log n) + O(m)$ *and length* $n + O(\log n) + O(m)$ *for which the boundary curve of* P_x *coincides with the line* $(0, t_0)$–$(1, t_1)$–$\ldots$–(k, t_k) *with* $O(\log n) + O(m)$ *precision: the distance between the set* P_x *and the set* $T = \{(i, j) \mid (i < k) \Rightarrow (j > t_i)\}$ *is bounded by* $O(\log n) + O(m)$.

(We say that the distance between two subsets $P, Q \subset \mathbb{Z}^2$ is at most ε if P is contained in the ε-neighborhood of Q and vice versa.)

Proof. For every i in the range $0 \ldots k$ we list all the sets of complexity at most i and size at most 2^{t_i}. For a given i the union of all these sets is denoted by S_i. It contains at most 2^{i+t_i} elements. (Here and later we omit constant factors and factors polynomial in n when estimating cardinalities, since they correspond to $O(\log n)$ additive terms for lengths and complexities.) Since the sequence t_i strictly decreases (this corresponds to slope -1 in the picture), the sums $i + t_i$ do not increase, therefore each S_i has at most $2^{t_0} \leqslant 2^n$ elements. The union of all S_i therefore also has at most 2^n elements (up to a polynomial factor, see above). Therefore, we can find a string of length n (actually $n + O(\log n)$) that does not belong to any S_i. Let x be a first such string in some order (e.g., in the lexicographic order).

By construction, the set P_x lies above the curve determined by t_i. So we need to estimate the complexity of x and prove that P_x follows the curve (i.e., that T is contained in the neighborhood of P_x).

Let us start with the upper bound for the complexity of x. The list of all objects of complexity at most k plus the full table of their complexities have complexity $k + O(\log k)$, since it is enough to know k and the number of terminating programs of length at most k. Except for this list, to specify x we need to know n and the sequence $t_0, \ldots, t_k$, whose complexity is m.

The lower bound: the complexity of x cannot be less than k since all the singletons of this complexity were excluded (via S_k).

It remains to show that for every $i \leqslant k$ we can put x into a set A of complexity i (or slightly bigger) and size 2^{t_i} (or slightly bigger). For this we enumerate a sequence of sets of correct size and show that one of the sets will have the required properties; if this sequence of sets is not very long, the complexity of its elements is bounded. Here are the details.

We start by taking the first 2^{t_i} strings of length n as our first set A. Then we start enumerating all finite sets of complexity at most j and of size at most 2^{t_j} for all $j = 0, \ldots, k$, and get an enumeration of all sets S_j. Recall that all elements of all S_j should be deleted (and the minimal remaining element should eventually be x). So, when a new set of complexity at most j and of size at most 2^{t_j} appears, all its elements are included in S_j and deleted. Until all elements of A are deleted, we have nothing to worry about, since A is covering the minimal remaining element. If (and when) all elements of A are deleted, we replace A by a new set that consists of first 2^{t_i} undeleted (yet) strings of length n. Then we wait again until all the elements of this new A are deleted, if (and when) this happens, we take 2^{t_i} first undeleted elements as new A, etc.

The construction guarantees the correct size of the sets and that one of them covers x (the minimal non-deleted element). It remains to estimate the complexity of the sets we construct in this way.

First, to start the process that generates these sets, we need to know the length n (actually something logarithmically close to n) and the sequence $t_0, \ldots, t_k$. In total we need $m + O(\log n)$ bits. To specify each version of A, we need to add its version number. So we need to show that the number of different A's that appear in the process is at most 2^i or slightly bigger.

A new set A is created when all the elements of the old A are deleted. These changes can be split into two groups. Sometimes a new set of complexity j appears with $j \leqslant i$. This can happen only $O(2^i)$ times since there are at most $O(2^i)$ sets of complexity at most i. So we may consider the other changes (excluding the first changes after each new large set was added). For those changes all the elements of A are gone due to elements of S_j with $j > i$. We have at most 2^{j+t_j} elements in S_j. Since $t_j + j \leqslant t_i + i$, the total number of deleted elements only slightly exceeds 2^{t_i+i}, and each set A consists of 2^{t_i} elements, so we get about 2^i changes of A.

Remark 6. It is easy to modify the proof to get a string x of length exactly n. Indeed, we may consider slightly smaller bad sets: decreasing the logarithms of their sizes by $O(\log n)$, we can guarantee that the total number of elements in all bad sets is less than 2^n. Then there exists a string of length n that does not belong to bad sets. In this way the distance between T and P_x may increase by $O(\log n)$, and this is acceptable.

Theorem 1 shows that the value of the complexity of x does not describe the properties of x fully; different strings of the same complexity x can have different boundary curves of P_x. This curve can be considered as an "infinite-dimensional" characterization of x.

Strings x with minimal possible P_x (Fig. 2, the upper curve) may be called *antistochastic*. They have quite unexpected properties. For example, if we replace some bits of an antistochastic string x by stars (or some other symbols indicating erasures) leaving only $K(x)$ non-erased bits, then the string x can be reconstructed from the resulting string x' with logarithmic advice, i.e., $K(x|x') = O(\log n)$. This and other properties of antistochastic strings were discovered in [24].

3.4   Optimality and Randomness Deficiency

In this section we establish the connection between optimality and randomness deficiency. As we have seen, the optimality deficiency can be bigger than the randomness deficiency (for the same description), and the difference is $\delta(x, A) - d(x|A) = K(A) + K(x|A) - K(x)$. The Levin–Gács formula for the complexity of pair ($K(u, v) = K(u) + K(v|u)$ with logarithmic precision, for $O(1)$-precision one needs to add $K(u)$ in the condition, but we ignore logarithmic size terms anyway) shows that the difference in question can be rewritten as

$$\delta(x, A) - d(x|A) = K(A, x) - K(x) = K(A|x).$$

So if the difference between deficiencies for some $(i * j)$-description A of x is big, then $K(A|x)$ is big. All the $(i * j)$-descriptions of x can be enumerated if x, i, and j are given. So the large value of $K(A|x)$ for some $(i * j)$-description A means that there are many $(i * j)$-descriptions of x, otherwise A can be reconstructed from x by specifying i, j (requires $O(\log n)$ bits) and the ordinal number of A in the enumeration. We will prove that if there are many $(i * j)$-descriptions for some x, then there exist a description with better parameters.

Now we explain this in more detail. Let us start with the following remark. Consider all strings that have $(i * j)$-descriptions for some fixed i and j. They can be enumerated in the following way: we enumerate all finite sets of complexity at most i, select those sets that have size at most 2^j, and include all elements of these sets into the enumeration. In this construction

- the complexity of the enumerating algorithm is logarithmic (it is enough to know i and j);
- we enumerate at most 2^{i+j} elements;
- the enumeration is divided into at most 2^i "portions" of size at most 2^j.

It is easy to see that any other enumeration process with these properties enumerates only objects that have $(i * j)$-descriptions (again with logarithmic precision). Indeed, each portion is a finite set that can be specified by its ordinal number and the enumeration algorithm, the first part requires $i + O(\log i)$ bits, the second is of logarithmic size according to our assumption.

Remark 7. The requirement about the portion size is redundant. Indeed, we can change the algorithm by splitting large portions into pieces of size 2^j (the last piece may be incomplete). This, of course, increases the number of portions, but if the total number of enumerated elements is at most 2^{i+j}, then this splitting adds at most 2^i pieces. This observation looks (and is) trivial, still it plays an important role in the proof of the following proposition.

Proposition 9. If a string x of length n has at least 2^k different (i,j)-descriptions, then x has some $(i*(j-k))$-description and even some $((i-k)*j)$-description.

Again we omit logarithmic term: in fact one should write $((i + O(\log n)) * (j-k+O(\log n)))$, etc. The word "even" in the statement refers to Proposition 8 that shows that indeed the second claim is stronger.

Proof. Consider the enumeration of all objects having $(i*j)$-descriptions in 2^i portions of size 2^j (we ignore logarithmic additive terms and respective polynomial factors) as explained above. After each portion (i.e., new $(i*j)$-description) appears, we count the number of descriptions for each enumerated object and select objects that have at least 2^k descriptions. Consider a new enumeration process that enumerates only these "rich" objects (rich = having many descriptions). We have at most 2^{i+j-k} rich objects (since they appear in the list of size 2^{i+j} with multiplicity 2^k), enumerated in 2^i portions (new portion of rich objects may appear only when a new portion appears in the original enumeration). So we apply the observation above to conclude that all rich objects have $(i*(j-k))$-descriptions.

To get the second (stronger) statement we need to decrease the number of portions (while not increasing too much the number of enumerated objects). This can be done using the following trick: when a new rich object (having 2^k descriptions) appears, we enumerate not only rich objects, but also "half-rich" objects, i.e., objects that currently have at least $2^k/2$ descriptions. In this way we enumerate more objects — but only twice more. At the same time, after we dumped all half-rich objects, we are sure that next $2^k/2$ new $(i*j)$-descriptions will not create new rich objects, so the number of portions is divided by $2^k/2$, as required. $\square$

Let us say more accurately how we deal with logarithmic terms. We may assume that $i, j = O(n)$, otherwise the claim is trivial. Then we allow polynomial (in n) factors and $O(\log n)$ additive terms in all our considerations.

Remark 8. If we unfold this construction, we see that new descriptions (of smaller complexity) are not selected from the original sequence of descriptions but constructed from scratch. In Sect. 6 we deal with much more complicated case where we restrict ourselves to descriptions from some class (say, Hamming balls). Then the proof given above does not work, since the description we construct is not a ball even if we start with ball descriptions. Still some other (much more ingenious) argument can be used to prove a similar result for the restricted case.

Now we are ready to prove the promised results (see the discussion after Example 1).

Theorem 2. *If a string x of length n is (α, β)-stochastic, then there exists some finite set B containing x such that $\mathrm{K}(B) \leqslant \alpha + O(\log n)$ and $\delta(x, B) \leqslant \beta + O(\log n)$.*

Proof. Since x is (α, β)-stochastic, there exists some finite set A such that $K(A) \leqslant \alpha$ and $d(x \mid A) \leqslant \beta$. Let $i = K(A)$ and $j = \log \# A$, so A is an $(i * j)$-description of x. We may assume without loss of generality that both α and β (and therefore i and j) are $O(n)$, otherwise the statement is trivial. The value $\delta(x, A)$ may exceed $d(x \mid A)$, as we have discussed at the beginning of this section. So we assume that

$$k = \delta(x, A) - d(x \mid A) > 0;$$

if not, we can let $B = A$. Then, as we have seen, $K(A \mid x) \geqslant k - O(\log n)$, and there are at least $2^{k - O(\log n)}$ different $(i * j)$-descriptions of x. According to Proposition 9, there exists some finite set B that is an $(i * (j - k + O(\log n)))$-description of x. Its optimality deficiency $\delta(x, B)$ is $(k - O(\log n))$-smaller (compared to A) and therefore $O(\log n)$-close to $d(x \mid A)$. □

In this argument we used the simple part of Proposition 9. Using the stronger statement about complexity decrease, we get the following result:

Theorem 3 ([43]). *Let A be a finite set containing a string x of length n and let $k = \delta(x, A) - d(x \mid A)$. Then there is a finite set B containing x such that $K(B) \leqslant K(A) - k + O(\log n)$ and $\delta(x, B) \leqslant d(x \mid A) + O(\log n)$.*

Proof. Indeed, if B is an $((i - k) * j)$-description of x (up to logarithmic terms, as usual), then its optimality deficiency is again $(k - O(\log n))$-smaller (compared to A) and therefore $O(\log n)$-close to $d(x \mid A)$. □

Note that the statement of the theorem implies that $d(x \mid B) \leqslant d(x \mid A) + O(\log n)$.

Theorem 2 and Proposition 7 show that we can replace the randomness deficiency in the definition of (α, β)-stochastic strings by the optimality deficiency (with logarithmic precision). More specifically, for every string x of length n consider the sets

$$Q_x = \{(\alpha, \beta) \mid x \text{ is } (\alpha, \beta)\text{-stochastic}\},$$

and

$$\tilde{Q}_x = \{(\alpha, \beta) \mid \text{there exists } A \ni x \text{ with } K(A) \leqslant \alpha, \ \delta(x, A) \leqslant \beta\}.$$

Then these sets are at most $O(\log n)$ apart (each is contained in the $O(\log n)$-neighborhood of the other one).

This remark, together with the existence of antistochastic strings of given complexity and length, allows us to improve the result about the existence of non-stochastic objects (Proposition 4).

Proposition 10 ([13, Theorem IV.2]). *For some c and for all n: if $\alpha + \beta < n - c \log n$, there exist strings of length n that are not (α, β)-stochastic.*

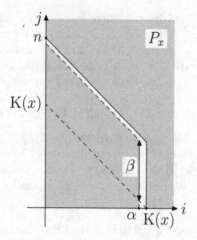

Fig. 3. Non-stochastic strings revisited. Left gray area corresponds to descriptions A with $K(A) \leqslant \alpha$ and $\delta(x, A) \leqslant \beta$.

Proof. Assume that integers n, α, β are given such that $\alpha + \beta < n - c\log n$ (where the constant c will be chosen later). Let x be an antistochastic string of length n that has complexity $\alpha + d$ where d is some positive number (see below about the choice of d). More precisely, for every given d there exists a string x whose complexity is $\alpha + d + O(\log n)$, length is $n + O(\log n)$, and the set P_x is $O(\log n)$-close to the upper gray area (Fig. 3).

Assume that x is (α, β)-stochastic. Then (Theorem 2) the string x has an $(i * j)$-description with $i \leqslant \alpha$ and $i + j \leqslant K(x) + \beta$ (with logarithmic precision). The set of pairs (i, j) satisfying these inequalities is shown as the lower gray area. We have to choose c in such a way that for some d these two gray are disjoint and even separated by a gap of logarithmic size (since they are known only with $O(\log n)$-precision). Note first that for $d = c'\log n$ with large enough c' we guarantee the vertical gap (the vertical segments of the boundaries of two gray areas are far apart). Then we select c large enough to guarantee that the diagonal segments of the boundaries of two gray areas are far apart ($\alpha + \beta < n$ with logarithmic margin). □

The transition from randomness deficiency to optimality deficiency (Theorem 2) has the following geometric interpretation.

Theorem 4. *The sets Q_x and P_x are related to each other via an affine transformation $(\alpha, \beta) \mapsto (\alpha, K(x) - \alpha + \beta)$, as Fig. 4 shows.*[5]

As usual, this statement is true with logarithmic accuracy: the distance between the image of the set Q_x under this transformation and the set P_x is claimed to be $O(\log n)$ for string x of length n.

---

[5] Technically speaking, this holds only for $\alpha \leqslant K(x)$. For $\alpha > K(x)$ both sets contain all pairs with first component α.

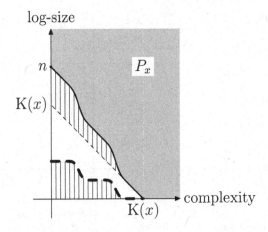

log-size

n P_x

$K(x)$

complexity

$K(x)$

Fig. 4. The set P_x and the boundary of the set Q_x (bold dotted line); on every vertical line two intervals have the same length.

Proof. As we have seen, we may use the optimality deficiency instead of randomness deficiency, i.e., use the set $\tilde{Q}_x$ in place of Q_x. The preimage of the pair (i, j) under our affine transformation is the pair $(i, i + j - K(x))$. Hence we have to prove that a pair (i, j) is in P_x if and only if the pair $(i, i + j - K(x))$ is in $\tilde{Q}_x$. Note that $K(A) = i$ and $\log \#A = j$ is equivalent to $K(A) = i$ and $\delta(x, A) = i + j - K(x)$ just by definition of $\delta(x, A)$. (See Fig. 4: the optimality deficiency of a description A with $K(A) = i$ and $\log \#A = j$ is the vertical distance between (i, j) and the dotted line.)

But there is some technical problem: in the definition of P_x we used inequalities $K(A) \leqslant i$ and $\log \#A \leqslant j$, not the equalities $K(A) = i$ and $\log \#A = j$. The same applies to the definition of $\tilde{Q}_x$. So we have two sets that correspond to each other, but their $\leqslant$-closures could be different. Obviously, $K(A) \leqslant i$ and $\log \#A \leqslant j$ imply $K(A) \leqslant i$ and $K(A) + \log \#A - K(x) \leqslant i + j - K(x)$, but not vice versa.

In other words, the set of pairs $(K(A), \log \#A)$ satisfying the latter inequalities (see the right set on Fig. 5) is bigger than the set of pairs $(K(A), \log \#A)$ satisfying the former inequalities (see the left set on Fig. 5). Now Proposition 8 helps: we may use it to convert any set with parameters from the right region into a set with parameters from the left region. □

Remark 9. Let us stress again that Theorem 2 claims only that the *existence* of a set $A \ni x$ with $K(A) \leqslant \alpha$ and $d(x|A) \leqslant \beta$ is equivalent to the existence of a set $B \ni x$ with $K(B) \leqslant \alpha$ and $\delta(x|A) \leqslant \beta$ (with logarithmic accuracy). The theorem does *not* claim that for *every* set $A \ni x$ with complexity at most α the inequalities $d(x|A) \leqslant \beta$ and $\delta(x, A) \leqslant \beta$ are equivalent (with logarithmic accuracy). Indeed, the Example 1 shows that this is not true: the first inequality does not imply the second one in general case. However, Theorems 2 and 3 show that this can happen only for non-minimal descriptions (for which the description

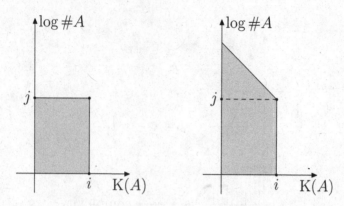

Fig. 5. The left picture shows (for given i and j) the set of all pairs $(K(A), \log \#A)$ such that $K(A) \leqslant i$ and $\log \#A \leqslant j$; the right picture shows the pairs $(K(A), \log \#A)$ such that $K(A) \leqslant i$ and $\delta(x, A) \leqslant i + j - K(x)$.

with smaller complexity and the same optimality deficiency) exists. Later we will see that all the minimal descriptions of the same (or almost the same) complexity have almost the same information. Moreover, if A and B are minimal descriptions and the complexity of A is less than that of B then $C(A|B)$ is small.

For the people with taste for philosophical speculations the meaning of Theorems 2 and 3 can be advertised as follows. Imagine several scientists that compete in providing a good explanation for some data x. Each explanation is a finite set A containing x together with a program p that computes A.

How should we compare different explanations? We want the randomness deficiency $d(x|A)$ of x in A to be negligible (no features of x remain unexplained). Among these descriptions we want to find the simplest one (with the shortest p). That is, we look for a set A corresponding to the point where the bold dotted line on Fig. 4 touches the horizontal axis. (In fact, there is always some trade-off between the parameters, not the specific exact point where the curve touches the horizontal axis, but we want to keep the discussion simple though imprecise.)

However, this approach meets the following obstacle: we are unable to compute randomness deficiency $d(x|A)$. Moreover, the inventor of the model A has no ways to convince us that the deficiency is indeed negligible if it is the case (the function $d(x|A)$ is not even upper semicomputable). What could be done? Instead, we may look for an explanation with (almost) minimal sum $\log \#A + |p|$ (minimum description length principle). Note that this quantity is known for competing explanation proposals. Theorems 2 and 3 provide the connection between these two approaches.

Returning to mathematical language, we have seen in this section that two approaches (based on $(i * j)$-descriptions and (α, β)-stochasticity) produce essentially the same curve, though in different coordinates. The other ways to get the same curve will be discussed in Sects. 4 and 5.

3.5 Historical Remarks

The idea to consider $(i * j)$-descriptions with optimal parameters can be traced back to Kolmogorov. There is a short record for his talk given in 1974 [16]. Here is the (full) translation of this note:

> For every constructive object x we may consider a function $\Phi_x(k)$ of an integer argument $k \geqslant 0$ defined as a logarithm of the minimal cardinality of a set of complexity at most k containing x. If x itself has a simple definition, then $\Phi_x(1)$ is equal to one [a typo: cardinality equals 1, and logarithm equals 0] already for small k. If such a simple definition does not exist, x is "random" in the negative sense of the word "random". But x is positively "probabilistically random" only if the function Φ has a value Φ_0 for some relatively small k and then decreases approximately as $\Phi(k) = \Phi_0 - (k - k_0)$. [This corresponds to approximate $(k_0, 0)$-stochasticity.]

Kolmogorov also gave a talk in 1974 [15]; the content of this talk was reported by Cover [10, Sect. 4, page 31]. Here $l(p)$ stands for the length of a binary string p and $|S|$ stands for the cardinality of a set S.

> 4. **Kolmogorov's H_k Function**
> Consider the function $H_k \colon \{0,1\}^k \to N$, $H_k(x) = \min_{p \colon l(p) \leqslant k} \log |S|$, where the minimum is taken over all subsets $S \subseteq \{0,1\}^n$, such that $x \in S$, $U(p) = S$, $l(p) \leqslant k$. This definition was introduces by Kolmogorov in a talk at the Information Symposium, Tallinn, Estonia, in 1974. Thus $H_k(x)$ is the log of the size of the smallest set containing x over all sets specifiable by a program of k or fewer bits. Of special interest is the value
>
> $$k^*(x) = \min\{k \colon H_k(x) + k = K(x)\}.$$
>
> Note that $\log |S|$ is the maximal number of bits necessary to describe an arbitrary element $x \in S$. Thus a program for x can be written in two stages: "Use p to print the indicator function for S; the desired sequence is the ith sequence in a lexicographic ordering of the elements of this set". This program has length $l(p) + \log |S|$, and $k^*(x)$ is the length of the shortest program p for which this 2-stage description is as short as the best 1-stage description p^*. We observe that x must be maximally random with respect to S — otherwise the 2-stage description could be improved, contradicting the minimality of $K(x)$. Thus $k^*(x)$ and its associated program p constitute a minimal sufficient description for x. $\langle \ldots \rangle$
> Arguments can be provided to establish that $k^*(x)$ and its associated set S^* describe all of the "structure" of x. The remaining details about x are conditionally maximally complex. Thus pp^{**}, the program for S^*, plays the role of a sufficient statistic.

In both places Kolmogorov speaks about the place when the boundary curve of P_x reaches its lower bound determined by the complexity of x.

Later the same ideas were rediscovered and popularized by many people. Kolmogorov himself gave a seminar talk [17] where he asked questions about (α, β)-stochasticity; papers [34, 46] were replies to his questions written by the participants of this seminar. (See also [47] for a survey of Vyugin's work and also his later paper [48] where similar questions were studied in terms of prediction). Koppel in [18] reformulates the definition using total algorithms. Instead of a finite set A he considered a total program P that terminates on all strings of some length. The two-part description of some x is then formed by this program P and the input D for this program that is mapped to x. In our terminology this corresponds to the set A of all values of P on the strings of the same length as D. He writes then [18, p. 1089]

Definition 3. The c-sophistication of a finite string S [is defined as]

$$\text{SOPH}_c(S) = \min\{|P| \mid \exists D \text{ s. t. } (P, D) \text{ is a } c\text{-minimal description of } \alpha\}.$$

There is a typo in this paper: S should be replaced by α (two times). Before in Definition 1 the description is called c-minimal if $|P| + |D| \leqslant H(\alpha) + c$ (here P and D are the program and and its input, respectively, H stands for complexity).

Though this paper (as well as the subsequent papers [19, 20]) is not technically clear (e.g., it does not say what are the requirements for the algorithm U used in the definition, and in [19, 20] only universality is required, which is not enough: if U is not optimal, the definition does not make sense), the philosophic motivation for this notion is explained clearly [18, p. 1087]:

The total complexity of an object is defined as the size of its most concise description. The total complexity of an object can be large while its "meaningful" complexity is low; for example, a random object is by definition maximally complex but completely lacking in structure.

⟨...⟩ The "static" approach to the formalization of meaningful complexity is "sophistication" defined and discussed by Koppel and Atlan [reference to unpublished paper "Program-length complexity, sophistication, and induction" is given, but later a paper of same authors [20] with a similar title appeared]. Sophistication is a generalization of the "H-function" or "minimal sufficient statistic" by Cover and Kolmogorov ⟨...⟩ The sophistication of an object in the size of that part of that object which describes its structure, i.e. the aggregate of its projectible properties.

A similar approach (models as two-part descriptions that consist of a total function and its argument) reappears in a later paper of Vitanyi [45, Sect. II, C].

One can also mention the formulation of "minimal description length" principle by Rissanen [33]; the abstract of this paper says: "Estimates of both integer-valued structure parameters and real-valued system parameters may be obtained from a model based on the shortest data description principle"; here "integer-valued structure parameters" may correspond to the choice of a statistical hypothesis (description set) while "real-valued system parameters" may correspond to the choice of a specific element in this set. The author then says that "by finding the model which minimizes the description length one obtains estimates of both the integer-valued structure parameters and the real-valued system parameters".

We do not try here to follow the development of these and similar ideas. Let us mention only that the traces of the same ideas (though even more vague) could be found in 1960s in the classical papers of Solomonoff [37,38] who tried to use shortest descriptions for inductive inference (and, as a side product, gave the definition of complexity later rediscovered by Kolmogorov [14]). One may also mention a "minimum message length principle" that goes back to [49]; the idea of two-part description is explained in [49] as follows:

> If the things are now classified then the measurements can be recorded by listing the following:
> 1. The class to which each thing belongs.
> 2. The average properties of each class.
> 3. The deviations of each thing from the average properties of its parent class.
> If the things are found to be concentrated in a small area of the region of each class in the measurement space then the deviations will be small, and with reference to the average class properties most of the information about a thing is given by naming the class to which it belongs. In this case the information may be recorded much more briefly than if a classification had not been used. We suggest that the best classification is that which results in the briefest recording of all the attribute information.

Here the "class to which thing belongs" corresponds to a set (statistical model, description in our terminology); the authors say that if this set is small, then only few bits need to be added to the description of this set to get a full description of the thing in question.

The main technical results of this sections (Theorems 1, 2, and 3) are taken from [43] (where some historical account is provided).

4 Bounded Complexity Lists

In this section we show one more classification of strings that turns out to be equivalent (up to coordinate change) to the previous ones: for a given string x and $m \geqslant C(x)$ we look how close x is to the end in the enumeration of all strings of complexity at most m. For technical reasons it is more convenient to use plain complexity $C(x)$ instead of the prefix version $K(x)$. As we have mentioned, the difference between them is only logarithmic, and we mainly ignore terms of that size.

4.1 Enumerating Strings of Complexity at Most m

Consider some integer m, and all strings x of (plain) complexity at most m. Let Ω_m be the number of those strings. The following properties of Ω_m are well known and often used (see, e.g., [8]).

Proposition 11

- $\Omega_m = \Theta(2^m)$ (i.e., $c_1 2^m \leqslant \Omega_m \leqslant c_2 2^m$ for some positive constants c_1, c_2 and for all m;
- $C(\Omega_m) = m + O(1)$.

Proof. The number of strings of complexity at most m is bounded by the total number of programs of length at most m, which is $O(2^m)$. On the other hand, if Ω_m is an $(m-d)$-bit number, we can specify a string of complexity greater than m using $m - d + O(\log d)$ bits: first we specify d in a self-delimiting manner using $O(\log d)$ bits, and then append Ω_m in binary. This information allows us to reconstruct d, then m and Ω_m, then enumerate strings of complexity at most m until we have Ω_m of them (so all strings of complexity at most m are enumerated), and then take the first string x_m that has not been enumerated. As $m < C(x_m) \leqslant m - d + O(\log d)$, the value of d is bounded by a constant and hence Ω_m is an $(m - O(1))$-bit number.

In this argument the binary representation of Ω_m can be replaced by its program, so $C(\Omega_m) \geqslant m - O(1)$. The upper bound $m + O(1)$ is obvious, since $\Omega_m = O(2^m)$. $\qquad \square$

Given m, we can enumerate all strings of complexity at most m. How many steps needs the enumeration algorithm to produce all of them? The answer is provided by the so-called *busy beaver numbers*; let us recall their definition in terms of Kolmogorov complexity (see [42, Sect. 1.2.2] for details).

By definition, the number $B(m)$ is the maximal integer of complexity at most m. It is not hard to see that $C(B(m)) = m + O(1)$. Indeed, $C(B(m)) \leqslant m$ by definition. On the other hand, the complexity of the next number $B(m) + 1$ is greater than m and at the same time is bounded by $C(B(m)) + O(1)$.

Note that $B(m)$ can be undefined for small m (if there are no integers of complexity at most m) and that $B(m+1) \geqslant B(m)$ for all m. For some m this inequality may not be strict. This happen, for example, if the optimal algorithm used to define Kolmogorov complexity is defined only on strings of, say, even lengths; this restriction does not prevent it from being optimal, but then $B(2n) = B(2n+1)$ for all n, since there are no objects of complexity exactly $2n + 1$. However, for some constant c we have $B(m + c) > B(m)$ for all m. Indeed, consider a program p of length at most m that prints $B(m)$. Transform it to a program p' that runs p and then adds 1 to the result. This program witnesses that $C(B(m) + 1) \leqslant m + c$ for some constant c. Hence $B(m + c) \geqslant B(m) + 1$.

Now we define $B'(m)$ as follows. As we have said, the set of all strings of complexity at most m can be enumerated given m. Fix some enumeration algorithm A (with input m) and some computation model. Then let $B'(m)$ be the number of steps used by this algorithm to enumerate all the strings of complexity at most m.

Proposition 12. The numbers $B(m)$ and $B'(m)$ coincide up to $O(1)$-change in m. More precisely, we have

$$B'(m) \leqslant B(m+c), \qquad B(m) \leqslant B'(m+c)$$

for some c and for all m.

Proof. To find $B'(m)$, it is enough to know m-bit binary string that represents Ω_m (this string also determines m). Therefore $C(B'(m)) \leqslant m + c$ for some constant c. As $B(m + c)$ is the largest number of complexity $m + c$ or less, we have $B'(m) \leqslant B(m + c)$.

On the other hand, if some integer N exceeding both m and $B'(m)$ is given, we can run the enumeration algorithm A within N steps for each input smaller than N. Consider the first string that has not been enumerated. Its complexity is greater than m, so $C(N) > m - c$ for some constant c. Thus the complexity of every number N starting from $\max\{m, B'(m)\}$ is greater than $m - c$, which means that $\max\{m, B'(m)\} > B(m - c)$. It remains to note that for all large enough m we have $m \leqslant B(m - c)$, as the complexity of m is $O(\log m)$. Thus for all large enough m the number $B'(m)$ (and not m) must be bigger than $B(m - c)$. Replacing here m by $m + c$ and increasing the constant c if needed, we conclude that $B'(m + c) > B(m)$ for all m. □

A similar argument shows that $B(n)$ coincides (up to $O(1)$-change in the argument) with the maximal computation time of the universal decompressor (from the definition of plain Kolmogorov complexity) on inputs of size at most m, see [42, Sect. 1.2.2]

The next result says how many strings require long time to be enumerated.

Proposition 13. After $B'(m - s)$ steps of the enumeration algorithm on input m there are $2^{s+O(\log m)}$ strings that are not yet enumerated.

We assume that the algorithm enumerates strings (for every input m) without repetitions. Note also that here B' can be replaced by B, since they differ at most by a constant change in the argument.

Proof. To make the notation simpler we omit $O(1)$- and $O(\log m)$-terms in this argument. Given Ω_{m-s}, we can determine $B'(m - s)$. If we also know how many strings of complexity at most m appear after $B'(m - s)$ steps, we can wait until that many strings appear and then find a string of complexity greater than m. If the number of remaining strings is smaller than $2^{s-O(\log m)}$, we get a prohibitively short description of this high complexity string.

On the other hand, let x be the last element that has been enumerated in $B'(m - s)$ steps. If there are significantly more than 2^s elements after x, say, at least 2^{s+d} for some d, we can split the enumeration in portions of size 2^{s+d} and wait until the portion containing x appears. By assumption this portion is full. The number N of steps needed to finish this portion is at least $B'(m - s)$. This number N and its successor $N + 1$ can be reconstructed from the portion number that contains about $m - s - d$ bits. Thus the complexity of $N + 1$ is at most $m - s - d + O(\log m)$. Hence we have

$$B(m - s - d + O(\log m)) > N \geqslant B'(m - s).$$

By Proposition 12 we can replace B' by B here:

$$B(m - s - d + O(\log m)) > B(m - s).$$

(with some other constant in O-notation). Since B is a non-decreasing function, we get $d = O(\log m)$. □

4.2 Ω-like Numbers

G. Chaitin introduced the "Chaitin Ω-number" $\Omega = \sum_k \mathbf{m}(k)$; it can also be defined as the probability of termination if the optimal prefix decompressor is applied to a random bit sequence (see [42, Sect. 5.7]).[6] The numbers Ω_n are finite versions of Chaitin's Ω-number. The information contained in Ω_n increases as n increases; moreover, the following proposition is true. In this proposition we consider Ω_n as a bit string (of length $n + O(1)$) identifying the number Ω_n and its binary representation.

Proposition 14. Assume that $k \leqslant m$. Consider the string $(\Omega_m)_k$ consisting of the first k bits of Ω_m. It is $O(\log m)$-equivalent to Ω_k: both conditional complexities $C(\Omega_k \mid (\Omega_m)_k)$ and $C((\Omega_m)_k \mid \Omega_k)$ are $O(\log m)$.

Proof. This is essentially the reformulation of the previous statement (Proposition 13).

Run the algorithm that enumerates strings of complexity at most m. Knowing $(\Omega_m)_k$, we can wait until less than 2^{m-k} strings are left in the enumeration of strings of complexity at most m; we know that this happens after more than $B(k)$ steps, and in this time we can enumerate all strings of complexity at most k and compute Ω_k. (In this argument we ignore $O(\log m)$-terms, as usual.)

Now the second inequality follows by the symmetry of information property. Indeed, since $C(\Omega_k) = k + O(1)$ and $C((\Omega_m)_k) \leqslant k + O(1)$, the inequality $C(\Omega_k \mid (\Omega_m)_k) = O(\log m)$ implies the inequality $C((\Omega_m)_k \mid \Omega_k) = O(\log m)$.

A direct argument is also easy. Knowing Ω_k and k, we can find the list of all the strings of complexity at most k and the number $B'(k)$. Then we make $B'(k)$ steps in the enumeration of the list of strings of complexity at most m. Proposition 13 then guarantees that at that moment Ω_m is known with error about 2^{m-k}, so the first k bits of Ω_m can be reconstructed with small advice (of logarithmic size; we omit terms of that size in the argument). □

There is a more direct connection with Chaitin's Ω-number: one can show that the number Ω_m is $O(\log m)$-equivalent to the m-bit prefix of Chaitin's Ω-number. Since in this survey we restrict ourselves to finite objects, we do not go into details of the proof here, see [42, Sect. 5.7.7].

---

[6] This number depends on the choice of the prefix decompressor, so it is not a specific number but a class of numbers. The elements of this class can be equivalently characterized as random lower semicomputable reals in $[0, 1]$, see [42, Sect. 5.7].

4.3   Position in the List Is Well Defined

We discussed how much time is needed to enumerate all strings of complexity at most m and how many strings remain not enumerated before this time. Now we want to study *which* strings remain not enumerated.

More precisely, let x be some string of complexity at most m, so x appears in the enumeration of all strings of complexity at most m. How close x is to the end, that is, how many strings are enumerated after x? The answer depends on the enumeration, but only slightly, as the following proposition shows.

Proposition 15. Let A and B be algorithms that both for any given m enumerate (without repetitions) the set of strings of complexity at most m. Let x be some string and let a_x and b_x the number of strings that appear after x in A- and B-enumerations. Then $|\log a_x - \log b_x| = O(\log m)$.

We may also assume that A and B are algorithms of complexity $O(\log m)$ without input that enumerate strings of complexity at most m.

Proof. Assume that a_x is small: $\log a_x \leqslant k$. Why $\log b_x$ cannot be much larger than k? Given the first $m - \log b_x$ bits of Ω_m and B, we can compute a finite set of strings B' that contains x and consists only of strings of complexity at most m. Then we can wait until all strings from B' appear in A-enumeration. After then at most 2^k strings are left, and we need k bits to count them. In this way we can describe Ω_m by $m - \log b_x + k + O(\log m)$ bits; however, Proposition 11 says that $C(\Omega_m) = m + O(1)$. Hence $\log b_x \leqslant k + O(\log m)$.

The other inequality is proven by a symmetric argument. $\square$

In this theorem A and B enumerate exactly the same strings (though in different order). However, the complexity function is essentially defined with $O(1)$-precision only: different optimal programming languages lead to different versions. Let C and $\tilde{C}$ be two (plain) complexity functions; then $\tilde{C}(x) \leqslant C(x) + c$ for some c and for all x. Then the list of all x with $C(x) \leqslant m$ is contained in the list of all x with $\tilde{C}(x) \leqslant m + c$. The same argument shows that the number of elements after x in the first list cannot be much larger than the number of elements after x in the second list. The reverse inequality is not guaranteed, however, even for the same version of complexity (small increase in the complexity bound may significantly increase the number of strings after x in the list). We will return to this question in Sect. 4.4, but let us note first that some increase is guaranteed.

Proposition 16. If for a string x there are at least 2^s elements after x in the enumeration of all strings of complexity at most m, then for every $d \geqslant 0$ there are at least $2^{s+d-O(\log m)}$ strings after x in the enumeration of all strings of complexity at most $m + d$.

Proof. Essentially the same argument works here: if there are much less than 2^{s+d} strings after x in the bigger list, then this bigger list can be determined by 2^{m-s} bits needed to cover x in the smaller list and less than $s + d$ bits needed to count the elements in the bigger list that follow the last covered element. $\square$

The last proposition can be restated in the following way. Let us fix some complexity function and and some algorithm that, given m, enumerates all strings of complexity at most m. Then, for a given string x, consider the function that maps every $m \geqslant C(x)$ to the logarithm of the number of strings after x in the enumeration with input m. Proposition 16 says that d-increase in the argument leads at least to $(d - O(\log m))$-increase of this function (but the latter increase could be much bigger). As we will see, this function is closely related to the set P_x (and therefore Q_x): it is one more representation of the same boundary curve.

4.4   The Relation to P_x

To explain the relation, consider the following procedure for a given binary string x. For every $m \geqslant C(x)$ draw the line $i + j = m$ on (i, j)-plane. Then draw the point on this line with second coordinate s where s is the logarithm of the number of elements after x in the enumeration of all strings of complexity at most m. Mark also all points on this line on the right of (=below) this point. Doing this for different m, we get a set (Fig. 6). Proposition 16 guarantees that this set is upward closed with logarithmic precision: if some point (i, j) belongs to this set, then the point $(i, j + d)$ is in $O(\log(i + j))$-neighborhood of this set. This implies that the point $(i + d, j)$ is also in the neighborhood, since our set is closed by construction in the direction $(1, -1)$.

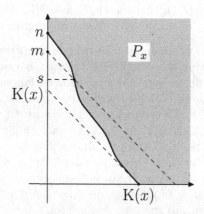

Fig. 6. For each m between $K(x)$ and n (length of x) we count elements after x in the list of strings having complexity at most m; assuming there is about 2^s of them, we draw point $(m - s, s)$ and get a point on some curve. This curve turns out to be the boundary of P_x (with logarithmic precision).

It turns out that this set coincides with P_x (Definition 3) with $O(\log n)$-precision for a string x of length n (this means, as usual, that each of the two sets is contained in the $O(\log n)$-neighborhood of the other one):

Theorem 5. *Let x be a string of length n. If x has a $(i * j)$-description then x is at least $2^{j-O(\log n)}$-far from the end of $(i + j + O(\log n))$-list. Conversely, if there are at least 2^j elements that follow x in the $(i + j)$-list then x has a $((i + O(\log n)) * j)$-description.*

Proof. We need to verify two things. First, assuming that x has a $(i * j)$-description, we need to show that it is at least 2^j-far from the end of $(i + j)$-list. (With error terms: in $(i + j + O(\log n))$-list there are at least $2^{j-O(\log n)}$ elements after x.) Indeed, knowing some $(i * j)$-description A for x, we can wait until all the elements of A appear in $(i + j)$-list (as usual, we omit $O(\log n)$-term: all elements of A have complexity at most $i + j + O(\log n)$, so we should consider $(i + j + O(\log n))$-list to be sure that it contains all elements of A). In particular, x has appeared at that moment. If there are (significantly) less than 2^j elements after x, then we can encode the number of remaining elements by (significantly) less than j bits, and together with the description of A we get less than $i + j$ bits to describe Ω_{i+j}, which is impossible.

Second, assume that there are at least 2^j elements that follow x in the $(i+j)$-list. Then, splitting this list into 2^j-portions, we get at most 2^i full portions, and x is covered by one of them. Each portion has complexity at most i and log-size at most j, so we get an $(i * j)$-description for x. (As usual, logarithmic terms are omitted.) □

Now we can reformulate the properties of stochastic and antistochastic objects. Every object of complexity k appears in the list of objects of complexity at most k' for all $k' > k$. Each stochastic object is far from the end of these lists (except, may be, for some k'-lists with k' very close to k). Each antistochastic object of length n is maximally close to the end of all k'-lists with $k' < n$ (there are about $2^{k'-k}$ objects after x), except, may be, for some k'-lists with k' very close to n. When k' becomes greater than n, then even antistochastic strings are far from the end of the k'-list. What we have said is just the description of the corresponding curves (Fig. 2) using Theorem 5.

4.5   Standard Descriptions

The lists of objects of bounded complexity provide a natural class of descriptions. Consider some m and the number Ω_m of strings of complexity at most m. This number can be represented in binary:

$$\Omega_m = 2^a + 2^b + \ldots,$$

where $a > b > \ldots$. The list itself then can be split into pieces of size 2^a, 2^b, ..., and these pieces can be considered as description of corresponding objects. In this way for each string x and for each $m \geqslant C(x)$ we get some description on x, a piece than contains x. Descriptions obtained in this way will be called *standard* descriptions. Note that for a given x we have many standard descriptions (depending on the choice of m). One should have in mind also that the class of standard descriptions depends on the choice of the complexity function and the enumeration algorithm, and we assume in the sequel that they are fixed.

The following results show that standard descriptions are in a sense universal. First let us note that the standard descriptions have parameters close to the boundary curve of P_x (more precisely, to the boundary curve of the set constructed in the previous section that is close to P_x).[7]

Proposition 17. Consider the standard description A of size 2^j obtained from the list of all strings of complexity at most m. Then $C(A) = m - j + O(\log m)$, and the number of elements in the list that follow the elements of A is $2^{j+O(\log m)}$.

This statement says that parameters of A are close to the point on the line $i + j = m$ considered in the previous section (Fig. 6).

Proof. To specify A, it is enough to know the first $m - j$ bits of Ω_m (and m itself). The complexity of A cannot be much smaller, since knowing A and the j least significant bits of Ω_m we can reconstruct Ω_m.

The number of elements that follow A cannot exceed 2^j (it is a sum of smaller powers of 2); it cannot be significantly less since it determines Ω_m together with the first $m - j$ bits of Ω. (In other words, since Ω_m is an incompressible string of length m, it cannot have more that $O(\log m)$ zeros in a row.)   □

This result does *not* imply that every point on the boundary of P_x is close to parameters of some standard description. If some part of the boundary has slope -1, we cannot guarantee that there are standard descriptions along this part. For example, consider the list of strings of complexity at most m; the maximal complexity of strings in this list is $m - c$ for some $c = O(1)$; if we take first string of this complexity, there are $2^{m+O(1)}$ strings after it, so the corresponding point is close to the vertical axis, and due to Proposition 16 all other standard descriptions of x are also close to the vertical axis. However, descriptions with parameters close to arbitrary points on the boundary of P_x can be obtained from standard descriptions by chopping them into smaller parts, as in Proposition 8. In that shopping it is natural to use the order in which the strings were enumerated. In other words, chop the list of strings of complexity at most m into portions of size 2^j. Consider all the full portions (of size exactly 2^j) obtained in this way (they are parts of standard descriptions of bigger size). Descriptions obtained in this way are "universal" in the following sense: if a pair (i, j) is on the boundary of P_x then there is a set $A \ni x$ of this type of complexity $i + O(\log(i + j))$ and log-cardinality $j + O(\log(i + j))$.

---

[7] In general, if two sets X and Y in $\mathbb{N}^2$ are close to each other (each is contained in the small neighborhood of the other one), this does not imply that their boundaries are close. It may happen that one set has a small "hole" and the other does not, so the boundary of the first set has points that are far from the boundary of the second one. However, in our case both sets are closed by construction in two different directions, and this implies that the boundaries are also close.

The following result says more: for every description A for x there is a "better" standard description that is simple given A (note that $d \geqslant 0$ in the following proposition and that optimality deficiency of B does not exceed that of A up to logarithmic term).

Proposition 18. Let A be an $(i * j)$-description of a string x of length n. Then there exists a standard description B that has parameters $C(B) \leqslant i - d + O(\log n)$ and $\log \#B \leqslant j + d + O(\log n)$ for some $d \geqslant 0$, and is simple given A, i.e., $C(B|A) = O(\log n)$.

Proof. If A has strings of length different from n, remove all those strings. In this way A becomes $(i * j)$-description for x with slightly larger i than before the removal and the same or smaller j. Now all the elements of A have complexity at most $m = i + j + O(\log j) = i + j + O(\log n)$, where the latter inequality holds, as after removal we have $j \leqslant n$. Consider the list of all strings of complexity at most m and the standard description B of x obtained from this list. As we know from Proposition 17, the sum of the parameters of this description is close to m (and therefore to $i + j$). We need to show that the size of B is large, at least $2^{j - O(\log n)}$ (recall that d in the statement should be positive). Why is this the case? Consider elements that appear after the last element of A in the list. There are at least $2^{j - O(\log n)}$ of them, otherwise the total number of elements in the list could be described in much less than m bits (that number can be specified by m, A and the number of elements after the last element of A). Therefore there are at least $2^{j - O(\log n)}$ elements in the list that appear after x, so B cannot be small.

Why B is simple given A? Denote the size of B by $2^{j'}$. Given A and m, we can find the last element of A, call it x', in the list of strings of complexity at most m. Chop the list into portions of size $2^{j'}$. Then B is the last complete portion. If B contains x', we can find B from m, j', and x' as the complete portion containing x'. Otherwise, x' appears in the list after all the elements from B. In this case we can find B from m and x' as the last complete portion before x'. Thus in any case we are able to find B from m, j', and x' plus one extra bit. □

For the same reason every standard description B of some x is simple given x (and this is not a surprise, since we know that all optimal descriptions of x are simple given x, see Proposition 9).

Proposition 18 has the following corollary which we formulate in an informal way. Let A be some $(i * j)$-description with parameters on the boundary of P_x. Assume that on the left of this point the boundary curve decreases fast (with slope less than -1). Then in Proposition 18 the value of d is small, otherwise the point $(i - d, j + d)$ would be far from P_x. So the complexities of A and the standard description B are close to each other. We know also that A is simple given B, therefore B is also simple given A, and A and B have the same information (have small conditional complexities in both directions).

If we have two different descriptions A, A' with approximately the same parameters on the boundary of P_x, and the curve decreases fast on the left of the corresponding boundary point, the same argument shows that A and A' have the same information. Note that the condition about the slope is important: if the point is on the segment with slope -1, the situation changes. For example, consider a random n-bit string x and two its descriptions. The first one consists of all n-bit strings that have the same left half as x, the second one consists of all n-bit strings that have the same right half. Both have the same parameters: complexity $n/2$ and log-size $n/2$, so they both correspond to the same point on the boundary of P_x. Still the information in these two descriptions is different (left and right halves of a random string are independent).

These results sound as good news. Let us recall our original goal: to formalize what is a good statistical model. It seems that we are making some progress. Indeed, for a given x we consider the boundary curve P_x and look at the place when it first touches the lower bound $i + j = \mathrm{C}(x)$; after that it stays near this bound. In other terms, we consider models with negligible optimality deficiency, and select among them the model with minimal complexity. Giving a formal definitions, we need to fix some threshold ε. Then we say that a set A is a ε-sufficient statistic if $\delta(x, A) < \varepsilon$, and may choose the simplest one among them and call it the *minimal ε-sufficient statistic*. If the curve goes down fast on the left of this point, we see that all the descriptions with parameters corresponding to minimal sufficient statistic are equivalent to each other.

Trying to relate these notion to practice, we may consider the following example. Imagine that we have digitized some very old recording and got some bit string x. There is a lot of dust and scratches on the recording, so the originally recorded signal is distorted by some random noise. Then our string x has a two-part description: the first part specifies the original recording and the noise parameters (intensity, spectrum, etc.) and the second part specifies the noise exactly. May be, the first part is the minimal sufficient statistic — and therefore sound restoration (and lossy compression in general) is a special case of the problem of finding a minimal sufficient statistic? The uniqueness result above (saying that all the minimal sufficient statistics contain the same information under some conditions) seem to support this view: different good models for the same object contain the same explanation.

Still the following observation (that easily follows from what we know) destroys this impression completely.

Proposition 19. Let B be some standard description of complexity i obtained from the list of all strings of complexity at most m. Then B is $O(\log m)$-equivalent to Ω_i.

This looks like a failure. Imagine that we wanted to understand the nature of some data string x; finally we succeed and find a description for x of reasonable complexity and negligible randomness and optimality deficiencies (and all the good properties we dreamed of). But Proposition 19 says that the information contained in this description is more related to the computability theory than to specific properties of x. Recalling the construction, we see that the corresponding standard description is determined by some prefix of some Ω-number, and is an interval in the enumeration of objects of bounded complexity. So if we start with two old recordings, we may get the same information, which is not what we expect from a restoration procedure. Of course, there is still a chance that some Ω-number was recorded and therefore the restoration process indeed should provide the information about it, but this looks like a very special case that hardly should happen for any practical situation.

What could we do with this? First, we could just relax and be satisfied that we now understand much better the situation with possible descriptions for x. We know that every x is characterized by some curve that has several equivalent definitions (in terms of stochasticity, randomness deficiency, position in the enumeration — as well as time-bounded complexity, see Sect. 5 below). We know that standard descriptions cover the parts of the curve where it goes down fast, and to cover the parts where the slope is -1 one may use standard descriptions and their pieces; all these descriptions are simple given x. When curve goes down fast, the description is essentially unique (all the descriptions with the same parameters contain the same information, equivalent to the corresponding Ω-number); this is not true on parts with slope -1. So, even if this curve is of no philosophical importance, we have a lot of technical information about possible models.

The other approach is to go farther and consider only models from some class (Sect. 6), or add some additional conditions and look for "strong models" (Sect. 7).

4.6   Non-stochastic Objects Revisited

Now we can explain in a different way why the probability of obtaining a non-stochastic object in a random process is negligible (Proposition 5). This explanation uses the notion of mutual information from algorithmic information theory. The mutual information in two strings x and y is defined as

$$I(x:y) = C(x) - C(x|y) = C(y) - C(y|x) = C(x) + C(y) - C(x,y);$$

all three expressions are $O(\log n)$-close if x and y are strings of length n (see, e.g., [42, Chap. 2]).

Consider an arbitrary string x of length n; let k be the complexity of x. Consider the list of all objects of complexity at most k, and the standard description A for x obtained from this list. If A is large, then x is stochastic; if A is small, then x contains a lot of information about Ω_k and Ω_n.

More precisely, let us assume that A has size 2^{k-s} (i.e., is 2^s times smaller than it could be). Then (recall Proposition 17) the complexity of A is $s + O(\log k)$, since we can construct A knowing k and the first s bits of Ω_k (before the bit that corresponds to A). So we get $(s + O(\log k)) * (k - s)$-description with optimality deficiency $O(\log k)$.

On the other hand, knowing x and k, we can find the ordinal number of x in the enumeration, so we know Ω_k with error at most 2^{k-s}, so $C(\Omega_k|x) \leqslant k - s + O(\log k)$, and $I(x:\Omega_k) \geqslant s - O(\log k)$ (recall that $C(\Omega_k) = k + O(1)$).

In the last statement we may replace Ω_k by Ω_n (where n is the length of x): we know from Proposition 14 that Ω_k is simple given Ω_n, so if condition Ω_k decreases complexity of x by almost s bits, the same is true for condition Ω_n.

Comparing arbitrary $i \leqslant n$ with this s (it can be larger than s or smaller than s), we get the following result:

Proposition 20. Let x be a string of length n. For every $i \leqslant n$

- either x is $(i + O(\log n), O(\log n))$-stochastic,
- or $I(x : \Omega_n) \geqslant i - O(\log n)$.

Now we may use the following (simple and general) observation: for every string u the probability to generate (by a randomized algorithm) an object that contains a lot of information about u is negligible:

Proposition 21. For every string u and for every number d, we have

$$\sum\{\mathbf{m}(x) \mid \mathrm{K}(x) - \mathrm{K}(x|u) \geqslant d\} \leqslant 2^{-d}.$$

In this proposition the sum is taken over all strings x that have the given property (have a large mutual information with u). Note that we have chosen the representation of mutual information that makes the proposition easy (in particular, we have used prefix complexity). As we mentioned, other definitions differ only by $O(\log n)$ if we consider strings x and u of length at most n, and logarithmic accuracy is enough for our purposes.

Proof. Recall the definition of prefix complexity: $\mathrm{K}(x) = -\log \mathbf{m}(x)$, and $\mathrm{K}(x|u) = -\log \mathbf{m}(x|u)$. So $\mathrm{K}(x) - \mathrm{K}(x|u) \geqslant d$ implies $\mathbf{m}(x) \leqslant 2^{-d}\mathbf{m}(x|u)$, and it remains to note that $\sum_x \mathbf{m}(x|u) \leqslant 1$ for every u.   □

Propositions 20 and 21 immediately imply the following improved version of Proposition 5 (page 11):

Proposition 22

$$\sum\{\,\mathbf{m}(x) \mid x \text{ is a } n\text{-bit string that is not } (\alpha, O(\log n))\text{-stochastic}\,\} \leqslant 2^{-\alpha + O(\log n)}$$

for every α.

The improvement here is the better upper bound for the randomness deficiency: $O(\log n)$ instead of $\alpha + O(\log n)$.

4.7   Historical Comments

The relation between busy beaver numbers and Kolmogorov complexity was pointed out in [12] (see Sect. 2.1). The enumerations of all objects of bounded complexity and their relation to stochasticity were studied in [13] (see Sect. 3, E).

5   Computational and Logical Depth

In this section we reformulate the results of the previous one in terms of bounded-time Kolmogorov complexity and discuss the various notions of computational and logical depth that appeared in the literature. (The impatient reader may skip this section; it is not technically used in the sequel).

5.1   Bounded-Time Kolmogorov Complexity

The usual definition of Kolmogorov complexity of x as the minimal length $l(p)$ of a program p that produces x does not take into account the running time of the program p: it may happen that the minimal program for x requires a lot of time to produce x while other programs produce x faster but are longer (for example, program "print x" is rather fast). To analyze this trade-off, the following definition is used.

Definition 4. Let D be some algorithm; its input and output are binary strings. For a string x and integer t, define

$$C_D^t = \min\{l(p)\colon D \text{ produces } x \text{ on input } p \text{ in at most } t \text{ steps}\},$$

the time-bounded Kolmogorov complexity of x with time bound t with respect to D.

This definition was mentioned already in the first paper by Kolmogorov [14]:

Our approach has one important drawback: it does not take into account the efforts needed to transform the program p and object x [the description and the condition] to the object y [whose complexity is defined]. With appropriate definitions, one may prove mathematical results that could be interpreted as the existence of an object x that has simple programs (has very small complexity $K(x)$) but all short programs that produce x require an unrealistically long computation. In another paper I plan to study the dependence of the program complexity $K^t(x)$ on the difficulty t of its transformation into x. Then the complexity $K(x)$ (as defined earlier) reappears as the minimum value of $K^t(x)$ if we remove restrictions on t.

Kolmogorov never published a paper he speaks about, and this definition is less studied than the definition without time bounds, for several reasons.

First, the definition is machine-dependent: we need to decide what computation model is used to count the number of steps. For example, we may consider one-tape Turing machines, or multi-tape Turing machine, or some other computational model. The computation time depends on this choice, though not drastically (e.g., a multi-tape machine can be replaced with a one-tape machine with quadratic increase in time, and most popular models are polynomially related — this observation is used when we argue that the class P of polynomial-time computable functions is well defined).

Second, the basic result that makes the Kolmogorov complexity theory possible is the Solomonoff–Kolmogorov theorem saying that there exists an optimal algorithm D that makes the complexity function minimal up to $O(1)$ additive term. Now we need to take into account the time bound, and get the following (not so nice) result.

Proposition 23. There exists an optimal algorithm D for time-bounded complexity in the following sense: for every other algorithm D' there exists a constant c and a polynomial q such that

$$\mathrm{C}^{t}_{D'}(x) \leqslant \mathrm{C}^{q(t)}_{D}(x) + c$$

for all strings x and integers t.

In this result, by "algorithm" we may mean a k-tape Turing machine, where k is an arbitrary fixed number. However, the claim remains true even when k is not fixed, i.e., we may allow D' to have more tapes than D has.

The proof remains essentially the same: we choose some simple self-delimiting encoding of binary strings $p \mapsto \hat{p}$ and some universal algorithm $U(\cdot, \cdot)$ and then let

$$D(\hat{p}x) = U(p, x)$$

Then the proof follows the standard scheme; the only thing we need to note is that the decoding of $\hat{p}$ runs in polynomial time (which is true for most natural ways of self-delimiting encoding) and that the universal algorithm simulation overhead is polynomial (which is also true for most natural constructions of universal algorithms).

A similar result is true for conditional decompressors, so the conditional time-bounded complexity can be defined as well.

For Turing machines with fixed number of tapes the statement is true for some linear polynomial $q(n) = O(n)$. For the proof we need to consider a universal machine U that simulates other machines efficiently: it should move the program along the tape, so the overhead is bounded by a factor that depends on the size of the program and not on the size of the input or computation time.[8]

Let $t(n)$ be an arbitrary total computable function with integer arguments and values; then the function

$$x \mapsto \mathrm{C}^{t(l(x))}_{D}(x)$$

is a computable upper bound for the complexity $\mathrm{C}(x)$ (defined with the same D; recall that $l(x)$ stands for the length of x). Replacing the function $t(\cdot)$ by a bigger function, we get a smaller computable upper bound. An easy observation: in this way we can match every computable upper bound for Kolmogorov complexity.

---

[8] This observation motivates Levin's version of complexity (Kt, see [21, Sect. 1.3, p. 21]) where the program size and logarithm of the computation time are added: linear overhead in computation time matches the constant overhead in the program size. However, this is a different approach and we do not use the Levin's notion of time bounded complexity in this survey.

Proposition 24. Let $\tilde{C}(x)$ be some total computable upper bound for Kolmogorov complexity function based on the optimal algorithm D from Proposition 23. Then there exists a computable function t such that $C_D^{t(l(x))}(x) \leqslant \tilde{C}(x)$ for every x.

Proof. Given a number n, we wait until every string x of length at most n gets a program that has complexity at most $\tilde{C}(x)$, and let $t(n)$ be the maximal number of steps used by these programs. $\square$

So the choice of a computable time bound is essentially equivalent to the choice of a computable total upper bound for Kolmogorov complexity.

In the sequel we assume that some optimal (in the sense of Proposition 23) D is fixed and omit the subscript D in $C_D^t(\cdot)$. Similar notation $C^t(\cdot|\cdot)$ is used for conditional time-bounded complexity.

5.2 Trade-off Between Time and Complexity

We use the extremely fast growing sequence $B(0), B(1), \ldots$ as a scale for measuring time. This sequence grows faster than any computable function (since the complexity of $t(n)$ for any computable t is at most $\log n + O(1)$, we have $B(\log n + O(1)) \geqslant t(n)$). In this scale it does not matter whether we use time or space as the resource measure: they differ at most by an exponential function, and $2^{B(n)} \leqslant B(n + O(1))$ (in general, $f(B(n)) \leqslant B(n + O(1))$ for every computable f). So we are in the realm of general computability theory even if we technically speak about computational complexity, and the problems related to the unsolved P=NP question disappear.

Let x be a string of length n and complexity k. Consider the time-bounded complexity $C^t(x)$ as a function of t. (The optimal algorithm from Proposition 23 is fixed, so we do not mention it in the notation.) It is a decreasing function of t. For small values of t the complexity $C^t(x)$ is bounded by $n + O(1)$ where n stands for the length of x. Indeed, the program that prints x has size $n + O(1)$ and works rather fast. Formally speaking, $C^t(x) \leqslant n + O(1)$ for $t = B(O(\log n))$. As t increases, the value of $C^t(x)$ decreases and reaches $k = C(x)$ as $t \to \infty$. It is guaranteed to happen for $t = B(k + O(1))$, since the computation time for the shortest program for x is determined by this program.

We can draw a curve that reflects this trade-off using B-scale for the time axis. Namely, consider the graph of the function

$$i \mapsto C^{B(i)}(x) - C(x)$$

and the set of points above this graph, i.e., the set

$$D_x = \{(i, j) \mid C^{B(i)}(x) - C(x) \leqslant j\}.$$

Theorem 6 ([2,6]). *The set D_x coincides with the set Q_x with $O(\log n)$-precision for a string x of length n.*

Recall that the set Q_x consists of pairs (α, β) such that x is (α, β)-stochastic (see p. 22).

Proof. As we know from Theorem 4, the sets P_x and Q_x are related by an affine transformation (see Fig. 4). Taking this transformation into account, we need to prove two statements:

- if there exists an $(i * j)$-description A for x, then

$$\mathrm{C}^{B(i+O(\log n))}(x) \leqslant i + j + O(\log n);$$

- if $\mathrm{C}^{B(i)}(x) \leqslant i + j$, then

  there exist an $((i + O(\log n)) * (j + O(\log n)))$-description for x.

Both statements are easy to prove using the tools from the previous section. Indeed, assume that x has an $(i * j)$-description A. All elements of A have complexity at most $i + j + O(\log n)$. Knowing A and this complexity, we can find the minimal t such that $C^t(x') \leqslant i + j + O(\log n)$ for all x' from A. This t can be computed from A, which has complexity i, and an $O(\log n)$-bit advice (the value of complexity). Hence $t \leqslant B(i + O(\log n))$ and $C^t(x) \leqslant i + j + O(\log n)$, as required.

The converse: assume that $C^{B(i)}(x) \leqslant i + j$. Consider all the strings x' that satisfy this inequality. There are at most $O(2^{i+j})$ such strings. Thus we only need to show that given i and j we are able to enumerate all those strings in at most $O(2^i)$ portions.

One can get a list of all those strings x' if $B(i)$ is given, but we cannot compute $B(i)$ given i. Recall that $B(i)$ is the maximal integer that has complexity at most i; new candidates for $B(i)$ may appear at most 2^i times. The candidates increase with time; when this happens, we get a new portion of strings that satisfy the inequality $C^{B(i)}(x) \leqslant i + j$. So we have at most $O(2^{i+j})$ objects including x that are enumerated in at most 2^i portions, and this implies that x has an $((i + O(\log n)) * j)$-description. Indeed, we make all portions of size at most 2^j by splitting larger portions into pieces. The number of portions increases at most by $O(2^i)$, so it remains $O(2^i)$. Each portion (including the one that contains x) has then complexity at most $i + O(\log n)$ since it can be computed with logarithmic advice from its ordinal number.   □

This theorem shows that the results about the existence of non-stochastic objects can be considered as the "mathematical results that could be interpreted as the existence of an object x that has simple programs (has very small complexity $K(x)$) but all short programs that produce x require an unrealistically long computation" mentioned by Kolmogorov (see the quotation above), and the algorithmic statistics can be interpreted as an implementation of Kolmogorov's plan "to study the dependence of the program complexity $K^t(x)$ on the difficulty t of its transformation into x", at least for the simple case of (unrealistically) large values of t.

5.3 Historical Comments

Section 5 has title "logical and computational depth" but we have not defined these notions yet. The name "logical depth" was introduced by C. Bennett in [7]. He explains the motivation as follows:

> Some mathematical and natural objects (a random sequence, a sequence of zeros, a perfect crystal, a gas) are intuitively trivial, while others (e.g., the human body, the digits of π) contain internal evidence of a nontrivial causal history. $\langle\ldots\rangle$
>
> We propose depth as a formal measure of value. From the earliest days of information theory it has been appreciated that information per se is not a good measure of message value. For example, a typical sequence of coin tosses has high information content but little value; an ephemeris, giving the positions of the moon and the planets every day for a hundred years, has no more information than the equations of motion and initial conditions from which it was calculated, but saves its owner the effort of recalculating these positions. The value of a message thus appears to reside not in its information (its absolutely unpredictable parts), nor in its obvious redundancy (verbatim repetitions, unequal digit frequencies), but rather is what might be called its buried redundancy — parts predictable only with difficulty, things the receiver could in principle have figured out without being told, but only at considerable cost in money, time, or computation. In other words, the value of a message is the amount of mathematical or other work plausibly done by its originator, which its receiver is saved from having to repeat.

Trying to formalize this intuition, Bennett suggests the following possible definitions:

> **Tentative Definition 0.1:** A string's depth might be defined as the execution time of its minimal program.

This notion is not robust (it depends on the specific choice of the optimal machine used in the definition of complexity). So Bennett considers another version.

> **Tentative Definition 0.2:** A string's depth at significance level s [might] be defined as the time required to compute the string by a program no more than s bits larger than the minimal program.

We see that Definition 0.2 consider the same trade-off as in Theorem 6, but in reversed coordinates (time as a function of difference between time-bounded and limit complexities). Bennett is still not satisfied by this definition, for the following reason:

> This proposed definition solves the stability problem, but is unsatisfactory in the way it treats multiple programs of the same length. Intuitively, 2^k distinct $(n+k)$-bit programs that compute same output ought to be accorded the same weight as one n-bit program $\langle\ldots\rangle$

In other language, he suggests to consider a priori probability instead of complexity:

> **Tentative Definition 0.3:** A string's depth at significance level s might be defined as the time t required for the string's time-bounded algorithmic probability $P_t(x)$ to rise to within a factor 2^{-s} of its asymptotic time-unbounded value $P(x)$.

Here $P_t(x)$ is understood as a total weight of all self-delimiting programs that produce x in time at most t (each program of length s has weight 2^{-s}). For our case (when we consider busy beaver numbers as time scale) the exponential time increase needed to switch from a priori probability to prefix complexity does not matter. Still Bennett is interested in more reasonable time bounds (recall that in his informal explanation a polynomially computable sequence of π-digits was an example of a deep sequence!), and prefers a priori probability approach. Moreover, he finds a nice reformulation of this definition (almost equivalent one) in terms of complexity:

> Although Definition 0.3 satisfactorily captures the informal notion of depth, we propose a slightly stronger definition for the technical reason that it appears to yield a stronger slow growth property $\langle\ldots\rangle$
> **Definition 1** (Depth of Finite Strings): Let x and w be strings [probably w is a typo: it is not mentioned later] and s a significance parameter. A string's *depth* at significance level s, denoted $D_s(x)$, will be defined as
> $$\min\{T(p)\colon (|p| - |p^*| < s) \wedge (U(p) = x)\},$$
> the least time required to compute it by a s-incompressible program.

Here p^* is a shortest self-delimiting program for p, so its length $|p^*|$ equals $\mathrm{K}(p)$.

Actually, this *Definition 1* has a different underlying intuition than all the previous ones: a string x is deep if *all programs that compute x in a reasonable time, are compressible*. Note the before we required a different thing: that all programs that compute x in a reasonable time are much longer than the minimal one. This is a weaker requirement: one may imagine a long incompressible program that computes x fast. This intuition is explained in the abstract of the paper [7] as follows:

> [We define] an object's "logical depth" as the time required by a standard universal Turing machine to generate it from an input that is algorithmically random.

Bennett then proves a statement (called Lemma 3 in his paper) that shows that his *Definition 1* is almost equivalent to *Tentative Definition 0.3*: the time remains exactly the same, while s changes at most logarithmically (in fact, at most by $\mathrm{K}(s)$). So if we use Bennett's notion of depth (any of them, except for the first one mentioned) with busy beaver time scale, we get the same curve as in our definition.

A natural question arises: is there a direct proof that the output of an incompressible program with not too large running time is stochastic? In fact, yes, and one can prove a more general statement: the output of a *stochastic* program with reasonable running time is stochastic (see Sect. 5.4); note that stochasticity is a weaker condition than incompressibility.

Let us mention also the notion of *computational depth* introduced in [4] (see also the later publications [3,5,27]). There are several versions mentioned in this paper; the first one exchanges coordinates in the Bennett's tentative definition 0.2 (reproduced in [4] as Definition 2.5). The authors write: "The first notion of computational depth we propose is the difference between a time-bounded Kolmogorov complexity and traditional Kolmogorov complexity" (Definition 3.1, where time bound is some function of input length). The other notions of computation depth are more subtle (they use distinguishing complexity or Levin complexity involving the logarithm of the computation time).

The connections between computational/logical depth and sophistication were anticipated for a long time; for example, Koppel writes in [19]:

⟨...⟩ The "dynamic" approach to the formalization of meaningful complexity is "depth" defined and discussed by Bennett [1]. [Reference to an unpublished paper "On the logical 'depth' of sequences and their reducibilities to incompressible sequences".] The depth of an object is the running-time of its most concise description. Since it is reasonable to assume that an object has been generated by its most concise description, the depth of an object can be thought of as a measure of its evolvedness.
Although sophistication is measured in integers [not clear what in meant here: sophistication of S is also a function $c \mapsto SOPH_c(S)$] and depth is measured in functions, it is not difficult to translate to a common range.

Strangely, the direct connection between the most basic versions of these notions (Theorem 6) seems to be noticed only recently in [6, Sect. 3], and [2].

5.4   Why so Many Equivalent Definitions?

We have shown several equivalent (with logarithmic precision and up to affine transformation) ways to defined the same curve:

- (α, β)-stochasticity (Sect. 2);
- two-part descriptions and optimality deficiency, the set P_x (Sect. 3);
- position in the enumeration of objects of bounded complexity (Sect. 4);
- logical/computational depth (resource-bounded complexity, Sect. 5).

One can add to this list a characterization in terms of split enumeration (Sect. 3.4): the existence of $(i*j)$-description for x is equivalent (with logarithmic precision) to the existence of a simple enumeration of at most 2^{i+j} objects in at most 2^i portions (see Remark 7, p. 20, and the discussion before it).

Why do we need so many equivalent definitions of the same curve? First, this shows that this curve is really fundamental — almost as fundamental characterization of an object x as its complexity. As Koppel writes in [18], speaking about (some versions of) sophistication and depth:

One way of demonstrating the naturalness of a concept is by proving the equivalence of a variety of prime facie different formalizations ⟨...⟩. It is hoped that the proof of the equivalence of two approaches to meaningful complexity, one using static resources (program size) and the other using dynamic resources (time), will demonstrate not only the naturalness of the concept but also the correctness of the specifications used in each formalization to ensure robustness and generality.

Another, more technical reason: different results about stochasticity use different equivalent definitions, and a statement that looks quite mysterious for one of them may become almost obvious for another. Let us give two examples of this type (the first one is stochasticity conservation when random noise is added, the second one is a direct proof of Bennett's characterization mentioned above). The first example is the following proposition from [40] (though the proof there is different).

Proposition 25. Let x be some binary string, and let y be another string ("noise") that is conditionally random with respect to x, i.e., $C(y|x) \approx l(y)$. Then the pair (x, y) has the same stochasticity profile as x: the sets Q_x and $Q_{(x,y)}$ are logarithmically close to each other.

Before giving a proof sketch, let us mention that an interesting special case of this proposition is obtained if we consider a string u and its description X with small randomness deficiency: $d(u|X) \approx 0$. Let y be the ordinal number of u in X. Then the small randomness deficiency guarantees that y is conditionally random with respect to x. Then the pair (X, y) has the same stochasticity profile as X. Since this pair is mapped to u by a simple total computable function, we conclude (Proposition 3) that the stochasticity profile of X is contained in the stochasticity profile of u (more precisely, in its $O(\log n + d(u|X))$-neighborhood). (More simple and direct proof of this statement goes as follows: if $\mathcal{U}$ is a description for X that has small complexity and optimality deficiency, we can take the union of all elements of $\mathcal{U}$ that have approximately the same cardinality as X; one can verify easily that this union also has small complexity and optimality deficiency as a description for u.)

The full statement of Proposition 25 would introduce some bound for the difference $l(y) - C(y|x)$ that is allowed to appear in the final estimate for the distance between sets. Recall also that we can speak about profiles of arbitrary finite objects, in particular, pairs of strings, using some natural encoding (Sect. 2.3).

Proof sketch. Using the depth characterization of stochasticity profile, we need to show that
$$C^{B(i)}(x, y) - C(x, y) \approx C^{B(i)}(x) - C(x).$$

Here "approximately" means that these two quantities may differ by a logarithmic term, and also we are allowed to add logarithmic terms to i (see below what does it mean). The natural idea is to rewrite this equality as
$$C^{B(i)}(x, y) - C^{B(i)}(x) \approx C(x, y) - C(x).$$

The right hand side is equal to $C(y|x)$ (with logarithmic precision) due to Kolmogorov–Levin formula for the complexity of a pair (see, e.g., [42, Chap. 2]), and $C(y|x)$ equals $l(y)$, as y is random and independent of x. Thus it suffices to show that the left hand side also equals $l(y)$. To this end we can prove a version of Kolmogorov–Levin formula for bounded complexity and show that the left hand side equals to $C^{B(i)}(y|x)$. Again, since y is random and independent of x, $C^{BB(i)}(y|x)$ equals $l(y)$.

This plan needs clarification. First of all, let us explain which version of Kolmogorov–Levin formula for bounded complexity we need. (Essentially it was published by Longpré in [23] though the statement was obscured by considering time bound as a function of the input length.)

The equality $C(x, y) = C(x) + C(y|x)$ should be considered as two inequalities, and each one should be treated separately.

Lemma

1. There exist some constant c and some polynomial $p(\cdot, \cdot)$ such that

$$C^{p(n,t)}(x, y) \leqslant C^t(x) + C^t(y|x) + c \log n$$

for all n and t and for all strings x and y of length at most n.
2. There exist some constant c and some polynomial $p(\cdot, \cdot)$ such that

$$C^{p(2^n,t)}(x) + C^{p(2^n,t)}(y|x) \leqslant C^t(x, y) + c \log n$$

for all n and t and for all strings x and y of length at most n.

Proof of the Lemma. The proof of this time-bounded version is obtained by a straightforward analysis of the time requirements in the standard proof. The first part says that if there is some program p that produces x in time t, and some program q that produces y from x in time t, then the pair (p, q) can be considered as a program that produces (x, y) in time $\text{poly}(t, n)$ and has length $l(p) + l(q) + O(\log n)$ (we may assume without loss of generality that p and q have length $O(n)$, otherwise we replace them by shorter fast programs).

The other direction is more complicated. Assume that $C^t(x, y) = m$. We have to count for a given x the number of strings y' such that $C^t(x, y') \leqslant m$. These strings (y is one of them) can be enumerated in time $\text{poly}(2^n, t)$, so if there are 2^s of them, then $C^{\text{poly}(2^n, t)}(y|x) \leqslant s + O(\log n)$ (the program witnessing this inequality is the ordinal number of y in the enumeration plus $O(\log n)$ bits of auxiliary information. Note that we do not need to specify t in advance, we enumerate y' in order of increasing time, and y is among first 2^s enumerated strings.

On the other hand, there are at most $2^{m-s+O(1)}$ strings x' for which this number (of different y' such that $C^t(x', y') \leqslant m$) is at least 2^{s-1}, and these strings also could be enumerated in time $\text{poly}(2^n, t)$, so $C^{\text{poly}(2^n, t)}(x) \leqslant m - s + O(\log n)$ (again we do not need to specify t, we just increase gradually the time bound). When these two inequalities are added, s disappears and we get the desired inequality. $\qquad\square$

Of course, the exponent in the lemma is disappointing (for space bound it is not needed, by the way), but since we measure time in busy beaver units, it is not a problem for us: indeed, $\text{poly}(2^n, B(i)) \leqslant B(i + O(\log n))$, and we allow logarithmic change in the argument anyway.

Now we should apply this lemma, but first we need to give a full statement of what we want to prove. There are two parts (as in the lemma):

- for every i there exists $j \leqslant i + O(\log n)$ such that

$$\mathrm{C}^{B(j)}(x, y) - \mathrm{C}(x, y) \leqslant \mathrm{C}^{B(i)}(x) - \mathrm{C}(x) + \varepsilon + O(\log n)$$

for all strings x and y of length at most n such that $\mathrm{C}(y|x) \leqslant l(y) - \varepsilon$;
- for every i there exists $j \leqslant i + O(\log n)$ such that

$$\mathrm{C}^{B(j)}(x) - \mathrm{C}(x) \leqslant \mathrm{C}^{B(i)}(x, y) - \mathrm{C}(x, y) + O(\log n)$$

for all strings x and y of length at most n;

Both statements easily follow from the lemma. Let us start with the second statement where the hard direction of the lemma is used. As planned, we rewrite the inequality as

$$\mathrm{C}^{B(j)}(x) + \mathrm{C}(y|x) \leqslant \mathrm{C}^{B(i)}(x, y) + O(\log n)$$

using the unbounded formula. Our lemma guarantees that

$$\mathrm{C}^{B(j)}(x) + \mathrm{C}^{B(j)}(y|x) \leqslant \mathrm{C}^{B(i)}(x, y) + O(\log n)$$

for some $j \leqslant i + O(\log n)$, and it remains to note that $\mathrm{C}(y|x) \leqslant \mathrm{C}^{B(j)}(y|x)$. For the other direction the argument is similar: we rewrite the inequality as

$$\mathrm{C}^{B(j)}(x, y) \leqslant \mathrm{C}(y|x) + \mathrm{C}^{B(i)}(x) + O(\log n)$$

and note that $\mathrm{C}(y|x) \geqslant l(y) - \varepsilon \geqslant \mathrm{C}^{B(i)}(y|x) - \varepsilon$, assuming that $B(i)$ is greater than the time needed to print y from its literary description (otherwise the statement is trivial). So the lemma again can be used (in the simple direction). □

This proof used the depth representation of the stochasticity curve; in other cases some other representation are more convenient. Our second example is the change in stochasticity profile when a simple algorithmic transformation is applied. We have seen (Sect. 2.3) that a total mapping with a short program preserves stochasticity, and noted that for non-total mapping it is not the case (Remark 3, p. 9). However, if the time needed to perform the transformation is bounded, we can get some bound (first proven by A. Milovanov in a different way):

Proposition 26. Let F be a computable mapping whose arguments and values are strings. If some n-bit string x is (α, β)-stochastic, and $F(x)$ is computed in time $B(i)$ for some i, then $F(x)$ is $(\max(\alpha, i) + O(\log n), \beta + O(\log n))$-stochastic. (The constant in $O(\log n)$-notation depends on F but not on n, x, α, β.)

Proof sketch. Let us denote $F(x)$ by y. By assumption there exist a $(\alpha * (C(x) - \alpha + \beta))$-description of x (recall the definition with optimality deficiency; we omit logarithmic terms as usual). So there exists a simple enumeration of at most $2^{C(x)+\beta}$ objects x' in at most 2^α portions that includes x. Let us count x' in this enumeration such that $F(x') = y$ and the computation uses time at most $B(i)$; assume there are 2^s of them. Then we can enumerate all y's that have at least 2^s preimages in time $B(i)$, in $2^\alpha + 2^i$ portions. Indeed, new portions appear in two cases: (1) a new portion appears in the original enumeration; (2) candidate for $B(i)$ increases. The first event happens at most 2^α times, the second at most 2^i times. The total number of y's enumerated is $2^{C(x)+\beta-s}$; it remains to note that $C(x) - s \leqslant C(y)$. Indeed, $C(x) \leqslant C(y) + C(x|y)$, and $C(x|y) \leqslant s$, since we can enumerate all the preimages of y in the order of increasing time, and x is determined by s-bit ordinal number of x in this enumeration.   □

A special case of this proposition is Bennett's observation: if some d-incompressible program p produces x in time $B(i)$, then p is $(0,d)$-stochastic, and p is mapped to x by the interpreter (decompressor) in time $B(i)$, so x is $(0+i,d)$-stochastic. (For simplicity we omit all the logarithmic terms in this argument, as well as in the previous proof sketch.)

Remark 10. One can combine Remark 4 (page 10) with Proposition 26 and show that if a program F of complexity at most j is applied to an (α, β)-stochastic string x of length n and the computation terminates in time $B(i)$, then $F(x)$ is $(\max(i,\alpha) + j + O(\log n), \beta + j + O(\log n))$-stochastic, where the constant in $O(\log n)$ notation is absolute (does not depend on F). To show this, one may consider the pair (x, F); it is easy to show (this can be done in different ways using different characterizations of the stochasticity curve) that this pair is $(\alpha + j + O(\log n), \beta + j + O(\log n))$-stochastic.

Let us note also that there are some results in algorithmic information theory that are true for stochastic objects but are false or unknown without this assumption. We will discuss (without proofs) two examples of this type. The first is Epstein–Levin theorem saying that for a stochastic set A its total a priori probability is close to the maximum a priori probability of A's elements; see [31] for details. Here the result is (obviously) false without stochasticity assumption.

In the next example [29] the stochasticity assumption is used in the proof, and it is not known whether the statement remains true without it: *for every triple of strings (x, y, z) of length at most n there exists a string z' such that*

- $C(x|z) = C(x|z') + O(\log n)$,
- $C(y|z) = C(y|z') + O(\log n)$,
- $C(x,y|z) = C(x,y|z') + O(\log n)$;
- $C(z') \leqslant I((x,y) : z) + O(\log n)$,

assuming that (x,y) is $(O(\log n), O(\log n))$-stochastic.

This proposition is related to the following open question on "irrelevant oracles": assume that the mutual information between (x,y) and some z is negligible. Can an oracle z (an "irrelevant oracle") change substantially natural properties of the pair (x, y) formulated in terms of Kolmogorov complexity? For instance, can such an oracle z allow us to extract some common information of x and y? In [29] a negative answer to the latter question is given, but only for stochastic pairs (x, y).

6   Descriptions of Restricted Type

6.1   Families of Descriptions

In this section we consider the restricted case: the sets (considered as descriptions, or statistical hypotheses) are taken from some family $\mathcal{A}$ that is fixed in advance.[9] (Elements of $\mathcal{A}$ are finite sets of binary strings.) Informally speaking, this means that we have some *a priori* information about the black box that produces a given string: this string is obtained by a random choice in one of the $\mathcal{A}$-sets, but we do not know in which one.

Before we had no restrictions (the family $\mathcal{A}$ was the family of all finite sets). It turns out that the results obtained so far can be extended (sometimes with weaker bounds) to other families that satisfy some natural conditions. Let us formulate these conditions.

(1)   The family $\mathcal{A}$ is enumerable. This means that there exists an algorithm that prints elements of $\mathcal{A}$ as lists, with some separators (saying where one element of $\mathcal{A}$ ends and another one begins).

(2)   For every n the family $\mathcal{A}$ contains the set $\mathbb{B}^n$ of all n-bit strings.

(3)   There exists some polynomial p with the following property: for every $A \in \mathcal{A}$, for every natural n and for every natural $c < \#A$ the set of all n-bit strings in A can be covered by at most $p(n) \cdot \#A/c$ sets of cardinality at most c from $\mathcal{A}$.

The last condition is a replacement for splitting: in general, we cannot split a set $A \in \mathcal{A}$ into pieces from A, but at least we can cover a set $A \in \mathcal{A}$ by smaller elements of $\mathcal{A}$ (of size at most c) with polynomial overhead in the number of pieces, compared to the required minimum $\#A/c$ (more precisely, we have to cover only n-bit elements of A).

We assume that some family $\mathcal{A}$ that has properties (1)–(3) is fixed. For a string x we denote by $P_x^{\mathcal{A}}$ the set of pairs (i, j) such that x has $(i * j)$-description *that belongs to* $\mathcal{A}$. The set $P_x^{\mathcal{A}}$ is a subset of P_x defined earlier; the bigger $\mathcal{A}$ is, the bigger is $P_x^{\mathcal{A}}$. The full set P_x is $P_x^{\mathcal{A}}$ for the family $\mathcal{A}$ that contains all finite sets.

For every string x the set $P_x^{\mathcal{A}}$ has properties close to the properties of P_x proved earlier.

Proposition 27. For every string x of length n the following is true:

---

[9] One can also consider some class of probability distributions, but we restrict our attention to sets (uniform distributions).

1. The set $P_x^{\mathcal{A}}$ contains a pair that is $O(\log n)$-close to $(0, n)$.
2. The set $P_x^{\mathcal{A}}$ contains a pair that is $O(1)$-close to $(C(x), 0)$.
3. The adaptation of Proposition 8 is true: if $(i, j) \in P_x^{\mathcal{A}}$, then $(i+k+O(\log n), j-k)$ also belongs to $P_x^{\mathcal{A}}$ for every $k \leqslant j$. (Recall that n is the length of x.)

Proof. 1. The property (2) guarantees that the family $\mathcal{A}$ contains the set $\mathbb{B}^n$ that is an $(O(\log n) * n)$-description of x.

2. The property (3) applied to $c = 1$ and $A = \mathbb{B}^n$ says that every singleton belongs to A, therefore each string has $((C(x) + O(1)) * 0)$-description.

3. Assume that x has $(i * j)$-description $A \in \mathcal{A}$. For a given k we enumerate $\mathcal{A}$ until we find a family of $p(n)2^k$ sets of size $2^{-k} \# A$ (or less) in $\mathcal{A}$ that covers all strings of length n in A. Such a family exists due to (3), and p is the polynomial from (3). The complexity of the set that covers x does not exceed $i + k + O(\log n + \log k)$, since this set is determined by A, n, k and the ordinal number of the set in the cover. We may assume without loss of generality that $k \leqslant n$, otherwise $\{x\}$ can be used as $((i + k + O(\log n)) * (j - k))$-description of x. So the term $O(\log k)$ can be omitted. □

For example, we may consider the family that consists of all "cylinders": for every n and for every string u of length at most n we consider the set of all n-bit strings that have prefix u. Obviously the family of all such sets (for all n and u) satisfies the conditions (1)–(3).

We may also fix some bits of a string (not necessarily forming a prefix). That is, for every string z in ternary alphabet $\{0, 1, *\}$ we consider the set of all bit strings that can be obtained from z by replacing stars with some bits. This set contains 2^k strings, if u has k stars. The conditions (1)–(3) are fulfilled for this larger family, too.

A more interesting example is the family $\mathcal{A}$ formed by all balls in Hamming sense, i.e., the sets $B_{y,r} = \{x \mid l(x) = l(y), d(x, y) \leqslant r\}$. Here $l(u)$ is the length of binary string u, and $d(x, y)$ is the Hamming distance between two strings x and y of the same length. The parameter r is called the *radius* of the ball, and y is its *center*. Informally speaking, this means that the experimental data were obtained by changing at most r bits in some string y (and all possible changes are equally probable). This assumption could be reasonable if some string y is sent via an unreliable channel. Both parameters y and r are not known to us in advance.

It turns out that the family of Hamming balls satisfies the conditions (1)–(3). This is not completely obvious. For example, these conditions imply that for every n and for every $r \leqslant n$ the set $\mathbb{B}^n$ of n-bit strings can be covered by $\operatorname{poly}(n)2^n/V$ Hamming balls of radius r, where V stands for the cardinality of such a ball (i.e., $V = \binom{n}{0} + \ldots + \binom{n}{r}$), and p is some polynomial. This can be shown by a probabilistic argument: take N balls of radius r whose centers are randomly chosen in $\mathbb{B}^n$. For a given $x \in \mathbb{B}^n$ the probability that x is not covered by any of these balls equals $(1 - V/2^n)^N < e^{-VN/2^n}$. For $N = n \ln 2 \cdot 2^n/V$ this upper bound is 2^{-n}, so for this N the probability to leave some x uncovered is less than 1. A similar argument can be used to prove (1)–(3) in the general case.

Proposition 28 ([44]). The family of all Hamming balls satisfies conditions (1)–(3) above.

Proof sketch. Let A be a ball of radius a and let c be a number less than $\#A$. We need to cover A by balls of cardinality c or less, using almost minimal number of balls, close to the lower bound $\#A/c$ up to a polynomial factor. Let us make some observations.

(1) The set of all n-bit strings can be covered by two balls of radius $n/2$. So we can assume without loss of generality that $a \leqslant n/2$, otherwise we can apply the probabilistic argument above.

(2) Clearly the radius of covering balls should be maximal possible (to keep cardinality less than c); for this radius the cardinality of the ball equals c up to polynomial factors, since the size of the ball increases at most by factor $n + 1$ when its radius increases by 1.

(3) It is enough to cover spheres instead of balls (since every ball is a union of polynomially many spheres); it is also enough to consider the case when the radius of the sphere that we want to cover (a) is bigger than the radius of the covering ball (b), otherwise one ball is enough.

(4) We will cover a-sphere by randomly chosen b-balls whose centers are uniformly taken at some distance f from the center of a-sphere. (See below about the choice of f.) We use the same probabilistic argument as before (for the set of all strings). It is enough to show that for a b-ball whose center is at that distance, the polynomial fraction of points belong to a-sphere. Instead of b-balls we may consider b-spheres, the cardinality ratio is polynomial.

(5) It remains to choose some f with the following property: if the center of a b-sphere S is at a distance f from the center of a-sphere T, then the polynomial fraction of S-points belong to T. One can compute a suitable f explicitly. In probabilistic terms we just change f/n-fraction of bits and then change random b/n fraction of bits. The expected fraction of twice changed bits is, therefore, about $(f/n)(b/n)$, and the total fraction of changed bits is about $f/n + b/n - 2(f/n)(b/n)$. So we need to write an equation saying that this expression is a/n and the find the solution f. (Then one can perform the required estimate for binomial coefficients.)

However, one can avoid computations with the following probabilistic argument: start with b changed bits, and then change all the bits one by one in a random order. At the end we hat $n - b$ changed bits, and a is somewhere in between, so there is a moment where the number of changed bits is exactly a. And if the union of n events covers the entire probability space, one of these events has probability at least $1/n$.   □

When a family $\mathcal{A}$ is fixed, a natural question arises: does the restriction on models (when we consider only models in $\mathcal{A}$) changes the set P_x? Is it possible that a string has good models in general, but not in the restricted class? The answer is positive for the class of Hamming balls, as the following proposition shows.

Proposition 29. Consider the family $\mathcal{A}$ that consists of all Hamming balls. For some positive ε and for all sufficiently large n there exists a string x of length n such that the distance between $P_x^{\mathcal{A}}$ and P_x exceeds εn.

Proof sketch. Fix some α in $(0, 1/2)$ and let V be the cardinality of the Hamming ball of radius αn. Find a set E of cardinality $N = 2^n/V$ such that every Hamming ball of radius αn contains at most n points from E. This property is related to *list decoding* in the coding theory. The existence of such a set can be proved by a probabilistic argument: N randomly chosen n-bit strings have this property with positive probability. Indeed, the probability of a random point to be in E is an inverse of the number of points, so the distribution is close to Poisson distribution with parameter 1, and tails decrease much faster that 2^{-n} needed.

Since E with this property can be found by an exhaustive search, we can assume that $C(E) = O(\log n)$ and ignore the complexity of E (as well as other $O(\log n)$ terms) in the sequel. Let x be a random element in E, i.e., a string $x \in E$ of complexity about $\log \#E$. The complexity of a ball A of radius αn that contains x is at least $C(x)$, since knowing such a ball and an ordinal number of x in $A \cap E$, we can find x. Therefore x does not have $(\log \#E, \log V)$-descriptions in $\mathcal{A}$. On the other hand, x does have $(0, \log \#E)$-description if we do not require the description to be in $\mathcal{A}$; the set E is such a description. The point $(\log \#E, \log V)$ is above the line $C(A) + \log \#A = \log \#E$, so $P_x^{\mathcal{A}}$ is significantly smaller than P_x. □

This construction gives a stochastic x (E is the corresponding model) that becomes maximally non-stochastic if we restrict ourselves to Hamming balls as descriptions (Fig. 7).

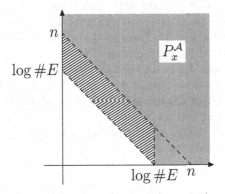

Fig. 7. Theorem 8 can be used (together with the argument above) to show that the border of the set $P_x^{\mathcal{A}}$ (shown in gray) consists of a vertical segment $C(A) = n - \log V$, $\log \#A \leqslant \log V$, and the segment of slope -1 defined by $C(A) + \log \#A = n$, $\log V \leqslant \log \#A$. The set P_x contains also the hatched part.

6.2 Possible Shapes of Boundary Curve

Our next goal is to extend some results proven for non-restricted descriptions to the restricted case. Let $\mathcal{A}$ be a family that has properties (1)–(3). We prove a version of Theorem 1 where the precision (unfortunately) is significantly worse: $O(\sqrt{n \log n})$ instead of $O(\log n)$. Note that with this precision the term $O(m)$ (proportional to the complexity of the curve) that appeared in Theorem 1 is not needed. Indeed, if we draw the curve on the cell paper with cell size $\sqrt{n}$ or larger, then it touches only $O(\sqrt{n})$ cells, so it is determined by $O(\sqrt{n})$ bits with $O(\sqrt{n})$-precision, and we may assume without loss of generality that the complexity of the curve is $O(\sqrt{n})$.

Theorem 7 ([44]). *Let $k \leqslant n$ be two integers and let $t_0 > t_1 > \ldots > t_k$ be a strictly decreasing sequence of integers such that $t_0 \leqslant n$ and $t_k = 0$.. Then there exists a string x of complexity $k + O(\sqrt{n \log n})$ and length $n + O(\log n)$ for which the distance between $P_x^{\mathcal{A}}$ and $T = \{(i,j) \mid (i \leqslant k) \Rightarrow (j \geqslant t_i)\}$ is at most $O(\sqrt{n \log n})$.*

We will see later (Theorem 8) that for every x the boundary curve of $P_x^{\mathcal{A}}$ goes down at least with slope -1, as for the unrestricted case, so this theorem describes all possible shapes of the boundary curve.

Proof. The proof is similar to the proof of Theorem 1. Let us recall this proof first. We consider the string x that is the lexicographically first string (of suitable length n') that is not covered by any "bad" set, i.e., by any set of complexity at most i and size at most 2^j, where the pair (i,j) is at the boundary of the set T. The length n' is chosen in such a way that the total number of strings in all bad sets is strictly less than $2^{n'}$. On the other hand, we need "good sets" that cover x. For every boundary point (i,j) we construct a set $A_{i,j}$ that contains x, has complexity close to i and size 2^j. The set $A_{i,j}$ is constructed in several attempts. Initially $A_{i,j}$ is the set of lexicographically first 2^j strings of length n'. Then we enumerate bad sets and delete all their elements from $A_{i,j}$. At some step $A_{i,j}$ may become empty; then we refill it with 2^j lexicographically first strings that are not in the bad sets (at the moment). By construction the final $A_{i,j}$ contains the first x that is not in bad sets (since it is the case all the time). And the set $A_{i,j}$ can be described by the number of changes (plus some small information describing the process as a whole and the value of j). So it is crucial to have an upper bound for the number of changes. How do we get this bound? We note that when $A_{i,j}$ becomes empty, it is refilled again, and all the new elements should be covered by bad sets before the new change could happen. Two types of bad sets may appear: "small" ones (of size less than 2^j) and "large ones" (of size at least 2^j). The slope of the boundary line for T guarantees that the total number of elements in all small bad sets does not exceed 2^{i+j} (up to a poly(n)-factor), so they may make $A_{i,j}$ empty only 2^i times. And the number of large bad sets is $O(2^i)$, since the complexity of each is bounded by i. (More precisely, we count separately the number of changes for $A_{i,j}$ that are first changes after a large bad set appears, and the number of other changes.)

Can we use the same argument in our new situation? We can generate bad sets as before 'and have the same bounds for their sizes and the total number of their elements. So the length n' of x can be the same (in fact, almost the same, as we will need now that the union of all bad sets is less than half of all strings of length n', see below). Note that we now may enumerate only bad sets in $\mathcal{A}$, since $\mathcal{A}$ is enumerable, but we do not even need this restriction. What we cannot do is to let $A_{i,j}$ to be the set of the first non-deleted elements: we need $A_{i,j}$ to be a set from $\mathcal{A}$.

So we now go in the other direction. Instead of choosing x first and then finding suitable "good" $A_{i,j}$ that contain x, we construct the sets $A_{i,j} \in \mathcal{A}$ that change in time in such a way that (1) their intersection always contains some non-deleted element (an element that is not yet covered by bad sets); (2) each $A_{i,j}$ has not too many versions. The non-deleted element in their intersection (in the final state) is then chosen as x.

Unfortunately, we cannot do this for all points (i,j) along the boundary curve. (This explains the loss of precision in the statement of the theorem.) Instead, we construct "good" sets only for some values of j. These values go down from n to 0 with step $\sqrt{n\log n}$. We select $N = \sqrt{n/\log n}$ points $(i_1,j_1),\ldots,(i_N,j_N)$ on the boundary of T; the first coordinates $i_1,\ldots,i_N$ form a non-decreasing sequence, and the second coordinates $j_1,\ldots,j_N$ split the range $n\ldots 0$ into (almost) equal intervals ($j_1 = n$, $j_N = 0$). Then we construct good sets of sizes at most $2^{j_1},\ldots,2^{j_N}$, and denote them by $A_1,\ldots,A_N$. All these sets belong to the family $\mathcal{A}$. We also let A_0 to be the set of all strings of length $n' = n + O(\log n)$; the choice of the constant in $O(\log n)$ will be discussed later.

Let us first describe the construction of $A_1,\ldots,A_N$ assuming that the set of deleted elements is fixed. (Then we discuss what to do when more elements are deleted.) We construct A_s inductively (first A_1, then A_2 etc.). As we have said, $\#A_s \leqslant 2^{j_s}$ (in particular, A_N is a singleton), and we keep track of the ratio

$$\text{(the number of non-deleted strings in } A_0 \cap A_1 \cap \ldots \cap A_s)/2^{j_s}.$$

For $s = 0$ this ratio is at least $1/2$; this is obtained by a suitable choice of n' (the union of all bad sets should cover at most half of all n'-bit strings). When constructing the next A_s, we ensure that this ratio decreases only by poly(n)-factor. How? Assume that A_{s-1} is already constructed; its size is at most $2^{j_{s-1}}$. The condition (3) for $\mathcal{A}$ guarantees that A_{s-1} can be covered by $\mathcal{A}$-sets of size at most 2^{j_s}, and we need about $2^{j_{s-1}-j_s}$ covering sets (up to poly(n)-factor). Now we let A_s be the covering set that contains maximal number of non-deleted elements in $A_0 \cap \ldots \cap A_{s-1}$. The ratio can decrease only by the same poly(n)-factor. In this way we get

$$\text{(the number of non-deleted strings in } A_0 \cap A_1 \cap \ldots \cap A_s) \geqslant \alpha^{-s} 2^{j_s}/2,$$

where α stands for the poly(n)-factor mentioned above.[10]

---

[10] Note that for the values of s close to N the right-hand side can be less than 1; the inequality then claims just the existence of non-deleted elements. The induction step is still possible: non-deleted element is contained in one of the covering sets.

Up to now we assumed that the set of deleted elements is fixed. What happens when more strings are deleted? The number of the non-deleted in $A_0 \cap \ldots \cap A_s$ can decrease, and at some point and for some s can become less than the declared threshold $\nu_s = \alpha^{-s} 2^{j_s}/2$. Then we can find minimal s where this happens, and rebuild all the sets $A_s, A_{s+1}, \ldots$ (for A_s the threshold is not crossed due to the minimality of s). In this way we update the sets A_s from time to time, replacing them (and all the consequent ones) by new versions when needed.

The problem with this construction is that the number of updates (different versions of each A_s) can be too big. Imagine that after an update some element is deleted, and the threshold is crossed again. Then a new update is necessary, and after this update next deletion can trigger a new update, etc. To keep the number of updates reasonable, we agree that after the update *for all the new sets A_l* (starting from A_s) *the number of non-deleted elements in* $A_0 \cap \ldots \cap A_l$ *is twice bigger than the threshold* $\nu_l = \alpha^{-l} 2^{j_l}/2$. This can be achieved if we make the factor α twice bigger: since for A_{s-1} we have not crossed the threshold, for A_s we can guarantee the inequality with additional factor 2.

Now let us prove the bound for the number of updates for some A_s. These updates can be of two types: first, when A_s itself starts the update (being the minimal s where the threshold is crossed); second, when the update is induced by one of the previous sets. Let us estimate the number of the updates of the first type. This update happens when the number of non-deleted elements (that was at least $2\nu_s$ immediately after the previous update of any kind) becomes less than ν_s. This means that at least ν_s elements were deleted. How can this happen? One possibility is that a new bad set of complexity at most i_s ("large bad set") appears after the last update. This can happen at most $O(2^{i_s})$ times, since there is at most $O(2^i)$ objects of complexity at most i. The other possibility is the accumulation of elements deleted due to "small" bad sets, of complexity at least i_s and of size at most 2^{j_s}. The total number of such elements is bounded by $nO(2^{i_s+j_s})$, since the sum $i_l + j_l$ may only decrease as l, increases. So the number of updates of A_s not caused by large bad sets is bounded by

$$nO(2^{i_s+j_s})/\nu_s = \frac{O(n2^{i_s+j_s})}{\alpha^{-s}2^{j_s}} = O(n\alpha^s 2^{i_s}) = 2^{i_s+NO(\log n)} = 2^{i_s+O(\sqrt{n\log n})}$$

(recall that $s \leqslant N$, $\alpha = \mathrm{poly}(n)$, and $N \approx \sqrt{n/\log n}$). This bound remains valid if we take into account the induced updates (when the threshold is crossed for the preceding sets: there are at most $N \leqslant n$ these sets, and additional factor n is absorbed by O-notation).

We conclude that all the versions of A_s have complexity at most $i_s + O(\sqrt{n\log n})$, since each of them can be described by the version number plus the parameters of the generating process (we need to know n and the boundary curve, whose complexity is $O(\sqrt{n})$ according to our assumption, see the discussion before the statement of the theorem). The same is true for the final version. It remains to take x in the intersection of the final sets A_s. (Recall that A_N is a singleton, so final A_N is $\{x\}$.) Indeed, by construction this x has no bad $(i*j)$-descriptions where (i,j) is on the boundary of T. On the other hand, x

has good descriptions that are $O(\sqrt{n \log n})$-close to this boundary and whose vertical coordinates are $\sqrt{n \log n}$-apart. (Recall that the slope of the boundary guarantees that horizontal distance is less than the vertical distance.) Therefore the position of the boundary curve for $P_x^{\mathcal{A}}$ is determined with precision $O(\sqrt{n \log n})$, as required.[11]   □

Remark 11. In this proof we may use bad sets not only from $\mathcal{A}$. Therefore, the set P_x is also close to T (and the same is true for for every family $\mathcal{B}$ that contains $\mathcal{A}$). It would be interesting to find out what are the possible combinations of P_x and $P_x^{\mathcal{A}}$; as we have seen, it may happen that P_x is maximal and $P_x^{\mathcal{A}}$ is minimal, but this does not say anything about other possible combinations.

For the case of Hamming balls the statement of Theorem 7 has a natural interpretation. To find a simple ball of radius r that contains a given string x is the same as to find a simple string in a radius r ball centered at x. So this theorem show the possible behavior of the "approximation complexity" function

$$r \mapsto \min\{C(x') \mid d(x, x') \leqslant r\}$$

where d is Hamming distance. One should only rescale the vertical axis replacing the log-sizes of Hamming balls by their radii. The connection is described by the Shannon entropy function: a ball in $\mathbb{B}^n$ of radius r has log-size about $nH(r/n)$ for $r \leqslant n/2$, and has almost full size for $r \geqslant n/2$. For example, error correcting codes (in classical sense, or with list decoding) are example of strings where this function is almost a constant for small values of r: it is almost as easy to approximate a codeword as give it precisely (due to the possibility of error correction).

6.3   Randomness and Optimality Deficiencies: Restricted Case

Not all the results proved for unrestricted descriptions have natural counterparts in the restricted case. For example, one hardly can relate the set $P_x^{\mathcal{A}}$ with bounded-time complexity (is completely unclear how $\mathcal{A}$ could enter the picture). Still some results remain valid (but new and much more complicated proofs are needed). This is the case for Propositions 8 and 9.

Let again $\mathcal{A}$ be the class of descriptions that satisfies requirements (1)–(3).

Theorem 8 ([44]).

- If a string x of length n has an $(i * j)$-description in $\mathcal{A}$, then it has $((i + d + O(\log n)) * (j - d + O(\log n)))$-description in $\mathcal{A}$ for every $d \leqslant j$.
- Assume that x is a string of length n that has at least 2^k different $(i * j)$-descriptions in $\mathcal{A}$. Then it has $((i - k + O(\log n)) * (j + O(\log n)))$-description in $\mathcal{A}$.

---

[11] Now we see why N was chosen to be $\sqrt{n/\log n}$: the bigger N is, the more points on the curve we have, but then the number of versions of the good sets and their complexity increases, so we have some trade-off. The chosen value of n balances these two sources of errors.

In fact, the second part uses only condition (1); it says that $\mathcal{A}$ is enumerable. The first part uses also (3). It can be combined with the second part to show that x has also $((i + O(\log n)) * (j - k + O(\log n))$-description in $\mathcal{A}$.

Though Theorem 8 looks like a technical statement, it has important consequences; it implies that the two approaches based on randomness and optimality deficiencies remain equivalent in the case of bounded class of descriptions. The proof technique can be also used to prove Epstein–Levin theorem [11], as explained in [31]; similar technique was used by A. Milovanov in [25] where a common model for several strings is considered.

Proof. The first part is easy: having some $(i * j)$-description for x, we can search for a covering by the sets of right size that exists due to condition (3); since $\mathcal{A}$ is enumerable, we can do it algorithmically until we find this covering. Then we select the first set in the covering that contains x; the bound for the complexity of this set is guaranteed by the size of the covering.

The proof of the second statement is much more interesting. In fact, there are two different proofs: one uses a probabilistic existence argument and the second is more explicit. But both of them start in the same way.

Let us enumerate all $(i * j)$-descriptions from $\mathcal{A}$, i.e., all finite sets that belong to $\mathcal{A}$, have cardinality at most 2^j and complexity at most i. For a fixed n, we start a selection process: some of the generated descriptions are marked (=selected) immediately after their generation. This process should satisfy the following requirements: (1) at any moment every n-bit string x that has at least 2^k descriptions (among enumerated ones) belongs to one of the marked descriptions; (2) the total number of marked sets does not exceed $2^{i-k}p(n)$ for some polynomial p. Note that for $i \geqslant n$ or $j \geqslant n$ the statement is trivial, so we may assume that i, j (and therefore k) do not exceed n; this explains why the polynomial depends only on n.

If we have such a strategy (of logarithmic complexity), then the marked set containing x will be the required description of complexity $i - k + O(\log n)$ and log-size j. Indeed, this marked set can be specified by its ordinal number in the list of marked sets, and this ordinal number has $i - k + O(\log n)$ bits.

So we need to construct a selection strategy of logarithmic complexity. We present two proofs: a probabilistic one and an explicit construction.

PROBABILISTIC PROOF. First we consider a finite game that corresponds to our situation. Two players alternate, each makes 2^i moves. At each move the first player presents some set of n-bit strings, and the second player replies saying whether it *marks* this set or not. The second player loses if after some moves the number of marked sets exceeds $2^{i-k+1}(n+1)\ln 2$ (this specific value follows from the argument below) or if there exists a string x that belongs to 2^k sets of the first player but does not belong to any marked set.

Since this is a finite game with full information, one of the players has a winning strategy. We claim that the second player can win. If it is not the case, the first player has a winning strategy. We get a contradiction by showing that the second player has a *probabilistic* strategy that wins with positive probability against any strategy of the first player. So we assume that some (deterministic) strategy of the first player is fixed, and consider the following simple probabilistic strategy: every set A presented by the first player is marked with probability $p = 2^{-k}(n+1)\ln 2$.

The expected number of marked sets is $p2^i = 2^{i-k}(n+1)\ln 2$. By Chebyshev's inequality, the number of marked set exceeds the expectation by a factor 2 with probability less than $1/2$. So it is enough to show that the second bad case (after some move there exists x that belongs to 2^k sets of the first player but does not belong to any marked set) happens with probability at most $1/2$.

For that, it is enough to show that for every fixed x the probability of this bad event is at most $2^{-(n+1)}$, and then use the union bound. The intuitive explanation is simple: if x belongs to 2^k sets, the second player had (at least) 2^k chances to mark a set containing x (when these 2^k sets were presented by the first player), and the probability to miss all these chances is at most $(1-p)^{2^k}$; the choice of p guarantees that this probability is less than $1/2^{-(n+1)}$. Indeed, using the bound $(1 - 1/x)^x < 1/e$, it is easy to show that $(1 - p)^{2^k} < e^{-(n+1)\ln 2} = 2^{-(n+1)}$.

The pedantic reader would say that this argument is not formally correct, since the behavior of the first player (and the moment when next set containing x is produced) depends on the moves of the second player, so we do not have independent events with probability $1 - p$ each (as it is assumed in the computation).[12] The formal argument considers for each t the event R_t: "after some move of the second player the string x belongs to at least t sets provided by the first player, but does not belong to any marked set". Then we prove by induction (over t) that the probability of R_t does not exceed $(1 - p)^t$. Indeed, it is easy to see that R_t in a union of several disjoint subsets (depending on the events happening until the first player provides $t + 1$ sets containing x), and R_{t+1} is obtained by taking a $(1 - p)$-fraction in each of them.

CONSTRUCTIVE PROOF. We consider the same game, but now allow more sets to be marked (replacing the bound $2^{i-k+1}(n+1)\ln 2$ by a bigger bound $2^{i-k}i^2\ln 2$) and also allow the second player to mark sets that were produced earlier (not necessarily at the current move of the first player). The explicit winning strategy for the second player performs in parallel $i - k + \log i$ substrategies (indexed by the numbers $\log(2^k/i), \ldots, i$).

The substrategy number s wakes up once in 2^s moves (when the number of moves made by the first player is a multiple of 2^s). It considers a family S that consists of 2^s last sets produced by the first player, and the set T that consists of all strings x covered by at least $2^k/i$ sets from S. Then it selects and marks some elements in S in such a way that all $x \in T$ are covered by one of the selected

---

[12] The same problem appears if we observe a sequence of independent coin tossings with probability of success p, select some trials (before they are actually performed, based on the information obtained so far), and ask for the probability of the event "t first selected trials were all unsuccessful". This probability does not exceed $(1-p)^t$; it can be smaller if the total number of selected trials is less than t with positive probability. This scheme was considered by von Mises when he defined random sequences using selection rules, so it should be familiar to algorithmic randomness people.

sets. It is done by a greedy algorithm: first take a set from S that covers maximal part of T, then the set that covers maximal number of non-covered elements, etc. How many steps do we need to cover the entire T? Let us show that

$$(i/2^k)n2^s \ln 2$$

steps are enough. Indeed, every element of T is covered by at least $2^k/i$ sets from S. Therefore, some set from S covers at least $\#T2^k/(i2^s)$ elements, i.e., $2^{k-s}/i$-fraction of T. At the next step the non-covered part is multiplied by $(1-2^{k-s}/i)$ again, and after $in2^{s-k} \ln 2$ steps the number of non-covered elements is bounded by

$$\#T(1 - 2^{k-s}/i)^{in2^{s-k} \ln 2} < 2^n(1/e)^{n \ln 2} = 1,$$

therefore all elements of T are covered. (Instead of a greedy algorithm one may use a probabilistic argument and show that randomly chosen $in2^{s-k} \ln 2$ sets from S cover T with positive probability; however, our goal is to construct an explicit strategy.)

Anyway, the number of sets selected by a substrategy number s, does not exceed

$$in2^{s-k}(\ln 2)2^{i-s} = in2^{i-k} \ln 2,$$

and we get at most $i^2 n2^{i-k} \ln 2$ for all substrategies.

It remains to prove that after each move of the second player every string x that belongs to 2^k or more sets of the first player, also belongs to some selected set. For tth move we consider the binary representation of t:

$$t = 2^{s_1} + 2^{s_2} + \ldots, \text{ where } s_1 > s_2 > \ldots$$

Since x does not belong to the sets selected by substrategies with numbers $s_1, s_2, \ldots$, the multiplicity of x among the first 2^{s_1} sets is less than $2^k/i$, the multiplicity of x among the next 2^{s_2} sets is also less than $2^k/i$, etc. For those j with $2^{s_j} < 2^k/i$ the multiplicity of x among the respective portion of 2^{s_j} sets is obviously less than $2^k/i$. Therefore, we conclude that the total multiplicity of x is less that $i \cdot 2^k/i = 2^k$ sets of the first player and the second player does not need to care about x. This finishes the explicit construction of the winning strategy.

Now we can assume without loss of generality that the winning strategy has complexity at most $O(\log(n + k + i + j))$. (In the probabilistic argument we have proved the existence of a winning strategy, but then we can perform the exhaustive search until we find one; the first strategy found will have small complexity.) Then we use this simple strategy to play with the enumeration of all $\mathcal{A}$-sets of complexity less than i and size 2^j (or less). The selected sets can be described by their ordinal number (among the selected sets), so their complexity is bounded by $i - k$ (with logarithmic precision). Every string that has 2^k different $(i * j)$-descriptions in $\mathcal{A}$, will also have one among the selected sets, and that is what we need.   □

As before (for the unrestricted case), this result implies that descriptions with minimal parameters are simple with respect to the data string:

Theorem 9 ([44]). *Let $\mathcal{A}$ be an enumerable family of finite sets. If a string x of length n has $(i * j)$-description $A \in \mathcal{A}$ such that $C(A|x) \geqslant k$, then x has a $((i - k + O(\log n)) * (j + O(\log n)))$-description in $\mathcal{A}$. If the family $\mathcal{A}$ satisfies the condition (3), then x has also a $((i + O(\log n)) * (j - k + O(\log n)))$-description in $\mathcal{A}$.*

This gives us the same corollaries as in the unrestricted case:

Corollary. Let $\mathcal{A}$ be a family of finite sets that satisfies the conditions (1)–(3). Then for every string x of length n three statements

- there exists a set $A \in \mathcal{A}$ of complexity at most α with $d(x|A) \leqslant \beta$;
- there exists a set $A \in \mathcal{A}$ of complexity at most α with $\delta(x, A) \leqslant \beta$;
- the point $(\alpha, C(x) - \alpha + \beta)$ belongs to $P_x^{\mathcal{A}}$

are equivalent with logarithmic precision (the constants before the logarithms depend on the choice of the set $\mathcal{A}$).

If we are interested in the uniform statements true for every enumerable family $\mathcal{A}$, the same arguments prove the following result:

Proposition 30. Let $\mathcal{A}$ be an arbitrary family of finite sets enumerated by some program p. Then for every x of length n the statements

- there exists a set $A \in \mathcal{A}$ such that $d(x|A) \leqslant \beta$;
- there exists a set $A \in \mathcal{A}$ such that $\delta(x, A) \leqslant \beta$

are equivalent up to $O(C(p) + \log C(\mathcal{A}) + \log n + \log \log \#A)$-change in the parameters.

7   Strong Models

7.1   Information in Minimal Descriptions

A possible way to bring the theory in accordance to our intuition is to change the definition of "having the same information". Although we have not given that definition explicitly, we have adopted so far the following viewpoint: x and y have the same (or almost the same) information if both conditional complexities $C(x|y), C(y|x)$ are small. If only one complexity, say $C(x|y)$, is small, we said that all (or almost all) information contained in x is present in y.

Now we will adopt a more restricted viewpoint and say that x and y have the same information if there are short *total* (everywhere defined) programs mapping x to y and vice versa. From this viewpoint we cannot say anymore that a string x and its shortest program x^* have the same information: for example, x may be non-stochastic while x^* is always stochastic, so there is no short total program that maps x^* to x because of Proposition 3.[13] Let us mention that if x and y have

---

[13] It is worth to mention that on the other hand, for every string x there is an almost minimal program for x that can be obtained from x by a simple total algorithm [40, Theorem 17].

the same information in this new sense, then there exists a simple computable *bijection* that maps x to y (so they have the same properties if the property is defined in the computability language), see [28] for the proof.

Formally, let us define the total conditional complexity with respect to a computable function D of two arguments, as

$$\mathrm{CT}_D(x|y) = \min\{l(p) \mid D(p,y) = x, \text{ and } D(p,y') \text{ is defined for all } y'\}.$$

(Note that D is not required to be total, but we consider only p such that $D(p,y')$ is defined for all y'.)

There is a computable function D such that CT_D is minimal up to an additive constant. Fixing any such D we obtain the *total conditional complexity* $\mathrm{CT}(x|y)$. In other way, we may define $\mathrm{CT}(x|y)$ as the minimal plain complexity of a total program that maps y to x.

We will think that y has all (or almost all) the information from x if $\mathrm{CT}(x|y)$ is negligible. Formally, we write $x \xrightarrow{\varepsilon} y$ if $\mathrm{CT}(y|x) \leqslant \varepsilon$ and we call x and y ε-*equivalent* and write $x \xleftrightarrow{\varepsilon} y$, if both $\mathrm{CT}(y|x)$ and $\mathrm{CT}(x|y)$ are at most ε.

Proposition 31. *If $x \xleftrightarrow{\varepsilon} y$ then the sets P_x and P_y are in $O(\varepsilon)$ neighborhood of each other.*

Proof. Indeed, if A is an $(i*j)$-description of x and p is a total program witnessing $x \xleftrightarrow{\varepsilon} y$, then the set $B = \{D(p,x') \mid x' \in A\}$ is an $((i+O(\varepsilon))*j)$-description of y. (We need p to be total, as otherwise we cannot produce the list of B-elements from the list of A-elements and p.) $\square$

7.2   An Attempt to Separate "good" Models from "bad" Ones

Now we have more fine-grained classification of descriptions and can try to distinguish between descriptions that were equivalent in the former sense. For example, consider a string xy where y is random conditionally to x. Let A be a model for xy consisting of all extensions of x (of the same length). This model looks good (in particular, it has negligible optimality deficiency). On the other hand, we may consider a standard model B for xy of the same (or smaller) complexity. It also has negligible optimality deficiency but looks unnatural. In this section we are interested in the following question: how can we formally distinguish good models like A from bad models like B? We will see that at least for some strings u the value $\mathrm{CT}(A|u)$ can be used to distinguish between good and bad models for u. (Indeed, in our example $\mathrm{CT}(A|xy)$ is small, while $\mathrm{CT}(B|xy)$ can be large.)

Definition 5. A set $A \ni x$ is an ε-*strong model* (or *statistic*) for a string x if $\mathrm{CT}(A|x) \leqslant \varepsilon$.

For instance, the model A discussed above is an $O(\log n)$-strong model for x. On the other hand, we will see later that, if y is chosen appropriately, then no standardbdescription B of the same complexity and log-cardinality as A is an ε-strong model for x, even for $\varepsilon = \Omega(n)$.

Strong models satisfy an analog of Proposition 8 (the same proof works):

Proposition 32. Let x be a string and A be an ε-strong model for x. Let i be a non-negative integer such that $i \leqslant \log \#A$. Then there exists an $\varepsilon + O(\log i)$-strong model A' for x such that $\#A' \leqslant \#A/2^i$ and $C(A') \leqslant C(A) + i + O(\log i)$.

To take into account the strength of models, we may consider the set

$$P_x(\varepsilon) = \{(i, j) \mid x \text{ has an } \varepsilon\text{-strong } (i * j)\text{-description}\}.$$

Obviously, we have

$$P_x(\varepsilon) \subset P_x = P_x(n + O(1))$$

for all strings x of length n and for all ε.

If the set $P_x(\varepsilon)$ is not much smaller than P_x for a reasonably small ε, we will say that x is a "normal" string and otherwise we call x "strange". More precisely, a string x is called (ε, δ)-*normal* if P_x is in δ-neighborhood of $P_x(\varepsilon)$. Otherwise, x is called (ε, δ)-*strange*.

It turns out that there are $\sqrt{n \log n}, O(\log n)$-normal strings with any given set P_x that satisfies the conditions of Theorem 1. On the other hand, there are $\Omega(n), \Omega(n)$-strange strings of length n. We are going to state these facts accurately.

Theorem 10 ([26]). *Let $k \leqslant n$ be two integers and let $t_0 > t_1 > \ldots > t_k$ be a strictly decreasing sequence of integers such that $t_0 \leqslant n$ and $t_k = 0$. Then there exists a string x of complexity $k + O(\sqrt{n \log n})$ and length $n + O(\log n)$ for which the distance between both sets P_x and $P_x(O(\log n))$ and the set $T = \{(i, j) \mid (i \leqslant k) \Rightarrow (j \geqslant t_i)\}$ is at most $O(\sqrt{n \log n})$.*

Proof. Consider the family $\mathcal{A}$ of all cylinders, i.e., the family of all the sets $\{ur \mid l(r) = m\}$ for different strings u and natural numbers m. Sets from this family have the following feature: if $A \ni x$ then A is an $O(\log n)$-strong model for x. Hence for all strings x we have $P_x^{\mathcal{A}} = P_x^{\mathcal{A}}(O(\log n))$.

By Theorem 7 and Remark 11 there is a string x of length $n + O(\log n)$ and complexity $k + O(\sqrt{n \log n})$ such that all sets $P_x, P_x^{\mathcal{A}}, T$ are $O(\sqrt{n \log n})$-close to each other. Hence all the three sets are close to the set $P_x^{\mathcal{A}}(O(\log n))$ as well. As the set $P_x(O(\log n))$ includes the latter set and is included in P_x, all the three sets are close to the set $P_x(O(\log n))$ as well. $\square$

The next theorem [40] shows that "strange" strings do exist.[14]

Theorem 11. *Assume that natural numbers k, n, ε satisfy the inequalities $O(1) \leqslant k \leqslant n$. Then there is a string x of length n and complexity $k + O(\log n)$ such that the sets P_x and $P_x(k)$ are $O(\log n)$-close to the sets shown on Fig. 8.*

---

[14] In this section we omit some proofs; see the original papers and the **arxiv** version of this paper.

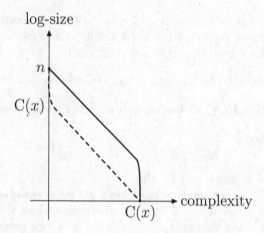

Fig. 8. The sets P_x and $P_x(k)$ for the strange string from Theorem 11, with $O(\log n)$-precision. The set P_x is to the right of the dashed line. The set $P_x(k)$ is to the right of the solid line.

Let $k = n/2$ in Theorem 11. Then the sets P_x and $P_x(n/2)$ are almost $n/2$-apart, since the point $(0, n/2)$ is in the $O(\log n)$-neighborhood of P_x while all points from $P_x(n/2)$ are $(n/2 - O(\log n))$-apart from $(0, n/2)$ (in l_1-norm). Thus the string x is $(n/2, n/2 - O(\log n))$-strange.

Recall that we have introduced the notion of a strong model to separate good models from bad ones. Indeed, there are some results that justify this approach. The following theorem by Milovanov (see [26] for the proof) states, roughly speaking, that there exist a string x of length n and a strong model A for x such that the parameters (complexity, log-cardinality) of every strong *standard* model B for x are $\Omega(n)$-far from those of A.

Theorem 12. *For all k there is a string x of length $n = 4k$ whose profile P_x is $O(\log n)$-close to the gray set shown on* Fig. 9 *such that*

- *there is an $O(\log n)$-strong model A for x with complexity $k + O(\log n)$ and log-cardinality $2k$ (that model witnesses the point $(k, 2k)$ on the border of P_x), but*
- *for every $m \geqslant C(x)$ and for every simple enumeration of strings of complexity at most m the standard model B for x obtained from that enumeration is either not strong for x or its parameters are far from the point $(k, 2k)$. More specifically, if B is an ε-strong model for x obtained from an enumeration provided by some program q, then $C(q) + |C(B) - k| + |\log \#B - 2k| + \varepsilon \geqslant \Omega(n)$.*

7.3   Properties of Strong Models

Once we have decided that non-strong descriptions are bad, it is natural to restrict ourselves to strong descriptions with negligible randomness deficiency (and hence negligible optimality deficiency).

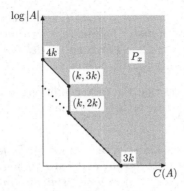

Fig. 9. The profile P_x of a string x from Theorem 12.

Consider some n-bit string x. Assume that A is an ε-strong description of x and the randomness deficiency of x in A is at most ε. Let u be the ordinal number of x in A with respect to some fixed order. Then $\mathrm{CT}(x\,|\,A, u) = O(1)$ and $\mathrm{CT}(A, u\,|\,x) \leqslant \varepsilon + O(1)$ (the latter inequality holds since $\mathrm{CT}(A\,|\,x) \leqslant \varepsilon$ and u can be easily found when x and A are known). As u is random and independent of A (with precision ε; note that $\mathrm{C}(u\,|\,A) \approx \mathrm{C}(x\,|\,A) \geqslant \log \#A - \varepsilon$), the sets $Q_{A,i}$ and Q_A are $\varepsilon + O(\log n)$-close (Proposition 25). On the other hand, the sets $Q_{A,u}$ and Q_x are $\varepsilon + O(1)$-close by Proposition 31. Thus we obtain the first property of strong models:

Proposition 33. If both $\mathrm{CT}(A\,|\,x)$ and $\log \#A - \mathrm{C}(x\,|\,A)$ are at most ε, then the sets Q_x and Q_A are $O(\varepsilon + \log l(x))$-close.

Assume that A is an ε-strong model for x with negligible randomness deficiency and ε; for simplicity we ignore these negligible quantities in the sequel. Assume that A is normal in the sense described above. Then the string x is normal as well. Indeed, for every pair $(i, j) \in P_x$ with $i \leqslant \mathrm{C}(A)$ the pair $(i, j - \log \#A)$ is in P_A (Proposition 25; note that x is equivalent to (A, u) and u is random with condition A) and hence there is a strong $(i * (j - \log \#A))$-description $\mathcal{B}$ for A. Consider the "lifting" of $\mathcal{B}$, that is, the union of all sets from $\mathcal{B}$ that have approximately the same size as A. It is a strong $(i * j)$-description for x.

It remains to consider pairs $(i, j) \in P_x$ where $i \geqslant \mathrm{C}(A)$. Then $i + j \geqslant \mathrm{C}(A) + \log \#A = \mathrm{C}(x)$. Hence the subset of A consisting of all strings x' whose ordinal number in A has the same $i - \mathrm{C}(A)$ leading bits as the ordinal number of x, is a strong $(i * j)$-description for x.

It turns out that for minimal models the converse is true as well. A model A for x is called $(\delta, \varkappa)$-*minimal* if there is no model B for x with $\mathrm{C}(B) \leqslant \mathrm{C}(A) - \delta$ and $\delta(x, B) \leqslant \delta(x, A) + \varkappa$.

Recall also that ε-sufficient statistic is a model whose optimality deficiency is smaller than ε.' This would be then the last sentence of a paragraph that starts with 'It turns out that for minimal models...'

Theorem 13 ([26]). *For some value $\varkappa = O(\log n)$ the following holds. Assume that A is an ε-sufficient statistic for an $(\varepsilon, \varepsilon)$-normal string x of length n. Assume also that A is a $(\delta, \varepsilon + \varkappa)$-minimal model for x. Then A is $(O((\delta + \varepsilon + \log n)\sqrt{n}), O((\delta + \varepsilon + \log n)\sqrt{n}))$-normal.*

The next theorem states that the total conditional complexity of any strong, sufficient and minimal statistic for x conditioned by any other sufficient statistic for x is negligible.

Theorem 14 ([39]). *For some value $\varkappa = O(\log n)$ the following holds. Assume that A, B are ε-sufficient statistics for a string x of length n. Assume also that A is an ε-strong and a $(\delta, \varepsilon + \varkappa)$-minimal statistic for x. Then $\mathrm{CT}(A|B) = O(\varepsilon + \delta + \log n)$.*

This theorem can be interpreted as follows: assume that we have removed some noise from a given data string x by finding its description B with negligible optimality deficiency. Let A be any "ultimately denoised" model for x, i.e., a minimal model for x with negligible optimality deficiency. Then $C(A|B)$ is negligible, as we have seen before. Hence to obtain the "ultimately denoised" model for x we do not need x: any such model can be obtained from B by a short program. Theorem 14 shows that any such *strong* model A can be obtained from B by a short *total* program.

7.4   Open Questions

1. Is the minimal strong sufficient statistic unique (up to ε-equivalence). More specifically, assume that A, B are ε-strong, ε-sufficient statistics for a string x of length n. Assume further that both A, B are $\delta, c(\varepsilon + \delta + \log n)$-minimal models for x. Is it true that $\mathrm{CT}(A|B), CT(B|A)$ are small in this case?
2. A similar question, but this time we do not assume that B is minimal. Is it true that $\mathrm{CT}(A|B)$ is small? (An affirmative answer to this question obviously implies the affirmative answer to the previous one.)
   Note that if, in these two questions, we replace total conditional complexity with the plain conditional complexity then the answers are positive and moreover, we do not need to assume that A, B are ε-strong (see Proposition 18 and the last two paragraphs on Page 37).
3. (Merging strong sufficient statistics.) Assume that A, B are strong sufficient statistics for x that have small intersection compared to the cardinality of at least one of them. Then it is natural to conjecture that there is a strong sufficient statistic D for x of larger cardinality (=of smaller complexity) that is simple given both A, B. Formally, is it true (for some constant c) that if A, B are ε-strong ε-sufficient statistics for x, then there is a $c\varepsilon$-strong $c\varepsilon$-sufficient statistic D for x with $\log \#D \geqslant \log \#A + \log \#B - \log \#(A \cap B) - c(\varepsilon + \log n)$ and $\mathrm{CT}(D|A), \mathrm{CT}(D|B)$ at most $c(\varepsilon + \log n)$? (A motivating example: let x be a random string of length n, let A consist of all strings of length n that have the same prefix of length $n/2$ as x, and let B consist of all strings of length n that have the same bits with numbers $n/4 + 1, \ldots, 3n/4$ as x. In this

case it is natural to let D consist of all strings of length n that have the same bits $n/4+1,\ldots,n/2$ as x, so that $\log \#D = \log \#A + \log \#B - \log \#(A\cap B)$.)

Acknowledgments. We are grateful to several people who contributed and/or carefully read preliminary versions of this survey, in particular, to B. Bauwens, P. Gács, A. Milovanov, G. Novikov, A. Romashchenko, P. Vitányi, and to all participants of Kolmogorov seminar in Moscow State University and ESCAPE group in LIRMM. We are also grateful to an anonymous referee for correcting several mistakes.

References

1. Adleman, L.M.: Time, space and randomness. MIT report MIT/LCS/TM-131, March 1979
2. Antunes, L., Bauwens, B., Souto, A,, Teixeira, A.: Sophistication vs. logical depth. Theory of Computing Systems. doi:10.1007/s00224-016-9672-6
3. Antunes, L., Fortnow, L.: Sophistication revisited. Theory Comput. Syst. **45**(1), 150–161 (2009)
4. Antunes, L., Fortnow, L., van Melkebeek, D.: Computational depth, In: Proceedings of the 16th IEEE Conference on Computational Complexity, pp. 266–273. IEEE, New York (2001). Journal version: Computational depth: concept and applications. Theoret. Comput. Sci. **354**(3), 391–404 (2006)
5. Antunes, L., Matos, A., Souto, A., Vitányi, P.: Depth as randomness deficiency. Theory Comput. Syst. **45**(4), 724–739 (2009)
6. Bauwens, B.: Computability in statistical hypotheses testing, and characterizations of independence and directed influences in time series using Kolmogorov complexity. Ph.D. thesis, University of Gent, May 2010
7. Bennett, C.H.: Logical depth and physical complexity. In: Herken, R. (ed.) The Universal Turing Machine: A Half-Century Survey, pp. 227–257. Oxford University Press, New York (1988)
8. Bienvenu, L., Desfontaines, D., Shen, A.: What percentage of programs halt? In: Halldórsson, M.M., Iwama, K., Kobayashi, N., Speckmann, B. (eds.) ICALP 2015. LNCS, vol. 9134, pp. 219–230. Springer, Heidelberg (2015). doi:10.1007/978-3-662-47672-7_18
9. Bienvenu, L., Gács, P., Hoyrup, M., Rojas, C., Shen, A.: Algorithmic tests and randomness with respect to a class of measures. Proc. Steklov Inst. Math. **274**, 34–89 (2011)
10. Cover, T.: Kolmogorov complexity, data compression and inference. In: Skwirzynski, J.K. (ed.) The Impact of Processing Techniques on Communications. NATO ASI Series, vol. 91, pp. 23–33. Martinus Nijhoff Publishers, Dordrecht (1985). doi:10.1007/978-94-009-5113-6_2
11. Epstein, S., Levin, L.: Sets have simple members. http://arxiv.org/abs/1107.1458, reposted as http://arxiv.org/abs/1403.4539
12. Gács, P.: On the relation between descriptional complexity and algorithmic probability. Theoret. Comput. Sci. **22**, 71–93 (1983)
13. Gács, P., Tromp, J., Vitányi, P.M.B.: Algorithmic statistics. IEEE Trans. Inf. Theory **47**(6), 2443–2463 (2001)
14. Kolmogorov, A.N.: Three approaches to the quantitative definition of information (in Russian). Prob. Inf. Trans. **1**(1), 4–11 (1965). English translation published: Int. J. Comput. Math. **2**, 157–168 (1968)

15. Kolmogorov, A.N.: Talk at the Information Theory Symposium in Tallinn, Estonia (then USSR) (1974) [As reported by Cover in his 1985 paper [10]]

16. Kolmogorov, A.N.: The complexity of algorithms and the objective definition of randomness. Summary of the talk presented April 16, 1974 at Moscow Mathematical Society. Uspekhi matematicheskikh nauk (Russian) **29**(4[178]), 155 (1974). http://mi.mathnet.ru/rus/umn/v29/i4/p153 (A short note in Russian)

17. Kolmogorov, A.N.: Talk at the seminar at Moscow State University Mathematics Department (Logic Division), 26 November 1981. [The definition of (α, β)-stochasticity was defined in this talk, and the question about the fraction of non-stochastic objects was posed.]

18. Koppel, M.: Complexity, depth and sophistication. Compl. Syst. **1**, 1087–1091 (1987)

19. Koppel, M.: Structure. In: Herken, R. (ed.) The Universal Turing Machine: A Half-Century Survey, pp. 435–452. Oxford University Press (1988)

20. Koppel, M., Atlan, H.: An almost machine-independent theory of program-length complexity, sophistication, and induction. Inf. Sci. **56**(1–3), 23–33 (1991)

21. Levin, L.: Randomness conservation inequalities; information and independence in mathematical theories. Inf. Control **61**(1), 15–37 (1984)

22. Li, M., Vitányi, P.M.B.: An Introduction to Kolmogorov Complexity and its Applications, 3rd edn. Springer, New York (2008)

23. Longpré, L.: Resource bounded Kolmogorov complexity, a link between computational complexity and information theory. Ph.D. Thesis, Department of Computer Science, Cornell University, TR 86–776 (1986)

24. Milovanov, A.: Some properties of antistochastic strings. In: Beklemishev, L.D., Musatov, D.V. (eds.) CSR 2015. LNCS, vol. 9139, pp. 339–349. Springer, Heidelberg (2015). doi:10.1007/978-3-319-20297-6_22

25. Milovanov, A.: Algorithmic statistic, prediction and machine learning. In: 33rd Symposium on Theoretical Aspects of Computer Science (STACS 2016), Leibnitz International Proceedings in Informatics (LIPIcs), vol. 47, pp. 54:1–54:13 (2016). doi:10.4230/LIPIcs.STACS.2016.54, http://drops.dagstuhl.de/opus/volltexte/2016/5755/

26. Milovanov, A.: Algorithmic statistics: normal objects and universal models. In: Kulikov, A.S., Woeginger, G.J. (eds.) CSR 2016. LNCS, vol. 9691, pp. 280–293. Springer, Heidelberg (2016). doi:10.1007/978-3-319-34171-2_20

27. Mota, F., Aaronson, S., Antunes, L., Souto, A.: Sophistication as randomness deficiency. In: Jurgensen, H., Reis, R. (eds.) DCFS 2013. LNCS, vol. 8031, pp. 172–181. Springer, Heidelberg (2013). doi:10.1007/978-3-642-39310-5_17

28. Muchnik, An.A., Mezhirov, I., Shen, A., Vereshchagin, N.K.: Game interpretation of Kolmogorov complexity. https://arxiv.org/abs/1003.4712

29. Muchnik, An.A., Romashchenko, A.: Stability of properties of Kolmogorov complexity under relativization. Prob. Inf. Trans. **46**(1), 38–61 (2010)

30. Muchnik, An.A., Semenov, A.L., Uspensky, V.A.: Mathematical metaphysics of randomness. Theoret. Comput. Sci. **207**(2), 263–317 (1998)

31. Muchnik, An.A., Shen, A., Vyugin, M.: Game arguments in computability theory and algorithmic information theory. https://arxiv.org/pdf/1204.0198.pdf

32. de Rooij, S., Vitányi, P.M.B.: Approximating rate-distortion graphs of individual data: experiments in lossy compression and denoising. IEEE Trans. Comput. **61**(3), 395–407 (2012)

33. Rissanen, J.: Modeling by shortest data description. Automatica **14**, 465–471 (1978)

34. Shen, A.: The concept of (α, β)-stochasticity in the Kolmogorov sense, and its properties. Soviet Math. Dokl. **28**(1), 295–299 (1983)
35. Shen, A.: Discussion on Kolmogorov complexity and statistical analysis. Comput. J. **42**(4), 340–342 (1999)
36. Shen, A.: Around Kolmogorov complexity: basic notions and results. In: Vovk, V., Papadopoulos, H., Gammerman, A. (eds.) Measures of Complexity: Festschrift for Alexey Chervonenkis, pp. 75–116. Springer, Switzerland (2015). doi:10.1007/978-3-319-21852-6_7. arXiv:1504.04955
37. Solomonoff, R.: A formal theory of inductive inference. Part I. Inf. Control **7**(1), 1–22 (1964)
38. Solomonoff, R.: A formal theory of inductive inference. Part II. Applications of the systems to various problems in induction. Inf. Control **7**(2), 224–254 (1964)
39. Vereshchagin, N.: Algorithmic minimal sufficient statistic revisited. In: Ambos-Spies, K., Löwe, B., Merkle, W. (eds.) CiE 2009. LNCS, vol. 5635, pp. 478–487. Springer, Heidelberg (2009). doi:10.1007/978-3-642-03073-4_49
40. Vereshchagin, N.: Algorithmic minimal sufficient statistics: a new approach. Theory Comput. Syst. **58**(3), 463–481 (2016)
41. Vereshchagin, N., Shen, A.: Algorithmic statistics revisited. In: Vovk, V., Papadopoulos, H., Gammerman, A. (eds.) Measures of Complexity: Festschrift for Alexey Chervonenkis, pp. 235–252. Springer, Switzerland (2015). arXiv:1504.04950
42. Vereshchagin, N., Uspensky, V., Shen, A.: Kolmogorov complexity and algorithmic randomness. In: MCCME 2013, 576 pp., Moscow (2013). http://www.lirmm.fr/~ashen/kolmbook.pdf (Russian version), http://www.lirmm.fr/~ashen/kolmbook-eng.pdf (English version)
43. Vereshchagin, N.K., Vitányi, P.M.B.: Kolmogorov's structure functions and model selection. IEEE Trans. Inf. Theory **50**(12), 3265–3290 (2004)
44. Vereshchagin, N.K., Vitányi, P.M.B.: Rate distortion and denoising of individual data using Kolmogorov complexity. IEEE Trans. Inf. Theory **56**(7), 3438–3454 (2010)
45. Vitányi, P.M.B.: Meaningful information. IEEE Trans. Inf. Theory **52**(10), 4617–4626 (2006). arXiv:cs/0111053
46. V'yugin, V.V.: On the defect of randomness of a finite object with respect to measures with given complexity bounds. SIAM Theory Prob. Appl. **32**(3), 508–512 (1987)
47. V'yugin, V.V.: Algorithmic complexity and stochastic properties of finite binary sequences. Comput. J. **42**(4), 294–317 (1999)
48. V'yugin, V.V.: Does snooping help? Theoret. Comput. Sci. **276**(1), 407–415 (2002)
49. Wallace, C.S., Boulton, D.M.: An information measure for classification. Comput. J. **11**(2), 185–194 (1968)

Lowness, Randomness, and Computable Analysis

André Nies$^{(\boxtimes)}$

Department of Computer Science, University of Auckland, Auckland, New Zealand
andre@cs.auckland.ac.nz

Abstract. Analytic concepts contribute to our understanding of randomness of reals via algorithmic tests. They also influence the interplay between randomness and lowness notions. We provide a survey.

1   Introduction

Our basic objects of study are infinite bit sequences, identified with sets of natural numbers, and often simply called sets. A lowness notion provides a sense in which a set A is close to computable. For example, A is *computably dominated* if each function computed by A is dominated by a computable function; A is *low* if the halting problem relative to A has the least possible Turing complexity, namely $A' \equiv_T \varnothing'$. These two notions are incompatible outside the computable sets, because every non-computable Δ_2^0 set has hyperimmune degree.

Lowness notions have been studied for at least 50 years [27,33,51]. More recently, and perhaps surprisingly, ideas motivated by the intuitive notion of randomness have been applied to the investigation of lowness. On the one hand, these ideas have led to new lowness notions. For instance, K-triviality of a set of natural numbers (i.e., being far from random in a specific sense) coincides with lowness for Martin-Löf randomness, and many other notions. On the other hand, they have been applied towards a deeper understanding of previously known lowness notions. Randomness led to the study of an important subclass of the computably dominated sets, the computably traceable sets [52]. Superlowness of an oracle A, first studied by Mohrherr [36], says that $A' \equiv_{tt} \varnothing'$; despite the fact that the low basis theorem [27] actually yields superlow sets, the importance of superlowness was not fully appreciated until the investigations of lowness via randomness. For instance, every K-trivial set is superlow [39].

Computable analysis allows us to characterise several randomness notions that were originally defined in terms of algorithmic tests. Schnorr [49] introduced two randomness notions for a bit sequence Z via the failure of effective betting strategies. Nowadays they are called computable randomness and Schnorr randomness. Computable randomness says that no effective betting strategy (martingale) succeeds on Z, Schnorr randomness that no such strategy succeeds quickly (see [11,40] for background). Pathak [44], followed by Pathak et al. [45] characterised Schnorr randomness: Z is Schnorr random iff an effective version of the Lebesgue differentiation theorem holds at the real $z \in [0,1]$ with binary expansion Z. Brattka et al. [6] showed that Z is computably random if and only

© Springer International Publishing AG 2017
A. Day et al. (Eds.): Downey Festschrift, LNCS 10010, pp. 738–754, 2017.
DOI: 10.1007/978-3-319-50062-1_42

if every nondecreasing computable function is differentiable at z. See Sect. 4 for detail.

From 2011 onwards, concepts from analysis have also influenced the interplay of lowness and randomness. The Lebesgue density theorem for effectively closed sets $\mathcal{C}$ provides two randomness notions for a bit sequence Z which are slightly stronger than Martin-Löf's. In the stronger form, the density of any such set $\mathcal{C}$ that contains Z has to be 1 at Z; in the weak form, the density is merely required to be positive. One has to require separately that Z is ML-random (even the stronger density condition doesn't imply this, for instance because a 1-generic also satisfies this condition). These two notions have been used to obtain a Turing incomplete Martin-Löf random above all the K-trivials, thereby solving the so-called ML-covering problem. We give more detail in Sect. 5; also see the survey [3].

In current research, concepts inspired by analysis are used to stratify lowness notions. Cost functions describe a dense hierarchy of subideals of the K-trivial Turing degrees (Sect. 6). The Gamma and Delta parameters are real numbers assigned to a Turing degree. They provide a coarse measure of its complexity in terms of the asymptotic density of bit sequences (Sect. 7).

The present paper surveys the study of lowness notions via randomness. In Sects. 2 and 3 we elaborate on the background in lowness, and how it was influenced by randomness. Section 4 traces the interaction of computable analysis and randomness from Lebesgue to the present. Section 5 shows how some of the advances via computable analysis aided the understanding of lowness through randomness. Sections 6 and 7 dwell on the most recent developments. A final section contains open questions.

2   Early Days of Lowness

Spector [51] was the first to construct a Turing degree that is minimal among the nonzero degrees. Sacks [47] showed that such a degree can be Δ_2^0. Following these results, as well as the Friedberg-Muchnik theorem and Sacks' result [48] that the c.e. Turing degrees are dense, the interest of early computability theorists focused on relative complexity of sets: comparing their complexity via an appropriate reducibility. Absolute complexity, which means finding natural lowness classes and studying membership in them, received somewhat less attention, and was mostly restricted to classes defined via the Turing jump. Martin and Miller [33] built a perfect closed set of computably dominated oracles. Jockusch and Soare [27] proved that every non-empty effectively closed set contains a low oracle. These constructions used recursion-theoretic versions of forcing. Jockusch published papers such as [25, 26] that explored notions such as degrees of diagonally noncomputable functions, and degrees of bi-immune sets. Downey's work in the 1980s was important for the development of our understanding of lowness. For instance, Downey and Jockusch [15] studied complexity of sets, both relative and absolute, using ever more sophisticated methods.

3   Randomness Interacts with Lowness

We begin with the following randomness notions:

weakly 2-random $\rightarrow$ ML-random $\rightarrow$ computably rd. $\rightarrow$ Schnorr rd.

Z is weakly 2-random iff Z is in no null Π_2^0 class. Section 5 will develop notions implied by weak 2-randomness, and somewhat stronger than ML-randomness.

Lowness can be used to understand randomness via the randomness enhancement principle [41], which says that sets already enjoying a randomness property get more random as they become lower in the sense of computational complexity. Every non-high Schnorr random is ML-random. A ML-random is weakly 2-random iff it forms a Turing minimal pair with $\emptyset'$. See [11,40].

Here we are mostly interested in the converse interaction: studying lowness via randomness. Let $K(x)$ denote the prefix free version of Kolmogorov complexity of a binary string x. The K-trivial sets were introduced by Chaitin [8] and studied by Solovay in an unpublished manuscript [50], rediscovered by Calude in the 1990s. Most of this manuscript is covered in Downey and Hirschfeldt's monumental work [11]. We say that A is K-trivial if $\exists b \forall n\, K(A \mid n) \leqslant K(n) + b$. By the Levin-Schnorr characterisation, Z is ML-random iff $\exists d \forall n\, K(Z \mid n) \geqslant n - d$. Since $K(n) \leqslant \log_2 n + O(1)$, this definition says that K-trivials are far from random. Each computable set is K-trivial; Solovay built a K-trivial Δ_2^0 set A that is not computable. This was later improved to a c.e. set A by Downey et al. [13], who used what became later known as a cost function construction.

An oracle A is called *low for a randomness notion* $\mathcal{C}$ if every $\mathcal{C}$-random set is already $\mathcal{C}$-random relative to A. K-triviality appears to be the universal lowness class for randomness notions based on c.e. test notions. A is K-trivial iff $A \in \mathsf{Low}(\mathsf{W2R}, \mathsf{CR})$, namely every weakly 2-random set is computably random relative to A. This was shown by the author [40, 8.3.14] extending the result that $\mathsf{Low}(\mathsf{W2R}, \mathsf{MLR})$ coincides with K-triviality [14]. As a consequence, for any randomness notion $\mathcal{D}$ in between weak-2 randomness and computable randomness, lowness for $\mathcal{D}$ implies K-triviality. For many notions, e.g. weak-2 randomness [14] and ML-randomness [39], the classes actually coincide.

Some of the K-trivial story, including the roles Downey, Hirschfeldt and the author have played in it, is vividly described in [42]. Background and more detailed coverage for the developments up to 2009 can be found in the aforementioned books [11,40].

4   Randomness and Computable Analysis

We discuss the influence exerted by computable analysis on the study of randomness notions. Thereafter we will return to our main topic, lowness notions.

Analysis and ergodic theory have plenty of theorems saying that a property of being well-behaved holds at almost every point. Lebesgue proved that a function of bounded variation defined on the unit interval is differentiable at almost every point. He also proved the density theorem, and the stronger differentiation theorem, that now both bear his name. The density theorem says that a measurable set $C \subseteq [0,1]$ has density one at almost every of its members z. To have density one at z means intuitively that many points close to z are also in C, and this becomes more and more apparent as one "zooms in" on z:

Definition 1. *Let λ denote Lebesgue measure on $\mathbb{R}$. We define the lower Lebesgue density of a set $C \subseteq \mathbb{R}$ at a point z to be the limit inferior*

$$\underline{\varrho}(C|z) := \liminf_{|Q| \to 0} \frac{\lambda(Q \cap C)}{|Q|},$$

where Q ranges over open intervals containing z. The Lebesgue density of C at z is the limit (which may not exist)

$$\varrho(C|z) := \lim_{|Q| \to 0} \frac{\lambda(Q \cap C)}{|Q|}.$$

Note that $0 \leqslant \underline{\varrho}(C|z) \leqslant 1$.

Theorem 2 (Lebesgue [31]). *Let $C \subseteq \mathbb{R}$ be a measurable set. We have $\underline{\varrho}(C|z) = 1$ for almost every $z \in C$.*

The Lebesgue differentiation theorem says that for almost every z, the value of an integrable function g at $z \in [0,1]$ is approximated by the average of the values around z, as one zooms in on z. A point z in the domain of g is called a *weak Lebesgue point* of g if

$$\lim_{\lambda Q \to 0} \frac{1}{\lambda(Q)} \int_Q g \, d\lambda \tag{1}$$

exists, where Q ranges over open intervals containing z; we call z a *Lebesgue point* of g if this value equals $g(z)$.

Theorem 3 (Lebesgue [30]). *Suppose g is an integrable function on $[0,1]$. Then almost every $z \in [0,1]$ is a Lebesgue point of g.*

Lebesgue [32] extended this result to higher dimensions, where Q now ranges over open cubes containing z. Note that if g is the characteristic function of a measurable set C, then the expression (1) is precisely the density of C at z.

In ergodic theory, one of the best-known "almost everywhere" theorems is due to G. Birkhoff: intuitively, given an integrable function g on a probability space with a measure preserving operator T, for almost every point z, the average of g-values at iterates, that is, $\frac{1}{n} \sum_{i<n} g(T^i(z))$, converges. If T is ergodic (i.e., all T-invariant measurable sets are null or co-null), the limit equals $\int g$; in general the limit is given by the conditional expectation of g with respect to the σ-algebra of T-invariant sets.

The important insight is this: if the collection of given objects in an a.e. theorem is effective in some particular sense, then the theorem describes a randomness notion via algorithmic tests. Every collection of effective objects constitutes a test, and failing it means to be a point of exception for this collection. Demuth [10] (see below) made this connection in the setting of constructive mathematics. In the usual classical setting, V'yugin [53] showed that Martin-Löf randomness of a point z in a computable probability space suffices for the existence of the limit in Birkhoff's theorem when T is computable and g is L_1-computable. Here L_1-computability means in essence that the function can be effectively approximated by step functions, where the distance is measured in the usual L_1-norm. About ten years later, Pathak [44] showed that ML-randomness of z suffices for the existence of the limit in the Lebesgue differentiation theorem when the given function f is L_1-computable. This works even when f is defined on $[0,1]^n$ for some $n > 1$. Pathak, Rojas and Simpson [45] showed that in fact the weaker condition of Schnorr randomness on z suffices. They also showed a converse: if z is not Schnorr random, then for some L_1-computable function f the limit fails to exist. Thus, the Lebesgue differentiation theorem, in this effective setting, characterises Schnorr randomness. This converse was obtained independently by Freer et al. [19], who also extended the characterisation of Schnorr randomness to L_p-computable functions, for any fixed computable real $p \geqslant 1$.

In the meantime, Brattka, Miller and Nies proved the above-mentioned effective form of Lebesgue's theorem [31] that each nondecreasing function is a.e. differentiable: each nondecreasing *computable* function is differentiable at every computably random real ([6], the work for which was carried out from late 2009). Later on, Nies [43] gave a different, and somewhat simpler, argument for this result involving the geometric notion of porosity. With some extra complications the latter argument carries over to the setting of polynomial time computability, which was the main thrust of [43].

Jordan's decomposition theorem says that for every function f of bounded variation there are nondecreasing functions g_0, g_1 such that $f = g_0 - g_1$. This is almost trivial in the setting of analysis (take $g_0(x)$ to be the variation of f restricted to $[0,x]$, and let $g_1 = f - g_0$). Thus every bounded variation function f is a.e. differentiable. For computable f, it turns out that ML-randomness of z may be required to ensure that $f'(z)$ exists; the reason is that the two functions obtained by the Jordan decomposition cannot always be chosen to be computable. Demuth [10] had obtained results in the constructive setting which, when re-interpreted classically, show that z is ML-random iff every computable function of bounded variation is differentiable at z. Brattka et al. [6] gave alternative proofs of both implications. For the harder implication, from left to right, they used their main result on computable nondecreasing functions, together with the fact that the possible Jordan decompositions of a computable bounded variation function form a Π_1^0 class, which therefore has a member in which z is random. See the recent survey [29] for more on Demuth's work as an early example of an interplay between randomness and computability.

5   Lebesgue Density and K-triviality

Can analytic notions aid in the study of lowness via randomness? The answer is "yes, but only indirectly". Analytic notions help because they bear on our view of randomness. In this section we will review how the notion of Lebesgue density helped solving the ML-covering problem, originally asked by Stephan (2004). This was one of five "big" questions in [34]. Every c.e. set below a Turing incomplete random is a base for ML-randomness, and hence K-trivial [23]. The covering problem asks whether the converse holds: is every c.e. K-trivial A below an incomplete random set? Since every K-trivial is Turing below a c.e. K-trivial [39], we can as well omit the hypothesis that A be c.e.

Computable analysis ⤳ randomness

Some effective versions of almost everywhere theorems lack a predefined randomness notion. In the context of Theorem 3, the statement that almost every point is a *weak* Lebesgue point will be called the weak Lebesgue differentiation theorem. We have already discussed the fact that the weak Lebesgue differentiation theorem for L_1-computable functions characterises Schnorr randomness. A function g is lower semicomputable if $\{x : g(x) > q\}$ is Σ_1^0 uniformly in a rational q, and upper semicomputable if $\{x : g(x) < q\}$ is Σ_1^0 uniformly in a rational q. Which degree of randomness does a point z need in order to ensure that z is a (weak) Lebesgue point for all lower (or equivalently, all upper) semicomputable functions?

Even ML-randomness is insufficient for this. Let $z = \Omega$ denote Chaitin's halting probability, and consider the Π_1^0-set $\mathcal{C} = [\Omega, 1]$. The real z is ML-random, and in particular normal: every block of bits of length k (such as 110011) occurs with limiting frequency 2^{-k} in its binary expansion. Suppose $z \in Q$ where $Q = (i2^{-n}, (i+1)2^{-n})$ for some $i < n$. If the binary expansion of z has a long block of 0 s from position n on, then $\lambda(Q \cap \mathcal{C})/|Q|$ is large; if z has a long block of 1 s from n on then it is small. This implies that $\lambda(Q)/|Q|$ oscillates between values close to 0 and close to 1 as Q ranges over smaller and smaller basic dyadic intervals containing z. So it cannot be the case that $\underline{\varrho}(\mathcal{C}|z) = 1$; in fact the density of $\mathcal{C}$ at z does not exist. This means that z is not a weak Lebesgue point for the upper semicomputable function $1_{\mathcal{C}}$.

We say that a ML-random real z is *density random* if $\underline{\rho}(\mathcal{C} \mid z) = 1$ for each Π_1^0 set $\mathcal{C}$ containing z. Several equivalent characterisations of density randomness are given in [35, Theorem 5.8]; for instance, a real z is density random iff z is a weak Lebesgue point of each lower semicomputable function on $[0, 1]$ with finite integral, iff z is a full Lebesgue point of each such function.

Randomness ⤳ lowness

The approach of the Oberwolfach group (2012) was mostly within the classical interplay of randomness and computability. Inspired by the notion of balanced randomness introduced in [18], they defined a new notion, now called Oberwolfach (OW) randomness [4]. A test notion equivalent to Oberwolfach tests, and easier to use, is as follows. A descending uniformly Σ_1^0 sequence of sets $\langle G_m \rangle_{m \in \omega}$, together with a left-c.e. real β with a computable approximation $\beta = \sup_s \beta_s$, form a *left-c.e.* test if $\lambda G_m = O(\beta - \beta_m)$ for each m. Just like in the original definition of Oberwolfach tests, the test components cohere. If there is an increase

$\beta_{s+1} - \beta_s = \gamma > 0$, then all components G_m for $m < s$ are allowed to add up to γ in measure, as long as the sequence remains descending. We think of first G_0 adding some portion of measure of at most γ, then G_1 adding some portion of that, then G_2 a portion of that second portion, and so on up to G_s.

The Oberwolfach group [4, Theorem 1.1] showed that if Z is ML-random, but not OW-random, then Z computes each K-trivial. They also proved that OW-randomness implies density randomness.

Analysis ⤳ randomness ⤳ lowness

Often the notions of density are studied in the context of Cantor space $2^{\mathbb{N}}$, which is easier to work with than the unit interval. In this context one defines the density at a a bit sequence Z using basic dyadic intervals that are given by longer and longer initial segments of Z. In the context of randomness this turns out to be a minor change. If z is a ML-random real and Z its binary expansion, then each Π_1^0 set $\mathcal{C} \subseteq [0,1]$ has positive density at z iff each Π_1^0 set $\mathcal{C} \subseteq 2^{\mathbb{N}}$ has positive density at Z iff Z is Turing incomplete, by a result in Bienvenu et al. [5]. Dyadic and full density 1 also coincide for ML-random reals by a result of Khan and Miller [28, Theorem 3.12].

Day and Miller [9] used a forcing partial order specially adapted to the setting of intermediate density to prove that there is a ML-random Z such that $\rho(\mathcal{C} \mid Z) > 0$ for each Π_1^0 class $\mathcal{C} \subseteq 2^{\mathbb{N}}$, and at the same time there is a Π_1^0 class $\mathcal{D} \ni Z$ such that $\rho(\mathcal{D} \mid Z) < 1$. Hence Z is incomplete ML-random and not Oberwolfach random. By the aforementioned result of the Oberwolfach group [4, Theorem 1.1] this means that the single oracle Z computes each K-trivial, thereby giving a strong affirmative answer to the covering problem.

Day and Miller also refined their argument in order to make Z a Δ_2^0 set. No direct construction is known to build a Δ_2^0 incomplete ML-random that is not Oberwolfach random. In fact, it is open whether Oberwolfach and density randomness coincide (see Question 17 below).

6   Cost Functions and Subclasses of the K-trivials

In this and the following section, we survey ways to gauge the complexity of Turing degrees directly with methods inspired by analysis. The first method only applies to K-trivials: use the analytical tool of a cost function to study proper subideals of the ideal of K-trivial Turing degrees. This yields a dense hierarchy of ideals parameterised by rationals in $(0,1)$.

The second method assigns real number parameters $\Gamma(\mathbf{a}), \Delta(\mathbf{a}) \in [0,1]$ to Turing degrees $\mathbf{a}$ in order to measure their complexity. These assignments can be interpreted in the context of Hausdorff distance in pseudometric spaces. In a sense, this second attempt turns out to be too coarse because in both variants, only the values 0 and $1/2$ are possible for non-computable Turing degrees (after a modification of the definition which we also present, the classes of sets with value 0 have subclasses that are potentially proper). However, this also shows that these few classes of complexity obtained must be natural and important. Although they richly interact with previously studied classes, they haven't as yet been fully characterised by other means.

Both approaches are connected to randomness through the investigations of the concepts, rather than directly through the definitions. We will explain these connections as we go along.

Cost functions

Somewhat extending [40, Sect. 5.3], we say that a *cost function* is a computable function

$$\mathbf{c}\colon \mathbb{N} \times \mathbb{N} \to \{x \in \mathbb{R} : x \geqslant 0\}.$$

For background on cost functions see [38]. We say that $\mathbf{c}$ is *monotonic* if $\mathbf{c}(x + 1, s) \leqslant \mathbf{c}(x, s) \leqslant \mathbf{c}(x, s + 1)$ for each x and s; we also assume that $\mathbf{c}(x, s) = 0$ for all $x \geqslant s$. We view $\mathbf{c}(x, s)$ as the cost of changing at stage s a guess $A_{s-1}(x)$ at the value $A(x)$, for some Δ_2^0 set A. Monotonicity means that the cost of a change increases with time, and that changing the guess at a smaller number is more costly.

If $\mathbf{c}$ is a cost function, we let $\underline{\mathbf{c}}(x) = \sup_s \mathbf{c}(x, s)$. To be useful, a monotonic cost function $\mathbf{c}$ needs to satisfy the *limit condition*: $\underline{\mathbf{c}}(x)$ is finite for all x and $\lim_x \underline{\mathbf{c}}(x) = 0$.

Definition 4 ([40]). *Let $\langle A_s \rangle$ be a computable approximation of a Δ_2^0 set A, and let $\mathbf{c}$ be a cost function. The* total $\mathbf{c}$-cost *of the approximation is*

$$\mathbf{c}(\langle A_s \rangle_{s \in \omega}) = \sum_{s \in \omega} \{\mathbf{c}(x, s) : x \text{ is least such that } A_{s-1}(x) \neq A_s(x)\}.$$

We say that a Δ_2^0 set A obeys $\mathbf{c}$ if the total $\mathbf{c}$-cost of some computable approximation of A is finite. We write $A \models \mathbf{c}$ (Fig. 1).

This definition, first given in [40], was conceived as an abstraction of the construction of a c.e. noncomputable K-trivial set in Downey et al. [13]. Perhaps

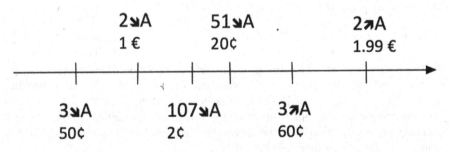

Fig. 1. Timeline illustrating the cost (in Euros) generated by an approximation of a Δ_2^0 set A for a particular cost function.

the intuition stems from analysis. For instance, the length of a curve, i.e. a C^1 function $f\colon [0,1] \to \mathbb{R}^n$, is given by $\int_0^1 \|f'(t)\| dt$. The "cost" of the change $f'(t)$ at stage t is the velocity $\|f'(t)\|$, and to have a finite total cost means that the curve is rectifiable.

The paper [38] also treats non-monotonic cost functions, where we define $\underline{c}(x) = \liminf_s c(x,s)$ and otherwise retain the definition of the limit condition $\lim_x \underline{c}(x) = 0$. Intuitively, enumeration of x into A can only take place at a stage when the cost drops. This is reminiscent of $\emptyset''$-constructions, for instance building a Turing minimal pair of c.e. sets. It would be interesting to define a pair of cost functions $\mathbf{c}, \mathbf{d}$ with the limit condition such that $A \models \mathbf{c}$ and $B \models \mathbf{d}$ for c.e. sets A, B imply that A, B form a minimal pair.

Applications of cost functions

Let β be a left-c.e. real given as $\beta = \sup_s \beta_s$ for a computable sequence $\langle \beta_s \rangle_{s \in \omega}$ of rationals. We let $\mathbf{c}_\beta(x,s) = \beta_s - \beta_x$. Note that $\mathbf{c}_\beta$ is a monotonic cost function with the limit condition. Modifying a result from [39], in [38] it is shown that a Δ_2^0 set A is K-trivial iff $A \models \mathbf{c}_\Omega$. Thus $\mathbf{c}_\Omega$ is a cost function describing K-triviality. This raises the question whether obedience to cost functions stronger than $\mathbf{c}_\Omega$ can describe interesting subideals of the ideal of K-trivial Turing degrees (being a stronger cost function means being harder to obey, i.e. more expensive).

By the "halves" of a set Z we mean the sets $Z_0 = Z \cap \{2n\colon n \in \mathbb{N}\}$ and $Z_1 = Z \cap \{2n+1\colon n \in \mathbb{N}\}$. If Z is ML-random and $A \leqslant_T Z_0, Z_1$ then A is a base for ML-randomness, and hence K-trivial. So we obtain a subclass $\mathcal{B}_{1/2}$ of the K-trivial sets, namely the sets below both halves of a ML-random. Bienvenu et al. [4] had already proved that this subclass is proper. Let $\mathbf{c}_{\Omega,1/2}(x,s) = \sqrt{\Omega_s - \Omega_x}$. In recent work, Greenberg, Miller and Nies obtained the following characterisation of $\mathcal{B}_{1/2}$.

Theorem 5 ([21], Theorem 1.1. and its proof). *The following are equivalent for a set A.*

(a) *A is Turing below both halves of some ML-random*
(b) *A is Turing below both halves of Ω*
(c) *A is a Δ_2^0 set that obeys $\mathbf{c}_{\Omega,1/2}(x,s)$.*

They generalise the result towards a characterisation of classes $\mathcal{B}_{k/n}$, where $0 < k < n$. The class $\mathcal{B}_{k/n}$ consists of the Δ_2^0 sets A that are Turing below the effective join of any set of k among the n-columns of some ML-random set Z; as before, Z can be taken to be Ω without changing the class. The characterising cost function is $\mathbf{c}_{\Omega,q}(x,s) = (\Omega_s - \Omega_x)^q$, where $q = k/n$. In particular, the class does not depend on the representation of q as a fraction of integers. By this cost function characterisation and the hierarchy theorem [38, Theorem 3.4], $p < q$ implies that $\mathcal{B}_p$ is a proper subclass of $\mathcal{B}_q$.

Following Hirschfeldt et al. [22] we say that a set A is *robustly computable* from a set Z if $A \leqslant_T Y$ for each set Y such that the symmetric difference of

Y and Z has upper density 0. In [21] it is shown that the union of all the $\mathcal{B}_q$, $q < 1$ rational, coincides with the sets that are robustly computable from some ML-random set Z.

Calibrating randomness notions via cost functions

Bienvenu et al. [4] used cost functions to calibrate certain randomness notions. Let $\mathbf{c}$ be a monotonic cost function with the limit condition. A descending sequence $\langle V_n \rangle$ of uniformly c.e. open sets is a $\mathbf{c}$-*bounded test* if $\lambda(V_n) = O(\underline{\mathbf{c}}(n))$ for all n. We think of each V_n as an approximation for $Y \in \bigcap_k V_k$. Being in $\bigcap_n V_n$ can be viewed as a new sense of obeying $\mathbf{c}$ that works for ML-random sets. Unlike the first notion of obedience, here only the limit function $\underline{\mathbf{c}}(x)$ is taken into account in the definition.

Solovay completeness is a certain universal property of Ω among the left-c.e. reals; see e.g. [12]. Using this notion, one can show that the left-c.e. bounded tests defined above are essentially the $\mathbf{c}_{\langle \Omega \rangle}$-bounded tests.

We now survey some related, as yet unpublished work of Greenberg, Miller, Nies and Turetsky dating from early 2015. Hirschfeldt and Miller in unpublished 2006 work had proven that for any Σ_3 null class $\mathcal{C}$ of ML-random sets, there is a c.e. incomputable set Turing below all the members of $\mathcal{C}$. Their argument can be recast in the language of cost functions in order to show the following (here and below $\mathbf{c}$ is some monotonic cost function with the limit condition).

Proposition 6. *Suppose that* $A \models \mathbf{c}$ *and* Y *is in the* Σ_3^0 *null class of ML-randoms captured by a* $\mathbf{c}$-*bounded test. Then* $A \leqslant_T Y$.

We consider sets A such that the converse implication holds as well.

Definition 7. *Let* A *be a* Δ_2^0 *set. We say that* A *is* smart *for* $\mathbf{c}$ *if* $A \models \mathbf{c}$, *and* $A \leqslant_T Y$ *for each ML-random set* Y *that is captured by some* $\mathbf{c}$-*bounded test.*

Informally, A is as complex as possible for obeying $\mathbf{c}$, in the sense that the only random sets Y Turing above A are the ones that are above A because A obeys the cost function showing that $A \leqslant_T Y$ via Proposition 6.

For instance, A is smart for $\mathbf{c}_\Omega$ iff no ML-random set $Y \geqslant_T A$ is Oberwolfach random. Bienvenu et al. [4] proved that some K-trivial set A is smart for $\mathbf{c}_\Omega$. This means that A is the hardest to "cover" by a ML-random: any ML-random computing A will compute all the K-trivials by virtue of not being Oberwolfach random.

In the new work of Greenberg et al., this result is generalised to arbitrary monotonic cost functions with the limit condition that imply $\mathbf{c}_\Omega$.

Theorem 8 (Greenberg et al., 2015). *Let* $\mathbf{c}$ *be a monotonic cost function with the limit condition and suppose that only* K-*trivial sets can obey* $\mathbf{c}$. *Then some c.e. set* A *is smart for* $\mathbf{c}$.

The proof of the more general result, available in [17, Part 2], is simpler than the original one. Since A cannot be computable, the proof also yields a solution to Post's problem. This solution certainly has no injury, because there are no requirements.

7   The Γ and the Δ Parameter of a Turing Degree

We proceed to our second method of gauging the complexity of Turing degrees with methods inspired by analysis. We will be able to give the intuitive notion of being "close to computable" a metric interpretation.

For $Z \subseteq \mathbb{N}$ the lower density is defined to be

$$\underline{\eta}(Z) = \liminf_n \frac{|Z \cap [0, n)|}{n}.$$

(In the literature the symbol ρ is used. However, the same symbol denotes the Lebesgue density in the sense of Definition 1, so we prefer $\underline{\eta}$ here.) Hirschfeldt et al. [24] defined the γ parameter of a set Y:

$$\gamma(Y) = \sup\{\underline{\eta}(Y \leftrightarrow S) \colon S \text{ is computable}\}.$$

The Γ operator was introduced by Andrews et al. [1]:

$$\Gamma(A) = \inf\{\gamma(Y) \colon Y \leqslant_T A\}.$$

It is easy to see that this only depends on the Turing degree of A: one can code A back into Y on a sparse computable set of positions (e.g. the powers of 2), without affecting $\gamma(Y)$.

We now provide dual concepts. Let

$$\delta(Y) = \inf\{\overline{\eta}(Y \leftrightarrow S) \colon S \text{ computable}\},$$

$$\Delta(A) = \sup\{\delta(Y) \colon Y \leqslant_T A\}.$$

Intuitively, $\Gamma(A)$ measures how well computable sets can approximate the sets that A computes in the worst case (we take the infimum over all $Y \leqslant_T A$). In contrast, $\Delta(A)$ measures how well the sets that A computes can approximate the computable sets in the best case (we take the supremum over all $Y \leqslant_T A$). Note that $A \leqslant_T B$ implies $\Gamma(A) \geqslant \Gamma(B)$ and $\Delta(A) \leqslant \Delta(B)$.

It was shown in [1] that $\Gamma(A) > 1/2 \leftrightarrow \Gamma(A) = 1 \leftrightarrow A$ is computable. Clearly the maximum value of $\Delta(A)$ is $1/2$. It is attained, for example, when A computes a Schnorr random set Y, because in that case $\underline{\eta}(Y \leftrightarrow S) = 1/2$ for each computable S. Merkle, Nies and Stephan have shown that $\Delta(A) = 0$ for every 2-generic A.

Viewing $1 - \Gamma(A)$ as a Hausdorff pseudodistance

For $Z \subseteq \mathbb{N}$ the upper density is defined by

$$\overline{\eta}(Z) = \limsup_n \frac{|Z \cap [0, n)|}{n}.$$

For $X, Y \in 2^{\mathbb{N}}$ let $d(X, Y) = \overline{\eta}(X \triangle Y)$ be the upper density of the symmetric difference of X and Y; this is clearly a pseudodistance on Cantor space $2^{\mathbb{N}}$ (that

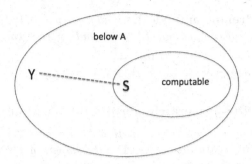

Fig. 2. Hausdorff pseudodistance $\sup_{Y \in \mathcal{A}} \inf_{S \in \mathcal{R}} d(Y, S)$.

is, two objects may have distance 0 without being equal). For subsets $\mathcal{U}, \mathcal{W}$ of a pseudometric space (M, d) recall the Hausdorff pseudodistance

$$d_H(\mathcal{U}, \mathcal{W}) = \max(\sup_{u \in \mathcal{U}} d(u, \mathcal{W}), \sup_{w \in \mathcal{W}} d(w, \mathcal{U}))$$

where $d(x, \mathcal{R}) = \inf_{r \in \mathcal{R}} d(x, r)$ for any $x \in M, \mathcal{R} \subseteq M$. Clearly, if $\mathcal{U} \supseteq \mathcal{W}$ then the second supremum is 0, so that we only need the first. The following fact, which is now clear from the definitions, states that $1 - \Gamma(A)$ gauges how close A is to being computable, in the sense that it is the Hausdorff distance between the cone below A and the computable sets (Fig. 2).

Proposition 9. *Given an oracle set A let $\mathcal{A} = \{Y : Y \leqslant_T A\}$. Let $\mathcal{R} \subseteq \mathcal{A}$ denote the collection of computable sets. We have*

$$1 - \Gamma(A) = d_H(\mathcal{A}, \mathcal{R}).$$

To interpret $1 - \Delta(A)$ metrically, we note that $1 - \delta(Y) = \sup_{S \in \mathcal{R}} d(Y, S)$. So we can view $1 - \Delta(A)$ as a one-sided "dual" of the Hausdorff pseudodistance:

$$d_H^*(\mathcal{A}, \mathcal{R}) = \inf_{Y \in \mathcal{A}} \sup_{S \in \mathcal{R}} d(Y, S).$$

For instance, for the unit disc $D \subseteq \mathbb{R}^2$ we have $d_H^*(D, D) = 1$.

Analogs of cardinal characteristics

The operators Γ and Δ are closely related to the analog in computability theory of cardinal characteristics (see [2] for the background in set theory). Both the cardinal characteristics and their analogs were introduced by Brendle and Nies in the 2015 Logic Blog [16], building on the general framework of an analogy between set theory and computability theory set up by Rupprecht ([46], also see [7]). We only discuss the versions of the concepts in the setting of computability theory.

Definition 10 (Brendle and Nies). *For $p \in [0, 1]$ let $\mathcal{D}(\sim_p)$ be the class of oracles A that compute a set X such that $\gamma(X) \leqslant p$, i.e., for each computable set S, we have $\underline{\eta}(X \leftrightarrow S) \leqslant p$.*

We note that by the definitions $\Gamma(A) < p \Rightarrow A \in \mathcal{D}(\sim_p) \Rightarrow \Gamma(A) \leqslant p$.

Definition 11 (Brendle and Nies). *Dually, for $p \in [0, 1/2)$ let $\mathcal{B}(\sim_p)$ be the class of oracles A that compute a set Y such that for each computable set S, we have $\underline{\eta}(S \leftrightarrow Y) > p$.*

For each p we have $\Delta(A) > p \Rightarrow A \in \mathcal{B}(\sim_p) \Rightarrow \Delta(A) \geqslant p$.

Collapse of the $\mathcal{D}(\sim_p)$ hierarchy for $p \neq 0$ after Monin

Definition 12. *For a computable function h, we let $\mathcal{D}(\neq_h^*)$, or sometimes $\mathcal{D}(\neq^*, h)$, denote the class of oracles A that compute a function x such that $\exists^\infty n \, x(n) = y(n)$ for each computable function $y < h$.*

This highness notion of an oracle set A was introduced by Monin and Nies in [37], where it was called "h-infinitely often equal". The notion also corresponds to a cardinal characteristic, namely $\mathfrak{d}(\neq_h^*)$ which is a bounded version of the well-known characteristic $\mathfrak{d}(\neq^*)$. The cardinal $\mathfrak{d}(\neq_h^*)$ is the least size of a set G of h-bounded functions so that for each function x there is a function y in G such that $\forall^\infty n[x(n) \neq y(n)]$. We note that $\mathcal{D}(\neq^*)$, i.e. the class obtained in Definition 12 when we omit the computable bound, coincides with having hyperimmune degree. See [7] for background, and in particular for motivation why the defining condition for $\mathfrak{d}(\neq_h^*)$ looks like the negation of the condition for $\mathcal{D}(\neq_h^*)$.

The proof of the following fact provides a glimpse of the methods used to prove that the $\mathcal{D}(\sim_p)$ hierarchy collapses.

Proposition 13. $\mathcal{D}(\neq^*, 2^{n!}) \subseteq \mathcal{D}(\sim_0)$.

Proof. Suppose that $A \in \mathcal{D}(\neq^*, 2^{n!})$ via a function $x \leqslant_T A$. Since $x(n) < 2^{n!}$ we can view $x(n)$ as a binary string of length $n!$. Let $L(x) \in 2^{\mathbb{N}}$ be the concatenation of the strings $x(0), x(1), \ldots$, and let $X \leqslant_T A$ be the complement of $L(x)$. Given a computable set S, there is a computable function y with $y(n) < 2^{n!}$ such that $L(y) = S$. Let $H(n) = \sum_{i<n} i!$. Since $x(n) = y(n)$ for infinitely many n, there are infinitely many intervals $[H(n), H(n+1))$ on which X and S disagree completely. Since $\lim_n n!/H(n) = 0$ this implies $\underline{\eta}(X \leftrightarrow S) = 0$. Hence $A \in \mathcal{D}(\sim_0)$.

We slightly paraphrase the main result of Monin's recent work [17]. It not only collapses the $\mathcal{D}(\sim_p)$ hierarchy, but also describes the resulting highness property combinatorially.

Theorem 14 (Monin). $\mathcal{D}(p) = \mathcal{D}(\neq^*, 2^{(2^n)})$ *for each $p \in (0, 1/2)$. In particular, $\Gamma(A) < 1/2 \Rightarrow \Gamma(A) = 0$ so only the values 0 and $1/2$ can occur when Γ is evaluated on incomputable sets.*

The proof uses the list decoding capacity theorem from the theory of error-correcting codes, which says that given a sufficiently large constant L, a fairly large set of code words of a length n can be achieved if one allows that each word of length n can be close (in the Hamming distance) to up to L of them. More precisely, independently of n, for each positive $\beta < 1$ there is $L \in \omega$ so that $2^{\lfloor \beta n \rfloor}$ codewords can be achieved. (In the usual setting of error correction, one would have $L = 1$, namely, each word is close to only one code word.)

Collapse of the $\mathcal{B}(\sim_p)$ hierarchy for $p \neq 0$ via a dual of Monin

Definition 15. *For a computable function h, we let $\mathcal{B}(\neq_h^*)$ denote the class of oracles A that compute a function $y < h$ such that $\forall^\infty n\, x(n) \neq y(n)$ for each computable function x.*

$\mathcal{B}(\neq^*)$, i.e. the class obtained when we omit the computable bound, coincides with "high or diagonally noncomputable" (again see, e.g., [7]). As a dual to Proposition 13 we have $\mathcal{B}(\sim_0) \subseteq \mathcal{B}(\neq^*, 2^{n!})$.

Theorem 16 (Nies). $\mathcal{B}(\sim_p) = \mathcal{B}(\neq^*, 2^{(2^n)})$ *for each $p \in (0, 1/2)$. In particular, $\Delta(A) > 0 \Rightarrow \Delta(A) = 1/2$ so there are only two possible Δ values.*

For a proof see again [17].

8 Open Questions

The development we have sketched in Sects. 4 and 5 has led to two randomness notions. The first, density randomness, was born out of the study of randomness via computable analysis. The second, OW-randomness, was born out of the study of lowness via randomness. We know that OW-randomness implies density randomness.

Question 17. Do OW-randomness and density randomness coincide?

One direction of attack to answer this negatively could be to look at other properties of points that are implied by OW-randomness, and show that density randomness does not suffice. By [35, Theorem 6.1] OW-randomness of z implies the existence of the limit in the sense of the Birkhoff ergodic theorem (Sect. 4) for computable operators T on a computable probability space $(2^{\mathbb{N}}, \mu)$, and lower semicomputable functions $g \colon X \to \mathbb{R}$. For another example, by [20] OW-randomness of z also implies an effective version of the Borwein-Ditor theorem: if $\langle r_i \rangle_{i \in \omega}$ is a computable null sequence of reals and $z \in \mathcal{C}$ for a Π_1^0 set $\mathcal{C} \subseteq \mathbb{R}$, then $z + r_i \in \mathcal{C}$ for infinitely many i.

Lowness for density randomness coincides with K-triviality by [35, Theorem 2.6]. Lowness for OW randomness is merely known to imply K-triviality for the reasons discussed in Sect. 3; further, an incomputable c.e. set that is low for OW-randomness has been constructed in unpublished worked with Turetsky.

Question 18. Characterise lowness for OW-randomness. Is it the same as K-triviality?

Section 7 leaves open several questions.

Question 19. Is $\mathcal{D}(\sim_0)$ a proper subclass of $\mathcal{D}(\neq^*, 2^{(2^n)}) = \mathcal{D}(1/4)$? Is $\mathcal{D}(\neq^*, 2^{n!})$ a proper subclass of $\mathcal{D}(\neq^*, 2^{(2^n)})$?

By Proposition 13 an affirmative answer to the first part implies an affirmative answer to the second. The dual open questions are:

Question 20. Is $\mathcal{B}(\neq^*, 2^{(2^n)}) = \mathcal{B}(\sim_{0.25})$ a proper subclass of $\mathcal{B}(\sim_0)$? Is it a proper subclass of $\mathcal{B}(\neq^*, 2^{(n!)})$?

Acknowledgement. Most of the research surveyed in this article was supported by the Marsden Fund of New Zealand.

References

1. Andrews, U., Cai, M., Diamondstone, D., Jockusch, C., Lempp, S.: Asymptotic density, computable traceability, and 1-randomness. Preprint (2013)
2. Bartoszyński, T., Judah, H.: Set Theory. On the Structure of the Real Line, p. 546. A K Peters, Wellesley (1995)
3. Bienvenu, L., Day, A., Greenberg, N., Kučera, A., Miller, J., Nies, A., Turetsky, D.: Computing K-trivial sets by incomplete random sets. Bull. Symb. Logic **20**, 80–90 (2014)
4. Bienvenu, L., Greenberg, N., Kučera, A., Nies, A., Turetsky, D.: Coherent randomness tests and computing the K-trivial sets. J. Eur. Math. Soc. **18**(4), 773–812 (2016)
5. Bienvenu, L., Hölzl, R., Miller, J., Nies, A.: Denjoy, Demuth, and density. J. Math. Log. **1450004**, 35 (2014)
6. Brattka, V., Miller, J., Nies, A.: Randomness, differentiability. Trans. AMS 368, 581–605 (2016). arXiv version at http://arxiv.org/abs/1104.4465
7. Brendle, J., Brooke-Taylor, A., Ng, K.M., Nies, A.: An analogy between cardinal characteristics, highness properties of oracles. In: Proceedings of the 13th Asian Logic Conference: Observation of strains, Guangzhou, China, pp. 1–28. World Scientific (2013). http://arxiv.org/abs/1404.2839
8. Chaitin, G.: Information-theoretical characterizations of recursive infinite strings. Theor. Comput. Sci. **2**, 45–48 (1976)
9. Day, A.R., Miller, J.S.: Density, forcing and the covering problem. Math. Res. Lett. **22**(3), 719–727 (2015)
10. Demuth, O.: The differentiability of constructive functions of weakly bounded variation on pseudo numbers. Comment. Math. Univ. Carolin. **16**(3), 583–599 (1975). (Russian)
11. Downey, R., Hirschfeldt, D.: Algorithmic Randomness and Complexity. Theory and Applications of Computability, p. 855. Springer, Heidelberg (2010)
12. Downey, R., Hirschfeldt, D., Nies, A.: Randomness, computability, and density. SIAM J. Comput. **31**(4), 1169–1183 (2002)
13. Downey, R., Hirschfeldt, D., Nies, A., Stephan, F.: Trivial reals. In: Proceedings of the 7th and 8th Asian Logic Conferences, Singapore, pp. 103–131. Singapore University Press (2003)
14. Downey, R., Nies, A., Weber, R., Yu, L.: Lowness and Π_2^0 nullsets. J. Symbolic Logic **71**(3), 1044–1052 (2006)
15. Downey, R.G., Jockusch Jr., C.G.: T-degrees, jump classes, and strong reducibilities. Trans. Amer. Math. Soc. **30**, 103–137 (1987)
16. Nies, A. (ed.): Logic Blog 2015 (2015). http://arxiv.org/abs/1602.04432
17. Nies, A. (ed.): Logic Blog 2016 (2016). cs.auckland.ac.nz/nies
18. Figueira, S., Miller, J.S., Nies, A.: Indifferent sets. J. Logic Comput. **19**(2), 425–443 (2009)

19. Freer, C., Kjos-Hanssen, B., Nies, A., Stephan, F.: Algorithmic aspects of Lipschitz functions. Computability **3**(1), 45–61 (2014)
20. Galicki, A., Nies, A.: Effective Borwein-Ditor theorem. In: Proceedings of CiE 2016 (2016)
21. Greenberg, N., Miller, J., Nies, A.: A dense hierarchy of subideals of the k-trivial degrees (2015)
22. Hirschfeldt, D., Jockusch, C., Kuyper, R., Schupp, P.: Coarse reducibility, algorithmic randomness. arXiv preprint arXiv: 1505.01707 (2015)
23. Hirschfeldt, D., Nies, A., Stephan, F.: Using random sets as oracles. J. Lond. Math. Soc. (2), **75**(3), 610–622 (2007)
24. Hirschfeldt, D.R., Jockusch, Jr. C.G., McNicholl, T., Schupp, P.E.: Asymptotic density and the coarse computability bound, Preprint (2013)
25. Jockusch Jr., C.G.: The degrees of bi-immune sets. Z. Math. Logik Grundlagen Math. **15**, 135–140 (1969)
26. Jockusch Jr., C.G.: Semirecursive sets and positive reducibility. Trans. Amer. Math. Soc. **131**, 420–436 (1968)
27. Jockusch Jr., C.G., Soare, R.I.: Π_1^0 classes and degrees of theories. Trans. Amer. Math. Soc. **173**, 33–56 (1972)
28. Khan, M.: Lebesgue density and Π_1^0-classes. J. Symb. Logic (to appear)
29. Kučera, A., Nies, A., Porter, C.: Demuth's path to randomness. Bull. Symb. Logic **21**(3), 270–305 (2015)
30. Lebesgue, H.: Leçons sur l'Intégration et la recherche des fonctions primitives. Gauthier-Villars, Paris (1904)
31. Lebesgue, H.: Sur les intégrales singulières. Ann. Fac. Sci. Toulouse Sci. Math. Sci. Phys. (3), 1, 25–117 (1909)
32. Lebesgue, H.: Sur l'intégration des fonctions discontinues. Ann. Sci. École Norm. Sup. (3), **3**(27), 361–450 (1910)
33. Martin, D.A., Miller, W.: The degrees of hyperimmune sets. Z. Math. Logik Grundlag. Math. **14**, 159–166 (1968)
34. Miller, J.S., Nies, A.: Randomness and computability: open questions. Bull. Symbolic Logic **12**(3), 390–410 (2006)
35. Miyabe, K., Nies, A., Zhang, J.: Using almost-everywhere theorems from analysis to study randomness. Bull. Symb. Logic **22**(3), 305–331 (2016)
36. Mohrherr, J.: A refinement of low_n, $high_n$ for the R.E. degrees. Z. Math. Logik Grundlag. Math. **32**(1), 5–12 (1986)
37. Monin, B., Nies, A.: A unifying approach to the Gamma question. In: Proceedings of Logic in Computer Science (LICS). IEEE press (2015)
38. Nies, A.: Calculus of cost functions. In: Cooper, B., Soskova, M. (eds.) The Incomputable: Observation of Strains: Journeys Beyond the Turing Barrier. Springer, Heidelberg (to appear)
39. Nies, A.: Lowness properties and randomness. Adv. Math. **197**, 274–305 (2005)
40. Nies, A.: Computability and Randomness, vol. 51 of Oxford Logic Guides, p. 444, Paperback version 2011 (2009)
41. Nies, A.: Interactions of computability and randomness. In: Proceedings of the International Congress of Mathematicians, pp. 30–57. World Scientific (2010)
42. Nies, A.: Studying randomness through computation. In: Randomness Through Computation, pp. 207–223. World Scientific (2011)

43. Nies, A.: Differentiability of polynomial time computable functions. In: Mayr, E.W., Portier, N. (eds.) 31st International Symposium on Theoretical Aspects of Computer Science (STACS 2014). volume 25 of Leibniz International Proceedings in Informatics (LIPIcs), pp. 602–613. Schloss dagstuhl-leibniz-zentrum fuer informatik, Dagstuhl, Germany (2014)

44. Pathak, N.: A computational aspect of the Lebesgue differentiation theorem. J. Log. Anal. 1, Paper 9, 15 (2009)

45. Pathak, N., Rojas, C., Simpson, S.G.: Schnorr randomness and the Lebesgue differentiation theorem. Proc. Amer. Math. Soc. 142(1), 335–349 (2014)

46. Rupprecht, N.: Effective correspondents to cardinal characteristics in Cichoń's diagram. Ph.D. thesis, University of Michigan (2010)

47. Sacks, G.E.: A minimal degree below $0'$. Bull. Amer. Math. Soc. 67, 416–419 (1961)

48. Sacks, G.E.: The recursively enumerable degrees are dense. Ann. Math. (2), 80, 300–312 (1964)

49. Schnorr, C.P.: Zufälligkeit und Wahrscheinlichkeit. Eine algorithmische Begründung der Wahrscheinlichkeitstheorie. Lecture Notes in Mathematics, vol. 218. Springer-Verlag, Berlin (1971)

50. Solovay, R..: Handwritten Manuscript Related to Chaitin's Work, p. 215. IBM Thomas J. Watson Research Center, Yorktown Heights, NY (1975)

51. Spector, C.: On the degrees of recursive unsolvability. Ann. Math. 2(64), 581–592 (1956)

52. Terwijn, S., Zambella, D.: Algorithmic randomness and lowness. J. Symbolic Logic 66, 1199–1205 (2001)

53. V'yugin, V.: Ergodic theorems for individual random sequences. Theor. Comput. Sci. 207(2), 343–361 (1998)

Erratum to: There Are No Maximal d.c.e. *wtt*-degrees

Guohua Wu[1](✉) and Mars M. Yamaleev[2]

[1] Division of Mathematical Sciences, School of Physical and Mathematical Sciences,
Nanyang Technological University, Singapore 637371, Singapore
guohua@ntu.edu.sg
[2] Institute of Mathematics and Mechanics, Kazan Federal University,
18 Kremlyovskaya Street, Kazan 420008, Russia
mars.yamaleev@ksu.ru

Erratum to:
Chapter "There Are No Maximal d.c.e. *wtt*-degrees" in:
A. Day et al. (Eds.):
Computability and Complexity, LNCS,
DOI: 10.1007/978-3-319-50062-1_28

The original version of this chapter contained an error. The spelling of Mars M. Yamaleev's name was incorrect. The original chapter was corrected.

Wu is partially supported by AcRF Tier 2 grants MOE2011-T2-1-071 (ARC 17/11, M45110030) and MOE2016-T2-1-083 from Ministry of Education of Singapore, and by AcRF Tier 1 grants, RG29/14, M4011274 and RG32/16, M4011672 from Ministry of Education of Singapore.
Yamaleev is supported by Russian Foundation for Basic Research (projects 15-41-02507, 15-01-08252), by research grant of Kazan Federal University, and by the subsidy allocated to Kazan Federal University for the project part of the state assignment in the sphere of scientific activities (project 1.2045.2014).

The updated original online version for this chapter can be found at
DOI: 10.1007/978-3-319-50062-1_28

A. Day et al. (Eds.): Downey Festschrift, LNCS 10010, p. E1, 2017.
DOI: 10.1007/978-3-319-50062-1_43

Author Index